The Matter with Things

THE MATTER WITH THINGS

*Our Brains, Our Delusions
and the Unmaking of the World*

Iain McGilchrist

· VOLUME TWO ·

What Then is True?

PERSPECTIVA PRESS

Copyright © 2021 Iain McGilchrist

᛭ First published in 2021 by
PERSPECTIVA PRESS
PO Box 75779, London, SW15 9HW
systems-souls-society.com

First paperback printing, 2023

All individual copyright information for material incorporated in this book can be found in the credits for the colour plates in volumes 1 & 2, and in the Acknowledgments.

The author and publishers have made every effort to trace the owners of copyright material reproduced in this book. In the event of any omission, please contact the publishers, who will make appropriate restitution in future editions.

Iain McGilchrist asserts his moral right to be identified as the author of this book.

All rights reserved. No part of this publication may be reproduced, stored in a retrieval system or transmitted in any form by any means, electronic, mechanical, photocopying, recording or otherwise, without the prior written consent of the copyright owner and the publisher, other than as permitted by UK copyright legislation.

ISBN (hardback) 978-1-914568-06-0
ISBN (ebook) 978-1-914568-05-3
ISBN (softcover) 978-1-914568-25-1

A CIP catalogue record for this book is available from the British Library.

Printed in Great Britain by TJ Books Limited, Padstow, Cornwall

᛭ PERSPECTIVA is a collective of scholars, artists, activists, futurists, and seekers operating as a registered charity based in London. We believe desirable futures for humanity depend upon forms of economic restraint and political cooperation that are beyond prevailing epistemic capacities and spiritual sensibilities. Our urgent one-hundred-year project is to develop an applied philosophy of education that cultivates forms of imaginative, aesthetic and emotional capacity to help usher in a world that is, at the very least, technologically wise and ecologically sound.

As part of our contribution to education, Perspectiva Press publishes books that offer soul food for expert generalists. The idea of an expert generalist is paradoxical but not oxymoronic. In theory, our very best philosophers, civil servants, political leaders, and writers are expert generalists – their defining skill is inclusive synthesis and their defining qualities are epistemic acumen and agility: knowhow with knowledge, having enough expertise in one domain to value different forms of understanding, and knowing how to integrate them while retaining curiosity towards whatever remains unfamiliar. And soul food is called for, and not merely brain food, because we seek to revitalise and reenchant the discourse about what it means to be human, in a time when that is all too often taken for granted or explained away.

Contents

VOLUME I

List of Figures	x
List of Plates	xiii
Acknowledgments	xv
Note to the Reader	xvii

• • •

INTRODUCTION		3

PART ONE: THE HEMISPHERES AND THE MEANS TO TRUTH

Chapter 1	Some preliminaries: how we got here	53
Chapter 2	Attention	67
Chapter 3	Perception	105
Chapter 4	Judgment	135
Chapter 5	Apprehension	181
Chapter 6	Emotional and social intelligence	193
Chapter 7	Cognitive intelligence	225
Chapter 8	Creativity	239
Chapter 9	What schizophrenia and autism can tell us	305
	CODA to Part I	371

• • •

PART TWO: THE HEMISPHERES AND THE PATHS TO TRUTH

Chapter 10	What is truth?	379

Science & truth

Chapter 11	Science's claims on truth	407
Chapter 12	The science of life: a study in left hemisphere capture	431
Chapter 13	Institutional science and truth	501

Reason & truth

Chapter 14	Reason's claims on truth	547
Chapter 15	Reason's progeny	571
Chapter 16	Logical paradox: a further study in left hemisphere capture	641

Intuition, imagination & truth

Chapter 17	Intuition's claims on truth	673
Chapter 18	The untimely demise of intuition	707
Chapter 19	Intuition, imagination and the unveiling of the world	753
	CODA to Part II	777

• • •

APPENDICES TO VOLUME I

Appendix 1	Hemisphere differences and creativity: an examination of Dietrich & Kanso's evidence	781
Appendix 2	Hoaxes are not confined to the field of science	793
Appendix 3	Why we should be sceptical of 'public science'	797
Credits for the colour plates in volume I		809

VOLUME II

PART THREE: THE UNFORESEEN NATURE OF REALITY

Chapter 20	The *coincidentia oppositorum*	813
Chapter 21	The One and the Many	843
Chapter 22	Time	881
Chapter 23	Flow and movement	945
Chapter 24	Space and matter	997
Chapter 25	Matter and consciousness	1037
Chapter 26	Value	1121
Chapter 27	Purpose, life and the nature of the cosmos	1167
Chapter 28	The sense of the sacred	1193
	CODA to Part III	1305
EPILOGUE		1309

• • •

APPENDICES TO VOLUME II

Appendix 4	What happens to time in depression	1337
Appendix 5	Hemisphere differences and morality: review of neuropsychological evidence	1343
Appendix 6	Hemisphere differences and beauty: review of neuropsychological evidence	1349
Appendix 7	Hemisphere differences and spirituality: review of neuropsychological evidence	1357
Appendix 8	The 'incompatibility' of science and religion, and the harmful nature of religion: some evidence to the contrary	1361
Bibliography		1377
Index of Topics		1561
Index of Names		1568
Credits for the colour plates in volume II		1579

List of Figures

VOLUME I

PART ONE

Fig. 1	Human brain, displaying the corpus callosum, engraving by Andreas Vesalius	20
Fig. 2	*All is vanity,* drawing by Charles Allan Gilbert, 1892	49
Fig. 3	Sagittal section of the human brain showing the midbrain or mesencephalon	57
Fig. 4	Drawings by patients exhibiting hemineglect	71
Fig. 5	Loss of the implicit in right hemisphere dysfunction	75
Fig. 6	Images copied by either hand of a commissurotomy subject (Bogen 1969)	82
Fig. 7	Drawings of targets from memory in subjects with unilateral deficits	99
Fig. 8	Assembling a mannequin in right hemisphere dysfunction (McFie *et al* 1950)	99
Fig. 9	Bi-stable image of stairs (Schröder 1858)	114
Fig. 10	Hemiprosopometamorphopsia (Bethlem Royal Hospital Archives)	125
Fig. 11	Alice-in-Wonderland syndrome: somatic effects (from Sir John Tenniel's illustrations for *Alice in Wonderland,* 1865)	126
Fig. 12	Alice-in-Wonderland syndrome: temporal effects (from Sir John Tenniel's illustrations for *Through the Looking-Glass,* 1871)	126
Fig. 13	The Rey-Osterrieth figure	152
Fig. 14	The 'smiley' sign (Regard & Landis 1994)	153
Fig. 15	The Rey-Osterrieth figure as copied by children aged 5 (Waber 1979)	165
Fig. 16	Drawings by a subject with autotopagnosia (Engerth 1933)	190
Fig. 17	Impaired emotional and social understanding in right hemisphere dysfunction (Baldo *et al* 2016)	209
Fig. 18	Literalistic versus metaphorical thinking (Myers *et al* 1981)	216
Fig. 19	'It was a dark day for Denis', drawing by Anita Klein, 1991	221
Fig. 20	Drawing by an artist before and after right hemisphere strokes (Schnider *et al* 1993)	262

Fig. 21	Change of style in artist following left hemisphere stroke (Blanke 2007)	267
Fig. 22	The nine dot problem	275
Fig. 23	The nine dot problem: solution	277
Fig. 24	*Melancholicus,* engraving by Gottfried Eichler, 1758–60	294
Fig. 25	*Corps Humain,* drawing by 'Katharina', 1965 (Collection de l'Art Brut, Lausanne)	346
Fig. 26	*The Air Loom,* engraving by John Haslam, based on a machine described by James Tilly Matthews (Haslam, 1810)	361
Fig. 27	Detail of Fig. 26, showing the recorder, Jack the Schoolmaster	362

PART TWO

Fig. 28	Caryokinetic figure in blastomere of a trout's egg (Thompson 1992)	483
Fig. 29	Transverse section of peripheral nerve	492
Fig. 30	Longitudinal section of peripheral nerve	492
Fig. 31	Transverse cut of bamboo	492
Fig. 32	Longitudinal cut of bamboo	492
Fig. 33	Flow diagram, after Peter Vamplew, 2014	519
Fig. 34	Chequerboard illusion	673
Fig. 35	The Monty Hall Dilemma	708
Fig. 36	The effect of expectation on perception	736

VOLUME II

PART THREE

Fig. 37	Geometric illustrations of Cusanus' concept of truth	948
Fig. 38	*Towards Pi 3.141552779,* drawing by Jason Padgett, 2011	949
Fig. 39	Early post-traumatic drawings of car wheel and balloon, by Jason Padgett (Brogaard *et al* 2013)	949
Fig. 40	Vortex in water	950
Fig. 41	*Black hole,* drawing by Jason Padgett, 2008	950
Fig. 42	Ripple in water	951
Fig. 43	*Water ripple,* drawing by Jason Padgett, c 2010	951

Fig. 44	Vortices in stream caused by uniform perpendicular obstruction (Douthat *et al* 1975)	962
Fig. 45	Detail of Fig. 44	962
Fig. 46	Effects of differing rates and duration of pulse on vortices in stream (*ibid*)	962
Fig. 47	Vortices in a stream in the wake of a regular rake of rods of rectangular cross section (*ibid*)	963
Fig. 48	Detail of Fig. 47	963
Fig. 49	Vortices formed when a stream of air impinges on a cube (*ibid*)	964
Fig. 50	Kármán vortex street (*ibid*)	964
Fig. 51	Vortices and contrary motion (Coats 2001)	966
Fig. 52	Corporal Trim's flourish, from *Tristram Shandy*, by Laurence Sterne (Bk IX, 1767)	977
Fig. 53	The plot structure of *Tristram Shandy* (Bk VI, 1761)	977
Fig. 54	Variations on the serpentine curve, by William Hogarth, 1753	977
Fig. 55	Golden ratio	1030
Fig. 56	Relation between the golden ratio and spiral	1030
Fig. 57	Fibonacci distribution of leaves; arrangement of leaves resulting from 137.5° compared with 135° separation (photo by Ron Knott; diagrams from Valladares *et al* 2004)	1031
Fig. 58	Harmonic oscillations	1108
Fig. 59	How continuous oscillations lead to discrete levels of energy (after Brooks 2016)	1108
Fig. 60	Oscillations in a string (*ibid*)	1109
Fig. 61	Canalised pathways of change within an epigenetic landscape (after Waddington 1957)	1176
Fig. 62	Pyramid of values according to Scheler	1324

List of Plates

VOLUME I

Plates 1–16 follow page 430.
Note that all attributions, permissions and copyright information can be found on p 809.

Plate 1	(a) Lobes of the brain (b) Diencephalon, basal ganglia and limbic system (c) Prefrontal cortex and limbic system (d) Optic pathways
Plate 2	(a) Superior longitudinal and uncinate fasciculi (b) Neural network of *Nematostella vectensis* (c) Comparative anatomy of neocortex in mouse, monkey and man
Plate 3	(a) Sagittal section of the brain of a dog and of a human brain (b) Simulation of palinopsia/akinetopsia
Plate 4	*The Garden of Earthly Delights*, by Hieronymus Bosch, oil on board, c 1505
Plate 5	(a) Simulation of hemiprosopometamorphopsia (b) *Head of a Woman*, by Pablo Picasso, oil on canvas, 1960
Plate 6	(a) *The Persistence of Memory*, by Salvador Dalí, oil on canvas, 1931 (b) The right temporoparietal junction in social interaction
Plate 7	Self-portrait sketches by Otto Dix, before and after RH stroke
Plates 8–9	Paintings and sketches by Lovis Corinth, before and after RH stroke
Plate 10	(a) *The Puppet-master*, undated watercolour, by a patient of the Bethlem Royal Hospital (b) *Self-portrait with Skeleton Arm*, by Edvard Munch, lithograph, 1895
Plate 11	*The Maze*, by William Kurelek, gouache on board, 1953
Plate 12	(a) Most significant areas of the brain for fluid intelligence (b) Brain activations during a creativity task, contrasting original and unoriginal subjects
Plate 13	(a) 'Horror graph' of cross-signalling between five intracellular signal transduction cascades (b) Eyes of a fly, a frog and a human (c) Norfolk Island pine (d) Eagle feathers
Plate 14	*Jacob's Ladder*, by William Blake, pen, ink and watercolour, c 1799–1807

Plate 15   *La belle captive,* by René Magritte, oil on canvas, 1948

Plate 16   (a, b) The chequerboard illusion
                   (c, d) The Wason test;
                   (e)   The Ebbinghaus illusion

VOLUME II

Plates 17–24 follow page 1166.
Note that all attributions, permissions and copyright information can be found on p 1579.

Plate 17   (a) Olfactory bulb of a dog, by Camillo Golgi, drawing 1875
               (b) Olfactory bulb of a mouse, by Santiago Ramón y Cajal, drawing, 1890

Plate 18   *Studies of Water Passing Obstacles and Falling,* by Leonardo da Vinci, ink and black chalk on paper, c 1508

Plate 19   (a) *Seated Old Man, and Studies and Notes on the Movement of Water,* by Leonardo da Vinci, ink on paper, c 1510
               (b) *Deluge,* by Leonardo da Vinci, black chalk on paper, c 1517

Plate 20   *The Great Wave off Kanagawa,* by Katsushika Hokusai, woodblock print, 1831

Plate 21   Diversity of lawful architective structure in snowflakes

Plate 22   (a)   Bi-stable image of woman's face and flower
               (b)   Inyo symbol
               (c)   Wabi-sabi/kintsugi ware
               (d, e) Fibonacci series in pine cones and sunflower

Plate 23   (a) Façade of the Villa Capra ('La Rotonda'), near Vicenza, designed by Andrea Palladio in 1570, finished 1592
               (b) Façade of The Rotonda, University of Virginia, designed by Thomas Jefferson, completed 1826

Plate 24   *Circle Limit IV,* by M. C. Escher, engraving, 1960

PART III

The Unforeseen Nature of Reality

> The universe is not to be narrowed down
> to the limits of the understanding, which has been men's practice
> up to now, but the understanding must be stretched and enlarged
> to take in the image of the universe as it is discovered.
> — SIR FRANCIS BACON[1]

> To say that metaphysics is nonsense *is* nonsense.
> — FRIEDRICH WAISMANN[2]

> In our description of nature the purpose
> is not to disclose the real essence of the phenomena
> but only to track down, so far as it is possible,
> relations between the manifold aspects of our experience.
> — NIELS BOHR[3]

---

1 Bacon 1858b, *Parasceve ad historiam naturalem et experimentalem,* aphorism §4.
2 Waismann 1968 (38; emphasis in original).
3 Bohr 1961 (18).

Chapter 20
The *coincidentia oppositorum*

A thing without oppositions *ipso facto* does not exist ... existence lies in opposition.
— C. S. PEIRCE[1]

It is the hallmark of any deep truth that its negation is also a deep truth.
— NIELS BOHR[2]

The heart's wave would never have risen up so beautifully in its cloud of spray, and become spirit, were it not for the grim old cliff of destiny standing in its way ...
— FRIEDRICH HÖLDERLIN[3]

According to an ancient Iroquois legend, the gradual fading of eternal power and light in the cosmos made necessary the activity of a creator god whose task was, for the sake of the whole universe, to bring into being the earth and all its creatures. His name in the Onondaga language, *De'haĕⁿhiyawă"khoⁿ'*, means He Grasps The Sky With *Both* Hands (my emphasis); and in the legend, he represents the power to remember one's higher identity in the midst of action in the world. He has, however, a twin brother who declares: 'I am not thinking about the place from where I came ... It is sufficient that my mind is satisfied in having arrived at this place ... This place will become exceedingly delightful and amusing to the mind ... I trust in the thing which my father gave me, a flint arrow, by which I have speech. This I will use perhaps to defend myself so that I will not think of that other place.' His name is *O'hā'ä'*, which means He Who Is Crystal Ice, He Who Is Flint; subsequently he is referred to simply as Flint. He represents 'evil in the form of forgetfulness, intentional forgetfulness of the higher identity'.[4]

He Grasps the Sky With Both Hands begins creating living creatures. Flint sees the animals that his brother creates and how good they are; and he is jealous. He gathers all his brother's animals together and puts them in a cave. Troubled by this, He Grasps The Sky With Both Hands tries to cut himself off from his brother. Flint then tries on his own to imitate his brother. He creates his own birds, flowers and fruits. His brother is more troubled than before. But his realisation is that it is only when Flint is cut off from his brother that he does wrong. So He Grasps The Sky With Both Hands rescinds the act of separating himself off from the evil, and returns to his brother to see what he has done.

---

1 Peirce 1931–60, vol 1, §457.

2 As quoted in Delbrück 1986 (167). A similar formulation – 'you can recognize a deep truth by the feature that its opposite is also a deep truth' – is sometimes attributed to physicist Frank Wilczek, because he included it in an account of Bohr's philosophy in his 2015 book *A Beautiful Question: Finding Nature's Deep Design*; but the context makes clear that he is (appreciatively) paraphrasing Bohr.

3 Hölderlin, *Hyperion*, 'Hyperion an Bellarmin VIII': » *Des Herzens Woge schäumte nicht so schön empor, und würde Geist, wenn nicht der alte stumme Fels, das Schicksal, ihr entgegenstände.* «

4 This account is derived from a landmark of Onondaga literature, dictated in 1900 by the Onondaga orator and priest, John Arthur Gibson, to his collaborator, JNB Hewitt, who was transcribing material for his *Iroquoian Cosmology* (1903, 1928). In this passage I am relying on a substantial report of the myth contained in Needleman 2003 (204–12), who in turn was dependent on Hewitt: all quotations here are verbatim from Needleman, including any emphasis. I am grateful to Robert Bringhurst for helping me understand the provenance of this account. Bringhurst points out that the names of the two brothers in current Onondaga orthography would be Tháęhya·wá?gih and Ohá·æ?. I am also grateful to the anthropologist Stefano Fait, who noticed 'a number of impressive parallels' between the story and the structure of our brains, for bringing this remarkable passage to my attention.

It turns out that Flint has created not birds, but flies and bats; not flowers, but thistles; not fruit, but thorns.

> Seeing this, the good brother embraces his brother's work, giving all that Flint has made their proper names (that is, assigning them their proper role in the scheme of things) and declaring, 'All this shall assist me. The flies shall assist me. The thistle will be food for small animals, the thorn will be food for game animals ...' The mind of Flint was gratified. But Flint goes on attempting to imitate the works of creation, and He Grasps The Sky With Both Hands comes to understand that it is right that he maintain a *small distance* from his brother, while at the same time keeping his attention upon him, neither letting him drift too far from his awareness, nor letting him blend with him. The good brother understands full well that Flint will forever attempt to destroy his rule.

He Grasps The Sky With Both Hands consults an 'Ancient One' who confirms this: Flint will aim to destroy his benign superintendence of creation.

He Grasps the Sky With Both Hands goes back to the most primitive source of being. From it he brings the light of the sun into the world. He starts to create human beings, a man and a woman. Into each he sees it is good that he should give some of his own life, his breath, his mind and his power of speech.

But all is not well. Seeing what his brother has done, Flint decides that he too can make human beings. Flint's experiments, however, result only in strange, anguished, misbegotten creatures that run from him and hide. So he turns to his brother for help.

> As Flint prepared to cross once again the narrow channel that separated him from his brother, he was startled to see that He Grasps The Sky With Both Hands had already crossed the water and was coming towards him. Flint greeted him, saying, 'I have come to meet you because I desire your aid in causing the human being to live'.
>
> He Grasps The Sky With Both Hands agreed and went to the place where the human being was. He Grasps The Sky With Both Hands took a portion of his own life and put it inside the human being. So also he took a portion of his own mind and enclosed it in the head of the human being. And so also a portion of his own blood and enclosed it inside the flesh of the human being. And so too did he take a portion of his own power to see and enclosed it in the head of the human being. So also he took a portion of his power to speak and enclosed it in the throat of the human being. Finally, he also placed his breath in the body of the human being. Just then the human being came to life, and he arose, and stood upon the earth present here.
>
> Turning to Flint, He Grasps The Sky With Both Hands spoke: 'I now have aided you in this matter. And now, I see that *this* human

being will become hostile to me. What will come to pass because of that?'

Flint quickly replied, 'Since both you and I took part in completing this human being, let both you and I have control over it. In that way you will have something to say concerning these human beings who will dwell on this earth.' He Grasps The Sky With Both Hands agreed to that, adding: 'That human being whom I alone created, who is the first human being to become alive on this earth – we shall call him real human being. And this human being whom you and I have now created and is now alive on this earth, we shall call him the hatchet maker, the bringer of strife.

In time, the moon is created, initially under the power of Flint and evil, but it eventually comes under the power of the good brother. The brothers depart the realm of this earth. But He Grasps The Sky With Both Hands, before he leaves, warns that there are two minds in human beings; and that if they pursue strife, rather than peace, they will end

> in the place where my brother dwells. And there you will see great suffering, and you will be famished, and you will be without liberty, and you will share the fate of my brother. I have confined him, and I have kindled a fire for him, and for this purpose I used his anger. This fire is hotter than any fire you have ever known; and this fire will burn eternally in that my brother even now desires to control all minds among human beings.

'Whichever mind you choose, you must obey it', he says. If mankind forgets, He Grasps the Sky With Both Hands will try to intervene twice on behalf of mankind, but

> if a third time it comes to pass that you forget, then you will see what will come to pass. The things upon which you live will diminish so that finally nothing more will be able to grow ... It will be my brother who will do all this, for he will be able to seduce the minds of all human beings and thus spoil all that I have completed. Now I leave the matter to you.

This extraordinary legend appears to me to be one of the most remarkable intuitions of the structure of mind and its influence on human destiny ever brought forth from the depth of the human imagination. There are many close parallels between its message and the account of hemisphere difference expounded in the course of this book, as will be obvious to the attentive reader.

Further, though a creation myth, it is one with an important difference. It is not merely a myth of a completed act of creation, dealing solely with origins, but an account which also looks forward: to creation as continuous. What's more, being an account of creation (of how worlds are brought into being) provided by a legend that intuits the structure of the mind, it concerns not simply 'the world' in a more limited sense, but the phenomenological world, the world that comes into being by the engagement of the human mind with whatever-it-is-that-exists-apart-from-ourselves: the only world we can ever know. At the risk of encroaching on the beauty and wisdom of myth, I will, with some misgivings, point to it from time to time as being a more vivid expression of an understanding of the world that also finds less vivid expression in the hemisphere hypothesis and in the efforts of various philosophers to put it in a more abstract form.

And there is another layer of meaning in the story: the dynamism of exchange between good and evil, the question of how close together they can become and how far apart they need to be.[5] I will turn to that in the final chapter of Part III.

5 Needleman 2003 (204).

THE GENERATIVE POWER OF OPPOSITES

There is so much that is unusual, and unusually apt, in this myth of generation. Yet another element is the need for two creative forces that are seen by one of those forces (Flint) as opposed, but are brought to work together by the other (He Grasps the Sky With Both Hands). All things arise from opposing, but in some form nonetheless related, drives or forces. Energy is always characterised by the coming together of *apparent* opposites – apparent because this is how we have conceived things left hemisphere fashion: as in the positive and negative poles of an electric circuit, the north and south poles of the magnet, or, in a quite different sense, the merging of male and female gametes in the origin of new life. To the imaginative mind, such a coming together of 'opposites' is, as Niels Bohr (above) suggests, a sign that we are at last approaching a deeper level of truth.

Bohr's greatest insight into the deep nature of the universe was that *contraria sunt complementa*: contraries *fulfil* one another. But it is also a timeless insight, to be found in one form or another in most ancient cultures of which I am aware. The most sophisticated of these, because of the detailed and refined exposition it has given rise to in China, is that of *yin* and *yang*, contrary forces that fulfil one another by their complementary nature (whose symbol, incidentally, Bohr placed at the centre of his coat of arms when ennobled by the Danish Government). And in that symbol the male and female principles are also implied. That the concepts may have become vulgarised is not a weakness in the concepts but, rather, in the cast of mind that does not measure up to them, much as a religion is not vitiated simply by the misguidedness of some of its followers. The idea of complementary opposites is, however, present at the beginning of the Western philosophical tradition, most notably in

Heraclitus. And as Nietzsche says, 'The world for ever needs the truth, hence the world for ever needs Heraclitus.'⁶

In one of his most penetrating observations, Heraclitus notes:

> They do not understand how a thing agrees at variance with itself: it is an attunement turning back on itself, like that of the bow and the lyre.⁷

It is the tension between the warring ends of the bow that gives the arrow the power to fly, as it is the tension in the strings of the lyre that gives rise to melody: this is what he meant by his saying 'war is the father of all things'. What looks like a waste of effort – pulling in opposite directions – is the essence of generative vitality. The word translated here as attunement is *harmoniē*.⁸ Harmony is, after all, the reconciliation of things that contend with one another. According to Charles Kahn, Heraclitus' most rewarding commentator, the word brings together three main ideas – the fitting together of surfaces that are 'true'; the reconciling of warring parties; and the accord of musical strings.⁹ Whatever is, therefore, brings together elements that are made to fit, and in a manner that is fitting; draws peace out of conflict; and gives birth to beauty out of this reconciliation.

Another fragment is equally pregnant, and of even greater density:

> Graspings: wholes and not wholes, convergent divergent, consonant dissonant, from all things one and from one thing all.¹⁰

(Incidentally, neither in this nor in the *harmoniē* fragment just quoted, does the word 'thing' occur in Greek: it is an interpolation that, significantly, the English language foists on the translator.)

The Greek word *syllapsies* – here translated 'graspings' – seems, again according to Kahn, to suggest several ideas: something 'grasped' (perhaps suggesting sudden comprehension); something that brings elements together; and fertility (Aristotle uses the word to mean the sexual generation of life). It is hard to overestimate the richness of this fragment. It says many things at once: that a deep understanding of the nature of reality comes in glimpses, or graspings – moments of insight; that, in that insight, all is neither simply single, nor simply manifold, neither simply whole, nor simply not whole, neither simply like nor simply unlike, each thing working with, and by the same token working against, the others; that the One and the Many bring one another forth into being, together *generating* the reality that has this structure at its core; and that despite (or, in light of all this, perhaps because of?) the nature of this multiplicity, all is held together in a *syllapsis*: the only word here not to be paired with an antithesis. And the whole saying is in itself a *syllapsis* – a gathering which, in its fertility, births a *syllapsis* – a moment of dawning insight in us.

Much of what we know of Heraclitus' sayings comes in reports at

---

6 Nietzsche 1962 (68). Heraclitus was, according to Karl Popper, 'a thinker of unsurpassed power and originality' (1994, 16).

7 Heraclitus fr LXXVIII [Diels 51, Marcovich 27] (trans C Kahn).

8 The usual word in Attic Greek is ἁρμονία (*harmonia*); Heraclitus uses the Ionic form ἁρμονίη (*harmoniē*). I have followed Kahn's preference for Heraclitus' usage, as I have in his preference for Heraclitus' Ionic σύλλαψις (*syllapsis*) over the Attic σύλληψις (*syllēpsis*).

9 Kahn 1979 (196–200).

10 Heraclitus fr CXXIV [Diels 10, Marcovich 25] (trans C Kahn).

second-hand, several of them from Plutarch. In one, Plutarch aims to clarify what Heraclitus is believed to have said:

> According to Heraclitus one cannot step twice into the same river, nor can one grasp any mortal substance in a stable condition, but by the intensity and rapidity of change it scatters and again gathers. Or rather, *not again nor later but at the same time* it forms and dissolves, and approaches and departs.[11]

Contemporary physics, as so often, confirms metaphysics. Thus, though it might be thought that a wave and a particle were mutually exclusive forms of manifestation – one possible at one moment, the other at another – light can be imaged as wave and particle simultaneously.[12]

This emphasis on simultaneity applies not just to individual experiences, but to whole categories of existence. Thus we are used to thinking of the individual and the general, the temporal and the eternal, the embodied and the disembodied as exclusive pairings; whereas they are not only inclusive, but – as it was possibly Goethe's greatest insight to see – are present *simultaneously* in one another. They are found, not by turning one's back on the supposed opposite, but by going more deeply into it. Thus the general is found *in* the individual, the eternal *in* the temporal, the spiritual *in* the embodied. This tension is creative, generative.[13]

A tension between opposites is at the heart of all creativity, not just in maths or science. For example, 'true poetry, complete poetry, consists in the harmony of contraries', wrote Victor Hugo: I will come to that shortly.[14] 'Good poetry is usually written from a background of conflict', wrote William Empson, himself no mean poet and the author of *Seven Types of Ambiguity* – seven types, in other words, of creative tension.[15] Donne, whose poetry delights in ambiguities and paradoxes, warned his friend Sir Henry Wotton that paradoxes are generative. 'They are nothings', he wrote, in a spirit of deliberate paradox, 'therfore take heed of allowing any of them least you make another'.[16]

Nothings, then, that *breed* – anticipating a richly suggestive paradox of Heidegger's: '*das Nichts selbst nichtet*'. In my view, the tetchy responses of analytic philosophers to Heidegger's words merely advertise their own limited thinking.[17]

The phrase is from Heidegger's 'Was ist Metaphysik?' a lecture delivered in 1929; it is sometimes translated as 'nothing noths'. There is no such verb in English, of course, but there was no such verb as *nichten* in German, either, until Heidegger decided to employ it. The phrase is notoriously difficult, but one way of understanding it that makes sense to me is this. Nothing, like Being, is no thing. Neither is it the mere *absence* of a thing: it is a subject of action, Heidegger

---

11 Heraclitus fr LI [Diels 91, Marcovich 40c³] (trans C Kahn).

12 Piazza, Lummen, Quiñonez *et al* 2015. Experiment shows an ensemble that has simultaneous wave-like and particle-like properties, but no specific particle was observed to behave as a particle and wave at the same time. But then the concept of a specific particle maybe simplistic: see Hut [undated], p 1006 below.

13 It is also what Kriti Sharma noticed about the interdependent nature of living organisms, both with one another and their environment: they and their environment are *simultaneously* constituted, not by a process of sequent alternation (see p 453 above).

14 Hugo, Preface to *Cromwell*, §60.

15 Empson 1949 (xiii).

16 Cited at Simpson 1948 (316).

17 For a perfect example of what I mean, see Carnap 1931.

implies (it positively 'noths'). There is nothing to which we can in any way liken nothing, but if there were anything, it would be, like Being, a process (in this case, a process of negation, whereas Being is a process of affirmation). Hence the formation of a verb from what looks like a noun. Nothing is not just some thing that fails to presence – an absence – but a certain irreducible element in whatever exists, on a par with, and complementary to, Being itself; not passive and sterile, but having energy of its own that takes part in the coming into being of whatever is. It is not an 'end', in the sense of the necessarily vacant termination of a linear process, but the other and necessary 'end' of the dipole that is Nothing and Being together.

If before there was Being there was Nothing, nothing could no longer be thought of as merely the *absence of something else*, because that implies, precisely, a something else: something outside itself to ground it, and there *was* no something else. Nothing can only ground itself in itself. It therefore fulfils itself through its noth-ing – the noth-ing of no-thing (as in Heidegger's 'nothing noths'), a self-denial: *das Nichts selbst nichtet* – nothing noths *itself*. (In Chapter 28 I will look at a creation myth that expresses this self-denial in an astonishingly vivid way.)[18] A negative × a positive just leads to another negative; but a negative × a negative leads to a positive. And so it is that Being begins to become.

Can anything co-exist with its *negation*? In Chapter 16 we saw that zero was first conceived by Brahmagupta in precisely this fashion: zero was a particular kind of coincidence of opposites, the presence of zero was the presence of x combined with the presence of its negative, -x. Notice also, at the *next* level, that zero's being a presence does not stop it being an absence; each is a different way of understanding the same element. Being and non-being coincide. As Hegel puts it, 'Being and nothing are the same; *but just because they are the same they are no longer being and nothing,* but now have a different significance'.[19]

Notice that they have been taken up into a synthesis that is doubly generative: not just of a new whole, which is what one sees at first; but of a transformation, for that very reason, of the elements that had gone to form the new synthesis. The process is reverberative between the 'parts' and the whole.

Although a thing and its opposite, or a thing and its negative, are customarily thought of as separate warring entities, they are mutually sustaining, inseparable, and intertwined. You cannot have heat without cold, or brightness without darkness. As Alan Watts pointed out, we often appear to believe we can keep the mountains and get rid of the valleys.[20] The opposite ends of a simple stick have to be present, not just at the same time, but as part of the same single phenomenon: a stick. A road must at the same time be the road up

18 See pp 1248–50 below. But a self-sacrificing God is a universal insight, and is most familiar in the Christian *mythos*.

19 Hegel 1969, §187 (emphasis in original).

20 Watts 1989 (35).

and the road down. Just as the road that is a road up is necessarily a road down, everything looks different depending on our viewpoint. A river changes, and in so doing remains the same. A boundary can become the pivotal point of connexion (since boundaries are necessary to give identity, and thereby to make *connexion* possible at all); or it can become the point of separation, where there is no longer reciprocal influence and interaction. It is equally one and the other: now you see it, now you don't. And depending on your attention you may help to *make* it one or the other. We have one such enigma at the core of our being, the corpus callosum: in connecting it separates, and in separating it connects. A passage in the Upanishads appears to express an intuitive awareness of this: 'In the space within the heart lies the controller of all ... He is the *bridge* that serves as the boundary to keep the different worlds *apart*.'[21]

The inhibitory action of the corpus callosum enables the human condition. Delimitation is what makes something exist. Friction, for example, the very constraint on movement, is also what makes movement possible at all. In its excess, true, we are immobilised; yet so we are in its absence. There is nothing to push against. Resistance can put the brakes on motion, or cause motion; it can prevent or cause change. It presents us with an obstacle, and thereby forces us to shift our point of view; it helps us shift the plane of focus, so that we see something new. In itself, resistance is neither necessarily good nor necessarily bad – just necessary.

Hegel was, in my experience, right to hold that

> antinomies ... appear in all objects of every kind, in all conceptions, notions and Ideas ... every actual thing involves a coexistence of opposed elements. Consequently to know, or, in other words, to comprehend an object is equivalent to being conscious of it as a concrete unity of opposed determinations. The old [Western] metaphysic, as we have already seen, when it studied the objects of which it sought a metaphysical knowledge, went to work by applying categories *abstractly* and *to the exclusion of their opposites*.[22]

The word 'abstractly' here could best be read as 'left hemisphere fashion'; hence naturally leading to the exclusion of opposites.[23] With this insight of Hegel's, Western philosophy caught up at last (both cumbersomely and belatedly) with what Heraclitus had already seen in the sixth century BC, and Taoism had been teaching for two thousand years.

For opposites to co-exist, they clearly cannot cancel or annul one another, but must rather give rise to something new: a form of *harmoniē*, as in Heraclitus' lyre.[24] But note that this harmony is not to be taken merely in the ordinary musical sense of concord, but in

---

[21] *Upanishads*, trans S Radhakrishnan, 1953 (27): *Brihadaranyaka*, iv, §4, 22 (emphasis added). Such boundaries are well-imaged by the ear drum, which separates inner from outer ear, but also connects them: its purpose is transmission of sound, and without it, the sound would not only not be transmitted, but could not arise. See Cilliers 2008 (48).

[22] Hegel 1975, Part I, §48; emphasis added.

[23] Throughout this passage, the 1874 edition had used the word 'contrary' for 'opposite' (*entgegengesetzt*). The revision made in 1892 seems to me preferable.

[24] See p 817 above.

the sense of *tonos*, tautness, from which we get our word 'tone'. If the opposing forces in lyre or bow simply annulled one another, the string would go slack – no 'tonus' – and nothing, no flight of notes, no arrow's flight – could come from either. This is also, by the way, what is intended by the Golden Mean: not a flabby compromise, but a position in which taut synergy produces a dynamic equipoise. For a good apple pie you need both tart apples and honey, both sourness and sweetness, not just apples that are bland.

Those people in whom balance is achieved merely by 'toning down to an unattractive equilibrium' are very different from those who achieve a living harmony, writes Schleiermacher:

> For this frequent phenomenon which so many value highly, we are not indebted to a living union of both impulses, but both are distorted and smoothed away to a dull mediocrity in which no excess appears, because all fresh life is wanting. This is the position to which a false discretion seeks to bring the younger generation ... Elements so separated or so reduced to equilibrium would disclose little even to men of deep insight ...[25]

We often represent the Golden Mean as a midpoint on an adynamic line, a position that understandably does not appeal to some people, especially perhaps the young, because they are still orienting themselves in the world by opposition to something. We must see the truth embodied in Heraclitus' image of the bow or lyre. William Blake wrote: 'Without contraries is no progression. Attraction and repulsion, reason and energy, love and hate, are necessary to human existence.' But he also wrote: 'There is a Negation, & there is a Contrary: The Negation must be destroyd to redeem the Contraries.' Negation here means what I would call annulment: it must be replaced by the energy of contraries, *neither* of which should be annulled, but both of which should be, by contrast, maximally present.

CONTRARY TO REASON?

The word 'coincidence' has fallen on hard times. 'What a coincidence!' we say when we meet someone who drives the same make and model of car. This is one reason I use the words *coincidentia oppositorum* in preference: the Latin phrase is owed to the fifteenth-century polymath Nicholas of Cusa, and I will come back to his use of it in a later chapter. It was later popularised through the work of Jung. What I want to emphasise for now is that 'coincide' here means more than that opposites happen to look like one another, even to cohere, to concur, or to be in accord, though those meanings are present, too: it means that they 'fall together', like the superposition of two images which, when projected on a screen, overlap precisely

[25] Schleiermacher 1994 (3–5). NB: this, the standard English translation, is taken from the third edition of Schleiermacher's text of 1831; this passage does not exist in quite this form in the 1799 edition.

to form a new image. That they can do so is not contrary to reason, though according to a narrow logic a thing and its opposite cannot both be true: in the machine sense it 'does not compute'.

Fortunately we understand far more than any machine (however much information it might hold) ever could. The presence or absence of contradiction, as the great mathematician Pascal said, is no basis on which to judge truth.[26] Kant, who synthesised rationalism and empiricism, was compelled to accept that there are important antinomies – that is to say, contradictory conclusions each of which seems an inevitable inference from logic and reason.[27] His four antinomies concern the infinity or otherwise of space and time; the existence or not of parts and wholes; the existence or otherwise of causal determinism; and the existence or not of a necessary being. These antinomies, if, as Kant believed, they are irresoluble, strike at the very root of any philosophical belief in a world that is both rational and mind-independent. However, another way of looking at antinomies is exemplified in the work of the philosopher and theologian Sergei Bulgakov, who does not see a problem so much as an insight. As he put it in his book *Unfading Light*: 'Antinomy does not imply a mistake in reasoning or the overall falsity of a given epistemological misconception, which can be clarified and thus eliminated. Entirely legitimate antinomies are inherent to reason.'[28]

Bulgakov was thinking of the realm of the infinite, a topic I approached in Chapter 16. But even in the finite realm, as Bergson points out, concepts 'generally go in couples and represent two contraries'. He is thinking of, to take a basic example, the way a tree is represented by science as it were from without, and the experience of the tree from within consciousness: the two coincide – overlap on the screen – perfectly, like your face superimposed on a radiograph of your skull, but mean quite different things:

> There is hardly any concrete reality which cannot be observed from two opposing standpoints, which cannot consequently be subsumed under two antagonistic concepts. Hence a thesis and an antithesis which we endeavour in vain to reconcile logically, for the very simple reason that it is impossible with concepts and observations taken from *outside* points of view to make a thing. But from the object, seized by intuition, we pass easily in many cases to the two contrary concepts; and as in that way thesis and antithesis can be seen to spring from reality, we grasp at the same time how it is that the two are opposed *and* how they are reconciled.[29]

Note we can begin with the description in terms of physics, but could never progress from that to the experiential tree; whereas we can begin with the experience and later incorporate within it the

---

26 Pascal 1976, §384 (Lafuma §177) : « Contradiction est une mauvaise marque de vérité. Plusieurs choses certaines sont contredites. Plusieurs fausses passent sans contradiction. Ni la contradiction n'est marque de fausseté ni l'incontradiction n'est marque de vérité » – 'Contradiction is a poor criterion of truth. Many things that are certain are contradicted; many things that are false pass without contradiction. Contradiction is not a sign of falsehood, nor the lack of contradiction a sign of truth.'

27 Equally Abelard, in the twelfth century, wrote a treatise called *Sic et Non*, which means literally 'Yes and No', in which he set out 158 propositions from theology which could be logically argued for or against. It is thought to have been intended to train students in logic.

28 Quoted in Florensky 2004, where the phrase *vpolne zakonomerniye* is translated as 'orderly'. I have substituted the more literal translation.

29 Bergson 1912 (39–40; emphasis added).

physics. The right hemisphere can incorporate the left's take, but the left cannot incorporate that of the right. The mechanistic vision can come only from experience, even if it is an experience from which much has been excluded; the experience of the tree can never emerge from the mechanistic vision.

On top of this, there is the reality that, from any one standpoint, only a partial vision is possible. The opposition of 'standpoints' is a way of speaking of difference wrought by context. As Whitehead put it, 'there are no whole truths; all truths are half-truths. It is trying to treat them as whole truths that plays the devil'.[30] Each truth conceals another, opposing, truth, that becomes apparent as soon as we move from the abstract to a real world context. Moreover we need *both* the vision that reveals separation *and* the vision that reveals union. According to the twentieth-century Zen master Shunryū Suzuki: 'We have to understand things in two ways. One way is to understand things as interrelated. The other way is to understand ourselves as quite independent from everything.'[31] Once again Whitehead comes to mind: 'To have seen it from one side only is not to have seen it.'[32]

The problem arises when we are so anxious to avoid contradiction that we are tempted to dismiss, or 'prove' to be wrong, one of the alternatives, both of which are needed. Many contemporary scientists and most analytic philosophers (it is intrinsic to their philosophical *credo*) are in servitude to this 'either/or' cast of mind. Alfred Kazin wrote that we should

> trust to the contradictions and see them *out*. Never annul one force to give supremacy to another. The contradiction itself *is* the reality in all its manifoldness … the more faithful [man] is to his perception of the contradiction, the more he is open to what there is for him to know … a contradiction that is *faced* leads to true knowledge.[33]

In a similar vein, Jacob Needleman wrote: 'Stay with the contradiction. If you stay, you will see that there is always something more than two opposing truths. The whole truth always includes a third part, which is the reconciliation.'[34]

Opposites genuinely *coincide while remaining opposites.* Some philosophies tend to collapse into the monism that opposites are identical; others into the dualism that opposites remain irreconcilable and are merely, at most, juxtaposed.[35] The important perception is that opposites not only co-exist, but give rise to and fulfil one another ('*sunt complementa*'), and are conjoined (like the poles of a magnet) without any intervening boundary, while nonetheless remaining distinct as opposites. And indeed the more intimately they are united, the more, not the less, they are differentiated.

30 Whitehead 1954 (19).

31 Suzuki 1999 (66).

32 Whitehead, as reported by his wife, Evelyn Wade, in Whitehead 1954 (19).

33 Kazin 2011 (217): entry for 2 June 1957 (emphases in original).

34 Needleman 2016 (20).

35 An example of the first might be the Hindu philosophy of Saṅkhāra; of the second, the biblical and Islamic doctrines of the transcendence of God: see Cousins 1978 (20).

THE CIRCULAR DIALOGUE OF LOVE WITH STRIFE

What we get when we become unaware of the neglected – that is to say, opposing – truths inherent in our position is extremism. We yield power to the dark side by ignoring it: by acknowledging it we free ourselves from its stranglehold.

It was the painter and architect Friedensreich Hundertwasser's view that 'the straight line leads to the downfall of mankind'.³⁶ This was not an idle reflection. He returned to it with a vengeance in 1985:

> In 1953 I realized that the straight line leads to the downfall of mankind. But the straight line has become an absolute tyranny. The straight line is something cowardly drawn with a rule, without thought or feeling; it is a line which does not exist in nature. And that line is the rotten foundation of our doomed civilization ... The straight line is atheistic and immoral. The straight line is the only sterile line, the only line which does not suit man as the image of God. The straight line is the forbidden fruit. The straight line is the curse of our civilization. Any design undertaken with the straight line will be stillborn. Today we are witnessing the triumph of rationalist know-how and yet, at the same time, we find ourselves confronted with emptiness. An aesthetic void, desert of uniformity, criminal sterility, loss of creative power. Even creativity is prefabricated. We have become impotent. We are no longer able to create. That is our real illiteracy.³⁷

No longer able to create: like Flint, with his trusty arrow.

Heraclitus' (considerably younger) contemporary Empedocles thought there were two opposing and equal forces, love (φιλότης) and strife (νεῖκος); in the presence of love only, or strife only, nothing could exist. These forces for union and for division, according to Empedocles, are present in the very stuff of all things, not just in their ultimate origin. And they are imaged as a circle, not a straight line.

Life is, after all, a dance to be celebrated, not a series of equations to be solved. And the dance, whatever may have evolved from it since, originated as a circle, based on a dipole that is not – and must never become – an opposition, but complementary, an act of consummation: that of man and woman. The circle represents both the finite and the eternal, since it has no end; it also represents that which moves and stays still. So Empedocles wrote:

> Thus insofar as they have learned to grow as one out of many,
> And inversely, the one separating again, they end up being many,
> To that extent they become, and they do not have a steadfast lifetime;
> But insofar as they incessantly exchange their places continually,
> To that extent they always are, immobile in a circle.³⁸

Yet linearity and circularity can co-exist, if what looks like a cir-

---

36 » *Die gerade Linie führt zum Untergang der Menschheit* « (from the 1954 catalogue of his first exhibition in Paris at the studio of Paul Facchetti, written in December 1953).

37 Hundertwasser 1985.

38 Empedocles, fr DK31: B 17.9–13.

cle from this point of view is actually a spiral, like an endless coiled spring viewed down its axis.

The idea of *complementarity* is foundational in Nature. So, for example, to turn one's back on the parts (the workings of the left hemisphere) and accept only the whole (the work of the right hemisphere) is not to 'get back to wholeness', because the whole is never an annihilation, but rather a subsumption, of the parts. The true whole exists precisely in this relationship, the tension between parts and an apparent whole. You recognise the earth's true essence, writes Schelling,

> only in the bond by virtue of which it eternally asserts its unity as the multiplicity of its things and, conversely, asserts this multiplicity as its unity. And it's not that you think that, apart from this infinity of things to be found in this earth, there is another one, which is the unity of these things; rather *the same* that is the multiplicity is also the unity, and *the same* that is the unity is also the multiplicity, and this necessary and indissoluble One of unity and multiplicity is in itself what you call its existence ... Existence is the conjunction of a being as One, with itself as a Many.[39]

The conjunction of the One and the Many is so important that it merits consideration in greater detail in the next chapter. It parallels that very necessary synthesis which is performed by the right hemisphere (if it is given freedom to do so) of its own awareness of wholeness together with the work of division which is provided by the left hemisphere. This is a synthesis that results in a new and enriched whole. Although it may sound as if this process is a linear sequence, it is in fact a continuous, reciprocal and immediate exchange, giving rise to a simultaneous fusion, yet always requiring the right hemisphere to oversee the process. This depends of course on the interhemispheric relationship functioning properly – and on Flint remaining under the superintendence of his brother.

Oppositions are the ground of energy, if properly seen, as in the tautness of the bowstring: the primal form of opposition is that between two opposed tendencies, one for division (strife) and one for union (love). Civilisations also flourish when they remember this, and fail when they forget. One of the strands of thought that flowed into and made the world view of the Renaissance so rich was the Kabbalah, a discovery of the humanists, spearheaded by the young polymathic scholar Giovanni Pico della Mirandola; and with it, the *coincidentia oppositorum* became almost an axiom of Renaissance thought, literature and art, imaginatively leavening the linear logic that had so deadened mediaeval scholasticism.

In the Kabbalah, the structure of human faculties takes the form of a tree with a right-hand side and a left-hand side; humanity's task

---

39 Schelling 1806 (56): » ... ihr wahres Wesen erkennst du allein in dem Band, kraft dessen sie ihre Einheit ewig als die Vielheit ihrer Dinge, und hinwiederum diese Vielheit als ihre Einheit setzt. Du stellst dir auch nicht vor, daß es außer dieser Unendlichkeit von Dingen, die in ihr befindlich sind, noch eine andere Erde gebe, welche die Einheit dieser Dinge ist, sondern dasselbe *was die Vielheit ist, dasselbe ist auch die Einheit, und was die Einheit ist, dasselbe ist auch die Vielheit, und dieses nothwendige und unauslösliche Eins der Einheit und Vielheit selbst in ihr nennst du ihre Existenz ... Existenz ist das Band eines Wesens als Eines mit ihm selbst als einem Vielem.* « (emphases in original).

is to integrate them, both laterally and vertically.[40] Specifically it is held that the mind is made up of two faculties: wisdom (*chochmah*) on the right, which receives the *Gestalt* of situations in a single flash, and understanding (*binah*), opposite it on the left, which builds them up in a replicable, step-by-step way. *Chochmah* and *binah* are considered 'two friends who never part', because you cannot have one without the other. *Chochmah* gives rise to a force for loving fusion with the other, while *binah* gives rise to judgment, which is responsible for setting boundaries and limits.[41] Their integration is another faculty called *da'at*, which is a bit like Aristotle's *phronesis*, or even *sophia* – an embodied, overarching, intuitive capacity to know what the situation calls for and to do it. What is more this tree is a true organism, each 'part' reflected in, and qualified by co-presence with, each of the others.

Schleiermacher thought that, inasmuch as the business of the cosmos is the differentiation of unity, every phenomenon comes from the working together of two opposing tendencies. Each of these emanations of the cosmos 'can only be actualized in two hostile yet twin forms, one of which cannot exist except by means of the other'. This includes the human soul, which, 'as is shown both by its passing actions and its inward characteristics, has its existence chiefly in two opposing impulses'. One impulse was towards individuation of the self, the other towards surrender of the self to union with the whole. He thought that different beings found different points along this continuum, so that 'the perfection of the living world consists in this, that between these opposite ends all combinations are actually present in humanity.' Yet, he asked, 'how are these extremes to be brought together, and the long series be made into a closed ring, the symbol of eternity and completeness?'[42]

Empedocles had taught that these two tendencies did form an endless *ring* of creation, as we have seen: 'Twofold is what I shall say: for at one time [the elements] grow to be only one out of many, at another time again they separate to be many out of one.'[43] What is united is to be divided, what divided to be united. The process involves cyclical returns. It follows from this cyclical nature that if you go far enough in any one direction you reach not more of what you desired but its opposite: go East and you eventually reach the West. And it follows that both of two opposites are simultaneously present, and need to be so, just as East and West are simultaneously present on the compass and need to be so, not just to navigate the world, but to have a world to navigate.

The discussion of the coincidence of opposites has so far been metaphysical, and may for that reason have seemed somewhat remote from daily life. But if I am right that this principle is generative of reality, it will also be generative as much at the level of everyday

---

[40] In this paragraph I draw on a private communication from Eric Kaplan, to whom I am grateful.

[41] These are known as *chesed* and *gevurah* respectively.

[42] Schleiermacher 1994 (3–5).

[43] Empedocles, fr DK 31: B 17.1–2

experience, both physical and mental. Let us turn for a moment to look at just such phenomena, beginning with the more physical, and moving on to the more psychical.

HORMESIS: UNEXPECTED CONSEQUENCES

Scientists involved in the Biosphere 2 project, which created the largest enclosed ecological system on earth, were puzzled by the fact that trees within the project repeatedly failed to achieve maturity before they fell over. Later, they realised that trees needed wind in order to grow strong. Exposure to winds causes the growth of 'stress wood', which is the core of the tree's strength and integrity. Winds also cause the root system to strengthen. Nietzsche got there before them, and immediately relates it to spiritual growth:

> Examine the lives of the best and most fruitful people and peoples and ask yourselves whether a tree which is supposed to grow to a proud height could do without bad weather and storms; whether misfortune and external resistance, whether any kinds of hatred, jealousy, stubbornness, mistrust, hardness, greed, and violence do not belong to the *favourable* conditions without which any great growth even of virtue is scarcely possible?[44]

In a similar physical phenomenon, repeated stress increases bone density and strength, and while excess stress can of course be damaging, too little is also harmful: the 'microgravity' environment in which astronauts operate would quickly lead to osteoporosis were steps not taken deliberately to counter the effect.

The name hormesis has been given to the phenomenon whereby a substance or process that is damaging to an organism above a certain level may have an opposite effect at lower levels. This reflects Paracelsus's dictum that all substances are poisons, depending only on the dose. Many notoriously poisonous plants may be medicinal in low doses: digitalis, a highly toxic extract of foxglove, has been used to treat heart failure since the eighteenth century; atropine, an extract of deadly nightshade, is a treatment for a seriously debilitating nervous disease, myasthenia gravis; arsenic has been used with some success to treat a host of illnesses for over two thousand years, and is still used in the treatment of syphilis and cancer.

Hormesis was first noted with respect to radiation, where the results can be dramatically opposed to one another. While radiation was clearly known to be a cause of cancer, it turned out that extremely low doses could stimulate DNA repair and delay cancer in laboratory animals.[45] Consonant findings have been found in humans, with very low exposure being superior to no exposure.[46] Some living organisms, including bacteria and plants, cannot grow without background-level radiation.[47] A review of the literature on radiation

44 Nietzsche 2001, §19 (43; emphasis in original).

45 Luckey 1991.

46 Vaiserman 2010.

47 Castillo, Schoderbek, Dulal *et al* 2015.

hormesis in 2019 concluded that 'practically all immune parameters are beneficially influenced by all forms of low-dose radiations.'[48]

The effect also applies to toxic chemicals in small amounts. Insecticides have been shown to stimulate growth of insects as well as plants, and fungicides can stimulate growth of fungi; low-level exposure to a stress agent that is harmful at higher levels can extend life in mice and fruit flies.[49] Dioxins are very widely known to be highly toxic, but have been shown in animal experiments at low doses to protect against liver tumours in rats.[50] Similar findings concerning dioxins have been confirmed in humans.[51] 'Hormesis', according to one group of experts, 'is *fundamental to evolution and highly generalizable*.'[52]

(I ought perhaps to emphasise that none of this should, of course, be interpreted under any circumstances as a reason not to be extremely watchful about pollution of the environment by radiation and chemicals. The point I am making *depends* on the indisputably toxic nature of the agents at anything other than minimal levels.)

In a more homely context, either complete physical inactivity or bouts of strenuous exercise increase the risk of infection, while moderate exercise strengthens the immune response.[53] Architects are now being asked to make designed environments somewhat more *difficult* to move around in the interests of human health. As Coleridge noted, 'the thing that causes instability in a particular state, of itself causes *stability* – as for instance wet soap slips off the ledge, detain it till it dries a little & it *sticks*'.[54]

Naseem Nicholas Taleb's book *Antifragile* is an exploration of the many ways in which by seeking to make a system less vulnerable, we succeed only in making it more so.[55] The mistake, he suggests, is that we seek to make it 'robust' and thereby inflexible; this leaves it open to catastrophic collapse under certain, non-vanishingly rare, circumstances. His message is that we should rather seek to make it 'antifragile', which paradoxically includes allowing certain vulnerabilities that make it capable of withstanding large shocks. Manageable small setbacks enable the whole to adapt, endure and evolve. Dinosaurs were robust, but when the right (or wrong) set of circumstances prevailed, they were doomed. The *locus classicus* is that of a type of forestry management so successful in avoiding small fires that the forest is vulnerable to total destruction in the (ultimately inevitable) event of a fire: in the natural course of things small-scale fires create firebreaks that, though they may locally appear to be a weakness, protect the future of the whole.

Turning to the more intangible realm of morality and human flourishing, we find the structure similar. Nothing good is achieved without a degree of adversity being overcome: health, resilience, courage, skill, knowledge, virtue and wisdom are no exceptions. As

---

48 Csaba 2019.

49 Son, Camandola & Mattson 2008; Hayes 2007; Le Bourg 2009.

50 Kociba, Keyes, Beyer *et al* 1978; Calabrese, Baldwin & Holland 1999.

51 Tuomisto, Pekkanen, Kiviranta *et al* 2006.

52 Calabrese & Mattson 2017 (emphasis added).

53 Radak, Chung, Koltai *et al* 2008.

54 Coleridge 1957, §1017 (emphases in original).

55 Taleb 2013.

Seneca is famed to have put it, 'a gem stone is not polished without friction, nor is a man without adversities'. It seems rational to strive to make life easy, to minimise effort: yet exerting oneself in a cause, neglecting one's ease for the sake of others, finding solutions to difficult problems, and overcoming resistance, all make for fulfilment, while a goal that is too easily achieved fails to satisfy; and what we come by too easily may lack meaning or value. In the words of the Midrash, 'one thing acquired through pain is better for man than one hundred things easily acquired.'[56]

FURTHER PROBLEMS WITH LINEAR THOUGHT

In Chapter 15, I reminded the reader of the Red Queen's advice to Alice. Given the circle of love and strife, the further we push single-mindedly in one direction, the more we achieve its opposite. Quoting Cicero, Hegel wrote: '*Summum jus summa injuria* [the extreme of the law is extreme injustice] which means that to drive an abstract right to its extremity is to do a wrong':

> In political life, as every one knows, extreme anarchy and extreme despotism naturally lead to one another. The perception of dialectic in the province of individual ethics is seen in the well-known adages: 'Pride comes before a fall'; 'Too much wit outwits itself'. Even feeling, bodily as well as mental, has its dialectic. Everyone knows how the extremes of pain and pleasure pass into each other: the heart overflowing with joy seeks relief in tears, and the deepest melancholy will at times betray its presence by a smile.[57]

A principle that is extended too far, without respect to the opposite that is always inherent in it, may turn into the very thing that is not only undesired, but is being denied. This is an understanding we would appear to have lost almost completely, and which, if we had kept it in mind, might have preserved us from many of the worst follies of contemporary culture.

In order to be natural, we must not try to be so; if we wish happiness, it is fatal to pursue it; freedom requires self-discipline; sometimes we must be cruel to be kind. Fear of pain leads to pain and fear: as Montaigne pointed out, 'anyone who is afraid of suffering suffers already of being afraid'.[58] So that they may be impervious to indoctrination, children must be indoctrinated in the principles of reason. If you want peace, you must prepare for war; to achieve freedom, you must bind it by laws; yet laws blindly applied or enforced with excessive zeal are a common source of injustice.[59] To which one well might add, as William James points out, that 'most human institutions, by the purely technical and professional manner in which they come to be administered, end by becoming obstacles to the very purposes which their founders had in view.'[60]

---

56 Midrash 1977, vol 1 (370).

57 Hegel 1975, Part 1, note to §81.

58 Montaigne 1991: Bk III, ch xiii, 'Of experience' (1243).

59 *ibid* (1216).

60 James 1909a (96).

On another level altogether, if depression does not destroy you, it can be a positive experience – one, as I too readily understand, associated with much suffering, but nonetheless positive in what it reveals to the sufferer. The poet William Blake, who said that 'without contraries there can be no progression', even thought that there must be sorrow in heaven if there was to be joy.[61] He also wrote:

> It is right it should be so;
> Man was made for Joy & Woe …
> Joy & Woe are woven fine,
> A Clothing for the Soul divine;
> Under every grief & pine,
> Runs a joy with silken twine.[62]

The imagination thrives on the implicit, and is deadened by the explicit. The explicit is single: the implicit is a coming together of opposites, and requires the simultaneous presence and absence of whatever is being gestured towards. We may become more aware of something if it is partially eclipsed, while a pure manifestation would not have achieved its end. If you are making a film, and wish to evoke the stillness of a warm day in the country, do you leave the soundtrack completely blank? No, you introduce the faint and intermittent buzzing of a fly. In an interior scene of tranquillity, the faint, slow, deep tick-tock of a grandfather clock intensifies both the peace and, as may seem oddly, the sense that time has come to a standstill. When the top of a mountain is hidden in cloud we appreciate its full immensity better than when the whole is in view. Limitation can intensify the sense of infinity. One of the best known poems in the Italian language is Leopardi's '*L'infinito*', in which he describes how a break of trees that partly obscures the view of an open landscape floods his mind with the sense of the infinite, and the stirring of the branches enhances the sense of stillness:

> Always dear to me has been this lonely hill,
>   and this line of trees which, from so much
>   of the furthest horizon, hides my view.
> Yet as I sit and gaze, in my thoughts
>   I conjure up boundless spaces far beyond it,
>   and superhuman silences,
>   and deepest quiet – until my heart
>   almost grows afraid. And as I hear the wind push
> Through the trees, I cannot help setting its sound against
>   that infinite silence; and the eternal envelops me
>   with the thought of seasons long past
>   and of the living present and all its sounds.
> Then in this immensity my thought goes under,
>   And sweet it is to me to drown in such a sea.[63]

---

61 In Crabb Robinson 1869, entry for 10 December 1825: 'Though he spoke of his happiness, he spoke of past sufferings, and of sufferings as necessary. "There is suffering in heaven, for where there is the capacity of enjoyment, there is the capacity of pain."' As quoted in Symons 1907 (259). Cf Nietzsche: 'What if pleasure and displeasure are so intertwined together that whoever *wants* as much as possible of one *must* also have as much as possible of the other – that whoever wants to learn to "jubilate up to the heavens" must also be prepared for "grief unto death"?' (Nietzsche 2001: Bk I, §12, 37–8). Nietzsche refers to Goethe's *Egmont*, Act III, sc ii.

62 Blake, 'Auguries of Innocence', lines 55–6 & 59–62.

63 Leopardi, 'L'infinito', trans Nigel McGilchrist, who comments: 'It is conventional to translate *siepe* as a hedgerow, which is exactly what it means in current Italian usage. But in older garden-design tracts *siepe* is often used to mean a "break of trees". Since the land drops sharply below the spot in Recanati where Leopardi is said to have composed the poem, only a line of tall trees rising from further down the slope, rather than a hedgerow, would effectively occlude his view.'

Pursuing something, then, can, for better or worse, lead to an outcome that seems paradoxical: we achieve the opposite. But we are also very often required to embrace opposites *at the same time*: as Kazin put it, 'The contradiction itself *is* the reality in all its manifoldness'. So Empson wrote: 'Extremely often, in dealing with the world, one arrives at two ideas or ways of dealing with things which both work and *are needed*, but which entirely contradict one another.'[64] We all experience this at every level from the most innocently trivial-seeming to the most sublime. We need universality and particularity, precision and flexibility, restriction and openness, freedom and constraint – simultaneously. Everything flows from the pairing. We 'lose ourselves', and consequently find ourselves, in music, dance, or contemplation of a beautiful painting or landscape.

However we look at it, the world forces us to acknowledge that opposites require to be satisfied together: no single goal can be successfully pursued without due acknowledgment, and indeed acceptance of, its contrary. To quote James again:

> Somehow life does, out of its total resources, find ways of satisfying opposites at once. This is precisely the paradoxical aspect which much of our civilization presents ... the way to certainty lies through radical doubt; virtue signifies not innocence but the knowledge of sin and its overcoming; by obeying nature, we command her, etc. The ethical and the religious life are full of such contradictions held in solution. You hate your enemy?—well, forgive him, and thereby heap coals of fire on his head; to realize yourself, renounce yourself; to save your soul, first lose it; in short, die to live.[65]

I have referred earlier to an innocence the other side of experience, a knowledge the other side of unknowing, a wisdom the other side of folly. 'The only simplicity for which I would give a straw', said the rather down-to-earth jurist, Oliver Wendell Holmes, 'is that which is on the other side of the complex – not that which never has divined it.'[66] Joseph Campbell writes:

> I think there are three states of being. One is the innocent expression of Nature. Another is when you pause, analyse, think about it ... Then, having analysed, there comes a state in which you're able to live as Nature again, but with more competence, more control, more flexibility.'[67]

This, I believe, reflects the proper reciprocal functioning of the two brain hemispheres.

In this sense, once more, the worse it gets, the better it gets. According to Jung,

> The grand plan on which the unconscious life of the psyche is constructed is so inaccessible to our understanding that we can never

---

[64] From Empson 1993 (46; emphasis added). Cf Schlegel 1991, §53: 'It's equally fatal for the mind to have a system and to have none. It will simply have to decide to combine the two'.

[65] James 1909a (98–9).

[66] Wendell Holmes 1961 (109).

[67] Campbell 1991 (10).

know what evil may not be necessary in order to produce good by enantiodromia, and what good may very possibly lead to evil.[68]

Enantiodromia means literally 'running in opposite directions'; Jung used the word to refer to the emergence, in the course of time, of the neglected opposite, the one of which we are unconscious. History is littered with examples of (at the time apparently) good actions that gave rise to ill effects, and (at the time apparently) disastrous events that yielded positive results. An instance: the elimination of wolves, the top predator in the ecosystem of Yellowstone Park in the 1930s, so unbalanced it that widespread destruction of the ecosystem ensued. Their reintroduction about 25 years ago has resulted in a tripling of the number of elk, their main prey; the sudden flourishing of the trees and shrubs that the elk ate (they formerly did not have to keep on the move and destroyed their own habitat); the return of beavers which used to be plentiful, but need willow trees to flourish; and the advance of many other species in what is called a 'trophic cascade'.[69] Nature testifies everywhere to enantiodromia.

Max Scheler wrote that 'it is originally our suffering of the resistance of the world to our vital drive impulses ... which sets everything going for us, including the very fact that one can be conscious of anything ... Everything stems from this original resistance, and the original subjective experience of suffering.'[70] This creative aspect of resistance is one to which I will need to return more than once. William James made a distinction between what he called the once-born, those who are naturally uncomplicatedly happy, and the twice-born, those who attain happiness only through enduring and surviving a state of abject misery. And he described the transformation of Tolstoy – a twice-born, if ever there was one – like this: 'The process is one of redemption, not of mere reversion to natural health, and the sufferer, when saved, is saved by what seems to him a second birth, a deeper kind of conscious being than he could enjoy before.'[71] Having suffered along the way *deepens* in some sense the experience. Viktor Frankl, who himself survived, and reflected on those who, like him, had survived, the death camps of the Nazis, recognised that suffering deepens meaning in life. In this sense, suffering is, obscurely, desirable, and yet absolutely not to be desired. *Contraria sunt complementa.*

THE ASYMMETRY OF THE *COINCIDENTIA OPPOSITORUM*

In the physics that makes the world possible, chaos and rigidity must be very finely balanced.[72] We need forces for stasis and conservation, and forces for flow and change: but they must work together. We can see this being played out in the task of balancing inherently

---

68 Jung 1953–79, vol 9i, §397.

69 This is a demonstration of the importance of certain 'keystone species' to a stable ecology. Such species are usually, but not inevitably, predators. The concept derives from ground-breaking work by the ecologist Robert Paine (1966).

70 Scheler 2008 (78).

71 James 1902 (157).

72 Scharf 2014.

contradictory or paradoxical elements such as being both united *and* distinct in spatial terms as well as both stable *and* changeable in temporal terms. However, as elsewhere, there is an asymmetry between the forces for division and the forces for union: ultimately they have to be united, not remain divided. The brothers must work together, as He Grasps the Sky With Both Hands sees, not against one another as Flint intends.

Actual occurrence involves the breaking of what is, considered in the abstract, symmetrical.[73] Small imbalances, differences among sameness, at all levels in nature make it work, starting with the initial inequality of matter and antimatter. The physicist Pierre Curie wrote in 1894 that

> ... certain elements of symmetry may coexist with certain phenomena, but they are not necessary. What is necessary is that certain elements of symmetry do *not* exist. *It is asymmetry which creates the phenomenon* ... The effects produced may be more symmetrical than the causes.[74]

(I will discuss asymmetry in the cosmos further in Chapter 24.) The principle for division and the principle for union need to be brought together, not divided. We need not *either* both/and *or* either/or, but *both* both/and *and* either/or.[75] We need not non-duality only, but the non-duality of duality and non-duality.

This is wittily encapsulated in a story I heard told by Rabbi Jonathan Sacks. A faithful man finds in the scriptures that Rabbi X said that a certain thing was true. Later he finds that Rabbi Y said that the very same thing was false. He prays for guidance: 'Who is right?' God answers: 'Both of them are right.' Perplexed, the man replies: 'But what do you mean? Surely they can't *both* be right?' To which God replies: 'All three of you are right.'[76]

In terms of the hemispheres it is once more not a symmetrical, but an asymmetrical, arrangement: not just between two dispositions (that of the left hemisphere and that of the right) towards the world, but between a disposition (that of the left) that sees the two dispositions as an antagonism that must ultimately lead to the triumph of one and the annihilation of the other, and a disposition (that of the right) that sees they need to be preserved together, neither being allowed to extinguish the other – *even though they are not of equal value*. One – the disposition of the right – overarches and takes into account the other; much as He Grasps the Sky With Both Hands not just protects Flint, but enables the fulfilment of Flint's contribution.

Here I want to turn to the remarkable book *Physical Spirituality: Changing the Paradigm* by Mike Abramowitz, a writer who approaches philosophical issues from a background in physics.[77] I can deal here only partially with the hypothesis of his book, but amongst

73 Stewart & Golubitsky 1992 (60).

74 Curie 1894 (400–1; emphasis in penultimate sentence in original): « ... certains éléments de symétrie peuvent coexister avec certains phénomènes, mais ils ne sont pas nécessaires. Ce qui est nécessaire, c'est que certains éléments de symétrie n'existent pas. C'est la dissymétrie qui crée le phénomène ... les effets produits peuvent être plus symétriques que les causes. »

75 For a penetrating insight into this logical 'paradox' in relation to the nature of Christ, see McCosker 2008 (esp 211).

76 This is a humorous retelling of a combination of two Talmudic passages (each only slightly different from the other). The original sources are from the Gittin and Eruvin, two volumes within the Talmud. Apparently a story similar to this is told in the 1971 film, *Fiddler on the Roof*. I am grateful to the late Rabbi Jonathan Sacks for this information.

77 The e-book (2018) is available free under a Creative Commons licence at physical-spirituality.neocities.org.

many points of interest he contrasts what he calls 'architective' with 'connective' interactions in physics.

Architective interactions use bonds to create persistent objects by disjunctive transformations: such are the chemical bonds that define molecular geometry. They are competitive for outcome, and result in something disappearing and something else taking its place: 'losing a contest is catastrophic for a bond'. Bonds are stable and resist change; but when they are forced to change, they do so 'stepping from one static architecture to another and in sequences of discrete reconfiguration events'. (They are characterised, in other words, by what I call 'stickiness', a core feature of the left hemisphere's take on the world.) Connective interactions, by contrast, allow the merging of entities, and are infinitely resolvable – waves being an example – where architectives are not. Connectives are 'fluid, highly susceptible to change, and do so in a smooth, unstepped motion.'

Architectivity and connectivity have a creative oppositional relationship. Architectivity constrains connectivity, but not vice versa. Yet 'architectivity attains its cosmic significance as a *contributor to the connectivity* of the cosmos rather than through its compulsive attempts to constrain connectivity.' At every scale, 'connectivity plays with whatever objects architectivity provides.' I cannot help seeing a similarity here with Whitehead's idea that potential 'plays with' – further creates by responding positively to – whatever it is that actualisation provides.

Architectivity is characterised by the fact that the motions of its constituent objects are always constrained, though due to this very fact their identity endures – or, equally, may be obliterated. None of this is true of connectivity. Architective aggregates exclude one another, can be defined and categorised with precision, display fixed hierarchies of rank, and can exercise precise control – in all of which respects, again, they form a contrast with connectives.

It will not have escaped the reader that there are clear parallels between what Abramowitz calls architectivity and the left hemisphere *modus operandi,* and what he calls connectivity and the right hemisphere *modus operandi*; and, moreover, with the idea that what the left hemisphere makes static and differentiated is taken up again into the right hemisphere's understanding of a living whole. I suspect that architectivity and connectivity reflect in their different ways, the same underlying structure. Both are necessary for creation.

Abramowitz also sees these patterns reflected, as do I, in the human world. Contemporary Western assumptions about reality are drawn, he says, from architectivity – for example,

> we unquestioningly accept that stasis is the natural 'rest state' of physical phenomena and that movement only arises when energy is imparted to an object that is otherwise naturally at rest.

But this, he argues, is at the very least an unwarranted assumption and may be profoundly misleading:

> a state of motion is at least equally entitled to being considered the natural state of things, one in which stasis only arises when constraints are imposed on objects that are naturally in motion.

Indeed: the evidence suggests a state of motion is more than equally entitled to be considered the 'natural' state.

More broadly, he writes that 'our cultures, history and traditions have been overwhelmingly shaped by an architective dominion'.[78] Cultures tend to accumulate architective knowledge, since it is the kind that can be

> precisely codified and stored, faithfully passed from generation to generation, built on and accumulated; while connective skills are not easily codified or handed down ... a personal lifetime of connective *nous* is usually buried with the individual.

Given this view, it is not surprising that he also contrasts 'purely connective religion' which tends to be fugitive, having less impressive power, but which has great spiritual depth – he mentions Sufism, Zen Buddhism, the Kabbalah, Tantric Hinduism and Christian mysticism – with architective religions possessed of dogmas, dominated by literalism, and enshrined in hierarchical structures. 'Rather', he writes, 'connective religions'

> see the parental myths as allegories pointing to a secret that is not knowable in any dogmatic sense and so must be alluded to by parable. This secret knowledge can only be attained by direct engagement with their spirits, so all indirect representations of them, including any iconic and dogmatic representations, even those of their parent religions, are considered to be a barrier to their direct revelation.

I have adverted to Abramowitz's intriguing insights into 'physical spirituality' not just because of the striking parallels with the hemisphere hypothesis – there are more that I have not had time to explore here – but because of his interesting perception that the balance between the two forces he describes depends on scale, in three respects.

First, bonds occur on a small scale, 'appear to become scarce at larger scales, and absent at very large scales. Connective interaction, on the other hand, is observable in abundance at every known scale.' Additionally, interactions based on physical shape, which is architective in nature, appear to be 'restricted to a window of scale residing between the molecular and the planetary ... It is not unreasonable to propose a window at a cosmic scale larger than the planetary in which all interaction is connective.'

---

78 That is certainly true of major civilisations. In at least some sustainable oral cultures, what gets passed down, stored, retained, includes myths with no fixed 'interpretation' – uncodifiable knowledge that is nevertheless essential for survival: knowledge that insists on developing, in the listener, the skill required for understanding it. (Thanks to Jan Zwicky.)

Second, although the connective spirit is manifest everywhere, 'its influence is comparatively small, especially inside pockets of architecture', because of the local dominance of architectivity wherever it is present.

Third, in the absence of such spatial limitations, 'connective phenomena offer an avenue of infinite subtlety in their capacity for infinite resolution and an avenue of infinite grandeur in their capacity for infinite extent. Infinity of resolution and extent are not available in architective contexts.'

Once again I would remind the reader that there is no reason for preferring an understanding from a narrow perspective to one from the broadest possible perspective: the phenomena the cosmos can give rise to ultimately tell us as much about the cosmos as the elements to which it may seem to be – but never can be – reduced.

Returning to the theme of the coincidence of opposites, I see here confirmation that fluidity at a cosmic level contains and does not annihilate, but creates out of, strong local forces that on their own tend away from flow and towards granularity. In the infinite case, and even in the very large case, the more yielding force of union most obviously takes up into itself and transforms the more locally prepotent forces for division, something which happens, though less obviously and with diminished power, even at the smallest scale.

We need both the architective and the connective. We need both division and union. We need both analysis and synthesis. We need both left and right hemispheres. This is clear. If I sometimes seem to emphasise the second unduly, it is for two reasons. In our reductionist culture it is the one of which we more urgently need reminding: a pragmatic reason. More importantly it is, like it or not, the 'senior partner' in their relationship: a metaphysical reason. The left hemisphere exists in service to the right. As long as this is respected, all goes well; when it is not, we court disaster. Each needs to be allowed its head: but at the end of the day the products of the first need to be taken up by the second into a newly enriched whole.

There is testimony to this master-servant relationship from many cultures and periods. Einstein wrote:

> Certainly we should take care not to make the intellect our god; it has, of course, powerful muscles, but no personality. It cannot lead, it can only serve; and it is not fastidious in its choices of a leader ... it is blind to ends and values. So it is no wonder that this fatal blindness is handed on from old to young and today involves a whole generation.[79]

On the topic of the equal necessity, but unequal status, of spiritual opposites, James reflects:

---

79 Einstein 1976 (260).

Looking back on my own experiences, they all converge towards a kind of insight to which I cannot help ascribing some metaphysical significance. The keynote of it is invariably a reconciliation. It is as if the opposites of the world, whose contradictoriness and conflict make all our difficulties and troubles, were melted into unity. Not only do they, as contrasted species, belong to one and the same genus, but *one of the species*, the nobler and better one, *is itself the genus, and so soaks up and absorbs its opposite into itself.*[80]

This rings true to my experience, too. Every adult human being must learn to accept the contradictions in himself or herself which we all inevitably embody; and learn even to embrace them. This acceptance and embrace is not just good for us in the sense that, while it does not change anything, it brings us to a position of reconciliation with ourselves: it does really effect a *change*. It helps us draw the venom of what Jung called the dark side. If we believe we must be only and always good and loving, paradoxically we give rise to the opposite of this in its 'most unbridled and perverse forms … Apparent contradictions within the human psyche are (as Jung later observed) mutually dependent relations.'[81]

Jung indeed observed that, not only do we often seem to act in what could be seen as far from our better interests – a phenomenon known to Plato and Aristotle as *akrasia* – but 'the self is made manifest in the opposites and the conflicts between them; it is a *coincidentia oppositorum*.'[82] In other words, the self is not just accidentally, frustratingly, and puzzlingly, contrary in its workings, but actually is itself a coincidence of opposites, and becomes apparent to us in and through those opposites. Montaigne, Western literature's most acute self-observer, testifies to this truth in almost every page of his *Essais*. There is certainly an argument that we are only aware of being a 'self' when there is some immediate and unresolved conflict. When we are absorbed in the flow of life, the fact of being a self – all 'self-consciousness' – retires from view.

THE SPECIAL ROLE OF ASYMMETRY IN THE LIVING WORLD

The union of union and division, of love and strife, in the living world is a recurrent theme of Goethe's:

> Dividing the united, uniting the divided, is the very life of Nature; this is the eternal systole and diastole, the eternal coalescence and separation, the inhalation and exhalation of the world in which we live, and where our existence is woven.[83]

In the generation of reality, this rhythmical movement forms a cycle, like the cardiac cycle: systole and diastole, the phases of con-

---

[80] James 1902 (388; emphasis in original).

[81] Drob 1997.

[82] Jung 1953–79, vol 12, §186.

[83] Goethe 1948–, vol 16 (199): Farbenlehre, §739: » Das Geeinte zu entzweien, das Entzweite zu einigen, ist das Leben der Natur; dies ist die ewige Systole und Diastole, die ewige Synkrisis und Diakrisis, das Ein- und Ausatmen der Welt, in der wir leben, weben und sind. «

traction and relaxation within the beat of the heart, the alternations of inspiration and expiration. But this is far from being a mechanical alternation, but alive and always responsive. Goethe's emphasis on *life* is important. And thinking of the eternal systole and diastole reminds me how forcibly I was struck by learning, while in training on the obstetric wards, that when the foetal heart is entirely regular, this is an emergency. A normal heartbeat is flexible, responsive, never regular: it is like a line of plainsong. When it becomes as regular as clockwork the life is, literally, going out of the body.[84]

All music involves – like plainsong – subtle variation of the flow, a degree, however slight, of *rubato*. There is sameness and there is difference within it. This is very marked in some forms of swing music, but it is also everywhere present in classical music, not just in performance but in its essence. Bach's music, for example, is the very image of difference within sameness, and sameness within difference – harmonically; contrapuntally; and rhythmically. Bach is in fact the greatest image of this I know.

It is the nature of all art, to establish a regularity and then depart from it – crucially, nonetheless, without imperilling the overall integrity of the work. There are many 'counterpoints' to this in poetry. All verse plays with departures from and returns to an established pattern of ictus that is heard behind the movement of the verse. Something similar could be said of the nature of rhyme, which is like a harmony that is stretched over time, rather than simultaneously perceived, as it is more usually thought of in music, and where the point is difference and sameness together.[85] Perhaps even more striking is the case of half-rhyme, so beautifully exploited in Wilfred Owen's visionary, spine-chilling poem, 'Strange Meeting', whose theme is indeed the 'other' that is not other: a dream in which he encounters a German soldier he has killed. The poem famously ends with the dead soldier's words to Owen:

> I am the enemy you killed, my friend.
> I knew you in this dark: for so you frowned
> Yesterday through me as you jabbed and killed.
> I parried; but my hands were loath and cold.
> Let us sleep now ...

It is interesting that the Russian word for 'other' or 'different', *drugoi*, has the same etymological root as the word for 'friend', *drug*.[86]

A consistent theme of this book is that to stand in relation to someone or something requires us to be close to, but sufficiently distinct from, the other: having what I call 'necessary distance'. Or half-rhyme. It is not enough that we should have the unison of sameness; we should also have the harmony of difference. The import of

---

84 In general the human heart beat, when it is in what is called sinus rhythm (that is to say, beating normally) is considerably variable, though regular: the interval between beats in any one sequence may vary from, say, around 0.85 to 1.35 beats per second. Loss of such variability is again a sign of stress and is not desirable. Note, this is quite distinct from an arrhythmia, which is normally undesirable, and in which the underlying regularity of beat structure is either temporarily or permanently lost.

85 I say 'more usually', because a melodic line also contains harmonies – ones that are not concurrent at a single moment.

86 Sakhno & Tersis 2008 (320): 'Etymologically one and the same word'.

the name *Sandokai*, probably the core text of Zen, is 'the harmony of difference and sameness'.

What we know of the physics of the universe emphasises the inescapable marriage of sameness and difference in the universe, and that this is central to its creative nature. Physicist David Oliver suggests that the laws of motion and emergent quantum randomness are like the warp and weft of a single fabric, order and randomness forming the creative whole:

> The quantum spontaneity of the universe is expressed in a law of nature, the Heisenberg Uncertainty Principle. This outstanding law declares 'there is no law.' This is not a paradox. We can more precisely state 'there is no law that completely fixes the outcomes of every physical interaction, every dynamic event.' It's the law. But much of the outcome is lawfully ordered and predictable. Nature is neither inevitably random nor completely lawful and predictable. Quantum spontaneity is only one-half the story. The other half is the regularity and predictability of the universe ... The uniquely quantum nature of the dynamic is that quantum dynamics produce both definite new states highly correlated with existing states and spontaneously random new states ... The Law of Motion captures the continuity and stability of motion while the quantum nips at its heels injecting wisps of novelty and new possibilities into every change of state.[87]

The balance between the repeatable and the unpredictable is maintained in genetics, as Denis Noble has emphasised: I will return to that in Chapter 27.

The brain is a beautiful expression of the need to combine sameness and difference, order and disorder. György Buzsáki and a colleague, Markku Penttonen, examined the relationship between different brainwave patterns ('neuronal oscillatory frequencies') in a rat's hippocampus.[88] They found there were three independently generated frequencies in widely different, but reliably predictable, ranges. Intrigued, they turned to a wide range of mammals. They discovered that there are similar 'bands' of frequencies, ranging from around 0.02 Hz to 600 Hz, generated by different brain structures. However, they were amazed to find that these frequencies were related, not by multiples of whole numbers, but by the natural logarithm base *e*, 2.71828. The point is that, since *e* is an irrational number – that is to say, it is never-ending and cannot be written as the simple ratio of two whole numbers – no frequency in the series can ever 'phase-lock' with any other frequency. What is more there were no gaps in this series of frequencies. 'In other words', writes Barbara Goodrich, 'the whole system of the brain is cooperating so as to permit the different frequencies *not* to entrain each other. This is understandable when we consider that a completely ordered,

87 Oliver 2019 (7).

88 Penttonen & Buzsáki 2003.

predictable system cannot itself predict or react to change very well.'[89] And she continues:

> In summary, Buzsáki views the mammalian organism as the most complex system of nature's devising, one which is built from elements relying on opposing forces, including opposing sodium and potassium ion flows, inhibitory versus excitatory neurons, and the predictability of individual oscillation frequencies interacting with the non-predictability of non-linear interactions among neurons kept in a metastable condition.[90]

Balance needs to be constantly disturbed and restored. Symmetry-breaking is everywhere in living organisms; it may be argued that all qualitative cellular transitions and cellular decision-making are forms of symmetry-breaking, and it is indeed 'fundamental to every physiological process'.[91] This echoes Schelling's perception, contained in a passage from his early treatise *On the World Soul* called 'Of the negative preconditions of life processes', in which he speaks of an equilibrium to life that *must* constantly be disturbed and re-established: lost and regained.[92] D'Arcy Thompson saw that,

> if the symmetry be ever so little disturbed, and the shape be ever so little deformed, then there will be forces at work tending to increase the deformation, and others tending to ... restore the spherical symmetry.[93]

Thus the structure of the system we begin with does not wholly determine its own result. Perfection can constitute a flaw. It may be for this reason that in traditional Chinese architecture, the last three tiles are always left off the roof, as the first great Chinese historian, Ssu-ma Ch'ien, records:

> Even heaven is not complete; that is why when people are building a house they leave off the last three tiles, to correspond. And all things that are under the sky have degrees. It is precisely because creatures are incomplete that they are living.[94]

There is also a necessity for slight imperfections in DNA transcription for there to be change and creativity: evolution.

Hegel believed that otherness was actually prior to sameness, sameness being merely an unusual subset of opposition in which the opposites happened to be wholly reconciled. Sameness is, indeed, sterile, and cannot give rise to anything: to this extent he must be right. For creativity there must be complementarity, a degree of resistance within oneness. This idea is bodied forth in the myth of the two brothers, and in the philosophical embrace of the *coincidentia oppositorum*.

---

89 Goodrich 2010 (emphasis in original).

90 *ibid.*

91 Li & Bowerman 2010.

92 Schelling 1798 (202ff): 'Von den negativen Bedingungen des Lebensprocesses'.

93 Thompson 1992 (334).

94 This passage comes from a chapter called 'The diviners by turtle shell and milfoil', which was added to the *Historical Records* (*Shih chi*) after Ssu-ma Ch'ien's death, by someone calling himself Mr Ch'u. It is a very early addition, however – probably of the first century BC. For the translation and an explanation of this passage I am indebted to the kindness of my erstwhile All Souls colleague, the consummate sinologist David Hawkes, who drew it to my attention, and encouraged – as many might not have done – my naïve interest in all things Chinese. I am told that Persian rugs also traditionally contain an imperfect stitch.

CONCLUSION

I have argued that, at the origin of everything, there lies a coincidence of opposites that is profoundly generative, indeed necessary for creation, and gives rise to all that we know; and that this coincidence of opposites is by no means contrary to reason. I have stressed that we must not be tempted, left hemisphere fashion, to resolve the necessary tension by pretending that one of the pairs of opposites either can safely be dispensed with, or is not real. Denying the concealed opposite is dangerous. The coincidence of opposites does not compromise their nature as opposites: rather they fulfil themselves through one another. At the foundation of everything is the opposition, recognised from Empedocles to Goethe, between Love and Strife: the opposite dispositions embodied by He Grasps the Sky With Both Hands and Flint. A harmonious world comes into being only if the forces of love and strife are unified: He Grasps the Sky allows Flint his degree of independence, but needs to retain a degree of oversight; without it, 'Flint will forever attempt to destroy his rule'. We need the union of division and union, of multiplicity and unity; the left hemisphere needs ultimately to act as servant to the right hemisphere master, since, unbridled, the left hemisphere is capable of destroying the world.

Moreover, what we think of as good may conceal harm, and what we consider harmful may bring something of great value. And I have suggested two geometrical images. We should be wary of linear models in our attempts to understand the world (except at the most minutely local level), and replace straight lines with helices, which incorporate an acknowledgment of the coincidence of opposites with the idea that there is always change and growth, not mere repetition, as the image of the circle risks suggesting. And, just as there is an asymmetry in the relationship of the hemispheres, there is an asymmetry in the *coincidentia oppositorum*. We need not just difference and union but the union of the two; we need, as I have urged, not just non-duality, but the non-duality of duality with non-duality; and we need not just asymmetry alone, or symmetry alone, but the asymmetry that is symmetry-and-asymmetry taken together. I will have more to say about the last of these particularly, in subsequent chapters. Now let us turn to one very particular opposition that is a conjunction: that between the one and the many.

Chapter 21 · The One and the Many

οὐκ ἐμοῦ ἀλλὰ τοῦ λόγου ἀκούσαντας ὁμολογεῖν σοφόν ἐστιν
ἓν πάντα εἶναι. ¶ It is wise, listening not to me but to the *logos*,
to agree that all things are one. ¶ It is wise, listening not
to me but to the *logos*, to agree that the one is all things.
—HERACLITUS[1]

Students achieving oneness will move ahead to twoness.
—WOODY ALLEN[2]

'In the making of every animal the presence of every other animal has been recognized', wrote John Muir:

> Indeed, every atom in creation may be said to be acquainted with and married to every other, but with universal union there is a division sufficient in degree for the purposes of the most intense individuality; no matter, therefore, what may be the note which any creature forms in the song of existence, it is made first for itself, then more and more remotely for all the world and worlds.[3]

Everything is part of one whole, connected to every other part by a matter of degree. But everything is also absolutely unique: has 'the most intense individuality'. The Anglo-Polish philosopher Leszek Kolakowski, following Bergson, relates this to the existence in all living things of some form of memory:

> A determinist states that, in the same conditions, the same phenomena occur. However, the same conditions can never, by definition, obtain in the life of the self, because each, artificially isolated, moment of its duration includes the entire past, which is, consequently, different for each moment. By contrast with the universe of abstract equations, the same situation never occurs twice in the being endowed with memory; since real time is absolutely irreversible, neither the same cause nor the same effect can ever reappear in experience.[4]

And in a more primitive sense, it has universal applicability, not just to the living. As Smolin says: 'no movement ever repeats. Looked at in enough detail, every event in the universe is unique ... the more detail we note, the more apparent it is that no event or experiment can be an exact copy of another.'[5]

In this chapter I want to consider the central paradox of the one and the many – or, rather, paradoxes, since there are many, not just one. I say 'central', because it takes us to the core of what it means to be at all.

---

1 Heraclitus fr XXXVI [Diels 50, Marcovich 26] (trans IMcG). The phrase is normally translated in the first sense, emphasising the unity of multiplicity: the Greek elegantly, and in the true spirit of Heraclitus, equally permits the second, emphasising the multiplicity of unity. And thus unifies division and union at a meta-level.

2 From a course description for Philosophy I in 'Spring Bulletin', *The New Yorker*, 29 April 1967 (37).

3 Muir 1875 (364). Nearly 200 years earlier, Leibniz wrote in a letter to Arnaud, dated September 1687: « *Toutes les substances sympathisent avec toutes les autres et reçoivent quelque changement proportionnel répondant au moindre changement qui arrive dans tout l'Univers ...* » – 'every substance is in sympathy with every other, and is to some degree altered in response to the least change that occurs in the entire universe ...'.

4 Kolakowski 1985 (20).

5 Smolin 2013 (215).

UNIQUENESS AND GENERALITY

Your identity, which in an important sense means that which distinguishes you from others, literally means sameness (from Latin, *idem*, the same; thence *identidem*, again and again). What makes you the same again and again – from moment to moment – is the very thing that makes you different from others. Internal sameness is a condition of external difference. As we have seen, in some kinds of right hemisphere dysfunction, continuity over time, the glue that holds the forms and patterns of the world together, is lost; thus your identity is lost. The logical result is that you are no longer unique and that you are at any instant *reproducible*. Hence we have the man who had eight 'doubles', each with a copy of his wife and children, each living in a replica of his city.[6] In this situation you have lost your identity because others are identical to you. This is why Capgras and Fregoli syndromes are manifestations of the same phenomenon, even though one involves seeing someone you don't recognise in someone you do, and the other involves seeing someone you do recognise in someone you don't. Loss of uniqueness is the core deficit in delusional misidentification. And that sense of uniqueness is dependent on the right hemisphere.[7]

The meeting of sameness and difference in identity may seem on the surface surprising. But uniqueness is always underwritten by some respect in which uniqueness does not apply. To know that anything is unique requires understanding the ways in which it differs from something else it might have been: you are a unique human being, and the generality of your being *a human being* is still there, but hidden in the particularity, if I simply call you 'unique'. If there were no general patterns at all, there would not be uniqueness, but mere chaos.

To see each thing as it really is requires a balancing act. On the one hand, we need to see it as unique: nothing that exists is ever the same as anything else. Yet one aspect of what it really *is* requires us to see where it fits into the context of everything else; and to see that, we need generalities. And to appreciate the relationship between uniqueness and generality means always to balance sameness and difference.

Numbers of things also intrinsically imply both sameness and difference. Referring to number, when applied to elements of experience, automatically invokes sameness and difference simultaneously. If we speak of two coins, we imply multiplicity, in that there are two, and singleness, in that they are coins (not, say, a coin and a button). In fact, for any two or more entities to be unlike, they must be like in some respect, and vice versa, so that, as Gerard Manley Hopkins put

---

[6] See p 96 above.

[7] I examine the evidence for this at pp 864–79 below.

it, in a Platonic dialogue on the nature of beauty, 'Likeness therefore implies unlikeness ... and unlikeness likeness'.[8]

This sycamore leaf in my hand is different from any other sycamore leaf that ever existed: it is unique. But it is also not just an interesting piece of abstract art, an irregular shaped, planar object of a certain colour and texture, but, precisely, a sycamore leaf, a certain type of thing, a leaf, and a certain type of leaf, that of a sycamore tree, which is a certain type of tree, a certain kind of living being, which means it has many predictable properties and functions, and has its own proper place in the world. If I ignore that, I scarcely know what it is at all. We need to see both the unique and the general *at once*. As a child, the excitement lies in discovering that not everything is unique, but that there are general categories giving shape to the world: 'Birdie!', 'Bunny!', 'Doggie!'. This gives pattern to experience. As adults we have become so used to this, that we have to make an effort in the opposite direction: the excitement comes only when we recover the uniqueness of what it is we contemplate.

Like identity, the word 'essence' also has a double life. In one sense it is what makes you absolutely you-and-not-someone-else: the very *essence* of you. Yet you are also, in *essence*, a human being. This was a distinction indirectly illuminated by the phenomenological philosopher Max Scheler, who contrasted here the child and the adult. Of Scheler's writing on this in *The Constitution of the Human Being*, John Cutting, who translated the English language version, comments: 'in childhood you are an empiricist, but in adulthood the essence squeezes out all new experience ... the essence brooks no opposition. This is how things are.'[9] This shift, from the unique presence to the general re-presentation is made palpable in the change with age of which Wordsworth complained: that the real, experienced presence of the mountains and lakes as a child was overwhelming – they were, in our terms, present 'in their very essence' to him, awe-inspiring and unique; while as an adult he could see them only as re-presented, now become the essential Mountain and the essential Lake. This different kind of essence has too much generalisation about it; it has pre-empted and squeezed out the other essence that is present only in experience. As Evelyn Underhill would say, the hare of reality was for him now already jugged.

Yet this essence also has its uniqueness of a kind, viewed at a different level. What functions as general at one level functions as unique at the next. Take the generalisation that the leaf is from a sycamore tree. Yet among sycamores, this particular sycamore tree is unique. And among trees, the type 'sycamore' is already unique. There is an essence, a proper form, of a sycamore. There is a 'thisness' to the type, as well as to the individual. It matters only at what

8 Hopkins 1963.

9 I am greatly indebted to John Cutting for his insights here. The quote comes from a personal communication. He deals with this at greater length in a published article (2018).

level you place the bar. To take Wordsworth's case, do we place it at the uniqueness of this experience of Ullswater here and now as it can only be when embodied and present? Or at the uniqueness of Ullswater, considered more generally? The uniqueness of the Lake District? The uniqueness of mountainous lakes? At one extreme, everything is alike; at the other, nothing is like anything else. In either case there is no pattern to reality. The trick is to find the level at which the richest patterns are revealed in the context in which one finds oneself.[10]

Normally, analytically, we think of difference and sameness as incompatible, like being both one and many. But they constantly interpenetrate one another and give life to one another. (As, in fact, do the One and the Many – always.)

BEAUTY, AND THE COINCIDENCE OF THE ONE AND THE MANY

Connexion always produces difference along with union. An interaction is, at the same time, *both* an act of coming together, *and* the process through which each party becomes more itself. In another observation of Hopkins',

> Beauty then is a relation ... and things which have a relation are near enough to have something in common, but not near enough to be one and the same.[11]

The coming together of sameness and difference makes relation possible; and, if, as I believe, everything exists *only in relation*, this 'coming together' must be essential – at the very ground of – all that is. Harmony is the instantiation, not just of sameness and difference, but of a special creative relationship between them: in an excess of either it disappears into mere unison or mere discord. And beauty is the experience of this harmony. Once again the trick is in finding the right level that gives the richest patterns.

In the Preface to *Lyrical Ballads*, Wordsworth describes 'the pleasure which the mind derives from the perception of similitude in dissimilitude. This principle is the great spring of the activity of our minds, and their chief feeder.'[12] In other words, he is ascribing a creative power to such a perception, stimulating what Coleridge called the Primary Imagination.

Hopkins continues his Platonic dialogue:

> — Then the beauty of the oak and the chestnut-fan and the sky is a mixture of likeness and difference or agreement and disagreement or consistency and variety or symmetry and change.
> — It seems so, yes.
> — And if we did not feel the likeness we should not feel them so beautiful, or if we did not feel the difference we should not feel

---

10 According to Nicholas of Cusa, the greatest philosophical and scientific mind of the fifteenth century, 'individual specimens are all that exist, but not only are they contracted images of God's oneness, but also contracted images of their types insofar as each more or less fulfils the possibilities of its nature.' This insight draws together the different levels of difference and sameness across the entire cosmos from the lowliest existing thing to God. See Miller 2017.

11 Hopkins *op cit* (103).

12 Wordsworth 1876, vol II (96).

them so beautiful. The beauty we find is from the comparison we make of the things with themselves, seeing their likeness and difference, is it not?

Harmony cannot exist without just the *right degree* of otherness within the whole. It is not, clearly, sameness alone or difference alone, nor sameness and difference together alone, but the nature of the *relationship between* sameness and difference: the betweenness. As in a thriving society, as in a healthy organism, as in anything beautiful, there is a proper relationship between the principles of differentiation and unity.

Beauty in music does not necessitate literal musical harmony, since the vagaries of a single melodic line also consist of the many relationships between notes, relations of qualified difference: a melody evokes harmonics that are spread over time. Some of the most beautiful sounds in the world are single lines – in folk music, in Bach's unaccompanied violin and cello music, and in Gregorian chant, for example. Indeed for the Greeks the term harmony in music was used, not of notes sounded at the same time, but 'only of notes that were heard successively': it was like rhyme, which is a kind of harmony that always creates tension between two or more moments in a temporal span. It was not until the Middle Ages that 'the word is applied to the concord between simultaneously sounding notes'.[13]   13 Begbie 2008 (89–90).

Yet literal harmony of this kind has a special richness. As an undergraduate, I looked forward to the moment in the church calendar when the choir of my then Oxford college would sing Tallis's setting of the Litany. It begins, in the full liturgical setting, very simply with many bars of repeated lines of plainchant: but then there comes a moment, when the choir takes a steep, and at that moment unexpected, plunge into homophony (chordal harmony), as restrained as it is profound. If ever one wanted confirmation of the mysterious power that harmony can add to melody, this was the moment. In retrospect it reminds me of the last three minutes of Tarkovsky's film *Andrei Rublev*, where, after three hours of unforgettable black and white cinematography, the film suddenly goes into colour as we contemplate the actual paintings that are the outcome of the suffering we have been witnessing for the last three hours. Colour too has its harmonies (and therefore its discords). But the harmony was richer for the plainsong; the colour for emerging from black and white.

As I mentioned in the previous chapter, Hegel made the important point that sameness and difference, unity and division, have themselves to be unified; this seems to me from experience clearly right. Note this suggests the ultimate priority of the principle of union over that of division, despite the necessary part played by division at one stage of the process.

SMALL DIFFERENCES CAN MAKE LARGE DIFFERENCES

Let us return to uniqueness. I love William James for many things, his unusual combination of intellect with wisdom, for one, and of intellect with humility, for another:

> An unlearned carpenter of my acquaintance once said in my hearing: 'There is very little difference between one man and another; but what little there is, is very important.' This distinction seems to me to go to the root of the matter.[14]

In a way this is the point I made, at the outset of this book, about hemisphere difference. Small differences can make large differences. It is manifestly true of individual people, but could also be applied to the differences between individual groups, including nationalities or the sexes, or those between a great and a minor artist. In face-to-face conversation, a minor change in the eyes, subconsciously detectable for only thousandths of a second, can make a huge difference to the direction of that conversation. As we encounter experience it is unique; yet as we represent it to ourselves, it becomes general. As Whitehead put it, 'we think in generalities, but we live in detail.'[15]

PARTS, WHOLES AND HEMISPHERES

The word 'detail' here is unfortunate, because it means the embodied and unique (rather than the general), which is not the same as 'detail' in the sense of a small part. So this prompts me to make an important distinction.

Each hemisphere deals in *parts* of a kind, and each deals in *wholes* of a kind; this is hardly surprising, because each has to negotiate and make sense of the world. But they do it in different ways, suggesting a different relationship between the 'parts' and the 'whole'. In the left hemisphere case, there are fragments, which must be put together to form an aggregate. In the right hemisphere case, there are wholes at any number of levels, in which parts can be distinguished. In the right hemisphere case things remain maximally diverse, yet unified; in the left hemisphere case things are minimally diverse, yet fragmented. A similar distinction can be made between the process of individu*ation*, an internal unfolding into complexity, which respects, never departs from, and enriches the whole; and mere individu*alism*, which fragments, atomises and destroys the whole. Heidegger is unusually clear on this, helped by the German language which distinguishes *Stücke* (fragments) from *Teile* (parts) – the word comes from *teilen*, to share or divide, and thus alludes to its *post factum*, retrospective not compositional, nature. He writes that

> the fragment [*das Stück*] is something entirely other than the part [*der Teil*]. The part shares and imparts itself [*teilt sich mit*] with [by,

---

14 James 1897, 'The importance of individuals': 255–62 (256–7).

15 Whitehead 1926a.

as, in] part of the organic whole [*das Ganze*]. It takes part in the whole, belongs to it. The fragment on the other hand is separated out and indeed is thus as fragment, as what it is, only as long as it is locked up in opposition to other fragments. It never shares and imparts itself in and as part of an organic whole.[16]

In societal terms, one might contrast the left hemisphere idea of an individual being, an idea that is, if you like, univalent (individual as defined over against the group); and a right hemisphere idea of an individual being, one that is bivalent, or reciprocal (individual as generated by, and in turn taking part in generating, the group).

LOVE IN THE TIME OF BUREAUCRACY

The drive behind categorisation is disconcerting when it results in too much emphasis on sameness. Then it speaks of alienation and the power to manipulate, both qualities associated with the left hemisphere rather than the right. It is how bureaucracies and commerce work in relation to individuals. It is a perhaps necessary evil, a tool that helps us organise knowledge towards a particular goal: as James also said, 'every way of classifying a thing is but a way of handling it for some particular *purpose*.'[17]

Mozart's opera *Don Giovanni* contains the most famous *exposé* of categorisation of all time. Here the Don's manservant Leporello reads from the list of his master's conquests, to Donna Elvira, his master's jilted lover:

> My dear lady, this is the list
> Of the beauties my master has loved,
> A list which I have compiled.
> Observe, read along with me.
>
> In Italy, six hundred and forty;
> In Germany, two hundred and thirty-one;
> A hundred in France; in Turkey, ninety-one;
> But in Spain already one thousand and three.
>
> Among these are peasant girls,
> Maidservants, city girls,
> Countesses, baronesses,
> Marchionesses, princesses,
> Women of every rank,
> Every shape, every age.
>
> With blondes it is his habit
> To praise their kindness;
> In brunettes, their faithfulness;
> In the white-haired, their sweetness.

16 Heidegger, 'Das Ge-stell'; in *Gesamtausgabe*, vol 79, 1994 (36); a still never officially translated lecture, here in the translation offered *en passant* in Levin 1999 (121).

17 James 1897, 'The sentiment of rationality': 63–110 (70; emphasis added). Bergson similarly thought that our ability to produce and understand generalities evolved purely for practical purposes, enabling us to classify our surroundings in such a way as to achieve maximum utility.

> In winter he likes fat ones.
> In summer he likes thin ones.
> He calls the tall ones 'majestic'.
> The little ones are always 'charming'.
>
> He seduces the old ones
> For the pleasure of adding to the list.
> His greatest favourite
> Is the young beginner.
>
> It doesn't matter if she's rich,
> Ugly or beautiful;
> If she wears a skirt,
> You know what he does.[18]

While the Don continues his wars of conquest, someone else, the servant who lies outside the circle of *eros*, matter-of-factly compiles the database – on his behalf. Quantity is, of course, the first essential: never a good starting place for love. Within it, all are carefully listed and neatly categorised by nationality, social status, height, build, age. Leporello could be seen as an obsessional lepidopterist, pinning another specimen for the Don's collection: but viewed another way he is a bureaucrat (he keeps detailed statistics) with excellent diversity credentials, as you see. He delights in recounting how the Don does not discriminate, but shows how he thereby falls into the category of those who are undiscriminating. The Don makes sure that while making no discriminations he has representatives of every discriminable category. Rather as modern bureaucracies devote considerable energy and attention to categories of people for the professed purpose of demonstrating that such categories are without foundation, he carves up his clientele so as to make sure he has them all in his list, but treats them all the same, whether they are rich, poor, thin, fat, old, young, dark or fair. The categories merely help him manipulate things better to his own ends: once the type is identified, he invariably applies the same tired formulae to make them conform. This is, of course, nothing to do with love. Such 'loving' devalues its objects, by substituting labels and categories for individually different living beings, makes them means rather than ends, and lends itself to greed, abuse and never-to-be-satisfied restlessness: Don Juanism.

Nietzsche called us, in the modern West, the Don Juans of cognition, hungrily acquiring what we take to be understanding, but acquiring only knowledge 'about' (*wissen*), not knowledge 'of' (*kennen*) – and passing everything through the mill of our categorising mentality.[19] He that would love, must love individuals, not generalities.

---

18 Da Ponte's libretto for *Don Giovanni*, Act I, sc v.

19 Nietzsche 1970 (234).

You can admire or be attracted to womanhood, or manhood, but you cannot *love* them.

Making things equal helps turn them into fodder for our own existing purposes. Nietzsche reflected a good deal on this in his notebooks, and put his finger on a number of important points. 'The entire apparatus of knowledge', he wrote, 'is an apparatus for abstraction and simplification – directed not at knowledge but at taking possession of things'.[20] He refers to 'the utilitarian fact that only when we see things coarsely and made equal do they become calculable and usable to us'.[21] This is the root of totalitarianism: we are the Don Juans of cognition.

Generalisation turns things into means to an already defined end, left hemisphere fashion. In this Nietzsche seized the essence of left hemisphere cognition: the will to power. And referring to the drive to render things equal – thus untroubledly the same – in order to be able to use them, he compares it to an amoeba engulfing and digesting its prey:

> In *our* thought, the essential feature is fitting new material into old schemas (= Procrustes' bed), making equal what is new [the LH trying to preserve its preferred way of looking at things at all costs] …[22]

And he carries on:

> All thought, judgment, perception, considered as comparison, has as its precondition … a "*making* equal". The process of making equal is the same as the process of incorporation of appropriated material in the amoeba …[23]

And further:

> the fundamental inclination to posit as equal, to *see* things as equal … corresponds exactly to that external, mechanical process (which is its symbol) by which protoplasm makes what it appropriates equal to itself and fits it into its own *forms and files* [the LH categorising what it needs to purposes of its own].[24]

Speaking of the left hemisphere's gross and approximate understanding that serves well its unsubtle purposes, he continues:

> the coarser organ sees much apparent equality; the intellect *wants* equality, ie, to subsume a sense impression into an existing series: in the same way as the body assimilates inorganic matter … the will to equality is the will to power – the belief that something is thus and thus (the essence of *judgment*) is the consequence of a will that as much as possible *shall be* equal.[25]

The lust for control lies behind the demand that all shall be equal

---

20 Nietzsche 1967, §503; emphasis in original in all quotations on this page.

21 *ibid*, §515.

22 *ibid*, §499.

23 *ibid*, §501.

24 *ibid*, §510.

25 *ibid*, §511. *Geist* can mean 'intellect' or 'spirit'; 'intellect' is here less confusing and I have accordingly substituted it.

to – reducible to – something else. The process is one, as Nietzsche makes clear, of triumph by reductionism: ingestion (appropriation by the left hemisphere), followed by digestion (lysis into parts).

First, uniqueness is lost in categorising: a triumph for sameness. The next step is to lose the uniqueness of the category – and the triumph for sameness is almost complete. To be a citizen of the world is to be a citizen of nowhere; to love everyone and no-one in particular is not to love. As Solzhenitsyn remarked in his Nobel acceptance speech:

> Were nations to disappear, we would be impoverished in exactly the same way as if all people suddenly became alike, with the same character and the same face. Nations are part of the wealth of the human race. Although generalised, they are its individuals. The smallest of them has its own special colours and hides in itself some special facet of God's design.[26]

Similarly Whitehead wrote that

> the first step in science and philosophy has been made when it is grasped that every routine exemplifies a principle which is capable of statement in abstraction from its particular exemplifications. The curiosity, which is the gadfly driving civilisation from its ancient safeties, is this desire to state the principles in abstraction. In this curiosity there is a *ruthless element which in the end disturbs*. We are American, or French, or English; and we love our modes of life, with their beauties and tendernesses. But curiosity drives us to an attempt to define civilisation; and in this generalisation we soon find that we have lost our beloved America, our beloved France, and our beloved England. The generality stands with a cold impartiality, where our affections cling to one or other of the particulars.[27]

Charitable schemes are not the same as, and no guarantee of, charity. We must deal with the actual and individual, not the theoretical and general. Blake famously wrote:

> He who would do good to another must do it in Minute
>   Particulars:
> General Good is the plea of the scoundrel, hypocrite & flatterer,
> For Art & Science cannot exist but in minutely organised
>   Particulars
> And not in generalising Demonstrations of the Rational Power.
> The Infinite alone resides in Definite & Determinate Identity ...[28]

'Generalising demonstrations of the rational power' are the left hemisphere's *modus operandi*. The last line here is striking for its paradoxical nature: 'The infinite alone resides in definite & determinate identity'. The point Blake is making is that we do not come

---

26 Solzhenitsyn 1973 (33).

27 Whitehead 1933 (181; emphasis added).

28 Blake, *Jerusalem*, ch 3, Plate 55, lines 60–4.

to understand or experience the infinite, or, for that matter, the eternal, by attempting somehow to transcend the finite or the temporal, but by immersing ourselves in them, in such a way as to pass into the infinite, manifest there where they are. The path to the infinite and eternal lies in, not away from – not even to one side of – the finite and the temporal. The infinite and eternal are manifest here, where we are, within the translucency of space and time. Blake saw that this forms a parallel to the relationship between the particular and the general, that we find the general manifest *in* the particular, not by turning our backs on it.

HOW THE GENERAL AND THE UNIQUE RELATE

On this relationship between the utterly unique and the whole world beyond, philosopher Jan Zwicky writes: '*This*ness is the experience of a distinct thing in such a way that the resonant structure of the world sounds through it. Each *this* focusses that resonant structure in a distinct way. But the structure so focussed is – of course – always the same. There is only one world.'[29]

The history of the cosmos looks like one of constant divergence into multiplicity and uniqueness, yet a uniqueness that is always subsumed within, and understood against the background of, a coherent whole. One hydrogen atom is more or less the same as another – though Smolin's point that 'every event in the universe is unique' stands. On the other hand, no one flower, tree, bird or animal, not even one snow crystal or lump of rock, is at all the same as any other, and whatever lives is constantly ramifying into ever-renewing newness. The effect of the evolution of matter, never mind that of life, is to diversify extravagantly; and even the function of genes is not just to fix, but also to vary.[30]

Life, in its essence, is a making new: a wholly superfluous, superabundant, self-overflowing – an exuberant, self-delighting process of differentiation into ever more astonishing forms, an unending dance, in which we are lucky enough to find ourselves caught up – not just, as the left hemisphere cannot help but see it, a series of survival problems to conquer. If reality is ultimately just an eternal, unchanging, perfect unity, as some philosophies seem to suggest, life is going the wrong way about making that clear. To the degree that we can discern any governing principle to the cosmos, it is not going to be parsimony. The One may be simpler than the Many, but the world as we know it emphasises individuation and multiplicity, not singleness and simplicity.[31] Connexions between relative samenesses (subatomic particles) produce great differences (the 10,000 things); the connexions between differences never produce sameness, but new wholes as they resonate with one another, forming new responsive relationships.

---

29 Zwicky 2014b, §55.

30 See pp 447ff above.

31 Mu 2009 (45): 'In sum, the *dao* is the ultimate source, unifying power, and fundamental principle of nature and the universe; it manifests itself through particular individual things'.

Shakespeare is as near to embodying the creative process as anyone we know; and Coleridge called Shakespeare 'myriad-minded', by which he meant that by a feat of the imagination he could feel his way into the depth of being of his characters, each of them entirely original, which meant that none of them was ever just a reproducible type but, like a real person, what we would call a complete 'one-off'. Not infrequently his characters' insistence on being who they were, not just what the stereotypes of the plot demanded of them, caused Shakespeare to change the story in order to accommodate their stubborn, vibrant thisness. Is not all creation, similarly, a work of myriad-mindedness, in which the stubborn resistant element too plays its part?

Thisness has a piercing vibrancy; generality is flat. 'We are pierced', writes Zwicky:

> The *this* strikes into us like a shaft of light. We are focussed by it and experience it as focussed: what is *this* is unique, it has an utterly distinct – and here notice the sense modality we reach for – flavour or fragrance. (What is important about the metaphor is that it recognizes the object as knowable but neither visible nor graspable.) Often the experience also includes an awareness of not being able to give an account of the *this* – we can point, but not say.[32]

As Hopkins found, too, flavour or fragrance are the modes of thisness: the sense of taste and smell are not only, as we have seen, the most finely discriminable of all sensory modes by a very long way, but they famously defy description in language (hence the common jokes at the expense of oenophiles), and are not decomposable into elements without their thisness being manifestly lost. Hopkins refers to

> that taste of myself, of I and me above and in all things, which is more distinctive than the taste of ale or alum, more distinctive than the smell of walnutleaf or camphor, and is incommunicable by any means to another man ... searching nature I taste self at but one tankard, that of my own being.[33]

It is possible to become so used to the left hemisphere, represented world, which deals in classes of things, and the rules they obey, that the embodied reality of what is never general, always unique, escapes us. Left hemisphere rules take over from right hemisphere insight. Yet it is the unrepeatable thisness of anything – at any rate, anything that is not mechanical or machine-made – that is its essence, its fulfilment and its value: even, one might say, its raison d'être. It is an invariable quality of everything we love: the particular essence of a person, a place, or a work of art, which cannot be replaced by

32 Zwicky 2014b, §53(a).

33 Hopkins 1963 (145–6).

anything else whatever, any more than can your best friend: if the work had not come into existence, we could not have imagined it, and there would be a hole in the universe where it should be.

This quality is present everywhere our attention allows, and it was this 'being always just itself and no other thing', that by definition defies definition, this *hæcceitas*,[34] that Hopkins celebrated in all his work, and most famously in the poem 'As kingfishers catch fire':

> As kingfishers catch fire, dragonflies dráw fláme;
> As tumbled over rim in roundy wells
> Stones ring; like each tucked string tells, each hung bell's
> Bow swung finds tongue to fling out broad its name;
> Each mortal thing does one thing and the same:
> Deals out that being indoors each one dwells;
> Selves – goes itself; *myself* it speaks and spells,
> Crying *Whát I do is me: for that I came.*

These opening lines also exemplify their subject: their style is highly idiosyncratic, and they take a bit of patient unpacking. But the message is clear. Each living creature (mortal thing) is unique; it is what it does; and doing it is *in itself* the purpose of the creature's existence. It is also, paradoxically, through each doing 'one thing and the same' that they become the many *different* beings the poem celebrates. Difference and sameness together, never just difference or just sameness.

Because of this manifest variety, general principles are less appropriate than they may customarily seem. 'One Law for the Lion & Ox is Oppression',[35] wrote Blake; and 'The eagle never lost so much time as when he submitted to learn of the crow'.[36] There is an uncanny echo of this in Sitting Bull's rejection of the white settlers: 'If the great spirit had desired me to be a white man, he would have made me so in the first place. He put in your heart certain wishes and plans, in my heart he put other and different desires. Each man is good in his sight. It is not necessary for eagles to be crows.'[37]

What Blake expresses is that each being has its proper role, its *telos*, and that the context created by its existence dictates that we cannot apply readymade rules derived from quite different contexts. In such a cosmos there is a place for the lion, the wolf and the tiger, as well as the ox, the lamb and the mouse, not by being comfortingly the same, but by being vitally different. We expect each lion or ox, say, to be reliably different from other animals, which means reliably similar to other lions or oxen: it is their essential difference that enables us to form any expectation at all (by their identity at one level they proclaim their non-identity at another). To apply the same laws to all situations indiscriminately is where the single-track

---

34 *Hæcceitas* (variously, also, *hæccitas* or *hæceitas*) was the term used by the mediaeval philosopher and theologian Duns Scotus to denote the *principium individuationis*, or formal principle of individuation – that which, for example, makes you not just 'a person' but the unrepeatable person you are. Hopkins was greatly influenced by Duns Scotus ('who of all men most sways my spirits to peace'), and invented the term 'inscape' to refer to the incarnate manifestations of this principle in the individual nature of every creature.

35 Blake, *The Marriage of Heaven and Hell*, Plate 24.

36 *ibid*, Plate 8, line 19.

37 Chief Sitting Bull 1883, quoted 1973 (see bibliography). There is a further layer of meaning: Sitting Bull was a Lakota Sioux, whose symbol is an eagle; and they and the Crow Indians were long-standing enemies. The Crow had sided with the US troops.

left hemisphere mindset leads. Context creates the necessity of discriminating what is proper to different contexts, since, when the context changes, everything within it changes.

According to R. G. Collingwood, our civilisation has neglected history, and therefore

> neglected to develop that kind of insight which alone could tell it what rules to apply, not in a situation of a specific type, but in the situation in which it actually found itself. It was precisely because history offered us something altogether different from rules, namely insight, that it could afford us the help we needed in diagnosing our moral and political problems.[38]

As we know, the right hemisphere, seeing more broadly, and more deeply, than the left is more sensitive to context.[39] So it is hardly surprising that it tends to see each situation as a potential 'one-off', not as one off the left hemisphere shelf. The right ventrolateral prefrontal cortex is particularly important for precisely what we are discussing, monitoring whether a course of action is appropriate in context, and preventing it from happening when it isn't.[40] It provides Collingwood's insight into the particular situation, rather than rules for 'such situations' in general.

A PLURALISTIC – BUT INTEGRATED – UNIVERSE

One way of looking at multiplicity is that it is the potential stored in oneness, as the potential for colour is 'stored' in white light. That potential multiplicity is constantly actualised over time. According to Antonio Negri, an expert on Spinoza (to whom the following distinction was particularly important), power can be thought of as having two forms: *potentia*, which is fluid, dynamic and constitutive; and *potestas* (the common Latin word for power), which fixes what is constituted. The former is, then, right hemisphere-congruent and the latter left hemisphere-congruent; and the former constantly gives way to the latter, as what *might* be becomes what *is* precipitated.[41] But because of the very uniqueness of what is, it is taken up again into the vibrancy of the now enhanced whole. Thus the hemispheres work together, from right to left to right, to maximum benefit.

The money in your bank account decreases as you spend it, as it is converted from potential into actual concrete individual elements in the world. Yet only by doing so can it realise its value. And potential individuates as it actualises: banknotes are effectively identical, whereas what they represent is infinitely diverse. As cells 'reproduce' themselves in the growing embryo, their potential becomes successively more limited: what is initially totipotential (has the potential to become any cell in the body) becomes specialised and yields to the actuality of a particular tissue. The potential takes part in what

38 Collingwood 1983 (101).

39 McElroy & Stroh 2013.

40 Chatham, Claus, Kim *et al* 2012; Aron, Robbins & Poldrack 2014.

41 The distinction is important in Spinoza. For a discussion, see Antonio Negri who writes: '*potestas* refers to power in its fixed, institutional or "constituted form", while *potentia* refers to power in its fluid, dynamic or "constitutive" form' (2004, 15). In English, our words 'potential' and 'power' tend to conflate the meaning; by contrast, French (*puissance/pouvoir*) and Italian (*potenza/potere*) do not.

becomes actual, just as the actual takes part in what, *after its appearance*, can be potential. (I shall have more to say about this relationship, in Chapter 22, on time, and Chapter 28, on the sacred.) The potential must be at least as great as the power it ultimately exhibits, possibly infinitely greater.

If one sees potential as energy, and the coming into being of the actual as that energy being consumed or transformed, this may be one of the possible meanings of another Heraclitean fragment: 'all things are requital for fire, and fire for all things, as goods for gold and gold for goods'.[42] In other words, energy ('fire') becomes matter, and matter becomes energy: but the gold for goods analogy adds specificity. The general becomes actual and the actual becomes general once more. On this requital, Blake wrote that 'Eternity is in Love with the Productions of Time'; but he also said 'The Ruins of Time build Mansions in Eternity'.[43] There is eternal, and eternally creative, reciprocity.

James referred to what he called 'a pluralistic universe' (the phrase became the title of his Hibbert Lectures, delivered in 1908, and published the following year). It seems to me, also, that the tendency of the world, the living world especially, is not towards oneness and sameness, but towards pluralism, difference and particularity – towards beings with a history: away from generalisation and equality, towards ever greater differentiation, relishing the uniqueness, the 'this-and-no-other-ness', of each being. Yet all this takes place as the enrichment of a whole from which it is never divorced, and to which, now enriched, it returns. According to the Genesis myth, God made the world by dividing – night from day, heaven from earth, the sea from the dry land, and so on. Division can be creative. Here it made a whole world. Before each cell replicates, the chromosome pairs divide and are actively drawn apart; later they re-form in new pairs to make a new cell. Life, as Shelley says, 'like a dome of many-coloured glass stains the white radiance of eternity'. But another way of looking at it is that the potential of the white light is actualised by the dome. As Newton was able to demonstrate, pure white light and the rainbow of colours are one and the same; the cosmos seems to me the prism through which the purity of white light is refracted. And yet, importantly, a prism not only splits light into colours, but can also recombine them into white light.

The left hemisphere tends to aggregate the reality-fragments it identifies into categories, within which what the right hemisphere sees as individuals become interchangeable. To that extent the left hemisphere is both a splitter and a lumper, the worst of both worlds, in which things are first artificially separated, and then artificially aggregated, by an effort of cognition. By contrast the right hemisphere sees already existing individual entities, each of them whole, and it

42 Heraclitus fr XL [Diels 90, Marcovich 54] (trans C Kahn).

43 Blake, *The Marriage of Heaven and Hell*, Plate 7, 'Proverbs of Hell'; and letter to William Hayley, 6 May 1800.

sees them as belonging in a contextual whole, from which they are not divided. It is, by contrast with the left hemisphere, neither a lumper nor a splitter.

DIFFERENTIATION WITHIN A NEXUS: THE CENTRAL NERVOUS SYSTEM

As it happens, this distinction lies behind an ancient and, to some degree, ongoing quarrel at the heart of neurology. In 1901, on the very first occasion of the awarding of a Nobel Prize, the Italian neuropathologist Camillo Golgi was put forward for the Prize for physiology and medicine. When he finally received it in 1906, he shared it with the other great neuropathologist of his era, Santiago Ramón y Cajal. You'd think that this would be the cause for common celebration of a shared major scientific advance; yet at the ceremony they did not speak to one another. What had happened?

Golgi developed the silver nitrate stain that has massively improved our ability to visualise nerve cells, and a version of which is still in use today; with its help, he identified the intracellular organelles still known as Golgi bodies. And it was Golgi's technique that enabled Cajal to do his detailed drawings. But they disagreed about the nature of what they saw. Golgi thought that the filaments of the nervous system were interconnected to form a unified network, or syncytium. Cajal believed that there were separations – what Sherrington, in 1897, first termed synapses – between the cells. As we know, Cajal proved to be right on this point. And yet there was truth of a kind in what Golgi saw, too. If one looks at his detailed drawings of nerves (see Plate 17[a]), it is clear that, for all he saw an ultimately connected structure, he also saw a highly differentiated one: equally Cajal (Plate 17[b]), for all he saw a highly differentiated structure, also saw an ultimately connected one.

They both saw *differentiation within union*. A flow, while single, is also differentiated within itself, producing pattern and local form without losing its unity (see Chapter 23). And if we conceive a living organism as a flow, rather than as a machine, all nerves are ultimately interconnected, as much by chemicals in the synapse as by fibres within each nerve. Yet it is the potential for both separation *and* connexion, made possible only by the synapse, that enables necessary differentiation between the elements that constitute the whole, so that the whole is properly articulated. Jean Pierre Flourens, the founder of experimental brain science in the early nineteenth century, proposed what he called a *sensorium commune*, or common sensorium:

> In the last analysis ... all of the essential and various parts of the nervous system have specific properties, proper functions, distinct effects, and in spite of this marvellous diversity of properties, of

functions and effects, they constitute nevertheless a unified system. When one point in the nervous system becomes excited, it excites all others ... There is community of reaction. Unity is the great reigning principle; it is everywhere; dominates everything. The nervous system is then only one single system.[44]

44 Flourens 1948 (139).

And yet it is effective as a single, unified system – which manifestly it is – precisely because of differentiation within. It is a resonant, not simply uniform, whole. If *every* nerve really always excited all the others, we could not function at all, any more than if *no* nerve excited any other. Each nerve communicates, potentially, with all others – yes; but either by facilitating and promoting, or by inhibiting and delaying. It is not that in one case there isn't, and in the other there is, communication. *Each is a form of communication.* The existence of synapses, which seems at first sight so cumbersome – why keep interrupting an electrical discharge, by insisting on gaps across which the message must be transmitted by propagating and receiving chemicals? – is in reality of crucial import. In terms of physiology, the advantage reveals itself in flexibility. But there is also metaphysical meaning to the arrangement.

What I mean is this. Since existence always must balance oneness and multiplicity, oneness and differentiation, and since the nervous system underwrites our experience of the world, we would expect it to reflect that structure. The synapse is an embodied metaphor of how every complex system, both in nature and in society, works. By its capacity either to strengthen or weaken connexions that are always present to some degree, it enables both independence and interdependence at the same time.

Disputes between neurological lumpers and splitters go on to this day – for example, between those who believe primarily in 'modules' and those who believe primarily in more global systemic complexes – quite unnecessarily, since each possesses half a truth that only offers its insight when reconciled, not at war, with the other. The moral? Neither a lumper nor a splitter be.

There is a tendency in the human mind to want to embrace either unity or multiplicity, but not both. According to Archilochus's distinction, 'a fox knows many things, but a hedgehog one important thing', a distinction made famous by the philosopher Sir Isaiah Berlin, who applied it to categorise (approximately) many great artists and thinkers as hedgehogs or foxes. Yet the intellect requires both. On the one hand, the knowledge of many things is of no use if it is not capable of being held together in a coherent framework; on the other, the single great thought requires unfolding and differentiation.

In marginalia to his copy of the works of Sir Joshua Reynolds, Blake wrote, in the same spirit as the passage I have already quoted:

'To Generalize is to be an Idiot; To Particularize is the Alone Distinction of Merit.'[45] The point is well taken: but, manifestly, it is also itself a generalisation. Reality is, as ever, two-sided.[46]

UNIQUE AND GENERAL AT THE SAME TIME: AN INSIGHT FROM ANGLO-SAXON VERSE

The sense of delight in characteristic difference and sameness is expressed in some Anglo-Saxon verses often referred to as *Maxims*, or *Gnomic Verses*. They are moving, powerful and largely neglected. They appear to have little in the way of structure, but could be thought of as existing simply to celebrate the thisness of everything that is, while conveying the fierce indomitable, ruggedness of the world their makers inhabited and of the proper human response to it. Yet what they do is quite complex.

Much turns on the rich ambiguity of a key verb, *sceal*, which is repeated nearly 90 times in the course of these four poems, each of them 60–70 lines long. The ancestor of our word 'shall' (and pronounced in a roughly similar way), *sceal* has a number of meanings. It does not predict the future, as it now generally does, but is somewhat closer to its still extant use in making a decree: 'no officer of the law *shall* under any circumstances accept a bribe'. But it goes much further than that: it honours the cosmic order, something hard for us to recognise given our chaotic vision of the world. That something *sceal* (be or do whatever it may be), is not only a statement of how things are, which would on its own be superfluous; but an acceptance and an affirmation of their being so, as well as an acknowledgment that it is right and proper that things should fulfil their nature in that way, a celebration *of* that fact, and at times an exhortation to do what is demanded of us as right and proper under the circumstances, so as to follow, and further, this flow of things. It has something of the quality of early Ancient Greek *logos* and Confucian Chinese *lǐ* about it.

It covers states of affairs about which we can do nothing – such as growth and decay, the seasons, the elements, or the behaviour of wild animals; as well as those about which we can do something – such as responding appropriately to circumstances that require generosity, magnanimity, skill, courage or fortitude. In other words, it suggests an order out of which we come and to which we respond, and to which we have a respons-ibility.

For examples of states of affairs we are powerless to affect:

> The tree *sceal* stand on the earth losing its leaves – the branches [*sceal*] grieve.[47]

> The ocean water *sceal* foam with salt, and the cloud and sea-flood [*sceal*] flow around each and every land, in mighty streams.[48]

---

45 A response to Edmond Malone's comment, 'generalising and classification is the great glory of the human mind'; in marginalia to Blake's copy, now in the British Library, of *The Works of Sir Joshua Reynolds* in 3 vols, Cadell & Davies, London, 1798, vol 1: 'An account of the life and writings of the author', by Edmond Malone (xcviii).

46 In the film *The Life of Brian*, the Christ-figure, Brian, expostulates with a crowd of followers: 'Look, you don't need to follow me, you don't need to follow anyone. You're all different.' In unison they reply: 'Yes, we are *all different*.' 'No, you don't understand. You are *all individuals*.' In unison they reply: 'Yes, we are *all individuals*.' Immediately after, one little bloke objects: 'I'm not'.

47 *Maxims* I A [Exeter Cathedral Library MS 3501, ff 88v–90r], lines 25–6.

48 *Maxims* II [British Library, MS Cotton Tiberius, B i, ff 115r–115v], lines 45–7.

For examples of those for which we bear some responsibility:

> A ship *sceal* be riveted …[49]
>
> A mast *sceal* be on the ship, the towering sail-yard; the sword *sceal* be on the breast, the noble iron …[50]
>
> The king *sceal* bestow rings in the hall …[51]
>
> An army *sceal* stand together, a glorious band of men: good faith *sceal* be in the warrior, and wisdom in the man.[52]

Sceal also celebrates the belonging of things:

> The bow *sceal* be for the arrow, they *sceal* be like companions together …[53]

yet without judgment, good and ill having their place together:

> A shield *sceal* be for the fighter, a cudgel for the robber; a ring *sceal* be for the bride, books for the scholar …[54]

and in a similar vein we are told that:

> The thief *sceal* go about his business in dismal weather.[55]

Most of all, however, *sceal* seems to say that this is no more, but also no less, than the playing out of the multiplicity of creation, whether that be (to mortal eye) for good or ill:

> The wild hawk *sceal* light upon the glove. The wolf *sceal* live in the forest, grim and alone; the boar *sceal* be in the woods …[56]
>
> The river *sceal* mix with the waves in the sea-flood … The fish *sceal* be in the water, propagating its kind … The bear *sceal* be on the heath, hoary and fearful. The rivers from the hills *sceal* flow down as grey as the sea … The woods *sceal* be on earth, blossoming and flourishing; the mountain *sceal* stand fast upon the green earth.[57]
>
> The birds above *sceal* sport in the wind; the salmon *sceal* dart in the pool.[58] The shower taken up by the wind of the heavens *sceal* come down on this world.[59]

Each thing, one might say, cries 'Whát I dó is me: for that I came'. For us, having pillaged and defaced the beauty of the earth, and squandered the richness of its living kinds, these verses have, I find, a terrible poignancy – us, the aftercomers their authors never foresaw.

Acceptance, although a virtue in every wisdom tradition since time began, is more a sin in modern Western society than a virtue. But do not imagine that the writer or writers of these verses, whoever they may have been, were simply prescriptive: we do not know everything, they seem to say:

---

[49] *Maxims I B* [Exeter Cathedral Library MS 3501, ff 90r–91r], line 23.
[50] *Maxims II*, lines 24–6.
[51] *Maxims II*, lines 28–9.
[52] *Maxims II*, lines 31–3.
[53] *Maxims I C* [Exeter Cathedral Library MS 3501, ff 91r–92v], lines 16–17.
[54] *Maxims I B*, lines 59–60.
[55] *Maxims II*, line 42.
[56] *Maxims II*, lines 17–19.
[57] *Maxims II*, lines 23–35.
[58] *Maxims II*, lines 38–40.
[59] *Maxims II*, lines 40–1.

There are as many ideas as there are men upon the earth – each of them has a mind of his own.[60]

60 *Maxims I* C, lines 30–1.
61 *Maxims II*, lines 61–6.
62 Cavill 1999 (48).

And again, speaking of the afterlife:

What is ordained to come is secret and dark – the Lord, the sustaining Father, alone knows. None ever return under these roofs, who may truly tell to men what the Lord's decree might be, nor speak of the seat of the victorious ones, where He himself dwells.[61]

The paradox is that these particularisations are at another level generalisations, and as generalisations they are particularisations. As Paul Cavill, a scholar of Old English, says, 'verbs in maxims can both generalise and particularise, often at the same time'.[62] By saying that something belongs in a special way to a special place or in a special role we are suggesting there is a pattern to things: that it is in the *nature* of a bear to be ferocious, and in the *nature* of a hare to be timid. They are thus different from one another (across kinds), yet the same as one another (within kind). It is proper for the wolf to roam in the forest as a predator, because that is its nature. Each wolf is individually different, but not so much that one might expect to encounter a beach-dwelling vegetarian wolf any time soon. This is a parallel to the already discussed case of the individual person: internal continuity is a condition of external difference. In a world without boundaries or patterns, although in one sense everything would be different, by the same token everything would be the same. We are what we are by virtue of our defining, delimiting (in each case, literally 'bounding') qualities, which nonetheless paradoxically liberate ('unbounding') us into being what we are: what we are is disclosed equally by what we are and are *not*. Which is why groups cease to cohere if they have no criteria of exclusion, one of the commonest observations in sociology.

It is not just that, as human beings, our individuality takes its nature and meaning from the groups to which we belong, and that they take their nature and meaning from the individuals that belong to them, so that they are inextricably intertwined and reciprocally generative. It's true of the non-human world, too, and, indeed, of every aspect of experience. If rabbits suddenly took the habit of sinking their teeth and claws into any passing creature, or eagles twittered as they pecked at grass seed, there would be no more rabbits or eagles. A degree of generalisation makes differentiation – types or species of beings or phenomena – possible. The more we break down barriers, the less differentiation we have. Meaning derives from the existence of, and a proper delight in, recognisable patterns.

This implies a probably unfashionable degree of essentialism. Of course, define essentialism rigidly enough and naturally it is bound to be wrong: we should never assent indiscriminately to any idea. But it is not the same as Don Juan-like categorisation. Women – and men – are a natural kind: countesses, peasants, blond and brunette, tall and short, are not. Both hemispheres categorise, in different ways, as I have explained. But what is involved can and should be as much a matter of *recognising* reliable patterns as *imposing* a rigid uniformity. Just because we don't want to pigeonhole doesn't mean we should deny the existence of pigeons. By doing so, we wilfully blind ourselves to the forms and patterns that are everywhere in the lived world, and which give it the beautiful, orderly, richly meaningful landscape it has, rather than that of a featureless desert, filled only with identical particles of sand scattered hither and thither by every gust that blows.

THE SORCERER'S APPRENTICE

In the Introduction I mentioned the relevance of the myth of the sorcerer's apprentice. The left hemisphere thinks reality is what it itself puts together, because that is all it knows – the theoretical construct in which it lives. For it, theory trumps life. If it therefore decrees that a state of affairs shall be a certain way, then reality will, it believes, bend to the decree. Since the left hemisphere uses language to label, this often involves a belief that changing the label will change the reality. The left hemisphere takes truth to be what it says on the piece of paper.

That there is no one fixed reality, and truth is not single or fully certain, does not mean that reality has been mysteriously abolished. That there is no one view that encompasses all truth about the world, does not mean that truth is made up, and can be whatever we want it to be. Because not all moral codes are exactly the same everywhere and at all times, does not mean that morality is a useful fiction. Because species are not fixed for all time, but flow, does not mean that I cannot tell the difference between a chaffinch and a hawk, or a lamb and a tiger, and expect reliably different behaviours from each of them. Because men and women's social roles may change from place to place and from time to time does not mean that there are no essential differences between men and women, starting in a host of embodied differences, which are neither trivial nor artificially encapsulated, and moving on to psychological differences, which may be less starkly obvious (to some), but since we are seamless beings, not mere disembodied psyches, are still neither trivial nor artificially encapsulated, and ultimately underwrite and inform their expression in social norms – though variations in how they are expressed may also be normal, given that contexts change. We may, of course, annotate the slate, but the slate is not blank.

The left hemisphere veers unstably between two unrealistic positions in what it sees as an *opposition*: either all is fixed or all must be formless flux. The right hemisphere, on the other hand, is capable of seeing that while nothing is fixed over long enough stretches of time, this does not mean that chaos ensues. There is identity over time. The mountain is flowing, yes – *always* flowing; but equally importantly it does so so slowly that from day to day, from millennium, even, to millennium, we can depend on its massive presence: 'The mountain *sceal* stand fast upon the green earth'. As Polixenes counsels in *The Winter's Tale*:

> Yet Nature is made better by no mean
> But Nature makes that mean; so over that art
> Which you say adds to Nature is an art
> That Nature makes ...
>
> ...This is an art
> Which does mend Nature – change it, rather – but
> The art itself is Nature.[63]

In other words, change is *always* happening organically, right hemisphere fashion, in accord with the thing itself, with its nature, with its flow, with the *tao*: not by abrupt disjunctive steps, by theory, by decree, by brute force, left hemisphere fashion.

LEVELS OF CATEGORIES

If every instance of anything was always seen as new, never quite the same as anything else (as is actually the case), we would have to start from the beginning with each encounter – which, with a leaf, for example, would be a huge waste of time, not to mention fatal with a snake. So each hemisphere does need to categorise to a degree; I have discussed the differences in strategy in Chapter 15. To remind you, the left hemisphere relies on there being a qualifying feature that makes whatever it is eligible for a certain category. If it ticks the box for that feature, it belongs to the category. 'Wears a skirt → I know what to do'. The right hemisphere, by contrast, categorises by 'family resemblance'.[64] In other words, you can see that the elements go together and have a likeness, but there is no one feature that each element has to display in order to be a member of the category. There is a likeness in the whole, in the *pattern*. 'No skirt, but definitely feminine.' Appropriately enough, this hemisphere difference in pigeonholing has been demonstrated in pigeons, and is in fact basic to any intelligent creature's knowledge of the world.[65]

However, these two ways of categorising have different effects on the uniqueness and individuality of their constituents. The right hemisphere way does less than the left to subsume the individual case in the category, and still less to substitute the category for the

---

63 Act IV, sc iv, lines 89–92, 95–7.

64 Laeng, Shah & Kosslyn 1999; Laeng, Zarrinpar & Kosslyn 2003; Gauthier, Behrmann & Tarr 1999; Gauthier, Tarr, Anderson *et al* 1999.

65 Laeng, Zarrinpar & Kosslyn *op cit*; Yamazaki, Aust, Huber *et al* 2007; Lux, Marshall, Ritzl *et al* 2004. Also see Halpern, Güntürkün, Hopkins *et al* 2005.

individual case. By contrast, as soon as the focus is on an abstraction, a feature possessed by all members of the group, everything else about them – what makes them unique – tends to recede. (Women become just men without penises.) This is no small issue, since we are constantly categorising. Standing certain instances together in a certain place in your mind because they seem similar leaves open the question of what is similar, and why, returning the mind to be attentive to their actualities. On the other hand, placing those instances firmly in a category according to their possession of a certain attribute inevitably elevates the attribute over the whole entity to which it belongs.

Many of our frustrations with what one might call 'machine thinking' in the modern world are caused by the way decisions are made on the basis of whether or not something ticks the boxes, not on a feel of the whole as something essentially unique, with family resemblances to other things that have been valued for working well. Decisions using the left hemisphere's type of categorisation can be made in the abstract – law-making; decisions using the right hemisphere's categorisation have to be made face to face, by experience – the skill of the judge in applying the law.

A further difference here between the hemispheres is that they categorise to a different degree of generalisation. The left hemisphere has highly generalised, overarching categories, while the right hemisphere has finer-grained, 'lower-level' ones.[66] As Stephen Kosslyn puts it, 'the right-sided subsystem comes to be more narrowly tuned, whereas the left-sided subsystem funnels a *range* of shapes into a *single* representation'.[67] What this means in real-life terms is that the left hemisphere certainly recognises 'birds', perhaps even 'waders' (handy feature: long legs), but as for the difference between a sandpiper and a snipe (both waders, neither of which actually have particularly long legs), it's over to the right hemisphere. Once again there is a tendency towards the abstract, the general and a part-wise approach in the left hemisphere (does it tick the boxes?), and towards the embodied, whole and unique in the right hemisphere. Uniqueness is, after all, the ultimate case of 'subordinate' categorisation: being in a category of one.

This ability for the right hemisphere to see uniqueness, *as well as* to understand a whole, and to remember, or manipulate complex 3-D structures in space, may partially explain why 'the right hemisphere is of much greater importance in facial discrimination and recall'.[68] Indeed, it has been suggested that faces may just be a special case of fine-grained ('subordinate') categorisation, at the point where it reaches the unique.[69] The face is an extraordinarily complex form that is always unique and yet almost identical to every other face in terms of its structure: another instance of James's carpenter's ob-

66 Grossman 1981; Laeng, Zarrinpar & Kosslyn 2003.

67 Kosslyn 1987 (emphasis added).

68 Meadows 1974.

69 Gauthier, Tarr, Moylan *et al* 2000.

servation. Despite this we are capable of recognising a single face in a crowd, at any angle, in varying light and wearing different facial expressions, even when moving at speed.

UNIQUENESS, FAMILIARITY AND THE HEMISPHERES

Let me explore the issue of uniqueness a little further, this time more explicitly in relation to the hemispheres. It is a very important one, not only for an understanding of the nature of the world each hemisphere perceives, but for our understanding of the world at large.

Elkhonon Goldberg and his colleagues have demonstrated that there is a reliable difference between the hemispheres in terms of their handling of fresh experience. They have shown that new experience of any kind – whether it be of real-life objects, sounds, skills or imaginary constructs – engages the right hemisphere. As soon as it starts to become familiar or routine, the right hemisphere is less engaged and eventually the 'information' becomes the concern of the left hemisphere only. This transfer of activity can be seen on imaging.[70] It is a finding we might logically expect, because new experience tends to come from the periphery of the field of attention, and that is the province of the right hemisphere – which is, after all, on the lookout for whatever unexpected is happening, at the moment that it happens; whereas the left hemisphere is concentrating on what is already identified as of interest, known and familiar, and at the centre of the attentional field – in order to grasp it. This means that to have knowledge of something as it is fresh to experience – while it is still unique, and before it has become just 'one of those things' – we rely on the right hemisphere. Once the left hemisphere becomes engaged the thing has lost its uniqueness and become familiar and available. The tendency for the left hemisphere to make hyperfamiliar judgments and to fail to grasp uniqueness was discussed at some length in Chapter 6. This difference is fundamental for any conscious being trying to make sense of experience; and so it is unsurprising that the same right-left difference in respect of newness is found in animals.[71]

Just as in most languages other than English we distinguish between types of knowing, referring to the different ways in which our two hemispheres 'take in' the world (see p 631 above), so there are also two kinds of familiarity, which reflect these different senses of knowing. Familiar comes from Latin, *familia*, meaning one's household, those who are close to one. One way of being familiar is to be like something else – the *type* is familiar. You might call it 'generalised' familiarity. It is familiar in the way that a *cliché* or an electronic 'icon' is familiar. It is worn, and somewhat lacking in life. This is the familiarity that leads things to the left hemisphere. The other is almost the opposite. It is familiar precisely because it is *not* like

---

70 Goldberg & Costa 1981.

71 Siniscalchi, Pergola & Quaranta 2013; Siniscalchi, Sasso, Pepe *et al* 2010; Siniscalchi, Sasso, Pepe *et al* 2011; see also congruent findings in Rogers & Andrew 2002.

anything else. You might call it 'unique' familiarity. This is the way in which your friends and family, your favourite music, is familiar – in all their difference and complexity. This is the familiarity that is bestowed by experience and the right hemisphere.

It's not so much a matter of certain generalised *types* of materials that distinguish left from right hemisphere involvement: instead, there's a gradient of relative hemisphere involvement across the board, depending on the degree to which the material is routine or not.[72] These differences are reflected in the way the two hemispheres relate to the autonomic nervous system, that aspect of the nervous system that is not under direct conscious control, and which regulates heart rate and respiration, blood pressure, digestive function and so on. The left hemisphere is more closely related to the parasympathetic system, which puts the body into a state of rest, whenever all is, in Nietzsche's terms, 'equal' – familiar and 'under control'. The right hemisphere is more closely related to the sympathetic system, which prepares the body to face the unknown, because it is on the alert – to whatever is not familiar, not already certain.[73]

Uniqueness presents particular problems for the left hemisphere's tool, language. Uniqueness brings everyday language to a standstill. Anything truly unique cannot be expressed in such language, which is why whatever is profound, personal, or sacred, if it is to be expressed in words, can be so expressed only in poetry, the language of the right hemisphere. In poetry, language subverts its normal tendency to precision and becomes rich with ambiguity, with potential meaning again; and through the rifts created in the enclosing veil of language the light once more streams in.

Reductionist thinking, more typical of the left hemisphere, to which uniqueness is opaque, holds that all can be accounted for by breaking things down to further, and yet further, entities. Uniqueness, however, halts analysis: it is a standing rebuke to our ever-ready categories. It cannot be accounted for in terms of its parts. That is what an in-dividual means – an entity that cannot be further *divided*, without ceasing to be what it is. Individuals are, after all, *Gestalt* wholes: that face, that voice, that gait, that sheer 'quiddity' of your friend, defying analysis into parts. Once you break everything down into parts – generosity, kindness, humour, brown hair, blue eyes, etc – you are lost in the realm of generalities only. You couldn't put her together from this information. To know her, you'd just have to meet her (quite a few times – it's a process). She is real, unique, and has extension in time (there is a history to her – she's as much a process as you are). Her analysed description is none of these things and has none of these qualities.

72 Marzi & Berlucchi 1977; Bever & Chiarello 1974; Martin, Wiggs & Weisberg 1997; Henson, Shallice & Dolan 2000; Gold, Berman, Randolph *et al* 1996; Shadmehr & Holcomb 1997; Haier, Siegel, MacLachlan *et al* 1992; Tulving, Markowitsch, Craik *et al* 1996; Bradshaw & Nettleton 1983.

73 Craig 2005; Oppenheimer, Gelb, Girvin *et al* 1992; Wittling, Block, Genzel *et al* 1998; Sullivan 2004.

'ALL THE BIRDS LOOK THE SAME'

We have seen that right hemisphere lesions can lead to loss of the ability to recognise a face.[74] Although prosopagnosia has specific neural correlates, it is an aspect of a much wider problem that the left hemisphere has in dealing with uniqueness. Numerous subjects with right hemisphere lesions have reported that they could no longer tell which street they were in, even if they had lived there all their lives: the houses had lost their individual characteristics and all looked the same.[75] Theodor Landis and colleagues describe 16 cases of people who showed inability to differentiate unique, familiar surroundings, their street, house or room, from general categories of streets, houses and rooms, while still being quite capable of finding them on a map.[76] All had posterior lesions of the right hemisphere. One said that he could

> 'logically' figure out the correct building but could not recognize it. What he did recognize were the small, distinctive features, such as the garage, mailbox, and doorway ...

Here the left hemisphere is doing its best to identify by parts – but what is required is a grasp of the whole. Another with a right hemisphere lesion reported, in addition to being unable to recognise houses and streets, an inability to recognise familiar handwriting, including her own, and the inability to recognise familiar pet animals. This confirms that we are dealing with a specific disorder of uniqueness, of recognising complex wholes, not primarily of topographical orientation.[77] For example, one subject with a right hemisphere lesion could no longer distinguish a peach from an apple.[78] A Swiss woman, with right hemisphere damage, whose hobby since childhood involved identifying 'all the birds in her country', poignantly lamented that 'all the birds look the same'.[79] A farmer, who had always known his cows by name, could no longer tell his cows apart after right hemisphere damage, although he could just about distinguish them from a horse.[80]

Landis and a colleague, Marianne Regard, reported two patients with posterior infarcts in opposite hemispheres. The patient with the right-sided lesion, therefore relying on her left hemisphere, could read what was written in letters she received, but could not say from looking at the writing who had written them (she also could not recognise faces); the patient with the left-sided lesion, relying on her right hemisphere, could identify the handwriting immediately, but was unable to read what was written, because of course the right hemisphere cannot easily read.[81] Again, this is not just about handwriting: a person with a left hemisphere lesion could not articulate

---

74 See pp 139ff above.

75 Eg, Bornstein 1963; Aoki, Hiroki, Bando *et al* 2003. Although all had right hemisphere lesions, interestingly not all had posterior lesions (at least one case was frontal) – see Yoshimura & Otsuki 2002.

76 Landis, Cummings, Benson *et al* 1986.

77 See, eg, Paterson & Zangwill 1945; McFie, Piercy & Zangwill 1950; Hemphill & Klein 1948; Hécaen & Angelergues 1962.

78 De Renzi, Faglioni, Scotti *et al* 1972.

79 Bornstein *op cit* (284 & 303–11).

80 Bornstein, Sroka & Munitz 1969.

81 Landis & Regard 1988.

what she saw in a painting, but she knew immediately it was by Van Gogh.[82] In fact people with left hemisphere damage were found to be actually *better than normal subjects* at recognising the style of a painter, whereas the right hemisphere-damaged were grossly inferior: when asked to categorise paintings by artist, they simply categorised them by subject matter.[83]

The link between, on the one hand, the left hemisphere and generalised abstraction, and, on the other, between the right hemisphere and embodied uniqueness, is demonstrated by some research by Tatyana Chernigovskaya, asking patients with temporary right or left hemisphere suppression to comment on paintings of natural scenes. The paintings chosen were by Corot (*Morning*; *Evening*; *Morning in Venice*; and *Windy Weather*); Monet (*Waterloo Bridge, the Effect of Fog*); Morland (*The Approaching Storm*) and Shishkin (*Before the Storm*). The researchers were interested in the accuracy and quality of subjects' understanding of the various natural phenomena (such as weather and time of day), and the way in which the subjects described the paintings.[84]

Under conditions of left hemisphere *suppression*, therefore with the right hemisphere dominant, subjects proved highly accurate in interpreting concrete specifics, such as weather conditions and time of day, better even than with both hemispheres functioning. Their responses were not only accurate but rapid and concise, often coinciding (unknown to them) with the actual name of the painting (eg, 'windy weather' or 'approaching storm'). Subjects were able to discriminate the important qualities of the painting – the precise use of colour, the nuanced distribution of light and shade, and so on. And they were able to give an emotional appraisal of the picture they were describing.

However, under conditions of right hemisphere *suppression*, the same subjects showed a 'drastic' reduction in their accuracy of recognition, an inability to synthesise a whole from the parts, and a failure to distinguish the important features of a painting. Many subjects showed an inability to point to any feature at all (one subject, for example, said of the Monet picture: 'I can't make out anything whatsoever – it's an abstract painting'; another, 'it could be the surface of the earth, or of some material, or another planet'; another, 'the surfaces seem to be concave or convex' (it is of interest that such illusions are sometimes found in schizophrenia).[85] In general their responses were 'sterile, vague, and rambling', frequently having nothing to do with the picture at all. A serious effort was made by some to *categorise* the painting (the genre, the 'school', the historic period or place of execution). No adequate emotional appreciation of the pictures was made.

82 Lhermitte, Chedru & Chain 1973.

83 Kaplan & Gardner 1989.

84 Černigovskaja 1993.

85 For comparisons with schizophrenia, see, eg, Jung, 'Schizophrenia': 1953–79, vol 3 (259).

Chernigovskaya comments that whereas the right hemisphere underwrites the relationship between a complex painting and elements of the real experiential world, the left fits the picture into a system of conventional categories, failing either to identify the relevant features, or to provide an overarching appreciative synthesis. And as Cutting comments:

> What we see here is that the right hemisphere's *modus vivendi* is to be concerned with the uniqueness of some object in the world. Its imagistic repertoire portrays a vivid scene with a time and a place, and its linguistic faculty further links a word with a definite thing. The left hemisphere, on the other hand, can barely provide images of anything definite, even though it can adequately construe what sort [category] of thing it is presented with.[86]

Thus, the left hemisphere seems biased to registering perceptual information more abstractly, perhaps even limited to generic, categorical information, such as prototypical features, and discarding unique information that does not generalise across the linguistic category.[87] As previously noted, generating unique rather than common responses shows greatest activation in the right temporal pole.[88]

Pallis reported in detail the case of a man with a right posterior cerebral artery infarct, who told him:

> I found out all faces were alike. I couldn't tell the difference between my wife and my daughters. Later I had to wait for my wife or mother to speak before recognizing them. My mother is 80 years old ... I have difficulty in recognizing certain kinds of food on my plate, until I have tasted or smelled them. I can tell peas or bananas by their size and shape ... I bought some copies of *Men Only* and *London Opinion* ['girlie' magazines]. I couldn't enjoy the usual pictures. I could work out what was what by accessory details, but it's no fun that way. You've got to take it in at a glance ... It's when I'm out that the trouble starts. My reason tells me I must be in a certain place and yet I don't recognize it. It has to be worked out each time.[89]

Pallis notes that his patient had difficulty, not just with human faces, but with animal faces and forms, too. He had to resort, left hemisphere fashion, to analysis by *parts*: 'A goat was eventually recognized by its ears and beard, a giraffe by its neck, a crocodile by its dentition, and a cat by its whiskers'. Most fascinatingly of all, while he had difficulties with faces, animals, food, pictures and places, he did not with matters of *utility*. 'He readily recognized individual items of cutlery, glassware, furniture, garden tools, or other objects in everyday *use*.'

---

86 Cutting 2015.

87 Evans & Federmeier 2007. See also Vuilleumier, Henson, Driver *et al* 2002; and Kosslyn 1987.

88 Asari, Konishi, Jimura *et al* 2008. See also McGilchrist 2009b (51ff *et passim*).

89 Pallis 1955.

SACKS'S DR P AND CONTEMPORARY
COGNITIVE SCIENCE

One is forcibly reminded of the case of Oliver Sacks's 'Dr P', the 'man who mistook his wife for a hat': his doing so was, of course, a consequence of inability to read or recognise faces. He knew the Platonic solids, and other regular bodies, like the back of his hand.[90] Such regular solids, being easily categorised and typical, are accessible to the left hemisphere in a way that complex, irregular and unique forms found in nature are not.[91]

It was quite another matter, however, when it came to faces. Dr P tried to understand them as if they were abstract puzzles: and, even when he knew they were faces, could not recognise their identity or understand their expression or intention:

> He did not relate to them, he did not behold. No face was familiar to him, seen as a 'thou', being just identified as a set of features, an 'it'.

In describing pictures, Sacks comments that Dr P failed to see the whole, seeing only details; he confabulated non-existent features; he exhibited partial left field neglect; and confused animate and inanimate, eg, his shoe for his foot. However Dr P himself 'seemed untroubled, indifferent, maybe amused': his chirpy insouciance is also in keeping with right hemisphere damage. When asked what is the matter, he replies – 'with a smile' – 'Nothing that I know of ...'

As Sacks reflects,

> abstract shapes clearly presented no problems. What about faces? ... He approached these faces – even of those near and dear – as if they were abstract puzzles or tests. Thus, there was formal, but no trace of personal, gnosis. And with this went his indifference, or blindness, to expression. A face, to us, is a person looking out – we see, as it were, the person through his *persona*, his face. But for Dr P. there was no *persona* in this sense – no outward *persona*, and no person within ...

Presented with a flower, and asked what it is, Dr P replies:

> 'Not easy to say.' He seemed perplexed. 'It lacks the simple symmetry of the Platonic solids, although it may have a higher symmetry of its own ... I think this could be an inflorescence or flower' ... He saw nothing as familiar. Visually, he was lost in a world of lifeless abstractions.
>
> Indeed, he did not have a real visual world, as he did not have a real visual self. He could speak about things, but did not see them face-to-face ... [Dr P] functioned precisely as a machine functions. It wasn't merely that he displayed the same indifference to the visual

90 Sacks 1986 (11ff).

91 Umiltà, Bagnara & Simion 1978. See also Blakeslee 1980; Deglin 1976.

world as a computer but – even more strikingly – he construed the world as a computer construes it, by means of key features and schematic relationships. The scheme might be identified – in an 'identikit' way – without the reality being grasped at all ...

His paintings, originally 'naturalistic and realistic, with vivid mood and atmosphere, but finely detailed and concrete', became 'far more abstract, even geometrical and cubist'. Finally, in the last paintings, the canvasses became 'mere chaotic lines and blotches of paint'. Dr P's wife calls Sacks a philistine because he cannot see the change as an increase in artistic *development* – renouncing the realism of his earlier years, he had, she thought, advanced into abstract, nonrepresentational art. '"No, that's not it", I said to myself ... This was not the artist ... but the pathology, advancing – advancing towards a profound visual agnosia, in which all powers of representation and imagery, all sense of the concrete, all sense of reality, were being destroyed'.

These many changes are all in keeping with right hemisphere impairment.[92] And there is one further detail that is interesting. His wife recounts that he can get dressed, eat and bathe, only while singing to himself: and, if he is interrupted or otherwise loses the thread, he grinds to a complete halt, 'doesn't know his clothes – or his own body ... He can't do anything unless he makes it a song.'

Music is, after all, something that flows, is integrating, has its own momentum or life, and is indeed a deep image of life itself. It was drawn on by Dr P as a lifeline: the one unifying force that he could summon from his dysfunctional right hemisphere.

Sacks does not in fact specify which hemisphere is damaged, as he was not able to follow up the case. But even if there were not here such a classic array of phenomena suggesting right hemisphere damage, there are two further clues.

The great Russian neuropsychologist Aleksandr Luria made a famous case study of a soldier, whom Luria called simply 'Zasetsky', who had sustained a penetrating injury to the left occipital lobe. Sacks recalls that the soldier might not have been able any longer to play games, but that the vividness of his imagination had remained unimpaired. He comments that Luria's patient and Dr P lived in worlds that were 'mirror images' of one another. And he remembered Luria's words: that his patient fought to regain his lost faculties 'with the indomitable tenacity of the damned' – whereas Dr P 'did not know what was lost, did not indeed know that anything was lost. But who was more tragic, or who was more damned – the man who knew it, or the man who did not?'[93]

The soldier with the left hemisphere damage was distressed, but lived in the real world, where he was aware of what he had lost, and

---

92 The left hemisphere may prefer abstract art (Coney & Bruce 2004), although a more recent study found no hemispheric difference in its appreciation (Nadal, Schiavi & Cattaneo 2018). More likely what one is seeing here is a form of a (classically right hemisphere) constructional apraxia combined with visual agnosia.

93 Sacks 1986 (10–16).

his imagination was intact. Dr P lived in a world that was its 'mirror image', the consequence of right hemisphere damage – and didn't know what it was he didn't know.

The other clue to hemispheric involvement lies in a postscript Sacks appended to his account of Dr P. He refers to having subsequently stumbled on an earlier report which he describes as 'indeed identical neuropsychologically and phenomenologically'.[94] This case, reported by Macrae and Trolle in 1956, had a left homonymous hemianopia, strongly suggesting a right parietal lesion, but unfortunately in the 1950s there was of course no CT or MRI, and x-ray investigations were inconclusive (the authors suggest it is safest to conclude damage to both parietal regions). However, this patient had what we now know to be a litany of problems dependent on damage to the right parietal region:

94 Macrae & Trolle 1956.

- no depth perception;
- 'tunnel vision' (2 degrees out of 360);
- inability to see, or visualise in his imagination, faces (including of himself, his family, his friends or the US president);
- inability to see, or visualise in his imagination, animals, or 'animate objects' of any kind (though he could recognise simple everyday objects *of use*, such as a bicycle, an iron, scissors, a kettle, a watch and a key);
- inability to tell a cow and dog apart, except by size;
- inability to imagine colours if they did not fit the stereotype (eg, he could imagine green grass, but not brown grass – when reminded of the brown grass of California, he could not visualise it);
- inability to recognise his wife in a public place unless she wore a striking identifying feature, such as a large hat;
- inability to recognise his own image (though he knew that he had three identifying features – black hair, a receding hairline and a black mole on one cheek – and this was the only way he could know it might be himself).

According to Macrae and Trolle, he also 'tended to belittle his defects of vision or explain them away'. Some feat, as I imagine you will agree, even for the left hemisphere's Mr Micawber.

Sacks concludes with a reflection which is highly germane to the thesis of this book – a reflection on the current state of cognitive science. He draws attention to the fact that human cognition is never just abstract and mechanical, but must be personal as well. As such, it involves not just calculating and categorising, but feeling and judging, and that this is *essential* to our humanity. If this is missing, says, Sacks, we become more like computers: in this we are not dissimilar to Dr P. And, to the extent that we eliminate feeling and judging, the personal, from the cognitive sciences, we 'reduce them to something as defective as Dr P – and we reduce our apprehension of the concrete and real':

> By a sort of comic and awful analogy, our current cognitive neurology and psychology resemble nothing so much as poor Dr P! We need the concrete and real, as he did; and we fail to see this, as he failed to see it. Our cognitive sciences are themselves suffering from an agnosia essentially similar to Dr P's. Dr P may therefore serve as a warning and parable – of what happens to a science which eschews the judgmental, the particular, the personal, and becomes entirely abstract and computational.[95]

Quite. These words, written in 1985, some 35 years ago, are truer now than ever, and apply not just to science, but to society as a whole. Sacks was able to make the connexion between the psychopathology of an individual psyche and that of a culture – something, as I have discovered, that seems to flummox more literal minds.

SHOULD YOU BE YOUR 'SELF'?

Finally I want to turn to a paradox that concerns not just the world experienced by consciousness but the nature of the individual consciousness doing the experiencing. Hopkins says that each thing 'selves'. Yet he was a priest; and, in every spiritual tradition, we are recommended to turn away from the self, even to cultivate a condition of 'no-self'. Is the self, then, something to celebrate, or to deny? And if it is the purpose of the lives of other creatures to fulfil that self, should it really be ours to stifle it? Is it a regrettable illusion, that it is our life's work to expunge; or is it, on the contrary, the purpose of our existence to find it, grow it and fulfil it? And what do spiritual masters mean when they say that the self is, even, an illusion?[96]

Of course even a spiritual master has a body that is distinct from other bodies, a birth, a life history and a death that is unique. We are separate physical entities in different locations in space and time. There are, as I have suggested, many senses in which we share our identity with others: we are social beings who co-create one another and the world. But however we become aware of and cultivate intersubjectivity, however empathic we become, there must always be differences in our bodily experiences and feelings, formed of a personal history, and we cannot dismiss those experiences and feelings as illusory because they are at the very ground of who we are. The tendency to deny reality to the realm of experience (which is always personal) is contagious and one soon finds oneself reaching the unfortunate conclusion that *everything* has to be an illusion –

95 Sacks *op cit* (19).

96 For a fuller treatment of the issues involved, see McGilchrist 2016.

including, presumably, our belief that everything is an illusion; and leading to the question who suffers from the illusion, if we don't exist, and what reality is it that we are deceived about. Indeed, given the illusion, why should we concern ourselves with anything at all?

Often what seems to be being claimed is not so radical. It is that the self in the sense in which many people (particularly now in the West) unreflectively conceive it is a misleading concept. The self, it is being claimed, is neither as separate from other selves, nor as static and unchanging, as it is often thought to be. These claims seem to me to be far more interesting than the more absolute, less finely articulated, claims, and have the considerable additional merit that they don't undermine the reality and urgency of what they help us see.

In the discourse of spirituality it is sometimes mistakenly assumed that the contradictions between 'no-self' and self can be resolved by declaring the first to be true and the second an illusion. However, by neglecting the other 'arm' of the dipole we create difficulties in understanding – difficulties that are not, however, overcome by merely recognising the need for both in some additive sense, rather through a synthesis that creates something new and beyond what was in either of the alternatives identified. 'We have to learn, so to speak, to get out of our own light', wrote Aldous Huxley; and yet 'we must not abdicate our personal, conscious self.'[97]

The apparent conflict between self and no-self parallels that between the One and the Many; and between changelessness and eternal flow on the other. When people say to me, 'All is One', I readily agree. 'And All is Many', I add: 'now what?' As Suzuki says, 'the secret of Sōtō Zen is "yes, but".'[98]

The claim that All is One is well-intentioned, but, it seems to me disastrous, because it is just *half* a truth. We sense that we are not as separate as our everyday manner of thinking implies, and that is wonderful. But the impulse to simplify causes problems – because the other equal truth is All is Many. I would suggest that this attempt to have it one way or the other comes from the left hemisphere's urge to resolve what it sees as a contradiction. In Buddhism, writes Jane Hirshfield, 'non-duality is not the negation of multiplicity in favour of some idea of the absolute; it is also not the nihilism so many Westerners think Buddhism to be.'[99]

'All is One' and 'All is Many' to the left hemisphere demands an either/or resolution. For the right hemisphere it is a differently structured problem, since, for it, what one might call differentiated wholes – not created by an effort of cognition, so much as by one of recognition – are all that there is. Precisely because the left hemisphere sees what amount to geometric abstractions, and categories, that are snatched from time and embodiment, its analytic bent leads to an abstract, eternally unchanging unity of perfect forms: all uniqueness

97 Huxley 1992 (58).
98 Suzuki 1999 (8).
99 Hirshfield 1998 (98); quoted in Zwicky 2014b, §16.

lost. By contrast, the right hemisphere sees a fractal or holographic world, a multitude of individually unique wholes, or *Gestalten*, that themselves form part of an ever greater *Gestalt*, which is filled with implicit differentiation, not just unitary.

When it comes to understanding the self, one can predict that each hemisphere will support a different version. The self as conceived by the left hemisphere, should be – and is – an entity that is relatively static, separate, fixed, yet fragmentary, a succession of moments, goal-orientated, with its needs at any moment perceived as essentially competitive (since others may similarly target the same resources), determinate, consciously wilful, circumscribed in the breadth and depth of what it sees, at ease with the familiar, certain and explicit, but less so with all that is fluid, ambiguous, and implicit, and unaware of the limitations of its own knowledge. The self as conceived by the right hemisphere should be – and is – more akin to a process than a thing, essentially fluid and less determinate, nonetheless forming a unique whole over time, aware that it is fundamentally inseparable from all else that exists, open to others and to experience, more concerned with co-operation than competition, less consciously wilful, more engaged in what one might call 'active passivity' (an open attendant disposition, in which one is ready to respond to what emerges), seeing the greater picture in space and time, and aware of the extent of its ignorance.[100]

In other words, the self as intrinsically inseparable from the world in which it stands in relation to others, the social and empathic self, and the continuous sense of self, with 'depth' of existence over time, is more dependent on the right hemisphere;[101] whereas the objectified self, the external self, and the self as an expression of will, is generally more dependent on the left hemisphere. This seems to me to reflect the distinction made by Jung between the self (here RH) and the ego (here LH), fulfilling different, but necessary, functions. For him, the self is the product of psychic integration over time and unites conscious and unconscious processes, while the ego is that part of the self identified with the conscious will, and which, though necessary in the earlier stages of development in order to anchor the growing individual in the world, comes to be transcended in the process of spiritual growth. In being transcended it is not abolished, but changes its nature by being taken up into a new whole where its role is altered. As Cynthia Bourgeault points out:

> The egoic selfhood does not go away; rather, it becomes a good servant. It's still a very useful tool for many of the functions we are called on to perform in this world. But it is now 'transcended and included'; we recognize that it is a modality of action and not the seat of our identity.[102]

---

100 See McGilchrist 2009b (87ff).

101 Fossati, Hevenor, Graham *et al* 2003; Keenan, Gallup & Falk 2003; Kircher, Senior, Phillips *et al* 2001; Decety & Sommerville 2003. Specifically the right dorsomedial frontal cortex is critical (Mega & Cummings 2001).

102 Bourgeault 2016 (53).

In its proper place, no longer master but faithful servant, it comes to perform a useful function. It can be redeemed: transcended but included. We would not be better for its non-existence. This is Flint as embraced by He Grasps the Sky With Both Hands.

It seems that we should not fail to fulfil our selves but precisely in doing so come to transcend the narrow sense of the self, the one that holds that we are radically distinct rather than intrinsically connected. When Emerson advised that 'the man who renounces himself, comes to himself',[103] he did not, surely, intend that the sense of the self should be lost: rather that it should be transformed. Provided that is understood, we need to nurture, not destroy the self. Indeed, in order to empathise with others and incorporate their experiences into our own – to forget ourselves in knowledge of them – *we require an intact sense of self*.[104]

Perhaps, then, All is One – and All is Many. That is a central insight of Bergson's philosophy and given expression in his view of time – a phenomenon which, as we have seen, the right hemisphere seems far more capable of understanding than the left.[105] William James shared Bergson's insight:

> It is the general conceptualist difficulty [as represented in the LH] of any one thing being the same with many things, either at once or in succession, for the abstract concepts of oneness and manyness must needs exclude each other ... The concrete pulses of experience [RH] appear pent in by no such definite limits as our conceptual substitutes for them [LH] are confined by. They run into one another continuously and seem to interpenetrate. *What in them is relation and what is matter related is hard to discern.* You feel no one of them as inwardly simple, and no two as wholly without confluence where they touch. There is no datum so small as not to show this mystery, if mystery it be. The tiniest feeling that we can possibly have comes with an earlier and a later part and with a sense of their continuous procession.'[106]

I earlier quoted Bergson's observation that we can move from an insight to analysis, but not from analysis to insight. The broad and flexible can see the value of being narrow and rigid at times, whereas the narrow and rigid, by definition, can only see the value of being narrow and rigid.

James is a pluralist, an articulate advocate for the reality of individuation in the face of the common philosophical drive for generality. Talking of what he calls the 'each-form', the result of individuation, which is brought about by evolution (evolution here in the broadest sense, not just biological or Darwinian evolution – though that would be included in it – but the progressive differentiated unfolding of reality through the agency of time), he not only rebuts the charge that such a process leads to loss of order, but

103 Emerson 2003 (109).

104 Johnstone, Cohen, Bryant et al 2015. See also Grattan, Bloomer, Archambault et al 1994. In psychopathological terms, the lack of a secure sense of the self is often associated with narcissism and lack of empathy.

105 For discussion, see Lawlor & Moulard-Leonard 2004.

106 James 1909a (281–2; emphasis added).

claims that, compared with the 'all-form' preferred by monists, the differentiated world brought about by time has a greater, and more fruitfully complex, order – one that is pregnant with possibility:

> Monism thinks that the all-form or collective-unit form is the only form that is rational. The all-form allows of no taking up and dropping of connexions, for in the all the parts are essentially and eternally co-implicated. In the each-form, on the contrary, a thing may be connected by intermediary things, with a thing with which it has no immediate or essential connexion. It is thus at all times in many possible connexions *which are not necessarily actualized at the moment* …

And, in a passage that seems to define not just phenomenological reality, but the very structure of the synaptic brain and nervous system, he continues:

> Our 'multiverse' still makes a 'universe'; for every part, tho it may not be in actual or immediate connexion, is nevertheless in some possible or mediated connexion, with every other part however remote, through the fact that each part hangs together with its very next neighbours in inextricable interfusion.[107]

RESOLVING THE ONE AND THE MANY

Schelling says that there is no higher revelation in all of science, religion or art, than that of the divinity of what he calls the 'All'; but this comes on the back of his recognition that each sphere of intellect and spirit – science, religion and art – sees something particular and special. In those ages, he warns, where we are mindful of this unity, a culture enjoys vigour, and vitality, and the fruits of the collaboration of the arts and sciences: but the price of losing that vision is the loss of everything we value. We struggle, he says, to put things together, adding grain of sand to grain of sand.[108] Unsurprisingly the left hemisphere, having dismantled the universe, it is at a loss to know how to put it together again.

So now we see the Many, but no longer the One. In Eastern thought (and the same can be found in Hegel, Heidegger and other thinkers of the Western tradition), there has always been an important dialectic between the figure and the ground in which it is set, between the distinguishable and the whole of which it is one element, between the specifiable (because limited) and the unspecifiable (because infinite) context which qualifies it. In the *Bhagavad Gita*, Lord Krishna advises Arjuna:

> When one sees Eternity in things that pass away and Infinity in finite things, then one has pure knowledge. But if one merely sees the diversity of things, with their divisions and limitations, then one has impure knowledge.[109]

---

107 *ibid* (324–5; emphasis added).

108 See p 1328 below.

109 *Bhagavad Gita*, 18: §20–1 (trans J Mascaró).

CONCLUSION

Whatever is general lies outside time and has no place in space. Whatever exists in time and space is *ipso facto* unique; though in it and through it one sees the general and the timeless, not as separate, but as another facet of the same entity. What is unique at one level is general at another; what is general at one level is unique at another. Thus, the pair of phenomena is coincident: '*not again nor later but at the same time* it forms and dissolves, and approaches and departs.'[110] We need both, and each gives rise to the other, not in sequence but simultaneously. Once again, the opposites that are indicated by the One and the Many, the unique and the general, remain opposite, while being nonetheless coincident; and hence generative.

Note that both the uniqueness of the individual case *and* the oneness of the whole are dependent for their appreciation on the right hemisphere. The left hemisphere substitutes membership of highly generalised categories for uniqueness, and then tries to achieve a sense of the whole by aggregating these categories. It is part of the unifying disposition of the right hemisphere to see similarity within difference, and part of its capacity for fine discrimination to see difference within similarity, whereas the isolating disposition of the left hemisphere sees similarity and difference as a simple opposition, at loggerheads with one another.

However, there is a role here for the left hemisphere – provided, as always, its contribution is in service to that of the right.

In these first two chapters of Part III, I have tried to illuminate the *processes* that bring into being, and give form or structure to, reality. In the subsequent chapters I will examine various *aspects* of that reality: time, flow, space, matter, consciousness, value, purpose and the sense of the sacred.

William James asked, in relation to the question of unity and multiplicity, why 'should the absolute ever have lapsed from the perfection of its own integral experience of things, and refracted itself into all our finite experiences?'[111] It's a very good question.

> Grant that the spectacle or world-romance offered to itself by the absolute is in the absolute's eyes perfect. Why would not the world be more perfect by having the affair remain in just those terms, and by not having any finite spectators to come in and add to what was perfect already their innumerable imperfect manners of seeing the same spectacle? Suppose the entire universe to consist of one superb copy of a book, fit for the ideal reader. Is that universe improved or deteriorated by having myriads of garbled and misprinted separate leaves and chapters also created, giving false impressions of the book to whoever looks at them? To say the least, the balance of rationality is not obviously in favour of such added mutilations. So this question

110 See p 818 above.

111 James *op cit* (120).

becomes urgent: why, the absolute's own total vision of things being so rational, was it necessary to comminute it into all these coexisting inferior fragmentary visions?[112]

112 *ibid* (118–9).

Asking this searching question invites another intimately related question: why the timeless could or should ever have given rise to time. In the chapter that follows, I will turn our attention to the question of the nature of time, and ask if the hemisphere hypothesis can help us come to a closer understanding of this mysterious element in experience.

Chapter 22 · Time

> Space and time can never be mere side-shows in philosophy. Their treatment must colour the whole subsequent development of the subject.
> —ALFRED NORTH WHITEHEAD[1]

> What, then, is time?
> —ST AUGUSTINE[2]

> I do not define time … as being well known to all.
> —ISAAC NEWTON[3]

'These questions above all force themselves on the speculative mind', wrote Schopenhauer in 1819:

> What is *time*? What is this being that consists in nothing but movement, without anything that moves? and, What is *space*, this omnipresent nothing, out of which no thing can emerge without ceasing to be something? … To think away time and space is completely impossible, while it is very easy to think away everything that appears in them. The hand can let go of everything, except itself.[4]

In the next few chapters I will be trying to come to some kind of understanding of this, in itself ungraspable, ground of our being in time and space, to grasp the hand that can let go of everything but itself. At any rate, as in the approach to all deep questions, at least to see more clearly what it is we now believe that is unlikely to be the case. And in this chapter it is specifically to time that I will turn.

Asking humans about time is like asking fish about water. In order to get a handle on something as tasteless, colourless and odourless as time, it is helpful to see what time is *not*. I earlier emphasised the consequences of distortions of time in schizophrenia: as the fish discovers the true value of water only when it is tossed up on the shore, we find out how necessary time is only when we lose all sense of it. Time is our home, and indeed death is our friend: the attempt to flee from either of them, as we do in the modern Western world, is not just doomed to fail, but to thwart a fulfilled and fulfilling life. I also believe that the philosophical positions that time is an illusion, or that it is in reality stationary (which amounts to the same thing), or is composed of moments or slices, entail philosophical mistakes with far-reaching consequences. They are, in effect, forms of delusion.

Before going any further, I should make it clear that naturally I do not pretend to do justice to a debate on the nature of time which has long preoccupied, and continues to preoccupy, philosophers and physicists, and, assuming time to be real, will carry on doing

---

1 Whitehead 1922 (222).

2 Augustine, *Confessions*, Bk XI, ch xiv, §17: 'Quid est enim tempus?'

3 Newton 1687.

4 Schopenhauer 2000, vol II (33): » Denn dem spekulativen Geiste drängen sich vor allen diese Fragen auf: Was ist die Zeit? was ist dies Wesen, das aus lauter Bewegung besteht, ohne etwas, das sich bewegt? – und was der Raum? dieses allgegenwärtige Nichts, aus welchem kein Ding herauskann, ohne aufzuhören Etwas zu seyn? … [Es bleibt die gänzliche] Unmöglichkeit Zeit und Raum hinwegzudenken, während man Alles, was in ihnen sich darstellt, sehr leicht hinwegdenkt. Die Hand kann Alles fahren lassen; nur sich selbst nicht. «

so for a long time to come. I have merely a different light to cast on the debate which might prove useful. What strikes me is that the debate centres on a nexus of issues that have definite hemisphere correlates: the difference between *representations* of experience and experience itself; between stasis and flow; between points of time and time as duration; between predictability and uncertainty; and between closedness and openness. If you have accompanied me so far in the book, you will know that, in each of these cases, the first is more associated with the left hemisphere and the second with the right. These elements, with the emphasis at times in different places, are commonly at the heart of many philosophical disagreements. What I can offer here is, as always, just one more perspective, but it is one on the basis of which two claims might be made.

One, the weaker one, is that these philosophical differences are to be expected, since they represent the difference in experience of the two hemispheres, and each provides a take on reality. Since our consciousness cannot be separated from the experience of time on which any debate is predicated, the division is natural.

But there is another, stronger, claim to be made, which depends on a view I have argued for throughout this book and its predecessor, namely that while each of the differing 'takes' on reality of the two hemispheres has value, their value is asymmetrical. If you are with me so far, you will want to see the need for both takes, but ultimately give priority to the take of the right hemisphere. I believe there is something other than the contents of our own minds to which each of us aims to be true – and the right hemisphere is, on any account we can advance, a better witness to that reality than the left. Because the left hemisphere is less perceptually based, more theoretically based, and because it tends to focus on what it brings to the foreground – in this case, time – without awareness of the background that alone makes a foreground possible – in this case, consciousness – the 'unreality of time' position looks like it just could be another mess the left hemisphere has got itself into.

There are further grounds for preferring the right hemisphere take that are particular to the nature of time. As I have already demonstrated (in Chapter 2), the right hemisphere is far more 'geared' to the appreciation of time than is the left – I will not repeat that evidence here. This particular asymmetry is not surprising since spans, whether in space or time, are better comprehended by the broad attention and more extensive working memory of the right hemisphere. It is also not surprising because the right hemisphere is the hemisphere that is more in touch with reality: as usual, when the left hemisphere cannot grasp something, its reaction is to deny its existence. And we see this in relation to time. We saw it at the phenomenological level in our exploration of schizophrenia and autism (Chapter 9); and at the metaphysical level in examining the paradoxes of Zeno (Chapter 16). In both cases time ceases to function. In both cases life and the world become incomprehensible and grind, in every sense, to a halt.

In what follows I will contrast two, in their own terms coherent, views of time – though, as I shall suggest, only one can be considered a view of time at all.

In one, time is a thing, to be grasped and – if only! – fixed, a commodity that diminishes as it is allowed to pass away; in the other, a process of becoming, to be inhabited and let go. In the first, time is frozen, and spatialised in its re-presentation by the analytical intellect and by language, delivering a world which is sliced and fragmented; in the second, it is forever presencing to our intuition and our embodied cognition, where it delivers a seamlessly integrated reality which cannot be divorced from value and from purpose. In the first, time is effectively denied, and life drained of meaning; in the second, both time – and, with it, life – are affirmed and celebrated.

TIME IS NOT A THING

A misunderstanding of time lies at the heart of modern culture, affecting what it means to be a living being; and it is to do with the reification of time. Conceptualising time immediately puts one on the outside of the experience, rather than being within it, the standpoint alone from which it can be understood. As a result, time is spatialised in our intellect and inevitably immobilised: and it becomes a thing. By the very act of conceiving it, we place *time itself already outside time*.

Time is no thing. In the thing-ridden left hemisphere world, it is a short step from being no thing to being nothing. But in reality time is more real than mere things could ever be. Time is adverbial, if it is anything that grammar recognises; an aspect of being (itself a verbal noun, or gerund) or of *Dasein* (also a gerund). Time is not separate from events or experience, and like love, which is also no thing, is revealed in and through events – it is itself an *aspect* of experience: it is, in other words, as I say, adverbial, not substantive. Friedrich Waismann says of time: 'The more we look at it the more we are puzzled: it seems charged with paradoxes ... But isn't the answer to this that what mystifies us lies in the *noun* form 'the time'? ... We are trying to catch the shadows passed by the opacities of speech.'[5]

It seems to me that Einstein is here stating a similar insight:

> Spacetime is not necessarily something to which one can ascribe a separate existence, independently of the actual objects of physical reality. Physical objects are not *in space*, but these objects are *spatially extended*. In this way the concept 'empty space' loses its meaning.[6]

5 Waismann 1968 (6).

6 Einstein 1961 (vii).

That is something I shall return to in a subsequent chapter when discussing space, but since space and time, however different they may and must be, are never entirely separable, we might be best to say of objects, similarly, that they are *temporally extended*, rather than existing *in time*. Time actually means something 'stretched': the Latin word for time, *tempus*, derives from a root meaning to 'stretch' in space like a string.[7]

Its status as a noun goes hand in hand with turning it into a thing. Abraham Heschel, in his classic meditation on time, *The Sabbath*, wrote:

> We are all infatuated with the splendour of space, with the grandeur of things of space. Thing is a category that lies heavy on our minds, tyrannizing all our thoughts ... Reality to us is thinghood, consisting of substances that occupy space; even God is conceived by most of us as a thing. The result of our thinginess is our blindness to all reality that fails to identify itself as a thing, as a matter of fact. This is obvious in our understanding of time, which, being thingless and insubstantial, appears to us as if it had no reality.[8]

And he continues, in a fascinating observation,

> there is no equivalent for the word 'thing' in biblical Hebrew. The word '*davar*', which in later Hebrew came to denote thing, means in biblical Hebrew: speech; word; message; report; tidings; advice; request; promise; decision; sentence; theme, story; saying, utterance; business, occupation; acts; good deeds; events; way, manner, reason, cause; but never 'thing'. Is this a sign of linguistic poverty, or rather an indication of an unwarped view of the world, of not equating reality (derived from the Latin word *res*, thing) with thinghood?[9]

As the reader knows, I think there is something the matter with things – or at least with the way in which we conceive them. Because even 'things' have changed their significance with the evolution of the West. Since the rest of this book is devoted to what I take to be the foundational elements of the world we know, and since time is the first great example of how thingness can lead us astray, now is the time to take a look at the nature of things.

Heschel's account of the word *davar* is far from being a comment on Hebrew alone. For the word 'thing' in English also derives from a remarkably similar array of uses. In Old Saxon it meant an assembly for judicial or deliberative purposes, conference, transaction, matter, or affair; and similar meanings are found in the evolution of the German, Dutch, Norwegian, Swedish and Icelandic terms for thing. The tenth-century Icelandic parliament, said to be the oldest of its kind in the world, was called the Althing. In Old English the first meaning of 'thing', dating from the seventh century, was a judi-

7 Allen 1880 (140).
8 Heschel 2005 (5).
9 *ibid* (7).

cial and deliberative assembly; hence, by the year 1000 AD, a *matter* (which contains the same ambiguity) brought before a court, and by extension an affair (literally the verbal phrase *à faire*, something 'to do') with which one was concerned. The root 'thing' can also mean a *cause*, both in the sense of the ground or reason for doing something, and in the sense of a cause that one pleads before a court, or a 'good' cause to which one is dedicated. Hence in French *chose*, and in Italian and Spanish *cosa* (from Latin, *causa*, a cause).

To begin with, then, a thing was not something fixed, isolated and inert, but something to be consensually decided; not certain (*certus* = already decided), but yet to be determined; not static, but moving – a process arising from the coming together of more than one party – just like Heschel's 'advice, request, promise'.

The Greek word for a thing, πρᾶγμα (*pragma*), meant originally an action, a deed (hence our 'pragmatic'), from πράττειν, to do, hence an 'affair' or matter of concern. Plato and Aristotle are responsible for its conversion into a concrete thing that is implicitly, or explicitly, 'real'. Until then such meanings were reserved to χρῆμα (*chrēma*), 'a necessity', and κτῆμα (*ktēma*), 'a possession'. The Latin word for a thing, *rēs*, also meant an issue or matter of concern, which is why formal letters are often headed *re* (originally *in rē*), meaning 'in the matter of' (whatever it may be).

In Chinese, the present word for a thing, *dōngxī*, literally means East-West, and though no one etymology has been established, it is thought that this might have originated, relatively late, in the market place, where material goods were traded across continents. However, an earlier word *wù shì*, contains the idea of 'matters' of an intangible kind (*shì*), and could also mean 'story'.[10]

What one sees in all this is that the concept of things seems to emerge historically from the nexus of deeds and affairs: things are then the elements of the picture that stand forth as being of particular concern to us. Even things were not always very thing-like, you might say, in the modern sense. They were in fact *reciprocal indeterminate processes that took place – in time*. Not only is time not a thing, but things are events in time.

This excursion into the history of our concept of things tends to support the view that I have argued for, that things are secondary properties of phenomena that emerge out of the web of experience, as 'objects' that attract our focussed (left hemisphere) attention. In fact an object is just that: it becomes an object by being the focus of a certain kind of foregrounding, isolating, immobilising attention (Latin *objectus*, 'thrown against'). It is what presents itself as useful to grasp. This is what I believe Nietzsche meant by saying that 'the "being" of things has been *inserted* by us (for practical, useful, perspectival reasons)'.[11]

---

10 I am indebted to Jeannemarie Gescher for this information. She points out that *wù shì* comes from the Southern Chinese tradition, considered more liberal than that of the North.

11 Nietzsche 2003b, §11 (212; emphasis in original).

THINGS ARE SECONDARY TO PROCESSES

Thing-ness sounds like a celebration of embodiedness, of earthiness – what am I saying? – of the *body*, of the *earth*. Not so. For a start, as that sentence demonstrates, it can be in the service of a form of mental abstraction. Thing-ness could even be seen as beginning in something as remote as a grammatical shift, which is less surprising when one considers that words help shape our conscious apprehension of the world. It is an accident, or rather a characteristic, of Greek grammar that gave us the definite article. This has the potential to turn adjectives into nouns. No longer are there only beautiful experiences – mountains, statues, poems or people that are beautiful (κάλοι). There is now such a thing as '*the* beautiful' (τὸ καλόν) – namely, beauty. This is not true of all languages: for example, there are very few abstract nouns in Japanese to this day, and no established method for their composition:

> Japanese does not get along well with abstract nouns. In fact, Japanese has far fewer abstract nouns than does English, and to a surprising degree. Ōno [the distinguished Japanese historian of language and philosopher, Susumu Ōno] reports that even abstract nouns for such basic concepts as right and wrong did not exist in old Japanese ...[12]

As a result, the Japanese have nothing that corresponds to the Platonic Idea, and in fact no true abstractions in general: they have never developed the dichotomy between the phenomenological world and the world of ideas.[13]

OK, you may say, but that just concerns abstract nouns. What about what we call concrete nouns? The mountains, statues, poems or people that were found to be beautiful? In what sense are they *not* things?

But as good a question as it is, in what sense *are* they things? A thing suggests permanence, and separation from what surrounds it. But that is all a matter of the timescale you happen to adopt. Entities that change fast we see as processes; entities that change slowly we see as objects. By any cosmic scale, the mountain that erupts, flows, sediments, then erodes has only boundaries of convenience, and permanence for a short while: as does, much more obviously, the statue, the poem or the person. It is really true that *everything flows*.

THE SPATIALISATION OF TIME

The conceptualising mind can deal with time only as a line in *space*. Julian Jaynes thought the spatialisation of time inevitable: 'You cannot, absolutely cannot think of time except by spatialising it', he wrote. 'Consciousness is always a spatialisation in which the dia-

---

12 Hasegawa 2012 (172). The reference is to Ōno 1978. In 1930s Japan, a silent movie created a sensation and achieved box office success, reportedly due in great part to its linguistically eccentric title, *Nani ga kanojo o sō saseta ka* (*What made her do it?*). It used familiar vocabulary and familiar grammatical structure, but it juxtaposed an abstract subject to the causative predicate, which just did not happen in normal Japanese.

13 Kawasaki 2002.

chronic is turned into the synchronic, in which what has happened in time is excerpted and seen in side-by-sideness.'[14] That is to say, automatically outside of time. Here Jaynes's 'consciousness' is the left hemisphere at work, which sees only time deprived of duration, projected at an instant as a single point, line or plane, become in his phrase synchronic – by contrast with diachronic time, duration, as given to the right hemisphere.

Speech is the mode of operation *par excellence* of the left hemisphere. In order to *speak about* time we are obliged to spatialise it, and this is so normal that most people are not even aware of the process and cannot imagine trying to avoid it. It obtrudes on our attention only when we encounter a culture that spatialises time in a different way.

Predominantly linear representations of time are a feature of Western societies, though the church calendar, with its recurring celebrations tied to the cycle of the seasons and their gifts, once acted as a living, unifying, counterpoint.[15] Time is no longer impressed with the circular shape that seasons – blossom, flower and fruit – and the cycle of the generations, born, marrying and dying within a community, made tangible. Circular time is collective time, time that binds us and the living world together, and its metaphysical representations go beyond the life of a single individual. Linear time is individual time and it is physically represented by an arrow flying ineluctably from the past to the future. Projected onto a society, it generates the myth of progress.[16] While both circular and linear representations of time can be found in most cultures, Chinese, Indian and Native American societies have predominantly circular representations of time. (This could be, and has at times been, misunderstood as meaning that they involve mechanistic repetition, which is not the case.)[17] Indeed most cultures use the metaphor of a spiral or wheel, based on the cyclical seasons, rather than positing that the Earth has a creation and a future ending, with 'progress' in between.

Circular time is the time of the body with its many rhythms, the time of nature's seasons, and of the rise and fall of civilisations. The shape of circular time made our relationship to one another and to the world porous, rather than, to use the term employed by Charles Taylor in *A Secular Age*, 'buffered': we were at home in the world, dwelling in it, rather than skating over it for a while on our simple linear path.

The effects of time on the surfaces of the world enabled us somehow to sink into it; they were *porous* to us. We no longer rest deeply in places any more but glide across their smooth, repellent surfaces, both the spatial ones and the temporal ones. The buildings we now make, though unlikely to last as long as the stone temples of

14 Jaynes 1993 (60).

15 Eg, Tordjman 2015; Le Guen & Balam 2012; and the Blackfoot (Kainai) elder Leroy Little Bear (personal communication).

16 See, esp, Wessels 2013.

17 Eg, Toynbee 1934–61: see Balslev 2014.

the Ancient World, seem to defy time, unable to age gracefully: their concrete, steel and glass either defiantly pristine or (eventually) dirty, scratched, covered in grime and ageing disgracefully. It is this aspect of modern architecture that the Finnish architect Juhani Pallasmaa has written about in *The Eyes of the Skin*, and resisted in his own work.[18]

18 Pallasmaa 2005.

19 Proust 1992 (68–9).

We have always loved old stone or wooden surfaces; and ultimately the pleasure we take in them is the tangible demonstration of the flow of time. Far from diminishments of the human, they are testimonies to what great things can be done by humans in keeping with time – though always ultimately respecting it, and yielding to it. We see there the triumph of process over stasis, of the *memento mori* over the man-made monument. And here, in his great prose poem on the subject of time, is Proust evoking the deep timelessness of the time-ravaged church at Combray:

> How I loved our church, and how clearly I can see it still! The old porch by which we entered, black, and full of holes as a colander, was worn out of shape and deeply furrowed at the sides (as also was the font to which it led us) just as if the gentle friction of the cloaks of peasant-women coming into church, and of their fingers dipping into the holy water, had managed by age-long repetition to acquire a destructive force, to impress itself on the stone, to carve grooves in it like those made by cart-wheels upon stone gate-posts which they bump against every day. Its memorial stones, beneath which the noble dust of the Abbots of Combray who lay buried there furnished the choir with a sort of spiritual pavement, were themselves no longer hard and lifeless matter, for time had softened and sweetened them and had made them flow like honey beyond their proper margins, here oozing out in a golden stream, washing from its place a florid Gothic capital, drowning the white violets of the marble floor …[19]

The stone becomes, through time's agency, something that flows, like the mountains that over aeons come to nothingness. The pleasure of ruins captures the imagination because our most monumental efforts to make things that stand still forever are seen to result ultimately in flow; because nature's vitality triumphs over the schemes of mankind; because the vestigial and dilapidated give rise to imaginative sublimity, as a mountain partially lost in mist seems more awe-inspiring than one in plain sight. All of which return us from the world of the left hemisphere to that of the right.

Above all, time is a way of precipitating out into infinitely various actuality the undifferentiated oneness from which the universe began; and it is this that I believe Blake meant by the wonderful line: 'Eternity is in love with the productions of time', which I have already quoted. Time, as Shelley might have said, 'like a dome of many-coloured glass, stains the white radiance of eternity', evoking the colours of the stained glass of Combray's church windows washing across the floor.

Movement is not possible without the existence of both space and time. Whether or not time or space could exist without motion, the least one can say is that they are inconceivable without it. If one tries to imagine what space would be like in a static universe, one cannot help one's point of view moving within it. And time in which there was no movement, and therefore no possibility of change, would have no meaning: it would be automatically 'outside time' (the German word for 'verb' is *Zeitwort*, a 'time-word'). Perhaps this displays only that our experience of time and space is bound up with our consciousness, and that we cannot transcend what our consciousness delivers: but since the very terms 'time' and 'space' are themselves attempts to describe aspects of experience, it is self-defeating to try to apply them to something we could never experience at all. They are both realms of potential in relation to the 'here' and the 'now' of experience, potential that is constantly, as I say, precipitated out into the actual through flow; and have meaning only in relation to the purposes of a conscious being that can *inhabit* them.

If one ceases to inhabit them, to be *present* within them, one has only *re*-presentations from the outside. It is this that is treacherous. In these representations space becomes two-dimensional, and time loses its single, only dimension, and becomes frozen.

THE REPRESENTATIONS OF LINEAR TIME

There is still, however, a place for the linear expression of time: we need to speak of yesterday and tomorrow. Once again, note: over the short term time appears linear, as the round horizon appears straight and the round earth flat – over the short term.

In linear time we normally think of the future as lying in front of us, as that 'space of time' into which we move forward, and the past as what we leave behind us. However, in some languages, for example the Peruvian language Aymara, or in Malagasy, the language spoken by the people of Madagascar, the future is in the space behind us – since we can't see it – and the past is in front of us, because it is visible, at any rate present to our mind's eye. These differences are expressed in the direction in which Aymara speakers gesture when referring to the future or past compared with Spanish speakers.[20]

In Chinese, there is both a vertical and horizontal time axis: left before right, as in the West, but also above before below. When asked to arrange pictures of actors in order of age, bilingual Mandarin and English speakers spontaneously arranged pictures of the American actor Brad Pitt in age order from left to right, but pictures of Jet Li, a Chinese film star, top to bottom.[21] It may be relevant that Chinese, Japanese and Korean script can be (and used only to be) written

20 For Aymara, see Núñez & Sweetser 2006; for Malagasy, see Lewis 2006 (61).

21 Miles, Tan, Noble *et al* 2011.

from top to bottom (and from right to left). When writing vertically, it is commoner for Chinese speakers to use Chinese numerals, and when writing horizontally to use Arabic numerals.[22]

One might ask why, when arranging the pictures of Brad Pitt horizontally, Chinese speakers do not arrange them right to left, rather than left to right. There may be two reasons. One is that although vertical writing was right to left, and therefore at one point Chinese horizontal writing was also right to left, the pressure to write horizontally was Western and the direction has since followed Western left-to-right convention. The other is that there may be an inbuilt sense of increasing magnitude as one moves into right hemispace: for instance, even three-day-old chicks already share this representation of numbers, consistently seeking lower numbers to the left of a target and larger numbers to the right.[23] This phenomenon has been put forward as the explanation of an intriguing finding that patients with left hemineglect (due to a right hemisphere lesion) fail to recall items in a story associated with the past, but remember items associated with the future: the researchers assume that this is because the past is represented in the left half of mental space, and the future in the right.[24] (It might have more to do with the possible predilection for the future in the left hemisphere and for the past in the right hemisphere: see p 920 below.) This 'timeline' representation survives right hemisphere damage, and so must be at least partly left hemisphere-dependent (no-one to date seems to have investigated whether the timeline survives left hemisphere damage).[25]

In Pormpuraaw, an Australian Aboriginal community, the direction of time relates to the direction of motion of the sun, the East-West axis, not to the direction the speaker happens to be facing at the time.[26] Time for the Yupno people of Papua New Guinea flows uphill and is not even linear. The past is always downhill, in the direction of the mouth of the local river, and the future is towards the river's source, which lies uphill – a rich metaphor indeed.[27]

Different languages tend to spatialise time differently in other respects, too. For example, English and Swedish speakers talk in a linear fashion about long and short periods, whereas Greek and Spanish speakers use images of volume, referring to big or small periods. This can be shown to have a subtle influence on the way they think about time.[28] But the differences are small compared with the overwhelming truth that in our imagination, always and everywhere, space trumps time. Representations of time depend on space more than representations of space depend on time.[29] Space is simply more easily conceptualised than time. Children produce spatial terms earlier than temporal ones: for example they use 'here' and 'there' long before they use 'now' and 'then'.[30] In schizophrenia, as we saw, time is spatialised, and one sign of this is that spatial adverbs tend

22 Li & Du 1987.
23 Rugani, Vallortigara, Priftis et al 2015.
24 Saj, Fuhrman, Vuilleumier et al 2013.
25 Pun, Adamo, Weger et al 2010.
26 Boroditsky & Gaby 2010.
27 Núñez, Cooperrider, Doan et al 2012.
28 Bylund & Athanasopoulos 2017.
29 Casasanto & Boroditsky 2008; Casasanto, Fotakopoulou & Boroditsky 2010.
30 Clark 1973.

to replace temporal ones: in schizophrenia, then, the hemisphere of representation trumps that of experience.

CAN WE AVOID THE SPATIALISATION OF TIME?

Jaynes thought, as we saw, that avoiding spatialisation of time was impossible. But is it? All our diagrammatic representations, it is true, represent the whole of time at an instant, at 'the same time', and as something in which points of time can be located. It is here that we need to turn to a proper understanding of Bergson. He gives us a very important clue which is in astonishing accord with the hemisphere hypothesis.

Bergson saw time as a fundamental reality distinct from space, not a series of 'instants', where each instant would correlate with a position occupied by an object moving along a path. Rather, time is, he held, like music, which unfolds seamlessly, and where each 'note' that we can identify is only understandable as part of a melody or musical sequence which is appreciated as whole, and where any one note, and those before and after, interpenetrate. They are not like the trucks of a goods train, 'a continuous line or a chain, the parts of which touch without penetrating one another', but like the flow of a river, what he calls a 'succession without distinction ... a mutual penetration, an interconnexion and organization of elements, each one of which represents the whole, and cannot be distinguished or isolated from it except by abstract thought'.[31]

31 Bergson 1910 (101).

Henri Bergson is someone I will draw on repeatedly in this chapter. Because he has been so neglected, many readers may know little about him, and what they do know about him might be inaccurate, if gathered at second-hand, which nowadays is mainly the case: hence the need for an aside on his significance.

When I first learnt, in my teens, about Bergson's most famous distinction, that between *temps* and *durée*, it was presented as a distinction between objective time (*temps*), as measured by the clock, which is regular, and subjective time (*durée*), which is widely variable. I felt rather surprised at the fuss about this, since there were surely no Nobel prizes for that one (and he did win a Nobel prize): every child knows that time goes more slowly in the dentist's waiting room. But that was not what Bergson was drawing our attention to. He was making a far more interesting distinction, one with immense, though not immediately discernible, consequences. Time, once it is measured – clock time – becomes spatialised as a succession of points: time as experienced is no such thing (though we *think* of it as such). In representing it we destroy its essence: and the consequences are, literally, world-shattering, as we shall see.

The eclipse of Bergson is interesting. He was the most famous philosopher of the first half of the twentieth century; in his day, he was better known, more prestigious and more influential than Einstein. His lectures were so well attended that a public lecture he gave in New York in January 1913 is credited with causing Broadway's first traffic jam. His influence on the Pragmatists, James and Dewey in particular, on the process philosophers Whitehead and Hartshorne, and on the phenomenological tradition of Husserl, Heidegger, Merleau-Ponty and Scheler has been far-reaching.[32] His thoughts on the nature of time are profound and of enduring interest. It is almost impossible to say anything useful about time without reference to him. Yet extraordinarily enough the entry on 'Time' in the *Stanford Encyclopedia of Philosophy* by Ned Markosian does not mention him. What happened here?

I believe there are several elements to his eclipse. First, and above all, he was a man ahead of his time, 'developing a theory of consciousness within a framework of physical thought which anticipated a crisis in classical physics yet unforeseen and other discoveries yet to come', principally those of quantum mechanics.[33] By the time those discoveries were made by physicists, his philosophical anticipation of them had been all but forgotten, and is only recently being rediscovered. Second, he developed a theory of direct perception and of memory long before such theories began to gain currency, though in recent years a version of his thinking has, belatedly, become a much-discussed – and much-needed – alternative to the idea that perception concerns only mental representations inside the head.[34] Then again, he was bold enough to posit an impulsion in living things which he called *élan vital*: unpopular with the mechanistic tradition in biology, by whom what Bergson intended has rarely been understood, it means nothing like what is often naïvely assumed: I will come to that in due course (see p 923 below). And finally there was the inconclusive debate with Einstein in Paris in 1922.

This debate continues to be a live issue among philosophers and physicists, with some maintaining that Bergson misunderstood Einstein, while others maintain that, on the contrary, Einstein misunderstood Bergson. On this question, if no other, it has been said that Einstein belonged to the school of Parmenides, and Bergson to the school of Heraclitus;[35] and we have already seen that that dispute is not resolved among philosophers. The physicist and historian of science Jimena Canales has written a fascinating investigation into what she calls the 'behind-the-scenes' correspondence between scientists whereby the view of Einstein came to be promoted at the expense of that of Bergson, despite Bergson's views being consonant with those of important physicists and mathematicians such as Poincaré, Lorentz, Michelson and de Broglie.[36] After describing how much both quantum mechanics and information theory have learnt from Bergson since the last war, she writes that,

---

32 As it was on Proust, who incidentally acted as best man at Bergson's wedding.

33 Robbins 2000.

34 See the work of, eg, Alva Noë, Riccardo Manzotti and others.

35 Canales 2016a (64).

36 Canales *op cit*.

across a growing number of disciplines, from psychology to cybernetics, scientists stressed the pertinence of Bergson's conclusions ... the vindication of Bergson's work by scientists strengthened in the late 1960s and continued for the rest of the century.

Canales quotes the philosopher Peter Gunter, editor of a collection of essays on Bergson by scientists and philosophers ranging from the physicist Louis de Broglie to the philosopher Milič Čapek: 'The charge of anti-scientific intention with which Bergson has been saddled is seen to be not merely misleading but radically false: Bergson's philosophy of intuition is the affirmation, not the negation of science.'[37]

In the tradition of 'both/and', it is perfectly possible for both Einstein and Bergson to be right. Bergson did not dispute Einstein's findings. Bergson was simply working at a different level, a subsequent level – that of what you take the findings to *mean*, which is always a matter of interpretation. He saw that Einstein's theory related to clocks, and measurement; but that what is measured is not the whole of time, because time is inextricably bound up with consciousness, and indeed at the core of what it is to be a human being. Einstein's relative time frames were in one sense absolute, in that they were consciousness-independent. His findings and theory were not *wrong* – just radically incomplete: they could not exhaust the true meaning and importance of time. Scientists sometimes imagine that their formulations *are* the reality which the formulations merely represent, whereas they can only ever be practical tools that help us manipulate one *aspect* of reality for a while. They are not wrong for being partial, like all truth. But the mistake is to take them for the absolute truth.[38]

However, that may not be the end of the story. Essentially, in the terminology of modern physics, Einstein believed in a 'block universe' (see pp 930–1 below), a universe in which all times coexist simultaneously. Einstein's theories could not be reconciled to quantum mechanics, a fact which vexed Einstein till his death; but Bergson's theory (like recent quantum field theory) could, and in the end, quantum field theory is the best description of physical reality we have. Einstein's 'relativity physics', according to Louis de Broglie, supporting Bergson,

> pushed the spatialisation of time and the geometrization of space to their extreme limits, because it is from this point of view the final development of classical physics ... in spite of the undeniable and admirable light which it brings to bear for us on many questions, relativity theory has not succeeded in interpreting phenomena in which quanta intervene; and in order to do so it was necessary to develop theories stranger than that of relativity. Today, it is certain

37 Canales 2016b (237–8). Amusingly, a zealous 'editor' of the Wikipedia entry for Eugène Minkowski has interpolated in italics the urgent warning: 'his book "Le Temps vécu" (*Lived Time*) is closely linked to Bergson's philosophy and his anti-scientific ideas, so all the book of Minkowski, including "psychopathological studies", is in doubt.' Even by Wikipedia standards this is prime tripe. First, there's no way an association with Bergson could invalidate a lifetime's careful observation of patients, even if Bergson was wrong about the nature of time, which in the opinion of many philosophers and physicists he was not. Second, Bergson is not anti-scientific in the least: it is hard to think of a philosopher of his age who had greater respect for, and interest in, physics and mathematics. Third, Minkowski was himself not in any way anti-scientific. And fourth, to support his anxious interpolation, the 'editor' refers to Canales's book, page 215. Unfortunately the 'Minkowski' referred to there is the Russo-German mathematician Hermann Minkowski, nothing to do with the French psychiatrist; and in any case the whole tenor of Canales's book is to challenge precisely this unreflective prejudice, and so presumably was not read by the 'editor' in question.

38 For further discussion, see Frank 2016.

that quantum theories penetrate into more profound strata of reality than all previous theories. The theory of relativity itself now appears to us as simply a macroscopic and statistical view of phenomena: it describes things approximately and in bulk and does not descend profoundly enough into the detailed description of elementary processes to allow us to perceive quantum discontinuities there. It is quantum physics, whose most advanced form is wave mechanics, which has enabled us to penetrate into the mysteries of elementary processes ...'[39]

According to de Broglie, Bergson's *Time and Free Will* (his doctoral thesis of 1889), 'antedates by forty years the ideas of Niels Bohr and Werner Heisenberg on the physical interpretation of wave mechanics'.[40] 'Space contains only parts of space, and at whatever point of space we consider the moving body, we shall get only a *position*', wrote Bergson.[41] In quantum mechanics, as de Broglie points out, a particle represents a point precisely located in geometric space with no motion, and a wave represents motion in a pure state with no spatial location.[42] As our uncertainty of a particle's position decreases and approaches a point in space, our uncertainty of its momentum necessarily increases, and vice versa. The measurement tends to project the particle to a point in space or on a geometric continuum, but to the very extent that the process succeeds, motion is excluded.[43] This is often seen as just one of a number of bizarre elements of quantum mechanics, but it could be seen as bound up with the very nature of time.

And this cannot be wholly separated from consciousness, and therefore from the brain. Bergson made a distinction between two forces in consciousness, each delivering a different and mutually incompatible take on reality, but between them constituting a normal person's mental world. Bergson called them *intellect* (sometimes *intelligence* – though it is not at all the same as intelligence as we now understand it) and *intuition* (sometimes *instinct* – though what Bergson called 'intuition' is more capable of disinterestedness and self-awareness than his 'instinct'). The terminology is potentially misleading, because 'intuition' has its intelligence, and 'intelligence' has its unconscious assumptions, but that is not important, as I will suggest. Just follow me for a few paragraphs.

What Bergson calls *intellect* sees things as separate from one another in space, and frozen in time – what he called the cinematographic representation of the world, a rapid succession of still frames. By contrast, *intuition* experiences directly the undivided flow of experience in time, in which no one thing can ultimately be separated from any other; intuition, according to Bergson, is even 'life itself'. Precisely because it is not a representation, but the very presence of life itself,

---

39 de Broglie 1969 (50).
40 de Broglie *op cit* (47).
41 Bergson 1910 (111).
42 de Broglie *op cit* (54).
43 Robbins 2000.

it cannot be *represented* in words without distorting its essence: only poetry, and ultimately music, could in any way succeed here.

The reason that an exact explication of Bergson's terminology is not important to readers of this book is that the deep distinction that is being made is quite astonishingly close to one with which they are already familiar. According to Kolakowski, perhaps his greatest expositor, Bergson

> argued ... that our analytic mind ... is not interested in reality as it truly is, only in its potential utility ... Our [analytic] intelligence is constructed in such a way as to be able to deal adequately with inert matter and to organise it according to the needs of life; it is primarily an organ of survival and of progress in technical skills. Its tendency is to reduce qualities to quantitative differences, new phenomena to old patterns, the unique to the repeatable and abstract, time to space.[44]

Here he makes (of course, unknowingly) at least four important hemisphere distinctions in one sentence. And Kolakowski continues:

> All the expressions of consciousness, our sensations, emotions, and ideas, 'display a double aspect: one is clear, precise, and impersonal; the other is confused, infinitely mobile, and inexpressible, because language cannot grasp it without immobilising its mobility'.[45] This inability of language to describe the "profound self", far from being a contingent defect, reveals the very nature of our linguistic apparatus: language, as part of our intelligence, is essentially a set of abstract signs; its task is to classify objects, to dissolve them into conceptual classes; uniqueness, unless it is an empirically unique collection of abstracts, is beyond its reach.[46]

Bergson's philosophical pronouncements amount to a clear description of each hemisphere, so we can now substitute the more scientifically correct and less confusing labels. Bergson also thought that, while 'intuition' (RH) was modelled along the same lines as life itself, 'intellect' (LH) was intrinsically incapable of understanding intuition, and indeed incapable of understanding life.[47] Subsequent neuropsychological research bears this out, as, once again, we have already seen, and will further see.

Bergson thought that our intellect (LH) permitted useful 'things' to be isolable in, and to stand forth from, the whole in which they inhere, because of what they unconsciously offer to the left hemisphere as opportunities to utilise them – and for that reason come into focus for it as 'things'.[48] In keeping with this, a 'particle' could be seen as the result of the intellect's attempt – the attempt of the left hemisphere – to grasp what the right hemisphere sees as a 'wave'.

Bergson refers to (what we now know to be) the takeover of the

44 Kolakowski 1985 (8 & 16).

45 Bergson 1970 (96).

46 Kolakowski *op cit* (18)

47 Bergson 1908 (179): « L'intelligence est caractérisée par une incompréhension naturelle de la vie » –'Intelligence is defined by a natural lack of understanding of life'.

48 This is similar to Nietzsche's insight that 'the "being" of things has been *inserted* by us (for practical, useful, perspectival reasons)': 2003b, §11[73] (212; emphasis in original).

right hemisphere's insights by the left hemisphere's representation. He calls this process 'converting intuition into symbols', thereby emphasising that the outcome of 'intellectualized time is space', and therefore, 'the intelligence [LH] works upon the *phantom* of duration, not on duration itself'.[49]

What Smolin says of time a century later corroborates Bergson precisely:

> If we confuse spacetime with reality, we're committing a fallacy, which can be called the fallacy of the spatialisation of time. It is a consequence of forgetting the distinction between recording motion in time and time itself. Once you commit this fallacy you're free to fantasise about the universe being timeless and even being nothing but mathematics. But, the pragmatist says, timelessness and mathematics are properties of representations of records of motion – and only that. They are not and cannot be properties of real motions. Indeed it's absurd to call motion 'timeless' because motion is *nothing but* an expression of time.[50]

TIME AS FLOWING PROCESS ONLY

'How can we help seeing', writes Bergson,

> that the essence of duration is to flow, and that the fixed placed side by side with the fixed will never constitute anything which has duration. It is not the 'states', simple snapshots we have taken once again along the course of change, that are real; on the contrary, it is flux, the continuity of transition, it is change itself that is real. This change is indivisible, it is even substantial [*il est même substantiel*].[51]

Hang onto that word 'substantial'. The word is carefully chosen.

Zeno's method involves destroying the fundamental nature of time and space by *atomising* them; and what Zeno shows most beautifully in the process is that atomism – the belief that we can analyse reality into distinct, separable, chunks without loss – does not apply to the real world. It applies only to the left hemisphere's representation of the real world: its diagram or map. In reality neither space nor time is atomistic. In fact, nothing is atomistic – not even atoms (as we shall see in Chapter 24). And that is something that the intuitive understanding of time from within experience, available to the right hemisphere, enables us to see.

What Zeno discovered was that, if you stop time's flow, and find *states* (by definition, 'static'), you make nonsense out of it. Because it doesn't just *have* flow, but *is* flow: stopping it therefore destroys its very nature. There is no flow without time, and there is no time without flow. As you will remember, Schopenhauer asked: 'What is *time*? What is this being that consists in nothing but movement,

---

49 Bergson 1912 (71); and 2007b (19; emphasis added).

50 Smolin 2013 (35; emphasis in original).

51 Bergson 2007a (6).

without anything that moves?'[52] Bergson had the same insight. For Bergson, there is always a priority of movement over the things that move; the things that move are an abstraction from the movement:[53]

> There are changes, but there are underneath the change *no things which change*: change has no need of a support. There are movements, but there is no inert or invariable object which moves: *movement does not imply a mobile*.[54]

[52] Schopenhauer 2000, vol II (33).

[53] Lawlor & Moulard-Leonard 2004.

[54] Bergson 2007e (122: in the original the whole passage is italicised).

[55] Bergson 2007d (105).

The sense that there must be things underlying the flow comes about because we imagine a change as being really composed of states, which are, however, only retrospective, left hemisphere representations. In the famous phrase of Yeats's I quoted in the Introduction, we cannot 'know the dancer from the dance'.

You cannot understand Bergson without 'getting' this insight; if you don't, he will seem to make little sense. Once one does, however, it becomes clear: simply the result of taking a long cool look at the nature of reality before the left hemisphere has taken that reality up and re-presented it as 'things set in motion'.

I shall want to look more fully into the concept of flow and its relationship with the nature of reality in the next chapter. But here, in discussing time, we cannot penetrate into Bergson's understanding of time without anticipating some of that discussion. So here goes.

Bergson asks that, instead of thinking that what we experience is a surface state covering something *else* that is real, 'maintaining with it a mysterious relationship of phenomenon to substance', we should see the *experience itself* as real: for intuition (the RH)

> will seize upon one identical change which keeps ever lengthening as in a melody where everything is becoming but where the becoming, being itself substantial, has no need of support. No more inert states, no more dead things; nothing but the mobility of which the stability of life is made.[55]

And he adds that 'a vision of this kind, where reality appears as continuous and indivisible', is the basis of an intuitively responsive and responsible philosophical understanding of reality. It is, in my view, the key.

Note this does not deny that there are elements in the flow that stand forth to us *as though* wholly separate: but insists that they are better seen as merely *distinct* (that is to say discernibly different, though ultimately not entirely divorced from one another or from the flow which they *are*). This insight is in line with traditions in East and West going back some thousands of years, but more recently lost from view.

It also reverses our customary post-Cartesian ontology – which is why it may be difficult, or even impossible, for some to see it at first. In our philosophy things are primary: how they come to be set in motion, and how they come to give rise to what we perceive, are then natural, but secondary, questions. Things are, as we say, *substantial*: that means literally that they 'stand under' (Latin, *sub-*, under, + *stare*, to stand) and provide necessary support, like a stage, on which the phenomena that we see and experience are 'played out'. Bergson's realisation was that this was back to front – or rather, upside down. Reality is what we experience – ever moving, changing, and continuous. Things, however, are secondary, static, products of perception which supervene on 'from above', not support 'from beneath', that field of flow.

What we perceive, he says, when we see something – and he was highly cognisant of developments in the physics of his day – is 'a series of extremely rapid vibrations. This alleged movement of a thing is in reality only a movement of movements.' And he continues:

> the truth is that there is neither a rigid, immovable substratum nor distinct states passing over it like actors on a stage. There is simply the continuous melody of our inner life, – a melody which is going on and will go on, indivisible, from the beginning to the end of our conscious existence. Our personality is precisely that. This indivisible continuity of change is precisely what constitutes true duration ...[56]

As with the continuity of personhood, we must avoid the dissolution of something that is seamless and in which all the 'parts' interpenetrate, into slices, or pieces, of something formerly alive. Once again the vital image is that of music, specifically melody (which, as you remember, is also appreciated largely or only by the right hemisphere):

> When we listen to a melody we have the purest impression of succession we could possibly have, – an impression as far removed as possible from that of simultaneity, – and yet it is the very continuity of the melody and the impossibility of breaking it up which make that impression upon us. If we cut it up into distinct notes, into so many 'befores' and 'afters', we are bringing spatial images into it and impregnating the succession with simultaneity: in space, and only in space, is there a clear-cut distinction of parts external to one another. I recognize moreover that it is in spatialized time that we ordinarily place ourselves ...[57]

Being absorbed in a piece of music, as opposed to standing back and analysing it, we perceive change itself, not *things* that change:

> This change is enough, it is the thing itself. And even if it takes time, it is still indivisible; if the melody stopped sooner it would no longer be the same sonorous whole, it would be another, equally indivisible.

[56] Bergson 2007e (124).
[57] Bergson *op cit* (124–45).

We have, no doubt, a tendency to divide it and to picture, instead of the uninterrupted continuity of melody, a juxtaposition of distinct notes. But why?

One reason might be that we are thinking of the efforts of the musical performer as discrete actions – moving an elbow, drawing breath – each player separately – and then added together, not as an embodied whole; and because we are tyrannised by the eye, seeing the music as if we were looking at notes placed next to one another upon an imaginary score, *reading* it: 'If we do not dwell on these spatial images, pure change remains, sufficient unto itself, in no way divided, in no way attached to a "thing" which changes'.[58] This accords, profoundly, with my experience.

In trying to help us shift our perception, Bergson repeatedly gives an accurate description of a number of differences between the right hemisphere's understanding and that of the left, though of course having no knowledge of hemisphere difference. Describing the narrowing of the attentional field that inevitably accompanies left hemisphere apprehension, he writes, extraordinarily enough, that 'the necessities of action tend to limit the field of vision … distinct perception is merely cut, for the purposes of practical existence, out of a wider canvas'; and he goes on:

> Our knowledge, far from being made up of a gradual association of simple elements, is the effect of a sudden dissociation: from the immensely vast field of our virtual knowledge, we have selected, in order to make it into actual knowledge, everything which concerns our action upon things; we have neglected the rest.

The matters that concern us become 'things' in the focus of our attention. We can *use* them. The rest of the picture, its background and context, the left hemisphere neglects. And then, with uncanny accuracy – remember this is 1911 – he continues:

> The brain seems to have been constructed with a view to this work of selection. One could say as much for perception. The auxiliary of action, it isolates that part of reality as a whole that interests us; it shows us less the things themselves than *the use we can make of them*. It classifies, it labels them beforehand; we scarcely look at the object, it is enough for us to know to which category it belongs. But now and then, by a lucky accident, men arise [great artists] whose senses or whose consciousness are less adherent to life. Nature has forgotten to attach their faculty of perceiving to their faculty of acting … It is therefore a much more direct vision of reality that we find in the different arts; and it is *because* the artist is less intent on *utilizing* his perception that he perceives a greater number of things.[59]

58 Bergson *op cit* (123).

59 Bergson *op cit* (113–4; emphasis added).

Focussing on what you currently think to be useful means that you inevitably see much less. 'Educationalists' and economists, take note.

Bergson sees the key move that philosophers – and all thinking people – need to make as 'turning this attention aside from the part of the universe which interests us from a practical viewpoint and turning it back toward what serves no practical purpose'. But, of course, by this he doesn't mean abandoning a sense of purpose. He means not being caught up in the *utilitarian* purposes of the left hemisphere; instead aiming to attend in a way that is not already committed or attached ('stuck' in the case of the left hemisphere) to a particular focus selected for its usefulness. 'This conversion of the attention', he says, 'would be philosophy itself.'[60] The mistake that is made by many traditional philosophers, he suggests, is to believe that freeing one's attention up in this way necessitates turning one's back on practical life, rather than, in fact, embracing it.[61] 'One should act like a man of thought', he wrote, in a memorable formulation, 'and think like a man of action.'[62]

Here I am aware of a number of strands that link the philosophical shift that Bergson exhorts on us with Eastern modes of cognition and perception. The emphasis on detachment from our usual busy concerns, yet without turning one's back on practical life; the ability to see the background as well as the foreground; to see ourselves within a bigger picture; to embrace what may seem initially paradoxical; above all, to sense flow and movement as the ground of existence; and to see that the whole is experienced as prior to its differentiation into the parts that we later identify as arising from it: it is not constructed from those parts.

One of the paradoxes is that change and permanence are not opposites (here we come back to the tree of Hegel – or the river of Heraclitus – each of which *persists* only by seamless *change*). Speaking of those who fail to see it, Bergson writes:

> Change, if they consent to look directly at it without an interposed veil, will very quickly appear to them to be the most substantial and durable thing possible. Its solidity is infinitely superior to that of a fixity which is only an ephemeral arrangement between mobilities … if change is real and even constitutive of reality, we must envisage the past quite differently from what we have been accustomed to doing through philosophy and language.[63]

If we succeed in doing so, the rewards are plentiful:

> what was immobile and frozen in our perception is warmed and set in motion. Everything comes to life around us, everything is revivified in us … grave philosophical enigmas can be resolved or even perhaps … need not be raised, since they arise from a frozen vision

---

60 Bergson *op cit* (115).

61 Bergson *op cit* (116).

62 Bergson 2011: « Je dirais qu'il faut agir en homme de pensée et penser en homme d'action. »

63 Bergson 2007e (125).

of the real and are only the translation, in terms of thought, of a certain artificial weakening of our vitality ...⁶⁴

The process of coming back to life, through engagement of what he calls 'intuition', is the exact inverse of the process of devitalisation which occurs in right hemisphere deficits, and in schizophrenia as we saw in Chapter 9. And his insight that often 'grave philosophical enigmas ... arise from a frozen vision of the real' is borne out by the discussion of paradox in Chapter 16.

With that necessary brief excursion into the nature of flow as it relates to time, I will now return to the topic of time itself.

TWO KINDS OF TIME – OR JUST ONE?

I have repeatedly referred to the importance of depth – in time, in space and in feeling; and to how they, like the sense of animacy, are delivered largely by the right hemisphere. Seeing things in relation to their past and future yields depth and brings them to life.

> Let us grasp afresh the external world as it really is, not superficially, in the present, but in *depth*, with the immediate past crowding upon it and imprinting upon it its impetus; let us in a word become accustomed to see all things *sub specie durationis*: immediately in our galvanized perception what is taut becomes relaxed, what is dormant awakens, what is dead comes to life again.⁶⁵

How we understand time influences what sense we make of the world: whether we cleave to the last couple of thousand years of Western philosophy, and its devitalising force, or see beyond what that model (fruitful in its way and for a time) taught us. Our understanding of time ramifies into everything that matters: especially into our understanding of life itself.

Merleau-Ponty saw the vital connexion between time and the conscious subject:

> There can be time only if it is *not completely deployed*, only provided that past, present and future do not all three have their being in the same sense. It is of the essence of time to be *in the process of self-production*, and not *to be*; never, that is, to be completely constituted. Constituted time, the series of possible relations in terms of before and after, is not time itself, but the ultimate *recording* of time, the result of its *passage*, which objective thinking always presupposes yet never manages to fasten on to.⁶⁶

What Merleau-Ponty calls 'constituted time' here is McTaggart's B-series time, in which you see the lot at once.⁶⁷ But as Merleau-Ponty hints, in what sense can that be said to be time at all? It is a 'record-

---

64 Bergson *op cit* (131–2).

65 Bergson 2007d (106).

66 Merleau-Ponty 1962 (482; emphasis on 'passage' in original, other emphases added).

67 See p 650 above.

ing', the result of its having all – past, present and future – already passed. It is time retrospected on and frozen: a re-presentation of a something, the nature of which 'something' is, in itself, quite different in kind from any possible representation; something whose very nature is already lost in the representation. Merleau-Ponty continues, speaking of this *'constituted'* time:

> It is spatial, since its moments co-exist spread out before thought; it is a present, because consciousness is contemporary with all times. It is a setting distinct from me and unchanging, in which nothing either elapses or happens.

And so he goes on:

> There must be *another true time, in which I learn the nature of flux and transience itself* ... time in short needs a synthesis. But it is equally true that this synthesis must *always be undertaken afresh*, and that any supposition that it can be anywhere *brought to completion* involves the negation of time.

Why a synthesis, and why must it always be begun afresh? Because if you treat the representation of time as itself a kind of time (even if 'constituted'), you are left with a conundrum. Time as conceived by the left hemisphere is a *thing*, that just *is*, once and for all, and can be broken down by analysis. For the right hemisphere, on the other hand, there is only change, forever coming into being and swirling away. Time, for the right hemisphere, is not something distinct from being, from reality flowing: it is always thus a *becoming*, never a something become.

And all this that we have heard from Bergson, from McTaggart, from Heidegger, from Merleau-Ponty was anticipated by the never-failing insight of Goethe, 200 years ago, that

> *Vernunft* is concerned with what *is becoming*, *Verstand* with what *has already become* ... The former rejoices in whatever evolves; the latter wants to hold everything still, so that it can utilise it.[68]

To remind you, *Vernunft* is 'reason' as the right hemisphere understands it; *Verstand* is 'rationality' as the left hemisphere understands it. Atomic moments only occur in retrospection and re-presentation: in the words of Sandra Rosenthal, 'movement from one interval to another is not a movement over discrete units but a spreading out of a continuous process of becoming other.'[69]

The left hemisphere's preoccupation with utility, as Goethe saw, leads it to seek the abolition of motion. 'Our action exerts itself conveniently only on fixed points', wrote Bergson:

> fixity is therefore what our intelligence seeks; it asks itself where the mobile is to be found, where it will be, where it will *pass* ... But

---

68 Goethe 1972, 545–73 (563; emphasis added): » *Die Vernunft ist auf das Werdende, der Verstand auf das Gewordene angewiesen ... Sie erfreut sich am Entwickeln; er wünscht alles festzuhalten, damit er es nutzen könne* «.

69 Rosenthal 1996 (556).

it is always with immobilities, real or possible, that it seeks to deal … Immobility being the prerequisite for our action, we set it up as a reality, we make of it an absolute, and we see in movement something which is superimposed.[70]

In other words, to the left hemisphere things appear simply static, and what has to be explained is how motion comes about on this static scene; whereas to the right hemisphere, everything flows, as it is immediately experienced, and what needs to be explained is how stability can arise from this flow. As the reader will remember, this issue was first raised in Chapter 12, where I contrasted the mechanical with the organic view of a living being.

WHEN TIME BREAKS DOWN: CARTESIANISM

If in reality 'everything flows', what could be more devastating than the loss of flow from one's world? Nothing, not even one's self-experience, would appear real any more.

The static and the timeless has been privileged in the West ever since Plato followed the path of Parmenides, not Heraclitus. With Descartes things went much further. Descartes deliberately excluded from his thought processes, like many subsequent philosophers in the analytic tradition, the kind of understanding available from the right hemisphere; and so he was obliged to see time as composed of static points or slices in need of a mysterious *stitching together*. According to the philosopher Charles Sherover, Descartes had 'problems with the very idea of temporal continuity, epitomized in his conviction that each moment is a somehow irreducible real self-enclosed atomic point in the structure of the universe, and is devoid of any sustaining continuity with any other moment'.[71] I quoted Descartes' words to the effect in Chapter 16; and his conclusion that 'the thing which endures may cease to be at any given moment.'[72]

Notice how precarious existence immediately comes to appear. Abstraction, as William James saw, generates 'a world which dies and is born anew at every instant', like the world which Descartes imagined.[73] I do not need to remind you that Descartes also believed that animals were merely automata,[74] and that human beings were only *assumed* not to be automata: 'if I look out of the window', he wrote, 'and see men crossing the square … do I see any more than hats and coats which could conceal automatons? I *judge* that they are men'.[75] His view of time is of a piece with the rest of his philosophy, and with the left hemisphere's world view. One of fragmentation, devitalisation and mechanism.

Descartes' philosophy famously helped to perpetuate a sundering of the natural union of mind and body, and his understanding of the ego was unaccommodating to the idea of intersubjectivity. In fact,

70 Bergson 2007a (5; emphasis in original); and Bergson 2007e (119–20).

71 Sherover 1989 (281).

72 Descartes: see refs at pp 662–3 above.

73 James 1909a (236–7).

74 Cottingham 1978 (552). Cottingham argues that this did not necessarily entail for Descartes that they had no feelings.

75 Descartes, 'Second Meditation': 2017 (26).

when one puts it all together, he experienced, through an effort of the analytic intellect, the same difficulties that are normally reserved to people with schizophrenia in the perception of time.

There is a clear connexion between the juddering sequence of static moments that replace flow and the alienation of the self, as well as the loss of a coherent narrative that might give it permanence. One patient reports:

> You are dying from moment to moment and living from moment to moment, and you're *different each time*.[76]

Once again we are reminded of Capgras syndrome: there is no continuity over time. In schizophrenia subjects do not, of course, lose the capacity to see difference. So, if no process of change is any longer experienced, difference can only mean a constant process of substitution of one isolated, static percept for the next. 'People's faces [are] changing', reports one patient: 'things on the ward aren't organized – they change all the time'; and another complains that the 'organisation of things' is 'different ... there was no-one I could recognize', they seemed 'changed'.[77] A change in appearance means a new 'thing' has been substituted: as a result everything becomes alien, unrecognisable.

WHEN TIME BREAKS DOWN: SCHIZOPHRENIA

In their fascinating, detailed explorations of patients who had unusual experiences of their own body – whether they were brain-damaged or those with so-called 'functional' illnesses, such as schizophrenia – neuropsychiatrists Henri Hécaen and Julian de Ajuriaguerra remark on those who had a problem with time that its effects are pervasive, and that it attacks one's whole sense of embodied being:

> The experience of lived time, the change that comes over personal time, appears to be at the core of the syndrome [of schizophrenia] ... No longer able to avail themselves of the integrating power of time, patients arrive at a condition (to use Minkowski's expression) of 'morbid dualism', in consequence of which things break down, even the *feeling of existing at all* ... Through loss of lived synchrony, which is to say loss of the dynamic element in vital phenomena and its replacement by the static, the *'mind-body synthesis' breaks down* ... the synthesis of one's personhood disintegrates *as the static replaces the dynamic* ...
>
> We have repeatedly found a disruption of the idea of time in subjects presenting with alterations of body awareness. Sometimes it is the feeling of continuity, of lived time, of the duration of time that is disturbed in these patients, and the synthesis of what it is to be 'myself' – which in the normal individual cannot be separated out into parts – seems to come undone ... The violation of the normal temporal assumptions wrenches our mental life out of the realm of

76 Chapman 1966.

77 Cutting 1997 (112–14).

what is automatic, and this disruption compels us to see our body more or less as a form that is distinct from our personhood ...[78]

The loss of the sense of time has been repeatedly singled out as the essential deficit, to be traced in all aspects of the psychopathology of schizophrenia. Pointing to its centrality, the contemporary phenomenologist Giovanni Stanghellini writes that

> temporality ... constitutes the bedrock of any experience ... The disintegration of time consciousness has obviously serious psychopathological consequences for the way one experiences the phenomenal world and relates to oneself.[79]

The psychiatrist Franz Fischer observed that every symptom of psychosis in schizophrenia can be related to the disturbance of time.[80] Hécaen and de Ajuriaguerra remarked in the 1950s that, in schizophrenia, 'the alteration in subjective time, the experience of lived time, seems at the core'.[81] Around the same time, the psychiatrist and philosopher Ludwig Binswanger described the way in which 'time virtually comes to a standstill' as the 'most characteristic feature' of the schizophrenic world.[82] And in his magisterial works from the 1920s and '30s, *La schizophrénie* and *Le temps vécu* (translated as *Lived Time*), Eugène Minkowski performed the richest and most profound examination of disrupted time, as the core phenomenological transformation in schizophrenia, ever conducted. It is astoundingly suggestive.

Fischer made a further observation, that the temporal thought of schizophrenics may become more saturated with internal space as the disease progresses.[83] Repeatedly, clinicians and psychopathologists have referred to a spatialisation of time in schizophrenia. What they mean by this is the substitution of the way we represent time – a succession of serial instants on a time line – for the way we experience it when we are living, and present within its flow; thereby killing it.

Making space temporal is an act of vivification; making time spatial an act of vivisection. Being depends on time, *Sein* on *Zeit*. With the loss of time the world becomes mechanical, and lifeless; ultimately it fragments and becomes incoherent. Mind (our thinking faculty) and body (our inhabited, intuitive experience) come apart. All this we saw clearly exemplified in the descriptions by patients in Chapter 9. Heidegger could not have known how clearly the connexion between time and being is demonstrated by the testimony of those who live with the devastating consequences of alienation from time.

Indeed time's being alien – 'an *alien time* seemed to dawn'[84] – is sometimes all a patient can say of it: its altered quality simply cannot be put into words.[85] Another patient says, '*new* time is being produced' and another that '*new time* was infinitely manifold and

78 Hécaen & de Ajuriaguerra 1952 (309 & 369–70):
« L'expérience du temps vécu, la modification du temps personnel, apparaît ainsi comme au centre du syndrome [de schizophrénie] ... Privés de ce pouvoir intégrateur du temps, les malades arrivent selon l'expression de Minkowski à un ‹ dualisme morbide › à partir duquel peut s'effondrer jusqu'au sentiment de l'existence ... Par la perte du synchronisme vécu qui est perte du dynamisme et entrée du statique dans les phénomènes vitaux, se rompt la ‹ solidarité organo-psychique › ... la synthèse de la personnalité se désagrège à mesure que le statique remplace le dynamique. » « À plusieurs reprises, nous avons trouvé une atteinte de la notion du temps chez les sujets présentant des modifications de la somatognosie. Tantôt c'est le sentiment de notre continuité, le temps vécu, la durée qui sont troublés chez les malades et la synthèse du moi semble se défaire en ses diverses constituantes évolutives non décomposables, chez l'individu normal ... L'atteinte des données temporelles désautomatise notre vie psychique et cette désautomatisation nous fait appréhender plus ou moins exactement notre corps en tant que forme séparée de notre personnalité. »

79 Stanghellini, Ballerini, Presenza et al 2016.

80 Fischer 1930b (469).

81 Hécaen & de Ajuriaguerra 1952 (309).

82 Binswanger 1987 (86). Thus in Sass 2017 (471, n43).

83 Fischer *op cit* (470). Fischer's views are however less clear-cut than Minkowski's better-known report of them: Minkowski 1968 (258).

84 Jaspers 1963 (86).

85 Mancini, Presenza, Di Bernardo et al 2014.

intricate and could hardly be compared with what we would ordinarily call time.'[86]

When it is possible to articulate the change in time in schizophrenia, it is that time ceases to flow, either in the sense that it has lost its fluent quality, or that it actually stands still. 'Time's standstill is never-ending, I live in a motionless eternity', says one patient. 'I see the clocks go round, but for me time does not pass.'[87] And another patient reports that 'time had failed and stood still ... thinking stood still; everything stood still *as if there were no more time.*'[88] This is a recognised symptom known as *Zeitstillstand*, here most vividly described – note that the consequence is that all meaningful existence is wiped out, with really nothing left:

> Time has come to a complete standstill for me, it seems ... I look at the clock – and if I look at it again, it's like an enormous stretch of time has passed, as if hours have gone by; but on the contrary it's been only a few minutes. It feels to me like an enormous period of time is passing. Yet time does not pass at all. I look at the clock but its hands are always stopped at the same position, they never move at all; they never go forward; I listen, to see if the clock has come to a halt, and I hear it ticking, but the hands are always in the same place. I don't think about my past ... Nothing would come to mind, *nothing* ... I could not make myself think about anything. I couldn't envisage anything at all in the future. The present does not exist for me when I am ill like this ... the past does not exist, the future does not exist.[89]

There is a failure of attunement of the psyche with life. I use the word 'attunement', since it avoids the suggestion that we are primarily dealing with a matter of understanding, or even of awareness; it is, rather, the inability to let oneself be entrained in time's flow – to move with the 'music of time'. Life becomes a thing to be observed, not an experience to be inhabited. 'Life is now a running conveyor-belt with nothing on it', says one patient:

> It runs on but is still the same ... I am now living in eternity ... outside everything carries on ... leaves move, others go through the ward but for me time does not pass ... *I wish I could run too, so that time might again be on the move* but then I stay stuck ... time stands still ... It is a boring, endless time. It would be fine to start again from the beginning and *find myself swinging along with the proper time*, but I can't ... *I get pulled back* ... Time is in collapse.[90]

This time is not just alien, but alienating – and boring. Note how the subject feels left out of the momentum of life. Time is the coherence-giving context in which we live. Not to have this is to disrupt many important facets of being, including the sense of belonging, of

---

[86] Jaspers *op cit* (82 & 87; emphasis in original).

[87] Tellenbach 1956a (12): »*Der Zeitstillstand ist unendlich, ich lebe in einer stehenden Ewigkeit. Ich sehe, daß die Uhren sich drehen, aber für mich vergeht die Zeit nicht* «. Also Jaspers 1963 (87).

[88] Jaspers *op cit* (87; emphasis in original).

[89] Muscatello & Giovanardi Rossi 1967 (784ff): 'Il tempo per me è fermo completamente, mi sembra ... Guardo l'orologio e mi sembra, se lo riguardo, che sia trascorsa una distanza di tempo enorme, come se fossero trascorse delle ore, mentre invece sono trascorsi pochi minuti. Mi sembra un tempo enorme che sta passando. Il tempo non passa mai, guardo l'orologio, ma le lancette sono sempre ferme nella stessa posizione, non si muovono mai, non vanno mai avanti; allora sento se l'orologio si è fermato, sento che batte, ma le lancette sono sempre ferme. Al passato non penso ... Non mi veniva in mente niente, niente ... non riuscivo a pensare a niente. Non riuscivo a prospettarmi niente, niente in futuro. Per me il tempo presente non esiste quando sto così male ... non esiste il passato, non esiste il futuro.'

[90] Jaspers *op cit* (87; emphasis in original).

connectedness, and of meaning to life. Life becomes lifeless – *boring*: a very modern concept (the word arose only in the eighteenth century, and is associated with a disengagement from the world which began with the Enlightenment).[91] And this lies at the heart of many problems experienced by subjects with schizophrenia – as well as homo sapiens in the twenty-first century – including a sense of the world as being remote or alien; a loss of emotional engagement with the world and others; and experiencing a sense of oneself, others and the world as disembodied mechanisms composed of parts, as we have seen.

As a result of the loss of the flow of time, writes psychiatrist Thomas Fuchs, there is

> a characteristic imbalance between lived time (or Bergson's *durée*) and static space. There is a weakening of the dynamic, flexible aspects of life and a corresponding *hypertrophy of the fixed, rational and geometrical* elements.[92]

From absence of time follows absence of motion: the flow congeals. The remarks of many patients refer to this explicitly: everything tends toward the condition of the snapshot, as we saw in Chapter 9. The fixed is graspable, whereas the flow of life never is:

> I look for immobility. I tend toward repose and immobilization. I also have in me a tendency to immobilize life around me. Because of this, I love immutable objects, things which are always there and which never change. Stone is immobile. The earth, on the contrary, moves; it doesn't inspire any confidence in me. I attach importance only to solidity. A train passes by an embankment; the train does not exist for me; I wish only to construct the embankment. The past is the precipice. The future is the mountain.[93]

Spatialisation of time. The same patient also said: 'I like immovable objects, boxes and bolts, things that are always there, which never change.'[94] Another schizophrenic patient, when asked, after a home visit, if she had been happy to see her mother, replied: 'There's movement: I don't much like that'.[95]

Superficially there are similarities between the experience of time in schizophrenia and depression. In reality, I believe they are different phenomena. I cannot pass over the issue, but I am wary of too great a digression in the argument, so I have relegated my reflections on their differences to Appendix 4.

I demonstrated earlier the extent to which our sense of time depends heavily on the right hemisphere. It is therefore not surprising that there are externally verifiable, measurable alterations in time perception in schizophrenia. A study aiming to uncover the neural

---

91 See McGilchrist 2009b (esp 336 & 400–3).

92 Fuchs 2013 (emphasis added).

93 Minkowski 1970 (279). See also Schreber 1955 (135): 'Apart from daily morning and afternoon walks in the garden, I mainly sat motionless the whole day on a chair at my table, did not even move towards the window, where by the way nothing was to be seen except green trees …'

94 This is the same individual reported at Minkowski 1987 (209).

95 Minkowski 2002 (145): « C'est du mouvement, je n'aime pas beaucoup ça ».

correlates of time distortion in schizophrenia found that in patients there was 'significantly lower activation of most right hemisphere regions' involved in time perception in normal subjects.[96]

As a result, though perceptual experiences are embedded in a continuous flow, schizophrenic subjects appear to prefer to process isolated stimuli.[97] They tend to perceive asynchronous stimuli as simultaneous (unless separated by longer than usual intervals),[98] and have difficulty discriminating temporal order.[99] They lose the ability to predict and follow the 'events flow', as if they were unable to experience the unfolding of a melody (which they often are).[100] Additionally schizophrenic subjects, again just like right hemisphere-impaired individuals, are impaired at judging time duration.[101]

With this vivid insight into the breakdown of time before us, let us return to an examination of its nature.

Even though it is from motion that we gain our sense of both space and time, the ultimate aim of the left hemisphere's focus has to be stasis. Yet everything comes into being only because of motion. 'There is', writes Bergson, speaking of what we now can see is the process of the left hemisphere,

> beneath these sharply cut crystals and this frozen surface, a continuous flux which is not comparable to any flux I have ever seen. There is a succession of states, each of which announces that which follows and contains that which precedes it. They can, properly speaking, only be said to form multiple states when I have *already passed them and turn back to observe their track*. Whilst I was experiencing them they were so solidly organized, so profoundly animated with a common life, that I could not have said where any one of them finished or where another commenced. In reality no one of them begins or ends, but all extend into each other.[102]

'Multiple states' exist only in retrospective representation: we saw this in considering logical paradoxes (Chapter 16). The world described by Bergson is one that is forever evolving, constantly creative, moving *forward* into newness: one of his greatest works is called *L'Évolution créatrice* (*Creative Evolution*).

This point seems simple enough but its implications are profound. William James shared the insight. Although, once it has come into being, anything can be taken apart, this process sheds

> no ray of light on the processes by which experiences *get made* ... to understand life by concepts is to arrest its movement, cutting it up into bits as if with scissors, and immobilizing these in our logical herbarium where, comparing them as dried specimens, we can ascertain which of them statically includes or excludes which other. This treatment supposes life to have already accomplished itself, for

---

96 Ortuño, Guillén-Grima, López-García *et al* 2011.

97 Lalanne, van Assche & Giersch 2012.

98 Lalanne, van Assche, Wang *et al* 2012; Foucher, Lacambre, Pham *et al* 2007; Giersch, Lalanne, Corves *et al* 2009; Schmidt, McFarland, Ahmed *et al* 2011.

99 Capa, Duval, Blaison *et al* 2014. These findings are interesting, too, because it seems to me that failures to observe temporal order might give rise to misperceiving causal influences, a common problem in delusions. If you seem to experience the thought 'Beckham shoots for goal' not immediately after, as is normal, but at the same time as, or just before, Beckham shoots for goal, you may believe that you influence Beckham to score the goal – therefore that events can, apparently, be brought about by thinking them.

100 EVENTS FLOW: Giersch, Poncelet, Capa *et al* 2015. MELODY: a significant and specific impairment of melody appreciation is found in schizophrenia (Kantrowitz, Scaramello, Jakubovitz *et al* 2014).

101 Wahl & Sieg 1980; Tysk 1984; Tracy, Monaco, McMichael *et al* 1998; Volz, Nenadic, Gaser *et al* 2001; Elvevåg, McCormack, Gilbert *et al* 2003; Carroll, Boggs, O'Donnell *et al* 2008; Lee, Bhaker, Mysore *et al* 2009; Waters & Jablensky 2009.

102 Bergson 1912 (11; emphasis added).

the concepts being so many views taken *after the fact*, are *retrospective* and *post mortem*.¹⁰³

The point James makes is that if we were capable only of reconstituting reality from what has already happened we would be trapped in a lifeless realm where imagination cannot penetrate. The left hemisphere always dealing with familiar re-presentations, there could only ever be assembled permutations of what is known. 'When you have broken the reality into concepts you never can reconstruct it in its wholeness. Out of no amount of discreteness can you manufacture the concrete.'¹⁰⁴

This has implications for determinism. Is the future just the unfolding of what was there all along in the past? Physical determinism is based on a belief in a mechanical universe. In Laplace's left hemisphere fantasy there was no room for free will.¹⁰⁵ But physics itself has done away with Laplace's universe. 'A most disturbing – or perhaps liberating – consequence of quantum physics is that the past no longer determines the future, at least not completely', writes physicist Richard Muller:

> limited ability to predict the future will remain a fundamental weakness of physics forever ... Despite arguments from classical philosophers, we now know that free will is compatible with physics; those who argue otherwise are making a case based on the religion of physicalism.¹⁰⁶

And he illustrates this from work he has repeatedly replicated as a particle physicist:

> we now know that two objects, objects that are completely identical, identical in every way, can behave differently. Two identical radioactive atoms decay at different times. Their future is not determined by their past or by their condition, their quantum physics wave function. Identical conditions do not lead to identical futures.¹⁰⁷

While this may look like a 'weakness' from one point of view, it is vital to the free nature of the creative process. To quote another physicist, David Oliver:

> There exist physical events the randomness of which cannot be dispelled by the most determined efforts to expose 'a deterministic cause' ... a stream of unpredictable information constantly enters the stream of definite information, a source of novelty, creativity, vitality.¹⁰⁸

Creativity is predicated on uncertainty. Strive for certainty and you kill creativity. As Smolin says:

103 James 1909a (239; emphasis in original; & 243–4; emphasis added).

104 ibid (261).

105 The distinguished eighteenth-century mathematician Pierre-Simon Laplace taught that the future could be predicted with certainty if one knew everything about the past.

106 Muller 2016 (9–10).

107 ibid (277).

108 Oliver 2019 (102 & 107).

On a personal level, to think in time is to accept the uncertainty of life as the necessary price of being alive. To rebel against the precariousness of life, to reject uncertainty, to adopt a zero tolerance to risk, to imagine that life can be organised to completely eliminate danger, is to think outside time. To be human is to live suspended between danger and opportunity.[109]

TWO KINDS OF POTENTIAL

In a crucial insight, Bergson distinguishes between two ideas of potential. In one, whatever it is that comes into existence is the actualisation of a pre-existing potential entity: there always was, among the possibilities of existence, a *Götterdämmerung*, and Wagner actualised it by writing it, as if he went to the cupboard where potentials are stored, dusted it off and set it upright in the living room of reality. In the other, there is no ghostly version hovering around waiting to be dusted off: Wagner wasn't selecting amongst the possible, which already in some sense pre-existed, but was truly *creating*.

To illuminate the distinction Bergson points to an asymmetry between possibility and impossibility. If I lock a gate, I know no-one can pass. If I open it again, I know someone is likely to pass, but not who it will be, or when. Thus, says Bergson, it is true that some things could come into existence, and some things could not. In that sense there is a field of the possible, and a field of the impossible, but they are not similarly structured. '*Hamlet* was doubtless possible before being realised', he writes,

> if that means that there was no insurmountable obstacle to its realisation. In this particular sense one calls possible what is not impossible; and it stands to reason that this non-impossibility of a thing is the condition of its realisation. But the possible thus understood is in no degree virtual, something ideally pre-existent ... from the quite negative sense of the term 'impossible' you pass surreptitiously, unconsciously to the positive sense. Possibility signified 'absence of hindrance' a few minutes ago: now you make of it a 'pre-existence under the form of an idea', which is quite another thing. In the first meaning of the word it was a truism to say that the possibility of a thing precedes its reality: by that you meant simply that obstacles, having been surmounted, were surmountable. But in the second meaning it is an absurdity, for it is clear that a mind in which the *Hamlet* of Shakespeare had taken shape in the form of [the] possible would by that fact have created its reality: it would thus have been, by definition, Shakespeare himself.[110]

It may be objected that a mind is unnecessarily posited, but is that not what *must* be posited in the case of something such as *Hamlet*? And of time? *Hamlet* without the mind of an author could not be *Hamlet*, even in potential: it would just be Bergson's lack of impossi-

[109] Smolin 2013 (xvi–xvii).

[110] Bergson 2007f (83).

bility, in the sense that an eternally running computer (quite a supposition in itself) would crank it out eventually, because it would crank out every combination of letters that is possible, like the proverbial ape with a typewriter. It would take many orders of time longer than the existence of our universe for the ape or computer to type it. But even when it did, the series of letters that composed *Hamlet* would not be, in any case, the play *Hamlet*.

IS POTENTIAL MORE IMPORTANT THAN ACTUALITY?

I want to sow here the seeds of an idea I will return to in the coming chapters: that, contrary to our usual way of thinking, potential is at least as important as, and (it may be argued) more important than, actuality. Admittedly, to a certain way of looking at it, it doesn't matter what the potential was: it's what actually happens that matters. This is undeniably true from within the frame of the actual. But the actual is smaller than, and tells us only one thing about, the field of potential. What is, through happenstance, actualised only poorly reflects one aspect of that potential. And if we are interested in the nature of reality, we are at least as interested in what it embraces as in what it has hitherto been limited to. It is possible that this is what Heidegger had in mind when he said that 'higher than actuality stands *possibility*.'[111] But there are a number of senses in which it appears to be true.

The unfolding of reality, in an image of Bergson's, is not like the unfolding of a lady's fan (he was writing in an era when this was a fashionable accoutrement) on which the same picture will inevitably be found, however fast or slowly one unfolds it, prefigured on the silk.[112] Philosophers, according to Bergson, have failed to grasp the positive nature of time, its necessary role in creation of all kinds, in producing what, prior to its coming into being, is potential yet inconceivable:

> They seem to have no idea whatever of an act which might be entirely new (at least inwardly) and which in no way would exist, not even in the form of the purely possible, prior to its realization. But this is the very nature of a free act. To perceive it thus, as indeed we must do with any creation, novelty or unpredictable occurrence whatsoever, we have to get back into pure duration ... Let us say then, that in duration, considered as a creative evolution, there is perpetual creation of possibility and not only of reality.[113]

Duration is *durée* – time as flow; not *temps* – a succession of instants like the already present folds of the lady's fan. But note that it seems to Bergson that the unfolding of things in time does not just, as it were, 'precipitate out' a pre-existing potential into reality, but actually is creative of *potential itself*: 'perpetual creation of possibility

111 Heidegger 1962, Int 11 (63); 1986 (38): » Höher als die Wirklichkeit steht die Möglichkeit «.

112 Bergson 2007a (9).

113 Bergson 2007a (8 & 10).

and not only of reality'. I will come to this shortly. Meanwhile let us consider the influence of what comes into being on what has already been, the past, since conventionally we see much of the potential of the present in the past. If they are seamless, then could the present influence the past?

T. S. Eliot made the point, in relation to a literary tradition, that new work in the tradition is not only formed out of the works that precede it, but in turn has an inevitable effect on how those works are henceforth experienced.[114] A similar point was made by Borges more dramatically: 'The fact is that every writer *creates* his own precursors. His work modifies our conception of the past, as it will modify the future.'[115] Since a work of art only fully exists in the experiencing of it, it could be argued that not just our way of experiencing it, but the underlying reality of a past work in the tradition has been changed by what comes after. Bergson seems at times to be saying that all experience is in this respect like the work of art: that what *is* only becomes fully what it *is* in the unfurling of the *whole* flow, that what is past only becomes what it really is once the future has unfurled. He gives this example: the Romanticism of which we now see aspects in prior Classical writers 'no more existed in Classical literature before Romanticism appeared on the scene than there exists, in the cloud floating by, the amusing design that an artist perceives in shaping to his fancy the amorphous mass.'[116] If Romanticism had never arisen, there would have been no prefigurings of it in Classicism.

At this point, Coleridge's idea of the Primary Imagination may be helpful. His conception, you remember, is that Imagination is the faculty by which we nurture reality into being, not the faculty whereby we *fashion an already existing reality*. According to Coleridge, that would be, by contrast, what fantasy – or as he would have put it, Fancy – achieves: Fancy can only rearrange already existing elements of experience. Imagination, however, brings into being something never before fully experienced. Temporality, it would seem, is bound up with this crucial distinction.

Imagination alone is creative, and can deliver the truly new. In *The Master and his Emissary* I likened the left hemisphere's 'fantasy' to those children's books with pages split into three, in which you can invent a new animal by putting together the head of a camel, the body of a seal and the legs of a goat. They 'create' apparent newness by simply recombining in a novel fashion, according to a formula, what is already known. Imagination, by contrast, strips the veil of familiarity from something that you thought you knew, so as to be truly revealed for the first time. At first glance, you might well think that nothing has changed, until the penny drops.

There is no possible formula for this. It's a quite different business in a range of respects. Bergson connects imagination, rightly in my

---

114 See, eg, Eliot 1941 (25–626): 'What happens when a new work of art is created is something that happens simultaneously to all the works of art that preceded it. The existing monuments form an ideal order among themselves, which is modified by the introduction of the new (the really new) work of art among them ... the relations, proportions, values of each work of art towards the whole are readjusted ... Whoever has approved this idea of order, of the form of European, of English literature will not find it preposterous that the past should be altered by the present as much as the present is directed by the past.'

115 Borges, 'Kafka and his precursors': 1964a (236; emphasis in original).

116 Bergson 2007a (12).

view, with the necessity of abandoning any piecemeal construction of reality. Though the piecemeal mode of thought has its uses at times, it cannot discern the truly new, which always exists as a new *Gestalt*:

> In psychology and elsewhere, we like to go from the part to the whole, and our customary system of explanation consists in reconstructing ideally our mental life with simple elements, then in supposing that the combination of these elements has really produced our mental life. If things happened this way, our perception would ... consist of the assembling of certain specific materials, in a given quantity, and we should never find anything more in it than what had been put there in the first place.[117]

As he says: 'The intellect [LH] combines and separates; it arranges, disarranges and co-ordinates; it does not create.'[118] On the other hand, it is the nature of time as duration, which the 'intellect' (the left hemisphere) simply does not understand, to make possible an 'unceasing creation, the uninterrupted upsurge of novelty'. Speaking of the Newtonian conception of time, Smolin writes that it 'brings no novelty or surprise. Change is just a rearrangement of the same facts.'[119] Similarly, the misconception of determinism, the predictability of the universe, depends on the left hemisphere's reducing of time to slices and wholes to parts. What Bergson calls 'our intelligence'

> imagines its origin and evolution as an arrangement and rearrangement of parts which supposedly merely shift from one place to another; in theory therefore, it should be able to foresee any one state of the whole: by positing a definite number of stable elements one has, predetermined, all their possible combinations.[120]

As in the child's 'novel' animal picture book. But not as in *War and Peace*.

CAUSATION

Built into the idea of causation is the notion of time. Conceived as it usually is, as a linear chain of events, a cause must precede its effect. Earlier causes later. But change does not always happen, it seems, in a *linear* fashion.

Consider quantum entanglement. What this means, conventionally, is that when two particles that have interacted or been in close enough proximity that they influence each other's basic properties, and these particles are subsequently separated, a change to one particle results in a complementary change to the other at exactly the same moment. Measuring a property, such as the position, momentum, spin, or polarisation of one particle, will instantaneously bring about a change in the quantum state *both* of the particle measured *and* of the other particle with which it is said to be 'entangled', in

117 Bergson 2007e (113).
118 Bergson 2007e (110).
119 Smolin 2013 (48).
120 Bergson 2007f (77).

a complementary fashion, so that it appears that the second one 'knows' instantly what happened to the first. If the first particle is in some respect 'up', the second will inevitably and instantaneously be found to be in the same respect, 'down'. The trouble is that the two particles may be separated by arbitrarily large distances, such that any information would have to have travelled many times faster than the speed of light – according to recent research at least 10,000 times faster – which is clearly impossible according to what we know, since nothing (assuming the special theory of relativity to be correct) can travel faster than light.[121]

The only thing that we can predict will *always* occur at *precisely* the same time as a given event is – itself! So what if there is really some sense in which what we conceive as two events are in reality two aspects of one and the same event?

We may be mistaking the particle for a distinct entity instead of seeing it as simply an aspect of a whole field – when you intervene at any one point in the field, you alter the configuration of the whole field.

In Plate 22[a], we see either a woman's face in three-quarter view or a flower and a butterfly. We can see one or the other, but not both at the same time: the one excludes the other. When a transformation occurs, it is not because the nose becomes a butterfly, which then causes the mouth to become a flower, which then causes the eye to become a leaf, and so on, each 'passing information' to the next: the shift happens instantaneously and as a whole.

It may be said that changes in consciousness are different in kind from changes in quantum physics. That is an assumption that may or may not be warranted, and as evidence of the inseparability of mind and matter accumulates it may be one we might need to re-evaluate. But in any case there are more homely examples that do not obviously involve consciousness. When I magnetise an iron bar, I don't create a north pole, which magically then 'tells' the south pole what it must become. They always have opposite polarity and any change in one instantaneously becomes a change in the other. The system as a whole reconfigures at one and the same time.

De Broglie noted that the essence of Heisenberg's uncertainty principle is that the projection of a *motion* to a *point* merely results in immobility – we have lost the motion.[122] Since Richard Feynman and Albert Hibbs demonstrated that the motion of a particle is continuous but not differentiable, the assumption of the differentiability (into particles) of spacetime has been called into question.[123] (In fact, according to the most successful current model in physics, quantum field theory, there are in any case no particles. When a field collapses and is quantised, it looks like a particle – but isn't.) As Stephen Robbins puts it, 'to say this in another way, the global evolution of

---

121 Ma, Herbst, Scheidl *et al* 2012; Yin, Cao, Yong *et al* 2013.

122 Robbins 2004 (783), referring to de Broglie 1969. This paragraph is based on my reading of Robbins, and at times follows approximately his wording.

123 Nottale 1996; Feynman & Hibbs 1965. Both cited in Robbins *op cit*.

---

124 Robbins *op cit* (783).

125 Bergson 1965 (116–7): « De quel objet, extérieurement aperçu, peut-on dire qu'il se meut, de quel autre qu'il reste immobile? Poser une pareille question, c'est admettre que la discontinuité établie par le sens commun entre des objets indépendants les uns des autres, ayant chacun leur individualité, comparables à des espèces de personnes, est une distinction fondée. Dans l'hypothèse contraire, en effet, il ne s'agirait plus de savoir comment se produisent, dans telles parties déterminées de la matière, des changements de position, mais comment s'accomplit, dans le tout, un changement d'aspect, changement dont il nous resterait d'ailleurs à déterminer la nature. » The standard translation by Paul and Palmer would appear to miss the meaning.

the matter-field over time ... cannot be treated as an infinitely divisible series of states'.[124] Bergson anticipated non-differentiable motion of a matter-field. 'How is it', he wrote, 'that we come to divide up the continuous extent of matter that we first perceive into so many *bodies*, each with its own substance and individuality?' It's not, he says, a matter of working out how each supposed part comes to change its position, but of how it is that there is a change in an aspect of the *whole* – 'a change, the nature of which would remain to be determined':[125]

> Without doubt, this continuity changes its appearance from one moment to the next: but why do we not simply note that the *whole* has changed, as if by the turn of a kaleidoscope? Why, at the end of the day, do we look in the movement of the whole for paths to be followed by bodies in motion? What we have is a moving continuum, in which everything both changes and at the same time remains: where do we get the idea of dissociating these two terms, permanence and change, in order to represent *permanence* by bodies and *change* by homogeneous movements in space? That does not come from our immediate intuition, and, if anything, it is still less a requirement of science; for science, on the contrary, sets out to rediscover the natural seams of the universe, where we have carved it artificially.[126]

Leibniz had already seen in 1671 that 'the essence of a body consists rather in motion [than extension]'.[127] As Stephen Robbins points out, what had looked like the movements of separate objects now become changes of state in a transformation of the whole. From this perspective, what Robbins calls 'primary memory' – how an 'instant' of the past that would be by definition non-existent comes to be connected to, and interact with, the 'instant' that is present – ceases to be a puzzle, and becomes a property of the field itself and of its 'melodic' flow.[128]

The principle of cause and effect breaks down at the quantum level: it appears that effects can be their own causes.[129] In relation to biology, Anjum and Mumford write:

> A problem, we argue, is that if one begins from the idea that causation has to be a relation between two discrete and 'static' events, then it may already be impossible to formulate a satisfactory theory of what causation is and how it works. The biological case makes this quite apparent. The assumption splits the world asunder into distinct, self-contained fragments, and then tries to find a relation that would stick them all back together again. It is not clear that any relation can do that job. Even if it could, would the world with which we are left – a Frankenstein-world of stitched together pieces – have all the features we require? Would it really be a world of continuity, fit for living biological processes?[130]

126 Bergson 1965 (117): « Comment morcelons-nous la continuité primitivement aperçue de l'étendue matérielle en autant de corps, dont chacun aurait sa substance et son individualité? Sans doute cette continuité change d'aspect, d'un moment à l'autre : mais pourquoi ne constatons-nous pas purement et simplement que l'ensemble a changé, comme si l'on avait tourné un kaléidoscope? Pourquoi cherchons-nous enfin, dans la mobilité de l'ensemble, des pistes suivies par des corps en mouvement? Une continuité mouvante nous est donnée, où tout change et demeure à la fois : d'où vient que nous dissocions ces deux termes, permanence et changement, pour représenter la permanence par des corps et le changement par des mouvements homogènes dans l'espace? Ce n'est pas là une donnée de l'intuition immédiate ; mais ce n'est pas davantage une exigence de la science, car la science, au contraire, se propose de retrouver les articulations naturelles d'un univers que nous avons découpé artificiellement. » The standard translation by Paul & Palmer is obscure and clumsy: I have tried to provide something more natural and comprehensible.

127 Leibniz, letter to Antoine Arnauld, late November 1671.

128 Robbins 2004 (783).

129 Oreshkov, Costa & Brukner 2012.

130 Anjum & Mumford 2018 (61–2).

The reality described by quantum field theory helps put one's finger on the problem. Linear causation may be a valid concept, but only at one, intermediate plane of focus, neither too minute (where there are quantum fields), nor too large (where one encounters complex 'systems' such as organisms, and ultimately a human mind). This is in keeping with what I suggested was the case when it came to looking at a physical entity as a mechanism: it works, but only at a quite specific scale, that of the microscopic detail. Any lower, and any higher, and it breaks down. Scale changes not just how much of some 'thing' you have, but what kind of a 'thing' it turns out to be.

THE SIGNIFICANCE OF SCALE

Linear causation is an artefact of our capacity, characteristic of the brain's left hemisphere, for shearing off from the attentional field all that does not concern us at this point – in space or time. It is like taking one arrow out of the diagram in Plate 13[a]. When one focuses on the brushstrokes in a portrait by Holbein, one can say that this application of paint here causes a darker line to appear in the face in the portrait; that the cause of the application was a movement of the painter's hand; and so on. But this, however often repeated, though it may give an apparently complete account of mechanisms involved, will never give an adequate account of how the portrait came into existence.

The Newtonian framework, says Smolin, 'is ideally suited to describe small parts of the universe, but it falls apart when we attempt to apply it to the universe as a whole ... Indeed as I will show in detail the very features that make these kinds of theories so successful when applied to small parts of the universe cause them to fail when we attempt to apply them to the universe as a whole.'[131]

One benefit of a *Gestalt* view is that causality no longer looks like it has to be the best way, let alone the only way, to think about the relationship between mind and brain.[132] There may, for example, be a form of entanglement between these different levels of reality involved, as physicist Richard Muller has suggested.[133] When two particles of matter are entangled, no one particle *causes* the other's behaviour; they are, it seems to me, best considered as aspects of one and the same event. During life it is possible that the spiritual and physical are entangled, neither *causing* the other, neither *depending* on the other for its existence, but their entanglement certainly depending on the co-existence of each: nonetheless each could exist separately, though then they would not exhibit the entanglement we recognise as the mind-brain relationship.

When I was discussing the distinctive features of organisms in Chapter 12, I suggested there were difficulties in applying linear causation to a waterfall or a tornado. In a flow, there are no clear

---

131 Smolin 2013 (xxiii–xxiv).

132 Cf Sass & Borda 2015 (480): 'There is, in any case, no reason to make the reductionist assumption that the causal direction must always be from brain to mind rather than the reverse (or even that causality is the best formulation of mind-brain relationships)'.

133 Muller 2016 (285ff).

trains of causation, with their one-to-one relationships of before and after, and a definable beginning and end. Schelling argued that mechanical cause-effect relations are abstractions from the reciprocal causation of self-organising processes.[134] An organic whole, in contrast to a mechanism, does not consist of a hierarchy of parts which exert control over other parts. Instead, it is a maximally responsive and transparent system in which changes and adjustments propagate *simultaneously* upwards, downwards and sideways, in the maintenance of the whole.

Instead of 'control' (LH), I suggest it is more appropriate and more fruitful to think in terms of 'communication' (RH).[135] 'It is assumed that we are both entitled and able to dismember the continuous flow of events', wrote F. C. S. Schiller,

> to dissever it into discrete stages, to distinguish certain elements in the infinitely complex whole of phenomena, and connect them with others as their causes or effects. But what if the Becoming of things be an integral whole, which could be understood only from the point of view of the whole? Would not the idea of causation be inherently invalid, just because it *isolates* certain factors?[136]

One way to think of a cause is 'that without which something would not be' – something which can, in a sense, take credit (or blame) for its coming about: this is approximately the root meaning of the word Aristotle uses for a cause, αἰτία (*aitia*). One can then distinguish different types of such 'without-which-nots', as, famously, does Aristotle, into four kinds: *material, formal, efficient,* and *final* causes. Take, for instance, a horseshoe: its *material* cause is metal; its *formal* cause is its characteristic shape; its *efficient* cause is the skill of the blacksmith; and its *final* cause (its purpose) is the preservation of a horse's hoof. Of these, the final cause, where there is one, has precedence over the others as the cause that best explains 'the reason why' there is a horseshoe.[137]

All this can be useful, in the way that analysis is, in making distinctions among apparent similarities. But it also sets up artificial barriers to seeing what is in front of you: that things are the way they are for a host of reasons that are literally endless and overlap. Borges again: 'There is no act that is not the coronation of an infinite series of causes and the source of an infinite series of effects.'[138] In causative chains, say that of the horseshoe, there are proximal (immediate) causes and ever more distal (remote) ones – the cause of the metal is a whole narrative – or *relation*, as we say – of events, involving ancient geology and modern mining and smelting at the very least. And so with the shape; and so with the skill of the blacksmith (a no doubt complex narrative, both personal and cultural); and the purpose of the *narrative* is to explain why there is a horseshoe at all, a narrative

---

134 See Gare 2018.

135 Ho & Popp 1994 (432).

136 Schiller 1891 (75–6).

137 Not everything has a final cause, however. Aristotle gives the example of an eclipse as something that has no final cause: *Metaphysics*, 1044^{b}12.

138 Borges 1964b (11).

which culminates in the reason it was made, to protect the horse's hoof – which is an instance of the future extending an influence back into the past. Chains of causation are conceived as working from the past towards the future: but the future, in the sense of internalised potential, pattern or *telos*, the drive that is in all living beings, may be as important a driver in the emergence of phenomena as the past.

Only a privative focus on detail delivers the clear cause and effect mechanism. Does the trumpeter cause a sound, or the trumpet – or is it the air blowing through it that causes the sound? Or the trumpeter's lungs, or the listener's ear? Or the instrument maker's skill? Or Handel, who wrote the music? Or all their parents for bringing them into being? Or the musical history of Europe? Or – the audience that *will* hear the performance tomorrow?

Here I am reminded of the work of Schelling, and his contention that, viewed correctly,

> the individual successions of causes and effects (that deceive us with the illusion of a mechanism) disappear, being infinitely small straight lines in the universal curvature of the organism, in which the world itself runs continually onward.[139]

Curvature: in which any starting point both influences and is influenced by its antecedents. Once again shape is a key to seeing the limitations of a certain kind of thought.

Bergson's all-important distinction between slices and flow opens our eyes to the distinction between a narrow view of causation as a linear sequence of discrete events, leading to determinism, on the one hand, and the autonomous reconfiguration of an indivisible, fluid form field or *Gestalt*, on the other.

THE PRESENT IS NEVER AN INSTANT BUT HAS THICKNESS

There is no point that is 'now', distinct from the immediate past and the immediate future – your consciousness straddles this supposed divide in a seamless way. There is no separate point of stasis in a stream: just longer and shorter stretches of flow. Awareness of the past always takes part in the present and the amount varies in extent depending on the type of attention you pay, the mood you are in, and other factors; it may extend back into the past, or away into the future, not just for seconds, but for days, or even years. It is only the deceptive habit of comparing time visually with space that makes us think of it as made of moments.

Whatever I am today includes my history over many decades, and the history of the culture and society that have formed me, and ultimately the history of humankind; all that partakes in, and is present in, exactly what I am *right now*. My body is history, and I am, at least

---

139 Schelling 1798 (x): » *Von dieser Höhe angesehen verschwinden die einzelnen Successionen von Ursachen und Wirkungen, (die mit dem Scheine des Mechanismus uns täuschen), als unendlich kleine gerade Linien in der allgemeinen Kreislinie der Organismus, in welcher die Welt selbst fortläuft.* «

in part, my body. Memory is stored everywhere in it. My body also anticipates the future, not just in the sense that I think about what is to come, but at an embodied, unconscious, even cellular, level an organism is acting *towards* certain outcomes, not merely passively responding to what has already happened. Equally my perceptions (what I am able to perceive), my interests (where I am directing my attention) are governed by the future, as well as the past: my aims and interests, where I project, and how I conceive, not just *my* future, but the future of us all. Bergson saw not only that the present has thickness, a thickness that is not circumscribed by any formula, or by any general prediction; but that the thickness varies and that this helps us to make sense of memory. The present is something 'thick, and furthermore, elastic, which we can stretch indefinitely backward by pushing the screen which masks us from ourselves farther and farther away'.[140] Elsewhere he writes:

> The distinction we make between our present and past is therefore, if not arbitrary, at least relative to the extent of the field which our attention to life can embrace. The 'present' occupies exactly as much space as this effort. As soon as this particular attention drops any part of what it held beneath its gaze, immediately that portion of the present thus dropped becomes *ipso facto* a part of the past. In a word, our present falls back into the past when we cease to attribute to it an immediate interest ...[141]

And this present, in as much as it includes elements of the past, does not do so by 'containing' a series of stills from our earlier life:

> An attention to life, sufficiently powerful and sufficiently separated from all practical interest, would thus include in an undivided present the entire past history of the conscious person, – not as instantaneity, not like a cluster of simultaneous parts, but as something continually present which would also be something continually moving: such, I repeat, is the melody which one perceives as indivisible, and which constitutes, from one end to the other – if we wish to extend the meaning of the word – a perpetual present, although *this perpetuity has nothing in common with immutability, or this indivisibility with instantaneity*. What we have is a present which endures.[142]

A present which endures. Compare the physicist Erwin Schrödinger: 'the present is the only thing that has no end.'[143] Thus, says Bergson, what is in need of explanation is not memory, but its 'apparent abolition'. He sees this as one of the primary functions of the brain (we might now say of the left hemisphere's conscious mind), to abolish memory: to lose 'irrelevant' experience, what is no longer of immediate *use*.[144]

According to Merleau-Ponty,

140 Bergson 2007d (106).
141 Bergson 2007e (127).
142 *ibid* (emphasis added).
143 Schrödinger 1964 (22).
144 Bergson *op cit* (128).

a past and a future spring forth when I reach out towards them. I am not, for myself, at this very moment, I am also at this morning or at the night which will soon be here, and though my present is, if we wish so to consider it, this instant, it is equally this day, this year or my whole life.[145]

These *tempora* do not, he says, need to be artificially synthesised: from the standpoint of experience, rather than any 'external' perspective, they seamlessly interpenetrate one another.

Though of course Merleau-Ponty does not intend that the past and the future should be thought of as symmetrical, there is a semi-automatic tendency to think of them as loosely symmetrical, in that they are both supposed to be equally 'unreal', and from the spatialised viewpoint they fall on a timeline with the past on one side and the future on the other side of the instantaneous 'now'. But they are *not at all* symmetrical, and this only points to the fallacy of the spatialisation and abstraction of time. Apart from the obvious experience that time's arrow moves only one way, so that the past is always growing while the future diminishes (unless each is infinite), the past has been 'passed' through the filter of being present, in the process acquiring embodiment, richness of human meaning and uniqueness. By contrast, the future is a theoretical projection, general, disembodied, and free to accept whatever meanings we care to throw at it. In that sense the future is *all theory*.

One would therefore expect the past to be of greater concern to the right hemisphere, and the future to the left hemisphere. Some evidence tentatively points to this being the case.[146] It is also clinically observable that in mania, in which the left frontal pole is dominant, subjects are orientated almost entirely towards the future, with completely unrealistic plans and projections, and seem shielded from thoughts of the past – they cut themselves off from their emotional history; whereas in melancholy, in which the right frontal pole is dominant, subjects are unable to free themselves from regretful thoughts of the past – of the bad that should not have happened and of the good that is lost – and shun the future.

TIME IS SHAPED BY OUR SENSE OF VALUE

This reminds us that our immediate concern, and the nature of our attention, alters time. A period that seems to pass particularly slowly at the time (eg, a period of depression) is in retrospect remembered as being relatively short, rather than relatively long: its emptiness prolongs it when experienced (presumably because we are too focussed on its passage, rather than taken up in engaged activity), but collapses it in retrospect (nothing 'there' to remember).

Why does time speed up as we get older?[147] A number of theories

---

145 Merleau-Ponty 1962 (489).

146 This is the argument of a paper by Joseph Dien (2008). See also: Saj, Fuhrman, Vuilleumier *et al* 2013. A good example of depression and time actually going backward can be found in Kloos 1938 (237). However, there are also occasional striking examples in schizophrenia: see, eg, case 5 ('Sche') and case 7 ('Ku') in Fischer 1929 (556 & 563): see also Saniga 2014 (it is exceedingly doubtful, however, that the case of disturbance of time in patient Jesse Watkins reported by RD Laing in Chapter 7 of *The Politics of Experience* represents schizophrenia).

147 There is an engaging book on the topic by Douwe Draaisma (2004).

have been put forward, but I think there are three principal factors that influence the speed with which time passes for most of us.

The first is the affect which attaches to the future compared with the affect which attaches to the past. In very early childhood, neither past nor future obtrude. Then, as the child grows, the future is increasingly longed for and is full of fantastic possibility, while the past is not, to any important extent, affect-laden: it does not draw us back. Hence time moves slowly in childhood, because, compared with our longing for the future, our 'shared' time seems to move too slow. In age, by contrast, the future is feared, and the past is highly charged with meaning – more than the present or future could be – and that draws us back: hence 'shared' time seems to us to be moving on too fast.

The second is the degree of absorption in the present moment. A child is completely taken up in play and directed towards his or her immediate goals: a child's time could therefore be said to be implicit. This implicit time is there for us, too, but normally overlaid by objectification. When we are absorbed, neither time nor our embodiment is in the foreground of our attention: when I am a lived body, I live time. But when I become consciously aware of my body as an object, time too comes to the foreground. When we are in the flow of life as Bergson remarked, 'there is neither a rigid, immovable substratum nor distinct states passing over it like actors on a stage. There is simply the continuous melody of our inner life.'[148]

While singing in a choir, or joining in a dance, or absorbed in a task requiring skill, or merely absorbed in thought, time appears absent – it no more obtrudes on our attention than does our body; though when finished, the sudden consciousness of what has happened in the world of 'shared' time makes time appear to have moved fast. Hence the folk wisdom that 'time flies when you are having fun'. These are the so-called flow experiences made famous by the Hungarian psychologist Mihály Csikszentmihályi, in which time effectively stands still; in which, to quote Thomas Fuchs, 'the sense of time is lost in unimpeded, fluid performance'.[149] As we get older we get less good at allowing ourselves to be absorbed in activities without keeping a watchful eye on time, 'looking before and after', as Hamlet says, and time becomes more pressing. It is less embodied, less *lived*: more like a fretful travelling companion. 'When we dwell in the past or become fixated upon the future', writes Glenn Parry, a psychologist with a special interest in Native American culture,

> we become unbalanced and can literally make ourselves 'time-sick'. Time-sickness is an artifact of the confusion about time. I believe it is actually a form of homesickness, because what we are missing, without realizing it, is a time when we lived in continuous connection with our homeland.[150]

148 Bergson 2007e (124).

149 Fuchs 2013. See also Csikszentmihályi 2014.

150 Parry 2015 (42).

That homeland is time's *flow*.

And the third factor that affects the speed of passing time is the degree of conformity of our experience to expectation. When unaccustomed things happen, and there is a break with routine, time slows down;[151] a weekend away from home, exploring somewhere new, can seem to have lasted a week. However, when everything is familiar, time speeds up. When we are young all is new: not so in age. This may be another good reason for practising mindfulness, which makes everything new once more.

CLOCK TIME *VS* LIVED TIME

Lee Smolin makes an observation that is interesting in the light of the hemisphere hypothesis. All established theories of physics, he points out, share a feature which makes them difficult to extend to the whole universe. For purposes of utility, one feigns to hold one part of reality fixed and invariable, when in reality it is not; and this allows one to observe an effect within a confined system which is allowed to move and change.

> Each divides the world into two parts, one that changes over time and the second assumed to be fixed and unchanging. The first part is the system being studied, whose degrees of freedom change with time. The second part is the rest of the universe; we can call it the background ... A distance measurement implicitly refers to the fixed points and rulers needed to measure that distance; a specified time implies the existence of a clock outside the system measuring the time ... The division of the world into a dynamical and a static part is a fiction, but it is an extremely useful one when it comes to describing small parts of the universe. The second part, assumed to be static, in reality consists of other dynamical entities outside the system being analysed. By ignoring their dynamics and evolution, we create a framework within which we discover simple laws ... The division of the world – into a dynamical part and a background that surrounds it and defines the terms with which we describe it – contributes to the genius of the Newtonian paradigm. But it is also what renders the paradigm unfit for application to the whole universe.[152]

The ruse is useful in helping us get a grasp, but at the expense of misrepresenting all of reality not encompassed in the system.

In his book on consciousness, *Out of My Head*, Tim Parks makes a nice observation: 'If there is one thing that is not a reliable measure of chronometric time, it is human experience. Or to put it another way: if there is one thing that is not a reliable measure of human experience, it is chronometric time.'[153] What is often called 'subjective' time is not unreal: it is the most real that time can be. Clock time is real, too, but it is just a much more restricted aspect of time. Einstein, in a moment of frustration, stated that 'the time of the

---

151 Pariyadath & Eagleman 2007; Tse, Intriligator, Rivest *et al* 2004; Ranganath & Rainer 2003; Ulrich, Nitschke & Rammsayer 2006.

152 Smolin 2013 (103–4).

153 Parks 2018 (75).

philosophers does not exist'.[154] This is no truer than that 'the time of the clockmakers, the measurers, does not exist'.

Minkowski refers to Bergson's concept of the *élan vital*. This concept has been comprehensively misunderstood, a misunderstanding that largely stems from a process of Chinese whispers, which is both the effect and the cause of people no longer troubling to read him. It is popularly imagined to refer to some hitherto undiscovered physical substance that magically confers life, much as scientists used to believe that a substance called phlogiston was responsible for the phenomenon of fire.

In fact Bergson refers to something quite different in kind, which in the rest of this chapter I will attempt to convey to the reader. Consciousness is always of something else: to use the jargon, it is said to be 'intentional', and therefore essentially both directed and dynamic. The *élan vital* is the drive inherent in consciousness, within something that is valued towards something that is valued. This is what Merleau-Ponty referred to as the 'intentional arc' of directed activity, 'the overriding temporal forms whereby our apprehension (eg, of a melody) and action (eg, speaking a sentence) takes place.'[155] Fuchs says of this basic energetic momentum of mental life that it

> is the root of spontaneity, affective directedness, attention and tenacious pursuit of a goal, which are characteristic of living beings generally, but it also lends the 'intentional arc' the tension and energy it needs.[156]

Jacques Ellul remarks that

> motion is the spontaneous expression of life, its visible form ... [Yet nowadays] motion is dissected into discrete aspects so that its form appears phenomenally, point by point. The immediate consequence of such analysis is that motion becomes completely disjoined from person and internal life ... Action is no longer a real function of the person who performs it; it is a function of abstract and ideal symbols, which become its sole criteria.[157]

Abstract and ideal symbols: clocks are like words, or like money. They take part in a structure that runs *parallel with* the real world, and each can carry out internal transactions in its own realm. But all of these – language, money and clock measurements – must at some point 'cash out' in the experience they parallel, where alone they have meaning. And in reality time is not regular; as Borges put it, 'Time can't be measured in days the way money is measured in pesos and centavos, because all pesos are equal, while every day, perhaps every hour, is different.'[158] To see only clock time is to miss most of its meaning. I am reminded of Blake's saying: 'The hours of folly are measur'd by the clock; but of wisdom, no clock can measure'.[159]

154 Canales 2016b (21).
155 Fuchs 2013.
156 ibid.
157 Ellul 1964 (330).
158 Borges, 'Juan Muraña': 2000 (53).
159 Blake, *The Marriage of Heaven And Hell*, plate 7, 'Proverbs of Hell'.

While asserting the all-pervasive reality of time, astrophysicist Michael Shallis observes:

> The techniques of instructional science cannot handle individual experience or admit to the quality of time. [Whereas] descriptive science can ... the fact that the experience of time is not quantifiable puts it into that arena of human perceptions that are at once richer and more meaningful than are those things that are merely quantifiable ... The lack of quantification of temporal experiences is not something that should stand them in low stead, to be dismissed as nothing more than fleeting perceptions or as merely anecdotal; rather that lack should be seen as their strength. It is because the experience of time is not quantifiable and not subject to numerical comparison that makes it something of quality, something containing the essence of being ...[160]

> Clock time is invented time, but man has been too gullible, he has ended up believing that his invention has an objective existence ... Objective time, clock time, exists because the mind invented clocks. That invention gave us a definition of an apparently objective time that we believe in too much ... [However] objective time has gone. It has gone in relativity, gone from the quantum world, gone in cosmology ... Only in the 'normal' world, which has been impoverished by our definitions and explanations which define poorly and explain little, does objective time still hold sway.[161]

In the immediate aftermath of her left hemisphere stroke, the neuroanatomist Jill Bolte Taylor found that she had to work out how to move. But once this had settled, she

> sensed the composition of my being as that of a fluid rather than that of a solid. I no longer perceived myself as a whole object separate from everything. Instead, I now blended in with the space and flow around me ... I felt truly at one with my body as a complex construction of living, thriving organisms.

Later that same day she found that, with her left hemisphere no longer functioning normally, she 'also lost the clock that would break my moments into *consecutive brief instances*'.[162]

Those who try to speak about time, find themselves, according to Waismann,

> pulled to and fro: 'I'm always in the present, yet it slips through my fingers; I am going forward in time – no, I am carried down the stream' ... *one* answer will never do.[163]

One reason we are pulled to and fro is that we try to combine an objectified image of time seen from the outside, with the experience of time as inhabited. Such an attempt is never completely successful.

160 Shallis 1982 (155–6).
161 ibid (198).
162 Taylor 2009 (42–3 & 68; emphasis added)
163 Waismann 1968 (5).

Bergson likens it to the impossibility of achieving the experience of walking around a town by compiling a potentially infinite number of still photographs from every conceivable angle; or, in poetry, of achieving the effect of the original by a translation. However much you carry on 'perfecting' it, then adding comments on your translation, and commenting on your comments, piling explanation on explanation, you can never quite, though the process be carried on to infinity, reach the simple effect of the original:

> the object and not its representation, the original and not its translation, is perfect, by being perfectly what it is. It is doubtless for this reason that the *absolute* has often been identified with the *infinite*.

The absolute has this in common with the infinite, that it can never be reached, so that progress is always asymptotic, like trying to approximate a curve using straight lines. Anything that is *sui generis*, incomparable to anything else and indecomposable into parts, shares this seemingly infinite quality. Unsurprisingly, then, time and movement share the characteristic:

> When you raise your arm, you accomplish a movement of which you have, from within, a simple perception; but for me, watching it from the outside, your arm passes through one point, then through another, and between these two there will be still other points; so that, if I began to count, the operation would go on for ever. Viewed from the inside, then, an absolute is a simple thing; but looked at from the outside, that is to say, relatively to other things, it becomes, in relation to these signs which express it, *the gold coin for which we never seem able to finish giving small change*. Now, that which lends itself at the same time both to an indivisible apprehension and to an inexhaustible enumeration is, by the very definition of the word, an infinite.[164]

TIME AS NECESSARY FOR INDIVIDUATION

In Chapter 9, I commented on the way that the loss or disruption of experienced time in schizophrenia had, among its many effects, that of devitalisation. Impermanence is vital, in the original sense of that word. 'Time is immediately given', writes Bergson. 'That is sufficient for us, and until its inexistence or perversity is proved to us we shall merely register that there is effectively a flow of unforeseeable novelty'.[165] The existence of time is inseparable from the evolving of difference and variety: therefore from freedom and creativity. Without time there is no principle of individuation, no human *being*, with its situation in the world, its capacity to feel, to act, and to relate to the cosmos at large.

'The continuing allure of Platonism', wrote Sherover,

164 Bergson 1912 (6–7; emphasis added).

165 Bergson 2007f (86).

may well be, as Heidegger once suggested, a sign of a continuing attempt to escape temporality. On any kind of Platonism, it is hard to see how one can explain why or how a completely atemporal realm of true 'reality' could or should bring a temporal order into being.[166]

There are two questions here: how could it, and why should it? On the first question, I am with those who would argue that time could not, and did not, arise from the atemporal; that time is an ontological primitive. On the second question, we are in the realm of speculation, but a starting point might be that only time can lend the *freedom* for individual elements of creation to come into being, and to have any meaning whatever:

> As absolute ... the world repels our sympathy because it has no history. As such, the absolute neither acts nor suffers, nor loves nor hates; it has no needs, desires, or aspirations, no failures or successes, friends or enemies, victories or defeats. All such things pertain to the world *qua* relative, in which our finite experiences lie, and whose vicissitudes alone have power to arouse our interest ... The doctrine on which the absolutists lay most stress is the absolute's 'timeless' character. For pluralists, on the other hand, time remains as real as anything, and nothing in the universe is great or static or eternal enough not to have some history.[167]

Indeed only 'in the world of finite multifariousness ... does anything really happen, only there do events come to pass.'[168] An immaculate vacuity is no contest for the vibrant imperfection of reality.

To have no history is not to exist; and to have a history of any kind is necessarily to exist in time. Time is the force field of individuation. Left hemisphere aspirations, embodied in much of the mainstream of Western philosophy for two thousand years following Plato, entail there being a single timeless unchanging unity, in which true creativity, individuation, and history come to be merely illusions, or at least fallings away from an ideal. Yet time is not simply a force for differentiation, but also for cohesion: it not only extends forwards and backwards along what seems like a line, but, as 'intersubjective time', also underwrites rhythms and cycles that keep different organic entities, and different levels of the same organic entity, entrained with one another. Rhythms, and rhyme, can exist only in time. Rhythms – regular vibrations and oscillations – are fundamental to the physics of the universe as much as to music, to brain function as much as to poetry, to nature's seasons as much as to the proper functioning of the body: science discloses them everywhere. And rhyme requires both separation in time and yet a binding together across time, sameness and difference, in which, as

---

166 Sherover 1989 (290 n).

167 James 1909a (47–9).

168 James *op cit* (50).

in a musical phrase, the past is made to resonate with the present. (I will have more to say about rhythm in Chapter 23.)

Borges expresses memorably the relation between time and individuation:

> To deny temporal succession, to deny the ego, to deny the astronomical universe, are apparent desperations and secret assuagements. Our destiny … is not horrible because of its unreality; it is horrible because it is irreversible and ironbound. Time is the substance I am made of. Time is a river that carries me away, but I am the river; it is a tiger that mangles me, but I am the tiger; it is a fire that consumes me, but I am the fire. The world, alas, is real; I, alas, am Borges.[169]

Borges is humorously, and humanely, pessimistic. But the words 'irreversible and ironbound' compel us to note that this view of time is applicable only when we retrospect. Nothing can undo time, true; but the pay-off is that we are free to choose in prospect – only because time exists. Perhaps it is the burden of freedom that proves 'horrible'.

According to Heidegger, the sense, the point, the very meaning of *Dasein* (each existing human being) is temporality: in his words, '*Der Sinn des Daseins ist die Zeitlichkeit*'.[170] The German word *Zeitlichkeit* is not so thoroughly technical and alien-sounding as temporality, something more like 'having the nature of time'. And so one might say that 'the meaning of human being is being *with* time'. The point is not the obvious one, that whatever we do we cannot escape the clutches of an external constraint called time, but that time is right at the core of our being: not an incidental, perhaps lamentable, aspect of the human condition, that in a better-ordered world would not – does not, according to Plato – exist. Heidegger's idea is not far from a saying of Merleau-Ponty: 'We must understand time as the subject and the subject as time', where, by 'the subject', Merleau-Ponty means human consciousness.[171] Our consciousness depends on time, and we humans have no meaning, and can find no meaning, outside time. That is one reason why Heidegger called *Dasein* the being-towards-death (*Sein zum Tode*).

In talking about time, Bergson sometimes uses the image of the colour spectrum, in which there is both continuity and maximum differentiation at once. What is more, none of the differentiations disrupts its continuity as a whole; the colours cannot be demarcated, merely indicated. As always, we need both differentiation *and* union, an idea so fundamental that one would anticipate its being built into the structure of the physical world both in space and in time.

I believe it is. Physics must speak of both waves and particles, or better still, as quantum field theory suggests, continuous fields and their ability to collapse instantaneously into quanta – discrete units.

169 Borges 1964b, 'New refutation of time' (186–7).

170 Heidegger 1986 (331).

171 Merleau-Ponty 1962 (490).

But, as we have seen, even discrete quanta, having interacted, appear linked in such a way that a change in one causes an instantaneous change in the other, even though the quanta are separated too far for information to have travelled between them in the available time. Perhaps more surprising to the reader may be to realise that physicists recently demonstrated that entanglement exists not just over space, *but over time*. They 'generated and fully characterized an entangled pair of photons that have never coexisted'.[172] These findings together are of considerable significance for a question that recurs throughout this book: how the parts relate to the whole.

My theme has been that the parts of 'things', most especially living organisms, are never isolated one from another. The finding of quantum entanglement in space, and now in time, underlines how fundamental this truth is to the structure of the world. As the researchers themselves point out, the apparent

> 'puzzle' of how parts fit with an overall whole presumes clear-cut spatial boundaries among underlying components, yet spatial non-locality cautions against this view. Temporal nonlocality further complicates this picture: how does one describe an entity whose constituent parts are not even coexistent?

Try a family, a culture, a civilisation. 'Nonlocality', Gleiser tells us, 'is an indelible feature of entanglement, and entanglement is an indelible feature of quantum mechanics.'[173] And of the living world.

TIME AND ETERNITY

As we have seen, Bergson remarked that 'that which lends itself at the same time both to an indivisible apprehension and to an inexhaustible enumeration is, by the very definition of the word, an infinite.'

Infinity in time is eternity. How can we relate time and eternity? If ever there were a topic that represents the gold coin for which we never seem able to finish giving small change, infinity must by definition be it.

I quoted Schrödinger's 'the present is the only thing that has no end.' There is a danger that any discussion of time and eternity leads to an infinite regress. Naturally we cannot conceive eternity except by means of time, and so all discussion of their relationship runs the risk of describing a snake that bites its own tail. Either time eats eternity or eternity swallows up time, depending on how you look at it. Merleau-Ponty exemplifies this problem when he prioritises time, by saying that 'the feeling for eternity is a hypocritical one, for eternity feeds on time. The fountain retains its identity only because of the continuous pressure of water.'[174] The French existential philosopher Louis Lavelle takes us more in the opposite direction – Schrödinger's direction – when he says that

---

172 Megidish, Halevy, Shacham *et al* 2013.

173 Gleiser 2014 (233).

174 Merleau-Ponty 1962 (492).

there is no true novelty other than the discovery, at every moment of time, of an eternity which delivers us from time ... Infinity is the negation of the end, and therefore of the way. It is itself the end, and the way. And the soul finds her equilibrium and her security only when she fixes her eyes on infinity *present here and now*, and has ceased to relegate it to an eternal beyond.[175]

So, the claim is, eternity is now and now is eternity. I feel the Zen saying 'Yes, but ...' coming over me. In this respect, I found this passage from the journals of Alexander Schmemann thought-provoking:

> It seems to me that eternity might be not the stopping of time, but precisely its resurrection and gathering. The fragmentation of time, its division, is the fall of eternity.

The fragmentation of time is exactly what we have been dealing with, here and in Chapter 16, and what Schmemann is saying is that this process not only denatures time but denatures eternity, that time and eternity are not enemies but stand or fall together. It is, I think, a powerful insight that to stop time would be to stop eternity. And he continues:

> The thirst for solitude, peace, freedom, is ... thirst for the transformation of time into what it should be – the receptacle, the chalice of eternity ... There are two irreconcilable types of spirituality: one that strives to liberate man from time; the other that strives to liberate time. In genuine eternity, all is alive.[176]

'These two ways of thinking, the way of time and history and the way of eternity and of timelessness', said Robert Oppenheimer in his Reith Lectures, 'are both part of man's effort to comprehend the world in which he lives. Neither is comprehended in the other nor reducible to it. They are, as we have learnt to say in physics, complementary views, each supplementing the other, neither telling the whole story.'[177]

My own take on eternity is dependent on another insight, that of the poet Blake, who famously wrote of holding 'infinity in the palm of your hand, and eternity in an hour'. For him, eternity was not a number that you reached by adding serially, unit by unit, left hemisphere fashion, but required a leap to a different understanding of reality altogether – as much a *qualitative*, as a quantitative, leap. Which is why infinity can fit in the palm of your hand, and eternity in an hour. Borges's remark that 'everything happens for the first time, but in a way that is eternal', expresses, I believe, a similar idea: eternity is not a thing, or amount, or measure, but a quality – a way of being: and everything that is present has – or can have – this quality.[178] Perhaps, one could say, eternity is more – like time itself – an adverb, not a noun.

175 Lavelle 1973 (79 & 190).
176 Schmemann 2000 (78).
177 Oppenheimer 1954 (69).
178 Borges 1999, 'Happiness' (441).

At the other end of the scale from the supposedly infinitely large, we can use the same insight to understand the supposedly infinitely small: the 'moment'. Although – and indeed precisely *because* – a moment can have no dimensions, it is not something atomistic. We may be encouraged to think in that way by the legacy of Aristotle, who likened 'the now' to a point. However, the 'moment' as referred to by Eastern sages such as Dōgen is entirely without dimensions, and is thus 'simultaneously unmeasurably brief and everlasting, always present'.[179] To quote Kierkegaard,

179 Raud 2012.

180 Kierkegaard 2000 (151).

> Thus understood, the moment is not properly an atom of time but an atom of eternity. It is the first reflection of eternity in time, its first attempt, as it were, at stopping time.[180]

SOME BRIEF REFLECTIONS ON THE DEBATE IN PHYSICS

This leads to consideration of time at the cosmic level. Scientists continue to argue about whether time exists or not. But most of these arguments, I believe, involve a fallacious choice: either there is left hemisphere time (Bergson's *temps*), time which has points, and measures the difference between points – clock time; or time does not exist, and in its place is the other classic left hemisphere conception, a Platonic, timeless, 'block universe', as physicists refer to it.

But these are not the only options. All things change, always: this the right hemisphere sees. 'Now' may be the only place we can inhabit, but it is not a point. Bergson's *durée* is something different: it flows. Flow is not, as we have seen, a series of slices, of train trucks, a succession of moments or points, gone as soon as they have come: nor is time. When you are properly in the flow *you do not experience time passing because you are flowing with it*. But it is there all the same in the flow.

I mentioned the Block Universe. There are three main contending models amongst physicists in relation to time:

> *The Block Universe* (*the large solid*): the past, present and future are all equally real; there is no privileged or moving present – every event is present at the time when it occurs, just as every *place* is a 'here'.
>
> *Presentism* (*the small sliver*): the universe consists of a succession of momentary (or very brief) phases; since only one of these presents is ever real, the past and future have no reality at all – they don't exist.
>
> *The Growing Block* (*the medium-sized solid*): the present and the past both exist, only the future is wholly unreal; since new 'slices'

of reality – new presents – are being created, and once created these remain in existence, the universe is constantly growing.[181]

His vision of a Block Universe led Einstein to claim that time did not exist:[182] indeed, Gödel thought that at last Einstein had bequeathed us a world without time.[183] But as the philosopher J. R. Lucas says: 'The block universe gives a deeply inadequate view of time. It fails to account for the passage of time, the pre-eminence of the present, the directedness of time and the difference between the future and the past' – in other words, just about everything that goes to constitute our idea of time.[184]

From a hemisphere point of view both Presentism and the Block Universe are typical left hemisphere constructs: either the whole thing is static, or it is made up of an effectively endless number of bits that are static.

As long as it does not depend on the acquisition of 'slices', the Growing Block seems more probable than the others, which just repeat the left hemisphere inabilities to deal with time: in the 'slices' version it, too, falls prey to the same problem. But why a 'block' at all? An improvement may be achieved by the model of cosmologist George Ellis, for whom time is a primary entity, one that can be understood by picturing an evolving universe, in which there is a growing 'volume' of spacetime. The surface of this volume can be thought of as the present moment. The surface represents the instant where 'the indefiniteness of the future changes to the definiteness of the past', thus 'spacetime itself is growing as time passes.'[185]

There is support from physics for the idea that time is more fundamental even than space. According to the Conformal Cyclic Cosmology (CCC) model put forward by Roger Penrose and Vahe Gurzadyan, in which universes succeed one another in time, transmitting information at the moment of transition, space is seen as 'essentially a geometrical construct', and the model 'relegates it [space] to an emergent phenomenon. Instead, it [the model] perceives the notion of time, and thus the flow of informational entropy, as more fundamental.'[186]

A further observation from physics is that the idea that time is an illusion is not consonant with the discovery that time's arrow cannot, after all, be reversed in the models of quantum mechanics and quantum field theory, the best models we have of the physical universe. As the Japanese theoretical physicist Satosi Watanabe first demonstrated, and as Andrew Holster has since confirmed, 'when the analysis is done correctly, it is clear that quantum mechanics is *time asymmetric* (*irreversible*). The probabilistic laws of quantum mechanics *simply do not hold of time-reversed quantum processes.*'[187] Dieter Zeh has also more recently argued this position in a series

---

181 Dainton 2014 (230–5).

182 Einstein, letter of March 1955, to Michele Besso's family following his death: 'People like us, who believe in physics, know that the distinction between past, present, and future is only a stubbornly persistent illusion' (quoted in Dyson 1979, 193). However, Einstein may be assumed to have been merely comforting the bereaved. According to Carnap: 'Once Einstein said that the problem of the Now worried him seriously. He explained that the experience of the Now means something special for man, something essentially different from the past and the future, but that this important difference does not and cannot occur within physics. That this experience cannot be grasped by science seemed to him a matter of painful but inevitable resignation' (Schilpp 1963, 37–8). On which, Smolin comments: 'Einstein's discontent comes down to a simple insight. A scientific theory, to be successful, must explain to us the observations we make of nature. Yet the most elemental observation we make is that nature is organised by time' (Smolin 2013, 91–2). We have choice about space, about where we move and the speed at which we do so, but not about time.

183 Yourgrau 2005.

184 Lucas 1990 (8).

185 See, eg, Ellis 2014. A very similar viewpoint is put forward in Richard Muller's fascinating book, *Now: The Physics of Time* (2016).

186 Currivan 2017 (193); Gurzadyan & Penrose 2016.

187 Holster 2014 (12; emphases in original).

of papers, as have a number of other physicists.¹⁸⁸ But in 1996, Ilya Prigogine had already noted that time's irreversibility was another aspect of the fruitful asymmetry at the core of things:

> it is precisely through irreversible processes associated with the arrow of time that nature achieves its most delicate and complex structures. Life is possible only in a nonequilibrium universe. Nonequilibrium leads to concepts such as self-organisation and dissipative structures ...¹⁸⁹

Not only is time obstinately real, but unfailingly creative. 'The only kind of universe that appears natural from the timeless perspective of the Newtonian paradigm is a dead universe in equilibrium', writes Smolin,

> ... obviously not the kind we live in ... But from the perspective of the reality of time, it is entirely natural that the universe and its fundamental laws be asymmetric in time, with a strong arrow of time that encompasses increases of entropy for isolated systems together with continual growth of structure and complexity.¹⁹⁰

TIME IS ESSENTIAL TO BEING IN THE WORLD

I began this chapter by quoting Schopenhauer's saying that to think away time and space is completely impossible. 'To think away time', writes Roger Scruton, 'is to think away myself. Indeed, it is to think away the whole observable world.'¹⁹¹ As for me, I consider time as real as anything I can know. To quote Smolin again:

> I no longer believe that time is unreal ... Not only is time real but nothing we know or experience gets closer to the heart of nature than the reality of time ... I will propose that time and its passage are fundamental and real and the hopes and beliefs about timeless truths and timeless realms are mythology ... Nothing transcends time, not even the laws of nature. Laws are not timeless. Like everything else, they are features of the present, and they can evolve over time.¹⁹²

Later he writes that, if we want to survive as a species,

> we have to understand the roots of the distinction between the artificial and the natural. These have a great deal to do with *time*. The false idea we have to put behind us is the idea that what is bound in time is an illusion and what is timeless is real.¹⁹³

This misconception underlies two thousand years of Western philosophical misjudgment.

Time is recalcitrant to thought in the left hemisphere sense of the word, and so I understand why it is tempting to 'think it away'. Yet a world without time is also unthinkable – literally inconceivable.

---

188 Zeh 2014; Haag 2014; Omnès 2014.

189 Prigogine 1997 (27).

190 Smolin 2013 (226).

191 Scruton 1996 (80).

192 Smolin *op cit* (xii–xiv). 'Time will turn out to be the only aspect of our everyday experience that *is* fundamental. The fact that it is always some moment in our perception, and that we experience that moment as one of a flow of moments, is not an illusion. It is the best clue we have to fundamental reality' (*ibid*, xxxi).

193 *ibid* (256).

For time is foundational to everything that exists. 'Were time to cease to exist, so would mountains and oceans cease to exist', wrote the thirteenth-century Zen master Dōgen.[194] Mountains flow, over time; oceans have tides, and flow. In other words, everything flows. Everything is extended in time.[195] According to Prigogine,

> ... the big bang was an event associated with an instability within the medium that produced our universe. It marked the start of our universe but not the start of time. Although our universe has an age, the medium that produced our universe has none. Time has no beginning, and probably no end.[196]

In an interesting examination of this issue, it has been argued at book length that even the conception of God as outside time must be mistaken.[197]

The essence of time is recalcitrant to thought because it is recalcitrant to language, as Augustine of Hippo most famously declared (I hope he may forgive my interpolations):

> Who even in thought can comprehend it, even to the pronouncing of a word concerning it? But what in speaking do we refer to more familiarly and knowingly than time? ... What, then, is time? If no one ask of me, I [RH] know; if I wish to explain to him who asks, I [LH] know not.[198]

The speaking hemisphere's re-presenting, rationalising tendency 'has the ability to isolate, immobilise and spatialise the flow of lived experiences inherent in *durée*, making them accessible to verbal description and analytic reflection',[199] but in the very fact of isolating and immobilising time it has lost the only elements of importance that it might have grasped: its universal nature and its motion. Language is a tool of convenience before it is anything else. As William James says: 'the concepts we talk with are made for purposes of *practice* and not for purposes of insight.'[200]

It may be said that here I am talking about experiential time, but that this gets us no nearer to the *reality* about time: the reality might be that it is an illusion after all. But for a scientific theory to be successful, it must explain the observations we make of nature; and 'the most elemental observation we make is that nature is organised by time'.[201] To regard time as a 'mistake' of experience is, in the words of philosopher Yuval Dolev, to remove 'all of experience from reality to the point of leaving empirical science, science based on experience, without trustworthy input to feed on ... discrediting experience to such an extent fatally cripples the very mechanisms through which the theory of relativity was arrived at to begin with.'[202] If we dismiss perceptual experience as a source of knowledge, we pull the rug from under the feet of science. And since we can only know

---

194 Dōgen Zenji 2007 (116).

195 Our English word *tide*, which used to mean 'time' (a usage preserved in words such as Yuletide) and by extension came to cover the tides of the sea, is cognate with German *Zeit*: both of these, as well as our more usual word *time*, come from a common root meaning to 'stretch out'. Time, therefore, as *durée*, not as *temps*.

196 Prigogine 1997 (6).

197 Mullins 2016.

198 Augustine, *Confessions*, Bk XI, ch xiv, §17.

199 Urfer 2001 (280).

200 James 1909a (290; emphasis in original).

201 Smolin 2013 (92).

202 Dolev 2016 (22).

anything about time by virtue of our consciousness, any attempt to know what it would 'really' be like outside of consciousness cannot succeed (indeed the very idea that reality is found by subtracting consciousness is fundamentally mistaken). It would be like trying to take a glimpse of what something looks like when it is not being looked at. What's more, time must be asymmetrical if science is to have a basis: 'Physics would not be possible at all', writes philosopher Michael Drieschner, 'if we did not start from this "direction" of time'.[203] As Merleau-Ponty says, the only thing that does not pass in time is the passing of time itself.[204]

If one espouses a view of the world as a flow, not as a collection of things, then all that exists is not just, inertly, being, but always 'be-coming'; and time and movement are bound up in that very concept. As it is written in the *Chuang Tzu*:

> There are no fixed limits
> Time does not stand still.
> Nothing endures,
> Nothing is final.
> You cannot lay hold
> Of the end or the beginning.
> …
> The game is never over,
> Birth and death are even,
> The terms are not final.[205]

Time is, in other words, according to this account, the condition of existence. And, as the rest of the passage from the *Chuang Tzu* goes on to say, in time our perspective changes, and what before seemed so important seems less, and what once seemed negligible now reveals itself to have untold depth. Above all, we cannot know what is to come – which is not a 'weakness' but a liberation. It opens up to possibility and is creative, not only of the future, as it unfolds, but of the present as we experience it – and even of the past as it is now unveiled to altered sight. Nothing is fixed; determinism is not determinable. Smolin again:

> A world without time is a world with a fixed set of possibilities that cannot be transcended. If, on the other hand, time is real and everything is subject to it, then there is no fixed set of possibilities and no obstacle to the invention of genuinely novel ideas and solutions to problems.[206]

In the absence of time, there can be no freedom, no evolution, no creation. Time is no tyrant, but the bringer of liberty.

The only surviving fragment of the pre-Socratic philosopher

203 Drieschner 2014 (135).
204 Merleau-Ponty 1962 (492).
205 *Chuang Tzu*, xvii, §1.
206 Smolin *op cit* (257).

Anaximander, born in the seventh century BC, already notes that there is a first principle, or ground of Being, beyond definition and without bounds (*apeiron*), 'out of which come into being the heavens and the worlds in them'; and that from this *apeiron,* 'the things that are come into being, and perish into the things out of which they come to be, according to necessity; for they give to each other justice and recompense for their injustice in conformity with the ordinance of time'.[207] Here are already hints of important ideas: that time grounds existence, since whatever happens does so in conformity with its ordinance; that 'the game is never over, birth and death are even'; that things arise not just from opposites, but from an element of *resistance*, which is bound up with time (compare Heraclitus' 'war is the father of all things'); and that the conflict nonetheless leads to resolution and reparation, in a never-ending flow.

Time's never-ending quality also ensures that 'the terms are not final'. Time is as much the creator as the destroyer: indeed ultimately more so, since what has been is never undone. '*Omnia mutantur*', wrote Ovid in the *Metamorphoses*, '*nihil interit*' – 'all things change, but nothing perishes'.[208] Aristotle quotes the poet Agathon's lines that 'One thing alone not even God can do, / To make undone whatever hath been done.'[209] Even the process of apparent destruction is merely part of the flow of creation. The entire business of life, from the tiniest single-celled organism to the whole sweep of evolutionary history, is one of constant flow, in which what we call 'things' appear and disappear; and yet it is one of constant creation and novelty, a present in which both past and future are importantly implicit. There are echoes of Hegel's image of the bud, blossom and fruit[210] in Yeats's poem, 'Among Schoolchildren', and they have a similar import:

> O chestnut tree, great rooted blossomer,
> Are you the leaf, the blossom or the bole?
> O body swayed to music, O brightening glance,
> How can we know the dancer from the dance?

So it is that, according to Hegel, the 'equal necessity of all moments constitutes alone and thereby the life of the whole'.[211] Each moment passes but no part of time is lost, each constituting part of the phenomenon as a whole. We do not regret the loss of each note of a melody as it is played: we do not regret the passing of each step in the dance. In the chapter of *The Principles of Psychology* which James devotes to the perception of time, he asserts that 'the knowledge of some other part of the stream, past or future, near or remote, is always mixed in with our knowledge of the present thing'.[212] So it is that each note of a melody is prepared by, and still contains the

---

207 Simplicius, *Commentary on Aristotle's Physics*, 24.13–21 (12A9 & 12B1).

208 *Metamorphoses*, bk 15, line 165.

209 Quoted by Aristotle, in *Nicomachean Ethics*, Bk VI, §2, 1139ᵇ10: μόνου γὰρ αὐτοῦ καὶ θεὸς στερίσκεται, ἀγένητα ποιεῖν ἅσσ' ἂν ᾖ πεπραγμένα.

210 See p 64 above. Yeats read Hegel widely: see Olsen 1983.

211 Hegel 1969 (68).

212 James 1890, vol 1 (606).

presence of, the notes that come before it, and prepares for and already contains an anticipatory experience of the notes that are to come. *If this were not the case, there would be no meaning whatsoever in the note at the instant it was heard.*

All is a flow, a pattern, in which each 'part' separated in time or space is implied in the whole – and therefore in the other 'parts'. And that includes each one of us.

TIME IN MODERNITY

The search for immobility and for the cessation of time's flow coincided with what I see as the left hemisphere's entry into the philosophical ascendant after Plato. It should therefore be particularly marked in modernism: and as Louis Sass points out, that search reaches its apotheosis in modernism. In this regard, he cites the cultural historian Roger Shattuck, who points to the time-machine fantasy of the early Dadaist Alfred Jarry as a symbol of the consciousness of the modernist age.[213] The fantasy machine achieves absolute immobility by a constant spinning motion, turning on its own centre, so as to escape the bounds of time – a sort of process of self-involution without end.

At a much more mundane level, is it not the fantasy of modern man – and woman – to defy time? Puritan diets, punitive exercise regimes, hair transplants, plastic surgery, cryogenics? An embargo on serious conversation about the fact that we are all born to die – quite soon? And that the meaning of life lies in its quality, not its quantity? Time and death are, it seems, our constant sources of anxiety and fear.

The idea of time as a thing leads ineluctably to the idea of time as a *resource*. This is in perfect synchrony with the left hemisphere's take on the world. The advent of spatialised clock time means that time appears to be something composed of chunks that can be counted – seconds, minutes and hours – *things*. 'Remember that TIME is Money', are the first words of Benjamin Franklin's 'Advice to a Young Tradesman'.[214] Hence time, like money, is 'stuff' that can be saved, wasted, or lost. As such it becomes something like money, merely a representation of something else that *is* real.

Lakoff and Johnson have done much to illuminate how the 'time as resource' metaphor changes our relationship with time. 'Cultures in which time is not conceptualised and institutionalised as a resource', they write, 'remind us that time in itself is not inherently resource-like. There are people in the world who live their lives without even the idea of budgeting time or worrying if they are wasting it.'[215] They quote an article from the 1984 *San Francisco Chronicle* which showed that the average weekly 'time theft' per employee

213 Shattuck 1988 (351–2); Sass 2017 (127).

214 Franklin 2004 (200).

215 See Lakoff & Johnson 1999 (161ff, esp 165).

amounted to 4 hours and 22 minutes. Time theft was, so the article claimed, America's 'number one crime' against business: the crime consisted in chatting with colleagues, dealing with personal matters in office time, extending lunch hours, and so forth.

Now we are all encouraged to see time as a resource, a thing, and therefore to rush our lives indiscriminately in order to pack 'more' into it. As a reaction to this there has arisen the Slow Movement, whose raison d'être, according to Norwegian philosopher Guttorm Fløistad, is that our most profound needs, those for closeness, care, love and appreciation of the good things in life, as well as of one another, depend on slowness in human relations: 'in order to master changes, we have to recover *slowness, reflection and togetherness.* There we will find real renewal.'[216] Pressure to acquire speeds things up; living in tune with the world moves things at the world's pace once again.

The urge to get more and more out of less and less time can reduce value overall: an apt example comes from scholarship. As Stefan Collini, who has written cogently about the plight of the modern university, puts it in his foreword to Berg & Seeber's *The Slow Professor: Challenging the Cult of Speed in the Academy*,

> 'Write more and publish less' is a valuable injunction, encouraging us to explore our thinking more, and only to publish when we are sure we have something worth saying ... we largely think by writing – or, rather, by trying to write and thus discovering that we don't quite know what we think. Similarly, rewriting is not chiefly a matter of buffing up already polished prose, but of coming to think a little more clearly and exactly. It all takes time.[217]

Economic pressures to do more faster may destroy quality and hence positively *waste* time.

Because we think of time as a 'thing' to be filled with other 'things' we foreground it (and of course spatialise it). We hasten always to pack more in, often seeking to do (but it cannot truly be done) as many things as we can at the same time. It leads us to feel we are always running against the clock, running after things and snatching them hastily, putting them in our little – always too little – bag of time.

Time, however, does not work like this. The more we hurry, the more it hurries, too. The more we try to do things at once, the less they mean, the less pleasurable they are, the less time we have, and the less we are alive. For we are never really *there*, but forever in the past or the future. Our mind is chasing another phantasm just as life offers us a potentially authentic experience: we munch it mindlessly, while watching TV, keeping an eye on our phone, half-listening to what our friends or children are saying, failing to notice what is

216 Di Nicola 2018.

217 Berg & Seeber 2016.

happening under our noses and unable to see that there is a problem except – that there is not enough time.

This is not your fault or mine. It is the way we feel obliged by our culture to live – if life it still is. Running away from mortality, we speed its arrival.

Time, however, expands, slows, and feels rich and companionable when we stop and simply attend to where we are. I am writing these words in rural Greece, somewhere I have always found time and life unhurried, with immensely healing properties. It is early November, the time of the olive harvest: the days are still warm and filled with sunlight. But over the valleys there comes the constant whining and braying of engines: just as leaves are no longer raked or swept in Britain, but blown noisily, ineffectually and extravagantly from place to place by petrol-driven, back-strapped blowers, the olives are now shaken from the branches by the flailing arms of a sort of strimmer. This means that the local farmer's olives are now gathered in in one morning by a gang of eight men, each armed with a machine. In the past it would have taken the family, men, women and children, three days to do the same work. How wonderful is that?

Well, it depends what you think life is about. Picking olives with friends and family was a companionable event. It involved singing and laughter. It brought together communities across the generations. It was work, but not in truth terribly hard work when the labour is shared. It would be punctuated with pauses to sit, chat, eat and drink. It had a meaning which is difficult to convey, surrounding the relationship between the often ancient trees, a proper reverence for them, their harvest, and its place in Greek culture, the process of gathering in something in the nature of a gift, in the peace of the autumn landscape, that would be stored and enjoyed over the whole coming year. It is also true that the olives were more carefully harvested, and there was less detritus – branches, leaves, odd plastic attachments from the flails – that got into the mash. But it's more about what it does to us and our relationship with nature than to the oil. A generative experience has been turned into a sort of violation: this was in fact the word used by the woman whose trees were being harvested in the village yesterday. Something the children would have remembered and hoped to repeat in their lives is gone. And so as to 'create time' – for what, exactly?

This is not a point about Greece, still less about olives; not even just country life – as it was before it was industrialised. It is a point about our attitude to time. The question always must be: what is our new way of doing things stopping us from experiencing (and how valuable or otherwise was that), and what are we doing with the time so 'retrieved' (and how valuable or otherwise is that)?

We have lost track of what seems, now, to us, like idleness: stillness, silence, peace, deep attention to what really matters. And this is quite compatible with work, if at the right pace. The title of one of the greatest 'works' of all Japanese literature, the *Tsurezuregusa* of Yoshida Kenkō, a fourteenth-century monk, translates into English as *Essays in Idleness*. One of its themes is a universal in Japanese and Chinese thought: the importance of transience to the meaning and beauty of all things.

> It is only after the silk wrapper has frayed at top and bottom, and the mother-of-pearl has fallen from the roller, that a scroll looks beautiful … If man were never to fade away like the dews of Adashino, never to vanish like the smoke over Toribeyama, but lingered on forever in this world, how things would lose their power to move us! The most precious thing in life is its uncertainty.[218]

[218] Kenkō 1967, §82 & §7. The first part is actually a phrase from the poet Ton'a, which Kenkō quotes here with approbation.

In its appreciation of whatever is irregular, unfinished to our hands, incomplete, always inevitably uncertain and ultimately transient, it both embodies much of the wisdom of the Orient, and singles out what it is that the right hemisphere alone can appreciate. To the left hemisphere these all seem like failings and fallings away from some illusory, timeless, perfection. How we strive, we in the West, to freeze the fleeting moment, to defy death and ageing, to make things regular and certain – deceiving ourselves, because none of these aims can be realised.

Two myths, one ancient and one modern, reveal that the ideal of permanence is a denial of everything we value. In the *Odyssey*, Homer gives his hero the opportunity to become immortal by remaining forever the lover of the beautiful goddess Calypso: but Odysseus is restless, and misses his wife and home on Ithaca. After seven years he is freed to go back to the everyday human world, with the implication that he must grow old and die. And in Julian Barnes's *A History of the World in 10½ Chapters*, the last chapter deals wittily with the consequences of living in what should be eternal bliss: his nurse-guardian angel confides that eventually, and often after only a few hundred years, all the residents of Heaven ask to go back to die.

The Greek island of Icaria hit the news when someone noticed that its inhabitants regularly lived to be over 100 years old. A man born there, but living in the US, was given six months to live, and decided to return to Icaria to pass his last days among the people he knew and loved. That was 50 years ago, and in his 90s he is still cultivating his vegetables. What is the secret of the island life? When everything else was taken into account, the main things that stood out were the habit of rising late (the doctor's surgery didn't even open till 11 am), taking at least one good daytime nap, working for a few hours in the open air growing food, and drinking two-thirds of a litre of wine a day. My own suspicion is that genes played a part, too – but the prescription for a good life is so good, why spoil it?

The *ancient* Greeks had no great love of work, either. They saw it as a necessity whose purpose was to clothe, shelter and feed the population, but that once that was out of the way, they could get back to the real business of life: namely, leisure. By this they did not mean just lazily lying about, nor, as leisure often seems to be nowadays, the frenetic pursuit of stimulation in a desperate attempt to fill the void so that we do not have to contemplate the emptiness of existence. No, it was in a way the opposite. A cultivation of stillness, the devotion of time and attention to what matters, and that has no ulterior purpose or instrumental 'point'– and is therefore in grave danger, nowadays, of being considered 'pointless': scholarship (which comes from the Greek word *scholē*, leisure), and the pursuit of insight, simplicity, beauty.[219] Leisure was the disposition of *receptive* openness to what *is*, all that we miss as we rush through life on our highway to the grave. In the German philosopher Josef Pieper's lectures, delivered just after the Second World War, and later published as *Leisure: the Basis of Culture,* he wrote: 'there can only be leisure, when man is at one with himself'.[220] We tend to overwork, he pointed out, as a means of self-escape, as a way of trying to justify our existence. Busy-ness, he contended, was the true laziness, a failure to engage fully and responsibly with oneself and the world.

Incidentally, leisure comes from Latin *licēre*, to have the opportunity or permission to do something. The German word for leisure, the word in Pieper's original title, *Muße*, is related to the modern German *müssen*, equivalent to our 'must', in the sense of having an obligation to do something. Originally, however, *müssen* meant to have the opportunity, the space, the time, the *leisure*, to do something; only latterly did it acquire the sense of obligation.[221] English 'must' took a similar route from permission to obligation, though perhaps even earlier.

Historically the attitude that 'busy-ness' was both abnormal and regrettable was far commoner than our own view of work as a way of life. I had a teacher who in his 40s left to farm a smallholding of 30 acres in Wales: he and his family were largely self-sufficient, with their own sheep, ducks and hens, their own corn and vegetables. When I went to stay, I was astonished to find how much leisure time there was, since I had anticipated it would be grindingly hard work. We put in about two or three hours in the morning, milling corn by hand, scything grass, chopping wood, and had the rest of the day free to read, talk and go for walks. When I expressed my astonishment, he explained that there were two busy periods a year:

---

219 The Latin word for business (literally, busy-ness) is *negotium*, from *nec* + *otium* – literally, 'no leisure'.

220 Pieper 1998 (30).

221 *Muße* is also 'the Muse' (Malsbary: Pieper 57 n 7). For more detail, see woerterbuchnetz.de/cgi-bin/WBNetz/wbgui_py?sigle=DWB&lemma=muessen. The transition seems to have been a consequence of negation, since 'must not' means 'does not have permission'.

a couple of weeks in the spring, during lambing, and a couple of weeks in the autumn, for harvest, when they and their neighbours all mucked in to help one another, and some of those days were very long. They were also associated with social gatherings across classes and generations, convivial festivals. Other than that there was no need for long hours. He recalled how William Langland berated the fourteenth-century peasantry, because of their soft lives, demanding long lunch breaks and refusing to eat bread made from unrefined flour. He explained that it was only with the eighteenth century that farming life became hard, and country people were expected to work punishingly long days: landowners had discovered greed, otherwise known as capitalism. They, and therefore all around them, were no longer content to produce just what was needed by the community, but to make the land yield more and more, to sell the excess and to become rich on the profits.[222] In this story, familiar to us today, there are some winners and many losers. And even the winners are, taking the broad view, losers, too.

'Faith in progress is deep within our culture', writes Juliet Schor. 'We have been taught to believe that our lives are better than those who came before us.' She rejects 'the notion that we work less than mediaeval European peasants, however poor they may have been'. And she reflects that 'the lives of so-called primitive peoples are commonly thought to be harsh – their existence dominated by the "incessant quest for food". In fact, primitives do little work. By contemporary standards, we'd have to judge them extremely lazy.' When we are asked to be thankful we no longer have to work as hard as our forebears, she says, the comparison

> conjures up the dreary life of mediaeval peasants, toiling steadily from dawn to dusk. We are asked to imagine the journeyman artisan in a cold, damp garret, rising even before the sun, labouring by candlelight late into the night. These images are backward projections of modern work patterns. And they are false. Before capitalism, most people did not work very long hours at all. The tempo of life was slow, even leisurely; the pace of work relaxed. Our ancestors may not have been rich, but they had an abundance of leisure. When capitalism raised their incomes, it also took away their time.

The working day had long breaks for food, and for sleep, which were universally expected. The Church calendar contained many holidays and festivals: 'all told, holiday leisure time in mediaeval England took up probably about one-third of the year. And the English were apparently working harder than their neighbours.' There were long vacations at Christmas, Easter and midsummer. Apparently even the *ancien régime* in France guaranteed 180 days holiday a year (that is,

222 Plato described this as *pleonexia* (greed): the behaviour of a tyrant and an example of lawlessness (*Republic*, 571^{b}5–7).

including Sundays). It was only with the advent of capitalism, and especially industrialisation, that lives became miserably hard.[223] It is capitalism that generated true poverty.

The philosopher Jacques Ellul points out that the fact that

> man until recently got along well enough without measuring time precisely is something we never even think about, and that we do not think about it shows to what a degree we have been affected by technique ... At most, life had been regulated since the fifth century by church bells; but this regulation really followed a psychological and biological tempo. The time man guided himself by corresponded to nature's time; it was material and concrete. It became abstract (probably toward the end of the fourteenth century) when it was divided into hours, minutes and seconds ...[224]

I mentioned that time becomes boring when we 'see' it in the left hemisphere sense – ie, fail to understand it. If we are readily bored, and quickly distracted by every sensation, what is our pathology?

It is known that people who are prone to boredom have little capacity to sustain attention,[225] and are also less accurate at judging the passage of time at multi-second intervals.[226] Attention-deficit/hyperactivity disorder (ADHD) is a right hemisphere deficit condition: amongst many other associations, subjects incidentally exhibit, as in right hemisphere damage, left-sided neglect,[227] and have thinning of the right frontal cortex.[228] Those with ADHD show deficits in right hemisphere attention.[229] Because the easily bored cannot sustain attention, they constantly make more errors, get less involved with the task, and find it takes longer to complete, creating a perfectly self-reinforcing circle; there is a clear two-way relationship between attention deficits and the experience of boredom.[230]

> When we are bored, our attitude toward time is altered, as it is in some dreamlike states. Time seems endless, there is no distinction between past, present and future. There seems to be only an endless present.[231]

We are, it seems, back to the left hemisphere condition of time without meaning: interminable – timeless – time. And yet we have never lived under such time pressure as we live now. For we are no longer able to be at ease in its flow; we are forever trying to row upstream.

Modern culture reflects many of the characteristics of left hemisphere domination: loss of focus, of history and continuity, disrupted attention, excess of detail, fragmentation, reification, loss of the embodied self, and so on. But we are also losing a sense of the importance of narrative. Louis Sass, pointing to a seminal work of

---

223 Schor 1992 (10 & 44–7). Schor provides extensive scholarly discussion and references (see *ibid*, 189–90, n7). The evidence on mediaeval conditions is more familiar than that regarding the *ancien régime*, on which she cites de Grazia 1962 (119).

224 Ellul 1964 (328–9).

225 Hamilton, Haier & Buchsbaum 1984.

226 Danckert & Allman 2005 (242); Jokic, Zakay & Wittmann 2018.

227 Braun, Delisle, Suffren *et al* 2013. This paper is an excellent overview of the evidence for right hemisphere deficits in ADHD. See also Panksepp, Burgdorf, Turner *et al* 2003; Braun, Archambault, Daigneault *et al* 2000; Geeraerts, Lafosse, Vaes *et al* 2008.

228 Almeida, Ricardo-Garcell, Prado *et al* 2010; Makris, Biederman, Valera *et al* 2007; Shaw, Lalonde, Lepage *et al* 2009.

229 Geeraerts, Lafosse, Vaes *et al* 2008.

230 Damrad-Frye & Laird 1989; Hamilton 1981; Wallace, Vodanovich & Restino 2003.

231 Wangh 1975 (541).

modernism, Joseph Frank's 'Spatial form in modern literature', sees modernism as denying time, preferring to dwell on descriptions of static objects rather than processes or actions.²³² Modernist art also, intriguingly, made concerted attempts specifically to disrupt narrative.

And not just in art, but in life. With increasing mobility ensuring that the continuity of local histories is disrupted, and identity politics masterminding the complete rewriting of a people's history in a manner hitherto confined to totalitarian regimes, we are becoming a society without a history, in a novel without a narrative. And breakdown of narrative risks loss of meaning, coherence and purpose, in a people as much as in a person.

CONCLUSION

My aim in this chapter has been not to argue in the abstract about the nature of time, which might not be particularly fruitful, but to suggest that there is a lead on this from what we know about the brain hemispheres. On this account, brain science harmonises with reason, with intuition from experience, and with imagination. Together they enable us to achieve an insight into which of the competing views of time is most likely to prove viable, fruitful in the longer run – thus passing the pragmatist test for truth.

In the time debate, I suggest, we should prefer a view that time is real to one that it is an illusion. We should back those who believe that it is essentially flow and cannot be fragmented; that there are no points in time but only duration; that its nature is incompatible with determinism; that the future is open, not closed; and that time is fundamentally asymmetrical. Of course I am not a physicist, but my understanding of the physics, for what it is worth, is that this position is coherent from a standpoint taken by (at least some) physicists.

I have tried also to suggest in this chapter that though we think of time as a thing, it is a quality of the process of life. Time is the process of creation, in fact – nothing less; and the process of creation is time. This insight is not just in Bergson, but in St Augustine's pregnant observation that 'the world is not made *in* time, but *with* time.'²³³ Meister Eckhart saw creation not as something that *has happened*, but what *is happening*, at each moment.²³⁴ Time is not an unfortunate incidental finding about the universe that may be an illusion, but the condition of its existence, and at the core of everything we can know.

Even if it were an illusion, we couldn't know that without time. As Rilke wrote eloquently in the *Sonnets to Orpheus*:

> Does Time, as it passes, really destroy?
> It may rip the fortress from its rock;
> but can this heart, that belongs to God,

232 Sass 2017 (126–7).

233 Augustine, *De civitate Dei*, Bk XI, ch vi: '*non est mundus factus in tempore, sed cum tempore.*' Borges somewhat misremembers this in his essay on time as '*non in tempore sed cum tempore Deus creavit cælum et terram*'.

234 See pp 1236–7 below.

> be torn from Him by circumstance?
> Are we as fearfully fragile
> as fate would have us believe?
> Can we ever be severed
> from childhood's deep promise?
> Ah, the knowledge of impermanence
> that haunts our days
> is their very fragrance.
> We in our striving think we should last forever,
> but could we be used by the Divine
> if we were not ephemeral?[235]

[235] Rilke, *Sonnets to Orpheus*, II, §27 (trans J May).

My question to you then is, how do you know that the world that passes is inferior to an imagined world in which nothing passed?

Chapter 23 · Flow and movement

The cause of coming-into-being of all things is the vortex.
—DEMOCRITUS[1]

Movement is reality itself.
—HENRI BERGSON[2]

I do not define … motion, as being well known to all.
—ISAAC NEWTON[3]

Let me tell you about the curious and extraordinary case of Jason Padgett. By his own account a down-to-earth, party-loving furniture salesman from Tacoma, Washington, he was brutally assaulted, in 2002, outside a nightclub in the early hours of the morning. Although he was concussed, scans at the time showed no obvious brain damage. But something remarkable revealed, nonetheless, a deep change in the functioning of his brain.[4]

From having no aptitude for, or interest in, geometry or mathematics, he suddenly developed a talent for abstract geometrical draughtsmanship which at first seemed to him to represent the meaning of π, and later to relate to equations such as $E = mc^2$ or $hf = mc^2$. He also developed obsessive-compulsive symptoms, which are of course widespread in the population, but which he did not have previously. They play a prominent role in autism, however, and he now developed a highly focussed, narrow preoccupation with stereotyped and restricted interests, inflexibility of routine and a persistent preoccupation with parts of objects – all of which are also typical of autism. Taken together, all of these features characterise, as we know, the phenomenological world of the left hemisphere, which prefers narrow focus, is relatively 'sticky' or inflexible, prefers rule-following, and is engaged by parts or fragments. And indeed the brain damage he was later discovered to have sustained, although bilateral, was more severe in the right hemisphere; he was told that 'because of this there was a possibility my left was compensating' by going into overdrive.[5]

In line with his reports, he was scanned while attempting to visualise mathematical formulae. During these attempts to visualise mathematical expressions geometrically, brain scans showed significant activity almost exclusively, unlike normal subjects, in the left hemisphere, especially in the parietal cortex. We would normally expect activation principally in the right lateral cortex: but for Padgett, 'experience of complex geometrical images emerging from mathematical formulas is restricted to the left hemisphere', according to the experimenters.[6] And, by way of confirmation, disruption of the

1 Democritus, according to Diogenes Lærtius, *Lives of the Eminent Philosophers*, Bk IX, §45.

2 Bergson 2007e (119).

3 Newton 1687.

4 Brogaard, Vanni & Silvanto 2013.

5 Padgett & Seaberg 2014 (210).

6 Brogaard *et al, op cit.*

left parietal cortex interfered with his ability to see the world in this way.[7] Half of the mathematical expressions shown to Padgett during the scans followed this pattern: the others failed to excite visual imagery altogether, one way or the other.

A note at this point. Although there has been a tendency to see this as a case of sudden, almost Einstein-like seeing into the heart of mathematical problems, this seems unlikely, and Jason himself does not make that claim. His actual maths skills remained fairly basic after the accident and he has, in his own words, 'about gone back to school to learn traditional mathematics' in the conventional fashion since.[8] Although he was drawing complex geometric shapes, he didn't initially see them as representing equations, and there is no evidence that they actually do. No physicist or mathematician to whom I have shown the drawings can see any connexion between the patterns and the mathematical formulae (eg, $hf = mc^2$) that often form their titles: and although Padgett & Seaberg refer to the formulae that *did not* induce imagery as 'nonsense formulas',[9] Berit Brogaard, in the report of her scanning study of Jason in *Neurocase*, is careful not so to describe them. Because it is not true. There is no single mathematical difference between the two lists, appended to Brogaard's study, of 15 formulae that induced imagery in Jason and 15 that failed to do so.[10] A physicist, who saw him making the drawings, urged him to get conventional mathematical training. Tim Chartier, a maths professor at Davidson, when asked, said that the drawings were remarkable, but commented that one had to be careful when using the word 'genius' and that Jason needed the help of a trained mathematician.[11] This does not mean, then, that he has found a sudden genius for maths, but rather developed an unusually perfectionistic attraction to creating diagrams that are self-referring – and entirely composed of straight lines.

That this has little to do with maths is not surprising since the visuo-spatial imagery accompanying mathematical problem-solving is usually associated with the right hemisphere, not the left. Bear in mind the distinction between abstract processing or *manipulation* of data, which is in the left hemisphere, and the *understanding* of its referents in the right. In keeping with this, release phenomena in the left hemisphere often show a disinhibited production of forms, not necessarily accompanied by meanings: procedures become more important than their real-world significance. Thus, in the realm of language a phenomenon known as hypergraphia (compulsive writing), or hyperlalia (compulsive utterance), sometimes combined under the term logorrhoea (excessive language), may follow right hemisphere damage: lacking the constraint of the right hemisphere, a meaningless hypertrophy of language may result, with a compulsive production of words devoid of clear reference.[12] It may be that

---

7 Unpublished study: I acknowledge with thanks personal communication from Professor Berit Brogaard on this matter.

8 Padgett 2014b.

9 Padgett *et al, op cit* (209).

10 Brogaard *et al, op cit* (appendix).

11 Padgett 2014a.

12 Yamadori, Mori, Tabuchi *et al* 1986; Frey & Hambert 1972; Arseni & Dănăilă 1977; Braun, Dumont, Duval *et al* 2004.

Jason's extraordinary, obsessional and painstaking drawings are a visual equivalent of this. The links that he makes to significant formulae are likely to be adventitious rather than essential.

In any case this looks like a release phenomenon, secondary to right hemisphere dysfunction, in which the left hemisphere's visual system has become overactive, and almost, if one can put it that way, *enriched*. After all, we have seen that an equivalent 'enrichment' of right hemisphere function sometimes accompanies damage to the left hemisphere, so why not the reverse? During scans, however, Jason's right hemisphere is not inactive, but *becomes inactive* when he conceives an equation as an image (the opposite of what one might expect in the normal subject). In Jason, then, here is someone whose visualisations, following trauma, seem to conform to the left hemisphere take on reality. Quite *how* true this is, though, has to be seen, as they say, to be believed. Listen to what he reports.

When he regained consciousness after the assault,

> things looked like individual picture frames coming in. And clouds moving – instead of looking smooth, they looked like little tangent lines in a spiral. Everything was discrete and chunky ... 'Raindrops, to me, they're these beautiful interference patterns', he said. 'They don't look like they're these smooth round ripples. They look like they're little tangent lines. So, again, the smoothness is gone from everything. Trees moving would be like an equation translating. Like, if you were to write an equation, and it translates, it makes graphs change.'[13]

The next morning, while running water in the bathroom, he noticed

> lines emanating out perpendicularly from the flow ... At first, I was startled, and worried for myself, but it was so beautiful that I just stood in my slippers and stared.[14]

And when he extended his hand out in front of him, it was like 'watching a slow-motion film', as if every slight movement was in 'stop-motion animation.'[15]

Where have we seen something very like this before? Indeed, all these elements, so beautifully gathered together in one example, suggest the phenomenology of the left hemisphere, bits and pieces of which we have seen in patients with right hemisphere damage. Wholes have become parts, discrete, 'chunky'. Elsewhere Jason describes everything he sees as having 'a pixelated look':[16] the wholeness of things broken down into tiny discrete units. In his book he writes: 'the house itself', that most familiar of lived spaces, 'seemed to fall away as a whole and become just a collection of shapes'.[17] Flow and smoothness in time has become disrupted, so that motion becomes

13 Padgett 2014b

14 Padgett *et al* 2014 (31).

15 Padgett 2014a.

16 Padgett 2014d.

17 Padgett *et al* 2014 (35).

[18] Padgett 2014d.
[19] Brogaard *et al, op cit* (567).
[20] Brogaard *et al, op cit*.
[21] Padgett 2014d.
[22] See p 918 above.

an indefinite number of individual 'frames', initially tending towards stasis, fragmented and slowed down. Later on, the disjunctive nature remained, but the speed normalised: he reported everyday vision as 'discrete picture frames with a line connecting them, but still at real speed',[18] like a ciné film in which the frames are each distinct, not smoothed into flow. It was as if 'someone is pressing the pause button on a video very quickly'.[19]

Similarly flow and smoothness in space have become disrupted, now an indefinite number of little lines. Specifically, curved lines have become reduced to an indefinite number of rectilinear ones. This is a very important feature for Jason. He reports that he experiences smooth contours as small tangent, and secant, lines. He is 'obsessed with drawing complex geometrical images using only straight lines'.[20] He says: 'There's no such thing as a perfect circle', because he can always see the edges of a polygon, however many-sided, that approximates the circle.[21]

Remember Schelling's description of causative chains:

> the individual successions of causes and effects (that deceive us with the illusion of a mechanism) disappear, being infinitely small straight lines in the universal curvature of the organism, in which the world itself runs continually onward.[22]

Remember, too, Bergson's description of the logician: 'Having in fact left the curve of his thought, to follow straight along a tangent, he has become exterior to himself. He returns to himself when he gets back to intuition.'

As mentioned, Nicholas of Cusa thought that truth is not a thing like other things, or put together from other things, but indivisible. The image he used to describe the problem is that the intellect moves in straight lines, whereas reality is curved, and the curve of reality can never be reached however many straight lines you use.

Interestingly Cusanus wrote a treatise on the impossibility of squaring the circle, a mathematical fact that was not proven until 1882.

Fig. 37. *Geometric illustrations of Cusanus' concept of truth: the circle represents truth, the polygon the intellect's never-fulfilled approach to truth*

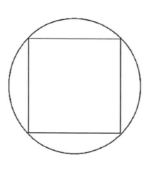
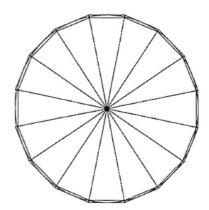

Though obviously stemming from a different period in intellectual history, Cusanus's view has its parallels in Schelling, Bergson and James, who all pointed to the importance of what they called intellectual intuition, rather than intellect *tout court*. The difference between straight lines and curves is not a simple geometric one, as Kepler expressed in a comment that should make us think. In his first work, the 1597 *Mysterium Cosmographicum*, he wrote, 'For this one fact, Nicholas of Cusa and others seem to me divine: that they attached so much importance to the difference between the Straight and the Curved.'[23] (As Hundertwasser was to say: 'The straight line is the curse of our civilization.')[24]

23 Kepler 1981 (92).

24 See p 824.

Let's return to Jason. Fig. 38 shows a circle as Jason sees it, constructed from 360 triangles, a polygon that can never (he repeated it with 720 triangles) become a circle:

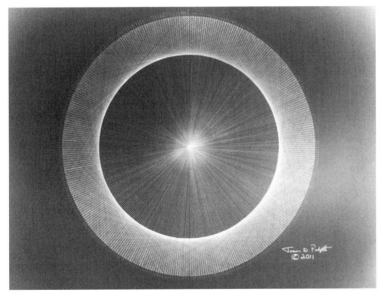

Fig. 38. *Towards Pi 3.141552779, drawing by Jason Padgett, 2011*

There was an immediate perceptual alteration following his injury, before he began his painstaking draughtsman-like diagrams. Here, from the early clinical setting, are his demonstrations of how he saw a circle (presumably this is a VW car wheel), approximated by tangents from the outside (with smaller inset instances of the same phenomenon); and an everyday object, a balloon, whose smooth contoured surface is broken up into numberless tangential straight lines (see Fig. 39).

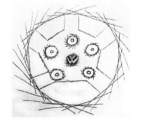

Fig. 39. *Early post-traumatic drawings of car wheel and balloon, by Jason Padgett* (from Brogaard *et al* 2013)

Five years before I read about Jason's accident, I described how:

> something that does not come into being for the left hemisphere is re-presented by it in non-living, mechanical form, the closest approximation as it sees it, but always remaining on the other side of the gulf that separates the two worlds – like a series of tangents

that approaches ever more closely to a circle without ever actually achieving it, a machine that approximates, however well, the human mind, yet has no consciousness, a Frankenstein's monster of body parts that never truly live ...[25]

Living, embodied forms, such as a tree, have become for Jason certainly something new and beautiful, but essentially abstract and inanimate, 'like an equation translating'. 'Numbers are an obsession, and I'm incapable of turning the fixation off. I can't climb stairs without counting them.'[26]

This might sound like typical OCD – and that is clearly an element – until one realises quite what he means by this. Everything in his world has not just latent (as is true for all of us), but appears to have *explicit*, abstract structure. The everyday, concrete and familiar has become something unusually abstract. Speaking of his morning coffee, he says: 'The perfect spiral is an important shape to me. It's a fractal. Suddenly, it's not just my morning cup of joe, it's geometry speaking to me.'[27] And: 'I see shapes and angles everywhere in real life', he says, 'from the geometry of a rainbow, to the fractals in water spiralling down a drain.'[28]

'Water spiralling down a drain' ... This point is worth pausing on. Because the water going down a drain is the best possible demonstration of a moving, ever-changing, ever-evolving, indivisible motion: a vortex. It is never, except artificially, frozen into a coherent image; and it is a constantly self-reforming asymmetrical structure (Fig. 40). Here, however, is Jason's beautiful, but very unusual, depiction (entitled 'black hole') of 'the pattern of lines I see overlaid on water going down the drain in the shower or the sink' (Fig. 41).[29]

On the facing page (Fig. 42) is a ripple in water, as we would normally see it: not uniform, but containing an inner *tension between opposites* – and beneath it (Fig. 43) is a water ripple[30] as seen by Jason.

Static, symmetrical, intricate in detail, composed entirely of straight lines – and skilfully made. But a ripple? For most of us a

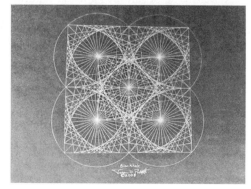

Fig. 40. *Vortex in water*

Fig. 41. Black hole, *drawing by Jason Padgett, 2008*

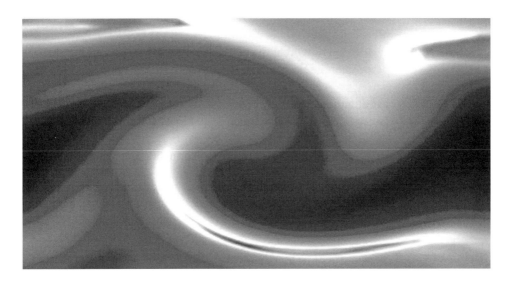

Fig. 42. *Ripple in water*

Fig. 43. Water ripple, *drawing by Jason Padgett, c 2010*

ripple is quintessentially moving, constantly evolving, never symmetrical, a fluid whole containing nothing but curved lines, if lines at all. Remember that the first moment of astonishment for him was perceiving the flow of water in the bath to have straight lines emanating from it. For Jason, 'water pours from the faucet in crystalline patterns'.[31] A beautiful word, crystalline, with its suggestion of stasis. Jason's picture is a splendid, fascinating abstraction, residing in the realm of the left hemisphere.

Its fractal structure nods in the direction of infinity, but, according to his interviewer, 'Padgett dislikes the concept of infinity, because he sees every shape as a finite construction of smaller and smaller units that approach what physicists refer to as the Planck length, thought to be the shortest measurable length.'[32]

But, as we have seen, neither extension in space nor time is actually comprehensible in terms of parts. A point is not a 'dot', because a dot already has a length, however short, and is therefore just a mini-line. A dot can't account for the introduction of length into a line, because it already presupposes what it is brought in to account for. But if it has no breadth at all – which is what differentiates an ethereal *point* from a worldly *dot* – it is in no better position to account for it either. A point is not 'part' of a line, and a line is not an aggregate of points. A straight line, however short, is not part of a circle; and a circle is not an aggregate of tangents. We end up having to define a line and a point by saying what they are *not*: a line is what has length, but no width or depth; and a point is what has neither length, nor width, nor depth.

In a picture of Jason's based on the so-called double slit experiment, he represents the ambivalence of his own work. In one world, the one alone he now perceives, there is no infinity, no flow, no wave, just finite, discrete shapes, composed of particles and straight lines, however complex; and yet at another level, there is also an awareness of the infinite, the flowing, the wave – hence the fascination for Jason and for us. Which of these two realities obtains may depend on which hemisphere is primarily attending to the phenomenon.

A few years ago I was asked to give the Vice-Chancellor's Lecture at an English university, and, contrary to my usual practice, I had written out my talk. In it I had incautiously included this passage: 'I sometimes think of the right hemisphere as what enables Schrödinger's cat to remain on reprieve, and the left hemisphere as what makes it either alive or dead when you open the box. It collapses the infinite web of interconnected possibilities into a point-like certainty for the purposes of our interaction with the world.' When I arrived, I was greeted by the Vice-Chancellor, who turned out, to my deep unease,

---

31 www.struckbygenius.com.

32 Padgett 2014c.

to be a physicist. I immediately started to have cold feet about what I had written, and made plans to omit the offending sentences. You can imagine my delight and surprise when he said, in so many words: 'I really enjoyed your book *The Master and his Emissary*. You know, I thought you could consider the right hemisphere as what enables Schrödinger's cat to remain on reprieve, and the left hemisphere as what makes it either alive or dead when you open the box. It collapses the infinite web of interconnected possibilities into a point-like certainty for the purposes of our interaction with the world.' So the passage remained … Since then I have come across this from physicist and philosopher Ruth Kastner: 'I too have thought that right brain corresponds to quantumland possibilities and left brain to spacetime actuals.'[33]

REALITY FLOWS, CONSCIOUSNESS FLOWS

James saw the entire universe as a seamless flow: 'its members interdigitate with their next neighbours in manifold directions, and there are no clean cuts between them anywhere.'[34] And he saw consciousness as having the same structure. In a famous passage he described, in a phrase that has gone into the language, what he called 'the stream of consciousness':

> Such words as 'chain' or 'train' do not describe it fitly as it presents itself in the first instance. It is nothing jointed; it flows. A 'river' or a 'stream' is the metaphor by which it is most naturally described. In talking of it hereafter, let us call it the stream of thought, of consciousness, or of subjective life.[35]

This *coincidentia* of the structure of the universe with the structure of our awareness of it is of profound significance. Not only is it not a coincidence, in the everyday sense, but it is not even, I suggest, a case of one modelling itself on, through or by, the other. They are, I suggest, aspects of one phenomenon. But here I am anticipating the argument of a later chapter.

Let us return, then, to Heraclitus, whose saying that one cannot step into the same river twice is familiar. Its power depends on the fact that the river has permanence: we call it the same river, because it is. Heraclitus points not to change only, but as much to permanence: flow which ever changes but ever remains. There is no *succession of things* involved in this change, because they always flow, *interpenetrating* one another. One of Plato's teachers, Cratylus, who was a pupil of Heraclitus, was right when he wittily added that we cannot step into the same river *once*, since at any instant it is already changing, and we too are in flux. And yet, like the river, it is only by changing that we can acquire the permanence we have.

33 Kastner 2017.

34 James 1909a (258).

35 James 1890, vol 1 (155).

Flow, then, is not primarily about change, since it is equally about persistence: I explored this coincidence of opposites in the chapter on the nature of the organism (Chapter 12), quoting Novalis.[36] Consciousness flows, the body flows; given this, it is hardly surprising that the evolution of the self is such a flow, too. At least the right hemisphere sees that this is the case: it is the right hemisphere, in particular the right dorsomedial prefrontal cortex (along with the right temporoparietal junction), that plays the critical role in sustaining the sense of a continuous self.[37] This is important because, in the left hemisphere's world, a self does not hang together. For Descartes, as we have seen, analysis reduced flow to a series of atomic points each devoid of any sustaining continuity with any other.[38] His radical doubt meant that he was sure of an existence at the 'now' point of his doubting (because *something* had to exist to be capable of doubt at all), but not of how he, René Descartes, should persist over time. I have also mentioned Derek Parfit, who, by following his distinctive analytic bent, concluded, remarkably enough, that persons have no continuous existence.[39]

Bergson, however, by seizing reality, as he put it, 'from within, by intuition [RH] and not by simple analysis [LH]', found that the one secure reality was our 'own personality in its flowing through time – our self which endures.' And, as I mentioned in the last chapter, he compared it to music, an idea I will return to in more detail.

So for both Bergson and James, the universe, consciousness, and the embodied self all flow. But there is a further point: change is accentuated when one sees '*things* that flow'; persistence when one sees the flow itself. And ultimately there is, they claim, just one incessant *flow*: though it may manifest as things flowing, the things arise on the surface of the flow, and do not constitute it. So why do we prioritise 'things'?

CONCEPTUAL THINKING DOES NOT FLOW

When we are absorbed in the natural business of living, our consciousness, as James says, forms a stream. He saw that in this stream

> no element *there* cuts itself off from any other element, as concepts cut themselves from concepts. No part *there* is so small as not to be a place of conflux. No part there is not really *next* its neighbours; which means that there is literally nothing between; which means again that no part goes exactly so far and no farther; that no part absolutely excludes another, but that they compenetrate and are cohesive; that if you tear out one, its roots bring out more with them; that whatever is real is telescoped and diffused into other reals; that, in short, every minutest thing is already its hegelian 'own other', in the fullest sense of the term.[40]

---

36 See p 444 above.

37 Fossati, Hevenor, Graham *et al* 2003; Keenan, Gallup & Falk 2003; Kircher, Senior, Phillips *et al* 2001; Decety & Sommerville 2003; Mega & Cummings 2001.

38 Sherover 1989 (281).

39 Parfit 1984.

40 James 1909a (271–2; emphases in original). One of James's eccentricities is spelling adjectives derived from proper names with a lower-case initial: I have preserved it here.

But when we inspect, focus on and therefore immobilise our conscious thought, the flow freezes. 'Concepts cut themselves from concepts': and according to Bergson, 'our concepts have been formed on the model of solids'.[41] (Concepts are in this respect, surprisingly, like *matter*.) They cannot interpenetrate; they are crystalline. This makes a kind of sense, because, Bergson continues, 'if the intellect [LH] has been made in order to utilise matter, its structure has no doubt been modelled upon that of matter'.[42] When it then attempts to grasp process, duration, the becoming of all that is, its very concepts externalise and freeze elements that are inseparable and dynamic in their essence. The flow is frustrated, being subjected to resistance. Equally James maintains that the very hard-and-fastness of categorical, conceptual thinking stops us seeing how things interpenetrate (its aim is to produce things that can be acted upon and manipulated, and therefore to *prevent* us from seeing it):

> A concept means a *that-and-no-other*. Conceptually, time excludes space; motion and rest exclude each other; approach excludes contact; presence excludes absence; unity excludes plurality; independence excludes relativity; 'mine' excludes 'yours'; this connexion excludes that connexion – and so on indefinitely; whereas in the real concrete sensible flux of life experiences compenetrate each other so that it is not easy to know just what is excluded and what not.[43]

What I think perhaps both James and Bergson underestimate – understandably, because they are rightly and admirably engaged in pushing back against the fragmenting nature of left hemisphere thought – is the value of resistance in differentiating that flow. This resistance is provided by conceptual thought.

LANGUAGE AND THING-WORDS

The tool of conceptual thought is, of course, language. As Kierkegaard put it, 'immediacy is reality and speech is ideality. Reality I cannot express in speech, for to indicate it I use ideality, which is a contradiction, an untruth.'[44] As we saw in Part II, Eastern traditions prize silence and mistrust language. Their philosophic utterances are largely apophatic, drawing attention to what is *not* the case. Much Eastern thought deliberately disrupts conceptual and linguistic thinking, pointing to its limitations: for example, 'the *tao* that can be named is not the real *tao*'. Such disruption could be seen as the whole aim of Zen, if having an overt aim were not in itself to subvert what Zen stands for. It brings to our awareness those things that must remain implicit if they are not to be denatured, that can never be clear when expressed in language, which is why Eastern thought so often seems to us what we call 'paradoxical'. By contrast, the aim

41 Bergson 1911a (ix).

42 Bergson 2007b (26).

43 James 1909a (253–4).

44 Kierkegaard 1967 (148).

for clarity resolves complexities into simple generalities, and what cannot be reduced in this manner is charged with being obscure, says Bergson:

> thus is explained the striking inferiority of the intuitive point of view in philosophical controversy ... Criticism of an intuitive philosophy is so easy and so certain to be well received that it will always tempt the beginner. Regret may come later ...[45]

Regret often *does* come later, if one has not by then become so narrow and rigid that one cannot see what it is one can no longer see. And, it was for this reason, as we saw in Part II, that many philosophers have asserted that philosophy must ultimately give way to poetry, or to music, which transcend conceptual thinking and everyday language.[46]

In using language we break up what is in reality inseparable, uncountable, immeasurable, into chunks: we substitute words for elements in experience, and these words, as Bergson puts it, 'ever after will cover them up; we then attribute to them the fixity, the discontinuity, the generality of the words themselves.'[47] That is certainly the danger. For James, 'events separated by years of time in a man's life hang together unbrokenly by the intermediary events. Their *names*, to be sure, cut them into separate conceptual entities, but no cuts existed in the continuum in which they originally came.'[48]

The simple act of putting events into language can have a fixative effect. But dealing in names, thing-words, changes further our perception of reality. Although nouns are not always 'object' words, and may denote actions, there is no doubt that some of the vitality is sucked out of language when we take the substitution of nouns for verbs too far. Contemporary Western language, in keeping with the left hemisphere bias of the culture, tends to 'nominalise' verbs. When we talk about an event, writes the philosopher Peter Simons,

> we typically nominalize a verb and form a derivative noun or noun-phrase: John's snoring last night, Luciano's rendering of 'Nessun dorma' in Madison Square Garden in 1987, Vesuvius's eruption in 79 A.D. all designate occurrents [events]. We are extremely adept at coining and using such nominalizations in both impromptu and routine ways.[49]

And we all recognise the tendency for bureaucratic and scientific prose to exploit nouns, especially abstract nouns, in place of verbs. In informal speech, we have now become so fond of nouns that we reach for them, in an unattractive trend, in place of verbs: to 'action' something, to 'inbox' someone, to 'architect' a plan, to 'dialogue', and so on – giving rise to the witticism that verbing weirds language.

---

45 Bergson 2007b (24).

46 See p 628 above.

47 Bergson 2007a (16).

48 James 1909a (285; emphasis in original).

49 Simons 2018 (50).

Chinese relies more on verb forms than does English; in a language such as Navajo, nouns are not necessary to a sentence and most information is expressed in verb forms. That both conceptual thought and language return us to the vision of reality as made up of bits, and thing-words then instantiate it, helps explain the one-sidedness of much Western philosophy and its historical lack of cogent defences of a right hemisphere understanding of the world, which for all the reasons we have already discussed, I hold to be truer to what is. 'Is it not possible for the syntax and grammatical form of language to be changed', mused David Bohm, 'so as to give a basic role to the verb rather than to the noun?'[50]

NECESSARY RESISTANCE

Yet, for all this, conceptual thinking is part of the evolution of self-conscious awareness. While it is certainly, and importantly, true, as James says, that 'when you have broken the reality into concepts you never can reconstruct it in its wholeness', fortunately we do not have to; unless, of course, we fall under the spell of language and abstract concepts to such an extent that we are no longer able to experience the intuitive wholeness from which, as secondary products, those concepts, and the things they denote, derive.[51]

That is a very substantial caveat in the world in which we now live. In some kinds of philosophical tradition this inability any longer to experience the intuitive wholeness is exactly what happens; and it so often happens nowadays that we need to be constantly alert to its effects, sceptical of it and on guard against its capacity to distort. The conceptual style of thought, James says, is simply

> superimposed for practical ends only, in order to let us jump about over life instead of wading through it; and if it cannot even pretend to reveal anything of what life's inner nature is or ought to be … we can turn a deaf ear to its accusations … We are so subject to the philosophic tradition which treats *logos* or discursive thought generally as the sole avenue to truth, that to fall back on raw unverbalised life as more of a revealer, and to think of concepts as the merely practical things which Bergson calls them, comes very hard.[52]

James mentioned a Hegelian other; and the thought takes us back to the need for something other than pure continuity and unity. There must be otherness, too, in some form within the flow. So it is that the very first passage I quoted from James above, the one about there being 'no clean cuts', begins '*without being one throughout*, such a universe is continuous'. That is the important point. 'Its members interdigitate with their next neighbours …'[53] So, note, there are still 'members', according to this account, even if they interdigitate.

50 Bohm 1980 (29).
51 James 1909a (261).
52 *ibid* (272–3).
53 *ibid* (258).

James rides the twin steeds of unity and multiplicity at once. And so do we all. It was James's life's work to enable us to see this more clearly, more beautifully, than Hegel, or any other philosopher, had ever done, or has done since.

In one of his essays, D. H. Lawrence reflected:

> What is it that man sees, when he looks at a horse? – what is it that will never be put into words? For a man who sees, sees not as a camera does when it takes a snapshot, not even as a cinema-camera, taking its succession of instantaneous snaps; but in a curious rolling flood of vision, in which the image itself seethes and rolls; and only the mind *picks out* certain factors which shall *represent* the image seen.[54]

Lawrence describes here perfectly the contrast between the seamless presence of experiential flow to the right hemisphere, and its representation in the left hemisphere; an inanimating takeover of the right hemisphere's animacy. Yet this left hemisphere intervention is not wholly misplaced. Indeed, it plays a crucial role in a certain kind of approach to reality if only by providing resistance. For, while there may well be a stream of consciousness, one aspect of consciousness, namely conceptual thought, as James realised, is not part of that flow. What he may have overlooked is that, to continue the metaphor, it provides rocks and stones in the stream: resistance to flow. That is its distinctive property. And for anything to 'arise' – rise up – tend, grow, change – there needs to be a degree of architective resistance within the connective flow.[55] In this the right hemisphere and left hemisphere are, as ever, complementary. And this architective resistance, causing something to endure for a while, manifests in mind as conceptual thought – much as, I shall later argue, it manifests in space as matter.

CONTINUITY AND DISCONTINUITY TOGETHER

We need, then, continuity and discontinuity together – to ride the *twin* steeds.[56] Here I would like once more to turn to Schelling, and his image of a stream, which is even more subtle and imaginative than the streams imagined by James or Bergson:

> Think of a stream, which is itself pure identity. Where it meets resistance, it forms an eddy. This eddy has no permanence, but is constantly disappearing and reappearing. Originally nothing in Nature is differentiated; all that she produces is at that point unseen and dissolved in the general productive potential. Only when there are points of resistance are Nature's products gradually precipitated out, emerging from the general identity. At every such point, the flow is broken up (so that productivity is destroyed), but at each moment the swell renews, and fills the sphere afresh.[57]

---

54 Lawrence 1992 (127–8; emphases in original).

55 See pp 834–6 above.

56 This is also in keeping with Pragmatism. Cf Rosenthal 1996: 'Pragmatic continuity can best be understood as having both unity and multiplicity' (559).

57 Schelling 1799a (30–1): » Man denke sich einen Strom, derselbe ist reine Identität, wo er einem Widerstand begegnet, bildet sich ein Wirbel, dieser Wirbel ist nichts Feststehendes, sondern in jedem Augenblick Verschwindendes, in jedem Augenblick wieder Entstehendes. In der Natur ist ursprünglich nichts zu unterscheiden; noch sind gleichsam alle Producte aufgelöst und unsichtbar in der allgemeinen Productivität. Erst wenn die Hemmungspunkte gegeben sind, werden sie allmählich abgesetzt, und treten aus der allgemeinen Identität hervor. – An jedem solchen Punkt bricht sich der Strom (*die Productivität wird vernichtet*), aber in jedem Moment kommt eine neue Welle, welche die Sphäre erfüllt. «

Resistance, according to Schelling, is *not a separate force*, but comes from the flow itself, and enables the unfolding of what in it is always potentially there; it precipitates the collapsing of its own potential into the actual. He is very clear that this resistance is part of the very same creative force that the resistance is there to resist. This is like the *coincidentia oppositorum*: some element of 'otherness' is incorporated within the force itself. All creation, he urges, is the result of this coincidence:

> A stream flows in a straight line forward as long as it encounters no resistance. Where there is resistance – a whirlpool forms. Every original product of nature is such a whirlpool, every organism. The whirlpool is not something immobilized, it is rather something constantly transforming – but reproduced anew at each moment. Thus no product in nature is *fixed*, but it is reproduced at each instant through the force of nature entire ... Nature as a whole co-operates in every product.[58]

The oneness is in the many-ness, and the many-ness in the oneness. 'This whole construction', he writes of the cosmos, 'therefore begins with ... a *dissonance*, and *must* begin this way.'[59] Dissonance is not external, in some sense, to the music, but itself part of the music. I will consider the analogy of flow with music shortly.

What he calls the 'products' of Nature – the fruits of Nature's generative quality, wherein its very essence lies – are never wholly distinct in themselves from Nature or from one another. They are only superficially 'thing-like'. Hence the comparison to eddies in a stream, which, although constantly inconstant, flowing on and yet renewing themselves, preserve for a while an identity distinguishable from the overall flow, but not separate from it, an identity which arises from the *very resistance which the water encounters within itself*.

For Schelling, then, 'things' did not have the meaning they have for us. They were more like processes, elements brought forth from Nature, but themselves, in turn, giving further process to Nature. He complained that, from our customary point of view, the primordial productivity of Nature disappears in the product: for the philosopher, he warns, the product must disappear in the productivity.[60]

What is more Schelling sees the very act of philosophising, which is nothing less than seeing into the nature of the world, as itself a liberating, creative act of Nature on behalf of Nature:

> To philosophise about Nature means to raise her up from the dead mechanism in which she seems to lie imprisoned, and, in freeing her, to bring her back to life, so to speak, and restore to her her own free unfolding – it means, in other words, to tear *oneself* away from

58 Schelling 2004, 'The unconditioned in nature': vol 1, i (18; emphasis in original).

59 Schelling 1994 (116; emphases in original).

60 Schelling 1799a (22): » in der gemeinen Ansicht verschwindet [*die ursprüngliche Productivität der Natur*] über dem Product; in der philosophischen verschwindet umgekehrt das Product über der Productivität. «

the common view, which sees in Nature only whatever happens to be – at best sees her activity merely as a *fact*, not the *activity* in itself.[61]

This is an account of philosophy reanimating the dead structure as perceived by the left hemisphere by attempting to regain the living process perceived by the right: what philosophy should be doing, as Schelling believes (and as do I), rather than the reverse process which it often serves in Anglo-American analytic philosophy, turning the living process into a dead structure. It is easy enough – indeed it happens automatically, and unconsciously, for us now – to see the mechanism and the parts; what is hard, without re-training one's attention, is to get back to seeing the living whole. To bring the corpse back to life.

Schelling held that there was a primordial energy, which he sometimes referred to as *Weltseele* ('world-soul'), which, like the stream, flows through the cosmos unimpeded until it encounters a difference within itself, a point of resistance. Then something is precipitated out of Nature, like the eddy in the current, a vortex. This 'product' of Nature is nothing mechanical or inert: it flows with vital energy itself, constantly renewing itself through the flow, which is both its cause and its effect, derived from this vital force's internal resistance. It is like a distinguishable pattern within the flow, precipitated out of the flow by the resistance in the flow. For Schelling, Nature continually evolves, due to the resistance that it carries within itself:

> Nature is originally pure identity – nothing to be distinguished in it. Now, points of inhibition appear, against which, as limitations to its productivity, nature constantly struggles. While it struggles against them, however, it fills this sphere again with its productivity.[62]

He also at times expressed the nature of being as a balance of contrary tendencies, a contractive force, 'gravity', and an expansive force, 'light'. All dynamic processes, he believed, could be seen as the interaction of these *contrary, but ultimately inseparable*, forces: a *coincidentia oppositorum*. This point is essential. For there to be a world at all, according to Schelling, the forces have to be part of one and the same process, otherwise either contraction would come to dominate, so that there would be no manifest universe, or else expansion would come to dominate, causing the universe to fly apart and dissipate at infinite speed. The final result would be the same in either case: there would simply be no world.[63] Only their being coextensive parts of one and the same process can guarantee their continual co-existence. A model for this is, once more, a magnet, whose opposing poles are inseparable from each other, even though they are opposites.[64]

This equilibrium of flow with resistance, of expansion with con-

---

[61] ibid (6): » *Philosophieren über die Natur heißt, sie aus dem todten Mechanismus, worinn sie befangen erscheint, herausheben, sie mit Freiheit gleichsam beleben und in eigne freie Entwicklung versetzen – heißt, mit andern Worten, sich selbst von der gemeinen Ansicht losreißen, welche in der Natur nur, was geschieht – höchstens das Handeln als Factum, nicht das Handeln selbst im Handeln – erblickt.* «

[62] Schelling 2004, 'The unconditioned in nature': I, i (18).

[63] See Bowie 2016.

[64] See, for example, Schelling, 'Of the magnet': 1988 (122–7).

traction, according to Schelling underwrites the continuing process of Nature as a whole:

> Visible Nature, in particular and as a whole, is an allegory of this perpetually advancing and retreating movement. The tree, for example, constantly drives from the root to the fruit, and when it has arrived at the pinnacle, it again sheds everything and retreats to the state of fruitlessness, and makes itself back into a root, only in order again to ascend. The entire activity of plants concerns the production of seed, only in order again to start over from the beginning and through a new developmental process to produce again only seed and to begin again. Yet all of visible nature appears unable to attain settledness and seems to transmute tirelessly in a similar circle.[65]

65 Schelling 2000 (2).

RESISTANCE BRINGS TO BIRTH BEAUTY AND COMPLEXITY

It is worth making a brief diversion at this point to take a look at resistance to flow and just how creative it is in its very nature. No obstruction in the path of a flow is required for it to become turbulent; *any inequality in speed or viscosity within the flow itself* can lead to turbulence. However in cases where there is a solid obstruction, the resulting flow exhibits surprising new qualities. From a straight flow in a rigid linear channel meeting a simple, smooth straight-edged obstruction such as a rod placed at right angles to the flow, the most extraordinary richness of design can emerge. It is an unfolding of potential that is completely unforeseen, until it manifests itself as eddies, as beautiful and strange as a plant emerging from a seed. It is the perfect image of multiplicity emerging from unity through resistance.

The images on the following pages (Figures 44–50) are all taken from laboratory experiments in which a simple flow was interrupted by a single, straight-edged, regular obstruction of some kind. The pattern of flow was made visible by pulses along a very thin platinum wire in the water, causing the release of tiny bubbles into the flow, which show up as white lines on a photographic plate.

The first image (Fig. 44) shows vortices in a water stream in the wake of a rod of circular cross section placed perpendicular to the stream. Fig. 45 is a detail from Fig. 44, which makes the resulting pattern more apparent.

The four images of Fig. 46 show an identical set-up: the vortices in a water stream in the wake of a rod of circular cross section placed perpendicular to the stream. However, different pulse rates and different durations have been employed in generating hydrogen bubbles. Each image has been reflected about the rod so as to incorporate a mirror reversal.

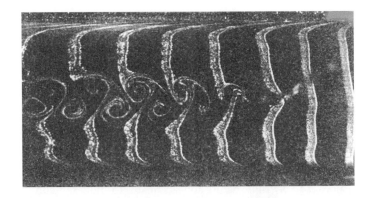

Fig. 44. *Vortices in a stream caused by uniform perpendicular obstruction*

(Figures 44–50 all from Douthat, Nagib & Fejer 1975)

Fig. 45. *Detail of Fig. 44*

Fig. 46. *Effects of differing rates and duration of pulse on vortices in stream*

CHAPTER 23 · FLOW AND MOVEMENT   963

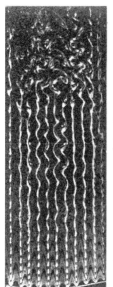

Fig. 47. Vortices in a stream in the wake of a regular rake of rods of rectangular cross section

Fig. 48. Detail of Fig. 47

The three images in Fig. 47 (previous page) are created by vortices in a water stream in the wake of a regular row or rake of rods of rectangular cross section. Fig. 48 is a detail from this series.

The next image shows vortices formed when a stream of air impinges on a regular cube:

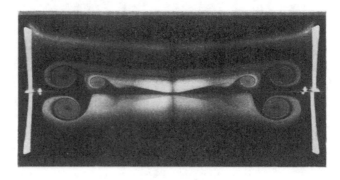

Fig. 49. *Vortices formed when a stream of air impinges on a cube*

And finally Fig. 50 shows the turbulence created behind a stick partially immersed in water. This kind of pattern, also seen in the first photographs above, is known as a Kármán vortex street.[66]

Leonardo da Vinci was fascinated all his life by water, its appearance and behaviour. He closely observed, in particular, whirlpools, vortices, turbulence and other complex flow phenomena. In fact he observed precisely the flow of water round an old man's stick partially immersed in a shallow stream (Plate 19[a]).

Leonardo observed that water often behaved like hair, something that is apparent from his detailed drawings in Plate 18: 'Consider the movement of the surface of water. It behaves like hair, which has two motions: one conforms to the weight of the mane, the others to the wandering of the locks. Likewise water has its eddying movements, one part of which follows the principal current, the other the random and reverse motion.'[67]

In another drawing, he observes the turbulence caused by a jet of water emptying from a rectangular opening straight into a pool, and at the same time the flow of hair (Plate 18).

Leonardo's late drawings of storms and deluges take things to the level where predictable patterns break down – as they do somewhat abruptly when turbulence passes a certain point.

That this degree of turbulence both has a life of its own and overwhelms human attempts to predict it is emphasised by the tiny village depicted near the foot of the drawing, caught up in the cataclysm (Plate 19[b]).

A close observer of the human body, Leonardo even understood the importance of vortices in the sinuses of Valsalva, at the root of the aorta, and their role in making possible the rhythmic opening and closing of the aortic valve, an observation that was confirmed

66 Schwenk 1976.

67 Royal Library (Windsor Castle): RL 12579r. As translated in Lugli 2016.

Fig. 50. *Kármán vortex street*

only in the twenty-first century using 4-D MRI.[68] Indeed spiral form is intrinsic to nature, as we have seen, and it is the spiral orientation of musculature in the heart and vessels which makes circulation – and therefore much of life – possible.[69]

THE CREATIVE POWER OF TURBULENCE

Turbulence is indeed a fascinating phenomenon. Vortices are absolutely fundamental in the structure of Nature, and almost all fluid flows occurring in nature, including all living organisms, are turbulent.[70] Turbulence is an expression of, and a consequence of, sustained energy flow, without which the vortex dies. As the mathematician Alexandre Borovik writes,

> The turbulent flow of a liquid consists of vortices; the flow in every vortex is made of smaller vortices, all the way down the scale to the point when the viscosity of the fluid turns the kinetic energy of motion into heat. If there is no influx of energy (like the wind whipping up a storm in Hokusai's woodcut), the energy of the motion will eventually dissipate and the water will stand still.[71]

For Hokusai's *Great Wave,* see Plate 20.

Turbulence is of unsurpassed complexity. According to the classic work on the subject, 'the mathematics of nonlinear partial differential equations has not been developed to a point where general solutions can be given. Randomness and nonlinearity combine to make the equations of turbulence nearly intractable.'[72] As a nice expression of quite how mind-boggling turbulence can be, Sir Horace Lamb, who had spent his life studying turbulent flows, quipped, in an address to the British Society for the Advancement of Science in 1932:

> I am an old man now, and when I die and go to heaven there are two matters on which I hope for enlightenment. One is quantum electrodynamics and the other is the turbulent motion of fluids. About the former I am rather optimistic.[73]

Richard Feynman described turbulence as the most important unsolved problem of classical physics.[74]

A tornado is a typical vortex: it contains contrary motions, cold air pushing down on the outside and hot air spiralling up on the inside. Vortices of all kinds both arise from and perpetuate such contrary motions. Most of us are not aware of how ubiquitous the vortex phenomenon is in nature. For example, it may account for how fish such as salmon and trout are able to rise up powerful columns of water that are pushing downwards, since the vortex has a core that moves as powerfully upward. (If a tornado can lift a house, a water current can lift a trout.) And they account for more 'everyday' phenomena, such as how fish are able to remain motionless in a flowing stream,

68 Bissell, Dall'Armellina & Choudhury 2014.

69 de Vries 2012 (311). Helical blood vessels permit a flow pattern that does not have stagnation points. The consequence of stagnation is atherosclerosis and, ultimately, death.

70 Tennekes & Lumley 1999 (1).

71 Borovik 2014 (32).

72 Tennekes *et al, op cit* (2).

73 Mullin 1989; Davidson 2004 (24).

74 Feynman, Leighton & Sands 1964.

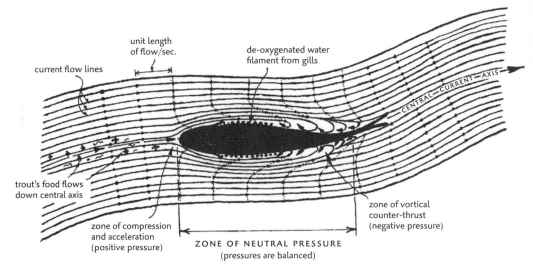

Fig. 51. *Vortices and contrary motion*
(Coats 2001, based on the researches of Viktor Schauberger)

75 Beal, Hover, Triantafyllou *et al* 2006.

76 Dabiri 2007.

77 FISH: Liao 2004: 'Comparisons with dead trout towed behind a cylinder confirm this intriguing observation that live trout may temporarily adopt the Kármán gait with no axial muscle activity, revealing paradoxically that at times fish can passively move against turbulent flow'. SWIFTS: Videler, Stamhuis & Povel 2004. BUTTERFLIES: Srygley & Thomas 2002. BASILISK LIZARDS: Dvořák 2014. JELLYFISH: Gemmell, Costello, Colin *et al* 2013. These principles were predicted by the work of Viktor Schauberger: see theartofnature.org/id20. html (a useful resource on the movement of fish, from which the illustration of trout movement is taken).

78 Tennekes *et al, op cit* (1–4).

without apparently consuming energy: they utilise the effect of vortices in the stream, often created behind rocks or stones, acting on their streamlined bodies. Even a dead trout can be shown to move passively upstream against the flow using such turbulence.[75] If none is present, fish will also create vortices themselves, using their gills and movement of the whole body. Vortices account for how many aquatic animals move: 'animals swimming in isolation and in groups are known to extract energy from the vortices in environmental flows, significantly reducing muscle activity required for locomotion.'[76] Vortices explain how swifts and butterflies fly, how basilisk lizards walk on water, and how the jellyfish comes to be the world's most efficient swimmer, using vortices to travel huge distances for the least possible expense of energy.[77]

Turbulence, then, has a number of familiar qualities that are suggestive of its creative potential, and I would say, perhaps not coincidentally, of its conformity to the metaphysics of the right hemisphere. First, all turbulent flows are irregular and therefore largely unpredictable; and yet not lawless, even if we cannot solve the equations. Second, every turbulent flow is different from every other, even though all turbulent flows have many characteristics in common; in this they represent multiplicity in unity and unity in multiplicity. Third, turbulence is a continuum phenomenon. Fourth, it is three-dimensional (two-dimensional flows are not generally turbulent). Fifth, it involves circular motion: it is rotational. Sixth, its characteristics depend on its environment: they are highly context-sensitive. Seventh, it exists in and as the transfer of energy. Further, one could say it produces a paradoxical union of opposites. And, most relevant of all, just as turbulence is not physically linear, it is not metaphysically linear: it is not only the *result* of resistance of flow, it is the *source* of resistance of flow, all the while itself *being* the flow.[78]

'WE ARE SIMPLY MORE ADVANCED WHIRLPOOLS'

Let us return with that hint to Schelling. He says that although we may want to start from imagining an infinite number of 'points of inhibition' in Nature, 'perhaps there is only *one* point of inhibition from which the whole of Nature develops itself'.[79] This one element of resistance is enough to bring into being the multiplicity of 'things'. I like this, because it seems to me that the element of resistance exists first at the level of there being a unity, and only becomes manifold points as the unity is diversified into multiplicity. It reflects that very fact.

This has implications for the relationship between Nature and the individual consciousness (Schelling assumes human consciousness, but the argument could be enlarged in view of the findings of contemporary science, to the consciousness of other beings). In Schelling's view each distinct consciousness arises as a vortex within an endless flow, an eddy in the stream. Compared with other aspects of, or 'products' of, Nature, 'we are simply more advanced whirlpools, more clearly articulated expressions of the absolute.'[80] For Schelling, the emergence of thinking subjects from nature is part of a process whereby an absolute subjective consciousness comes to know itself. The process of consciousness is creative not just of what it comes to know but of itself, which are ultimately one and the same thing: 'What in us *knows*', says Schelling, 'is the same as what *is* known.'[81] As I suggested in the first chapter, creativity is always self-creation as well as other-creation. This self-and-other-creation, according to Schelling, is the business of the cosmos.

To summarise, one could say that, in Schelling's view, (1) there are no static things, only processes; that (2) there is an intrinsic otherness within unity, that is not separate from unity, but gives rise, by virtue of its resistance, to the multiplicity of phenomena; that (3) these phenomena, the 'products of Nature', are also processes, not things, and are not in any sense separate from Nature – rather an unfolding of the potential in Nature; and that (4) their being precipitated out of the primordial energy is the very process of creation. From my reading of him, this potential cannot be of Bergson's first kind, where it is all prefigured, like the painting on the lady's fan, just waiting to be revealed: it is of his second kind, where even the *Weltseele*, the world-soul, itself does not know what is to come, since it gets to know itself only *through* the process of creation whereby simultaneously it and the world come into being together – not again as two distinct events just happening to happen simultaneously, but simultaneous because we are seeing *one and the same* process, just from two different standpoints.

Let's now turn briefly to look at the nature of the two great ex-

[79] Schelling 2004, 'The unconditioned in nature', 1, i (18).

[80] Bowie 2016.

[81] Schelling 1994 (130).

emplars of flow – music and water – and, through them, at what in Chinese is discovered everywhere in the cosmos, the principle of flow known as *lĭ*.

MUSIC AND PHILOSOPHY

Music is extraordinary and unaccountable. Of music, the philosopher Josef Pieper wrote:

> Not only is music one of the most amazing and mysterious phenomena of all the world's *miranda*, the things that make us wonder (and, therefore, the formal subject of any philosopher ...). Not only has it even been said, and rightly so, that music may be nothing but a secret philosophising of the soul ... yet, with the soul entirely oblivious that philosophy, in fact, is happening here ... Beyond that, and above all, music prompts the philosopher's continued interest because it is by its nature so *close to the fundamentals of human existence*.[82]

Music speaks 'of the essence' of life, where the other arts 'speak only of the shadow', thought Schopenhauer.[83] Nietzsche went so far as to say that 'without music life would be a mistake.' And he continued: 'Germans even think of God as singing *Lieder*.'[84] My favourite remark in this regard comes from the Franco-Romanian philosopher Emil Cioran, whose general outlook on life can be inferred by the astute from the title of one of his better-known books, *The Trouble With Being Born*. In an interview in *Newsweek* in December 1989 Cioran asserted that 'without Bach, God would be a completely second-rate figure', and that 'Bach's music is the only argument proving the creation of the Universe cannot be regarded as a complete failure.'[85]

Music deals with that realm of experience, of such vital importance and yet so hard to express in concepts or language, which would and should be the proper object of philosophy if philosophy were only capable of dealing with it. Suzanne Langer, in her classic book *Philosophy in a New Key*, wrote perceptively:

> There are certain aspects of the so-called 'inner life' – physical or mental – which have formal properties similar to those of music – patterns of motion and rest, of tension and release, of agreement and disagreement, preparation, fulfilment, excitation, sudden change, etc.[86]

Note that she emphasises dynamic elements, what is always becoming, where analytic philosophy accentuates always the static, what has become.

In listening we do not stand stock still on the bank of the stream with a flow gauge and a clipboard in hand, but move together with and entrained by the flow. Nothing, according to Basil de Sélincourt, is 'more forced in music than a suggestion that time is passing while we listen to it ... our own continuity must be lost in that of the sound

---

82 Pieper 1990 (39).

83 Schopenhauer 2000, vol I (257).

84 Nietzsche 1889, 'Sprüche und Pfeile': §33: » *Ohne Musik wäre das Leben ein Irrthum. Der Deutsche denkt sich selbst Gott liedersingend* «.

85 Cioran 1989.

86 Langer 1942 (228).

to which we listen.'[87] We are carried with and in the flow that is both ourselves and the music *at once*. We cannot tell, to coin a phrase, the dancer from the dance.[88]

What music is also able to do is to express persistence in a changing flow, even though such a flow appears on the surface to involve its opposite, impermanence. This is Bergson's 'indivisible continuity of change' which is 'precisely what constitutes true duration.'[89] Of this, Langer writes:

> The elements of music are moving forms of sound; but in their motion nothing is removed. The realm in which tonal entities move is a realm of pure duration ... It is not a period – ten minutes or a half hour, some fraction of a day – but is something radically different from the time in which our public and practical life proceeds. It is completely incommensurable with the progress of common affairs ... Such passage is measurable only in terms of sensibilities, tensions, and emotions; and it has not merely a different measure, but an altogether different structure from practical or scientific time ... *Music makes time audible, and its form and continuity sensible.*[90]

Music is as different from the separate notes seen on the score as life is from the language that aims to throw its net around it; when, in Robert Graves's phrase, the 'cool web of language winds us in'. (As an aside, score-dependent musicians rely more on the left hemisphere, improvising musicians on the right hemisphere.)[91] In music, as in the living world, change is permanent, stasis is transitory. 'What is perceived', writes Thomas Fuchs, 'is not a sequence of discrete tones but a dynamic, self-organising process which integrates the tones heard to create a melody'. *Self*-organising, note: it is an 'automatic synthesis, not one actively performed by the subject'.[92] In other words we have to escape the effortful sense of constructing something if we are to allow a flow simply to be. There must not be two elements here, but one. We must be actively *receptive* in relation to it, not actively expressive – as also in prayer and meditation. This image from music is a perfect example of a philosophical insight into life that is otherwise hard to express explicitly. And Fuchs sees that our lives as social beings must belong to something that is best expressed as a dance or a piece of music, if we are to enmesh, engage, connect:

> It is a procedural knowledge in the sense of being accessible only in contact with others, and of being organised temporally: as a feeling for the rhythm of action and reaction, for the *crescendo* and *decrescendo* of a sequence of behaviour, for the 'dancing steps' of the interaction. Implicit relational knowledge is a 'musical' memory – one is able to hear the 'undertones', the 'music' that plays inaudibly in the interaction with the other. Also, when we speak of the sense of 'tact', it points to the relevance of rhythm and synchronisation

87 de Sélincourt 1920: quoted in Langer *op cit* (110).

88 Cf Eliot, 'The Dry Salvages', v: 'you are the music while the music lasts'.

89 Bergson 2007e (124).

90 Langer 1953 (109–10; emphasis in original).

91 Harris & de Jong 2015.

92 Fuchs 2013.

for the intercorporeal sphere. Common sense is also a feeling for the proper timing.[93]

In the performance of mediaeval choral music there was a pulse, or *tactus* (the Latin word for touch), a measure based on the beat of the human heart, that was developed, propagated, sensed and shared through bodily contact, singers often touching one another's shoulders with their hands.

Single pulses appear to interrupt flow, having the nature of points, much as is a single beat on a drum. It is perhaps not accidental that the potentially disjunctive element in music, rhythm, is underwritten commonly by the left hemisphere, while the potentially conjunctive elements of music, harmony and melody, are usually dealt with by the right hemisphere. Together they go to make the structure of music's flow, a union of division and union: however, within this flow, rhythm is wholly transformed and *becomes itself a unifying force*. This transmogrification of an element of differentiation ultimately into a force for union is essential to the nature of creation. To quote Dewey: individuality is a 'phase, though a decisive and outstanding one, of a process having continuity'.[94]

The everyday business of describing the world uses a structure of symbols – language, and in particular nouns – that leads us to believe that there are non-unique things, and that our representations of reality are the reality itself. By transcending language one may see the world as unique wholes *that themselves together constitute unique wholes at a higher level*, and so on without limit. As William James remarked:

> The essence of life is its continuously changing character; but our concepts are all discontinuous and fixed, and the only mode of making them coincide with life is by arbitrarily supposing positions of arrest therein. With such arrests our concepts may be made congruent. But these concepts are not *parts* of reality, not real positions taken by it, but *suppositions* rather, notes taken by ourselves, and you can no more dip up the substance of reality with them than you can dip up water with a net, however finely meshed.[95]

Reality, like the river, is a flow, which only seems to be composed of discrete drops when we try – and fail comprehensively – to catch it in the net of language: the bits that we do catch, the drops from the net, are an *artefact of our process of investigation*. No net, no drops. Similarly modern physics tells us that the entities that we discover when we probe the subatomic world are shaped by the process we use to investigate it – famously so, in the case of particles and waves. What seems to be fundamental is pattern and relationship, not the semi-distinct entities that are patterned and related. Yet immediately

---

93 Fuchs 2001.

94 Dewey, 'The philosophy of Whitehead': 1988 (135).

95 James 1909a (253).

we sense that our everyday language is inadequate to what is meant: some differentiation is necessary for there to be anything out of which a relationship can be constituted. And yet what is differentiated can never be separate, because it is what it is *only in relationship* to everything else. This is the very essence of music.

WATER AND PHILOSOPHY

If life without music would be a mistake, life, of any kind, without water would be impossible. Schelling clearly intuited that water was a potent metaphor or symbol for the nature of reality, as indeed it is in Oriental philosophy, especially Taoism, which constantly recurs to the image of flowing water. In his treatise *On the World-Soul*, which Schelling wrote when he was only 23, he speaks of an equilibrium to life that must constantly be disturbed and re-established. Although he does not use this image, what comes to mind for me is the way in which the acrobat can balance on a tightrope only by virtue of its constant capacity for disequilibrium: if it were rigid he would fall off. And in the same passage where Schelling makes this claim, he names oxygen and hydrogen, the elements of water, as the two principles of life in the animal body, acting as 'weights on the lever of life', the local disequilibrium of each allowing for the maintenance of life itself at the scale of the organism as a whole.[96]

Yet there are aspects of the nature of water that even Schelling did not explore. John Herschel deduced in 1830 that 'the solid, liquid and aeriform states of bodies are merely stages in a process of *gradual* transition from one extreme to the other', which need not necessarily be separated by 'sudden or violent lines of demarcation'.[97] In other words bodies have solid, liquid and gaseous forms – palpable differentiations: but the forms are not necessarily abruptly distinct, but seamlessly continuous. This insight, another example of relative discreteness within a relative continuum, came from studying water.

Nothing seems more intuitive than the nature of water; yet water is a peculiar substance. Felix Franks, a physical chemist who devoted his life to studying the properties of water, concluded that 'of all the known liquids, water is probably the most studied and least understood.'[98] Because of its hydrogen-bonded character, water is a liquid that flows at ambient temperatures, unlike almost any other substance.[99] 'Not for nothing is the science of flow called *hydrodynamics*', writes Philip Ball in his fascinating exploration of water's nature. 'Water is not, of course, the only substance that forms a liquid, but very few do so for the conditions of temperature and pressure under which life is comfortably conducted.'[100] And yet it has a high degree of internal 'structure', brought about by the 'stickiness' that can 'bind the H_2O molecules into a dynamic, ever-changing labyrinth'.[101] Elaborating on the oddness of water, Ball writes:

96 Schelling 1798 (204–6).
97 Ball 2000 (148; emphasis added).
98 Franks 1972 (18).
99 Other examples would include decane and mercury.
100 Ball *op cit* (x).
101 *ibid* (168).

The hydrogen bond, then, is what sets water apart from other liquids. But it doesn't immediately explain why water is *so* odd – why it is denser than ice, why it is central to life, why it has such a capacity to absorb heat and so forth. Can't we think of water as being just like any other liquid, except more strongly bound? Not a bit of it.[102]

'It is very tough to make a good computer model of an H_2O molecule', he writes, in a phrase that invites one to ponder: surely, of all molecules involved in life, water – two hydrogen atoms and one of oxygen – looks like one that should be rather easy to model. But the difficulty arises 'because the way a hydrogen bond forms as two water molecules approach one another is so complex. The crucial difficulty is that *hydrogen bonds form cooperatively*, not independently.'[103] This difference, that makes computer modelling so difficult, expresses something intrinsic, I suggest, to life itself.

Life depends on water because of its very peculiarity. As Philip Ball explains, 'when it comes to water, the refined tools of the theory of simple liquids are often woefully inadequate. It becomes like chiselling with a screwdriver or banging in nails with a mallet – you can just about get the job done, but it doesn't always look very convincing or elegant.'[104]

Water has, to use Abramowitz's terminology, architectivity at one temperature as well as connectivity at another.[105] Ball writes:

> A continuing mystery about dendritic snowflakes is why all six of their branches seem to be more or less identical. The theory of dendritic growth explains why the side branches will develop at certain angles, but it contains no guarantee that they will all appear at equivalent places on different branches, or will grow to the same dimensions; indeed, these branching events are expected to happen at random. Yet snowflakes can present astonishing examples of coordination, as if each branch knows what the other is doing.[106]

I reported earlier something similar about the formation of deer antlers,[107] but from this account of water, it would seem not to be a matter confined to the world of the living. The single flake comes into being as a single *Gestalt*.

Unlike almost every other liquid, when water freezes, rather than contracting and becoming more dense, it *expands* and becomes *less* dense. If this were not the case, the world would be a very different place:[108]

> if the ocean bottom waters were frozen, [oceanic] circulation could not take place, with the result that northern countries would be much colder ... Hot water cools very slowly since it must lose a lot of heat for its temperature to fall significantly. So our hot water tanks and our baths stay hot for a long time. Water's large heat capacity means

---

102 *ibid* (158; emphasis in original).
103 *ibid* (165; emphasis added).
104 *ibid* (154).
105 See pp 834–6 above.
106 *ibid* (177).
107 See p 462 above.
108 Ball *op cit* (140–1).

that warm ocean currents can carry a phenomenal amount of heat. The Gulf Stream, which ultimately keeps Northern Europe warmer than Labrador (at the same latitude) by carrying heat from tropical South America northwards across the Atlantic Ocean, bears with it every day twice as much heat as would be produced by burning all of the coal mined globally in a year.[109]

One of the difficulties in modelling water is, according to Ball, precisely that the models start from stasis, not from flow: a problem to which the reader will by now be alert. 'The fundamental difficulty with all these models', he writes, 'is that they try to force a static description on a changing scene.'[110] And he goes on to elaborate:

> We don't have the technologies yet for tracking huge numbers of water molecules as they play out their tactics in the real liquid. Although it is now possible to take frame-by-frame snapshots of molecular processes over the incredibly short timescales – perhaps a few trillionths of a second – during which they occur, it may never be feasible to monitor directly and simultaneously the environments and trajectories of thousands of individual molecules in a complex and messy system like a liquid.[111]

'Messy': one might be forgiven for hearing there, if it were not for Ball's delight in its 'messiness', the left hemisphere of the Western scientist speaking. To which what we don't understand and can't neatly predict is, of course, a mess.

THE ESSENTIAL PRINCIPLE OF FLOW: LǏ

However to the Eastern mind, though processes may not follow programmatic rules, they are far from being any kind of mess. An ancient Chinese concept that stems from Confucianism is that of *lǐ*.[112] It indicates a formal principle in all things, that is considered, together with *ch'i* (a vital force or energy, also written *qi*), ontologically prior to the cosmos itself, and to have given rise to it. This idea of generation as an energetic force entering into a receptive form is itself deeply generative.

According to Joseph Needham, *lǐ* indicates,

> the order and pattern in Nature, not formulated Law. But it is not pattern thought of as something dead, like a mosaic: it is dynamic pattern as embodied in all things living, and in human relationships, and in the highest human values.[113]

And it is not just in the living, as I understand it. *Lǐ*, as the ordering principle in the world, is something like 'reason', according to Alan Watts, but not in the now normal, Platonic, sense of that word.[114] It is perhaps more like what Heraclitus called the *logos*. At that stage

109 *ibid* (143–4).
110 *ibid* (164).
111 *ibid*.
112 This concept is central to Taoism, and was originally applied to the markings in jade.
113 Needham 1954–98 (1956), vol 2 (558).
114 Watts 1975 (15).

logos had not yet come to mean 'reason' in the rather limited modern sense, instead meaning the common principle that makes complexity, beauty and meaningful order arise in place of chaos, both in the living world and what we consider the non-living; and gives rise to a fittingness, or rightness, or dignity, in human affairs where they arise. *Lǐ* is closely related to the idea of the *tao*, the flowing formal principle of the cosmos.

Lǐ can manifest in a wide range of patterns and intrinsically beautiful, irregular regularities found throughout Nature. David Wade, in his excellent short book *Li: Dynamic Form in Nature*, distinguishes, inevitably arbitrarily, 24 types of such pattern, which he illustrates from, amongst other things, animal markings, shells, leaves, films, crystals, eggs, insect wings, sand, webs, stalks, rock formations, soap films, cabbages, clouds, lichens, asphalt, ceramics, tree-barks, ice, parched earth, dried paint, grasses, ferns, coastlines, agates, brains, bone, particles of soot, streams, bacterial growths, and static electricity. *Lǐ* can refer to the markings in jade, but is present above all in the patterns of flowing water. 'We see those patterns of flow memorialized, as it were', writes Watts, 'as sculpture in the grain in wood, which is the flow of sap, in marble, in bones, in muscles. All these things are patterned according to the basic principles of flow.'[115] In the West, the technique of 'marbling' paper, used widely by bookbinders in the past to decorate the edges, endpapers and even covers of books, makes use of 'rather sophisticated flow phenomena in viscous fluids.'[116] In other words, *lǐ*.

Viewed as a guiding pattern that lies behind phenomena, yet importantly without controlling or predicting their expression in any one instance, *lǐ* has many of the same qualities expressed in the idea of the *élan vital*. It is the creative core of existence. Lao Tzu says of the *tao*, the underlying natural order of the universe:

> The great *tao* flows everywhere, to the left and to the right;
> it loves and nourishes all things, but does not lord it over them.[117]

This last phrase is important. In other words it does not control, force into conformity with its will, direct or have power over what is essentially free. Its openness is more in accord with the way of being of the right hemisphere than that of the left. To know is also not to pin down: 'knowing is not a smash-and-grab raid on the object but what [Heidegger] calls a being-with, a concern, a not-having-power-over.'[118] The power of *lǐ* is permissive, not privative: freeing up into possibility, not closing down into certainty. And as we have seen turbulence, which is what all natural flow is, and what the *tao* is, resists precise formulation, and therefore precise prediction.

Of the patterns representing *lǐ* gathered in his book, Wade writes:

---

115 Watts [undated].

116 Douthat, Nagib & Fejer 1975.

117 Watts 2006 (128). In Chinese philosophy, left and right have special significances, very much in line with the hemisphere hypothesis: see McGilchrist 2009b (457). That the *tao* is in both seems to me to relate to Schelling's perception that the element of resistance to flow is part of the flow itself, a part that makes it what it is.

118 Eggert 1999.

'The aesthetic appeal that links each of these examples ... derives from the attractive combination of geometry and pure chance.'[119] This coming together of order with disorder is highly reminiscent of the pattern of such musical forms as jazz, *raga*, and *fado* that I invoked as metaphors of existence in the first pages of this book. Even what we call 'things' are non-random, but also non-determined, patterns: *Gestalten*.[120]

RHYTHM

Nothing more strongly entrains the entire body to participate in its flow than musical rhythm. I mentioned a few pages back that rhythm might be considered punctate, and yet it turns what might have been isolated elements into a continuous, unbreakable flow, with a past and a future strongly implied in, and bound up inextricably with, its present. Rhythm could be thought of, as it undoubtedly is wherever it exists (and it exists everywhere) in Nature, as a highly ordered series of obstructions to flow, from within flow, and forming part of flow. We have seen the beautiful, regular, yet irregular, vortex formations that are caused by pulsations in a flowing liquid: oscillations are present, in some form, often extremely rapid though they may be, throughout all the processes of the cosmos. Note that an interruptive or resistive element is turned into something that does not weaken, but rather strengthens, the sense of flow it ostensibly opposes.

I also mentioned the left hemisphere's role in rhythm. But there are qualifiers, one exception being that syncopated or cross rhythms are better dealt with by the right hemisphere.[121] And the left hemisphere is also, as one might imagine, unable to deal with long sequences:

> a healthy right hemisphere appears to be more suited for the discrimination of longer rhythms than a healthy left hemisphere. This effect was seen only for the 'longer' rhythms and not for the 'shorter', probably because the latter are too simple to discriminate for an effect to be detected.[122]

The same is true of visual rhythmic patterns: as the number of elements increases, sequential processing, typical of the left hemisphere, 'becomes inefficient and the right hemisphere takes over, representing the rhythm as a single temporal pattern'. As the experimenters here conclude, 'the finding that longer rhythms yield right hemisphere superiority points to the possibility that rhythms can be processed as Gestalts, ie, whole patterns.'[123]

A *Gestalt* is not mechanically decomposable, but a pattern that is recognisable despite having no simple formula. In this it is like *lǐ*. Here Whitehead has some interesting observations to make on rhythm:

119 Wade 2007 (8).

120 Whitehead 1919 (195): 'An object is a pattern'.

121 Professional musicians are an exception: see discussion of evidence and possible cause in McGilchrist 2009b (75).

122 Alpherts, Vermeulen, Franken *et al* 2002.

123 Ben-Dov & Carmon, 1984.

A rhythm involves a pattern and to that extent is always self-identical. But no rhythm can be a mere pattern; for the rhythmic quality depends equally upon the differences involved in each exhibition of the pattern. The essence of rhythm is *the fusion of sameness and novelty*; so that the whole never loses essential unity of the pattern, while the parts exhibit the contrast arising from the novelty of their detail. A mere recurrence kills rhythm as surely as does a mere confusion of differences.[124]

This is the essence of the phenomenon known as *rubato*, where slight slowings or accelerations in tempo are detectable, and bring expressive life to the music; but subliminal rhythmic changes are always present in human music, which is why mechanical rhythms are so dull, and even offensive to the ear. Remember that in the beating of the heart – and not just the foetal heart, but the adult heart – mechanical regularity is a warning sign.[125]

Whitehead believed there was a close identification of rhythm with 'the causal counterpart of life; namely, that wherever there is some rhythm there is some life, only perceptible to us when the analogies are sufficiently close. The rhythm is then the life.'[126] And where there is life, there is rhythm. Elsewhere he observes that what he calls 'the Way of Rhythm pervades all life and indeed all physical existence. This common principle of rhythms is one of the reasons for believing that the root principles of life are, in some lowly form, exemplified in all types of physical existence.' And he notes that these rhythms form unending cycles, in which the end of one cycle initiates the seamless beginning of the next.[127]

THE FLOWING LINE

We have often had reason to return to the difference between a curving motion and a rectilinear motion. Curvature, as I have suggested, is more characteristic of the intellectual world of the right hemisphere, in which opposites can be reconciled, in which the direct approach may for many purposes be inferior to the indirect, and in which continuous variation is united with coherent pattern. Curves are complex, like turbulence: straight lines are simple. Curves are also more sympathetic to the natural world, in which there are no straight lines.[128] In particular, one might say that curves have a living quality that straight lines lack.

In relation to that, I might mention a reader who wrote to me to say that he started getting tattoos a few years ago, beginning with a tree shape on his left shoulder. He quickly found that he wanted the left side of his body to be characterised by natural and organic forms and his right side by geometric and linear designs. It wasn't until he read *The Master and his Emissary* that it struck him

---

124 Whitehead *op cit* (198; emphasis added).

125 Dekker, Crow, Folsom *et al* 2000. And see p 838 above.

126 *ibid* (197).

127 Whitehead 1929b (16–17).

128 See McGilchrist 2009b (387 & 509, n129). Since writing those words, more than one person has written to me to point out that light rays follow a straight path. I suppose if one were to be pedantic one would say that they follow what appear to be straight lines, but lines which in reality follow the curvature of space. However, accepting the point does not do much to negate the general truth.

that he had 'naturally gravitated to a left/right brain way of thinking about my tattoos and literally embodied that idea with ink on my skin'.

Sterne's Corporal Trim demonstrates freedom with a flourish of his stick (Fig. 52).

And Tristram Shandy gives the storyline of Book V as something that looks in places fortuitously like a deranged electrocardiogram, with plenty of vital variability:

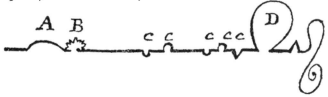

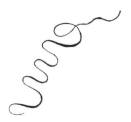

Fig. 52 (above). *Corporal Trim's flourish, from* Tristram Shandy, *by Laurence Sterne (Bk IX, 1767)*

Fig. 53 (left). *The plot structure of* Tristram Shandy *(Bk VI, 1761)*

The connexion between curved lines and life was made by William Hogarth in his *Analysis of Beauty*, in which the 'Line of Beauty and Grace' is depicted as a slender serpentine double curve (see Fig. 54).[129]

Alberti had made the same point in his treatise *On Painting*, in the fifteenth century, where he finds the beauty of such a line embodied in the depiction of a woman's flowing hair, or a horse's mane, which should 'swirl' or 'wave in the air while it imitates flames', rising 'sometimes in this or that direction'.[130] And Herder, in his treatise on sculpture, writes of 'the beautiful line that constantly varies its course': it is never 'forcefully broken or contorted, but rolls over the body with beauty and splendour; *it is never at rest but always moving forward*'.[131]

There is indeed something fearful – like the tiger's symmetry – about the straight line, which has so come to dominate our world in the last hundred years. I mentioned Hundertwasser's view that 'the straight line leads to the downfall of mankind'.[132] Bergson seems to have seen that what he called intellect could only approximate the living curve that intuition could reach directly. Like the tangents that approximate a circle without ever reaching it, he thought that analytic thinking has to try to approximate the findings of intuition by a series of short straight lines that zigzag around its path, making an endless series of corrections, where intuition enters the line at one stroke.[133]

Spirals and vortices bring order into being out of chaos. Early Celtic and Teutonic cosmologies held that life emerges from a vortex, an idea also common to many pre-Socratic Greek philosophers. Anaximander moved away from the idea that there was some single primordial substance – such as Thales' water or, later, Heraclitus' fire – towards the idea that there was an indefinable something (*apeiron*) that expressed itself in an eternal motion, which brought everything

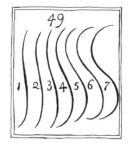

Fig. 54 (above). *Variations on the serpentine curve, by William Hogarth. 'Strictly speaking, there is but one precise line, properly to be called the line of beauty, which in the scale of them is number 4'.*

(from *Analysis of Beauty*, 1753 [49])

129 It has been suggested that an analysis of 'Hogarth curves' shows that his preferred curve, number 4, embodies the Golden Ratio more closely than any of the others (Zhu, Luo, Huang et al 2018).

130 Alberti 2013 (67).

131 Herder 2002 (40; emphasis in original).

132 See n 36, p 824 above.

133 Bergson 2007d (89).

into being by a reciprocation of opposites.¹³⁴ A vortex is such a tension between opposing forces, specifically centripetal and centrifugal forces, and thus it both separates and combines. Later Anaxagoras and Empedocles agreed with Anaximander that this motion is a vortex. Indeed according to Democritus, 'the cause of coming-into-being of all things is the vortex'.¹³⁵

Is movement more generally, then, foundational?

THE PRIMACY OF MOTION OVER STASIS

At some deep level, reality requires both elements of motion and (relative) stasis, the wave and the particle, the continuous and discrete. However, if the continuous and the moving is best appreciated by the right hemisphere, the discrete and static is best represented by the left hemisphere. It would not be surprising, therefore, if we found that they are not quite symmetrical in ontological status, with the continuous being the ontological primary – albeit in reciprocal relation with the discontinuous.

This primacy of union over division, however necessary division might be, is reflected in the fact that one can move from an extended whole in space or time to parts (though losing almost everything on the way), but not from the parts to the whole. This point of view, which we have associated with Bergson, has been advanced by many philosophers, and was anticipated by Leibniz. Indeed Leibniz denied that anything real could ever be an aggregate of parts:

> I believe that where there are only entities through aggregation, there will not even be real entities … To be brief, I hold as axiomatic … that what is not truly *one* entity is not truly one *entity* either …¹³⁶

Similarly Leibniz remarks that mathematical points, 'even an infinity of points gathered into one, will not make extension.'¹³⁷ So extension in space and extension in time cannot be constituted by aggregation of points or parts. Both of these ideas are expressed more eloquently by Bergson and by James, but my point is that they are not confined to a 'Bergsonian' or 'Jamesian' view: they are there in the European tradition well over 200 years previously, and are mathematically correct according to Leibniz, who invented the calculus.

And since a point is conceived as static, no matter how many of them you have, you can never get from a point or points to motion. 'How can one fail to see', says Bergson, 'that the essence of duration is to flow and that one static element stacked on another will never result in anything that has duration?'¹³⁸

However, you may say, a flow, while it cannot be made up (prospectively) from static parts, could have recognisable 'regions' in it once it has come into being. So when Bergson says: 'We shall think of all change, all movement, as being absolutely indivisible',¹³⁹ he

---

134 Simplicius, *Commentary on Aristotle's Physics*, Diels-Krantz A9.

135 DEMOCRITUS, according to Diogenes Lærtius, *Lives of the Eminent Philosophers*, Bk IX, §45. For ANAXAGORAS, see fr 9 & 12; for EMPEDOCLES, see fr B35; for EARLY CELTIC & TEUTONIC, see McCormack 2012 (1).

136 Leibniz, letter to Antoine Arnauld, 30 April 1687; emphases in original.

137 Leibniz, letter to Bartholomew Des Bosses, 30 April 1709.

138 Bergson 1969 (11): « Comment pourtant ne pas voir que l'essence de la durée est de couler, et que du stable accolé à du stable ne fera jamais rien qui dure? »

139 Bergson 2007e (118).

does not mean that it cannot be retrospectively divided, but that such a division can apply only to the representation, not to the entity itself, whose essence is change, motion, flow. His point is that the representation has lost the very element – duration or extension – that is the essence of what it tried to capture. For, once time – or motion – is *retrospected* on, it is *no longer time or motion*. In attempting to account for a pure process of change analytically, we can do so only by reducing it to parts time and again, until at last we reach something that no longer changes. But, at the very moment we exclaim 'Finally, I have it!', we have, by definition, only immobilities; and from them no change can ever emerge. The change we were trying to account for must forever elude our *grasp*.

> If change, which is evidently constitutive of all our experience, is the fleeting thing most philosophers have spoken of, if we see in it only a multiplicity of states replacing other states, we are obliged to re-establish the continuity between these states by an artificial bond; but this immobile substratum of mobility [see note], being incapable of possessing any of the attributes we know – since all are changes – recedes as we try to approach it: it is as elusive as the phantom of change it was called upon to fix.[140]

As for metaphysicians' attempts, from Zeno onwards, to account for motion and change, Bergson says that 'of change they retained what does not change, and of movement what does not move.'[141]

All motion requires both a following and a headwind: it requires resistance, some degree of friction. If everything moved in the same direction at the same speed in a frictionless universe, this would effectively be stasis: motion is always relative to something. Since there is no framework by which one element can be granted fixity, it is thus always a matter of both, or all, parties moving, never just one.

A defining quality of all living things is that they move purposively – growth being a form of motion; and we have seen that in order merely to remain in existence they need constantly to change. Aristotle recognised that locomotion is only one of four types of movement, the other three being growth, decay and change of state (metamorphosis), all of which are present in plant life.[142] Animals, unlike most plants, are locomotive, and spend a great deal of their often costly energy in moving, sometimes over very large distances. Motion is important to continuing life, if nothing else. But it is more important even than that, as I shall suggest: an irreducible facet of being itself.

Earlier I considered a particular kind of motion which followed logically from the discussion of time, namely the motion of flow, and how it represents both continuity and discontinuity *together*. This is, I believe, of some philosophical importance. Now I want

---

140 Bergson 2007e (130). This, the standard English version, has 'this immobile substratum of IMMOBILITY', which seems wrong, and is. The original reads '*ce substrat immobile de la* MOBILITÉ': Bergson 1911b (34).

141 *ibid* (117).

142 Aristotle, *Physics*, III.1: 201ª10–14.

to broaden the discussion to consider the grounding role of movement in general.

First, though, we should explore the role played by motion in grounding all aspects of our mental life.

MOTION AND COGNITION

Our knowledge is derived from action and interaction with the world. Before we have language or concepts, we are navigating and manipulating the world, solving problems, making decisions and coming to an understanding of, not just the world, but ourselves – purely through moving about in it.

In a letter to *Nature* in 1887, Francis Galton takes issue with the philologist Max Müller's idea that thinking depends on language.[143] He makes two points that are relevant here. One needs to be mentioned only in passing. In suggesting that prowess at chess owes nothing to language, he remarks: 'I myself cannot conceive that the names – king, queen, &c. – are of any help in calculating a single move in advance. For the effect of many moves I use them mentally to record the steps gained, but for nothing else.' What he points to is that we mainly do not think in language, but *record* thought, after the event, in language. It is another example of the retrospective, rather than prospective, aspect of left hemisphere function. I have to say that, though Galton is certainly right about certain kinds of thought, it is my experience that some branches of thought become fully formed only through their elaboration in language – so, while language is neither necessary nor sufficient for thought, it nonetheless often plays an important part in the process, even if its effect can also be obstructive.[144] It provides 'necessary resistance'.

His other point is more telling. Using examples from his favourite activities – which include both inventing and mending 'mechanical contrivances', tools and mechanisms; playing billiards and chess; indulging in fencing, climbing and scrambling – he makes it clear that bodily movement itself, not language, is the key element in most of our thoughtful interactions with the world. And he carries on to emphasise that even reasoning emerges from motion. He refers to Binet's *La psychologie du raisonnement*, published the previous year, which had emphasised 'the important part played by visual and motile as well as audible imagination in the act of reasoning'; as well as to numerous instances of thought depending on an active, embodied engagement with a scene, involving *movement* – certainly not on any form of words.[145]

That was 130 years ago. The idea that thought depended on bodily movement did not fit with the scientific prejudices of the age. The close relationship between embodiedness and cognition is now appreciated, but conventionally has not been taken into account in the

---

143 Under the heading 'Thought without words', the letter was published in the issue for 12 May 1887, 36, 28--9; and a further response to Müller's reply is dated 2 June 1887, 36, 101. Both are worth reading in their entirety, and can be accessed at archive.org/stream/naturejournal36londuoft#page/100.

144 This point is considerably elaborated in McGilchrist 2009b (esp in Chapter 3).

145 Galton refers to Binet's work (incorrectly) as *Sur Le Raisonnement*.

analytic tradition of philosophy, which has been happy to entertain, for instance, accounts of 'brains in vats' as useful ways to arrive at an understanding of the human mind.

It is only in recent years that we have begun to take notice of, and learnt very much more about, the role of the body in cognition. The argument that the contribution of literal and explicit thinking is minimal compared with largely metaphorical, unconscious thinking, rooted in the body, was put forward by George Lakoff & Mark Johnson in their *Metaphors We Live By* and *Philosophy in the Flesh*; and Guy Claxton has marshalled an impressive array of evidence that takes the argument further, in his *Intelligence in the Flesh*. 'We think through the body', writes Matt Crawford. 'The boundary of our cognitive processes cannot be cleanly drawn at the outer surface of our skulls, or indeed of our bodies more generally. They are, in a sense, distributed in the world that we act in.'[146]

I'd like to consider motion from a number of different angles, each of which may tell us something different about its significance. First, I will look at what brain structure and function, both normal and abnormal, can tell us about the part played by motion in different aspects of our experience. That will lead naturally to a consideration of the relationship between motion and various aspects of the psychology of phenomenal experience: thinking, feeling, perceiving, and social intelligence. From there I will venture into some reflections on motion in ontological terms. What I aim to show is that motion is at the core of every aspect of our experience, and of our ability to make sense of it, in a way of which we are normally unaware, because our analytic intellect cannot deal with it; and that motion is foundational to existence and stillness merely the limit case of motion, not stillness primary, and motion some form of aberration or disturbance of the foundational inertia.

MOTION AND COGNITION: IN THE BRAIN

Mind and body are inseparably connected. The conventional divide between thinking and motor function simply can no longer be supported. The dichotomy is, in fact, entirely misleading, as has been confirmed in a number of interesting ways. And some of the evidence, once again, comes from diseases of the nervous system.

We know that just thinking about a certain activity motivates parts of the brain connected with performing the action, and causes subtle changes in tone of the relevant musculature – so much is thinking bound up with bodily action. Just reading action words activates motor systems; words, such as 'kick' or 'lick', change activity in those areas of the brain having input to the legs and mouth, respectively.[147] Equally, the converse is true: movements of the hands affect the processing of words for actions involving the hands, and the same is

---

146 Crawford 2015 (51).

147 Hauk, Johnsrude & Pulvermüller 2004. See also Barsalou 2008; Fischer & Zwaan 2008; Pulvermüller 2005; Jirak, Menz, Buccino *et al* 2010; Gallese 2008; Boulenger, Shtyrov & Pulvermüller 2012; Aziz-Zadeh & Damasio 2008; Cappa & Pulvermüller 2012.

true of the feet.[148] (Reading nouns with strong taste, smell or sound associations has been shown to activate the relevant sensory brain regions.[149]) When we comprehend sentences, we internally simulate the state of the world described in them.[150] The term 'motor cognition' has been adopted to cover those aspects of cognition involved in planning, executing, understanding and imagining bodily movements and actions.[151]

The close connexion between thinking and moving is evident, too, in pathology. Some of the same proteins and genes are involved in both classically 'motor' diseases, such as motor neurone disease, and in classically 'cognitive' diseases, such as frontotemporal dementia.[152] In what used to be thought of as purely movement disorders, we now recognise that aspects of cognition are affected.[153] For example, motor neurone disease, as its name suggests, has been conventionally thought of as affecting only the neurones of the peripheral nervous system, ones that convey motor commands to muscles. To a large extent this is correct; but more recently physicians have become aware that patients with motor neurone disease have word-finding difficulties.[154] In fact clear language deficits can be detected in about half of patients with motor neurone disease.[155]

What is significant is that in these conditions it is the retrieval of *action words* that is principally affected. According to the neurologist Thomas Bak, a leading figure in the world of embodied cognition pathology, every single reported patient with motor neurone disease and cognitive impairment that has been tested on generating or processing nouns versus verbs, showed a more pronounced deficit for verbs, with a difference in performance of up to 50%.[156] Similar word retrieval problems affecting specifically verbs have been repeatedly found in other movement disorders such as Parkinson's disease,[157] progressive supranuclear palsy[158] and cortico-basal degeneration,[159] as well as in other rarer conditions.[160] We saw earlier that disrupting the speech motor cortex disrupts comprehension of action words.[161]

That is not, however, because verbs are simply the first to go whenever there is cognitive impairment – indeed the reverse is generally true. Though children acquire nouns before verbs, once acquired verbs are tenacious: in cognitive impairment adults tend first to lose not verbs but nouns. While it is true that some forms of dementia spare nouns, some spare verbs, and some spare neither, nouns are generally more vulnerable than verbs. The first description of someone who had lost the use of verbs was made in 1744 by the philosopher Giambattista Vico: 'There is a good man living among us who, after a severe apoplectic stroke, utters nouns but has completely forgotten verbs.'[162] By an odd coincidence, the first description of the converse, loss of noun retrieval, was made in the following year,

---

148 Shebani & Pulvermüller 2013.

149 Eg, Kiefer, Sim, Herrnberger et al 2008; Barrós-Loscertales, González, Pulvermüller et al 2012; González, Barrós-Loscertales, Pulvermüller et al 2006. For review, see Moseley & Pulvermüller 2014.

150 Zwaan 2004.

151 Jeannerod 2006.

152 Bigio 2012; Bak 2013.

153 Brown, Lacomblez, Landwehrmeyer et al 2010; Bak 2010; Bak 2013.

154 Eg, Bak *supra*. See also Liepmann's description of the (quite different) limb-kinetic apraxia, where jerkiness is not part of the clinical picture, but in which he observes the connexion between movement and imagination: 'the person with damage to kinaesthetic concepts returns to the stage of one who has not yet learned a certain manipulation. Movements are influenced in a general way by the ideational concept. He can move his limbs to an extent according to imagined paths, but there is a lack of fineness and certainty of movement, which is the result of practice ... the movement ... relates to the ideational concept or correct execution as a helpless copy relates to the original, the finer details of which are completely absent in the copy.' Lange 1988 (153–4), describing a patient of Westphal's (Westphal 1907).

155 Taylor, Brown, Tsermentseli et al 2013.

156 Bak, O'Donovan, Xuereb et al 2001.

157 Péran, Rascol, Démonet et al 2003; Boulenger, Mechtouff, Thobois et al 2008; Rodriguez-Ferreiro, Menéndez, Ribacoba et al 2009.

1745, by the botanist and physician Carl Linnæus: as far as we know, he had no knowledge of Vico's case.¹⁶³

The cognitive deficits that accompany motor diseases may be an inalienable consequence of the close connexion between action and thought. I alluded in Chapter 12 to the well-known Hebbian formula in neuroscience that 'what fires together wires together' – that is to say, repeated use of a particular neuronal pathway causes structural changes which further facilitate its future use. Since we now have evidence that neurodegeneration may spread along the reinforced synaptic connexions created by this process, Bak has coined the phrase 'what wires together *dies* together'.¹⁶⁴

It seems clear, however, that the key difference is not a grammatical one – nouns versus verbs, as such – but an experiential one – in as much as nouns tend to suggest objects, and verbs tend to suggest action. The brain's connexions between thought and motion are mediated through imagined experience, not through linguistic rules.

MOTION AND COGNITION:
WHAT THE CEREBELLUM CAN TEACH US

Until the last 50 years, the cerebellum, or 'ancient brain', which lies below and behind the cerebrum, or 'new brain', was thought to be more or less confined to co-ordinating motor behaviour. Its name means the little brain – it has only about 8–10% the volume of the cerebrum¹⁶⁵ – yet it contains an astonishing 80% approximately of all the brain's neurones.¹⁶⁶ This is particularly striking since, as the number of neurones increases, the number of connexions does so exponentially. It seems unlikely that this vast complexity, so much greater than that of the cerebrum – what we normally think of as our 'brain' – is devoted entirely to finessing motor control. (It is of no little interest, incidentally, that while consciousness can be sustained by just a fraction of the normal volume of the cerebrum, the entire cerebellum, with its vastly greater interconnective potential, cannot sustain consciousness on its own.)

In recent years, we have come to learn that the medial cerebellum, in particular the central part known as the vermis (literally, from its appearance, the 'worm'), plays a key role in emotional and social interaction. Abnormalities of the cerebellar vermis,¹⁶⁷ and the medial cerebellum more generally,¹⁶⁸ have been linked to autism and similar affective and social disorders.¹⁶⁹ There is a particularly close relation between movement and social cognition, since thought, feeling and social connectedness depend on, and in turn reinforce, shared resonances of movement, both in action and perception.¹⁷⁰ At the same time, principally lateral cerebellar regions have been linked to a variety of more purely cognitive deficits.¹⁷¹

158 Daniele, Giustolisi, Silveri *et al* 1994.
159 Silveri & Ciccarelli 2007.
160 Bak, Yancopoulou, Nestor *et al* 2006.
161 See p 106 above.
162 Vico 1971 (505) / 1948 (137). See also Denes & Dalla Barba 1998.

163 VICO: see Denes & Dalla Barba 1998, referring to Vico 1971 (137); LINNÆUS: Östberg 2003, referring to Linnæus 1745/1943 (trans in Viets 1943).
164 Bak & Chandran 2012.
165 Henery & Mayhew 1989.
166 Herculano-Houzel 2009.
167 Courchesne, Yeung-Courchesne, Press *et al* 1988; Bauman & Kemper 1985.
168 Muratori, Cesari & Casella 2001.
169 Tavano, Grasso, Gagliardi *et al* 2007; Schmahmann 2010.
170 Cook 2016.
171 Schmahmann *op cit*.

There are close links between the cerebellum and all parts of the brain.[172] We now know that projections from cortical areas direct to the cerebellum and to the pons (the part of the brainstem that forms the 'bridge' between the rest of the brain and the cerebellum) are concerned, not only with motor functions, but with an array of non-motor functions, including spatial awareness, spatial memory, higher-order visual processing and language.[173] In relation to time intervals, the cerebellum is important for their *perception*, not only, as was once thought, for the execution of motor responses to them.[174] There are also projections from many areas in the prefrontal cortex – the most lately evolved part of the brain – to this 'ancient brain', projections which are critical for what we conventionally consider the highest-order processes, including complex reasoning, judgment, attention and working memory.[175] Pontine nuclei exchange information of motivational and affective significance with association areas throughout the brain in many modalities.[176] It is now beyond doubt that an important role is played by the cerebellum in the acquisition of higher-order cognitive, affective and social skills.[177] And in an earlier chapter I referred to its possible role in creativity.[178]

In fact we had reason to suspect that this might be the case nearly two centuries ago, but the information was overlooked because it did not fit with the pervasive assumption that motor skills must be separate from cognitive and emotional skills. The first description of intellectual deficits in a patient with cerebellar agenesis is by Combettes in 1831, who reported the case of an 11 year-old girl with serious cognitive, linguistic impairments whose cerebrum appeared normal at autopsy, though she completely lacked a cerebellum.[179] Since then a handful of other cases (14 to date) of cerebellar agenesis have been described.[180] Non-motor impairments caused by cerebellar disease or dysfunction may involve visuo-spatial construction, conceptual understanding and organisation of sequential behaviour,[181] and may also include perseveration, distractibility, lack of mental flexibility and disturbances of grammar, as well as deficits in working memory. Such a persistent pattern of interlinked executive, visuo-spatial, linguistic, and affective impairments is now known as 'cerebellar cognitive affective syndrome'.[182]

In the late 1960s, the developmental psychologist James Prescott developed the hypothesis, based in part on his observations of experimental monkeys, that stimulation by movement was an important element in normal emotional development.[183] What if abnormal patterns of emotional behaviour – being withdrawn, apathetic, autistic or hyperactive, for example – which all had motor aspects, could be at least in part attributable to dysfunction of the cerebellar regulatory system?

---

172 Schmahmann 1991.

173 Schmahmann 2010. I mentioned that in patients with motorneurone disease, there are problems with retrieving verbs of action: there is activation in the right posterolateral and midline cerebellum, which can be visualised on brain imaging, when subjects are asked to generate verbs for nouns, a process conventionally believed to be confined to 'lexical processing areas' of the neocortex (Fiez & Raichle 1997). See Vigliocco, Vinson, Druks *et al* 2011, for a sensible and comprehensive review of the separate debate about whether nouns and verbs are processed independently in the brain.

174 Ivry & Keele 1989

175 Schmahmann *op cit*.

176 ibid.

177 Tavano, Grasso, Gagliardi *et al* 2007.

178 See p 278 above.

179 Combettes 1831.

180 Velioğlu, Kuzeyli & Özmenoğlu 1998; Yu, Jiang, Sun *et al* 2015; Gelal, Kalaycı, Çelebisoy *et al* 2016.

181 Botez-Marquard & Botez 1993.

182 Schmahmann *op cit*.

183 Prescott 1970.

To Harvard neurologist Jeremy Schmahmann, it seemed possible that whatever the cerebellum did for motor control, it might also do for behaviours outside the motor domain.[184] In the same way that the cerebellum regulates the rate, force, rhythm and accuracy of movements, it might regulate the speed, capacity, consistency and appropriateness of mental or cognitive processes. He called this the 'dysmetria of thought' hypothesis, referring to the Greek-derived term (dysmetria) conventionally applied to the ataxic gait or disproportionate and uncoordinated movements associated with cerebellar lesions. Symptom complexes in the cognitive and affective domain, he reasoned, could also be seen as reflecting exaggeration or diminution of responses to the environment, similar to those in the motor domain.[185] Dysmetria of thought, then, would according to this hypothesis, eventuate in disjunctive and disproportionate elements of social interaction.[186]

A range of motor, cognitive and affective disorders in patients with focal lesions in the cerebellum, and subcortical regions such as the basal ganglia or thalamus, are similar to those that are caused by lesions of the cerebral cortex, though often with qualitative differences.[187] In *The Master and his Emissary*, I described, for example, how widely different, but experientially coherent, phenomena across the motor, affective and cognitive realms could be elicited by minute variations in the positioning of an electrode in the subthalamic nuclei, themselves tiny structures only millimetres in diameter, deep below the cortex.[188] What one sees there, as in the cerebellum, is that complex phenomena at the highest level are already coherent wholes – ready-to-go, as one might say – at this apparently low level, well below conscious awareness. There the motor, affective and cognitive elements are *inextricably linked in ways that have human meaning*, and do not need to be combined or 'assembled' mechanically by any higher process. We are used to thinking that the higher functional level must unite what is separate at a lower level; but it is just as likely that what is united at a lower level is separated at a higher level. As I suggested in *The Master and his Emissary*, the 'binding problem' – how all those little 'modules' we have identified get 'put together' into one more or less seamless whole – may be an artefact of our chosen epistemology.[189]

My point in this excursus has not been so much to explore the contributions of the cerebellum in and of themselves, intriguing though recent research in the area may be, but to illuminate the inextricable alliance in the brain *at every level* – phenomenological, physiological and neuroanatomical – between motor control and all aspects of thought, including not just perception but complex cognition and emotional and social regulation.

184 Schmahmann 1991.
185 Schmahmann 2010.
186 *ibid.*
187 Schmahmann & Pandya 2008.
188 McGilchrist 2009b (225–56).
189 McGilchrist 2009b (225).

MOTION AND COGNITION:
SCHIZOPHRENIA AND AUTISM

I mentioned that some theories of autism implicate the cerebellum; and we have seen that there motion, cognition and emotion are closely related. It is informative to take a closer look at abnormalities of movement in autism and schizophrenia – *not* in an attempt to localise any one 'function', be it in cerebellum or cerebrum, but purely to demonstrate how these aspects of experience vary together, wherever they may be found in the brain.

Organisms move as a whole, and their movement is characteristically fluid: machines work by action of part on part, and their motions typically lack organic fluidity. And in a sense what the schizophrenic subject, and to some extent the autistic subject, experiences is something like a mechanisation of time, experience and the body.

I mentioned that schizophrenic subjects have an altered sense of time, but I did not particularly emphasise one highly characteristic aspect of it, the jerkiness of its tempo. Like right hemisphere-damaged subjects, they experience the time-lapse (*Zeitraffer*) phenomenon (see pp 79–80): one subject, for example, explicitly reported that 'time wasn't moving slowly, but in jerks'.[190]

'When I move quickly it's a strain on me', reports another patient. 'Things go too quickly for my mind. They get blurred and it's like being blind. It's as if you were *seeing a picture one moment and another picture the next*.'[191] Life itself becomes a series of snapshots, and, as Sass reflects, there is 'the feeling that there are gaps between one moment and the next', as under a strobe light.[192] A patient of Jaspers reported that

> he only saw the space between things; the things were there in a fashion but not so clear; the *completely empty space* was what struck him.[193]

One of Minkowski's patients actually 'conceived of the idea of putting a buffer day between the past and the future. Throughout this day I try to do nothing at all. I will go for forty-eight hours without urinating'.[194] Another makes the explicit connexion between his awareness that if there are 'instants', there must be gaps, an 'emptiness' – a 'nothing' – in between the instants:

> I feel like in the movie *Groundhog Day*: time and again I wake up, and the same things happen again and again. That's how I feel – like in a dream ... All other people live a normal life, but for me, it's different, it's like *cut–cut–cut* ... I look at an entity, and I look at other entities, and there is emptiness in between, there is *nothing in between*.[195]

This disjointed aspect of time experience passes over into cognition. In many schizophrenics, there are characteristically jerky con-

---

190 Cutting & Silzer 1990.

191 McGhie & Chapman 1961 (emphasis added).

192 Sass & Pienkos 2013c.

193 Jaspers 1963 (82; emphasis added).

194 Patient quoted by Minkowski 1970 (279).

195 Fuchs 2013 (emphasis added).

nexions (or lack of them) in thought and its expression, known as 'knight's moves', after the right-angle turns made by a knight in chess. There the flow of thought, like the flow of time, is no longer continuous: there is a loss of 'a smooth train of thought'.[196] And sometimes it stops altogether in a common symptom of schizophrenia known as 'thought block', when a line of thought suddenly disappears in mid-train. No motion at all: 'nothing in between'.

By contrast, depression tends towards right hemisphere dominance; and as Sass & Pienkos report 'many depressed patients in fact experience an increase in the number of thoughts. These are experienced as co-existing simultaneously in an unpleasant continuous flooding.'[197]

The connexion between disjunctive thought processes and a disjunction of the experienced world is made clear by one of Minkowski's patients: 'the ideas which he had entertained seemed … to have appeared in isolation, with no links between them', and as a result 'his whole life seemed to have evolved in fits and starts. It was not a continuous line, supple and elastic, but one that was broken in several places.'[198]

Inevitably this discontinuity further passes into bodily motion, both its perception and execution. Remarkably, schizophrenic subjects often perceive smooth biological movement itself as jerky, discontinuous – and even *unnatural*. In other words, machine-like.[199] And both the way we experience and perceive movement are related reciprocally to our capacity for emotional and *social* cognition. Deficits in the one area cause deficits in the other.[200]

On testing, schizophrenic subjects have difficulty distinguishing human from random movement, or normal from abnormal biological motion.[201] They are impaired in the processing of *coherent* motion, which requires ability to see the whole. In keeping with this, local motion processing may be relatively intact, but global motion processing lost.[202] It is, however, at the global level that one must understand the fluid movements of a living body.

The loss of smoothness in the perception of movement appears to be particularly associated with a disturbed sense of self: patients who report this loss also endorse statements such as 'I have difficulty in forming my own opinion' and 'I have lost my own self'.[203] This is in part due to the loss of the coherence-giving *Gestalt*, the sense of a life and person as an embodied whole enduring over time.

Thinking and moving, perceiving and acting, are bound up with and reflected in one another. Walking has been long considered an aid to fluent thinking – the philosophical school founded by Aristotle was known as 'peripatetic', from the habit of walking while discussing philosophy. There is an observable and reciprocal relation between objectively fluid movement on the one hand, and subjectively fluid

196 Kim, Takemoto, Mayahara *et al* 1994 (433).

197 Sass & Pienkos 2013b (103–30).

198 Minkowski 1987 (206).

199 Cook, Blakemore & Press 2013; and see Kretschmer *op cit*. As also discussed in Chapter 9.

200 Shiffrar 2011; Cook 2016.

201 Kim, Park & Blake 2011.

202 Chen, Nakayama, Levy *et al* 2003.

203 Kim *et al* 1994 *op cit* (432).

cognition and creativity on the other.²⁰⁴ Since both subjects with schizophrenia and those with autism have an impaired sense of fluid time, we should expect them also to display loss of fluidity in cognition, difficulties in perception and interpretation of motion – and a loss of fluidity in their observed bodily movements. And they do exhibit all three: flow has been replaced, across all domains, by the jerkiness of 'successive segments' of motion or time.

At times, indeed it can leave the subject rooted to the spot: 'I get stuck, almost as if I am paralysed at times. It may last for a minute or two but it's a bit frightening.'²⁰⁵ I mentioned the tendency to immobility in schizophrenic subjects in Chapters 9 and 22. According to Minkowski, 'their posture and behaviour bear the stamp of this morbid immobility. It shows itself in their stereotyped movements, which are a kind of perpetual repetition of one movement.'²⁰⁶ And it is there in thought, too: so-called 'thought perseveration', an obsessive repetition of insignificant thoughts or mental images; 'thought echo', a feeling that one's thoughts become automatically and involuntarily repeated; and *Gedankenlautwerden* (literally, 'thoughts become audible'), the feeling that one's thoughts are already anticipated and uttered out loud an instant before one 'has' them.

Kretschmer noticed what he called a 'jerkiness of motor tempo' in schizophrenia spectrum disorders.²⁰⁷ There was, he thought, 'a peculiar military stiffness in expression and movement, as an inherited peculiarity in schizoid families', affecting them 'somatically and psychically alike'.²⁰⁸ He related it to the progression of the condition itself. He called affective psychoses 'circular' psychoses, because they moved in cycles, whereas schizophrenic psychoses come in jerks. Something had got out of order in the inner structure. The whole structure might collapse inside, he observed, or perhaps only a few slanting cracks might appear. But in the majority of cases there remained, he thought, something that never got patched up.²⁰⁹

Changes of temperament in such subjects, too, are, he says, 'abrupt, jagged': 'schizoids vary between tenacious and jerky' in their style.²¹⁰ If schizophrenia is an example of bodily motion being dislocated in parallel with emotional and social cognition, one might expect Kretschmer's 'cycloid', that is to say, affective, psychoses (for whom the right hemisphere and its sense of time are relatively intact) to show normal motor fluency. Kretschmer reports that in them 'motor expressions and movements are well-rounded, fluid and natural.'²¹¹

A literal jerkiness of body movement is already found in the prodrome to schizophrenia (ie, before the illness is manifest).²¹² Similar disturbances of fluidity of motion have been found in autism. Both Leo Kanner, who first described classical autism, and Hans Asperger, after whom Asperger's syndrome is named, noted motor abnormalities such as 'sluggish' reflexes, 'clumsy' gait and an absence, from an

---

204 Slepian, Weisbuch, Pauker *et al* 2014; Slepian & Ambady 2012.

205 McGhie & Chapman 1961 (106).

206 Minkowski 1987 (200).

207 Kretschmer 1925 (174 &175).

208 *ibid* (175).

209 *ibid* (147).

210 *ibid* (176–7).

211 *ibid* (134).

212 Lin, Chuah, Mohan *et al* 2011.

early age, of anticipatory postures when being picked up.[213] Autistic children who have never been treated with medication display 'a stiffer gait', in which the 'usual fluidity of walking' has been lost.[214] Individuals with autism display instability, during both standing and walking, have impairments in gait, posture, balance, speed and coordination, as well as fine motor control (often evidenced in handwriting), and they move 'more jerkily', and with unusual acceleration.[215] The degree to which this is the case correlates with the severity of autism, and with a tendency to perceive the motion of living things as not just hard to interpret, but as positively unnatural.[216] Some subjects report a lack of 'feedback' from the body: some describe lacking awareness of their own facial expressions, movements or the position of their limbs in space.[217] One person with autism writes:

> I think the fluidity of access to various places in my brain is dependent upon neurological movement between places ... Sometimes my speaking is hindered, other times my thinking, and sometimes my physical movement. The hardest is when thinking is not working smoothly. When that happens, I have to line up one thought at a time, *like train cars*. I like it much better when my thoughts do not have to be methodically lined up, but are more *fluid* ...[218]

Like train cars: an evocative description of the difference between explicit serial connexion of parts and the flow of the whole.

Observe the fluidity of the natural walking movement of someone from a non-industrialised culture, compared with our generally awkward, angular, graceless way of moving. Smoothness of motion has disappeared from modern life: the strobe and the break-dance – are they in some ways reflections, even celebrations, of this? Whether that be the case or not, the fragmentation of modern culture, with its abrupt nonsensical transitions, sometimes accidental, as with juxtapositions on television, and sometimes contrived deliberately, as in postmodern art, mimics this loss of flow.

MOTION AND COGNITION: THE HEMISPHERES

Distinct regions of the right hemisphere are given over to perception of motion in general, of biological motion, and, separately, of human motion.[219] I have mentioned that the right posterior superior temporal sulcus is crucial to the perception of others' bodies and bodily motion, and is particularly strongly activated during the perception of eye or mouth movements, body language, and biological motion of all kinds.[220] Suppressing the right superior temporal sulcus causes a loss of ability to discriminate living motion, whereas suppressing the same region in the left temporal lobe produces no significant result.[221]

The same right hemisphere region is involved in assessing the

213 Kanner 1943; Kanner 1968; Asperger 1944.

214 Nobile, Perego, Piccinini *et al* 2011.

215 Eg, Ghaziuddin & Butler 1998; Jansiewicz, Goldberg, Newschaffer *et al* 2006; Noterdaeme, Mildenberger, Minow 2002; Rinehart, Tonge, Iansek *et al* 2006.

216 See note 198 above.

217 Eg, Blackman 1999; Hale & Hale 1999; Williams 1996, 1999 & 2003.

218 Judy Endow (Facebook, 25 January 2009), quoted in Donnellan, Hill & Leary 2013 (emphasis added).

219 Han, Bi, Chen *et al* 2013.

220 Allison, Puce & McCarthy 2000; Grossman, Battelli & Pascual-Leone 2005; van Kemenade, Muggleton, Walsh *et al* 2012; Hein & Knight 2008; Redcay 2008; Carrington & Bailey 2009.

221 Grossman, Battelli & Pascual-Leone 2005; van Kemenade, Muggleton, Walsh *et al* 2012.

goals and intentions underlying another's movements – again, motion, emotion and meaning are closely linked.[222] The closely associated (and physically proximate) right fusiform gyrus is critical for reading facial expressions.[223] Independently it has been shown that if right hemisphere attention is compromised, we lose the intuitive understanding of motion.[224] And we have seen that the perception – and the enactment (since they are inseparable) – of human social relations is also highly dependent on the right hemisphere.

Tying in with the findings in autism and schizophrenia, a review confirms the findings of at least six earlier studies, showing that motor stability and coherence of gait and posture were more affected by right hemisphere deficits than by left.[225] According to Coslett & Heilman, 'the right hemisphere is dominant for motor activation or intention'.[226]

MOTION AND COGNITION:
THE EMERGENCE OF THE PHENOMENAL WORLD

Analytic thought and language tend to immobilise the world. John Cutting notes that the phenomenological reduction envisaged by Max Scheler, a thought experiment whereby 'the real world and everything of it which pertained to a human being was envisaged as cancelled out', is strikingly close to the world as experienced by the schizophrenic subject, and he details eleven respects in which this could be said to be the case.[227] The one that interests us for now is what Cutting calls 'adynamy'.

This is not *just* the abolition of motion, which we have already seen very clearly is an aspect of the schizophrenic world, but the abolition, too, of the connexions that make sense of the world. For the subject, nothing any longer happens by chance; and at the same time causal relations are misunderstood, seeing them where they cannot be, and failing to see them where they must be. Chance and causation both depend, in different ways, on motion, and in schizophrenia the subject abjures motion, strives to eliminate chance, and forms disrupted causal – or in truth non-causal – connexions. This is just like the 'phenomenologically reduced' subject described by Scheler, who 'not only loses any sense of accident or chance in their experience but any sense of movement as well'. Cutting quotes Scheler:

> There disappear all causal interconnections because reality itself is the basis of all such causes. The result of this is that the reduced world is a perfectly adynamic world – nothing has an effect on anything else any more.[228]

Thus, as Cutting expresses it, the world is 'frozen in time, *because nothing causes anything to move*'.[229] We saw many accounts of this from the mouths of patients themselves in Chapter 9.

Wittgenstein repeatedly remarks on the way that stopping acting

---

222 Lee, Gao & McCarthy 2014; Shultz, Lee, Pelphrey *et al* 2011; Saxe, Xiao, Kovacs *et al* 2004; Pelphrey, Singerman, Allison *et al* 2003; Shultz & McCarthy 2012; Shultz & McCarthy 2014; Pelphrey, Morris & McCarthy 2004.

223 Puce, Allison, Gore *et al* 1995; Haxby, Horwitz, Ungerleider *et al* 1994; Kanwisher, McDermott & Chun 1997; Sergent, Ohta & MacDonald 1992.

224 de Vito, Lunven, Bourlon *et al* 2015.

225 Chen, Novak & Manor 2014; Titianova & Tarkka 1995; Cassvan, Ross, Dyer *et al* 1976; Pérennou, Bénaim, Rouget *et al* 1999; Pérennou, Mazibrada, Chauvineau *et al* 2008; Genthon, Rougier, Gissot *et al* 2008.

226 Coslett & Heilman 1989 (274); see also Heilman & van den Abell 1979.

227 Cutting 2018. Note that Scheler's reduction is quite different from that developed by Husserl, which according to Cutting, 'was merely a judgement, on Husserl's part, of what a human being would experience of some object (an apple tree in blossom being his example) if the reality of the actual object were "bracketed"'.

228 Scheler 2008 (78).

229 Cutting & Musalek 2016 (emphasis added).

and engaging with the world in order to reflect on it makes things appear alien. His thrust as a philosopher is to help us get on with things, to 'move about around things and events in the world instead of trying to delineate their essential features'.[230]

Perception is certainly altered by movement and action in the world. Perception appears on the face of it to be a passive process (in an age in which the camera has become the false analogy for perception). Yet it is an active process in which we go to meet phenomena. There is widespread evidence that the motor and visual systems are intrinsically linked and mutually influence each other. Observations and perceptions affect movements, and movements affect observations and perceptions. Sometimes the process may unduly affect judgment: for example, observers lifting boxes perceive the weight of an object being lifted by an actor in a way that depends on the weight of the box that they themselves are lifting,[231] and walking affects one's judgments of the speed of another walker.[232] Measurement, then, may sometimes be better from a static position (measurement depends largely on the left hemisphere which tends towards stasis); but learning how to move, or react, or to understand the *meaning* or *purpose* of a movement or action, depends on reciprocity of action and observation. In general this process aids rather than distorts perception.[233] Disrupting motor regions temporarily (using transcranial magnetic stimulation) distorts judgments about actions;[234] and patients with motor deficits have deficits in action recognition.[235] But that is just the tip of the evidence iceberg: motion and perception are clearly and deeply intertwined.[236]

The idea that perception is altered by, and may even be secondary to, activity is familiar from the *Gestalt* theory of perception, according to which we are co-operatively constructing the aspects of the world that we see. It was already there in the thinking of John Dewey, who, in a now famous essay on the nature of the 'reflex arc', wrote in 1896 that

> upon analysis, we find that we begin not with a sensory stimulus but with a sensori-motor coordination ... and that in a certain sense it is the movement which is primary, and the sensation which is secondary, the movement of body, head and eye muscles determining the quality of what is experienced. In other words, the real beginning is with the act of seeing; it is looking, and not a sensation of light.[237]

The so-called perceptual 'stimulus' and motor 'response' cannot be considered separately, outside the context of their interaction, though Dewey hints that indeed the motor element – normally seen as the response – may be primary. Perception is an active, not a passive process – or better, it is a profoundly interactive process. Movement lies behind, and in, every one of our senses.

230 van der Merwe & Voestermans 1995.
231 Hamilton, Wolpert & Frith 2004.
232 Jacobs & Shiffrar 2005.
233 Wohlschläger 2000; Brass, Bekkering & Prinz 2001.
234 Pobric & Hamilton 2006.
235 Buxbaum, Kyle & Menon 2005; Pazzaglia, Smania, Corato *et al* 2008; Negri, Rumiati, Zadini *et al* 2007.
236 I explore the intersubjective and embodied nature of perception in McGilchrist 2009b (esp 161–9 & 265–6). For a book-length examination of the topic, the reader is referred to Claxton 2016.
237 Dewey 1896 (358–9).

This idea has gathered further scientific backing in recent years. The Colombian neuroscientist Rodolfo Llinás has argued, starting from the examination of simple marine invertebrates such as the sea squirt, that the capacity for motion underlies all knowledge:

> What I must stress here is that the brain's understanding of anything, whether factual or abstract, arises from our manipulations of the external world, by our moving within the world and thus from our sensory-derived experience of it.[238]

Similarly neuroscientist György Buzsáki claims that perception is founded on motion and cognition, not motion and cognition founded on perception. He regards activity 'as not only interwoven with perception but *prior* to perception, prior both in terms of evolution and in terms of initiating processes within and outside the organism that result in the organism's perceiving.'[239] In relation to the evolutionary claim, he points to some primitive sea animals that are capable only of a rhythmic movement of cilia to bring in nutrients, with no (presumed) perceptual abilities at all.[240]

And those creatures that have perceptual abilities need to act, after all, even if only to organise and interpret their perceptions. Perception is an active process of going to meet the world – interactive, exploratory – 'something we *do*.'[241] 'Seeing', as Kevin O'Regan & Alva Noë put it, 'is a way of acting. It is a particular way of exploring the environment.'[242] This is another way of saying that attention changes the world.

Cognition, too, in as much as it can be separated from perception, involves movement. By this I do not mean simply that, as Galton first pointed out, we often think in and by acting, though that is an important truth. Nor that we often think better when in motion, as the peripatetic school of Aristotle believed (especially, from my experience, when driving). On this topic Nietzsche was characteristically expressive:

> *On ne peut penser et écrire qu'assis* [one can think and write only when seated] (G. Flaubert). – Now I've got you, you nihilist! Sitting on your arse is precisely the *sin* against the Holy Ghost. Only those thoughts that come in *walking* have any value.[243]

Heidegger, whom George Steiner called an indefatigable walker in unlit places, would have agreed.

Nor is it simply that, when we do think in language, we are using adaptations of the motor system, though there is an evolutionary relationship, memorialised in the neuroanatomy of the brain, between speech and movement of the right hand, and it has even been claimed that syntax derives from ambulation.[244]

No, what I am getting at here is best articulated by Schelling, who

---

238 Llinás 2001 (58–9).

239 I am indebted to Barbara Goodrich's interesting paper (2010) from which this quotation, and these examples, are taken.

240 Buzsáki 2006 (ix & 30).

241 *ibid* (228).

242 O'Regan & Noë 2001.

243 Nietzsche 1889, 'Sprüche und Pfeile': §34: » *Damit habe ich dich, Nihilist! Das Sitzfleisch ist gerade die* Sünde *wider den heiligen Geist. Nur die ergangenen Gedanken haben Wert* « (emphases in original).

244 See McGilchrist 2009b (111–26 *et passim*). On syntax, see Vowles 1970.

made the point that every assertion is inseparable from the set of circumstances that gave rise to it. There is no genuine proposition (not even, presumably, this one), he asserted, that has universal validity, such that it could be separated from what he calls the movement (*die Bewegung*) – the active engagement with the world – out of which it arises: 'Movement', he says, 'is the essence of knowledge; take away this vital element, and it dies like fruit stripped from the living tree.'[245]

Emotion, too, very clearly implies motion. The very word tells us that; but every emotion we experience is founded on a concern or care for some aspect of the world, implying a tendency towards action of some kind. The capacity to move and be moved by others may be fundamental to our conscious being.[246] James Prescott was surely right to think that emotional development depended on movement. Ethics, the experience of moral values, quite clearly implies motion, too, for such values are similarly rooted in active dispositions towards others and the world at large, eventuating in actions, and inevitably tied to emotions. Just to have a body at all, according to Merleau-Ponty, implies motion: 'My body', he says, 'appears to me as an attitude directed towards a certain existing or possible task'.[247] James sums it up well: 'The fact is that there is no sort of consciousness whatever, be it sensation, feeling, or idea, which does not directly and of itself tend to discharge into some motor effect.'[248]

THE ONTOLOGY OF MOTION

'Change, flux, or becoming is the Absolute', wrote the Indian philosopher Balbir Singh. 'What we ordinarily call a thing is itself a process, a ceaseless coming to be and passing away.'[249] Is it in fact possible that all 'bodies' – animate and otherwise – depend for their nature on motion?

> We say that movement is composed of points, but it comprises, in addition, the obscure and mysterious passage from one position to the next. As if the obscurity was not due entirely to the fact that we have supposed immobility to be clearer than mobility and rest anterior to movement! As if the mystery did not follow entirely from our attempting to pass from stoppages to movement by way of addition, which is impossible, when it is so easy to pass, by simple diminution, from movement to the slackening of movement, and so to immobility![250]

I quoted Schelling just now on the relation between motion and thought. But his insight into their inseparability went further and deeper. At a second level, our knowledge of anything is embedded in movement since, as Fichte too had seen, consciousness is not a fact, or a thing – but an *act*. It is not just that it is a flowing movement, as

245 Schelling 1856–61a (208): » *Die Bewegung ist aber das Wesentliche der Wissenschaft; diesem Lebenselement entnommen, sterben sie ab, wie Früchte vom lebendigen Baum getrennt.* «

246 Ellis & Newton 2012.

247 Merleau-Ponty 1962 (137). Merleau-Ponty 1945 (134–5): « *mon corps m'apparaît comme posture en vue d'une certaine tâche actuelle ou possible* ».

248 James 2008 (83).

249 Singh 1987 (10).

250 Bergson 1912 (50–1).

James saw: it is that movement unites me with the world. Motion is intrinsic to the betweenness, and the directedness, of consciousness itself.[251] Consciousness is always consciousness *of* something, reaching out and going to meet something beyond the self; not a self-enclosed Cartesian theatrical display, but a reverberative process, already aimed towards the real, living world – *out of which it also comes*. Creation of other is also creation of self; knowledge of other is also knowledge of self. And in consciousness Nature, according to Schelling, is coming to create and to know itself – in more conventional terms one might say both as 'subject' and as 'object', except that in this very process that duality of subject and object reveals itself to be false, and comes to be transcended.

Movement then is not only implied in each and every aspect of mental life, as we have seen, but in the body and in consciousness. It is essential to Nature, both animate and inanimate – to all existing things from waves and particles onwards. Schelling, unlike Fichte, however, did not see consciousness as the ground of everything that exists: that role was occupied in his philosophical system by Nature, out of which consciousness arose and from which it was inseparable. Nature herself, existence itself, *was*, according to Schelling, motion. '*Being itself* = absolute activity', he wrote in the opening pages of his *First Outline of a System of the Philosophy of Nature*.[252]

If we accept that being itself is motion, immobility becomes merely a fictional representation imposed by the mind for the purposes of calculation, as Bergson realised: 'there never is real immobility, if we understand by that an absence of movement. Movement is reality itself.'[253] Here again Bergson seems to share an insight, not just with Schelling, but with Leibniz. Thus Bergson says:

> How could the moving object *be* in a point of its trajectory passage? It *passes through*, or in other terms, it *could be there*. It would be there if it stopped; but if it should stop there, it would no longer be the same movement we were dealing with ...[254]

Compare Leibniz: 'Whatever moves is never in one place, not even in an infinitesimal instant'.[255] This is a version of the point that Zeno made in in the Paradox of the Arrow, but with the opposite conclusion: not that motion cannot exist, but that a dimensionless point or instant is a mental fiction. Something moves, says Hegel, in a similar vein,

> not because at one moment it is here and at another there, but because at one and the same moment it is here and not here, because in this 'here', it at once is and is not. The ancient dialecticians must be granted the contradictions that they pointed out in motion; but it does not follow that therefore there is no motion, but on the contrary, that motion is *existent* contradiction itself.[256]

---

251 Elias 1991 (33–4; emphasis in original): '"Reason", "mind", "consciousness" or "ego" ... all give the impression of substances rather than functions, of something at rest rather than in motion ... [Yet] they are functions which – unlike those of the stomach or bones, for example – are directed constantly towards *other* people and things'.

252 Schelling 1799b (5): » *Das Seyn selbst = absoluter Thätigkeit* « (emphasis in original).

253 Bergson 2007e (119).

254 ibid (emphases in original).

255 Leibniz, letter to Antoine Arnauld, early November 1671. Cf Kolakowski 1985 (14–15): 'The moving arrow never *is* at any certain point, and we can understand this easily if, instead of starting with space, we take the fact of movement as the original, irreducible reality'.

256 Hegel 1969 (440).

This was more than a hundred years before Heisenberg. An entity, Bergson would say, is never in a point in space, it is only in a movement. In fact it only *is* a movement. Could motion, then, be said to be in some sense foundational? Like time and space, motion cannot be derived from anything else. You can never get from stasis to motion, as Aristotle was obliged ruefully to recognise.[257] You have to start with motion itself. It is in that sense foundational, in that it cannot be derived. But does it form the foundation, ontologically speaking, for other aspects of experience?

Space and time are not just abstract concepts, but aspects of lived experience. They are not things, even disembodied things. Nowadays in the West they are almost always replaced, as soon as mentioned – so lightning fast that we can scarcely be aware of it – by abstractions; and abstractions are always secondary developments from lived experience. And not just for the layman: 'the method of freezing time has worked so well that most physicists are unaware that a trick has been played on their understanding of nature'.[258] Our education teaches us not just to think of space and time as abstractions, but, because of our tendency to *privilege abstractions*, to see them as primary – and movement as secondary. I suggest that movement is as foundational as space and time. Each requires the other. Space is the *potential* for something to move within it; time is the *potential* for something to change within it.[259] Both become actualised in flow. To attempt to negate motion, then, threatens to undermine any means we might have of approaching and understanding reality.

Yet to the left hemisphere rest is the primary state. Rest is, as far as it is concerned, how things *really* are, and motion is something adventitious that is superadded. Its efforts are always directed, as far as possible, towards the eradication of motion. As we have seen, analysis, the forte of the left hemisphere, ultimately yields stasis. 'The separation of thought and action, theory (*theoria*) and practice, is a fundamental requirement of the form of rationality for which Descartes is arguing', writes the philosopher D. M. Levin.[260] And with those words I am reminded of how Roland Kuhn's schizophrenic patient Franz Weber 'wanted activity of any sort to be replaced by knowledge'.[261]

Analysis wants measurement, and measurement begins the process of immobilisation and fragmentation. Yet it is never quite equal to what it measures. As Claudio Ronchi puts it: 'Once a unit is given, our mind can conceive of its measurement only through a sequence of contiguous and not overlapping units that cover the object in defect or excess.'[262] Analysis also starts its attempts to reconstruct the world from stasis, though motion can, of course, never be reached by aggregation of static elements. '*Fixed concepts can be extracted by our thought from the mobile reality*', said Bergson, '*but there is no*

[257] See p 648 above.

[258] Smolin 2013 (30).

[259] I am grateful to John Cutting for drawing my attention to the fact that Scheler said just this: see Scheler 1973, 288–356 (esp 331, 333 & 341).

[260] Levin 1999 (37).

[261] See p 351 above. From Cutting 2002 (161). I am indebted to Cutting's account of the patient (160–2). The account is taken from Kuhn 1952.

[262] Ronchi 2014 (132).

263 Bergson 2007c (160; emphasis in original).
264 Bergson 2007b (22).
265 James 1909a (236).
266 trans D Jenkins & Y Moriguchi.
267 Suzuki 2011.

means whatever of reconstituting with the fixity of concepts the mobility of the real.'²⁶³ And this hemisphere difference is what he was alluding to, unknowingly, when he said: 'intelligence starts ordinarily from the immobile ... intuition starts from movement'.²⁶⁴ James writes:

> you cannot make continuous being out of discontinuities, and your concepts are discontinuous. The stages into which you analyse a change are *states*, the change itself goes on between them. It lies along their intervals, inhabits what your definition fails to gather up, and thus eludes conceptual explanation altogether.²⁶⁵

What this means is that analytic thinking will never be up to the job of providing a complete account of reality, and what it lacks will be hardly trivial, but what is in every sense vital to the project of philosophy.

≈

In Oriental religions and in early Western philosophy, before the time of Plato, we used to know that change is something very deep, not just something that 'happens to be' – or worse, happens not to be at all. In 1212, Kamo no Chōmei wrote the poetic meditation which is familiar to generations of Japanese schoolchildren, known as *Hōjōki*, which encapsulates the Buddhist idea of *mujō*, or impermanence:

> The flowing river never stops and yet the water never stays the same. Foam floats upon the pools, scattering, re-forming, never lingering long. So it is with man and all his dwelling-places here on earth.²⁶⁶

Shunryū Suzuki reinforced the absolutely foundational nature of this insight:

> That everything changes is the basic truth for each existence. No one can deny this truth, and all teaching of Buddhism is condensed within it ... Without realising how to accept this truth you cannot live in this world. Even though you try to escape from it, your effort will be in vain. If you think there is some other way to accept the eternal truth that everything changes, that is your delusion. This is the basic teaching of how to live in this world. Whatever you may feel about it, you have to accept it.²⁶⁷

Of course if Heraclitus did really say 'everything flows', his saying has the advantage over 'everything changes' (which it is often understood to mean) in that it says not only that things change, but Bergson-like, there is no succession of things (time slices) involved in change, because they flow, interpenetrating one another – and therefore *remain*. And not only that, but, again Bergson-like, there are not things that flow, but there is just – flow, which manifests as things flowing; it's the flowing that is the ultimate reality.

Chapter 24 · Space and matter

Something unknown is doing we don't know what – that is what our theory [of the physical world] amounts to.
—SIR ARTHUR EDDINGTON[1]

I do not define ... space ... as being well known to all.
—SIR ISAAC NEWTON[2]

Since Einstein we speak of an entity called spacetime. From one perspective, however, there may not be an entity, or even separate entities, involved at all. I have mentioned his suggestion that we should speak of things being 'spatially extended' rather than 'in space'. Space and time are not containers in which we live, but aspects of being.

'It is interesting that in some mythologies, Time appears as a god, a mythic *person* of sorts', observes Jan Zwicky, 'but, to the best of my knowledge, space is never so represented.'[3] Time speaks profoundly to the human condition in a way that space, however fundamental it might be, simply does not. Time is relentless, like another being's will, where space is pliable and may be fashioned, though not without limits, to our own. Time is emotive; space is bland. Time is always single. 'I venture to say only one true integer may occur in all of physics', says physicist David Tong. 'The laws of physics refer to one dimension of time. Without precisely one dimension of time, physics appears to become inconsistent.'[4] By contrast space is multiple: it has between three and 11 or even more dimensions, depending on whose theory you are inclined to embrace. Time is irreversible: space open to endless revision. Time creates and corrodes: space lends (temporary) permanence to what is. Writing, for example, was invented to give permanence in space to the fleetingness of thought. Consciousness exists in time, but not in space.[5] For elements that are often conflated, their qualities could hardly be more different.

Max Scheler points to the dynamic and living quality of time, compared with the adynamic and comparatively inanimate quality of space. 'Life', he wrote,

> is an event or a process. It can only be defined functionally and dynamically. No structural definition is sufficient and there is no way one can grasp its meaning by invoking some spatial arrangement of its parts ... Life is a sort of being, which can only be properly got to grips with by emphasising its coming-to-be ... A spatial arrangement is a pre-requisite of the inorganic. Within living creatures events are not arranged along any spatial dimension, but along temporal.[6]

---

1 Eddington 1928 (291; emphasis in original).

2 Newton 1687.

3 Zwicky 2014b, §73 (emphasis in original).

4 Tong 2012.

5 Cf McGinn 1995, for whom consciousness 'seems not to be the *kind of thing* that falls under spatial predicates. It falls under temporal predicates ... but it resists being cast as a regular inhabitant of the space we see around us and within which the material world has its existence' (220; emphasis in original).

6 Scheler 2008 (300–1).

Yet, despite this, space has the means to be generative: it is the *potential* for motion, and gives rise to form, which is what we see, precipitated out of potential, in space. In the project of differentiation – of species, of individuals and in all the creative flow towards generating new, unique creatures, events and circumstances – space is *required*, otherwise all would collapse into unity.

While time shows us that aspect of reality which is always incomplete, space shows us that aspect of reality which has been achieved. Both have their place. Thus Whitehead:

> Apart from space, there is no consummation. Space expresses the halt for attainment. It symbolises the complexity of immediate realisation. It is the fact of accomplishment. Time and space express the universe as including the essence of transition [time] and the success of achievement [space]. The transition is real, and the achievement is real. The difficulty is for language to express one of them without explaining away the other.[7]

It is possible to see time as the leading edge of a waveform which actualises what we experience out of the potential of space as it moves forward. The question why the arrow of time can't go backwards is a mistake due to the spatialisation of time. You might as well ask why a stream doesn't flow backwards.

WHAT IS FORM?

It is worth distinguishing between structure and form. The word 'structure' comes from Latin *struere*, to build, and implies construction: putting parts together to produce an edifice (or a mechanism). 'Form', on the other hand, has no such implications and refers to the overall shape, or *Gestalt*, of something that may be constantly changing, even if, indeed, changing in order to stay the same.[8] Forms are usually evidenced in matter, and matter in forms: no-one has seen matter without form. It is sometimes said that no-one has seen form without matter, but that seems to me wrong: I can see forms of light, and I can see imaginary or abstract forms, with the mind's eye, or in certain states of mind with the bodily eye. But matter is inconceivable without form. So it seems to me – and here I am in agreement with Plato and Aristotle – that form is more fundamental than matter. This is also, according to Schrödinger what physics confirms:

> The habit of everyday language deceives us and seems to require, whenever we hear the word 'shape' or 'form' pronounced, that it must be the shape or form of *something* ... when you come to the ultimate particles constituting matter ... they are, as it were, pure shape, nothing but shape.[9]

---

7 Whitehead 1938 (139–40).

8 See Chapter 12.

9 Schrödinger 1951 (21; emphasis in original).

Form is the coming together of an essentially static, receptive potential (space) with an essentially motivating, 'informing' energy (force).[10]

That receptivity is what Buddhists call emptiness: 'the very essence of the phenomenon, emptiness, produces form and needs to manifest itself in certain superficial ways', ie, as the phenomenal and material world we experience.[11] Emptiness is the commonest theme underlying the metaphorical expressions of the classic Buddhist text, the Diamond Sutra.[12] *Śūnyatā*, the Sanskrit word translated as emptiness, derives from *śvi*, which denotes hollowness and swelling, as of a womb, or of a seed as it expands.[13] It is the opposite of emptiness as it is normally understood in the West, namely, a condition of inertia and sterility: rather it is one of limitless, undifferentiated potential.[14] As often, there are parallels in quantum field theory. Space is never in fact empty: 70% of the entire universe is dark energy, we are told, and 25% dark matter: far greater in extent than all the 'things' we could see. Even a vacuum is full of energy fields. Space could be seen as a property of fields of *energy* – fields, therefore, of potential creation. Fields do not need a substance (Latin *sub-* + *stans*, that which 'stands under') in which to exist: space does not 'carry' or support the oscillating magnetic fields: they support themselves. As James puts it, 'the directly apprehended universe needs, in short, no extraneous trans-empirical connective support, but possesses in its own right a concatenated or continuous structure' – or form, as I would prefer to put it.[15]

The forms that are found in Nature are the result of motion, and embodied movement, not stasis; similarly, movements found in Nature enact forms, not structures. The great biologist and mathematician D'Arcy Thompson saw form as inseparable from the energy involved in the processes which generate it.[16] We have already seen that many flows in Nature are vortices, and that self-organising and self-promulgating patterns of complexity and beauty – fractals, spirals, *lǐ*-formations – are everywhere in the world, both organic and inorganic. The spiral is an expression of dynamism (DNA is the 'betweenness' of two spirals), where the circle is an expression of stasis.[17] In the spiral, the end point of each turn does not quite match the beginning. And so we have, suddenly, because of this symmetry-breaking 'mis-step', something that is mobile, three-dimensional, endlessly generative, while never being wholly predictable (because always moving onward into a new realm of space, not residing always in the old one). It replaces something atemporal, two-dimensional, repetitive, and entirely regular, namely a circle. All the same, viewed down the axis of the spiral it still has the eternally unchanging quality of the circle – particle-like: though viewed from the side it has an

---

10 Some of my thinking here was influenced by the work of the mycologist, Alan Rayner: see Rayner 2017. Incidentally, at first glance, this is the opposite of the way in which the origins of the world are expressed according to the creation myth in Genesis, where it is written that 'the earth was without form, and void; and darkness was upon the face of the deep: and the Spirit of God moved upon the face of the waters.' But what was real at that stage was the Spirit of God, and the world was merely a receptive space ('void') without any identity: it was the creative force of the Spirit of God that brought form to it, and thereby brought it into being. So for what it is worth, there seems to me to be no contradiction here.

11 Lu & Chiang 2007 (344).

12 Gao & Lan 2018 (237).

13 Watson 2014.

14 Gao & Lan 2018 (237).

15 James 1911b (xii–xiii).

16 See, eg, Thompson 1992 (16).

17 For an interesting sidelight on the relationship between the structure of DNA and the golden ratio, see Chen, Huang & Sun 2014.

oscillatory or vibratory movement, wave-like, changing, progressing and alive. Fractals, though quite different in nature, have this in common with spirals, that they generate *difference that is also* a kind of *sameness*.

Intellectual life also has more of the spiral than the straight line or the circle about it. I have pointed out that many important truths cannot be expressed explicitly or arrived at linearly, but must be, so to speak, taken by stealth. They must be disclosed from a number of different perspectives that converge, rather like following a spiral path around them. This is how the best philosophy, it seems to me, is done, and it is what Donne referred to in the passage I quoted in the Coda to Part 11. Like Blake's ladder that circles its way to Heaven, and thus to spiritual truth, Donne sees truth, both spiritual and intellectual, as being achieved by a path that is spiral-like. Since the Enlightenment this important truth has been neglected.

Galileo said that 'the book [of the universe] is written in mathematical language, and the symbols are triangles, circles and other geometrical figures, without whose help it is impossible to comprehend a single word of it.'[18] I would venture a caveat. The book of the universe is one of pattern: and mathematics is just one way of looking at those patterns. The patterns come first in water, earth, fire, and air: maths comes afterwards, abstracting those patterns. And I'd enter another caveat. The patterns are not just the regular, afore-knowable, patterns of Platonic philosophy – triangles, circles and so on. Many are such that we have no mathematical expressions to describe them.[19] In the words of Whitehead, 'beyond all questions of quantity, there lie questions of pattern, which are essential for the understanding of Nature.'[20]

Perhaps Galileo was a little seduced by the exhilarating feeling of being one of a select group of *cognoscenti*. It is, after all, perfectly possible to comprehend more than 'a single word', or even two or three, of the book of the universe without reading the mathematical language at all. The Buddha might stand as an example. Reading the maths adds something both beautiful and very useful, but it is not essential, except to one kind of understanding; and, as with everything, opening one channel of understanding risks closing another. Not being able to see *that* is part of the trouble with the modern West.

THE NATURE OF DEPTH

Let me consider a neglected aspect of space: its depth. In examining the question it is impossible to divorce the literal from the metaphorical. What is it and where do we turn to find it?

Isaiah Berlin wrote that 'the notion of depth is something with which philosophers seldom deal':

---

18 Galilei 1623.

19 See p 965 above.

20 Whitehead 1938 (195).

Nevertheless it is a concept perfectly susceptible to treatment and indeed one of the most important categories we use … There is no doubt that, although I attempt to describe what … profundity consists in, as soon as I speak it becomes quite clear that, no matter how long I speak, new chasms open. No matter what I say I always have to leave three dots at the end … I am forced in my discussion, forced in description, to use language which is in principle, not only today but for ever, inadequate for its purpose …[21]

21 Berlin 1999 (102–4).

'New chasms open': the attempt to clarify depth leads back into the abyss. Discouraging as these words are for those who expect precise answers to deep questions, they pose a challenge: 'one of the most important categories we use' is seldom touched on, and yet is 'perfectly susceptible to treatment' – though after demonstrating that it is almost impossible to clarify, Berlin gives few clues as to how to set about treating it. So what is depth, and where do we turn to find it? What do we mean when we call a person, a perception, a philosophy, a work of art, or an experience, deep?

What is clear is that depth cannot easily be replaced by other more approachable concepts. When we call something deep, we do not mean that it is just beautiful, important, special, complex or difficult to understand – though it may be any, or all, of those things. In fact the simultaneous sense of something that is charged with ultimate meaning and yet (perhaps for that very reason) hard to put into words, gets close, as Berlin hints.

The history of 'depth' since ancient times reveals that the spatial and the spiritual senses of the term have always co-existed, as I will show, so that it is not clear which is literal and which metaphorical. Faced with the important moments where conventional language breaks down, we turn to metaphor; and the metaphor of depth, *par excellence*, is the sea. As soon as the word 'deep' was first used in the English language, over a thousand years ago, it was already a word for the sea itself: the deep. Its counterparts, too, *bathus* (βαθύς) in Greek, and *profundus* in Latin, were from the very start applied to the sea. We call things that defeat our everyday grasp unfathomable. The sea, after all is both creator and destroyer, the source of life and the end of it, the buoyant force, the upholder, and the endless space in which we drown. It is what connects land to land and yet divides each from the other. It changes and yet in changing remains the same.

Etymologically, both *bathus* and *profundus* refer to the lowest part or foundation of whatever it is we know. But they had a range of other implications. They could be applied to a cliff, or mountain – a 'deep' cliff or 'deep' mountain – of awe-inspiring proportions (the words are importantly connected with the feeling of awe). In other words the idea of depth was not yet limited in direction, and could be measured up, as well as down. And in the third dimension: it also, importantly, referred to the sheer *solidity* of existence in lived space. (Latin *solidus* comes from the same Proto-Indo-European root *sol-* as, for example, the Greek *holos*, whole, and Latin *salvus*, healthy: and of course 'whole', 'healthy', 'hale', and 'holy' have common roots). It implied three-dimensional 'thickness'. Solid bodies are distinguished by their depth, by contrast with their representations, which lack it. In classical Greek, a body of soldiers could be described as so many ranks 'deep', impressing in us the sense of standing over against a forceful presence. And then the terms *bathus* and *profundus* were used to describe something 'far', remote, almost lost to our knowledge, alerting us to other meanings of deep that we still recognise: deep in the forest, deep in rural France (the phrase '*la France profonde*' rightly has its own Wikipedia page); and by extension, deep in sleep, deep in learning, deep in age.

Time also has this quality of depth that both invites our exploration, and yet signals that we can never fully succeed. Shakespeare's 'dark backward and abysm of time': far distant times are mysterious ('dark'), a realm which stands 'in back' of us somewhere, out of sight – and are unfathomably deep (an 'abysm'). New chasms open.

The word *bathus* was often used to imply *fertility* and *richness*: 'deep' soil and its fruits, fields 'deep' with grain, forests 'deep' with trees, a woman's 'deep' and luxuriant hair. This is important because when we speak of depth we never imagine vacuity – indeed we use those words as antonyms. It is a depth that breeds, that has potential, that is generative of something other than the familiar: *śūnyatā*.

And so there is depth of mind, of wisdom (a use already present in Æschylus, the first great dramatist of Ancient Greece). When we imported the word 'profound' from French in the early fourteenth century, it was first used in the metaphorical sense, referring to the mind, since we already had 'deep' for the purely physical sense, and it became literal only later; while 'a deep', moving in the opposite direction, by the seventeenth century was no longer just the sea, but, according to the OED, could signify a 'secret, mysterious, unfathomable, or vast' region of thought, feeling, or being. Depth is also a quality of what we mean by soul, a quality not captured by any of our workaday categories of cognition or emotion. What is deep is profound, awe-full, mysterious, solemn, not to be confined. Such epithets conjure the vast realms of the unconscious mind, that we conventionally think of as lying below our superficial cognition, and in doing so go *deep*.

Conscious thought, according to James and Bergson, is relatively superficial, if practically useful, when compared with intuition and experience. 'Thought deals ... solely with surfaces', writes James:

it can name the thickness of reality, but it cannot fathom it, and its insufficiency here is essential and permanent, not temporary ... Instead of being the only adequate knowledge, it is grossly inadequate, and its only superiority is the practical one of enabling us to make short cuts through experience and thereby to save time.[22]

22 James 1909a (250 & 252).

23 Tillich 1958.

We need the short cuts, but we should not mistake them for the knowledge from experience through which they cut.

We expect deep truths to cohere (even if one of those truths is that coherence is not necessary for truth). We would be very surprised to learn that the deep truths of different cultures did not largely overlap. It is as though our differences might be spread out on the circumference of a sphere, but as we dig deeper we reach aspects of the same core. We speak naturally of spheres, not lines, grids, squares, cubes, or rhomboids, of existence.

Depth, then, may characterise what is tangible, solid and reliable, as foundations are, but also what is indescribable, ineffable, awe-inspiring and beyond reach, as are the mysteries of a solemn ritual. In doing so it re-enchants the realm of matter. What is deep has abundance, fertility, much packed within, like a great poem or piece of music, though what that is cannot be known in full. It reaches not just beyond our viewpoint, but beyond what we can grasp. To believe that all is within our grasp, is surely a failure of imagination, one that, I contend, leads in turn to a failure to apprehend the deeper meaningfulness of existence.

The question of the nature of depth, like that of the nature of truth, has a peculiar urgency for us now. Is there, in fact, anything left in our world that we would be prepared to describe as not just clever or innovative, but deep? I resonate with Paul Tillich's judgment: 'The decisive element in the predicament of Western man in our period is his loss of the dimension of depth.'[23]

A number of critiques of the impact of technology on the Western mind have appeared in recent years: Nicholas Carr's account of 'What the Internet Is Doing to Our Brains' bears the accusation in its title, *The Shallows*. Similar admonitions have been published, among them neuroscientist Susan Greenfield's *Mind Change*, subtitled 'How Digital Technologies Are Leaving Their Mark On Our Brains'. The clever have scoffed and cavilled; the wise have paid (not uncritical) heed. Since the brain is *always* changed by experience, it would indeed be a miracle if something as powerful and unprecedented as digital technology did not leave a substantial mark on our brains. In fact it could not possibly be otherwise. It is therefore not only reasonable, but of vital importance, to ask what the nature of that imprint might be. Some of it might be positive, but not all is likely to be: and it would be worth noticing what is happening, very fast, to human beings. We might be modelling single, selected and simple aspects of our being (however complex-looking in IT terms), and projecting them outwards into the environment so that the feedback loop reinforces in ourselves the impoverished model of ourselves we externalise in computers. And don't forget that the further into this process we allow ourselves to be drawn, the less we can see any more what it is that we are missing.

I believe that we are indeed losing depth, but because of a whole change of heart and mind that has come over us in the last 200 years or so, only part of which can be said to be caused by technology, and of which technology is at least as much an effect as a cause. Technology is one expression of the desire for power and control over the world, which is of course the primary motivation of the left hemisphere, in which it repudiates the right hemisphere on which we rely for our sense of depth in every sense of the word.

THE PRIMACY OF RELATIONSHIP

In the presence of depth we enter into a relationship with what we see, one that embodies nonetheless, as one might expect, a tension between opposites. This is true of depth in emotion, depth in time, and depth in space. Take the example of spatial depth: on the one hand, unlike a flat, two-dimensional screen which rebuffs our approach and draws attention to itself as an object of observation, depth draws us into the possibility of connexion. On the other hand, and by the very same token, it makes us aware of separation. This together-yet-apartness operates also in the 'depth' of time, where, through memory, we are one with, and yet, by definition, sundered from, whatever it is we remember. And so it is with emotional depth: we are one with, and *not* one with, the object of our feelings. Depth is, in other words, the *union* of differentiation and union. An element of uncertainty, of unknowability, of what is other, is always present in depth: and that, too, is an important indicator of its being real.

One of the most important words for space in the Japanese language is *ma*, a word that has no exact English translation. It is, as far as I believe I understand it, something similar to what I would call 'betweenness'. It refers to both time and space. It suggests the idea of an interval, an interstice; and yet a span, and a relationship. It does not refer to an objective space, but to a sense of the space in the consciousness of an observer. Without *ma*, nothing can fulfil itself. It also, like *śūnyatā*, suggests the generative potential of space, a potential place where something can come into being: '*ma* is filled with nothing but energy and feeling.'²⁴ It is written with the ideograms for a door and the sun, so might be thought of as the crack

24 Canning [undated].

through which the light gets in. Of it, the influential graphic designer Alan Fletcher writes:

> Mallarmé conceived poems with absences as well as words. Ralph Richardson asserted that acting lay in pauses ... Isaac Stern described music as 'that little bit between each note – silences which give the form'. Franz Kafka warned that 'the Sirens have a still more fatal weapon than their song, namely their silence ... Someone might possibly have escaped from their singing; but from their silence, certainly never'... The Japanese have a word (*ma*) for this interval which gives shape to the whole. In the West we have neither word nor term. A serious omission.[25]

In Japanese the word for a fool, *manuke*, means someone lacking in *ma*, which, given that we Westerners haven't got *ma* at all, ought to bring us up short. *Ma* is the depth that is expressed by reticence, silence, what is *not* said; it is a space that is the opposite of empty. This is reminiscent of the lines from the *Tao Te Ching*:

> Thirty spokes meet in the hub,
> but the empty space between them
> is the essence of the wheel.
>
> Pots are formed from clay,
> but the empty space within it
> is the essence of the pot.
>
> Walls with windows and doors form the house,
> but the empty space within it
> is the essence of the home.[26]

The ways in which distinct (but never wholly separate) entities are related are more important than the *relata*, the things so related. That applies whether the relation be one of conjunction or disjunction. Indeed, though this may sound paradoxical at first, I believe the relations are primary, and form the bedrock of our experience, out of which emerge, secondarily, the elements that we *retrospectively* see as 'things related'. When I say retrospectively, I do not mean that we have actively to reflect on experience at some future point to see this, but that it comes about through the left hemisphere's re-presentational gaze. Here I am again with the mathematician Poincaré, who, while stressing that science must tell us *something* of reality, insisted that 'the things themselves are not what it can reach, as the naïve dogmatists think, but only the relations between things. Outside of these relations *there is no knowable reality*.'[27]

One way of making sense of this is that relationships must be

---

25 Fletcher 2001 (369).

26 *Tao Te Ching*, §11 (translation untraced).

27 Poincaré 1913a (28; emphasis added).

primary, since entities *become what they are only through their situation in the context of multiple relations*. All entities are essentially interconnected, changing, flowing, ungraspable: their thingness is an emergent property of the field. 'It isn't the atoms and molecules that are at the hard core of reality', writes biophysicist Don Mikulecky; 'it is the relations between them and the relations between them and things called processes which are at the core of the real world.'[28]

As I was nearly finished writing this book, which was originally to have been called *There Are No Things* (a title which I eventually rejected because of its unfortunate suggestion that I might subscribe to a nihilistic brand of postmodernism), a physicist interested in my work sent me a piece by his colleague Piet Hut, Professor of Astrophysics at the Institute for Advanced Study, Princeton, called: 'There are no things'. Most strikingly it expresses in terms of quantum fields my long-held philosophical belief that relations are prior to *relata*, one for which I had despaired of finding a sympathetic hearing. Hut quotes his colleague David Mermin's maxim: 'correlations have physical reality; that which they correlate does not.'[29] He comments: 'the verbs are verbing all by themselves without a need to introduce nouns ... properties are all there is. Indeed: there are no things.'[30] And Mermin names Niels Bohr, Lee Smolin, Carlo Rovelli and Gyula Bene as adopting a similar position.[31] One consequence of this is that, according to Mermin, 'if physical reality consists only of correlations, nothing physically real ever changes *discontinuously*.'[32]

SPACE AND THE HEMISPHERES

As in the case of time, the right hemisphere has the advantage in comprehending depth in space.[33] I have reviewed the aspects of visual perception at which the right hemisphere is superior (see pp 107ff), and many of the remarkable spatial distortions that ensue on right hemisphere injury earlier (see pp 118ff), and so I will keep here to a few essentials.

In general, spatial organisation is right hemisphere-based, specifically involving the right hippocampus.[34] EEG activity in the right neocortex suggests it plays a special role in spatial navigation.[35] 'Left-lateralized spatial memory strategies, relying on serial order, are not efficient if not accompanied by right-brain spatial functions': note that the left hemisphere produces, as with time, 'serial order' of points, places or slices, so that the 'right brain is crucial for mental rotations, necessary for spatial updating'.[36] Spatial integration across sensory modalities, large-scale spatial orientation, spatial memory and navigation skills are all preponderantly reliant on the right, not left, temporoparietal cortex.[37] Subjects with right hemisphere damage, with or without neglect, do very poorly at following a map, which involves converting abstract information into an

---

28 Mikulecky 2000 (419).

29 Mermin 1998.

30 Hut [undated].

31 Mermin 1998.

32 *ibid*.

33 Durnford & Kimura 1971; Nikolaenko & Egorov 1998.

34 Maguire, Gadian, Johnsrude et al 2000; Abrahams, Pickering, Polkey et al 1997; Burgess, Maguire & O'Keefe 2002; Iglói, Doeller, Berthoz et al 2010; Maguire, Frackowiak & Frith 1997; Bohbot, Jech, Růžička et al 2002; Squire, Ojemann, Miezin et al 1992.

35 Specifically, gamma oscillations: Jacobs, Korolev, Caplan et al 2010.

36 Belmonti, Berthoz, Cioni et al 2015.

37 Dieterich & Brandt 2018.

---

38 Palermo, Ranieri, Boccia et al 2012.

39 Brunyé, Holmes, Cantelon et al 2014.

40 Banich & Federmeier 1999; Hellige & Michimata 1989; Kosslyn, Koenig, Barrett et al 1989; Laeng 1994; Rybash & Hoyer 1992; Baumann, Chan & Mattingley 2012.

41 Kosslyn 1987; Kinsbourne 1993 (6).

42 For space, see eg, pp 81–7; for time, see eg, pp 75–81; and for emotion, see eg, pp 87–9 & 197ff. See also McGilchrist 2009b (60, 64, 77–9, 181–3, 447); and, eg, Liotti & Tucker 1994; and Ross, Homan & Buck 1994.

43 The case of a Mexican artist has been reported in detail (Orjuela-Rojas, Sosa-Ortiz, Díaz-Victoria et al 2017). After resection of a neurocytoma that affected the right thalamo-parietal connexions, he suffered an impairment of the ability to create

[note continues on facing page] ▸

understanding of the experienced three-dimensional world: those with left hemisphere damage, by contrast, reveal no deficit.[38] Navigation efficiency and spatial memory can be significantly improved by stimulating the right, but not left, temporal region.[39]

A distinction has repeatedly been made between the right hemisphere's capacity to identify the location of something in the depth of space, its 'co-ordinates', so to speak, and the left hemisphere's strategy of categorising something as 'above' or 'below' something else.[40] The right hemisphere's organisation of space depends more on this sense of where things *are*, nearer or further 'from me' – on depth, in other words.[41]

The left hemisphere has a problem dealing with depth, whether that be depth in space, in time or in emotion.[42] And, as far as spatial transformations go, it is not just that three-dimensional depth is lost and objects appear flattened; they also become more distanced, more generalised, more stylised, less fluid, more symbolic and more geometric. This strange visuo-spatial world is sometimes dramatically obvious after a right hemisphere stroke. The sense of overall shape, the *Gestalt*, may be lost, reduced to an aggregate of details without form; and with that the vital flow is lost. In general, drawings by the isolated left hemisphere show important characteristics of a virtual, rather than a real world image, or, as we might now say, an 'icon', rather than what that icon stands for in the lived world. They become 'views from nowhere'.[43]

In keeping with my hypothesis that the left hemisphere is orientated to a theoretical, rather than experiential, world view, the processing of nonsensical and non-real-world images activates preferentially the left hemisphere, unlike most other imagery.[44] The left hemisphere is, when compared with the right, at home dealing with distorted, non-realistic, fantastic – ultimately artificial – images.[45] This may be because they invite analysis by parts, rather than as a whole. But there are independent hints (no more than that) that the left hemisphere may have a positive bias towards whatever is bizarre or non-existent.[46]

Our brains have a remarkable early facility for negotiating space. Elizabeth Spelke, along with her colleagues, introduced blind, or blindfolded, toddlers into a room with objects in four locations. They had the children walk between them on a specific path, then asked them to use another path to move one object to another – putting a toy on a chair, for example. The children proved strikingly adept at the task.[47] Our spatial sense of the world is both deep and intuitive, and very largely right hemisphere-dependent.[48]

It is only later that we develop abstract ways of representing space. The mathematician Henri Poincaré wrote:

perspective in his paintings and omitted shapes only on the left side of his paintings. Before the procedure, his paintings evinced 'a play of light and shadow': he achieved the representation of a complex volume using different depths of shading, and suggesting a light source coming from the right side of the picture. For example, in a landscape painting 'each shape is associated with a shadow that depends on the position of the light source. This is the case with the rock in the foreground, the house, and the mountains, creating not only three dimensionality but also giving credibility to the painting … [After the procedure there was] no efficient perspective, the brush stroke is loose and imprecise, there are no figurative details (birds, trees, flowers, clouds) in the left space, but distribution of colour throughout the work remains. There is no strong light source; thus, the figures, including the tree that casts no shadow, lack volume … The origin of light is ambiguous throughout the image, constantly *changing in an arbitrary way*' (emphasis added). This is the difference between a real, unique, embodied view and a 'view from nowhere', which, as discussed both in this book and *The Master and his Emissary*, is remarkably like the view taken, both literally and metaphorically, by the left hemisphere.

44 Yomogida, Sugiura, Watanabe et al 2004.

45 Laeng, Shah & Kosslyn 1999; Zaidel & Kasher 1989.

46 Cutting 1997 (162–3); Wexler 1986.

47 Landau, Gleitman & Spelke 1981.

48 Chiron, Jambaqué, Nabbout et al 1997.

Another frame which we impose on the world is space ... the space our senses could show us differs absolutely from that of the geometer ... We therefore conclude that the first principles of geometry are only conventions; but these conventions are not arbitrary ...[49]

The mapping and measuring of space, as Poincaré points out, though different entirely from our experience of space, is by no means arbitrary, otherwise it would scarcely be of any use. Indeed its very plausibility is why it can be deceitful: like spatialised time, it is easy for us to be taken in by it. But, as always, its goal is utility, not fidelity. It provides the map, rather than access to the territory. Colin McGinn warns us that

> what we need from space, practically speaking, is by no means the same as how space is structured in itself. I suspect that the very depth of embeddedness of space in our cognitive system produces in us the illusion that we understand it much better than we do.[50]

SCHIZOPHRENIA AND THE APPRECIATION OF SPACE

In Chapter 9 I spoke about the spatial distortions that commonly occur in schizophrenia, and which closely parallel those found in right hemisphere dysfunction. One patient describes the external world as 'like a two-dimensional transparency, something like an architect's drawing or plan'.[51] The loss of spatial depth spontaneously reported by schizophrenic subjects is corroborated by laboratory research.[52] The following quotation from one of Chapman's schizophrenic patients brings out the relationship between three elements, each associated with right hemisphere deficit states – lack of depth, piecemeal composition of both space and motion, and impaired judgment of distance:

> I see things flat ... It's as if there were a wall there and I would walk into it. There's no depth, but if I take time to look at things I can pick out the pieces like a jigsaw puzzle ... Moving is like a motion picture ... If you run you receive the signals at a faster rate. The picture I see is literally made up of hundreds of pieces. Until I see into things I don't know what distance they are away.[53]

Merleau-Ponty pointed to the fact that only something that has depth can have different facets, or aspects in the true sense. Without depth all it has is *parts* or *pieces*.[54]

The reducing of space to something without depth is expressed by one schizophrenic subject: 'Space oppresses me – for me it is no longer there'. He describes his experience in such a way as to compare what has happened to space with what has happened to time: 'the lack of the feeling for space, the *pokiness of the moment*'.[55] For

---

49 Poincaré 1913a (29).

50 McGinn 1995 (230).

51 Sass 2017 (270).

52 Kantrowitz, Butler, Schechter et al 2009; Schechter, Butler, Jalbrzikowski et al 2006; Phillipson & Harris 1985.

53 Chapman 1966 (230).

54 Murata 1998 (298), referring to Merleau-Ponty 1969.

55 Fischer 1930a (244–5): » *Der Raum bedrängt mich. Er ist für mich nicht da ... Das mangelnde ‚Raumgefühl', die Engigkeit der Augenblicke* « (emphasis added).

Tatossian's patient Hélène Jacob, there is a collapse of space, of the near and the far, both being equal.[56] The world is like a canvas, and only things painted on it can lend a degree of reality to space, since 'space that is materially empty is totally empty, empty of being. Thus to her the sea seems like a sort of hole in the "canvas of the world".'[57]

The sea – the *deep* – has become merely a hole, a nothing.

The effect of non-differentiation – near is far and far is near, small is large and large is small – is that 'no space' can equally be read as 'all space'. These veerings between extremes come from the loss of betweenness, a loss in which everything and nothing, omnipotence and impotence, coincide.[58] It is only relationships that can give real meaning to these terms. Jaspers notes the frequency of descriptions of infinite space by schizophrenic subjects, quoting one who said:

> Space seemed to stretch and go on into infinity, completely empty. I felt lost, abandoned to the infinities of space … it seemed the complement of my own emptiness … the old physical space seemed to be apart from this space, like a phantom.[59]

The 'old physical space' was something quite other than this space, which is 'completely empty'. Another patient says:

> In my mind's eye I thought I saw below the pale blue evening sky a black sky of horrible intensity. Everything became limitless, engulfing … I knew that the autumn landscape was pervaded by a second space, so fine, so invisible, though it was dark, empty and ghastly. Sometimes one space seemed to move, sometimes both got mixed up …'[60]

Whether limitless or empty, this space lacks depth.

The collapse of spatial planes destroys normal boundaries. A patient complains:

> When I look in a mirror, I no longer know if I'm seeing myself there in the mirror, or if it's that I'm in the mirror, seeing myself here. I'm standing between two mirrors, so that an endless series of 'seeing myself' results, which bewilders me. [NB: here as elsewhere lack of boundaries leads to an infinite series.] If I see another person in the mirror, I'm incapable of distinguishing him from myself. Or worse still, the very difference between myself and another real person is lost. When I'm watching TV, I no longer know if it's me speaking, there on the box, or I'm the one here, hearing what's spoken. I no longer know if the inside is turned out, or the outside turned in. It's as if the foundations of my being are destroyed. Might there not be two of me?[61]

Another abnormal aspect of schizophrenic space perception is spatial dislocation. It has been shown that auditory information about the spatial surroundings is also highly right hemisphere-dependent.[62] And so it is not surprising that a schizophrenic patient reports:

---

56 Tatossian 2014a (78).

57 ibid (76).

58 See p 367 above.

59 Jaspers 1963 (81).

60 ibid.

61 Kimura 1994 (194): » Wenn ich in einen Spiegel sehe, weiss ich nicht mehr, ob ich hier mich dort im Spiegel sehe oder ich dort im Spiegel mich hier sehe. Stehe ich zwischen zwei Spiegeln, dann entsteht eine unendliche Kette des Mich-selbst-sehens, was mich verwirrt. Sehe ich einen anderen im Spiegel, so vermag ich ihn nicht mehr von mir zu unterscheiden. In einem schlechteren Befinden geht auch der Unterschied zwischen mir selbst und einem wirklichen anderen verloren. Im Fernsehen weiss ich nicht mehr, ob ich dort im Fernsehapparat spreche oder das Gesprochene hier höre. Ich weiss nicht, ob sich das Innere nach aussen kehrt oder das Äussere nach innen. Mir kommt es vor, wie wenn der Boden meines Seins untergeht. Ob es nicht zwei Ichs gibt? «

62 Dietz, Friston, Mattingley et al 2014.

I have noticed a lot recently that I seem to get a little mixed up about where sounds are coming from. Often I have to check up if someone speaks to me and several times I thought someone was shouting through the window when it was really the wireless at the front of the house.[63]

And another:

I've had difficulty in tracing where sounds are coming from although I am not deaf. If the wireless is on, for example, I know the wireless is there but sometimes I feel that the sounds are coming from behind my back.[64]

Thus it is that perceptual distortions in space feed paranoid beliefs, and paranoid states lead to perceptual distortions of space. Schizophrenia, as we have seen, as well as autism, obsessive-compulsive disorder and anorexia, are all characterised by an essential preference for stasis and sameness over motion and change.

Morbid symmetry

Before leaving schizophrenia and what it might tell us about right hemisphere dysfunction, a word about an important concept that we would normally think of as purely spatial, but has another meaning in physics: symmetry. In mathematics and physics, the term refers not just to spatial symmetry about an axis, but to any procedure which one can perform on an object and leave it unchanged. It also signifies independence from contingency – in other words, universality: if a law obeys symmetry, it is universally applicable. All these meanings ally it with the realm of stasis, of universals, of simple, ideal forms: the left hemisphere. Newtonian mechanics are conventionally said to obey symmetry. The word means 'equal measure' (Greek *sym-* + *metron*) and, outside of physics and maths, equal measure can mean two different things: identity (that is to say, unison), simplicity, predictability; or harmony, the coming together of disparate elements in a new balanced whole.

As schizophrenic subjects are attracted to immobility, so they can become obsessed with symmetry. I have spoken about the 'morbid geometrism' displayed by many schizophrenic subjects.[65] One aspect of this is an excessive need for symmetry. Schizophrenics deal with regular and symmetrical stimuli more easily than ones with complex or asymmetrical patterns: it is the meaning of symmetry as simplicity, predictability, not that of harmony of disparate elements, that is alluring here.[66] Minkowski writes of one schizophrenic patient that a 'mania for symmetry' took hold of him.[67] In Chapter 9 we heard of another of Minkowski's patients, that 'life has no regularity or symmetry, and for that reason I manufacture my own reality.'[68] It may

---

63 McGhie & Chapman 1961 (105).

64 *ibid.*

65 See p 352 above.

66 Knight, Manoach, Elliott *et al* 2000; Silverstein, Bakshi, Chapman *et al* 1998.

67 Minkowski 1987 (207).

68 *ibid* (208).

be for similar reasons that autistic subjects, obsessive-compulsive subjects, and subjects with anorexia nervosa (a condition which has some overlapping symptoms with each of the other syndromes) also have an abnormally elevated preference for symmetrical patterns,[69] which is thought to be because such patterns are more readily graspable, and involve less complex transitioning between part and whole. Strikingly one sufferer from anorexia nervosa refers to 'the horrid soft place of inner *asymmetry* for which there are many vile words: gut, viscera, abdomen'.[70] It is of particular interest that, when obsessive-compulsive and anorexic patients are stressed by being shown pictures of messy rooms or asymmetrical arrays, areas of the right hemisphere that become active in normal subjects are hypoactive in patients.[71]

In my discussing of time and schizophrenia, I emphasised that changes in the depth perception of time affect the experienced world as a whole: the consequences go well beyond the obviously temporal. Similarly when there are changes in the depth perception of space, the consequences affect the quality of experience as a whole and in ways one might not immediately have anticipated.

LOSS OF SPATIAL DEPTH BRINGS LOSS OF LIFE

Insight into the effects associated with altered depth perception comes from a series of experiments in which it was suggested to hypnotic subjects that they experienced either intensification or attenuation of depth. The trials were carried out in young men, verified hypnotic-suggestive subjects, who were free from mental illness. The advantage of this kind of study is that it is naturalistic and the resulting paper reports the actual experiences of the subjects, often verbatim.

When primed that the dimension of depth was 'gone', one subject 'at once showed marked schizophreniform behaviour with catatonic features.'[72] Another reported that 'colours, shapes, and sounds all seemed less intense. He reported a loss of sensitivity to touch. He became bored, withdrawn, and hostile … [and] felt that his environment had become alien and the people around him dehumanized.'[73] The experimenters comment that 'the *no depth* condition seems accompanied by a general dulling of perceptual experience. The crucial variable here in determining the schizophreniform response seems to be an increased insubstantiality [lack of *solidity*] of all objects in the environment, including the self.'[74]

By contrast,

> when the dimension of depth was expanded, a psychedelic state resulted similar to that described by Huxley in *The Doors of Perception*. Lines seemed sharper, colors intensified, everything seemed to have a place and to be in its place, and to be aesthetically satisfying. The

---

69 AUTISM: Perreault, Gurnsey, Dawson *et al* 2011; OCD: Evans, Orr, Lazar *et al* 2012; Mataix-Cols, Nakatani, Micali *et al* 2008; Silverstein, Matteson & Knight 1996; AN: Matsunaga, Kiriike, Iwasaki *et al* 1999; Matsunaga, Miyata, Iwasaki *et al* 1999.

70 Private communication from the patient.

71 RIGHT THALAMUS AND RIGHT INSULA IN OCD: Gilbert, Akkal, Almeida *et al* 2009; RIGHT PARIETAL LOBE AND RIGHT PREFRONTAL CORTEX IN AN: Suda, Brooks, Giampietro *et al* 2014.

72 Aaronson 1967 (247).

73 ibid (249–50).

74 ibid (251).

hand of God was manifest in an ordered world ... Colours seemed intensified, lines more distinct, and sounds crisper.[75]

One subject, primed with the suggestion of expanded depth, reported that 'riding in a car was like taking a wonderfully exhilarating roller coaster ride to everywhere. The landscape was at once a gargantuan formal garden and a wilderness of irrepressible joyous space. Even now, I feel dumb-struck and preposterous in trying to describe this perceptual miracle which has somehow been given me.'[76] The experimenters report that 'the usual perception of objects in the environment as things in themselves, independent of their surroundings, seems replaced by a perception of objects as being in *interaction with* their surroundings and with the *active* properties of the space around them.'[77] In other words the world showed itself to be resonant.

The 'lack of depth' condition, then, produces schizophreniform features in a normal subject: hostility, a sense of alienation, and boredom, coupled with a dulling of perception, and a sense of reality as insubstantial. Meanwhile the 'increased depth' condition produces a sense of oneness with things, vividness of perception, a sense of awe, and, a sense, not of isolated objects, but of things 'in interaction with one another and their surroundings'. We know that when the right hemisphere is suppressed, so is the sense of depth; here, when the sense of depth is suppressed, it is as if so were the right hemisphere.

MODERNISM, DEPTH AND THE LEFT HEMISPHERE

I will be coming to the physics of space very shortly, but following the discussion of schizophrenia and space it's worth considering briefly the parallels once again between schizophrenia and modernism.[78]

In his essay on 'The thing', Heidegger writes of modernity, again confirming its similarities to schizophrenia:

> What is happening here when, as a result of the abolition of great distances, everything is equally far and equally near? What is this uniformity in which everything is neither far nor near – is, as it were, without distance? Everything gets lumped together into uniform distancelessness. How? Is not this merging of everything into the distanceless more unearthly than everything bursting apart?[79]

Although we are taught that perspective in painting was first adopted by Giotto, and first demonstrated by Brunelleschi in 1415, it was known to both the Greeks and Romans, and that knowledge was subsequently lost in the Dark Ages and had to be rediscovered.[80] Harmony in music is the equivalent aurally of spatial depth: it is interesting that both harmony and perspective came in with the early Renaissance and departed in the twentieth century.

---

75 ibid (247 & 249).

76 ibid (249).

77 ibid (251; emphasis added).

78 See Chapter 9.

79 Heidegger 1975c (164).

80 Gombrich 1960 (112). According to Vitruvius, Agatharchus wrote a treatise on perspective in the fifth century BC: see Vitruvius, 1914, intro to Bk VII (198).

Modernist painting is notable for its insistence on the flatness of the canvas. Clement Greenberg described the flatness of modern painting as the goal of its progression. He argued that the essential and unique element in Modernist painting is its flatness: 'the Cubist counter-revolution eventuated in a kind of painting flatter than anything in Western art since before Giotto and Cimabue – so flat indeed that it could hardly contain recognizable images.'[81]

Lack of depth is a key to understanding modernism more generally. In a strange way, the art, literature and thought of the period takes a perverse pride in the avoidance of depth, proffering in its place a knowing celebration of shallowness. Sophisticated as she is, Elfriede Jelinek was nonetheless expressing, whether she was aware of it or not, a problem at the core of modern culture when she published her manifesto, 'I want to be shallow'. In theatre, the effect of reality is achieved through great artifice: without it, its productions seem merely artificial. 'The machinery, in other words, is hidden', as Jelinek writes. The actor produces 'nuances of expression'. By contrast, Jelinek continues,

> I prefer not to have anything on stage that smacks of this sacred bringing to life of something divine ... nobody should ever be able to say of them that something quite different is going on inside of them, something that one can read only indirectly on their faces or their bodies ... Perhaps I just want to exhibit activities which one can perform as a presentation of something, but without any higher meaning ... A fashion show, because on that occasion one could also send out the clothes by themselves. Get rid of human beings who could fabricate a systematic relationship to some invented character! Like clothes, you hear me? ... But now to our collaborators: How can we remove these dirty marks, these actors, from the theatre, so that they won't pour themselves from their zip-lock packages all over us? I mean: that they won't overwhelm us with their fluids! ... Let's simply remove them from the inventory of our life! Let's flatten them out into celluloid! ... They drop out of our body perception and turn into *surfaces* that move before our eyes ...[82]

The fastidious turning away from the messiness of bodily existence, and the inebriated embracing of the abstract left hemisphere representational world, flattened, as Jelinek puts it, into celluloid, makes her manifesto the quintessence of a certain aspect of modernism. In its sophisticated way, modernism, only half mockingly, embraces an aspect of popular modern culture: its unabashed superficiality.

All is not what it seems here, though. One driving force behind the emphasis on surfaces has, since Nietzsche, been an *affirmation* of phenomenal experience, as against the view that phenomena merely obscure an ideal Kantian 'reality' hidden behind them. The

[81] Greenberg 1960.

[82] Jelinek 1983 (102; emphasis added).

message seems to be, what you see *is* what you get, and not another thing. And yet any resultant cultivation of superficiality seems no more revealing, indeed less so, than what it replaces.

The problem here is the fruitless switching between the two parts of a false dichotomy. The structure of this problem is similar to that of the subjective/objective dilemma. The distinction we should make in contemplating that dilemma is not between subjective, on the one hand, and objective, on the other, but between the view that subjective and objective are two opposed entities, on the one hand, and the view that they form a continuous dipole (like the magnet), on the other. They are distinct, but not at war; not discordant, but concordant. The more subjective view is not something we must (even if we could) abandon, but by means of which we are enabled to see beyond it to what we call a more objective view. In the case before us here, the distinction is not between the surface, on the one hand, and the depth, on the other, but between the view that artificially separates surface from depth, on the one hand (seeing surface as opaque and obscuring a reality that lies beyond), and that which takes the surface as what one sees *into* in order to find the depth, lying just where it is, in full view: there is no discernible *line* between surface and depth.

This latter view is what I call semi-transparency, or translucency, and in brief I believe it helps clarify many mistaken dichotomies. Thus we see the infinite not by turning away from the finite, but by looking *into* it: a looking-into in which the finite is not just a means to the ends of infinitude – hence I say *semi*-transparency, because the finite is precious in its own right, and worthy of the eye's delay as it passes through.

We see the general not by turning away from the particular, but by looking intently at it so as to see *into* it, whereby the value of the particular is not in any way negated, but taken up (*aufgehoben*) into something greater beyond. Similarly, I suggest, we find the soul not by turning away from the body, but by embracing it in a way that spiritualises the body; and we find the sacred not by turning away from the world, but by embracing it, in a move that sanctifies matter. The soul is both in and transcends the body, as a poem is in and yet transcends mere language, a melody in, yet transcends, mere sound, a painting in, yet transcends, the merely frescoed wall.[83]

The left hemisphere is 'either/or' in its persuasion, the right hemisphere 'both/and'. Thus the apparently paradoxical way in which what language and the left hemisphere set up as opposites nonetheless cohere and enrich one another is something the right hemisphere is more capable of accepting. Which in turn is why we need the take of both hemispheres, but for the right hemisphere to be the master of both.

---

83 Cf Octavio Paz, from 'The Consecration of the Instant': 'What characterizes the poem is its necessary dependence on the word as much as its struggle to transcend it.' Quoted in Zwicky 2014a, §108b.

Apart from loss of depth, there are a range of other stylistic features of twentieth-century art that mimic right hemisphere dysfunction. I wrote about this in *The Master and his Emissary*, and referred to cubism, pointillism, surrealism and collage techniques, amongst others.[84] I therefore recognised this in a recent examination of spatial distortions following right hemisphere damage:

> Interestingly, the distortions created by painters of the Cubism period, characterized by an asymmetry of objects and body representations, a specific enlargement or reduction of parts of space, or even by complex distortions of 3D space are analogous to those classically reported in right-brain-damaged patients (unilateral spatial neglect, hyperschematia, constructional apraxia).[85]

[84] McGilchrist 2009b (414–6).

[85] Rode, Vallar, Chabanat *et al* 2018.

ASPECTS OF THE PHYSICS OF SPACE

Here I want to shift from the phenomenology of space, and its metaphysics, to the physics. Physics and metaphysics are always implied in one another.

I have no pretensions to understand the complexities of modern physics. Reassuringly, even those – perhaps particularly those – who, unlike me, have spent years studying them make clear that neither they nor anyone else really understand the various paradoxical models that contend for a place in our understanding of the fundamentals of the physical world, and that a belief that you do understand them is more or less a guarantee that you don't. But where they say that from experience, I say it out of sheer ignorance. In what follows, I have therefore depended on the more accessible accounts given by a number of physicists, among them David Tong, Michael Green, Carlo Rovelli, Henry Stapp, Jim Al-Khalili, Brian Greene, Rodney Brooks and Marcelo Gleiser. It goes without saying that any misunderstandings are my own.

It might seem better not to risk saying anything, but I think physics and metaphysics cannot be treated wholly separately without grave losses on both sides – and what are we non-physicists supposed to do about that? Importantly I do not claim to *found* anything on physics: I simply observe occasions on which physics seems to me to have reached conclusions similar to those reached independently from a philosophical and neurological standpoint. It seems to me likely that neurology, philosophy and physics should all be approaching a similarly structured reality, albeit from different paths, and therefore seeing different aspects of the same whole.

So, though I don't mind in the least being told I am mistaken, I will carry on and do my best to summarise very briefly why I think that the discussions in neurology and philosophy that have flowed from the hemisphere hypothesis and filled this book so far, can illuminate, and in turn be illuminated by, certain issues in physics. These issues are:

1. To what extent are the physical foundations of reality knowable?
2. If so, what do they look like? In particular, are they discrete or continuous?
3. What clues do we have about the nature of mass and matter?
4. Is asymmetry built into the foundations of reality?

1 · *Are the physical foundations of reality knowable?*

Physics is Pragmatist in its approach to truth. It does not claim to describe reality, but to offer models that work. To quote Bohr: 'There is no quantum world. There is only an abstract quantum physical description. It is wrong to think that the task of physics is to find out how nature is. Physics concerns what we can say about nature.'[86] The models enable us to make predictions with greater or lesser degrees of accuracy. The elements in the model are not observable directly, but are implicit in the effects that we observe within the model.

We can measure no aspect of a particle without in some fashion interacting with it, and that interaction changes the particle's nature. This becomes dramatically obvious at the quantum level: for anything to be literally seen requires the object of our interest to be struck by a photon. That collision alters the system that is observed. With particles smaller than can be seen by light, we may use electromagnetic waves with much smaller wavelength, but such waves have inevitably greater energy, and interfering with a particle with a high energy beam changes the particle more radically. The German theoretical physicist Pascual Jordan went so far as to say: 'Observations not only *disturb* what has to be measured, they *produce* it!' Of the electron, he continued, 'we compel it *to assume a definite position* ... We ourselves produce the results of measurement.'[87] This is the (by now well-known) observer effect: that by the mere fact of observing a phenomenon we cannot but change that phenomenon, or even, as Jordan suggests, bring it about. Observation, it is thought, collapses what behaves like a wave into what behaves like a particle.

As to precise knowledge, a number of factors operate. The more closely one pins down one measure (such as the position of a particle), the less precise another measurement pertaining to the same particle (such as its momentum) must become – Heisenberg's Uncertainty Principle. It is intrinsically impossible to know the values of all of the properties of the system at the same time. Further, elements of matter are neither isolated nor certain, but interconnected and probabilistic. David Tong, Professor of Theoretical Physics at Cambridge, draws attention to a further point:

---

86 As quoted in Petersen 1963 (12).

87 As quoted in Jammer 1974 (161; emphasis in original). Cf Smolin 2013 (81): 'We can manipulate quantum particles in experiments and talk about how they respond to being measured. But we cannot visualise what goes on in the absence of our manipulation of nature.'

Physical quantities are not integers but real numbers – continuous numbers, with an infinite number of digits after the decimal point. The known laws of physics, *Matrix* fans will be disappointed to learn, have features that no one knows how to simulate on a computer, no matter how many bytes its memory has. Appreciating this aspect of these laws is essential to developing a fully unified theory of physics.[88]

He points to the fact that 40 years of efforts to simulate the Standard Model on a computer have so far failed. 'To perform such a simulation, one must first take equations expressed in terms of *continuous* quantities and find a *discrete* formulation that is compatible with the bits of information in which computers trade ... It remains one of the most important, yet rarely mentioned, open problems in theoretical physics.'[89] This suggests not necessarily that the simulation could never be made, but more fascinatingly, that even if it could, it would involve translating the inherently continuous nature of reality into the inherently discrete nature of machine processes – a problem which might or might not prove intractable.

And he continues:

> Scientists are not entirely sure what to make of our inability to simulate the Standard Model on a computer. It is difficult to draw strong conclusions from a failure to solve a problem; quite possibly the puzzle is just a very difficult one waiting to be solved with conventional techniques. But aspects of the problem smell deeper than that. The obstacles involved are intimately tied to mathematics of topology and geometry ... [it may be] that the laws of physics are not, at heart, discrete. We are not living inside a computer simulation.

In the attempt to describe quantum reality, ordinary language simply breaks down. As Bohr put it: 'We must be clear that, when it comes to atoms, language can be used only as in poetry. The poet, too, is not nearly so concerned with describing facts as with *creating images* and *establishing mental connections*.'[90] Another twentieth-century physicist (and, incidentally, a logical positivist), Philipp Frank, was of the opinion that 'even the statements of Newtonian physics cannot really be formulated in common-sense language, but in the relativity and quantum theories the impossibility becomes obvious.'[91]

If reality is such that our knowledge is intrinsically, not accidentally, *incomplete*; if it is intrinsically, not accidentally, *uncertain*; if it is intrinsically inexpressible in everyday language, requiring exceptional, non-denotative, highly metaphoric, 'poetic' use of language to get beyond the limits of language; if we must deal not with facts but with connexions; if entities are never wholly separable from other entities; if the process of a knower coming to know is interactive or

88 Tong 2012.

89 Tong *op cit.*

90 Conversation recorded in Heisenberg 1971, '"Understanding" in modern physics': 27–42 (41; emphasis added). It is widely cited incorrectly as from 'Discussions about language (1933)'.

91 Frank 1958 (60).

reverberative, each changing the other – not distanced, inert and owing nothing to the presence, and possibly the nature, of the one who comes to know; and if any attempt to model it reduces what is continuous and moving, to what is static and discrete – if all of this is true, it is clear which hemisphere will be better suited to discerning it. Once more the right hemisphere's take seems more veridical.

2 · Fields or particles?

James's insight was that what he called a pluralistic universe 'possesses in its own right a concatenated or continuous structure'. Expanding this idea, he wrote that 'the relations between things, conjunctive as well as disjunctive, are just as much matters of direct particular experience, neither more so nor less so, than the things themselves. The generalized conclusion is that therefore the parts of experience hold together from next to next by relations that are themselves parts of experience.'[92]

A hundred years or more later, what do we know about this insight? A good deal, I submit, that would bear out both his and Bergson's view of the universe: that it is neither wholly discretised, though discretised enough to generate individuals; nor wholly continuous, though continuous enough to be ultimately unified, and that the discreteness arises secondarily out of the continuity.

'Physicists routinely teach that the building blocks of nature are discrete particles such as the electron or quark', writes Tong:

> That is a lie. The building blocks of our theories are not particles but fields: continuous, fluidlike objects spread throughout space ... The objects that we call fundamental particles are not fundamental. Instead they are ripples of continuous fields ... Deep down, the theory is not quantum. In systems such as the hydrogen atom, the processes described by the theory mould discreteness from underlying continuity.[93]

'All particles are waves in a universally distributed continuous shared field that envelops each and all of us: values in the field change with space and time,' says Tong.[94] Wave and particle are two modes of being of the same field phenomenon: this makes possible the coming together of union and division, of continuity (the wave) with discreteness (the particle) within a single uniting phenomenon (the field). Louis de Broglie proposed that the dipole of wave and particle is universal: everything that moves – which is another way of saying everything – has some aspects of a wave and some aspects of a particle.[95]

This is where it is useful to think in terms of quantum field theory (QFT). In QFT there are no particles; there are *only* fields.[96] But those fields are not the same as classical fields: they can become,

---

92 James 1911b (xii–xiii).

93 Tong 2012.

94 Tong 2016.

95 Smolin 2013 (81).

96 Joseph Needham, the great historian of science in China, comments that 'atomism in the physico-chemical sense never played any role of importance in traditional Chinese scientific thinking, which was wedded to the ideas of the continuum and action at a distance': 1969 (224).

emergently, quant*ised*. The quantum of a field is what was conventionally conceptualised as a particle. According to QFT, the continuity of the field is more foundational than the particle-like quantum: the wave-like property of a field collapses into what behaves like a particle, and thus the 'particle' is an emergent property of the field. 'I, among many others', writes Tong, 'think that reality is ultimately analogue rather than digital. In this view, the world is a true continuum. No matter how closely you zoom in, you will not find irreducible building blocks.'[97] In another context he remarked: 'At a fundamental level is nature discrete or continuous? I see no evidence whatever for discreteness. All the discreteness we see in the world is something which emerges from an underlying continuum … Quanta are emergent … they are not built into the heart of Nature.'[98]

Not everyone would agree with Tong, of course – not everyone would agree with any point of view, especially in contemporary physics. But as he points out, such a view has a long pedigree. Maxwell and Faraday were the first to think in terms of fields. He draws attention to Nobel Prize winner Wilhelm Ostwald, who, 100 years ago,

> pointed out that the laws of thermodynamics refer only to continuous quantities such as energy. Similarly, Maxwell's theory of electromagnetism describes electric and magnetic fields as continuous. Max Planck, who would later pioneer quantum mechanics, finished an influential paper in 1882 with the words: 'Despite the great success that the atomic theory has so far enjoyed, ultimately it will have to be abandoned in favour of the assumption of continuous matter.'[99]

More recently, in the 1980s, a 'celebrated theorem' proved that 'it is impossible to discretise the simplest kind of fermion … We can handle all kinds of hypothetical fermions but not the ones that actually exist.'[100] Fermions could be considered the 'building blocks of matter': electrons, protons and neutrons are all types of fermion. And yet, according to Tong, they can't be discretised.

Many hold that no theory compares with quantum field theory for the range of phenomena with which it is compatible. For example, quantum mechanics (QM) and relativity are mutually incompatible, something that famously distressed Einstein. But QFT *is* compatible with relativity, an important point in its favour. According to Rodney Brooks, QFT is

> a theory that has produced more precise agreement with experiment than anything before, that encompasses all forces and all matter, that unifies QM and relativity, that resolves Einstein's Enigmas, and that reintroduces common sense into physics … Common sense is choosing, wherever possible, the simplest, most intuitively satisfying explanation *that is consistent with our observations*. This is not the same thing as accepting intuition blindly …[101]

---

97 Tong 2012.
98 Tong 2016.
99 Tong 2012.
100 The theorem referred to is the Nielsen-Ninomiya theorem: Tong 2012.
101 Brooks 2016 (location 371).

I began this chapter with some reflections on reasons for a non-physicist to regret the conflation of space with time. In QFT, says Brooks,

> gravity is caused by a force field – not curvature, and ... contrary to popular belief, QFT is compatible with general relativity ... In QFT, space is the same three-dimensional 'Euclidean' space that we intuitively believe in, and time is the same time that we intuitively believe in.[102]

In other words QFT does not embrace the idea that spacetime is an entity, the curvature of which causes gravity. At the end of the day, spacetime may not exist. 'In physics and, more generally, in the natural sciences, space and time are the foundation of all theories', writes George Musser:

> Yet we never see spacetime directly. Rather we infer its existence from our everyday experience ... But the bottom-line lesson of quantum gravity is that not all phenomena neatly fit within spacetime. Physicists will need to find some new foundational structure ...[103]

In string theory, which, as mentioned in Chapter 11, is another popular, though not generally accepted, or even clearly defined, theory, the 'whole collection of apparently disconnected particles are in fact unified in the sense that they correspond to the many ways in which a single object, a string, can vibrate', according to Michael Green, former Lucasian Professor of Mathematics at Cambridge.[104] A string is a single entity in constant motion which, like the vibrating string of a musical instrument, yet has the potential to vibrate at different frequencies – which pleasingly unites continuity and discreteness in the very same event.

It is indisputable that there are phenomena that come in *chunks*. These are what used to be called particles. The trouble is that the term is misleading, conjuring up distinct concrete entities with known co-ordinates in space and time. Philipp Frank wrote that

> in quantum theory the term 'particle' is employed as a thing which has no precise position and velocity, and so is clearly incompatible with the full common-sense meaning of this word. I once asked Niels Bohr whether it would not be practical to eliminate the term 'particle' completely from quantum theory. Bohr agreed that one could do so in the interest of unambiguity.[105]

So, even if there are no particles as such, it is not as if continuity is everything: Tong himself argues not that there are no quanta, but that they are *emergent* from continua. Let us explore that a little further, since one of our metaphysical themes has been the need for the

---

102 Brooks *op cit* (location 2593).

103 Musser 2018.

104 Green 2014.

105 Frank 1958 (64).

union of discreteness and union – albeit with this asymmetry, that the union is prior ontologically to the discreteness, and embraces it.

Whereas QM posits a world of particles, and its equations give the probability that the particle is at a given point, QFT posits a world of fields, and its equations give the strength of the field at a given point. Frank Wilczek, a theoretical physicist and mathematician who was a Nobel Laureate, and Professor of Physics at MIT, puts it like this: 'The Core theory, which summarises our best current understanding of fundamental processes, is formulated in terms of quantum fields ... [What we take to be particles] appear as secondary consequences; they are localised disturbances in the primary entities – that is, in quantum fields.'[106] This is metaphysically familiar from James and Bergson.

Particularly fascinating from our point of view is that quanta may overlap each other, nonetheless each one maintaining its separate identity. The field quantum 'lives a life and dies a death of its own', as Brooks puts it. 'In that sense, *and that sense only*, field quanta resemble particles.'[107] Physicist Art Hobson puts it this way: 'An electron is nothing like, say, a tiny pea. An electron is simply an energy increment of a spread-out matter field.'[108] Yet the electron is a real quantum of the field, that has an independent existence to a degree: for example, 'it may be held in place temporarily (or even indefinitely) by the electrical attraction of a nucleus, but given enough energy it is perfectly capable of going off on its own.'[109] In that way, it is more like an event within a temporal and spatial continuum than a thing in time and space: something we can talk about as if discrete, which in a sense it is, yet acknowledging that it is not wholly disjunct from everything else that exists.

Again the need to combine two apparently opposite characteristics, without collapsing them into one or the other; the recognition of continuity and yet of individual entities within the continuity arising from it and unified by it; the reclamation of events from the realm of thingness; all this forms an exact parallel with the right hemisphere's take, which unites its own fundamental *Gestalt* perception with the left hemisphere's secondary representation in particulate form. Furthermore it is deeply consonant with the worldview of Schelling.

3 · *What can we know about the nature of mass and matter?*

Matter is whatever occupies space and has mass. Or rather 'exists spatially', since space should not be thought of as a container for 'things'. But it must have mass. And what is mass? Fascinatingly, mass is the tendency of an entity to *resist* – changes in course or speed. This should resonate, since, from a purely metaphysical point of view, I have been arguing that, though flow is generative, nothing comes

106 Wilczek 2008 (236).

107 Brooks 2016 (location 262; emphasis added).

108 Hobson 2007 (312).

109 Brooks 2016 (location 1881).

into existence except by means of *resistance* to flow. The recalcitrance of mass gives rise to the possibility of enduring form.

Photons, particles of light, have no mass: when, however, they split into an electron and a positron, matter and antimatter, they lose energy and gain mass. And when an electron and a positron recombine they lose mass once more, which becomes energy again in the form of light. Light gives rise to mass and the material world we see. Without light we could see nothing: without mass there would be nothing to see.[110]

Mass is not the same as weight, though weight is another fascinating concept distinct from resistance: the tendency of all bodies to approach one another. On earth it is, of course, the pull exerted by the earth's gravitational field on all material things. Thus it could be said that while mass comes about because of an attachment to – a drive to make cohere and endure – some particular form, weight comes about because of an attraction *between* existing material forms.[111] The first causes a new event to persist, for a while, in the face of instability; the second causes the coming together of what so persists, so that this creative achievement can grow, can realise further potential. Gravity converts a resistant or disjunctive force, namely mass, into a propulsive or conjunctive force, namely weight.

In classical physics mass is a measure of inertia. Similarly in QFT the effect of mass is

> to slow down the speed at which a field evolves and propagates, so mass plays the same inertial role in QFT as it does in classical physics. But this is not all that mass does. This same term also causes the field to oscillate, and the greater the mass, the higher the frequency ... the energy of a quantum is represented by oscillations in its field intensity: the more energy, the faster the oscillations.[112]

Mass and energy are 'interpotential', as the most famous of all equations, $E=mc^2$, proclaims. Increasing mass increases the speed of oscillations, and that in turn increases energy. Might it be legitimate to think of mass as the price paid in energy for making things endure? Or as the way energy manifests in the so-called 'emptiness' (more truly the infinite potential) of space, so as to bring about phenomena that have some thickness in space and duration in time? Bohm called matter 'as it were ... condensed or frozen light'.[113]

In case my account here seems to privilege physics too much over philosophy, this relationship between matter and consciousness – that matter is continuous with mind, but forms relatively persistent elements in it – was an insight of Peirce's:

> in obedience to the principle, or maxim, of continuity, that we ought to assume things to be continuous as far as we can, it has been urged

---

110 Life is bound up with energy fields that emit light. 'Organisms and tissues spontaneously emit measurable intensities of light, ie photons in the visible part of the electromagnetic spectrum (380–780 nm), in the range from 1 to 1,000 photons per second per square centimetre, depending on their condition and vitality. It is important not to confuse ultraweak photon emission from living systems with other biogenic light-emitting processes such as bioluminescence or chemiluminescence ... Ultraweak photon emission also supports the understanding of life sustaining processes as basically driven by electromagnetic fields' (Schwabl & Klima 2005).

111 Cf Schopenhauer 1974, vol 1 (99): 'Matter is that which persists and endures'.

112 Brooks 2016 (location 2748).

113 Bohm 1986 (44).

that we ought to suppose a continuity between the characters of mind and matter, so that matter would be nothing but mind that had such *indurated habits* as to cause it to act with a peculiarly high degree of mechanical regularity, or *routine*.[114]

According to Brooks,

a quantum field may spread over great distances, but it always maintains its separate identity. In QFT, what we call electrons, protons, etc, are really quanta of a field.[115]

Not everyone is easily reconciled to this. The physicist Abraham Pais, for example, felt that this was unsatisfactory. 'In electrodynamics', he wrote,

the continuous field appears side-by-side with the material particle as the representative of physical reality. This dualism, though disturbing to any systematic mind, has today not yet disappeared … the successful physical systems that have been set up since then represent rather a compromise between these two programs, and it is precisely this character of compromise that stamps them as temporary and logically incomplete …[116]

Yet if aesthetics is to be the guide – and one cannot dispute that for many scientists and mathematicians it has proved, historically, a reliable one – for me there is even greater beauty in the idea that reality can appear both discrete and continuous depending on how it is observed; and that nonetheless continuity has ultimate priority. *Pace* Pais, this may be neither an unsatisfactory 'compromise', nor an instance of 'dualism'. Rather it may be a way of describing two aspects of a single event. After all, as we have already heard from Tong, 'it is impossible to discretise the simplest kind of fermion', and electrons, protons and neutrons are all fermions. So quanta are in one sense discrete and in another not discrete.

What makes quanta behave like particles is that only one field quantum can occupy a given state at a time (due to the so-called Pauli Exclusion Principle). This immediately imports the necessary degree of discreteness. But what we think of as qualities of 'solid' matter, such as mass, charge and energy, are actually, again, properties of fields. 'Taking the existence of all these transmutations into account, what remains of the old ideas of matter and substance? The answer is *energy*', writes Pauli. 'This is the true substance, that which is conserved; only the form in which it appears is changing.'[117]

The true *substance* is energy, which appears in changing forms. This is entirely in keeping with the view that reality is a dynamic, ever-changing, flowing process, not an assemblage of things; a view

114 Peirce 1931–60, vol 6, §227 (emphasis added).

115 Brooks 2016 (location 1732).

116 Pais 1982 (289 & 463).

117 Pauli 1954 (14).

which is to be found in the wisdom traditions of East and West, and is given in the world as delivered to us by the right hemisphere, and not the left.

Here is Brooks again:

> The difference [between fields and particles] lies in how we choose to perceive reality – the picture in our mind – and this difference is immense. A particle is a localised bounded object. It has edges, or perhaps, as some think, it is only a point. A particle exists *in* space and around it is empty space. A field, on the other hand, is something that exists everywhere as a property *of* space. Its intensity may be small, but it is never zero. Even in areas where there are no quanta, there is a small amount of field called the vacuum field. In QFT there is no such thing as empty space. While particles move through space like baseballs, fields are diffuse and spread-out; they exhibit interference and diffraction. Nevertheless, when a spread-out quantum is finally absorbed, it collapses into one location.[118]

This collapse seems to occur only when 'measurement' takes place.[119] But, as David Mermin points out, this whole concept of measurement privileges one very limited and specialised kind of interaction – that with a measuring device – over the broad concept of interaction in general. It is 'interaction [that] produces correlation', correlation in this context meaning the tendency of two systems to interact so as to produce altered behaviour in each. And he argues that 'the very much broader concept of correlation ought to replace measurement in a serious formulation of what quantum mechanics is all about.'[120] He argues, in fact, that 'all correlations are real and objective – not just those produced by a measurement', and that it is micro-interactions that cause the collapse: 'it is simply a bad habit not to grant micro-micro ... -micro correlations as much objective reality as the traditional emphasis on measurements has granted to micro-macro correlations.'[121]

This bears on the question of how quantum collapse could have occurred prior to the advent of an 'observer'. Would it nonetheless have required consciousness within the correlation? That might not be as odd as it seems, since consciousness may be prior, ontologically, to matter. The implications of that I shall leave till the next chapter.

Richard Feynman championed a 'particle only' theory of photon and electrons. Pleasingly, in 1949 Freeman Dyson showed that Richard Feynman's particle diagrams gave the same results as the field theory approach that was championed by Julian Schwinger and Shin'ichirō Tomonaga. However, according to Wilczek, when Feynman realised his theory was mathematically equivalent to the usual theory, instead of his finding it exhilarating, 'it crushed his deepest hopes ... he gave up when, as he worked out the mathemat-

---

118 Brooks 2016 (location 1259–78; emphasis in original).

119 Manning, Khakimov, Dall *et al* 2015; Ma, Kofler, Qarry *et al* 2013.

120 Mermin 1998. John Bell took the view that the word 'measurement' had had 'such a damaging effect on the discussion, that I think it should now be banned altogether in quantum mechanics' (Bell 2004, 216).

121 Mermin *op cit*.

ics of his version of quantum electrodynamics, he found the fields, introduced for convenience, taking on a life of their own. He told me he lost confidence in his programme of *emptying space*.'[122] The fields came to life in a way the particles did not. Space is certainly not empty. As David Bohm wrote:

> What we call empty space contains an immense background of energy, and ... matter as we know it is a small, 'quantized' wavelike excitation on top of the background, rather like a tiny ripple on a vast sea.[123]

Physicist Robert Oerter writes:

> Wave or particle? The answer: both, and neither. You could think of the electron or the photon as a particle, but only if you were willing to let particles behave in the bizarre way described by Feynman: appearing again, interfering with each other and cancelling out. You can also think of it as a field, or wave, but you had to remember that the detector always registers one electron, or none – never half an electron, no matter how much the field has been split up or spread out. In the end, is the field just a calculational tool to tell you where the particle will be, or are the particles just calculational tools to tell you what the field values are? Take your pick.[124]

Is there an argument that, importantly, we should *not* take our pick? Might we not need both? At least, let me rephrase that, since it would seem that the word 'particle' causes more trouble than it's worth: we might need both continuity and discreteness together in the one picture. And that seems to be exactly what we have. This is not an unfortunate conclusion whereby we reach a compromise, or can't yet decide (not enough to go on), or reluctantly accept a 'dualism'. It is cause for deeply disconcerted awe. It reflects, or constitutes at another level, everything that the philosophy and neurology on which this book is based would lead us to believe about existence. And it has ramifications well beyond physics into the ontology of the world, both inanimate and animate.

Another marvellous element in this picture is that the fields come with *degrees* of independence. The quantum is as discrete as it gets: a separate, indivisible element, arising in a field, that, for a while, has a distinct existence, as Brooks puts it, 'living a life and dying a death of its own'. Then there are self-fields. These live a dependent existence, being created by a source to which they are permanently attached. According to Brooks, examples include the electric field around an electron or proton, and the strong and weak fields around a nucleon. And then, universally present at a higher – or lower – level (depending on how you look at it), is the vacuum field:

> The equations of quantum field theory do not permit the field strength ever to be zero. Even in regions where there are no quan-

122 Wilczek 2008 (84 & 89; emphasis added).

123 Bohm 1980 (191).

124 Oerter 2006 (218).

ta or self-fields, there is a background field called the vacuum field. The vacuum field is especially important in the case of the Higgs mechanism.[125]

And even in a vacuum there is entanglement. You might think that entanglement depended on the existence of particles, but no. Entanglement is also a matter of *degrees* of independence, *degrees* of cohesion, and it does not concern particles so much as it concerns fields. 'Even in a vacuum, with no particles around', writes George Musser in *Nature*, 'the electromagnetic and other fields are internally entangled. If you measure a field at two different spots, your readings will jiggle in a random but coordinated way ... The big realization of recent years – and one that has crossed old disciplinary boundaries – is that the relevant relations involve quantum entanglement. An extra-powerful type of correlation, intrinsic to quantum mechanics, entanglement seems to be *more primitive than space*.'[126]

Nonlocality is, according to Marcelo Gleiser, Professor of Physics at Dartmouth College, 'an indelible feature of entanglement, and entanglement is an indelible feature of quantum mechanics.'[127] Or QFT. It may even be prior, as Musser suggests, to the space in which it is manifest. Nor is it confined to the microscopic level. It has been demonstrated in the macroscopic realm, in entanglement within diamond crystals,[128] and has also been demonstrated in biological systems,[129] as well as between two entities that never co-existed in time.

That was in Chapter 22; and there I reflected that if it makes sense – and I have no idea if it does – to view the interaction that gives rise to the observation of entanglement as a sudden shift in a whole *Gestalt* (right hemisphere take), such as a field, rather than a change in one discrete entity causing a change in another discrete entity sequentially (left hemisphere take), entanglement would no longer require faster than light communication over unlimited distances between particles. Every part of the cosmos would be necessarily connected in some form to every other part. This is, in fact, the import of Bell's theorem. Oddly enough, and beautifully enough, the quest to reduce reality – at last! – to its elementary particles merely returns the searcher to wholeness.

We are too reluctant to let go of thingness. If we think of particles as things, they seem separate and fixed, whereas they are connected and in flux. 'Isolated material particles are abstractions', wrote Niels Bohr, 'their properties being definable and observable only through their interaction with other systems.'[130] 'We customarily say', wrote David Bohm,

> 'one elementary particle acts on another', but ... each particle is only an abstraction of a relatively invariant form of movement in the whole field of the universe. So it would be more appropriate to

---

125 Brooks 2016 (location 2738).
126 Musser 2018 (emphasis added).
127 Gleiser 2014 (233).
128 Chalmers 1995. See Ringbauer, Duffus, Branciard *et al* 2015; Ma, Zotter, Kofler *et al* 2012; Megidish, Halevy, Shacham *et al* 2013; Lemos, Borish, Cole *et al* 2014.
129 Shi, Kumar & Lee 2017.
130 Bohr 1961 (56–7).

say, 'Elementary particles are on-going movements that are mutually dependent because ultimately they merge and interpenetrate'. However, the same sort of description holds also on the larger-scale level. Thus, instead of saying, 'An observer looks at an object', we can more appropriately say, 'Observation is going on, in an undivided movement involving those abstractions customarily called "the human being" and "the object he is looking at"'.[131]

In other words, as Mermin put it, not so much measurement, perhaps not even observation, but interaction.

Dynamic relationships are prior to static entities. Einstein saw that fields were of primary importance. 'It needed great scientific imagination', he wrote, 'to realise that it is not the charges nor the particles but the *field in the space between* the charges and the particles that is essential for the description of physical phenomena'.[132] Once again it is relationships that are more significant than the *relata*. Betweenness is more essential than the 'things' that exhibit it. As Gleiser points out, photons are

> massless bundles of pure energy. Physics was thus proposing that something could exist without mass, that things could exist without being material. Since what exists defines physical reality, the new physics suggested that reality could be immaterial. Energy is more fundamental than mass, more essential.[133]

So energy is ontologically prior to matter. To me that makes perfect sense philosophically speaking. Matter is a condensate riding on the sea of energy, as Bohm indicated it was.

Fields may not only precede 'particles' ontologically, they may have done so historically. Immediately (10^{-30} secs) after the Big Bang there were no particles – for those who still recognise particles – but there were already quantum fields. Quantum fields were everywhere in the universe. In a vacuum, again, there can be no particles, but the fields are still there: and they are moving and differentiated (ie, are something like fluctuating, shape-shifting blobs). The Heisenberg Uncertainty Principle dictates that there is never such a thing as stasis: there are always what are called 'quantum vacuum fluctuations' – complex patterns of flow. They are real, measurable, and indispensable. David Tong describes a vacuum as 'the simplest thing we can imagine in the entire universe and it is astonishingly complicated.'[134]

What emerges here for me is further confirmation that we need both discreteness and continuity, both the knowable and the unknowable, the graspable and the ungraspable *together*; that the second in each pairing is ontologically prior to the first; and that here, as elsewhere, betweenness is the most important element. In terms of the manner in which the hemispheres deliver the world this means that there are

131 Bohm 1980 (29).
132 Einstein 1971 (244).
133 Gleiser 2014 (168).
134 Tong 2016.

elements delivered by each, but that the take of the right hemisphere must be both foundational and overarching, grounding and fulfilling the process in which what the left hemisphere offers is intermediary.

It also suggests to me that matter is part of the process of creative resistance within the flow that precipitates individual phenomena into existence, for a while.

Bergson saw the way we discretise matter as something that the part of our consciousness devoted to using the world (a perfect description of the left hemisphere) brings about for its own purposes: carrying out 'the double work of solidification and of division which we effect on the moving continuity of the real in order to obtain there a fulcrum for our action, in order to fix within it starting-points for our operation ...'[135] He saw it as analogous to the way in which we discretise time – again for our purposes. In realising that this was, nonetheless, neither the only way in which matter can be thought of, nor the most fundamental way in which it can be thought of, since it draws too harsh a distinction between the static and dynamic characteristics of matter, 'Bergson arrived', according to philosopher Peter Gunter, 'at insights closely resembling those of quantum physics.'[136] Only Bergson got there first.

If continuity is ontologically prior to discontinuity, and there is a need to understand discontinuity as arising on the background of continuity; if nonetheless each has to be given a place alongside the other without the tension being resolved by the dissolution of one or the other; if both independence and entanglement are matters of degree, rather than absolute; if relationship is prior to *relata*; all this conforms better to the take on the world of the right hemisphere, and will be better understood by it than by the left.

It is the quantum vacuum fluctuations that give rise to inequalities in the cosmic background radiation. And this asymmetry or inequality in the cosmic background radiation was necessary for the universe to come into being at all. Without it, the quantities of matter and antimatter would have 'cancelled one another out'. According to Gleiser, 'a universe that began with equal amounts of matter and antimatter would quickly evolve into a void filled mostly with radiation, a world nothing like what we see around us.'[137] Which leads to the fourth question, the one about the status of asymmetry.

4 · *Is asymmetry built into the foundations of physical reality?*

Already, in the 1870s, Pasteur was writing sentences that would have anguished his Enlightenment forebears. 'The universe as a whole', he wrote,

> is asymmetrical, and I have come to believe that life, as it is manifest to us, is a function of the asymmetry of the universe ... Without any doubt, I repeat, if the basic principles of life are asymmetrical, it is

135 Bergson 1911c (280).
136 Gunter 1969 (45).
137 Gleiser 2014 (249).

because asymmetrical forces of the cosmos preside over their unfolding … Life is dominated by the effects of asymmetrical forces whose enveloping cosmic existence we sense intuitively. I would even say that living species are primordially, in their structure, in their external forms, functions of the cosmic asymmetry.[138]

Before carrying on to look at the asymmetry of the cosmos, let us pause to consider what we mean by symmetry and asymmetry.

We saw that the greater meaning of symmetry is that an operation leaves something unchanged. (Our everyday use of the term is just a special case of that, whereby reflecting an object through 180° leaves it unchanged.) To this extent it is sterile. Asymmetry, by contrast, is generative – in more than one sense. First, asymmetrical operations produce not the same outcome, but something new. Second, even considered in its more everyday sense, asymmetry can be seen as the coming together of two *different* elements on either side of the axis so as to form a new coupling – the object that is asymmetrical. And, third, asymmetry also always implies symmetry – the symmetry that is broken. Asymmetry can be spoken of only where the potential for symmetry exists, but is gainsaid. Not everything that is not symmetrical is asymmetrical (the contents of my car's glovebox are not asymmetrical, just a mess). Whereas the façade of the Villa Rotonda is deliciously asymmetrical in its apparent symmetry (compare Plates 23[a] & [b]). It is this simultaneous affirmation, yet denial, of sameness that is so fruitful.

A similar structure can be discerned in another important phenomenon that we have already seen is highly generative: that of metaphor. The metaphor 'x is a y' (Churchill was a bulldog) works only if simultaneously 'x is not a y' (Churchill was not, actually, a bulldog). What counts is the coming together of an equivalence that is asserted, with one that is denied. However, though x not being a y is necessary, it is clearly not sufficient: Churchill is also not a bookcase. There must be potential proximity, for the spark, so to speak, to jump the gap. So it is with asymmetry. Metaphor and asymmetry each set up a *tension*, and the energy lies in this tension.

Symmetry is sometimes approximated by living things, which on closer inspection are, however, like the brain, not truly symmetrical, and are constantly changing and moving. They suggest symmetry, yet deny it. Symmetry-breaking, as cell biologists Li and Bowerman put it, is 'fundamental to every physiological process. In fact, it may be argued that all qualitative cellular transitions and cellular decision-making are forms of symmetry breaking'.[139] Symmetry-breaking is what makes morphogenesis possible at the single-cell level, or in multicellular structures – and equally importantly, given the important role of motion in all living things, it is essential for generating

---

138 Pasteur 1874; in Vallery-Radot 1922, vol 1 (361, 375 & 377): « L'univers est un ensemble dissymétrique et je suis persuadé que la vie, telle qu'elle se manifeste à nous, est fonction de la dissymétrie de l'univers … Sans nul doute, je le répète, si les principes immédiats de la vie sont dissymétriques, c'est que, à leur élaboration, président des forces cosmiques dissymétriques … La vie est dominée par des actions dissymétriques dont nous pressentons l'existence enveloppante et cosmique. Je pressens même que toutes les espèces vivantes sont primordialement, dans leur structure, dans leurs formes extérieures, des fonctions de la dissymétrie cosmique. »

139 Li & Bowerman 2010.

body axis and direction, and enabling cell motility. Particularly interesting are asymmetric divisions that lead to different cell fates. For instance stem cells are undifferentiated cells that divide both to renew themselves *and* to give rise to more specialised cell types. Many cases exist in which one daughter cell maintains the stem cell characteristic while the other daughter is differentiated.[140]

The golden ratio ϕ (*phi*) 1:~1.618 is asymmetrical, though universally recognised for its beauty.[141]

140 Klar 2002.

141 See pp 595–6 above, and pp 1159–61 below.

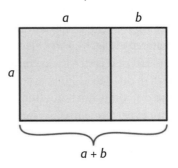

Fig. 55. *Golden ratio*

It is also generative in a number of respects. Starting from a rectangle whose sides are in the ratio 1: ~1.618, adding a square produces another rectangle, also in the golden ratio, that includes the first. Or, moving in the opposite direction, within any 'golden rectangle' a square can be inscribed which produces another golden rectangle. And so *ad infinitum*.

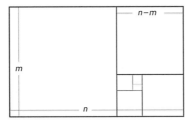 

Fig. 56. *The relationship between the golden ratio and spiral*

This process yields a spiral that is found throughout nature. The 'golden spiral' can be seen in, for example, the form of nautilus shells, ammonites, the way hurricanes unfurl, and how spiral nebulae come into being.

Generation means diversification, and yet the 'golden rectangle' is simultaneously *unifying*, with the 'parts' and the whole always in the same relation. It is a spatial expression of the union of diversification and unification.

Another way to approach ϕ is through the Fibonacci series, which, like the golden ratio, is ubiquitous in Nature. The Fibonacci series is generated by adding each number to its predecessor and making the result the next member of the series, thus: 1, 1, 2, 3, 5, 8, 13, 21, 34, 55, 89 … The ratio of any member of the series to that which comes immediately before it approximates to the golden ratio (approx 1.618), the approximation getting more and more accurate the further one continues the series, converging on ϕ at infinity. The sequence is generative because it is asymmetrical, open-ended and non-repeating – not symmetrical, closed and repetitive. The spiral disposition of leaves round the stem of a plant in many species is based on the Fibonacci series. When 360, the number of degrees in a full circle, is divided by ϕ, it yields the number of degrees by which neighbouring leaves on the stem are separated, which closes on approximately 222.5° (or 137.5°, when looked at in the opposite direction of rotation, since the numbers add up to 360°). This has the consequence that no leaf ever completely blocks light from another. Even small variations from this 'golden angle' cause the property to be lost: varying the angle only from 137.5° to 135° produces a repeating pattern, offering just eight positions in place of the infinitely non-repeating set of positions using the 'golden angle'. What is particularly intriguing is that this orderly and beautiful way of achieving its end was not purely the fallout of necessity, since other, less elegant solutions could have done a reasonable job of ensuring access to light.[142] In other words, the omnipresence of the Fibonacci spiral in Nature is not just a matter of utility (though it encompasses that along the way), but of structural principle that is both beautiful and generative.

So it is that the spiralling in pine cones and sunflower discs follows the Fibonacci series (see Plates 22[d] & [e]). The ratio of leftward to rightward spirals in the sunflower obeys the golden ratio, demonstrating exactly 55 leftward spirals and exactly 89 rightward spirals, just as the Fibonacci series dictates: and in the pine cone there are 8 rightwards spirals and 13 leftwards spirals.

Perfect symmetry is an abstraction from reality. Any symmetrical system is highly vulnerable to small inequalities. These then magnify to produce patterns that are less highly symmetrical, yet regular and familiar. Mathematician Ian Stewart gives the example of a flat sandy desert that is highly symmetrical. It takes only a few grains of sand to pile up for some reason and that causes eddies in the wind, which causes more sand to shift and so on – resulting in rhythmically undulating dunes, which have a lesser degree of symmetry. Any part of a perfectly flat surface – the flat desert – could theoretically be slid across itself by any amount and it will not appear changed; but once there are dunes, you'd have to slide the dunes by specific distances, in order to make the new positions of the dunes coincide with the old ones. Breaking symmetries of a certain kind in a similar way leads to irregularly regular ribbed patterns forming – eg, the stripes of a zebra: 'there's a hidden unity: the processes take different physical forms, but the patterns are universal'.[143] Stewart describes shapes arising 'through a collective process in which the

142 See Niklas 1988; and Valladares & Brites 2004.

143 Stewart 2018.

137.5°

135°

Fig. 57. (*Top*) Typical Fibonacci distribution of leaves, viewed from above; (*centre and bottom*) arrangement of leaves resulting from 137.5° compared with 135° separation (photo: © Ron Knott; diagrams from Valladares & Brites 2004)

entire pattern jostles around until everything fits together'. In other words the whole is involved in a reverberation, generating its own rhythmic pattern or *Gestalt*, not reductive to 'parts' that are specified by a unidirectional linear process.

Dendritic, or branching form (see Plate 21), such as one sees in frost on a window pane, or in a plant such as a fern, is fractal. There is a main stem, with smaller stems coming off it, and ones smaller still branching off them, and so on. Interestingly, this has to do with symmetry-breaking:

> The reason for this kind of shape is that as something grows (ice crystals, plant), as it passes some critical size, the smooth edge becomes unstable and starts to get bumpy. Then the bumps grow, obeying much the same rules, and the process repeats. It's actually another case of symmetry-breaking, analogous to ripples forming in a flat plane, but here the ripples form along a flattish edge. So again we see mathematical unity in very different physical systems.[144]

We see unity, yet we also see discreteness. As symmetry is broken, and sand dunes arise, or waves rise up from a flat sea, or a string vibrates, there is *discreteness*, though the surface of land or sea, or the string, is in truth *continuous*. The form produces regularities; and those very regularities are a testimony to there being harmony. This is how unique things come into being at all: otherwise there would be only undifferentiated unison. *We* are the notes of this polyphonic motet.

With that, let us return to Pasteur and the nature of the cosmos. I remarked that his Enlightenment forebears would have been anguished, because for them asymmetry was an obvious flaw: perfection must be symmetrical. Symmetry is a prominent feature of the architecture, the music, the poetry and even of the prose style of the Age of Reason. The thought that at the core of things there was *asymmetry* would have been enough to trigger a mass outbreak of *felo de se* amongst these decent, orderly folk.

Pasteur's astonishing intuition was that an asymmetry in the very stuff of the universe gave rise to, and was reflected in, the asymmetries of living things. In fact we now know that it is to the asymmetry between electromagnetic and weak forces at low temperatures – they are symmetrical at high temperatures – that we owe the existence of life at all: 'our existence', writes physicist Frank Close, 'is critically dependent on nature having broken symmetry as the universe cooled.'[145] Not only are living beings subtly, but manifestly and importantly, asymmetrical in their form – notably, of course, in their neural networks or brains – but the stuff of the cosmos appears to be so, too. Earlier I mentioned that Pasteur's younger contemporary, the

144 Stewart 2018.

145 Close 2000 (152).

physicist Pierre Curie, emphasised that *'it is asymmetry which creates the phenomenon.'* And his intuition, too, has since been shown to be right by subsequent developments within physics.

Tsung-Dao Lee and Yang Chen-Ning were the first to point out that there is no theoretical or experimental basis for the assumption of left-right symmetry in matter; for this they received a Nobel Prize after Chien-Shiung Wu, in 1956, demonstrated violation of parity (absence of left-right symmetry) in the case of the weak interaction.[146] From that point it was established that the universe has handedness, or chirality ('chiral' just means 'handed', from Greek, χείρ, a hand) – that is to say has asymmetrical properties, in the same way that a right-hand glove and a left-hand glove are mirror images of one another, but the images have different properties: you cannot put your right hand into the left-hand glove. Similarly a helix can spiral clockwise or anticlockwise: the two spirals are mirror images, but are not identical and have different properties that are very important in chemistry. Asymmetry and handedness are at the core of reality.

'A chiral theory is delicate', writes Tong:

> Subtle effects known as anomalies are always threatening to render it inconsistent. Such theories have so far resisted attempts to be modelled on a computer. Yet chirality is not a bug of the Standard Model that might go away in a deeper theory; it is a core feature. At first glance, the Standard Model, based on three interlinking forces [the electromagnetic, weak, and strong interactions], seems to be an arbitrary construction. It is only when thinking about the chiral fermions that its true beauty emerges. It is a perfect jigsaw puzzle, with the three pieces locked together in the only manner possible. The chiral nature of fermions in the Standard Model makes everything fit together.[147]

Chirality is everywhere in the cosmos. Radioactivity is chiral; polarised light is chiral; biological molecules are chiral (amino acids are mostly left-handed; sugars right-handed). Left-handed amino acids always give rise to right-handed helical proteins, while right-handed amino acids give rise to left-handed proteins. A mixture of both forms cannot form helices at all. DNA is always a right-handed helix.

Different-handedness, giving rise to two different enantiomers (the 'left-hand' and 'right-hand' versions) of a molecule, may lead to different phenomena at the human sensory level. For example, the smell of caraway seed flips into the smell of spearmint depending on whether left-handed or right-handed forms of the molecule carvone are involved. Of more consequence is that bioactive molecules, such as many medications, are effective in the case of only one enantiomer; a racemic (50:50) mix may not just be half as effective, but ineffective, as each may tend to interfere with the other.[148]

---

146 Brooks 2016. Fermions (one of the two basic classes of particles, which includes, among others, electrons, protons, neutrons, neutrinos and quarks) that spin in a counter-clockwise direction feel the weak nuclear force, and those that spin in a clockwise direction do not.

147 Tong 2012.

148 One mechanism for this is that neuronal receptors have built-in chirality, and when a drug is designed to 'lock on' to a receptor and activate it, the wrong enantiomer may block the receptor, but without activating it. This is somewhat like the way in which the wrong key, if it is like enough to the one that opens the lock, may jam the lock and prevent it from opening. This probably explains why, for example, the S (or left-hand) enantiomer, escitalopram, is a much more effective antidepressant than the racemic drug, citalopram.

Chirality is a form of fundamental asymmetry: and yet, once again, it invokes a symmetry – mirror symmetry – too. The metaphor of the hand from which it takes its name makes this clear.

It has become apparent from a number of disciplines that dynamic systems depend on what is often expressed as a balance between opposing and complementary forces: between chaos and order, stability and flexibility. As John Polkinghorne puts it,

> The physical world as science discerns it, is one in which order and disorder interlace and fertilize each other. The creative interplay of chance (happenstance – the occurrences which are the seeds of novelty) and necessity (lawful regularity which sifts and preserves the novelties thrown up by happenstance) lies at the root of all the fruitful history of the universe ... Physical process is not merely mechanical but it has an inherent openness which makes this a world of true becoming.[149]

Complete order would prevent adaptive changes; unconstrained disorder would render self-organisation impossible.[150] The so-called 'edge of chaos', where chaos and order are maximally present to one another, is the most fruitful condition of an open system, including the creative potential of the human brain.[151] As Nassim Nicholas Taleb argues in his book *Antifragile*, a certain vulnerability is bound up with the potential to survive: the attempt to do away with it leads to extinction. However, these conjunctions of opposites are not expressive of symmetry, but of the complementarity of elements that are fundamentally asymmetrical. This in turn helps one see that what is needed is both symmetry *and* asymmetry – themselves an asymmetrical pairing – *together*. It is neither the breaking of the symmetry alone, nor (alone) the symmetry that is broken, but the new breaking, in conjunction with the old symmetry, the combination of the two, that is generative.

This generative asymmetry of symmetry and asymmetry is imaged in the way that the sex chromosomes in men and women are asymmetrical, not just different. It is not just that the X and Y chromosomes themselves are an asymmetry, but that the female pair (XX) is symmetrical and the male pair (XY) is asymmetrical. Together they are generative.

The idea of *yin* and *yang*, from which all things arise, has apparent symmetry, but in fact refers, respectively, to the shady side and to the sunny side of a hill, which are not symmetrical. Note that like the poles of the magnet the two sides of a hill cannot be separated, so are not dual. The Japanese equivalent of the *taijitu* symbol, the *inyo* symbol (where *in* = *yin* and *yo* = *yang*) captures more clearly than the *taijitu* the idea of asymmetry combined with symmetry, and at the same time the way in which the energy of *yo* (*yang*) sits

---

149 Polkinghorne 2008.

150 See, eg, Theise & Kafatos 2016.

151 Pierre & Hübler 1994; Wotherspoon & Hübler 2009; Bilder & Knudsen 2014; Korn & Faure 2003.

within the receptive space of *in* (*yin*). Again the two elements are asymmetrical, as is their positioning in relation to one another (see Plate 22[b]).

In fine, a world which obeyed symmetry could not exist: that would be the left hemisphere fantasy of stasis and timelessness in which there is no place for creation, none for anything new or unknown. If, as theoretical physicist and Nobel Laureate Philip Anderson put it, symmetry means 'the existence of different viewpoints from which the system appears the same',[152] then at the heart of physics lies the *difference* of asymmetry. Oddly enough, those two points of view – one in which things are momentarily 'static', discrete, at least partly measurable, the other in which they are forever in motion, continuous, and indeterminate – happen to have their counterparts in the modes of consciousness of the two hemispheres of the asymmetrical brain. And in physics they are not symmetrically important, in that one is capable of embracing the other, just as in the case of the brain's hemispheres.

For that to be more than a coincidence would suggest that either the brain is so made as to reflect the seemingly paradoxical nature of reality, or its structure and function themselves *mould* such a reality, since quantum field theory implies that nothing takes the form that it does wholly independently of consciousness. Each hypothesis is fascinating. But perhaps it is not a matter of having to choose one or the other. To quote Heisenberg: 'The same organizing forces that have shaped nature in all her forms are also responsible for the structure of our minds.'[153] And brains.

[152] Anderson 1972 (394).

[153] Heisenberg 1971, 'Atomic physics and Pragmatism': 93–102 (101).

Chapter 25 · Matter and consciousness

> Mental and physical events are, on all hands, admitted to present the strongest contrast in the entire field of being. The chasm which yawns between them is less easily bridged over by the mind than any interval we know.
> —WILLIAM JAMES[1]

> As we live, we are transmitters of life.
> —D. H. LAWRENCE[2]

'The existence of consciousness is both one of the most familiar and one of the most astounding things about the world', writes the philosopher Thomas Nagel. 'No conception of the natural order that does not reveal it as something to be expected can aspire even to the outline of completeness.'[3] I agree on both counts. And when we ponder the relationship between matter and consciousness, our thoughts naturally turn to that between brain and mind.

While no-one could possibly dispute the existence of an intimate relationship between brain and mind, the nature of that relationship remains highly disputable. Lucretius was probably the first to argue that, since mental faculties are affected, first by the maturation, and then by the degeneration, of the brain; since alcohol and drugs alter experience; since epilepsy and head injuries alter consciousness; and since memories appear to be stored in the brain; it follows that mind is dependent on the brain for its existence. This same set of observations lies behind the popular belief that matter gives rise to mind. Now, in the era of brain scanning, in which it might seem that mental events can be visualised through their brain correlates, it is even harder for many to resist such a conclusion.

But do we know that matter can give rise to consciousness? This is merely an assumption. When a TV set malfunctions, it can distort the image or sound it relays in a large number of ways, depending on where in the system the malfunction lies. To an engineer, the nature of the distortion may be a clue to the location of the problem, as the nature of brain pathology is to the neurologist. To an observer from another planet, it might prove impossible to tell that the TV set did not generate, but merely transmitted, its material: pull the plug and the show ceases to exist.

The intimacy of the relationship between two parties has in itself nothing whatever to say about its nature. In the history of the cosmos, matter might give rise to mind, or mind to matter; or each might equally give rise to the other interdependently; or they might run in parallel, perhaps because they are different aspects of some

[1] James 1890, vol 1 (134).
[2] Lawrence, 'We are transmitters'.
[3] Nagel 2012 (53).

ultimately unified phenomenon. When it comes to the brain, the intimate relation between brain activity and states of mind cannot in itself help distinguish between theories of emission, transmission, and permission as its basis. In other words, the same findings are equally compatible with the brain *emitting* consciousness, *transmitting* consciousness, or *permitting* consciousness. (The latter two options are similar, except that permitting substitutes the idea of a constraint that is creative, fashioning what it allows to come into being, for the merely passive idea of transmission.)

I am going to argue that it is the last of these possibilities – permission – that is the most convincing. The prejudice in favour of the most bizarre – the emissive option – in which consciousness is some form of secretion of the brain, stems from the mistaken belief that, while we may not understand consciousness, we do at least understand matter. But we don't. Even if we did, this would be a quite astonishing conjuring trick, since no-one has the slightest idea of a mechanism by which consciousness could emerge from unconscious matter; in any case matter evanesces as we look at it more closely and turns out to be every bit as inscrutable as consciousness itself. The atom has the curious property that while from a distance it has a blurry self-consistency, it does not become clearer, but more indistinct, as you zoom in, so that there is less and less to see, until it evades your grasp entirely. And its nature changes by the mere fact of being observed. All this is fine, but it renders impossible the role in which materialists wish to cast matter: the solid mind-independent basis of mind. It is not just brains, but the tiniest particles of matter, we discover, that cannot be considered independent of consciousness. 'Matter loses solidity more and more under the steady scrutiny of relentless rationality', as geneticist Mae-Wan Ho notes.[4]

It is clear that we cannot continue to consider matter as straightforward in any way that helps us ground reality in its apparently comforting tangibility. Nearly three-quarters of a century before Murray Gell-Mann and George Zweig discovered the existence of quarks, the philosopher F. C. S. Schiller wrote that 'the connection of the scientific conception of matter with the hard matter of common experience has become fainter and fainter, as science is compelled to multiply invisible, impalpable and imponderable substances in the "unseen universe", by which it explains the visible.'[5]

The great neurophysiologist and Nobel Laureate Sir Charles Sherrington reflected on mind and matter:

> For myself, what little I know of the how of the one does not, speaking personally, even begin to help me toward the how of the other. The two for all I can do remain refractorily apart. They seem to me disparate; not mutually convertible; untranslatable the one into the other.[6]

[4] Ho 1994 (189).

[5] Schiller 1891 (297–8).

[6] Sherrington 2009 (312). Cf James 1890 (as in note 1 above).

Like many a profession of unknowing (all statements about consciousness must be hedged round with such unknowing) this statement carries with it a tincture of positive knowledge. It is already an important perception that the difference lies in the *how*, not the *what*. As soon as we escape the spell of thingness, and see that what we are dealing with is different modes of being, rather than different entities, we have begun to make an inroad into unknowing.

Faced with Sherrington's feeling of the incommensurateness of mind and matter, which I think we must all, to some extent, share, we have several options. These can be roughly stated as: (1) to deny the existence of consciousness; (2) to deny the existence of matter; (3) to accept that they each exist but are totally distinct; (4) to assert that they each exist but are the same; or (5) to entertain the possibility that they are distinct phenomena that reflect different aspects of a nonetheless importantly indivisible reality.

WHAT DO WE MEAN BY CONSCIOUSNESS?

When we speak of consciousness, what are we talking about? The word literally means 'knowing *with*' (Latin *con-*, with, + *scientia*, knowledge): it is therefore in essence, not a thing, but a betweenness.

Within that definition it is obvious that consciousness can mean a number of different things in different contexts, and I do not want to spend a lot of time on territory that has been endlessly disputed, and at times even a little clarified, elsewhere.[7] My aim here is as much as possible to avoid misunderstandings, rather than to provide definitions of what is intrinsically not amenable to definition; and I do not want to set up sharp distinctions where such distinctions themselves merely provide ground for fruitless contestation.

When I use the word 'consciousness', I refer very broadly to all that we might call 'the experiential'. This covers all the activities that go on, for each of us, as we say, *un*consciously and *pre*consciously, as well as consciously; but could not go on without what is conventionally referred to as subjectivity, or inwardness, of some kind. According to one source, 'our psychological reactions from moment to moment ... are 99.44% automatic.'[8] You don't have to buy the precision to get the point. Only a tiny part of our psychological – our mental – life enters our full consciousness so as to form a subject of reflection, that of which we are aware we are aware.

That of which we are unaware of being aware can easily pass into reflexive awareness, and vice versa, as whatever it is comes into, or slips out of, the focus of attention. We should not think of two realms, spatially, as we are wont to do (the unconscious lying *underneath* the conscious), any more than the part of a stage-set on the left lies in a different realm from the part on the right when the spotlight happens

7 This area is well covered by, eg, Zeman 2004.

8 Bargh 1997 (243).

to bring one forward and allow the other to recede into darkness. Nor does that part of experience of which we are reflexively aware always have the same qualities. The same set of apparently objective circumstances can appear very differently when our awareness of them changes. William James pointed up the need to avoid conceptualising hard and fast categorisations, when he wrote that 'our normal waking consciousness, rational consciousness as we call it, is but one special type of consciousness, whilst all about it, parted from it by the filmiest of screens, there lie potential forms of consciousness entirely different',[9] including a kind of consciousness that particularly interested him (and interests me) in which opposites are experienced as reconciled.

And it is not as if what goes on outside the spotlight is inferior stuff. I discriminate, reason, make judgments, find things beautiful, solve problems, imagine possibilities, weigh possible outcomes, take decisions, exercise acquired skills, fall in love, and struggle to balance competing desires and moral values all the time without being reflexively aware of it. Note that these are not just *calculations*, but rely on my whole embodied being, my experience, my history, my memory, my feelings, my thoughts, my personality, even – dare I say it? – my soul: 'psyche' in the broadest sense. It might be asked if it is really then 'I' that am responsible for these activities. Yet, regardless of how you choose to phrase it, all this activity still goes on, and in so doing helps constitute my world, who I am, and the quality of my experience. Sooner than place the Cartesian ego first, I would place first and foremost thoughts, feelings, imaginings and responses to the world, embodied and embedded in the world – experience – and allow the ego to take a back seat. As Lichtenberg said, '*Es denkt*': it thinks. Thoughts take place in what I call the 'field of me'. Even Descartes derived the ego from experience, not experience from an ego. What can more conservatively be claimed is that something about me – in the 'field of me' – *permits* these particular activities: and that something is what, in the broad definition, I am calling my consciousness. When I say the 'field of me' do not imagine for an instant that I mean something alien, impersonal and abstract, onto which an illusion of 'me' is plastered, but something that is at the core of me, something I can convey best in the words, once more, of Gerard Manley Hopkins: 'that taste of myself, of I and me above and in all things, which is more distinctive than the taste of ale or alum, more distinctive than the smell of walnutleaf or camphor, and is incommunicable by any means to another man ... searching nature I taste *self* at but one tankard, that of my own being.'[10]

There is, in the physicist Wolfgang Pauli's terminology, 'a psyche long before there is consciousness.'[11] We do not know the extent of the inner world of other creatures, but as Schiller pointed out, 'even in highly developed minds judging [the making of judgments in

---

9 James 1902 (388). Often misquoted as 'flimsiest' of screens.

10 Hopkins 1963 (145–6); see p 854 above.

11 Letter of Pauli to Léon Rosenfeld, 16 April 1952; in von Meyenn 1996 (610–11).

self-awareness] is a relatively rare incident in thinking, and thinking in living, an exception rather than the rule, and a relatively recent acquisition.' He reframes such explicit cognitive activity as not the apogee, nor even the norm, of consciousness, but as a somewhat regrettable lapse. The 'elaborate and admirable organisation' that results in habit is so efficient that 'thought, therefore, is an abnormality which springs from a disturbance. Its genesis is connected with a peculiar deficiency in the life of habit.' Self-conscious thought 'becomes biologically important' when, due to the peculiarities of a particular case, 'the guidance of life by habit, instinct, and impulse breaks down'. He continues:

> Thinking, however, is not so much a substitute for the earlier processes as a subsidiary addition to them. It only pays in certain cases, and intelligence may be shown also by discerning what they are and when it is wiser to act without thinking … Philosophers, however, have very mistaken ideas about rational action. They tend to think that men ought to think all the time, and about all things. But if they did this they would get nothing done, and shorten their lives without enhancing their merriment [how marvellous that a philosopher could then say such things!]. Also they utterly misconceive the nature of rational action. They represent it as consisting in the perpetual use of universal rules, whereas it consists rather in perceiving when a general rule must be set aside in order that conduct may be adapted to a particular case.[12]

This is reminiscent of Whitehead's admonition, quoted earlier, that 'operations of thought are like cavalry charges in a battle – they are strictly limited in number, they require fresh horses, and must only be made at decisive moments.'[13] It is clearly far more efficient to limit the contents of awareness, rendering responses wherever possible automatic. To quote William James: 'It is a general principle in psychology that consciousness deserts all processes where it can no longer be of use.'[14] What begins as effortful and self-conscious becomes effortless and intuitive, as when a surgeon, pilot or chess player acquires skill, according to 'a general pattern in which things move from initially conscious to gradually being more automatic.'[15] By that point, the unconscious knowledge is rich, sometimes considerably richer than conscious knowledge. We have already seen that unconscious processes are often (though by no means always) superior to conscious thought.[16] Self-consciousness may not be necessary, and may even interrupt, the ability to integrate sensory information, to assess emotion and motivation (including the integration of one's mental states with that of others – we rapidly and unconsciously weigh what others know, want and see) and to achieve a global 'take' on a situation.[17] And decisions which require the integration of many sources of information are better made un-

12 Schiller 1929 (197–8).
13 Whitehead 1911 (61).
14 James 1890, vol 2 (496).
15 Baumeister, Masicampo & Vohs 2011 (336).
16 See Chapters 17–19. Also de Neys 2006.
17 Frith & Metzinger 2016; Samson, Apperly, Braithwaite et al 2010.

consciously, since conscious deliberation interferes with the process of assigning appropriate weight to different sources of information.[18]

My point is that included in the broader concept of consciousness as experientiality is much material of which I am not often, or perhaps in some cases ever, reflexively aware. If that suggests that the subject/object divide is not as robust as we assume, and that our perceptions, thoughts and feelings are not located only 'inside' our heads, that is something we may have to accept; and I will return to it again later in the chapter.

CAN WE DENY CONSCIOUSNESS ALTOGETHER?

Attempts to resolve the 'hard' problem of how consciousness relates to matter by denying the existence of consciousness are surely the least promising of all – even less promising than attempts to resolve it by denying the existence of matter. We are told by some prominent biologists, skilled undoubtedly in their chosen field, that, once you take science seriously, consciousness and subjectivity are seen to be an illusion. But they've got it back to front. What science actually seems to tell us (since I take it that physics is closer to examining the nature of the world – that is what 'physics' actually means – than molecular genetics) is that, once you take science seriously, you see that *the notion of science as distinct from consciousness is an illusion*. I expect those of us who are still conscious are more likely to think the physicists right on this one.

I am with analytical philosopher Galen Strawson here, as often on the topic of consciousness, in his opinion that 'this particular denial is the strangest thing that has ever happened in the whole history of human thought, not just the whole history of philosophy.' For, as he puts it, 'experience is itself the fundamental given natural fact; it is a very old point that there is nothing more certain than the existence of experience.'[19]

To claim that consciousness is non-existent is self-exploding, since it requires consciousness both to make, and to make sense of, the claim: and to state that consciousness exists, but is an illusion, is no better, since an illusion requires a consciousness in which such an illusion might occur. Some philosophers 'are prepared to deny the existence of experience', writes Strawson:

> At this we should stop and wonder. I think we should feel very sober, and a little afraid, at the power of human credulity, the capacity of human minds to be gripped by theory, by faith. For this particular denial is the strangest thing that has ever happened in the whole history of human thought, not just the whole history of philosophy. It falls, unfortunately, to philosophy, not religion, to reveal the deepest woo-woo of the human mind. I find this grievous, but, next to this

---

18 See, eg, Dijksterhuis, Bos, Nordgren *et al* 2006; Dijksterhuis & Nordgren 2006; Levine, Halberstadt & Goldstone 1996; Engel & Singer 2008.

19 Strawson 2006 (4). Cf James 1897, 'The will to believe': 1–31 (15): 'There is but one indefectibly certain truth, and that is the truth that pyrrhonistic scepticism itself leaves standing, – the truth that the present phenomenon of consciousness exists'.

denial, every known religious belief is only a little less sensible than the belief that grass is green.[20]

'The capacity of human minds to be gripped by theory...' The reader knows by now which hemisphere backs theory in the face of experience. In any case, it is absurd, Strawson continues, to reject the idea of there 'seeming' to be experience:

> The phenomenon of there seeming to be experience – the phenomenon we're supposing to be an illusion – can't exist unless there really is experience. Daniel Dennett tries this move. He proposes that 'there is no such thing' as phenomenology: 'There seems to be phenomenology ... but it does not follow from this undeniable, universally attested fact that there really is phenomenology.' In fact it does follow, for the reason I've just given: for there to seem to be phenomenology is for there to be phenomenology. When it comes to experience, you can't open up the is/seems gap.[21]

If we seem to have the experience of sunlight on a bowl of strawberries, that means we have the experience of sunlight on a bowl of strawberries. 'Dennett and his kind find themselves at one with many religious believers', Strawson comments on another occasion:

> This seems at first ironic, but the two camps are deeply united by the fact that both have unshakable faith in something that lacks any warrant in experience. That said, the religious believers are in infinitely better shape, epistemologically, than the Dennettians ...[22]

COULD CONSCIOUSNESS BE REDUCED TO ANYTHING ELSE AT ALL?

If it is true that consciousness is 'the fundamental given natural fact', it clearly follows that it cannot be reduced to something more fundamental. But *is* consciousness – the experiential – fundamental? Schrödinger thought so: 'Consciousness cannot be accounted for in physical terms. For consciousness is absolutely fundamental. It cannot be accounted for in terms of anything else.'[23] And others, as we shall see, were with him.

I am not inclined to demur. We know about the experiential directly – from experience; while we assume the non-experiential only indirectly – from experience. Similarly it cannot be denied that matter is something disclosed to me by my mind: I do not know that mind is something disclosed to me by matter – it might be, or it might not. Strawson again:

> If we ask what evidence there is for the existence of non-experiential concrete reality, the answer is easy and mathematically precise. There

20 Strawson 2008 (55).

21 Strawson 2013.

22 Strawson 2008 (55).

23 Schrödinger 1984 (334).

is zero evidence. There is zero observational evidence for the existence of any non-experiential concrete reality. Nor will there ever be any. All there is is one great big, wholly ungrounded, wholly question-begging theoretical intuition or conviction: we don't see most of the matter around us doing things that we think of as showing signs of experientiality, so we conclude we know it's not experiential, and that it's ridiculous to think otherwise. This appears to be the great foundation of the wildly anti-naturalistic naturalism *de nos jours*.[24]

In other words, a form of anthropomorphism operates in reverse: 'doesn't do what humans do with their consciousness, so can't have it'. Sometimes I am asked, 'surely you can't think a mountain has awareness?' I feel like replying, 'but how would you expect a mountain to behave if it did have awareness? Mow the lawn, drink a beer and go to Sainsbury's?' The idea that consciousness is an ontological primary looks much less bizarre in any other culture than our own. We don't have the luxury of adopting a common-sense attitude here, because our alternatives are (a) either consciousness does not exist at all (see above), or (b) it was there all along, in everything. Which is the more absurd?

Strawson continues: 'The experiential/non-experiential divide, assuming that it exists at all, is the most fundamental divide in nature (the only way it can fail to exist is for there to be nothing non-experiential in nature).'[25] Here Strawson alludes to his view, shared by some other Anglo-American analytic philosophers such as Christian de Quincey,[26] that the divide can be denied only if, effectively, panpsychism is true (that there is nothing that has no inwardness of some kind, nothing that is wholly *non-experiential* in nature). Panpsychism has a long and venerable history in philosophy both Eastern, as might be anticipated, and Western, which might not.[27]

To say that experience exists, 'but is really just something whose nature can be fully specified in wholly non-experiential, functional terms', is to deny its existence.[28] What would it mean for non-experientiality (matter as conventionally conceived) to give rise to experience? This would seem to involve a miracle. Strawson disposes of the argument that it is no different from the 'emergence' of liquidity from molecules of H_2O, and other similar examples. 'For any feature Y of anything that is currently considered to be emergent from X', he writes, 'there must be something about X and X alone in virtue of which Y emerges, and which is sufficient for Y.'[29] Thus there are obvious, well-known, features of water molecules that predict their fluid motion over one another – in other words, their 'wateriness': nothing needs to be added for this to arise. However, there is no feature of matter as conventionally conceived that explains how it could possibly on its own give rise to consciousness.[30] Stating that matter 'just does' give rise to consciousness is to invoke what

---

[24] Strawson 2013.

[25] Strawson 2006 (18).

[26] De Quincey 2010.

[27] Such ideas are hardly confined to Eastern traditions: see Skrbina (2017) for a scholarly and detailed account of the long tradition of such ideas in Western thinking.

[28] Strawson 2008 (54).

[29] Strawson 2006 (13 & 18).

[30] Cf Nagel 2012 (56): 'Harmless emergence is splendidly illustrated by the example of liquidity, which depends on the interactions of the molecules that compose the liquid. But the emergence of the mental at certain levels of biological complexity is not like this. According to the emergent position now being considered, consciousness is something completely new.'

is called brute emergence, and 'brute emergence is by definition a miracle every time it occurs.'[31]

Notice that I say 'matter as conventionally conceived'. Because if you understood matter in a radically different way, as part of a wholly experiential cosmos, you would have solved the mind/matter problem. But that possibility is open only to those who are willing to entertain that matter is sufficiently different from our usual conception of it that it no longer has the explanatory power relied on by reductionists. According to J. B. S. Haldane, if consciousness were not present in matter, such emergence would be 'radically opposed to the spirit of science, which has always attempted to explain the complex in terms of the simple … if the scientific point of view is correct, we shall ultimately find them [signs of consciousness in inert matter], at least in rudimentary form, all through the universe.'[32]

Once again William James delineates the problem most sharply:

> The demand for continuity has, over large tracts of science, proved itself to possess true prophetic power. We ought therefore ourselves sincerely to try every possible mode of conceiving the dawn of consciousness so that it may *not* appear equivalent to the irruption into the universe of a new nature, non-existent until then. Merely to call the consciousness 'nascent' will not serve our turn. It is true that the word signifies not yet *quite* born, and so seems to form a sort of bridge between existence and nonentity. But that is a verbal quibble. The fact is that discontinuity comes in if a new nature comes in at all. The *quantity* of the latter is quite immaterial. The girl in 'Midshipman Easy' could not excuse the illegitimacy of her child by saying, 'it was a very small one'. And consciousness, however small, is an illegitimate birth in any philosophy that starts without it, and yet professes to explain all facts by continuous evolution. *If evolution is to work smoothly, consciousness in some shape must have been present at the very origin of things.*[33]

How could consciousness have emerged *de novo* from matter? It does not matter how much you invoke recursive loops and meta-processes. They only work once you have already got consciousness: they can't themselves account for consciousness. The explanatory gap is unbridgeable, certainly by such means. Philosopher of mind Nakita Newton writes that 'phenomenal consciousness itself is *sui generis*. Nothing else is like it *in any way at all*'.[34] Of course, she is consummately right. It comprehensively fails the H_2O-to-liquid test. Other philosophers have expressed similar views. For example, Colin McGinn puts it with customary vividness: 'You might as well assert that numbers emerge from biscuits or ethics from rhubarb.'[35] Searching for an explanation in molecular biology appears to be a category mistake (unless consciousness is already assumed there, in which case it is not an explanation). So Gunther Stent writes:

---

[31] Strawson 2006 (18).

[32] Haldane 1932 (113).

[33] James 1890, vol 1 (148–9; emphases in original).

[34] Newton 2001 (48; emphasis in original).

[35] McGinn 1993 (60).

Searching for a 'molecular' explanation of consciousness is a waste of time, since the physiological processes responsible for this wholly private experience will be seen to degenerate into seemingly quite ordinary, workaday reactions – no more and no less fascinating than those that occur in, say, the liver.[36]

Even Sherrington came to the conclusion that 'we have difficulty in assigning the lower limit of the mental. It may therefore be that its distribution extends to all organisms, and *even further* … it is as though the elementary mental had never been wanting.'[37] Neuroscientists V. S. Ramachandran and Colin Blakemore conclude that 'consciousness, like gravity, mass, and charge, may be one of the irreducible properties of the universe for which no further account is possible.'[38] Physicists agree. According to Heisenberg, 'if we go beyond biology and include psychology in the discussion, then there can scarcely be any doubt but that the concepts of physics, chemistry, and evolution together will not be sufficient to describe the facts.'[39] This is very similar to Bohr's insight that 'consciousness must be part of nature, or, more generally, of reality, which means that, quite apart from the laws of physics and chemistry, as laid down in quantum theory, we must also consider laws of quite a different kind.'[40] The great mathematician and physicist von Neumann confirmed that 'it is inherently entirely correct that the measurement or the related process of the subjective perception is a new entity relative to the physical environment and is not reducible to the latter. Indeed, subjective perception leads us into the intellectual inner life of the individual, which is extra-observational by its very nature.'[41] And in similar vein, Adam Frank, Professor of Astronomy at the University of Rochester, New York, writes that we must entertain the 'radical possibility that some rudimentary form of consciousness must be added to the list of things, such as mass or electric charge, that the world is built of.'[42]

There is, to put it conservatively, a good chance they are right. It may be irritating to some to face the fact that after several thousand years of ratiocination and experimentation we are arriving at truths that were anciently known to philosophers and sages, East and West, though it is exciting and perhaps reassuring to have them confirmed by elaborate experimentation.

I must expect at this stage a group of reductionists (themselves rarely physicists) who think of themselves as 'sceptics' to treat the very invocation of physics in support of the status of consciousness as something they can dismiss with scorn: derisive phrases such as 'quantum flimflam' are bandied around. Physicists themselves do not, for the most part, see it that way. To quote Bohr, who frequently made connexions between an understanding of consciousness and the world,

---

36 Stent 1968 (395).

37 Sherrington 2009 (354 & 266; emphasis added). Cf Waddington 1961 (121): 'Are we not forced to conclude that even in the simplest inanimate things there is something which belongs to the same realm of being as self-awareness?'

38 Ramachandran & Blakemore 2001 (176).

39 Heisenberg 1958 (95).

40 As recorded in Heisenberg 1971, 'The relationship between biology, physics and chemistry': 103–16 (114).

41 von Neumann 1955b (418).

42 Frank 2017.

in our description of nature the purpose is not to disclose the real essence of the phenomena but only to track down, so far as it is possible, relations between the manifold aspects of our experience.[43]

I believe the passion with which the 'sceptics' dismiss such relations is itself neither scientific nor rational. Its very vehemence leads one to suspect that they may know there is something here so important, and so hard to reconcile with their philosophical position, that they are worried that people will spot that the 'sceptics' are wearing no clothes. (By the way, is it more sceptical, or on the contrary more gullible, to believe that consciousness does not exist, or that it can arise with a handwave out of matter?) So we must remember, when predictable noises are heard, that the materialists are entitled to their point of view, but that the attempt to rule quantum physics out of court does not wash, and should be politely refused. To quote the great physicist John Bell: 'to restrict quantum mechanics to be exclusively about piddling laboratory operations is to betray the great enterprise. A serious formulation will not exclude the big world outside of the laboratory.'[44]

[43] Bohr 1961 (18).
[44] Bell 2004 (217).
[45] Bohr 1961 (56–7).
[46] Schiller 1891 (69).

CAN WE DENY MATTER ALTOGETHER?

Matter itself is an abstraction which no-one has ever seen: we have only seen elements of the world to which we attribute the quality, within our consciousness, of being material. It both substitutes an idea for an experience (which is a kind of event) and, in doing so, produces something static, no longer in process: no longer an experience, now a *thing*. According to Bohr, 'isolated material particles are abstractions, their properties being definable and observable only through their interaction with other systems.'[45] Materialism derives the only thing we undeniably know, the concreteness of experience, from an unknown abstraction: matter.

A number of philosophers had already come to this conclusion. Thus F. C. S. Schiller wrote in 1891:

> It appears that all we know of matter is the forces it exercises. Matter, therefore, is said to be unknowable in itself ... And yet it is perhaps hardly astonishing that a baseless abstraction should be unknowable in itself. And matter certainly is such an abstraction. For all that appears to us is *bodies*, which we call material. They possess certain more or less obvious points of resemblance, and the abstraction, 'matter', is promptly invented to account for them.[46]

And later:

> All the sensible qualities of matter are due to forces, gravitative, cohesive, propulsive, chemical, electrical, or to motions (like heat, sound, light, etc), or 'motive forces'. Matter itself, therefore, is left as the un-

known and unknowable substratum of force ... It is not required to explain the appearance of anything we can experience, and is merely a metaphysical fiction designed to provide forces with a vehicle.[47]

But, as we saw in the previous chapter, energy needs no substratum. Above all, matter cannot be called on to bring about the 'annihilation of the mind by means of one of [the mind's] own abstractions.'[48]

In a 2017 essay entitled 'Minding matter: the closer you look, the more the materialist position in physics appears to rest on shaky metaphysical ground', Frank eloquently expresses both the mystery of consciousness and the puzzling reluctance of many biologists to entertain theories that venture beyond viewing consciousness as a result of processing in the brain. 'Materialists appeal to physics to explain the mind', he writes, 'but in modern physics the particles that make up a brain remain, in many ways, as mysterious as consciousness itself.' And he continues:

> Some consciousness researchers might think that they are being hard-nosed and concrete when they appeal to the authority of physics. When pressed on this issue, though, we physicists are often left looking at our feet, smiling sheepishly and mumbling something about 'it's complicated'. We know that matter remains mysterious just as mind remains mysterious, and we don't know what the connections between those mysteries should be. Classifying consciousness as a material problem is tantamount to saying that consciousness, too, remains fundamentally unexplained.[49]

Moreover, the observer cannot be excluded from the reality observed. To quote physicist Paul Davies: 'You can't do away with the observer and that's a fact.'[50] On which topic, Frank continues that 'putting the perceiving subject back into physics seems to undermine the whole materialist perspective ... A theory of mind that depends on matter that depends on mind could not yield the solid ground so many materialists yearn for.'

All this may be importantly true, but I think we should no more conclude that we can deny matter than that we can deny consciousness. Neither way out of our dilemma is at all satisfactory. What, then, can we say about matter? Strawson uses the word physicalism to refer to a reality which is not antithetically divided into matter and mind, but incorporates both elements as indivisible. From this perspective he writes: 'It's not just that we don't definitely know the nature and limits of the physical. We definitely don't know the nature or limits of the physical. Physics may tell us a great deal about the structure of physical reality', he writes:

---

47 *ibid* (272).
48 *ibid* (271).
49 Frank 2017.
50 Davies 2006.

but it seems that it can't tell us anything about the intrinsic nature of reality in so far as its intrinsic nature is more than its structure ... It's plain that the human science of physics can't fully characterise the nature of concrete reality, even in principle ... On many matters, such as experience, physics is simply silent. If you're not clear on this limitation, you have no idea what physics is. This isn't New Age anti-scientism, it's hardnosed physicalism.'[51]

If asked my view, I would say that matter appears to be an element within consciousness that provides the necessary resistance for creation; and with that, inevitably, for individuality to arise. All individual beings, including ourselves, bring forms into being and cause them to persist: each of us is not, ultimately, any one conformation in matter, but, Ship of Theseus-like, the conformation itself, the morphogenetic field, which requires matter in order to be brought into being, but, once existent, persists while matter comes and goes within it.[52] Remember Schopenhauer: 'Matter is that which persists and endures.'[53] Since the development of new such form-fields in consciousness seems to be part of the creative process of the cosmos, the lending of persistence (for a while) may be matter's peculiar role.

That may raise more questions than it answers, but it is something we can come back to. Meanwhile, as Frank puts it, 'rather than trying to sweep away the mystery of mind by attributing it to the mechanisms of matter, we must grapple with the intertwined nature of the two.'[54]

ARE MATTER AND CONSCIOUSNESS ONE AND THE SAME?

The problem for Descartes was how mind, *res cogitans*, could interact with matter, *res extensa*. That each affects the other is amply confirmed by modern physics, if by nothing else (I will come to other evidence in due course). So the idea that they are so separate that they cannot interact can safely be dismissed. Are they perhaps, then, the same?

Whitehead thought that the 'sharp division between mentality and nature has no ground in our fundamental observation', and that 'we should conceive mental operations as among the factors which make up the constitution of nature.'[55] Certainly matter is not complete without the incorporation of consciousness. The theoretical physicist and mathematician Eugene Wigner pointed out that 'it was not possible to formulate the laws of quantum mechanics in a fully consistent way without reference to the consciousness.'[56] Physicist and philosopher Carl von Weizsäcker went further: 'consciousness and matter are different aspects of the same reality.'[57]

51 Strawson 2013.
52 See pp 662–3 above.
53 Schopenhauer 1974, vol I (99).
54 Frank 2017.
55 Whitehead 1938 (214).
56 Wigner 1970 (172).
57 von Weizsäcker 1980 (250).

Such ideas go back a long way in philosophy before modern physics. Schopenhauer, for example, held that

> a consciousness without object is no consciousness at all ... although materialism imagines that it postulates nothing more than this matter – atoms for instance – yet it unconsciously adds not only the subject, but also space, time, and causality, which depend on special determinations of the subject ... the intellect and matter are correlatives, in other words, the one exists only for the other; both stand and fall together; the one is only the other's reflex. They are in fact really one and the same thing, considered from two opposite points of view.[58]

That might seem odd, simply because matter doesn't behave like consciousness. However ice doesn't behave like water: it is solid, comes in chunks, is opaque, and so hard it can split your head open, while water is fluid, seamless, transparent and can be so dispersed that we cannot see it or touch it at all. Ice is a 'phase' of water. Could matter be a 'phase' of consciousness?

At the quantum level, are consciousness and matter like those creatures that are neither plant nor animal, or both plant and animal? Is a quark material, or an aspect of consciousness – or both, or neither? In a letter to his colleague Abraham Pais, Wolfgang Pauli speculated that a science of the future would refer to such a basic reality as neither psychic nor physical, but somehow both of them and somehow neither of them.[59] He suggested that the mental and the material domains of basic reality should be understood as complementary aspects under which this reality can appear.[60] This suggests that we have to give up a conception of matter that excludes consciousness and a conception of consciousness that precludes matter, even though we may be able to conceive them distinctly only one at a time. Mass and energy are interconvertible: the brain is a manifestation as mass, the mind a manifestation as energy.

'Bohr's principle of complementarity is the most revolutionary scientific concept of this century and the heart of his fifty-year search for the full significance of the quantum idea', wrote the theoretical physicist John Archibald Wheeler.[61] Complementarity – the principle that objects have certain pairs of complementary properties which cannot be observed or measured simultaneously – is most commonly thought of in relation to position and momentum, but has further applications within physics. Extending the concept of complementarity somewhat, Pauli writes:

> the only acceptable point of view appears to be the one that recognises *both* sides of reality – the quantitative and the qualitative, the physical and the psychical – as compatible with each other, and can embrace them simultaneously ... it would be most satisfactory of

---

58 Schopenhauer 2000, vol II (15–16).

59 Letter of Pauli to Abraham Pais, 17 August 1950; in von Meyenn 1996 (152): » *Meine persönliche Ansicht ist die, daß in einer zukünftigen Wissenschaft die Realität weder ‚psychisch‘ noch ‚physisch‘ sein wird, sondern irgendwie beides und irgendwie keines von Beiden.* «

60 Pauli 1952 (164).

61 Wheeler 1963 (30).

all if physis [physical nature] and psyche could be seen as complementary aspects of the same reality.[62]

Pauli did not intend that to be seen as a merely technical point. He advanced the opinion that the 'issue of complementarity within physics naturally leads beyond the narrow field of physics to analogous conditions of human knowledge'.[63] And Bohr would have agreed: 'Bohr's preeminent concern was to extend the idea of complementarity beyond physics.'[64]

That contraries fulfil one another (or *contraria sunt complementa*, as Bohr's coat of arms declared) appears odd to us only because of the pervasive and restrictive influence on the Western capacity for thought of Plato and Aristotle. In a memoir of Bohr at the time of his death, his colleague the physicist Léon Rosenfeld wrote:

> I had occasion to discuss Bohr's ideas with the great Japanese physicist [Hideki Yukawa], whose conception of the meson with its complementary aspects of elementary particle and field of nuclear force is one of the most striking illustrations of the fruitfulness of the new way of looking at things that we owe to Niels Bohr. I asked Yukawa whether the Japanese physicists had the same difficulty as their Western colleagues in assimilating the idea of complementarity and in adapting themselves to it. He answered 'No, Bohr's argumentation has always appeared quite evident to us'; and, as I expressed surprise, he added, with his aristocratic smile, 'You see, we in Japan have not been corrupted by Aristotle.'[65]

To which, Rosenfeld adds: 'If Yukawa had also mentioned Plato, his epigram would have given a complete characterization, which it would be difficult to make more pregnant, of the significance of Bohr's contribution to philosophical thought.'

All this is in keeping with Strawson's position that there is nothing '*merely* physical' about the physical:

> If everything that concretely exists is intrinsically experience-involving, well, that is what the physical turns out to be; it is what energy (another name for physical stuff) turns out to be. This view does not stand out as particularly strange against the background of present-day science, and is in no way incompatible with it.[66]

So far, so good. But should we conclude that matter and consciousness are simply one? They may be instantiations or aspects of one entity or process, but they are surely different. And the observer and the observed, though they are each affected by the other, must remain at least partly distinct. Yet in the quantum world we cannot assume that any of these are ultimately separate. The world I *know* is the world as *I* know it. According to Werner Heisenberg:

62 Pauli 1955 (208 & 210).

63 Pauli 1950 (79) (trans H Atmanspacher & H Primas).

64 Atmanspacher & Primas 2006 (21).

65 Rosenfeld 1963 (47).

66 Strawson 2008 (8).

Natural science does not simply describe and explain nature; it is a part of the interplay between nature and ourselves; it describes nature as exposed to our method of questioning. This was a possibility Descartes could not have thought, but it makes the sharp separation between the world and the I impossible.[67]

Planck puts it like this: 'As every act of research measurement has a more or less causal influence on the very process that is under observation, it is practically impossible to separate the law that we are seeking to discover behind the happening itself from the methods that are being used to bring about the discovery.'[68] Sir James Jeans had this to say in an address to the British Association for the Advancement of Science, of which he was President:

> There is, in fact, no clear-cut division between the subject and object; they form an indivisible whole which now becomes nature. This thesis finds its final expression in the wave-parable, which tells us that nature consists of waves and that these are of the general quality of waves of knowledge, or of absence of knowledge, in our own minds.[69]

This suggests two important conclusions. The first is that the way in which we approach Nature governs what we find (science gives not an account of nature *tout court*, but an account of nature 'as exposed to our *method of questioning*'). Thus, to quote Pauli, it may even depend on the *nature* of the individual observer:

> Modern microphysics turns the observer once again into a little lord of creation in his microcosm, with the ability (at least partially) of freedom of choice and fundamentally uncontrollable effects on that which is being observed. But if these phenomena are dependent on how (with what experimental system) they are observed, then is it not possible that there are *also* phenomena ... that depend on *who* observes them (ie, on the nature of the psyche of the observer)?'[70]

The second is this: that the conclusion we should draw is – *not* that all that we can encounter are *representations* of something we cannot know – but the precise opposite: that we do actually deal with reality and know it, just with an aspect of it that *we partly call forth ourselves by our approach*. The fact that we play a part in its being what it is does not make it unreal.

In a variation of a thought experiment known as the Frauchiger-Renner experiment, Matthew Leifer has shown that there is no single outcome of a given measurement that's objectively true for all observers: the results of measurements depend on the perspective of the observer.[71] This bears out Wheeler's dictum that 'quantum states are not physical objects: they exist only in our imagination ... reality may be different for different observers.'[72] But note that noth-

---

67 Heisenberg 1958 (75).
68 Planck 1932 (95).
69 Jeans 1934.
70 Letter of Pauli to CG Jung, 23 December 1947; in Meier 2001 (32–3).
71 Frauchiger & Renner 2018; Ananthaswamy 2018.
72 Peres 2005 (514).

ing in all of this makes what we take to be reality 'really' an illusion. Einstein is credited with having said that 'reality is merely an illusion, albeit a very persistent one'. There is no evidence that I know of that Einstein ever said any such thing, nor is such a statement in keeping with, for example, his distress at not being able to square relativity with quantum mechanics. Whether Einstein said anything like it or not, the remark is deeply mistaken. If you believe matter is the only reality, and you then learn that matter as you think of it is illusory, you will conclude that reality is illusory. But it is not. It is matter, *as we think of it*, that is an illusion. And there is more to reality than matter. It was your thinking that misled you.

IF MATTER AND CONSCIOUSNESS ARE DIFFERENT, IN WHAT RESPECT?

How on earth might consciousness – immaterial and lacking extension in space as it is – emerge from matter, which is very clearly both material and extended in space? Since, as Colin McGinn reflects, this 'looks more like magic than a predictable unfolding of natural law', he suggests 'the following heady speculation: that the origin of consciousness somehow draws upon those properties of the universe that antedate and explain the occurrence of the big bang ... If so, consciousness turns out to be older than matter in space, at least as to its raw materials.'[73] That would be one very important difference.

In that form it is indeed a speculation, but so is the idea that matter precedes consciousness; by contrast, that consciousness precedes matter is an idea that has an ancient lineage, and more than a little, I shall suggest, going for it. Matter could be born of consciousness without either being the same as, or wholly distinct from, the other. And if true, a form of asymmetry familiar to the readers of this book would operate: mind and matter being aspects of the same thing, but that not of itself making them equal.

Schrödinger points out that though what we see depends on consciousness, consciousness itself is nowhere to be seen.[74] We look for it in the field, and do not see it, not realising it is itself the ground on which we stand to search: as the eye sees itself nowhere in the picture of the world it brings into being. Perhaps, then, we should expect matter to be visible within consciousness, but hardly consciousness within matter, since consciousness would have to see itself therein. This would agree with Schrödinger's conclusion: 'Mind has erected the objective outside world of the natural philosopher out of its own stuff. Mind could not cope with this gigantic task otherwise than by the simplifying device of excluding itself – withdrawing from its conceptual creation. Hence the latter does not contain its creator.'[75]

A long roll call of the most distinguished physicists would support the view that the originary 'stuff' of the universe is consciousness.

[73] McGinn 1995.

[74] Schrödinger 1967a (122): 'We are faced with the following remarkable situation. While the stuff from which our world picture is built is yielded exclusively from the sense organs as organs of the mind, so that every man's world picture is and always remains a construct of his mind and cannot be proved to have any other existence, yet the conscious mind itself remains a stranger within that construct, it has no living space in it, you can spot it nowhere in space. We do not usually realise this fact, because we have entirely taken to thinking of the personality of a human being, or for that matter also that of an animal, as located in the interior of its body. To learn that it cannot really be found there is so amazing that it meets with doubt and hesitation, we are very loath to admit it'.

[75] *ibid* (121).

Thus Max Planck was famously asked whether he thought consciousness could be explained in terms of matter and its laws. 'No', he replied. 'I regard consciousness as fundamental. I regard matter as derivative from consciousness. We cannot get behind consciousness. Everything that we talk about, everything that we regard as existing, postulates consciousness.'[76] It is worth noting that the interviewer prefaces his piece with the remark: 'In my interview with him Professor Planck replied to all my questions with a quite remarkable lack of hesitation. It would seem that his ideas on these subjects are now definitely formed, or else that he thinks with remarkable rapidity – probably both suppositions are true.' Thirteen years later, and three years before he died, Planck went further:

> As a physicist, and therefore as a man who has spent his whole life in the service of the most down-to-earth science, namely the exploration of matter, no one is going to take me for a starry-eyed dreamer. After all my exploration of the atom, then, let me tell you this: there is no matter as such. All matter arises and exists only by virtue of a force which sets the atomic particles oscillating, and holds them together in that tiniest of solar systems, the atom ... we must suppose, behind this force, a conscious, intelligent spirit. This spirit is the ultimate origin of matter.[77]

Eugene Wigner concurs: 'It will remain remarkable, in whatever way our future concepts may develop, that the very study of the external world led to the scientific conclusion that the content of consciousness is the ultimate universal reality.'[78] This theme is picked up by the geneticist Ho. When 'reductionist, atomistic science is pursued to its logical conclusion', she writes, 'and pushed to its very limits, it can only undermine the basis on which it was built. For everywhere, it reaffirms the unity of nature in which the knowing being is inextricably embedded, compelling us towards a new knowledge system and a new way of knowing that is at the same time very old.'[79]

A passage of the great astronomer, physicist and mathematician Sir Arthur Eddington deserves quotation in full:

> The universe is of the nature of a thought or sensation in a universal Mind ... To put the conclusion crudely – the stuff of the world is mind-stuff. As is often the way with crude statements, I shall have to explain that by 'mind' I do not exactly mean mind and by 'stuff' I do not at all mean stuff. Still this is about as near as we can get to the idea in a simple phrase. The mind-stuff of the world is, of course, something more general than our individual conscious minds; but we may think of its nature as not altogether foreign to the feelings in our consciousness ... *It is the physical aspects of the world that we have to explain.*

---

76 Planck 1931a. A revealing story: I found this missing from Max Planck's Wikiquote page. Looking up the editorial discussion page (en.wikiquote.org/wiki/Talk:Max_Planck), I found the following: 'Dear Editors and Readers, I have removed the following citation [there follows this quote *verbatim*]. Rational [*sic*]: I checked this citation in its source – that is, on page 17 in the January 25, 1931 issue of the newspaper *The Observer* in the article entitled "Interviews with Great Scientists – VI. – Max Planck", in the digital archive of *The Guardian* and *Observer*, and this citation has proved to be false ... The file I attached to my letters contains a detailed reasoning.' However, another reader wrote: 'Dear all. I took the liberty of double checking in *The Observer*, 25 January, 1931, p.17 ...The interview by J. W. N. Sullivan does indeed include the said quote, verbatim, in the second column! (Copy available on request.) I therefore request that the quote be reinstated on the page ...' It had not been reinstated. I checked for myself and found the quotation, *verbatim*, just where the previous commentator had described, except that it was in column 3. I therefore reinstated it.

77 Planck 1944: » Als Physiker, also als Mann, der sein ganzes Leben der nüchternen Wissenschaft, nämlich der Erforschung der Materie diente, bin ich sicher frei davon, für einen Schwarmgeist gehalten zu werden. Und so sage ich Ihnen nach meiner Erforschung des Atoms dieses: Es gibt keine Materie an sich! Alle Materie entsteht und besteht nur durch eine Kraft, welche die Atomteilchen in Schwingung bringt und sie zum winzigsten Sonnensystem des Atoms zusammenhält ... so müssen wir hinter dieser Kraft einen bewussten, intelligenten Geist annehmen. «

And, in a phrase that is in direct opposition to the aspirations of materialist reductionism: 'Our bodies are more mysterious than our minds.'[80]

Eddington continues:

> Consciousness is not sharply defined, but fades into sub-consciousness; and beyond that we must postulate something indefinite but yet continuous with our mental nature. This I take to be the world-stuff ... We have only one approach, namely, through our direct knowledge of mind. The supposed approach through the physical world leads only into the cycle of physics, where we run round and round like a kitten chasing its tail and never reach the world-stuff at all ... It is difficult for the matter-of-fact physicist to accept the view that the substratum of everything is of mental character. But no one can deny that mind is the first and most direct thing in our experience, and all else is remote inference – inference either intuitive or deliberate.'[81]

Another classic statement comes again from Sir James Jeans, also a physicist, astronomer and mathematician:

> The stream of knowledge is heading towards a non-mechanical reality; the universe begins to look more like a great thought than like a great machine. Mind no longer appears to be an accidental intruder into the realm of matter ... we ought rather hail it as the creator and governor of the realm of matter.[82]

Many modern physicists speak in a similar vein, though perhaps less eloquently. Sir Roger Penrose declares: 'I think that matter itself is now much more of a mental substance'.[83]

Astronomical physicist Richard Conn Henry writing in *Nature* avers that 'the Universe is entirely mental ... and we must learn to perceive it as such'.[84] Elsewhere he expands on this theme, and goes further:

> Non-local causality is a concept that had never played any role in physics, other than in rejection ('action-at-a-distance'), until Aspect showed in 1981 that the alternative would be the abandonment of the cherished belief in mind-independent reality; suddenly, spooky-action-at-a-distance became the lesser of two evils, in the minds of the materialists. Why do people cling with such ferocity to belief in a mind-independent reality? It is surely because if there is no such reality, then ultimately (as far as we can know) mind alone exists. And if mind is not a product of real matter, but rather is the creator of *the illusion of* material reality (which has, in fact, despite the materialists, been *known* to be the case, since the discovery of quantum mechanics in 1925), then a theistic view of our existence becomes the only rational alternative to solipsism.[85]

78 Wigner 1961/1983 (169).
79 Ho 1994 (181).

80 Eddington 1928 (276–7; emphasis added).
81 *ibid* (280 & 281).
82 Jeans 1930 (137).
83 Penrose & Clark 1994 (17–24).
84 Henry 2005.
85 Henry & Palmquist 2007 (emphases in the original).

(In Chapter 28 I will return to what may or may not be meant by theism here.)

Astrophysicist Bernard Haisch states: 'It is not matter that creates an illusion of consciousness, but consciousness that creates an illusion of matter.'[86] And according to Henry, 'that is correct physics: it is not controversial in the *slightest* degree that there is no reality; this has been demonstrated in both theory and experiment'.[87] Presumably, once again, it depends on what you mean by reality. While it is not the way it seems till it seems it to you, must we conclude that it is invented *ex nihilo* by your looking at it? Surely not. There are constraints on what it can be, otherwise everything would be as true as everything else, and thought, speech, action and even existence itself would become pointless.

The danger of naïve idealism

There is a sort of suffocation induced by the view that we alone create reality. Planck says 'all ideas of the form of the outer world are ultimately only reflections of our own perceptions',[88] and something in me partly consents, partly recoils. There is a wonderful passage at the end of Eddington's book *Space, Time and Gravitation*, that I cannot forbear quoting:

> The theory of relativity has passed in review the whole subject-matter of physics. It has unified the great laws, which by the precision of their formulation and the exactness of their application have won the proud place in human knowledge which physical science holds to-day. And yet, in regard to the nature of things, this knowledge is only an empty shell – a form of symbols. It is knowledge of structural form, and not knowledge of content. All through the physical world runs that unknown content, which must surely be the stuff of our consciousness. Here is a hint of aspects deep within the world of physics, and yet unattainable by the methods of physics. And, moreover, we have found that where science has progressed the farthest, the mind has but regained from nature that which the mind has put into nature. We have found a strange foot-print on the shores of the unknown. We have devised profound theories, one after another, to account for its origin. At last, we have succeeded in reconstructing the creature that made the foot-print. And Lo! it is our own.[89]

While appreciating the deep wisdom of this image, I think that we are not alone in a dead universe. It is that reciprocity that I miss here, stirring as the passage may be. It is important to note that while we help bring the world about, we are constrained by something other than ourselves – and at that, not a little. According to Henry Pierce Stapp, an American mathematical physicist who worked closely with Heisenberg, Pauli and Wheeler, 'in the quantum world the observing processes of acquiring empirical knowledge must disturb, or perhaps

---

86 Haisch 2006 (137).

87 Henry 2008; referring to Gröblacher, Paterek, Kaltenbaek *et al* 2007 (emphasis in the original). See more recently, eg, Manning, Khakimov, Dall *et al* 2015.

88 Planck 1960 (53).

89 Eddington 1920 (200–1).

even bring into existence, the values that we observe. By virtue of Heisenberg's discovery, the process of our acquiring knowledge about the material aspects of nature cannot merely reveal already existing values. The process of our acquiring knowledge injects our mental aspects in an essential way into the process that determines "what we will find if we look".[90] In other words, the process is genuinely both creative and undetermined.

On the other hand, it seems to me that there is an electron we did not actually put there. When all's said and done, what seems most true to experience is the rather modest view expressed in Wheeler's famous pronouncement: 'This is a participatory universe.'[91] The phrase is somewhat in line with Bohr's reflection that 'the new situation in physics has ... forcibly reminded us of the old truth that we are *both onlookers and actors* in the great drama of existence.'[92] Both. Or as Stapp puts it,

> the new theory elevates our acts of conscious observation from causally impotent witnesses of a flow of physical events determined by material processes alone to irreducible mental inputs into the determination of the future of an evolving psycho-physical universe. In this orthodox quantum mechanical understanding of the world our minds matter.[93]

Wheeler goes further:

No theory of physics that deals only with physics will ever explain physics. I believe that as we go on trying to understand the universe, we are at the same time trying to understand man. Today I think we are beginning to suspect that man is not a tiny cog that doesn't really make much difference to the running of the huge machine but rather that there is a much more intimate tie between man and the universe than we heretofore suspected ...[94]

While, of course, sympathetic to this, I find the emphasis on human life rather than life more generally (to be fair, characteristic of its period) jars slightly. And he rephrased it here:

Modern quantum theory, the overarching principles of twentieth-century physics, leads to quite a different view of reality, a view that man, or intelligent life, or communicating observer participators are the whole means by which the very universe is created: without them, nothing.[95]

What seems to me indisputable is that there is a relationship of reciprocity. That for me would be a touchstone of whether we are on the right path. In such a relationship both – or all – parties are changed. The reciprocal truth of the observer changing what is observed is that what is observed changes the observer. This was a view

90 Stapp 2017 (9).
91 Wheeler 1990 (5).
92 Bohr 1961 (119; emphasis added).
93 Stapp 2017 (x).
94 Wheeler 1973, as quoted in Chandos 2015.
95 Wheeler 2006.

espoused by Goethe, and not just in the obvious ways you might imagine. According to him, we literally grow faculties. He held that an object properly contemplated generates in the beholder the faculty proper to its own perception: 'Every new object, well contemplated and clearly seen, opens up a new organ within us'.⁹⁶ Contemplation of the world in a spirit of openness and humility fundamentally enlarges our being, where dogma and complacency simply narrow it.

Equally it enables the greater reality of the cosmos – whatever it may be – to fulfil itself through us.

A theme of our explorations has been the status of things. Things, it seems, emerge from our descriptions of experience: they do not constitute it. Whether a thing enters our world or not depends on the scale at which it is seen, or sought, as stem cell researcher and complexity theorist Neil Theise demonstrates:

> Ant colonies are a good example: from afar, the colony appears to be a solid, shifting, dark mass against the earth. But up close, one can discern individual ants and describe the colony as the emergent self-organization of these scurrying individuals. Moving in still closer, the individual ants dissolve into myriad cells … The cell as a definable unit exists only on a particular level of scale. Higher up, the cell has no observational validity. Lower down, the cell as an entity vanishes, having no independent existence. The cell as a thing depends on perspective and scale: 'now you see it, now you don't', as a magician might say.⁹⁷

And as Theise and a colleague, physicist Menas Kafatos, put it elsewhere: 'No single scale of observation can reveal the whole; at the moment selection is made of a scale of observation, the features of other levels of scale are hidden from view.'⁹⁸ This intrinsic incompleteness of any one point of view is a theme of this book, wonderfully imaged in the Japanese Zen garden, Ryōan-ji, referred to earlier.⁹⁹

It may be objected that, whether we see something or not, it still exists. But what that tells us is that a 'thing' is a category within our thought. I am reminded here of Ruskin's observation about seeing things clearly. 'We never see anything clearly', he writes; 'what we call seeing a thing clearly, is only seeing enough of it to make out what it is; this point of intelligibility varying in distance for different magnitudes and kinds of things'.¹⁰⁰ If Nietzsche was right in saying that 'A thing = its qualities',¹⁰¹ and qualities change unrecognisably with scale, so do things come and go from experience depending on how they are observed. As the perspective shifts so do the *Gestalten*.

Theoretical physicist and mathematician Freeman Dyson noticed something interesting about scale: how seeing a *mechanism* is a feature of a certain scale of vision. 'For the biologists', he wrote,

---

96 Goethe 1988 (39): »*Jeder neue Gegenstand, wohl beschaut, schließt ein neues Organ in uns auf*«.

97 Theise 2005; Theise 2006; Kurakin 2011.

98 Theise & Kafatos 2016.

99 See p 550 above.

100 See McGilchrist 2009b (181–2).

101 Nietzsche 2003b (73).

'every step down in size was a step toward increasingly simple and mechanical behaviour':

> A bacterium is more mechanical than a frog, and a DNA molecule is more mechanical than a bacterium. But twentieth-century physics has shown that further reductions in size have an opposite effect. If we divide a DNA molecule into its component atoms, the atoms behave less mechanically than the molecule. If we divide an atom into nucleus and electrons, the electrons are less mechanical than the atom.[102]

Once one reaches quantum level, events cannot be separated from the consciousness of the observer, and

> the laws leave a place for mind in the description of every molecule … In other words, mind is already inherent in every electron, and the processes of human consciousness differ only in degree but not in kind from the processes of choice between quantum states which we call 'chance' when they are made by electrons.[103]

It is sometimes objected that quantum effects 'wash out' when a system is viewed at the level of everyday. Not so, according to Stapp:

> The oft-heard claim that 'quantum mechanics is not relevant to the mind-brain problem because quantum theory is only about tiny things', is absolutely contrary to the basic quantum principles … Quantum mechanics is explicitly designed to cover 'big' systems, and by becoming 'big' a quantum system does not become classical! Indeed, the fact that quantum mechanics is explicitly designed to cover big things is important to the solution of the mind-brain problem.[104]

IS CONSCIOUSNESS, THEN, NOT JUST IN US BUT IN EVERYTHING THAT EXISTS?

Physicalism, according to Strawson, 'entails panexperientialism or panpsychism. All physical stuff is energy, in one form or another, and all energy … is an experience-involving phenomenon.'[105] Heraclitus already saw that 'mind is common to all things'.[106] Indeed he said that all things are full of soul, which is perhaps even more astute.[107] Soul (*psuchē* or psyche) in Heraclitus is limitless (not material or extended in space), as is Anaximander's *apeiron*, the unbounded origin of all things. Soul, too, according to Heraclitus scholar Charles Kahn, is for Heraclitus a first principle, or *archē*.[108] In Heraclitus' philosophy, all things are requital for fire, and fire for all things – which implies that matter and energy are aspects of one and the same entity (something that science has discovered relatively recently: $E = mc^2$).[109] Energy, in the form of fire, and psyche, play a similar role in his philosophy:

---

102 Dyson 1979 (248).

103 Skrbina 2017 (256).

104 Stapp 2017 (13).

105 Strawson 2006 (25).

106 Heraclitus fr XXXI [Diels 13, Marcovich 23d]: ξυνόν ἐστι πᾶσι τὸ φρονέειν (trans IMcG).

107 Heraclitus fr XXXV [Diels 45, Marcovich 67]: this is the interpretation offered by Kahn in his commentary (Kahn 1979, 129).

108 Kahn 1979 (128).

109 Heraclitus fr XL [Diels 90, Marcovich 54].

each is universal, each is all-inclusive. And of the soul, he wrote, 'You will not find out the limits of the soul whatever way you travel, so deep is its *logos*.'[110]

After Heraclitus, Empedocles too gave it as his opinion 'that all things have thought (*phronēsis*) and a share of mind (*noēma*).'[111] The Roman philosopher Epicurus argued that will could not arise *de novo*, and therefore that, since Nature exhibits will in her creatures, the atoms from which everything was made had to possess a kind of will themselves. Nature embodies drives, but drives themselves are not material. Will must therefore be present in some form in the fundamental principles of the universe. Much later Spinoza and Leibniz were panpsychists, as were probably Goethe, and his contemporary, the philosopher Herder, who considered the forces in plants and stones as analogous to the soul, each endowed with a different degree of consciousness.[112]

Schopenhauer was also a kind of panpsychist. 'Among people untrained in philosophy', he wrote, 'there still exists the old, fundamentally false antithesis between mind and matter.'[113]

His argument, that we do not understand matter so long as we imagine it to be simple 'stuff' devoid of mind, is based on observations that (for obvious reasons) do not rely on quantum physics, but apply to classical physics too, and are therefore deserving of special consideration:

> You believe you perceive dead [ie completely passive] material, void of all qualities, because you suppose you can truly understand everything which you are able to trace back to a *mechanistic* effect. But as physical and chemical effects are avowedly incomprehensible to you so long as you cannot trace them back to mechanistic effects, so are these mechanistic effects themselves – that is to say, modes of expression proceeding from weight, impenetrability, cohesion, hardness, inflexibility, elasticity, fluidity, etc – just as mysterious as these others, indeed as mysterious as thought in the human head.[114]

And he anticipates Bertrand Russell's point that from physics we know nothing of the physical world other than its mathematical structure. 'What is really comprehensible through and through in mechanics', writes Schopenhauer,

> extends no further in any account of an effect than what is purely mathematical; it is limited, that is to say, to determining its spatial and temporal qualities ... In short: all ostensible mind can be attributed to matter, but all matter can likewise be attributed to mind; from which it follows that the antithesis is a false one.[115]

Here is Russell for comparison: 'we know the intrinsic character of the mental world to some extent, but we know absolutely nothing of

---

110 Heraclitus fr XXXV [Diels 45, Marcovich 67] (trans IMcG).

111 Empedocles [Diels 31.B 110] as at Barnes 1987 (163).

112 See Nisbet 1970 (11); also Skrbina 2017 (113).

113 Schopenhauer 2004 (212).

114 *ibid* (212–3). Hollingdale has 'dead, ie completely passive material void of all qualities'. I have added punctuation to make it easier to read.

115 *ibid* (213).

the intrinsic character of the physical world'.[116] What Russell means by this is, of course, from physics alone (or what Schopenhauer calls 'mechanics'). Hence his remark that 'physics is mathematical, not because we know so much about the physical world, but because we know so little: it is only its mathematical properties that we can discover. For the rest our knowledge is negative.'[117] Again he speaks of our knowledge *from physics*. Sir Arthur Eddington made a similar point in a passage I quoted earlier (see p 1056).

But since it is only such a kind of thinking, from which all human understanding has been stripped away and reduced to maths, that would make us think that consciousness and matter were sundered in the first place, it is the reductionist's unwarranted assumption that consciousness does not enter into matter that we must address.

Amongst philosophers since Schopenhauer who were in some respect panpsychist one would have to include Peirce, James, Dewey, Bergson, Whitehead and Hartshorne amongst others. Even the sceptical Russell addressed the issue of mind and matter in a spirit that is open to panpsychism:

> My own feeling is that there is not a sharp line, but a difference of degree; an oyster is less mental than a man, but not wholly un-mental. And I think 'mental' is a character, like 'harmonious' or 'discordant', that cannot belong to a single entity in its own right, but only to a system of entities ... inanimate matter, to some slight extent, shows analogous behavior [to memory] ... The events that happen in our minds are part of the course of nature, and we do not know that the events which happen elsewhere are of a totally different kind. The physical world ... is perhaps less rigidly determined by causal laws than it was thought to be; one might, more or less fancifully, attribute even to the atom a kind of limited free will.[118]

The eminent evolutionary biologist Sir Julian Huxley wrote that the relation between mind and matter is so close that

> mind or something of the nature as mind must exist throughout the entire universe. This is, I believe, the truth. We may never be able to prove it, but it is the most economical hypothesis: it fits the facts much more simply ... than one-sided idealism or one-sided materialism.[119]

And in his introduction to Teilhard de Chardin's masterwork *The Phenomenon of Man* he wrote:

> evolutionary fact and logic demand that minds should have evolved gradually as well as bodies and that accordingly mind-like properties ... must be present throughout the universe ... we must infer the presence of potential mind in all material systems, by backward

---

116 Russell 1927 (307).

117 *ibid* (163).

118 Russell *op cit* (209, 306 & 311).

119 Huxley 1942 (141).

extrapolation from the human phase to the biological, and from the biological to the inorganic.[120]

In biology the issue is no less pressing, and perhaps more pressing, since the issue of emergence of mind is not purely historical, located in some, for practical purposes theoretical, primaeval past, but immediate and present every time a new life comes into being. If the two cells, sperm and egg, out of which the embryo proceeds, are to be considered mindless and yet a newborn infant is not, at what point in this process does mind, brutishly, emerge? And why just there? And what is to one side of it? To quote again the mother of the illegitimate baby in *Mr Midshipman Easy*: 'if you please, ma'am ... it was *such a little one*.'[121] There is no difficulty with 'little' for those who believe consciousness was there all along; but a problem for those who don't, because, however little, its coming about at all still requires a miracle.

About 30 years after Haldane, the biologist Bernhard Rensch, an architect of the mainstream Modern Synthesis theory of evolution, asserted that, just as there is a blurring of categories when we examine the evolution of one life form to another at the level of microorganisms and cells, the stark division between living and non-living systems is blurred, and a mistaken distinction likely carries over to the boundaries of conscious experience as well.[122]

In sum, it seems that (1) mind and matter have a close relationship; that (2) we cannot logically dismiss the existence of consciousness; and (3) ought to be unwilling to dismiss the existence of matter; that (4) they are not so distinct that they cannot interact; that (5) neither are they identical; and yet (6) may be aspects of one and the same reality. Nonetheless (7) they are not equal, in that there is reason to believe that consciousness is prior ontologically to matter.

CONSCIOUSNESS AND LIFE

Up to this point I have concerned myself mainly with consciousness in the universe as a whole, both animate and inanimate. But animacy, otherwise known as life, is different from inanimacy. Where there is animacy, both consciousness and matter equally *evolve* much faster than they would do in its absence; or, to use a both more familiar and less appropriable terminology, they *become* faster than in its absence. I recently came across this quote from Bohm: 'when one analyses processes taking place in inanimate matter over long enough periods of time, one finds a similar behaviour [to living processes]. Only here the process is so much slower...'[123] And what gives us the impression of increasing complexity in evolution is the way consciousness, more than matter, *becomes* – increasingly complex.

The re-admission of the observer's consciousness into the de-

120 Huxley 1959a (16).
121 Marryat 1836 (20).
122 Skrbina 2017 (249–50).
123 Bohm 1957 (163).

scription of the cosmos is a change of unequalled significance in the history of science since its banishment in the seventeenth century. In a theme that should be familiar to my readers, that exile enabled us to become hugely, indisputably, powerful; but at the price of a lack of understanding of what it is we had power over. 'An adequate scientific theory of reality', writes Stapp, 'ought to accommodate *all* the regularities of human experience. This includes not only the results of experiments pertaining to astronomical, terrestrial, and atomic physics, but also to the experiences of normal everyday life.'[124] Until the advent of quantum mechanics, the most important fact of nature, the existence of consciousness, was banished from the scientific world-picture, in which everything was supposed to be accounted for in non-experiential terms. But its enforced repatriation by quantum physics, though unreservedly to be welcomed, will scarcely have more than a technical effect on science, as long as science still assumes that all can, ultimately, be reduced to physics and chemistry.

Physics can't, as Russell points out, tell you anything about reality except the mathematics of its structure; and it can't begin to help with what cannot be measured. Even in what can be measured, the observer, as we have seen, plays a key role, but where measurement is excluded the observer is all-important. The phrase 'the observer', by the way, already seems to refer to a third party. But it doesn't. The observer is *you and me*. And observation describes not an 'I–It' but an 'I–Thou' relationship. We are not distinct from – *over against* – Nature: we emerge out of, live within, and return to Nature. In a sense we are Nature herself reflecting on Nature. Seeing this does not exclude the adoption of other perspectives, when required; but this particular perspective cannot be excluded or circumvented. Thus Whitehead:

> For natural philosophy everything perceived is in nature. We may not pick and choose. For us the red glow of the sunset should be as much part of nature as are the molecules and electric waves by which men of science would explain the phenomenon ... We are instinctively willing to believe that by due attention, more can be found in nature than that which is observed at first sight. But we will not be content with *less*.[125]

There is a dialogue between our own consciousness and the aspects of the world we experience. For example, the experience of listening to music is a betweenness. Is it just out there, on its own? Clearly not. Is it, then, just in my brain? Clearly not. It exists only when outer and inner come together: that is, it lies in the betweenness. Experience – mind – is always a betweenness. And I believe all reality is like this. Outer and inner should not be separated; if they

[124] Stapp 2017 (6; emphasis in original).

[125] Whitehead 1920 (29; emphasis added).

are, they become separately inert, since everything arises from their being together. In Goethe's wonderful words: *Natur hat weder Kern / Noch Schale, / Alles ist sie mit einemmale* ('Nature has neither kernel nor shell: she is everything at once').[126] Apparently Wittgenstein was so taken with this expression of the transcendence of dualism that he considered using it as an epigraph to the *Investigations*.[127] This transcendence of dualism is also the core insight of Schelling's philosophy.

Science can legitimately add to our experience – that's why we value it – but it cannot legitimately diminish our experience simply on the grounds that it doesn't know how to deal with much of it. Whitehead continues that science must not imply that we 'add in' what we understand and perceive, as though it were somehow extraneous to the 'scientific facts'. We should not be asked to imagine that we *add* the greenness of grass: 'what is given in perception is the green grass. This is an object which we know as an ingredient in nature.' He argues against what he calls

> the bifurcation of nature into two systems of reality, which, in so far as they are real, are real in different senses. One reality would be the entities such as electrons which are the study of speculative physics. This would be the reality which is there for knowledge; although on this theory it is never known.[128]

The other sort of reality, the part that is in reality directly known, would be seen as merely the 'byplay of the mind' – a 'dream'. In this scheme our 'conjecture', our theory, about reality would take precedence over experience. *Wissen* would triumph over *kennen*. This is what Whitehead called the *fallacy of misplaced concreteness*. It operates wherever the theory, the map, displaces the terrain that is mapped, and is taken for the reality. It is a fallacy to which, as we have seen, the left hemisphere is peculiarly susceptible, since the left hemisphere is better at producing models than engaging with reality – and models are inevitably simpler (therefore more prone to convey a false sense of total understanding) than the reality they model. Whitehead insists that

> a cosmology should above all things be adequate. It should not confine itself to the categoreal notions of one science, and explain away everything that will not fit in. Its business is not to refuse experience but to find the most general interpretive system.[129]

However, the lack of acknowledgment of the place of psyche within the framework of physics, which Schrödinger, Planck, Heisenberg, Pauli and Bohr all lamented, no longer obtains. Things are moving on. According to Stapp, it is crucial to appreciate that orthodox quantum theory is intrinsically a *psychophysical* theory, 'fundamentally

[126] Goethe, 'Allerdings: dem Physiker', lines 15–17.
[127] Baker & Hacker 2005 (30).
[128] Whitehead 1920 (30).
[129] Whitehead 1929b (69).

a causal weaving together of our streams of conscious experiences, described in psychological terms, with a theoretical representation of the physical world described in mathematical language.'[130] In 1938 Jung, who collaborated intensely with Pauli over a period of nearly 30 years, wrote to another colleague:

> though the methods of modern physics are different from those of psychology their fundamental ideas are not. I would not be surprised if one day we saw a far-reaching agreement between the basic formulations of psychology and physics. I am convinced that if the two sciences pursue their goals with the utmost consistency and right into the ultimate depths of man they must hit upon a common formula.[131]

In Chapter 12, I tried to convey the sheer scale and complexity of organisation, responsiveness to the outer and inner environment, decision-making and apparent purposiveness conveyed by a single cell, let alone millions – or trillions – of them acting in concert. I say 'apparent', because it is intrinsically impossible to be certain that some experientiality, some awareness, some purposive drive is being manifested. However it is, for the same reason, intrinsically impossible to be certain that it is not; and anyone watching cells at work, and engaging with what they can see, will have to do violence to their intuition and imagination, and simultaneously indulge in what Whitehead called 'brilliant feats of explaining away', in order to convince themselves that no purposeful responses are to be seen there.[132] And given the *intrinsic* undecidability of the matter, one way or the other, why would one even *want* to believe that? Let alone be so sure one is right as to turn the idea into an orthodoxy that no scientist must deny, on pain of expulsion from the scientific 'community'. (Science has its excommunications.) Here is cell biologist Bruce Lipton:

> Each cell is an intelligent being that can survive on its own, as scientists demonstrate when they remove individual cells from the body and grow them in a culture … these smart cells are imbued with intent and purpose; they actively seek environments that support their survival while simultaneously avoiding toxic or hostile ones. Like humans, single cells analyse thousands of stimuli from the microenvironment they inhabit. Through the analysis of this data, cells select appropriate behavioural responses to ensure their survival. Single cells are also capable of learning through these environmental experiences and are able to create cellular memories, which they pass on to their offspring.[133]

It seems to me, given everything else we know, that the explanatory burden is on those who would have mindless mechanisms give rise to sentient beings to tell us how, why and at what point they do so.

130 Stapp 2007 (2).
131 Letter of Jung to Oscar Hug, 24 May 1938; in Adler & Jaffé 1973, vol I (246).
132 Whitehead, 1929a (17). This issue was discussed in Chapter 12, and will be further discussed in Chapter 27.
133 Lipton 2005 (7).

We have seen that one aspect of nature, both animate and inanimate, is its tendency to repeat patterns at different levels of the same whole: fractality. When one compares the single cell to the whole organism to which it belongs – in other words, over the most colossal scale – one finds that common to each are the needs to coordinate information, to exchange energy with the surroundings, to have a functioning system of internal distribution, to have the ability to repair and protect itself, to exert 'border control', and to have the capacity to carry on producing itself. In fact, zooming outwards from the human individual to society, rather than inwards to the single cell, one finds the same fractal pattern – the same needs – much as one might expect. Those needs and drives that we know at one level as explicit are implicit at other levels; there is no absolute, merely a relative, difference in kind as the scale changes. Lipton continues:

> the biochemical mechanisms employed by cellular organelle systems are essentially the same mechanisms employed by our human organ systems. Even though humans are made up of trillions of cells ... there is not one 'new' function in our bodies that is not already expressed in the single cell. Virtually every eukaryote [cell that includes a nuclear membrane] possesses the functional equivalent of our nervous system, digestive system, respiratory system, excretory system, endocrine system, muscle and skeletal systems, circulatory system, integument (skin), reproductive system, and even a primitive immune system, which utilises a family of antibody-like 'ubiquitin' proteins.[134]

This pattern, then, transcends scale and seems to be even universal in living entities.

ARE NEURONES, LET ALONE BRAINS, NECESSARY FOR AWARENESS?

In Chapter 12, I quoted cell biologist James Shapiro: 'Living cells do not operate blindly: life requires cognition at all levels.'[135] I argued that single cells behave intelligently, dealing with unforeseen events, for which they could not be 'programmed', in ways that are original and highly adaptive. In the same chapter I quoted microbiologist Brian Ford to the effect that living cells exhibit 'considerable ingenuity ... our conviction that these phenomena become manifest only through cell communities is a fundamental misconception. Ingenious, perceptive and intelligent behaviour is apparent in a single living cell.' And, he continues, to the cells in the body, performing their complex daily tasks, 'the brain is ordinarily an irrelevance.'[136]

So where *do* brains fit in?

It seems that neither complexity of cell aggregation, nor, more specifically, neuronal complexity, is sufficient, or even necessary, for

---

134 *ibid.*

135 Shapiro 2011 (7).

136 Ford 2017 (280–5).

137 Part of the variance in numbers is due to differences in methodology, and part to do with sex differences – 'the total glial cell numbers were 27.9 billion in females and 38.9 billion in males, and the total number of neocortical neurons 21.4 billion in females and 26.3 billion in males [a 22% increase], providing a total neuron and glial cell number of 49.3 billion for females and 65.2 billion for males [a 32% increase]': Pelvig, Pakkenberg, Stark *et al* 2008; 'this region [medial temporal lobe] showed 34 % more neurons in men than in women: 525.1 million against 347.4 million': Oliveira-Pinto, Andrade-Moraes, Oliveira *et al* 2016. (Strictly speaking this is not correct: it shows, for this brain region, 34% fewer neurons in women than in men, 51% more neurons in men than women.) Interestingly, treatment of female-to-male sex change patients with testosterone causes increase, and treatment of male-to-female sex change patients with oestrogen and anti-androgens causes decrease, in cortical thickness across the brain (Zubiaurre-Elorza, Junque, Gómez-Gil *et al* 2014; Pol, Cohen-Kettenis, van Haren *et al* 2006). The lower overall estimates of 16 and 69 billion come from Suzana Herculano-Houzel's research group: Azevedo, Carvalho, Grinberg *et al* 2009. The higher estimates are more usual in most other sources both before and since Herculano-Houzel's study: eg, Carassiti, Altmann, Petrova *et al* 2018; Pelvig, Pakkenberg, Stark *et al* 2008; Andersen, Korbo & Pakkenberg 1992.

awareness. In humans, the presence of massive neuronal complexity can be insufficient for awareness; and, conversely, awareness may be fully present in its almost total absence. Though it is often argued that, in some unspecified way, consciousness emerges as a function of scale and complexity, there are further considerations that compound the already discussed formidable problems surrounding any idea of brute emergence. For example, there are thought to be an astonishing 16–26 billion neurons in the cerebral cortex; but there are a staggering 69–101 billion neurons, thus approximately four times as many, in the cerebellum.[137] These include Purkinje cells, some of the largest and most profusely connected cells in the nervous system.[138] Yet the cerebellum is completely incapable of supporting self-awareness – what we normally call waking consciousness. Moreover there are cases of viable human beings for whom the cerebellum is wholly absent.[139] In any case, complexity is not always advantageous: a recent paper entitled 'Optimal degrees of synaptic connectivity' claims that 'sparse connectivity is sometimes superior to dense connectivity,'[140] a finding that helps to explain the remarkable fact that the 'mean number of neurons reaches a maximum at 28 weeks of gestation and then declines by approximately 70% to achieve a stable number of neurons around birth.'[141] Note: not *to*, but *by*, 70%. It also accords with the neglected but important principle that there is an optimal level of everything, however good it may seem up to a point; and that inhibition can be as creative as excitation, and absence as presence.

As Annaka Harris points out, 'just because the cerebellum is not responsible for the part of my brain that governs language or for the flow of consciousness that I consider to be "me," we can still speculate that it's *another region* of consciousness, just as we can speculate that a worm or a bacterium might be conscious.'[142] Indeed it could be argued that the entire body, every single cell, contributes to consciousness in the normal state, as long as one bears in mind that consciousness has levels.[143] Certain meanings of consciousness would be excluded in the absence of cortical function: principally reflexive self-awareness.

Or so we assume. In a *Science* article by Roger Lewin, provocatively entitled 'Is your brain really necessary?', John Lorber, Professor of Paediatrics at Sheffield University, reports:

> There's a young student at this university ... who has an IQ of 126, has gained a first-class honours degree in mathematics, and is socially completely normal. And yet the boy has virtually no brain ... When we did a brain scan on him, we saw that instead of the normal 4.5 centimetre thickness of brain tissue between the ventricles and the cortical surface, there was just a thin layer of mantle measuring a millimetre or so. His cranium is filled mainly with cerebrospinal

138 Herndon 1963.

139 See p 984 above.

140 Litwin-Kumar, Harris, Axel et al 2017.

141 Rabinowicz, de Courten-Myers, Petetot et al 1996.

142 Harris 2019 (38; emphasis in original).

143 See esp, Panksepp 2007.

fluid ... I can't say whether [he] has a brain weighing 50 grams or 150 grams, but it's clear that it is nowhere near the normal 1.5 kilograms...

Professor Patrick Wall, of University College London, commented that 'scores of similar accounts litter the medical literature and they go back a long way.'[144] Indeed a more recent report of four children aged between 5 and 17 with hydranencephaly, a condition in which the cerebral hemispheres are very largely or totally absent and the space filled with cerebrospinal fluid, demonstrated that, despite each having minimal or practically non-existent cortex, they all nevertheless possessed discriminative awareness: they had functional vision, could orient themselves, could distinguish familiar from unfamiliar people and environments, showed toy preferences, could interact socially, could respond to and discriminate pieces of music, and demonstrated not just awareness of their own body, but appropriate affective responses to others, as well as associative learning. One passed the 'mirror test', supposedly a test of self-awareness, that very few species pass.[145] All of which invites the question whether neuronal complexity is necessary for awareness at all, since the remaining areas of diencephalon and brainstem amount to only 6–10% of the neuronal mass of the normal brain. Another survey of hydranencephaly concludes:

> Most cortical areas are simply missing in hydranencephaly, and with them the organized system of corticocortical connections that underlie the integrative activity of cortex and its proposed role in functions such as consciousness[146] ... The evidence and functional arguments reviewed in this article are not easily reconciled with an exclusive identification of the cerebral cortex as the medium of conscious function.[147]

Let us start at the bottom of the evolutionary heap – with slime moulds. Slime moulds, of course, have no neurones whatever. I have already discussed the complexity of their behaviour.[148] This involves being aware of the environment in such a way as to act 'intelligently' on preferences, initiate sequences of behaviour that are coordinated and flexible, learn from experience and memorise, and produce what appear to be a complex series of transformations in the whole organism carried out at the level of individual cells, which at some stages act independently, but at others in the service of higher orders of structure of which the individual cells appear to be aware. Awareness of the environment is exhibited at several different levels, but together.[149]

The behaviour of plants is vastly various, and can at times reach much greater degrees of complexity: awareness in what we see as the parts, and in what we see as the whole, appears so seamless as to

---

144 Lewin 1980 (1232).

145 Shewmon, Holmes & Byrne 1999.

146 Baars, Ramsøy & Laureys 2003; Sporns, Tononi & Edelman 2000.

147 Merker 2007.

148 See pp 482–3 above.

149 See, eg, Alcantara & Monk 1974.

be perhaps aspects of one larger field of awareness. Again, of course, plants have no neurones at all. Can plants learn, remember and make decisions?

The Venus flytrap can tell the difference between rain drops and the arrival of a small insect, at which it snaps shut. Clever. But can plants learn from experience? Wilhelm Pfeffer was one of the first to demonstrate in the 1870s that plants can learn, through studying the behaviour of the sensitive plant *Mimosa pudica*.[150] More than 50 years ago Holmes and Gruenberg demonstrated that *Mimosa* can discriminate between stimuli, in this case a finger and a drop of water. The plants initially reacted to drops of water by closing the leaflets self-protectively: with repetition they learnt there was no need to do so. Then, if the leaflets were touched with a finger, the plants responded by closing again, even though they no longer responded to a drop of water. If the habituated response to water had been due to fatigue, they would not have responded to the finger.[151] Rather, they could tell the difference between a familiar and unfamiliar stimulus and responded appropriately. There is an energy cost to closing, so a plant needs to weigh the relative risk of closing or remaining open quite carefully. It turns out that plants are more willing to take risks at lower light levels: when kept in artificial lighting conditions, they respond as appropriate to the diurnal light cycle in the world outside the lab.[152]

The most comprehensive study of habituation of *Mimosa* was reported in 2014 by Monica Gagliano and colleagues.[153] She confirmed earlier findings by vertically dropping plants in their pots, which initially responded by closing, and then habituated (stopped reacting). Being dropped in a pot is, needless to say, not a normal historical experience for *Mimosa*. More fascinating is that plants are able to learn to make utterly new connexions which could not be foreseen in nature. Pea plants were grown in an environment where they craved light; experimentally a light would turn on for periods of time in one or another arm of a Y-shaped 'maze' above the plant. Although the side was randomly varied, for one group of seedlings air from a fan coming down the same arm as the light always preceded it; for another group, the set-up was similar, except that the air always came down the arm *opposite* to the one in which the light would subsequently appear. Having been trained over a three-day period, the plants equally learnt to predict the light from the stream of air and turned appropriately towards whichever side, in their experience, would deliver light – whether that was the same arm as the air stream, or the opposite arm.[154]

Plant biologists Stanisław Karpiński and Magdalena Szechyńska-Hebda describe what they call the 'life wisdom' of plants, that is, their

150 Pfeffer 1873 & Pfeffer 1900.
151 Holmes & Gruenberg 1965: as cited by Abramson & Chicas-Mosier 2016.
152 Gagliano, Renton, Depczynski *et al* 2014. I am grateful to Rebecca Hosking for drawing this body of research to my attention.
153 *ibid*.
154 Gagliano, Vyazovskiy, Borbély *et al* 2016.

capacity to integrate and simultaneously read and act on stimuli of all kinds, and prioritise their responses.[155] They can 'store and use information from the spectral composition of light for several days or more to anticipate changes that might appear in the near future in the environment, for example, for anticipation of pathogen attack.' To do this they need to know how to use memorised information, and to project forward; in this they adopt an approach which the authors liken to military strategy. 'Different groups of chloroplasts and cells in the same leaf under identical constant and stable light, temperature and relative humidity condition have different opinions' on what to do in such conditions, and try out different possibilities, 'like different military exercisers during peace time'. However, cells and leaves exposed to a stressful excess of light have 'one clear opinion': to use a particular strategy as much as possible, 'like in a real war zone'. For example, when a plant detects excess of light, it will orientate its leaves so that they are side-on to its source, in such a way as to limit damage. Plants, they conclude, as many other plant biologists have done, 'can actually think and remember.'[156]

Plants can discriminate more sustained signals from transient background noise,[157] using both spatial and temporal cues to identify relevant information.[158] Indeed they respond to roughly 15 environmental factors acting simultaneously to different extents and each modulating the other.[159] According to philosopher Michael Marder, the result is 'infinite variations of selectively variable responses.'[160] Flexibility is an important and sophisticated feature. A plant forms enduring memories of past events, recognises stimuli, and then generalises from what it has learnt to new contexts.[161] However, it must also be able to select what is relevant to memorise, to focus on 'the important information in its environment, while filtering out stimuli or events that, over time, have repeatedly proven to be irrelevant and innocuous', and thus to select those memories it most needs in order to 'modify the timing, quality or quantity' of its response to the prevailing circumstances.[162] Discernment is 'a key feature of intelligence', writes Marder, and he gives the example of a plant's ability to differentiate between damage from insect feeding and a mechanically induced wound, and respond appropriately.[163] Plants assess when to bloom, weighing different factors, not just reacting automatically to, say, temperature.[164] 'If consciousness literally means being "with knowledge", then plants fit the bill perfectly', says Marder.[165]

Plant memory is not an abstract representation of, eg, light, but a capacity to draw on 'the actual photon energy absorbed in excess by some leaves to improve the chances of survival for the whole plant ... in the future.'[166] Plants, at least according to Marder, also attend – and indeed pay different *types* of attention. 'Plant attention is active, rather than merely contemplative', and, according to him, plants exhibit focal attention, and attention to the periphery, as well as both selective attention and sustained attention.[167] According to František Baluška, Professor of Cellular and Molecular Botany at Bonn University, 'it is well known that stress perceived at one site is rapidly communicated through the whole plant body.'[168] Plants communicate internally from one focus of attention to other tissues not directly affected by the stimulus, or coordinate the many different attentive foci, 'each of them singling out a vital piece of information about environmental conditions – often, by way of parallel processing, as in the case of leaf photosensitivity.'[169]

Plants are by no means the immobile and passive entities they have been taken to be. They actively 'forage' for light, water and nutrients. 'Foraging behaviours in plants are highly selective. They are accompanied by attention to numerous environmental factors, foremost among them resource availability and the presence or absence of competitors', writes Marder.[170] Such movement 'isn't just adaptive behaviour, it's anticipatory, goal-directed, flexible behaviour', writes one researcher, Paco Calvo. And he points out that plants 'do things so slowly, that they can't afford to try again if they miss.'[171] Although generally rooted in place, they develop and reform themselves both above and below ground. Animals usually move quickly; mountains, although they move, do so slowly; plants are somewhere in between. Time-lapse photography reveals how coherently, and purposively, plants appear to move. When animals intend something, they move their muscles; when plants intend something, they grow and change their form.[172] Vines grow more rapidly and redirect their growth once they sense a nearby anchor. Plants detect and react to different sounds, bending root tips toward a sound source[173] and can detect the presence of water by sound, as well as by sensing moisture directly: their detection is susceptible to being disrupted by environmental vibration and noise.[174] 'Intact growing root apices can, under appropriate circumstances, perform crawling-like searching movements, which closely resemble the type of behaviour of a lower animal', write Baluška and colleagues.[175]

The importance of the root tip is something Darwin addressed in one of his last works, *The Power of Movement in Plants*, published in 1880. 'It is hardly an exaggeration', he wrote, 'to say that the tip of the radicle … having the power of directing the movements of the adjoining parts, acts like the brain of one of the lower animals.'[176] He drew attention to the way in which plants explore possibilities through what he called 'circumnutation', a slowly spiralling movement affecting all parts of the plant, including the roots. The turns would be generally slightly asymmetric, describing ellipses:

167 Marder 2013.
168 Baluška, Mancuso, Volkmann *et al* 2009.
169 Marder *op cit*.
170 ibid.
171 Geddes 2021. The research report is: Raja, Silva, Holghoomi *et al* 2020.
172 Marder 2012.
173 Gagliano, Mancuso & Robert 2012.
174 Gagliano, Grimonprez, Depczynski *et al* 2017.
175 Baluška, Mancuso, Volkmann *et al* 2009.
176 Darwin 1880 (573). Cf Charles Darwin in a letter to Asa Gray of 22 October 1872: 'The point which has interested me most is tracing the *nerves!* which follow the vascular bundles. By a prick with a sharp lancet at a certain point, I can paralyse one-half the leaf, so that a stimulus to the other half causes no movement. It is just like dividing the spinal marrow of a frog: – no stimulus can be sent from the brain or anterior part of the spine to the hind legs; but if these latter are stimulated, they move by reflex action. I find my old results about the astonishing sensitiveness of the nervous system (!?) of Drosera to various stimulants fully confirmed and extended': Darwin 1887, vol III (322; emphasis in original).

If we look, for instance, at a great acacia tree, we may feel assured that every one of the innumerable growing shoots is constantly describing small ellipses; as is each petiole, sub-petiole, and leaflet … If we could look beneath the ground, and our eyes had the power of a microscope, we should see the tip of each rootlet endeavouring to sweep small ellipses or circles, as far as the pressure of the surrounding earth permitted. All this astonishing amount of movement has been going on year after year since the time when, as a seedling, the tree first emerged from the ground.[177]

This enables trees, and plants in general, to perceive, and to respond by appropriate motion to their perceptions. 'Their world-construction is accomplished in common', writes Marder.[178] For example, 'in a drought, specimens of *Pisum sativum* communicate the onset of adverse environmental conditions through biochemical messages emitted by the roots to other pea plants unaffected'.[179] He further argues that because plants communicate with one another, defend their health, and make decisions, among other things, they may well have some sense of self, too.[180]

Evolutionary ecologist Monica Gagliano insists that plants are intelligent: 'When I talk about learning, I mean learning. When I talk about memory, I mean memory.' Despite having no nervous system they behave like intelligent beings. She says that if plants can summon knowledge about an experience repeatedly – as was the case with the potted plants that stopped curling their leaves after they learnt they would come to no harm – then plants are clearly able to remember and learn from experience.[181] Philosopher Michael Marder concludes: 'Plants are definitely conscious, though in a different way than we, humans, are.'

While it would be improbable that all scientists agree with these voices, since the orthodoxy of current science is set firmly against such ideas, there are increasing numbers who do.[182] It is by now established that trees form communities of a kind, communicate with and support one another as they grow, warn one another of impending threats, and even share nutrients with those who are sick or struggling.[183] The ecologist Suzanne Simard discovered that two tree species she was studying, the Douglas-fir and paper birch, shared what are called mycorrhizal networks, complex underground networks of fungi that connect individual plants, and are capable of transferring water, carbon, nitrogen, and other nutrients and minerals. The species were engaged 'in a lively two-way conversation.' In the summer months, when the Douglas-fir needed more carbon, the birch sent more carbon to the fir; and at other times when the fir was still growing but the paper birch needed more carbon because it was leafless, the fir sent more carbon to the birch. The two species were in fact interdependent.[184] Apparently Douglas-fir 'mother trees'

---

177 Darwin 1880 (558). Circumnutation can now be seen with the help of time-lapse photography: eg, www.youtube.com/watch?v=nmIYAX6Tw6w.

178 Marder 2012.

179 Falik, Mordoch, Quansah *et al* 2011.

180 Livni 2018.

181 ibid.

182 For an insight into some of those who don't, see an opinion piece by Lincoln Taiz and colleagues in *Trends in Plant Science*, which states that 'there is no evidence that plants require, and thus have evolved, energy-expensive mental faculties, such as consciousness, feelings, and intentionality, to survive or to reproduce' (Taiz, Alkon, Draguhn *et al* 2019). Of course this already makes numerous questionable assumptions. Taiz told *The Guardian* newspaper: 'Our criticism of the plant neurobiologists is they have failed to consider the importance of brain organisation, complexity and specialisation for the phenomenon of consciousness.' That, too, makes unnecessary assumptions. But most importantly it fails to do justice to much of the material in this chapter suggesting that, whatever our theory may dictate, the relationship between brains and consciousness is not transparent at all. As Monica Gagliano responded, 'If we think we already know how things are and fail to continuously question our own assumptions, but construct our claims on a system of beliefs we are dearly attached to, then we are in deep trouble and miss the opportunity for true scientific discovery to occur … Miserably, this opinion piece seems yet another missed opportunity, one that makes strikingly no headway towards a better scientific understanding of what consciousness is' (Sample 2019).

are able to distinguish between their own kin and a neighbouring stranger's seedlings:

> Simard found that the mother trees colonized their kin with bigger mycorrhizal networks, sending them more carbon below ground. The mother trees also 'reduced their own root competition to make room for their kids', and, when injured or dying, sent messages through carbon and other defence signals to their kin seedlings, increasing the seedlings' resistance to local environmental stresses.[185]

None of all this is as surprising as it appears unless one makes the mistake of believing that only human beings and perhaps a handful of what we call 'higher' animals have awareness. It seems that plants, like animals, use complex calcium-signalling networks and 'neurotransmitters'. The stimulation of a plant cell leads to changes that result in an electrical signal – similar to the reaction caused by the stimulation of nerve cells in animals – and 'just like in animals, this signal can propagate from cell to cell, and it involves the coordinated function of ion channels including potassium, calcium, calmodulin, and other plant components.'[186] What has been called plant neurobiology, according to Baluška and colleagues, 'neatly closes the gap between animals/humans and plants ... Both animals and plants are non-automatic, decision-based organisms. Should Charles and Francis Darwin have witnessed these unprecedented discoveries, they would surely have been pleased by them.'[187]

Gagliano and colleagues conclude that 'brains and neurons are just one possible, undeniably sophisticated, solution, but they may not be a necessary requirement for learning.'[188] What about single cells? I have written already about the way in which cells are described as making decisions based on assessing a multitude of sources of information. We saw that amoebas and slime moulds are able to learn from experience and exhibit apparently purposeful behaviour – sometimes to a level that astonishes any observer whose mind is not firmly closed to the possibility that there is a form of awareness here, different of course in nature and extent from our own. Is this magic? Not at all. Though it remains awe-inspiring, it is not in conflict with what we know about cell life. To quote Lipton, a large number of recent scientific studies have confirmed that

> 'invisible forces' of the electromagnetic spectrum profoundly impact every facet of biological regulation. These energies include microwaves, the visible light spectrum, extremely low frequencies, acoustic frequencies, and even a newly recognised form of force known as scalar energy. Specific frequencies and patterns of electromagnetic radiation regulate DNA, RNA, and protein syntheses; alter protein shape and function, and control gene regulation, cell division, cell differentiation, morphogenesis (the process by which cells assem-

183 See Tudge 2006; and Wohlleben 2017.

184 Simard 2016.

185 Livni 2018.

186 Chamovitz 2012b (68–9).

187 Baluška, Mancuso, Volkmann *et al* 2009.

188 Gagliano, Renton, Depczynski *et al* 2014 (69–70).

189 Lipton 2005 (81).

190 *ibid* (100).

191 Darwin 1881 (98).

ble into organs and tissues), hormone secretion, and nerve growth and function.[189]

In doing so, 'signals released by cells into the environment allow for a coordination of behaviour among a dispersed population of unicellular organisms. Secreting signal molecules into the environment enhanced the survival of single cells by providing them with the opportunity to live as a primitive dispersed "community".'[190] All of which is perfectly in keeping with the way in which plants are able to exhibit responses that require coordination within and between organisms.

The question of what it is like to be such a single-cell creature is impossible to answer. Its level of awareness can be neither finally affirmed nor denied. Do plants have experientiality? I believe it is both a reasonable and an intuitively compelling assumption that they do. It may be said, quite correctly, that a computer can be programmed so as to learn, remember and make decisions. But that capacity comes from a human being's cleverly externalising an image of such faculties, faculties which come from consciousness whose origins we do not know, and creating a system that appears 'intelligent'. There is no need to suppose experientiality – inwardness – driving this process. In plants, however, we have what look like human faculties appearing in something that was never programmed, but 'programmes' itself – making its 'programme' in the act of making itself. This is extraordinary. Plants appear to be driven to survive, adapt, and help other members of their species just as we are. Even single cells have such drives.

Coming further up the evolutionary tree, it is worth quoting Darwin on earthworms. He notes that in their assessment of how to solve problems relating to different sizes of leaves and the different shapes of their burrows, and how to get the one into the other, they 'act in nearly the same manner as would a man under similar circumstances'; and he continues, having dismissed either instinct alone or simple exhaustive trial and error,

> One alternative alone is left, namely, that worms, although standing low in the scale of organization, possess some degree of intelligence. This will strike every one as very improbable; but it may be doubted whether we know enough about the nervous system of the lower animals to justify our natural distrust of such a conclusion.[191]

Fish injected with a substance thought to cause pain will preferentially swim into water containing a pain-killing substance, even though it is in other respects an 'unpreferred environment': they will not swim there if it does not contain such a substance. 'They made a choice they'd not normally make', writes philosopher Peter

Godfrey-Smith, 'and they made it in a situation where the idea of a more painful or less painful *environment* would be quite novel to them: evolution could not have set them up with a reflexive reaction to this situation.'[192] Crabs, and some shrimps, nurse injured limbs and groom injured areas. 'You can still doubt that these animals feel anything, yes. But you can doubt that about your next-door neighbour. Scepticism is always possible, but a case is being built here. These results do provide support for a view of pain as a basic and widespread form of subjective experience, one present in animals with very different brains from ours.'[193]

Godfrey-Smith contrasts what he calls the 'latecomer' and 'transformer' views of experience – in other words, the view that consciousness somehow emerged late in evolution versus the view that consciousness was never absent but was transformed during evolution.[194] And he points out that if you are an adherent to the 'latecomer' story, not only do you have the problem attaching to the magic trick of brute emergence, but you have to perform the magic trick several times over. He establishes to his satisfaction, and I believe convincingly, that crabs, octopuses, and cats have consciousness. 'By the Cambrian', he continues,

> the vertebrates were already on their own path (or their own collection of paths), while arthropods and molluscs were on others. Suppose it's right that crabs, octopuses, and cats all have subjective experience of some kind. Then there were at least three separate origins for this trait, and perhaps many more than three.'[195]

Indeed like octopuses, other creatures such as honeybees and spiders have perceptual constancy, that is to say, the tendency to recognise from a percept that varies in its shape, brilliancy and size a nonetheless constant, familiar object.[196] Honeybees and spiders can perform elementary arithmetical calculations.[197] Crows can remember individual human faces for several years after a single encounter, respond appropriately to different people and are capable of solving new logical problems consisting of up to eight steps, including avoiding deliberate distractor 'tools' that do not function.[198] Ravens can plan flexibly, barter and teach their offspring how to use tools.[199] As birds have no neocortex, clearly cortical processing cannot be a requirement for higher-order cognition.[200] Although birds have more neurones than expected from their small brain weights,[201] their absolute neurone count is still low compared to cortical neurone numbers in primates. Yet in, for example, abstract numerical competence and visual memory tasks pigeons reach performance levels comparable to those of macaques,[202] and in some cognitive tasks they can outperform humans;[203] corvids can outperform pigeons, primates and humans at some cognitive tasks.[204] While affirming

192 Godfrey-Smith 2018 (94): referring to Sneddon 2011.
193 *ibid* (95).
194 *ibid* (92–5).
195 *ibid* (97).
196 *ibid* (100).
197 Howard, Avarguès-Weber, Garcia *et al* 2019; Rodríguez, Briceño, Briceño-Aguilar *et al* 2015; Nelson & Jackson 2012.
198 Marzluff, Miyaoka, Minoshima *et al* 2012; Gruber, Schiestl, Boeckle *et al* 2019; Taylor, Elliffe, Hunt *et al* 2010.
199 Kabadayi & Osvath 2017.
200 Güntürkün & Bugnyar 2016; Jarvis, Güntürkün, Bruce *et al* 2005.
201 Olkowicz, Kocourek, Lučan *et al* 2016.
202 Güntürkün, Ströckens, Scarf *et al* 2017.
203 Letzner, Güntürkün & Beste 2017.
204 Wright, Magnotti, Katz *et al* 2017.

that brain size increases in more intelligent creatures, Darwin rejects the idea that there need to be elaborate neuronal ganglia for consciousness and cognition:

> It is certain that there may be extraordinary mental activity with an extremely small absolute mass of nervous matter: thus the wonderfully diversified instincts, mental powers, and affections of ants are generally known, yet their cerebral ganglia are not so large as the quarter of a small pin's head. Under this latter point of view, the brain of an ant is one of the most marvellous atoms of matter in the world, perhaps more marvellous than the brain of man.[205]

Before we conclude that it is absurd to suppose that other organisms, perhaps far removed from us in terms of evolutionary history, have awareness, let us remember that the detached post-Enlightenment view of life as mechanical, and of consciousness as something we must not make the mistake of attributing to any creature other than ourselves, on the basis that to do so is to make assumptions we cannot validate, is both historically anomalous and illogical. Historically anomalous, because such a view would never have been accepted by Greek or Roman, Chinese or Indian philosophers, or our own, until Descartes. Illogical, because to assume that they do not have awareness is also an assumption we cannot validate, but which, unlike its alternative, does violence to every other human faculty.

And it is an assumption with a far from glorious history. Descartes affirmed that 'in my view pain exists only in the understanding. What I do explain is all the external movements which accompany this feeling in us; in animals it is these movements alone which occur, and not pain in the strict sense.'[206] This idea made it easier to excuse animal experimentation without anaesthesia, which persisted for hundreds of years. I was shocked to learn from an anaesthetist during my medical training that human infants were operated on well into the 1980s without anaesthetics, because, unable to verbalise their pain, they were clearly not capable of feeling it.[207] Their screams and cries were like those of animals, creakings of a machine. This practice was perpetuated, and given credence, by an influential paper by the psychologist Myrtle McGraw, published in 1941, which promoted the idea that infants do not experience pain.[208] It was not until 1987 that the American Academy of Pediatrics issued a declaration that it was no longer ethical to perform surgery on preterm babies without anaesthetics; a declaration in which they confirmed that 'hospitalized newborns, from preemies to babies up to 18 months of age, have been routinely operated upon without benefit of pain-killing anaesthesia. This has been the practice for decades.'[209] However babies' brains react in 18 out of 20 regions that adult brains do to pain,[210] and it is

---

205 Darwin 1871, vol I (145).

206 Letter of Descartes to Marin Mersenne, 11 June 1640: in Descartes 1984–91, vol 3 (148).

207 Anand & Hickey 1987.

208 McGraw 1941: 'it is reasonable to assume that the sensori-motor reactions of the newborn infant do not extend appreciably beyond the level of the thalamus.'

209 Chamberlain 1991.

210 Goksan, Hartley, Emery *et al* 2015.

deeply perverse to deny that they experience suffering, as I believe it is to deny outright the possibility of suffering to any living being.

Let us beware of what Sydney Brenner calls Occam's Broom, a device that helps one sweep under the carpet any findings that cast doubt on the current paradigm. Brenner, who won a Nobel Prize for his work in molecular biology wrote:

> Molecular biology has been a great leveller and has made thinking unnecessary in many areas of modern biology ... So powerful are contemporary tools for extracting answers from nature that pausing to think about the results, or asking how one might find out how cells really work, is likely to be seen as a source of irritating delay to the managerial classes, and could even endanger the career of the questioner ... I found that many people were applying what I called Occam's Broom, which was used to sweep under the carpet any unpalatable facts that did not support the hypothesis ... The orgy of fact extraction in which everybody is currently engaged has, like most consumer economies, accumulated a vast debt. This is a debt of theory and some of us are soon going to have an exciting time paying it back – with interest, I hope.[211]

SO WHY BRAINS AT ALL?

There can be no doubt that brains serve an important role in the evolution of higher levels of consciousness, though no-one knows how. There are tantalising clues that what enables consciousness to cohere in the brain may be aspects of quantum field theory, but this area lies beyond (my) non-specialist grasp. According to Matthew Fisher, Professor of Physics at the University of California at Santa Barbara, the nuclear spin of phosphate atoms could serve as rudimentary quantum bits (so-called 'qubits') of information in the brain, since such phosphate atoms, bonded with calcium in Posner molecules (clusters of nine calcium atoms and six phosphorus atoms), can prevent coherent neural 'qubits' from collapsing into decoherence (non-quantum states) for long enough to enable the brain to function somewhat like a quantum computer.[212] As far as entanglement goes (see pp 913–14 above), it was always said that entanglement was difficult enough to observe under laboratory conditions and at very low temperatures, never mind at room temperature and in the soggy environment of the brain, but this now seems to be much less serious an objection than it was once thought to be. Quantum entanglement could coordinate processing in many different parts of the brain simultaneously.[213] According to Fisher, 'if we find that quantum mechanics is in operation cognitively, then it could be a necessary component of consciousness.'[214]

Another implication of entanglement is that it might underlie

---

211 Brenner 1997.

212 Johnson 2018. On Posner molecules, see: Swift, van de Walle & Fisher 2018.

213 Fisher 2015.

214 Johnson *op cit*.

the collapse of the wave function, which is not a *mechanical* effect of consciousness on the system; it could be the outcome of the observer's brain and the observed system becoming entangled in consciousness.²¹⁵ Quantum theory, which Stapp uncontroversially describes as a 'hugely successful rational approach that yields validated predictions of high accuracy ... involves, in an essential way, the causal participation of the minds of us observers, while classical mechanics strictly bans any such effect of mental realities on the world of matter.'²¹⁶ It differs from classical physics in attributing a key role in the actualisation of one particular potential (out of a range of potentials) to the agency of the observer. 'What a person's brain does', he writes,

> can, according to the quantum theory, be strongly influenced by a nonlocal causal process connected to the person's conscious choices and mental efforts. Consciousness can play a nonredundant causal role in the determination of our actions: it can play the very role that we intuitively feel that it plays. Quantum theory allows your mind and your brain to co-author your physical actions.²¹⁷

Note 'co-author': since brain and mind interact, each will be causative of our behaviour. And as Denis Noble says: 'Agency is central to understanding life.'²¹⁸

Remember Stapp's statement that 'quantum mechanics is explicitly designed to cover "big" systems, and by becoming "big" a quantum system does not become classical.'²¹⁹ Quantum effects are involved in photosynthesis, as well as migration in birds through magnetoreception.²²⁰ Physicists Andreas Albrecht and Daniel Phillips explain that 'the quantum nature of fluctuations in the gasses and fluids around us can lead to a fundamental quantum basis for probabilities we care about in the macroscopic world.' They assert that 'the randomness in collections of molecules in the world around us has a fully quantum origin ... We expect that all practical applications of probabilities can be traced to this intrinsic randomness in the physical world', so that, eg, 'the outcome of a coin flip is truly a quantum measurement (really, a Schrödinger cat)'. Unpredictability is a physical reality embedded in the fabric of the quantum world. And it starts to make itself felt not in one or two rare circumstances, but in those of everyday. To give some idea of this, they take the case of ricocheting billiard balls, the perfect image of Laplacian determinism, and show by calculation 'the number of collisions after which the quantum spread is so large that there is significant quantum uncertainty as to which billiard takes part in the next collision.' Several million, perhaps? No, just *eight*.²²¹

Based on von Neumann's orthodox mathematical formulation of quantum mechanics, Stapp describes a two-stage process, likened

---

215 'Consciousness does not mechanically cause the wave function to collapse or influence physical particles. Rather, the observer's brain and the observed system are synchronously entangled' (Wallace 2007, 82).

216 Stapp 2017 (1–2).

217 Stapp 2009 (242).

218 Noble [undated].

219 Stapp 2017 (13).

220 Lambert, Chen, Cheng *et al* 2013.

221 Albrecht & Phillips 2014. I am grateful to Professor David Anthony Oliver for bringing this paper to my attention.

to a 'questioning and response', between the observer and the rest of nature, in which one part of the process involves freedom from determination. As a result of the 'question asked' by the observer there is a 'proliferation in the brain of representations of many different possible immediate courses of action', which then collapse into one single actuality as a result of the interactive process.[222] In mechanistic terms, Stapp hypothesises that the necessary dynamic uncertainty is introduced by the quantum-indeterminable action of single ions (eg, calcium ions) passing through ion channels in neuronal membranes and resulting in the release (or not) of neurotransmitters at the synaptic cleft. 'Thus the pertinent-for-us essence of quantum mechanics is the causal dynamical linkage that quantum mechanics specifies between our conscious thoughts and our atomic-particle-based brains.'[223] This quantum uncertainty is played out at each of the quadrillions of synapses active in the brain – rather more than eight. Stapp's model 'makes consciousness causally effective, yet it is compatible with all known laws of physics, including the law of conservation of energy'.[224]

'The empowering message of quantum mechanics', he concludes,

is that the empirical data of everyday life, and also our intuitions, are generally veridical, not delusional; and hence that our mental resolves can often help bring causally to pass the bodily actions that we mentally intend. The role of our minds is to help us, not to deceive us as the materialist philosophy must effectively maintain.[225]

That consciousness interacts with matter, an insuperable problem in the seventeenth century, is no longer insuperable, since matter is already intrinsically a field that interacts with a field of consciousness. Additionally fields affect matter 'at a distance' all the time: gravity, accepted by Newtonian mechanics, is just one instance, and since Newton's time we have become familiar with electromagnetic forces and other types of force fields, which, while not tangible, cause rearrangements in the form of matter. It is not incomprehensible that such fields of force in consciousness affect matter, and it would perhaps be harder to account for if they did not.[226] In any case it is not disputed that observation changes the nature of matter, and not just in some 'incidental' fashion.

At the end of the nineteenth century Schiller had written of action at a distance that 'the objection to it seems nothing but the survival of the primitive prejudice that all action must be like … a tug-of-war. If metaphysics had been consulted, it would have been obvious that no special medium was required to make interaction possible between bodies that *co-exist*, seeing that their co-existence is an ample guarantee of their connexion and of the possibility of their interaction.'[227] Here again Schiller's metaphysics seem to have

222 'Process 1 first selects, from the Process-2-generated continuum of potentialities, a particular perception that 'might' occur. Then 'nature' chooses, subject to the statistical Born Rule, either to accept the possibility selected by the observer, and then actualize the global consequences of that acceptance, or actualize the global consequences of rejecting the observer's proposal' (Stapp 2017, 24).

223 Stapp 2017 (13).

224 Stapp 2009 (23).

225 Stapp 2017 (12).

226 An intriguing theory developed by Jonathan Schooler, Professor of Psychological and Brain Sciences at the University of California, Santa Barbara, and Tam Hunt, a philosopher of mind at the UCSB's Department of Brain and Cognitive Sciences, builds on the non-controversial idea that matter and energy are oscillating fields, and indeed that all existing things, however static they may seem, are 'just vibrations of various underlying fields'. Hunt (2018) points to the fact that these oscillations and vibrations are entrained – come into 'synch' – in closely coupled entities: he gives the examples of light emissions from neighbouring fireflies, neuronal firing, and the rotation of the moon in relation to its orbit. He and Schooler propose what they call a 'resonance theory of consciousness', adopting a panpsychist position, and positing that 'particular linkages that allow for macro-consciousness to occur result from a shared resonance among many micro-conscious constituents'. This thesis is consonant with Sheldrake's idea of morphogenetic fields.

227 Schiller 1891 (71–2; emphasis in the original).

anticipated modern physics. Princeton physics philosopher Hans Halvorson has concluded that a form of 'superentanglement' links every aspect of everything in the universe.[228]

Nor is there any lack of everyday evidence that changes wrought in the mind have material effects. The evidence that mind and body are inextricably interrelated, with reciprocal effects on one another, is an everyday experience for a physician, yet the philosophical implications of that fact are seldom recognised. Note not just *brain* and body (viz, 'body and body'), but *mind* and body. The belief that a substance will produce a cure (placebo) – or harm (nocebo) – is a potent predictor that it will do so even if the substance is inert.[229] Although it is one of the most familiar and best attested phenomena in medicine (and one of the most reliable and effective), the mechanism by which the placebo operates has been little researched, for, I suspect, three main reasons: there is no money in it for drug companies – perhaps the reverse; it is an embarrassment to the reductionist materialist mainstream in biological research; and there is little chance of a mechanism being found any time soon. Recent reviews of the phenomenon demonstrate the chasm open between the silence of neuroscientists when contemplating the interaction between mind and brain, and their fluency when on home ground, dealing with the brain and body as a closed system.[230] Attempts to 'find consciousness in the brain' are inevitably driven by this home ground fluency, and leave the problem of the chasm unaddressed.

Hypnosis would appear to present similar difficulties, since simple suggestion can produce a whole range of physiological changes, even causing, or alternatively preventing, an allergic skin reaction.[231] Equally hypnosis can in many cases help clear skin conditions (and of course help treat many other conditions).[232] Autosuggestion and the enacting of imaginary scenarios have measurable effects on body functions. A report of the US National Research Council on behalf of the military found that mental practice was almost as effective as physical practice in the acquisition of motor and other skills.[233] None of all this would surprise a psychiatrist. Indeed something as well-validated as cognitive therapy[234] could not work, according to the materialist hypothesis, unless beliefs were able to change the brain, as they evidently do.[235] The dogma that while matter can affect mind, mind cannot affect matter is irrational and baseless. The brain can be altered by changes in the mind, which accords with an array of neuroscientific findings, just as the mind can be altered by changes in the brain.

A recent case report of a woman with multiple personality disorder is interesting. The woman exhibited a variety of dissociated personalities ('alters'), some of which claimed to be blind. On EEG, the brain activity normally associated with sight was not present when a

---

228 Musser 2015 (139).

229 For an overview of the phenomenon in practice, see Brown 2013.

230 Geuter, Koban & Wager 2017; Požgain, Požgain & Degmečić 2014.

231 Ikemi & Nakagawa 1962. The paper reports in detail a study of suggestion in 57 boys aged between 15 and 18 who were touched with a leaf of 'lacquer tree' or 'wax tree', which are both known to cause contact dermatitis, on one arm, and with a chestnut tree, which is non-allergenic, on the other. Amongst a subgroup of 13 that were known to have had previous marked reactions to the poisonous leaves, subjects hypnotised to believe the harmless leaf was poisonous, and the poisonous leaf was harmless, in every case produced a clear reaction to the harmless leaf, and in only one case a reaction to the poisonous leaf. This study is difficult to replicate for ethical reasons, but needs replication. A highly sceptical review of reported cases up to 1963 concludes that, 'even though the majority of these reports are grossly lacking in controls, experimental design, etc, and are subject to alternative explanations, the author concludes that skin anomalies have been produced by suggestion in some instances. Additional studies of psychogenic vascular changes add credence to the possibility of central control of these phenomena' (Paul 1963).

232 Shenefelt 2017.

233 Druckman & Swets 1988 (esp 61–70).

234 See, eg, Butler, Chapman, Forman *et al* 2006.

235 See, eg, Porto, Oliveira, Mari *et al* 2009; Yang, Oathes, Linn *et al* 2017; and Collerton 2013.

blind alter was 'in control', even though her eyes were open: when a sighted alter assumed control, the usual brain activity returned. The authors note that 'visual evoked potentials were absent in the blind personality states but were normal and stable in the seeing states. A switch between these states could happen within seconds.' The authors of the report 'assume a top-down modulation of activity', of the brain by consciousness.[236]

The most contentious area is that of so-called 'paranormal' phenomena, such as extra-sensory perception, telepathy, psychokinesis, near-death experiences, remote viewing, presentiment, etc. Partly because of my lack of a thorough enough familiarity with the evidence, and partly because it risks triggering the descent of a red mist in otherwise rational and well-disposed readers, I will not deal with it: to do so fairly would surely demand a book-length review of the evidence, and one I could not write. However, for the sake of the progress of science, as much as anything else, nothing should be denied fair scientific investigation, and the difficulty seems to be that to accord it such would require years of hard work that very few people can afford, since the price would be exclusion from the 'serious' scientific community. Those who have genuinely accepted the challenge with an open mind are worth reading: they do not generally reach the conclusion that it is safe to dismiss the evidence. Such are the writings of the systematic, analytic philosopher C. J. Ducasse;[237] and a recent review by neuroscientist Mario Beauregard and colleagues which reaches the conclusion that there is a convergence of evidence suggesting that we should take what they call post-materialist hypotheses seriously.[238] Clearly the danger is that one defines as 'normal' only the things that one happens to believe now: and everything else becomes, by definition, abnormal, paranormal, supernatural, irrational.

Paul Nuñez, a biomedical engineer who specialises in the physics of brain function, writes that

> an appreciation of the grand conceptual leap required in the transition from classical to quantum systems may give us some vague feeling for how far from current views neuroscience may eventually lead. Such humbling recognition will perhaps make us especially sceptical of attempts to 'explain away' (that is with tautology) data that do not merge with common notions about consciousness, such as multiple conscious entities in a single brain, hypnosis, and so on.[239]

Social psychologist Roy Baumeister, looking at the question at a different level from physicists, writes, in a review:

> The evidence for conscious causation of behaviour is profound, extensive, adaptive, multifaceted, and empirically strong. However, conscious causation is often indirect and delayed, and it depends

---

236 Strasburger & Waldvogel 2015. As an aside, there are elevated levels of perfusion in the left temporal lobe in such dissociative identity disorders: Schlumpf, Reinders, Nijenhuis et al 2014. In a detailed study of two cases of multiple personality disorder, both 'were in a state of relative left hemisphere [EEG] activation across all cerebral regions and task conditions': Flor-Henry, Tomer, Kumpula et al 1990.

237 Ducasse 2006.

238 Beauregard, Trent & Schwartz 2018. See also Cardeña, Lynn & Krippner 2017. Certain particular findings stand out: the capacity for intention to influence non-living systems, as determined by Radin, Michel & Delorme 2016; the possibility of verifiable detailed vision of some kind in blind subjects, including the congenitally blind (hence my rider, 'of some kind'), during near-death and out-of-body experiences, as reported in an impressively balanced assessment of the evidence by Ring & Cooper 1997 (both Ring & Cooper and Beauregard et al dispose convincingly of the suggestion that NDEs are merely a side-effect of hypoxia, hypercarbia or another brain distress state); prescience, as determined by Mossbridge, Tressoldi & Utts 2012.

239 Nuñez 1995 (158).

on interplay with unconscious processes. Consciousness seems especially useful for enabling behaviour to be shaped by non-present factors and by social and cultural information, as well as for dealing with multiple competing options or impulses. It is plausible that almost every human behaviour comes from a mixture of conscious and unconscious processing.[240]

Almost all actions evolve out of a nexus of wishes, desires, emotions, imaginings, intuitions and explicit acts of cognition, each playing a part at some stage. Some of these elements will be conscious, though most unconscious, and of course they may come and go from conscious awareness at different times. In the broader usage of consciousness employed so far, both self-awareness and awareness at any level are included. These streams emerge out of 'the field of me'.

The brain and free will

I am sometimes asked when I wrote a particular book. I might reply with the year of publication. But when did I begin to write it – and when did I decide to write it? When I typed the long since discarded first words? When I started writing down disparate thoughts that tended to its topic? When I, much earlier, started reading about related matters? When I first had certain insights as a teenager that inspired me to follow certain paths of thought and research? When I had that conversation with a friend? There is no answer to this question. Much of the work was done unconsciously, but it would be normally accepted that a book has an author or authors, and that author in this case was me. It may be thought that this is a special case, but it is only a clear one. When and how did I decide to go for a walk today rather than tomorrow? I can't unravel things to reveal a moment when I did. There are many possibilities and they collapsed into a certain single event at a certain moment when I put on my boots. I am both my unconscious and my conscious mind. When the decision comes it comes out of, or at least, through me.

The classic experiments are those carried out by Benjamin Libet in the 1980s. He asked volunteers, who were monitored by an EEG, to press a button 'at random', and record the moment at which they made the decision to act. Since the 1960s it had been known that there is a characteristic event in the EEG called the 'readiness potential' (*Bereitschaftspotential*), which was thought to mark the beginning of the train of events leading to an action. What he found was that the readiness potential was evident about half a second *before* the subject decided to act: spooky! This led to all kinds of wonderful claims, such as that 'I' am an illusion, or that my brain makes decisions in which 'I' play no part, but for which 'I' later ignorantly take credit. Estimates by experimental subjects of when they think they

[240] Baumeister, Masicampo & Vohs 2011 (331).

decided to carry out an action indicate only the timing of when they came to have a belief about when they became aware of a conscious immediate intention to act. In the famous Libet experiments, subjects were instructed at the outset that they would be making choices to move a finger, and so they and their brains were at all times in a state of meta-readiness to *decide when to make that decision*.

There are many other problems with the Libet experiments. Recording the 'moment' of a decision is intrinsically difficult and doing so while reading a clock accurately makes it harder to be sure one knows what one is recording at exactly which point. Moreover, it turns out that the readiness potential occurs frequently without heralding action, and its apparent prominence may be the consequence of studying only cases in which, by definition, it did. It may signal the brain gathering evidence for or against action, rather than initiating action. In fact it is certainly not simply causative of anything at all. It may be an artefact of a highly artificial situation in which subjects have been told to resist all obvious reasons for movement and produce a random event (therefore, unusually, coming to rely on a symmetry-breaking event of some kind). Furthermore in a study of monkeys tasked with choosing between two equal options, the monkey's choice correlated with brain activity before the monkey had even been presented with options.[241] With a control condition of periods of inaction, scrutiny of the EEG alone could not distinguish between action and non-action conditions. In a modified version of the Libet experiments, subjects showed a readiness potential when they decided *not* to move.[242] In yet another variation, in which subjects were asked to press one of two buttons in response to images on a computer screen, they showed a readiness potential even before the images came up on the screen, suggesting that it was not related to deciding which button to press.[243] And, to crown it all, the readiness potential is found in an area of the brain, the supplementary motor area (SMA), which is normally activated by imagining, not initiating, movements (which is associated with more posterior cortical regions).[244] Neither Hans Helmut Kornhuber nor Lüder Deecke, who discovered the *Bereitschaftspotential*,[245] nor Benjamin Libet, believed they had disproved free will – which is just as well, because they hadn't. This has not stopped 'science has proved we have no free will' getting firmly lodged in the popular imagination. The design of Libet's experiment was so poor that its renown could not have lasted as long as it has except for its conclusion being in keeping with the orthodox materialist dogma. By contrast some of the experiments on psychical phenomena are amongst the most stringently designed and meticulously executed in the field of psychology, but because of their clear indication of an effect of consciousness on matter, which though quantitatively small has been repeatedly demonstrated to

241 Maoz, Rutishauser, Kim *et al* 2013. See also Gholipour 2019.

242 Trevena & Miller 2010.

243 Herrmann, Pauen, Min *et al* 2008.

244 Roland, Larsen, Lassen *et al* 1980; Desmurget, Reilly, Richard *et al* 2009. For an excellent brief overview, on which I have drawn here, see Taylor 2019.

245 Kornhuber & Deecke 1965.

be of statistically very high significance, they have been ignored or dismissed rather than falsified, because their conclusions question the consensus view.[246]

Above all, 'I' am not confined to my reflexive awareness of my decisions. Most of our decisions to act have little reflexive awareness to them at all. But, significantly, many of our more important decisions to act do. That such reflexive awareness is only sometimes required is not to say that when required it is not effective. Much of the time I remain warm without lighting a fire; that doesn't mean that when I do light a fire it is ineffective in keeping me warm. And, when required, reflexive awareness does not have to be unmixed with unconscious thoughts. As Baumeister says, 'any evidence that conscious thoughts are themselves the results of other causes (presumably including unconscious processes and brain events) is irrelevant.' The immediate cause of movement is *always* an action potential in a motor axon, which is unconscious. 'The proper question is whether the conscious processes can play any causal role' in that chain.[247] And they manifestly can.

Mozart was quite unaware of where his music came from; Poincaré was taken by surprise by his sudden revelation of Fuchsian functions. Neither came as acts of will. Yet we do not suppose that Mozart and Poincaré could have done what they did had they been zombies, beings without consciousness. In the sense examined by Libet, consciousness is just the individual's attentional spotlight; but it passes over a massively larger field of awareness, existing at many levels. It's good that it's on somewhere; but it doesn't have to be shining on our chosen place of interest for us to be there. And if we are doing many things simultaneously – imagine playing an organ fugue, requiring the working of ten fingers and two feet – the searchlight having only one focus, it would be logically impossible for it to be on them all. (In any case, if the performance is to be a success, it must be on none of them.) That does not prevent them being *our* actions, rather than suddenly becoming something 'decided' on by individual neurones themselves.

I am working my way towards the question what consciousness is for. But first, we should return to the nature of the role played by the brain in consciousness. It might give us a clue.

I started by suggesting that everything we know about the brain would be compatible with it acting in an emissive, transmissive or permissive capacity. I then gave some reasons for doubting that the brain somehow generates consciousness. The engineer and inventor Nikola Tesla is said to have said: 'My brain is only a receiver. In the universe there is a core from which we obtain knowledge, strength, inspiration. I have not penetrated into the secrets of this core, but I know that it exists.' Of course, his hunch may be wrong,

---

246 See, eg, Radin 2018 (ch 6).

247 Baumeister, Masicampo & Vohs 2011 (333). For a useful discussion of the fallacies of 'willusionism', see Mele 2009.

but it expresses a healthy scepticism about the idea that consciousness *starts out* in neurones. Paul Nuñez writes that 'whole brains or special parts of brains might behave like antenna systems sensitive to an unknown physical field or other entity that, for want of a better name, may be called Mind.'[248] When Strawson says 'there is a lot more to neurons than physics and neurophysiology record (or can record)' – a view with which I wholly concur – he nicely leaves the door open as to what that unrecordable element might be.[249] There can be no question that reality is more than we can see or measure or know precisely. But this does not mean a speculative free-for-all. The principles of Pragmatism apply: only what answers best to experience as a whole will pass. What does experience suggest?

PERMISSION

'How do you know but ev'ry Bird that cuts the airy way, is an immense world of delight, clos'd by your senses five?', wrote Blake.[250] An interesting phrase, because we normally think of the senses as openings on the world. I suggest that the function of the brain is to create by permission, in other words by acting as a kind of filter. This includes the idea of transmission but adds a further element. Consciousness is sculpted: by saying 'no' to some things it enables others to stand forward into being, as Michelangelo's hand caused his David to come into being by a process of discarding stone from the formless block and allowing other stone to remain. Cells are the most primitive life forms, and they already define themselves by an exercise of coming into being through permission. The cell membrane is a (highly active) *semi*-conductor: it conducts some things across, while keeping some things out.[251] This is what makes it possible for the cell to exist, to persist, and to embody a purpose. The membrane both says 'no' and does not say 'no'. It doesn't make the cell happen, but permits it to come into being – it is its ground of possibility. In a vastly complex organism such as the human being, what we see on the small scale is fractally expressed on a much larger scale. Lipton writes:

> While it is the job of the membrane in a single cell to be aware of the environment and set in motion an appropriate response to the environment, in our bodies those functions have been taken over by a specialised group of cells we call the nervous system. It is not a coincidence that the human nervous system is derived from the embryonic skin [the ectoderm], the human counterpart of the cell's membrane.[252]

This already casts the idea of the nervous system as a permissive and non-permissive system – a filter of sorts. With this in mind, here is James:

[248] Nuñez 2010 (274).
[249] Strawson 2006 (7).
[250] Blake, *The Marriage of Heaven and Hell*, Plate 6, 'A memorable fancy'.
[251] Lipton 2005 (60).
[252] *ibid* (58).

when we think of the law that thought is a function of the brain, we are not required to think of productive function only: we are entitled also to consider permissive or transmissive function. And this the ordinary psycho-physiologist leaves out of his account.

Invoking Shelley's timeless image of life as a 'dome of many-coloured glass that stains the white radiance of eternity', James implies that consciousness is given its particular unique quality in each living being by a process of filtering or shaping, carried out by the brain, of the formless, featureless condition in which it otherwise would exist; much as the air passing through his vocal chords is limited and restricted in such a way as to 'shape it into my personal voice'.[253] Limitation is intrinsic to creation. Elsewhere he actually used the metaphor of sculpture: 'the mind ... works on the data it receives very much as a sculptor works on his block of stone'.[254]

Schiller had the same intuition. 'Matter is not that which *produces* consciousness', he wrote, 'but that which *limits* it and confines its intensity within certain limits: material organisation does not construct consciousness out of arrangements of atoms, but contracts its manifestation within the sphere which it permits.'[255] In fact he saw the brain as 'admirably calculated machinery for regulating, limiting and restraining the consciousness which it encases'. And he has this interesting reflection to make:

> if, eg, a man loses consciousness as soon as his brain is injured, it is clearly as good an explanation to say the injury to the brain destroyed the mechanism by which the manifestation of consciousness was rendered possible, as to say that it destroyed the seat of consciousness. On the other hand, there are facts which the former theory suits far better. If, eg, as sometimes happens, the man after a time more or less recovers the faculties of which the injury to his brain had deprived him, and that not in consequence of a renewal of the injured part, but in consequence of the inhibited functions being performed by the vicarious action of other parts, the easiest explanation certainly is that after a time consciousness constitutes the remaining parts into a mechanism capable of acting as a substitute for the lost parts.[256]

This is a point that has not been much discussed. How does consciousness as a whole have a sense of functions that have gone missing and try to re-instigate them – somewhere – in the brain? Schiller's point is worthy of serious debate, if nothing else.

Bergson, similarly, thought that 'if consciousness is not a function of the brain, at least the brain maintains consciousness fixed in the world in which we live; it is the organ of attention to life'. The perception that attention was central is of critical importance. He thought, like Schiller, that in the case of brain injury, 'the mecha-

---

253 James 1992 (1110).

254 James 1890, vol 1 (288).

255 Schiller 1891 (295; emphases in original).

256 *ibid* (295–6).

nism is thrown out of gear', so that its properly limiting function is deficient, and certain aspects of thinking are no longer capable of being shaped into being. When functioning well, the brain 'canalises, and also limits, the life of the mind.' The part the brain plays 'is that of shutting out from the field of our consciousness all that is of no practical interest to us, all that does not lend itself to our action.'[257]

Interestingly, suppression of the left medial frontal cortex has been found experimentally to increase significantly a subject's ability to influence the numerical output of a random event generator. An arrow moving at random on the screen, as the output of the random number generator dictated, was influenced to make significantly more movements in the 'contralesional' direction, namely towards the right. The authors conclude: 'The medial frontal lobes may act as a biological filter to inhibit *psi* [phenomena that are anomalous according to contemporary orthodoxy] through mechanisms related to self-awareness.'[258] Feelings of self-transcendence follow *damage* to either left or right parietal region.[259] Philosopher Bernardo Kastrup, who also supports the filter hypothesis, offers evidence from 10 different kinds of psychological phenomena.[260] One study he reports, of mediums writing during a trance, compared with writing under normal conditions, showed marked reduction of brain activity in key regions such as the frontal lobes and hippocampus.[261] When the written material was later scored for complexity, however, the material written under trance scored consistently higher than material produced without trance. One would conventionally expect more complex material to be associated with increased activity in the frontal and temporal lobes, not decreased activity.

There is also a phenomenon known as 'terminal lucidity', whereby, shortly before death, patients suffering from severe and sometimes incapacitating, psychiatric and neurological disorders, and who may be dull or unconscious, and have been so for years, exhibit an unexpected return of mental clarity and memory, with normal, or unusually enhanced, mental abilities, including considerable elevation of mood and unaccustomed spiritual expression. In about half of cases this happens in the last 24 hours of life, and in almost 90% within a week of death.[262] 'It's as if the damaged brain prevents the person from consciousness, but then as the brain finally begins to die, consciousness is released from the grasp of the degenerating brain.'[263] Examples include case reports of patients suffering from brain abscesses, tumours, strokes, meningitis, dementia or Alzheimer's disease, schizophrenia, and affective disorders.[264] This is not a rare phenomenon, but is estimated to occur in 6–40% of cases.[265]

Most compelling is the evidence of people who report complex 'near death' experiences at a time when the EEG shows brain activity to be absent.[266] People who are judged clinically dead and are resus-

257 Bergson 1920 (77).

258 Freedman, Binns, Gao *et al* 2018.

259 Urgesi, Aglioti, Skrap *et al* 2010.

260 Kastrup 2014 (esp 42–50).

261 Peres, Moreira-Almeida, Caixeta *et al* 2012.

262 Nahm, Greyson, Kelly *et al* 2012. In another study 100% of cases died within nine days: Lim, Park, Kim *et al* 2018.

263 Greyson 2019.

264 Nahm *et al, op cit*; Lim *et al, op cit*; Nahm & Greyson 2009.

265 Macleod 2009; Nahm & Greyson 2009.

266 For a detailed overview, see Holden, Greyson & James 2009.

citated or revived after a brief interval with memories of what they have experienced, according to Bruce Greyson, who has studied the phenomenon,

> typically report exceptional mental clarity, vivid sensory imagery, a clear memory of their experience, and an experience more real than their everyday life; all of this occurring under conditions of drastically altered brain function under which the materialist model would say that consciousness is impossible ... There's a sense of the person's thoughts being much faster and clearer than usual, and finally there was a life review or panoramic memory where their entire life seems to flash before them ... Typical emotions reported during the NDE [near death experience] include an overwhelming sense of peace and well-being, a sense of cosmic unity or being one with everything, a feeling of complete joy and a sense of being loved unconditionally ...

And he continues:

> One of the things about NDE that interests me as a psychiatrist are the profound after-effects that people report, a consistent change in values that don't fade over time. Near death experiencers report overwhelmingly that they're 'more spiritual' after the experience, they have more compassion for others, and a greater desire to help others, a greater appreciation for life, and a stronger sense of meaning or purpose in life ... An analysis of their medical records shows that mental functioning is significantly better in those people who come closest to death. Many NDErs experience a panoramic life review, not just visual images, but elaborate events, sometimes the entirety of that person's life.[267]

Such experiences are reported in many cultures, and have been commented on since at least the fifteenth century.[268] In a British study, involving a five-year retrospective and a one-year prospective study, the majority (70%) of caregivers had witnessed 'end-of-life experiences' in their patients just before death;[269] and the number may be greater, since previous unawareness of the phenomenon in staff and relatives alike may lead to under-reporting. These experiences differ from drug-induced hallucinations, and delirious states, and occur in clear consciousness.

Returning to Bergson, he saw memory, as well as consciousness, as being filtered by the brain:

> it does not serve to preserve the past, but primarily to mask it, then to allow only what is practically useful to emerge through the mask. *Such, too, is the part the brain plays in regard to the mind generally.* Extracting from the mind what is externalisable in movement, inserting the mind into this motor frame, it causes it to limit

---

267 Greyson 2019.

268 Fenwick, Lovelace & Brayne 2007; Lim *et al, op cit.*

269 Brayne, Lovelace & Fenwick 2008. See also Fenwick, Lovelace & Brayne 2010; and Macleod 2009.

270 Bergson 1920 (71; emphasis added).

271 Schiller 1891 (295–6).

272 Penfield & Perot 1963: examples include: (a) 'It was like being in a dance hall, like standing in the doorway – in a gymnasium – like at the Kenwood Highschool ... If I wanted to go there it would be similar to what I heard just now'; (b) 'Some crazy things ran through my mind; I was younger, at school. I was playing with a polo bat.' When asked, he said he remembers doing this when going to school at about the age of 10; (c) 'Oh, I had the same very, very familiar memory in an office somewhere. I could see the desks. I was there and someone was calling to me, a man leaning on a desk with a pencil in his hand'; (d) 'I feel as though I was in the bathroom at school'; (e) She had heard 'Frankie and the neighbourhood noises'. She was asked whether it seemed to her to be a memory and she replied, 'Oh no, it seemed more real than that.' She thought she was looking into the yard and saw as well as heard the boy; (f) 'I hear my brother talking.' When asked what he was saying, she said he was talking to one of his schoolmates; (g) While the electrode was held in place the patient said, 'Something brings back a memory, I could see Seven-Up Bottling Company – Harrison Bakery.'

273 Such might be the case of Franco Magnani, memorialised by Oliver Sacks (1992); or of 'HK', who suffered a serious haemorrhage as a neonate, which left him blind, but with almost total autobiographical recall (Ally, Hussey &

[note continues on facing page] ▶

its vision, but also it makes its action efficacious. This means that the mind overflows the brain on all sides, and that cerebral activity responds only to a very small part of mental activity.²⁷⁰

In view of the way in which the left hemisphere deliberately narrows focus ('to limit its vision'), and prioritises utility ('makes it action efficacious'), this seems, as so often with Bergson, to be an almost uncanny case of the mind intuiting brain function from careful introspection.

Similarly Schiller thought that the brain functioned so as to limit memory: 'this will serve to explain not only the extraordinary memories of the drowning and the dying generally, but also the curious hints which experimental psychology occasionally affords us that nothing is ever forgotten wholly and beyond recall.'²⁷¹ As the stimulation by electrodes of brains exposed at surgery by Wilder Penfield illustrate, some kinds of memories, at least, appear to be retained intact in the brain, yet under normal circumstances most are not available, for what Schiller, James and Bergson see as practical reasons.²⁷² Occasionally brain injury may result in exceptional access to memory.²⁷³ Memories that are no longer accessible may nonetheless influence behaviour and can sometimes be accessed under hypnosis. That the brain needs to filter at least its own output in order to function properly is not controversial: less is often more.²⁷⁴

This idea that I have expressed as permission, or 'filtering', is remarkably consonant with Stapp's explanation of the interaction of consciousness with Nature, whereby a set of potential outcomes is narrowed to one actual outcome. One way of putting this is that potentiality is collapsed into the actual through its interaction with consciousness, and that that is what we experience.²⁷⁵

WHAT IS CONSCIOUSNESS OF?

Consciousness always has a correlate – what we are conscious *of*. What should we take the objects of consciousness to be? I suggest that they are just what they seem to be – not representations inside a closed mind of something unknowable outside it. On this, again, Strawson is admirably clear:

> What is it to be a real realist about experience? One way to answer this is that it's to continue to take colour experience, say, or taste experience or pain experience to be exactly what one took it to be, quite unreflectively, before one did any philosophy ... However many new and astonishing facts real realists about experience learn from psychologists – facts about 'change blindness' or 'inattentional blindness' – their basic general understanding of what colour experience or pain experience is remains the same as it was before

Donahue 2013). It is thought possible that deficits in the left temporal lobe may underlie the condition, as it may other savant syndromes: 'the anatomic substrate for the savant syndrome may involve loss of function in the left temporal lobe with enhanced function of the posterior neocortex' (Hou, Miller, Cummings et al 2000). Interestingly the original 'Rain Man' was able to read extremely rapidly, simultaneously scanning one page with the left eye and the other page with the right eye: he had no corpus callosum (Treffert 2009). Cf the case of CS Peirce, who reportedly could simultaneously write a question with one hand and answer it with the other (see Stigler 1992 [63]).

274 See, eg, Chrysikou, Weber & Thompson-Schill 2014; and Shimamura 2000.

275 'The first phase selects a possible next subjective perception on the part of the observer. This 'possible/potential' next perception defines a corresponding brain correlate, which has, according to the theory, a certain statistical weight. The second phase of Process 1 then reduces the material universe to two parts, one that definitely contains this brain correlate and the other that definitely does not contain this brain correlate, and it "actualises" either one part or the other. This choice made by nature between the two parts accords with a certain statistical rule known as the "Born Rule". This Born-Rule choice is the (unique) place where "an element of chance" enters into the quantum dynamics. The preceding choice of a possible next perception reflects the history and the felt values of the observer, and is identified with what the observer feels is his or her personal subjective choice of what physical property of the observed system to probe or inquire about. No element of chance is ascribed to this choice made by an observer of a particular possible probing action' (Stapp 2017, 8).

they did any philosophy. It remains, in other words, entirely correct, grounded in the fact that to have experience is already to know what it is, however little one reflects on it. To taste pineapple, in Locke's old example, is sufficient, as well as necessary, for knowing what it's like to taste pineapple ... I'm talking about the lived character of experience. With this in place, one can say the following. If, as a physicalist naturalist, you think that naturalism or physicalism gives you any good reason to give an account of experience that is in any way deflationary or reductionist relative to the ordinary pre-philosophical understanding of experience, then you have gone wrong.[276]

This is not the same as saying that we cannot be deceived. The earth, we now know, is not flat. While experience is not negotiable, the interpretation of it is up for grabs. Imagination is involved. In the words of Lakoff and Johnson,

> as embodied, imaginative creatures, *we never were separated or divorced from reality in the first place.* What has always made science possible is our embodiment, not our transcendence of it, and our imagination, not our avoidance of it.[277]

According to geneticist Ho, there is

> no mismatch between knowledge and our experience of reality. For reality is not a flat impenetrable surface of common-sensible literalness. It has breadths and depths beyond our wildest imagination. The quality of our vision depends entirely on the extent our consciousness permeates and resonates within her magical realm. In this respect, there is complete symmetry between science and art. Both are creative acts of the most intimate communion with reality.[278]

If one accepts that awareness is a fundamental element in whatever is, rather than a late emergent property from insensate matter, qualia (subjective qualities of experience) are not a 'hard problem' to solve, but the foundational nature of all existence.[279] Calling himself a conscious realist, Donald Hoffman considers

> conscious experiences as ontological primitives, the most basic ingredients of the world. I'm claiming that experiences are the real coin of the realm. The experiences of everyday life – my real feeling of a headache, my real taste of chocolate – that really is the ultimate nature of reality.[280]

This is because we cannot get behind experience to some fantasy of an objective (unexperienced) world, since experience is always involved in reality. But as I explained in the Introduction, this does not mean that there is no reality, just that we are part of its constitution.

This would also be Bergson's position. '*There is a reality*', he wrote, '*that is external and yet given immediately to the mind.* Common-sense

---

276 Strawson 2013. This is highly reminiscent of an equally vibrant passage in another philosopher, Bryan Magee: 'Ordinary human beings not engaged in philosophizing take it for granted that a so-called material object is the sum total of its observable characteristics. If you say to someone, 'How do you know I am holding a leather glove in my hand?' he might well say with impatience, 'Well, don't be silly. I can *see* it. And here' – reaching out 'I can touch it and feel it. And' – taking it and raising it to his nose – 'I can tell from the smell that it's leather.' And so on and so forth. And when he had gone through every attribute of the glove that was accessible to his senses he would think there was nothing left he could say about it. It would not normally occur to him to suppose that the real glove, the glove-in-itself, was an invisible, ineffable substratum that was for ever inaccessible to his observations even though it sustained the incidental characteristics he observed. He would, on the contrary, take the glove to *be* the sum of the characteristics he could observe' (Magee 1998a, 123–4; emphasis in original).

277 Lakoff & Johnson 1999 (93; emphasis in original).

278 Ho 1994 (182).

279 See, eg, Theise & Kafatos 2016.

280 Hoffman 2016.

is right on this point, as against the idealism and realism of the philosophers.'[281] (Schelling, another touchstone for me on this path, was one of the first modern Western philosophers to transcend that dichotomy.) It is what I think the Renaissance philosopher and polymath Pico della Mirandola was getting at when he said, referring to Sufism, that man is 'the living union (as the Persians say), the very marriage hymn of the world'.[282]

SO WHAT IS CONSCIOUSNESS FOR?

Built into the idea of the brain as permissive, or filtering, is, once again, the idea of resistance. Sculpting occurs through an impediment to the otherwise free flow of whatever it may be – in this case, of consciousness. In *The Master and his Emissary*, I touched on the philosophy of Heidegger's contemporary Max Scheler, whereby the experiential world is the product of two metaphysical forces or drives, which he called *Geist* and *Drang*. *Drang* is primary, and is a more elemental force, *Geist* a secondary, comparatively rarefied, intellectual force. 'While there is no simple equation between the right hemisphere and Scheler's *Drang* and the left hemisphere and his *Geist*', I wrote,

> I believe this nonetheless illuminates an important element both in how the hemispheres relate to one another, and in how they together relate to whatever it is that exists apart from ourselves. The relationship between the hemispheres is *permissive* only. The right hemisphere can either fail to permit (by saying 'no') or permit (by not saying 'no'), aspects of Being to 'presence' to it. Until they do so, it does not know what they are, and so cannot be involved in their being as such prior to their disclosure. Subsequent to this, the left hemisphere can only fail to permit (by saying 'no'), or permit (by not saying 'no') aspects of what is 'presented' in the right hemisphere to be 're-presented': it does not know what the right hemisphere knows and therefore cannot be involved in its coming into being as such.[283]

You will also remember that, in its genesis, the brain becomes more powerful by shedding neurones, and 'pruning' connexions; that a primary function of the corpus callosum is to inhibit, and probably the single most important function of the frontal lobes is to inhibit; and that the human brain has proportionately more inhibitory neurones than that of any other species.

Now let us go back to the philosophy of Schelling. His notion of *Weltseele* (world-soul) is of a primordial energy, not unlike Scheler's *Drang*, though more all-encompassing. This stream of energy or world-soul flows in the cosmos, like a vital current, unimpeded until it encounters an asymmetry, an inequality, a difference within itself, which acts as a point of *resistance*. By this resistance something is

281 Bergson 1912 (65; emphasis in original).

282 della Mirandola 1956 (4).

283 McGilchrist 2009b (197; emphasis in original: for more detail, see 197ff & 230–1).

precipitated out of the otherwise even flow, much as an eddy, a vortex, arises in a stream of water. This 'product' of Nature is nothing mechanical or inert: it flows with vital energy itself, constantly renewing itself through the flow, which is both the product's cause and its effect. Individual human consciousnesses, according to Schelling, themselves arise from just such a vortex. Since everything that comes into being arises from a similar vortex within the flow, no one thing is absolutely distinct from anything else that exists, nor from the flow at large. But each has a special role, which is, arising out of the *Weltseele*, to achieve an expression of the *Weltseele*, such that it – the (always already) conscious ground of existence – can come to *know itself*. According to Schelling scholar Andrew Bowie, Schelling 'endeavours to explain the emergence of the thinking subject from nature in terms of an "absolute I" coming retrospectively to know itself in a "history of self-consciousness" that forms the material of the system'.[284] The consciousness of living beings is an expression of this primordial force through which it discovers the (potentially infinite) variety of which it is capable, and which it already is in essence. Such consciousness is the unpacking, the unfolding (literally, the ex-plication) of that force's implicit order, without ever being a lifeless 'product' or mechanism, *natura naturata*: rather it is something in process and possessed of life of its own, *natura naturans*.

The physicist Henry Stapp underlines the way in which quantum theory lays the accent on potential, change and becoming – *events* – rather than actualisation, states and being – *things*:

> The aspects of nature represented by the [quantum] theory are converted from elements of *being* to elements of *doing*. The effect of this change is profound: it replaces the world of *material substances* by a world populated by *actions*, and by *potentialities* for the occurrence of the various possible observed feedbacks from these actions. Thus this switch from 'being to 'action' allows – and according to orthodox quantum theory demands – a draconian shift in the very subject matter of physical theory, from an imagined universe consisting of causally self-sufficient mindless matter, to a universe populated by allowed possible physical actions and possible experienced feedbacks from such actions. A purported theory of matter alone is converted into a theory of the relationship between matter and mind.[285]

The evolutionary biologist Julian Huxley saw something very similar to this. 'Man', he said, 'is that part of reality in which and through which the cosmic process has become conscious and has begun to comprehend itself.'[286] And Pierre Teilhard de Chardin, perhaps more strictly in Schelling's spirit (though I am not aware that he read Schelling) wrote: 'Man discovers that *he is nothing else than*

---

284 Bowie 2016.

285 Stapp 2007 (20; emphases in original).

286 Huxley 1959b (236).

*evolution become conscious of itself.*²⁸⁷ Taking this forward again into the world of quantum physics, Bohm notes that 'there is no need to regard the observer as basically separate from what he sees nor to reduce him to an epiphenomenon of the objective process. More broadly one could say that, through the human being, the universe has created a mirror to observe itself.'²⁸⁸ In Thomas Nagel's words, 'each of our lives is a part of the lengthy process of the universe gradually waking up and becoming aware of itself'.²⁸⁹

In this light, the question 'what is consciousness for?' appears to be based on a false premise. Consciousness is nothing to our purpose; we are to the purpose of consciousness. We are part, more properly, of a purpose – that of self-knowledge – which lies fulfilled within itself, not requiring buttressing from without by any justification in terms of utility.

Philosophy faces its most fundamental challenges at the interface between consciousness and matter: all the great questions are there. Why is there something rather than nothing? Why is that 'something' complex, beautiful and orderly, rather than meaningless, ugly and chaotic? Why is it perfectly, exquisitely, calibrated so as to make the existence of matter possible? And life, and sentient beings, possible? And how is it that we are equipped to understand its complexity, beauty and order? As physicist David Deutsch puts it, 'The extraordinary thing is not that there are laws but that we can understand them. Why should we be able to understand them? It's almost as if, whenever we land on a new and unknown planet the inhabitants come up and speak to us in English. Why should the laws of the universe be understandable by us? Well, either it's a fantastic coincidence or there's some deep reason why it had to be that way.'²⁹⁰

'Physicalism', writes Whitehead scholar Matt Segall, referring to what I call scientific materialism,

> is the idea that the universe is fundamentally composed of entirely blind, deaf, dumb – DEAD – particles in purposeless motion through empty space. For some reason, these dumb particles follow the orders of a system of eternal mathematical laws that, for some reason, the human mind, itself made of nothing more than dumb particles, is capable of comprehending. On this definition of physicalism, 'life' and 'consciousness' are just words we have for epiphenomenal illusions with no causal influence on what happens. 'Life' is a genetic algorithm and 'consciousness' is a meme machine, in Dawkins' and Dennett's terms. We are undead zombies, not living persons, on this reading of physicalism.²⁹¹

By contrast, according to Thomas Nagel, 'the inescapable fact that has to be accommodated in any complete conception of the universe',

287 Teilhard de Chardin 1959 (221; emphasis in original).
288 Bohm 2002 (389).
289 Nagel 2012 (85).
290 Deutsch 2006.
291 Segall 2018.

is that the appearance of living organisms has eventually given rise to consciousness, perception, desire, action, and the formation of both beliefs and intentions on the basis of reasons. If all this has a natural explanation, the possibilities were inherent in the universe long before there was life, and inherent in early life long before the appearance of animals. A satisfying explanation would show that the realisation of these possibilities was not vanishingly improbable but a significant likelihood given the laws of nature and the composition of the universe. It would reveal mind and reason as basic aspects of a nonmaterialistic natural order.[292]

If one is certain that consciousness emerged from matter, then this is indeed a conundrum, and I agree with Nagel's conclusions. But to me even for the *possibilities* of consciousness, and all the rest, to be inherent, the actualities must have been present, otherwise we are back to what one might call the Midshipman Easy problem. Nagel points up the trap for reductionism, since, if consciousness exists (and it does) and it cannot emerge (and it cannot), this implies that consciousness was there all along. So he continues: 'since conscious organisms are not composed of a special kind of stuff, but can be constructed, apparently, from any of the matter in the universe, suitably arranged, it follows that this monism will be universal. Everything, living or not, is constituted from elements having a nature that is both physical and nonphysical – that is, capable of combining into mental wholes.'[293]

OPEN AND CLOSED SYSTEMS

Here I would make a reflection on the relationship between consciousness and life. 'Thought and matter have a great similarity of order', says Bohm, 'in a way, nature is alive, as Whitehead would say, all the way to the depths. And intelligent. Thus it is both mental and material, as we are.'[294]

I have quoted from Schrödinger's *What Is Life?* before. It is often read as reassuringly decomplexifying life, by approximating it to physics; but as the biophysicist, mathematician and systems theorist Robert Rosen realised, what many readers have not noticed is that what Schrödinger was actually doing was complexifying physics, by approximating it to life. In Rosen's words, though Schrödinger was one of the outstanding theoretical physicists of all time,

> one of the striking features of his essay is the constantly iterated apologies he makes, both for his physics and for himself personally. While repeatedly proclaiming the 'universality' of contemporary physics, he equally repeatedly points out (quite rightly) the utter failure of its laws to say anything significant about the biosphere and what is in it. What he was trying to say was stated a little later, perhaps even more vividly, by Albert Einstein. In a letter to Leo Szilard, Einstein

292 Nagel 2012 (56).
293 *ibid* (57).
294 Bohm 1982 (39).

said, 'One can best feel in dealing with living things *how primitive physics still is*'.[295] Schrödinger and Einstein were not just being modest; they were pointing to a conundrum about contemporary physics itself, and about its relation to life.[296]

What Rosen suggests is that Schrödinger, for the first time, had realised, not just that the study of matter might teach us about organisms, but that 'organisms teach new lessons about matter in general'.[297] All living things involve open systems. An open system is not just a closed system with a little bit added on, but a wholly different kind of system, nothing like the artificially isolated, predictable, closed systems, at or near equilibrium, assumed by classical physics. What Schrödinger saw was that life was not a marginal phenomenon within classical physics, but rather classical physics, with its 'assumption of excessively restrictive closure conditions', conservation laws and similar highly atypical features, was a special, limiting case of the physics of open systems such as, but by no means confined to, life. Such irreducibly complex systems have the capacity to generate and maintain stable patterns that, though partially constrained, are neither rigid nor wholly predictable, and thus never wholly capable of being objectified or computed. And they are everywhere in the universe – in stars and galaxies, as well as in what we normally call life. David Oliver sees something very like this dynamism, and the capacity for self-organisation, in the realm of the inorganic:

> Life is the outstanding example of dynamism, but not all dynamism is life. Beyond life, the geological structures of the planet send and receive seismic signals which in turn induce tectonic motion. Before life, the physics and chemistry of organic and amino acids and amphiphilic compounds were sufficiently evolved to self-assemble bilayers separating one set of chemicals from another – an important precursor to the cell. Inorganic nature is rich in self-organization in which simple atoms and molecules interact through their force fields to create complex abiotic chemical cycles and structures: dynamism outside of life, much of which is a prelude to life.[298]

When it comes to approaching complex systems, one cannot seek an understanding only by going downwards to subsystems, by analysis. This is where reductionism is impotent. When it comes to open systems, in order to understand what is going on 'inside' the system, we need to look equally upwards and outwards to the system's environment. A system cannot be fully known if its subsystems are not known; yet knowing a subsystem requires knowledge of the larger system to which it belongs. There is a recursive loop here.

Reductionist science, Rosen claims, is based on three ideas, each of which is easily challenged. The first is that simplicity is generic,

295 Letter of Einstein to Leo Szilard, 12 July 1947; as at Clark 1984 (34–5; emphasis added).

296 Rosen 2000 (7).

297 *ibid* (20).

298 Oliver 2019 (4).

and complexity rare. The second is that these simple, generic systems are independent of context. And, third, that you get from simple to complex along a continuous gradient by merely adding simple, context-free systems together. This last leads to the assumption that 'the analysis of more complex systems is merely a matter of inverting the accretions that produced them.' Collectively, he continues, these assumptions

> serve to make biology unreachable from contemporary physics, which is based on them. At the same time, qualifying any of them would profoundly modify the conceptual basis of contemporary physics itself. This is why the phenomena of biology embody a foundations-crisis for contemporary physics, far more profound than any that physics has yet seen. Indeed it is more accurate to invert the presumptions. Namely, there is a sense in which complex systems are far more generic than simple, context-independent ones. Moreover, analysis and synthesis are not simple rote operations, nor are they in any sense inverses of one another. In short, the entire identification of *context-independence* with *objectivity* is itself far too special and cannot be retained in its present form as a foundation for physics itself.[299]

What Rosen suggests is that such a model is not only not generic, but is 'far too restricted and specialised to accommodate things like organisms':

> Biology is not simply a special case, a rare and overly complicated anomaly, a non-generic excrescence in the world of objective things. To the contrary, it is that world itself that is non-generic, and it is organisms which are too general to fit into it. This too counts as an objective fact, and it is one with which (contemporary) physics *must* come to terms, if it indeed seeks to comprehend all of material nature within its precincts.[300]

The question 'How does life emerge from inanimacy?' is, like 'How does consciousness emerge from matter?', a question based on premises that are themselves questionable. As matter is a special case of consciousness, inanimacy should be seen, according to Rosen, as a special, indeed the limiting, case of life. C. S. Peirce would have agreed: acknowledging his debt to Schelling, he also held that matter was 'mere specialized and partially deadened mind'.[301] Our idea of life comes from every living species, not just the most primitive ones, *each* telling us something about the nature of life. As Whitehead asks, 'why construe the later forms by analogy to the earlier forms. Why not reverse the process?'[302] Higher, biological, forms of organisation of matter tell us about simpler ones, as much as simpler ones are supposed to tell us about what life is – and *more*.[303] The structure

---

299 Rosen *op cit* (36; emphasis in original).

300 *ibid* (37; emphasis in original).

301 Peirce 1955 (339).

302 Whitehead 1929b (11).

303 Cf Bergson 2007f (74): 'Why must we speak of an inert matter into which life and consciousness would be inserted, as in a frame? By what right do we put the inert first?'

of the universe is, according to Bohm, 'much more reminiscent of how the organs constituting living beings are related, than it is of how parts of a machine interact.'[304]

'Science is taking on a new aspect which is neither purely physical, nor purely biological', Whitehead says elsewhere. 'It is becoming the study of organisms. Biology is the study of the larger organisms; whereas physics is the study of the smaller organisms.'[305] The rigid partition between organic and inorganic breaks down; and when it does so one consequence is that, since the more complex can't be constructed out of the less complex by mere summation of simple components, but the less complex *can* be constrained out of the more complex by setting enough limiting conditions, it is at least as logical, if not more so, to see the inanimate as a limit case of the animate, as to see life as a peculiar subset of inanimacy. Here is Rosen again:

> organisms, far from being a special case, an embodiment of more general principles or laws we believe we already know, are indications that these laws themselves are profoundly incomplete. The universe described by these laws is an extremely impoverished, non-generic one, and one in which life cannot exist. In short, far from being a special case of these laws, and reducible to them, biology provides the most spectacular examples of their inadequacy. The alternative is not vitalism, but rather a more generic view of the scientific world itself, in which it is the mechanistic laws that are the special cases.[306]

In other words, organicism, like consciousness, may be a foundational principle of the cosmos: the mechanisms we wish to base our understanding on may be no more than simplified mental abstractions from the whole. Note that this is not vitalism, which requires the addition of an extraneous element to matter, but its precise opposite: nothing needs adding to matter, since matter is nothing like what it is projected to be – namely, simple, inert until set in motion, and separate from consciousness. Schelling had already pointed out that the problem is to account not for motion in Nature, but for whatever is resting, or permanent; since permanence 'is a limitation of its own activity': stasis is the limit case of life.[307]

Because this may be a step too far for some minds priding themselves on their rationalism – there is of course nothing irrational about what Rosen is suggesting, just unusual by the standards of present day scientistic assumptions – I'd point out that the great Enlightenment *philosophe* Diderot, writing in 1769, had this to say:

> All is in perpetual flux … every animal has something of the human; every mineral has something of the plant; every plant has something of the animal. There is nothing circumscribed in nature … Each form has its own proper fortune or misfortune. From the elephant

304 Bohm 1980 (175).

305 Whitehead 1926b (129).

306 Rosen 2000 (33–4). See also Waddington 1961 (121): 'something must go on in the simplest inanimate things which can be described in the same language as would be used to describe our self-awareness.'

307 Schelling, 'The unconditioned in nature': 2004, I, i (17).

to the flea ... from the flea to the sensitive and living molecule, the origin of everything, there is not a point anywhere in the whole of nature that does not suffer or take delight.[308]

A NEW WAY OF SEEING THINGS

So much, for now, of life. Let us return to consciousness, and see what happens if we set aside for a moment the idea that it arises from matter in a random universe (none of which seems very likely in any case, for reasons some of which I have touched on), and instead think along the lines that I have indicated. I am not claiming to prove anything at all – in this area the very idea of proof is inappropriate; and I cannot expect to convince the reluctant. I want merely to offer an alternative *Gestalt* or world-picture that, to me at least, accommodates some of these questions better than other world-pictures I have explored. The reader must decide for himself or herself whether it resonates with experience and intuition better or worse than reductionist materialism.

If matter is a product of primal consciousness, and is the creative element of resistance within it which causes 'things' – becomings and processes – to be precipitated out as differentiated entities in space and time, this requires no further purpose than itself. It is the never-ending coming into being of the cosmos, in which the primal consciousness comes to understand itself, and even fulfil itself, through the realisation of what would otherwise remain unknowable potential, in a constant act of creation. There is no sundering between the 'positive' processive part of the whole and the 'negative' resistant part of the whole; they are not a duality, but complementary and mutually necessary aspects of the same entity, just as the vortex is caused *by* the flow, from an element *in* the flow which partly *impedes* the flow – and *is* the flow for that time and that place. To use the time-honoured Chinese formulation, it is thus that the One and the 10,000 Things can be seen to be one.

If there is anything in this idea, it helps us approach the thorny questions differently. That is all – it is not designed to compel consent, which in these matters cannot in any case be compelled (if it could, philosophy would have come to an end a long time ago). But if the material cosmos is an emanation or projection of a grounding consciousness it will as a matter of course have the necessary, apparently fine-tuned, conditions to come into existence; it will naturally have the qualities of order, beauty and complexity because it issues from a consciousness that, like us, is attuned to and gives rise to such elements; it will naturally produce conscious beings, and the conscious beings will naturally be able to speak its language, since they are generated by it. Of course this does not answer the unan-

---

308 Diderot 1875 (138–40):
« Tout est en un flux perpétuel ... Tout animal est plus ou moins homme; tout minéral est plus ou moins plante; toute plante est plus ou moins animal. Il n'y a rien de précis en nature ... Chaque forme a le bonheur et le malheur qui lui est propre. Depuis l'éléphant jusqu'au puceron ... depuis le puceron jusqu'à la molécule sensible et vivante, l'origine de tout, pas un point dans la nature entière qui ne souffre ou qui ne jouisse. »

swerable question, why there is something rather than nothing. It can do no more than postulate that the grounding consciousness is intrinsically creative and that part of its self-realisation is the realisation of the cosmos: something, not nothing.

Of course this is to make assumptions, but it is impossible not to make assumptions. The standard materialist position makes assumptions of its own, eg, that all is entirely random and meaningless; that nothing exists apart from matter; or that if consciousness exists, it comes about secondarily at some point in evolution out of something fundamentally alien to consciousness; that the order and beauty and 'apparently' purposive drive in things cannot be explained, as neither can our capacity to appreciate and understand more than a very small part of them (see Chapter 26), and that these are all remarkable coincidences (whenever we land on a new and unknown planet the inhabitants come up and speak to us in English). On either my outline or this one, order still emerges out of chaos and things naturally complexify without outside intervention; the difference is that in the materialist paradigm all that is inexplicable. So is the fine-tuning of the cosmos. Attempts to explain it result in the extravagant postulate that there are an infinite number of universes, so that eventually one like this is bound to come about by chance. Occam wept.

As an alternative to the view that teleology is of the stuff of the universe, as Nagel points out, there is the view that

> this universe is not unique, but that all possible universes exist, and we find ourselves, not surprisingly, in one that contains life. But that is a cop-out, which dispenses with the attempt to explain anything … One doesn't show that something doesn't require explanation by pointing out that it is a condition of one's existence. If I ask for an explanation of the fact that the air pressure in the transcontinental jet is close to that at sea level, it is no answer to point out that if it weren't, I'd be dead.[309]

309 Nagel 2012 (95, n9).

And it does not address the question why life and its necessary conditions *exist at all*.

The grounding consciousness is not deterministic. It has none of the characteristics of an omnipotent and omniscient engineering God constructing and winding up a mechanism. It is in the process of discovering itself through its creative potential (one thing we all know directly from our own experience is that consciousness is endlessly creative). According to philosopher Arran Gare,

> life is self-organizing and has to be appreciated as such. Schelling rejected both appeals to a creator of Nature and to a life-force to account for life. 'You *destroy* all idea of *Nature* from the very bottom, as soon as you allow the purposiveness to enter her from without, through a

transfer from the intelligence of any being whatever.'[310] 'Life-force', Schelling went on, 'is a completely self-contradictory concept.'[311]

It may sound as if his *Weltseele*, the energetic creative force, is just such a life force. But it is not. What Schelling is fending off is the notion that Nature, or the ground of Being, needs a life force adding in, extrinsically, so to speak, to make what happens happen, the effect of which addition would be actually to rob Nature of its *intrinsic* creative potential. This also applies, as he says, to 'purposiveness', and it is perhaps easier to see the problem there.

Many discussions about whether life or the cosmos more generally display purpose founder on an important distinction between two meanings of purpose that are at loggerheads with one another. This can be seen most easily by speaking about works of art or people. If a piece of music is thought of as having an extrinsic purpose, such as to help you relax, it is misunderstood and diminished. If mindfulness is thought of as a way to make bankers more efficient at generating profits, it is misconceived and degraded. If a person is thought of as having a purpose, he becomes a mere tool. But this doesn't exclude a different meaning of purpose. Each of these – the music, the spiritual practice, the person – has a kind of purpose which is fulfilled by its mere being itself (this is not simply to say that they are valued: see Chapter 27). Importing the extrinsic kind of purpose immediately destroys the intrinsic kind of purpose. A watchmaker, omniscient and omnipotent engineering God imports extrinsic purpose and is an idea of this destructive kind. So would be a 'life force' that was brought in: it would deny the vitality of the cosmos *per se*.

This puts me in mind again of Nagel, speaking of theistic conceptions:

> To make the possibility of conscious life a consequence of the natural order created by God while ascribing its actuality to subsequent divine intervention would then seem an arbitrary complication.[312]

The *Gestalt* I offer has a number of other things going for it, I believe. It chimes with, and I believe makes best sense of, Eddington's view that 'we have found a strange foot-print on the shores of the unknown ... we have succeeded in reconstructing the creature that made the footprint. And Lo! it is our own.'[313] At the same time it removes some of the unintended implications of solipsism that are troubling in that formulation. In fact it importantly suggests how and why the realism/idealism divide and the objective/subjective divide can be, and are, transcended. The key point is that the ground of consciousness has given rise to something other than itself wholly out of itself, that is *nonetheless not wholly determined by the ground of consciousness*. There is an element that cannot in the nature of things

---

310 Schelling 1988 (34).
311 *ibid* (37); in Gare 2011.
312 Nagel *op cit* (95).
313 See p 1056 above.

be predicted: there is room for freedom. After only eight collisions of the billiard balls, we are already in – not accidentally, but *essentially* – unpredictable territory. The process of discovery that this makes possible depends on there being distinctness but not complete separation: the unity of difference with union.

I think it also gives us a new potential take on the so-called binding problem: how the various modules of brain function result in a single sense of one experiencing self. If consciousness does not have to be 'put together' from modules in such a way as to result in a unified consciousness, but is instead permitted or moulded as a whole by the brain, the famously unsolved binding problem does not arise. Just a tiny example, but some nice research on infant development by Elizabeth Spelke suggests this may be right. Piaget had contended that in babies the five senses operated separately, and that 'intermodal perception' – the capacity to recognise, for instance, that the clattering sound you just heard and the pot lid you just saw your mother drop are part of the same event – developed later, during toddlerhood. In other words, a 'put-together-the-modules' theory. Spelke was sceptical of the Piaget line; in the mid 1970s, she devised an experiment to test it. She showed four-month-old infants two films, on side-by-side screens, while playing the soundtrack for just one of them. Babies focussed on the screen for which the soundtrack was appropriate, correctly matching the picture and sound.[314] To quote Giulio Tononi and Christof Koch, 'consciousness is unified: each experience is irreducible to non-interdependent subsets of phenomenal distinctions'.[315]

This is quite compatible with hemisphere differences, as Koch demonstrates when he says, 'because both the speaking *and* the mute hemispheres carry out complex, planned behaviours, both hemispheres will have conscious percepts, even though the character and content of their feelings may not be the same'.[316] It is not that consciousness is fragmentary and must be integrated if there are to be individual beings; it's that consciousness is integral and must somehow be divided if there are to be individual beings. The whole business of consciousness limiting itself by embracing the stickiness of matter (remember, matter slows energy down and makes its forms persist longer than they otherwise would), is to produce differentiation, individuation, thisness, actuality precipitated out of a sea of potential. The process of individuation involves sculpting, filtering – however one wants to put it, delimiting and distinguishing – parts of the seamless whole. Thus the brain needs two streams of consciousness, one in each hemisphere, but they are like two branches of a stream that divide round an island and then reunite.

A stumbling block for the scientific reception of any such model is that science is not well disposed towards what cannot be localised,

314 Spelke 1976.

315 Tononi & Koch 2015.

316 Koch 2004 (287–94).

sensed nor measured, such as is mind (contemporary physicists may be, perforce, an exception). Nonetheless mind remains embarrassingly real. Nobel Prize-winning physiologist George Wald thought that 'the stuff of which physical reality is composed is mind-stuff. It is Mind that has composed a physical universe that breeds life, and so eventually evolves creatures that know and create ... In them the universe begins to know itself.'[317] This is entirely in keeping with the model I am recommending for consideration. But, echoing Yukawa's words, Wald reflected: 'Let me say that it is not only easier to say these things to physicists than to my fellow biologists, but easier to say them in India than in the West.'[318] He continues: 'Mind is not only not locatable, *it has no location*. It is not a thing in space and time, not measurable; hence – as I said at the beginning of this paper – not assimilable as science.' He put forward the hypothesis that

> Mind, rather than being a very late development in the evolution of living things, restricted to organisms with the most complex nervous systems – all of which I had believed to be true – that Mind instead has been there always, and that this universe is life-breeding because the pervasive presence of Mind had guided it to be so. That thought, though elating as a game is elating, so offended my scientific possibilities as to embarrass me. It took only a few weeks, however, for me to realize that I was in excellent company. That kind of thought is not only deeply embedded in millennia-old Eastern philosophies, but it has been expressed plainly by a number of great and very recent physicists.[319]

IS CONSCIOUSNESS WHOLLY PERSONAL?

What is the relationship between consciousness as a whole and the individual consciousness?

The twelfth-century Muslim philosopher Averroes believed that only one intellect exists for the whole human race, in which every individual participates to the exclusion of personal immortality. 'A human being is a part, limited in time and space, of the whole that we call the Universe', wrote Einstein. 'He experiences himself and his feelings as cut off from the rest – an optical illusion of his consciousness.'[320] Other physicists have embraced similar notions. Bohm proposed human participation in 'a greater collective Mind in principle capable of going indefinitely beyond even the human species as a whole.'[321] Such a collective mind parallels the collective unconscious posited by Jung, and which for him made sense of the existence of archetypes that seem given, rather than acquired, in the human psyche. In most cultures there is a belief in consciousness that transcends the locality of the body in space and time. Schrödinger

---

317 Wald 1984.
318 Wald 1989.
319 *ibid* (emphasis in original).
320 Letter of Einstein to Robert Marcus of the World Jewish Congress, who had recently lost a son, 12 February 1950 (retranslated from his handwritten draft): » *Ein Mensch ist ein räumlich und zeitlich beschränktes Stück des Ganzen, was wir ‚Universum' nennen. Er erlebt sich und sein Fühlen als abgetrennt gegenüber dem Rest, eine optische Täuschung seines Bewusstseins.* «
321 Bohm 2002 (386).

put it like this: 'consciousness is a singular of which the plural is unknown.'[322]

The year before he died, James wrote:

> Out of my experience, such as it is (and it is limited enough) one fixed conclusion dogmatically emerges, and that is this, that we with our lives are like islands in the sea, or like trees in the forest. The maple and the pine may whisper to each other with their leaves ... But the trees also commingle their roots in the darkness underground, and the islands also hang together through the ocean's bottom. Just so there is a continuum of cosmic consciousness, against which our individuality builds but accidental fences, and into which our several minds plunge as into a mother-sea or reservoir.[323]

And in *A Pluralistic Universe*, published in the same year: 'Mental facts do function both singly and together, at once, and we finite minds may simultaneously be co-conscious with one another in a superhuman intelligence.'[324] What we consciously experience as individuals is, for James, selected by our brains from a much larger field of consciousness. This would also be compatible with Scheler's view that when we share thoughts and feelings, this does not describe the linking of two locations, one in my head and one in yours, but a literal sharing, outside of any such spatial considerations. Perhaps we should say not that thoughts and feelings are in us, but we in them.

'Sentience was never our private possession', writes David Abram. 'We live immersed in intelligence, enveloped and informed by a creativity we cannot fathom':

> The world we inhabit ... is a sensitive sphere suspended in the solar wind, a round field of sentience sustained by the relationships between the myriad lives and sensibilities that compose it. We come to know more of this sphere not by detaching ourselves from our felt experience, but by inhabiting our bodily experience all the more richly and wakefully, feeling our way into deeper contact with other experiencing bodies, and hence with the wild, inter-corporeal life of the earth itself.[325]

Suppose that I am right and that everything is ultimately part of one consciousness, that individual consciousnesses are never wholly separate from the whole – much as vortices in the stream, or waves in the sea, are visible, measurable and truly distinguishable, but not separate, from the body of water in which they arise – then the individual correctly perceives a self, but a self that is connected to the whole: wholly a self, and wholly part of the whole. To see the eddy as a separate entity from the river, or the wave as something 'additional to' the sea, causes the self to become unstable: either wholly

322 Schrödinger 1967b (89).
323 James 1909b (589).
324 James 1909a (292).
325 Abram 2011.

localisable or lost in the whole, not part and whole together. A better image perhaps of the proposed relation of the individual consciousness to the whole is that of a cell membrane that has protrusions or 'evolutions' in it. Inside any such villous protrusion, one would seem to be completely bounded by membrane on all sides – except right at the foot of it, where the interior of the villus is seen to be seamlessly merged with the cytoplasm of the cell as a whole.

Our consciousness goes out to meet other consciousnesses, where it can merge without loss, but rather with enrichment, of self. Heidegger captured this when he wrote to Hannah Arendt of the 'sweet burden' of love, that 'we become what we love and yet remain ourselves.'[326] More stringently, Schrödinger conveyed it as the impossibility of seeing the whole at once from any one viewpoint: 'This life of yours which you are living is not merely a piece of the entire existence, but is in a certain sense the *whole*; only this whole is not so constituted that it can be surveyed in one single glance.'[327] Ryōan-ji again.

I believe that what exists is a locally differentiated, but ultimately single, field of potentiality, which is constantly actualising itself. Thus all is one, and all is many. Each differentiation is, however, also a *Gestalt* that is complete in itself – a new whole, not a fragment. This seems to me the essence of creation – differentiation of something that is not thereby destroyed in its unity, but enriched, as with the unfolding of something hitherto implicit into a new more explicit order. So what we see as distinct is distinct, but it does not have to be effortfully 'connected' to anything else, because it is never wholly separate. Here I find Wordsworth's formulation, 'in Nature everything is distinct, yet no thing defined into absolute, independent singleness', simple, beautiful and true.[328] I see matter as a special case, or 'phase', of consciousness, which is the primal 'stuff' out of which the universe is made.

IMPLICATIONS OF THE HEMISPHERE HYPOTHESIS

What if any are the implications of the hemisphere hypothesis here? And how legitimate is it to make comparisons between hemisphere differences and the findings of physics?

Types of attention to matter matter

First of all, if our experience is mediated for us by a brain that pays attention to the world in two broadly different ways, the world will, as I have tried to show throughout this book, have apparently paradoxical qualities, sometimes conforming to the qualities revealed by one and sometimes to the qualities revealed by the other. We have seen how this applies in the realm of the everyday, but it applies just as much in the realm of the very small or large. In fact it is particu-

---

326 Letter of Martin Heidegger to Hannah Arendt, 21 February 1925; in Ludz 2004 (4–5).

327 Schrödinger 1964 (22).

328 Wordsworth 1876, vol II (122).

larly evident that it must apply to the interaction of attention with physical systems at the quantum level, since the role of attention there is explicit in a way that in the everyday setting is conventionally glossed over – though it is just as real there. Quantum theory focusses our attention on the role of attention.

Second, we have already seen that Pauli raised the question whether there were phenomena 'that depend on *who* observes them (ie, on the nature of the psyche of the observer)'. It is not incoherent to believe that qualitative differences – which it may be assumed have at least something to do with the quality of attention – may affect what is observed to happen. And a main source of difference in the quality of attention is which hemisphere is primarily engaged in attending. As Bohm explains, in doing so seeming to refer to hemisphere differences in attention:

> Man's general way of thinking of the totality, ie his general world view, is crucial for overall order of the human mind itself. If he thinks of the totality as constituted of independent fragments, then that is how his mind will tend to operate, but if he can include everything coherently and harmoniously in an overall whole that is undivided, unbroken and without a border (for every border is a division or break) then his mind will tend to move in a similar way, and from this will flow an orderly action within the whole.[329]

And he continued, more explicitly referencing hemisphere differences, and their relationship to the thriving of a culture:

> My suggestion is that at each stage the proper order of operation of the mind requires an overall grasp of what is generally known, not only in formal, logical, mathematical terms, but also intuitively, in images, feelings, poetic usage of language, etc. (Perhaps we could say that this is what is involved in harmony between the "left brain" and the "right brain"). This kind of overall way of thinking is not only a fertile source of new theoretical ideas: it is needed for the human mind to function in a generally harmonious way, which could in turn help to make possible an orderly and stable society ... however, this requires a continual flow and development of our general notions of reality.[330]

This leads naturally to the question of the legitimacy of drawing parallels between psychology and physics. There is a consensus among many physicists that what they are talking about must be applicable outside of the narrow confines of physics. According to Heisenberg, one of Bohr's oft-repeated 'quotations' was 'clarity is gained through breadth'.[331] (If it is indeed a quotation, he may have been thinking of Heraclitus' dictum: 'men who love wisdom must be good enquirers into many things indeed'; or of Plotinus: 'all things

329 Bohm 1980 (xi).

330 *ibid* (xi–xiv).

331 Heisenberg 1971, 'Elementary particles and Platonic philosophy': 237–47 (246).

are full of signs, and it is a wise man who can learn about one thing from another'.)³³² Such breadth is itself the forte of the right hemisphere. Pauli wrote that 'this business of complementarity within physics leads naturally beyond the narrower field of physics to analogous situations to be found in the usual conditions of human knowledge ... Indeed the relationship between subject and object has paradoxical qualities, which form a far-reaching analogy to the relationship between the means of observation and the system being observed, as we encounter it in quantum physics.'³³³

Bohr thought that 'the necessity of considering the interaction between the measuring instruments and the object under investigation in atomic mechanics corresponds closely to the peculiar difficulties, met with in psychological analyses, which arise from the fact that the mental content is invariably altered *when the attention is concentrated on any single feature of it*.' He contended that this analogy 'offers an essential clarification of the psycho-physical parallelism.'³³⁴ Elsewhere he compared the contrast between wave and particle to the contrast between the 'continuous onward flow of associative thinking and the preservation of the unity of the personality'. This, he thought,

> exhibits a suggestive analogy with the relation between the wave description of the motions of material particles ... and their indestructible individuality. The unavoidable influence on atomic phenomena caused by observing them here corresponds to the well-known change of the tinge of the psychological experiences which accompanies any direction of the attention to one of their various elements.³³⁵

Planck wrote that

> Natural science and the intellectual sciences cannot be rigorously separated. They form a single interconnected system and if they are touched at any part the effects are felt through all the ramifications of the whole, the totality of which is forthwith set in motion ... It would be absurd to assume that a fixed and certain law is predominant in physics unless the same were true also in biology and psychology ... the nature of any system cannot be discovered by dividing it into its component parts and studying each part by itself, since such a method often implies the loss of important properties of the system. We must keep our attention fixed on the whole and on the interconnection between the parts.³³⁶

Note that both Bohr and Planck pass from the analogy between the psychological and the physical to aspects of the psychological world that mirror established hemisphere differences. Thus Bohr refers to capacity for flow and interpenetration on the one hand

---

332 Plotinus, *Enneads*, Bk II, Tractate III, §7 (trans AH Armstrong).

333 Pauli 1961 (15–16): »Dieser Sachverhalt der Komplementarität innerhalb der Physik führt in natürlicher Weise über das engere Gebiet der Physik hinaus zu analogen Situationen bei den allgemeinen Bedingungen der menschlichen Erkenntnis ... In der Tat hat die Beziehung zwischen Subjekt und Objekt paradoxe Eigenschaften, die eine weitgehende Analogie zu Beziehung zwischen Beobachtungsmittel und beobachtetem System haben, wie wir ihr in der Quantenphysik begegnen.«

334 Bohr 1933 (emphasis added). A version of this paper had been published in Bohr 2011 (3–12). I have drawn on both versions.

335 Bohr 1961 (99–100).

336 Planck 1936 (31–3).

and distinctness on the other. Planck refers to seeing the whole and seeing the parts. Something similar is seen in Heisenberg, who drew a parallel between complementarity in physics and different modes of attention to music. 'We realise that the situation of complementarity is not confined to the atomic world alone; we meet it … when we have the choice between enjoying music and analysing its structure.'[337]

From 1932 until his death in 1958, Pauli collaborated with Jung in conceptualising the unconscious mind as the psychological analogy of the field concept in quantum physics. An interesting thought arises, namely that it is specific focussed attention to quantifying detail that causes collapse of the wave function, or, put differently, causes the field to become quantised.[338] Such attention is characteristic of the brain's left hemisphere, which requires the 'either/or' aspect of experience to stand forward: what is ambivalent becomes univalent, what was flowing becomes fixed, what was diffusely connected to and embedded in the context becomes arbitrary. The right hemisphere could be seen as the field of potential in which things remain fluid, non-localised, implicit and interconnected. To the extent that the right hemisphere is more identified with the unconscious mind and the left hemisphere with the self-reflexively aware mind, this corroborates Pauli's insight into Jung.

Asymmetry of continuity and discontinuity

Jeans, too, hinted at the way in which complementarity must exist at the mental level, noting that, since the observation that the electron did not obey the particle model, 'there had been a progress towards the truth', in which

> each step was from particle to waves, or from the material to the mental; the final picture consists wholly of waves, and its ingredients are wholly mental constructs … the cumulative evidence of various pieces of probable reasoning makes it seem more and more likely that reality is better described as mental than as material … There is no longer a dualism of mind and matter, but of waves and particles … The two members of this dualism are no longer antagonistic or mutually exclusive; rather they are complementary. We need no longer devise elaborate mechanisms, as Descartes and Leibniz did, to keep the two in step, for one controls the other – the waves control the particles, or in the old terminology the mental controls the material.[339]

If reality is mental, as so many physicists repeatedly stress, and has two primary modes with mutually incompossible but nonetheless complementary features, both of which are needed, this constitutes a remarkable parallel between the complementarity of the physicists

[337] Heisenberg 1958 (155).

[338] See p 1018 above.

[339] Jeans 1942 (200, 202 & 204).

and the complementarity of the two modes of attention that, I believe, underwrite our paradoxical mental life. Jeans here also corroborates the idea that, within the complementarity, one of the parties is ontologically prior to the other: mind to matter, wave to particle.

That makes sense to me, because while a particle symbolises discreteness, a wave symbolises both discreteness and union. There is one aspect of a wave that I have not touched on so far: the possibility of a standing wave. A standing wave is one which persists, by maintaining harmonised oscillations that do not cancel one another out: this requires a whole number of oscillations. Think of a string, such as that of a musical instrument, that is fixed at either end. When it is plucked, it must make a whole number of troughs and peaks: it can make 1, 2, 3, 4 etc, but not, for example, 1½, or 2¼ peaks:

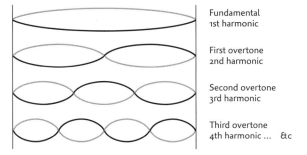

Fig. 58. *Harmonic oscillations*

If one thinks of electronic orbits as waves that are standing, they can do so only if the number of oscillations is, similarly, a whole number, otherwise at the point where the wave completed a circuit it would not coincide with its origin:

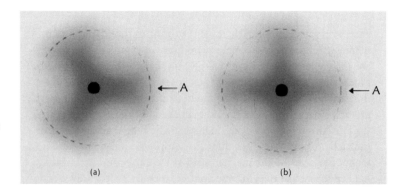

Fig. 59. *How continuous oscillations lead to discrete levels of energy. Each diagram is a 'snapshot' taken at a single instant. This is not intended to be a realistic picture of a hydrogen atom: it is designed only to show how the field picture, with its oscillations, leads to discrete levels of energy.*

(adapted from colour version in Brooks 2016 [location 1888])

Fig. 59 shows simplified patterns of oscillation of the electron field in a hydrogen atom. The dashed circle shows the electron's path as a waveform, circling the nucleus: bolder dashes and shading represent peaks, while fainter dashes and shading represent troughs. In (a) there are three peaks of the wave, corresponding to three zones of field intensity; in (b), with a higher frequency of oscillation, there are four peaks. These patterns change with time as the fields oscillate. As the wave, or field intensity, 'travels' round the nucleus, it must meet its starting point (A) on its return. This becomes obvious if one thinks again of the oscillations of a string (see Fig. 60). In (a) three cycles, and in (b) four cycles, are possible: in (c) it is clear that 3¼ cycles are not.

'This was de Broglie's great insight', writes Brooks: 'if the electron in an atom is made of waves then the number of waves must be an integer, and the corresponding frequencies must be discrete. And since the frequency of oscillation is related to the energy of the electron field, the energy states must be discrete.'[340] What this implies is that the waveform, while wholly continuous, incorporates necessary discreteness – *seamlessly*. This is beautiful.

And de Broglie has this to say:

> In physics, as in every other branch of knowledge, the problem of continuity and discontinuity has existed at all times: for in this science, as elsewhere, the human mind has always manifested two tendencies at once antagonistic and complementary. On the one hand, there is the tendency which tries to reduce the complexity of phenomena to the existence of simple elements indivisible, and capable of being counted; a tendency whose analysis of Reality seeks to reduce it to a dust-cloud of individuals. On the other hand, there is the tendency, based on our intuitive notion of Time and Space, which observes the universal interaction of things and regards every attempt to disengage definite individual entities from the flux of natural phenomena as artificial …

This is an uncannily accurate description of the most quintessential hemisphere differences. And he goes on:

> What gives a special interest to the question of the continuous and the discontinuous in modern Physics is the fact that, during the last few years, it has arisen in a particularly clear-cut and also novel form. More definitely than ever before, the need has been realised of effecting a synthesis of the two opposed points of view, while at the same time the very real difficulties raised by this problem have led physicists to discuss questions which pass beyond the proper technique of their science, and merge with the general problems of philosophy.[341]

In a later passage from the same book, de Broglie writes:

> In the psychological, ethical and social sphere an uncompromisingly rigid definition or argument often leads away from, rather than towards, Reality. It is true that the facts tend to assume a certain order within the framework supplied by our reason; but it is no more than a tendency, and the facts invariably overflow if the framework is too

340 From Brooks 2016 (location 1896).

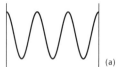
(a)

(b)

(c)

Fig. 60. *Oscillations in a string:* (a) *and* (b) *are possible, while* (c) *is not*

(adapted from Brooks 2016 [location 1888])

341 de Broglie 1939 (217–8).

exactly defined ... Even in the most exact of all the natural sciences, in Physics, the need for margins of indeterminateness has repeatedly become apparent – a fact which it seems to us, is worthy of the attention of philosophers, since it may throw a new and illuminating light on the way in which the idealisations formed by our reason become adaptable to Reality.[342]

Again one sees a physicist sensing that while classical Newtonian physics had a way of representing reality as particulate, mechanical and precisely known (conforming to the left hemisphere's take), a view that is instead based on waves and fields, interconnectedness (entanglement) and intrinsic uncertainty (conforming to the take of the right hemisphere) has proved itself to be more veridical.

The unique and the contextual

And reality has further qualities that again require a right hemisphere, not left hemisphere, world-picture. 'Consciousness is specific: each experience is the particular way it is', write Tononi and Koch.[343] And yet by means of the unique consciousness – never by ignoring, circumventing, or worse still, opposing that uniqueness – we can arrive at that which *transcends* the personal. It is not an opposition but a dipole. And – in another right hemisphere-oriented characteristic – quantum phenomena are *contextual*. One cannot speak of 'independent' outcomes without the kind of attention used to examine contextual phenomena.[344]

What we have here is physicists, like philosophers before them, such as James and Bergson, intuiting by introspection the bipartite nature of the mind. James, like Bergson, always decades ahead of his time, had described, in 1890, what he indeed called 'split' modes of consciousness, 'which coexist but mutually ignore each other and share the objects of knowledge between them, and – more remarkable still – are complementary'.[345] Bergson promulgated, from the 1880s onwards, his understanding of the need for two opposed although complementary ways of knowing: a useful, but unfaithful, representation of the world, on the one hand, and a veridical apprehension of the deep nature of reality that is of limited utility, on the other.[346] This is again a precise intuition of how the two hemispheres of the brain are required to work together.

The conformity of matter to mind here is, of course, only to be expected. Our capacity to understand the world will bear the imprint of the brain we use to do so. According to Eddington,

the subjective laws are a consequence of the conceptual frame of thought into which our observational knowledge is forced by our method of formulating it, and can be discovered *a priori* by scruti-

342 *ibid* (281–2).
343 Tononi & Koch 2015.
344 Theise & Kafatos 2016.
345 James 1890, vol 1 (206).
346 Bergson 1911a (342–3).

nising the frame of thought as well as *a posteriori* by examining the actual knowledge which has been forced into it. The characteristic form of the fundamental laws of physics is the stamp of subjectivity.[347]

It might be asked, then, whether it is that we see the world as having this structure because our brains have the structure they do, or that our brains are so constructed because the world they have evolved to deal with has this structure. If I am right that matter arises from consciousness as a way of Nature coming to know herself, and if I am right that consciousness and the cosmos co-evolve, each furthering the becoming (the evolution) of the other, the question is based on a false premiss. Here von Weizsäcker expressed it well: 'We can only understand nature if we think about her, and we can only think because our brain is built in accordance with nature's laws'.[348] He also said, wisely, that 'all our thinking about nature must necessarily move in circles or spirals',[349] suggesting that the linear mode of approach typical of the left hemisphere will not answer to the structure of nature.

Earlier in this book, I suggested that philosophy cannot escape being moulded by the particular cast of mind of the individual philosopher. It is of some interest that faced with the problems created by assuming a meaningless, mechanical universe – such as the existence of consciousness in it, the extraordinarily fine-tuning of the cosmos required to permit life, and the role of consciousness in determining outcomes in physical reality – several of the commonly touted responses from scientists and philosophers correspond uncannily closely to delusions occurring in schizophrenia and right hemisphere-deficit states, in other words in conditions of left hemisphere prepotency. In pointing this out, I am not, of course, ascribing pathology to those who espouse such theories. Rather, I am illustrating how every point of view, pathological or not, inevitably has its hemispheric predilections, and it is worth being aware of its genealogy. Those of the current establishment thinking in philosophy and materialist science are for the left hemisphere, to the virtual exclusion of all else: so when faced with phenomena hard to accommodate, the solutions offered would be likely to illustrate that hemispheric style. And they do.

The world as representation

The favoured theory of consciousness these days is internalism, the belief that what we experience is a representation inside the head of whatever 'really' exists outside it, as if projected on a screen and, presumably watched by – what? whom? This entails that reality is what the mathematics of physics describes, and that my sensory ex-

347 Eddington 1939 (104).

348 See Heisenberg 1971, 'Elementary particles and Platonic philosophy': 237–47 (244).

349 *ibid.*

perience of the world is an illusion. Thus the theoretical skeleton world of physics and maths is more real than the empirical world. This is straight left hemisphere stuff; and, as Tim Parks and Riccardo Manzotti note, it carries with it an immediately recognisable tone of condescension towards, even of moral superiority when compared with, the supposedly naïve rest of us:

> I recall now how the neuroscientist Christof Koch claims that our experience of color is 'a con job'. David Eagleman talks about vision being the result of 'fancy editing tricks'. The philosopher Eric Schwitzgebel accuses us of being 'ignorant and prone to error' about everything we see. A moral nuance is smuggled into the debate, as if there were something shabby and lazy in the way we see the world. It feels like we're being blamed, or condescended to, for not perceiving things as science thinks we ought to.[350]

[350] Manzotti & Parks 2020 (21).

This superior, complacent tone is characteristic of one who does not know what it is he doesn't know. I have already given my view that there is no reason to dub our full experience of the world an illusion, while ascribing to the very partial mathematical description of the world the reality. I also believe internalism is a mistake, and that consciousness is located not inside us, but in a non-spatial 'betweenness' created by our attention and the object of our attention. It is, therefore, always a partial revelation – and partly, also, a creation of the act of experiencing – but our part in it does not negate its reality: that *is* reality. Reality is always *coming into* being. A true *presencing* of something, not just a *re-presentation*. But that is, as the reader will know, what we would predict the left hemisphere would make of experience.

Taken altogether, then, that in the discussion of consciousness, the Western mainstream favours a view of reality as a representation rather than a presencing; favours the abstract and disembodied over the sensory and embodied; and believes it has superior knowledge to the rest of us; all could be predicted from the dominance of the left hemisphere's take in the world of cognitive science and philosophy, as elsewhere. That there is a 'problem' of consciousness at all is due to this left hemisphere-mindedness. The right hemisphere has no problem, in that it is already engaged with reality; what it encounters, the *Gestalt* it perceives, is then taken up by the left hemisphere, which now wants to be certain, at a meta-level, what this perception, this *Gestalt*, 'really is'. And since the left hemisphere doesn't recognise the existence of what is 'outside' itself, we end up with the dogma of internalism. Once again it sees consciousness as a thing, rather than a process of connexion, or rather a betweenness, out of which such 'things' as we and the world we inhabit co-arise.

Zombies

Then, another fascinating phenomenon. The 'hard problem' gives rise in some minds to the reconceiving of apparently human subjects as zombies, a popular topic of current philosophical debate; in others to doubting the difference between people and machines, a widespread and even automatic assumption of modern neuroscience and cognitivist philosophy. This goes beyond playing with ideas. That we are effectively no different from zombies or machines is to some a revealing insight: similar conclusions are common in, indeed characteristic of, schizophrenia.[351] An example I have already quoted is scarily close to some current philosophical positions: 'I'm actually deluding myself into thinking I could think … I was actually searching my memory bank … non-mechanical thinking? I can't conceive of that any more.'[352]

Most people who ever lived, and most people alive now around the world, would correctly consider these assessments of the human condition to be a sign, not of wise insight, but of madness. In the world of philosophy, they first showed up in the mind of Descartes, who found he had no means of disproving that the people he could see from his window were automata; and they have proved hard to dislodge from Western thinking ever since. Those who have followed the argument so far will know why that could not have avoided being the case, given the prevailing cast of mind.

Giovanni Stanghellini wrote a book about how the schizophrenic mind becomes possessed by such thoughts.[353] It is easy to see: R. D. Laing reports a schizoid patient who saw his wife as a mechanism:

> She was an 'it' because everything she did was a predictable, determined response. He would, for instance, tell her (it) an ordinary funny joke and when she (it) laughed this indicated her (its) entirely 'conditioned', robot-like nature …[354]

This again, in its assumption of determinacy and empty mechanistic 'behaviour', reflects what is hardly even a parody of a certain not uncommon scientific position.

Amongst other considerations, this is a natural consequence of the particular cast of mind enjoined on those who engage in science in the modern West. (Sophisticated science was successful, without espousing any such mindset, in China and the Middle East long before modern Western science took off in the seventeenth century.)[355] In the current scientific paradigm, one of the key elements is the rigorous adoption of the third-person perspective. A corollary of this is the conceiving of mental life by cognitive scientists schematically from without, not phenomenologically from within. The ultimate case of this has to be the absurd dogma of behaviourism, now luckily

351 See pp 344–5 above.
352 Sass 2017 (280).
353 Stanghellini 2004.
354 Laing 1990 (49).
355 As Joseph Needham's monumental *Science and Civilisation in China*, and Stephen Gaukroger's *The Emergence of a Scientific Culture: Science and the Shaping of Modernity,1210–1685*, make evident.

356 Mormann, Dubois, Kornblith et al 2011; Yang, Bellgowan & Martin 2012; Martin & Weisberg 2003; Castelli, Happé, Frith et al 2000.

357 See pp 96–7 & 140 above.

358 Smolin 1999 (44–5).

---

359 See Rees 2015.

360 Hawking 1989 (126). This quote is *not* to suggest that Hawking himself favoured fine-tuning – he didn't; but that was because he believed apparent fine-tuning could be explained by inflation. That hypothesis, in turn, is far from straightforward, however. 'The difference between good and bad [inflation] hinges on the precise shape of the potential energy curve, which is controlled by a numerical parameter that could, in principle, take on any value whatsoever. Only an extremely narrow range of values could produce the observed temperature variation. In a typical inflationary model, the value must be near 10^{-15} – that is, zero to 15 decimal places. A less fine-tuned choice, such as zero to only 12 or 10 or eight decimal places, would produce bad inflation: the same degree of accelerated expansion (or more) but with a large temperature variation that is inconsistent with observations ... Both sets of configurations are rare, so obtaining a flat universe is unlikely overall. Penrose's shocking conclusion, though, was that obtaining a flat universe without inflation is much more likely than with inflation – by a factor of 10 to the googol (10^{100}) power! ... In 2008 Gary W. Gibbons of the University of Cambridge and Neil G. Turok of the Perimeter Institute for Theoretical Physics in Ontario showed that an overwhelming number of extrapolations have insignificant amounts of inflation. This conclusion is consistent with Penrose's. Both seem counterintuitive because a flat and smooth universe is unlikely, and inflation is a powerful mechanism for obtaining

[note continues on facing page] ▶

---

no longer, as it was, *de rigueur*, in which interiority was altogether denied. But behaviourism has been replaced by something not obviously more human: cognitivist psychology, which is effectively psychology as a machine might try to 'understand' it.

Discriminating the animate from the inanimate depends heavily on the right hemisphere, especially on the right temporal lobe.[356] And so does understanding of an instance as properly unique, not merely one of a series: which takes us to two other, interestingly related, theoretical positions, the 'multiverse' hypothesis and the so-called 'many worlds interpretation' (MWI) of quantum mechanics. Both mimic reduplicative phenomena, which outside of philosophy and science are typically due to right hemisphere, typically right temporal lobe, damage.[357]

Endlessly reduplicative phenomena

The multiverse hypothesis suggests that the explanation for the unimaginably intricate interrelationship of highly precise factors necessary to permit the evolution of life in the cosmos just happening to be present together and to the right extent is that, as long as you keep multiplying universes indefinitely, eventually you are bound to end up with one like this. It is worth setting the probability in context, because it shows that the number of such universes would have to be effectively infinite. Here is astrophysicist Lee Smolin:

> We must understand how it came to be that the parameters that govern the elementary particles and their interactions are tuned and balanced in such a way that a universe of such variety and complexity arises. Of course, it is always possible that this is just coincidence. Perhaps before going further we should ask just how probable is it that the universe created by randomly choosing parameters will contain stars. Given what we have already said it is simple to estimate this probability ... The answer, in round numbers, comes to about one chance in 10^{229}. To illustrate how truly ridiculous this number is, we might note that the part of the universe we can see from earth contains about 10^{22} stars which together contain about 10^{80} protons and neutrons. These numbers are gigantic, but they are infinitesimal compared to 10^{229}. In my opinion, a probability this tiny is not something we can let go unexplained. Luck will certainly not do here; we need some rational explanation of how something this unlikely turned out to be the case.[358]

That's just to get to a universe that could contain stars. It does not get us anywhere *near* life or consciousness (if you hold that they must, somehow, arise out of inert matter).

Indeed, to get to a universe at all, the ratio (known as *N*) of the gravitational force to the electrical force would have to be almost exactly 1: 10^{36}, otherwise nothing could exist. Then, the value of nuclear efficiency (ε) – the percentage mass of the nuclear constituents that is converted to heat when the nuclear constituents react via nuclear fusion to form heavier nuclei – is 0.007: if it had had a value of 0.006 there would be no other elements than hydrogen; if it had a value of 0.008, protons would have fused in the Big Bang, leaving no hydrogen: in either case, no universe.[359] As Stephen Hawking (who nonetheless did not believe in fine-tuning: see note), observed: 'If the rate of expansion one second after the Big Bang had been smaller by even one part in a hundred thousand million million, the universe would have recollapsed before it ever reached its present size'.[360] Had it been just slightly larger, the universe would have flown apart without creating galaxies. Even a change of 0.4% in the strength of another force, the nucleon-nucleon force, would make carbon-based life impossible, 'since all the stars then would produce either almost solely carbon or oxygen, but could not produce both elements'.[361] And so on through a number of constants that have to be almost exactly what they in practice turn out to be for there to be something rather than nothing. Legendary astrophysicist Fred Hoyle famously remarked: 'A commonsense interpretation of the facts suggests that a super-intellect has monkeyed with physics, as well as with chemistry and biology, and that there are no blind forces worth speaking about in nature'.[362] Astrophysicist Paul Davies's view similarly is that 'there is for me powerful evidence that there is something going on behind it all. It seems as though somebody has fine-tuned nature's numbers to make the Universe ... the impression of design is overwhelming'.[363] And for Einstein, the cosmos showed evidence of 'an intelligence of such superiority that, compared with it, all the systematic thinking and acting of human beings is an utterly insignificant reflection'.[364]

To get to life, things get worse – much, much worse. Wilczek wrote that

> it is logically possible that parameters determined uniquely by abstract theoretical principles just happen to exhibit all the apparent fine-tunings required to produce, by a lucky coincidence, a universe containing complex condensed structures [such as is life]. But that, I think, really strains credulity.[365]

Wolfgang Pauli wrote the following to his fellow physicist Niels Bohr:

> In discussions with biologists I met large difficulties when they apply the concept of 'natural selection' in a rather wide field, without being able to estimate the probability of the occurrence *in an*

the needed smoothing and flattening. Yet this advantage appears to be completely offset by the fact that the conditions for starting inflation are so improbable. When all factors are taken into account, the universe is more likely to have achieved its current conditions without inflation than with it ... [In inflation] rogue regions spawn new rogue regions, as well as new islands of matter – each a self-contained universe. The process continues ad infinitum, creating an unbounded number of islands surrounded by ever more inflating space. If you are not disturbed by this picture, don't worry – you should not be. The disturbing news comes next ... In an eternally inflating universe, an infinite number of islands will have properties like the ones we observe, but an infinite number will not. The true outcome of inflation was best summarized by Guth: "In an eternally inflating universe, anything that can happen will happen; in fact, it will happen an infinite number of times."... What does it mean to say that inflation makes certain predictions – that, for example, the universe is uniform or has scale-invariant fluctuations – if anything that can happen will happen an infinite number of times? And if the theory does not make testable predictions, how can cosmologists claim that the theory agrees with observations, as they routinely do?... The oft-cited claim that cosmological data have verified the central predictions of inflationary theory is misleading, at best ... Highly improbable conditions are required to start inflation. Worse, inflation goes on eternally, producing infinitely many outcomes, so the theory makes no firm observational predictions' (Steinhardt 2011).

361 Oberhummer, Csótó & Schlattl 2000.

362 Hoyle 1982 (16).

363 Davies 1988 (203).

364 Einstein 1999 (29).

365 Wilczek 2006.

empirically given time of just those events, which have been important for the biological evolution. Treating the empirical time scale of the evolution theoretically as infinity they have then an easy game, apparently to avoid the concept of purposiveness. While they pretend to stay in this way completely 'scientific' and 'rational', they become actually very irrational, particularly because they use the word 'chance', not any longer combined with estimations of a mathematically defined probability, in its application to very rare single events more or less synonymous with the old word 'miracle'.[366]

In fact, if the chances of getting a universe with stars are 10^{-229}, only what is effectively an infinity of universes will do if it is life we want. In an infinite number of universes, by definition, everything that can happen must happen, over and over again: problem solved! But just to invoke infinity as the 'explanation' of anything – because whatever it is you can imagine must happen repeatedly in infinity – is not only not a cogent explanation, it is not an explanation at all. It is also intrinsically unverifiable, which means that it is a matter of faith. 'Of course', writes Paul Davies, 'one might find it easier to believe in an infinite array of universes than in an infinite Deity, but such a belief must rest on faith rather than observation.'[367] Deity or not, 'we cannot observe any of the properties of a multiverse ... as they have no causal effect on our universe ... The hypothesis that a multiverse *actually* exists will always be untestable,' writes astrophysicist Luke Barnes.[368] There is nothing wrong with that – unless one claims it is science, rather than faith.[369]

While it is easy to talk about the concept of such an 'infinite', that is not the same as accepting the reality of anything infinite in this sense. David Hilbert famously argued that 'the infinite is nowhere to be found in reality. It neither exists in nature nor provides a legitimate basis for rational thought ... The role that remains for the infinite to play is solely that of an idea ... which transcends all experience and which completes the concrete as a totality'.[370] Were it to exist, the consequences for science and philosophy would be devastating, since in a world where everything is true, nothing is true. As cosmologist and mathematician George Ellis and his colleagues have pointed out, multiverses offer not an explanation, but an infinite regress. A coherent description of a multiverse is possible only through the existence of certain regularities of structure and origin: 'a multiverse consisting of completely causally disconnected universes is a problematic concept.'[371] And, if there are regularities, we are faced with the same problems that apply in the universe we have: 'What would explain the existence of an ensemble, and its specific properties? Why should there be this particular ensemble, rather than some other one? Why this multiverse with these prop-

---

366 Letter of Pauli to Niels Bohr, 15 February 1955; in Meyenn 2001 (105). And he continued with a shrewd observation: 'I found for instance H.J. Muller very characteristic for this school of biologists (see also his recent article 'Life' in "Science", issue of January 7, 1955, which certainly contains very interesting material), but also our friend Max Delbrück. With him this is combined with vehement emotional affects and a permanent threat to run away which I interpret as obvious signs of overcompensated doubts.'

367 Davies 1983 (174).

368 Barnes 2011 (559; emphasis in original).

369 For intrinsic unverifiability, see Ellis 2011. For a detailed cosmological examination of the multiverse claim and its coherence or otherwise, see Lewis & Barnes 2016; and Barnes 2011. Barnes shows that all proposed alternatives to 'fine-tuning' in the accepted sense require fine-tuning of their own, and to a similar extent.

370 Hilbert 1984 (201).

371 Ellis, Kirchner & Stoeger 2004.

erties rather than others? What endows these with existence and with this particular type of overall order? What are the ultimate boundaries of possibility – what makes something possible, even though it may never be realised?'[372] To quote theoretical physicist Paul Steinhardt:

> Scientists proposed the multiverse as a way of resolving deep issues about the nature of existence, but the proposal leaves the ultimate issues unresolved. All the same issues that arise in relation to the universe arise again in relation to the multiverse. If the multiverse exists, did it come into existence through necessity, chance or purpose? That is a metaphysical question that no physical theory can answer for either the universe or the multiverse.[373]

On this situation the mathematician Martin Gardner commented wryly:

> The stark truth is that there is not the slightest shred of reliable evidence that there is any universe other than the one we are in. No multiverse theory has so far provided a prediction that can be tested ... As far as we can tell, universes are not as plentiful as even *two* blackberries. Surely the conjecture that there is just one universe and its creator is infinitely simpler and easier to believe than that there are countless billions upon billions of worlds, constantly increasing in number and created by nobody. I can only marvel at the low state to which today's philosophy of science has fallen.[374]

'Simpler and easier to believe' because it is a more rational deduction. Still, my purpose here is not to debate the correctness of the multiverse theory – I will have more to say about it in a later chapter – but to draw attention to its interesting structure.

The 'many worlds' hypothesis has a similar structure. It supposes that every time there is quantum collapse, leading to an event occurring one way rather than another (eg, Schrödinger's cat being either dead or alive, but not both), what actually happens is that both outcomes occur: the universe splits, and in one new universe the cat is alive, while in the other the cat is dead. Such moments of collapse lie behind all phenomena everywhere for all time – every single change in the cosmos – and the numbers expand exponentially with each universal split, so the number of universes being created every nanosecond by changes in my body alone is beyond all conception and beyond all measure, and might as well be considered infinite.

What lies behind this less than parsimonious fantasy is the nature of the quantum state as now conceived. It is not the case that a quantum system is actually in one or other of the possible states, and it's just that we cannot know which; the system is in some sense per-

372 Stoeger, Ellis & Kirchner 2004.

373 Steinhardt 2011.

374 Gardner 2003 (9).

mitted to be in all states at once. This leads to the problem of where all the other states go when the wave function collapses. According to physicist Max Tegmark, 'the act of making a decision' – a measurement generating a particular outcome – 'causes a person to split into multiple copies'.³⁷⁵ But as Philip Ball suggests, this is incoherent. 'What can it mean to say that splittings generate copies of me?'

> In what sense are those other copies 'me?' ... Tegmark waxes lyrical about his copies: 'I feel a strong kinship with parallel Maxes, even though I never get to meet them. They share my values, my feelings, my memories – they're closer to me than brothers.' But this romantic picture has, in truth, rather little to do with the realities of the Many Worlds Interpretation. The 'quantum brothers' are an infinitesimally small sample cherry-picked for congruence with our popular fantasies. What about all those 'copies' differing in details graduating from the trivial to the utterly transformative?³⁷⁶

The phrase 'feeling a strong kinship – closer to me than brothers', perhaps intended to humanise this inhuman vision, has for me the chilling, reverse, effect. And it seems hard to know how to make the picture coherent. Under these circumstances there can be 'neither facts nor a you who observes them', as Ball points out:

> It says that our unique experience as individuals is not simply a bit imperfect, a bit unreliable and fuzzy, but is a complete illusion. If we really pursue that idea, rather than pretending that it gives us quantum siblings, we find ourselves unable to say anything about anything that can be considered a meaningful truth ... Its implications undermine a scientific description of the world far more seriously than do those of any of its rivals ... it destroys any credible account of what an observer can possibly be.³⁷⁷

Strikingly, a schizophrenic patient of Bin Kimura's reported that 'whenever one perceives anything, it bifurcates: at that instant the world is split in two ... I think it's plausible that there are other 'I's.'³⁷⁸ Thus he thought that when he had cleaned his teeth, there was one world in which he had, and one in which he hadn't. This is almost exactly a 'many worlds' theory. Fantasies of multiple copies of the self are sometimes the left hemisphere's attempt to make sense of a changing world in the presence of right hemisphere dysfunction (see p 96): remember, for example, the man who had eight copies of himself and his wife and children in eight different cities where all the circumstances of his life were somehow replicated, though subtly different; or the patient who had 50 'twinnies'. Again, let me emphasise – I am absolutely not suggesting that the MWI is evidence of a delusional mind. But in light of the way in which many paradoxes

---

375 Tegmark 2003.

376 Ball 2018.

377 ibid.

378 Kimura 1994 (194): » Wenn man irgend etwas erkennt, dort verzweigt es sich; in diesem Augenblick wird die Welt entzweigerissen ... Ich glaube, dass es auch andere Ichs wahrscheinlichkeitsmässig gibt. «

illuminate the clash between left hemisphere and right hemisphere ontology it seems worth pointing out that such resolutions of the paradoxical character of the quantum world, and the multiverse response to the appearance of rightness in the cosmos, have structure that speaks of left hemisphere epistemology and ontology.[379]

CAN WE TALK ABOUT THE TOPIC AT ALL?

I am aware that talk about consciousness and its relation to reality is foolhardy. The topic defies language. However, for that very reason, it is not irrational to push the bounds of language and thought, and if I seem to have done so here, perhaps I may be forgiven. I'd just say that I do so less than most alternative theorisers. I do not claim to know what experience *is* any more than anyone else, except that it is the condition on which I know anything at all. And yet we all understand it directly. It is what we know (*kennen*) better than anything at all, and yet know (*wissen*) least of all about. It is therefore difficult to discuss, since neither language nor reason are well adapted to it. Language is already at a remove from direct experience. Bryan Magee captures the difficulties:

> this direct experience which is never adequately communicable in words is the only knowledge we ever fully have. *That* is our one and only true, unadulterated, direct and immediate form of knowledge of the world, wholly possessed, uniquely ours. People who are rich in that are rich in lived life. But the very putting of it into words translates it into something of the second order, something derived, watered down, abstracted, generalised, publicly shareable. People who live most of their outer or inner lives in terms that are expressible in language – for example, people who live at the level of concepts, or in a world of ideas – are living a life in which everything is simplified and reduced, emptied of what makes it lived, purged of what makes it unique and theirs.[380]

I quoted Bohr as saying: 'We must be clear that, when it comes to atoms, language can be used only as in poetry. The poet, too, is not nearly so concerned with describing facts as with *creating images* and *establishing mental connections*.'[381] That means the language of science alone is not going to deliver an understanding of ultimate reality. As Sam Matlack puts it: 'if poetry is necessary for talking about the foundations of physical reality, this should both elevate the importance of poetry and help to disabuse us of the idea that we can exclude the more personal, parochial, poetic forms of language and still truly apprehend reality. Far from making poetic speech a mere means of translating a scientific message, talking about the constitution of the physical world must be poetic in some way.'[382]

---

379 A further sidelight on this comes from David Mermin's model, in which there are only correlations – no correlata (see p 1006 above). He quotes Christopher Fuchs, Professor of Physics at Boston, as suggesting that the distinction between the 'many worlds interpretation' and the 'correlations without correlata' position of Mermin is 'most succinctly expressed by characterizing many worlds as correlata without correlations' (Mermin 1998). In other words, all betweenness, or configuration of the *Gestalt*, with no entities to be fixed, in the first case; and all fixed entities, but no betweenness, or *Gestalt*, arising in the second. This too reflects the difference between right hemisphere and left hemisphere takes on reality, respectively.

380 Magee 1998a (98).

381 See p 1017 above (emphasis added).

382 Matlack 2017.

Language is no match for the topic. And reason cannot reach the depths of experience, either. 'We have to confess', writes James,

> that the part of [mental life] of which rationalism can give an account is relatively superficial. It is the part that has the *prestige* undoubtedly, for it has the loquacity, it can challenge you for proofs, and chop logic and put you down with words. But it will fail to convince or convert you all the same ... if you have intuitions at all, they come from a deeper level of your nature than the loquacious level which rationalism inhabits.[383]

Strawson would probably agree: 'discursive thought is not adequate to the nature of reality: we can see that it doesn't get things right although we can't help persisting with it ... the nature of reality is in fundamental respects beyond discursive grasp.'[384] Marcelo Gleiser puts it with panache: 'Unless you are intellectually numb, you can't escape the awe-inspiring feeling that the essence of reality is unknowable.'[385]

Rather famously the psychologist Stuart Sutherland wrote in the *International Dictionary of Psychology*: 'Consciousness is a fascinating but elusive phenomenon; it is impossible to specify what it is, what it does, or why it evolved. Nothing worth reading has been written on it.' I cannot hope to have been an exception to this rule.

---

[383] James 1902 (73; emphasis in original).

[384] Strawson 2006 (28).

[385] Gleiser 2014 (193).

Chapter 26 · Value

> Where in the Schrödinger equation do you put the joy of being alive?
> —EUGEN WIGNER[1]

> The whole universe appears as an infinite storm of beauty…
> —JOHN MUIR[2]

> Truth, outlasts the Sun –
> —EMILY DICKINSON[3]

What life brings, I would maintain, is not consciousness, then – which, as I have argued, is present from the beginning – but the coming into being of the capacity for *value*: thus, a mountain cannot value, though it can have value for creatures, like ourselves, who value. And it is not just we, but all living creatures, that for the first time are able to recognise value. Life vastly enhances the degree of responsiveness of, to and within the world.

The customary way to think of values is to see them as piggybacking on and arising out of our consciousness: a human invention. An alternative view is that values are not invented but discovered and disclosed, and it takes life to discover and disclose them: that they declare themselves in and through the responses of living beings to the world and the world's response to them. 'Value', writes Thomas Nagel, 'is not just an accidental side-effect of life; rather, there is life because life is a necessary condition of value.'[4] Valuing depends on a relationship; only in its appreciation is value fulfilled. But it is not we who originate the possibility of truth, or goodness, or the beauty of the cosmos. We help fulfil them (or not). I see value as intrinsic to the universe; and the possibility of appreciating and responding to value – therefore fulfilling its potential – as one reason for the cosmos having evolved life.[5] Indeed, life could be seen as the very process of the cosmic consciousness continually both discovering and furthering its beauty, truth, and goodness; both contemplating and (not separately but in the same indivisible act) bringing them further into being: a process.

This is not surprising if awareness is foundational to the universe, rather than arising from it late in the day. And this is a point on which Schelling and contemporary science could be seen as coming together. As Theise and Kafatos put it, 'the universe is non-material, self-organizing throughout, comprised of a holarchy of complementary, process-driven, recursive phenomena. *The universe is both its own first observer and subject.*'[6] Though they make no reference to Schelling, or even to his philosophical era, this is uncannily close,

1 Wigner, as quoted in Dyson 2004 (294)

2 Muir 1915 (5).

3 Dickinson, fr 1455 (in the Harvard *variorum* edition).

4 Nagel 2012 (123).

5 The kabbalists regarded elements such as will, wisdom, love, and compassion, which form part of the *sefirot*, not simply as aspects of the human mind but as 'the very elements of the world itself' (Drob 1997).

6 Theise & Kafatos 2016 (emphasis added).

200 years on, to the structure of the world envisaged in Schelling's philosophy.

What are values? 'There is something in common between truth, beauty, and goodness', writes Andrew Steane, Professor of Physics at Oxford: 'they each make demands on us, and also fulfil us, and also leave us thirsty for more.'[7] Values evoke a response in us and call us to some end. They are what give meaning to life: such things as beauty, goodness, truth – and purpose. Science can tell us what their brain correlates may be, but cannot help us understand their nature. It can, though, help us misunderstand them. This is for two main reasons.

One is that science starts where values leave off. Having for good reasons done its best to exclude from its view of the world any consideration of value, it finds no value in the world to consider. Purporting to eschew value in its workings, it implies that values are secondary phenomena that are, so to speak, painted on afterwards to suit human predilection, and, to that extent, obscure the primary reality. This, I will argue, is back to front; and contrary to the day-to-day practice of science, which so clearly implies value, if only that of truth, at its core.

The other reason is that when science turns its gaze directly on values, it immediately begins to account for them in terms of something else assumed to be more fundamental. But for ultimate values there can be no such thing, much as there can be no such thing in the case of consciousness. In an age when it is widely thought that science alone can answer our questions, values may therefore become overlooked – and even devalued. Not a few readers for example, may be surprised by my including value alongside time, space, motion, consciousness and matter as a constitutive element of reality. Yet I believe it is as foundational as consciousness. It is, I suggest, at least worth seeing what follows if we abandon our usual assumptions on the matter.

CONCERNING TRUTH

An earlier chapter addressed the nature of truth, and the difference between the take on it of the right and left hemisphere. Moreover the theme of this entire book has concerned the comparative reliability – the truth – of what each hemisphere permits us to encounter. So I will make relatively few observations here.

In this, and each of the subsequent sections of this chapter – on goodness and beauty – I am making two principal claims. The main claim is that value, whether it is truth, goodness or beauty, is not, as our culture has come to regard it, an 'add-on', a human invention, some sort of extra that is not intrinsic to the nature of the cosmos, but is, rather, itself constitutive of the cosmos and is discovered by,

[7] Steane 2018 (139).

and disclosed in, the encounter of life (and not just human life) with whatever it is that exists. The attendant claim is that that encounter is best served – indeed, served only – by the right hemisphere, optimally when it is assisted by the left; and if, on the contrary, the left hemisphere usurps the right hemisphere and 'goes it alone', it will not only fail to comprehend what is true, good or beautiful, but, by misconceiving it, help to destroy it.

With that in mind, let us look at the nature of truth.

Science is a product of a human aspiration which is difficult to explain in purely scientific terms: a desire to know, and to acquire knowledge for its own sake, not just for the sake of some useful purpose. Of course science has values of its own, just as it has very real value in itself. Values are what command our allegiance. Contemporary science seems to me governed by one overarching value, and an unquestioned assumption that governs how that value is interpreted. The value is, of course, truth: a value that is essential, timeless, and of the highest importance. By contrast, the assumption is adventitious and culture-bound: that truth requires maintaining an ideal of meaninglessness – or to put it more positively, the avoidance at all costs of implying meaning in the world. A piece of work may still be science if it is wrong – science advances necessarily error by error; but it will not be accepted as science if it imputes meaning to the processes of the cosmos or of life. For that would have the unwelcome implication that meaning is not something we need to *invent*, if, for emotional reasons, we feel we must; but something we need, perhaps urgently, to *discover*, unless, for emotional reasons, we feel we must not. I do understand the reaction against premature attribution of meaning, which is often irrational. But not all attribution of meaning is premature, and it would be in itself irrational to deny meaning if it were present. We need to keep an open mind.

Truth carries within it the whole purpose of science, and gives meaning to its activities. However, science will not admit anything that is not empirically verifiable – yet the value of truth, like all value, is incapable of empirical proof. It is, instead, quite correctly, assumed. But that leaves the door open to an important question. Where does the overwhelming intuition – for that is what it is – of the overriding import of truth come from?

It cannot come from utility. Some untruths might have greater utility than the truth. Suppose, for example, it were true that people derived great comfort in their darkest moments from a belief in an afterlife, which also encouraged them, more than they might otherwise, to lead a moral life and do good to others. Suppose you felt strongly that they were wrong because life, according to you, is pointless, matter is all that exists, and the cosmos is meaningless. My question is, why would your truth be so much more important than this human benefit?[8] *I think truth matters more than utility, because I see truth as an ultimate value, irreducible to anything else:* that is part of my view of the cosmos as pregnant with meaning, soliciting our allegiance, rather than a place where it is good enough to get by with comfortable lies. For me, values are part of its very fabric, not optional adornments. But why would those who do not share such a belief in ultimate value think so? In a meaningless universe, without ultimate values, shouldn't we just maximise happiness? Where does this idea of a transcendent truth that surpasses all other considerations, including those of the greater happiness of others, come from? Or is the defence that *ultimately* only truths can make the world a happier place? How on earth could we know that? In any case, do you believe that scientists for the most part really value truth only instrumentally, rather than for and in itself?

Why for that matter believe in reason as a guide to truth, let alone the only guide to truth? I believe it helps us approach truth because I believe the cosmos has meaning, I believe reason is an aspect of its coherence, which both grounds it and which it in turn grounds, and I believe we are an expression of that cosmos. But if rationality is simply a derivative of evolution in a meaningless universe, why trust it to disclose reality rather than be simply a useful tool developed through evolution for getting by, regardless of any deeper relationship to truth? Indeed it has been argued by evolutionary psychologists that reason evolved purely as a useful weapon for winning arguments, without regard for truth.[9] Useful assumptions are not always truthful; equally, true assumptions are not always of practical use.

What, then, once again, is truth? And – a question jesting Pilate did not even stay to ask – where does it come *from*? My view, as laid out earlier in the book, is that truth (cf *troth*) is an act; one of trust in, or faithfulness towards, whatever *is*. It characterises the proper relationship between consciousness and the world. It is therefore not a function of some other value. Nonetheless it does imply that being faithful – though not blindly so – has value in and of itself; and that the 'something else' to which we are faithful also has intrinsic value, perhaps goodness or beauty, or the faith would be blind. In other words, rather than closing down on a single foundational element in a causal chain, we find this process leading in the opposite direction, to a web of interconnectedness that we cannot by any means get behind, or beneath, in which values cohere and sustain one another. This web of values is foundational, underwriting the meaning of our actions – including those of the reductionist, though he won't be aware of it. Kant believed not in moral values because there was a God, but in God because there were moral values: not in a rule-engendering Nobodaddy in the sky, that we had better obey, but in an ultimate moral force in the universe to which we are intrinsically

---

[8] 'The author Philip Pullman, one of Dawkins's friends, recalled a discussion the two once had about what you might tell a terminally ill child about death. Dawkins "was very unwilling, or seemed to be, to say that it would be OK at that point to tell a fairytale about heaven", Pullman said' (Elmhirst 2015).

[9] See p 557 above.

attracted. Truth is a moral value, like beauty, and goodness. Only our familiarity with truth, beauty and goodness makes us take them for granted; but they didn't have to exist at all. We miss their essentially mysterious – indeed essentially good, beautiful and true – nature.

Not all values are fundamental in this way. In particular utilitarian values are not: they are derived from the value of pleasure. But some, like beauty and goodness – and indeed meaning and purpose, as I shall later suggest – are not derivable in this way. Even if they led to suffering we would be right to hold them as non-negotiable, and indeed to hold them in reverence. To value such values.

Are values useful *for* something else, or to be prized for their own sake? Are good deeds only a useful investment in the expectation of future help from others? Is beauty merely a scheme for selecting healthy mates? Is a love of truth simply what gives us power over nature? Are courage and heroism either blind risk-taking or bids for admiration? The point at issue here is a very familiar one, so I shall not labour it – how instrumentality limits and degrades whatever it touches, be that art, nature or human beings themselves. Values are not just validated by the outcomes they achieve: they are inseparable from our deepest emotional experience. This does not make them suspect: rather they become suspect precisely when emotions have not played a sufficiently large part in their application, since emotions can take into account a host of implicit considerations that abstract argument would miss. What passes through my mind in making a judgment is not only, or even to any great extent, what passes through it consciously and explicitly: the judgment is a distillate of ever-growing experience. And when we disregard that fullness of experience we dehumanise what we describe and rob it of the value we are intent on pinning down.

The Victorian poet Coventry Patmore wrote a poem about truth. In it he writes of watching the tides come and go on the shore, and reflects:

> For want of me the world's course will not fail;
> When all its work is done, the lie shall rot,
> The truth is great, and shall prevail,
> When none cares whether it prevail or not.[10]

10 Patmore, 'Magna est veritas'.

Truth is not a human invention. It is possible to construe both truth and falsehood as having meaning only in relation to assertions in language, but that is to miss their depth – and what is more their essential asymmetry. There will be truth when we are no longer around to see it, but there will be no falsehood: the lie shall rot. It takes a human to lie.

In the words of Emily Dickinson,

> Truth — is as old as God —
> His Twin Identity
> And will endure as long as He
> A Co-Eternity.[11]

The fascination with and love for truth is something deep in us, which in science and philosophy responds to the world with wonder and excitement, as if exploring ever deeper an enchanted realm. Truth is not a thing to be possessed, however immaterial, but a path to follow, a process. 'It is not the possession of truth', according to Max Planck, 'but the success which attends the seeking after it, that enriches the seeker and brings happiness to him'. He was anticipated by the eighteenth-century German philosopher Lessing:

> The true value of a man is not determined by his possession, supposed or real, of truth, but rather by his sincere exertion to get to what lies behind the truth. It is not possession of the truth, but rather the pursuit of truth by which he extends his powers and in which his ever-growing perfectibility is to be found. Possession makes one *passive, indolent, vain* – If God held enclosed in his right hand all truth, and in his left hand the ever-living striving for truth, although with the qualification that I must for ever err, and said to me 'choose', I should humbly choose the left hand and say 'Father, give! pure truth is for thee alone.'[12]

Note that Lessing's advice is to choose the hand that does the bidding of the right hemisphere. Psychopaths – and this is surely fascinating in itself – lie quite gratuitously and needlessly, *not* just instrumentally, so as to achieve a purpose; it is almost as if they reverence the lie. They care nothing for truth; and they cannot love or trust. They also have severe right hemisphere dysfunction.[13]

Truth, as Patmore suggests, is a value that exists whether we recognise it or not. I will argue that this is true of other values, such as beauty, goodness and purpose. Truth also exemplifies another aspect of value: that, as Goethe suggested, what we at first only dimly perceive calls forth in us the faculty whereby it *can* better be perceived, a faculty, what is more, that is susceptible of development through practice, and may not function at all if not practised at all. That faculty, however cognitive it may become (at some later stage and for some purposes), is based in emotional intelligence and practical wisdom (*phronēsis*), which it is the business of a sympathetically lived life to nurture. A value can be calibrated cognitively, but it itself is first *perceived* pre-cognitively, much as we perceive colour or a musical tone directly, not as a cognitive elaboration. In German the word for perception is *Wahrnehmung*, literally truth-taking: Scheler

---

[11] Dickinson, fr 836 (in the Harvard *variorum* edition).

[12] Lessing 1979 (trans HB Garland 1937: 171).

[13] Narayan, Narr, Kumari *et al* 2007; and see p 1346 below.

introduced the word *Wertnehmung*, literally value-taking, as a parallel. His point was twofold: that value is a primal phenomenon, like colour – it speaks for itself; and, related to this, that it is a *Gestalt* and therefore cannot be decomposed into bits or parts, or steps or slices, but must be taken in as a whole.

There is an argument that love, too, is a value. But is it not, you may object, an emotional experience? If so, in this it is not different from other values. It might be objected that it is, surely, value that leads us to love something: love is *based* on value. This is true; but equally true is that value is based on love. Love is not merely a product of value, but also its foundation. Thus Scheler quotes Goethe:

> 'One knows nothing save what one loves, and the deeper and more complete that knowledge, the stronger and livelier must be one's love – indeed passion' … He repeated this thought in countless variations all his life.[14]

Wisdom has many facets that distinguish it from our common conception of knowledge in the modern West, and one of these is that a true understanding requires a certain disposition of the mind towards its object. While it is decidedly not one of excessive attachment, it cannot be that of complete indifference, either. There is a sense in which an open affection for its object is as much a requirement for a deep understanding, as it is a product of it. Emerson reflected that 'love is fabled to be blind, but to me it seems that kindness is necessary to perception.'[15]

True understanding in other words already *presupposes* a connexion, rather than being the prerequisite of such a connexion. The order of precedence is wrong. Knowledge cannot be confirmed by some external criterion that is not itself already an object of knowledge: we can know 'from the inside' only, not 'from the outside'. For there is no such outside. A hermeneutic circle is involved, which means access to knowledge cannot be made certain and definite, but requires a step of faith to get going at all. I say 'step', because the normal expression 'leap of faith' makes it sound potentially random, whereas random is the last thing it is. It is no more random than one's willingness to trust an outstretched hand that enables one to cross a stream.[16]

The incoherent attempt to see things from the outside means that there is something excessively cold and alien about the Western idea of knowledge. Consider the three greatest acknowledged influences on Western thought of the last 150 years (and they are by no means the most icily objective of thinkers you could hope to find): Darwin, Marx and Freud.

Here is Freud writing to his colleague the psychoanalyst Oscar

---

14 Scheler 1992, 'Love and knowledge' (147). The quote comes from Goethe's letter to Friedrich Heinrich Jacobi, 10 May 1812: » *Man lernt nichts kennen, als was man liebt, und je tiefer und vollständiger die Kenntniß werden soll, desto stärker, kräftiger und lebendiger muß Liebe, ja Leidenschaft seyn.* «

15 Emerson, 'Journal C', entry for 10 April 1837: 1965 (294).

16 Cf Polkinghorne (2008): 'Faith involves an act of commitment, but that commitment is motivated; it is a leap into the light and not into the dark'.

Pfister about his patients: 'I must tell you that in private life I have no patience at all with lunatics'.[17] As Freud's colleague Sándor Ferenczi wrote, 'one learned from [Freud] and from his kind of technique'

> various things that made one's life and work more comfortable: the calm, unemotional reserve; the unruffled assurance that one knows better; and the theories, the seeking and finding of the causes of failure in the patient instead of partly in ourselves ... and finally the pessimistic view, shared only with a few, that neurotics are a rabble [*Gesindel*], good only to support us financially and to allow us to learn from their cases: psychoanalysis as a therapy may be worthless.[18]

Then there is Karl Marx: what is one to say of this undoubtedly clever, but unattractive, self-obsessed tyrant? Certainly not that he loved his life's professed project, the improvement of the lot of the proletariat. He described the peasants as 'troglodytes',[19] the workers as 'those asses', 'the rabble', 'the mob': and his acolyte Engels wrote to him, clearly anticipating approval, that 'the people are of no importance whatever'.[20]

And then there is poor Darwin, who, I believe, really did love the natural world that he observed, as evidenced by his consistent appreciation of its beauty; though even here the process of scientific detachment almost drove it out of him: as he wrote to his cousin William Fox,

> I am at work at the second volume of the Cirripedia, of which creatures I am wonderfully tired. I hate a Barnacle as no man ever did before, not even a sailor in a slow-sailing ship.[21]

His mind seemed to him to have become 'a kind of machine for grinding general laws out of large collections of facts': the capacity to wonder and to love seemed to have been lost. 'A scientific man', you will remember, 'ought to have no wishes, no affections, – a mere heart of stone.' This makes an interesting contrast with Thoreau's approach to Nature, which, however, as he concedes, would not issue in a communication to the Royal Society.[22]

It is not science *itself*, but something about the Western approach to understanding the world that demands this loss, and induces the antipathy for one's subject evidenced by each of these otherwise important thinkers. The Japanese word *kansatsu*, used in scientific accounts of experiments where we would use the word 'observe', has implications of a relationship, and is closer in meaning to the word 'gaze', a word which we use only when we are in a state of rapt attention in which we lose ourselves, and feel connected to the other: indeed the syllable *kan* in *kansatsu* implies a 'one-body-ness' with the object of gaze.[23]

---

17 Freud to Oskar Pfister, letter of 21 June 1920: in Freud 1961 (330–1).

18 Ferenczi, diary entry, 4 August 1932; in Dupont 1995 (185–6).

19 Schwarzschild 1948 (194).

20 *ibid* (297): Engels, letter to Karl Marx, 3 December 1851.

21 Letter from Darwin to WD Fox, 24 October 1852; in Darwin 1887, vol I (384–5).

22 See above, p 620 for Darwin and p 1209 for Thoreau.

23 Kawasaki 1992, quoted by Ogawa 1998 (153).

Not that all of the Western tradition demands that understanding should be based on a loss of connexion with its 'object'. Thus Pascal:

> instead of saying, when speaking of human things, that one must know them before one can love them, which has become a proverb, the saints say on the contrary, speaking of divine things, that one must love them in order to know them, and that no one attains the truth except by means of love ...[24]

Similar sentiments have been expressed since the time of St Augustine, who said, 'one may not enter into truth except by means of love';[25] and St Anselm, whose saying *credo ut intelligam* – 'I believe in order that I might understand' – reverses the normal Western assumption.[26] As I have already pointed out, the modern notion of belief as propositional is not what is intended here: belief is not holding a proposition, but a disposition, an openness to trust, *in order that* one may experience, and therefore *know*. Conviction will come, if it comes at all, from experience – never from trading propositions: this is *credo* as *cor do* ('I give my heart').[27]

Such recommendations are not, however, confined to religious contexts, in which they clearly have a particular application. Nor are they as paradoxical as they may seem to the modern mind. For loving is not something that 'just happens', but something that like other kinds of understanding, can, and must, be learnt. So Nietzsche says, emphatically, that '*One must learn to love*'; and he compares it to achieving an understanding of a piece of music. At first, he says, one must '*learn to hear* a figure and melody at all,' and be open to it, attend to it as a life in itself; then one needs effort and 'good will to *tolerate* it in spite of its strangeness, to be patient with its appearance and expression, and kindhearted about its oddity.'[28]

Only thus do we come, he says, to a point of becoming

> its humble and enraptured lovers ... But this is what happens to us not only in music. That is how we have *learned to love* all things that we now love. In the end we are always rewarded for our good will, our patience, fairmindedness, and gentleness with what is strange; gradually it sheds its veil and turns out to be a new and indescribable beauty. That is *its thanks* for our hospitality.[29]

Not to love is not fair-mindedness, but an unfairness in itself: a bias against. We cannot know anything without attending to it, and the nature of that attention alters what we find: so to avoid bias, our task is not to adopt a peculiarly alienating form of attention, but to be aware of how we attend. We need 'necessary distance', yes, but this is neither a closeness that blinds us, nor a distance that alienates. It is *that* for which we should strive.

---

24 Pascal 1964: « *au lieu qu'en parlant des choses humaines on dit qu'il faut les connaître avant que de les aimer, ce qui a passé en proverbe, les saints au contraire disent en parlant des choses divines qu'il faut les aimer pour les connaître, et qu'on n'entre dans la vérité que par la charité ...* ».

25 Augustine, *Contra Faustum*, Bk XXXII, ch xviii: '*Non intratur in veritatem nisi per caritatem.*'

26 Anselm, *Proslogion*, ch 1.

27 See p 386 above.

28 Nietzsche 1974b §334 (262: emphases in original).

29 *ibid*.

And love is not just for the sake of the other – or even primarily for the sake of the other. It is rewarding, because it enables us to see something beautiful which without it we cannot understand. 'Understanding and loving are inseparable', wrote Erich Fromm: 'if they are separate, it is a cerebral process and the door to essential understanding remains closed.'[30] Max Scheler asserted that 'before he is an *ens cogitans*, or an *ens volens*, man is an *ens amans*': that before he is a being that thinks, or a being that wants, he is a being that loves.[31]

Scheler also cites Pascal as saying words to the effect that love must first *disclose* what reason then may judge. And he refers to

> the common, and as far as I can see, specifically modern bourgeois judgement, prevalent since the Enlightenment, that 'love makes one blind', that all true knowledge of the world can rest only on *holding back* the emotions and simultaneously ignoring differences in value of the objects known.[32]

It is not that Scheler thinks there is no place for detached analysis in certain kinds of knowledge. Clearly there is. His point is that this cannot be the first stage, but must be a subsequent stage, of acquiring knowledge, not the ground of the process. If one mistakes this first step, one sets off already on the wrong path to understanding. Whatever we see as value-free, according to Scheler, can be so only through an *already achieved process of abstraction*, whereby we have set aside the value-quality which, he claims, is not given after – not even at the same time as – but *before* the object fully discloses itself.

The philosopher Warren Heiti writes: 'It is not value or meaning which is peculiar; on the contrary, it is the allegedly value-free fact which is "utterly different" from anything else in the universe.'[33] He is drawing here, not on Scheler, but on the writings of Simone Weil; and he goes on to point out that there is nothing mystical about the idea, which is, for example, central to the psychologist James Gibson's widely accepted idea of affordances, which we have already encountered: those aspects of the environment that reveal directly and immediately its meaning for good or ill in the animal (or human)'s first single, whole, perception of it. Certain affective responses at least, such as whether we are attracted to something or not, occur before any cognitive processes: this is a phenomenon known as the primacy of affect.[34] Affect may too readily be equated with emotion. Emotions are certainly part of affect, but are only part of it. Something much broader is implied: a way of attending to the world (or not attending to it), a way of relating to the world (or not relating to it), a stance, a disposition, towards the world – ultimately a 'way of being' in the world. The point is that for the world to 'presence' to us we already have to have adopted a disposition of our consciousness toward it, and the disposition determines the

---

30 Fromm 1994 (193).

31 Scheler, *Ordo Amoris*: 1973 (110–1). See also Maria Scheler 1957.

32 Scheler 1992, 'Love and knowledge' (147). Cf Pascal 1910 (425): 'Let us not therefore exclude reason from love, since they are inseparable. The poets were not right in painting Love blind; we must take off his bandage and restore to him henceforth the enjoyment of his eye'.

33 Heiti 2018 (273).

34 See Zajonc, Pietromonaco & Bargh 1982; and Zajonc 1984. See also McGilchrist 2009b (184–6 & 491, nn26 & 29).

value, *including the situation in which we preclude it having value*. Such a stance is already *value-driven*.

Scheler called value the first 'harbinger' of the particular nature of anything:

> while the object itself remains indistinct and unclear, its value may be distinct and clear already. Whenever we apprehend a situation, at the same moment, at a glance, we apprehend the unanalysed whole and, in this whole, its value.[35]

Of this, the philosopher Guido Cusinato writes,

> Value, then, is what announces the phenomenon and gives direction to the further unfolding of its expression. The nuances of value in an object are the primary elements that reach us, and, so to speak, the medium through which the form and meaning of the object come into being and declare themselves.[36]

This does *not* mean that such value-ception is invariably correct – just that it is always there, immediately. Clear and correct perceiving is a learnable art: a skill like any other that comes with experience. And the acquisition of greater skill may not negate, so much as go some way to explicating (= unfolding), the original judgment. Thus Bergson says:

> Where would the difference be between great art and pure fancy? If we reflect deeply upon what we feel as we look at a Turner or a Corot, we shall find that, if we accept them and admire them, it is because we had already perceived something of what they show us. But we had perceived without seeing.[37]

By contrast the scholar/scientist, according to Scheler, is so used to imagining that the process works in reverse – valueless perception first, value merely 'painted on' later – that he doesn't notice what he is doing:

> in the scholar this kind of abstraction can become so habitual, so much his 'second nature', that he is in fact inclined to regard the value-free entity of both psychic and natural phenomena not only as more fundamental *in esse* than their value-qualities but even as preceding them in order of perception. Consequently he casts about, on this false assumption, for some kind of 'yardstick' or 'norms' which might restore value-distinctions to his value-free entity.[38]

The process described by Scheler – the casting about for some kind of yardstick – is the left hemisphere's response to something it does not properly understand: 'let's account for it in terms of something else, something more accessible which we can measure'. Indeed, what I have been describing so far in this chapter is the difference

---

35 Scheler 1916 (13): » Sein Wert schreitet ihm gleichsam voran; er ist der erste ‚Bote' seiner besonderen Natur. Wo er selbst noch undeutlich und unklar ist, kann jener bereits deutlich und klar sein. Bei jeder Milieuerfassung erfassen wir z.B. zugleich zunächst das unanalysierte Ganze und an diesem ganzen seinen Wert «.

36 Cusinato 2012 (140ff): » Der Wert ist also das, was das Phänomen verkündet und der weiteren Entwicklung seines Ausdrucks Richtung gibt. Die Wertnuancen eines Gegenstandes sind das Primäre, was von ihm auf uns zukommt, und gleichsam das Medium, in dem das Bild und die Bedeutung des Objekts aufgehen und sich kundtun. «

37 Bergson 2007e (112).

38 Scheler 1960 (86).

between the left hemisphere take and the right hemisphere take on value (including truth). For the left hemisphere, value is something we *invent*; which is *separate* from and, as it were, painted onto the world; and whose function is *utility*. For the right hemisphere, on the other hand, value is something *intrinsic* to the cosmos; which is *disclosed* and responded to in a pre-cognitive take on the *Gestalt*; and is not, other than incidentally, in service of anything else.

This leads naturally to thoughts about the calculus of utility which operates in a certain kind of ethics. So let us turn, now, to what it means to be good.

CONCERNING GOODNESS

Here, in keeping with what has gone before, I suggest that the good is, like other values, part of the nature of a conscious cosmos, not some sort of human 'add-on' divorced from its constitution, as our Western culture has come to think of it; and furthermore that we depend on our right hemisphere for this constitutive good to be disclosed to us. If instead we rely on our left hemisphere we not only fail to apprehend it, but are led ultimately to destroy it.

The dominant approach to ethics in our culture is utilitarianism, the belief that what is good is so because it issues in utility (see Chapter 18). Given that the governing value of the left hemisphere is to aid manipulation of a creature's environment, this is exactly what we would expect, as the reader will recognise, if the left hemisphere were trying to give an account of what goodness might be. This move also turns whatever intimations of goodness and badness it receives from the right hemisphere into representations it can both control and measure; it substitutes for the complexity of reality a series of cause and effect mechanisms, with the ultimate focus on outcomes; it suggests that such outcomes can be assessed by calculation (the greatest happiness of the greatest number); and, in keeping with its lesser emotional and social intelligence, removes the essential interiority of morality and replaces it with externalities of the kind it prefers. It then calls itself objective, thereby implicitly trumping all competitor theories, and installs itself in university departments across the world.

We know that experimentally induced left hemisphere predominance (whether through left hemisphere activation or right hemisphere suppression) in a normal subject leads towards utilitarianism. Perhaps, champions of utilitarianism might argue, the right hemisphere is just not a very good judge of morality? Fortunately we know a good deal about the disposition of each hemisphere when it comes to questions of morality, and the picture is clear. I have reviewed earlier the evidence that the right hemisphere is the principal substrate for social and emotional understanding (see Chapter 6). Furthermore the right temporoparietal junction is of critical importance in the maintenance of a coherent sense of one's own embodied self;[39] and the right inferior parietal lobe and right insula are crucial for a sense of owning one's actions (both may be areas of dysfunction in schizophrenia, causing subjects to believe that their own thoughts, feelings and actions belong to other people).[40] Since each of these elements – social and emotional understanding, and a robust sense of the self as agent, taking responsibility for his or her actions – are in turn central to morality, it is not surprising that a mass of research of differing kinds suggests strongly that the right hemisphere is more important for morality, too.[41]

In Chapter 15 I alluded to the fact that with either the right temporoparietal junction or the right dorsolateral prefrontal cortex suppressed, normal subjects can be induced to misjudge a situation, attending only to the consequences, so that an accidental poisoning gets to be considered morally more culpable than a failed murder attempt.[42] The right hemisphere is superior to the left hemisphere at 'theory of mind'; the capacity to understand what another person is thinking, an essential component of moral evaluation unless one is purely utilitarian in outlook. Both right hemisphere damage and frontal lobe damage are independently associated with more utilitarian judgments.[43] The abnormally high rate of utilitarian judgments observed in frontal brain-injured patients with deficits in emotional response suggests that their decisions are mostly *cognitive, intentional and conscious*, unaided by emotion.[44] Brain-injured patients tend to assess personal moral dilemmas based solely on cognitive criteria, 'conscious abstract reasoning processes and cost-benefit analysis'.[45] Five studies specifically examining lateralisation of lesions, in particular of prefrontal lesions, and adequate response to emotionally demanding moral dilemmas have concluded that right hemisphere involvement is the critical factor.[46] 'Normal judgments of morality require full interhemispheric integration of information critically supported by the right temporal parietal junction and right frontal processes'.[47] In moral decision-making, then, the right hemisphere is the more important: it takes into account intention and context.

As one might expect, an atypical pattern of moral judgments has independently been observed in patients with deficits in emotional understanding:[48] specifically, subjects with emotional blunting make more utilitarian judgments.[49] The tendency to adopt a calculating and utilitarian approach in judging moral issues is more marked in those with reduced aversion to harming others,[50] lower trait empathy,[51] higher psychoticism (which is itself characterised by reduced empathy and emotional blunting),[52] a greater sense of the meaninglessness of life and greater Machiavellianism.[53] It is also characteristic of the moral thinking of psychopaths,[54] as well as subjects

39 Tsakiris, Costantini & Haggard 2008.

40 Farrer, Franck, Georgieff *et al* 2003; Chaminade & Decety 2002; Karnath & Baier 2010.

41 See n 47 below.

42 See p 602 above.

43 Koenigs, Young, Adolphs *et al* 2007 (in this study the orbitofrontal lesions were, as one might expect, more right-sided, though that is apparent only on inspecting the scan data).

44 Anderson, Barrash & Bechara 2006.

45 Martins, Faísca, Esteves *et al* 2012.

46 Tranel, Bechara & Denburg 2002; Young, Cushman, Adolphs *et al* 2006; Mendez & Shapira 2009; Demaree, Everhart, Youngstorm *et al* 2005; Martins *et al, op cit*.

47 Miller, Sinnott-Armstrong, Young *et al* 2010.

48 Eslinger 1998; Gazzaniga, Ivry & Mangun 2008; Greene & Haidt 2002; Mendez & Shapira 2009.

49 Koenigs, Young, Adolphs *et al* 2007. In this study the orbitofrontal lesions were, as one might expect, more right-sided, though that is apparent only by inspecting the scan data.

50 Cushman, Gray, Gaffey *et al* 2012.

51 Choe & Min 2011.

52 Wiech, Kahane, Shackel *et al* 2013.

53 Bartels & Pizarro 2011.

54 Koenigs, Kruepke, Zeier *et al* 2012.

with schizophrenia.[55] The fact that frontal lesions result in anger and frustration in specific circumstances appears to be related to a difficulty in processing the social emotions, such as compassion, shame and guilt, that are closely linked to moral values, and these are better understood by the right hemisphere.[56]

Apologies to any philosopher who takes offence here, but these are findings that are hard to dismiss: what you make of them is up to you. As one research group points out, however, the widespread assumption in academic philosophy departments that utilitarianism is the appropriate framework by which to evaluate moral judgment and that individuals who endorse non-utilitarian solutions to moral dilemmas are committing an error, is a curious one. It leads to the 'counterintuitive conclusion that those individuals who are least prone to moral errors also possess a set of psychological characteristics that many would consider prototypically immoral.'[56]

In a number of experiments examining moral responses, those that were designated utilitarian were often driven, 'not by concern for the greater good, but by a calculating, egoist, and broadly amoral outlook', and 'were strongly associated with … primary psychopathy, rational egoism, and a lenient attitude toward clear moral transgressions'. Researchers repeatedly found 'associations between "utilitarian" judgment and antisocial and self-centred traits, judgments and attitudes'.[58] Moreover, it was not just that such decisions tended to be associated with a certain type of personality, but that they might have widespread untoward consequences:

> The kind of no-nonsense, tough-headed and unsentimental approach to morality that makes it easier for some people to dismiss entrenched moral intuitions may also drive them away from a more impartial, all-encompassing and personally demanding view of morality, and might even lead some to skepticism about morality itself.[59]

Utilitarianism's stock-in-trade are scenarios designed to force unpalatable choices in an attempt to make us aware of our 'irrationality'. These are amusing, but have a number of problems associated with them. Here are a couple of well-known thought experiments:

> A trolley is hurtling down a track towards five people. You are on a bridge under which it will pass, and you can stop it by putting something very heavy in front of it. As it happens, there is a very fat man next to you – your only way to stop the trolley is to push him over the bridge and onto the track, killing him to save five. Should you proceed?

> A doctor has five patients who are awaiting an organ transplant and who will die without it. Unfortunately, as of yet, there are no available organs. An innocent person visits the doctor for a check-up and he happens to be a perfect match for all five of the patients. The doctor

[55] McGuire, Langdon & Brüne 2014; Johnson 1960.

[56] Martins *et al, op cit*. See also Adolphs 2001; Adolphs, Baron-Cohen & Tranel 2002; Damasio, Grabowski, Bechara *et al* 2000; Eslinger 1998.

[57] Bartels & Pizarro 2011.

[58] Kahane, Everett, Earp *et al* 2015 (200, 206).

[59] *ibid* (207).

is tempted to take the innocent person's organs, thereby causing him to die, so as to save the five patients. In doing this, he saves the lives of five persons by killing one. Should he do it?

That people calling themselves moral philosophers can seriously debate whether it might be right for the doctor to act in this way, and even in some cases conclude that he should – or even *must* – suggests that there is something very wrong with the way we do moral philosophy nowadays. If the road leads you to the wrong destination, take a different road.

The goodness or badness of what I do to you hangs on many things not involved in the calculus of thought experiments: what, for example, it does to me as a spiritual, emotional, cognitive, and physical whole, to be the perpetrator (for example, brutal treatment of animals also brutalises us), and what happens to the world at large in which we are inevitably embedded (a world in which doctors can't be trusted not to cannibalise you for spare parts) – neither of which considerations can be ultimately isolated from one another or from the impact on the victim. There are consequences that could in theory be taken into account, though they rarely are in practice because of the difficulty in knowing how much weight to give them. How do we know what guilt the person who pushes the fat man will feel later – perhaps for years? What if the man does not die, and does not even stop the trolley, but ends up horribly injured? It's no good stipulating certainties here – that he must die, and that he must stop the trolley – and then hoping to deduce what our responses would be were such an unlikely situation to arise in real life. One reason that most people tend to look askance at utilitarian views of morality is that things are *intrinsically uncertain* in this world – something the right hemisphere is much better at understanding and accepting – and, on something so important as the ethics of killing, it is better to be guided by an intuition than to attempt to make a calculus.

As this suggests, the kind of thought experiments that seem to challenge our intuitions are not without their problems. First of all, in most thought experiments, the situation is unrealistically circumscribed, and taken out of context. What people say on a survey or in a seminar is not necessarily what they actually do when faced with the real-life experience. Often they cannot know how they would react until they have to do so. It is not to be expected that intuitions and judgments drawn from living and validated in life will apply to bizarre and extreme cases. Saying to people that they should somehow blind themselves to certain aspects of the thought experiment, or to assume that 'other things are equal', when in real life they couldn't, cannot be a good way of assessing our moral principles in practice. Moral judgments are made by human intuitions that include *everything* we know from experience, and on an understanding of what would really be involved in the cases we are asked to respond to: we can't just 'exclude' certain factors or 'fix' others at will. Once you stipulate that we live in a world where you can know things that in real life we cannot know – such as that the fat man will stop the trolley – you are not testing our intuitions fairly. In moral dilemmas we are not asked to act like omniscient beings, but like truly human beings. And the one precludes the other.

Such thought experiments seem to presume that we are machines that have no significant past or future, and do not ramify into the world around us. In a way which suggests an imbalance in favour of left hemisphere-dominant processes, the fashionable cognitivist thought experiments are too tightly circumscribed: in depth of time, space and emotion. The right hemisphere sees human individuals as existing over time, not just at a moment in time, beings therefore who have a past and a future held together by a coherent narrative on which significant actions could have an indelible impact. It sees them never as isolated entities, but rather as embedded in a complex web of relationships – as social beings, in other words, within a culture. And it sees them as not just cognitive mechanisms, but as fully embodied and emotionally complex, with all that that entails. This makes us behave in 'irrational' ways, such as not maximising our benefits. (People may perform a task less well and with less commitment if given greater material rewards; and children's intrinsic interest and motivation can be sapped by turning a task into an explicit means to an extrinsic reward.)[60]

A further important problem is that they are not exercises in exploring what *might* be valid moral motivations and conclusions, since that is *assumed to be known at the outset*. Decisions are judged correct only in as far as they conform to the utilitarian calculus: all other answers are assumed to be fallaciously swayed by 'irrational' considerations. But we are feeling beings, and emotion and reason are not so easily separated. That is not because we are foolish, but because we have the capacity to be wise, since emotion and reason contribute to, and help constitute, one another.

> To see the philosophical enterprise in general, and the utilitarian project in particular, as engaged in a war to the death with intuition is evolutionarily misconceived, philosophically sloppy and tactically foolish. Philosophy, if it's not to be pure intellectual onanism, needs to deal with the substrate (intuitive humans) that it has, not the substrate that would, if it existed, fit more neatly in its equations. If utilitarianism wants a war with intuitions (and it does), it's doomed: to embark on it is the conceit of Canute. Philosophy departments everywhere have embarrassingly wet feet; many are drowning ...[61]

---

60 Deci 1971; Ariely, Gneezy & Loewenstein, 2009; Lepper, Greene & Nisbett 1973.

61 McGee & Foster, forthcoming.

And what would it do to humanity – what would already have happened to humanity – if we really accepted that moral decisions could be made purely on a calculus?

From this perspective, utilitarianism could be seen as not so much a *kind* of moral philosophy, as an *alternative* to moral philosophy, in subjects with a disconnexion from the moral sense, for whom a purely cognitive approach is all that is left; rather as echolocation is not so much a kind of sight – though it may grow in time to feel like one – as an aid to orientation in the blind. Except that utilitarianism is much less reliable than echolocation. This is partly because it can at times lead to what are rightly perceived as revolting – or 'repugnant' – conclusions; partly because it is forced to attempt an unconvincing reduction of all values to pleasure or utility; and partly because it involves the 'turtles all the way down' problem.[62] In other words, if you ask me to demonstrate why it is wrong to torture children for fun, the fault is not mine if every answer I give leads you to a further calculation, and a further question, so that consequently no answer can ever satisfy you.[63] The fault lies in the mind that posed the question. Can we really make a calculus of the pleasure gained by perhaps a large number of sadistic paedophiles against the suffering of one innocent child? The calculus of pain and pleasure not only fails us as a guide to goodness, but is in itself morally wrong as a way to approach the situation. There comes a point where one has to say 'certain things are just wrong: if you can't see it for yourself, I can't help you'. Not all moral issues are like torturing children, and not many of them are independent of context, but this would certainly be one. As Charles Foster and Andrew McGee put it,

> in the case of the moral certainties, there is nothing we can appeal to that is more certain than the belief itself. This claim is a *rational* claim; it is about what rationality would require in order for these beliefs to be susceptible to questioning. Far from it being the case, then, that we admit beliefs in our moral framework that escape the tribunal of reason, we instead claim that reason itself shows some beliefs to be beyond question. Some beliefs are beyond question in the sense that we just wouldn't know what it means to raise doubts about them.[64]

Moreover, as they point out, morality is not just reducible to evolutionary utility. Many of our commonly accepted rules and intuitions include beliefs that have no obvious evolutionary explanation; for example, that we should look after our elderly parents (it not being clear how caring for them, and keeping them alive and healthy, would have helped us propagate our genes, while the practice is a far from negligible drain on what might be scarce resources, and delays responsiveness in an emergency).

---

62 A perhaps apocryphal story has an old lady approach William James after a lecture on cosmology and put forward her view that the world rests on the back of a turtle. When James politely enquired what the turtle rested on, she rebuked him: 'You can't catch me out that easily, you know, Professor James – it's turtles all the way down'. But there are many versions based on the same idea, which derives from an obvious question prompted by ancient Indian cosmology.

63 Cf Vygotsky 1986 (252): 'Behind every thought there is an affective-volitional tendency, which holds the answer to the last "why" in the analysis of thinking'.

64 McGee & Foster, forthcoming.

Given the extent and importance of the evidence of hemisphere differences concerning morality, and so as not to disrupt the argument, I have included a review of the evidence in Appendix 5, which the reader is encouraged to consult. There are three aspects to the matter. First, further evidence confirms that the right hemisphere is much more important for reaching moral judgments: bizarre judgments arise when the person has to rely on left hemisphere input. Second, it demonstrates that the right hemisphere is more involved than the left in inhibition, an essential aspect of moral behaviour. And, third, it demonstrates the right hemisphere's involvement in promoting prosocial – and the left hemisphere's in promoting antisocial – behaviour. In the words of David Hecht, of University College London, 'moral and immoral thinking are associated with activity in the right hemisphere and left hemisphere, respectively.'[65]

I have more than once touched on the left hemisphere's tendency towards deterministic thinking. It is therefore of interest that a belief in determinism also leads, independently, to antisocial attitudes and behaviour, including increases in deceitfulness, aggressive behaviour, and selfishness, lower achievement levels, and an increased susceptibility to addiction.[66]

A moral act is the expression of a moral being. It is not just about, or even mainly about, an outcome. What we call a morally good action is not a thing, but the result of the disposition of a morally good being towards the world. The great mediaeval philosopher and theologian Meister Eckhart had this to say:

> People should not worry so much about what they have to do; they should consider rather what they are. If people and their ways were good, their deeds would shine brightly. If you are righteous, then your deeds will be righteous. Do not think to place holiness in doing; we should place holiness in being, for it is not the works that sanctify us, but we who should sanctify the works.[67]

I am aware that, though Eckhart makes clear what is important, there is at least a *partially* self-referring loop here, because one part of discriminating what a good person is comes from knowledge of his deeds: 'by their fruits ye shall know them'. But that this circularity should be a problem, rather than part of the solution, is a consequence of the analytic quest to 'get to the bottom' of this thing called goodness, as though it were a mystery that had to be solved, and once we could string out a chain of reasoning in relation to it we would have captured it – and could calculate it. But morality is a nexus, not a chain. The disposition of mind (or of the whole being, or soul) that led to an action is more important than the nonetheless clearly important consequences of the action – at least those that could be foreseen; and along with that the intuitive sense of our non-local,

---

65 Hecht 2014.

66 Vohs & Schooler 2008; Baumeister, Masicampo & DeWall 2009; Stillman, Baumeister, Vohs et al 2010; Vohs & Baumeister 2009.

67 Eckhart, 'Of the value of resignation: what to do inwardly and outwardly', *The Talks of Instruction*, §4, 2009 (489).

distributed being among the society to which we belong. All goes together, each as much a receiver from, as a contributor to, this nexus, in a way that looks like Escher's famous portrayal of hands drawing hands that draw hands into being: not like a piece of string.[68]

T. H. Huxley exemplifies the misunderstanding of moral choice from the left hemisphere viewpoint. He claimed he would have preferred to be a machine, provided it thought and acted correctly, as though morality were purely about specifiable, explicit and programmable outcomes (left hemisphere fashion), rather than about a disposition towards the world (right hemisphere fashion):

> I protest that if some great Power would agree to make me always think what is true and do what is right, on condition of being turned into a sort of clock and wound up every morning before I got out of bed, I should instantly close with the offer.[69]

By definition, such 'thinking what is true' and 'doing what is right' could not be either, since neither truth nor moral 'rightness' is a thing we possess, but lies in a relationship to the particular circumstances in the world, in a choice with consequences. Huxley was a fascinating, contradictory character, in some senses repellent (his attitudes to the animal world, for example), but vulnerable, and in that respect sympathetic. He describes himself with a degree of revulsion at his own embodiment, as having 'a very pale, thin, lanky, ugly body ... with dreadfully long hair ... and a generally neglected style of attire'.[70] His love of the mechanical leaves one to speculate as to what psychopathology lay behind his personal philosophy of life:

> As I grew older, my great desire was to be a mechanical engineer, but the fates were against this and, while very young, I commenced the study of medicine under a medical brother-in-law. But, though the Institute of Mechanical Engineers would certainly not own me, I am not sure that I have not all along been a sort of mechanical engineer ... I am now occasionally horrified to think how very little I ever knew or cared about medicine as the art of healing. The only part of my professional course which really and deeply interested me was physiology, which is the mechanical engineering of living machines ...[71]

Utilitarianism tends to lead to the overvaluing of individualistic pleasure and individualistic determination, otherwise known as autonomy. While each is a reasonable enough goal, each needs to be tempered with other considerations, since unmitigated pursuit of either is not only bad for society but bad for the individual. A pleasure-filled life is not the same as a happy life, and a happy life is not the same as a meaningful life.[72] Being happy is a matter of 'feeling good', which might, of course, be a sign of enlightenment; but it could sig-

---

68 MC Escher, 'Drawing Hands' (1948): see McGilchrist 2009b (134).

69 Huxley 1893 (192–3).

70 Desmond A, entry for Huxley in the *Oxford Dictionary of National Biography*, 2004.

71 Huxley 2015 (25).

72 Baumeister, Vohs, Aaker et al 2013.

nify ignorance, thoughtlessness, insensitivity, lack of insight into oneself, lack of empathy for others, and a healthy bank balance. That could well do the trick – as would being fortunate enough to be born with a serene temperament. As Wittgenstein said, or should have said if he didn't, 'I don't know why we are here, but I'm pretty sure that it is not just in order to enjoy ourselves'. There are many other things that call to us and motivate us – in other words, other values.

Happiness cannot in any case successfully be pursued, since it comes as a by-product of forgetting oneself; and the attempt tends to lead, not to fulfilment, but to the pursuit of languid pleasure, in a process of diminishing returns known as the hedonic treadmill. Empty feelings of elation, such as are experienced in hypomanic episodes or with the help of alcohol or drugs, do not produce enduring happiness but more often its opposite. Ideally, hedonic happiness should be matched by eudaimonic well-being. Eudaimonia refers to the fulfilment resulting from a life well lived: this is virtue ethics, not just *feeling* good, but *being* and *doing* good and feeling fulfilled as a consequence. The Epicureans, who have lent their name to a form of hedonism, nonetheless advised moderation and the cultivation of modest needs and desires. Their goal was not hedonistic pleasure, which is fickle, transitory, and dependent on circumstance, but eudaimonic pleasure, which is more stable, tends to endure, and is less dependent on fate. It is no coincidence that happiness is about the present moment, independent of other moments, whereas meaning links events across time, thus integrating past, present, and future.

We need both, not just one, and they are not, once again, of symmetrical importance. People who report being happy but have little or no sense of meaning in their lives have the same gene expression patterns as people who are enduring chronic adversity, such as loneliness, bereavement, or poverty.[73] Meaning comes from an orientation to something bigger than the self. As Jung noted, 'the least of things with a meaning is worth more in life than the greatest of things without it.'[74]

A certain degree of self-respect is important; but the cult of self-esteem always seemed to me overrated. Now we have proof. Social psychologist Roy Baumeister, a promoter of the value of self-esteem in the 1980s, discovered that it doesn't deliver on its promise. In fact the cult of self-esteem at all costs (as opposed to 'if truly deserved'), seems too often to lead to mediocrity and an insufferable self-conceit. 'After all these years,' Baumeister commented, 'my recommendation is this: forget about self-esteem and concentrate more on self-control and self-discipline. Recent work suggests this would be good for the individual and good for society, and might even be able to fill some of those promises that self-esteem once made but could not keep'.[75] A better approach, Baumeister and his team have concluded,

---

73 Their bodies go into threat mode, activating a stress-related gene pattern that leads to an increase in the activity of pro-inflammatory genes and a decrease in the activity of genes involved in anti-viral responses: Fredrickson, Grewen, Coffey *et al* 2013; but see Coyne 2013, and a reply to Coyne: Cole & Fredrickson 2013.

74 Jung 1962 (75).

75 Baumeister 2005.

would be to boost self-esteem as a reward for ethical behaviour and worthy achievements.

Autonomy is another unusual preoccupation of our age. Too little is very obviously a social ill, and so it is certainly a worthy aim. But it may conflict with other important values, requiring us to qualify its application. There is a false antinomy between personal fulfilment and the fulfilment of others: we constitute society and society constitutes us. In a healthy society, the needs of self and others are as much as possible harmonised, and neither should be allowed to tyrannise the other. The main casualty of autonomy as a principle for reaching moral conclusions is a proper concern for the impact of an action on how we view humanity at large – on human dignity, to use an unfashionable term. For a deeply thought-provoking look at how this plays out in the courts of law, I recommend Charles Foster's *Choosing Life, Choosing Death: The Tyranny of Autonomy in Medical Ethics and Law*:

> The universal law of the medical autonomists is not geographically universal. In fact it is only to be found in a relatively small, highly educated part of the West. There is a wider point: autonomy itself (as opposed to the universal liberal law at its heart) is a Western idea – mysterious to and frowned upon by those outside the West.[76]

Which does not of course make it wrong, but may alert us to what it is that we are failing to take into account in our attempts to harmonise competing claims, a necessity for good moral decisions.[77]

Autonomy may be a culturally local phenomenon, but it is not as if the majority of traditional moral values are primarily the product of enculturation. They lie much deeper than that. Most social codes embody what are known as deontological principles, that is, ideas concerning duty and obligation (the word 'deontology' is derived from the Greek word for duty, δέον, not from the Latin word for God, *deus*). These lead to judgments made on the basis of the intrinsic rightness or wrongness of an action, quite separately from its consequences – principles to which we are duty-bound to cleave. With few exceptions children do not have to be taught that murdering and torturing are wrong – they feel it intuitively: the exceptions should alert us to a serious problem. Indeed monkeys will starve themselves for days rather than shock another monkey.[78] Such self-denying responses, according to one of the researchers, are 'observable throughout the animal kingdom'.[79] They come from an intuitive interaction between our minds and the world at large. The social behaviour of freely acting animals is too complex to fit the mechanistic scheme of behaviourists. Animals in social groups are not constantly fighting for supremacy, but expend a great deal of energy in making sure that the group as a whole is peaceful and

76 Foster 2009 (11).

77 Foster cites these wise words of Lord Justice Hoffman: 'there is no morally correct solution which can be deduced from a single ethical principle like the sanctity of life or the right of self-determination. There must be an accommodation between principles, both of which seem rational and good, but which have come into conflict with each other': *Airedale NHS Trust v Bland* [1993] AC 789, 827 (Hoffman LJ). On the narrowness of Western morality compared with that of other countries, see Jonathan Haidt (2013).

78 Masserman, Wechkin & Terris 1964.

79 Masserman 1960.

successful. Altruism is common among animals of all kinds, and cannot always be explained away as kin selection: numerous species are known to take life-threatening risks to rescue other animals, not even those of their own species. Non-relatives help each other and form alliances that can properly be called friendships; and primates earn respect within a group in which they display kindly behaviour.[80] And, ultimately, animal bodies could be seen as societies of socially self-sacrificing bacteria, numbered in trillions.

The only reason we don't acknowledge all this more readily is that we have been taught to think of nature as merely a blood-drenched battle, rather than a narrative in which co-operation and competition play important roles *together*. Humans are intuitively co-operative, as well as competitive. When we act intuitively we are most often gracious and generous – it is further reflection that makes us selfish and greedy.[81] According to psychologists Jamil Zaki and Jason Mitchell,

> rather than requiring control over instinctive selfishness, prosocial behavior appears to stem from processes that are intuitive, reflexive, and even automatic. These observations suggest that our understanding of prosociality should be revised to include the possibility that, in many cases, prosocial behavior – instead of requiring active control over our impulses – represents an impulse of its own.[82]

If life were not essentially collaborative, it would not be possible at all. As biologist Colin Tudge puts it:

> Taken all in all, and in most circumstances, collaboration is the best survival tactic – and so we should expect Darwinian natural selection to favour collaborativeness. In the case of animals, which interact largely through physical contact, with all the hazards that this entails, we would expect natural selection to favour sociality: the ability to get along with others of one's own kind at a *personal* level (and sometimes with other creatures of different kinds). We would also expect that as evolution proceeds, the nature and the scope of the social relationships would become more and more intricate and more subtle. On the whole, this is precisely what we do find.[83]

Joshua Greene, whose view is that 'moral judgment is just a brain process – that's precisely why it's possible for these researchers to influence it using electromagnetic pulses on the surface of the brain', and whose stated aim is to take morality and 'break it down in mechanical terms' so as to dismiss the existence of a soul,[84] found to his surprise that people's default position is honesty, not deception; and that honesty is not a matter of exercising will-power but of being effortlessly disposed to behave honestly. It is *deceit* that (with the exception of psychopaths) requires an effort. He and his fellow researchers found this 'somewhat surprising', having conducted a

---

80 See Tudge 2013 (121–3).

81 Rand, Greene & Nowak 2012.

82 Zaki & Mitchell 2013.

83 Tudge *op cit* (99: emphasis in original).

84 Quoted in Hamilton 2010.

survey to assess its *a priori* plausibility, in which people showed themselves inclined, as he was, to the more cynical view.[85] And, of course, it should be obvious that altering moral judgment by altering the brain does not show that it is 'just a brain process', any more than any conceivable outcome of Greene's experiment could either deny or affirm the existence of the soul.

Incidentally, I think it is fair to say that the prevailing cast of mind in reductionist science, whether biological or psychological, is effectively cynical. It takes the view that, for instance, where there is altruism it must be covert selfishness; that we are maximisers of our self-interest; that we are blind mechanisms. Philosopher David Stove writes:

> There is a perennial human type to whom this belief [that no one ever acts intentionally except from motives of self-interest] is peculiarly and irresistibly congenial. It is almost never a woman. It is the kind of man who is deficient in generous or even disinterested impulses himself, and knows it, but keeps up his self-esteem by thinking that everyone else is really in the same case. He prides himself on having the perspicacity to realise, what most people disguise even from themselves, that everyone is selfish, and on having the uncommon candour not to conceal this unpleasant truth.[86]

Cynicism, the negative belief that self-interest drives human behaviour, is associated with worse health outcomes, poorer psychological well-being and poorer economic well-being.[87] According to a survey of several hundred thousand people, most people believe the stereotype that cynics are smarter. However, on formal testing, people with high levels of cynicism have repeatedly been found to be less intelligent,[88] and have lower educational levels.[89] (This is one case in which a stereotype is false.) Intelligence and education help detect and avoid deceit in the first place, contributing to a more positive view of human nature, and a greater inclination to trust. Studies using trust games show that people typically earn more if they are willing to trust strangers.[90] Following subjects over time corroborates this, suggesting that 'cynical individuals earn lower incomes due to their ineptitude for cooperation, and cynicism might therefore be not that smart in terms of financial success.'[91] Competent adults tend to have cynical viewpoints only when it is warranted by a given social situation. Less competent individuals are more likely to be cynical across the board. Cynicism appears to be a coping strategy by the cognitively less gifted to avoid being duped by others.[92]

If I ask myself why I do certain, in some sense altruistic, things, the answer that seems most apt is 'because I don't want to live in a world where …' I don't want to live in a world where no 'rational' person voted, or made efforts to conserve energy, just because their

85 Greene & Paxton 2009 (12509).

86 Stove 2006 (115–6).

87 Chen, Lam, Wu et al 2016; Everson, Kauhanen, Kaplan et al 1997; Haukkala & Uutela 2000; Smith 1992; Stavrova & Ehlebracht 2016.

88 Solon 2014; Mortensen, Barefoot & Avlund 2012; Barnes, De Leon, Bienias et al 2009; Carl 2014; Carl & Billari 2014; Hooghe, Marien & de Vroome 2012; Oskarsson, Dawes, Johannesson et al 2012; Sturgis, Read & Allum 2010. However, it may be that 'those individuals whose personalities include such dark traits as Machiavellianism, narcissism, and psychopathy are neither brute dullards nor evil geniuses on average' (O'Boyle, Forsyth, Banks et al 2013).

89 Haukkala 2002; Stavrova & Ehlebracht 2018.

90 Fetchenhauer & Dunning 2010.

91 Stavrova & Ehlebracht 2016.

92 Stavrova & Ehlebracht 2019; Wilson, Near & Miller 1996.

contribution made no significant difference; I don't want to live in a world where we turn our back on the weak, the suffering and the needy, because they are not productive; I don't want to live in a world where we *always* counted the cost before engaging in acts of helping others. This acknowledges the fact that every decision we make is not just a response to a known and certain world, but is part of co-creating that world for what it is.

One response to these problems is deontology, with its focus on duty and obligation. But what are we to do when such duties conflict, as they often do in real life? Where our duty to save a life means causing an innocent person harm? Or only some lives can be saved? Any schema is compromised: deontology on its own is unable to help when duties conflict; utilitarianism cannot on its own provide justice for the few against the tyranny of the many. Both run the risk of encouraging an attitude that is too rigid and rule-based towards morality. 'There is no such thing possible as an ethical philosophy dogmatically made up in advance', wrote William James, 'there can be no final truth in ethics … until the last man has had his experience and said his say.'[93]

If we need to temper the calculus, how, and on what authority, other than some intuitive sense, are we to make that compromise? As philosopher David Misselbrook asks,

> How can one say that either [utilitarianism or deontology] is a theory that allows one to accurately understand moral principles if one then has to fix the argument in order to ensure the right result? And what faculty within us is able to tell that a particular outcome is wrong if we cannot deduce that by the use of our right moral model? … When John Rawls feels the need to balance the deontological perspective with the utilitarian perspective, where does the need to balance the two come from? By what method of moral reasoning have we discerned that either was not in balance in the first place? How will I know the balanced result when I see it?[94]

It seems to me that we need a better approach, one that takes seriously enough the inner perspective of the doer, not just the externally observed deed; and the *whole* of the doer's inner self, not just their attachment to principles or calculations. Indeed one based on *attitude* itself, and founded on the ontology of the right hemisphere. There is no escape from some dependency on intuition, a disclosure of something of the nature of the good, mediated by the right hemisphere. Such an approach must focus on being, more than doing. According to such an account, what is primarily of importance is the moral character of the actor as a whole, bodied forth in actions in particular contexts. Its concern is not so much with outcomes alone, as attitudes, emotional and social understanding, sensitiv-

---

93 James 1897, 'The moral philosopher and the moral life': 184–215 (184).

94 Misselbrook 2015; referring to Rawls 1971.

ities to context, personality and the choices that issue from them, as well as what it means to be a fulfilled human being balancing many competing considerations through the exercise of practical wisdom. Such an approach is called virtue ethics.

It is the strength and the weakness of virtue ethics – for no one approach can answer every objection – that there are no rules, leaving matters to judgment. Devoid of procedures that could ideally lead to the one correct answer, the left hemisphere is lost. But that is the very nature and purpose of judgment; it introduces a concept that is capable, precisely, of going beyond rules. The claim is not that it is impregnable, merely that it is better than any other option. The price of certainty is absurdity; the prize of uncertainty is wisdom.

As Misselbrook puts it,

> virtue ethics is ethics for grown-ups living in a complex world, with far more than 50 shades of grey. As Beauchamp and Childress remark, 'we see disunity, conflict and moral ambiguity as pervasive aspects of the moral life. Untidiness, complexity and conflict are unfortunate features of communal living ...[95]

We now see all over the world examples of social groups and individuals attempting to do what they have worked out is the right thing in the abstract, and hence doing it for the wrong reason – with catastrophic results. In the words of Sartre, 'evil is the systematic substitution of the abstract for the concrete'.[96] From atheistic totalitarian regimes to religious extremists to 'social justice warriors', the perpetrators of callous crimes comfort themselves that they are acting out of noble motives, but are blind to their own inhumanity. Even at a much higher level of self-sublimation than evidenced in these displays of narcissism, theoretical right beliefs provide no guarantee of good-heartedness. Kretschmer makes a fascinating observation, that strikes me as true to experience, namely that schizoid personalities often display a striving

> after the theoretical amelioration of mankind, after schematic, doctrinaire rules of life, after the betterment of the world, or the model education of their own children, often involving a stoic renunciation of all needs on the part of the individuals themselves. Altruistic self-sacrifice in the grandest possible style, especially for general impersonal ideals (socialism, teetotalism), is a specific characteristic of many schizoids ... on an average they themselves are surpassed by the cyclothymes in natural warm kind-heartedness towards individual men, and patient understanding of their peculiarities.[97]

On which matter Colin Tudge writes:

> The [Christian] commandments may seem simply to be deontological, but in practice for the most part they are an exercise in virtue

95 Misselbrook *op cit*; Beauchamp & Childress 1979.

96 Sartre 1963 (213).

97 Kretschmer 1925 (163).

ethics ... what matters most in all religious ethics is the underlying attitude: and the attitude that all the great religions demand of us is always the same. *All* preach personal humility, and all teach what the Buddhists call compassion and Jesus called love. I suggest that these two – personal humility and compassion (and particularly compassion) – are indeed the most fundamental notions or feelings that underpin all moral codes, of everyone, whether they deem themselves to be 'religious' or not. We could (and I believe should) add a third: the sense of reverence towards all life and towards the universe as a whole.[98]

In summary, the utilitarian approach, strongly linked to the left hemisphere, emphasises isolated events viewed from the outside; the virtue ethics approach, closer to the take of the right hemisphere, emphasises dispositions, processes and relationships viewed from the inside. It is hardly surprising that research demonstrates that seeing the bigger picture, including 'intellectual appreciation of contexts broader than the issue, sensitivity to the possibility of change in social relations, intellectual humility and search for a compromise between different points of view', leads to outcomes in which individual and societal interests are best harmonised, and are thus likely to command consent more widely and for longer.[99] And every aspect of this – seeing the big picture, sensitivity to broader contexts, intellectual humility, and the capacity for compromise – is better served by the right hemisphere than the left.

But my point is not just that the right hemisphere is superior to the left hemisphere with regard to ethics. That is the case: but there is a more important point. According to the left hemisphere's model of reality, it is the author of all its experience, so that goodness, like truth, is its own invention. In intuiting, by contrast, what is good, the right hemisphere makes room for the idea that something that is not just its own invention, but part of the order of things, is being disclosed to us. It creates the disposition (of humility, love, and reverence) that allows it to respond to the good that is, I suggest, in some form constitutive of the cosmos, as is the consciousness that makes possible its apprehension, its intuition and its disclosure in the world.

CONCERNING BEAUTY

The dominant contemporary account of what the world is made of has a bit of a problem with beauty: it doesn't know what to make of it. It recognises that everyone talks about beauty, and that for many people beauty is terribly important to their lives. But how is it to be fitted in to an account that regards the cosmos as a meaningless, materialist affair functioning in a broadly deterministic manner of cause followed by effect? One ploy is to recruit beauty to the ranks

---

98 Tudge 2013 (234–5; emphasis in original).

99 Grossmann, Brienza & Bobocel 2017.

of the one value which such a materialist account embraces, namely the value of utility. Specifically, it sees beauty as a necessary cog in the machinery of evolution: beauty serves as a means of ensuring sexual attraction and therefore the continuation of the evolutionary process. Problem solved.

Except that this doesn't solve the problem. In particular it begs the very first question: where does beauty come from? Natural selection and sexual selection cannot be the answers; they are the answers to a quite different question, namely, '*given* beauty, how might it be used to advantage?' Having described how bright colours and other adornments in flowers and animals, can be 'largely attributed to the agency of selection', as they clearly can, Darwin twice makes the same puzzled observation:

> How the sense of beauty in its simplest form – that is, the reception of a peculiar kind of pleasure from certain colours, forms and sounds – was first developed in the mind of man and of the lower animals, is a very obscure subject … how the sense of beauty in its simplest form was first acquired, – we do not know.[100]

100 Darwin 1872 (162 & 414).

In this, as so often, he was, and remains, entirely right.

Moreover beauty in nature, and especially in humans, is not simply what evolutionary psychology would predict. Of course, it goes without saying that it plays an important role in reproductive success, since reproduction, more than any other element in life, requires attraction. To take the most obvious example, the colours of flowers are more attractive to insects that will help them propagate. But what is it that attracts and why? There are many aspects of the beauty of the forms of plants and trees – their elegant, delicate or majestic shapes and forms – that go well beyond any such mechanism, are probably not apparent to the insect eye at all, and convey no additional advantage. So why are they so beautiful to humans, who do not share a common ancestor with insects? Why aren't they a matter of indifference to us? The colours and forms, and the sweet scents, of plants are extraordinarily beautiful to humans, but having a sense of their beauty serves no utility, and may even, in the case of attractive poisonous plants, be fatally deceptive. Why is the aesthetic sense of birds so similar to our own, not just blatant or tawdry? Birds could be ugly to us and it really wouldn't matter to our survival or theirs. Nor can it be that all living things are considered equally beautiful, as the case of bats, toads and spiders, despite their no doubt sincere aficionados, attests (and, for an ugly bird, just try the unfortunate shoe-billed stork). Why do we think there is a kind of beauty in the finest horse that could never be matched by the finest donkey, however attractive? Both animals have been useful to man in roughly the same degree.

Please note, I am absolutely not arguing that the beautiful colours of the flowers and birds, and all the other beauties of nature, were created by an engineering God for human delight. (I shall have more to say against the concept of an engineering God later.) As Darwin says,

> Were the beautiful volute and cone shells of the Eocene epoch, and the gracefully sculptured ammonites of the Secondary period, created that man might ages afterwards admire them in his cabinet? Few objects are more beautiful than the minute siliceous cases of the diatomaceæ: were these created that they might be examined and admired under the higher powers of the microscope?

Clearly not: but what, then, *was* their beauty for?

Though Darwin himself accepted that animals and birds have a sense of beauty, neo-Darwinians deny that there is anything there other than the decoding of an informational signal about reproductive health, or a lure to copulation. The philosophical arguments here are complex, and perhaps one cannot be definite on the issue.[101] But one does not need to be, in order to see that reductionists have a dilemma. If they are forced to conclude that animals have an aesthetic sense that is over and above utilitarian functionality – a disposition to find certain colours or forms in themselves attractive, so that that disposition could be then harnessed by utility in the first place – how do they account for it? This difficulty is avoided by denying it. But if they deny that animals have any such sense, then the appreciation of beauty, since it clearly exists in humans, must have *arisen* in humans; in which case why does it have so little to do with promoting *human* survival? Beauty in people is not at all the same as sexiness: there are many people who are beautiful without being at all sexually attractive, and many who are sexually attractive without being at all beautiful. More generally, epicene looks in men, not a sign of fertility, robustness or hierarchy dominance, have over a very long period been considered particularly beautiful by both sexes.[102] Equally, ideal feminine beauty is not a matter of vaunting the most exaggerated symbols of fertility. If it is a basic instinctual quality geared solely to reproductive success, why would it have changed in humans – and changed in an era when human survival was more precarious?

In landscape, it is indisputably true that sometimes the kind that is highly productive or offers shelter is considered beautiful, but by no means always. Often, in fact, wild and inhospitable places, mountains, oceans and deserts, have ravished the human spirit with their beauty. There is nothing less likely to fill the heart with longing than a Lincolnshire field full of cabbages. You may think this is a late, Romantic, judgment, and it is very well-known that some

---

101 For a useful and clear, if limited, discussion, see Welsch 2004.

102 See, eg, Greer 2003.

landscapes we now consider beautiful once inspired fear and awe; but sublimity is part of beauty, and the beauty of wild landscapes, mountains, gorges and forests, was celebrated from as early as the Han dynasty, spanning the third century BC to the third century AD, in China; Eucherius describes the desert of the Middle East as *speciosa* ('strikingly beautiful') in the fifth century AD;[103] the Inuit have ancient words for the sheer beauty of the frozen wastes of their homeland; and so on.[104] All this has nothing whatever to do with the cultural phenomenon of Romanticism in eighteenth- and nineteenth-century Europe.

And that is only the tiniest beginning. There's the beauty of a minor third, never mind of a Schubert piano sonata; of an elegant chess move; of a Zen gravel garden; of snow on a mountain top; of Euler's equation $e^{i\pi} + 1 = 0$; of a crucifixion by Cimabue. Of course, I do not say that someone somewhere could not find a way to defend the idea that they are all by-products of reproductive signalling, because someone somewhere can always be found to defend any point of view, however silly. But what on earth lies behind such an attempt?

In the last chapter I supported the view that it is more rational, and better in keeping with science, to suppose that matter arose out of consciousness than consciousness out of matter, and that what we experience is the way in which the cosmos becomes not only aware of itself, but becomes more itself. A logical view, then, is that value is the proper object of consciousness; that beauty is foundational, like truth and goodness, and is, like them, an aspect of the cosmos revealed to us *by* conscious life; and that it has been requisitioned by a massively important biological drive to its service, as just one of the many elements of human experience in which beauty plays a crucial part. I suspect that the appreciation of beauty is one of the things life is for: it is not surprising, then, that it plays a part in bringing life about. And just as human consciousness reveals goodness to be more than animal empathic sensibilities are able to encompass, without denying their reality, human consciousness reveals the sense of beauty to be more than an aid to procreational choices, without disputing that that is one of its roles.

If, then, utility fails to provide the prevailing materialist account of the cosmos with an adequate explanation for the phenomenon of beauty, how should that account regard beauty? D. H. Lawrence wrote that 'science has a mysterious hatred of beauty, because it doesn't fit in the cause and effect chain'.[105] His point was not, of course, that scientists cannot appreciate beauty in their subject, but that scientific culture was and perhaps still is, committed to unweaving the rainbow by instrumentalising beauty: beauty becomes only the small part of it that can be revealed through the lens of functional use, what Lawrence calls the 'cause and effect chain'. Moreover he

103 Eucherius (c 428 AD), *De laude eremi*, §38.

104 Pelly 2016.

105 Lawrence 1950 (14).

was reacting to the obvious fact that, though Darwin constantly expressed astonishment at the beauty of nature (the word *beautiful* or *beauty* appears on average on every seventh page in *The Origin of Species*, and on every second page in *The Descent of Man*), there is no place for such language in the literally an-aesthetic writings that form the bulk of modern science literature.

In short, the prevailing account of the cosmos washes its hands of the phenomenon of beauty. Beauty is merely an extra, an add-on, something surplus to the requirements of the model. 'Surplus' is perhaps the most appropriate word to describe this view because it reflects a common attitude that the appreciation and cultivation of beauty are a sort of luxury that can be afforded only once we have satisfied our most basic needs for survival. In any event, beauty is not central to anything that matters: it is a marginal affair.

But this account simply does not accord with what we know about the attitude to beauty of people with infinitely less of a material surplus than we modern Westerners enjoy.[106] Receptiveness to, and valuation of beauty seem so deep in us that it could be said to be a primal instinct. Introduced by the remark of a stone-knapping expert that early flint tools were the work of 'creative' people who wanted to 'make beautiful objects, not just functional objects', Neil MacGregor, then Director of the British Museum, commented about the making of a series for the BBC called *A History of the World in 100 Objects*:

> One of the greatest discoveries for me was that actually as soon as we start making things we start making beautiful things, that it looks as though even one and a half million years ago we want things to be beautiful, we want them to be complicated and we want that just as much, apparently, as wanting them to be fit for purpose.[107]

Even the development of the use of metals seems to have been motivated primarily by their beauty rather than, as assumed, by their utility. According to the historian of technology Robert James Forbes, metal was at first prized for its fascinating beauty, and only subsequently found to have use.[108] Anthropologist Christopher Hallpike concurs: 'The gold, copper, and iron our ancestors would have found in their pure or "native" state', he writes,

> would have been useless for practical purposes, being softer and blunter than flint, and also required the development of a whole new technology. So why did our Neolithic ancestors bother with metals at all? ... It is only because gold and copper are beautiful and rare that people initially treasured them, and were sufficiently motivated to explore their properties further ... It was man's aesthetic sense, his love of self-decoration, and his desire to own rare and precious objects that was responsible for the early development of metal-

106 Social worker Susanne Sklar writes: 'I was not surprised when impoverished single mothers in the slums of Tijuana, Mexico told an American foundation that what they most wanted for their undernourished and ill-clad children was a school that was beautiful' (Sklar 2007).

107 *A History of the World in 100 Objects*, programme 3, BBC Radio 4, 2010.

108 Forbes 1971 (10).

109 Hallpike 2008 (195).

110 Hallpike 2011 (150: emphasis in original).

111 It seems to me that this is due mainly to a loss of the sense of harmonic proportion. My choice of date refers to Britain, the first nation to industrialise: this may coincide with the date at which the first builders and designers brought up entirely in an industrialised urban society there began to influence architecture and artefacts. My sense is that, since the process of industrialisation began in other nations a little later, the birth of ugliness in other parts of the world may have been correspondingly delayed slightly. It is also worth noting that the Georgian period (that preceded this date in Britain) is particularly recognised for its beautiful architecture and artefacts, so the effect is stark. Naturally I do not intend to imply anything so absurd as that all beauty disappeared from design: just that inadvertent ugliness in design made its first significant appearance at around this point in history, and grew.

lurgy and ... of glass as well, long before their practical possibilities became obvious.[109]

And he points out that 'it was often the stimulus of *non-practical motives*, such as aesthetics, intellectual curiosity, magic and religion, pride and status, or entertainment', that led to functional developments 'that they would never have thought of in the ordinary work of daily life, or in relation to material needs.'[110] And as to beauty being only for the affluent, it is arguable that ugliness came into human life only with affluence. In fact, I'd venture to say that there are few artefacts or buildings that we know of prior to 1830 that would be generally considered ugly.[111] Such ugliness depends on the externalisation of a certain aspect of human thinking – one that has no intuitive understanding of proportion or *Gestalt*, but is focussed on parts, and intent on utility.[112]

It seems, then, that beauty is an irreducible element in experience, and more fundamental than utility. Indeed it is particularly perverse to attempt to subordinate beauty to utility since one of the distinguishing features of beauty is that, as Kant pointed out, it pleases us disinterestedly. Exactly what marks it out, he says, is its purposiveness without presenting any purpose (*Zweckmäßigkeit ... ohne Vorstellung eines Zwecks*).[113] We *contemplate* the beautiful:

> nor is this contemplation, as such, directed to concepts, for a judgment of taste is not a cognitive judgment (whether theoretical or practical) and hence is neither *based* on concepts, nor directed to them as *purposes* ... the liking involved in taste for the beautiful is disinterested and free, since we are not compelled to give our approval by any interest, whether of sense or of reason.[114]

Leibniz called beauty a 'disinterested love';[115] Burke, a form of love that is 'different from desire'.[116] And this is not a sentiment confined to the Enlightenment. Marcus Aurelius said that 'anything in any way beautiful derives its beauty from itself, and asks nothing beyond itself'.[117] In this it is not different from other values. Seneca, as is well known, asserted that virtue is its own reward; and Spinoza that not happiness, but virtue itself, was the prize of virtue.[118] Values are not instrumental. That X has an effect Y does not make Y the purpose of X. For example, beauty relieves pain;[119] but while that is a sign of its power, it does not explain the existence of beauty, or indicate that its purpose is analgesia.

In the words of Emily Dickinson,

> Beauty — be not caused — It Is —

And she continues,

112 According to Kluge's *Etymological Dictionary of the German Language*, the word *Gestalt* may be derived from the compound Old High German word *ungistalt* (Middle High German *ungestalt*), meaning 'disfigured.' Thus the *Gestalt* is what is lost when something is ugly. I am not in a position to pass philological judgment, but I like that.

113 Kant 1987, sect I, Bk I, iii, §17, Akademie, vol 5 (236).

114 Kant 1987, sect I, Bk I, i, §5, Akademie, vol 5 (209–10: emphases in original).

115 Leibniz 1996, Bk II, ch xx, §5 (163).

116 Burke 1887, vol 1 (166).

117 Marcus Aurelius, *Meditations*, Bk IV, §20.

118 Seneca, *On the Happy Life*, IX, 4: '*virtus ipsa pretium sui*'; Spinoza, *Ethics*, V, §42: '*beatitudo non est virtutis præmium, sed ipsa virtus*'.

119 de Tommaso, Sardaro & Livrea 2008.

Chase it, and it ceases —
Chase it not, and it abides —[120]

Since there is no formula for beauty, any more than there is for truth or goodness, it cannot be commanded, but must be wooed. And it is certainly not just about things that make us feel comfortable and safe. 'Beauty', wrote Rilke, 'is nothing but the onset of terror we can only just bear, and which we admire in awe because it serenely disdains to destroy us.'[121] In experiencing it – as in experiencing truth, goodness and a sense of purpose – we are aware of being in the presence of something greater than ourselves.

And, though it is said that beauty is in the eye of the beholder, that is only half the story: the eye must also *discover* the beauty that it could never create. Much as I argued that moral values, within certain limits, are universal, beauty is also more universal than we have been taught to think. Cross-cultural agreement is remarkable, even though the forms taken by beauty in art may differ widely in style.[122] Westerners have no difficulty in responding to Japanese or Chinese aesthetics, nor do Far Easterners have difficulty in responding to Western aesthetics. Norwegians acculturated to a Western musical tradition make precisely the same associations between particular emotions and particular musical intervals as are made in Ancient Indian music – a radically different musical tradition.[123] Generally Westerners and Easterners agree on what is beautiful in their respective cultures, though there are bound to be some differences. For me the most moving demonstration of the cultural universality of beauty is in a sequence made for French television, in which Amazonian tribesmen are exposed to images of the modern Western world, and their reactions recorded. They express revulsion at almost everything – our treatment of our elderly, of animals and trees, our violation of the moon and our terribly destructive wars. In just one sequence, however, the tone changes dramatically. They are shown a clip of Maria Callas singing Bellini's 'Casta diva', and are asked what they think. The men are rapt. For the first time, someone stands up and moves towards the camera in order to speak. He is a young tribesman: 'This music is not our culture. We do not know what it means. We can only watch and listen. But we are touched by it.' And an elder continues: 'I find it overwhelming. Without understanding her, we sense that there is something sacred there.'[124]

On a more banal level, agreement between individuals *from* different cultures about facial attractiveness *in* different cultures 'is one of the best-documented and most robust findings in facial attractiveness research since the 1970s'.[125] Nor do children have to learn cultural norms in order to make beauty judgments. According to psychologist Stephen Ceci,

---

120 Dickinson, fr 516 (in the Harvard *variorum* edition).

121 Rilke, *Duino Elegien*, Erste Elegie: » Denn das Schöne ist nichts | als des Schrecklichen Anfang, den wir noch grade ertragen, | und wir bewundern es so, weil es gelassen verschmäht, | uns zu zerstören «.

122 See McGilchrist 2009b (419–21).

123 Mithen 2005 (91); Oelman & Loeng 2003 (393–6).

124 www.youtube.com/watch?v=eafOkWXjqjc&feature=youtu.be at 06:35.

125 Little, Jones & DeBruine 2011. See also Langlois, Kalakanis, Rubenstein *et al* 2000; and Cunningham, Roberts, Barbee *et al* 1995.

starting at seventy-two hours after birth, babies prefer to look at faces that have been rated as attractive over faces that have been rated as average or less attractive. It doesn't happen fifteen minutes after birth. It takes three days looking at faces that they form this sort of prototype of what a face should look like. But starting at seventy-two hours, scientists have shown that babies reliably prefer to look at faces that have been rated as attractive. The one caveat, though, is that they're not necessarily proportionate or symmetric faces.¹²⁶

I suggest that 72 hours is too short a time to arrive at a 'prototype' from experience. And of course you cannot expect a baby to prefer faces at 15 minutes: it has got to learn to see in daylight and become minimally oriented first. The beautiful face is not a 'prototype' or a stereotype, but an archetype. Only once or twice in a lifetime does one see the archetype of the beautiful female or male face realised, and then the experience is one of unforgettable awe.

Beauty is also constantly referred to by mathematicians, for whom its only consequence is delight: I discussed this in relation to mathematical intuition in Chapter 19 – how often mathematicians and physicists were convinced of the rightness of a conclusion by its beauty, even though at the time they could not see why it must be correct, and perhaps even knew of evidence against it that only later was found to be mistaken. Music and mathematics are often compared: 'mathematics is *the music of reason*', according to mathematician Paul Lockhart.¹²⁷ The question arises whether this beauty is an adventitious quirk, or fundamental to the structure of reality: is maths, and its beauty, discovered, or invented? Astronomer Mario Livio writes: 'Since the laws of nature have the elements of beauty engraved in them, it should come as no surprise that aesthetic principles played a major role in the shaping of our thinking about the origin of the universe.'¹²⁸ Here is the account of Paul Dirac (who, by the way, called himself an atheist):

> It seems to be one of the fundamental features of nature that fundamental physical laws are described in terms of a mathematical theory of great beauty and power, needing quite a high standard of mathematics for one to understand it. You may wonder: Why is nature constructed along these lines? One can only answer that our present knowledge seems to show that nature is so constructed. We simply have to accept it. One could perhaps describe the situation by saying that God is a mathematician of a very high order, and He used very advanced mathematics in constructing the universe. Our feeble attempts at mathematics enable us to understand a bit of the universe, and as we proceed to develop higher and higher mathematics we can hope to understand the universe better.¹²⁹

126 BABIES LESS THAN 72 HOURS-OLD: Slater, Quinn, Hayes et al 2000; Slater, Bremner, Johnson et al 2000; 2–8 MONTH-OLD BABIES: Langlois, Roggman, Casey et al 1987; 3–6 MONTH-OLD BABIES: Samuels & Ewy 1985; 6 MONTH-OLD INFANT PREFERENCES FOR ATTRACTIVE FACES GENERALISE ACROSS SEX, RACE AND AGE: Langlois, Ritter, Roggman et al 1991; 5–15 MONTH-OLD BABIES PREFER ATTRACTIVE PREMATURE INFANT FACES: Van Duuren, Kendell-Scott & Stark 2003; EFFECT EXTENDS BEYOND CONSPECIFICS: Quinn, Kelly, Lee et al 2008; NOT BASED ON SYMMETRY: Samuels, Butterworth, Roberts et al 1994.

127 Lockhart 2009 (37: emphasis in original).

128 Livio 2000 (43).

129 Dirac 1963.

130 Thompson 1992 (1096–7).

131 I am indebted to Julian Marshall for this insight.

132 Peirce 1931–60, vol 5, §130.

133 Tolstoy 2009 (100).

In a similar vein, D'Arcy Thompson pronounced that 'the harmony of the world is made manifest in form and number, and the heart and soul and all the poetry of natural philosophy are embodied in the concept of mathematical beauty.'[130] Anciently there was a perception that natural form is *intrinsically* beautiful. The word 'cosmos' was Pythagoras's term and it originates in a stem that connotes order, harmony and beauty (of which the basest reflection only remains in our word 'cosmetic'). I have earlier discussed the Chinese concept of *lǐ* (the form that is everywhere in the cosmos and is beautiful and right); the word *rūpa* in Sanskrit means form, but its deeper meaning is beauty;[131] and the Latin word *formosus*, from *forma*, a form, means beautiful (cf modern Italian and Portuguese *formoso*, and Spanish *hermoso*).

At the core of beauty is the capacity to lead us to truth, but also to destruction: its consequences are as often disruptive to, as confirmatory of, our proper goals. Of course, one might make beauty one's *only* goal, but in doing so one risks making a monster of oneself – callous without, desolate within. For it seems that beauty alone, though it addresses itself to the soul like little else, is not enough to sustain the soul, which requires also goodness and truth. If beauty were simply truth, and truth beauty, with due respect to Keats, we wouldn't have two very different concepts. They need their potential both for independence and for interdependence: each aspect of the relationship is real.

C. S. Peirce had this to say about the relation between goodness and beauty:

> An ultimate end of action ... must be a state of things that reasonably recommends itself in itself aside from any ulterior consideration. It must be an admirable ideal, having the only kind of goodness that such an ideal can have; namely, aesthetic goodness. From this point of view the morally good appears as a particular species of the aesthetically good.[132]

If it is true that goodness is a particular species of beauty, there is still certainly no equation here, since manifestly the beautiful is not always a species of good. To quote Tolstoy, 'it is amazing how complete is the delusion that beauty is goodness'.[133]

In Appendix 5, I assess the left hemisphere's contribution to the appreciation of goodness, and in doing so considered three elements: its contribution to an understanding of what goodness, in the abstract, *is*; its role in making particular *judgments* of goodness in particular cases; and its immediate *value-ception* of goodness. In each of these the left hemisphere was inferior to the right hemisphere because of its tendency to rationalise and make explicit: that is not just my opinion – we have seen the evidence that this is so.

With beauty, things are similar, yet slightly different. The left hemisphere's understanding of what beauty *is* tends, once again, toward a rationalised, retrospective account in terms of utility. But specific judgments of beauty, and the sense of beauty, are harder to put into words, with the expectation that the left hemisphere will be less likely than it was in the case of goodness to be seduced into making *particular* judgments on the basis of principles of utility. So how do the hemispheres compare when it comes to appraising beauty?

Starting from first principles, the essence of beauty is harmony, including its judicious violations: appreciation, therefore, not of things, but the relations *between* things that are simultaneously similar but different. This is a strength of the right hemisphere. Furthermore, given that in art the ambiguous and unexpected are important aspects of aesthetic preference;[134] and given that the sense of beauty depends on understanding implicit meaning and gives rise to a response that one cannot entirely separate from emotion; and given that the beautiful, and the sublime, can never be made explicit; and given that beauty is the most embodied of all values; there seems to be a *prima facie* case that the right hemisphere is going to play an important role in aesthetics. Interestingly, Scott Thybony reports that 'the Navajos have two ways of looking at the landscape. One's with hard eyes and the other's with soft eyes. Hard eyes are used when looking for things like game, water, pop machines. Soft eyes are used to take in the beauty of the scene.'[135] This sounds very like the sharply defined gaze of the grasping left hemisphere (hard eyes) for acquisition, contrasted with the broad, open, receptive gaze of the right hemisphere (soft eyes) for appreciation of beauty.

There are a number of further characteristics of beauty that suggest it would be better understood by the right hemisphere than the left. Roger Scruton comments about beauty's freedom from all goal-directed designs that 'one sign of a disinterested attitude is that it does not regard its object as one among many possible substitutes'.[136] The mind sees the beautiful as irreplaceable and *unique*: no two people are beautiful in the same way, whereas in, for example, a commercial enterprise they become fungible.

Another right hemisphere characteristic is that it pleases without *concepts*.[137] This is akin to Scheler's view that beauty's appeal is pre-cognitive. Scruton makes a distinction between what he calls the immediate, sensory, intuitive character of the experience of beauty and 'the way in which an object comes before us' more deliberately.[138] This looks like a distinction between the right hemisphere's appreciation of beauty as *zuhanden*, and the left hemisphere's attempt to deal with it once it becomes *vorhanden*.[139]

Yet another right hemisphere characteristic is its *Gestalt* nature. My argument about goodness was that our sense of it does not come

134 Berlyne 1970; Humphrey 1973; Martindale & Moore 1988; Zeki 2001; Seashore & Metfessel 1925.

135 Thybony 1997 (12).

136 Scruton 2009 (27–8).

137 Kant 1793: Part 1 'Critique of aesthetic judgement'; Second Moment, §9; Fourth Moment, §22.

138 Scruton *op cit* (26).

139 See McGilchrist 2009b (153–4); and earlier reference to Heidegger's terminology on p 340.

from the summation of component elements, but from a sense of the whole. Similarly with beauty. 'Recall the queasy feeling', writes Scruton, 'that ensues, when – for whatever reason – you suddenly see a *body part* where, until that moment, an embodied person had been standing. It is as though the body has, in that instant, become *opaque*.'[140] Beauty is a matter of seeing through the surface to the depth, seeing through the parts to see the whole. In Pope's famous phrase, ''Tis not a lip or eye we beauty call, / But the joint force and full result of all.'[141] And it was this that lay behind my argument, in an early book called *Against Criticism*, that works of art are *Gestalten*, which the analytic, explicit, abstracting process of criticism fails to account for and destroys. 'No one component of a film can have any meaning in isolation', wrote the great film director Andrei Tarkovsky: 'it is the film that is the work of art. And we can only talk about its components rather arbitrarily, dividing it up artificially for the sake of theoretical discussion.'[142]

Yet another reason for supposing right hemisphere preponderance is beauty's link with the imagination, and its intrinsic ineffability. 'Things are pretty', wrote Emerson,

> graceful, rich, elegant, handsome, but, until they speak to the imagination, not yet beautiful. This is the reason why beauty is still escaping out of all analysis. It is not yet possessed, it cannot be handled ... It instantly deserts possession ...[143]

And his compatriot Emily Dickinson wrote in her lapidary fashion: 'The Definition of Beauty is / That Definition is none —'[144] This ineffability signals not just that it is beyond language, but beyond limitation, as poets have often attempted to convey:

> Beauty is everlasting
> and dust is for a time ...[145]
>
> A thing of beauty is a joy forever ...[146]
>
> Estranged from Beauty — none can be —
> For Beauty is Infinity —[147]

It also cannot be *used* in the service of knowledge, if knowledge is understood as *wissen*, rather than *kennen*. Art is how humans have expressed their sense of beauty, but it is the route to an understanding of what cannot appear as an object of knowledge. Rather we come to acquaint ourselves with something unique that may be understood only by experience. So, Tarkovsky says,

> Some say that art helps man to know the world, like any other intellectual activity. I don't believe in this possibility of knowing ...

---

140 Scruton *op cit* (48).
141 Pope, *An Essay on Criticism*, Part II.
142 Tarkovsky 1986 (114).
143 Emerson, 'Beauty': 2004 (161).
144 Dickinson, fr 988 (in the Harvard *variorum* edition).
145 Moore, 'In Distrust of Merits'.
146 Keats, 'Endymion'.
147 Dickinson, fr 1474 (in the Harvard *variorum* edition).

Knowledge distracts us from our main purpose in life. The more we know the less we know; getting deeper our horizon becomes narrower. Art enriches man's own spiritual capabilities, and he can then rise above himself ...[148]

And elsewhere,

The allotted function of art is not, as is often assumed, to put across ideas, to propagate thoughts, to serve as an example. The aim of art is to prepare a person for death, to plough and harrow his soul, rendering it capable of turning to good.[149]

When it comes to discriminating the relative contribution of each hemisphere to our appreciation of beauty, there are almost insurmountable difficulties. How does one attempt to measure the correlates of beauty, as science inevitably must? To start with, aesthetic appreciation involves perception, emotion, intuition and cognition, in any of several modes or more together, and to differing degrees in different cases and in different individuals; emotions of pleasure, disgust, sadness, joy, or awe; and draws on responses to form, colour, sound, action, memory, associations with bodily sensations, sexual responses, action preparations, and much more. Though musical appreciation is generally speaking more reliant on the right hemisphere, sad music activates the right hemisphere more greatly and happy music the left – irrespective of beauty.[150] Findings can also be skewed by the choice of comparator: with the ugly? the bland? the merely symmetrical? And with the nature of the stimuli: are we looking here at faces, landscapes, abstract patterns, nudes, or what? Moreover, when subjects are asked to form a judgment, rather than merely experience, this radically alters which parts of the brain are involved.[151] And then there are differences between naïve and trained subjects (one might assume that the untrained subjects would rely more on intuitions and the trained subjects on rules, but it is just as likely to be the other way round.)[152] A few of the undoubted difficulties in knowing what it is that one is measuring here are outlined in a paper by Marcos Nadal and colleagues.[153]

What's more, the sexes differ in their brain responses (see below), and there are, for either sex, differences between early (~300 milliseconds) and late (~600 milliseconds) responses. Jacobsen and Höfel have put forward a two-stage model of aesthetic preference.[154] During the first stage, they suggest, at around 300 milliseconds, an initial impression is formed. This is associated with midline, thus non-lateralised, frontal activity. The second stage, a deeper aesthetic evaluation, begins close to 600 milliseconds and is related to widespread right hemisphere activity.[155] Thus, much may depend simply on timing in any experimental setting: this is virtually impossible

148 Tarkovsky 1983.

149 Tarkovsky 1986 (42).

150 Mitterschiffthaler, Fu, Dalton et al 2007.

151 Liu, Brattico, Abu-Jamous et al 2017.

152 Hadravová 2019.

153 Nadal, Munar, Capó et al 2008.

154 Jacobsen & Höfel 2003. See also Brattico, Jacobsen, De Baene et al 2003; Jacobsen & Höfel 2001.

155 Nadal et al, op cit.

to assess using fMRI which has only gross time resolution – a few seconds at best – though it is possible using EEG.

And then there's the fact that as people age and get more experienced, they acquire different kinds of preferences. Furthermore, different aspects of the same hemisphere seem to be involved with different qualities involved in the object of appreciation. One study compared popular music with 'artistic' (so-called classical) music: 'activation of right putamen tracked the aesthetic ratings of popular music, whereas the right medial prefrontal cortex tracked the aesthetic ratings of artistic music ... artistic music activated theory of mind areas ... when compared with popular music. And these areas also tracked aesthetic ratings of artistic music but not those of popular music.' 'Artistic' music, the authors conclude, involves intelligence and social cognition to a greater extent than popular music.[156]

Given all of that, it would be a minor miracle if there was consistency. There can be no one circumscribed 'beauty centre': too much is going on. As David Bentley Hart points out,

> Beauty is something other than the visible or audible or conceptual agreement of parts, and the experience of beauty can never be wholly reduced to any set of material constituents. It is something mysterious, prodigal, often unanticipated, even capricious. We can find ourselves suddenly amazed by some strange and indefinable glory in a barren field, an urban ruin, the splendid disarray of a storm-wracked forest ...[157]

The situation looks unpromising. As one paper puts it, 'rational reductionist approaches to the neural basis for beauty ... may well distil out the very thing one wants to understand'.[158]

Readers by now will not be surprised to learn that our receptiveness to and appreciation of the aesthetic is, with all such caveats, highly dependent on the right hemisphere. Not everyone would necessarily accept such an assessment, so readers interested in following the evidence can read it in an appendix and reach their own conclusions (Appendix 6). Suffice it to say here that lesion studies, where alone one has naturalistic accounts of subjects describing, specifically, a loss of the sense of beauty, are almost conclusive (92%) in favour of the right hemisphere as the locus of beauty appreciation; studies of appreciation in normal subjects confirm this tendency; and even studies of explicit *linguistic* judgment of an aesthetic nature, despite a built-in tendency favouring the left hemisphere, offer broad support. Given the aforementioned complexities involved in designing studies in normal subjects, this degree of cohesion is highly significant. As if to confirm the picture, it turns out that the intact brain as a whole has 'aesthetic preferences that are very close to the ones of the right hemisphere'.[159]

---

156 Huang, Huang, Luo *et al* 2016,. On the relationship between classical music and intelligence, see, eg, Kanazawa & Perina 2012; also Dutton 2013.

157 Hart 2013 (279–80).

158 Conway & Rehding 2013 (3–4).

159 Lotman & Nikolaenko 1983.

There is reason, I suggest, to believe that beauty is neither 'surplus' to what matters in the cosmos, nor a human invention: it would appear to be, rather, a constitutive element in the cosmos to which the right hemisphere is particularly attuned, an aspect of reality which it discloses. If that were the case, and given that I have throughout this book presented evidence that the right hemisphere is a more reliable guide to the nature of things than the left, we would expect that beauty had qualities that could neither be reduced to utility, nor were conformable to the assumptions of the left hemisphere's model of reality. I have already alluded to some of these qualities: beauty's relational nature; its affinity for the ambiguous and unexpected; its implicit and embodied nature; its dependence on appreciation of the unique; its *Gestalt* nature; its capacity to please without concepts, its close link with the imagination, and its intrinsic ineffability.

But there are further attributes of beauty that strongly suggest it is an insight into the deep nature of the cosmos, not just a mental construct. On first principles, we would expect a preference for the perfect over the imperfect, and the symmetrical over the asymmetrical. (Certainly this would be the case if beauty was *just* an aid to mate selection.) What we find, though, is that beauty often attends a coupling of symmetry with asymmetry, of perfection with imperfection: and that these couplings are themselves asymmetrical, as in the case of the brain hemispheres, one element being capable of incorporating its opposite, while the other cannot.

Let us take again the example of the golden ratio, also known as *phi* (ϕ), which I mentioned in Chapters 15 & 24, where I referred to the way in which it incorporates asymmetry in its symmetry. Its value is approximately 1.618.[160] It occurs in the structures of the Ancient Egyptians,[161] and is thought to have been known to the Greeks and Romans, as well as being present in Eastern cultures, such as those of the Japanese, Chinese and Indian civilisations, who also recognised the Fibonacci series to which it is related.[162]

The role of ϕ in aesthetics is not, however, undisputed. Though Leonardo knew it as 'the divine proportion', and undoubtedly painters, architects and designers have used and continue to use it as a reference point for beautiful proportion for centuries, it has been the target of recent, predictable, 'debunking', probably because it is often known as 'divine', appears to be threaded through nature in ways we cannot fully fathom, and might suggest intrinsic beauty and order in the cosmos (all red rags to the Nothing Buttery bull). There is a serious point to be made on behalf of such exercises: namely, that if you do enough comparisons of measurements you are likely to come up with an approximation to ϕ *somewhere*. That may be true, but it ignores the evidence that, for example, in many beautiful faces ϕ can be found in around 10 places, but rarely in less

160 See p 1030 above.

161 Badaway 1965 (24ff); Pile 2005 (29).

162 See, eg, Noguchi & Murota 2013; Singh 1985.

beautiful faces.¹⁶³ Questioning it is entirely legitimate, but for critics to be taken seriously they'd need something that passes for a logical argument, not just, a grumpy assertion that 'it's bullshit', in the words of one hapless 'debunker'.¹⁶⁴

The aspiring debunker's arguments are not impressive. For example, ϕ, it is objected, is, like π, an irrational number, the full expression of which has an infinite number of digits after the decimal point: ergo, it cannot exist in the real world. Clearly this can carry no weight: we don't dispute the existence or usefulness in the real world of π on these grounds. Moreover, in the real world no-one could possibly appreciate the difference made by a few extra million decimal places of accuracy. Then, it is said, only approximations to ϕ tend to show up in the real world. See previous. But repeated measurements come out so close as to make mere dismissal look embarrassingly simple-minded. To take an example: the classic ϕ ratio in the human body (that between overall height and the distance of navel from the ground) observed by Leonardo's contemporaries comes out in populations of Germans, Indians and Italians as astonishingly close to 1.618 to three decimal places.¹⁶⁵

It is claimed that, if there is no evidence that artist X spoke about ϕ, this must mean it would be wrong to find it in X's work. Clearly, however, (1) that we have no record of the artist speaking about it doesn't mean he didn't speak about it, especially if we are talking of hundreds of years ago; (2) not speaking about it doesn't mean he didn't know about it; but (3) why must he have known about it in order to use it in his work? Why should only explicit use count as use? While many cases of the use of ϕ are deliberate, many cases in art and architecture are equally likely to be intuitive rather than calculated. But this does nothing to rob them of interest or discredit them: rather the opposite, since it seems that artists find the golden section by intuition, not merely by following a formula. One ardent debunker, Keith Devlin, dismisses averages that converge on ϕ, as though only single cases count. He puts up straw-man arguments, such as that ϕ is claimed (by whom?) to be all that matters in assessing beauty, or that it must always be present in all instances of beauty, or that it can be used as a 'diagnostic test' for beauty. He exhibits grotesque distortions of known faces, claiming that they are the result of applying ϕ; but this is so silly, and so far from any attempt to engage with the issue in good faith, that one is pushed to find a charitable explanation, other than that he assumes his audience is stupid. For a far more balanced assessment, that acknowledges pitfalls, and presents well-sourced evidence, I suggest an overview that took 20 years in the making, and is continually updated online, by Gary Meisner.¹⁶⁶

The brain correlates of the appreciation of the golden ratio have

163 Meisner 2015.

164 Brownlee 2015.

165 Davis and Altevogt investigated the anthropometric proportions described by Leonardo's contemporaries (ratio between height and the distance of navel from the ground) in 459 German and Indian subjects. In the Germans, they found the average to be the golden ratio precisely to three decimal places: 1.618 (100% of ϕ) with a range from 1.610 (99.5%) to 1.637 (101.2%). In the Indians, they found an average of 1.626 (100.5% of ϕ), with a range between 1.623 (100.3%) and 1.634 (100.9%), also very close to the golden ratio (Davis & Altevogt 1979). A more recent study found the same value of 1.626 (100.5% of ϕ), with a range of 1.573 (97%) and 1.671 (103.3%) in 20 Italian subjects (Iosa, Morone, Bini *et al* 2016). If this degree of precision were encountered in any other biological context, it would be considered compelling. For a review of other possible instances of ϕ in the human frame and other human phenomena, see Iosa, Morone & Paolucci 2018.

166 Meisner 2015.

not been widely researched. Two studies now show that the 'two key regions' that respond to 'the canonical (φ) proportion' are both in the right hemisphere – the right occipitotemporal and right parietal areas. When sculptures exhibiting the golden ratio were presented, these areas repeatedly responded, but when the proportions of the sculptures were subtly altered they did not.[167] It has separately been reported that the right hemisphere is attracted to figures drawn according to the golden ratio, as well as to relatively square shapes, while the left hemisphere seems to be more attracted by shapes that are tall and narrow.[168] (Perhaps not coincidentally, this corresponds to the shape of most modern urban buildings.)

Beauty is closely linked to balance and harmony. In turn the presence or absence, or subtle variation, of symmetry affects our assessment of beauty.[169] Symmetry perception requires global processing, so might on first principles be expected to engage the right hemisphere preferentially. And, indeed, the assessment of symmetry or asymmetry is specifically right hemisphere-dependent – 'a cognitive function lateralized to the right hemisphere for most of the population'.[170] This is also the case in blind people for tactile symmetry.[171] Strikingly, a patient with occipital and only mild right parietal damage had normal spatial attention, but no ability to discriminate symmetry.[172]

Symmetry is independent of beauty, though it often plays a covert, if not overt, role. As I have pointed out, asymmetry depends on the underlying possibility of symmetry, and registers the tension between its absence and potential presence. Japanese art has long exalted asymmetry and imperfection in art,[173] an ideal that is intimately linked to the philosophy of Zen Buddhism.[174] It also idealises incompleteness and brokenness, both of which suggest something in process, not finished, as well as austerity and modesty. Importantly this imperfection is said to arouse a sense of serene melancholy and a spiritual longing in the one who sees it (not mere pleasure or comfort): in this it seems to me very like most great music.[175] This is the principle behind the most important concept in Japanese aesthetics, that of *wabi-sabi*, that beauty is transient, that imperfections connect us with nature and the work of the maker, that what is empty is richer than what is full. There is an art form called *kintsugi* (or *kintsukuroi*), in which broken objects are repaired with lacquer, sometimes gold, in a way that draws attention to the flaws, rather than attempting to hide them. The results are sometimes ravishing (see Plate 22[c]).

According to Robert Wicks, 'traditional Japanese aesthetics is an aesthetics of imperfection, insufficiency, incompleteness, asymmetry, and irregularity.'[176] Note these qualities also distinguish the beauty of natural processes from the beauty of mechanical artefacts.

167 Di Dio, Macaluso & Rizzolatti 2007; Iwasaki, Noguchi & Kakigi 2018.

168 di Ludovico 2010, vol 1 (239); referring to the research of Lotman and Nikolaenko 1983.

169 Ramachandran & Hirstein 1999.

170 Bertamini, Silvanto, Norcia et al 2018. See also, for similar assessments: Bertamini & Makin 2014; Brysbaert 1994; Wilkinson & Halligan 2002; Di Dio, Canessa, Cappa et al 2011; Wright, Makin & Bertamini 2017; Wright, Makin & Bertamini 2015; Corballis & Roldan 1974; Bona, Cattaneo & Silvanto 2015; Bona, Herbert, Toneatto et al 2014; Prete, Fabri, Foschi et al 2017; Makin, Rampone, Wright et al 2014; Jacobsen, Schubotz, Höfel et al 2006; Verma, van der Haegen & Brysbaert 2013.

171 Bauer, Yazzolino, Hirsch et al 2015.

172 Vecera & Behrmann 1997.

173 Saito 1997.

174 Koren 2008 (15).

175 Juniper 2003. On music and melancholy, see McGilchrist 2009b. There is a stronger affinity between the right hemisphere and the minor key, as well as between the left hemisphere and the major key (Suzuki, Okamura, Kawachi et al 2008). The pre-Socratic philosopher Gorgias wrote that 'awe [*phrike*] and tearful pity and mournful desire enter those who listen to poetry', and at this time poetry and song were one (Gorgias, *Encomium*, §9).

176 Wicks 2005 (88).

Compare this observation from English composer Julian Marshall on the sublime and its beauty:

> it's usually writings where 'brokenness' is a feature where, for me, beauty shows up. In other words I rarely find a 'one way' ticket to the sublime does it – that's like a quick fix. It's where there is a journey through something incomplete that puts me in touch with a deeper, truer experience of beauty...[177]

Nor is symmetry necessary to beauty in the living. A number of studies have found facial attractiveness to be unrelated to symmetry;[178] and some have found that asymmetry is actually preferred.[179] Symmetry runs the risk of looking uncanny and mechanical. However, the exact relationship between beauty and symmetry in the human face is still open to debate.[180]

There is something just a little grandiose about symmetry: it speaks of perfection, complete control, something finished and outside of time – and hence outside of life. 'Even heaven is not complete.'[181]

So far we have talked of where symmetry/asymmetry is *assessed*, but not of where it is *preferred*. Symmetry is better assessed, but not necessarily better preferred, by the right hemisphere. One interesting finding is that symmetry is more *favourably* assessed in the centre of the field than in peripheral vision: this is the part of the visual field in which left hemisphere attention is concentrated.[182] People with conditions that are associated with right hemisphere deficits, such as schizophrenia, autism and anorexia nervosa, have an unusual predilection for, at least simple, symmetry. It was also, of course, the hallmark of the movement known, with appropriate *hubris*, as the Enlightenment. Compare these two images (Plates 23[a] & [b]), one of a Renaissance façade by Palladio, the other of an Enlightenment façade by Jefferson, both beautiful in their way, both probably ultimately derived from the Roman Pantheon.

Extremely prominent in the Palladian façade are elements of life that are clearly asymmetrical: in the statues, in the central plaque, and in the 'balancing' (itself asymmetrical) of male and female forms. These set up a wonderful counterpoint with the symmetry of the core. By comparison the Jefferson building seems somewhat adynamic and devitalised, though undoubtedly pleasing, in its perfect symmetry, up to a point. It is appropriate that a clock, a device that spatialises and measures time, occupies the central position in the pediment; in Palladio's façade it is an (asymmetrical) crest depicting an animal.[183]

Symmetry is about relations between parts, but asymmetry is about the relations between symmetry and its absence, something

---

177 Personal communication.

178 Farrera, Villanueva, Quinto-Sánchez *et al* 2015 (387–96); Van Dongen 2014; Zaidel & Hessamian 2010.

179 There is a 'biological trend away from perfect symmetry in primates consequent to adaptive evolutionary alteration favouring functional asymmetry in the brain, perception, and face': Zaidel & Deblieck 2007. See also Swaddle & Cuthill 1995; Zaidel & Cohen 2005. Even when the symmetry is recognised as expressing health, it may still be thought less beautiful: Zaidel, Aarde & Baig 2005.

180 For an opposite view, see, eg, Little, Jones & DeBruine 2011.

181 See p 840 above.

182 Rampone, O'Sullivan & Bertamini 2016.

183 I am grateful to Nigel McGilchrist for elucidating that it is a goat standing on its hind legs. *Capra* is the Italian for a goat, and this was the armorial bearings of the Capra family, who owned the Villa Capra at the point of its completion.

still deeper. This relationship between a thing and its absence is central to Japanese aesthetics – and not to Japanese aesthetics alone. In his classic work, *In Praise of Shadows*, Jun'ichiro Tanizaki writes:

> We find beauty not in the thing itself but in the patterns of shadows, the light and the darkness, that one thing against another creates … Were it not for shadows, there would be no beauty … I would call back at least for literature this world of shadows we are losing. In the mansion called literature I would have the eaves deep and the walls dark, I would push back into the shadows the things that come forward too clearly …[184]

Beauty is, Tanizaki suggests, resistant to the bright beam of the intellect's spotlight: bright light can kill beauty, as photographers and painters well understand. And beauty depends on contrast. If everything attracts us equally, nothing in particular attracts us. As Ruskin says:

> beauty deprived of its proper foils and adjuncts ceases to be enjoyed as beauty, just as light deprived of all shadow ceases to be enjoyed as light. A white canvas cannot produce an effect of sunshine; the painter must darken it in some places before he can make it look luminous in others …[185]

The very existence of beauty, goodness and truth requires that some things be more beautiful, better, and truer, than others; they depend for their existence on their dipolar nature. Theologian Hans Urs von Balthasar wrote:

> Since the beautiful comprises both tension and its release, and reconciliation of opposites by their interaction, it extends beyond its own domain and necessarily postulates its own opposite as a foil. The sublime has to be set off by the base, the noble by the comic and grotesque, even by the ugly and the horrible, so that the beautiful may have its due place in the whole, and that a heightened value may accrue from its presence.[186]

The most beautiful hymn in praise of erotic love in modern times – perhaps ever – Strauss's *Der Rosenkavalier*, depends, to some extent, on the vulgar crassness of Baron Ochs, much as Shakespeare's tragedies require their burlesque scenes.

The beautiful is not the same as the sublime,[187] yet it is closer than it has often been made out to be. The essential element in the sublime is not merely something large, but something whose limits, like a mountain top that is lost in cloud, are unknown: it is both there and not there, never fully knowable, and more vital for not being subject to pinning down. It is what aesthetician Patrick

---

184 Tanizaki 1977 (30 & 42).

185 Ruskin 1863, vol 3, Part IV, ch iii, §14 (34).

186 von Balthasar 1989, 95–126: 'Revelation and the beautiful' (105).

187 See McGilchrist 2009b (363).

Colm Hogan calls 'targeted absence', by which he means the use of vagueness or ellipsis to foster aesthetic response, by invoking 'the interrelation of presence and absence'.[188] Elsewhere he comments on both sublimity and beauty:

> Specifically, the emotions at issue in feelings of beauty and sublimity appear to be primarily attachment, on the one hand, and a profound sense of isolation, on the other ...[189]

This highly creative bringing together of distance and union, of negation with affirmation, of absence with presence, has, of course, been a theme of this book. The only neuroscientific study to date to have examined specifically the sublime, as distinct from the beautiful, found, as one might expect, that the two areas of cerebral cortex specifically involved were in the right hemisphere: the right caudate and right putamen, both areas we have come across before in relation to the beautiful.[190] The emotion associated with the sublime is awe, which is effectively what Rilke had in mind in his account of beauty that I quoted – another reason to consider any sharp distinction artificial.

I stressed earlier that I was not suggesting that the beautiful colours of the flowers and birds, and all the other beauties of nature, were created by an engineering God for human delight. Rather it seems to me that we, as they, are the manifestations of an intrinsically beautiful cosmos, and to borrow again Deutsch's formulation, this explains why everywhere we go 'the inhabitants come up and speak to us in English'. Beauty is our native tongue.

It is suggestive that changes of behaviour are, according to American neuroscientist Bud Craig,

> actually initiated by activity in the right anterior insula, followed by the anterior cingulate, rather than the cingulate initiating behavior.[191] Other functional imaging studies of 'free won't' or error awareness suggest that the left anterior insula is predominantly involved in monitoring, rather than generating, behaviours.[192]

Craig refers to 'this complementary pairing – the right side leading and the left side following'.[193] I say 'suggestive', because three important roles of the right, specifically anterior, insula that have emerged are: appreciation of beauty, emotional understanding, and initiating a change of behaviour.[194] By beauty we are moved to move.

≈

Of the two claims outlined towards the beginning of this chapter, one is susceptible of proof and the other isn't.

Claims concerning the role of the right hemisphere in the disclosing of value can be put to the test. They are not claims of an 'all

---

188 Hogan 2017 (89).

189 Hogan 2013 (319).

190 Ishizu & Zeki 2014.

191 Sridharan, Levitin & Menon 2008.

192 Klein, Endrass, Kathmann et al 2007; Brass & Haggard 2007.

193 Craig 2010.

194 *ibid*: 'The insula, and especially the anterior insula, is known to play a critical role in emotional processing'.

or nothing' kind – no claims about human life can be; but I believe the evidence suggests that these particular claims are very likely to be correct.

However, the main claim is not of this kind. It concerns how we understand ourselves, the world and our relation to it. The prevailing dominant account of a meaningless, purely material cosmos, supplied by the reductionist strategy of the left hemisphere, fails to make sense of value, whether that be truth, goodness or beauty, just as it fails to make sense of consciousness. Its answer in every case is the same: that they must be emanations of that purely material cosmos – emanations of matter – that exist purely to further utility. It seems to me that, if you believe that, you will believe anything; you might even end up believing that consciousness is an illusion.

Neither this point of view nor any alternative to it can be proved, so our best recourse is to apply science, reason, intuition and imagination to experience. It seems to me that the reductionist account is contrary to scientific findings, unreasonable, counterintuitive, and shows a complete refusal to exercise intelligent imagination: all the hallmarks of its birth in the left hemisphere. The result is that values themselves become devalued. Beauty, morality and truth have been downgraded, dismissed or denied. If you want to see the consequences, you need do no more than look around you.

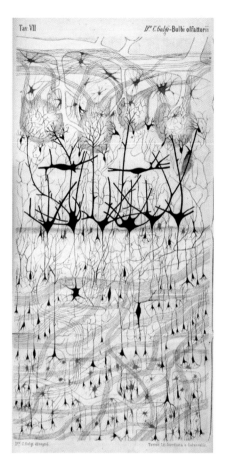 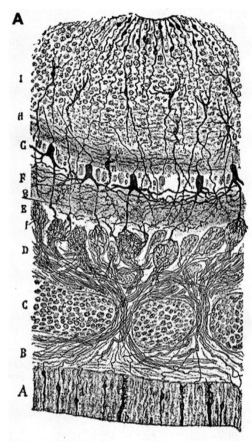

PLATE 17[a] · Olfactory bulb of a dog, drawing by Camillo Golgi, 1875
(Golgi 1875)

PLATE 17[b] · Olfactory bulb of a mouse, drawing by Santiago Ramón y Cajal, 1890
(Ramón y Cajal 1890)

Camillo Golgi and Santiago Ramón y Cajal, the two most eminent neuropathologists of the nineteenth century, saw the same need for simultaneous differentiation and union within the central nervous system, but differed over how this was achieved.

See pp 858–60.

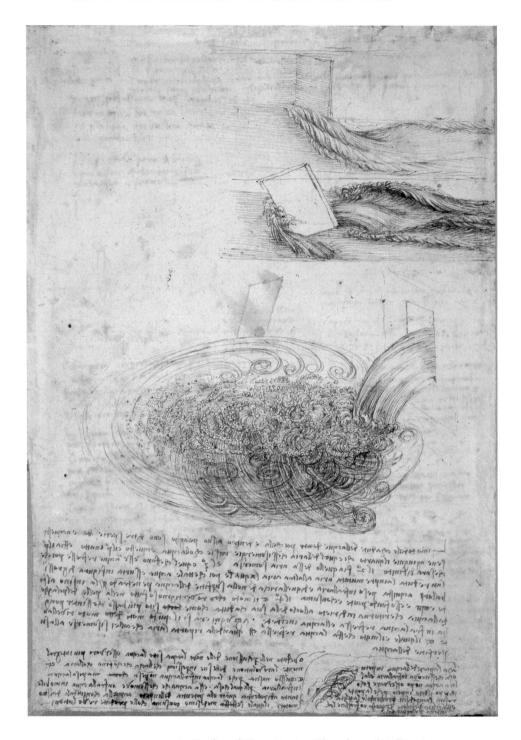

PLATE 18 · *Studies of Water Passing Obstacles and Falling*, by Leonardo da Vinci, ink and black chalk on paper, c 1510–13
See p 964.

PLATES 18 & 19   1166·*iii*

PLATE 19[a] · *Seated Old Man, and Studies and Notes on the Movement of Water,* by Leonardo da Vinci, ink on paper c 1510

PLATE 19[b] · *Deluge,* by Leonardo da Vinci, black chalk on paper, c 1517–18
See p 964.

PLATE 20
The Great Wave off Kanagawa, by Katsushika Hokusai, woodblock print, 1831

Both wave and crystal forms of water are individually entirely unique and unpredictable, yet are in no sense disorderly or chaotic, but fractal and lawful in their morphology.

See p 965.

PLATE 21· Diversity of lawful architective structure in snowflakes
See pp 833–6 & 971–3.

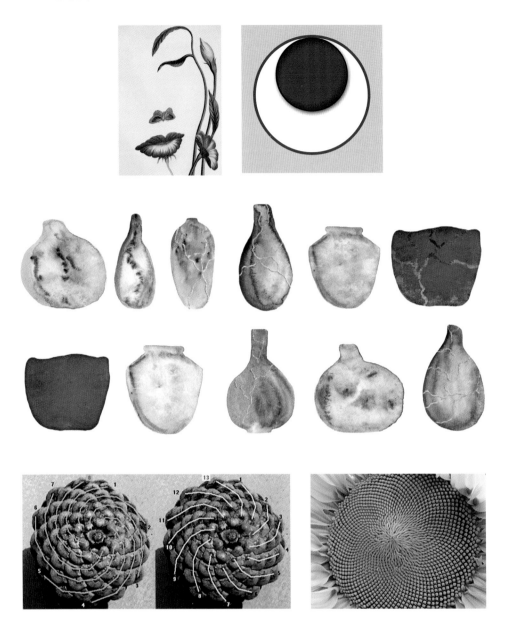

PLATE 22[a] · Bi-stable image of woman's face and flower
See p 914.

PLATE 22[b] · Inyo symbol
See p pp 1034–5.

PLATE 22[c] · Wabi-sabi / kintsugi ware
See p 1161.

PLATE 22[d] & [e] · Fibonacci series in pine cones and sunflower
See p 1031.

PLATE 23[a] · Façade of the Villa Capra ('La Rotonda'), near Vicenza, designed by Andrea Palladio in 1570, finished 1592

See p 1162.

PLATE 23[b] · Façade of The Rotonda, University of Virginia, designed by Thomas Jefferson, completed 1826

See p 1162.

PLATE 24 · *Circle Limit IV* (also known as 'Heaven or Hell', or 'Angels and Devils'), by M. C. Escher, engraving, 1960

'If only you knew the things I have seen in the darkness of night', remarked Escher (Ernst 2007: 20). Many of Escher's drawings involve being able to reverse figure and ground. The left hemisphere tends to focus on what stands out against the ground at the expense of the ground itself; the right hemisphere is better able to see both. The flexibility of view that Escher's drawings demand tends to rely on right hemisphere perceptual capacity (as with so-called 'bi-stable percepts').

See pp 1299–1302.

Chapter 27

Purpose, life and the nature of the cosmos

> Not every end is the goal. The end of a melody is not
> its goal, and yet if a melody has not reached its end,
> it has not reached its goal. A parable.
> —FRIEDRICH NIETZSCHE[1]

> The teleology of the universe is directed to the production of beauty.
> —ALFRED NORTH WHITEHEAD[2]

Nietzsche's 'parable' points up a number of ways in which we can talk at cross-purposes about purpose: I will explore a few in what follows. It's an important topic – that of purpose – because it is close to the whole question of the meaning or meaninglessness of life. The point is not that there should be a readily expressible answer to the purpose of life, or the cosmos, and we'd better set out to pin down what it is: that would be to mistake the nature of such a purpose. It is perfectly coherent to discern a purposefulness in what one sees without being able to delimit in language (outside perhaps of poetry) what that purpose 'is', especially if that purpose is not purely instrumental. John Dewey thought the 'deepest problem of modern life' is that we have failed to integrate our beliefs about the world with our beliefs about value and purpose.[3] One of the reasons is that nowadays we ask science to answer questions it is not equipped to answer, and its answer to the question whether there is purpose in the world is a resounding 'no' – which is hardly surprising, since it excludes purpose from its considerations from the outset.

It is also much easier to ignore purpose if, in accordance with the left hemisphere's dictates, we focus on detail to the exclusion of the whole picture. Narrow views, which are intrinsic to molecular biology, encourage focus on local mechanisms and effectively would not be in a position to perceive purpose if there was one. They provide a description which, while appearing complete in its own terms, misses crucial insights which are hard to accommodate in its model. But naturally that cannot be a reason to ignore them; indeed, for that very reason they must not be ignored.

WHAT IS PURPOSE?

Purpose means different things; and it is a matter of observation – and hardly likely to be a coincidence – that those differences accord with two differing dispositions towards the world that the reader will by now recognise.

[1] Nietzsche 1913, §204.
[2] Whitehead 1933 (341).
[3] Dewey 1984 (204).

The first difference is whether purpose is thought of as being extrinsic and instrumental – eg, the purpose of a photocopier lying in the copied sheet that emerges from it – or, on the contrary, intrinsic and entirely fulfilled in the process itself – as might be a dance, or a poem: by no means pointless, but the point lying within, not like the copied sheet, without, the process that brought it into being.

The second difference is that between a narrowly determined and mechanistic account of purpose, on the one hand, and a largely undetermined and free account, on the other.

The third is a matter of scale: the difference between a narrowly localised and a broad, overarching view of purpose.

These three distinctions map, in the same order, onto each other: extrinsic, deterministic and narrowly focussed according to one view; intrinsic, free and attending to the broad *Gestalt* according to the other.

NON-INSTRUMENTALITY

To deal with the first: a machine has an extrinsic purpose – not its own, but that of the person who made it. It exists purely to achieve an end: another being's end. By contrast, a melody has an intrinsic purpose: it's hardly pointless, but it has no other point than itself. This is the distinction made by James Carse between finite games that come to an end when their goal is achieved, and infinite games that do not issue in an 'outcome' – or do so only incidentally.[4]

Everything of ultimate importance forms part of what George Steiner referred to, in a wonderful phrase, as 'the sovereignly useless'.[5] Such things have intrinsic purpose: trying to find an extrinsic purpose fails and, in the process, devalues them: it feels like a betrayal. Gifts and selfless actions are obvious examples; but celebrations, music, art, poetry, drama, liturgy, meditation, philosophy, spirituality, leading a good life, or ultimately just being the human being you are, are all examples. These are not pointless – about as far as one can get from that; but seeing them as purposed to anyone's gain, whether yours or another's, destroys the essence for which they are prized. They become instrumentalised. Here one sees the first and most obvious hemispheric distinction: that between a strategic purpose, as the left hemisphere sees it, designed to issue in utility; and purpose as the celebration of life, creativity, difference and uniqueness, as the right hemisphere takes it to be.

This takes us back to the discussion of beauty, and my emphasis on the inadequacy of utility as providing even a start on its essential nature. Beauty is gloriously superfluous and unnecessary: its nature is that of a gift. Hart writes that beauty

> is not simply this or that aspect of its composition, not simply its neurological effect, not simply its clarity or vividness or suggestive

---

4 See p 396 above.

5 Steiner 1978a (16).

associations, and so on; it is not even just the virtuosity of its execution or the mastery exhibited in its composition. Rather, it is all of these things experienced as sheer fortuity. I may be speaking of something that escapes exact definition here, but it seems clear to me that the special delight experienced in the encounter with beauty is an immediate sense of the utterly unnecessary thereness, so to speak, of a thing, the simple gratuity with which it shows itself, or (better) gives itself. Apart from this, even the most perfectly executed work of art would be only a display of artisanal proficiency or of pure technique, exciting our admiration but not that strange rapture that marks the most intense of aesthetic experiences. What transforms the merely accomplished into the revelatory is this invisible nimbus of utter gratuity ... the beautiful presents itself to us as an entirely unwarranted, unnecessary, and yet marvellously fitting gift.[6]

The machine fulfils an extrinsic purpose only after abandoning its stable state, that of something naturally static. A single cell, by contrast, has no extrinsic purpose, but nonetheless is extremely active in pursuing its intrinsic purpose, its own continuance and development, ie, the persistence of its stable state as something that is naturally flowing. Even as part of a multicellular organism its purpose continues to be intrinsic, though not, unless something is badly wrong, at the *expense* of the organism as a whole, but in harmony with the organism – not doing so is what makes cancer cells so very unusual. (In some important cases, such as occur in the immune system, the cell's own purposes are actually subordinated to that of the organism – self-sacrificial cells: I will come to that later.) There is an alignment of aims, as a citizen in a properly functioning society is not just a tool of that society, but flourishes in it, while contributing to the flourishing of society as a whole.[7]

Most discussions of purpose in life or the cosmos founder on this distinction. If you understand purpose to mean extrinsic purpose, you invent an engineering God who made the universe as an infinitely complex mechanism to serve some unknown end of his own. Such a God is just a projection of the left hemisphere's fantasy of endless power to manipulate – a divine left hemisphere, detached from the cosmos and running the show according to a foreordained plan. Such a God belongs in a mechanical universe, a museum of clockwork, and that is the last thing we are dealing with. If, like me, you can't, understandably, believe in such a God, you might jump to the conclusion that this infinitely complex 'mechanism' has simply no purpose. But that is just to make the same error, that of conceiving purpose only in extrinsic terms: as if the only alternatives were the purposes of an engineering God, or a cosmos without purpose.

6 Hart 2013 (282–3).

7 See Chapter 12.

OPENNESS

The second difference is this: between a purpose that is closed down to a predetermined mechanism and one that is free and largely undetermined. 'The success of Darwinian natural selection', wrote Popper,

> in showing that the *purpose or end* which an organ like the eye seems to serve may be only apparent has been misinterpreted as the nihilist doctrine that all purpose is only apparent purpose, and that there cannot be any end or purpose or meaning or task in our life.[8]

To some extent this is an extension of the previous distinction, since there will usually be a predetermined mechanism only where there is a clear extrinsic purpose: back to engineering. But that is not because there *could* not, in principle, be a detailed plan of, eg, how to write a poem, or how to lead a good life – there could: it's just that it would be worth rather little. We recognise that such detail is inimical to achieving the valued end. A degree of *freedom* necessary to the process would have to be sacrificed; and having too prescriptive a plan will prevent one achieving the end at all. The poem becomes doggerel; the good life Pharisaic.

Thus pursuing non-mechanical purposes, bearing in mind something one values and wants to bring about, requires a sense of direction, yes – but a necessary, patient, withdrawal from control over the detail. The process must be *flexible*. As James says, where the undetermined plays a part, 'the system of other things has no positive hold on the chance-thing ... [it] says hands off! coming, when it comes, as a free gift, or not at all.' Note that here again, the nature of a gift is invoked.

Here, then, is the second hemispheric distinction: between what is controlled, rigid, with focus on detail, and what is flexibly responded to, with appreciation of the bigger picture. The purpose is overarching, not contained in the detail. A deterministic finalism is just the flipside of mechanism. A controlling God whose creatures are the tools of his purpose; the biological determinism of 'robot vehicles blindly programmed to preserve the selfish molecules known as genes'; and the Laplacian determinism of the cosmic billiard table; all come from the same place. William Paley, the eighteenth-century cleric and philosopher, was, after all, a utilitarian. That left hemisphere mindset is clear from his tell-tale model: that the universe is a mechanism, like a watch; and a watch needs a watchmaker.[9] As mentioned, Dawkins says that 'people like me might be labelled neo-Paleyists, or perhaps "transformed Paleyists"': Paleyists without God, still believers in a watchmaker, though blind.[10]

---

8 Popper 1978 (342: emphasis in original).

9 It is of some interest that a Jesuit priest (Roger Boscovich) may have proposed a fully deterministic universe more than half a century before Laplace: it is after all the logical conclusion to draw from the existence of an engineering ('deistic') God.

10 Dawkins 1998 (16). *The Blind Watchmaker*, is, of course, the title of Dawkins's book published in 1986.

SCALE

And third, the matter of scale. Scale matters far more than is often realised, especially by those who have a mechanistic or reductionist outlook. We have seen examples of what is predictable at the local level being ultimately unpredictable at a higher level (collisions of billiard balls); and examples of the reverse, what is unpredictable at the local level being predictable at a higher level (numbers of dog bites). Purposes are equally dependent on scale. A creature with a purpose can belong to a whole that has none, or vice versa. That a creature has local 'purposes' is a separate issue from whether it has a purpose. Thus a lioness hunts with a purpose – a local and instrumental purpose, lunch – which is not to say that hunting is her greater purpose, or even that she has an overall purpose at all (though I think she has one). That Y is the effect of cause X does not make Y the purpose of the whole in which the cause and effect chain operate. Copulation results in the propagation of genes, without that necessarily being the purpose of the whole in which this cause and effect chain operates: life. Turning the engine of a Nazi staff car makes the car capable of motion, but that cannot be assumed to be therefore the purpose of the whole event in which this particular cause and effect chain operates. There's far more to both than meets the myopic eye. On this, again Andrew Steane is good:

> Perhaps the purpose of a lioness is to dissipate entropy as quickly as possible (I doubt it). Perhaps the purpose of a lioness is to produce more lioness genes (I doubt it). Perhaps the purpose of a lioness is to kill, eat, and copulate (I doubt that, too). Perhaps the purpose of a lioness is to be a lioness (this seems to me to be on the right track). Perhaps the purpose of a lioness is to express whatever good a lioness is capable of expressing (I am inclined to pick this one). Perhaps the purpose of a lioness is to express an aspect of what the verb *to be* can signify.[11]

11 Steane 2018 (94).

I'd be inclined to pick the last one, ultimately. Everything that exists could be seen as an unfolding of the potential within being, and a re-enfolding of it again into a now enriched whole. At the local level there are only what look like *extrinsic* purposes. At a larger level one sees a picture that modifies the value and purpose of the detail. And at the global level one sees *intrinsic* purposes; purposive goings-out to meet the world, guided by attractors, but not to achieve utility. They are gloriously, magnificently, sovereignly useless. This is like the mother's relationship to her children – or, if she is an artist, to her creation. There are no blue-prints: there is a clear tendency or purpose towards creation in this universe, but one of a flexible and completely non-instrumental kind.

In Chapter 12, I showed that a complex system that is intrinsically unpredictable can nonetheless contain short linear chains that are very largely predictable. Kolakowski, expanding on Bergson, writes: 'The mechanistic view is applicable to some fragments which we artificially cut out from the world for practical purposes: it can be applied neither to the universe as a whole nor to the phenomenon of life.'[12]

I would suggest that whatever creative energy underwrites the unfolding of the phenomenal universe is continually active and involved in that universe; that the future is tended towards, but not closely determined; rather it is open, evolving, self-fulfilling. This means that it seems 'purposeless' to some; richly 'purposeful' to others. To me, a universe with tendencies towards beauty, complexity, and the rich unfolding of uniqueness is already teleological. It is a verb with many adverbs, not just a matter of nouns chasing nouns.

Though purpose may be more or less apparent at different levels, the process of life cannot, in reality, be broken down in this way. This is because the purposes of an organism are shared with a group of other organisms, or with the 'environment' (which is another way of saying the same thing). Creatures are not the passive playthings of necessity, but determine their environment as much as the environment determines them.[13] The idea of a niche, separate from an animal whose niche it is, is nonsensical. The animal shapes the niche to its purposes; the niche in turn shapes the animal to its own. Every part of the world is a potential niche for something. Living beings equally shape and are shaped by a world that is coming into being alongside them in a reciprocal, and reciprocally paced, process, not negotiating a path across its already pre-formed surface.[14] The environment of the rain forest has accommodated the most staggering, teeming variety of creatures of every conceivable kind, not just one 'solution to rain-forest living' that works best. All creatures and their environment together are interdependent. Interdependence means not just interaction, but mutual constitution. In keeping with this free model of evolution, both Peirce and Whitehead, as well as, among others, the physicists Paul Dirac, John Archibald Wheeler and Richard Feynman, preferred the idea of habits in Nature, that themselves evolve, to the idea of universal laws.[15]

DARWIN AND PURPOSE

Darwin is often thought to have done away with purpose. Brought up to believe not only in an engineering God but in one who was benevolent, too, Darwin was genuinely and appropriately troubled by his observation of cruelty in nature, just as he was genuinely and

---

12 Kolakowski 1985 (56).

13 See Lewontin, Rose & Kamin 1984 (274–5).

14 See Ingold 2006.

15 Smolin 2013 (xxv–xxvi).

appropriately awestruck by his observation of nature's beauty. He did not exclude that there was what he called 'design' (purpose) at a higher level, while the lower level remained unplanned. In a famous passage from a letter to Asa Gray, the most important American botanist of the nineteenth century, he wrote:

> I cannot persuade myself that a beneficent & omnipotent God would have designedly created the Ichneumonidæ with the express intention of their feeding within the living bodies of caterpillars, or that a cat should play with mice. Not believing this, I see no necessity in the belief that the eye was expressly designed. On the other hand I cannot anyhow be contented to view this wonderful universe & especially the nature of man, & to conclude that everything is the result of brute force. I am inclined to look at everything as resulting from designed laws, with the details, whether good or bad, left to the working out of what we may call chance.[16]

When Gray reviewed Darwin's legacy for *Nature*, in June 1874, he wrote: 'Let us recognise Darwin's great service to Natural Science in bringing back to it Teleology; so that, instead of having Morphology *versus* Teleology, we shall have Morphology wedded to Teleology.'[17] Darwin replied: 'What you say about teleology pleases me especially, and I do not think anyone else has ever noticed the point. I have always said that you were the man to hit the nail on the head.'[18] (Much earlier, in a letter to Jeffries Wyman, Professor of Anatomy at Harvard, Darwin had written: 'No one other person understands me so thoroughly as Asa Gray. If ever I doubt what I mean myself, I think I shall ask him!').[19]

His son, Francis Darwin, wrote that 'one of the greatest services rendered by my father to the study of Natural History is the revival of Teleology';[20] and the idea was corroborated by Thomas Huxley, 'Darwin's bulldog':

> perhaps the most remarkable service to the philosophy of Biology rendered by Mr Darwin is the reconciliation of Teleology and Morphology, and the explanation of the facts of both, which his views offer ... it is necessary to remember that there is a *wider teleology* which is not touched by the doctrine of Evolution, but is actually based upon the fundamental proposition of Evolution ... The teleological and the mechanical views of nature are not, necessarily, mutually exclusive.[21]

A wider teleology: a broader understanding of purpose. This is an important point: there may be several factors operating at different levels. It is noticeable that those who believe in purpose rarely, if ever, deny the existence or importance of chance, or of mechanism

---

16 Letter of Darwin to Asa Gray, 22 May 1860; in Darwin 1887, vol II (312).

17 Gray 1874 (81).

18 Letter of Darwin to Asa Gray, 5 June 1874; in Darwin 1887, vol III (189).

19 Letter of Darwin to Jeffries Wyman, 3 October 1860; in Darwin 1993 (405).

20 Darwin 1887, vol III (255).

21 Darwin 1887, vol II (80: emphasis added).

as an almost universal causative factor at the local level; while those who believe in a mechanistic materialism rarely allow the possibility of there being purpose at any level, or anything other than mechanism involved.[22] This may be part of the 'either/or' mentality that is attracted to mechanistic belief systems in the first place, and to their focus on detail.

The word 'evolution' – which, incidentally, appears for the first time in Darwin only in the sixth and final edition of *On the Origin of Species* – means a spinning out, an unfolding of what is latent within. As we have seen, Bergson, to whose philosophy the concept of evolution was central, points out that this can be conceived in two ways: as like the unfolding of a lady's fan, on which the painting is fixed and the process merely brings what was already there to our eyes; or as a genuinely creative act in which what is brought 'out' is hitherto not just unknown but unknowable, because the evolution actually brings it into being for the first time.

It seems that Darwin's ambivalence about purpose hinges on this difference: the first kind – limited, wholly pre-designed and finished – he cannot accept; the second – neither micro-controlled nor wholly random – he can. Writing to William Graham, the author of a book entitled *The Creed of Science*, in 1881, less than a year before he died, Darwin notes that the book was of great interest, and that, even though there were points in it with which he could not agree, 'nevertheless you have expressed my inward conviction, though far more vividly than I could have done, that the universe is not the result of chance.'[23] And a month later, he wrote to his friend T. H. Farrer these words: 'If we consider the whole universe, the mind refuses to look at it as the outcome of chance – that is, without design or purpose. The whole question seems to me insoluble.'[24] (There is a stark contrast between Darwin's intellectual humility and the attitude of his neo-Darwinian disciples: to quote, 'the more one thinks the more one feels the hopeless immensity of man's ignorance').[25]

I would love to build a bridge to Richard Dawkins and those of his mind here: would they be willing to consider a description of Nature as something that discovers what it is in the process of becoming what it is, and the point and purpose of which lies in itself: a free, exuberant creation, not a micro-controlled one? No algorithm, no programme, no robots: just an endless self-discovering act of creation?

COMPATIBILITY OF RANDOMNESS WITH PURPOSE

It is often assumed that if events were random that would exclude the possibility of purpose; but it doesn't at all. Randomness is something purpose can work with. It introduces variation into a system: the *very stuff which purpose requires in order to act*. 'Blind stochastic-

---

22 As Gilson points out, 'normally, mechanism excludes finalism [teleology], but finalism does not exclude mechanism. On the contrary, it necessarily implicates it' (Gilson 2009, 125).

23 Letter of Darwin to William Graham, 3 July 1881; in Darwin 1887, vol I (316).

24 Letter of Darwin to TH Farrer, 28 August 1881; in Darwin 1903, vol I (219).

25 ibid.

ity [randomness] is a misconceived idea as it has been developed in evolutionary biology', writes biologist Denis Noble:

> Far from proving that evolution is necessarily blind, randomness is the clay from which higher level order can be crafted ... Organisms can and do harness stochasticity in generating function ... Randomness and functionality necessarily co-exist at different levels ... If one focusses only narrowly on the bottom end, blind chance can then seem to be the sole determinant of variation even when, in fact, the variation is directed in response to environmental challenges.[26]

It's a matter of scale again. Noble's point was succinctly put by Popper: 'the selection may be from some repertoire of random events, without being random in its turn. This seems to me to offer a promising solution to one of our most vexing problems, and one by downward causation.'[27]

When describing events as locally random, we are blind to intelligibility within the larger framework, much as we forget that the slightest departure from precisely equal weight, size and density in a pair of dice will spoil their ability to produce chance outcomes: randomness here, as it seems to us paradoxically, is the product of ultra-high specification.[28] We can have randomness only because almost everything is not random. Randomness is the limit case of order (one that is strictly speaking impossible fully to achieve); not order the limit case of randomness. Producing chance outcomes is a necessary part of the freedom in Nature's purposiveness, not a negation of it. New life is achieved by *imperfections* in transcription. Too much disorder and there is no structure for purpose to express itself in: too little disorder and there is nothing to enable purpose to express itself with. There must be the possibility of adaptive change. Creative self-organisation implies both a degree of constraint and a degree of freedom.

At the level of the human individual, too, the fact that much is unconstrained is perfectly compatible with purpose. It doesn't have to be one or the other exclusively. For example, I can have the clear purpose to start a family, but when, where, with whom and by what means not only *might* be, but cannot help being, to a considerable degree, matters of chance. But without that purpose, I might never put myself in the way of its happening at all.

C. H. Waddington, the most interesting philosophical biologist of the last century, proposed the existence of what he called 'chreodes' (literally, 'useful roads'), possible paths of development which might be represented by a landscape of valleys, divided by ridges, which canalise streams.[29] If a stream flowing down a valley is diverted, it will usually return to the centre of the valley further down. A very large diversion, however, might result in the stream flowing over a ridge

26 Noble 2017.

27 Popper 1978.

28 Gilson 2009 (xx).

29 See Waddington 1957; and Gare 2017.

into a different valley. The nature and degree of the buffering is represented by the steepness and height of the ridges and the difficulty the stream would have in flowing to a different valley. The point at issue here is that, however randomly water may fall, the landscape it encounters brings about a far from random set of outcomes.

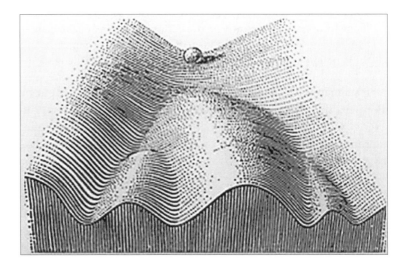

Fig. 61. *Canalised pathways of change within an epigenetic landscape*
(after Waddington 1957)

But there is something more active involved, too. A concrete example of the necessary fruitful relation between purpose and indeterminacy comes from a cell's response to threat. Cells actively promote mutations under certain circumstances, and this process begins not from DNA, but merely uses DNA as a resource. Faced with the need for a new antigen, the mutation rate in part of the genome can be accelerated by as much as 1,000,000 times. According to Noble, 'so far as we know, the mutations occur randomly. But the location in the genome is certainly not random. The functionality in this case lies precisely in the targeting of the relevant part of the genome.'[30]

Earlier I gave a number of examples of how organisms do not wait around for chance to save them from extinction, but both greatly accelerate, and appear to select, new mutations, so that they can recover something as complex as flagella within as little as a few days, rather than over the many millennia that chance mutation would require.[31] We now know that even egg and sperm do not just meet randomly, but both are actively selective.[32]

To quote Kolakowski:

> The evolutionary process teems with dead ends, failures, half-baked projects, and circuitous routes; nature proceeds somewhat gropingly, often trying several roads before it finds the right one. But it is driven constantly by an inherent tendency, and to uncover this tendency would be to understand the life of the universe.[33]

30 Noble 2017.
31 See pp 467–8 above.
32 Levitan 2017.
33 Kolakowski 1985 (56).

PURPOSE AND LIFE

What evidence might suggest that purpose is a real element in the cosmos, at any rate in the realm of life? Is purpose intrinsic in the fabric of reality, or a happy local accident from time to time, or an invention of the human mind in order to blot out the true vacancy of life?

Before addressing the question, I'd suggest that, where two assumptions fit the data, and neither is open to proof or disproof, both may be to an extent correct (as with purpose and randomness). Certainly, in the absence of proof either way, it is just irrational to rule out the intuitively more likely assumption. That is not to claim that intuition is an infallible guide: see Chapters 17 and 18. Quite how counterintuitive and, indeed, how unlikely, sheer randomness is as a model, I will turn to next. But then science itself does not in practice behave as though the universe were a random occurrence; and 'philosophy destroys its usefulness when it indulges in brilliant feats of explaining away'.[34]

UNADULTERATED CHANCE AS SOMEWHAT UNLIKELY

We have seen that mutations are not purely random. When it comes to the larger narrative of biology, I have already quoted Pauli writing to Bohr, to the effect that biologists use the word 'chance' in an irrational way.[35]

We talked about the improbability of there being a universe which contains stars and planets in Chapter 25. The existence of life, in the absence of purposiveness – in fact, in the absence of tendencies in the cosmos to overwhelmingly unlikely ends – takes improbability to a whole new level. All life, however simple, is dependent, and as a minimum, on systems which can replicate DNA and translate it into functional proteins. These systems are vastly complex. Eugene Koonin, Senior Investigator at the National Center for Biotechnology Information, and a recognised expert in the field of evolutionary and computational biology, points to a problem here:

> The origin of the translation system is, arguably, the central and the hardest problem in the study of the origin of life, and one of the hardest in all evolutionary biology. The problem has a clear catch-22 aspect: high translation fidelity hardly can be achieved without a complex, highly evolved set of RNAs and proteins but an elaborate protein machinery could not evolve without an accurate translation system.[36]

The point here, to put it in the simplest possible terms, is that, for the evolution of life forms to get started, a certain minimum level of accurate transcription must be achieved (known as the Eigen

---

34 See p 1065 above.

35 See pp 1115–6 above.

36 Wolf & Koonin 2007. The words are repeated in a book of which Koonin is sole author (2011).

threshold). Below this level of fidelity, there is too little stability. However, the achievement of this level of fidelity demands the evolution of a still more complex system to have been *already* achieved. This makes no sense according to our conventional schemata.

There is what Koonin calls 'staggering complexity inherent even in the minimally functional translation system', leading to 'the dramatic paradox of the origin of life':

> in order to attain the minimal complexity required for a biological system to get on the Darwin-Eigen spiral, a system of a far greater complexity appears to be required. How such a system could evolve, is a puzzle that defeats conventional evolutionary thinking, all of which is about biological systems moving along the spiral …[37]

According to Koonin, despite adopting a model in which he 'assumes a deliberately inflated rate of RNA production', the probability that a coupled translation-replication emerges by chance in a single observable universe is $< 10^{-1018}$ – that is, less than 1 in a number expressed as 1 followed by 1,018 zeros.[38] Impossible to imagine; but to put it in context, the entire number of subatomic particles in the observable universe is estimated to be a mere 10^{86} – by comparison, so infinitesimally small as to be practically non-existent. Thus, effectively, if chance alone is allowed to operate, the first step on the path to life is so improbable as to be an impossibility in a single universe.

But don't imagine that Koonin's argument is that therefore there is likely to be purpose, because in contemporary science that cannot be entertained on pain of excommunication. In reply to a reviewer's comments on a paper in which he puts forward his hypothesis, he states: 'There is no teleology at all involved in my approach in this paper. No teleology. It is the opposite of teleology.' After all teleology is, he says, 'a non-scientific concept'.[39] But of course. Teleology is a philosophical concept, though it is perfectly compatible with science.

The only way to avoid it is to posit an infinite number of universes: obviously, in an infinite number of universes everything happens, not because of any purpose, but by exhausting all the possibilities. An infinity of monkeys and typewriters means that every book that ever has existed or could exist gets written. What exactly, though, is the *scientific* status of the concept of an infinite number of universes, for which there is no shred of evidence? Choosing an explanation here is not a matter of science, but of philosophy.[40]

The multiverse 'solution' is a gambit whereby you can never lose an argument, but by the same token never win one either, because everything both is and is not true. It leaves reality untouched. And, by the way, if there is a universe for everything, there will also be one in which there is intrinsic purpose, so the possibility you most fear is wrapped up in your own supposition. (Unfortunately there

---

37 ibid. The Darwin-Eigen spiral, or cycle, is a series of steps allowing a succession of increases in replication accuracy followed by an increase in the size of the genome. Every time the accuracy of replication is increased, it allows an increase in the amount of information coded. This, in turn, allows for the selection of an increase in accuracy, which allows more coding, which allows selection for more information, and so on. Koonin's point is that to start this process requires something that can only be generated once the process is already under way.

38 Koonin 2007 (19: appendix 1).

39 Koonin 2007 (11).

40 Hart has this to say about fine-tuning: 'Certainly all of the cosmos's exquisitely fine calibrations and consonances and exactitudes should speak powerfully to anyone who believes in a transcendent creator, and they might even have the power to make a reflective unbeliever curious about supernatural explanations. But, in the end, such arguments also remain only probabilistic, and anyone predisposed to explain them away will find plentiful ways of doing so …': Hart 2013 (39). The best known such critic of fine-tuning is Victor Stenger, who contends that 'the claim of fine-tuning of the parameters of physics is equivalent to a claim that our languages have been fine-tuned so they have grammatical rules that are highly unlikely to have occurred naturally … No fine tuning; units are chosen for convenience and have no fundamental significance': Stenger 2013 (48 & 50). This seems to me to mistake the representation for that which is represented. The units could have been otherwise, of course; but they refer to realities that are independent of the units, just as two languages will use

[note continues on facing page] ▶

must also be one in which there is extrinsic purpose, that of a cosmic engineer with a Meccano set.) Added to which, as astrophysicist Marco Bersanelli points out, 'we end up with a picture of reality that is dangerously similar to a materialised projection of the set of logical possibilities of our human mind. This may appear even more rigidly *homo-centred* than the anthropic flavour that multiverse speculations seek to remove.'[41] Dangerously similar to a materialised 'projection of the set of logical possibilities of our human mind': precisely. In other words, the left hemisphere in action again.

In that respect it is like the idea of an engineering God. According to Paul Davies, 'the general multiverse explanation is simply naïve deism dressed up in scientific language. Both appear to be an infinite unknown, invisible and unknowable system. Both require an infinite amount of information to be discarded just to explain the (finite) universe we live in.'[42]

To be clear, there can be, in the nature of things, no certain answer to the question of purpose versus multiverse – not just now, but ever.[43] The only court before which to bring such a question, like all the most important questions in life, is that of such wisdom as we may have acquired through experience. So reader, your choice: which seems to you the more reasonable? That there is, overall, purpose involved in the way the universe works (implying a conscious cosmos – though, importantly, not entailing an engineering God or a detailed plan, ideas that I explicitly reject); or that what is acknowledged to be essentially impossible by chance is explained by positing an infinite number of universes, in which, by definition, absolutely anything and everything, however improbable, must happen, over and over again for ever? Reductionists, straining at a gnat, swallow a camel. Apart from its intrinsic improbability, and its breath-taking extravagance, the second option means not just the end of science but of all forms of meaning, since if everything happens, nothing *happens* (the root *hap* means chance – what chances to happen rather than not): nothing and everything is equally real, or unreal.[44] The first, by contrast, is not only far easier to accept, but constitutes the beginning, rather than the end, of meaning, and is perfectly compatible with science, as is obvious, if from nothing else, from the fact that very few of the great scientists on whose work we have long depended believed the universe was purposeless. 'An assertion such as "it has no goal or purpose" is a religious response, not a scientific deduction', writes Steane: 'That is, such an assertion is not one that is handed to us by the data, it is one that we generate'.[45]

Teleological beliefs are not the result of indoctrination in the dogmas of Western culture – though their rejection is. Such beliefs are present from an early age, exist in cultures widespread across the globe, are present in the educated and uneducated alike, and are no

different words to describe the same table and chair. It doesn't matter what symbols, concepts or terms are used: the universe would not have come into existence for there to *be* humans and their symbols, concepts and terms without the 'six numbers' of Martin Rees having the values they do. To take a homely example, a value that is clearly *not* one of those numbers, the freezing point of water is 'chosen for convenience' as 0°C, a 'number which has no fundamental significance'; it could just as easily have been, say, 32°, as it is Fahrenheit. But so what? If water didn't have the freezing point it does – however you measure it and whatever you call it – the seas might never have given rise to life.

---

41 Bersanelli 2011 (204: emphasis in original).

42 Lennox 2011 (53).

43 See Ellis 2011: 'All the parallel universes lie outside our horizon and remain beyond our capacity to see, now or ever, no matter how technology evolves. In fact, they are too far away to have had any influence on our universe whatsoever. That is why none of the claims made by multiverse enthusiasts can be directly substantiated.'

44 See astrophysicist Marco Bersanelli: 'Nothing really happens in a world where everything always goes on infinitely often … We need a concept of infinity in which genuine novelty can happen': Bersanelli *op cit* (205 & 214).

45 Steane 2018 (91).

less pervasive in science graduates than humanities graduates.[46] I think those who would dismiss them should explain how they just know teleology has to be mistaken before doing so.

The instinct to reject the sort of God and the sort of design that the machine model implies is surely right; yet Dawkins and his disciples continue to espouse it, just substituting the random installation of a blind algorithm for a divine plan. In doing so they have the least plausible of all worlds: a mechanistic universe, but one without a mechanic. If I believed the universe was mechanistic, I hope I'd have enough insight to be troubled by the question 'How is it that the universe is such that it *is* so highly ordered that it appears to all observers to have design, given there are no purposes?' Does that not in itself suggest a purpose at a meta-level? Does not a universe in which there are purposes imply purposiveness is latent in that universe?

AGAINST PURPOSE

Life, it would seem, delights in difference. Life is in essence the drive to differentiation, a process that happens many orders of magnitude faster and more extensively in life than anything in the realm of the inanimate. Indeed it has been suggested that we start from unrelieved uniformity: 'all particles in the world are fundamentally identical, ie, belong to the same species. Different masses, charges, spins, flavours, or colors then merely correspond to different quantum states of the same particle, just as spin-up and spin-down do.'[47] Inanimate substances – at a level beyond mere particles – can be differentiated, however. Iron is different from quartz. But a vast acceleration and simultaneous increase in the complexity of this differentiation is the hallmark of all living things: no two blades of grass are ever the same.

Does any of this, however, demonstrate purpose? Some may object that, as biologist Allen Orr puts it, 'if nature is trying to get somewhere, why does it keep changing its mind about the destination?'[48] There are several answers to this question.

First, there *is* no 'destination' to an infinite game, which does not thereby become pointless, but whose purpose lies, not in a heretofore describable end, but, precisely in the continued process itself, wherever it leads.

Second, if you wish to use the 'destination' metaphor, it must be with the proviso that the 'destination' is in principle unknowable. It is coherent to be trying to get somewhere and to change tack, even if the destination is known – but especially if it isn't foreknowable. Watch an animal purposefully exploring its environment: its purposiveness is evidenced precisely by the fact that it moves towards, and then backs off from, certain directions and turns in another. If I write a poem, I have a feel of it at the outset, but do not know the completed poem before writing it: there will be forays and retractions. Would you ask me why I didn't just write the poem straight down and be done with it? The system is open not closed: it must be *flexible*, not rigidly pre-planned. Creation is always like this, and life is one huge act of creation. Yet it would be a basic error to suppose that that meant that the process was random – that the animal had no sense of what it was about, the poet had no sense of what it was that was coming into being. Earlier I used the example of someone whose declared purpose was to settle down and have children: would you ask her why, if that's the case, she had so many false starts along the way – a way that nonetheless resulted in her achieving her aim? Humans certainly have drives, but their every act is not determined by them, or even in line with them: nor are drives the same as purposes.

Although a drive implies a purpose, a purpose is not identical to a drive, because a single purpose is compatible with a number of sometimes conflicting drives. Thus a purpose that leads to complex and beautiful life may include a drive to compete and a drive to co-operate: indeed it not simply tolerates, but positively requires, them both. And given the free nature of the process, some outcomes we deprecate as evidence of 'bad design' – eg, instances of ruthless competition – are merely evidence that the process is working itself out freely, as it must.

Any path of free creation not only will, but must, err. Blindness at the everyday level is entirely compatible with a purpose overall. But to deny all goals, values or drives in the wider cosmos leaves one able to do no more, when asked why the universe tends to order, complexity and beauty, than shrug one's shoulders and say that they just emerged at random. And that would in itself be no small miracle, suggesting a very special universe.

I have mentioned that life is no respecter of the law of parsimony. One reason often put forward for doubting purpose in the universe is its size relative to the earth: we seem so insignificant. Physicist Robert Park rejects the possibility of purpose thus:

> If the universe was designed for life, it must be said that it is a shockingly inefficient design. There are vast reaches of the universe in which life as we know it is clearly impossible: gravitational forces would be crushing, or radiation levels are too high for complex molecules to exist, or temperatures would make the formation of stable chemical bonds impossible … Fine-tuned for life? It would make more sense to ask why God designed a universe so inhospitable to life.[49]

49 Park 2009 (11).

However, both the scale and the rate of expansion of the universe are *exactly* what would be expected if life were to have any chance of existing in it anywhere at all.⁵⁰ A smaller more compact universe, or one in which the cosmological constant (which governs the rate of expansion) was only very slightly different from the one that exists, would render life impossible.⁵¹ 'Only a universe as big as ours could have lasted the fifteen billion years that are needed to make men and women', to quote theoretical physicist John Polkinghorne. 'It's a process that can't be hurried.'⁵² The existence of the rest of the universe is simply an affront to our sense of power – the value of the left hemisphere, that of instrumentality. We have no control over the vastness of the universe, so it seems alien and our impotence in the face of it triggers a feeling of insignificance. But the size of the universe is entirely irrelevant. Whether things in our world have value or not is a question *wholly* independent of the size of the universe. I recall a remark of comedian Peter Cook's: 'As I looked out into the night sky, across all those infinite stars, it made me realise how unimportant ... they are.'

WARNING: PURPOSE MAY DAMAGE YOUR HEALTH

Instinctual behaviours are characteristically purposeful. I touched on instinctual behaviours in an earlier chapter on embodied and intuitive knowledge. They are deeply embedded drives that do not have to be learnt, and are not consciously pursued. Yet they are supremely purposeful: how to build a nest, how to fly, where to migrate, how to weave a cobweb, the steps of a mating ritual – how to do a thousand things of vital importance to survival. Once again, I say, this has nothing to do with an engineering God. Some of the best examples of purposiveness in nature also happen to be among the best arguments against such a God: if an engineer, then hardly benevolent. Purpose may not further what look to us like benign projects. Let's take a look.

Resistance to chloroquine in the malaria parasite is hardly desirable from the human perspective, and is hard to bring about: it depends on the co-occurrence of simultaneous mutations the chances of which are vanishingly small.⁵³ Even when allowing for compensatory mechanisms, 'genetic mutation rates for complex organisms such as humans are dramatically lower than the frequency of change for a host of traits, from adjustments in metabolism to resistance to disease', according to Michael Skinner, a professor of biological science at Washington State University.⁵⁴ There are tendencies acquired in an organism's own lifetime, and embodied in epigenetic changes, that can markedly accelerate processes in the organism's favour, in a way inconceivable by random genetic mutations alone.

Nature makes no judgment on the 'virtue' of her creations. Some of the most ingenious improvisations in Nature are occasioned by

---

50 See, eg, Piran, Jimenez, Cuesta *et al* 2016.

51 In theory, according to cosmologist Alan Heavens (2016), Director of the Imperial Centre for Inference and Cosmology at Imperial College London, either the constant should be hundreds of orders of magnitude higher than it appears to be, or it should be zero, in which case the universe wouldn't accelerate. But this would disagree with what astronomers have observed: 'The small – but nonzero – size of the cosmological constant is a real puzzle in cosmology'.

52 Polkinghorne 2008.

53 Summers, Dave, Dolstra *et al* 2014.

54 Skinner 2016.

parasites and viruses. Even the humble 'winter flu' virus has the effect of making people more sociable, increasing the chances of spread.[55] (Fortunately it doesn't seem to have the same effect on the host's wish to visit the office.) A type of carpenter ant (*Camponotus leonardi*), which is extraordinarily sophisticated – it forages and communicates food sources to comrades, carves wooden galleries to live in, and indulges in farming (the ants corral and protect aphids in order to get a sweet substance called honeydew from them, which they achieve by stroking the aphids with their antennae) – nonetheless meets its nemesis in the form of a lowly fungus, the so-called zombie ant fungus (*Ophiocordyceps unilateralis*), which exists in rainforests in Asia, Africa and South America.[56] While the ant forages on the forest floor, it is infected with a spore, which takes several days to develop inside the ant's body. The fungus then takes over the nervous system of the ant and makes it behave entirely against its nature. The ant climbs, always to about the same height, about 25 centimetres up a tree, to a site with exactly the right amount of humidity for the fungus to grow. The fungus then makes its host do something contrary to its nature, to bite onto vegetation, before killing it: the ant clamps onto the underside of a leaf with its mandibles, hanging above its colony, and dies. Within 24 hours, threads of fungus burst out of the corpse. Finally, a stalk thrusts up out of the ant and starts to shower spores onto the rainforest floor, where they, in turn, infect more ants.

The so-called kamikaze horsehair worm (*Paragordius tricuspidatus*) has a similarly ingenious life story. First, the tiny larva is eaten by the larva of another insect, such as a mosquito or mayfly: once this emerges, a cricket will gobble it up. The horsehair worm then develops inside the cricket. However, its final stage of development needs to take place in water. Since crickets don't swim, and don't tend to live near water, the worm has a problem. It solves this by hijacking the cricket's nervous system, impelling it to jump – against its nature – into water. Once the cricket drowns, the horsehair worm is free to emerge and reproduce.[57] Now it's got a mechanism, it works well: but how did it survive long enough, as a species, to develop the necessary ability to compel a cricket towards water by sheer chance, given that it needs water to reproduce from the outset? It would fall at the first fence.

And probably the best known of all is the case of toxoplasmosis (a condition caused by the parasite *Toxoplasma gondii*, which can infect humans). This single-celled parasite mainly infects rats and mice, but needs to be eaten by a cat so that it can reproduce. Infected rats and mice bizarrely lose their fear of cats and will even seek them out, making the cat's job easy and the parasite's work done.[58] It can then, as an *encore*, get into the human central nervous system

[55] Reiber, Shattuck, Fiore *et al* 2010.

[56] Andersen, Gerritsma, Yusah *et al* 2009.

[57] Thomas, Ulitsky, Augier *et al* 2003.

[58] Vyas, Kim, Giacomini *et al* 2007.

where its presence is believed to cause behavioural and personality changes, and has been implicated in the genesis of schizophrenia.[59] Who said purpose was all good, clean fun?

Cancerous tumour cells can suppress apoptosis, a mechanism that ensures the death of excessive cells where appropriate, 'rewire' metabolism, so as to promote the proliferation of blood supply to the tumour, and evade immune responses, in such a way as to promote tumour growth at the expense of the organism.[60]

I bring up these examples because I do not want to be thought of as basing the idea of purpose on a simple idea of an avuncular engineer, which, like Darwin, I cannot accept. But, like Darwin, I am staggered by the ingenious improvisations everywhere in evolution, whether to human sensibilities they be elevating or disturbing. Darwin's writing is full of rapturous appreciation of Nature: in *On the Origin of Species*, he writes 'we see beautiful adaptations everywhere and in every part of the organic world'.[61] Speaking of the *orchis pyramidalis*, for example, and the adaptation of its parts, Darwin confesses 'I never saw anything so beautiful … the beauty of the adaptation of parts seems to me unparalleled.'[62]

LIFE TENDS TO COMPLEXITY: WHY?

Aristotle 'thought he had come across the truth of nature from the moment when he had perceived its beauty', writes Gilson,

> Not so much aesthetic beauty, such as that of light and colours or forms; but first of all and above all the intelligible beauty, which consists in the apperception by the mind of the order which rules the structure of the forms and presides over their relations.[63]

This is remarkably like Darwin's reaction to beauty, order, and fittingness in nature – and they are there too in what we call the inanimate universe. But why? Saying that things are such that they 'just do' naturally behave, as physicists and mathematicians constantly attest, in an orderly and elegantly satisfying fashion, begs the question *why*, unless order, beauty and complexity are natural tendencies. You might object that in fact the general tendency is for order to dissipate; but just as the fact of death doesn't relieve you of the necessity to account for life, the existence of decay doesn't relieve you of the necessity to account for the existence of order – nor to account for the tendency, not just of life, but of all complex open systems to buck the trend by unceasingly creating vastly complex order, and in the case of life speeding up the process immeasurably through evolution.

When we look back in cosmic history, says Smolin, 'we see a universe evolving from less to more structured, from equilibrium to complexity':

---

59 Fuller Torrey & Yolken 2003; Yolken, Dickerson & Fuller Torrey 2009.

60 See, eg, for a review, Fouad & Aanei 2017.

61 Darwin 1859 (61).

62 Letter of Darwin to JD Hooker, 12 July 1860; in Darwin 1887, vol III (263); letter of Darwin to JD Hooker, '1861'; *ibid* (265).

63 Gilson 2009 (24).

64 Smolin 2013 (221–2).

65 Watanabe 1969 (70 & 75): 'In quantum mechanics, a physical state changes according to two completely different kinds of processes. First, the state changes gradually with the course of time and following a certain differential equation termed a wave equation. This process is causal and reversible. Second, the state changes abruptly the moment the observer makes a new observation of the system. This process is statistical and irreversible. Quite naturally, the renewal of the investigation changes the knowledge of the state, and hence the state itself. It can be demonstrated that entropy is not in the least changed by the first process. (This is intimately bound up with the fact that the process of change is reversible.) On the other hand, entropy changes rapidly in the second process, and always so as to increase. This means that the act of observation itself increases the entropy: a shocking situation in the eyes of classical physicists since an objective physical measure is now closely dependent on the intervention of the observing subject. The principle of entropy is no longer a simple physical law, but must be considered as the law of the development of knowledge. In the new mechanics, entropy measures, so to speak,

[note continues on facing page] ▶

This is certainly not the picture a naïve application of the second law of thermodynamics would suggest. The second law says that isolated systems increase their randomness, becoming more disordered and less complex and structured as time moves forward. This is the opposite of what we see happening in the history of our universe, in which complexity increases as structures form on many scales, with the most intricate structures being the most recent. Evolving complexity means time. There has never been a static complex system.⁶⁴

As he goes on to explain, this does not abnegate the second law of thermodynamics, which applies to closed systems near to equilibrium, and even if a (non-closed) system is increasing in complexity at one point in space it is decreasing elsewhere: 'as long as the increase in entropy caused by heating a dust grain somewhere in space is greater than the decrease of entropy caused by forming a molecular bond, the long-term outcome is in agreement with the second law.' The great theoretical physicist Satosi Watanabe suggested that entropy is not indeed a general law of the universe, and that 'to speak of the entropy of the entire universe is to risk passing beyond the boundaries of science ... the increase of entropy applies only to what is observed, and consequently not at all to the observer ... The observing self is thus always excluded from the field to which the law of entropy applies.'⁶⁵ Since the observing self is a functional part of every known system in science, that is not an insignificant exception.

Claude Bernard, who believed that determinism is 'the only possible scientific philosophy', nonetheless wrote that 'the materialist doctrine is just as imprecise, in that general agents of a physical nature capable of causing the appearance of vital phenomena in isolation do not explain their orderliness, harmony and integration.'⁶⁶ Must there not, then, be a tendency towards such outcomes?

Just as values are foundational in the cosmos, in my view, so is purpose. Leibniz wrote in a letter to de Volder, professor of philosophy and mathematics at Leiden, 'it may be said that there is nothing in the world except simple substances, and, in them, perception and appetite'.⁶⁷ Nothing 'just is' where it is – in this universe everything *tends* somewhere: appetite means literally (Latin, *ad- + petere*) a seeking, in other words, a tendency.

I discussed how conflicted biology is about teleology in Chapter 12: the mistress the biologist cannot live without, but with whom he is unwilling to be seen in public. There I was concerned not with the purposefulness of life, but the purposes of individual creatures. Contemporary physics, too, has found teleology hard to ignore because of a change in the way we conceive probability. Probability, in a Newtonian universe, described our incomplete state of knowledge about something that does not now exist, though at some point in

the degree of precision of our knowledge ... the principle of entropy applies only to the object observed by the observer. This involves two important consequences. In the first place, to speak of the entropy of the entire universe is to risk passing beyond the boundaries of science ... Secondly, the increase of entropy applies only to what is observed, and consequently not at all to the observer ... The observing self is thus always excluded from the field to which the law of entropy applies'.

66 Bernard 1885 (344): « *La doctrine matérialiste est tout aussi inexacte, en ce que les agents généraux de la nature physique capables de faire apparaître les phénomènes vitaux isolément n'en expliquent pas l'ordonnance, le consensus et l'enchaînement.* »

67 Leibniz, letter to Burchard de Volder, 30 June 1704.

the future it might do. However, according to Heisenberg, 'the probability wave ... meant more than that':

> it meant a *tendency* for something. It was a quantitative version of the old concept of 'potentia' in Aristotelian philosophy. It introduced something standing in the middle between the idea of an event and the actual event, a strange kind of physical reality *just in the middle between possibility and reality*.[68]

Potentia is something present and taking part in being. In the Newtonian world, probability is inert and concerns something non-existent; *potentia*, by contrast, is something that implies movement, a 'tendency towards' whatever may be, and is something that already exists, now, in a world of fields and forces that already tend to shape the future. In Newtonian probability, the uncertainty lies in our knowledge; in quantum probability the uncertainty lies in the world we aim to know, and is an objective aspect of it. 'The objective probabilities of quantum mechanics exist even though there is nothing to be ignorant of', writes David Mermin:

> They express correlations in the absence of correlata. To avoid such linguistic traps it would be better to speak not of 'probabilities' but of 'propensities' or 'dispositions', or to eschew all talk of probability in favour of talk about correlation.[69]

LIFE AND THE PURPOSES OF THISNESS

Returning to biology, I suggest that living beings are neither purposeless nor random, but have, as well as goals of their own, intrinsic purpose. It is not that 'Nature has a purpose' of which individual creatures are the passive victims: Nature's purposiveness includes and is predicated on the freedom of her creatures, each of which expresses its purpose, which ahead of time she does not know, through her. Every creature, in Hopkins' phrase, cries 'What I do is me: for that I came.' Note, '*for that*': the purpose is right there, in the very thisness, quiddity, *hæcceitas*, of each unique being. We are grateful for the existence of things *in themselves*, for no other reason. Life is a dance – a celebration; and it has the nature of a gift. It is neither a mechanism to be exploited nor a problem to be solved.

One might query whether claiming that, eg, 'a lion has a purpose' says any more than that we value the lion. That in itself is no mean thing, since, as I have said, it looks like the purpose of the emergence of life from consciousness is to enable the recognition of value. But I'd answer that the lion has that purpose *whether we value it or not*, indeed whether beings capable of valuing it existed or not. I think that one of the many things that human awareness brings to the party (granted all the disadvantages of the species) is that it helps

---

68 Heisenberg 1958 (42: emphasis added).

69 Mermin 1998.

fulfil that intrinsic purpose – of the production of untold richness of difference – by its greater capacity to appreciate it. (The greater the shame if we fail to do so.) 'Born to blush unseen', the flower still has a purpose, or purposes: being; continuing-in-being; and continuing-in-being-by-reproduction. Once *seen*, however, its purpose is more richly fulfilled – its resonances are wider. Incidentally, as Leon Kass says, in relation to continuing in being, 'very few people have noticed that this [Darwinian] non-teleological explanation of change not only assumes but even depends upon the immanent teleological character of all living organisms. The desire or tendency of living things to stay alive and their endeavour to reproduce, both of which are among the minimal conditions of Darwinian theory, are taken for granted and unexplained.'[70] Why make so much effort, embrace so many sacrifices?

A point worth pondering. After all, by far the best strategy for *persisting in being* is to avoid being alive at all. How is it that life ever took off? As Whitehead observes,

> life itself is comparatively deficient in survival value. The art of persistence is to be dead. Only inorganic things persist for great lengths of time ... The problem set by the doctrine of evolution is to explain how complex organisms with such deficient survival power ever evolved. They certainly did not appear because they were better at that game than the rocks around them. It may be possible to explain "the origin of *species*" by the doctrine of the struggle for existence among such organisms. But certainly this struggle throws no light whatever upon the emergence of such a general type of complex organism, with faint survival power.[71]

And once they arrived, why did organisms further evolve towards creatures with vastly lower survival prospects? A tree can live a thousand years. A human life is on average 70–80 years, and until recently considerably less. Individual examples of some actinobacteria, however, are thought to be over a million years old, and still going strong.[72] In the survival stakes we lose hands down to a monocellular organism.

As I have often had occasion to remark, creativity depends on a degree of resistance. It seems to need to overcome adversity. 'The life energy has to overcome the obstacles erected not only by matter but by life forms as well', writes Kolakowski:

> the results usually turn against the principle of creativity itself, as in human efforts of expression: once produced, each form sticks to its identity and resists further changes. But the original force never sleeps. The species which focus their efforts on building defensive shelters and shells usually close the road to progress, whereas those

70 Kass 1988 (261).

71 Whitehead 1929b (2–3; emphasis in original).

72 Johnson, Hebsgaard, Christensen *et al* 2007. Nor are they dormant, but in a metabolically active state, relying on DNA repair.

that take greater risks and buy flexibility of movement at the price of weakened armour, prove to be winners in terms of evolution.⁷³

As Taleb points out, being robust is weaker than being antifragile. While Occam's razor must be observed in pursuing human explanations, there is no *lex parsimoniæ* – no principle of parsimony – governing the natural order. It is an extravagantly creative, inventive, superabundant, self-delighting process. As von Bertalanffy observed:

> If life, after disturbance from outside, had simply returned to the so-called homeostatic equilibrium, it would never have progressed beyond the amoeba which, after all, is the best adapted creature in the world – it has survived billions of years from the primaeval ocean to the present day. Michelangelo, implementing the precepts of psychology, should have followed his father's request and gone into the wool trade, thus sparing himself lifelong anguish although leaving the Sistine Chapel unadorned.⁷⁴

Whitehead, speaking of the determination with which humans set about modifying their environment, wrote: 'the explanation of this active attack on the environment is a three-fold urge: (i) to live, (ii) to live well, (iii) to live better. In fact the art of life is *first* to be alive, *secondly* to be alive in a satisfactory way, and *thirdly* to acquire an increase in satisfaction.'⁷⁵

All that we have considered here seems to me to speak of mind, not just in the sense of awareness, but of ingenuity and purpose. As Kass says, 'do we really understand what we are claiming when we accept the view that a mindless universe gave rise to mind?'⁷⁶

PURPOSE IN MIND AND TIME

One aspect of mind in the cosmos would be the ability to 'see' beyond the immediate, not just to register this moment, but to have some ability to situate it in a trajectory over time. There are elements in evolution which appear to require some such oversight of the general path, though there is no tendency to interfere in the detail: elements which seem to take greater account of the broader view in time and space than the highly focal vision of mechanism would allow. If Shapiro is right that all life has cognition, this is not strange: it is to be expected.

As Bruce Charlton argues, there may be many reasons for evolution 'preferring' something *once it exists*, without it being at all obvious how that something would have come into existence in the first place in order to be subsequently preferred. His point is similar to Whitehead's. An everyday example of it is the existence of sexual reproduction. In the short-to-medium term, as he points out, needing to find a suitable member of the opposite sex with whom to repro-

---

73 Kolakowski 1985 (59).

74 von Bertalanffy 2003 (192).

75 Whitehead 1929b (5: emphasis in original).

76 Kass 2004 (287).

duce greatly reduces the chances of reproductive success, and at the same time halves the number of potential reproductive units. Once established, it works – but at the outset there would be grave competitive disadvantages without their being offset by the advantages that come only further down the line. How this happened remains, as he puts it, 'utterly unclear.'[77] In evolution there are many instances with a similar structure: that is to say, an initial development in a process that will one day reach a desirable goal, is either neutral, or actually negative, and therefore unlikely to be selected for.

Similarly there are advantages in altruism from a point of view of the group, once established, but very few for the first altruists:

> In a multi-cellular organism, the dividing component cells are constantly being naturally selected for neoplastic (eg cancerous) change – such that they cease to cooperate with and contribute to the organism, and instead exploit it as a 'host' environment.[78] How, then, did multicellular organisms evolve the many integrative systems (eg nervous, paracrine, hormonal and immune systems) designed to impose cooperation of specialized cells and suppress non-functional and actively parasitic (eg mutated) cell variants; bearing in mind that all such integrative systems are themselves intrinsically subject to neoplastic evolution (as well as loss of function from cumulative damage)?[79]

Why, Charlton asks, does development of multicellular organisms not 'utterly collapse into a chaos of ever smaller and faster replicating, more mutually exploiting purposeless entities, as we observe it does not?' The net effect of natural selection, as he points out, is to break down the major transitions of evolution before they can be established, unless the tendency is overcome by a purposive ('and indeed cognitive'), integrative and complexity-increasing tendency, which is able to avail itself of a longer-term view: an intentional arc, to use Merleau-Ponty's phrase.

Charlton deduces that there must be a mechanism of teleological intent, either in the form of a hitherto unknown source of information built into the cell's own structure, or in the form of a field that acts on the whole form, much as Sheldrake has proposed. He gives a couple of familiar examples:

> the mechanisms of cell-suicide or apoptosis – such that if a cell experiences a mutation that may endanger the organism … the cell destroys itself (for the good of the whole organism) …
> Some types of motile white blood cells such as macrophages (which resemble free living amoebae) will kill themselves in the process of defending the organism against microorganism invasion … and this purpose is apparently built into them in terms of their core functionality.[80]

77 Charlton 2016.
78 Charlton 1996.
79 Charlton 2016.
80 ibid.

He concludes that there is, in general, considerable altruism built in at the cellular level of a multicellular organism. This seems undeniable. I would incline towards there being something that operates not just at the level of the single cell but at the level of the whole organism (and of course these ideas are not mutually exclusive: both may be required). I showed in Chapter 12 a number of examples of situations in which the whole organism seems to become aware of a problem in a perhaps isolated part of it, and responds not only as a whole (rather than just locally), but sometimes with an entirely novel strategy for which it could never have been 'programmed'.[81]

The conventionally proposed solutions to problems of altruism, 'such as inclusive fitness/kin selection and various types of reciprocal benefit'[82] answer a different question, namely, why altruism, *once established*, is advantageous and worth sustaining, not how it came into being in the first place. To prevent 'the amplification of selfish, short-termist, parasitic variants and lineages (which are immediately advantageous, and much more strongly selected for)', observes Charlton, requires some countervailing tendency that favours long-term cohesion, and the survival and reproduction of the group. This sounds to me like purposefulness.

Such a tendency is a sign that things are better thought of as being attracted towards certain goals, rather than pushed blindly forwards by a mechanism from behind. Blind processes can't 'see' ahead to let a future aim guide a present development. James makes the point thus:

> Romeo wants Juliet as the filings want the magnet; and if no obstacles intervene he moves towards her by as straight a line as they. But Romeo and Juliet, if a wall be built between them, do not remain idiotically pressing their faces against its opposite sides like the magnet and the filings with the card. Romeo soon finds a circuitous way, by scaling the wall or otherwise, of touching Juliet's lips directly. With the filings the path is fixed; whether it reaches the end depends on accidents. With the lover it is the end which is fixed, the path may be modified indefinitely.[83]

I am reminded, and not for the first time, of Dupré's analogy of asking a friend to buy a loaf of bread: specifying the goal is vastly more likely to produce the desired result than specifying the individual actions to be taken to get to the shop. One ought to take into consideration how often certain common outcomes are arrived at through different processes of exploration. Evolution is full of examples whereby a similar end is repeatedly arrived at by different means, sometimes even in related, or the same, organisms.[84] Simon Conway Morris demonstrates that evolution converges: the same patterns of outcome keep attracting evolution. This doesn't look like aimlessness or lack of purpose.

---

81 See especially Barbara McClintock, on p 463 above.

82 See, eg, Ridley 1996.

83 James 1890, vol 1 (7).

84 See p 455 above.

Consciousness is, in its essence, always disposed *towards something*; its nature is to be a reaching out. Similarly, always reaching out beyond itself, it favours, or tends towards, its own greater complexity. Consciousness can not only purpose the creation of utility, an end that is now in sight, but seems able to purpose something which, more like a work of art, can be known only once it is achieved. Here again a longer trajectory of sight or insight than the immediate, local cause and effect feedback mechanism seems to be required. I come back to Bergson, as interpreted by Kolakowski:

> Life, then, is a continuous process in which the original drive divides itself into a growing variety of forms, but retains a basic direction. It has no goal, Bergson says, in the sense that human actions have goals, in other words, no one can anticipate its future course, which is more similar to an artistic creation than to the operation of a machine.[85]

Teleology in nature, then, is not just a matter of purposing solutions to local problems, but a tendency towards greater complexity and higher levels of awareness. And this requires a sense of the trajectory of time. At the simplest level, Prigogine comments:

> Figuratively speaking, matter at equilibrium, with no arrow of time, is 'blind', but with the arrow of time, it begins to 'see'. Without this new coherence due to irreversible, nonequilibrium processes, life on earth would be impossible to envision ...[86]

Is denial of purpose coherent? At the beginning of Part II, I quoted Bacon: 'The universe is not to be narrowed down to the limits of the understanding, which has been men's practice up to now, but the understanding must be stretched and enlarged to take in the image of the universe as it is discovered.' We have been very successful in applying a methodology, that of ever further analysis to mechanisms, that strips the universe of any coherence at a higher level. As Whitehead observed:

> The brilliant success of this method is admitted. But you cannot limit a problem by reason of a method of attack. The problem is to understand the operations of an animal body. There is clear evidence that certain operations of certain animal bodies depend upon the foresight of an end and the purpose to attain it. It is no solution of the problem to ignore this evidence because other operations have been explained in terms of physical and chemical laws. The existence of a problem is not even acknowledged. It is vehemently denied.[87]

And he continues: 'Scientists animated by the purpose of proving that they are purposeless constitute an interesting subject for study.' Thomas Nagel picks up the implications of proving oneself the mere product of randomness: 'Evolutionary naturalism implies that we

85 Kolakowski 1985 (58).
86 Prigogine 1997 (3).
87 Whitehead 1929b (12).

shouldn't take any of our convictions seriously, including the scientific world picture on which evolutionary naturalism itself depends.'[88]

I fear that an over-simplified vision is, as usual, the problem here. In the view of Nobel Prize-winning physicist Charles Townes,

> People are misusing the term intelligent design to think that everything is frozen by that one act of creation and that there's no evolution, no changes. It's totally illogical in my view. Intelligent design, as one sees it from a scientific point of view, seems to be quite real ... design could include evolution perfectly well. It's very clear that there is evolution, and it's important. Evolution is here, and intelligent design is here, and they're both consistent ...[89]

It was Descartes' ambition, through science, to 'render ourselves, as it were, the masters and possessors of nature'.[90] I believe that a significant motivation for the resistance to the idea of purpose is the implied affront to our sense that there must be no theoretical limits to the power we may aspire to have over nature: and as the philosopher Étienne Gilson comments, there is no place for final causality in such a scheme.[91] Only we are allowed to have purposes; the rest must submit passively to ours.

CONCLUSION

In this chapter I suggested that the concept of teleology is quite compatible with science, and that denying it leads to improbable claims, to a loss of substance, as Ernst Mayr suggested, and a failure to ask the right questions about *how* purpose works as a force in Nature. I also suggest that there has been confusion about what is meant by purpose, and a fear that it implies an engineering God, which it does not. This brings me to a turning point in the course of the book.

All that I have covered so far in Part III suggests that the cosmos is likely to be differently constituted from the way we have come to believe. That leaves unaddressed an issue of a quite different order: the mystery *that* the cosmos should exist at all, and exist in the way I have accounted for it up to the end of this chapter. This is an order of questions that it is beyond science's reach to answer, although philosophical or theological approaches to these questions can and should be informed by whatever science can tell us that might prove relevant; and whatever each of reason, intuition and imagination at any time can tell us. It is in that spirit, and to this different order of questions, that I turn in the next chapter.

---

[88] Nagel 2012 (28).
[89] Townes 2005.
[90] Descartes 1994 (87).
[91] Gilson 2009 (23).

Chapter 28 · The sense of the sacred

> As a human being one has been endowed with just enough intelligence to be able to see clearly how utterly inadequate that intelligence is when confronted with what exists. If such humility could be conveyed to everybody, the world of human activities would be more appealing.
> —ALBERT EINSTEIN[1]

> The first gulp from the beaker of knowledge estranges us from God, but at the bottom of the glass God is waiting for him who seeks.
> —CARL FRIEDRICH VON WEIZSÄCKER[2]

> By love He can be caught and held, but by thinking never.
> —THE AUTHOR OF *THE CLOUD OF UNKNOWING*[3]

> For every thing that lives is holy, life delights in life …
> —WILLIAM BLAKE[4]

THE GROUND OF BEING

For me, and for many philosophers historically, the deepest question in all philosophy – both the most important, and the hardest to answer – is why there should be something rather than nothing.

And close on its heels comes the question why that 'something' turns out to be complex and orderly, beautiful and creative, capable of life, feeling and consciousness, rather than merely chaotic, sterile, and dead. It is not a matter of opinion, but a fact if ever there was one, that, somehow or other, this 'something' has within it the capacity to give rise to Bach's *St Matthew Passion*. Any attempt at understanding the cosmos needs to take that stark fact into account. There is a parallel between the false view that we are separate from and over against Nature (encapsulated in the disastrous idea of Nature as the 'environment') and the idea that we are separate from and over against the cosmos. This cannot be true, for the same reason in either case. We were born out of, and return to, the one and the other. It therefore makes no sense to set us up as proud, lonely, tragic figures, struggling against Nature, trying to subdue her, or struggling defiantly to bring love, goodness and beauty into a hostile cosmos. Any love, goodness and beauty we can bring come out of Nature and out of the cosmos in the first place: where else can they possibly come from?[5]

Moreover how is it that, as Deutsch pointed out, the universe is intelligible? Not only that, I might add, but how is intelligence possible unless intelligence is in some sense 'part of the stuff of the

---

[1] Letter of Einstein to Queen Elizabeth of Belgium, 19 September 1932; in Dukas & Hoffmann 1979 (48).

[2] » *Nach einem alten Satz trennt uns der erste Schluck aus dem Becher der Erkenntnis von Gott, aber auf dem Grunde des Bechers wartet Gott auf den, der ihn sucht.* « Usually, and probably wrongly, attributed to Heisenberg: it comes from von Weizsäcker 1948 (124: very generally cited as 152, despite the book having only 138 pages).

[3] Anon, *The Cloud of Unknowing* (fourteenth century) 1961 (60).

[4] Blake, 'America: a Prophecy', Plate 8.

[5] Cf Mary Midgley 2006 (45: emphasis in original): 'This cosmos is, after all, the one that has produced us and has given us everything that we have. In what sense, then, is it *hostile*? Why this drama?' It is itself a myth generated by the Cartesian view of mind over against the world, the view that Schelling's philosophy transcended.

universe'? I am not speaking of information-processing, but of *understanding*, which is what intelligence literally means: a phenomenon to which we are so close that we hardly perceive its uniqueness. Taking it for granted is a rather big assumption: don't let its familiarity lead to underestimating its magnitude. If we are invited to assume (as we are) that the cosmos consists of mindless matter, the problem arises of how such an *un*intelligent cosmos is constructed so that intelligence might arise *de novo* in it, engage in the business of comprehension, and find it comprehensible. If that doesn't trouble you, you are missing something: intelligence looks disquietingly like an intelligent idea rather than a property of mindless matter. I believe this was what Einstein meant by his famous words: 'The eternal mystery of the world is its comprehensibility ... The fact that it is comprehensible is a miracle.'[6] However, let's set these questions aside for now, and focus on the more fundamental question: why there should be anything at all, rather than nothing.

The question is frequently misunderstood as 'what was the first element in the chain of causation that led to the existence of the universe?' But that is an entirely different question (and there may be no such first element at all, since for all we know the universe may have no beginning in time). It is not a question of a temporal cause in a sequence, one lying on the same plane as the sequence itself, but of an ontological cause, underlying and sustaining any such sequence. In other words, not 'what was it that set some process in motion at a point in time?', but, rather, 'how does it come about that there is a process, or motion, or a point in time, at all – now or ever?' The answer to this question is of an altogether different order, and must lie on a plane different from, and deeper than, everything else. The question cannot be answered in terms of a physical entity or process, because that already presupposes what we are questioning – why there are physical entities and processes. The proper object of this question is that which underwrites, timelessly and eternally, whatever is: in other words, the ground of Being. This is, I believe, what Wittgenstein was referring to when he wrote that 'the mystical is not *how* the world is, but *that* it is.'[7] Similar reflections have been made by philosophers at all times and in all places since the very beginnings of philosophy.

Wittgenstein, among many others, saw a mystery here; others only something so self-evident as to warrant no further discussion. Wittgenstein had foreseen that, too. 'God grant the philosopher insight into what lies in front of everyone's eyes', he wrote.[8] A deep mystery may be clothed as a commonplace that, for that very reason, we can scarcely see at all. And what eagerly presents itself to us as obviously the case may be a deception. In the months before the outbreak of the First World War, Evelyn Underhill observed that:

---

6 Einstein 1936a (351) and 1936b (315): » *Das ewig Unbegreifliche an der Welt ist ihre Begreiflichkeit ... die Welt unserer Sinneserlebnissen [ist] begreifbar, und dass sie es ist, ist ein Wunder.* «

7 Wittgenstein 1961, §6.44.

8 Wittgenstein 1984 (63e) & 1998 (72e).

the surface-self, left for so long in undisputed possession of the conscious field, has grown strong and cemented itself like a limpet to the rock of the obvious ... building up from a selection amongst the more concrete elements offered it by the rich stream of life, a defensive shell of 'fixed ideas'. It is useless to speak kindly to the limpet. You must detach it by main force. That old comfortable clinging life, protected by its hard shell from the living waters of the sea, must now come to an end.[9]

9 Underhill 1914 (60–1).

This surface-self, the possessor of the conscious field, attached to its fixed ideas, which it sees as resoundingly obvious, will be recognisable to the reader of this book as an expression of the left hemisphere's *modus operandi*. For the surface-self, the limpet on the rock of the obvious, there is no mystery about Being: it is simply self-evident. But we must disturb its complacency.

First, however, I should enter an important caveat, without which my purpose may be misunderstood. In what follows I have of course no final answers to any of the big questions. But I believe we must not, under any circumstances, cease to be mindful of these questions, even while we know there can be no definite answers. Having ready answers means you don't understand; understanding here means never letting go of the questions. Unknowing will turn out to be a sign not of weakness, but of wisdom.

To many people speaking of a ground of Being is entirely pointless: after all, what grounds the ground of Being? And so on, *ad infinitum*. But that is to misunderstand. In speaking of the ground of Being, we do not provide an answer, but draw attention to a problem. The point is not to make a question go away – 'well, that's that sorted, then!' – but rather to place it centre stage, and allow the light, in time, to dawn.

I am certainly not attempting to argue for the existence of God: no argument for, or against, the existence of God can possibly succeed. Moreover, as you will see, I believe the concept of God to be fraught with difficulties. Whether it can be rehabilitated in some form is a question I will come to, but I suggest that for now the reader put it out of mind, with all its unhelpful baggage, which has a tendency to get between us and seeing something important. In particular in what follows I am not trying to revamp some version of a cosmological or ontological 'argument' for the existence of God. I am merely indicating that, whatever we choose to call it, there is almost certainly more here than we have words for, or can expect ever to understand using reason alone. Such an expectation would itself be irrational. The proper response to this realisation is not argument, but awe. To be human, in my view, is to feel a deep gravitational pull towards something ineffable, that, if we can just for once get beyond words and reasons, is a matter of experience, and to which we reach out, silently, though not without misgivings; something outside our conceptual grasp, but nonetheless present to us through intimations that come to us from a whole range of unfathomable experiences we call 'spiritual'. This has been true of humanity the world over and throughout time, and is true now as much as ever; no advance in science can have anything to say about it one way or the other. To think that it could is to misunderstand science as much as spirituality.

Almost a defining characteristic of the left hemisphere is that it has no sense of the limits of its own understanding: it doesn't know what it is it doesn't know. It operates inside a framework, within which all questions are referred back, and all answers form part of a reassuringly familiar schema; if they don't, they are simply pronounced nonsense. But it doesn't see the bounds of its own world view; in order to do that, it would have to see there is something beyond the bounds – and that is something it cannot do. The right hemisphere, on the other hand, is turned outwards, attentive to whatever comes to it, without attempting to make it conform to the familiar, or to the uses of everyday language. It alone therefore 'purges from our inward sight the film of familiarity which obscures from us the wonder of our being'.[10] It has the capacity to express what it perceives, but this can be done only indirectly, through metaphor and myth. To quote Whitehead: 'The difficulty of philosophy is the expression of what is self-evident ... Our understanding outruns the ordinary usage of words. Philosophy is akin to poetry.'[11] Understanding poetry depends heavily on the right hemisphere.

Nothing is less true than that we understand something only when we express it in language. On the contrary, language at times places a barrier between us and understanding, substitutes a crabbed expression for a living reality, and pretends to 'explain' it – away. Thought often far outstrips language: it has been a recurrent theme in this book, and I explored much of the evidence in *The Master and his Emissary*.[12] The philosopher Bryan Magee again hits the nail on the head:

> How does one say the *Mona Lisa*, or Leonardo's *Last Supper*? The assumption that everything of significance that can be experienced, or known, or communicated, is capable of being uttered in words would be too preposterous to merit a moment's entertainment were it not for the fact that it has underlain so much philosophy in the twentieth century ...

And he continues in words I have already quoted, 'direct experience which is never adequately communicable in words is the only knowledge we ever fully have'.[13]

Things that can be understood only by direct experience can be

---

10 Shelley 1921.
11 Whitehead 1938 (68).
12 See McGilchrist 2009b (ch 3).
13 Magee 1998a (98).

spoken of only indirectly: and conversely what is talked about directly is usually experienced only indirectly, because in the process of articulation it has inevitably become a re-presentation – something *other* than what we experience. Nietzsche thought that distortion was not just a limitation of language, but of its essence. And it is true that the more important something is, the harder it is to grasp in language. 'Most events are unsayable, occur in a space that no word has ever penetrated, and most unsayable of all are works of art', wrote Rilke, 'mysterious existences whose life endures alongside ours, which passes away'.[14]

Not just language, but the thinking to which it is allied, has strict limits. Deep intuitions can flourish only when there is enough space granted by *not* knowing, in the recognition that conventional 'wisdom' does not apply. What we take to be ob-vious may prove an ob-stacle; it may ob-trude on, ob-fuscate, ob-struct, ob-scure, and ultimately ob-literate the truth. In fact the meaning of the word 'obvious' (Latin, *ob-* against, + *via* way) is that which stands 'in the way' – in both the good sense, that it is what we first encounter on the highway of our cognition, and the bad sense, that it impedes our path. We must always be alert to precisely what our customary way of thinking leaves out. This is not just true of poetry: 'If the study of science teaches one anything', writes Polkinghorne, 'it is not to take everyday thought as the measure of all that is.'[15]

Language is a tool that was evolved for everyday use. In philosophical thinking of all kinds, according to Whitehead, we wrestle with 'the difficulty of making language express anything beyond the familiarities of daily life … the struggle of novel thought with the obtuseness of language': one of the problems of philosophy, according to him, was the 'uncritical trust in the adequacy of language', for 'in philosophy linguistic discussion is a tool' – it is a useful servant – 'but should never be a master'.[16]

Speaking of the ground of Being, the Zen monk Shunryū Suzuki writes: 'The true source, *ri*, is beyond our thinking: it is pure and stainless. When you describe it, you put a limitation on it. That is, you stain the truth or put a mark on it.'[17] In the *Analects* of Confucius it is written: 'The Master said, does Heaven speak?'[18] Famously Lao Tzu tells us that 'the *tao* that can be named is not the eternal *tao*'.[19] In the Eastern tradition, then, there are many such statements of the impossibility of capturing the source of all things in language. What is interesting is that they are also plentiful, as I shall demonstrate, in the Western sacred – though not secular – tradition. That very opposition, sacred versus secular, however, is in itself typically Western. It is a sign of a significant rupture that never existed in the same degree in other cultures, and should never have existed in our own. It is the result of the progressive sequestration of

14 Rilke 2011 (21).

15 Polkinghorne 2008.

16 Whitehead 1933 (153 & 293).

17 Suzuki 1999 (52).

18 Confucius, *Analects*, XVII, §19.

19 Lao Tzu, *Tao Te Ching*, §1.

the sacred, whatever is by definition of the highest possible value, from the centre, to its present ghetto on the margins, of Western mental life.

DENYING THE GROUND OF BEING

Since what cannot be expressed, and must always remain implicit, is antithetical to the left hemisphere's way of being, the immediate reaction from those approaching the problem from a purely analytical perspective is to deny the topic of the ground of Being altogether: the question or questions are said to be non-questions, or to make no sense. To me, while such a conclusion may well follow from the narrow premisses of such an approach – a quick trip round the hall of mirrors confirms whatever we already believe – it does not measure up in any way to the experience of the simple fact of Being, if one allows oneself truly to face it, and does not succeed in sidestepping it by a convenient intellectual reflex.[20] Nor is it in line with my reading of the ancient wisdom literature of East or West, which I am not able to dismiss just because a perhaps fashionable school of contemporary philosophy does not agree with it.

Attempts to dispense with the problem follow a number of approaches. First, it might be said that, although everything we know is sustained in its being by the existence of other existences, in the case of the universe, there is an exception – ultimately there must just be an uncaused cause, or an ungrounded ground. But this is to affirm a unique ground of Being, not to negate it. The philosopher Jakob Boehme actually refers to the fount of all being as the 'unground' (*Ungrund*) – a ground to be understood in a sense like no other. In whatever way we recast it, we are faced with an exception, and an exception that cannot be rationalised so as to be safely packed away again within our familiar categories.

Another is to rely on brute 'emergence'. Being simply – somehow, it is claimed – emerged out of Nothing: nothing here to explain. But by this strategy the existence of a miracle is not denied, but confirmed: one miracle is simply recast as another. This type of argument is familiar to us from our discussions of consciousness, and it is no more successful here than it was there. Just as the emergence of consciousness out of matter wholly devoid of consciousness is a miracle, brute emergence of Being from non-Being is either the miracle of all miracles or a straight impossibility. Does it happen at a stroke? A miracle. Or does it truly *emerge*? An impossibility: for it falls prey to the 'Midshipman Easy problem', of which it is surely the most extravagant case.[21] Being is not something that can happen gradually, beginning with a *very little bit* of Being that *slightly* is. The mystery remains.

It has also been suggested that there is no mystery because there

20 Waismann: see p 556 above.

21 See p 1045 above.

is nothing here to explain, since the sum of all the electrical charges, angular momenta and energy quanta in the cosmos is zero: thus, it is claimed, absolutely nothing has unfolded into absolutely nothing. But as Rupert Shortt points out, this is like arguing that the Taj Mahal doesn't amount to anything, because it contains the same volume of stone as was quarried from the ground to make it.[22] Others have suggested that Being might have emerged from a quantum vacuum; but a quantum vacuum is not Nothing. It, manifestly, has being: it is a physical entity and obeys the laws of physics. We saw that the physicist David Tong describes it as 'the simplest thing we can imagine in the entire universe and it is astonishingly complicated.'[23]

Yet another argument that there is nothing here to explain asserts that the laws of physics alone are enough to account for things coming into existence. There are numerous problems with this line of reasoning. For a start, physical theories claiming to show how the universe could spontaneously arise from nothingness invariably turn out, like the quantum vacuum case, not to be that at all. They involve, as David Bentley Hart points out, 'transition from one physical state to another, one manner of existence to another, but certainly not the spontaneous arising of existence from nonexistence.' Such an event is impossible within physics.[24] Existence is 'most definitely not a natural phenomenon', he continues; 'it is logically prior to any physical cause whatsoever; and anyone who imagines that it is susceptible of natural explanation simply has no grasp of what the question of existence really is.'[25] To repeat, my point is not that I can explain existence: I can't. I merely question the claim that there is nothing here to explain.

Further, the identification of so-called laws does not answer anything. What grounds the laws, if there is absolutely Nothing? We are concerned with that which is prior to and underlies all systems, underwrites all so-called laws, and everything whatsoever. And even if there is something – namely, laws – how could these laws themselves unfold the infinitely complex universe? Laws cannot cause anything to happen. They are merely a description of an observed regularity in phenomena. What causes the observed regularity remains unspecified and is unaltered by being labelled a law. What's more, laws have to operate on something. Even if some handwaving were to 'grant' (ie pass over) initial conditions and a set of laws to operate on them, there are two questions which the Newtonian paradigm will never be able to answer: 'why these laws?' and 'why these initial conditions?'[26] Thus the universe may well operate according to the principles of natural selection, as has been claimed; but it cannot be *explained* by a process such as natural selection, since natural selection operates according to principles ('laws') that natural selection cannot itself explain.[27] So Wittgenstein declared: 'At the basis of the

22 Shortt 2019b (37–8).

23 Tong 2016.

24 Berry 2000 (29): 'If at last "all tangible phenomena" are empirically reduced to the laws of physics, then we will merely have completed a circle. We will have arrived again at the question that preceded Genesis: where did the physical come from? And physics of course can have no answer.'

25 Hart 2013 (18).

26 On this, see Smolin 2013 (97–8).

27 *ibid* (xxvii): 'Nor do the laws of nature wait, mute, outside of time for the universe to begin. Rather the laws of nature emerge from inside the universe and evolve in time with the universe they describe.'

whole modern view of the world lies the illusion that the so-called laws of nature are the explanations of natural phenomena.'²⁸

TALKING ABOUT THE GROUND OF BEING

Being, then, is mysterious. The problem is that if we are to say anything about it, we still need some sort of placeholder, within language, for all those aspects of Being that defy direct expression, but which we sense are greater than the reality which language is apt to describe, almost certainly greater than whatever the human mind can comprehend. If we don't have such a placeholder they will disappear from our awareness; yet *what* that placeholder signifies must not, above all at first, be tied down too tightly – if indeed it ever can be. In this it is rather like learning a language from experience only, without a grammar book or dictionary: in such a process what a word means must be initially left open, and narrowed only with deeper and repeated acquaintance. A drive for precision at the outset becomes the enemy of understanding. 'The pretension, under such conditions, to be rigorously "scientific" or "exact" in our terms', writes James, 'would only stamp us as lacking in understanding of our task.'²⁹

What we need, in fact, is a word unlike any other, not defined in terms of anything else: a sort of *un*-word. This is no doubt why in every great tradition of thought – and perhaps beyond that, in every language of every people – there *is* such an un-word. It holds the place for a power that underwrites the existence of everything – the ground of Being; but, as I shall suggest, it holds a place for more than that, otherwise some such phrase as 'ground of Being' would itself be enough. To Heraclitus it was the *logos*; to Lao Tzu the *tao*; to Confucius *li*; in Hinduism *Brahman*, and to the Vedic tradition *ṛta*; in Zen *ri*; to Arabic peoples, since pre-Islamic times, *Allah*; to the Hebrews YHWH.³⁰ And in the Western tradition it is known as God.

But what a host of troubles lies in that name ... There is a very good reason why in most traditions there is a prohibition on use of the un-word. It is inevitably a mendacious re-presentation of ultimate truth – that is to say, an idol. And the word God is obfuscated and overlaid with so many unhelpful accretions in the West that it is not surprising that people recoil from this idol. It's not just that, obviously, God is not some old man sitting on a cloud, but that very much else that is often believed, or at any rate assumed by atheists to be believed by theists, badly gets in the way of an understanding.

Here is the dilemma, and why I speak of an un-word: if we have no word, something at the core of existence disappears from our shared world of awareness; yet if we have a word, we will come to imagine that we have *grasped* the nature of the divine, pinned it down and delimited it, even though by the very nature of the divine this is something that can never be achieved. So we find, as we might expect, that

---

28 Wittgenstein 1961, §6.371. And he continues: 'the view of the ancients is clearer in so far as they have a clear and acknowledged terminus, while the modern system tries to make it look as if everything were explained.' §6.372.

29 James 1902 (39). Cf Cusanus 1985, I, 2, §8: 'someone who desires to grasp the meaning must elevate his intellect above the import of the words rather than insisting upon the proper significations of words which cannot be properly adapted to such great intellectual mysteries'.

30 Heraclitus' *logos* is not to be confused with the very different *logos* of post-Platonic philosophy. See discussion at pp 627 & 973–4 above.

there are parallels in the Western tradition to Lao Tzu's admonition that the *tao* is unnameable and inapprehensible. In Judaism the name of God is the Unnameable Name. 'God is something that cannot be and, at the same time, has to be spoken of', wrote Kolakowski. 'No logical devices will make this tension vanish.'[31]

And what is true of language is a symbol of what is true in thought. 'There never was nor will be a man who has certain knowledge about the gods and about all the things I speak of', wrote Xenophanes. 'Even if he should chance to say the complete truth, yet he himself knows not that it is so.'[32] In Buddhism, Hinduism and each of the monotheistic religions, Islam, Judaism and Christianity, there is an ancient and powerful apophatic tradition, or *via negativa*, which holds that all positive assertions about God will be false: we must approach God by clearing away untruth, as we reveal the statue by discarding stone from the primal block. So St Augustine warns that 'if you understand, it is not God you understand'.[33] That does not mean, however, that the attempt is pointless, or that we are absolved from making the attempt. As the polymath Nicholas of Cusa, whose greatest work is entitled *On Learned Ignorance*, put it: 'the deeper we know our unknowing, the nearer we are to truth'.[34]

Odd as it may seem, and though it is impossible to avoid the form of words 'God's existence', it is even arguable, without denying God, that God does not exist. The twelfth-century Jewish philosopher Maimonides, for example, says that God 'exists, but not through existence.'[35] If you are frustrated by this deeply paradoxical form of words, I can understand; but do not succumb to the temptation to dismiss it. His point is an important one: that to predicate existence of God is to mistake the divine nature. God cannot be said to 'exist', a word which in its origins means to 'stand forth' (Latin, *ex-* 'out', + *-sistere*, reduplicative of *stare*, 'to stand') in the way that a *thing* stands forth for us against the ground of our already existing field of vision. God *is* that ground. God is above all not a thing alongside other things – even one equipped with ultra-special powers. God simply *is* – in a use of the verb that requires that we understand God both to *have* Being and to *be the ground* of Being at one and the same time. In similar manner, the Origin (*Ursprung*) referred to by the philosopher Jean Gebser is the source from which all springs, but it is also that which itself springs forth.[36] And again Suzuki says, reflecting on the *Sandokai*, once more using the image of water:

> In the darkness the branching streams flow everywhere, like water. Even when you are not aware of water, there is water. Water is inside our physical body and in plants too; there is water all over. In the same way the pure source is everywhere. Each being is itself pure source, and pure source is nothing but each being. They are not two things ... The stream is pure source, and pure source is the stream.

31 Kolakowski 1985 (33).

32 Xenophanes fr 34 (Diels' notation, trans Burnet).

33 Augustine, '*Si enim comprehendis, non est Deus*': Sermon 117.

34 Cusanus 1985: '*Et quanto in hac ignorantia profundius docti fuerimus, tanto magis ipsam accedimus veritatem*' (*De docta ignorantia* I, 3).

35 Maimonides 1919, Bk I, ch lvii.

36 Mahood 1996.

The pure source is flowing all over, even though you don't know it. This 'don't know' is what we call 'dark', and it is very important.[37]

And in the Sufi tradition, the poet Rumi writes: 'Truly it is the water; that which pours; and the one who drinks – all three become one when your talisman is shattered. That oneness you can't know by reasoning.'[38]

In connexion with this, the theologian and philosopher Paul Tillich also makes what at first sight is a puzzling observation about the way we think, which emphasises that we misunderstand all aspects of being that are not thing-like when we try to render them as things. 'Nothing, perhaps, is more symptomatic of the loss of the dimension of depth', he says,

> than the permanent discussion about the existence or non-existence of God – a discussion in which both sides are equally wrong, because the discussion itself is wrong and possible only after the loss of the dimension of depth. When in this way man has deprived himself of the dimension of depth and the symbols expressing it, he then becomes a part of the horizontal plane. He loses his self and becomes a thing among things.[39]

The point I believe Tillich is making is that we cannot rightly speak of God's existence (or non-existence) for several, loosely associated, reasons. One is that to speak in such a way is to imagine God as a thing in the world alongside other things, such as bus passes and mountain bikes, just much, much grander. As a result, the field of possibilities has become two-dimensional, which Tillich refers to as the loss of the dimension of depth (he also implies, in my view rightly, that, in this loss, our lives, too, become lacking in depth). Another is that it makes Being ontologically prior to God, as though existence was an *attribute* of God: but Being cannot 'validate' God, since it is God, if there is a God, that 'validates' Being. As the late fifth-century mystic Dionysius the Areopagite says, God is 'the cause of being for all, but is itself nonbeing, for it is beyond all being.'[40] But the most fundamental reason that 'the discussion itself is wrong' is that it suggests that invoking God is an *explanation* on the same level as, and as a potential alternative to, invoking the laws of thermodynamics. John Gray refers to

> the 'God debate' – a tedious re-run of a Victorian squabble between science and religion ... the idea that religion consists of a bunch of discredited theories is itself a discredited theory – a relic of the nineteenth-century philosophy of Positivism.[41]

Science helps inform philosophy and philosophy helps ground science. The quest of science must tend always beyond what cur-

---

37 Suzuki 1999 (52–3).

38 Rumi, *Mathnawi* II: 717–8; from 1999 (111). This formulation parallels the Christian doctrine of the Trinity.

39 Tillich 1958.

40 Dionysius the Areopagite (also known as Pseudo-Dionysius), *The Divine Names*, ch 1, §1: 1920 (53).

41 Gray 2018 (9).

rent science sees – and not only within science. It has a role in leading on to metaphysics. Aristotle implied that, in addition to each science studying its own peculiar subject matter, the sciences 'have a further function as leading to a goal outside themselves, namely the discovery of what they logically *presuppose*.'[42] (I have addressed the fallacy that science makes no presuppositions in an earlier chapter.) Aristotle called this foundation *sophia*, wisdom, or *theologikē*, discourse about God. He thought that the subject of metaphysics was the ground of Being, and that therefore metaphysics was the 'first philosophy', not in the sense that it was what came first in a sequential process, but that it lay deepest – probing the very nature of the cosmos – and was therefore in fact the *last*, temporally speaking, in the process. All other knowledge led to this end: 'the person who studies it', wrote the philosopher R. G. Collingwood, 'will be doing what in all his previous work he was preparing himself to do.'[43] Science is not just a technique, but, rightly conceived, the groundwork for illuminating something lying beyond itself. This is why philosophy needs science, and science needs philosophy. 'The First and Last Science is therefore the science of that which stands as ultimate logical ground to everything that is studied by any other science', continues Collingwood. 'The ordinary name for that which is the logical ground of everything else is God.'[44]

[42] Collingwood 1940 (5–6: emphasis added).
[43] *ibid* (9).
[44] *ibid* (10).
[45] Wittgenstein 1981 (74e): entry for 8 July 1916.
[46] McCabe 2010 (128).

WHY 'GOD'?

I can easily understand someone saying 'I can see why we need to go beyond what science can tell us. Very few people now believe that science can answer all our questions. And you have already suggested that metaphysics and science when properly understood sustain one another in different ways. But why bring God into it? God is no more an explanation of how things came about than the ground of Being. Where, after all, does God come from? We are still as ignorant. Theology gets us no further than philosophy.' I am sympathetic, particularly because the weight of history attached to the word makes it hazardous to invoke the word God.

It is perfectly true that invoking God does not *explain* anything. But, importantly, that is not its purpose. The recognition of God is not an answer to a question: it is to fully understand the question itself. In this spirit Wittgenstein wrote: 'To believe in a God means to understand the question about the meaning of life.' And he continued: 'To believe in a God means to see that the facts of the world are not the end of the matter. To believe in God means to see that life has a meaning.'[45] The point of invoking God is to ensure that we do not lose sight of the deepest of life's enigmas. 'When we speak of God', writes theologian Herbert McCabe, 'we do not clear up a puzzle; we draw attention to a mystery.'[46] When the word disappears from our vocabulary, we don't abolish that mystery; we just cease to recognise that it is there. We no longer know what it is we do not know. There is nothing shameful in not knowing: the human mind is inevitably characterised by its ignorance more than by its limited understanding. But the deeper ignorance is when we choose to put out of mind what it is we do not know, and pretend to know what we never can.

Yet the word God cannot easily be substituted *merely* by Being. That is immediately to substitute something else that has its own, more limited, meaning; whatever word we use, it must have no other referent whatsoever. To quote Plotinus: it refers to 'a nobler principle than any thing we know as Being; fuller and greater; above reason, mind and feeling; *conferring* these powers, not to be confounded with them.'[47] Moreover we have lost the sense of how mysterious Being in itself is. To some it is impossible to get beyond the view that 'is' is just a verbal copula. It is hard for us to awaken ourselves again to the essential strangeness of existence – to recover, as Heidegger thought we must, the 'radical astonishment' of being.[48] If such an idea means nothing to you, that could, indeed, be a rebuff to it; but it might be the best demonstration of the truth of Heidegger's assertion that our problem is that we have lost sight of, forgotten – even forsaken – Being.[49]

Being being a mystery, it is therefore better expressed as an object of awe than as an object of knowledge. Numerous philosophers, among them Plato and Aristotle, as well as Goethe and Heidegger, have been moved to observe that philosophy begins in wonder. It also often ends in wonder, as Goethe reflected, awe being a better gauge of one's commerce with truth than certainty.[50] Similarly, Whitehead thought that, while 'philosophy begins in wonder ... at the end, when philosophic thought has done its best, the wonder remains.'[51] According to Josef Pieper, wonder is not just what initiates, but what sustains philosophy. He calls wonder 'the abiding, ever-intrinsic origin of philosophizing. It is not true to say that the philosopher ... ever "emerges from his wonder" – if he *does* depart from his state of wonder, he has ceased philosophizing.'[52] Knowing does not dispel awe, though ignorance can banish it.

That awe and wonder are the end as well as the beginning of philosophy is one reason why God may be a better name than just 'the ground of Being' for this creative mystery. A phrase like 'the ground of Being', too, may have its conventional, cultural baggage – in this case, its presumed dullness. It could serve only as long as we see Being as having already something unfathomable about it – somewhat of the nature of God. But that is precisely what modern Western culture does not entertain.

So, providing we remain appropriately sceptical about language,

---

47 Plotinus, *Enneads*, Bk v, Tractate IV, §14 (trans S MacKenna & BS Page).

48 The phrase in this form is George Steiner's: 1978b (27).

49 Thus he spoke of *Seinsvergessenheit* and *Seinsverlassenheit*.

50 Eckermann 1970 (296): entry for 18 February 1829.

51 Whitehead 1938 (232).

52 Pieper 1998 (106).

we not only *can* use a term other than ground of Being, but, it seems to me, we *must*. Metaphysical argument can take us some of the way, but it deals only with the what, not the how. Even the rather abstract question 'why should there be anything at all?' is not, after all, just an intellectual puzzle. It is a fundamental question – *the* fundamental question – for human beings; and we miss the point if we suppose that it is a matter for abstract reasoning alone.

In a wonderful passage Schelling writes about how we should prepare ourselves for an understanding of any subject:

> First and foremost, any explanation should do justice to what is to be explained, not devalue it, explain it 'away', diminish it or mutilate it, simply so as to make it easier to grasp. The question is not 'what view must we adopt so as to explain the appearances in a way that accords neatly with some philosophy?', but precisely the opposite: 'what philosophy do we need if we are to measure up to our object, and be on a par with it?' It is not how the phenomenon must be turned, twisted, skewed or stunted, if need be, so as to be explicable according to principles which we have already resolved never to go beyond. The question is 'in what way must *we* broaden our thinking so as to get a hold on the phenomenon?'

And he goes on:

> But he who refuses, for whatever reason, to broaden his thinking in this way should at least be honest enough to count the phenomenon amongst those things (which, when all's said and done, are for all of us plenty enough) that he does not understand; rather than drag it down and degrade it to the level of his own conceptions; and, if he is incapable of raising himself up to the level of the phenomenon, at least to stop short of holding forth about it in wholly inadequate terms.[53]

I believe that, in the necessary process of achieving a fit between our understanding and what there is to be known, our present materialist culture has disregarded Schelling's advice, and contracted the scope of what we allow to exist to our limited understanding, rather than enlarging our understanding to meet the scope of what exists. We are like someone who, having found a magnifying glass a revelation in dealing with pond life, insists on using it to gaze at the stars – and then solemnly declares that if only people in the past had had such a wonderful magnifying glass to look through, they'd have known that, on closer inspection, stars don't actually exist at all.

I want to emphasise that there is a distinction between something beyond our means of grasp and something beyond our means of knowing. To follow the *via negativa* is an intellectual process in which we recognise the inadequacy of our conceptions of God,

53 Schelling, 'Seventh Lecture': 1856–61b (137): » *Bei jeder Erklärung ist das Erste, daß sie dem zu Erklärenden Gerechtigkeit widerfahren lasse, es nicht herabdrücke, herabdeute, verkleinere oder verstümmle, damit es leichter zu begreifen sey. Hier fragt sich nicht, welche Ansicht muß von der Erscheinung gewonnen werden, damit sie irgend einer Philosophie gemäß sich bequem erklären lasse, sondern umgekehrt, welche Philosophie wird gefordert, um dem Gegenstand gewachsen, auf gleicher Höhe mit ihm zu seyn. Nicht, wie muß das Phänomen gewendet, gedreht, vereinseitigt oder verkümmert werden, um aus Grundsätzen, die wir uns einmal vorgesetzt nicht zu überschreiten, noch allenfalls erklärbar zu seyn, sondern: wohin müssen unsere Gedanken sich erweitern, um mit dem Phänomen in Verhältniß zu stehen. Wer aber aus was immer für einer Ursache vor einer solchen Gedanken-Erweiterung Scheu trüge, der sollte, anstatt die Erscheinung zu seinen Begriffen herabzuziehen und zu verflachen, wenigstens so aufrichtig seyn, sie in die Zahl der Dinge zu setzen, deren es für jeden Menschen noch immer sehr viele gibt, in die Zahl der Dinge, die er nicht begreift; und wenn er unfähig ist, sich selbst zu dem den Erscheinungen Gemäßen zu erheben, sollte er wenigstens sich hüten, das ihnen völlig Unangemessene auszusprechen* « (emphasis in original).

but it is not at all the same as just giving up hope of knowing God. Though we may approach a subject of knowledge apophatically, not by asserting what is the case but by seeing what is *not* the case, there will nonetheless be intimations, affirmative signs within the field of our unknowing, or we would not even be able to see that there is something there beyond our current knowing to know. The knowing is a process that knows no end.

It is argued that our knowledge of science, too, has an apophatic structure – and it does: approximating truth by stripping away falsehoods.[54] But in relation to the divine, unlike spacetime, there is also a realm of spiritual gnosis that does not apply to physics, since God is far more accessible to heart and soul than to intellect. Plotinus puts it, I think, exceptionally well:

> We do not, it is true, grasp it by knowledge, but that does not mean that we are utterly void of it ... Those divinely possessed and inspired have at least the knowledge that they hold some greater thing within them though they cannot tell what it is; from the movements that stir them and the utterances that come from them they perceive the power, not themselves, that moves them ...[55]

Take those 'placeholder' terms – *logos, li, tao, ṛta*, and so on. The place they hold is not nearly filled by the mere idea of a ground of Being. They suggest much more: a response to the second question with which this chapter began – why does Being take the creative, complex, orderly, beautiful, intelligible – *vital* – form that it does? And, though arising in different cultures, what they suggest is remarkably consonant. They suggest a co-ordinating principle in the universe which is evidenced in order, harmony and fittingness; a principle that is not only true, but the ultimate source of truth. This principle applies to all 'levels' of existence and therefore wraps within itself the human soul. Speaking of *ṛta*, for instance, Raimon Panikkar writes that it can be seen as the order behind the manifest world, the harmony among all aspects of manifestation, 'each of which obeys its own level'. *ṛta* is in the nature of things: 'Man being an aspect and expression of this order has within him a reflection thereof'.[56]

The point is that we are engaged with it through the whole experience of being alive: through our way of meeting the world not just in the intellect but in the heart and mind. It engages us in the form of a question that is so much more than the kind inviting a merely analytic reply: the question 'what does it mean to be alive?'

The consequence of speaking about God rather than merely about the ground of Being is not only that it keeps in our field of attention the ineffable mystery of existence, as McCabe points out – that to which Heidegger exhorts us to respond with radical astonishment; it also alerts us to the inadequacy of a response couched primarily in

---

54 Eg, Heller 2011 (222: emphasis in original): 'We collect information from inside a given spacetime (by following the behaviour of geodesics in it) to learn something about the way its structure breaks down. The *apophatic* character of our knowledge is mitigated by tracing vestiges of what we do not know in the domain open to our investigation'.

55 Plotinus *op cit*.

56 Panikkar 2001 (350–1).

terms of propositions parried back and forth in the cut and thrust of argument. What the term 'God' requires of us is not a set of propositions about what cannot be known but a disposition towards what must be recognised as beyond human comprehension. The primary response, therefore, is not intellectual. It is awe and wonder – not mere curiosity, which motivates us to find out more information, more knowledge (valuable as that is), but wonder at the immensity of what we must recognise we can never know. Yet that very wonder is what increasingly we lack.

'As civilization advances, the sense of wonder declines', wrote Abraham Heschel:

> Such decline is an alarming symptom of our state of mind. Mankind will not perish for want of information; but only for want of appreciation. The beginning of our happiness lies in the understanding that life without wonder is not worth living. What we lack is not a will to believe but a will to wonder. Awareness of the divine begins with wonder. It is the result of what man does with his higher incomprehension. The greatest hindrance to such awareness is our adjustment to conventional notions, to mental clichés. Wonder or radical amazement, the state of maladjustment to words and notions, is therefore a prerequisite for an authentic awareness of that which is.[57]

[57] Heschel 1955 (46).

A DISPOSITION TOWARDS THE DIVINE: SUSPENDING THE LEFT HEMISPHERE

Heschel's terms will ring bells for readers who have followed the hemisphere hypothesis that underlies this book. 'Conventional notions' and 'mental clichés' are exactly what are produced when the left hemisphere takes on the role of master and believes in the sufficiency of its re-presentations of reality, when it traps us in a hall of mirrors of its own creation, and tells us that's all there is. Equally, 'higher incomprehension' is a grand way of speaking of the right hemisphere's openness to what it doesn't already know, which it can't (in any conventional way) make sense of, but whose reality it intuits and does not deny. It will come as no surprise, then, that a disposition towards God is largely dependent on the right hemisphere, the hemisphere that we already know brings us closer to truth than the left. (I explore the evidence for significant right hemisphere dependency in Appendix 7.)

But, before we look further at this disposition, it's worth noting where we've got to and how we have got here. We've reached the point, in this discussion, where the left hemisphere can do no more for us – at least, not for the time being. We're faced with the unavoidable fact of the existence of the ground of Being (whatever we may choose to call it) and of the fact that we cannot treat it as we treat manifestations of Being itself: that's to say, manifestations of which the left hemisphere can make re-presentations so that we may control them. Going beyond mere acknowledgment of the ground of Being's existence to see what is 'there', so to speak, is beyond the left hemisphere's capacity.

But here's the thing: note that it is the left hemisphere and its reasoning processes that have brought us to this point, the point at which its own limitations become evident. Put another way, the left hemisphere as servant has provided a vital role: drawing attention to the limits of its service. You have to be able to think clearly in order to see that there are limits on thinking. Here we may recall Pascal: 'the ultimate achievement of reason is to recognize that there is an infinity of things which surpass it. It is indeed feeble if it can't get as far as understanding that.'[58] And although the left hemisphere cannot do more for us now, I shall argue that its service will be needed again once the right hemisphere has done its work of gleaning 'an authentic awareness of that which is' by means of a necessary 'mal-adjustment to words and notions' (in Heschel's marvellous phrase). Thereafter the helpful left hemisphere will be useful again but it will need to keep to its place: as servant it will be invaluable, but as master it will be disastrous.

As far as humanity is concerned, the summit of knowing is knowing that you do not know. Such was the expressed conclusion of, to name only a few, the Buddha, Confucius, Socrates, St Paul, Montaigne and the Indian saint, Swami Ramdas. Wisdom is so far from being knowledge in the usual sense, of knowing many 'things', that one of the only pieces of advice offered by wisdom traditions is to value not-knowing.

Not-knowing, however, is not the same as ignorance: it is what is left before us when ignorance is left behind. There is ignorance prior to knowledge, and there is not-knowing, once knowledge has been outlived. The essence is encapsulated in a remark attributed to 'a Master' by the poet Charles Wright: 'for knowledge, add; for wisdom, take away'.[59]

In an earlier chapter I alluded to the form of the spiral, as opposed to the circle: arriving where you started from, in one sense, and yet knowing it for the first time. As there is an un-knowing the other side of knowing, which is far superior to both knowing and ignorance, there is an innocence the other side of experience, which is superior both to experience and naivety; and a wisdom the other side of folly, which is superior both to folly and common sense. This insight lies behind the well-known Zen saying, sometimes attributed to Dōgen: 'Before I sought enlightenment, the mountains were mountains and the rivers were rivers. While I sought enlightenment, the mountains were not mountains and the rivers were not rivers.

---

58 Pascal 1976, §267 (Lafuma §188): « La dernière démarche de la raison est de reconnaître qu'il y a une infinité des choses qui la surpassent. Elle n'est que faible si elle ne va jusqu'à connaître cela ».

59 Wright 1995 (59); quoted in Zwicky 2014b, §87(b). Cf Eckhart, 'Sermon on the Fourth Sunday after Trinity', 1958 (194): 'God is not found in the soul by adding anything, but by a process of subtraction'.

After I attained enlightenment, the mountains were mountains and the rivers were rivers.'

More down to earth, perhaps, but no less worth listening to, is Thoreau. 'It is only when we forget all our learning that we begin to know', he writes:

> I do not get nearer by a hair's breadth to any natural object so long as I presume that I have an introduction to it from some learned man. To conceive of it with a total apprehension I must for the thousandth time approach it as something totally strange. If you would make acquaintance with the ferns you must forget your botany. You must get rid of what is commonly called *knowledge* of them. Not a single scientific term or distinction is the least to the purpose, for you would fain perceive something, and you must approach the object totally unprejudiced. You must be aware that *no thing* is what you have taken it to be. In what book is this world and its beauty described? Who has plotted the steps toward the discovery of beauty? You have got to be in a different state from common. Your greatest success will be simply to perceive that such things are, and you will have no communication to make to the Royal Society.[60]

This allowing something to 'presence' is surprisingly hard for us, since as Evelyn Underhill put it, for us the 'hare of reality is always ready-jugged'. It involves eschewing knowledge *about*, in order to have knowledge *of*, something; the relinquishing of serial processes that claim to help us understand, in order that we can see the thing at once; the abjuring of the need to narrow down or fix one's object.

In the *Diamond Sutra*, it is said that knowing is seeing, but seeing is not knowing.[61] By this I understand that true knowing, *understanding*, is not a matter of accumulating facts, but a form of perception in which one at last sees into the depth of things as it were 'at once', and recognises them for what they are, no longer overlaid by our projections – something like the process described by Thoreau. At last we *see* them. At the same time seeing things in the normal sense, of resting our eyes on their surface, is not to know them at all. This distinction also lies behind Josef Pieper's remark that in the modern age 'man's ability to *see* is in decline.'[62]

The idea of unknowing may sound – well, negative. But it is the secret to the greatest creative power. 'To know truth, one must get rid of knowledge', said Lao Tzu: 'nothing is more powerful and creative than emptiness, from which men shrink.' This emptiness goes well beyond our normal understanding, because there is an emptiness, a 'nothing', a living 'nothing' very different from what we generally now think of as nothing (which is dead), a *no thing on the other side of something*. I have mentioned the Buddhist concept of emptiness, or *śūnyatā*: like a receptive womb there needs to be a place for the new understanding, the new wisdom to grow.[63] Fullness is where

---

60 Thoreau 1906, vol 12 (371): entry for 4 October 1859 (emphases in the original).

61 Gao & Lan 2018 (254).

62 Pieper 1990 (31).

63 See p 999 above.

there is no room for anything to grow. And since everything *is* a growing and becoming, and nothing that exists is *not* growing and becoming – when there is no room to grow there *is*, in that fullness, truly nothing.

When our assumptions flood in to the available space, they drive out the new before it has a chance to take root. We must first say 'no' to the *ob*-vious, that which stands in the way; we must get Underhill's limpet off the rock. Saying 'yes' depends on something to say 'yes' to, but whatever it is has not yet come into being. We therefore say 'no' to what is already known. Thus, in creation, 'no' comes first.

At this point we must suspend the action of the left hemisphere, for it would at once try to see all this as a matter of propositional knowledge – and ultimately that is not to be dismissed, since the existence of a divine or sacred realm is not contrary to reason: it's just that at this stage the left hemisphere is not up to doing the intuiting in the first place. The right hemisphere, however, *is* better capable of engaging with this whatever-it-is as something experienced and lived; as something relational, and reciprocal, in nature. It starts with the advantage that it 'believes', so to speak, that there is a 'whatever-it-is' out there with which to engage. 'Belief' is not a matter of coming to terms with arcane propositions, but nothing other than a disposition towards the world *as if God exists*, in order to open the *possibility* of an encounter with whatever the word 'God' designates. And it is *only* through the encounter that we can know – not through argument, or any amount of thinking in the abstract. One can sit on the brink for a lifetime waiting to learn how to swim, but without getting into the water one can never learn to swim at all. 'Seek not to understand so that you may believe', wrote St Augustine, 'but to believe so that you may understand.'[64] Without having sincerely made that attempt, it is impossible for anyone to know what it is they are denying. To say 'it's obviously not true' is to make a mistake at the very first step.

There are *prima facie* reasons to suppose that an understanding of the divine is sustained largely by the right hemisphere. In metaphysical terms, it requires being open to something 'Other', something not already familiar, not part of the self-consistent system in which one operates; not ignoring – or simply not seeing – whatever does not fit the accepted paradigm. As Wendell Berry writes, 'the incompleteness of a system is rarely if ever perceptible to those who made it or to those who benefit from it.'[65] The left hemisphere is effectively a closed system, dealing with the comfortably known and familiar, in which everything refers internally. Adopting this mindset makes it hard to 'see' what is meant by the divine. To bring things closer to our age, there is a partial parallel in computer programming called the

---

64 Augustine, *Tractates on the Gospel of St John*: XXIX, §6 (trans P Schaff): '*Ergo noli quærere intelligere ut credas, sed crede ut intelligas*'.

65 Berry 2000 (35).

Blub paradox. This indicates that a programmer who is accustomed to using a certain language (call it 'Blub') will realise its comparative strengths, but will be blind to its weaknesses. This is because writing in a language means thinking in that language: typically, according to computer scientist Paul Graham, programmers are 'satisfied with whatever language they happen to use, because it dictates the way they think about programs.'[66]

This should immediately alert us to a problem, since by common consent no adequate language for God exists. Hence the conviction that religious thinking is delusional. 'The habit of religion will always be derided by the atheist as an exercise in ever-increasing self-deception', writes Jonathan Gaisman, in an essay called 'The devout sceptic', which I recommend. 'Of course, precisely the same can be said about the habit of seeing the world in purely materialist terms. All mental habits lead – as is obvious – to habituation.'[67]

The right hemisphere is better at accepting uncertainty and limits to knowledge. An understanding of the divine must rely on indirect and metaphorical expression, not direct and literal expression; it must tolerate ambiguity; and it has to be at ease with accepting that both of what, on the surface, appear to be contradictory elements might be true – in other words, it must be receptive to what we call paradox. It must see continuous processes, rather than a succession of isolated events, a process of becoming, not simply the fact of being; it must be able to apprehend 'betweenness', a web of relationships, not just an assemblage of entities – and that what is of primary importance in the web is the relationships, not the entities related. It involves appreciating a *Gestalt*, not a construction of parts; entering into an 'I–Thou', not just an 'I–It' relationship, with its subject; sustaining emotional depth; and seeing that spirit and body are not distinct, or opposed, but discernibly different aspects of the same being. In epistemological terms, it requires knowing in the sense of *kennen* more than *wissen*; valuing 'active receptivity', as well as *not* doing and *not* knowing. It involves sustaining attention, and stilling the inner voice, as in prayer and meditation. In ethical terms, it places a value on empathy; on the consensual rather than individualistic self; and on vulnerability. It must recognise, not deny, the dark side to human consciousness, and be capable of understanding that good may, despite everything, emerge from suffering.[68] All of this (as only the reader who has accompanied me so far through the book will understand) in one way or another and to some degree suggests that the right hemisphere will play the key role. In this it is not different from other areas of life, since the proper relationship is always that of Master and emissary.

It may seem odd to cite the mediaeval mystic Meister Eckhart

66 Graham 2010 (176).

67 Gaisman 2018.

68 On this I find the words of Hermann Hesse, writing just after the First World War, thought-provoking: 'All the good in a man, for which he is praised or loved, is merely good suffering, the right kind, the living kind of suffering, a suffering to the full … From suffering springs strength, from suffering springs health': 'Zarathustra's return' (1919; in 2018, 96 & 98).

in support of an association between what we now know to be the right hemisphere and the approach to the divine; but, like James and Bergson so many centuries later, he seems to have intuited a distinction between two modes of understanding, one of which is more skilful here than the other. Consider these quotes from Eckhart in the light of what we have learnt:

> The active intellect [LH] ... cannot entertain two images together, it has first one and then the other. [But] ... if God prompts you to a good deed ... whatever good you can do takes shape and presents itself to you together in a flash [RH], concentrated in a single point.[69]

Here he could be seen as contrasting the serial concatenations of the left hemisphere with the immediate intuitive *Gestalt* formation more characteristic of the right.

Then he says: 'When the intellect discerns true being it descends on it, comes to rest on it, pronouncing its intellectual word about the object it has seized on [LH].' But, says Eckhart, it can never do what it longs to do, namely to say 'this is this, it is such and not otherwise'. It carries on in 'questioning and expectation; it does not settle down or rest, but labours on, seeking, expecting and rejecting ... Thus there is no way man can know what God *is*. But one thing he does know: what God is *not*. And this a man of intellect will reject.'[70] Eckhart seems to be telling us that the left hemisphere is unhappy until it can put words to experience and say with certainty what it is. While it can never succeed in this, it might have acquired an understanding by apophasis; but it can't understand that, and rejects the apophatic path.

Another passage bears comparison:

> Intellect peeps in and ransacks every corner of the Godhead, and seizes on the Son in the Father's heart and in the ground, and sets him in its own ground. Intellect forces its way in, dissatisfied with wisdom or goodness or truth or God Himself. In very truth, it is as little satisfied with God as with a stone or a tree. It never rests ...[71]

The 'intellect' (aka the left hemisphere here) is intrusive, overbearing, disrespectful, appropriative, insatiable, relentlessly striving and wilfully going about its business. And note the suggestions of curiosity rather than wonder: it peeps in, ransacks, seizes, forces its way in, is never satisfied, never resting ... it treats the divine according to its own conceptions ('sets him in its own ground'), and as a thing alongside other things, something commensurate with a stone or a tree. When it *does* see God, it doesn't recognise the fact, but goes on with its rampaging.

It is not surprising that Eckhart tried to image the opposite of

---

69 Eckhart, Sermon (Pfeiffer) III: 2009 (49). Wherever I have quoted from Eckhart in this chapter I have compared the translations by Evans with those by Walshe, and have in each case used whichever seemed to me clearer. I have retained Pfeiffer's 1857 numeration (also used in the earlier Evans translation) for ease of reference and comparison.

70 ibid (50).

71 Eckhart, Sermon (Pfeiffer) XLII: 2009 (237).

---

72 See p 775 above.

73 See p 1202 above.

74 In Walshe's translation of this Sermon (Pfeiffer IV) this passage is translated thus: 'The only name it has is "potential receptivity" which certainly does not lack being, nor is it deficient, but it is the *potential* of receptivity in which you will be perfected ...': 2009 (56: emphasis in original). Modern German texts, in different translations, give *die Möglichkeit des Empfangens* or *eine vermögende Empfänglichkeit*. Thus Quint, 1955: » *Ihr Name besagt nichts anderes als eine Empfänglichkeitsanlage, dies (indessen) durchaus nicht des Seins ermangelt oder entbehrt, sondern eine vermögende Empfänglichkeit, worin du vollendet werden sollst.* « This is no doubt correct, but gives an impression of greater abstraction than the Middle High German original:

[note continues on facing page] ▶

this as *un*doing, *un*knowing, *un*saying, darkness, emptiness and silence – something like the 'fertile night' that Chargaff referred to.⁷² (It is only at dusk that the owl of Minerva spreads her wings. Notice that the leaders of the Enlightenment called themselves '*les lumières*', and the leaders of the attack on religion judge themselves to be the 'Brights': as Chargaff pointed out, the brighter the light, the less you see; and as Suzuki noted, 'this "don't know" is what we call "dark", and it is very important'.⁷³) We switch out the light in order to see the stars. In one sermon Eckhart expands on the meaning of darkness. 'You cannot do better than to place yourself in darkness and unknowing', he says; and he imagines a bystander asking him: 'But what is this darkness and unknowing? And what is its name?' To this he replies: 'I can only call it a loving and open receptiveness, which however in no way lacks being: it is a receptive potential by means of which all is accomplished.'⁷⁴ This suggests the fertility of union between a creative principle and a receptive womb-like space (female principle) in which something is to grow.

In other words the darkness is not merely negative, but the active opening of a field of potential, what I have called *active receptivity*: the mode of the right hemisphere's attention.

In another passage Eckhart seems to distinguish two necessary modes of consciousness, one directed upwards towards the divine, the other obscuring reality by its representations of it in images and words:

> The soul has something in her, a spark of intellect, that never dies; and in this spark, and at the apex of the mind we place the 'image' [*bilde*] of the soul. But there is also in our soul a knowing directed towards externals, the sensible and rational perception which operates in images [*glîchnisse*] and words to obscure this from us.⁷⁵

Elsewhere, he warns us of re-presentations, the left hemisphere's stock in trade: eg, 'God is not seen except where he is seen spiritually, free of all images [*glîchnisse*].'⁷⁶

The personal accounts of Jill Bolte Taylor and Steve McKinnell, both of whom experienced a left hemisphere stroke, reveal that they felt an increase in spirituality, a new interest in meditation and an increased feeling of empathy following their stroke: according to Bolte Taylor, if she had to pick one word to describe the intent of her right hemisphere, she 'would have to choose *compassion*.'⁷⁷

Looking more closely at religious practices, we see this pattern demonstrated vividly. Music is very largely right hemisphere-dependent.⁷⁸ So, it seems is meditation, especially the kind known as mindfulness meditation. Bhante Henepola Gunaratana, a Bhuddist monk who is an acknowledged master of meditation and mindful-

» *Was ist das dünsternuzz und das unwizzen oder wie sind sein nam? Sein namen sint nicht anders dann ein mynnecliches enpfencklicheit die dann zümal wesens nich mangelt. Es ist auch ein müglich enpfencklicheit durch die es alles volbracht ist.* « (Steer, Klimanek & Löser 2002, 427–8.) This to me has a quite different energy. First of all, unlike Quint (and Walshe following him), it speaks expressly about 'unknowing', *das unwizzen*, which is lost in subsequent translations into modern German and English. The word *mynneclich* is also passed over in both. Its root is *minne*, meaning primarily love (as in *die Minnesänger*, the wandering minstrels who sang of courtly love), but also remembrance. Then there is the richness of *enpfencklicheit*, which suggests, to different degrees, a complex of meaning including, along with receptivity, a sensitivity or inclination towards, attunement to, susceptibility for, a responsiveness, openness and alertness to, as well as a vulnerability and tenderness towards its object.

75 Eckhart, Sermon (Pfeiffer) VII: 2009 (73).

76 Eckhart, Sermon (Pfeiffer) IX: 2009 (86).

77 McKinnell 2017; Taylor 2009 (170: emphasis in original).

78 See McGilchrist 2009b (72–5).

ness, writes in a classic text on mindfulness practice (I have italicised and numbered the more than 20 phrases that, with remarkable specificity in many cases, differentiate hemisphere styles of awareness):

> Before you first become aware of something, there is a fleeting instant of *pure awareness just before you conceptualize the thing* [1], *before you identify* [2] *it* ... you experience a softly *flowing* [3] moment of pure experience that is *interlocked with the rest of reality, not separate from it* [4]. Mindfulness is very much like what you see with your *peripheral vision* [5] as opposed to the *hard focus* [6] of normal or central vision. Yet this moment of soft, unfocused, awareness contains a *very deep* [7] *sort of knowing that is lost as soon as you focus your mind* [8] and *objectify* [9] *the object into a thing* [10] ... It stays forever in the present, *perpetually on the crest of the ongoing wave of passing time* [11] ... Mindfulness is *participatory observation* [12]. The meditator is both *participant and observer at one and the same time* [13] ... Mindfulness is *not an intellectual awareness* [14]. It is just awareness. The mirror-thought metaphor breaks down [15] here. It is *objective, but it is not cold or unfeeling* [16]. It is the wakeful *experience of life* [17], an alert participation in the *ongoing process* [18] *of living* ... Mindfulness is *a process, but it does not take place in steps* [19] ... it does not *categorise* [20] ... it does not *aim* [21] at anything ... you can't develop mindfulness *by force* [22]. Active teeth gritting willpower won't do you any good at all. As a matter of fact, it will *hinder progress* [23] ...[79]

79 Gunaratana 2011 (132–46).
80 Earle 1981.

The relation of this strong phenomenological account with, principally, the right hemisphere is borne out by a range of neuroimaging studies, an account of which can be found in Appendix 7. Spiritual practices such as meditation are designed purposely to transcend typical left hemisphere reactions to perceived events: in the tradition of such practices the fact that verbal, analytical thought is antithetical is often expressed in the form of a *warning*.[80]

WHAT IS IT LIKE?

So many things that are very real cannot be conveyed by language; and if they can't even be put into words, how can they possibly be argued about in a way that helps determine meaning? Ideas, the native currency of the left hemisphere, can be discussed at arm's length; direct awareness, the native currency of the right, has no arm's length about it: it is immediate in experience. It's where we finally get into the water and learn to swim.

Communication of such direct awareness can only be an account of personal experience: in recognition of this, I shall do my best to provide some indication of my own. Though I am someone whose left hemisphere, as the reader may have noticed, is far from inactive, I can attest that certain kinds of direct experience, unmediated by reflection, have been for me the undoubted wellspring from which life draws its meaning. So here I will briefly suspend my usual mode of discourse so far in this book, and attempt to provide a brief account of such direct engagement with what, both then and now, seems to me of the utmost importance.

Perhaps I should start by saying that neither of my parents, and none of my grandparents, was at all religious – rather the opposite. They did not take me to church, and, nearing the end of my life, I still don't attend one. Am I a religious person? Not in a conventional sense, then. Yet in a sense that is as important to me as it is hard to express, I surely am. I didn't get there by believing propositions, but from experience. And don't imagine that what I mean by experience is anything *outré* or supernatural. No visions or voices. It has all been entirely, quintessentially, natural – starting indeed from Nature herself.

It struck me from an early age that there was more going on than my senses were able to encompass: that nothing 'super' needed to be added to the 'natural' for it to invoke wonder. I suspect that if you rigorously disattend to that intuition, it will sooner or later go away. However I did not, and it didn't. It seemed to me that there was something 'beyond', in some sense, that drew me forward; something I had intuitive acquaintance with but could say almost nothing about, except that it seemed both real and beautiful. Later I discovered that it was akin to the feeling of being in love, a yearning for something that can only ever be known in part, but given as much in the *process* of discovery as in the never-completed discovery itself. I felt that the world, and especially the natural world, was far from inert and unresponsive, but rather that I was already deeply connected to something inconceivably great, awe-inspiring and vivifying. I saw it in landscapes of hills and fields, in the cliffs and the sea, in the light of day-skies and the stars at night; in the flowing of streams, and the life of living creatures. In fact it seemed to me that Nature in all her forms, including those we call inanimate, was alive. I contrasted what I experienced with what I knew – or thought I knew: and I could only shake my head in disbelief at what my culture was teaching me to think.

Intellectual debate in the culture in which I grew up seemed dominated by what all my life has seemed to me a terribly impoverished philosophy, that of reductive materialism – just one model among many, and an outlier in the history of humanity; and this formed the starkest possible contrast with the world I encountered, not just in Nature, but in music, poetry and painting; and friendship, and, later, in love. At age 13, my first hearing of Tallis's *Lamentations*, of Beethoven's late piano sonata in A flat major, and Schubert's C major quintet, were each like an epiphany. As I explored the music of Tallis's contemporaries in England and across Europe, and the other late sonatas and quartets of Beethoven, Mozart's chamber music and the last sonatas of Schubert, I knew without question that the world was far greater, and far more mysterious, than the schema of it I was being offered by the popular voice of science. I still cannot describe such great works in terms other than those of the sacred and the ensouled.

At the same time I was reading poets avidly, especially Donne, Herbert and their contemporaries; I discovered the poetry of John Clare, not a religious poet in any explicit sense, but a delighted and awestruck observer of Nature, and Gerard Manley Hopkins. Another epiphany – I can specify exactly when and where it occurred – was the moment I heard read, as if *for the first time*, a poem (I thought of as) long familiar to me, Wordsworth's Tintern Abbey ode – one indeed that as a teenager I had learnt by heart: it induced a transfiguration of the accustomed world, and signalled the beginning of a lifetime's devotion to the poet.[81] From every side I saw how clearly the edifice of Western civilisation – its literature, its poetry, its architecture, its painting, its music – at all times and in all ages was to a large extent an expression of a spiritual impulse.[82] And one we have largely lost.

As the reader may well point out, it has surely also been an expression of *eros*. And I concur wholeheartedly. The celebration of *eros*, the force of life itself, is often a celebration of the sacred – Sappho's lyrics and Petrarch's sonnets, for example, come immediately to mind. Indeed erotic love may be the closest that many of us in our secular age get to experiencing the sense of something sacred.[83] To compare the spiritual and the erotic is not to denigrate the soul, but to recognise what erotic love can be: while clearly distinct phenomena, they have common facets, not least of which being their capacity to induce awe, to transform one's vision of the world, and to forge a sense of connexion to something much greater beyond both space and time. Some may be inclined to be cynical here, but I think those who have experienced what I mean will understand. Plato was right to conceive our longing for communion with whatever is real as an expression of the impulse of *eros*.

What underlies and unites all these aspects of experience for me is the conviction of a direct and reciprocal engagement with whatever-it-is that is the ground of Being, and which we call God. I know of no better description of such direct experience and of the conviction it elicits than Wordsworth's:

> A presence that disturbs me with the joy
> Of elevated thoughts; a sense sublime
> Of something far more deeply interfused,
> Whose dwelling is the light of setting suns,

---

81 It is no coincidence that Wordsworth was Whitehead's favourite poet, as he is mine. 'The deepest impact on Whitehead came through the poetry of Wordsworth, which he studied throughout his life. According to his daughter's testimony, he would read *The Prelude* almost daily "as if it were the Bible, poring over the meaning of various passages"' (Wyman 1956, 283).

82 Cf Rupert Spira (2002): 'Sacred art is work that comes from a deep desire to explore the true nature of our experience, or from an intuition of it. So if we are trying to find out who we really are and what the world really is, it makes no sense to predefine or limit either … Sacred art takes us beyond these limitations, because it is inspired by that which is beyond them. If a work of art is inspired by these limitations it will only lead the viewer back to them. We could say that a work of art is like a pathway; it bears the signature of its origin. The senses are the medium through which we travel this path … But true art comes from transparency, not from feelings of isolation, separation or despair. Of course, such feelings often precipitate the sort of openness and sensitivity that are the origin of creativity, but it is important to make this distinction, because without it we may think that these feelings are themselves the source of creativity and we end up with an art form that celebrates the hopeless, banal, vacuous despair of a culture whose paradigm has lost sight of the sacred, of any true enquiry into the nature of reality …'

83 Jason Cowley (2018) on the philosopher Bryan Magee: 'He describes sex as an "other-worldly" experience and likens its effects to that of great music, "the deepest we can penetrate into a world other than this world, the world beyond appearances"'.

> And the round ocean and the living air,
> And the blue sky, and in the mind of man:
> A motion and a spirit, that impels
> All thinking things, all objects of all thought,
> And rolls through all things.[84]

In summary, my religious disposition, if that indeed is what it is, has resulted from a largely private lifelong exploration of the experience of being alive, guided by meditative reading of the spiritual texts of different cultures, experiencing holy places in different lands, encounters with human beings who seemed to me to be deeply spiritual people, sporadic attendance at rituals of great beauty, a lifelong celebration of art, and poetry, but of music above all; and love; and long communing with the astonishing beauty of the natural world.[85] All this, coupled to an abiding sense – intensified in proportion as I came to understand more – of how very little we can possibly hope to understand of all that exists.

Over my lifetime, I have repeatedly and increasingly become aware of the way in which such direct experience is vulnerable to diminution and dismissal in the light of the limited vision of the left hemisphere. This was in part the motivation for writing my first book, *Against Criticism*, and, in a sense, of my change of career to medicine; and of whatever has followed from it.

For those who have hung on to the reality of direct experience, and not let it become overshadowed by a representation that has none of its living qualities, we should reclaim the word 'expert' – literally, one who has experienced. According to the accounts of such experts, this vision is transformative. It induces, their example tells us, a humility before the greatness of the cosmos, and how little we understand it; compassion for others and ourselves; and reverence towards the living world. It allows us to acknowledge that there is something way before, behind, above, and beyond our selves; that that something is not inert or remote, but 'speaks' to us and calls to us to respond, and that we feel the need to do so with seriousness, reverence and gratitude. And that is what gives meaning to life.

In short, creation and the mystery of what lies behind it become *sacred*; and the disposition that sees it thus is what is meant by a religious disposition. It is a disposition that perceives depth. Except where such a religious disposition becomes perverted, as all too often it does (as I shall explain shortly), it is the exact opposite of the disposition towards creation of the left hemisphere, which sees itself as master – detached, confident, domineering, wanting to have control over creation and either disregarding the mystery of Being or recruiting it to its own purposes. That is the disposition that now dominates our world, and, unhappily, no longer in the West alone.

---

84 Wordsworth, 'Tintern Abbey'.

85 Sir Thomas Browne (*Religio Medici*, 1, xvi): 'Thus there are two Books from which I collect my Divinity; besides that written one of God, another of his servant Nature, that universal and publick Manuscript, that lies expans'd unto the Eyes of all, those that never saw him in the one, have discovered him in the other ... all things are artificiall; for Nature is the Art of God'.

More than 60 years ago, Tillich already reflected that

> modern man is neither more pious nor more impious than man in any other period. The loss of the dimension of depth is caused by the relation of man to his world and to himself in our period, the period in which nature is being subjected scientifically and technically to the control of man. In this period, life in the 'dimension of depth' is replaced by life in the horizontal dimension. The driving forces of the industrial society of which we are a part go ahead horizontally and not vertically ... He transforms everything he encounters into a tool: and in doing so he himself becomes a tool. But if he asks, a tool for what, there is no answer ...[86]

Our view, says Tillich, has become superficial, lacking in depth, all on one plane. In Chapter 2 we saw that this loss of depth in time, in space, in emotion, and in understanding is one consequence of right hemisphere suppression. Our take on the world has become, Tillich says, that of instrumentalism, of transforming everything into a tool: the main characteristic of the left hemisphere. And we end by instrumentalising ourselves, for a purpose unknown.

He continues:

> If the dimension of depth is lost, the symbols in which life in this dimension has expressed itself must also disappear. I am speaking of the great symbols of the historical religions in our Western world, of Judaism and Christianity. The reason that the religious symbols became lost is not primarily scientific criticism, but it is a complete misunderstanding of their meaning ... The first step toward the non-religion of the Western world was made by religion itself. When it defended its great symbols, not as symbols, but as literal stories, it had already lost the battle ... If the symbol of the Fall of Man, which points to the tragic estrangement of man and his world from their true being is transferred to the horizontal plane, it becomes a story of a human couple a few thousand years ago in what is now present-day Iraq. One of the most profound psychological descriptions of the general human predicament becomes an absurdity on the horizontal plane ...[87]

Yet there never was a time in which the story of the Fall seems more pertinent – indeed prophetic.

RELIGION

I may have given the impression that the sense of the sacred is something isolated within the person, if not within the mind, and at the same time that it is purely transcendent of time and place. Yet as well as personal, the sense of the sacred is inevitably shared and communal; as well as being an inner realisation, it is realised externally in the visible, tangible, world; as well as being transcendent,

---

86 Tillich 1958.

87 Tillich *loc cit*.

it is immanent, having to do with the thisness of things in time and place, not just with abstract generalities. That is why there is not just my disposition or your disposition, but such a thing as religion.

What, though, is religion? Religion, as James points out at the start of *The Varieties of Religious Experience*, means so many things. However, there are common elements we can point to.

The principal way in which humanity has felt compelled both to express a sense of, and to make contact with, the divine is through music. And in this it seems to me that it has succeeded so immediately and so indubitably that language is scarcely needed. Both abstract and at the same time deeply, powerfully, wholly, embodied; both timeless and situated in time; both personal and universal; both particular and beyond all particularity; taking us into realms that declare themselves despite being utterly beyond language. Music unifies these apparent contraries. And it is also a feature of music in every known culture that it is used to communicate with whatever is by definition above, beyond, 'Other than', ourselves.[88] It forms the bridge: that between human and divine, and that between human and fellow human, and is at the heart of religious worship everywhere. Music exists entirely in the betweenness of tones; and religion exists – or rightly understood exists – in the betweenness of human beings, out of which, as with musical notes, something far greater than the sum of its parts emerges.

Many misunderstandings of spiritual and religious teachings come from a narrow adherence to the view that language, if sufficiently carefully analysed, will reveal truths. There is wisdom in lines from the ninth-century Chinese poem *Bǎojìng Sānmèi*, considered one of the masterpieces of Zen literature: 'The meaning is not in the words, but it responds to the inquiring impulse.'[89]

Religion, at its best, is a cultural expression of that enquiring impulse; of an awareness of and openness to a God or gods; of a context that transforms our understanding of the world, and which enables this sense to be shared and celebrated with others; in other words, it involves community, in space, but also over time. Indeed, it helps to bind a community together: that is what religion means (from Latin *religare*, to bind). It makes tangible the betweenness, the relational nature of existence. And in this respect, if no other, it is hard to replace. What there is to be known is reciprocally bound up with the way that we attempt to know it, something science generally glosses over. The way we choose to attempt to know anything has moral implications, a point I have repeatedly emphasised. The myths of religions convey truths that are absent from everyday thought and language, and speak directly to us at the deepest level of our understanding of life itself.

Secular gatherings, by contrast, do not remind us of what our lives

---

88 Nettl 1983. See also McGilchrist 2009b (77).

89 Suzuki 1999 (17). *Bǎojìng Sānmèi* is also widely known by its Japanese title, *Hōkyō Zanmai*.

mean *sub specie æternitatis*, but merely confirm further our everyday views. This probably explains the enduring desire for religious ceremonies of birth, marriage and death among those who are not regular attenders at a place of worship.[90] In fact one of the reasons for having religion is constantly to remind us of a broader context; a moral order; a network of obligations to other humans, to the earth, and to the Other that lies beyond. Extending beyond our lives, that is, in space and time, yet rooted firmly in places, spaces, practices, here and now. A religion forms the bridge between worlds, which is the purpose of metaphor – and the purpose of ritual, which is metaphor embodied. One of the beautiful things about many religions, especially perhaps Hinduism, but also certainly in some traditions within the monotheistic religions – those I know of include Eastern (so-called Orthodox) Christianity, and Judaism – is that there are brief prayers of only a sentence or so, gestures, beautiful small rituals, that sanctify the familiar routine actions of daily life, and set them within the perspective of the infinite, of which we so easily lose sight. And inculcate a habit of reverence and gratitude towards the world: of seeing the sacred in every part of what is given.

I earlier referred to Confucius' advice that a ruler needs weapons, food and, above all, trust, in order of ascending importance. As if to confirm this, in commenting on prehistoric sacred rituals, archaeologist Clive Gamble comments that beliefs could not have been

> something special and separate, but an integral part of social life, pervading all activities ... I think that belief systems at this stage could have been almost as important as having enough people to defend your territory, or even controlling a food supply: because shared beliefs would have allowed people to connect across social universes much larger than the local social group.[91]

Trust depends on shared beliefs; religion is the manifestation of that trust, and the embedding of it into the fabric of daily life. Religion embodies awareness of God in the world through deeply resonant myths, narratives and symbols, enacted in rituals, conducted in holy places, that parallel the cyclical passage of time. In doing so it exists as a repository of the accumulated wisdom of good men and women, so that each living being does not have to 're-invent the wheel' but can benefit from common insights. While religions differ, particularly in their more superficial representations, their insights are for the main part congruent across time and across the world, a sort of 'perennial philosophy'. A religious life expresses, as I have suggested, a *disposition* towards the world that has consistently the same qualities: humility, compassion, reverence. Of James, Jonathan Rée writes:

---

90 Habermas commented on an atheist friend who nonetheless wanted a church burial: 'the enlightened modern age has failed to find a suitable replacement for a religious way of coping with the final *rite de passage*.' On this, Stanley Fish (2010) observed: 'The point can be sharpened: in the context of full-bodied secularism, there would seem to be nothing to pass on to, and therefore no reason for anything like a funeral.'

91 MacGregor 2018 (10).

The cynicism of modern atheists was as alien to him as the gullibility of traditional believers. Religion, as far as he was concerned, meant regarding the world with reticence, tenderness and love, and in that sense he still regarded himself as religious.[92]

It is the opposite of cynicism or trivialisation. Out of a religious disposition, at its best, arises a harmony between deeds and words, manifest in a certain way of being. It is an allegiance to certain values; a synthesis of rational beliefs with valuable intuitions; of faith with doubt. Since it is the disposition that matters, it cannot – as neither can virtue ethics – be identified with words alone, or deeds alone, or beliefs alone, though all these form a part. It is not just the 'what' of the parts, but the 'how' of the whole, that matters.

According to James, a religious disposition 'favours gravity, not pertness':

> it says 'hush' to all vain chatter and smart wit … There must be something solemn, serious, and tender about any attitude which we denominate religious … The divine shall mean for us only such a primal reality as the individual feels impelled to respond to solemnly and gravely, and neither by a curse nor a jest.[93]

Which is not at all to say, of course, that the religious person should abjure fun or humour. The point is about respect, something we're not strong on these days. Wisdom is often closely allied to humour, a truth enshrined in Zen Buddhism and Sufi stories, as well as in the Talmud, and it has its place: just not in the solemnities of ritual (if only we still remembered this simple fact – it's once again the importance of context). James's point is that religion prefers wisdom to cleverness. Our contemporary culture is not obviously well disposed towards gravity, the cessation of chatter, thoughtfulness, acknowledgment of the depth of our unknowing. It prefers cleverness to wisdom.

The way in which one equips oneself to understand religious truths is not – fairly obviously, but I'm afraid it still needs saying – by the scientific method. God is not a force in physics that we have not yet discovered. Propositional beliefs are what science has to offer. Yet propositional belief, while indisputably valuable, is the *least* that religion has to offer. Indeed we are not dealing primarily with propositions at all, certainly not with a simple body of propositions the truth of which could in principle be determined in the same way that the date of the Big Bang, or the number of bonds in a carbon atom, can be determined. Religion offers deep, imaginative archetypal truths about the human condition that cannot easily be expressed in any other way, never mind in the sort of prose you might expect

92 Rée 2019.

93 James 1902 (37–8).

in a science text book. And such truths are primarily experiential, although they may have cognitive aspects. In order to understand, you not only may, but *must*, try for yourself. Knowledge is of many kinds. Science is a matter of *wissen*, knowing facts; religion a matter of *kennen*, knowing by experience. (I am coming to believe that the limited nature, and many of the confusions, of the Anglo-American tradition in philosophy is in part due to the fact that in English we have only one word for 'know'.) Science is – at least purports to be – purely a matter of cognition. Religion is about the whole business of human being, human existence. Cognition alone will not do for that. As Wittgenstein saw, 'even if all *possible* scientific questions be answered, the problems of life have still not been touched at all.'[94]

To quote Jonathan Rée, 'if there are religious truths, they are more like truths of love than truths of science: they depend on facts that will not come to pass unless we go half way to meet them.'[95] On this point I cannot help thinking of a story told me by a Jewish friend. There was a very poor, but good, rabbi whose life would have been very much more comfortable if he had money; and so he prayed repeatedly to God: 'Please just let me win the lottery'. And his prayer never seemed to be answered. One day he was at prayer as usual, when God said to him, 'Look, Manny, meet me halfway: *buy a ticket*'. Understanding any spiritual truth depends on at least buying a ticket. Knowledge of this universe in which we live must be participatory. If you are not prepared to participate, or to take any risks, love will never be part of your life. Risk and vulnerability are of love's essence. And love – as you will know if you have made the experiment and experienced it – opens aspects of reality that would otherwise be concealed from you.

Faith can never be certain: it follows that doubt is a necessary part of faith. Because of the prominence in news bulletins of religious fanaticism – something utterly different from what I'm talking about, and to which I'll return – those who have never tried to find out more about religion imagine that doubt is the opposite of faith. It is rather its inalienable companion. Faith and doubt are a living dipole. Faith is neither certain nor blind, but a sense of allegiance born of experience.

It is significant that if you ask people, certainly in Britain, if they are religious, a large majority say they are not; if you ask people if they believe that there is more to reality than the realm of material things described by science, an even larger majority say that they do. I suspect that many of what I would see as the most honest kind of believers call themselves agnostic; they are quite understandably frightened away from God by the off-putting and sometimes thoroughly left hemisphere way in which religions these days in the West present themselves, having allowed themselves to stray from their

---

94 Wittgenstein 1961, §6.52.

95 Rée *op cit.*

origins into something that the left hemisphere, they believe, will surely understand. And by the idea that something called 'blind faith' is involved. There is nothing blind about faith, but there is nothing certain about it, either. It is like trusting the outstretched hand that helps you ford the stream: you see the stream, you see the hand; you do not blindly step, but step you must.

'ONE MUST KNOW WHEN TO STOP':
RELIGION AND THE PROBLEM OF LANGUAGE

There are, of course, many religions. But one of the striking things about those religions with their different placeholder 'un-words' for the divine – *logos*, *li̇̌*, *tao*, *r̥ta* and so on – is that none of these 'un-words' suggests a thing, but always a process: a dynamic source of energy, often imaged as fire or water – or, at another level, as love and life. Though *r̥ta* is the ultimate foundation of everything, 'this is not to be understood in a static sense … It is the expression of the primordial *dynamism* that is inherent in everything.'[96] Like *li̇̌*, like *logos*, and like the *tao*, it is creative energy, and it flows. God is a verb.[97] That these 'un-words' for the divine should share a conception of it as a process of flow is further evidence that it is the right hemisphere which, so to speak, divines the divine. Nothing more economically expresses the deep distinction between a view of the cosmos, on the one hand, as a specifiable assemblage of distinct entities (LH) and, on the other as an unnameable, undivided flow (RH), than this insight from the *Tao Te Ching*, warning us against division and the naming of parts:

> The Tao is *forever* undefined …
> Once the whole is divided, the parts need names.
> There are already enough names.
>
> One must know when to stop.
> Knowing when to stop averts trouble.
> Tao in the world is like a river flowing home to the sea.[98]

The left hemisphere never 'knows when to stop'. And so we *do* use more than 'enough names' when trying to express a response to the divine – indeed the response almost excites linguistic extravagance – and the trouble is that those names are then taken to refer to distinct, individually analysable components of some sort of divine 'system'. The words 'God', the 'divine', the 'holy' and the 'sacred', both in their origin and in their usage, have never been wholly separable, nor should they be, since they refer to one coherent experience of the world and our response to it. They refer not to distinct *entities*, but to a *relationship*. And if one is willing to use the term sacred, or holy, one cannot do so without implying relationship with the

96 Panikkar 2001 (351: emphasis added).

97 This is the title of a book on the Kabbalah by Rabbi David Cooper (1997). 'God is a verb' is also, I should point out, the title of a slightly ghastly poem by Buckminster Fuller, which understands the phrase in its own particular way.

98 *Tao Te Ching*, §32 (trans G-F Feng & J English: emphasis added).

divine. Blake uses such words interchangeably: what, however, we can be clear about is that when he says that all living things are holy, or that the human form is divine, he does not mean that they are merely beautiful, wonderfully intricate, or to be admired in some other way, but are part of an intrinsically divine cosmos.

This is not just another case of how thinking in terms of things leads us astray, but the ultimate case. The words designate not some unfamiliar thing to relate to, but a new relationship with the familiar, manifested throughout: an allegiance, which is what faith (from Latin, *fides*, as in 'fidelity') means.

Throughout I have contended that dynamic relationships are not only more important than the entities related, but odd as it must undoubtedly sound, ontologically prior to them – so that what we call 'things' arise out of the web of interconnectedness, not the web out of the things. This has long been my intuition, and recent advances in physics, to some of which I have adverted, seem to corroborate it at a different level (see Chapter 24). It also recalls the Vedic image of Indra's net, which encompasses the universe, and in which all that exists is connexion; at each point of intersection there is a jewel, like the drops of dew on a spider's web. In each jewel all the others are reflected, something akin to a hologram.

If there are, then, 'already enough names', if the divine is perceived as an unnameable, undivided flow, and if relationship to and within that flow marks the nature of the right hemisphere's engagement with whatever-it-is that it perceives as divine, then it is to be expected that what religions *primarily* do in response should itself be consistent with the mode of the right hemisphere: they engage in acts of worship, ceremonies, rituals that celebrate the sacred. But human beings feel the need to speak about their experiences, and religions, being about collective, shared experience, feel the need to find a common language for that experience. So, despite the warning about 'names', religion cannot, unfortunately, escape the problem of language.

We saw, in previous chapters, that time as experienced cannot be represented and still be time; that space cannot be represented and still be space; that flow and motion cannot be represented and still be motion and flow. These elements are not further clarified by language, but rather become something else in its clutches – a derogation of what can be known only through direct experience. Typically the left hemisphere deals with each of them, as it deals with everything, through representation; so, for us to gain a fuller understanding of any of them, involves reversing that process. We need to be liberated from our familiar *idea* of it, and allow whatever it is – time, space, or flow – to regain experiential depth. We must prize the limpet off the rock of the obvious. I have attempted to do that in some of the preceding chapters.

It is no different with God. Attempts to deal with the divine, not in experience, but in language other than poetry, too easily lead to the substitution of a re-presentation for a living experience – and are doomed.

We *must*, then, have recourse to metaphor, as the philosopher Paul Ricoeur makes clear: we have no other option.[99] To do so is not an evasion, as some atheists seem to believe, but the inevitable result of being aware of what one is dealing with. We have seen that, according to the philosopher Whitehead, philosophy is akin to poetry. And we have heard from Bohr that physics can be described only in language that has the nature of poetry: this does not render physics somehow unreal – rather the opposite. He was equally clear about the need for poetry in dealing with the divine. The fact that spiritual traditions through the ages have spoken in images, parables, and paradoxes, Bohr said, 'means simply that there are no other ways of grasping the reality to which they refer. But that does not mean that it is not a genuine reality.'[100] If you are a biologist you can get by (while, nonetheless, mistaking the nature of what you deal with) in a comfortingly familiar metaphysical world. Physicists, by contrast, spend much of their time grappling with the ultimate nature of reality, and must take in their stride the failures of analytic rationalising, classical mechanics and ordinary language. As a result, contemporary physicists often say things that the great wisdom traditions of the world have been declaring for thousands of years; and since one of the findings of contemporary physics is, indeed, that 'inner' and 'outer', consciousness and matter, are not separate, it is only to be expected that the truths they discover in the physical world will be in harmony with those disclosed in spiritual traditions. And so Bohr continued, dealing with another point sometimes made by opponents of such traditions, namely that there are differences between the ways in which those traditions express themselves:

> I can quite understand why we cannot speak about the content of religion in an objectifying language. The fact that different religions try to express this content in quite distinct spiritual forms is no real objection. Perhaps we ought to look upon these different forms as complementary descriptions which, though they exclude one another, are needed to convey the rich possibilities following from man's relationship with the central Order ...[101]

That is because what we are dealing with is not subject to being defined; it can be approached only tentatively through different analogical metaphors or myths, which *should*, if they are true, be various in nature, because a single, simple analogy can never be right. (There are, separately, problems in finding analogies of any kind whatsoever to something that is utterly *sui generis*).[102] The point is,

99 See Paul Ricoeur's *La métaphore vive* (1975), translated as *The Rule of Metaphor* (1977) – the title thereby losing its metaphoric power.

100 As recorded in Heisenberg 1971, 'Science and religion': 82–92 (88).

101 *ibid* (89). He even went on to say of divine intervention, a concept on which I have to acknowledge the extent of my ignorance, that 'we quite obviously do not refer to the scientific determination of an event, but to the meaningful connexion between this event and others or human thought. Now this intellectual connexion is as much a part of reality as scientific causality; it would be much too crude a simplification if we ascribed it exclusively to the subjective side of reality' (91).

102 Cusanus 1985: I, 1, §4: 'Every inquiry proceeds by means of a comparative relation, whether an easy or a difficult one. Hence, the infinite, qua infinite, is unknown; for it escapes all comparative relation.'

the difficulty lies not in *finding* the right words: the difficulty lies in there *being* no right words, and so when we use words carefully we must always be both saying and un-saying.

Even poets, after all, struggle with the limitations of language. Rilke wrote wonderfully about this problem, how language destroys the immediacy of experience, how it neutralises awe, and even tries to encroach on the divine. For language, he says, no mountain is miraculous anymore; the 'garden and property' of language reach *right up* to the very border with God:

> The words of men make me very afraid.
> They make everything sound so terribly clear:
> This is called 'dog', and that is called 'house',
> And here's the beginning, and there is the end.
>
> I fear their intentions, their playful sneers,
> They know all that was and ever will be;
> Now there's no mountain can fill them with awe;
> Their estate reaches right to the border with God.
>
> I want to resist them and shout: 'Keep away!
> I so love to hear the singing of things.
> You touch them – and they fall as silent as stone.
> You're destroying the life of things for me.'[103]

The single most dramatic demonstration of the inadequacy of prose to speak of the divine comes from the life of St Thomas Aquinas, by common consent one of the greatest of theologians that ever lived. His massive analytical work the *Summa theologiæ*, the most comprehensive account of the Christian faith ever written, is a masterpiece of scholarship and philosophical thought, and he is often spoken of in the same breath as Aristotle and Plato. On the 6 December 1273, the feast of St Nicholas of Myra, he had a mystical experience while saying mass, after which, having written at a furious pace without ceasing for 35 years, he decided he could not continue. His secretary Reginald of Piperno naturally urged him to resume. Aquinas replied:

> I adjure you by the living almighty God, and by the faith you have in our order, and by the love you bear me, that you never reveal in my lifetime what I am about to tell you ... Everything that I have written seems to me like chaff compared to those things that I have seen and which have been revealed to me.[104]

He stopped in the middle of Part III of the *Summa*, at the point where he happened to be writing about penance, and never wrote another word.[105]

It is not that language and rational thought here are not valuable:

---

103 Rilke, from *Mir zur Feier: Gedichte* (composed 1897, published 1899: trans IMcG): Ich fürchte mich so vor der Menschen Wort. | Sie sprechen alles so deutlich aus. | Und dieses heißt Hund und jenes heißt Haus, | und hier ist der Beginn und das Ende ist dort. || Mich bangt auch ihr Sinn, ihr Spiel mit dem Spott, | sie wissen alles, was wird und war; | kein Berg ist ihnen mehr wunderbar; | ihr Garten und Gut grenzt grade an Gott. || Ich will immer warnen und wehren: Bleibt fern. | Die Dinge singen hör ich so gern. | Ihr rührt sie an: sie sind starr und stumm. | Ihr bringt mir alle die Dinge um.

104 Bartholomeus de Capua, in D Prümmer OP (ed), *Fontes vitæ S Thomæ Aquinatis*, Toulouse [undated: foreword, however, dated 1911], §79, 377: 'ego adiuro te per Deum vivum omnipotentem et per fidem quam tenetis ordini nostro et per caritatem quam michi stringeris quod ea que tibi dixero nulli reveles in vita mea ... Omnia que scripsi videntur michi palee respectu eorum que vidi et revelata sunt michi.' I have preserved the unorthodox mediaeval Latin spelling.

105 As one might expect, it has been suggested that Thomas could have had a stroke. However, this is wholly unsubstantiated. Aquinas was not aphasic, and his intellectual and speech abilities remained unaltered until his death, which makes a left hemisphere stroke, as an explanation of his sudden aversion to using words, unlikely; a right hemisphere stroke would hardly have opened his eyes to what it is that language fails to capture. See Charlier, Saudamini, Lippi et al 2017.

they are. But they are there to be struggled with, and finally, having been found wanting, let go. The struggle was not wasted effort. In Hegel's metaphor, the beauty of the bud is sacrificed, necessary as it was, once the flower comes into being; and still more the flower must be sacrificed once the fruit comes into being; yet each has its place, and all were valuable, as part of the greater creation as a whole. So words have their place, but only up to the borders with God.

And, beyond language, what is to be understood by the idea of God passes through a familiar process, that of unlearning what we thought we knew. Freeman Dyson describes the three stages whereby students come to terms with quantum mechanics. First they learn the 'tricks' – how to calculate and get the right answers – a trouble-free stage. Then they come to a painful awareness that they have no understanding of what they have been doing, and feel confused and abashed. Then 'unexpectedly, the third stage begins':

> The student suddenly says to himself, 'I understand quantum mechanics', or rather he says, 'I understand now that there isn't anything to be understood'. The difficulties which seemed so formidable have mysteriously vanished. What has happened is that he has learned to think directly and unconsciously in quantum-mechanical language. He is no longer trying to explain everything in terms of prequantum conceptions.[106]

There is a clear parallel here with the attempt to understand the idea of God by the conventional terms of the left hemisphere: we can't do it with 'pre-quantum' terms.

As Meister Eckhart says, 'Since it is God's nature not to be *like* anyone, we have to come to the state of being *nothing* in order to enter into the same nature that God is'.[107] Of those who no longer know enough to experience what Einstein called 'rapturous amazement', Bohr reportedly said: 'Thinking they know things when they know only words, they will not know their ignorance and will never wonder.'[108] The space held open by words, in particular the un-word that is the divine name, must never be closed too tightly. Yet, paradoxically, without words we may, as a culture, if not as individuals, forget what everyday language obscures from our vision.

The thirteenth-century mystic, Marguerite Porete, in her book *Le Miroir des âmes simples et anéanties* – translated as *The Mirror of the Simple Souls Who Are Annihilated [and Remain Only in Will and Desire of Love]* – gives 12 'names' for the soul, the last of which is *Oubli* (Forgetting). Our souls, she says, readily experience and understand the divine, but as readily let it pass from mind.[109] I believe it to be important that we recognise what has happened and aim to recover some of what we have lost. We need what Plato called *anamnesis* (unforgetting), or what Heidegger calls remembering of Being.

106 Dyson 1958 (78).

107 Eckhart, Sermon (Pfeiffer) XLII: 2009 (74: emphasis in original).

108 Quinn 2002 (219).

109 Porete 1927 (26). 'Forgettelle is her name, for it is her manner much to comprehend and soon to forget.'

There is no doubt that we are again in an area where language and reason are inadequate to the task, but that does not mean there is nothing to be conveyed. 'Whoso is unable to follow this discourse', said Meister Eckhart, 'let him never mind. While he is not like this truth he shall not see my argument.'[110] In similar vein Plotinus said: 'To any vision must be brought an eye adapted to what is to be seen, and having some likeness to it. Never did eye see the sun unless it had first become sunlike.'[111] That might sound irritatingly like a rhetorical ploy, but I believe there is, nonetheless, clearly truth in it: we must meet things on their own terms, not our preconceived ones, if we are to see them at all. This is the point so beautifully made by Schelling above. Kolakowski puts it well, speaking of the contradiction apparent in saying that 'God is ineffable':

> The adjective 'concrete' is abstract, the adjective 'incommunicable' is communicable, the adjective 'unique' is general, and to utter the word 'intuition' is not itself an act of intuition. We cannot get rid of the barriers of language when we try to convey to others something that language is intrinsically not designed to deal with; we can use it none the less to produce various hints, metaphors, or aesthetically powerful images, in order to awaken in other people the faculty of intuition which, even if dormant, is a part of the universally human endowment.[112]

BRING BACK THE LEFT HEMISPHERE: THEOLOGY

Once we have acknowledged the limitation of language we are able, paradoxically, to bring back that most language-dependent, articulate and clear-thinking of our capacities, the left hemisphere. And we need to do so, even if only to confirm its limitations. That is to say, if we are to try to think and speak at all systematically about what a religious disposition may have had disclosed to it – if we are, to take a most basic example, to grasp at an *intellectual* level the primacy of metaphor and myth in understanding – we need the clarity, the ordering, the trying-to-make-sense-of-things that are the left hemisphere's forte. Within religion the left hemisphere helps us supplement ritual with theology. But the left hemisphere must remain the servant: what it 'serves up' must maintain respect for the provisionality, the ultimately incomprehensible mysteriousness of what the right hemisphere has had disclosed to it. Bringing back the left hemisphere here conforms to what I have consistently argued is the optimal mode of human understanding – reliance solely neither on the left hemisphere nor on the right, but rather on a process in which all that is to be known must initially 'presence' to the right hemisphere (we have no other access); then be transferred to the

---

110 Eckhart, Sermon (Pfeiffer) LXXXVII: 1931–47, vol 1 (221).

111 Plotinus, *Enneads*, Bk I, Tractate VI, §9 (trans S MacKenna & BS Page). Eckhart also said of the knowledge of God: 'The eye with which I see God is the same with which God sees me. My eye and God's eye is one eye, and one sight, and one knowledge, and one love.' Sermon (Pfeiffer) XCVI: 1931–47, vol 1 (240).

112 Kolakowski 1985 (33).

left hemisphere so as to gain expression through re-presentation; and that re-presentation returned to the right hemisphere where it is either recognised for its consonance with the initial presencing and subsumed into a new *Gestalt*, or rejected. It is no different with the presencing of the divine. It's just that it is going to be more difficult, not least because it will once again present us with paradox.

As always we need the hemispheres to co-operate in fulfilling their proper roles. A religion inevitably formalises its philosophy and its ethical teaching, whether or not that is in the spirit of those whose teaching forms its foundation. For this there needs to be an optimal balance between autonomy and community, as in any healthy social group. Metaphysical speculation demands the bringing to bear of reason in the abstract, which is largely a left hemisphere specialisation: such a process does, however, need to know its limitations. The development of a moral code, based on a body of propositional knowledge, the establishment of a body of conventionally approved texts, the elaboration of celebratory rites and customs, require a balance of hemispheric specialisations, but are clearly to a considerable extent left hemisphere-dependent. The same applies to the propagation of good works through institutions of different kinds involved in government, in education and in the care of the sick. All of this requires not just inspiration, imagination and vitality, but formal procedures providing cohesion and permanence; and this in turn relies to a large extent on a proper interhemispheric balance.

In the rabbinic tradition, two kinds of teaching are distinguished, one literal and legalistic, the other metaphorical and imaginative. These are referred to as *halakhah* and *aggadah*, respectively. According to the Midrash, a body of early rabbinic commentary on scripture, when God promised his people corn and wine, the corn was *halakhah*, the wine *aggadah*.

Abraham Heschel has this to say of these concepts, in a passage I quote at length, because the correspondences with the phenomenology of the two hemispheres, as described in this book, are so many and so striking:

> *Halakhah* represents the strength to shape one's life according to a fixed pattern; it is a form-giving force. *Aggadah* is the expression of man's ceaseless striving that often defies all limitations. *Halakhah* is the rationalisation and schematisation of living; it defines, specifies, sets measure and limit, placing life into an exact system. *Aggadah* deals with man's ineffable relations to God, to other men, and to the world. *Halakhah* deals with details, with each commandment separately; *aggadah* with the whole of life, with the totality of religious life. *Halakhah* deals with the law; *aggadah* with the meaning of the law. *Halakhah* deals with subjects that can be expressed lit-

erally; *aggadah* introduces us to a realm that lies beyond the range of expression. *Halakhah* teaches us how to perform common acts; *aggadah* tells us how to participate in the eternal drama. *Halakhah* gives us knowledge; *aggadah* gives us aspiration. *Halakhah* gives us the norms for action; *aggadah*, the vision of the ends of living. *Halakhah* prescribes, *aggadah* suggests; *halakhah* decrees, *aggadah* inspires; *halakhah* is definite; *aggadah* is allusive ...

Halakhah, by necessity, treats with the laws in the abstract, regardless of the totality of the person. It is *aggadah* that keeps on reminding that the purpose of performance is to transform the performer, that the purpose of observance is to train us in achieving spiritual ends ...

Halakhah thinks in the category of quantity; *aggadah* is the category of quality. *Aggadah* maintains that he who saves one human life is as if he had saved all mankind. In the eyes of him whose first category is the category of quantity, one man is less than two men, but in the eyes of God one life is worth as much as all of life. *Halakhah* speaks of the estimable and measurable dimensions of our deeds, informing us how much we must perform in order to fulfil our duty, about the size, capacity, or content of the doer and the deed. *Aggadah* deals with the immeasurable, inward aspect of living, telling us how we must think and feel; how rather than how much we must do to fulfil our duty; the manner, not only the content, is important.

And Heschel draws interesting conclusions. They are both necessary, he emphasises: '*Halakhah* without *aggadah* is dead, *aggadah* without *halakhah* is wild ... There is no *halakhah* without *aggadah*, and no *aggadah* without *halakhah* ... Our task is to learn how to maintain a harmony between the demands of *halakhah* and the spirit of *aggadah*.' Nonetheless, not only is one greater than the other, but, as with the Master and his emissary, the one that should be subservient has come to dominate:

> To reduce Judaism to law, to *halakhah*, is to dim its light, to pervert its essence and to kill its spirit. We have a legacy of *aggadah* together with a system of *halakhah*, and although, because of a variety of reasons, that legacy was frequently overlooked and *aggadah* became subservient to *halakhah*, *halakhah* is ultimately dependent upon *aggadah*. *Halakhah*, the rationalisation of living, is not only forced to employ elements that are themselves unreasoned, its ultimate authority depends upon *aggadah* ...[113]

113 Heschel 1997 (175ff).

With varying degrees of success a healthy balance is maintained between these hemispheric forces, provided, as Heschel says, *halakhah* (the viewpoint of the left hemisphere) plays a supporting, not the lead, role. In this it is no different from the rest of life. That is how we come to have functioning religious traditions.

A THEOLOGICAL SPECULATION: PANENTHEISM

Niels Bohr thought that simple truths were consistent, but deep truths paradoxical in nature: and he greeted the emergence of a paradox with the words 'now we have some hope of making progress.'[114] I have explored the paradox of the One and the Many in Chapter 21. It is central to an understanding of the divine, because the divine is widely held in many cultures to be transcendent (beyond the world) and undivided, yet at the same time immanent (in the world) and present in all things. As Lord Krishna says to Arjuna of the wise man: 'When he sees me in all and he sees all in me, then I never leave him and he never leaves me': those 'with spiritual vision … worship me as One and as Many, because they see that all is in me.'[115]

As always we need to resist choosing one truth only and ignoring the other; rather, we must see how the greater truth may hold both together.

I referred to the image of Indra's net in which each point of connexion is linked to every other one and indeed reflects the whole net in itself. This web could be thought of as encompassing all things in the cosmos. So is this complex, creative, interconnectedness itself what we mean by God? That would certainly be one version of pantheism, the belief that all things are God, and God is all things. This, however, would be a God that was wholly immanent, as though the world could speak of nothing beyond itself. To me, that is a lack almost as unsatisfactory in its own way as its opposite pole, deism, whereby God is wholly transcendent – 'a transcendent engineer on sabbatical leave', as such a God has been described;[116] remote and unknowable, like the God envisaged by James Joyce's Stephen Dedalus, 'invisible, refined out of existence, indifferent, paring his fingernails'.[117]

Relationship seems to me the central element that is lost when things collapse into either pole in this way. Let's go back to Wordsworth. The passage I quoted earlier seems to me the expression in poetry of pan*en*theism, a belief that all things are *in* God, and God *in* all things:

> … A motion and a spirit, that impels
> All thinking things, all objects of all thought,
> And rolls through all things.

And Wordsworth continues:

> Therefore am I still
> A lover of the meadows and the woods
> And mountains; and of all that we behold
> From this green earth …

[114] Moore 1966 (140).

[115] *Bhagavad Gita*, ch 6 v 30 and ch 9 v 15, trans J Mascaró.

[116] Panikkar 2004 (162).

[117] Joyce 2000 (233).

For me, this answers, better than pantheism, to my sense of there being something flowing, life-giving, creative, responsive, awe-inspiring and sacred *in* 'all that we behold *from* this green earth' – note, not just '*on* this green earth' – but which can never be reduced to what can be seen or fully known; something that both inspires from within the world (is immanent) and embraces it from without (is transcendent); that takes into itself and owns what is good, together with what we are minded to oppose or reject as evil, without thereby becoming equated with such good or evil, even in part; something indeed having no parts; and being both *immediately* knowable and completely unknowable at the same time – as if 'hiding in plain sight'. Moreover, engaging us by love: 'therefore am I still a lover ...' Of course language breaks down here: so either you will find this absurd, or its meaning so transparent that it hardly needs saying at all.

Keith Ward, a formerly atheist analytic philosopher who, following a religious experience, became a theologian and went on to become Regius Professor of Theology at Oxford, points out that the belief that 'God is certainly greater than but *includes* the universe', is not just his own view, but one 'very widely held amongst Christian theologians now', as well as being an ancient Hindu belief.[118] This of course does not make it true, but it does indicate that a similar intuition can be found in widely diverse traditions. It also resonates with a beautiful and (formerly) well-known phrase from the New Testament: 'in him we live, and move, and have our being'.[119]

Panentheism is a theological category, a re-presentation by the left hemisphere of something intuited by the right. And although professional theologians might shudder at the idea, an intuitive form of panentheism is animism. As Tim Ingold points out, animism is not, as anthropologists used to suppose, a belief system (such as is panentheism): that would be to impose our left hemisphere mentality on it. It is not, he says, 'a way of believing about the world but ... a condition of being in it'. As he puts it,

> people do not universally discriminate between the categories of living and non-living things. This is because for many people, life is not an attribute of things at all. That is to say, it does not emanate from a world that already exists, populated by objects-as-such, but is rather immanent in the very process of that world's continual generation or coming-into-being.

The significance of this is that in almost all of Southeast Asia, China, Tibet, Japan, the Pacific Islands, Central and South America, Africa, and the circumpolar North – indeed in almost all parts of the world apart from Western Europe, the Middle East, and non-native North America – animism is the normal way to see the world. Animism, according to Ingold,

---

118 From 'The anthropic universe' 2006: www.abc.net.au/radionational/programs/scienceshow/the-anthropic-universe/3302686 (see transcript).

119 Acts 17:28. I am grateful to Nick Spencer for pointing out that this phrase appears to originate with Epimenides' *Cretica*, where it refers to Zeus.

could be described as a condition of being alive to the world, characterised by a heightened sensitivity and responsiveness, in perception and action, to an environment that is always in flux, never the same from one moment to the next. Animacy, then, is not a property of persons imaginatively projected onto the things with which they perceive themselves to be surrounded. Rather ... it is the dynamic, transformative potential of the entire field of relations within which beings of all kinds, more or less person-like or thing-like, continually and reciprocally bring one another into existence. The animacy of the lifeworld, in short, is not the result of an infusion of spirit into substance, or of agency into materiality, but is rather *ontologically prior to their differentiation.*[120]

This way of being in the world seems to me so much more sophisticated than the way we normally carry on our lives in the West nowadays that it is no surprise that most Westerners can't begin to understand it. And what they fail to understand, they are, with the left hemisphere's characteristic arrogance, inclined to treat dismissively – as something 'primitive' or 'childish', to be 'outgrown'. It has nothing to do with propositional beliefs, but is a matter of direct perception: animists see a world full of spirits not because they are trying to *explain* it, and come up with theories, which science will later 'correct'; but because they perceive immediately that the natural world is not separate from them and is charged with holiness – not something, then, that as it were happened to fall across their path,[121] as if it were a puzzle to be fretted over and *solved*, but full of individual living entities that are experienced directly as part of the same living whole as ourselves; individuals, like us, within a differentiated, but never wholly divided, unity.[122]

Panentheism sees everything as sacred, not just for our *use*. As John Muir put it:

> No dogma taught by the present civilization seems to form so insuperable an obstacle in the way of a right understanding of the relations which culture sustains to wildness as that which regards the world as made especially for the uses of man. Every animal, plant, and crystal controverts it in the plainest terms. Yet it is taught from century to century as something ever new and precious, and in the resulting darkness the enormous conceit is allowed to go unchallenged.[123]

The Greeks, the Inuit, the Penan, the Chinese, the Indians, the intellects of the Western Middle Ages and of the Renaissance, the Australian Aboriginals, the Romantics, the Navajo, the Romans, the Blackfoot and the modern Japanese – and countless others – all thought, or think, that there is something speaking to us in nature. If we alone suddenly can't hear it, in the West in the twenty-first century, how do we know it's we who are right?

---

120 Ingold 2006 (10: emphasis added). As Schelling understood, the distinction between humanity and the world comes very late in our history: it would be anachronistic to think of concepts which we owe to the furthest past of human history as the result of efforts of a subject to make sense of the world by holding a theory about it.

121 Not coincidentally, Paley likened the world to just such an object fallen in one's path: 'In crossing a heath, suppose ... I had found a watch upon the ground, and it should be inquired how the watch happened to be in that place ...'

122 Men 1997, vol 2.

123 Muir 1875 (364).

PROCESSES AND EVENTS

I have referred to Goethe's insight into the distinction between the two kinds of reason, *Verstand* (LH) and *Vernunft* (RH), before: '*Vernunft* is concerned with what *is becoming*, *Verstand* with what *has already become* ... The former rejoices in whatever evolves; the latter wants to hold everything still, so that it can utilise it.'[124] In one of his conversations with Eckermann, he commented on the relation between these two faculties and our capacity to understand Nature, or the God of Nature:

> Mere rationality (*Verstand*) cannot reach as far as Nature; man must be capable of raising himself up to the heights of intuitive reason (*Vernunft*), to touch the Godhead which expresses itself in the fundamental phenomena, physical as well as moral, and which lies behind them and from which they proceed. But the Godhead works in the living, not in the dead; it is in the becoming and the changing, not in what has become and is fixed. So it is that intuitive reason, in her inclination towards the Divine, has to do only with what is living and becoming, rationality with the already fixed and become, so that it may make use of it.[125]

I mentioned Jean Gebser's idea – already there in ancient Zen writings – that the divine is the deepest of springs, yet also that which continually springs forth from it; God, then, as an eternal process. Process theology is, put very simply, the belief that the divine is misconceived as purely a static entity outside time (though that is an accepted aspect of divinity), and is, at least in some important aspects, better seen as a process within time, an eternal Becoming rather than merely an eternal Being – though it is that, too. We need Heraclitus' insight that 'by changing, it stays the same'. In the Introduction, I mentioned that flow continually creates newness, while remaining itself stable: changing and yet not changing.

Process theology is a natural counterpart or companion to panentheism, since it, too, implies that God is in everything without being reducible to the sum of everything: the spring and that which comes forth from the spring.

There is a common prejudice that process theology is a modern invention: perhaps it is in its current forms (there are many). But it has resonances with a host of traditions including pre-Socratic philosophy, Taoism, Buddhism (including Zen), Judaism, mediaeval Christian mysticism, the philosophies of Hegel and Schelling, the writings of Wordsworth and Coleridge, and the philosophies of James, Bergson and Whitehead, among others – indeed Whitehead is the figure generally credited with originating process theology as such.

---

124 See p 902 above.

125 Eckermann 1889–96, vol 6: entry for 13 February 1829. The contrast that Goethe is getting at here between *Verstand* and *Vernunft* could be rendered as the contrast between Bergson's *intellect* and *intuition*, the Latin *ratio* and *intellectus*, Greek *dianoia* and *nous* – and, effectively, LH and RH. The slightly unsatisfactory phrase 'intuitive reason' is the best I can do to capture the combination of reasoning and intuiting expressed by *Vernunft*. » *Der Verstand reicht zu ihr nicht hinauf, der Mensch muss fähig sein, sich zur höchsten Vernunft erheben zu können, um an die Gottheit zu rühren, die sich in Urphänomenen, physischen wie sittlichen, offenbaret, hinter denen sie sich hält und die von ihr ausgehen. Die Gottheit aber ist wirksam im Lebendigen, aber nicht im Toten; sie ist im Werdenden und sich Verwandelnden, aber nicht im Gewordenen und Erstarrten. Deshalb hat auch die Vernunft in ihrer Tendenz zum Göttlichen es nur mit dem Werdenden, Lebendigen zu tun, der Verstand mit dem Gewordenen, Erstarrten, dass er es nutze.* «

It is also suggested by the hemisphere hypothesis. For what the left hemisphere knows is secondary to whatever the right hemisphere knows. What the left hemisphere sees as finished, perfect, single, abstract, detached, motionless, beyond space and time, is virtual only; a reduced re-presentation at an instant outside time of what according to the right hemisphere is always evolving, ever both self-differentiating and self-unifying, and involved with the process of creation and what it creates.

Throughout Part III, I have emphasised a number of positions that are not the norm in our culture (though they are, I believe, accepted by many contemporary physicists); the primacy of motion over stasis, and the importance, in particular, of flow; the reality of time as an expression of that extended flow, not a series of linear moments; consciousness and matter as not simply irreconcilable, leaving us with the problem of how to get consciousness out of matter, but in reality aspects of one another, in which consciousness is nonetheless primary; and the world having purposiveness without reductive, preordained purposes. As the reader will have observed, all of these themes are consistent with what I have to say in this chapter.

Similarly the idea that relationships and processes are more fundamental than things and 'states' has been a main theme of this book, so I will deal with this only briefly here, emphasising, though, for the first time how it is expressed within the world's great religious traditions, and in spiritual thinking to the present day.

Where, then, do I see it?

In the Biblical story of Moses and the burning bush, God is said to have declared, in most English translations, 'I am that I am'. Apparently the simplest and most direct translation of the Hebrew words, *ehyeh asher ehyeh*, is 'I will be who (or, that which) I will be'. In the *Zohar*, a body of kabbalistic texts, the appellation 'I will be' (*ehyeh*) is applied to the highest of the emanations of the infinite by which the cosmos is constantly sustained and created, known as *Keter*. This indicates, according to Sanford Drob, '*Keter*'s limitless potential, and its wilful movement toward a future.'[126] Here I find a close parallel to the view of a God and cosmos that are purposeful, yet undetermined, and *each* in the process of becoming what they are. (If God is in the cosmos and the cosmos in God, as in panentheism, in such a way that they cannot be wholly separated – though never merely equated – this would naturally follow.)

In the account of the creation contained in the opening of the Book of Genesis, after each act of creation, it records that God looked and '*saw* that it was good'.[127] To me this speaks of an encounter with something new, of something free and hitherto undetermined – of veritable creation; not just, as Bergson put it, the unfolding of a fan.[128] What it suggests is that God did not know already that it was good

---

126 Drob 1997. *Keter* is the highest of the *sefirot*.

127 These words are repeated after every day's work, except for the second day. But as has been pointed out, on this day, uniquely, God did not create, but rather divided what had already been created.

128 See p 911 above.

(the Hebrew word can also be translated 'beautiful') without having to see it. It speaks of something free, and *Other*. (The realm of the right hemisphere is relatively speaking, that of the uncertain and undetermined, the left that of the certain and determined; moreover the right hemisphere is the one that preferentially relates to what is Other, the left hemisphere to what is already known and has been internalised. And, in Chapter 8, I related the evidence that, unsurprisingly therefore, *creativity* involves principally the right hemisphere.)

Turning to the tradition of Christianity, philosopher John Lucas points out, in a fascinating paper entitled 'Begotten not made', that the phrase used in St John's Gospel to express the incarnation, the word 'was made' flesh, departs from the sense of the original Greek ἐγένετο, which is much better translated as 'became'.[129] 'The word was made flesh' suggests a single, discrete, wilful, act of 'being done to'; by contrast, ὁ λόγος σὰρξ ἐγένετο (*ho logos sarx egeneto*) suggests 'a natural inherent process' of becoming. The intended meaning is that the *logos*, the universal origin, ground or reason, *became embodied*.

'The distinction between *becoming* and *being made*', Lucas reflects,

> continues through the Middle Ages down to the present day. Duns Scotus was a 'becoming' thinker. He sought explanations, arguing for the existence of God as the ultimate explanation, with the universe evolving in an intelligible way ... William of Ockham, by contrast, was a 'maker'. He emphasised the importance of the will, and the power of an omnipotent God to do whatever he liked. We can see him as the spiritual progenitor of the Atomists' point-particles placed in a uniform and featureless space wherever God wanted them to be according to the inscrutable counsels of omnipotence.

Here Lucas contrasts, in the works of two loosely contemporary mediaeval philosophers, an evolutionary flow, on the one hand, a process of Becoming, in which all comes to be individuated (Duns Scotus was also the originator of the term *hæcceitas* to express the importance of the unique individual element), though remaining part of one whole; and an account, on the other hand, of the wilful exercise of omnipotence, arbitrarily disposing things as 'point-particles ... in uniform and featureless space'. It is impossible not to recognise a right hemisphere-congruent way of conceiving reality – 'becoming', a process – contrasted with a left hemisphere-congruent way – 'was made', literally (*factum*), a deed or fact.

The greatest of the mediaeval mystics, Meister Eckhart, is associated with what has been called a 'metaphysics of flow'.[130] For him Being, which emanates from the loving God, must flow, because, like love, it constitutes a continual movement toward the Other. Moreover, it is a Becoming: according to Eckhart, God (in his in-

---

129 Quite why this happened reveals the treachery of language. 'Why had the Authorised Version plumped for "was made"? Because in Latin, "I become" is *fio*, whose past tense is *factus sum*', writes Lucas. The point is that the translators of the Authorised Version of the Bible where possible followed earlier translations, which in turn had been made from the Latin Vulgate. The paper was privately communicated to me, and is not, as far as I know, published: before he died Lucas gave me permission to quote from it and to make it available on my website.

130 Radler 2010.

131 Eckhart, Sermon (Pfeiffer) XXVI: 1931–47, vol 1 (76). I gratefully acknowledge Professor Charlotte Radler's help.

carnate form as Christ) streams forth from the Father's heart endlessly into the God-loving soul: 'he is being born anew unceasingly … And this same birth today in the God-loving soul delights God more than his creation of the heavens and earth.'[131] Eckhart is said to have said: 'We are all meant to be mothers of God, for God is always needing to be born.'[132] According to him, then, God also depends on us for his Becoming.

The idea that God is love, or even the 'word' (*logos*), suggests that ultimately what is primary is relationship: a word exists only in the betweenness of utterance and audition, which has the same structure as love. Love is an experience always in process, never a thing or anything like a thing. The mediaeval mystic Mechthild of Magdeburg spoke about the 'flowing light', the 'flowing fire' of God's love, which 'never stands still and always flows effortlessly and without ceasing in so sweet a flood.'[133]

All of this is more consonant with the right hemisphere's take on the world than that of the left: it depends on process not stasis; it places the emphasis on relationship not on entities that must be related *post factum*; it embraces, rather than flees from, paradox. If God is Becoming, at least as known from within creation – which is all we can know – this has important consequences. Oddly (since it might seem to do the opposite) it has the effect of removing God from time (in the sense of *temps*, though not of *durée*: see Chapter 22). For if God were to be the first cause, in some sequential sense of that word 'first', or a creator God that 'started things off', God would be inevitably situated as performing a function in time (if not, God could not act as the temporal first cause). Instead, a God that is endlessly Becoming is already situated in an eternal Now; and we are back to the true meaning of the first cause – not a cause on a timeline, but an ongoing ground of Being. So Meister Eckhart says: 'God is in all things … God is creating the whole world now, this instant … Where time has never entered and no form was ever seen, at the centre, the summit, of the soul, there God is creating the whole world.'[134]

In a not dissimilar vein the early seventeenth-century philosopher and mystic Jakob Boehme says: 'It is an everlasting beginning. It begins itself perpetually and from eternity to eternity, where there is no number; for it is the Unground.' According to Boehme, what he calls the Unground is an eternal Nothing, which yearned for Something: out of which there arose *Becoming*. Longing, the right hemisphere's awareness of God,[135] was (according to Boehme) the Unground's drive to create. This Becoming is a constant longing *and* its fulfilment in a something that is evermore coming into Being.[136] This idea shares resonances with Buddhism; and with the

132 I cannot trace the origin of this exact phrase, though it is frequently repeated in anthologies, and expresses well what Eckhart never formulated quite so succinctly. The closest would seem to be in Sermon (Pfeiffer) LXXXV: 'God has been begetting his only-begotten Son and is giving birth to him … now and eternally, he, like any woman, being brought to bed in every virtuous soul who has embarked on the interior life … What does it profit me the Father's giving his Son birth unless I bear him too? God begets his Son in the perfect soul and is brought to bed therein that she may bring him forth in all her works': 1931–47, vol 1 (216). There are also parallels in 'The nature of the soul … want[s] to have God begotten in her': Sermon (Pfeiffer) XX: 1931–47, vol 1 (65); and 'He *must* be born in you' (Sermon 24a [no Pfeiffer equivalent], 2009 (160). I am grateful to Professor Bernard McGinn, Dr Rebecca Stephens and Fr Richard Woods OP for their advice here.

133 Mechthild of Magdeburg 2012.

134 Eckhart, Sermon (Pfeiffer) LXVI: 1931–47, vol 1 (164); and cf 'God makes the world and all things in this present now': Sermon (Pfeiffer) LXXXIII, *ibid* (209).

135 See below; and lecture by this author, entitled 'Wanting and longing' delivered at Heythrop College, 7 March 2017: www.youtube.com/watch?v=lSiKK2x4LJg (accessed 15 March 2021).

136 Boehme 1920: 'The Fourth Text', §9 and §1. Rückert's lines addressed to his beloved, and set to music by Schubert in 'Du bist die Ruh', come to mind: » *Die Sehnsucht du, | Und was sie stillt.* «

Wordsworthian idea of Nature, both as she is in herself and in our understanding and response to her, as, in a wonderful phrase, 'something evermore about to Be'.

And Nature, and the one unifying spirit that Wordsworth believed we encounter in Nature, *flows*. He felt himself swept up in a flow of vital energy when in moments of sublimity 'my blood appeared to *flow* / With its own pleasure, and I breathed with joy'. So Emerson wrote that 'in the presence of nature, a wild delight *runs* through the man, in spite of real sorrows'. Wordsworth's sense of a spirit that impels, and '*rolls* through all things' is close to Emerson's description of rapture occasioned by Nature, when 'I am nothing; I see all; the currents of the Universal Being *circulate* through me; I am part or particle of God.'[137]

What Wordsworth expressed more poetically, his friend and literary companion Coleridge expressed more philosophically. The best of Coleridge's thinking, in my view, lies not in his more abstract, Schellingian disquisitions, but in the many astute observations he made throughout his life on Shakespeare, whose works were for him like the productions of Nature herself. (Kant, in whose work Coleridge was immersed, thought that only artists could attain to genius, since they participated intuitively in the very creativity of organic Nature: they brought new forms into being in their wholeness, not simply putting them together from distinct parts.) In his reflections on Shakespeare, Coleridge crystallised some philosophical insights that we may recognise.

The first might be this. By contrasting the works of Shakespeare with those of his lesser contemporaries, Coleridge is able to illuminate the relationship between a true creator (in this case Shakespeare) and his creation. He does this through a contrast between the nature of the 'dead', static, finished, concatenated products of a controlling, ever obvious, will – the God of deism; and, on the other hand, the 'living' creations of an animating, in some sense hidden and unknowable spirit, which are themselves unknown ahead of time, and come seamlessly into being, as if generating themselves within the flow – God as the *tao*. This is the mediaeval philosophical distinction, later taken up by Spinoza, between, respectively, *natura naturata* and *natura naturans*. In the works of Beaumont and Fletcher (Shakespeare's more workaday contemporaries), says Coleridge, 'you will find a well arranged bed of flowers, each having its separate root, and its position determined aforehand by the *will* of the gardener'. In Shakespeare, by contrast, 'all is growth, evolution, γένεσις [genesis] – each line, each word almost, begets the following – and the will of the writer is an interfusion, a continuous agency, no series of separate acts.'[138]

In one of his lectures, Coleridge says:

---

137 Emerson 1836 (13).

138 From a marginal note by Coleridge, contrasting Shakespeare with Beaumont and Fletcher, made between 1815 & 1819 in a copy of *The Dramatic Works of Ben Jonson, Beaumont and Fletcher*, 1811: in Raysor 1936 (88–9: emphasis in original).

The form is mechanic when on any given material we impress a predetermined form, not necessarily arising out of the properties of the material – as when to a mass of wet clay we give whatever shape we wish it to retain when hardened – The organic form on the other hand is innate, it shapes, as it developes itself from within, and the fullness of its development is one & the same with the perfection of its outward Form. Such is the Life, such is the form. Nature, the prime genial artist, inexhaustible in diverse powers, is equally inexhaustible in forms.[139]

Earlier I remarked on the generous profligacy of Nature, and its inversion of the *lex parsimoniæ*. The literary scholar A. D. Nuttall observed of Shakespeare that his works may be 'likened to Ockham's beard, golden, luxuriant, not yet subdued (happily) by the famous razor.'[140]

Then, second: we have seen that, in *living* systems, it is by changing that things remain the same – permanence and impermanence, durance and process, are aspects of one and the same phenomenon. So Coleridge remarks that what Shakespeare gives us is a union of 'the liveliest image of succession with the feeling of simultaneousness'.[141] 'In Shakespeare', he writes elsewhere, 'there is neither past nor future, but all is permanent in the very energy of nature.'[142] Process and durance together.

Thirdly, he sees that this process involves the all-important union of division with union; maximal individuation, that does not disrupt, but rather enriches, the whole. So, in Shakespeare, 'the play is a *syngenesia*', he writes, invoking an image from botany suggestive of individuation within a unified whole.[143] Of Shakespeare's characters he writes, 'each has indeed a life of its own and is an *individuum* of itself, but yet an organ to the whole.'[144] The creative force is still entirely itself, even when it is expressed in the most *individuated* form: thus, 'Shakespeare becomes all things, yet forever remaining himself;'[145] he is enabled 'to become by power of Imagination another Thing – Proteus, the river, a lion, yet still the God felt to be there.'[146] In this last phrase, the analogy between the very greatest of human creations and the divine creation of the cosmos is made explicit in Coleridge's language. These expressions, if applied to the relation of God to his creation, rather than Shakespeare to his, are pure panentheism.

And finally, in an analogy with the hiddenness of God, who is both wholly revealed and wholly concealed in his Creation, Coleridge writes of Shakespeare's work that it is 'all Shakespere, & nothing Shakespere'.[147] These thoughts were later to be echoed by T. S. Eliot: 'The world of a great poetic dramatist is a world in which the creator is everywhere present, and everywhere hidden.'[148] He was, of course, thinking of Shakespeare.

139 Coleridge 1987, vol II (495).
140 Nuttall 2007 (181).
141 Coleridge 1817, vol II (20).
142 Coleridge 1856, lecture IX (107).
143 OED: 'If ... the filaments ... are free and distinct, but the anthers are connected together, so as to form one body, then your plant will be found in the class syngenesia': taken from Jean-Jacques Rousseau's *Letters on the Elements of Botany*, as translated by Thomas Martyn, 1794 (94).
144 Coleridge 1987, vol II (151).
145 Coleridge 1817, vol II (22).
146 Coleridge 1987, vol I (69).
147 Coleridge 1962, §2086.
148 Eliot, 'The three voices of poetry', in 1957 (102). I owe this connexion to Perry 1999.

Ultimately we have to reconcile the tendency to flux with a tendency to stasis, change with (some degree of) permanence. The worlds of the right hemisphere and the left have to be brought into fruitful conjunction. Here too Coleridge saw what was needed:

> To *reconcile* therefore is truly the work of the Inspired! This is the true *Atonement* – i.e. to reconcile the struggles of the infinitely various Finite with the *Permanent*.[149]

Atonement is literally 'at-one-ment': reconciliation of apparent incompatibles.

These philosophical positions would have been familiar to Schelling, from whom Coleridge learnt much. I have discussed Schelling's philosophy, as it relates to the creative process in Nature, in a previous chapter.[150] Matthew Segall notes that 'Nature was no mere appearance for Schelling, but rather the living ground and visible body of an *eternally incarnating* divinity.'[151] Not a divinity that had incarnated once, note, but one that is always incarnating itself in the evolving cosmos. For Schelling, and it is a position to which I subscribe, the imagination is not, as for Kant, a faculty that creates merely the best we can manage as a *re-presentation* of the world; nor is it making the world up from scratch. It is collaboratively allowing the world to *presence*, bringing the world into existence; and if it is the case that the soul is not separable from the God that is the ground of all that is, this is entirely in keeping with the imagination helping to constitute the world as it really is. This is remarkably similar to Eckhart's deep insight that 'the nature of God ... is to give birth,' and that the birth happens in the soul of each one of us.[152]

Whitehead wrote that 'the elucidation of meaning involved in the phrase "all things flow" is one chief task of metaphysics'.[153] I follow him in seeing processes and relationships, not things themselves, as the fundamental realities.

Does this tell us something about the nature of God? Whitehead thought so. Moreover he saw God as the principle that made possible, and was expressed in, the newness of creation, the presence of order in complexity, and of purpose, within the cosmos. His view of God's interaction with the cosmos is dialectical, in that God and the world fulfil each other and bring each other into being. The one and eternal becomes many and ceaselessly changing, just as the many and ceaselessly changing become one and eternal. In the words of Schelling once more: 'Existence is the conjunction of a being as One, with itself as a Many.' This is not, then, as Whitehead crucially recognised, two processes, but two *facets of a single process*:

> It is as true to say that God creates the World, as that the World creates God ... Neither God, nor the World, reaches static completion. Both are in the grip of the ultimate metaphysical ground, the

149 Coleridge 1962, §2208 (emphases in original).

150 See pp 958–9 *et passim* above.

151 Segall 2018 (emphasis added).

152 Eckhart, Sermon (Pfeiffer) XII: 2009 (101).

153 Whitehead 1929a (208).

creative advance into novelty. Either of them, God and the World, is the instrument of novelty for the other.[154]

God, truth, and infinity are all processes, not things; comings into being, not entities that are already fixed. All three seem to me, however, like rivers, to combine, stability with flux. 'All things flow'; but 'by changing, a thing remains the same'.[155] Ultimately Being and Becoming are aspects of the same thing.[156] It's just that our culture emphasises Being to the exclusion of Becoming. However, as usual, there is an asymmetry: they are not equal. In the philosophy of Whitehead, the divine is Becoming, and Becoming is even more fundamental than Being. In keeping with this, Wolfgang Pauli described the quantum theory he helped to establish, not as a theory of being, but as 'a theory of *becoming*'. He saw what he called the 'interplay' between possibility and actuality as 'dialectical'.[157] And the theoretical physicist David Bohm said, even more clearly, that

> in my scientific and philosophical work, my main concern has been with understanding the nature of reality in general and of consciousness in particular as a coherent whole, which is never static or complete but which is an *unending process of movement and unfoldment* ...[158]

What Bohm called the Implicate Order unfolds itself continually. But he also believed it was then re-enfolded.

Oddly enough, we find ourselves here close to the perceptions of a much earlier figure, one that we have repeatedly encountered in Part III, Nicholas of Cusa. He tried to explain his understanding of the relationship between God and his creation through a metaphor of *un*folding from inside outwards (*explicatio*, from Latin *ex-*, out), and *en*folding, from outside inwards (*complicatio*, from Latin *com-*, with, + *plicare*, to fold).[159] According to this formulation, while all beings are an 'unfolding' of God in time and space, they are at the same time 'enfolded' in the undifferentiated oneness of God, their divine source. In terms of the hemisphere hypothesis, we recognise the same process by which we come to understand anything whatsoever: what is at first *im*plicate is taken up by the right hemisphere, then *ex*plicated by the left hemisphere, and then the products of that explication re-enfolded or reintegrated in its more *com*plicated form by the right hemisphere's vision of the whole once more. This is how the hemispheres, when they work well together, co-operate in giving us insights into the depths of reality.

POTENTIALITY AND ACTUALITY

According to Whitehead, the World converts potentiality into actuality; what the World makes actual, God takes back into a field

---

154 ibid (348 & 349).

155 Heraclitus fr LII [Diels 84a, Marcovich 56A]. See p 431 above.

156 Thus Cusanus saw in God the *coincidentia* of motion and rest, 'not as two but as above duality and otherness' (Cusanus 1978b, §74).

157 Pauli 1994 (48: emphasis added). In a footnote to the remark about 'theory of becoming', he adds: 'One may, following [the Swiss mathematician and philosopher] F Gonseth, denote the interplay of the two aspects as "dialectical".'

158 Bohm 1980 (ix).

159 The terms he used were, respectively, *complicatio* & *explicatio*. This is not dissimilar to an idea in Eckhart, Sermon (Pfeiffer) VII: 2009 (87): 'How marvellous to be without and within, to embrace and be embraced, to see and be seen, to hold and be held – that is the goal, where the spirit is ever at rest, united in joyous eternity.'

of receptive potentiality, one so shaped as to draw what has been created onwards towards further fulfilment in actualisation. It is because of this reciprocal, yet opposite, motion, that God and the World ceaselessly bring something into being, the world drawing actuality out of possibility, and God responding to that actuality with further possibility.[160]

This idea is already there in early Renaissance theology. In *On the Summit of Contemplation* (*De apice theoriæ*), Cusanus calls God *posse ipse*, that is, Possibility Itself. But he also used the term *possest* for God. *Possest* is, as Bond puts it,

> a play on words, a coincidence of *posse* ('can') and *est* ('is'), the Can, the Possibility that at the same time Is, the Can-Is, which only God can be.[161]

This is another way of expressing a similar insight to Whitehead's, since what Whitehead calls God and World, Cusanus saw as different aspects of the divine (which is why it can be all three – *posse*, *est* and *possest* – at once). I do not believe that Whitehead would have dissented from this. He did not see God and World as wholly separate entities, but as two distinct aspects of a seamless process, the process of creation.[162]

Such ideas as these are hardly peripheral or unorthodox in the history of Christianity. Thomas Aquinas thought of God as an infinite potential, attracting things to their fulfilment. Yet in doing so God is not seen as determining, engineering or controlling, though neither is God merely passive. From this perspective, God is seen as the ultimate good who attracts all things to their flourishing, the possibility that is most fulfilling for them, but does not *compel* them to take that path: they have the freedom to respond for better or for worse. This is like a lover, who by virtue of love draws whatever emerges in the loving relationship towards a greater fulfilment in love, but cannot in any way enforce such an outcome.

In the last hundred years such ideas have taken various forms that reflect the same basic structure. Jung's idea of the collective unconscious, while certainly not equivalent to the concept of *posse ipse*, is nonetheless an attempt to articulate a sense of an absolute origin that also guides the individual life: we arise from its common and timeless field, yet are independent of it, while being nonetheless in constant relation to it, drawn towards it and influenced by it, as if by a gravitational field. We are free to heed or not to heed (and at present, we seem to be heeding very little). The form or field of potential – God, *Ein-sof* in the Kabbalah, the collective unconscious to Jung – draws something out of the world to meet itself.[163] This is Whitehead's constant creative advance into the newness of being: Becoming.

---

160 See Whitehead 1929a (341–9): and, for discussion, Oomen 2015. This is not to be seen as two separate processes, one attributable to God, the enrichment of potential; and the other to the World, the enrichment of actualisation. Rather, the World is that aspect of God that gets to be actualised, towards which God strives just as much as to ever-renewed potential. Cf Whitehead 1933 (357): 'We must conceive the Divine Eros as the active entertainment of all ideals, with the urge to their finite realization, each in its due season. Thus a process must be inherent in God's nature, whereby his infinity is acquiring realization.'

161 Bond 1997 (58).

162 Similarly in the Kabbalah, *Ein-sof* includes within itself, although distinct, the *sefirot*, which are modes or attributes through which God's Being is manifest in the cosmos.

163 *Ein-sof* should be understood as God prior to any self-manifestation: *Ein* is a negative, *sof* means grasp. This is conventionally translated as the Infinite. Is it possible that it also suggests that which cannot be grasped (or indeed that which itself *does not grasp*)?

This relationship is put at its most dramatic by Meister Eckhart who says that God depends on the human soul as much as the human soul depends on God. Equally in the Kabbalah, according to Drob,

> just as humanity is dependent for its existence upon *Ein-sof*, *Ein-sof* is dependent for its actual being upon humanity. The symbols of *Ein-sof*, *Shevirah* (rupture) and *Tikkun* (repair) thus express a coincidence of opposites between the presumably opposing views that God is the creator and foundation of humanity and humanity is the creator and foundation of God.[164]

Returning once more to Whitehead, he did not see the process as something remote, abstract and mathematical – as though the world and God had no part in the values that are embedded and embodied in the process – but one characterised by compassion (literally, 'suffering with'): 'What is done in the world is transformed into a reality in heaven, and the reality in heaven passes back into the world … In this sense, God is the great companion – the fellow-sufferer who understands.'[165] Christianity is, above all, the religion that speaks of vulnerability and love, in the image of a God that cared for creation in such a way as to be unable not to suffer in and alongside it. Whitehead nonetheless thought Christianity had erred by presenting God as a divine ruler, whose outstanding characteristic is power: he preferred what he called 'the brief Galilean vision of humility', characterised by love.[166] There are hemispheric implications here, too, that are too obvious to need pointing out.

HOLARCHIES

I quoted physicist David Mermin's saying that 'correlations have physical reality; that which they correlate does not'. In explaining this he writes:

> The physical reality of a system is *entirely contained* in (a) the correlations among its subsystems and (b) its correlations with other systems, viewed together with itself as subsystems of a larger system.[167]

All of reality is a network in which our attention artificially isolates different, but cognate, sets of immediate relations at each level with which it interacts. This, in turn, is not unlike Koestler's idea of what he called a holarchy. Theise and Kafatos, writing of the phenomenon in biology, say:

> in fact, the whole exists not at any single level of scale, nor in a hierarchy of systems, but, to use Koestler's term, as a holarchy, a holistic (quantum-like) superposition of all levels of scale, [as imaged in] our bodies which are comprised of human and non-human cells. At the nanoscopic scale, cells themselves disappear from view to reveal atoms and molecules self-organizing in aqueous suspension.

164 Drob 2009 (133).
165 Whitehead 1929a (351).
166 ibid (342–3).
167 Mermin 1998 (emphasis added).

> No single scale of observation can reveal the whole; at the moment selection is made of a scale of observation, the features of other levels of scale are hidden from view.[168]

Cusanus was fond of the saying *quodlibet in quolibet*: 'each thing is in every thing'. According to Cusanus scholar C. L. Miller,

> just as God is present to each creature that stands as a contracted image of the divine, so the universe as a macrocosm is present to each creature or constitutive part as microcosm. In that way, each natural thing is an image of the collective whole. But since this collectivity is made up of interrelated parts, each thing is also the totality of its connections with everything else. 'Each thing is in each thing' because each is an image reflecting the oneness of the whole and thus of all other individuals that are the interrelated parts of that whole.[169]

This gives reality an essentially nested structure, in which what looks like a part at one level, is a whole at another. Thus the subatomic particle is to the atom, as the atom to the molecule, as the molecule to the compound – to the organelle, to the cell, to the tissue, to the organ, to the body, to the family, to the community – and so on up to the whole earth, and the cosmos beyond. As one moves the plane of focus, one *Gestalt* comes into being and another is relinquished. This ex-plicates (and com-plicates, in Cusanus's sense) an insight that is part of the perennial philosophy, and is encapsulated in the Vedic saying: 'as is the individual, so is the universe; as is the universe, so is the individual'.[170] It also recalls the image of Indra's net, since in each jewel all the others are reflected. And according to the kabbalists, since the microcosm perfectly mirrors the macrocosm, the *sefirot* are 'not only the dimensions of the *universe*, but also the constituent elements of the *human mind*.'[171]

This draws attention to the fact that the universe is one whole and yet is seen in every part: if one looks deeply into the particular, one sees the universal, one of Goethe's most important insights. Or as Whitehead put it: 'you cannot abstract the universe from any entity ... so as to consider that entity in complete isolation ... In a sense, every entity pervades the whole world.'[172]

Referring to the way in which, until the last couple of centuries, we used to understand everything in the cosmos by analogy, Bohm commented: 'The human being was a microcosm of the cosmos, so that he had implicitly in him the possibility of understanding it. The general view before our modern times was more favourable to wholeness – in Europe as well as in the East.'[173] The mediaeval idea that we are microcosms, that we reflect the universe, is curiously close to the idea of fractality, or the hologram – we are images of the

---

168 Theise & Kafatos 2016.
169 Miller 2017.
170 'Yatha pinde tatha brahmande, yatha brahmande tatha pinde.'
171 Drob 1997 (emphasis added).
172 Whitehead 1929a (28).
173 Bohm 1989.

very forms and processes of the universe. That the cosmos is fractal or holographic is indeed not an idea alien to modern physics.[174] It is also a striking feature of the philosophy of the three thinkers in German intellectual history who were most myriad-minded in their interests and abilities, in their 'universal genius': Nicholas of Cusa, Leibniz and Goethe.[175]

THE HIDDEN AND THE MANIFEST

In some places, Cusanus images God's relation to the world by using the metaphor of light. We cannot see light if there is nothing to reflect it (space looks dark, even though light constantly passes through it). When there is a reflection, we see not light, but the object that reflects it. Light itself we cannot see: yet in its absence we can see nothing.[176] The created world is a direct reflection of God in creation: we do not 'see' God, but without God we would 'see' nothing. If we take this metaphor from the realm of the senses to that of the intellect and the soul, we are made to understand that there is no being without Being, yet we do not perceive Being itself: rather we perceive its reflection in the world, and everything we do perceive is perceived only because of that reflection. The light is in itself invisible, a darkness to our intellect. This may be, I suggest, part of what mystics mean when they say that God is everywhere visible and nowhere visible, and what Goethe meant by saying that nature is a 'holy open secret' (*heilig öffentlich Geheimnis*).[177] The idea that the hidden and the manifest are dipolar characteristics of one reality is well expressed in the *Tao Te Ching*:

> Ever desireless, one can see the mystery.
> Ever desiring, one sees the manifestations.
> These two spring from the same source but differ
>   in name; this appears as darkness.
> Darkness within darkness.
> The gate to all mystery.[178]

I mentioned in Chapter 20 that the phrase *coincidentia oppositorum* comes from Cusanus. Though generally cited as found in his treatise *On Learned Ignorance* (*De docta ignorantia*), this would appear not to be correct. It occurs several times, however, in *The Vision of God* (*De visione Dei*): for example, Nicholas writes

> Hence I observe how needful it is for me to enter into the darkness, and to admit the coincidence of opposites, beyond all the grasp of reason, and there to seek the truth where impossibility meeteth me …

And, particularly movingly, to my way of thinking,

---

[174] See, eg, Currivan 2017.

[175] See, eg, Beck 1969 (58).

[176] William Earle, describing what he calls the Technical Philosopher of the mid-twentieth century, writes that he 'is the point of pure negativity, an eye which would like to see pure light but *cannot because of visible things*' (Earle 1960, 374: emphasis added).

[177] Goethe, 'Epirrhema'. Hadot has written well of Goethe and this 'open secret': see Hadot 2006, esp ch 9, 'Isis has no veils', 247–61. One of Hadot's phrases is especially apt: 'if Isis is without veils, it is because she is entirely form, that is, entirely veil; she is inseparable from her veils and her forms' (259).

[178] *Tao Te Ching*, §1 (trans G-F Feng).

I begin, Lord, to behold Thee in the door of the coincidence of opposites, which the angel guardeth that is set over the entrance into Paradise.[179]

In these passages Cusanus suggests the need to enter into a darkness of unknowing which is also the door to Paradise – itself a coincidence of opposites – accessible only to whomever is able to abjure *finally* the attempt to apprehend God through reason (at least in as much as that is of the kind that precludes the possibility of 'opposites' being true). God is beyond the familiar domain where opposites hold sway: in God they are reconciled. Thus it is that the darkness is in reality a form of extreme brightness. In the mystical literature, God is often described in these terms. Dionysius the Areopagite, in his *Mystica theologia*, wrote of 'the superessential Radiance of the Divine Darkness ... the superessential Darkness which is hidden by all the light that is in existing things'.[180] Much later, in the seventeenth century, the mystical poet Henry Vaughan, whose work is impregnated with imagery of light and dark (the title of his great collection of poems *Silex Scintillans* means 'the flint that yields a spark') wrote: 'There is in God (some say) / A deep, but dazzling darkness ...'[181]

GOD AS COINCIDENTIA OPPOSITORUM

In his *The Hunting of Wisdom* (*De venatione sapientiæ*) Cusanus alludes to Dionysius and endorses his view that 'opposites are to be affirmed and denied of God at the same time'.[182] And in *The Vision* he notes that we affirm of the incarnate God 'most true contradictories. For you are creator and likewise creature, the attracting and likewise the attracted, the infinite and likewise the finite.'[183] God is not simply oneness, according to Cusanus, but 'oneness to which neither otherness nor plurality nor multiplicity is opposed'.[184] Unity and multiplicity are finally reconciled in God.

When one reaches the infinite, opposites coincide and the opposition is resolved (any 'arc' of an infinitely large circle would be a straight line: any straight line infinitely prolonged would become a circle). In Buddhism, one is constantly reminded that true fulfilment is emptiness. In the Kabbalah, *Keter*, the utmost crown of the *sefirot*, represents emptiness in the infinite. 'At infinity thoroughgoing coincidence occurs', writes Bond referring to Cusanus. However:

> The coincidence of opposites provides a method that resolves contradictions without violating the integrity of the contrary elements and without diminishing the reality or the force of their contradiction. It is not a question of seeing unity where there is no real contrariety, nor is it a question of forcing harmony by synthesizing resistant parties. Coincidence as a method issues from coincidence as a fact or condition of opposition that *is resolved in and by infinity*.[185]

---

179 *De visione Dei*, ch ix, §36, lines 1–3: '*Unde experior, quomodo necesse est me intrare caliginem et admittere coincidentiam oppositorum super omnem capacitatem rationis et quærere ibi veritatem, ubi occurrit impossibilitas*'; and ch x, §40, lines 1–2: '*Unde in ostio coincidentiæ oppositorum, quod angelus custodit in ingressu paradisi constitutus, te, domine, videre*' (translations from Cusanus 1960, 43 & 46).

180 Dionysius 1923 (9 & 12). Cf Thomas Merton's 'Prayer before Midnight Mass', Christmas 1941: 'Your brightness is my darkness. I know nothing of You and, by myself, I cannot even imagine how to go about knowing You. If I imagine You, I am mistaken. If I understand You, I am deluded. If I am conscious and certain I know You, I am crazy. The darkness is enough.'

181 Henry Vaughan, 'The Night'.

182 Cusanus 1978a, §67.

183 Cusanus 1988, §92. See Hopkins undated.

184 Cusanus 1985, §76.

185 Bond 1997 (22: emphasis added).

Neither opposite is excluded. Only their mutual exclusion is excluded.¹⁸⁶

The dipolar relationship between being and becoming in theory should – and it seems to me in practice does – apply to infinity. The mathematical physicist Edward Nelson distinguished potential and actual infinity. He described a feeling of 'loathing' and oppression when thinking about what he called 'actual infinity'. Following Whitehead he called it 'the antithesis of life, of newness, of becoming – it is finished.'¹⁸⁷ 'Perfect' literally means completed – done with. The achieved infinite of the left hemisphere is unreal (can never be achieved) and is lifeless; the constantly becoming, processual, infinite of the right hemisphere is real and life-giving. Note that this is an *inversion* of what we normally hold, namely, that potential means unreal, and actual means real; something I touched on in Chapters 21 and 22.

It seems to me, that, as usual, when it comes to the nature of Being, or God, we are easily attracted to, and too readily comforted by, the idea of a fully achieved perfection, rather than one of open dialectical creativity, continually both expressive and receptive. When God is analogous with Nelson's 'actual' infinity, he is re-presented – literally no longer present – and therefore, if one can put it that way, for the time being dead.

Another important figure in the rich literature of mediaeval mysticism is the fourteenth-century Flemish writer Jan van Ruysbroeck. 'God in the depths of us', he said, 'receives God who comes to us: it is God contemplating God.'¹⁸⁸ God comes to know himself in the Other: we come to know ourselves in God: the whole comes to know itself through the part, but the part comes to know itself through the whole. And he goes further: 'We behold that which we are, and we are that which we behold; because our being, *without losing anything of its own personality*, is united within the Divine Truth.'¹⁸⁹ We are fully ourselves, and our union with God does nothing to negate (rather fulfils) that reality; so that if, through the purity of soul of which Ruysbroeck is speaking, we are enabled to see God, we see the reality of the soul. This is remarkably similar to Schelling's 'What in us *knows*, is the same as what *is* known.'¹⁹⁰ For him, for James, for Bergson, for Whitehead, and for others, a conscious creative cosmos comes to know itself through its never-ending coming into being. And as I suggested in the first few pages of the Introduction to Part I, creativity is always also self-creation: discovery of the self as well as of the other.

That the Creation is fulfilled in the process of differentiation and yet wholly at one with the God of creation is implicit in pantheism: 'the more we understand individual things, the more we understand God', wrote Spinoza.¹⁹¹ As Roger Scruton comments on Spinoza the

186 McCosker 2008 (46).

187 Nelson 2014 (76).

188 Quoted in Kingsland 1927 (94).

189 Ruysbroeck J, 'De calculo', *The Sparkling Stone*, c1340: quoted thus at Underhill 1911 (vi: emphasis added). On page 423 of the same book, it is quoted slightly differently: 'We behold that which we are, and we are that which we behold; because our thought, life and being are uplifted in simplicity and made one with the Truth which is God.'

190 Schelling 1994 (130: emphases in original).

191 Spinoza 1910, v, §24 (214).

pantheist: 'the distinction between the creator and the created is not a distinction between two entities, but a distinction between two ways of conceiving a single reality'.[192] Pan*en*theism, however, permits something further: the possibility that God has a relationship not just with the divine self, but with something Other; and this, it seems to me, is the drive behind there being a creation at all. For me, this is further grounds for preferring panentheism to pantheism. We need immanence, yes, which pantheism offers; but we need the union of transcendence with immanence, which only some form of panentheism encompasses. Yet again, we need union, but we need that to be the *union of division with union*.

REPORTING BACK TO THE RIGHT HEMISPHERE

There was a lot of the left hemisphere's struggling with reality in the last few pages, and there's nothing wrong with that. But, as we have seen, the occupational hazard of the left hemisphere is to believe that its necessarily narrower re-presentations constitute the reality. The danger applies here too: we are tempted to treat panentheism as a left hemisphere intellectual re-presentation, a source of argument, rather than something intuited by the right hemisphere as a source of wonder. We should resist the temptation to take it as gospel – which is why I talked about a 'speculative' theology of panentheism. There are no certainties here. In a well-known saying, attributed to Eugene Gendlin: 'We think more than we can say. We feel more than we can think. We live more than we can feel. And there is much else besides.' But the attempt to use language is not irrational: the struggle, as I say, is not wasted effort. It is merely not enough.[193]

It has been a consistent argument of this book that the left hemisphere is a fine servant but a disastrous master, and that its results need 'returning' to the right hemisphere for assessment of their validity, and reintegration into a now enriched vision of the world. It is their consistency with what the right hemisphere has intuited with greater richness, but less of an argumentative structure, that is required. The most we can hope for is some sense that an analytic theology might at least seem true to – faithful to – the intuited perceptions of the right hemisphere expressed, as they tend to be, in terms of metaphor and myth. How does panentheism fare?

One myth that seems to me strikingly consonant not only with panentheism but, even more extraordinarily, with the hemisphere hypothesis of this book is to be found in the Lurianic Kabbalah. (I should say that I was not brought up in the Jewish tradition and knew nothing whatever of the literature until the last few years: this makes the correspondences with the hemisphere hypothesis and the thesis of this book all the more striking to me.)

When the first Being, *Ein-Sof*, made the world, there were three

---

192 Scruton 1996 (78).

193 According to Joseph Campbell (1991), his teacher the Indologist Heinrich Zimmer used to say: '"The best things can't be told" – because they transcend thought. "The second best are misunderstood" – because those are the thoughts that are supposed to refer to that which can't be thought about ... and one gets stuck with the thoughts. "The third best are what we talk about"'. (I have rationalised the punctuation.)

phases. What was *Ein-Sof*'s first act? To stretch out a hand and make something happen? No – it was to *withdraw*. To make a place for something other than *Ein-Sof* to be: an act of self-abnegation. This phase is known as *tzimtzum*, contraction.

In the space that results, there lie the *sefirot*, emanations of *Ein-Sof*, vessels prepared to receive the divine light. Sparks of light from the divine Being land on the vessels made to receive them, but the vessels cannot contain them, and they shatter. This phase is known as *shevirat ha-kelim*, the shattering of the vessels.

The third phase is known as *tikkun*, repair, in which the sparks of light are gathered up again into the Godhead, and the fallen, shattered pieces, still retaining sparks of light, are rebuilt as stronger and more beautiful vessels.

It is important, incidentally, that humanity is called on to play a central role in *tikkun*. Thus again we are not just passive responders to a divine cosmos, but are part of bringing it about – or declining to bring it about – through our free will:

> Shards from the shattered vessels attached themselves to sparks of divine light and were scattered throughout the cosmos. These kernels of entrapped divine energy are to be found everywhere, and especially within the human soul. According to Luria, each man and woman is enjoined to 'complete creation' by liberating and raising the sparks within his or her own soul and environment and reconstructing the sefirot in a new, more complete and stable form which reflects the image of both God and humanity.[194]

194 Drob 1997.

195 ibid.

Though not conventionally panentheistic, the panentheist character of this *mythos,* with its sense of God as both transcendent and immanent, hardly needs pointing out. In the Kabbalah, the idea of God being other than and yet one with his creation is made both more explicit and more particular. *Ein-Sof* in the kabbalistic literature is that aspect of God which is infinite. The *sefirot* are the emanations from that infinite *Ein-Sof* which sustain creation: the highest of these is *Keter*, which we have encountered, and the lowest is *Malchut*, or kingship. 'If the goal of creation is the actualization of what exists only potentially within *Ein-Sof*', writes Drob, '... then the sefirah *Malchut* is the very fulfilment of the divine plan.' Thus '*Ein-sof* finally comes to know Himself in an "other".'[195]

This extraordinarily redolent *mythos* appears to intuit the way in which I believe everything comes into being – not just in the cosmos at large, but for us, phenomenologically, through the interaction of the hemispheres. The first 'act' is not the making of something happen, but the open receptive attentional field offered by the right hemisphere in which all new experience begins. It creates a space for something to be: *tzimtzum*. The attempt is then made to 'pour' whatever is received by the right hemisphere into the various categories, verbal or otherwise, that the left hemisphere brings to bear on it, but these prove inadequate to contain the meaning that was there in the first manifestation or 'presencing', and they break down. Analysis and language, in other words, reach their limits: *shevirat hakelim*. The meaning has to be returned to the right hemisphere to be 'restored' by understanding it as a whole again. This synthesis engenders a new, richer, wholeness: *tikkun*. I have compared this with the process whereby we learn a piece of music: initial receptivity; fragmentation and analysis in the pursuit of technical proficiency; followed by a new synthesis in which the previous phase is entirely banished from mind. Yet, crucially, no phase of the process is dispensable: every phase, including fragmentation and analysis at one stage, the shattering of the inadequate vessels, played its part. As in the kabbalistic *mythos*, the process is dialectical.

Of course there are any number of creation myths, many of them bearing little relationship to either panentheism or the hemisphere hypothesis. So to repeat what cannot be repeated too often: the fact that a panentheist theology finds resonance in a Lurianic creation myth does not make it 'true' in the sense that our left hemisphere has come to use the word. There can be no certain truth in speaking of the divine. But there is resonance, and the test is whether it answers to experience. It answers to mine.

There is, further, more than a little similarity here with Hegel's view of how something becomes an object of thought, an idea. 'The idea as a process runs through three stages in its development', he says. 'The first form of the idea is life: that is, the idea in the form of immediacy [RH]. The second form is that of mediation or differentiation; and this is the idea in the form of knowledge [LH].' And finally, the result is 'unity enriched by difference ... the third form of the idea, the absolute idea [RH synthesis]', which Hegel says is 'at the same time the true first', having 'a being due to itself alone'.[196] It is the 'true first', and has a being 'due to itself alone', because it is how the idea first becomes real and alive to us, and this is not just a summation of the two preceding processes, therefore *containing* them, but a *transformation* of them into something new. There may be a parallel here with the idea of a 'coherent superposition' in quantum physics, which is something in itself distinct from its components, as its components are one from another.[197]

There may also be a similarity here, at least in part, with Cusanus's dialectic of *explicatio*, or unfolding of the implicit into the created world; and *complicatio*, its enfolding in God. In this *mythos*, the infinite enters into the finite, the transcendent becomes immanent: and thus the finite is equally taken up into something greater than itself. These phases are not merely historical, but continuous and

196 Hegel 1975, Part 1, §215.

197 Zukav 1979 (284). In Christology, Christ is not simply man, nor simply God, nor both, nor neither.

eternal. This takes us back to Eckhart's idea of a continuous incarnation. Such constriction of the divine within the finite is a central theme of the Christian incarnation: in the words of the carol, 'For in this rose contained was / Heaven and earth in little space; / *Res miranda*'. And so Donne, addressing the Virgin, marvels at 'immensitie cloysterd in thy deare wombe'.

Here again we see the creative nature of negation. The Chabadic understanding of *tzimtzum* is that in order to *say* anything at all, one must select one thing out of all that one knows: thus the totality of one's knowledge remains, yet the utterance is simple and single. In other words a limiting or filtering process is required for something to come into being; but that does not negate the 'immensitie'. Immediately parallels come to mind: it is that limiting of potential which causes the single observable event, when the wave function collapses into a particle; James reminds us that it is the limiting constriction of his vocal cords that shapes the limitless air into 'my personal voice'; and it is in this sense that I have argued that the brain constrains the field of consciousness to cause personal experience, and allow us the freedom to influence the course of events. The fourteenth-century Persian mystic poet Hafiz wrote: 'I am a hole in a flute that the Christ's breath moves through.'[198] (Note that for Hafiz there was no insuperable divide – quite the opposite – between Christ and Mohammed.)

Creative negation is not only imaged in the first phase of divine contraction, *tzimtzum*, but also in the existence of the constraining vessels in the second phase: paradoxical as it may seem, such constriction is the only way that God, or *Ein-Sof*, may enter into, be received by, and understood by, his creation. This has deep resonances with the Christian doctrine of what is called kenosis, the self-renunciation, or 'self-emptying' of the divine nature in the act of incarnation.

THE HEMISPHERES AND OMNIPOTENCE AND OMNISCIENCE

There is a further aspect of divine constraint that I want to suggest. This relates to the attributions of omnipotence and omniscience.

As conceived by the left hemisphere, a religion is likely to emphasise power and certainty above all. There is little room in its monolithic structures for the freshness of creation and the vulnerability that alone makes love possible. To me, this is a misunderstanding of the nature of religion.

For what it is worth, I do not believe in a God of love who is also omnipotent and omniscient. I am encouraged by the fact that once again William James was there before me. In a letter to a friend, he wrote: 'The "omnipotent" and "omniscient" God of theology I regard as a disease of the philosophy-shop.'[199]

---

198 Hafiz 2002 (153). I am told that Ladinsky's translations are somewhat free.

199 James, letter to Charles Strong, 9 April 1907; in H James 1920, vol 2 (269).

But I recognise the likelihood of my being misunderstood. Not to be omniscient and omnipotent sound like failings in a God. I cannot believe that God is either omniscient or omnipotent: but I also think God is not '*not* omniscient', and not '*not* omnipotent'. It's that the terms just don't apply. (Similarly it would be equally baseless to claim that God is green or not green. The terms are misconceived.)

To know, according to the left hemisphere, is already to have something fixed and represented in memory – pinned down. To have power, similarly, is to have the ability to interfere in the course of things and manipulate an intended outcome: that is, after all, the left hemisphere's raison d'être. They are both aspects of the need to *control*: but if there is to be veritable creation, creation must be *not* wholly under the creator's control. We are thinking in the wrong way, if we think like this about God. For neither power nor knowledge is only of this kind.

God is not in a left hemisphere sense, but in a right hemisphere sense, all-knowing and all-powerful. Knowledge, as understood by the right hemisphere, is a process of openness and receptivity in which two entities progress ever closer to one another through experience. *Kennen*, not *wissen*. In this sense, God alone has knowledge of everything, whereas we have knowledge of only that limited part of reality that we can encounter. If God were to know everything, in the sense of 'knowing the facts', God would be importantly limited, because then Creation could no longer be truly free and with that the possibility for love – which depends on the free will of a true Other – would be lost. As Max Scheler has it, 'even God is not party to an unconscious ecstatic foreknowledge of His own coming-to-be'.[200]

And power? Power as understood by the right hemisphere is *permissive*: creative power, the power to allow things to come into being, precisely by underwriting the existence of a creative field, but *not* interfering and manipulating within it. Not making things happen according to *fiat*, but allowing things to grow. That is true creation. I am reminded of the story of the Tibetan Buddhist monk who carried on serenely praying as Chinese soldiers ordered his fellow monks out of the monastery at gun point. A soldier roughly prodded him with the butt of his rifle, and told him to get up and get out, shouting 'Don't you know that I have the power to kill you?' The monk looked up and replied, 'Don't you know that I have the power to let you?' Semantics, you may say. But I think it is much more.

There is an ambiguity to the account of the creation in Genesis. Each act of creation is initiated by God saying 'let' something come into being. This can be seen as a command, exercising controlling power, but as I have already suggested this is at odds with the idea of a free creation which God 'saw' for the first time as it came into being. I would suggest that it could also be seen as evocative – not

---

200 Scheler, 'Supplementary remarks: ideas are not pre-existing things'; in 2008 (408).

even permitting, but actually calling something into being. This idea is enshrined in other creation myths, such as that of the Mayan people: the primal gods also speak the world into being – they say 'mountains', and mountains emerge, 'as if the mountains were there in the primordial world all along and were revealed, little by little, as the clouds parted.'[201]

In a slightly different and more profoundly paradoxical take on God's constraint the twelfth-century kabbalist Azriel of Gerona argues that God must have both infinite and finite powers, since if one supposes 'that [*Ein-Sof*] has unlimited power and does not have finite power, then you ascribe imperfection to his perfection.'[202] This is not just sophistry. An all-embracing power embraces limitation; a creative force requires an element of resistance. Jordan Peterson refers to a commentary on the Torah: 'Imagine a Being who is omniscient, omnipresent and omnipotent. What does such a being lack? The answer? *Limitation*.' And he continues:

> If you are already everything, everywhere, always, there is nowhere to go and nothing to be. Everything that could be already is, and everything that could happen already has. And it is for this reason, so the story goes, that God created man. No limitation, no story. No story, no Being.[203]

And no Becoming. I would say that it is for this reason that God created – full stop. And if God has initiated a process that generates what is genuinely new, genuinely free – a process of truly creative evolution – why would God destroy it by omniscience and omnipotence? That the divine mind contains all possibilities, does not imply knowledge of which particular possibilities will be actualised.

God is not like a human agent performing acts of *will*. The sun does not will to shine, nor can we will it to shine: it always shines, and it is only the presence of cloud that obscures it. We need, then, to be in a state of highly active passivity, or 'active receptivity'[204] – what Freya Stark calls, in an even better phrase, 'fearless receptivity'.[205] So it is that Meister Eckhart says:

> Do not imagine that God is like a human carpenter, who works or not as he likes, who can do or leave undone as he wishes. It is different with God: as and when God finds you ready, He has to act, to overflow into you, just as when the air is clear and pure the sun has to burst forth and cannot refrain.[206]

This is remarkably like Scheler's understanding of human awareness of reality in general, that it is a process of saying no, or *not* saying no. We do not know what we are saying 'yes' to until we encounter it, so the most we can do is decide not to say 'no' to whatever it is that is coming into being.

---

201 Tedlock 1996 (65–6 & 226). This insight comes from Tedlock's Mayan teacher Andrés Xiloj. I take this account from Thomas Alexander (2016), where he contrasts it with what he sees as the more imperative account of creation in Genesis.

202 Dan 1986 (90): quoted in Drob *op cit*.

203 Peterson 2018 (343).

204 See, eg, McGilchrist 2009b (173ff & 230–1).

205 Stark 2013 (107). I am indebted to Patrick Curry for drawing this phrase to my attention.

206 Eckhart, Sermon (Pfeiffer) IV: 2009 (58).

God, however, unlike us, can say 'yes'. And to say 'yes' to everything includes *saying 'yes' to 'no'* – limitation – which may explain the existence of sin. But equally, if the spirit of evil says 'no' to everything – Goethe's Mephistopheles says '*Ich bin der Geist der stets verneint*' ('I am the spirit that always negates') – this must include *saying 'no' to 'no'*, which is why the force for negation has the potential to be turned to good.[207] This aligns itself closely with the *coincidentia oppositorum* theme of Chapter 20.

In the *Tao Te Ching*, it is said that 'being and non-being produce each other'.[208] The Chinese is notoriously such that it cannot be pinned down to just one interpretation. In this word-form it seems peculiarly abstract. The insight behind this saying, it seems to me, is one that I have touched on repeatedly; that creation is the precipitation of something out of unlimited potential into limited actuality, which then inevitably interacts further with potential, in such a way that potentiality influences what is further actualised. In other words, there is a continuous reciprocity or calling-forth between the potential and the actual, the unbounded and the bounded, in Whitehead's terms between God and the World, each helping to shape the other. This meaning is perhaps more apparent in another translation: 'what is and what is not create each other'.[209]

Yet another translation, which refers more obviously to hemispheric interplay, is 'the hidden and the manifest give birth to each other'.[210] Note, not just that the hidden gives rise to the manifest (since the left hemisphere draws on the right), but that the manifest at the same time gives rise to the hidden (the right hemisphere receives what the left hemisphere produces, taking it up again into an implicit whole). Goethe's 'holy open secret' is, like the divine, both everywhere manifest and everywhere hidden. Of *Ein-Sof*, which unifies within itself being and nothingness, Azriel remarks that it 'is the essence of all that is *concealed* and *revealed*'.[211] And for Luria, according to Sanford Drob,

> God does not create the world through a forging ... of a new, finite, substance, but rather through a contraction or concealment of the one infinite substance, which prior to such contraction is both 'Nothing' and 'All'. Like a photographic slide, which reveals the details of its subject by selectively filtering and thus concealing aspects of the projector's pure white light (which is both 'nothing' and 'everything'), *Ein-sof* reveals the detailed structure of the finite world through a selective concealment of its own infinite luminescence. By concealing its absolute unity *Ein-sof* gives rise to a finite and highly differentiated world.[212]

This is Shelley's 'dome of many-coloured glass', staining the white radiance of eternity. And the sense in which negation is creative is

---

[207] This truth is recognised by Mephistopheles in the preceding lines, in which he announces himself as » *Ein Teil von jener Kraft, | Die stets das Böse will, und stets das Gute schafft* « (Part of that power which wills only evil, but forever works to the good').

[208] *Tao Te Ching*, §2 (trans Y Wu).

[209] *Tao Te Ching*, §2 (trans S Senudd).

[210] *Tao Te Ching*, §2 (trans JCH Wu).

[211] Azriel, 'The explanation of the ten Sefirot', 1966 (emphasis added). See also Scholem 1987, 423.

[212] Drob 2000.

bodied forth at various levels in the structure of the brain: the creation of the brain by the paring away of neurones on a colossal scale even before birth; the continual pruning of synapses over a lifetime; the inhibitory relationship between frontal and posterior cortex, and between the cortex and subcortical regions; the unveiling of savant skills through damage to the brain or through selective suppression of parts of the frontal cortex; and the mutually inhibitory relationship between the hemispheres. The brain, in other words, is a process of sculpture – indeed, auto-sculpture. And the creation of *individual* consciousness comes about, I believe, as I have explained, by the permissive action of the brain.[213]

Meister Eckhart says that God negates negation.[214] Indeed he says that One Itself is the negation of negation (*unum ipsum negatio negationis*):

> Here Eckhart says that the unity appropriated to the Father is nothing other than 'the negation of negation which is the core, the purity, the repetition of the affirmation of existence' ... God as *negatio negationis* is simultaneously total emptiness and supreme fullness.[215]

As mentioned, the philosopher and Christian mystic Jakob Boehme called the ultimate ontological ground of the cosmos the 'unground' (*Ungrund*). And we saw that, according to the Lurianic Kabbalah, the primal ground of Being (*Ein-sof*), brought about the created cosmos by an act of withdrawing, or self-abnegation, known as *tzimtzum*. According to the *Tao Te Ching*, the *tao* is 'the deep source of everything. It is nothing, and yet in everything'[216]. And according to mathematician and philosopher William Byers, the ancient Indian idea of *śnya*, which is behind the Buddhist philosophy of emptiness,

> has two self-consistent but mutually incompatible meanings. For other civilisations (such as the ancient Greek), the idea of 'nothing' has only the negative connotation of absence. It is a triumph of Indian civilisation that it manages to look at nothing as this self-contradictory concept, give it a coherent meaning and even symbolic representation.

In other words, the Indian invention of the concept of zero, on which modern science and mathematics is dependent. Zero thus contrives to be both a non-entity and an entity:

> the power and importance of a concept such as 'zero' may be proportional, not to its properties of harmony and consistency, but to the inner contradiction to which it gives form and by so doing resolves in some way. Thus the important thing about 'zero' is precisely its inner contradiction.[217]

---

213 See Chapter 25.

214 Eckhart, Sermon (Pfeiffer) c: 2009 (467).

215 McGinn 2001 (84 & 93–4). The passage referred to is from Eckhart's 'Expositio sancti Evangelii secundum Iohannem', in Christ, Decker, Koch *et al* 1994, vol 3 (485).

216 *Tao Te Ching*, §4

217 Byers 2007 (103–4).

The relation between Being and Nothing is something Hegel repeatedly confronted: 'The truth of Being and of Nothing is accordingly the unity of the two: and this unity is Becoming ... Becoming is the unity of Being and Nothing.' And, moreover, 'Being and Nought are empty abstractions ... Becoming is only the explicit statement of what Being is in its truth.'[218] If one thinks of Zeno's paradox of the arrow, the options Zeno considered were for the arrow to 'be there' at some point, or 'not to be there' – being or nothing. But the *flight* of the arrow is a becoming: it really *is* becoming. It is never just there absolutely (being) or just not there at all (nothing), which are purely theoretical abstractions, merely *post hoc* representations of the living flow in such a way as to be grasped by the left hemisphere.

Hegel went on to emphasise that Becoming is not only the unity of Being and Nothing, but 'also inherent unrest – the unity ... which, through the diversity of Being and Nothing that is in it, is at war within itself.' By contrast brute Being, 'all that "is there and so"', as he puts it, is 'one-sided and finite.'[219] And essentially dead. Elsewhere he wrote that it was 'one of the fundamental prejudices of logic as hitherto understood and of ordinary thinking' that contradiction was less essential than sameness (identity). However, if one had artificially to separate them, and to say which was more essential,

> contradiction would have to be taken as the profounder determination and more characteristic of essence. For as against contradiction, identity is merely the determination of the simple immediate, of *dead* being; but contradiction is the root of all movement and *vitality*; it is only in so far as something has a contradiction within it that it moves, has an urge and activity.[220]

I quoted earlier Schelling on the cosmos: 'this whole construction begins with ... a *dissonance*, and *must* begin this way.'[221] It is the tension between what one is and what one has it in one to become that gives rise to purpose. It is the tension in the bow, the tension in the string of the lyre, that gives it power to let the arrow fly, to let the music live.

Azriel of Gerona, writing in the twelfth or thirteenth century, was perhaps the first kabbalist to articulate clearly the doctrine of *coincidentia oppositorum*. He makes the point that the nature of the *sefirot*, the emanations from the divine that sustain creation, is a synthesis of every thing with its opposite: 'for if they did not possess the power of synthesis, there would be no *energy* in anything.'[222]

IS GOD UNKNOWABLE?

Can we know God? 'In theological matters negations are true and affirmations are inadequate', wrote Cusanus.[223] On the other side of the 'step across' from the everyday realm in which we are contained by left

---

218 Hegel 1975, Part 1, §88.
219 *ibid*.
220 Hegel 1969, §956 (emphasis added).
221 See p 959 above.
222 Azriel 1966 (94: emphasis added).
223 Cusanus 1985, I, 26, §89.

hemisphere thinking, whatever is encountered – not just God – can be expressed only in terms of what it is *not*. The language-defined left hemisphere sees the world as the opportunity to *do*, to manipulate. It can't deal adequately with whatever refuses such manipulation. It can handle such elements only by expressing them as the negation of something it has already, to its satisfaction, established. It is as if life were referred to as undeath, giving in itself no idea of what life is like – except that it is not death. So we say the *in*finite, we call things *in*divisible, we refer to *non*-quanta – we simply have no positive terms in the realm of 'beyond': all is defined by what it is not. I find that in itself revealing. According to Cusanus, whose thinking about God was couched in terms of mathematics and geometry: 'A finite line is divisible, and an infinite line is indivisible; for the infinite, in which the maximum coincides with the minimum, has no parts.'[224] (Division, as we have seen, is the operation of the intellect reducing things to parts.) Cusanus saw God as like the circumference of a circle that even an infinite number of straight lines can only approximate, but never finally achieve (see Fig. 37, p 948). It is not just language, but mathematics, that declares its own limitations here.

Ein-Sof, literally means 'no limitation', therefore 'the infinite'. And at times the name is reduced simply to *Ein* ('Nothing').[225] The source of Being, then, is also No-thing. The mediaeval Spanish kabbalist Joseph Gikatilla had this to say:

> The depth of primordial being is called Boundless. Because of its concealment from all creatures above and below, it is also called Nothingness. If one asks, 'What is it?', the answer is, 'Nothing', meaning: 'No one can understand anything about it'. It is negated of every conception. No one can know anything about it – except the belief that it exists. Its existence cannot be grasped by anyone other than it. Therefore its name is 'I am *becoming*'.[226]

Or 'I will be what I will be'. The freedom is in the unknowing and undetermined becoming of God, but therein lies also the freedom of God's creation. In the Kabbalah, the cosmos and God are, as in so many of the great *mythoi* of the world, reflected in one another: as above, so below. And in the human soul, too, there is this salvific emptiness. The existence of human free will is the ultimate expression of *tzimtzum*, the 'standing back' of God.[227]

All that matters most to us can be understood only by the indirect path: music, art, humour, poems, sex, love, metaphors, myths, and religious meaning, are all nullified by the attempt to make them explicit. This is the right hemisphere's preferred territory; and, beyond that, the right hemisphere is more capable of holding together apparently conflicting positions (as the left hemisphere would see them); it's more at ease with the idea that both of two viewpoints

---

224 *ibid*, I, 17, §47.

225 Eg, *Zohar* III: 288b.

226 Matt 2009 (67): from Gikatilla, *Sha'are Orah*, 44a–b (emphasis added).

227 I am indebted to Noah Lubin for this insight.

are necessary. Such ambiguities and 'conflicts' of meaning or reasoning lie at the core of the mystical religious tradition, which is notoriously reliant on paradox to convey truths that transcend our everyday understanding. And we are now discovering the same necessity in science.

Mystery does not imply muddled thinking. On the other hand, thinking you could be clear about something which in its nature is essentially mysterious *is* muddled thinking. Nor does mystery betoken a lack of meaning – rather a superabundance of meaning in relation to our normal finite vision. Philip McCosker puts it well, taking a homely example that speaks to each one of us:

> I know I was born and will die and that knowledge is mysterious, in the way that knowing that one is in love is mysterious: the mysteriousness of both cases does not primarily or easily lie in a lack of knowledge, for it is the knowledge itself that is mysterious.

It represents, he says, a 'tense conjunction of knowing and unknowing'.[228]

This transfers our attention from knowing in an intellectual, above all linguistic sense, to knowing in the sense of experience. Most people who believe in the divine would say that their belief is a matter of experience rather than ratiocination, and that the experience is hard indeed to communicate – like attempting to communicate the taste of pineapple to someone who has never tasted it – but nonetheless carries conviction for the one that experiences. To quote the Psalms: 'O taste and see how gracious the Lord is'.[229] If you want to know how pineapple tastes, you have to eat it: and obviously the deep analogy implied here is with the nature of love.

In Chapter 26, I suggested that love is not just a product of value, but might be considered itself a value; and that, like the other values I have discussed, such as truth, beauty, goodness, and purpose, it is foundational. By that I mean that it does not emerge from something else, but is an irreducible aspect of the cosmos, and in that sense an 'ontological primitive'. Here we approach an area in which, it could be argued, we may find it more reasonable to speak of the divine than not. If we accept that, then such values become no longer separate individual primitives, substantive elements, but adjectival, qualities – but of what? Of the one ontological primitive, namely God. Rather oddly, it seems more acceptable among philosophers (and Iris Murdoch is a case in point) to speak of 'the Good' than of God, as though it were a matter of education, if not of good manners, to steer clear of any whiff of divinity. At the same time quite how Murdoch's idea of the Good differs from God is unclear. Substantivizing qualities was arguably an accident of the Greek language's invention of the definite article – no longer having only the adjective καλός, it

---

228 McCosker 2008 (3–4).

229 Psalm 38: 4.

became possible to speak of a noun, τὸ καλόν. One then has a number of 'unmoored' (no pun intended) properties to account for. Derek Parfit was a 'naturalist' in the sense that he did not believe in aspects of reality that go beyond a scientific account in the broad sense; yet he took an objectivist view of values, which he saw as irreducible properties. Again it remains unclear what, for him, sustains such values ontologically. While invoking God does not, as I say, answer our questions, it is part of a picture – a *Gestalt* – that makes more sense to me as a whole than a *Gestalt* that avoids the divine. But I readily accept that here we reach, as so often, a point at which language hinders rather than furthers understanding. How are we to express the *sui generis* in language, and what does it add if we do so? We can only keep returning to experience.

This problem of how to comprehend a whole that is *sui generis* is not a form of special pleading, and it is not just a problem in approaching the divine. It has a precise analogy in physics. 'It remains a great temptation to take a law or principle we can successfully apply to all the world's subsystems and apply it to the universe as a whole', writes Smolin.

> To do so is to commit a fallacy I will call the *cosmological fallacy*. The universe is an entity different in kind from any of its parts. Nor is it simply the sum of its parts. In physics, all properties of objects in the universe are understood in terms of relationships or interactions with other objects. But the universe is the sum of all those relations and, as such, it cannot have properties defined by relations to another, similar entity.[230]

Nicholas of Cusa made the same point in the fifteenth century, in respect of infinity. It is intrinsically unknowable because there is no comparison: 'Therefore, every inquiry is comparative and uses the means of comparative relation ... Hence, the infinite, qua infinite, is unknown; for it escapes all comparative relation.'[231]

Here the account of artists, poets and composers seems to me of interest. The poet Gregory Orr writes:

> One of the terms we poets use in our considerable effort to avoid religious and spiritual terminology is 'beautiful'. Of course no one can define the word, or everybody defines it differently, and yet we believe in it. Beauty is an article of faith among poets. I think many of us are trying to sidestep religion, and beauty is a word we use to do that.[232]

Beauty speaks directly to us of something beyond, an idea that is at the core of all Hopkins' poetry. When we are in the presence of Taverner's Kyrie 'Leroy', or one of Fra Angelico's annunciations (he painted at least four), it is not just beauty we experience but the

230 Smolin 2013 (97).
231 Cusanus 1985, I, 1, §3.
232 Kaminsky & Towler 2011 (281–2).

strangeness of another realm. 'The encounter with the aesthetic', writes George Steiner,

> is, together with certain modes of religious and of metaphysical experience, the most 'ingressive', transformative summons available to human experiencing. Again, the shorthand image is that of an Annunciation, of 'a terrible beauty' or gravity breaking into the small house of our cautionary being. If we have heard rightly the wing-beat and provocation of that visit, the house is no longer habitable in quite the same way as it was before. A mastering intrusion has shifted the light (that is very precisely, non-mystically, the shift made visible in Fra Angelico's Annunciation).'[233]

NEGATION AS CREATIVE

Such experience is the only positive knowledge we can have. But, like the *tzimtzum* of *Ein-sof* in creating the cosmos, there needs to be an emptying out, a receptive space so as to make a place for it to live: a primary act of negation.

I have repeatedly emphasised the creative role of reciprocal inhibition in the brain. And I have discussed Heidegger's saying '*Das Nichts selbst nichtet*', and the relation between Nothing and Being. In the Kabbalah, '*Keter*'s very negativity is what brings all of the succeeding *sefirot* into being. This negativity is, in fact, the essential manifestation of the primal will.'[234]

Moreover, the *sefirot* comprise ten powers or principles; and for the cosmos to be at all, both the principle of love (*chesed*), and the power of restraint (*gevurah*), are required. At one level they are opposites, but, at another, each is vitally needed for the fulfilment of the other. It is only with the restraint of *gevurah*, which is made evident in the phase of creation called *tzimtzum* (divine withdrawal), that finite creatures can subsist *without being reabsorbed into Ein-Sof*. Neither principle alone could sustain creation. It is only through their tension and complementarity, one dividing, the other uniting, that a world can come into being at all.[235] Human flourishing, too, depends on remembering this wisdom. To be secure, children need boundaries and discipline as well as acceptance and unconditional love; for the mature adult, there needs to be a proper balance between self-acceptance and self-criticism, the taking of responsibility and self-forgiveness: permission and constraint.[236]

Although it might have seemed puzzling at the time, I have throughout emphasised that some degree of resistance, of negativity, is necessary to creation: Heraclitus' 'war', Hegel's antithesis. The divine creation is the archetype of creation, so we must expect it to apply there, too: in *tzimtzum*, and in the Christian idea of the Incarnation. I suggest that this is why it involves the recalcitrance

---

[233] Steiner 1991 (143).
[234] Drob 1997.
[235] ibid.
[236] Drob himself (*ibid*) makes this connexion.

of matter, as well as the freedom of the spirit; processes of constriction and permission as much as of liberation and generation; division and differentiation as well as union and the formation of new wholes. It is out of the conjunction of 'all is one' and 'all is not one' that everything arises. I find this in Kolakowski:

> … the original eruption of creative energy must begin by erecting obstacles to its further expansion: in order to organise conditions for a freedom other than his own, God had to produce matter, in which he subsequently finds an eternal foe; matter is both the condition of the movement of life and a resistance to be overcome.[237]

Negation is not just a limit on freedom, but is *itself* liberating. Too ready a foreclosure on certainty destroys the suspended space in which creation alone can come to being: we need Keats's 'negative capability'.[238] Wordsworth knew this and much of his poetry depends on the suggestion of possibility by negatives and comparatives. His most famous short poem begins with both: 'Earth hath *not* anything to show *more* fair'; but it is everywhere in his work, opening up to possibility by negating what our mind first grasps hold of: 'thoughts of *more deep seclusion*', 'with *some uncertain* notice, as *might seem* of *vagrant dwellers* in the *houseless* woods', and so on. Absences, comparatives. Comparisons open up possibility by suggesting what is *not* the case: this is not what you might think 'deep' to mean, the grammar seems to say, but something still 'more deep'.

And negation can imply a whole background, a whole depth in time, that affirmation could never achieve so economically. How is that possible? Pushkin's wonderful long poem *Yevgeny Onegin* begins: 'When my uncle, a man of principles, fell ill *ne v shutku* …' *Ne v shutku* means 'not in jest'. So, 'seriously', then? It won't do at all: 'fell seriously ill' happens to convey a truth, but the least important part of the truth. Because the point is that the uncle was a tedious old hypochondriac, a valetudinarian who was always crying wolf and keeping the family at his bedside, cooling their heels: and, then, at last he really was ill. Only the formulation '*not* in jest' conveys all of this picture in one little phrase.

The constant emphasis on the ungraspability of God may have left the impression that God is something remote, intellectual, and abstract. But everything is paradoxical here: because God is also the least of all these things. One of my favourite sayings, because it answers to experience exactly, is from the palaeontologist and priest Pierre Teilhard de Chardin:

> By means of all created things, without exception, the divine assails us, penetrates us and moulds us. We imagined it as distant and inaccessible, whereas in fact we live steeped in its burning layers.[239]

237 Kolakowski 1985 (64).

238 See p 589 above.

239 Teilhard de Chardin 1960 (99). Cf Stephen Batchelor 2016 (231): 'The mystical does not transcend the world but saturates it'.

240 Genesis 28:16.

241 Eckhart, Sermon (Pfeiffer) LXXXII: 1931–47, vol 1 (206).

You cannot get less remote, intellectual or abstract than that. The divine is not a realm transcending life, but an aspect of life itself. When Jacob awoke from his dream, he realised that God was 'in *this* place, and I knew it not.'[240] Once again, it is a matter of seeing exactly where we are, but with different 'eyes' – or, in terms of the hemisphere hypothesis, from a point of view from which the right hemisphere is not excluded.

There is, furthermore, no conflict between reason and religion. Belief is not the antithesis, but the complement, of reason; not the opposite of knowledge, but its inevitable basis – and its outcome.

Though a belief in God or otherwise cannot be a matter of argument, there is nonetheless virtue in having a sort of scaffolding in place, even though it cannot reach heaven. The scaffolding won't do the job, but it will be reassuring to those who are wont to arrive at truths by erecting scaffolding (or so they believe). People have to start from where they are at the time, and many can't get past the first hurdle: that they believe the idea of God is an affront to the rational mind. But, though there are many paths that may lead to God, I can't believe any of them does so in a coercive fashion – one that leads someone to say, 'OK, I give in, there is a God'. That you must always be free to choose, and free to doubt, seems to me part of the deal. (If you believe we are not free, then beliefs, including that one, don't matter, since they are merely predetermined.) What does love mean, to the lover or the one that is loved, if it is compelled?

I have throughout this book suggested that the cosmos is not an unfolding of something already present in its origins, but a free process of true and original creation, not foreknown, not even to a God, if there is one. And similarly there was no necessity for there to be a God (though given that there is anything at all, it arguably becomes highly probable). There was no necessity for there to be anything at all. Being is mysteriously unnecessary. As Meister Eckhart put it, being is God's idiosyncrasy.[241]

On the lines of what might be possibly persuasive, rather than coercive, let me put to you my version of Pascal's Wager. Pascal, one of the world's greatest mathematicians and theorists of probability, contended, you will remember, that it is not just wise, but rational, to believe; because if there is no God, faithfulness does no harm, and possibly some good; whereas if there is a God, and you don't know him, you're lost. As a teenager, I used to be troubled by the question whether one can will what one believes; but since I came to see that belief is dispositional, not propositional, that difficulty was overcome. Pascal advised us not to begin from propositions themselves but from adopting a way of life that embodies the disposition we seek to emulate: practical, embodied wisdom of the highest degree.

Logic dictates that one takes into account not just the likelihood of an outcome, but the cost of failing to have prepared adequately for it should it occur. Since the cost would be extraordinarily high – we don't know how high that would mean, but we know the reality of suffering – and since God is certainly not a vanishingly unlikely possibility, what is more logical? Moreover, to decide is not just a matter of weighing propositions: I have all along emphasised that truth in this area is something one can find out *only* by experience, by seriously acting 'as if'.

So what is McGilchrist's Wager? For me, Pascal's doesn't quite cover the bases. That's because I don't think this is necessarily a matter of 'either/or'. If you accept that God is in process, as is the cosmos, there is an important third option, much more significant than either of the other two. With Pascal's Wager, there just *is* a state of affairs, which we either recognise or do not; we cannot play any part in its coming about. But if it is true that the cosmos depends on us to become what it is, there are three possibilities, not two. Either – as with Pascal – there just is a God, and all depends on our recognising him; or – as with Pascal – there just isn't a God, in which case nothing is lost by believing; or (*per* McGilchrist) if God is an eternal Becoming, fulfilled as God through the response of his creation, and we, for our part, constantly more fulfilled through our response to God; then we are literally partners in the creation of the universe, perhaps even in the *becoming of God* (who is himself Becoming as much as Being): in which case it is imperative that we try to reach and know and love that God. Not just for our own sakes, but because we bear some responsibility, however small, for the part we play in creation (and indeed how 'big' or 'small' we cannot know: the terms are derived from our limited experience of a finite world).[242]

Since the atheist cannot be logically certain of his lack of belief, and since one cannot rule out that the consequences of being mistaken about this are as great as anything can be – for the atheist, an encounter with God in whatever happens after death might be seen as the ultimate Black Swan event – the rational path is clear. Pascal's limited Wager still applies; but if the nature of reality is not already fixed, but, rather, evolving, participatory, reverberative, it is both rational and important to open your mind and heart to God, in order to bring whatever it is evermore into existence.

Of course, the nature of reality may be unalterable, and the universe may be mechanistic, as some claim. In such a case, it would still be better to believe, because of the good effects of belief locally.[243] Some will say, 'but I can't take steps to *believe*: it all seems to me like a childish fairy tale – it's just not true'. I am immensely sympathetic to this, because I, too, believe truth has intrinsic, not just instrumental, value. But my question, then, is this. Bearing in mind what is at stake, if you think people simply cling to religion for comfort,

242 The physical universe of measurement is incommensurate with the universe of consciousness. So Eckhart: 'The least of the powers of my soul is wider than the expanse of heaven': Sermon (Pfeiffer) XXIX: 2009 (178).

243 That is of course a matter of debate, but I believe the effects of religion overall outweigh whatever is alleged against it, a topic that would require, and has had, dispassionate book-length treatment: see, eg, Shortt 2019a, entitled *Does Religion Do More Harm Than Good?*

why is it that your truth is so much more important than their happiness? I think you are right to believe that it is. But why would *you* think so? In a meaningless universe, shouldn't utility alone count? Where does the sacrosanct nature of truth come from, and what does it matter, in an intrinsically amoral universe?

Here I think of the saying of Meister Eckhart: 'God loves the soul so deeply that were anyone to take away from God that divine love of the soul, that person would kill God.'[244] And in another place: 'God loves my soul so much that his very life and being depend upon his loving me, whether he would or no.'[245] In some sense God's very existence, says Eckhart, expressing it, of course, in such a way as to create maximum effect, depends on loving us (as does ours on loving God). Could it make sense to consider that this union might even be ontologically prior to those that in it are unified – the relation prior to the *relata*? Love is a relationship. That would mean God and the soul do not produce, but are manifestations of, the same love; the logical consequence being that if you take away the love of the soul you take away God, as well as taking away the soul.

A hungering for certainty and the desire to over-clarify are allied to literalistic thinking. They can afflict believers and atheists alike. Too great a need for precision crystallises our thinking, fixes our path, too early – and thus leads us astray. 'He who thinks greatly must err greatly,' wrote Heidegger.[246] The need to be right at every step often deceives and leads astray. Instead we need to be more alert and attentive, like a tracker or hunter. 'Everything here is the path of a responding that examines as it listens', he wrote. 'Any path always risks going astray ... Stay on the path, in genuine need, and learn the craft of thinking, unswerving, yet erring.'[247] Never to err is for the gods alone.

Approach to the divine involves living with a 'cloud of unknowing'; but there is also a 'cloud of knowing', which leads many people astray: 'the pride connected with knowing and sensing lies like a blinding fog over the eyes and senses of men', as Nietzsche says.[248] They fail to attend to that sense that is in all of us, because they already think they know for certain it can't be right. That is their faith.

Unknowing could be thought of as what happens when we transcend analysis. 'We experience more than we can analyse', wrote Whitehead. 'For we experience the universe, and we analyse in our consciousness a minute selection of its details.'[249] Analysis *on its own* not only fails to see most of the picture – just 'a minute selection of its details' – but removes the connexions between what we take to be 'things', and deceives us into thinking that this is how we come to know what they are: it is the very connexions, not the things, that constitute reality. Analysis also tends to focus attention on the immediate and salient – the obvious – ignoring the background. According

---

244 This is how it is normally rendered. However, Eckhart writes: 'If anyone were to rob God of loving the soul, he would rob Him of His life and being, or he would kill God, if one may say so; for the self-same love with which God loves the soul is His life ...' Sermon (Pfeiffer) XLII: 2009 (234). This translation follows the German more closely than the Evans [1857] translation I have used elsewhere: » *Der daz gote benæme, daz er die sêle niht enminnete, der benæme im sîn leben und sîn wesen, oder er tôte got, ob man daz sprechen sölte; wan diu selbe minne, dâ mite got die sêle minnet, daz ist sîn leben.* «

245 Eckhart, Sermon (Pfeiffer) V: 1931–47, vol 1 (62). See also: 'God can no more do without us than we without Him, for even if we were able to turn from God, God still could not turn from us ... he is of such nature and essence that he *must* give. He who would deprive God of this would deprive him of his own being, of his very life': Sermon (Pfeiffer) XI: 2009 (98).

246 Heidegger 1975a (9).

247 Heidegger 1975c, Epilogue to 'The thing' (186).

248 Nietzsche 2012 (4).

249 Whitehead 1938 (121).

to Takahiko Masuda & Richard Nisbett, who have written widely on culturally divergent thinking styles, whereas East Asians tend to view the world by attending to the entire field and the relations among objects, Westerners tend to view the world analytically, focussing on the attributes of distinct, salient objects. East Asian cultures, preferring holistic thinking, require 'the active suppression of analysis, and *not* attending solely to focal stimuli, but rather actively attending to both focal and contextual information'.[250] Notice that holistic thinking is already a 'both/and'. It is its nature to include its opposite. There is nothing wrong in itself with analysis; it is often a useful staging-post, but should never become the journey's end. Taking a point of view that is unbalanced towards analysis *only*, we tend to see the choices as *either* details *or* whole – and have, as could be predicted, chosen details.

Sequential analysis will never succeed in revealing truth in areas such as that of the sacred and divine. It would be like trying to tell whether the sun is shining by listening for the sound it makes. C. L. Miller warns the reader that 'Cusanus's thought has to be viewed as a whole, for it works more by correspondences and parallels between the domains he is interested in expounding than in a linear fashion or by direct argument.' This is despite the fact that nowadays the common prejudice is that a scientist and mathematician such as he was would be sure to construct analytic arguments. However his intelligence was synthetic, as much as it was analytic, and it was precisely this that enabled him to have the many insights, across so many fields of knowledge, that he had (see Appendix 8).

Reason is wholesome, and by no means leads necessarily to atheism. But it remains true that by focussing too much on reason we miss all the things that can't be reasoned about, or precisely expressed – only alluded to. One might make a distinction between what is irrational (against reason) and what is 'suprarational' (beyond reason). Music might act as an everyday example of something very real, possessed of deep meaning, and not irrational, but suprarational. 'The whole world lies, so to speak', says Schelling, 'in the nets of rationality and reason; but the question is, *how* exactly it got into these nets, since there is obviously something other, and something *more*, than mere reason in the world – indeed something that actually strives outwards, beyond its bounds.'[251]

How *did* it get into those nets? My answer is contained in the second half of *The Master and his Emissary*. It involves, at some level, hemisphere imbalance in the history of the Western World. But let me turn now to what happens to the sense of the sacred when the left hemisphere ceases to be a faithful servant, and takes it upon itself to be Master.

250 Masuda & Nisbett 2001 (emphasis in original). See also Miyamoto, Nisbett & Masuda 2006; Masuda, Wang, Ito *et al* 2012.

251 Schelling, 'Hegel': 1966 (163: emphases in original): » *Die ganze Welt liegt gleichsam in den Netzen des Verstandes oder der Vernunft, aber die Frage ist eben, wie sie in diese Netze gekommen sei, da in der Welt offenbar noch etwas anderes und etwas mehr als bloße Vernunft ist, ja sogar etwas über diese Schranken Hinausstrebendes.* «

LUCIFER REBELS: WHEN THE LEFT HEMISPHERE ASSUMES MASTERY

When the left hemisphere acts as servant its assistance is often invaluable. We have seen in this chapter its role in helping fashion coherent propositions out of an intuitive disposition, that of wonder at the mystery of Being. But in the end it is the right hemisphere that understands. As Heschel put it, *halakhah* and *aggadah* are both needed, but *aggadah* grounds *halakhah*. We saw in Part 1 how the left hemisphere is less reliable, by comparison, even in the mundane exercise of navigating our way through daily life. And we have also seen what happens when the servant assumes the role of master, and the left hemisphere arrogates to itself, as it so often tries to do, the right hemisphere's role, that of understanding.

In relation to understanding the divine, the left hemisphere usurps the role of the right in two apparently opposed, but remarkably similar ways. In one it appropriates the tentative speculations of uncertain theology, faith that is always imbued with doubt, and turns them into dogmas and unbending moral codes of 'true' or fundamentalist religions. In the other, it denies everything. It excludes from its hall of mirrors all that the right hemisphere intuits about the ground of Being and declares there is no God. Both are rejections of the critically important need to remain open. Both are forms of Lucifer's rebellion.

Religious dogmatism

The first is a transformation of religion, whereby it becomes a matter of certainties, in the process, at the outset, losing what is arguably its central quality. Matters of great subtlety that defy language are replaced by linguistic doctrines of great complexity. The Christian doctrine of the Trinity, for example, an ancient mystical concept with pre-Christian origins,[252] became the ground – or perhaps more accurately the pretext – for disputes about the relationship between the persons of the Trinity that drove the Western and Eastern churches apart for over a thousand years.[253] I do not mean to suggest that in such superficially superficial disputes there are not serious intellectual issues worthy of exploration and debate; but that an obsession with truth-as-correctness and a focus on detail – both characteristic of the left hemisphere's approach – result in a loss of the more important vision enshrined in the Christian *mythos* of which the Trinity forms part. The proper part played by the doctrine of the Trinity is explained by the theologian Jürgen Moltmann in his Gifford lectures:

> The trinitarian concept of creation binds together God's transcendence and his immanence. The one-sided stress on God's transcendence in relation to the world led to deism, as with Newton. The

---

252 Sinclair 1876 (382): 'It is generally, although erroneously, supposed that the doctrine of the Trinity is of Christian origin. Nearly every nation of antiquity possessed a similar doctrine. St Jerome testifies unequivocally, "All the ancient nations believed in the Trinity"'. An image of the Trinity that helps me is that of a book. What is the book? Is it what was present in the mind of its writer? Or the tangible volume on the table in front of me? Or what goes on in the mind of the receptive reader? Clearly it is each and all.

253 I refer to the '*filioque* clause'. The original creed, compiled in the fourth century by the First Council of Constantinople, asserts that the Holy Spirit emanates from God the Father. The Western Church from the eleventh century added the phrase *filioque*, 'and from the Son'.

one-sided stress on God's immanence in the world led to pantheism, as with Spinoza. The trinitarian concept of creation integrates the elements of truth in monotheism and pantheism. In the pan*en*theistic view [my emphasis], God, having created the world, also dwells in it, and conversely, the world which he has created exists in him ... the de-divinisation of the world has progressed so far that the prevailing view of nature is totally godless, and the relationship of human beings to nature is a disastrous one. This means that today we have to find an integrating view of God and nature which will draw them both into the same vista. It is only this that can exert a liberating influence on nature and human beings alike.[254]

[254] Moltmann 1993 (98).

[255] McGilchrist 2009b (314–29).

The extraordinary power of the Christian *mythos* lies in its central idea of incarnation – the intimate relationship between consciousness and matter, and the core idea of panentheism. Yet focus on minutiae coupled with a need to be right (both left hemisphere tendencies) explains the fissiparous nature of religious groupings – and, for that matter, identity-conscious groups anywhere in politics: immortalised in *The Life of Brian* as the dispute between the People's Front of Judaea and the Judaean People's Front.

When the left hemisphere predominates, the space of unknowing in which spiritual life flourishes comes to be replaced by dogma; openness by contention; tolerance by self-righteousness; forgiveness by stigma; orderliness by legalism. There emerge steep hierarchies. Fundamentalism insists that the truth lies in a written word, a holy book; whatever wisdom the book enshrines no longer seen as the work of variously inspired, yet fallible, humans, but of a divine hand; taken out of its historical setting and viewed as absolute; conferring on its adherents the possibility to be finally right, and those who doubt them unquestionably wrong. Truth indeed changes its nature, and becomes simplistic, literal, stateable and knowable, explicit and abstracted from context. The body becomes no longer the best image of the human soul, in Wittgenstein's phrase, but the soul's prison and antagonist. Representations come to replace the living presence they purport to represent (see the commentary on the Reformation in *The Master and his Emissary*).[255] Codes of conduct lose contact with their primary source in life and become bafflingly recondite: the Talmudic tractate *Shabbat*, for instance, identifies 39 categories of activity prohibited on the Sabbath day, one of which includes drawing the bones out of fish, though drawing fish off the bone may be permitted as long as the fish is to be eaten immediately, and no utensil was used in the process (there are many fine points of interpretation). Again, I mean no disrespect to the pious, but simply to illustrate a trend towards the triumph of form over content.

Fascinatingly, since poetry, music and humour are all strongly right hemisphere-dependent – and (not incidentally) effective discouragements to dogmatism and self-righteousness – we find that the poetry of metaphor, along with music, dance and laughter are all abjured once fundamentalism rules. The Reformers, and later the Puritans, laid waste to art and sculpture on a tragic scale, disapproved of music in worship, pissed (literally) on the sacred, took metaphorically intended language literally, and wrote tracts against laughter and comedy, which when the Puritans came to rule England in the mid-seventeenth century, they outlawed. Fundamentalist Islam is currently engaged in destroying irreplaceable masterpieces of sculpture and architecture, and is similarly recalcitrant to metaphor, has disapproved of music for centuries,[256] and disapproves of laughter.[257] (Of course, Islam has its own wonderful history of religious poetry, music and dance in Sufism; the richness of poetry, art, and music in the history of Christianity is equally obvious.) And so it is that what should be a glorious spontaneous expression of humility, gratitude, awe, and vitality, becomes one of Pharisaic vaunting, regulation, right-thinking and fearful discouragement.

This happens whenever fundamentalism raises itself up, because fundamentalism is itself an expression of left hemisphere thinking. It has nothing to do with religion in itself: the triumph of left hemisphere thinking is demonstrable in wholly secular contexts such as, for example, Sovietism, under which jokes were an imprisonable offence, art (once it had done its necessary work as propaganda) was to be abolished, and all music had to be in service to the state to be accepted. Extreme in its view, black and white in its judgment, angry and self-righteous in its disposition, it seeks control and power, the raison d'être of the left hemisphere, not understanding, which is the raison d'être of the right. (Not dissimilar fundamentalist mindsets – especially in identity politics – are readily identifiable in our own age and in the West.) Religion of this kind is coercive: so intent on imposing control that it will resort to all sorts of wickedness and cruelty in order to do so.

The Christian religion is unusual for its metaphysically complex creed, which unfortunately leads straight into the territory of the left hemisphere. No other religion seems to expect such a level of assent to what are cast superficially as improbable intellectual propositions. But that is because they have been translated by the left hemisphere-minded out of *mythos* and into *logos*, giving the impression that a religious disposition involves having to believe six impossible things before breakfast. Although as a teenager I therefore tended to dismiss its tenets as incomprehensible and possibly nonsensical, with living I have come to see them as intuitive insights, misrepresented to me as if they were something to evaluate like a chemistry experiment.

Legalism is an unattractive element in all three of the monotheistic

---

256 For an illuminating account, see Shiloah 1997.

257 Roy 1994 (83).

religions. I see it as the inevitable reaction of the left hemisphere to an essentially right hemisphere phenomenon, an attempt to make what are necessarily uncertain, but nonetheless profound, insights into the nature of Being, certain and declarative: 'let's codify, make the rules explicit, clarify our terms, rationalise the subtleties out of existence, and make it all *cognitive*.' If you want a lasting institution, that is what you may feel forced to do. Fortunately in Christianity, Judaism and Islam, there is an equally strong belief in the importance of praxis, a rich *mythos* and a profound mystical literature, which counterbalance all this. Learning here is not about facts. 'Probably most worthwhile learning', writes Charles Foster, 'is actually *anamnesis*: unforgetting.'[258] In other words, as Plato taught, being brought to a place where we once again recognise what our daily habits of thought have prevented us from seeing and have clouded over.

The need for certainty is a sign of mental imbalance, and nothing is a greater waste of time than debating with someone who doubts everything, on the grounds that only certainty is admissible. Similarly with the need always to be right. The Germans have a word for such a person, *ein Rechthaber*. Offering, if need be, yet more evidence of hemisphere imbalance, the left hemisphere tendency to black and white views is more common among the technically minded, which is why engineers are over-represented amongst religious fundamentalists and radicals.[259]

Czesław Miłosz quotes 'an old Jew of Galicia':

> When someone is honestly 55% right, that's very good and there's no use wrangling. And if someone is 60% right, it's wonderful, it's great luck, and let them thank God. But what's to be said about 75% right? Wise people say it is suspicious. Well, and what about 100% right? Whoever says he is 100% right is a fanatic, a thug, and the worst kind of rascal.[260]

The smaller the question, the clearer the answer. Expecting clear answers to big questions is to be thinking too small. Informed opinion and judgment are all each of us has to go on when dealing with any of the big questions about the world. And that's no bad thing by any means; it's much more intelligent than thinking everything is a matter of fact. An opinion is always a choice of some part of a whole, concealing some things, as any viewpoint must, as well as revealing others. Being sceptical is not a mindless matter of rejecting as wholly false something that cannot be proved, but which might contain truth; any more than it is to accept something indiscriminately because it is theoretically correct without testing it on the business of living. It is to run a course between uncritical rejection and uncritical acceptance. The word 'sceptical' ultimately comes from the Greek word *skopos*, a watchman: the verb *skeptesthai* means

258 Foster 2012 (159). As mentioned previously, the Greek word for truth, *aletheia*, means un-forgetting; see pp 386 & 776.

259 Gambetta & Hertog 2009. Reductionists tend to rely heavily on engineering metaphors when attempting to understand living organisms.

260 Quoted in the frontispiece of *The Captive Mind* (Miłosz 1955).

to keep a look out, to pay vigilant attention, so as to be in a position to know for yourself.

Indeed, though belief in God is sometimes presented as credulity, and its rejection as scepticism, the reverse is equally true. To some people, believing in the reality of something that cannot be seen or measured is credulity; to others, believing that there is only what can be seen and measured is credulity. Choose your scepticism. As Kierkegaard pointed out, it can come about that the unreasonable sceptic 'precisely out of fear of being deceived is thereby deceived.'[261]

The approach to God requires imagination: but, as you know from Chapter 19, I take imagination to be our only means of approaching reality of any kind, *a fortiori* that of God. It is certainly not a guarantor of truth – there isn't any; but its absence *is* a guarantor of failure – failure to properly understand truths of any kind, including those of science. Imagination is what enables us to forge the link between our individual experience and something beyond, that is, ultimately, the universal – and by that very process the universal becomes no longer beyond the confines of experience. Imagination is always extending from the dead realm of language into the living world that it alone can bring forth for us. But imagination is proscribed once religion is misunderstood by the left hemisphere.

Militant atheism

I mentioned a second form in which the left hemisphere, assuming mastery, treats the divine. When the unaided left hemisphere comes to confront something which it doesn't understand and cannot imagine how to approach, it turns it into something it could understand – and then either accepts it or rejects it. We have seen accepting it: perverted religion. Now let's look at rejecting it: atheism.

On the closeness of the phenomena Jonathan Rée is acute. Speaking of the militant atheist A. C. Grayling, 'quite considerably a left-hemispheric creature' by his own description,[262] he writes:

> Militant atheism makes the strangest bedfellows. Grayling sees himself as a champion of the Enlightenment, but in the old battle over the interpretation of religious texts he is on the side of conservative literalist fundamentalists rather than progressive critical liberals. He believes that the scriptures must be taken at their word, rather than being allowed to flourish as many-layered parables, teeming with quarrels, follies, jokes, reversals and paradoxes. Resistance is, of course, futile. If you suggest that his vaunted 'clarifications' annihilate the poetry of religious experience or the nuance of theological reflection, he will mark you down for obstructive irrationalism. He is, after all, a professional philosopher, and his training tells him that what cannot be translated into plain words is nothing but sophistry and illusion.

[261] Kierkegaard 1988 (119).
[262] Grayling 2009/10.

The distinction between believers and unbelievers may be far less important than Grayling and the New Atheists like to think. At any rate it cuts right across the rather interesting difference between the grim absolutists, such as Grayling and the religious fundamentalists, who think that knowledge must involve perfect communion with literal truth, and the sceptical ironists – both believers and unbelievers – who observe with a shrug that we are all liable to get things wrong, and the human intellect has a lot to be modest about. We live our lives in the midst of ambiguities we will never resolve. When we die our heads will still be filled with a few stupid certitudes mixed in with some more or less good ideas, and we are never going to know which are which. There is no certainty, we might say: so stop worrying about it.[263]

In his *Atheist Manifesto*, Sam Harris, perhaps like Grayling, shows a similar preference for fundamentalists:

> Although it is easy enough for smart people to criticize religious fundamentalism, something called 'religious moderation' still enjoys immense prestige in our society, even in the ivory tower. This is ironic, as fundamentalists tend to make a more principled use of their brains than 'moderates' do.[264]

Atheism and religious fundamentalism have a common quality that would be puzzling if we could not see that both stem from the same place in that brain. 'Religious beliefs', continues Harris, 'to be beliefs about the way the world is, must be as evidentiary in spirit as any other. For all their sins against reason, religious fundamentalists understand this; moderates – almost by definition – do not.'[265] What counts as 'evidentiary' is not discussed. Neither is the word 'principled': which principles? Those of the left hemisphere only, of course. Why? Because they are *obviously* right. I am reminded of the limpet stuck to the rock of the obvious: 'Atheism is not a philosophy; it is not even a view of the world; it is simply a refusal to deny the obvious.'[266] The thought crosses one's mind that more philosophy might help here. 'A little philosophy inclineth man's mind to atheism', wrote Bacon already in 1597; 'but depth in philosophy bringeth men's minds about to religion.'[267]

The significant divide is, as usual, not based on the 'what', but the 'how': not between atheism and belief at all, but between those who approach the world literally and dogmatically, and those who approach the world with a richer understanding of metaphor and a capacity to tolerate uncertainty, be they agnostics or believers in a divine cosmos.

To the left hemisphere's mindset many things must seem obvious. Among them would be: that anything true can be expressed in everyday language; that all truths must agree with one another;

---

263 Rée 2013. Grayling certainly loves certitude. In a piece entitled 'Probably a ridiculous caveat', about the 'atheist bus campaign' (to which the last sentence of Rée's piece above obliquely refers), Grayling wrote: 'I was not happy about the word "probably" in the slogan "There is probably no God". I would question the rationality of anyone who thought that there is probably no Father Christmas, or probably no fairies at the bottom of the garden, etcetera, and since such beliefs and beliefs in the gods of Olympus and Ararat and all other religions are on a par, there is no "probably" about it' (Grayling 2009). The campaign, backed by Richard Dawkins and The British Humanist Association, took place in 2008–9, and involved advertisements on public transport stating: 'There's probably no god. Now stop worrying and enjoy your life.' The assumption that God must interfere with the enjoyment of life is odd (though God might consider really spoiling your fun in an afterlife, I suppose). As Harold Bloom put it, 'It's no fun being an atheist' (Bloom 2005). The evidence appears to be, by contrast, that religious people are very considerably happier, healthier, both physically and mentally, less lonely, more fulfilled and longer-lived than others (see Appendix 8; also Klein, Keller, Silver *et al* 2016). Not, of course, that that is a sufficient reason for believing.

264 Harris 2005.

265 ibid.

266 ibid.

267 Bacon, 'Of Atheism', 1902 (38–41: 40).

that paradox is a sign of sloppy thinking; that what is not precise or certain just requires more thought to become precise or certain; that metaphor is a tool of obfuscation; that it makes no sense that not-knowing can sometimes lead one closer to truth than knowing, not-naming lead closer than naming; that everything can be understood if it can just be reduced systematically to its parts; that context can't alter the nature of an entity, which is something it has in *itself*; that dealing with the representation of something is just as good as dealing with its presence, except quicker and more efficient; that the living is in reality just mechanical; that believing in a God is, to quote Bertrand Russell, like believing 'that between the Earth and Mars there is a china teapot revolving about the sun'.[268] And so such people are not going to be able to get to first base when it comes to approaching the divine. Nor, if it comes to that, with physics – or with almost any academic discipline, with the exception of Anglo-American analytic philosophy.

What is encountered spiritually can be conveyed best by poetry, drama, ritual, image, narrative, music – by means, in other words, that are implicit, embodied, and contextually rich. They are resonant rather than declarative. The inherent weakness of the analytic method applied to theological matters has been aptly described as 'cognitive *hemianopia*'.[269] This suggests that only half the visual field, and therefore one hemisphere's view, is being taken into account.

As William James noted more than a century ago, 'the more fervent opponents of Christian doctrine have often enough shown a temper which, psychologically considered, is indistinguishable from religious zeal'.[270] Militant atheism is marked by the same left hemisphere characteristics of intolerance, self-righteousness and a refusal to accept that all positions, without exception, make assumptions that are open to question. It has, too, its own rituals, its denunciations, and its sacred texts. 'The Brights', as a group of them like to call themselves, sound uncomfortably like the Chosen Few.[271]

George Steiner, in his Massey Lectures, observed that 'the political and philosophic history of the West during the past 150 years can be understood as a series of attempts – more or less conscious, more or less systematic, more or less violent – to fill the central emptiness left by the erosion of theology.'[272] Central emptiness is a vacuum, and nature abhors it: something else will find its way to occupying the space formerly occupied by God, but we can't do away with the space itself. Steiner noted that these attempts are characterised by familiar claims to be able to explain everything, by canonic texts delivered by a founding genius, and by a zealous orthodoxy, amongst other things. He calls these attempts 'a kind of *substitute theology*':

> They are systems of belief and argument which may be savagely anti-religious, which may postulate a world without God and may deny

---

268 Russell 1952.

269 Stump 2010 (25).

270 James 1902 (35).

271 Although some people who do not identify as atheists want to include themselves among 'Brights', the history of the term clarifies its import. According to Wikipedia, Paul Geisert, who coined the term 'Bright' and co-founded the Bright Movement, is a one-time Chicago biology teacher, professor, entrepreneur and writer of learning materials. In deciding to attend the Godless Americans March on Washington in 2002, Geisert disliked the label 'godless' because he thought it would alienate the general public to whom that term was synonymous with 'evil'. He sought a new, positive word that might become well-accepted and improve the image of those who did not believe in the supernatural, in the same way that the term 'gay' did. A few weeks later, Geisert came up with the noun 'bright' after brainstorming lots of ideas. He then ran into another room and told his wife: 'I've got the word, and this is going to be big!' Their self-identified numbers include Richard Dawkins, Steven Pinker and Daniel Dennett: en.wikipedia.org/wiki/Brights_movement.

272 Steiner 2004 (2).

an afterlife, but whose structure, whose aspirations, whose claims on the believer, are profoundly religious in strategy and in effect … we will recognise in them not only negations of traditional religion (because each of them is saying to us, look, we don't need the old church any more – away with dogma, away with theology), but systems which at every decisive point show the marks of a theological past … [and] are very much like the churches, like the theology, they want to replace.[273]

One of the myths of atheists is that science and religion are incompatible. Belief in God cannot be in conflict with science, because such belief is not part of a competition between differing reasons or explanations. God is rather seen as the ground of there being such things as reasons or explanations at all. But that is not something of which the natural *modus operandi* of the left hemisphere can make sense, so one should expect confusion here.

In an earlier chapter I quoted Einstein to the effect that when it comes down to one's grounding assumptions in physics 'there leads no logical path, but only intuition, supported by being sympathetically in touch with experience.'[274] I commented that this is a beautiful expression of the purest Pragmatism. It might be noted that the approach to truth that I find most compelling, that of the Pragmatists, underlies not only the best way to deal with the everyday world, but both the best science and the best sacred traditions, much as practitioners in either of these fields might not welcome the insight into their relationship.

There is, put simply, more to religion than science can possibly prove or disprove.[275] For this reason none other than the atheist's ally 'Darwin's bulldog', T. H. Huxley, wrote

> of all the senseless babble I have ever had occasion to read, the demonstrations of these philosophers who undertake to tell us all about the nature of God would be the worst, if they were not surpassed by the still greater absurdities of the philosophers who try to prove that there is no God. [276]

Huxley would have had little time, then, for Jerry Coyne's *Faith vs. Fact: Why Science and Religion Are Incompatible*, a book notable not just for its 'no holds barred' title, but because it has prestige among the New Atheists and may be considered representative of their thinking.[277]

On the first page of his book Coyne declares that 'science and religion … are *competitors* at discovering truths about nature' (my emphasis), which is hardly promising, since religions are not explanatory theories about the world but, in Wittgenstein's phrase, 'forms of life'.[278] To call them competitors at the very outset is to assume the conclusion you were supposed to be arguing towards, commonly

273 *ibid* (4–5: emphasis in original).

274 See p 420 above.

275 Cf Schiller 1891 (268): 'It is a mistake to suppose that all things require to be proved, for proof is an activity of thought, and thought does not constitute the whole of consciousness. A fact may be as surely attested by feeling or will, as by the most rigorous demonstration, and ultimately all demonstration rests on such self-evident facts'.

276 Huxley 1893 (245).

277 Steven Pinker calls it 'timely and important' and 'superb', and says that it is 'clear and gripping, and should be read by anyone interested in the tension between science and religion'; Richard Dawkins calls it 'outstandingly good', and says 'it's hard to see how any reasonable person can resist the conclusions of his superbly argued book'; and Sam Harris calls it 'a profound and lovely book [that] should be required reading at every college on earth'.

278 Hudson 1968 (58).

known as begging the question. For him a religious text such as the Bible can be read only literally and he is simply uncomprehending of those who suggest it might be read otherwise. He writes:

> One of the most common arguments against such literalism is this: 'The Bible is not a textbook of science'. When I see that phrase, I automatically translate it as, 'The Bible is not entirely true', for that is what it means.[279]

Since, according to Bohr, 'when it comes to atoms, language can be used only as in poetry', it may be that Coyne not only misunderstands the nature of religion, but misunderstands the nature of the 'textbook of science'.[280]

The literalism of the fundamentalist Christian is paralleled by the literalism of the fundamentalist atheist; and here Coyne is a beautiful example. Coyne assumes that there cannot be (at least superficially) incompatible truths. Throughout this book I have emphasised that there are often 'incompatible' truths, and that the intelligent thing to do is not to force an 'either/or' (the left hemisphere's only gambit) but accept a 'both/and' (as the right hemisphere understands), since, importantly, they may be truths on different levels, truths of different kinds, or, since all truths can be partial only, may both be needed to see a fuller picture. And that leads to the question what we mean by true. For some that will be, naturally, 'obvious', but they might be better not to advertise the fact.

'In this book', Coyne writes early on in *Faith vs Fact*, 'I will avoid the murky waters of epistemology by simply using the words "truth" and "fact" interchangeably'.[281] But if one is really going to avoid the murky waters of epistemology to that extent, might it not have been better to save oneself the trouble of writing the book?[282] As readers will know, such an equation simply renders the exercise worthless, since absolutely central to all debate in this area are such epistemological questions as: what does it mean to know, and how do we come to know anything? The answers are not obvious. As you can imagine, then, Coyne is impatient of theologians and philosophers who think there is more to discuss here; their fine distinctions and desire to deal with things in a manner not wholly cut and dried are interpreted by him as merely the wriggling of those who want to escape the rigour of his shiny new logic.

'My claim is this', Coyne might well have written: 'science and the arts are incompatible because they have different methods for getting knowledge about reality, have different ways of assessing the reliability of that knowledge, and, in the end, arrive at conflicting conclusions about the universe.' But actually that's not what he wrote. He wrote the very same words, except that where I have 'the arts', he had 'religion'.[283] However, if the structure of the argument is

---

[279] Coyne 2016 (55).
[280] Niels Bohr: see p 1017 above.
[281] Coyne *op cit* (29).
[282] See the argument of Chapter 15 throughout.
[283] Coyne *op cit* (64).

correct, it must follow that science and the arts are also incompatible, and that we had better be campaigning to do away with one of them.

Why should we consider literal truth superior to, rather than just different from, metaphoric truth? We need both and they have different proper applications. They are not in conflict. It may be that, ultimately, literal truth is merely a special case, the limit case, of metaphorical truth, as actuality is a special case, the limit case, of potentiality; and that making the difference into a dichotomy is a product of modern Western ways of thinking. *Mythos* was considered anciently truer than *logos*, and not by simple people either, but by sophisticated people who had a different outlook on the world.[284] According to Christianity, God is not separable from the creation and is *in* the tangible world, not just spectating from outside; since the Christian *mythos* concerns divine instantiation in the world, it would seem fitting that this was an instantiation in every possible meaning of the term – literal as well as metaphorical. Since we don't know much, this is not something we can dogmatise about. I certainly don't know. I feel like echoing Bohr here: 'I wish Einstein would stop telling God what he can and can't do with his dice'. Faith, like true science, is not static and certain, but a process of exploration that always has in sight enough of what it seeks to keep the seeker journeying onward.

Sir Francis Bacon wrote that 'they are ill discoverers that think there is no land when they can see nothing but sea.'[285] To which I would add, they are ill discoverers that think the land they are set out to seek will be like the land they have left behind.

In a different world, one where such things are more certain, Coyne knows already that dialogue with theists is a waste of time. 'Anything useful', he writes, 'will come from a monologue – one in which science does all the talking and religion the listening. Further, the monologue will be constructive for only the listener.'[286] The image is of the religious person as the naughty child, corrected by the prissy headmistress. I can't immediately bring to mind a better expression of the left hemisphere's arrogance: nothing outside its hall of mirrors could possibly have any merit. Anyone trying to see whether there is anything beyond it needs to be hauled back in to be re-educated: there is nothing beyond what it knows.

But, in circumstances such as these, instead of saying 'I'll do the talking round here', it might prove a better plan to enter into a tentative conversation with those who see what it is you don't. It's a well-tried path; and it means you stand a chance of gaining insight into something about which you pride yourself on knowing nothing.

Here is Popper warning science – not religion – of the danger he saw:

[284] See pp 625–8 above.
[285] Bacon 1857 (355).
[286] Coyne *op cit* (257).

I am on the side of science and of rationality, but I am against those exaggerated claims for science that have sometimes been, rightly, denounced as 'scientism'. I am on the side of *the search for truth*, of intellectual daring in the search for truth; but I am against intellectual arrogance, and especially against the misconceived claim that we have the truth in our pockets, or that we can approach certainty.[287]

And here is the astrophysicist Carl Sagan, an atheist, on the importance of accepting differing kinds of intelligence and knowing:

> If we know only one kind of life, we are extremely limited in our understanding even of that kind of life. If we know only one kind of intelligence, we are extremely limited in knowing even that kind of intelligence ... broadening our perspective, even if we do not find what we are looking for, gives us a framework in which to understand ourselves far better. I think if we ever reach the point where we think we thoroughly understand who we are and where we came from, we will have failed. I think this search does not lead to a complacent satisfaction that we know the answer, not an arrogant sense that the answer is before us, and we need to do only one more experiment to find out.[288]

Understanding that there are different kinds of knowing – without which a project such as Coyne's book is a massive exercise in missing the point – is not in itself a matter of religious belief at all. It is rather the outcome of having reflected a little on experience and of having reached some degree of insight into what it is we can say we know. Thinking you know prevents you knowing, since knowing is always a process, not a state, and one that never ends. 'One can only become a philosopher', wrote Friedrich Schlegel, 'not be one. As soon as one thinks one is a philosopher, one stops becoming one.'[289]

All in all, the history of the supposed 'conflict' between science and religion makes fascinating reading.[290] But it has done great harm to clear thinking. It has entrenched a false antithesis: an entirely material view of the world as the only alternative to 'superstition'. This has done immeasurable harm. According to Hart,

> the materialist metaphysics that emerged from the mechanical philosophy has endured and prevailed not because it is a necessary support of scientific research, or because the sciences somehow corroborate its tenets, but simply because it determines in advance which problems of interpretation we can all safely avoid confronting.[291]

In other words, it narrows the field to those questions that such a mechanistic 'philosophy' is equipped to answer. But, if we are astute, we should try to be aware of what is being excluded from the conversation – that of which we are otherwise unaware that we are

---

[287] Popper 1978 (emphasis in original).

[288] Sagan 2006 (221.

[289] Schlegel 1991, §54.

[290] See, eg, Russell 1991; and Ronchi 2014.

[291] Hart 2013 (65).

unaware – and why. Modern physics has forced much on our attention that was formerly, mistakenly, disregarded by mechanists.

On the matter of God, the evolutionary biologist Stephen Jay Gould put it rather clearly:

> To say it for all my colleagues and for the umpteenth millionth time (from college bull sessions to learned treatises): science simply cannot adjudicate the issue of God's possible superintendence of nature. We neither affirm nor deny it; we simply can't comment on it as scientists.[292]

One can sense his frustration: why did it have to be said umpteen millions of times? Because scientism has all the qualities of the worst kind of religion (with luck, this remark will confirm itself by being appropriately anathematised). Things over which science has no jurisdiction are a threat, not to real scientists, only to adherents of scientism, whose motive is power.[293] Scientism, according to Roger Scruton, is 'a bid to reassemble the complex matter of human life, at the magician's command, in a shape over which he can exert control. It is an attempt to *subdue* what it does not understand'.[294] Good scientists should be doing their utmost to distance themselves from the sorcerer's apprentices of this world. Scientism damagingly makes science appear craven, foolish, power-hungry. That is a terrible sleight on something – science – which we haven't learnt how to value properly.

In a famous speech celebrating the legacy of Darwin, Karl Popper had this to say:

> It is important to realize that science does not make assertions about ultimate questions – about the riddles of existence, or about man's task in this world. This has often been well understood. But some great scientists, and many lesser ones, have misunderstood the situation. The fact that science cannot make any pronouncement about ethical principles has been misinterpreted as indicating that there are no such principles.[295]

Science can't pronounce on moral principles, and it can't answer the question of God's existence one way or the other. That doesn't mean there is a *conflict* here between science and God: that would only be possible if, on the contrary, it *could* answer the question. Peter Medawar, whom Richard Dawkins himself described as 'a giant among scientists',[296] unequivocally affirmed 'the existence of questions that science cannot answer, and that no conceivable advance of science would empower it to answer'.[297] Top of that list has to be, *pace* Dawkins and Coyne, the existence of God: to think otherwise is to commit a category mistake.

Many of the greatest scientists, of course, would have been hor-

292 Gould 1992.

293 Although some scientists dismiss 'scientism' as a contentless term, or 'boo word', it is not: McGilchrist 2013.

294 Scruton 2013 (46: emphasis in original).

295 Popper 1978.

296 Dawkins 2003 (198).

297 Medawar 1988 (66).

rified that the New Atheists should claim science rendered religion, effectively, hogwash. That so many scientists, past and present, have succeeded in combining the practice of science with religion is relevant to Coyne's assertion that they are incompatible. Instead of taking a hint from this, Coyne brushes it aside as evidence of the human tendency to be inconsistent. But that won't quite do: many scientists, including the very greatest of scientists, undeniably go out of their way to talk of their religious beliefs and their scientific beliefs in the same breath, and even in terms of one another. (I have included some examples, drawn from the greatest among scientists, in Appendix 8.) According to a book-length study of the beliefs and characteristics of Nobel Laureates,[298] overall only 10.5% described themselves as 'atheist, agnostic, freethinker or otherwise nonreligious at some point in their lives'. What is striking, however, is that while the figure reaches as high as 35% for Laureates in literature, the figures for science are 8.9% in physiology/medicine, 7.1% in chemistry, and 4.7% in physics. Since Nobel prizes are a twentieth-century invention, that means that, as far as we know, 95.3% of the most acclaimed physicists of the twentieth and twenty-first centuries were consistent theists – still more striking when one realises that agnostics and 'freethinkers' counted, for the purposes of this exercise, as non-believers.

A very one-sided, not to say downright inaccurate, account of the role played by religious institutions in the development of science has gone unchallenged into pop culture. Yet undoubtedly a vast body of scientific research and development occurred under the aegis of the Church, which both supported and encouraged it. The account has been not just one-sided, but subject in some instances to deliberate misdirection. For example, do you believe that people thought the world was flat until Columbus's voyage? Or that Copernicus 'dethroned' the world by demonstrating that it was not at the centre of the universe? You may be surprised to learn that these are nineteenth-century inventions (see Appendix 8 for further details).

Before leaving the New Atheists, it's worth noting another aspect of their outlook that is highly characteristic of the left hemisphere when it assumes the role of master: anger. We saw this tendency of the left hemisphere in Chapter 6. Here it manifests itself in their attitude to anyone who has the effrontery not to share their views. A representative example is Richard Dawkins, who once described religion on Twitter as 'an organised licence to be acceptably stupid'. In a subsequent moment of illumination he tweeted: 'How dare you force your dopey unsubstantiated superstitions on innocent children too young to resist? How DARE you?'[299] Of course, if he is referring to the sort of dogmatic, fundamentalist religion conjured up by the left-hemisphere-as-master and discussed in the previous

---

298 Shalev 2002 (57–61). Religious beliefs are broken down into 25 categories, though these include Buddhism (1.2%) and neo-paganism (0.2%), which though treated as religions, are not conventionally associated with the idea of God. (However, on Buddhism, see Shaku 2017 (25): 'At the outset, let me state that Buddhism is not atheistic as the term is ordinarily understood. It has certainly a God, the highest reality and truth, through which and in which this universe exists.') On everyday scientists, see 2009 Pew Research Center poll: 48% describe themselves as 'atheist, agnostic or nothing in particular'.

299 Elmhirst 2015.

section, then I can see his point; though I have no particular interest in intervening in a civil war within Lucifer's camp. But his anger seems directed more widely. He seems to believe that unless 'forced' to think differently, children (and, one imagines, people in general) would naturally incline to the atheist position that Dawkins champions. He seems to take it as read that this is the human default position: hence 'how DARE you?'

Fortunately some necessary intellectual content can be added to this 'debate' by research on children's spirituality, and human spirituality in general. Evidence on this and a number of other matters, such as the relationship between religious belief and health, is also to be found in Appendix 8. In summary, it demonstrates that children naturally express themselves in spiritual terms, irrespective of their upbringing; and that religious belief in general is shaped by, but not originated by, culture. The research raises a number of questions. If religious beliefs arose evolutionarily to enable us to account for what we cannot explain, why are they invoked in some types of cases only, and our ignorance accepted in others? If, on the other hand, religion is a product of cultural indoctrination, why does it surface in the face of the strongest cultural discouragement (for example, in an atheist totalitarian state)? One answer might be that we have a religious sense because there is something there in the world (as with all the other senses) in response to which the appropriate 'perceptive organ' arose. There are, of course, other answers that don't involve such a conclusion, but they don't have the field to themselves, from a purely rational point of view.

The research shows that experiences of the divine are far from uncommon, and when they do occur, often do so in formerly non-believing subjects; have a benign and sometimes life-transforming impact; and are quite unlike hallucinations and delusions in quality, effect, and subsequent evaluation by the individuals who experienced them. While it has been argued that theism is a product of mental style and culture, it is just as arguable that atheism is a combination of personality type, mental style and culture – one that promotes atheism as smart and religion as naïve. And given the commonalities between right hemisphere deficit conditions and autistic spectrum disorders, it is relevant that autism-spectrum disorders make belief in God less likely (in a series of four studies, neurotypical (viz, 'normal') subjects were 10 times as likely as those on the spectrum strongly to endorse the idea of God).[300]

The evidence is also that religious belief has a dramatic positive impact on both psychological and physical health, certainly comparable with, if not superior to, most 'lifestyle' changes known to modern medicine to have a beneficial effect; and is associated with prosocial behaviour (Appendix 8).

[300] Norenzayan, Gervais & Trzesniewski 2012. See Appendix 8 for further discussion of this point.

301 Commenting on Coyne 2016, according to the latter's publisher's blurb.

302 Muller 2016 (287).

303 Gaisman 2018.

304 For Carse on finite and infinite games, see p 396 above.

Sam Harris writes that the 'honest doubts of science are better – and more noble – than the false certainties of religion.'[301] But this is a false antithesis. Honest doubts are by definition better than false certainties; and though the imputation is that religion is falsely certain and science honest in its doubt, it is easy to find examples of the precise reverse. Only the worst of religion is falsely certain, and only the best of science is honest in its doubt. I can think of certain dogmas of materialist science, just as I can of religious dogmas, that are false certainties that impede its access to truth; the oppressive intolerance for ideas that do not conform to the narrow orthodoxies of modern biology is so well documented that the idea that nobility, honesty and self-questioning characterise science alone is laughable. Equally the genuinely exploratory process involved in an openness to the idea of God raises as many questions as it answers, just as does the genuinely exploratory process of a truly open science.

Problems began only when science started to categorise whole swathes of experience as inadmissible evidence. This began relatively recently, not earlier than the mid-nineteenth century. What can be measured was alone henceforth real. This is obviously untrue: love, merely to take one example, is both real and unmeasurable. 'Anyone who claims, "if it can't be measured, if it can't be quantified, it isn't real" is not without religion', writes Richard Muller, professor of physics at Berkeley.[302] The demand is a typical left hemisphere misunderstanding. What is more, it is demonstrably founded in neither science nor reason. 'The atheist's essential case that the only truths that we should accept are those for which there is an empirical or scientific basis', writes Jonathan Gaisman, 'does not itself have an empirical or scientific basis. The notion that the only things worthy of belief are those which are objectively verifiable is not itself an objectively verifiable belief.'[303]

Invoking God draws attention to the intrinsic limits of knowing; invoking science tends to result in the opposite – to a belief that all will one day be known, that there are only problems to be solved, and ones for which we are getting ever nearer the solution: 'the answer is before us'. It would be easy to reply that science, too, is a process, and in one sense that is obviously true. But that would be to fail to grasp a difference. Science is, in Carse's terms, a finite game; philosophy and religion are infinite games.[304] Science is convergent on a target, that becomes – theoretically, if not in practice – more certain as it is approached, and the point of the process lies in achieving that target; yet philosophy and religion are divergent, as the nature of the target becomes less certain as it is approached, and the point of the process lies in itself. For science knowledge is things; for philosophy and religion knowledge is a process of understanding *what these things mean*.

So should we conclude that science and religion are separate, but nonetheless compatible, modes of understanding the world? Gould famously thought they were: he called them 'non-overlapping magisteria'.³⁰⁵ This might seem in contrast to Julian Huxley's saying that 'it is no longer possible to maintain that science and religion must operate in thought-tight compartments ... The religiously-minded can no longer turn their backs upon the natural world ... nor can the materialistically-minded deny importance to spiritual experience and religious feeling.'³⁰⁶ But it is not. Huxley, like Gould, did not deny the validity, on its own terms, of either form of understanding: what he emphasised was that each should take into account the other – a quite different matter. Both Gould and Huxley's thoughts on the topic have, however, been widely ignored. This is a calamity.

Just as naïve materialists do great damage to science by their over-reaching claims of access to 'the truth' – even to sole access to truth – on her behalf, so do misguided religious figureheads and their lay 'supporters' to religion, when they don't know enough to see what it is they don't know. Freed from such damaging accretions, however, science and religion have much in common: both being projects that, when carried out in a spirit of humility, are potentially beautiful and good. They are both, at their best, an honest, dignified reaching after truth, of different kinds, though never finally in possession of it. They should be able to honour one another, and work together towards the common goal of understanding the world better. If they can't, something has gone badly wrong.

Science has its own assumptions and beliefs – that's not a criticism, it could not be otherwise. It is an approach to the world we very much need. The question is just whether it is the *only* one we need. After all, why should it be? 'What we see of the world', writes physicist Marcelo Gleiser,

> is only a sliver of what's 'out there' ... Like our senses, every instrument has a range. Because much of Nature remains hidden from us, our view of the world is based only on the fraction of reality that we can measure and analyse. Science, as our narrative describing what we see and what we conjecture exists in the natural world, is thus necessarily limited, telling only part of the story ... We strive toward knowledge, always more knowledge, but must understand that we are, and will remain, surrounded by mystery.³⁰⁷

A secular construction of the human world in the West, with little or no place in it for the spiritual, has become the default view among the deracinated, intellectual classes, who prioritise autonomy over most other values. As Peter Berger explains:

> There exists an international subculture composed of people with Western-type higher education ... that is indeed secularised beyond

305 Gould 1997.
306 Huxley 1959a (26).
307 Gleiser 2014 (xiii–xiv).

measure. This subculture is the principal 'carrier' of progressive, Enlightenment beliefs and values. While its members are relatively thin on the ground, they are very influential, as they control the institutions that provide the 'official' definitions of reality, notably the educational system, the media of mass communication, and the higher reaches of the legal system ... I can only point out that what we have here is a globalised *élite* culture.[308]

A proper scepticism, it should be remembered, applies not just to discerning the limits of intuitions, as it almost always seems to do nowadays, but to discerning the limits of analysis, as it rarely, if ever, does any more. 'Over the last century', writes Hart,

> Anglo-American philosophy has for the most part adopted and refined the methods of 'analytic' reasoning, often guided by the assumption that this is a form of thinking more easily purged of unexamined inherited presuppositions than is the 'continental' tradition. This is an illusion. Analytic method is dependent upon a number of tacit assumptions that cannot be verified in their turn by analysis: regarding the relation between language and reality, or the relation between language and thought, or the relation between thought and reality's disclosure of itself, or the nature of probability and possibility, or the sorts of claims that can be certified as 'meaningful', and so on. In the end, analytic philosophy is no purer and no more rigorous than any other style of philosophizing. At times, in fact, it functions as an excellent vehicle for avoiding thinking intelligently at all; and certainly no philosophical method is more apt to hide its own most arbitrary metaphysical dogmas, most egregious crudities, and most obvious flaws from itself, and no other is so likely to mistake a descent into oversimplification for an advance in clarity. As always, the rules determine the game and the game determines the rules.[309]

Trying to understand religion purely analytically is hardly more likely to be effective than trying to understand Schubert by doing a statistical breakdown of the frequency of particular notes, their lengths, and so on. One really might as well rewire the amplifier into a food processor.

'Though religion will live on in the minds of the unlettered, in educated circles faith is entering its death throes', writes Richard Dawkins. 'Symptomatic of its terminal desperation are the "apophatic" pretensions of "sophisticated theologians", for whose empty obscurantism Coyne reserves his most devastating sallies.'[310] Dawkins is clearly more easily impressed than I was: Coyne's sallies, like many of his own, seemed to me, if I am honest, somewhat less than devastating, and to show that he had not fully understood what he was talking about. And Dawkins must surely know better than to call the apophatic path a symptom of what he calls 'terminal desperation'. Its Christian origins lie in the second century AD, and it has been

308 Berger 1999 (10).
309 Hart 2013 (46).
310 Front matter of Coyne 2016.

an important tradition throughout a period of nearly two thousand years in the West in which religion was in no way defensively embattled against atheists. It is recognised as a path to wisdom in every major tradition of both East and West – in Buddhism (Zen is the world's greatest expression of it), Hinduism, Judaism, and Islam as well as Christianity, for millennia – and has nothing to do with 'pretensions' of any kind. Nor is it aimed to deflect his kind of cavils, whatever he might like to imagine. Indeed Meister Eckhart, a representative of the tradition in the thirteenth century, risked everything to express himself in this way, and has the distinction of being the only mediaeval theologian to be tried for heresy: it was feared by the Dominican authorities that what he said could 'easily lead simple and uneducated people into error'.[311] Given that, on this evidence, even those in 'educated circles' can no longer begin to understand what Eckhart was talking about, it looks to me as though they might have had a point.

There is of course more to say, some of which is to be found in Appendix 8.

311 McGinn 2001 (1 &14).

312 Shogaku Shunryū Suzuki-rōshi, 'There must always be differences in our bodily experiences and feelings', delivered Sunday, 30 March 1969; included, in edited form, in Suzuki 2002 (59).

WHY IT MATTERS

Perspective alters what we see. The left hemisphere is rigid, but we need to think flexibly, so as not to get trapped into one world view and imagine it is everything. According to Suzuki:

> In Japanese, we call someone who understands things from just one side a *tamban-kan*, 'someone who carries a board on his shoulder'. Because you carry a big board on your shoulder, you cannot see the other side. You think you are just an ordinary human, but if you take the board off, you will understand, 'Oh, I am Buddha, too. How can I be both Buddha and an ordinary human? It is amazing!' That is enlightenment.[312]

There is something much too small about a world in which we are isolated from the divine; in a 'disenchanted' world, as it has been called, one that has no place for the sacred, we ourselves loom, imaginatively, far too large, as if, occupying too much of the screen. At the same time we see ourselves conceptually as diminished, because as soon as we pan out, we see ourselves dwindle to a pointless speck in a barren cosmos. A religious cast of mind sets the human being and human life in the widest context, reminding us of our duties to one another, and to the natural world that is our home; duties, however, that are founded in love, and link us to the whole of existence. The world becomes ensouled. And we have a place in it once more.

Max Scheler called man the *ens amans*, the being whose defining feature it is, that it is capable of love. (I'd say that we are not alone in this, but the point stands.) If there were not already enough reasons to reject the eighteenth-century image of 'man the machine', this is another. No machine is capable of love. Yet, love is creative and curative: as we have seen, according to traditional Judaic teaching, *tikkun*, repair, is brought about by human acts of compassion in the world, much as the redemption which Christ symbolises is realised and renewed in the acts of kindness and mercy of each one of his followers. There are sparks of divine fire still burning in the fragments of the vessels out of which we rebuild the world.

I keep coming back to Blake's amazing line, 'Eternity is in love with the productions of time.'[313] And, you can be sure of it, what Blake meant was nothing cosy. He was the least cosy of poets, and one of the most insightful, that ever lived. This love he speaks of is not merely some tasty, nutritious and comforting *olla podrida*. It certainly does add flavour and substance to life, and at times is a great comfort; but it is, as it is the nature of love to be, spiritually demanding, capable of shaking you to the core: the *mysterium tremendum et fascinans*.[314] According to the great philologist Max Müller,

> Soul is the Gothic *saivala*, and this is clearly related to another Gothic word, *saivs*, which means the sea. The sea was called *saivs* from a root *si* or *siv*, the Greek *seio*, to shake; it meant the tossed-about water, in contradistinction to stagnant or running water. The soul being called *saivala*, we see that it was originally conceived by the Teutonic nations as a sea within, heaving up and down with every breath, and reflecting heaven and earth on the mirror of the deep.[315]

Evelyn Underhill pointed out that the mystics 'all knew, as Richard of [St] Victor said, that the Fire of Love *burns*.[316] We have not fulfilled our destiny when we have sat down at a safe distance from it, purring, like overfed cats.'[317] We need what the Czech phenomenological philosopher and human rights activist Jan Patočka called 'shakenness', so as to be fundamentally open to transcendence, to the radical astonishment of Being. Not to be astonished is not to be truly alive. It is something of which we have intimations, intimations that fill us with a healthful fear and awe; fear in the sense of reverence, not timidity. So C. S. Lewis said:

> I do not see how the 'fear' of God could have ever meant to me anything but the lowest prudential efforts to be safe, if I had never seen certain ominous ravines and unapproachable crags. And if nature had never awakened certain longings in me, huge areas of what I can now mean by the 'love' of God would never, so far as I can see, have existed.[318]

Yet in Eckhart's words, 'God is nearer to me than I am to my own self.'[319] God situates us firmly in the cosmos. Where we have been taught to see human consciousness as 'an anomalous tenant of an

---

313 Blake, *The Marriage of Heaven and Hell*, Plate 7, 'Proverbs of Hell'.

314 Otto 1917.

315 Müller 1885, vol 1, 391–448, 'The theoretical stage, and the origin of language' (434–5).

316 Underhill is referring to *The Four Degrees of Passionate Charity*, also translated as *The Four Degrees of Violent Love*, written in or around the year 1170 by Richard of St Victor, a Scottish philosopher and mystic who was one of the Victorines, a group of scholars gathered round the Abbey of St Victor in Paris.

317 Letter of Underhill to Marjorie Robinson, 13 October 1909; in Poston 2010 (162).

318 Lewis 1988 (23–4).

319 Sermon (Pfeiffer) LXIX: 1931–47, vol 1, 171.

alien universe', that consciousness is reframed as 'the most concentrated and luminous expression of nature's deepest essence.'[320] Alan Watts puts the matter well:

320 Hart 2017.

321 Watts 1989 (8–9: emphasis in original).

322 See p 1167 above.

> This feeling of being lonely and very temporary visitors in the universe is in flat contradiction to everything known about man (and all other living organisms) in the sciences. We do not 'come into' this world; we come *out* of it, as leaves from a tree. As the ocean 'waves', the universe 'peoples'. Every individual is an expression of the whole realm of nature, a unique action of the total universe.[321]

Religion takes seriously both the thisness of the individual, at one extreme of the scale, and the fate of the cosmos at the other, and shows them to be part of one whole. It is the absence of this integrative perspective that Dewey called the 'deepest problem of modern life'.[322]

This book has had a consistent message: that the right hemisphere is a more reliable guide to reality than the left hemisphere. In Part I, we saw that it has a greater range of attention; greater acuity of perception; makes more reliable judgments; and contributes more to both emotional and cognitive intelligence than the left. In Part II, we saw that the right hemisphere is responsible for, in every case, the more important part of our ability to come to an understanding of the world, whether that be via intuition and imagination, or, no less, via science and reason. In Part III, I have suggested that the right hemisphere's capacity to deal with what we call 'paradox' is greater; that its understanding of space, motion, and time is deeper and more resonant with the findings of contemporary physics (and all philosophy other than the purely Anglo-American analytic tradition, itself a left hemisphere venture); and that it contributes more to important aspects of consciousness, including the appreciation of values such as goodness, beauty, and truth. So the least strong claim one might make is that the spiritual and divine will be misapprehended if one brings to bear on them only the process for which the left hemisphere is best equipped – analysis to parts, following of procedures, and the presentation of results in language: a process that prioritises the known, the certain, the fixed, the partial, the explicit, the abstract, the general, the quantifiable, the inanimate and whatever is 're-presented', over the unknown, the uncertain, the flowing, the living and the implicit, the whole, the contextual, and whatever is of unique quality; and all that 'presences' to us, before it has been represented through rationalisations in language. A stronger claim is that, not just here, but in general, we should prefer, wherever possible, an approach that can be identified with the habits of mind of the right hemisphere – a claim which I think, on the evidence I have presented, entirely justified.

One way of thinking about religion is that it instantiates the 'form of life', to return to Wittgenstein's term, of the right hemisphere. In another era there was no way of knowing that that's what it was, of course, but it took the form of a deep intuition about the division of the human spirit, one which has a very long history. The reason that so much time, expense and skilled artistry was devoted to religion, from the most ancient, prehistoric epochs, when lives were shorter and resources scarcer, arose from the understanding that what we now know to be the 'form of life' of the left hemisphere, though obvious and easily expressible, belongs to the banality of everyday, and is far less important than the realm of what is comparatively hidden, but in which everything that matters to us and that gives meaning to life resides. If we should forget that, we would have forgotten how to be fully human. Without such instantiation in the fabric of a culture, it might easily be forgotten: that, then, was the role of religion.

And now it seems we *have* forgotten. In the absence of a tradition that embodies the values that the left hemisphere can't see, we arrive at conclusions about ethical and metaphysical matters of the greatest significance in a manner that might result from putting everything we cared about in the hands of a mediocre bureaucracy, or having to bring it before a law court in a degenerate regime where the art of legal judgment had given way to the slavish following of numbered statutes – or running it through a computer. Meanwhile the very stuff of life ebbs away.

The raison d'être of the left hemisphere is control and calculation. Importantly *we* are not exempt from being the objects of control and calculation in a culture in which all is controlled and calculated. We have the illusion of being in control, whereas there is in truth very little we can control; rather we are controlled, in what Adorno called *die verwaltete Welt*, the 'administered world', one where a new form of total control has taken root in the form of administration – a self-legitimising bureaucracy.[323] It is not that bureaucrats are our new slave-masters; for they are themselves as much enslaved by their bureaucracy as the rest of us. They are, like us, in the grip of the disenchanting drive to control. As Tillich says, we become tools, but when asked for what purpose, there is no reply.[324]

This drive is bigger than us, and it is driving us towards destruction. Thus historian of ideas Paul Bishop says, in a fascinating article on which I here draw, that disenchantment is 'a consequence and the effect of *being controlled by others*'.[325] Referring to a radio discussion between Adorno, Horkheimer and Kogon in September 1959, he writes:

> In this broadcast Adorno argues that, as the Austrian writer Ferdinand Kürnberger put it, 'life no longer lives' (*das Leben lebt nicht*). 'There is', Adorno maintains, 'no longer any life in the sense in which

323 Greisman & Ritzer 1981.

324 See p 1218 above.

325 Bishop 2012 (emphasis added).

we all use the word life', because there has been 'a transition of the entire world, of life as a whole, to a system of administration, to a particular kind of control from above'. In particular, he emphasizes that, in its most recent forms, bureaucracy serves 'to rationalize the irrational', whilst the system of planning conceals a complete lack of real planning.

Only the human soul is left to resist this process, being, unlike cognition, beyond 'administrative control'; it is that part of us that intuits the divine, is in touch with unconscious knowledge and wisdom, and resists the banality of the desacralised, if not actually desecrated, world. As such it is a threat to the administered world, which responds with an otherwise inexplicable wish to crush and exterminate it. Everything that is not explicit, and therefore not reducible to material terms that can be predicted and administered, is to be done away with, as in Freud's declaration 'where id was, there shall ego be'. For Jung, on the other hand, 'the unconscious is the residue of *unconquered* nature in us, just as it is also the matrix of our unborn future'.[326]

According to French philosopher Bertrand Vergely, 'our world is not merely disenchanted, it is in the grip of a malaise', one that is linked to our incapacity for wonder.[327] We are *curious* about things, until we know enough to control them; but that is not the same as wonder at all – rather its opposite. We do not permit ourselves to be awestruck or amazed by what it is we come to see. Dennis Quinn writes: 'Among those defective in wonder are the inveterately curious'.[328] There is, of course, good in curiosity, especially in the young, but it confines one to the smaller questions: no-one, as he points out, says 'I am curious to know the nature of God', or 'I am curious about the meaning of life'. And in reality the lust here is appropriative, to know for myself, so as to have a better *grasp*; whereas in wonder there is an acknowledgment of what must forever lie beyond our grasp, an insight that comes with experience. The curious materialist, by contrast, assumes that, theoretically at least, everything can be definitively known: there are no things in heaven or earth that are not knowable in his philosophy.

Awe and wonder involve an encounter, and are other-directed emotions, in which the ego takes a back seat. On the relationship between this and the seeking after knowledge of all kinds, not least in science, the philosopher Mary Midgley wrote with great insight:

> Wonder involves love. It is an essential element in wonder that we recognize what we see as something we did not make, cannot fully understand, and acknowledge as containing something greater than ourselves. This is not only true if our subject-matter is the stars; it is notoriously just as true if it is rocks or nematode worms ...

326 Freud 1933 (112); Jung, 'Psychological types' (1923): 1953–79, vol 6 (510–23: emphasis added).

327 Vergely 2010 (112).

328 Quinn 2002 (25).

Knowledge here is not just power; it is a loving union, and what is loved *cannot just be the information gained*; it has to be the real thing which that information tells us about ... First comes the initial gazing, the vision which conveys the point of the whole. This vision is in no way just a means to practical involvement, but itself an essential aspect of the whole. On it the seeker's spirit feeds, and without it that spirit would starve.[329]

In an itself wonderful lecture entitled 'The work of wonder', Patrick Curry reminds us that 'wonder, like the relationships it consists of, is wild; so the correct attitude to it is therefore unpossessive, respectful and indeed reverent. It follows that anything that is being mastered and managed must be something else':

> Wonder, enchantment, astonishment, delight, joy – these are experiences that are not, and cannot be, simply willed into existence or manufactured on demand. They are not under our control, not something we do but something that happens to us (or doesn't). Indeed, trying to use them, to do something purposeful with them, destroys them. Their only use inheres in their use-lessness; their point is precisely in their point-lessness. They can only do their work, whatever that is, if we don't try to put them to work. So whatever else wonder may be, it is not itself instrumental but sufficient unto itself ... It cannot be tested, evaluated, improved, rolled out or developed, and there can be no system or method (let alone methodology) to achieve it – even if the goal is re-enchantment! In fact, that attempt would not only be arrogant but insidious, since it would require tacitly replacing wonder with a biddable simulacrum, complete with targets, outcomes, benchmarks, assessments, impacts and other instruments of modern management; in other words, a grievous betrayal.[330]

It would be, in short, the betrayal of the right hemisphere Master into the hands of the left hemisphere emissary.

'All humanity, including scientists, have to learn to live with mystery; and with mystery should come humility. (Shouldn't it?)' So writes Colin Tudge.[331] Greek tragedy concerns the effects of *hubris*, the vain delusion whereby man sees himself as being god-like. The result is inevitably catastrophic. Indeed our term 'catastrophe' (from Greek *katastrephein*, to overturn) refers specifically to the dramatic downfall of the victim of *hubris* in Greek tragedy. This ancient insight would appear to be of immediate relevance today.

The importance of humility in the face of the divine is because it is what is proper to divinity, not something demanded by a supposedly petulant God. That would be to misunderstand humility as abasement. And it may prove ultimately that the faithful person's attitude of humility comes *from* the divine element in himself

---

329 Midgley 1989 (41: emphasis added). Cf Paul Hawken, Commencement Address to the Class of 2009, University of Portland, 3 May 2009: 'Ralph Waldo Emerson once asked what we would do if the stars came out only once every thousand years. No one would sleep that night, of course. The world would create new religions overnight. We would be ecstatic, delirious, made rapturous by the glory of God. Instead, the stars come out every night and we watch television.'

330 Curry 2018. See further Curry 2017.

331 Tudge 2013 (156).

(otherwise known as the soul) and is aimed *at* the divine in himself, and is therefore the divinity properly attending to the divinity. Ruysbroeck's 'God in the depths of us receives God who comes to us: it is God contemplating God.' This has nothing whatever to do with a supposed master-slave relationship. Instead it is, like awe and wonder, not an abasement but an *ecstasis* (from Greek *ek*, out, + *stasis*, standing), a standing outside oneself while still being oneself. We are both united with something greater than ourselves, in which we share, and simultaneously aware of the separation, in which one feels one's smallness: the union of division and union. (The great is not recognised by the limited, not because the great is not great, but because the limited is limited: 'No man is a hero to his valet; not, however, because the hero is not a hero, but because the valet is – the valet'.)[332]

Lao Tzu said: 'the un-wanting soul sees what's hidden, and the ever-wanting soul sees only what it wants.'[333] Want requires fulfilment. On the other hand, wonder leads to longing of a spiritual kind, which is fulfilled in itself, 'an unsatisfied desire which is itself more desirable than any other satisfaction', in Lewis's words. 'I call it Joy, which is here a technical term and must be sharply distinguished both from happiness and from pleasure.'[334] It is a kind of desire that 'is itself desirable' and is therefore 'the fullest possession we can know on earth.'[335] According to him it is what we sense we are made for, but which we can never know in more than intimations – intimations that are nonetheless authentic and beautiful, in such a way as to lead the soul onward:[336]

> The books or the music in which we thought the beauty was located will betray us if we trust to them; it was not *in* them, it only came *through* them, and what came through them was longing. These things ... are not the thing itself; they are only the scent of the flower we have not found, the echo of a tune we have not heard, news from a country we have never yet visited.[337]

It involves a sense of transcendence in which one is transported to somewhere where one is in touch with things beyond the everyday world, that is nonetheless familiar – a sort of long-lost home, as Novalis is supposed to have said.[338]

Even Emil Cioran, who described God's creation as a total failure were it not for the music of Bach, described that same music as a 'ladder of tears on which our longings for God ascend.'[339] We see God through the beauty of this world, but we must not stop there. Hence the importance of that sense of the loss of something beloved and beyond.

---

332 Hegel 1977, §665. More precisely: 'No man is a hero to his valet, not, however, because the hero is not a hero, but because the valet is – the valet, with whom the hero has to do, not as a hero, but as a man who eats, drinks, and dresses'.

333 *Tao Te Ching*, §1 (trans UK Le Guin).

334 Lewis 2017 (19).

335 Lewis 2017 (205).

336 Lewis 2015 (150–1).

337 Lewis 1988 (20).

338 Berlin 1999 (104): 'When Novalis was asked where he thought he was tending, what his art was about, he said "I am always going home, always to my father's house" ... attempts to go back, to go home to what is *pulling* and *drawing* him, the famous *Sehnsucht* of the romantics, the search for the blue flower, as Novalis called it.'

339 Cioran 1949: « échelle de larmes sur laquelle gravissent nos désirs de Dieu ».

Religion or spirituality: a matter of societal flourishing?

Religion can undoubtedly give rise to fanatical behaviour, oppression and cruelty. The evidence is undeniable. But in this it is not unique: so, alas, can any human structure conferring power, since humans are deeply flawed beings. Atheist regimes have been around for only a hundred years or so, compared with the millennia of religion: the evidence is that atheist regimes, such as those of Lenin, Stalin, Mao and Pol Pot, have committed the most extensive atrocities in human history. If we were hunter gatherers we would not need formalised religion. But since we settled, grew our own crops, and built and dwelt in cities, organised social structures, inevitably conferring power, have been unavoidable. And secular ones alone are simply not adequate: that way far too much is left out of our picture of the world. Religion has equal power to do great good, both through its charitable actions, and, even more, by its capacity to bring before the eyes of the mind another realm than the one that is so 'obvious' to our gaze – that is 'too much with us', as Wordsworth put it; to provide a common moral focus that is not merely pragmatic or politically motivated; to enrich life with beauty and pattern; to revitalise us, celebrate our being in the world for now, and transform what might otherwise seem like meaningless life for the better.

John Gray, in his *Seven Types of Atheism*, makes the point that the values that we hold to be 'obviously', even rationally, correct have not seemed so and do not still seem so to societies with other nonetheless coherent codes of behaviour.[340] Jürgen Habermas sees Christianity playing a tacit role in modern society far beyond that of a mere precursor or a catalyst. He, like many others, views our contemporary allegiance to freedom of life, social solidarity, emancipation, individual morality of conscience, justice, and human rights as directly derived from the Judaeo-Christian tradition. The fact that our civilisation has not yet completely fallen apart is a demonstration not that the tradition can be dispensed with, but that it continues, for a while, to be rooted in our psyche. Of its legacy, he says, 'to this day, there is no alternative to it. And in light of the current challenges of the post-national constellation, we continue to draw on the substance of this heritage. Everything else is just idle postmodern talk.'[341] We are enjoying the afterglow of a fire we are doing our best to extinguish: when it goes out it cannot be rekindled within our lifetime.

Rupert Read links the current ecological crisis to a decline in a collective belief system which was provided by Christianity. 'Yes, anti-clericalism has brought us to where we are today: a declining culture, bereft of values, and heading rapidly towards complete self-destruction.' Political liberalism of the Rawlsian type, he holds, is 'a severe form of tacit anti-clericalism, a profoundly-anti-religious

340 Gray 2018.

341 Habermas 2006 (151).

fundamentalism', in which everything whatever is tolerated, provided that it tolerates (and in practice even if it doesn't) everything else: 'toleration has turned to indifference ... In contrast, we need to start to de-indifference our world. We need to seek truths that can re-unite us, that can revive community, not a mere glorified alleged *modus vivendi* that keeps us separate in our private worlds, while the public world declines to ruin.'[342]

This is all the more difficult since a corollary of this form of liberalism is, as Stanley Fish puts it, that

> religion must give up the spheres of law, government, morality and knowledge; reason is asked only to be nice and not dismiss religion as irrational, retrograde and irrelevant. The 'truths of faith' can be heard but only those portions of them that have secular counterparts can be admitted into the realm of public discourse.[343]

This will not provide us with a moral compass. We turn to science, perhaps, for help; but our faith in science is naïve, since science, as Habermas saw, has no capacity of itself 'to provide reasons, aside from the reason of its own keeping on going, for doing [what it does] and for declining to do it in a particular direction because to do so would be wrong.'[344] The way of being of animist cultures issued in a form of life that endured and flourished for millennia: it led to a certain harmony between man and Nature. There is nothing wholly good or bad; and science, for all its wonderful discoveries, has lost us that capacity. Science is now used to many fine ends; but, alas, inevitably much of the time in the service of the subjugation and destruction of Nature.

Under these circumstances is it enough to be 'spiritual, but not religious'? That it is hard to know what such a thing means or entails may indeed represent one of its attractions, and I understand only too well why conventional religion is hard to accept for many people these days. I share many of those feelings. But the one feeling that I don't share is the idea that I can't learn anything from a tradition; that my limited rationalising on the basis of my limited experience in one time and place on the planet is enough. Indeed even such a belief is an outcome of a culture – just one that is in meltdown. Only I can take responsibility for what I believe, hence Habermas's point that individual conscience is a key element in Christianity. But I cannot possibly penetrate to the core of the enigma of life by my own efforts. Nor can I wilfully invent myths or rituals without their being trivial and empty. This is why we have traditions of art, philosophy and, above all, religion. The fetishisation of novelty and the repudiation of history are reflections of a capitalist culture that depends on dissatisfaction with what we have and the constant seeking after new 'improvements' in order to fuel demand. It is not

[342] Read 2011 (99).
[343] Fish 2010.
[344] *ibid.*

only false but obviously immoral in a number of respects. A culture (and the point of religion is to embody the ethos of a culture) is of critical importance for a society's survival. Cultures are living; but precisely because of that they can be killed. A plant can be flexibly trained, but it cannot be avulsed from its roots and still live. And if our culture dies, so will we who live in it.

In 2014, the RSA in London instigated a series of workshops, debates and events dealing with the question of the nature of spirituality, in a project entitled Spiritualise. The report makes interesting reading.[345] One element in it is striking: groups of young people from all over Britain were asked to enter a competition to 'design a service, product, environment, or communications campaign that addresses spiritual needs in contemporary contexts'. Perhaps the left hemisphere terminology contained the seeds of its own failure; but the results showed that 'while the brief was about using design to help reconceive the spiritual in the modern age, most submissions suggested the students did not significantly differentiate between spirituality and wellbeing ... the general impression was of a range of good designs that didn't really speak of the spiritual'.

Smith and Denton reporting on the spiritual lives of American teenagers found a common belief that, as they wryly put it, God was 'something like a combination Divine Butler and Cosmic Therapist',[346] who was available on demand but undemanding.[347] This has been popularly characterised as 'benign whateverism'.[348] Its core is that we should try to be nice, kind, respectful and responsible, and by doing so achieve a state of 'feeling good, happy, secure, at peace.'[349] Worse things might certainly be believed; but this is not enough to support a civilisation, inspire great art, induce fidelity, inculcate sanctity, motivate self-sacrifice, or lead us to insights into the nature of existence.

When Solzhenitsyn asked himself what had given rise to the catastrophic brutalities of the twentieth century, his conclusion was that men had forgotten God. In a speech given in 1983, he repeated: 'If I were called upon to identify briefly the principal trait of the entire twentieth century, here too, I would be unable to find anything more precise and pithy than to repeat once again: men have forgotten God.' More than this, a positive 'hatred of God', he thought, was the principal driving force behind the philosophy and psychology of Marxism-Leninism: 'militant atheism is not merely incidental or marginal to Communist policy; it is not a side effect, but the central pivot.'[350] The hatred of God is indeed a fascinating phenomenon, one more and more evident in our time – and not just in political philosophies, but in the vox pop of media scientists. Lucifer – 'the Bright' – cannot bear the imputation of anything higher than he.

Jonathan Gaisman, in his reticent fashion, addresses the issue of

---

345 Rowson 2014. Formerly the Royal Society of Arts, the RSA now styles itself by its initials only, referring to its history as 'the royal society for arts, manufactures and commerce' and commitment to 'a future that works for everyone. A future where we can all participate in its creation' (www.thersa.org/about).

346 Smith & Denton 2005 (165).

347 assets.ngin.com/attachments/document/0042/5177/NationalStudy Yout_Religion.pdf (accessed 15 March 2021).

348 ibid.

349 Smith et al, op cit (164).

350 Solzhenitsyn 1983.

religion or spirituality. 'It is open to anyone to retain a personalised and entirely individual faith', he says,

> but for the majority the natural course at this stage is to reach out to an established tradition, which is most likely to be that in which they were brought up. This is for several reasons. First, like the followers of any other predilection, religious people typically form groups. Second, those groups provide a collective enhancement of the experience shared, as anyone who has been part of a theatre audience will attest. Third, all religions concern themselves with relations between human beings, and prescribe rules of conduct towards one's fellows, so that it is natural for people of faith to reach out to others. Fourth, and perhaps most important, the profound human desire for sacred spaces and rituals is best served by religions whose lengthy past has brought such places and actions into being, polished and veneered by the observances of succeeding generations.[351]

Religion performs a role of incomparable importance, whether one believes in it or not, which is why, presumably, it attracts such strong, and strongly opposed, feelings. Ten years before he died, William James wrote in a letter to a friend: 'I myself invincibly do believe, that, although all the special manifestations of religion may have been absurd (I mean its creeds and theories), yet the life of it as a whole is mankind's most important function.'[352] I have found that James was rarely wrong. The intellectually wrought specifics are going to be approximate at best: the disposition of the soul is everything.

This great turning of our backs on the sacred began with the Enlightenment. Already in the eighteenth century Schiller prophetically lamented what Weber would later call, in a famous phrase, 'the abolition of the sacred'. If the words sacred and holy still mean anything to you, then your world must contain the divine. As Blake's saying 'all living things are holy' reveals, for him the world was divine throughout, since to the imagination everything lives. Nowadays, of course, we react to such ecstatic insights with distancing gestures of irony: we are clever. But these are the ways in which we kill the soul. As Friedrich Schlegel declared already, 27 years before Blake died, 'what gods will rescue us from all these ironies?' He foresaw what James referred to as 'pertness', vain chatter and smart wit.

As we have seen, according to Goethe (and Plotinus before him), aspects of the world call forth in us, if we are open and attentive, the faculties that are needed to respond to them. The faculty to perceive the divine is no exception. Indeed that faculty is what we mean by soul. Soul does not exclude feeling or intellect or imagination, but it is not nearly exhausted by them. Though natural, it can be developed or stunted. Keats, who was wise beyond his years, called this world a 'vale of Soul-*making*'.[353] We grow a soul – or we can snuff it out. It is the most important purpose of a culture – any culture – to ensure

351 Gaisman 2018.

352 Letter of James to Frances Rollins Morse, 12–13 April 1900; in H James, 1920, vol 2 (127).

353 Letter of Keats to George and Georgiana Keats, 14 February – 3 May 1819: 1970 (249: emphasis added).

that such faculties are aided to grow: the invocation of archetypal symbols, the practice of rituals, and the deployment of music and holy words in the approach to the divine have been universal across the world over time. It is only very recently that this universal practice has been abandoned. If you are convinced that in principle you know and can account for everything, you will see only what you think you know. You will never give yourself a chance to know what it is you might not know. And as Colin Tudge points out, 'we cannot know how much we don't know, unless we are already omniscient.'[354]

The business of life then becomes like a dance watched by a deaf person: puzzling, pointless and somewhat absurd. Death becomes just the meaningless end of a life itself without meaning. Goodness becomes mere utility, and suffering just frustration of utility. Eros becomes just lust; longing just want; sleep and dreams an inefficiency that we should do away with if we could; art a toy; the natural world a heap of resource; and wonder merely a measure of our failure, rather than, as I believe it to be, a measure of our insight.

We are now in the grip of an obsession with human power and the subjugation of nature, at bottom of which lies our infatuation with technology: the power to manipulate. We have subordinated ends to means. We emphasise self at the expense of others, our rights rather than our duties, what we have rather than who we are, the material rather than the spiritual, and vaunt the reach of the unaided human intellect. In such a world God is a nuisance. 'Men despise religion', wrote Pascal. 'They hate it and are afraid it may be true.'[355]

When our society generally held with religion, we might indeed have committed many of the same wrongs; but power-seeking, selfishness, self-promotion, narcissism and entitlement, neglect of duty, dishonesty, ruthlessness, greed, and lust were never *condoned* or actively and openly encouraged – even admired – in the way they sometimes are now. In other words, we have lost all shame. And that can't help but make a difference to how we behave.

Pride and arrogance, believing we know it all, are the opposite of the religious disposition of humility, reverence and compassion. And without them, neither we, nor the whole far greater, astonishing, living world, over which for better or worse we now have the power we so much craved, can thrive. It is pride that will destroy us, and quickly. With so much going for us, rising educational standards, better healthcare, public welfare and humane and stable government, what could be against us? We ourselves. Pride was always considered the greatest of the 'seven deadly sins', and it may in the past have proved difficult for many to see why. But the evidence is all around us now; and it is there in the epic narrative of one of the greatest poems of our language, *Paradise Lost*, for all to read.

Let no-one think for a moment that I can exculpate myself from

---

354 Tudge 2013 (156).

355 Pascal 1976, §187 (Lafuma §12): « Les hommes ont mépris pour la religion. Ils en ont haine et peur qu'elle soit vraie. »

any of this. I find it all as hard as anyone; but sometimes one has to say what one sees, whether one feels a hypocrite or not.

And when we die – when anything dies – what then? Of course we can't know, but as I put forward in the chapter on consciousness, I believe each being is an individuating, and individuated, actualised expression of the potential of consciousness as a whole, which is never ultimately divisible from it. The attachment to self, which is a necessary concomitant of the individuating process while life lasts, at death must be relinquished, but the achievement of thisness, an unfolding or explication of that always unique aspect of the whole, is not swallowed up and lost, but is reincorporated into the now enriched whole. Division and union are unified, without loss of either necessary element of the dipole. Such a view of death seems to me coherent with the rest of what I have put forward in this book.

It is in dealing with death that one is most forcibly made aware of how we have yielded, hands down, to the forgetting of Being. One of the few occasions on which *at last* modern man might be able to grasp the enormity of existence is in the contemplation of death. Yet this is just what we ignore. It is a commonplace that while the Victorians did not talk about sex, they were open about death; we do not talk about death, but are clinically explicit about sex. Unfortunately for us, being open about something robs it of its power, while hiding increases it.

The clergy are no exception to this 'costly aversion of the eyes from death'. Perhaps, like some doctors, some priests – strangely enough – just don't know how to deal with suffering and would prefer everyone to be mindlessly happy. At the funerals of friends and loved ones, they are so anxious that people should not be uncomfortable, or in touch with their natural and healthful feelings of grief, that all sense of a solemn mystery is sacrificed.

When my brother and I had to bury our parents, we asked, mainly for our sakes, for the time-honoured words of the 1662 Book of Common Prayer to be said, just as they are written, since at such times there is something important and moving in the subsumption of the individual life in the common lot of mankind, the ritual of saying the same solemn words that would have accompanied our parents' parents, and their parents' parents, to the grave. Curiously, it is the universality and anonymity that are comforting in the burial service. (There should be no funeral oration on the life of the deceased. That is quite proper, but on another occasion: for now it is the universal, not the particular, that must be faced.) In any case, at each occasion a different priest assured us that that was what he would do; and in neither case could he bring himself to do so. They managed bits of the rite, it is true; the bits that were consonant with their decent, progressive views. But they omitted anything that seemed to them unreasonably gloomy, or might have suggested that something of consummate importance was at stake. The ancient words no longer sit well in their kindly modern mouths, and are better not said: they come from an age when people still really lived, and really died, seeing life and death *sub specie æternitatis*. Great words, deep words – 'He cometh up, and is cut down like a flower; he fleeth as it were a shadow, and never continueth in one stay. In the midst of life we are in death …' – are read, if at all, in a tone of voice more appropriate for a children's bedtime story. Yet the point of hearing these words is to remind the living, just for once, of the context in which they lead their lives: something not entirely trivial. Or the supplication: 'Thou knowest, Lord, the secrets of our hearts; shut not thy merciful ears unto our prayers; but spare us, Lord most holy, O God most mighty, O holy and most merciful Saviour, thou most worthy judge eternal, suffer us not at our last hour, for any pains of death, to fall from thee.' Words, I suppose, never more to be heard: too powerful, too gripping, too real – there's not enough irony there for our taste. Perhaps it's that the language is cast in terms that suggest retribution, in which I no more believe than do the clergy; and hence they avoid them. But that's not the only way to understand what is being said. In death no-one really knows what we face. But if we are not separate from the rest of creation, the gaining, or regaining, of insight into the whole, and our part in it, might well – it is not too hard to believe – give rise to suffering. The judgment feared comes, then, more from within the individuated soul than from without; the mercy sought comes more from the divine whole than from within. And according to the Christian *mythos*, mercy is ours if desired. The cry for God's blessing or mercy is ancient, and does not at all depend on judgment in the sense of damnation – more on the sense of fallenness, somehow falling short of something we intuit we might be, and of which we are always aware. Which of us if honest does not feel this? Hence humility, hence reverence, hence compassion for self and others: hence the prayer.

And finally there is the mechanical disposal of the body in a modern crematorium. The burning *ghats* would be quite another matter: it's the clinical, mechanised efficiency invoked. And the loss of the grave: in both senses.

Perhaps the clergy no longer know what to believe – I can understand that. But if, whatever that strange thing one calls one's consciousness may be, it does not end at death – and nobody can be certain that it does – but persists in some form at which we can only guess, there is a real chance that an afterlife might be worse, not just better. I have experienced places in the universe that are real, but not reassuring. Faith is not simply comforting, but puts before one the full tremendousness of being, its meaning and its consequences. If one is a materialist, then there is no meaning in death, apart from the negation of life. Such people live in the comforting thought that life is meaningless, and has only personal consequences or none; then death, too, must be meaningless, and the quicker its acknowledgment is over, and the less grief there is, the better.

But if life has meaning, death has meaning, and if death has meaning, so does life. They interpenetrate one another, and to be anaesthetised to this is to be deceived. 'The fact, and only the fact, that we are mortal, that our lives are finite, that our time is restricted and our possibilities are limited, this fact is what makes it meaningful to do something, to exploit a possibility and make it become a reality, to fulfil it', wrote Viktor Frankl. And he continued:

> Death is a meaningful part of life, just like human suffering. Both do not rob the existence of human beings of meaning but make it meaningful in the first place. Thus, it is precisely the uniqueness of our existence in the world, the irretrievability of our lifetime, and the irrevocability of everything with which we fill it – or fail to fill it – that give significance to our existence.[356]

It may, actually, matter how one lives one's life, because we may play a part in the coming into Being of whatever is, and we cannot separate ourselves from whatever is, perhaps for ever. *Something depends on our way of being, and it is not just we ourselves.* In a world where no-one can avoid the experience of suffering we know that it is a real part of consciousness: suffering is a central element, not just in Christianity, but in Buddhism, and no doubt in most, if not all, religions. So the *how* of life, not just the *what* – its mere existence or non-existence, huge as that is – matters: it has a value and price we cannot fully conceive.

Similarly one life and all life are reflected in one another, as we would see if we looked into the gems of Indra's net. Life requires death; death is the friend of life, not its foe. Goethe wrote: 'all that is to persist in being must dissolve to nothing'.[357] According to historian of religion Mircea Eliade, myths from all over the world convey the mutual sustenance of death and life: 'in countless variants', there is a widespread myth that 'the world and life could not come to birth except by the slaying of an amorphous Being.'[358] In one of Goethe's greatest poems, he writes: '"Die – and so continue into being!" As long as you fail to see this, you are no more than a forlorn guest on the dark earth.'[359] The Latin word *homo* is related to *humus*, the earth, and to *humilis*, humble; and man is 'made of the dust of the earth'. Another ancient word for man is the Sanskrit *marta*, 'he who dies', cognate with Latin *mortalis*. I understand the Christian belief in the redemption of death through God's own suffering to mean that death is not an end, but plays a part – like the intermediate phase of

356 Frankl 2019 (50–1).

357 From 'Eins und Alles':
» *Wirkt ewiges lebendiges Tun ... Es soll sich regen, schaffend handeln,* | *Erst sich gestalten, dann verwandeln;* | *Nur scheinbar steht's Momente still.* | *Das Ewige regt sich fort in allen:* | *Denn alles muß in Nichts zerfallen,* | *Wenn es im Sein beharren will* « – 'An eternal living force ... that cannot but move, cannot but act to create, first to form, and then to transform, itself; its moments of rest are only apparent. The eternal presses onward in all things: for all that is to persist in being must dissolve to nothing.'

358 Eliade 1978 (205–7).

359 From 'Die selige Sehnsucht':
» *... so lang du das nicht hast, Dieses: Stirb und werde! Bist du nur ein trüber Gast Auf der dunklen Erde.* « Goethe's expression is compressed and has foxed translators. My translation is not quite literal, but I believe it is as close as can be managed while capturing Goethe's meaning.

destruction, of fragmentation, of the shattering of the vessels – in the greater story of repair and restoration; a story that is both mine and not mine, taking place in the immensity of a living cosmos where the part and the whole are as one, yet without the loss of the meaning of the part that is each one of us.

Or so it seems to me.

EVIL

Ein-sof, the infinite God, is a 'unity of opposites', reconciling within itself even those aspects of the cosmos that are opposed to or contradict one another:

> *Sefer yetzirah*, an early (third to sixth century) work which was of singular significance for the later development of Jewish mysticism, said of the *Sefirot* that 'their end is imbedded in their beginning and their beginning in their end'.[360]

The idea of the eternal return, founded on an eternal cycle of life and death, creation and destruction, each requiring the other for its fulfilment, is imaged in the ancient Egyptian sacred image of the *ouroboros*, the serpent eating its own tail, which surrounded the cosmos and kept it alive. But the foundational work of Kabbalah, the *Zohar*, takes the image imaginatively into a new realm: 'The Holy One, Blessed be He, has curled a serpent around the realm of holiness'.[361] According to the *Zohar*, there is no path to holiness except by way of the serpent – the shadow side. The journey always involves an encounter with something you have to be courageous enough to withstand.

The 'problem of evil' is as ancient as humanity. It seems that we need resistance in order for moral issues to stand forth or for moral goals to be achieved. Too easy a life leads us to complacency. The Duke in *As You Like It*, banished from court to the forest of Arden, finds that

> Sweet are the uses of adversity,
> Which, like the toad, ugly and venomous,
> Wears yet a precious jewel in his head;
> And this our life exempt from public haunt,
> Finds tongues in trees, books in the running brooks,
> Sermons in stones, and good in every thing.[362]

George Herbert, too, believed that excessive comfort dulls our ability to seek God. In his poem 'The Pulley', he describes how God bestowed on humanity all the blessings that he could, though deliberately withholding rest, for fear that we would then remain satisfied with what we knew, and never seek its source beyond, in

---

360 Drob 2009 (130–1).

361 *Zohar* II: 173b: in Lachower & Tishby 1991, vol 2 (468).

362 Shakespeare, *As You Like It*, Act II, sc I, lines 14–19.

God's self. From that point of view there is something in the idea that – in accord with the *coincidentia oppositorum* – the better it gets, the worse it gets; and the worse the better.

The recalcitrance of the world seems to me, incidentally, also a guarantee that it is to some degree independent of us, not wholly determined by us. While we cannot ever be absolutely sure that the world is not simply a fabrication of our own minds, the element of resistance, if nothing else, holds an important clue. If there were no mind-independent world, the fact that wishing does not always make it so would be hard to explain.

Resistance in nature is the cause of suffering, but, by the very same token, of creativity. According to Paul Cilliers, a philosopher of complex systems, 'for self-organisation to take place, some form of competition is a requirement'.[363] Self-organisation means thriving: adapting, creating, fulfilling potential and surviving. In Nietzsche's famous phrase, that which does not kill us makes us stronger.

The existence of suffering in nature is every bit as real as the existence of exultant life and beauty: we may not focus on either to the exclusion of the other. And it is one of the oldest and most pressing problems in philosophy and theology. It seems that everything exists only as a dipole, a coincidence of opposites. Ruskin thought that beauty required the existence of ugliness, much as Blake thought there must be sadness in heaven if there was to be joy;[364] Nietzsche called happiness and misfortune 'two siblings and twins who either grow up together – or … *remain small* together!'[365] They are like the opposites that Heraclitus and Anaximander saw as not just an incidental *aspect* of existence, but the core of existence itself: the structure of the world conceived as counterpoint. (See Plate 24.)

Ursula Le Guin's searing short story, 'The ones who walk away from Omelas',[366] describes a city of extraordinary beauty, whose people lead lives of delight, great wholesomeness and vitality. They are a morally and politically sophisticated society, technologically not advanced, but living peacefully in a glorious and celebratory harmony with Nature, and with one another, without either rulers or slaves. However, those who grow up there come to learn that there is a dark secret on which the very life of the community depends: a child who is kept prisoner in a lightless basement, where it leads a squalid and desperate existence:

> Some of them understand why, and some do not, but they all understand that their happiness, the beauty of their city, the tenderness of their friendships, the health of their children, the wisdom of their scholars, the skill of their makers, even the abundance of their harvest and the kindly weathers of their skies, depend wholly on this child's abominable misery.

363 Cilliers 1998 (94–5).

364 For Ruskin, see p 1163 above. For Blake, see p 830 above.

365 Nietzsche 2001, §338 (192: emphasis in original).

366 Le Guin 1973. The idea was prompted by a famous passage in William James's lecture of 1891, 'The moral philosopher and the moral life' (James 1897, 184–215).

Some go to visit in order to confirm for themselves the truth; others prefer to rely on what they are told. Some are able to put out of mind what they have seen, and accept it as the necessary price to be paid for the welfare of the community at large. But some who go to visit are haunted by what they know; one day, sooner or later, they become the ones who walk away from Omelas. Whither they walk we do not know; or whether they are fortunate in finding a place whose beauty and goodness is bought at a less costly price. LeGuin's concluding words are equivocal:

> The place they go towards is a place even less imaginable to most of us than the city of happiness. I cannot describe it at all. It is possible that it does not exist. But they seem to know where they are going, the ones who walk away from Omelas.

If there is a force for good, there must be a force for evil. I don't flinch from such a conclusion. I think I have seen forces both for good and evil at work during my lifetime, and I think, that seeing them as forces, having dynamism, initiative, and purpose, is to me more truthful than seeing them as inert abstractions of the mind – abstract nouns, as is our way, derived from adjectives we use to describe behaviour we happen to like and behaviour we happen not to like. I think as our civilisation breaks down, we will all be able to identify these drives more clearly. But, if the thesis of this book is right, though they may be polar opposites, they cannot be wholly separate. And they will not be wholly equal.

As many mystical writers have intimated, the divine must be capable of encompassing in some form its opposite, while not, nonetheless, being diminished by this inclusion.[367] But does divine containment also have the power to embrace, to constrain, and to set limits to whatever it may be: even to change its nature? If so, is this not a trick? As a teenager, wrote Jung, 'Faust struck a chord in me and pierced me through in a way that I could not but regard as personal. Most of all, it awakened in me the problem of opposites, of good and evil, of mind and matter, of light and darkness'.[368] The problem of opposites is central to Faust because it was a recurring source of fascination and inspiration to Goethe, too; both Goethe and Jung saw that opposites extended far enough reach the same point.

I experience both good and evil as real, and see them as necessary opposites; but while evil can, goodness knows, locally overwhelm good, it cannot subsume good into itself. The goodness of love can embrace its opposite; the evil of hate cannot. As we saw in Chapter 20, opposition has it in itself to become a union of opposites, but only by the forces of union, not by those of opposition.

The easiest way to deal with evil is to deny its existence. In a

---

367 Sikka 1994 (426): 'Evil is thus nothing merely negative. It is positive discord, arising when the part strives to be the whole, when a particular will opposes itself to the universal will by seeking to be the creative ground of all reality. It then seeks to usurp the place of God, who is that ground, and does so by perversely asserting its own existence in opposition to, rather than in harmony with, the existence of other natures and other beings. It is not limitation that is evil, then, but the refusal, on the part of a finite being, to accept limitation'.

368 Jung 1961 (235).

society that is unwilling to countenance the existence of a moral force of any kind, and sees good and bad as merely labels we use to indicate degrees of societal approval, this recommends itself as the easiest path. For most of us, evil is just behaviour we disapprove of more strongly, and the term is thought better avoided, as it has no explanatory power.

Yet I cannot see this as satisfactory when we are faced with the dynamism of sheer wanton cruelty, the lust to destroy and cause suffering, that delights in power for its own sake. It can't be denied that it exists. And power is the key: power is a form of *energy*.

In contemplating totalitarianism and its atrocities, such as the Nazi death camps, Hannah Arendt speaks of 'radical evil', by which she means a form of wrongdoing which is not captured by other moral concepts. Central to it is the making of humans into things, devitalised automata, living corpses without will or spontaneity. 'According to Arendt a distinctive feature of radical evil is that it isn't done for humanly understandable motives such as self-interest, but merely to reinforce totalitarian control and the idea that everything is possible.'[369] In other words, the self of the evil-doer, too, is destroyed, in the service of a power greater than itself, the ideal of which is simply – power. Power to do anything and everything possible. We have that lust within us.

For those who recognise a moral force for good, derived from God, the problem is different, and more acutely embarrassing. If God is omnipotent, why evil? The traditional Neoplatonic answer is that evil is mere privation of good. This seems to me an intellectual gambit only. Christians especially must take into account that Christ, according to the gospels, was tempted by Satan; and taught his followers to pray continually to be delivered from evil – not just to avoid not being quite good enough. If your belief in God stems from the refusal to deny an experienced reality, I suspect denying the reality of evil in experience is like putting the telescope to one's blind eye. Moreover, intellectual gambit that it may be, it does not really work; since why, if God is omnipotent, is privation possible? To me it makes more sense to deny God's omnipotence, of which I have no experience, than deny either God or evil, of both of which I do.

One way of thinking of this (it is hardly original) is that a divine principle of love needs something Other to love, since love is essentially directed outwards; that that Other must be free to respond, since a love that is compelled is not love; and that this necessarily means that the Other must be free to reject the love that is proffered. This seems to me necessarily true, if such a divine principle of love exists. To reduce the living spontaneity and freedom of a being to mere automaticity of behaviour would be, after all, to bring about,

[369] Calder 2018.

precisely, Arendt's idea of 'radical evil'. God's powers are limited here essentially, rather than accidentally, much as no God could be expected to make 2 + 2 =17. It can be argued that in God's creation there is a kenosis, a deliberate renouncing of power, much as *tzimtzum* is the self-limiting act of *Ein-Sof*. But at the end of the day, renunciation of power, a setting of limits, is still what it says on the tin.

The drive to evil comes from somewhere – and I do think it is a drive. In the Kabbalah, mankind has both an innate tendency to good (*yetzer hatov*) and an innate tendency to evil (*yetzer hara*).[370] Here the same logic applies, that humanity must be free to choose. But the tendencies are built into the way things are. No-one seems keen to say why.

I admire the serenity of those Buddhists who are able to say 'not two', and see the whole concern about good and evil as a misunderstanding caused by making false distinctions; but I find it difficult to share this serenity (especially in the face of the Holocaust, and a myriad other examples of innocent suffering). And while I indeed believe the endpoint is right, that the two eventually come together, and that we cannot know what good may flow from what we now call evil, or what evil may flow from what we now call good – Jung's enantiodromia – my response to 'not two' is 'yes, but ... two'; as my response to 'two' is 'yes, but ... not two'. To my way of thinking, something here is at risk of being short-circuited. The *coincidentia oppositorum* involves *both* the union *and* separation of good and evil. It is not possible to get round that. I am reminded again of Escher's 'Angels and Devils' (Plate 24.)

In Taoism, good follows from absolutely abjuring human will and following the spontaneous flow of Being: bad occurs only when humans strive against the *tao*. It is left unspoken, however, why humans alone, being themselves inevitably an expression of the *tao* like all other creatures, strive against the *tao* and exert an opposing will. But, though there is here no God as such, rather a creative force that grounds and orders the cosmos, this is not really different in its structure from the Christian *mythos* of evil – Milton's *Paradise Lost* being perhaps the most psychologically sophisticated expression of it ever written, an account written after Milton had gone totally blind, and during the bloodiest civil war in England's history: he probably understood his subject. His Satan was formerly Lucifer, the most beautiful of all angels, not an evil creation. What makes Satan evil is a disposition: pride, envy, resentment, anger and a delight in exercising power to assail the divine order. Satan seduces Adam and Eve to eat the fruit of the tree of the Knowledge of Good and Evil. It is said that the meaning of the Hebrew words translated as good and evil 'mean precisely the useful and the useless, in other words, what

---

370 According to the *Zohar* (1: 205a), God 'made a Right and a Left for the ruling of the world. The one is called "good", the other "evil", and he made man to be a combination of the two.'

is useful for survival and what is not'³⁷¹ – which happens to characterise the value structure of the left hemisphere (and of modern society): utility above all else. And, for us, utility has become the Good. It leads us onwards towards the abyss.

The Kabbalah is once again an invaluable resource. Luria held that the divine principle of the cosmos is both *Ein-sof* (that which is without end) and *Ayin* (absolute nothingness). He also held that creation is both an emanation and a contraction (*tzimtzum*); that *Ein-sof* is both the creator of the world and itself created and completed through the spiritual, ethical and 'world-restoring' acts of humanity (*tikkun*, repair); and, finally, that the *sefirot* both are the originating elements of the cosmos and are only fully realised when the cosmos is displaced and shattered (*shevirat ha-kelim*, the breaking of the vessels).³⁷² This is redemption: it is from the divine sparks within the shards of the vessels that this greater unity, this fulfilment of creation, this redemption, is achieved, so that 'a world that is alienated from and then reunited with God is superior to one that had never been alienated or divided at all':

> this return to the primal unity is all the more exalted for having passed through the dichotomies and multiplicities of a finite world; for such a restored unity is not simply a restoration of the original divine oneness, but is actually the completion and perfection of *Ein-sof* itself. According to the Kabbalists, it is incumbent upon humankind to recognize and even facilitate the distinctions within the finite world, while at the same time, through an appreciation of the coincidence of opposites, to comprehend the unity of all things.³⁷³

Returning to the Iroquois myth with which I began Chapter 20, both brothers are sons of heaven. One, He Grasps the Sky With Both Hands, remembers his divine origins: the other, Flint, does not. He relies on language and tools to do as he wishes ('I trust in the thing which my father gave me, a flint arrow, by which I have speech'). He Grasps the Sky With Both Hands can, on his own, create the panoply of life; Flint can create only monsters. His attempts to do so result in deformities, which his brother accepts and does his best to redeem. Flint's creatures do not live, until He Grasps the Sky With Both Hands puts some of his own life force into them. Flint is proud, envious and delights in his own power; his brother realises that to constrain this power, he must keep his brother, who represents 'evil in the form of ... intentional forgetfulness of the higher identity', close to him, so that as much good as possible can come from the work of both of them – but not too close, since Flint ultimately desires his brother's destruction.³⁷⁴ There is, on the one hand, the real human being, made by He Grasps the Sky With Both Hands alone,

---

371 Watts 1963 (9). See also Wilber 2001 (30).

372 Drob 2000. The resonances with the Japanese concept of *kintsugi* are impossible to ignore (see p 1161 and Plate 22[c]). That one is a spiritual concept and the other an aesthetic concept should not, of itself, perturb a reader who has accompanied me this far.

373 ibid.

374 This is also the message of *I Ching*: that the Creative and the Receptive must work together, but always with the Receptive serving the Creative. 'For the Receptive must be *activated and led* by the Creative; then it is productive of good. *Only when it abandons this position and tries to stand as an equal* side by side with the Creative, does it become evil. The result then is opposition to and struggle against the Creative, which is productive of evil to both.' *I Ching* 1968 (11), commentary on hexagram 2: K'un. I am grateful to Stephen Lowy for bringing this, and the other references to *I Ching* that follow, to my notice.

perhaps like Adam and Eve before the fall, perhaps like the soul, in Cottingham's phrase, pointing to 'the better selves we are meant to be';[375] and, on the other, there is the hatchet maker, the bringer of strife – the human beings we are. He Grasps the Sky With Both Hands warns us that Flint's anger has created a fire that 'will burn eternally in that my brother even now desires to control all minds among human beings.' We must struggle to avoid it. He Grasps the Sky With Both Hands will come to our aid – twice: but

> if a third time it comes to pass that you forget, then you will see what will come to pass. The things upon which you live will diminish so that finally nothing more will be able to grow ... It will be my brother who will do all this, for he will be able to seduce the minds of all human beings and thus spoil all that I have completed. Now I leave the matter to you.

I have nothing to add.

375 Cottingham 2020 (153).

Coda to Part III

As I maintained at the outset of this book, complex as the concept of truth may be, it is indispensible: some things are truer than others. The first two Parts of this book have been devoted to exploring how we could maximise the chances of encountering truth, given the contributions made by each brain hemisphere; and, Part III, devoted to how on adopting such an approach the cosmos would appear to be constituted.

In Part I, I showed that in relation to the main portals of access to knowledge – attention, perception, judgments formed on those perceptions, social and emotional intelligence, cognitive intelligence and creativity – while both hemispheres contribute, the right hemisphere is in every case superior to the left. Only in the business of grasping something in order to use it is the left hemisphere superior. I also showed how we can now *recognise* what is contributed to our synthesised world-picture by each hemisphere, and thus, *ceteris paribus*, how much weight to accord to the contribution. That is an advance in understanding in itself.

In Part II, I considered each of four main pathways to understanding: science, reason, intuition and imagination. My conclusion was that we should never rely on one pathway alone, and that, where appropriate and possible, all four should be brought to bear. I also concluded that in each case, while again both hemispheres make a contribution, it is the contribution made by the right hemisphere that is ultimately of greater importance in arriving at a full understanding. In practice we in the contemporary West tend to restrict ourselves to one, sometimes two, of these pathways at any one time, and, moreover, often neglect the more important right hemisphere contribution to those on which we do choose to rely.

In Part III, I have looked at the 'stuff' of which the cosmos is made – time, space, motion, matter, consciousness, etc – as well as why we might see it as divine in nature. And its apparently paradoxical nature. I suggested that paradox generally resulted from the clash between the ways in which the right and left hemisphere construe reality. 'One way to unify things that appear different is to show that the apparent difference is due to the difference in the perspective of the observers', writes physicist Lee Smolin. 'A distinction that was previously considered absolute becomes relative. This kind of unification is rare and represents the highest form of scientific creativity. When it is achieved, it radically alters our view of the world.'[1] When one applies the lens of the hemisphere hypothesis to important issues in philosophy, including the philosophy of science, many can be

1 Smolin 2008 (22).

seen, I suggest, as reflecting, at one level, a conflict between the two ways of looking at the world that are typical of the two hemispheres.

The thinking of our culture is still dominated by the superannuated, eighteenth-century model of mechanistic reductionism, despite the discoveries of quantum physics and an increasing understanding of complex systems, neither of which is compatible with it. Such a way of thinking posits that the cosmos is a purely material machine, in principle fully comprehensible by analysis into its parts. It is determined, rather than freely creative. And the whole farrago is without any meaning or purpose. So it is said.

This view is far removed from what physics tells us of the reality which it is thought to describe. But – here is the beauty of the alternative view I am recommending – this conventional view does not have to be wholly dismissed. In fact it *must* not be. It is not without its elements of truth – just very limited ones by comparison with the understanding of the right hemisphere. A map contains some truth, but unless you can interpret it in the light of real-world experience, it is useless. And although a map depends on the world, the world does not depend on a map. The map is a special case of the world, by which it is included, not the world a special case of the map. This is an important point because it is sometimes assumed that if you can break the world down, by the same process you could build it up. As the theoretical physicist Philip Anderson once put it, 'the ability to reduce everything to simple fundamental laws does not imply the ability to start from those laws and reconstruct the universe'.[2]

As always there is an asymmetry: the right hemisphere's take is in no sense a special case of the left hemisphere's, but that of the left hemisphere is a special case of the right's, which already includes it.

I argue that the reductionist paradigm has had a good run for its money, but its time is over. We need a new one: like any paradigm, the new one can be no more than a working model, but I believe it is one that makes a better fit with everything we know from experience, by which I mean all that can be learnt from science and reason at their best, in combination with intuition and imagination at theirs. It is a paradigm supported by the argument and evidence in this book, from brain science and from physics, and by several increasingly important schools of thought, especially those of the Pragmatists, process philosophers and phenomenologists, as well as ancient bodies of wisdom, of both East and West, that have never been of greater consequence than they are today. That confluence alone is striking. This paradigm is in accord with the understanding of the brain's right hemisphere, which, I hold, has evolved to be our primary means of comprehending the world, while the left hemisphere has evolved to execute procedures that serve our will in manipulating the world. On such a basis, even if this alternative

2 Anderson 1972 (393).

paradigm did not seem a much better fit with our knowledge and experience, it would be *prima facie* an account less likely to deceive us. But it does seem a better fit. And the right hemisphere also has the capacity to unite what the left hemisphere offers with its own understanding, in a way that the left hemisphere cannot.

What is this right hemisphere understanding? In line with Bohr's insight, it holds that the deep truths about reality are likely to appear initially paradoxical. What look like things – with inevitable suggestions of stasis, certainty and fixity – are processes; and all such thing-like processes are interdependent with others, bringing one another into being. Everything is changed by context; strictly speaking the context is everything else that exists or existed, but for practical purposes we mean by it that part of the net of interrelations which, due to proximity or potency, has the greatest part to play in constituting whatever it is on which we are focussed at the time. All parts may be considered as wholes in relation to some lesser parts, and all wholes may be considered as parts in relation to some greater whole. The whole which a part goes to make up tells us as much about the nature of the part as the parts reveal about the nature of the whole. Relations are not secondary to *relata*, and it may be argued that *relata* may be secondary to relations, as the nodes in a web are secondary to the intersection of the threads (think Indra's net):

> According to the prevailing theory, as the universe evolved after the Big Bang, matter became distributed in a web-like network of interconnected filaments separated by huge voids. Luminous galaxies full of stars and planets formed at the intersections and densest regions of the filaments where matter is most concentrated.[3]

In quantum field theory, the absolutely primary elements of reality are force fields, that is to say relationship and process, not things. Moreover things emerge from processes in the 'Standard Model':

> In the Standard Model of Particle Physics, which is the best theory we have so far of the elementary particles, the properties of an electron, such as its mass, are dynamically determined by the interactions in which it participates ... No longer are there absolutely 'elementary' particles; everything that behaves like a particle is, to some extent, an emergent consequence of a network of interactions.[4]

And, of course, consciousness takes part in their being. There is absolutely no reason to believe matter to be ontologically prior to consciousness: there is only a prejudice that now needs to be retired.

The cosmos is in process, one in which the potential that is infolded within being is constantly unfolding into actuality and then being re-infolded into the now enriched whole, in a cycle that endlessly returns. But to see the bigger picture, one has to see not only

---

[3] Stephens 2020, referring to Burchett, Elek, Tejos *et al* 2020.

[4] Smolin 2013 (xxx).

the actual, but the potential through and behind the actual. This means having always before one's eyes the bigger *Gestalt* from which the actual comes and to which it returns; being able to remember, while in the field of actuality, the potential out of which it arises and to which it contributes its part, for a time. The first thing we learn about the twin brothers is that He Grasps the Sky With Both Hands does not forget his higher identity in the midst of action in the world; while his brother Flint declares: 'I am not thinking about the place from where I came ... It is sufficient that my mind is satisfied in having arrived at this place.'

I have also suggested that whatever creative energy underwrites the unfolding of the phenomenal universe is continually active and involved in that universe. This is a teaching which in one form or another is at the core of almost every ancient mythology, every religious tradition, has literal truth in terms of physics, and is true to a Whiteheadian vision: that of the world and a creative dynamism forever bringing one another into being.

This also touches on what, in *The Master and his Emissary*, I referred to as the ontological asymmetry of the hemispheres. In doing so I quoted Blake's saying that 'Energy is the only life and is from the Body; and Reason is the bound or outward circumference of Energy'. Our power to conceptualise, then, according to Blake, is parasitic on and derives any vitality it may seem to have from being the very boundary, the limit place of that Energy. The right hemisphere, being the primary mediator of experience, from which the bloodless, conceptualised, re-presented world of the left hemisphere derives, is never remote from the creative energy of the body, the emotions and life: it is involved in the world, though aware of there being much beyond. He Grasps the Sky With Both Hands brings the human form to life with life of his own, giving it of his own blood, his own consciousness and his own breath. The left hemisphere, He Who Is Crystal Ice, He Who Is Flint, is 'satisfied in having arrived at this place'. He has got his words and his arrows. He has forgotten where he derives his energy from. And so He Grasps the Sky With Both Hands says of their joint creation, the hatchet maker, the bringer of strife: 'I see he will become hostile to me ... What will come to pass because of that?'

Epilogue

You live in a deranged age – more deranged than usual –
because despite great scientific and technological advances,
 man has not the faintest idea of who he is or what he is doing.
  —WALKER PERCY[1]

The great departs; the small approaches.
Thus heaven and earth do not unite,
and all beings fail to achieve union.
Upper and lower do not unite,
and in the world, states go down to ruin.
  —I CHING[2]

Strength lies in improvisation. All the decisive blows
are struck left-handed.
  —WALTER BENJAMIN[3]

I often feel sympathy for novelists and crime-writers, who must be aware that there is a type of reader who will leap straight to the final chapter to discover whodunnit even before anything has been dunn. It destroys, of course, one of the pleasures of reading the book. Authors of books such as this one are not immune from the same problem, and in our case it is even worse. Readers who turn first to the final chapter often do so as a substitute for reading the book as a whole, and inevitably come away with a much cruder idea of what the book suggests. Final chapters, epilogues, and other 'last words' can contain none of the subtleties, complexities or indeed the caveats, of the book itself. Yet some sort of concluding chapter is almost always necessary.

However, I have to disappoint the casual browser. Let me state clearly that what follows is in no sense even an approximate 'summary' of the book's argument. It isn't a summary at all.

I feel my loyalty is to those readers who have accompanied me from the beginning and have arrived at this point in the conventional manner. I therefore see paying my debt to them as my showing how what, at times, might have seemed somewhat remote from the concerns that press on us – discussions of the nature of time, space and movement, for example – nonetheless bear directly upon our experience in the world today. In the book I have put forward for the reader's consideration a form of cosmology – and, in the simplest sense, a philosophy of life: all of that was best expressed where I took time to do so, and to express it less carefully now would be both superfluous and counterproductive. This epilogue, then, presents not a backward glance at where we have been, but a projection

---

1 Percy 1983 (76).

2 Hexagram 12: P'i – Stagnation (the exact opposite of hexagram 11: T'ai – Peace). Commentary on the Decision: 'Evil people of the time of standstill do not further the perseverance of the superior man. The great departs; the small approaches. Thus heaven and earth do not unite, and all beings fail to achieve union. Upper and lower do not unite, and in the world, states go down to ruin. The shadowy is within, the light without; weakness is within, firmness without; the inferior is within, the superior without. The way of the inferior is waxing, the way of the superior is waning' (trans R Wilhelm).

3 Benjamin 1995 (650).

forward into the plight of our contemporary world, and how that philosophy may be of relevance here.

Those who read only what follows rather than the book as a whole may be tempted to see here the conventional 'conclusions' which a final chapter might provide. If so, they might well protest that the 'conclusions' are nothing new. In one sense they are right – there is nothing new under the sun. All I hope to do in these last pages is hint at how our exploration of the brain and its contributions to mind can help us see things we might not otherwise see, and which indeed we are no longer permitting ourselves to see; see new connexions between otherwise disparate phenomena; and see what we *do* see in a different light. Whether I have succeeded or not, however, can be judged, I would suggest, only by those who do, in fact, read the book.

~

It is often said that we are experiencing a crisis of meaning. Not coincidentally, far more of us than ever before in the history of the world live divorced from Nature, alienated from the structures and traditions of a stable society, and indifferent to the divine. These three elements have always been what have provided us with an overarching sense of belonging: our relations with the living world, with one another, and with a divine realm. (They also turn out from contemporary research, unsurprisingly, to be the three elements that most determine one's happiness and fulfilment as a human being.)[4] Alas, they are no longer open to us: we have seen to that. We exist in the world, of course, but we no longer *belong* in this world – or any world worthy of the name. We have unmade the world. This is entirely new in the history of humanity and it is impossible to exaggerate its significance.

Each of these divorces has come about very swiftly – in a mere 250–300 years – the twinkling of an eye in relation to the age of humanity. And they have each come about because of an access of hubris. Nature has become mere resource; the divine mere superstition; and the unruly complexity of life can, we believe, be simply rationalised, ironed out, and subjected to our conscious control, by technology, by bureaucracy and, where necessary, by law. We know far more, we think, than people of other ages and cultures; indeed we pretty much know it all. However, I suggest, no people that ever lived has understood so little.

In the ninth-century Chinese classic, *The Secret of the Golden Flower*, it is written that 'the conscious mind is like a violent general of a strong fiefdom controlling things from a distance, until the sword is turned around.'[5] The sinologist Thomas Cleary comments: 'Zen Buddhism traditionally describes the mechanism of delusion as

---

[4] For the living world, see pp 1325–6 below; for one another, see McGilchrist 2009b (esp 434–6); for a divine realm, see Appendix 8, section 4.

[5] *The Secret of the Golden Flower*, Bk 11, §5: 1991 (14).

mistaking the servant for the master. In the metaphor of this passage, the general is supposed to be a servant but instead usurps authority.'⁶

In both the Zen and Taoist traditions, the narrowly circumscribed conscious mind, according to Cleary, 'is supposed to be a servant of the original mind' – original here meaning the ontologically prior and deeper-lying mind, on which the 'conscious mind' depends for understanding. When 'the sword is turned around ... the original mind retrieves command over the delinquent conscious mind'.⁷

In a subsequent passage Cleary adds, unknowingly, but precisely, describing the way in which the two hemispheres work best together (the interpolations in square brackets are of course mine):

> Intuition belongs to the original spirit; intellect belongs to the conscious spirit. The essence of Taoism is to refine the conscious spirit [LH] to reunite it with the original spirit [RH] ... self-delusion occurs when the servant has taken over from the master; self-enlightenment takes place when the master is restored to autonomy in the centre.

As he points out, this is an image of 'an ideal relationship between the original spirit as the source of power and the conscious spirit as a subordinate functionary':

> In this way the intellect [LH] functions efficiently in the world without that conscious activity inhibiting access to deeper spontaneous knowledge through the direct intuition of a more subtle faculty [RH].⁸

Why is the sword said to be turned around? Because the highest achievement of the analytic intellect – and this only very rarely happens – comes when it knows when to stop: how to turn its power, where necessary, on itself, so as to see its proper limits and to abide by them. To quote Heidegger once more, 'The evil and thus keenest danger is thinking itself. It must think against itself, which it can only seldom do.'⁹

I dare to hope that this book may aid in one of those rare instances of the intellect's becoming aware of its own limitations; coming once more to play the invaluable role of servant, rather than pretending to be the Master, without having any of the necessary insight into, or wisdom about, what it is doing.

In *The Master and his Emissary* I laid out, first, the neuropsychological grounds of the hemisphere hypothesis and its philosophical consequences; and, then, what I could see happened to a civilisation when its ethos, instead of encouraging the proper working together of the hemispheres, began to favour a very particular outlook, one that can readily be shown to conform to the mode of operation of the left hemisphere alone. I did this by reviewing the major turning

6 Commentary on the above: 1991 (78). (Cleary correctly uses the less familiar Chinese term 'Chan' for what the reader will probably know by its Japanese appellation, 'Zen'. I have changed the term in the interests of a readier understanding.) I am grateful to Professor Kenneth Wilson for bringing this passage to my attention.

7 ibid.

8 *The Secret of the Golden Flower*, 'Translator's afterword': 1991 (138–9).

9 Heidegger 1975a (8).

points in the history of ideas in the West through the lens of that hypothesis, which to me provided a grave warning. I was, and am now still more, fearful that unless we *radically* change the path we are pursuing we cannot survive – certainly as a civilisation, and perhaps as a species. In the last chapter of that book I asked the reader to imagine what the world would look like if I were right that we had more or less confined ourselves to seeing it from the very narrow, highly skewed, standpoint of the left hemisphere. Few readers have needed much prompting to recognise in it the world where we live now.

Any reader of this book will, by now, be very familiar with the consequences of the view made possible by the left hemisphere. So I will not need to spend much time on them: the most obvious of these can be listed fairly succinctly. In each case we achieve the opposite of what we intend.

OUR LEFT HEMISPHERE WORLD: 'ARROGANT DRAGON WILL HAVE CAUSE TO REPENT'

We have lost the sense of the broad picture, and the feeling for the whole, resulting in despoliation of the planet, exploitation of poorer nations, and extinction of the way of life of indigenous peoples. We have for some time now disregarded, and latterly traduced, the ancient wisdom of our cultures, and simultaneously acted with contempt for future generations. Social cohesion is breaking down in the face of atomistic individualism; and we are experiencing an epidemic of depression, anxiety and loneliness. Context is absolutely *essential* to the understanding of every utterance and every deed; yet we neglect the importance of context, so that words and deeds are ripped from their anchors in life and subject to comprehensive misunderstanding – all of which is exacerbated by the Procrustean effect of soundbite culture, and by the fragmentation of our attention. We sit in judgment on other times and other places, the art, literature and history of the past, and other people's cultures; yet we have no idea how very weird our own values are when set in context. We neglect all that is implicit, which is nine-tenths of what matters, and which can't be made explicit without destroying its meaning and value: this is even affecting our ability to understand tone of voice and to read faces. The 'what' has come to triumph over the 'how', the goal over the means, the brute 'this must be done' over the manner in which it is done. There is a growth of machine-like inflexibility, loss of judgment and discretion, and a culture of petty rules that strangle initiative and affront our humanity. We are witnessing the triumph of black and white judgments, especially in the 'culture wars', where there is no vestige of subtlety in our thinking,[10] no patience for the complex, and often little or no empathy, but rather anger and self-righteousness. We are newly beset by a tyranny

[10] Rigid views, regardless of political orientation to right or left, are associated with extreme cognitive inflexibility (Zmigrod, Rentfrow & Robbins 2020).

of literal-mindedness – affecting our capacity to understand metaphors, humour, and irony, which increasingly are being driven out of public converse and out of our lives. We have replaced the living and unique by simple categories everywhere: a tick-box mentality. We have substituted quantity for quality – and 'productivity' for creativity (another aspect of the 'what' killing off the 'how'). We have an insatiable need for control – approaching total control, not just of action, but now of thought, through surveillance by both state and global capitalism. We are engaged in a war on the body, since it resists our free fantasies of limitless possibility, even if only by ensuring that one day we die; simultaneously we deny the existence of the soul, which is not something that exists over against embodiment, but is enmeshed with it. We kid ourselves that frenetic activity is a sign of life, not the desperation of people who have no idea what they are doing or why they are doing it, locked into a competition they cannot see how to escape; and which is destroying the world. And all this is accompanied by a blind optimism, born of the denial of consequences. Every aspect of this reflects, as the reader will recognise, a prepotency of the left hemisphere's *modus operandi*. In Parts I and II, we saw how this leads, literally, to delusion; and in Part III, by revisioning the world, I drew attention to how we are unmaking it.

We are out of touch with reality, to which the right hemisphere's world-picture would still give us access, if only we didn't dismiss it. In our world, the real, possibly cataclysmic, threats we are facing are in effect denied, while in their place new, hitherto unimagined, 'threats' to our sensitive natures are invented, which then proceed to take up a disproportionate part of our attention. It puts me in mind of a patient I looked after who, following a right hemisphere stroke, was not the least bit concerned that the whole left side of her body was paralysed – indeed, like many right hemisphere-damaged patients, she denied that it was so; but she was absolutely convinced that she was being victimised by a patient in the next bed, who had, she thought, contrived to steal her magazines and subtly poison her food.

One of the themes of this book has been that when we don't truly understand what we are doing and why we are doing it, life appears 'paradoxical': we set out to achieve one end and reach its exact opposite. Indeed, because of our unusual world view, we are constantly in a state of surprise about the way our plans don't work out. (When I say 'we', I acknowledge that many of my readers will object that they are no part of this; that is good, but you must then accept that you are the dissidents on whom nothing less than our future depends.)

As a society, we pursue happiness and become measurably less happy over time. We privilege autonomy, and end up bound by rules to which we never assented, and more spied on than any people since the beginning of time. We pursue leisure through technology, and discover that the average working day is longer than ever, and that we have less time than we had before. The means to our ends are ever more available, while we have less sense of what our ends should be, or whether there is purpose in anything at all. Economists carefully model and monitor the financial markets in order to avoid any future crash: they promptly crash. We are so eager that all scientific research result in 'positive findings' that it has become progressively less adventurous and more predictable, and therefore discovers less and less that is a truly significant advance in scientific thinking. We grossly misconceive the nature of study in the humanities as utilitarian, in order to get value for money, and thus render it pointless and, in this form, certainly a waste of resource. We 'improve' education by dictating curricula and focussing on exam results to the point where free-thinking, arguably an overarching goal of true education, is discouraged; in our universities many students are, in any case, so frightened that the truth might turn out not to conform to their theoretical model that they demand to be protected from discussions that threaten to examine the model critically; and their teachers, who should know better, in a serious dereliction of duty, collude. We over-sanitise and cause vulnerability to infection; we over-use antibiotics, leading to super-bacteria that no antibiotic can kill; we make drugs illegal to protect society, and, while failing comprehensively to control the use of drugs, create a fertile field for crime; we protect children in such a way that they cannot cope with – let alone relish – uncertainty or risk, and are rendered vulnerable. The left hemisphere's motivation is control; and its means of achieving it alarmingly linear, as though it could see only one of the arrows in a vastly complex network of interactions at any one time. Which is all it can.

If these paradoxes surprise us it is because we have not thought far enough ahead in time or broadly enough in space: we take a small part of the complex for the whole. The awareness coming from the right hemisphere can embrace that of the left, but not the other way round. When the hemispheres are working together under the unifying influence of the right hemisphere, the effect is not purely additive, but transformative. However, since the left hemisphere not only 'takes in' less, but understands what it takes in less well, our almost exclusive reliance on it, the servant, in contemporary Western culture, is a problem of some considerable proportions. Indeed, like civilisations before us, which drifted further and further to the outlook of the left hemisphere, we would appear to be engaged in committing suicide, intellectual and moral – if not indeed literal; excluding whole aspects of reality, resulting in a version of the world that 'computes' as far as the left hemisphere is concerned, but is grossly impoverished and lacking in meaning. One that is, in sum, more fit for a computer than a human being.

The focus of our thought is constrained because we think in terms of linear systems, when linear thinking is inadequate to the task. The result is what David Bohm called 'sustained incoherence', and its manifestations are all around us. He linked it to what he calls 'thought', which is clearly a description of left hemisphere thinking, even down to the denial:

> There is another major feature of thought: *thought doesn't know it is doing something and then it struggles against what it is doing*. It doesn't want to know that it is doing it. And it struggles against the results, trying to avoid those unpleasant results while keeping on with that way of thinking. That is what I call *sustained* incoherence.[11]

Later he has this to say: 'If there is sustained incoherence, it just keeps on going in spite of the fact that there is evidence which would show that it's incoherent. Now, we could say that an intelligent response on seeing incoherence would be to stop it, to suspend it and begin to look out for the reason for the incoherence and then to change it. But I say there is a defensive incoherence'.[12] In other words the left hemisphere, above all else, doesn't want to hear why it might have got things wrong, as we have seen in this book time and time again.

I see widespread evidence of this strange mentality in corporations, governments, health systems and education – everywhere that management 'culture' holds sway – that when things go wrong it is never that we have been travelling in the wrong direction, or have gone too far in what may once have been the right direction, only that we have *not gone far enough*. This links to the Dunning-Kruger effect: the less you know, the smarter you think you are. A further finding by Dunning and colleagues, however, helps make clear the relationship with the left hemisphere mindset, not only because of its relatively blinkered vision, but because of its preference for simple linear algorithms and procedures that 'logically must' lead to a certain outcome. Those who have such procedures think they must be in the right, even when the outcome ought to compel them to the opposite conclusion. Once they are committed to their theory of how things work, drawing attention to its glaringly obvious failure in the real world leads not to a flicker of doubt, but to a rise in confidence and redoubled efforts along the same line.[13]

Once more the *I Ching* offers insight:

> The person who ... keeps on striving when the maximum of influence has already been achieved, knowing only how to press forward and not how to retreat, isolates himself from the human sphere and loses his success. For what is complete cannot endure, and what

11 Bohm 1994 (10–11: emphases in original).

12 *ibid* (55).

13 Williams, Dunning & Kruger 2013: 'In 6 studies, participants completed tests involving logical reasoning, intuitive physics, or financial investment. Those more consistent in their approach to the task rated their performances more positively, including those consistently pursuing the wrong rule. Indeed, completely consistent but wrong participants thought almost as highly of their performance as did completely consistent and correct participants. Participants were largely aware of the rules they followed and became more confident in their performance when induced to be more systematic in their approach, no matter how misguided that approach was.'

is pushed to the limit ends in misfortune. Thus the last line says, 'Arrogant dragon will have cause to repent'.[14]

Although our current public discourse often implies the opposite, nothing is ever simply yes or no, but always a matter of degree: 'yes, but'. Or, in another Zen saying I love: 'not always so'. Carved into the stone of the ancient temple of Apollo at Delphi, along with the injunction to 'know thyself', was the epitome of wisdom: *mēden agan*, 'nothing in excess'. (I didn't fully understand that in my teens; I do now.) Any principle that is extended too far, without respect to the opposite that is always inherent in it, may turn into the very thing that is being avoided and denied: its dark side. This is especially the case if there is no civilised dialogue between opposing camps. One extreme will simply flip into the other.

Moreover, bureaucracy is an unwitting accomplice in generating extremism. The answer to every ill now is to appoint, often at public expense, an officer, and an office, whose job is to see only one side of a question. Once appointed, these 'tsars' and their tsardoms will not do themselves out of a job by saying, at any point whatever, there is little to be done, since a balance has now been achieved. They will press on regardless, and without any obvious end in sight, finding ever more minute infractions of their principles to deal with by means of further regulation: a hugely wasteful and destructive process, leading to the incubation of resentment.

And nobody dares wind them up, for fear of giving the 'wrong message'.

Disregard for the future is matched by disregard for the past and its hard-won wisdoms. 'No self is of itself alone', wrote Schrödinger: 'The "I" is chained to ancestry by many factors ... This is not mere allegory, but an eternal memory.'[15] Because we are part of a living whole in time and space, we have responsibilities to our ancestors, to one another now, and to future generations, something that what we call 'simple' peoples and 'simple' cultures have always known and taken into account. Ironically, with globalisation, the world has become contracted in space and time to the infinitely thin slice which represents just 'me', here and now.

A society is not derived from an assemblage of individuals, but individuals derive from the organism that is a society. 'The individual actually grows', wrote the social philosopher Norbert Elias, 'from a network of people existing before him into a network that he helps to form: the individual person is not a beginning and his relations to other people have no beginnings.' And he compares it to a conversation where 'the questions of one evoke the answers of the other and vice versa', something that 'can be understood – like the figure of a thread in a net – only from the totality of the net-

---

[14] Wilhelm & Wilhelm 1995 (75).

[15] Schrödinger, writing in July 1918; quoted in Moore 1992 (113). The most famous expression of this is Burke's vision of society as 'a partnership not only between those who are living, but between those who are living, those who are dead, and those who are to be born': *Reflections on the Revolution in France* (1790), in Burke 1887, vol 3, 231–563 (359).

work'.¹⁶ For this reason, as Viktor Frankl observed, 'individuality can only be valuable when it is not individuality for its own sake, but individuality for the human community.'¹⁷ Social understanding, empathy, connectedness, being concerned more with duties than rights, seeing the need for self-discipline, not self-indulgence – all this the reader will know is also far better understood by the right hemisphere than the left.

Is it not the case, though, it may be objected, that technology, the left hemisphere's extension of its power, is helping to bring one world together? 'Poisonous nonsense.' That was the answer given by Waldo Frank, an American historian of mixed race, in an essay called 'The central problem of modern Man', published in 1946. Pointing to air travel, television and radio as the media of such supposed 'one-worlding', he noted, with remarkable foresight, that with them

> come voices, visions, wills, that are products of, and spokesmen for, disintegration; come special and sectional pulls; come egoisms both individual and collective ... the crowded, noisy ignorance of the modern world is less easily invaded by knowledge than was the silence of illiterate epochs. But the peril is more positive ... Each faction of the unassembled body of integration, each with a part of truth but standing alone, asserting itself alone, multiplies disintegration; causes a pendulum swing from one partial extreme to another.

One wonders what he would say if alive today. But to me what makes his comment so astute is the comparison he makes. Unlike a single cell, or an organism (as the reader of Chapter 12 will remember vividly), to which a society is often compared, we lack any distinct sense of an overall form, or *Gestalt*. So he continues:

> ... the reason is, of course, the want of pre-existent balance, the organism's way of digesting all it absorbs, using and rejecting according to its *form*. We thus find ourselves in a vicious circle: because we lack the organic sense to begin with, we misuse or deform the trends in our world that should move us toward it.¹⁸

Those who have accompanied me on the trip so far will recognise in North American native myths, ancient Judaic wisdom and Chinese philosophical literature analogies to the myth enshrined in the title of *The Master and his Emissary*. Versions of this insight into the over-reaching nature of mere left hemisphere 'intellect' exist, in fact, in many cultures. (I put the word intellect in inverted commas because, as the reader will know, intellect must not be confused with intelligence, which it often lacks.) The insight is present in Rumi;¹⁹ it is present in the Jewish legend of the *Golem*; and versions are also found in Sanskrit literature including the *Mahabharata*. All of these, including *The Secret of the Golden Flower*, were unknown to

16 Elias 1991 (32–3).
17 Frankl 2019 (55).
18 Frank 1946 (emphasis in original).
19 According to Idries Shah, 'it also appears in Rumi's work and again and again in oral legends of Jesus, of which there are a large number.' In a tale of the Sufi dervishes, Isa (Jesus) is besieged by some followers who beg from him the secret name by which Isa restored the dead to life. Not being fitted for such knowledge, they 'make a trial' of the Word on a heap of bleached bones in the desert: immediately a ravenous wild beast comes to life and tears them to shreds (Shah 2016, 138–9).

me when I wrote that book. Yet these ancient insights were exactly where neurological research has led me.

Nor are these 'merely' ancient insights. Here is the neuropsychologist Stuart Dimond speaking: 'Language immediately makes a claim for itself to be a *prince* among the mental processes, and to be the exclusive medium of mental function, denying often by its existence the presence of other modes of thought.'[20] As David Bohm also saw, our conscious thought processes are not, as they proclaim themselves to be, neutral, but motivated by the values, such as they are, of the left hemisphere:

> One of the obvious things wrong with thought is *fragmentation*. Thought is breaking things up into bits which should not be broken up ... and at the same time we are trying to establish unity where there isn't any [through categorisation] ... Thus we have false division and false unification. Thought is always doing a great deal, but it tends to say that it hasn't done anything, that it is just telling you the way things are ... that 'you' are inside there, deciding what to do with the information. But I want to say that you don't decide what to do with the information. The information takes over. *It runs you.* Thought runs you. Thought, however, gives the false information that you are running it, that you are the one who controls thought, whereas actually thought is the one which controls each one of us. Until thought is understood – better yet, more than understood, *perceived* – it will actually control us; but it will create the impression that it is our servant, that it is just doing what we want it to do.[21]

Making 'thought' perceivable to our thinking has been one major aim of this book.

Bohm carries on to say that 'thought is creating divisions out of itself and then saying that they are there naturally ... thought divides itself from feeling and from the body.' Once again he unwittingly describes the left hemisphere, cut off from emotion, the body, and all other forms of knowledge; not seeing what it is it cannot see; power-hungry, and seductive as the serpent with its air of innocence offering fruit from the tree of 'knowledge'.[22] But his astonishingly clear insight is that, as I too believe, we are in the grip of something bigger than us that tells us it has our interests at heart in order the better to control us. And it has been devastatingly successful. As the theory of complex systems shows, within such a system, while every single part has its effect on the whole, no single part can determine, or necessarily influence, or even perceive, where the whole is nonetheless tending.

In 1922, Rilke wrote in one of his *Sonnets to Orpheus*:

> When the machine presumes to inhabit our minds, rather than be at our command, it threatens everything we have achieved. That the lovelier hesitancy of the master's hand may be no longer manifest,

[20] Dimond 1979c (216; emphasis added).

[21] Bohm 1994 (3–6; emphases in original).

[22] That is to say, knowledge as understood by the left hemisphere: see p 1252 above.

it cuts the stone for its more resolute work to an unyielding line. It tarries nowhere, so that we might escape from it, just for once – and leave it be, oiling away in some quiet factory. It is life – it thinks it knows best, as it orders and makes and destroys with the same determination.[23]

This sounds a lot like Bohm's 'thought' that thinks it knows best, pretends to serve, but is quietly in control: and indeed that 'the machine presumes to inhabit our minds, rather than be at our command' is a perfect description of the left hemisphere usurping the right. What Rilke's poem as a whole makes clear is that 'the machine' is a drive, not just a thing: a drive that is implacably opposed to the human, and (as the rest of the poem declares) to all that is vulnerable, beautiful, subtle, holy or awe-inspiring. It destroys. And as before, in the very act of giving up our freedom, we imagine we are the ones that are in control.

Analysis has the capacity to take to pieces (literally, from Greek ἀναλύειν, to 'undo', 'break up' or 'dissolve') what already exists: but, like Flint in the Iroquois legend, it is impotent to bring things into being on its own.[24] I earlier quoted Jan Zwicky on the analytic process: 'Carried too far too often, we lose the sense that something is amiss when the patient exhibits no life. We come to take pride in our cases of polished bone.' Life or vitality comes to us from the body and the unconscious, not from the exercise of analytic reason; much as the gardener does not *make* the plants, or *make* them grow, but merely trims and tends the natural vigour that comes from elsewhere – in this case out of the earth and the seed. D. H. Lawrence referred to 'the true unconscious, where our *life* bubbles up in us, prior to any mentality';[25] and elsewhere, in speaking of his embodied existence, whether sick or well, 'I am alive, alive to the depths of my soul, and in touch somewhere with the vivid life of the cosmos'.[26] It is hard to feel alive when you have done everything possible to deny the existence of your embodiment, as man or woman – which is flesh and blood, not merely dry bones, however well-polished.

And then we will die, of course; but does that necessarily mean we were ever really alive? It seems to me that, in Kürnberger's phrase, life itself no longer lives. We are, more than anything else, *devitalised*: another predictable consequence of left hemisphere domination. Mechanism has triumphed over organism.

Because of our lack of sense of the cohesion of a dynamic whole, we end up acting in bad faith. Not for the first time, it is Erwin Chargaff who puts it most vividly:

> Our time is cursed with the necessity for feeble men, masquerading as experts, to make enormously far-reaching decisions ... You can stop splitting the atom; you can stop visiting the moon; you can stop

---

23 Rilke, *Die Sonette an Orpheus*, II, 10; and it carries on: 'But existence is still enchanted for us; it wells up in a hundred places. A play of pure forces, that no-one can touch who does not kneel and adore. Words falter, in delicacy, before the ineffable ... And music, ever new, builds from the most tremulous stones, in space that can never be made useful, its divinely hallowed home.' » *Alles Erworbene bedroht die Maschine, solange | sie sich erdreistet, im Geist, statt im Gehorchen, zu sein. | Daß nicht der herrlichen Hand schöneres Zögern mehr prange, | zu dem entschlossenern Bau schneidet sie steifer den Stein. || Nirgends bleibt sie zurück, daß wir ihr ein Mal entrönnen | und sie in stiller Fabrik ölend sich selber gehört. | Sie ist das Leben, – sie meint es am besten zu können, | die mit dem gleichen Entschluß ordnet und schafft und zerstört. || Aber noch ist uns das Dasein verzaubert; an hundert | Stellen ist es noch Ursprung. Ein Spielen von reinen | Kräften, die keiner berührt, der nicht kniet und bewundert. | Worte gehen noch zart am Unsäglichen aus ... | Und die Musik, immer neu, aus den bebendsten Steinen, | baut im unbrauchbaren Raum ihr vergöttlichtes Haus.* «

24 See pp 814–5 above.

25 Lawrence 1980 (207; emphasis added).

26 Lawrence 1961 (202).

using aerosols; you may even decide not to kill entire populations by the use of a few bombs. But you cannot recall a new form of life ... The world is given to us on loan. We come and we go; and after a time we leave earth and air and water to others who come after us. My generation, or perhaps the one preceding mine, has been the first to engage, under the leadership of the exact sciences, in a destructive colonial warfare against nature. The future will curse us for it.[27]

27 Chargaff 1976

28 See p 1140 above.

29 DEPRESSION: Wood & Joseph 2010; MORTALITY: see meta-analysis of 62 papers studying a total of 1,259,949 subjects by Martín-María, Miret, Caballero *et al* 2017; and more recently Becchetti, Bachelet & Pisani 2019; LATERALISATION: while hedonic happiness shows no significant lateralisation, eudaimonic happiness is, as expected, more dependent on the right hemisphere (Costa, Suardi, Diano *et al* 2019; Lewis, Kanai, Rees *et al* 2014).

30 Hofstede 2011.

In the second half of *The Master and his Emissary* I aimed to show how we have succumbed twice in the West (at the end of the Greek and Roman civilisations) and are now succumbing for the third time, to the temptation to see the world only through the eyes of the left hemisphere emissary. In the past this has coincided with the over-reaching of an empire, as today, and the collapse of a civilisation, as I fear awaits us tomorrow. But never has the hold of the left hemisphere on us been more complete than it is today. Its form of attention to the world and way of being in it confront us wherever we look. The sword must be turned around if we are to survive.

THE PURSUIT OF HAPPINESS

Let me ask one simple question. Has our weird inversion of values made us happier?

We are the world's most relentless ever seekers after hedonic happiness: pursuing personal pleasure. This only succeeds in setting us on 'the hedonic treadmill', whose results are a restless, endless yearning for more of something we know not what, a craving that can never be fulfilled.[28] Eudaimonic happiness, by contrast, is the result of leading a more 'other-centred', rather than self-centred life, of self-restraint – in short, living unostentatiously what has traditionally been called a virtuous life; it gives rise to a sense of fulfilment and vitality. By contrast with hedonic happiness, it protects against depression, and lowers all-cause mortality: it is also more dependent on the right hemisphere.[29]

The pursuit of hedonic happiness depends upon a set of values that our society has taken to be axiomatically superior. These are the values of individualism: personal autonomy, control, choice. But the evidence is that they do not provide a pathway to a happy life – rather the reverse. In fact at the heart of all this is what has come to be known as the 'vulnerability paradox'. Cultures can be characterised by four main measures, according to the most widely adopted criteria in cross-cultural psychology, those of Geert Hofstede: greater or lesser inequalities of power, greater or lesser autonomy, more or less structured codes of behaviour, and more or less fluid roles for men and women.[30] To these he later added measures of longer-term (self-denying) or shorter-term (self-gratifying) orientation. The seeming paradox lies in the fact that countries that score 'best' (as the prevailing climate of opinion would see it) in terms of these factors – that is to say societies with lower power inequalities, greater autonomy, less structured codes of behaviour, and 'low masculinity' – have higher rates of mental illness and suicide. That such an outcome should be labelled 'paradoxical' by professional psychologists is a measure of just how unquestioning has become our commitment to values which do in fact turn out to cause harm: if we were not so blindly committed to them, there would be no paradox, simply the observation that the values that underpin the pursuit of hedonic happiness result in higher rates of mental illness and suicide.

Satisfaction and happiness ratings in all modern societies that have been researched have declined as they have become more prosperous and more 'Westernised'.[31] (The decline has been particularly steep among women: over the 35 years between 1970 and 2005 women's happiness declined both absolutely and relative to men.[32]) The suicide prevalence in lower-income countries is lower than in higher-income groups, and in modernised societies higher than in traditional societies. In Britain, while suicide rates have always been and remain higher for men (roughly three times those for women), the rate of increase in suicide is steepest in young women between 10 and 24 years of age.[33] This is in line with findings in other Western countries.[34]

Psychologist Jean Twenge studied rates of psychopathology in adolescents, relying on serial contemporaneous assessments using the same objective assessment tool and meeting stringent standards, over the period from 1938 to 2007. By these means she avoided the problems of retrospective accounts, of changing diagnostic fashions and of new patterns in willingness to seek help. She found that there were between five and eight *times* as many students that met a common cut-off for psychopathology in the latest cohort compared with the earliest, and this may be an underestimate because many recent subjects were already stabilised on an antidepressant, a possibility that did not exist for the earliest cohorts.[35]

A major 2018 survey, which charts social isolation using a common measure known as the UCLA Loneliness Scale, shows, according to Arthur Brooks, that

> loneliness is worse in each successive generation. In the 'siloed', or isolated, worlds of cable television, ideological punditry, campus politics and social media, people find a sense of community in the polarized tribes forming on the left and right in America. Essentially, people locate their sense of 'us' through the contempt peddled about 'them' on the other side of the political spectrum. There is profit to be made here. The 'outrage industrial complex' is what I call the industries that accumulate wealth and power by providing this simulacrum of community that people crave – but cannot seem to find in real life.[36]

31 See McGilchrist 2009b (434ff) for references.

32 Stevenson & Wolfers 2009.

33 Report of the Office of National Statistics, 'Suicides in the UK: 2018 registrations', 3 September 2009.

34 Rates of depression differ markedly between cultures, probably by as much as 12-fold, and such differences in rates of depression appear to be linked to the degree of stability and interconnectedness within a culture (Weissman, Bland, Canino *et al* 1996). In *The Master and his Emissary*, I cited by way of illustration the fact that rates of psychological disturbance in Mexican immigrants to the USA start at a low level, but increase in proportion to the time spent in the US. The lifetime prevalence of any mental disorder in one large study was 18% for Mexican immigrants with less than 13 years in the US, 32% for those with more than 13 years, but only for those born in the US did it approximate, at 49%, the national rate for the whole US (Vega, Kolody, Aguilar-Gaxiola 1998). What is particularly striking is that, in a Swedish study, refugees are no more likely to attempt suicide than non-refugee migrants, and both have markedly lower rates of attempted suicide than the native population. Only with progressive acculturation do they begin to approach levels in Swedish natives (Hollander, Pitman, Sjöqvist *et al* 2020; Björkenstam, Helgesson, Amin *et al* 2020: see also Norredam, Olsbjerg, Petersen *et al* 2013; Di Thiene, Alexanderson, Tinghög *et al* 2015. Similar findings obtain in the US (Nasseri & Moulton 2011) and Canada (Kliewer & Ward 1988).

35 Twenge, Gentile, DeWall *et al* 2010.

36 Brooks 2018.

Findings from the Cigna US Loneliness Index 2018 include: when asked how often they feel like no-one knows them well, more than half of the respondents surveyed said they feel that way always or sometimes; approaching half of respondents felt that their relationships were not meaningful; more than half of adults aged 18–22 identify with 10 of the 11 feelings associated with loneliness.[37]

There is clear contemporary research evidence that supports the positive correlation between suicide and individualism.[38] Durkheim had already observed in the late nineteenth century that the loss of the sense of belonging and having a secure role in a society – what he called *anomie* – and an individualistic casting off of well-defined societal values, norms, and goals – which he called *égoïsme* – are prominent concomitants of suicide.[39] Indeed, if you had set out to destroy the happiness and stability of a people, it would have been hard to improve on our current formula: remove yourself as far as possible from the natural world; repudiate the continuity of your culture; believe you are wise enough to do whatever you happen to want and not only get away with it, but have a right to it – and a right to silence those who disagree; minimise the role played by a common body of belief; actively attack and dismantle every social structure as a potential source of oppression; and reject the idea of a transcendent set of values.

And there is a further point to make. Something noticeable has happened to our emotional range, which also illuminates how our values have been corrupted. It strikes me that, as a culture, we are losing the capacity for sorrow. This might sound like a good thing, but is far from being so. Indeed it may even be a sign that we are losing our humanity. Sorrow is a normal part of life, and nothing like anxiety and depression (in which, contrary to popular belief, a capacity for sorrow may be diminished). We have lost that sense of deep connexion and communion, eliciting feelings of longing, tenderness and compassion, and which is more prevalent than any other sense in the musical traditions of the whole world. Music, like great works of tragedy, and like the rites of a religion, acknowledges sorrow and redeems it by taking it up into something with a capacity to heal. In our world, in place of sorrow or sadness, we have anger, resentment and self-righteous indignation. Sorrow and sadness depend on connexion; anger, resentment and self-righteousness on alienation. Sorrow leads to insight; anger to blindness. As the reader will know a capacity for sorrow is closely connected to a capacity for empathy, and both are heavily dependent on the right hemisphere; whereas anger, like denial, is heavily dependent on the left hemisphere.[40] Our public expressions in art, in films, and in the stories and myths

---

[37] www.cigna.com/assets/docs/newsroom/loneliness-survey-2018-full-report.pdf (accessed 15 March 2021).

[38] Webster Rudmin, Ferrada-Noli & Skolbekken 2003; Greenberg, Carey & Popper 1985; Kearl & Harris 1981; Lester 1997; Rangaswami 1996; Sinha 1988; Stack 1993.

[39] Durkheim 1951.

[40] See p 197 *et passim* above. Interestingly, schizophrenic subjects, who show a pattern of right hemisphere deficits, can recognise happiness or anger, but not sorrow, in another's face: Bonfils, Haas & Salyers 2019; Bonfils, Ventura, Subotnik *et al* 2019.

we espouse showcase conflict, self-assertion, violence, aggression, torture and horror – or alternatively a sentimental and unremitting positivity – but little in the spectrum of sorrow, or tenderness, certainly when compared with other times and other cultures. This is part of our overvaluing of the rhetoric of power, which often seems to be the only value that gets discussed, whether it be in relation to political values, societal discourse or even the critique of works of art. With it goes a dereliction of the value of what is vulnerable.

Given this rhetoric, it is hard for us to realise quite how much we close the door on when we are too brittle to allow ourselves to remain open to life as it is, not as we imagine it ought to be. In the face of all that we are destroying, we have lost, it seems, the capacity to mourn. Perhaps acknowledging the loss would be, for many of us, simply too painful.

THE INVERSION OF VALUES

I have suggested that values should be considered ontological primitives (ie, irreducible and foundational); that, of course, we may observe and esteem any one value or not, but that we do not manufacture them. Rather we recognise them, respond to them and help them grow – or not; and that in doing so, as in everything, the hemispheres see them differently. Max Scheler thought there was a hierarchy of values, with those of pleasure and utility – the values of utilitarianism and the left hemisphere – at the lowest level, and rising by stages to that of the holy or sacred, which he considered the highest: a value which I suggest is incomprehensible to the left hemisphere. In between were, first, the *Lebenswerte* or values of 'life', such as courage, magnanimity, nobility, humility and loyalty;[41] and then the *geistige Werte*, the values of mind or spirit, such as beauty, goodness and truth – which I have suggested are better understood by the right hemisphere. In the world we live in, the reductionist narrative is that holiness is a deceit employed by clergy to maintain power and privilege; that beauty is a tool of sexual selection; goodness and truth tools of social cohesion; the *Lebenswerte* merely opportunities for simple-minded individuals to indulge in self-sacrifice for the benefit of the group; and that all that really matters is pleasure and utility – the only values that do not require affective or moral engagement with the world. In other words, in a thoroughly cynical assessment of what it means to be human, we have inverted Scheler's hierarchy and exalted the individual ego. Instead of his *ens amans* we have *homo economicus*. This inversion is just as predicted from a culture marked by left hemisphere dominance. And it has rendered many virtues, including, but not confined to, beauty, goodness and truth, fugitive species.

41 There are almost insuperable difficulties in localising the correlates of *Lebenswerte*. I know of only one, an ingenious experiment involving subjects with a phobia for snakes bringing a live snake in close proximity to their head while in a scanner – a reasonable proxy for courage – which found that the brain correlates were principally in the right hemisphere (Nili, Goldberg, Weizman *et al* 2010).

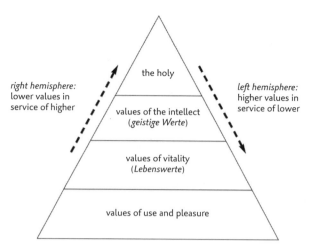

Fig. 62. *Pyramid of values according to Scheler*

The left hemisphere's raison d'être being power and control, it naturally puts values of utility and hedonism, those of the lowest rank in Scheler's pyramid, first. I may be wrong, but it is my distinct impression that there has been a decline in courage, loyalty and humility in our society – indeed in all behaviour that carries its costs upfront, rather than concealing its sting in the tail; speaking the truth takes courage, and it would seem that those in the institutions of government, science and the universities would rather conform than confront untruth. The powerhouses of intellect, the universities, have lost their nerve, and become passive, conformist and feeble – and excessively bureaucratic. Perhaps because of this I can't help noticing that many of the most interesting ideas in science, in politics and in philosophy these days come from outside the institutions – something I say only with the greatest of regret, as someone who always saw himself as a proud beneficiary of their ancient traditions, and who wished to further those traditions. And along with the loss of courage to speak the truth, there has been an undeniable withdrawal from the beautiful and the sacred. All of this combines to reinforce a loss of sense of purpose and direction – or 'the faintest idea of what we are doing', in Walker Percy's phrase. Hence the crisis of meaning that it is, by now, a commonplace that we face.

Considered just from the left hemisphere's narrowly utilitarian point of view, its blind destruction of the three elements that are most important in securing our well-being – closeness to the natural world, social cohesion and embrace of the divine – is self-defeating. That these elements are so vital to well-being is not my opinion, but as near a fact as we can get. I deal with the evidence regarding the extraordinary range of beneficial effects on health, both physical and mental, and on resilience, happiness and fulfilment, associated with religious belief and observance in Appendix 8. I dealt with the similar evidence regarding social cohesion in *The Master and his Emissary*.[42] But there is equally impressive evidence attaching to the effects of our being close to Nature. A multitude of studies testify that spending time in Nature is conducive to improved physical and mental health on a wide range of indicators, to a sense of well-being and happiness, and to improved attention, cognition and social functioning; while deprivation can lead to the opposite, including anxiety, anger, and frustration, problems with attention and cognition, and the psychological and physical concomitants of stress.[43] The irony is that these three aspects of life, which can be fully appreciated only by transcending the narrowly utilitarian view, nonetheless ought, on purely rational grounds, to be embraced even by those whose values go no further than the wholly utilitarian. The *non*-utilitarian value of religious observance, social cohesion and closeness to Nature would have seemed a foregone conclusion in any other culture than our own: and now we know the utilitarian value is also overwhelming. Are we, suddenly, so 'Bright' after all?

The notes for page 1325 appear overleaf ▶

In summary, two things stand out. The first is that, although we are inextricably part of it, derive our whole existence from it, and are ourselves incomprehensible when sundered from it, we have abandoned Nature. Nature – not some abstract, eviscerated, bureaucratic entity known as 'the environment', which exists to be managed and exploited; not some technical set of mechanisms called an ecosystem; but Nature, whose meaning is that-which-is-about-to-be-born, and is feminine – and what's more a goddess; Nature, that, like Kali, is wild, and gives life and destroys – or rather does not destroy, but transforms one being into another;[44] Nature, that is our mother and our healer and our home, as well as our ultimate fate; Nature, that we are reviling and doing our best to devastate – is the great whole to which we belong. All the elements of the left hemisphere insurrection can, individually and together, be seen as an attack on Nature – and, with it, on the body; and hence on life itself.

'Let us allow Nature to play her part', wrote Montaigne, sceptic that he was: 'she understands her business better than we do.'[45] Nowadays we are sceptical about Nature, because too often she refuses to behave according to our new rules. But scepticism of one position merely leads to certainty about another unless scepticism is equally applied to its opposite; total scepticism of both leaves us impotent; a vestige of scepticism about scepticism re-engages us with a possible balanced solution.

The other thing that stands out is that we have abandoned even the pretence of seeking wisdom, instead intent on seeking power. Wisdom and humour are both expressions arising from the shared suffering involved in acquiring one of the greatest flowers of life, a sense of proportion. They are both important elements in turning us away from aggression and towards healing. But, for us now, humour

NOTES TO PAGE 1325:

42 See McGilchrist 2009b (436–8).

43 Just walking in a forest has been shown reliably to reduce blood pressure, lower blood cortisol, strengthen immunity, and promote muscle relaxation, as well as bringing about improvements in subjective well-being, confidence and social relationships: Ives, Abson, von Wehrden *et al* 2018; Tsunetsugu, Park, Ishii *et al* 2007; Hansen, Jones & Tocchini 2017; Ideno, Hayashi, Abe *et al* 2017; Chang, Hammitt, Chen *et al* 2008; Berman & Anton 1988; Polenz & Rubitz 1977; Richardson, Cormack, McRobert *et al* 2016; Capaldi, Dopko & Zelenski 2014; Pretty 2004; Bowler, Buyung-Ali, Knight *et al* 2010; DE Aldous 2007. In one study of subjects with major depressive disorder, the effects of locale on cognitive-behavioural therapy were compared; only 5% of the control (no-treatment) group achieved remission over a four-week period, while 21% achieved remission with CBT delivered in a hospital. However, this compared with 61% of those treated in a forest setting (Kim, Lim, Chung *et al* 2009). Disconnexion from nature may lie behind a number of psychological ills of modern urban life, especially when combined with the exhaustion induced by constant artificial stimulation. Spending time in nature heals stress and causes restoration after mental fatigue: see Berto 2014 for a review. In particular, see Hartig, Evans, Jamner *et al* 2003; Ulrich, Simons, Losito *et al* 1991. It also reduces anger, aggression and fear: Ulrich 1979; Ulrich, Simons, Losito *et al* 1991; Kuo & Sullivan 2001. By contrast environments devoid of natural elements can produce anxiety, anger, and frustration: Coss 1991; Ulrich 1993. Scenes with natural landscapes produce more positive emotional effects than urban scenes with inanimate objects: Lohr & Pearson-Mims 2006. Hospital patients have better outcomes on a wide range of measures if their windows overlook trees rather than a manmade environment: Ulrich 1984; Diette, Lechtzin, Haponik *et al* 2003; Miller, Hickman & Lemasters 1992.

Prisoners' health is better if their cell windows include views of farmland and trees: Moore 1981. Natural settings increase performance on tasks requiring attention and cognitive processing: in an urban setting attention is constantly aggressively captured, requiring immediate directed attention, even if only so as to avoid traffic: Berman, Jonides & Kaplan 2008; Felsten 2009; Tennessen & Cimprich 1995; Berto 2005; Kaplan & Kaplan 1989. This causes fatigue, an impoverished capacity for global attention and increased stress. It also leads ultimately to the inability to focus, with more frequent performance errors, an inability to plan, loss of social civility and increased irritability: Kaplan 1995. A study of girls living in the same housing complex found that being in or near a natural setting increased self-control and self-discipline, improved the ability to concentrate and to delay gratification, and decreased impulsiveness: Taylor, Kuo & Sullivan 2002. Interestingly blue and green space, but blue space in particular, embodies important therapeutic qualities for older adults, which may explain why so many retire by the sea: Finlay, Franke, McKay 2015.

44 See Goethe, p 1297, n 356 above.

45 Montaigne, 'On experience' (trans M Screech): « *Laissons faire un peu à nature: elle entend mieux ses affaires que nous.* »

is dying along with wisdom: the left hemisphere Puritans are seeing to that. Puritanism always was the enemy of *both*; and it has historically often been associated with aggression, destruction and a ferocious need to control – with power, in other words, above all else.

A member of the Swiss Parliament, Lukas Fierz, recalls as a boy meeting his celebrated neighbour, Carl Gustav Jung. In the course of conversation,

> Jung told us about his encounter with a Pueblo chief whose name was 'Mountain Lake'. This chief told him, that the white man was doomed. When asked why, the chief took both hands before his eyes and – Jung imitating the gesture – moved the outstretched index fingers convergingly towards one point before him, saying 'because the white man looks at only one point, excluding all other aspects'.

Many years later, Dr Fierz, who is a physician and a founding member of the Green Party in Switzerland, recalls that a significant adversary of that movement was a successful industrialist and self-made billionaire:

> I asked him what in his view was the reason for his incredible entrepreneurial and political success. He took both hands before his eyes and moved the outstretched index fingers convergingly towards one point before him, saying 'because I am able to concentrate on only one point, excluding all other aspects'. I remember that I had to swallow hard two or three times, so as not to say anything ...[46]

The secret of success – of a kind; and the formula for ruin.

SO WHAT SHOULD WE DO?

I am convinced that it will do little good to tackle these problems piecemeal, while essentially retaining the mindset that caused them to arise in the first place. Of course at some stage the piecemeal method is necessary, even vital – the left hemisphere, in other words, acting properly as the useful servant, not the master. But such efforts are of no avail unless they form part of a greater project that involves radically revising our whole concept of the reality of which we are a part, and of who we are within it. Only then do we stand a chance of being possessed of a different set of values and of changing our behaviour accordingly. Without deep-rootedness and appropriate soil a tree will not stand. So, if we want to 'tackle climate change' we most certainly do have to take whatever steps are necessary to stop destroying the rain forests. But if we go on thinking of the extraordinary richness and beauty of abundant life in them merely in terms of what it can do for us, what it is 'worth' to us in terms of utility, which ultimately translates as economic value, we might as well forget trying to save ourselves, and allow ourselves to sink.

46 Dr Lukas Fierz, private communication. In keeping with my view that many of the phenomena of modernism reflect the culture's broader imbalance in favour of the left hemisphere (how could they avoid doing so?), Le Corbusier stated: « L'homme marche droit parce qu'il a un but ; il sait où il va. Il a décidé d'aller quelque part et il y marche droit. » – 'Man walks in a straight line because he has a goal; he knows where he is going. He has made up his mind to go to some particular place and he goes straight to it' (1922, 33).

As long ago as 1805 Schelling already saw the root of our malaise. There is no higher realisation in all of science, he believed (and he was well versed in the physics and chemistry of his time) and none higher in art, philosophy, or religion, than that of the sacredness of what, for want of language, he refers to as the 'All'; and that in those ages where we are mindful of this unity, a culture enjoys vigour, and vitality, and the fruits of the collaboration of the arts and sciences. But

> whenever the light of that revelation faded, and men came to see things not in relation to the All, but as distinct from one another – not in their unity, but in their disjunction – and in the same spirit came to conceive themselves as in isolation and estrangement from the All; then one sees science, lost in the desert expanses, struggling to take meagre steps in the furtherance of knowledge, grain of sand by grain of sand, so as to put the universe together. And with it one sees the beauty of life vanish; and the spread of a savage war of opinion over the most essential and important things, while everything falls apart in a mass of details.[47]

Like Schelling, I believe that without an overarching understanding of the 'All' of which we are a part, however tentative and incomplete it must necessarily be, we are bound to go on acting in such a way that we lose everything we value – or all that, when in our right minds, we value. The left hemisphere has dismantled the universe and is unable to put it back together again. Without a radically different understanding we just can't carry on. That is why I have written this book.

Behind my attempt to offer a truer account of the reality in which we belong, there lie three fundamental questions.

The first of these is: 'In our explorations of reality, how can we know what to trust – that is to say, to believe to be true of it – at least in the sense of truer than any of the alternatives?' Addressing that was the burden, in different respects, of Parts I and II of the book, and my answer was given at length in those pages.

The second was this: 'What account of reality then emerges?' That was the burden of Part III. There I attempted to provide an account of what Schelling calls the 'All' that is richer and truer than the narrowly left hemisphere account which has been our lodestar for the last three hundred years or more, and which has brought us to this pass. It is in no sense a new account. Readers of the book will be well aware of the debt it owes to the wisdom of ancient myths and to the insights of philosophers over the centuries, many of whom have occupied a dissident place on the fringes of the dominant culture. What I hope I have done, however, is to bring fresh support to their insights from what the hemisphere hypothesis tells us – and, incidentally, from what recent, exciting, developments in the philosophy

47 Schelling 1985, vol 3 (629): » Wo das Licht jener Offenbarung schwand, und die Menschen die Dinge nicht aus dem All, sondern aus einander, nicht in der Einheit, sondern in der Trennung erkennen, und ebenso sich selbst in der Vereinzelung und Absonderung von dem All begreifen wollten: da seht ihr die Wissenschaft in weiten Räumen verödet, mit großer Anstrengung geringe Fortschritte im Wachsthum der Erkenntniß, Sandkorn zu Sandkorn gezählt, um das Universum zu erbauen; ihr seht zugleich die Schönheit des Lebens verschwunden, einen wilden Krieg der Meinungen über die ersten und wichtigsten Dinge verbreitet, alles in Einzelheit zerfallen. «

of both biology and physics are confirming. The hemisphere hypothesis demonstrates very clearly, I believe, that the prevailing account of the 'All' is the product of an unbalanced, untempered and characteristically domineering, go-it-alone left hemisphere view of the world, one that is false, impoverished and dangerous; and that a form of attention to the world that allows into the picture what the right hemisphere discerns produces a truer, richer account of the 'All', without which we shall perish.

The world, I suggest, is far different from the way it has been generally understood – with notable, mainly disregarded, exceptions – over Western history, certainly recent history. It is a world in which relationships are ontologically primary, foundational; and 'things' a secondary, emergent property of relationships. It is one where matter is an aspect of consciousness, not consciousness an emanation from matter. It is one in which there is a natural process of individuation, but one the aim of which, far from disrupting wholeness, is to enrich it. It is one where opposites are not as far as possible removed from one another but tend to coincide. Change and motion are the universal norm, without disrupting stability and duration. It is one where nothing is wholly determined, though there are constraints, and nothing is wholly random, though chance plays an important creative role. Indeed the whole cosmos is creative; it drives towards the realisation of an infinite potential. It is one to which we are profoundly connected – out of which we arise and to which we again revert; Nature being our specific home in that cosmos from which we come and to which in time we return. It is one that absolutely cannot be properly understood or appreciated without imagination and intuition, as well as reason and science: each plays a *vitally* important role. It is neither purposeless nor unintelligent, but simply beyond our full comprehension, though we are given enough insight to kindle wonder and awe, and the impetus to try to understand what little we can. It is an expression of energy and form, form within energy and energy within form, each both 'challenging' and 'liberating' the other to further creative realisation. It's more a dance than an equation. And at the core is something we call the divine, which is itself forever coming into being along with the world that it forms, and by which, in turn, it too is formed.

And now: the third question, which follows from this, since we are part of the All: 'We, then – who *are we*?'

On what grounds I base my answer to this last question, only the reader of this book will understand; but what I say is this.

Ultimately of course we cannot really know fully who we are. But we can know what we are not. We are definitely not the lonely, isolated, predatory egos we have been taught to think of ourselves as being; hurled into an alien universe, mere accidents of cosmic history, whose lives, like the cosmos itself, are pitifully devoid of meaning, purpose and value. We are not the playthings of necessity. Nor in our dealings with Nature are we merely detached observers, manipulative, tinpot gods – in our dealings with one another forever doomed to fight with and destroy one another. Power and pleasure are not our only values, nor by any means our greatest values, nor do they on their own satisfy us, add beauty and goodness to existence, or help save us from ourselves. The boxes into which we force reality are a mental expedient, since no two things and no two people are ever really the same in any quality whatsoever, except the fact of their existence. This is central, because all is, once it is properly seen, unique; and it is the business of living to unfold that unending potential for uniqueness. At the same time nothing and no-one is ever wholly disconnected from anything else: the uniqueness is not in any way opposed to that connectedness, but vitally dependent on it, and in turn *constitutive* of it.

We are temporarily material entities, capable, we do not know how or why, not just of awe before creation, but of playing a part in creation itself; beings that emerge out of the original consciousness, eddies in a seamless flow that embraces everything that is and was and will be; for a while distinct, but never wholly separate from the flow, since we *are* for a while that flow, wherever it finds itself. We are embedded in the cosmos that gives rise to us: the least that can logically be said is that there is something about this cosmos that makes it such that we and all our capabilities, all our achievements and creations, for good or ill, can arise. They simply cannot come from anywhere other than the cosmos in which we belong. What is wonderful about us is not our pitiful lust for power, our self-absorption and our armour-plated invulnerability, but precisely our capacity to be vulnerable, to wonder, and to love: which alone makes what we most value possible.

This is all very well, someone might say, but it is a myth. That is of course true, but it doesn't deliver the hoped-for, knock-out blow. Just as there is no option to think without metaphor, there is no such thing as not having a myth: those who think they don't have a myth have merely bought into the prevailing myth of the time – in our case, the myth of the machine – so thoroughly that they are not even aware of it. Our only option is to espouse a worse or, one hopes, a better myth. As the reader will know, I have argued that myths, like metaphors, are by no means lies that are *invented*, but aspects of truth that are *discovered*, and form an inevitable, indeed crucial, part of the search for that truth. Myths oversee – or underwrite – what we are capable of seeing. The nature of the attention that we bring to bear on the world, and the values which we bring to the encounter, change what we find; and in some absolutely non-trivial sense, change what it is. At the same time, the encounter, as is always the way with encounters, changes who we are. Thus in addressing the question 'who are we?', the attention we pay and the values we hold both contribute to the answer which we receive.

The aim of this book has been to address the issue of values by, first, attempting to provide a truer account of the reality we inhabit and, then, of our identity within it. The answers, such as they are, that have emerged cannot be stated simply – a fact which explains what the reader may feel has been the inordinate length of the book. But the experience of readers of *The Master and his Emissary*, as they have reported and continue to report it to me, encourages me to believe that changing the whole way in which we see our lives, our experience, our world, can be achieved by such means.

The account we give of ourselves helps determine our values, and hence our behaviour: and, since how we behave is central to whether we could ever save ourselves and our world from the current tragic state of affairs, all this matters profoundly. We need the best myth we can have. I offer a myth that I believe, if lived, will be found truer than the reductionist one peddled in the market place. One that promotes much needed healing, not further destruction. That is all. But on that 'all' everything depends.

Remember Thoreau's words: it's not what you look at that matters, it's what you see. We need to bring ourselves to see a new *Gestalt*, one in which what seem like fragments of knowledge form part of a coherent whole. This book has covered a lot of ground; that was inevitable if I was to show that there are deep accords between patterns of phenomena arising in what are conventionally thought of as widely disparate areas of knowledge – neurology, philosophy, physics: a consilience (literally, 'a jumping together'), to use a fashionable word, that is together more revealing than its constituents could ever be apart. In reality they are never wholly separable. But I want to emphasise again: it is not just a matter of putting together ever more facts, but of developing an understanding. Neurology *matters*. It is not just a technical exercise in electronics, whereby we discover more data about the 'circuitry'. It means absolutely nothing, unless it helps us better understand human minds and lives: who *we are*. Philosophy *matters*. It is not just a procedural business of logic-chopping and point-scoring. It means absolutely nothing, unless it helps us to a wiser view of human experience and the nature of human being. Physics *matters*. It is not just a tool for manipulating the world, but for understanding it. All three help answer the questions of what the world is and who we are, on which all else depends.

≈

It will not surprise the reader that I have spent much time in the company of Roger Sperry, though I was never lucky enough to meet him. Like me he started out in the humanities, and of course he went on to be the person who most deeply investigated the significance of the bipartite human brain. Sperry was not merely a neuroscientist but a philosopher of considerable discernment, many of whose insights came from the research into hemisphere difference which won him the Nobel Prize. Reflecting on the ills of Western society nearly half a century ago, he wrote that 'what it comes down to is that modern society discriminates against the right hemisphere.'[48] He later amplified this observation:

> By evolutionary time standards, the fate of life on our planet has suddenly, and quite abruptly, come to rest on an entirely new form of security and control, based on the machinery of the human brain. The older, noncognitive controls of nature that have regulated events in our biosphere for hundreds of millions of years, the forces of nature that lifted life from the amoeboid to the human level and created man, are no longer in command. Modern man has intervened and now superimposes on nature his own cognitive brand of global domination ...
>
> Any attempt to attack directly the overt symptoms of our global condition – pollution, poverty, aggression, overpopulation, and so on – can hardly succeed until the requisite changes are first achieved in the underlying human values involved. Once the subjective value factor has been adjusted, corrections will follow readily in the more concrete features of the system ... Among the vast complex of forces that influence and control the brain and behaviour of man, the factor of human values stands out as a universal determinant of all human decisions and actions ...[49]

This book is my best attempt to address Sperry's concern and to suggest how 'the factor of human values' he sees as all-determining might be shifted from its current, catastrophic, condition.

Let me be clear: replacing the right hemisphere in its proper position, one in which the left hemisphere works with it, but serves it, rather than aims to control it, is not to wish for some dream of softness, lack of rigour and ease. Quite the opposite. This is not an easy path. It makes demands on us, far more than the selfish, complacent vision of the left hemisphere has ever done: which just adds to the reasons we don't want to change. As the reader has seen, the right hemisphere has more of a grasp of moral values, is apt to take responsibility rather than blame others, is more apt to inhibit our first impulse to what is simple and easy, is more intelligent, and complicates our simple mechanistic vision by its insight. And for the umpteenth time, as the reader of this book knows, we do *not* achieve any of this by turning our backs on reason and on science.

---

48 Sperry 1973 (209).

49 Sperry 1974b (7 & 9).

As I have explained, the right hemisphere makes the most valuable contribution to both. If anything reason and science are worryingly diminished in our world: reason, by its narrowness, become unreasonable; science, by its narrowness, unscientific. Both have become tarnished by dogma. We should expect not less, but more and better of both; it is dogma we must avoid at all costs. Dogma is the besetting sin of the age; and if one wanted one it would be hard to find a better expression of the left hemisphere's take on the world than dogma.

Towards the end of his life Einstein was asked if he had any regrets. Reportedly he replied: 'I wish I had read more of the mystics earlier in my life.'[50] I have been forcibly struck by the remarkable similarities in the wisdom enshrined in writings coming out of a breadth of traditions – Hindu, Taoist, Buddhist, Christian, Hebrew, Islamic, those of the ancient cultures of North America, or those of ancient Greece – that I have encountered; and I would be surprised, on that basis, if there were not a great number more, not just texts but whole traditions, of which I remain ignorant, that could be added to such a body of wisdom.

Scheler wrote that 'every finite spirit believes either in a God or in an idol.'[51] Having abandoned God, our idol is – *ourselves*, as 'God': the gaining of human power over every manifestation of Nature, so as to subdue it in the furtherance of a plan that came to us, viewed from a historical perspective, barely minutes ago, and which we have therefore had no opportunity properly to evaluate, and whose outcome is fully unknown, although we can already see the most ominous signs that mean we can't possibly go on living the way we do now. 'Nothing is more useful than power, nothing more frightful', wrote Heschel. 'We have often suffered from degradation by poverty, now we are threatened with degradation through power.'[52]

Despite being deeply suspicious of a great deal of 'God talk' myself, and while fully acknowledging the problematic nature of the very word *God*, I feel our repudiation of God is not a wise move. It is easy to misunderstand what cultures wiser than ours were trying to express by speaking of God; still easier to reject the idea of God entirely. But easy is not enough. It is our duty to do the more difficult thing: to find out the core of wisdom in this ill-understood, though universal, insight, for that there is such an inestimably valuable core seems to me more credible than anything else I know.

I have said what I can. If you wish more, I can do no better than repeat the words of the seventeenth-century priest and physician, Angelus Silesius:

> Friend, that is surely enough. And should you want to read more,
> Then go and become yourself the words, and yourself the Being.[53]

---

50 Fox 2011 (1).

51 Scheler 1954 (261): » *Jeder endliche Geist glaubt entweder an Gott oder an einen Götzen* «. It may be argued that Buddhism acknowledges no God; once again I would quote Shaku (see p 1278, n 297 above): Buddhism 'has certainly a God, the highest reality and truth, through which and in which this universe exists.' This is also the God of the Taoists.

52 Heschel 2005 (3).

53 Angelus Silesius, 'Beschluss', from *Cherubinischer Wandersmann*, 1657: » *Freund, es ist auch genug. Im Fall du mehr willst lesen, | So geh und werde selbst die Schrift und selbst das Wesen.* «

Appendices to Volume 11

↜ *Appendix 4* · What happens to time in depression

This evidence should not be read in isolation from its context in the overall argument of Chapter 22.

As with schizophrenia, the whole range of existential changes encountered in depression can also be conceived in terms of a disturbance of time.¹ And that means equally in terms of the body, since the one affects the other. For example, when I am absorbed in the flow of experience, enjoying what I am doing, my body is in the background, as that *through which* I experience the world – it is the lived body, *der Leib*; and time, too, is in the background, that *in which* I live, think and act intuitively. They do not obtrude, as things, objects of experience – that is to say, not subject to the Gorgon stare of the left hemisphere.

When, by contrast, I am outside the flow of experience – because, for example, I am bored, stressed or in pain – my body is foregrounded, and becomes focussed on as a *thing* in the world, *ein Körper*; and time, too, becomes my focus, some *thing* I observe anxiously from the outside, constantly looking at my watch, only too aware of something I call 'lateness'.² This change in the nature of attention changes the experience of time and the body.

In our ordinary experience of time, past, present and future are entirely seamlessly connected. In our awareness there are, at one and the same time, yesterday's pleasure or pains, and tomorrow's hopes or fears, inextricably bound up with one another, as each note of a tune inevitably brings with it its relation to those parts of the tune that are technically past (no longer detectable by a recording device, but crucially alive to the mind) and those that are to come (which the mind inevitably anticipates). The philosopher Matthew Ratcliffe writes:

> Even if you haven't heard [a melody] before, there is a sense of roughly what will come next, as illustrated by the surprise you feel when a note is out of tune. This expectation also shapes how a present note is experienced. Notes that have just passed are experienced in the present but *as* having passed, and *as* that out of which the present has arisen.³

More importantly, no part of the melody makes any sense or exhibits any power on its own. It works as a whole or not at all. This

---

1 See Ratcliffe 2015 (148).

2 *ibid* (150).

3 *ibid* (151). These ideas derive from Bergson.

process of fulfilment and anticipation is dynamic: as certain possibilities are actualised, others appear. It is inseparable from the experienced 'flow' of time.

Though the experience of time in schizophrenia and depression may seem superficially similar – for example, in both it may be experienced as slowed down – I believe the two are distinct. As one might expect for a right hemisphere-dominant condition, depressives are able to judge time duration accurately when specifically tasked to do so – in fact, they are actually better than normals.[4] This may be part of a generally greater realism in depression. However, personal time is another matter. The sense of time passing is slowed down for depressive subjects by contrast with the time everyone else is living – of which, unlike in schizophrenia, the subject is only too aware.[5]

In schizophrenia both body and time have become insubstantial, airy, unreal, virtually non-existent. The overall feel of the schizophrenic world has been aptly described as 'peculiarly insubstantial, evanescent, and hovering'.[6] In depression, by contrast, time and the body become oppressively real: more substantial, heavier and darker than usual.

To the depressive patient the body is hardly insubstantial: it is heavy, solid, sluggish, but very real. Its functioning appears to slow down (and literally does so, including the autonomic functioning of, for example, the bowels), but the body remains an ever-present focus of concern, not something from which the patient has become alien. Quite the reverse: it is the very sense of bodily effort, the lack of energy, motivation and drive, that make action so difficult; it is the feeling of 'going nowhere' which ensures that the body is all too present. As Sass observes, 'whereas the person with schizophrenia feels *detached* from his body, the melancholic feels somehow over-identified with it.'[7] With remarkable insight Shakespeare makes the melancholic Hamlet complain:

> O that this too too solid flesh would melt,
> Thaw, and resolve itself into a dew!

For the depressive subject, the lived body is inescapable, yet lacks its usual grace to move in accordance with the flow of the world; whereas, for the schizophrenic subject, the lived body has simply evanesced, and one is left with a somewhat bizarre scientific exhibit.

Images of 'being weighed down' and 'being in a state of darkness' are the two most familiar metaphors produced in depression.[8] Many melancholics describe an actual sense of pressure on the body, particularly around the area of the heart ('heavy heart'); and similarly in depression colour is literally experienced as darker – in particular,

---

4 Kornbrot, Msetfi & Grimwood 2013.

5 Mundt, Richter, van Hees *et al* 1998; Lewis 1932; Wyrick & Wyrick 1977; and Kitamura & Kumar 1982.

6 Schmidt 1987 (115).

7 Sass & Pienkos 2013b (122).

8 Jackson 1986 (383–404).

as greyer and bluer, the recovery from depression being accompanied by a sense of the return of the vividness of colour and light.⁹ (Goethe observed that blue is 'on the negative side', noted its 'affinity with black', and remarked that 'the appearance of objects seen through a blue glass is gloomy and melancholy'.)¹⁰

So it is with time. For the schizophrenic subject, time has ceased to have meaning, or even existence; while for the depressive subject, time is ponderous, inescapable and oppressive. It is the *contrast between* the present weight and lethargy of both body and time, and the quickness and lightness of body and time when one is in the flow, that is so overwhelming to the depressed subject. *And time must be present for such a comparison even to take place.* Remember how Zingerle's right hemisphere stroke patients had no sense of how their bodies or they themselves could ever have been otherwise?¹¹ Such a comparison is not generally made by the schizophrenic subject, for whom neither body nor time, as normally understood, exist at all. The melancholic is quite able to 'feel that I cannot feel'; by contrast the schizophrenic can scarcely 'feel that I cannot feel'.

There is a difference between the almost metaphysical statement of the schizophrenic that there is no longer any future, present or past, and the depressive's saying that 'for me there is no future'; 'my past is all gone'; 'I cannot live in the present'. Behind those statements are implied 'I know that the future exists and other people have a future', 'the past is an oppressive loss', and 'I no longer feel myself to be present as other people do'. In schizophrenia there *is no* time; in depression, there is an all too oppressive time, that has slowed or stopped. Schizophrenics live outside time, and are supremely indifferent to it; whereas depressives are involved with temporality, obsessed with the loss of the past and afraid of what the future will bring. In depression it is not that the future doesn't exist, but that it is hard to imagine one's own future, because one is still frozen in the painful experience of the past, without any possibility of redemptive change. This may be the opposite (as according to the hemisphere theory it should be) in mania, where all is geared towards an unrealistically brilliant future. 'In melancholia, the past becomes more determining, and the future less open; in mania it is the reverse'.¹² In depression, time is freighted with minatory meaning, and it is this that stops us moving or becoming: we no longer move, but the world *does* continue to move.

Memory of the past is a *living* process, not a *thing*, and neither fictional nor insubstantial; stored in the body and substantiating who we are from moment to moment. It is not an inert record, like a computer data bank, but part of the flow that constitutes what Merleau-Ponty calls the body-subject. A person with no past and no

9 This is primarily from my own experience and that of my patients, but there is research that confirms it: Bubl, Kern, Ebert et al 2010; Bubl, Tebartz Van Elst, Gondan et al 2009. See also McGilchrist 2017 (2–5). As we know, depression is associated with the right hemisphere, and with the colour blue. There is a possibility that the left hemisphere prefers red, a colour favoured by people who are manic or hypomanic: Pettigrew 2001 (94); and McGilchrist 2009b (63 & 474).

10 Goethe 2006, §§779–84 (171).

11 See pp 75–6 above.

12 Sass & Pienkos 2013c (140).

narrative would be a contradiction in terms: which is why schizophrenic subjects sometimes deny their existence.[13]

Humans being complex, these distinctions between schizophrenia and depression are not absolute and invariable: in particular there are mixed, so-called schizo-affective, conditions, and some chronic schizophrenic subjects would also qualify as depressed. But the distinction is, I think, nonetheless immediately recognisable to anyone working with psychotic patients.[14]

Cutting writes perceptively:

> Schizophrenics talk a lot about death, and are prone to delusions which fit the conventional category of nihilistic delusions. But what they mean by 'being dead' is completely different from what depressives mean by this, as is what they mean when they talk about 'nothingness'. The schizophrenic generally intends 'dead' or 'nothingness' as *never-alive thingness*; the depressive generally intends these as the *dying of living beings*.[15]

This distinction between two superficially similar, but radically different, states goes to the core of the two conditions. In one, the left hemisphere-dominant condition, there never was life, only mechanism; in the other there was all too clearly life, and it is this loss that is lamented. Those who see the life that is being lost, the world that is dying, grieve; others cannot see how we could ever have seen the world and ourselves as anything other than timeless mechanisms.

One distinguishing feature of depression is the excessive willingness to accept responsibility, as in delusional guilt; and a distinguishing feature of schizophrenia is the inability to accept responsibility, as in delusions of control of thoughts, feelings and actions by *others*. It is as though the depressed subject is excessively attuned to community, whereas the schizophrenic is insensitive to community altogether.[16] Hence Bleuler's designation of 'autism' (self-involvement) as a cardinal feature in schizophrenia. This would seem to be a classic left/right difference, since the sense of social connectedness is highly dependent on the right hemisphere, especially a network involving the frontal lobe and the right temporoparietal junction.[17]

Connexions to others are burdensome to the depressive subject in numerous ways. The inability to feel for loved ones what one normally feels causes distress and guilt. At the same time the suffering of others is felt more acutely. There is not a lack, but an excess, of empathy. It becomes impossible to take part in the usual social nexus that gives meaning to life, but this is not a matter of indifference: it is a constant source of pain and a sense of falling away from the lived world. Awareness of the effect of one's own mental state

---

13 See Chapter 9.

14 For further discussion of the asymmetrical, non-dipolar, relationship between schizophrenia and depression and how they may relate, see my discussion of the matter in relation to the philosophy of Max Scheler: McGilchrist 2009a. Here I part company from Cutting, who sees a more symmetrical, dipolar, relationship between the conditions. Since the right frontal pole of the brain is in a dynamic equilibrium with its own (right) posterior cortex, on the one hand, and with the contralateral (left) frontal pole, on the other, depression is usually associated with hypofunction of either of those two areas, and its resolution with recovery of function in them. In hemisphere terms, therefore, the situation with regard to depression is not straightforward. For a more thorough exploration of this complex area in terms of laterality, see McGilchrist 2009b (62–4).

15 Cutting 2002 (158; emphasis added).

16 Tatossian 1997 (56 & 62). And see O'Connor, Berry, Lewis *et al* 2007.

17 Plate 6(b).

on others adds to the burden. One's sense of being poisoned seeps into the surroundings and contaminates them, too, with poison. And in some cases subjects feel that they have committed terrible crimes against others, of which they could not possibly be capable. Cutting and colleagues note:[18]

> The depressive, furthermore, is someone who is more attuned to communal values than the sane, unlike the schizophrenic, who has lost this facility.[19] This takes the form of a hypersensitivity to the morals of society.[20]

In general one might say that in schizophrenia, values tend to be absent, so their fulfilment or otherwise is not an issue; whereas in depression they are *present*, but accompanied by an acute awareness that now they are, for the depressed individual, painfully incapable of fulfilment.[21]

[18] Cutting, Moreira, Naudin *et al* 2017.

[19] Tatossian *op cit* (56).

[20] *ibid* (62).

[21] For further discussion, see McGilchrist 2009a.

← *Appendix 5* · Hemisphere differences and morality: review of neuropsychological evidence

This evidence should not be read in isolation from its context in the overall argument of Chapter 26.

There are a number of aspects to this. One is the superior involvement of the right hemisphere in making moral judgments. Another is its greater involvement in inhibition, an essential aspect of moral behaviour. And most intriguing is the right hemisphere's involvement in promoting prosocial – and the left hemisphere's in promoting antisocial – behaviour.

First, moral judgments. 'The right hemisphere plays an important role in our ability to tell right from wrong', writes David Hecht of University College London, in an excellent review of the literature in this area on which, amongst other sources, I draw here.[1] Assessing the morality of actions, moral reasoning, and promoting prosocial norms all depend on the proper functioning of the right hemisphere. Investigation of a split-brain patient shows that the right hemisphere is 'not only necessary, but also sufficient, for intent-based moral judgement.'[2] Hecht continues:

> Collectively, the studies with healthy and brain damaged participants suggest that the right hemisphere is involved, to a relatively greater extent than the left hemisphere, in mediating empathy and compassion. The left hemisphere, on the contrary, is more involved in mediating anti-social emotions and mental states (eg, gloating and justifying a crime).

Activity in the right posterior superior temporal cortex is correlated with altruistic tendencies.[3] The right dorsolateral prefrontal cortex 'not only participates [in] a rational cognitive control process, but also integrates emotions generated by contextual information', a process which is 'decisive' in reaching moral judgments.[4] When unpleasant moral and non-moral stimuli were compared, 'the most striking findings' were increases in activations in the right hemisphere (medial orbitofrontal cortex, medial frontal gyrus and surrounding the posterior superior temporal sulcus).[5]

Our sense of fairness, too, it seems, is underwritten by the right hemisphere, particularly again by the right dorsolateral prefrontal cortex, and with suppression of this area, we act more selfishly.[6] In males (at least), empathic responses are shaped by valuation of

1 Hecht 2014.

2 Steckler, Hamlin, Miller *et al* 2017.

3 Tankersley, Stowe & Huettel 2007; Montague & Chiu 2007; Morishima, Schunk, Bruhin *et al* 2012. However, things are more complex than they might seem, since the right prefrontal cortex is also differentially activated when people make independent-minded, rather than interdependent-minded, assertions (Wang, Peng, Chechlacz *et al* 2017). This may be because of the right hemisphere's contribution to the sense of the self as moral agent, which becomes more obvious when thinking or acting independently.

4 Tassy, Oullier, Duclos *et al* 2012; Greene, Nystrom, Engell *et al* 2004.

5 Moll, de Oliveira-Souza, Eslinger *et al* 2002.

6 Knoch, Pascual-Leone, Meyer *et al* 2006.

other people's social behaviour, such that they empathise with fair opponents while favouring the punishment of unfair opponents. In particular, they are willing to make sacrifices in order to punish unfair behaviour, so-called altruistic punishment, which benefits the group at the cost of the individual.[7] This behaviour is right hemisphere-mediated, as are judgments of fairness generally, including the unfairness of not punishing unfairness.[8]

The right hemisphere tends to make moral judgments by reference to the intention of the doer (as in deontology or virtue ethics), the left hemisphere by reference to the consequences of the deed (utilitarianism): and 'normal judgments of morality require full interhemispheric integration of information critically supported by the right temporal parietal junction and right frontal processes'.[9] Issues of morality depend on being able to understand what is going on in another person's mind.[10] This is the faculty known as 'theory of mind', which is impaired in individuals on the autistic spectrum. Such individuals find it difficult to pick up indirect expressions, hints, tone of voice, irony, facial expressions, body language and all the other ways we tend to infer what is going on in another person's mind – which may be very different from what they say. Right hemisphere damage impairs each of these indirect expressions and their interpretation; unsurprisingly, then, damage to the right frontal lobe makes us less capable of seeing another's point of view, and impairs insight into our own mental processes.[11] More generally, right hemisphere damage tends 'to predispose the individual to misinterpret complex information and integrate irrelevant stimuli into false beliefs',[12] suggesting that moral judgment is likely to depend to a large extent on the right hemisphere. There are a number of lines of evidence that suggest that this is indeed the case.

Then, inhibition. The right hemisphere – far from its popular caricature as the 'let-it-all-hang-out' hemisphere – is responsible for inhibition, the necessary counterpart to emotional, intellectual and spiritual health.[13] 'One of the striking aspects of the studies reviewed', writes Marie Banich, a cognitive neuroscientist at the University of Colorado,

> is the clear lateralization of function, with right prefrontal regions differentially engaged as compared to left prefrontal regions across most aspects of inhibitory control. As of yet, the underlying reason for this rather dramatic degree of lateralization remains unclear ...[14]

It seems to me that the key role of the right hemisphere in making sure that a response is appropriate in context implies just such a capacity to inhibit. Interestingly, in the light of the greater contribution made to intelligence by the right hemisphere, inhibitory control is more efficient in those of higher intelligence.[15] Similarly there is an association between adopting 'a global mindset' – another right hemisphere connection – and greater self-control in the presence of temptation.[16] Since we know that such a mindset is associated with taking longer-term views,[17] this may operate by bringing to mind the consequences of one's current decisions and actions *in the long run*; or simply of *broader* consequences of one's actions at any one moment. The right frontal lobe's capacity to inhibit our natural impulse to selfishness means that it is also the area on which we most rely for self-control and the power to resist temptation.[18]

And then there is the issue of prosocial behaviour – or otherwise. A sizeable body of research indicates that the right hemisphere is more prosocial and the left hemisphere more antisocial in its style.[19] Feeling others' pain predicts prosocial behaviour,[20] and feeling others' pain is largely right hemisphere-dependent.[21] According to the research of psychologists Julius Kuhl and Miguel Kazén, power-seeking, instrumental, means-to-ends motivations are associated with the left hemisphere; prosocial, relational, co-operative and non-instrumental motivations with the right hemisphere.[22] 'At a conscious level', writes neuropsychologist Allan Schore, 'the left side of the brain concerns itself primarily with power motives, while the right side of the brain is steeped in affiliation drives.'[23]

An assessment from medical records, by clinicians blind to brain scan data, of the extent of criminal, aggressive, or sexually deviant behaviour, alienation from family, friends or employers, financial recklessness, and abnormal emotional and social responses exhibited by patients with frontotemporal dementia found that undesirable behaviour was an early presenting symptom in 11 out of 12 right-sided cases, but in only 2 out of 19 left-sided cases.[24] A study of temporal lobe atrophy reached a similar conclusion, namely that aggressive and antisocial behaviour followed atrophy of the right, but not of the left, temporal lobe.[25] Violent sexual sadists have been repeatedly shown to exhibit right temporal deficits.[26]

This is in line with the fact that motivation research 'consistently shows that individuals with a propensity to engage attentional functions of the right hemisphere ... show more socially adaptive behaviour than individuals with a propensity to engage attentional functions of the left hemisphere'.[27]

There are positive anatomical and physiological relationships between the right hemisphere and prosocial personality traits. Affiliativeness (the desire for benign social connectedness),[28] agreeableness,[29] and a tendency to trust other people[30] all implicate the right hemisphere – the last of these, essential for the formation of and sustaining of rewarding social relationships, is associated with activity in the right ventrolateral prefrontal cortex (which lies adjacent to an area, the right ventromedial prefrontal cortex, that is

16 Fujita & Carnevale 2012; Chiou, Wu & Chang 2013.

17 Malkoc, Zauberman & Bettman 2010.

18 Knoch & Fehr 2007; Alonso-Alonso & Pascual-Leone 2007.

19 Kuhl & Kazén 2008; Quirin, Meyer, Heise *et al* 2013; Quirin, Gruber, Kuhl *et al* 2013. For review, see Hecht 2014.

20 Hein, Lamm, Brodbeck *et al* 2011.

21 Jackson, Brunet, Meltzoff *et al* 2006.

22 Kuhl *et al, op cit*.

23 Schore & Marks-Tarlow 2019 (155).

24 Mychack, Kramer, Boone *et al* 2001. See also Mendez & Shapira 2009.

25 Chan, Anderson, Pijnenburg *et al* 2009.

26 Langevin, Ben-Aron, Coulthard *et al* 1985; Litman 2004; Hucker, Langevin, Dickey *et al* 1988.

27 Schultheiss 2018.

28 Whittle, Yücel, Fornito *et al* 2008.

29 Nestor, Nakamura, Niznikiewicz *et al* 2013; Kapogiannis, Sutin, Davatzikos *et al* 2013.

30 Yanagisawa, Masui, Furutani *et al* 2011.

well-known to be dysfunctional in psychopaths). And a disposition towards gratitude – being appreciative of and thankful for the kindness of others – correlates with the volume of the right inferior temporal cortex.[31] By contrast, the volume of the left lateral orbital gyrus correlates with Machiavellianism – the tendency to manipulate other people in order to control and exploit them for one's own benefit.[32] Patients with damage mostly in their left hemisphere reveal greater levels of warmth, sociability and agreeableness, whereas right hemisphere-damaged patients tend to try to control and dominate: trust, straightforwardness, altruism, compliance, modesty, and tender-mindedness correlate positively with right orbitofrontal, and negatively with left orbitofrontal, volume.[33] Moreover, studies of lesion patients show that right hemisphere damage can lead to sexual aggression, physical assaults and acquired psychopathy.[34] When normal subjects were engaged in mental simulations of immoral acts, there was 'a remarkable shift' in their brain activity towards the left hemisphere.[35] As Hecht comments, 'these studies suggest that moral and immoral thinking are associated with activity in the right hemisphere and left hemisphere, respectively.'[36]

Guilt is often a motivator for prosocial behaviour: only psychopaths have no capacity for guilt. The level of guilt felt by subjects correlates with activation of the right frontal, specifically right orbitofrontal, cortex.[37] Suppressing right prefrontal cortex facilitates lying and reduces the levels of guilt subjects feel about their deceit.[38] Incidentally, psychopaths exhibit thinning in the right frontal and temporal cortex, specifically in grey matter, when compared to normal controls.[39] I mentioned earlier that there is, as Werner Scheid astutely observed, a 'pointer of guilt' that is turned inwards on the self in depression (a right frontal predominant state), whereas in paranoia (a left hemisphere predominant state) the finger of guilt is turned firmly outward, pointing at others: our condition is always someone else's responsibility, never our own.[40] This also appears to be a defining characteristic of modern society.

I have already mentioned that anger, aggression and hostility are associated with a relatively greater activity in the frontal left hemisphere compared with the right.[41] This has been confirmed repeatedly by scanning and EEG studies, including an EEG study of extremely violent offenders with long-term prison sentences, compared with normal subjects;[42] by the finding that enhancing left hemisphere activation in normal subjects gives rise to higher levels of anger and aggression;[43] and by neuroanatomical studies which found that aggression is associated with a larger left orbitofrontal cortex,[44] and a reduced right anterior cingulate cortex.[45] Psychopathy is associated with a hemisphere imbalance steeply skewed towards the left hemisphere.[46]

31 Zahn, Garrido, Moll et al 2014.
32 Nestor, Nakamura, Niznikiewicz et al 2013.
33 Rankin, Rosen, Kramer et al 2004; Sollberger, Stanley, Wilson et al 2009.
34 Mendez 2009; Mendez, Chen, Shapira et al 2005; Tranel, Bechara & Denburg 2002.
35 Cope, Schaich Borg, Harenski et al 2010.
36 Hecht 2014.
37 Wagner, N'Diaye, Ethofer et al 2011; Amodio, Devine & Harmon-Jones 2007; Matsuda, Nittono & Allen 2013.
38 Karim, Schneider, Lotze et al 2010.
39 Müller, Gänssbauer, Sommer et al 2008; Müller, Sommer, Döhnel et al 2008; Yang, Raine, Colletti et al 2009.
40 Scheid 1934. Kretschmer made similar observations.
41 Everhart, Demaree & Harrison 2008; Harmon-Jones 2003; Harmon-Jones & Allen 2003; Harmon-Jones & Sigelman 2001; Harmon-Jones 2004; Hiraishi, Haida, Matsumoto et al 2012; Kubo, Okanoya & Kawai 2012; Peterson, Gravens & Harmon-Jones 2011; Verona, Sadeh & Curtin 2009.
42 Harmon-Jones 2007; Keune, van der Heiden, Várkuti et al 2012.
43 Hortensius, Schutter & Harmon-Jones 2012; Peterson, Shackman & Harmon-Jones 2008.
44 Boes, Tranel, Anderson et al 2008.
45 Gansler, McLaughlin, Iguchi et al 2009.
46 Hecht 2011.

It may be neither possible nor desirable to avoid conflict altogether, but whether that conflict turns out to be creative or destructive seems to depend on hemisphericity even more than motivation. Inhibition is largely associated with the right hemisphere, as described, and experimental measures of inhibition have been independently verified as indicating attention to the left visual field (right hemisphere).[47] Even where subjects are motivated by power rather than affiliation, those with right hemisphericity on visual field testing show more socially adaptive behaviour, predicting 'management success as reflected in high organizational clarity and team morale', rather than a 'self-aggrandizing leadership style'.[48] And even where subjects are motivated by affiliation rather than power, those with left hemisphericity on testing evince 'high levels of physical and psychological partner abuse'.[49] As psychologist and neuroscientist Oliver Schultheiss writes,

> this is consistent with McGilchrist's hypothesis, based on a large neuropsychological literature, that individuals who get stuck in a left hemisphere information processing mode (as opposed to those with a right hemisphere-favouring mode and/or fluid interhemispheric information exchange) have a peculiar maladaptive mindset ... characterized, for instance, by reality distortion through denial of expectation-violating information, an inability to deal with ambiguity, a tendency to view and treat others as mere tools for the advancement of one's interests and goals, and a profound lack of empathy.[50]

47 Schultheiss, Riebel & Jones 2009.

48 McClelland & Burnham 2003.

49 Mason & Blankenship 1987; Schultheiss, Riebel & Jones 2009.

50 Schultheiss 2018.

↩ *Appendix 6* · Hemisphere differences and beauty: review of neuropsychological evidence

This evidence should not be read in isolation from its context in the overall argument of Chapter 26.

Dahlia Zaidel states: 'No current research would suggest that the right hemisphere specializes in aesthetic judgment.'[1] Let's have a look.

Seeing the wood for the trees will be difficult, because both hemispheres are inevitably involved at some point, in some way, in aesthetic appreciation. For this reason, looking at brain scans is like trying to interpret a painting of Monet's by getting closer to the coloured dots: too much conflicting detail that can be understood only by standing back. In teasing out which is the more important hemisphere, it is the effects of cerebral lesions on the appreciation of beauty that should be particularly revealing: if lesions in one hemisphere have a consistently much greater impact than lesions in the other, that would be highly informative about hemisphere difference. Since appreciation of beauty is one of the least manifest impairments after a stroke, doesn't fit neatly into any of the standard neurological categories, and is seldom if ever enquired about by clinicians, it is not often reported, and still less often recorded in print. However, as it happens, David Fischer and colleagues at Harvard Medical School assembled four cases of, specifically, the loss of appreciation of visual beauty – three from a review of the literature, and one of their own.[2]

The first was the case of a man who reported that 'he could no longer become emotionally or sexually aroused by visual stimuli, and that his visual world had become drab and uninteresting':

> Once an assistant city planner, he was no longer able to appreciate subtle aesthetic differences between buildings. He had ceased hiking because he found natural scenery dull, 'all the same'. He complained bitterly about his loss of emotional reaction to viewing pretty girls or erotic visual stimuli. He had cancelled his subscription to *Playboy*, and had withdrawn from heterosexual contact because he found it difficult to become aroused by visual cues.

It was also noted that he approached 'complex visuoperceptual tasks in a piecemeal, feature-analytic manner that greatly compromised speed and accuracy'.[3]

The second concerned a woman who 'especially appreciated aesthetics; she was fond of painting aquarelles and enjoyed taking care

[1] Zaidel 2013.

[2] Fischer, Perez, Prasad *et al* 2016.

[3] Bauer 1982. Interestingly, he also 'described three unsatisfying sexual encounters since his accident in which he had been unable to achieve erection until he began talking himself into arousal by verbal-auditory means'.

of the many flowers that grew in her garden'. Following her stroke she complained of

> a lack of emotive reaction elicited by visual stimuli that formerly aroused a powerful feeling of well-being or contentment: 'I loved flowers so much before ... Their charm doesn't enter my mind anymore. Looking at the landscape through the window, I see the hills, the trees, the colours, but all those things cannot convey their beauty to me ... Everything looks ordinary, indefinite. I feel indifferent about it. What I lack is feeling.'[4]

The third was a middle-aged man who complained that

> Flowers to me have lost their essence, I fail to see them as part of nature. They have become almost synthetic, artificial, I seem to lack a kind of knowledge, no, it's not really a knowledge, rather a certain clarity to see nature itself. I fail to see the flower in all its authenticity ... Just as with flowers, there is also an emptiness to landscapes. I cannot appreciate them, I cannot grasp the beauty of nature. I lack a kind of lucidity, a lucidity in my vision that would normally allow me to appreciate it; its colours, the temperature of its colours so to speak. I cannot think of a right word to explain it. I just cannot enjoy that sense of beauty that nature brings.[5]

And the last is their own case, a man who following a stroke noted a 'sense of visual estrangement', his feelings being cut off from what he saw, and whose surroundings did not 'feel real', which he attributed to a sense of emotional disconnectedness from his visual perceptions.

Note 'ordinary', 'all the same', failing to arouse 'emotive reaction', 'artificial', lacking 'authenticity'. All of this – together with 'a piecemeal, feature-analytic manner' – suggests right hemisphere disruption. So what did the authors find? Their principal finding was that 'all four lesions specifically overlapped with the expected trajectory of the right inferior longitudinal fasciculus'. This an important white matter tract connecting the temporal and occipital cortex, and involved in the internal coherence of right hemisphere function, so a very significant finding. Meanwhile Massimo Marianetti had found decreased perfusion of the right, but not left, medial temporal lobe caused by drug toxicity in a subject with 'visual hypoemotionality' similar to that described in these cases.[6]

It is notable that the condition is 'frequently associated with prosopagnosia',[7] and other right-lateralising signs such as *déjà vu*, as indeed it was in some of these cases.

By contrast, a man with a glioma of the *left* temporal lobe, and a consequent right-sided visual field defect – thus indisputably involving limitations of the left hemisphere's perceptual system – saw

4 Habib 1986.

5 Sierra, Lopera, Lambert *et al* 2002; and Lopera & Ardila 1992.

6 Marianetti, Mina, Marchione *et al* 2011.

7 ibid.

flowers as 'extraordinarily beautiful', odours 'intensified', his vision 'remarkably keen'.[8]

It is notable that Minkowski describes a schizophrenic subject who is reminiscent of Sacks's 'Dr P' (see discussion, pp 871–4 above). He had lost *Gestalt* perception. Of him, Minkowski reports that his mind

> decomposed every object that it met. The clock ... was not just a clock but an assemblage of instruments of torture – cogs, key, hands, pendulum, etc. Every object that he saw was like the clock ... The essential values of an object or another being – such as the aesthetic value – could not be appreciated by him; he was unable to adopt the appropriate attitude. 'You see these roses?' he asked me. 'My wife would say that they are beautiful but, as far as I can see, they are just a bunch of leaves and petals, stems and thorns.[9]

Let us turn to music. Loss of the ability to appreciate music is called amusia. I have reviewed the lateralisation evidence in *The Master and his Emissary*. Most cases of acquired amusia in which language is spared are due to right hemisphere lesions, and cases are more severe where there is right hemisphere involvement.[10] In congenital amusia, structural MRI studies point to an extended network involving temporal and frontal lobes principally of the right hemisphere;[11] and a functional MRI study of individuals with congenital amusia revealed reduced activity in the right inferior frontal gyrus to small pitch changes.[12]

However, one needs to distinguish music perception and recognition from the aesthetic sense, the ability to see and feel its beauty. Some patients, with classical amusia, have impaired recognition of music despite a preserved affective response to it.[13] Others, by contrast, who exhibit 'musical anhedonia', have technical understanding and can recognise, perceive and play music normally – they just cannot appreciate it. There are seven such cases reported in the literature.

A conductor who sustained a right putaminal haemorrhage 'found himself unable to have any emotional experience while listening to music':

> He could not elicit interest from any type of music, even his favourite genre ... Before the illness he felt great joy and 'chills' when the harmony sounded complete ... He felt that the sound was dull and lacked freshness, though he could acoustically recognise that the harmony was correct in its physical sound properties ...[14]

A 24 year-old amateur guitarist experienced a right temporoparietal haemorrhage from an arteriovenous malformation and found that 'aesthetic pleasure for the musical world had completely vanished':

8 Cushing 1922; and Horrax 1923 (apparently reporting the same patient).

9 Minkowski 2004 (136).

10 See McGilchrist 2009b (74ff); and Marin & Perry 1999; Stewart, von Kriegstein, Warren *et al* 2006; Särkämö, Tervaniemi, Soinila *et al* 2010; Jafaria, Esmaili, Delbari *et al* 2017.

11 Hyde, Zatorre, Griffiths *et al* 2006; Hyde, Lerch, Zatorre *et al* 2007; Mandell, Schulze & Schlaug 2007.

12 Hyde, Zatorre & Peretz 2010.

13 Peretz, Cagnon & Bouchard 1998.

14 Satoh, Kato, Tabei *et al* 2016.

On hearing pieces played on the piano, he complained, 'my perception is changed ... it's flat, it's no longer 3-dimensional; it's only on two planes ... there's no emotion ...' His difficulties increased as the presented compositions became more complex: '... this is even worse: I can distinguish the different instruments, but I can't perceive the whole ... in jazz pieces, the relationships between the accompaniment and the soloist escape me.'[15]

An elderly retired teacher, who from his youth had been an enthusiastic lover of classical music, found, after a right temporoparietal lobe stroke, that he

> could not elicit interest in any music, even his favourite music or artists. He described the music as dull and lacking freshness. He thought that this symptom might be caused by listening to recorded music by the stereo and would disappear if he listened to music at a live concert. At a concert of his favourite singer, he felt that, as was expected, the quality of sound was much superior to the recorded one. But he could not experience any emotional response.[16]

A 43-year-old amateur musician was found to have an ischaemic lesion in the right superior temporal lobe, and showed both musical anhedonia and amusia. He complained spontaneously of a loss of the feelings he had been used to experience when listening to music previously.[17]

Another man, following right temporal stroke, experienced

> 'loss of pleasure' or 'loss of aesthetic appreciation' in listening to music: the patient claimed it was difficult for him to express verbally this loss of 'aesthetic pleasure', but, in any case, this feeling was distinct from acoustic distortions and just perceived as 'aesthetic'.[18]

A piano teacher, 10 years after a stroke affecting the right frontoparietal and temporal region, complained that 'he no longer appreciates music as he used to in the past, and spontaneously reports that he seems to have lost the capacity to "conceive the whole".'[19]

And finally there is the case of a middle-aged radio announcer, which is interesting because his loss followed a *left*-sided infarct, involving the insula and extending into the frontal lobe and amygdala. However in his case it was only, and quite specifically, an emotional response to Rachmaninov preludes that was lost: otherwise his sense of musical enjoyment was preserved.[20] So not really a loss of the aesthetic sense? Indeed, some might say that the left hemisphere stroke improved his musical taste.

In six out of the seven recorded cases, then, the deficit was right-sided; in all cases depression had been excluded. It is interesting that two of the cases spontaneously mentioned that the problem was in conceiving or perceiving the whole; that two referred to dullness

---

15 Mazzoni, Moretti, Pardossi et al 1993.

16 Satoh, Nakase, Nagata et al 2011.

17 Hirel, Lévêque, Deiana et al 2014: « le patient rapportait spontanément une perte d'intérêt pour la musique, et notamment une perte des émotions qu'il ressentait auparavant lors de l'écoute de la musique. »

18 Mazzucchi, Marchini, Budai et al 1982.

19 Judd, Arslanian, Davidson et al 1979; as quoted in Samson & Zatorre 1994 (296): « il se plaint de ne plus apprécier la musique comme dans le passé et rapporte spontanément qu'il semble avoir perdu la faculté de 'créer une gestalt'. »

20 Griffiths, Warren, Dean et al 2004.

and a loss of freshness; and that one referred to flatness and a loss of three-dimensionality – all recognised right hemisphere deficits.

To summarise, of the 12 known reported cases of specific loss of the feeling of beauty, the key lesion was right-sided in 11: and in the single case of an *increase* in the feeling of beauty it was left-sided.

Now let us turn to experiments conducted in normal subjects. In the majority of these studies an explicit judgment, most typically articulated in language, is involved – both of which factors skew things towards the left hemisphere. The results are more mixed, but nonetheless still show a right hemisphere preponderance.

According to one recent paper, 'a substantial body of research suggests that the right hemisphere may have a greater role than the left hemisphere in a number of perceptual processes that may be central to perceiving attractiveness', including attention to the whole, symmetry judgments and perceptual acuity.[21] A study of appreciation of beauty in art in experts found that the right hemisphere was preferentially involved.[22] In another study, activity in the right, but not left, lateral occipital region was directly and positively correlated with aesthetic evaluation of artistic images.[23] Disrupting the right lateral occipital cortex using transcranial magnetic stimulation 'decreased aesthetic appreciation of representational artworks'.[24] Complexity in aesthetic judgments activated right lateral fronto-orbital and right inferior frontal cortex.[25] The three most significant areas of activation in aesthetic, as compared with pure symmetry, judgments were all in the right hemisphere (anterior cingulate cortex, frontomedian cortex, and inferior frontal gyrus).[26] In three other studies, activations in aesthetic judgments were either bilateral or right-sided.[27] However in one study of experts, activations in aesthetic judgments were either bilateral or left-sided.[28] In a series of four experiments, the left side of a stimulus (the part appreciated by the right hemisphere) was repeatedly and robustly shown to be more important than the right side in aesthetic judgments.[29] Filmed dance movements that were rated more beautiful produced increased activity in the right premotor cortex.[30] Another study of response to images of varying degrees of beauty showed that the default mode network was activated: the researchers found 'significantly greater activation in the right ventral striatum for the most highly recommended images'.[31] (This is of special interest given that inability to experience aesthetic pleasure in schizophrenia is associated with reduced activation in the very same area.)[32] In this study, participants reacted to pictures during scanning; afterwards they were asked to score the intensity of their experience numerically according to a list of general verbal concepts representing categories of emotional reaction. Higher scores were associated with greater activity in the left hemisphere, as well as in the right superior temporal gyrus.

21 Rodway, Schepman, Crossley et al 2019.

22 Fudali-Czyż, Francuz & Augustynowicz 2018.

23 Lacey, Hagtvedt, Patrick et al 2011.

24 Cattaneo, Lega, Ferrari et al 2015.

25 Jacobsen, Schubotz, Höfel et al 2006.

26 ibid.

27 Lacey, Hagtvedt, Patrick et al 2011; Kirk 2008; Kirk, Skov, Hulme et al 2009.

28 Kirk, Skov, Christensen et al 2009.

29 Rodway, Schepman, Crossley et al 2019.

30 Calvo-Merino, Jola, Glaser et al 2008.

31 Vessel, Starr & Rubin 2012.

32 Dowd & Barch 2010.

A study of face perception found that activating the right, but not the left, dorsolateral prefrontal cortex increased attractiveness of the faces, irrespective of the sex of the faces or viewers. Identical stimulation over the same site did not affect other facial characteristics such as age: 'overall, our data suggest that the right dorsolateral prefrontal cortex plays a causal role in explicit judgment of facial attractiveness.'[33] In another study of facial beauty 'activity in the right orbitofrontal cortex increased linearly as a function of attractiveness'.[34]

Directing attention in such a way that 'the majority of the picture [is] located within the left visual field and analyzed by the right hemisphere' increases aesthetic preference.[35] And this holds in advertisements and even abstract paintings.[36]

I mentioned sex differences. In one study, judgments of beauty, whether of art or natural scenes, were confined to the right parietal lobe in men, but in women responses were bilateral. Nonetheless, the response was also greater in the right hemisphere for females.[37] (Injuries to the right posterior parietal lobe can affect the perception of selective aspects of art in either sex.)[38]

The most comprehensive meta-analysis of neuro-aesthetic processing in explicit judgment tasks to date considered 93 neuroimaging studies, involving only positive judgments, across four sensory modalities: sight, sound, taste and smell. 'The results demonstrate that the most concordant area of activation across all four modalities is the right anterior insula.'[39] In a separate study, a lesion of the right anterior insula affected aesthetic judgment of ambiguous artworks.[40] Not only do patients with right hemisphere disease display autonomic hyporeactivity,[41] but arousal is specifically associated with the right insula, the location that is singled out here.[42]

Not all explicit judgment studies, however, favour the right hemisphere: three studies found bilateral activations with no significant lateralisation.[43] And turning to those that find a significant *left* hemisphere contribution to explicit judgment tasks, the most significant is Nadal's study which showed that both male and female participants liked representational paintings more when presented in the right visual field, favouring the left hemisphere: there was no effect for abstract paintings.[44] However, Coney and Bruce found the opposite: a right visual field preference for abstract paintings, but not for representational paintings.[45] In either case the right visual field effect is contrary to expectation from much of the rest of the literature. Two fMRI studies report left sensorimotor cortex activations associated with explicit subjective judgments of ugliness.[46] Compared to reading prose, beauty in poetry increased activity in left hemisphere areas in Chinese subjects listening to Chinese poetry.[47] (This contrasts with the finding that English-speaking subjects showed

---

33 Ferrari, Lega, Tamietto et al 2015.

34 Tsukiura & Cabeza 2011.

35 Friedrich, Harms & Elias 2014. See also: Christman & Pinger 1997; Mead & McLaughlin 1992; Beaumont 1985.

36 Hutchison, Thomas & Elias 2011; McDine, Livingston, Thomas et al 2011.

37 Cela-Conde, Ayala, Munar et al 2009.

38 Bromberger, Sternschein, Widick et al 2011.

39 Brown, Gao, Tisdelle et al 2011.

40 Boccia, Barbetti, Piccardi et al 2017.

41 Heilman, Schwartz & Watson 1978; Morrow, Vrtunski, Kim et al 1981: 'non-dominant hemisphere damage suppresses arousal altogether'.

42 Craig 2005.

43 Mizokami, Terao, Hatano et al 2014; Calvo-Merino, Urgesi, Orgs et al 2010; Cupchik, Vartanian, Crawley et al 2009.

44 Nadal, Schiavi & Cattaneo 2018.

45 Coney & Bruce 2004.

46 Kawabata & Zeki 2004; Di Dio, Macaluso & Rizzolatti 2007.

47 Gao & Guo 2018.

specifically right-sided activations relating to both reading poetry and the emotional response to it, a significant finding since almost all verbal tasks activate the left hemisphere.)[48] As the authors point out, the Chinese subjects were asked to give an explicit judgment, unlike the case of Bohrn *et al* (see below),[49] who deliberately separated the *experience* of beauty from explicit *judgment*, and found the key area in experience to be the right caudate. Both hemispheres, but more left than right, were activated in a task in which subjects viewing artwork were told to 'experience the mood of the work and the feelings it evokes, and to focus on its colours, tones, composition, and shapes' in an explicit fashion.[50] I suspect the explicit focus on a procedure is important here. And finally a 3% improvement in aesthetic ratings in an explicit judgment task occurred after stimulation to left dorsolateral prefrontal cortex (in something so variable and hard to measure, it is not clear that a 3% change is significant).[51] Oddly the experimenters did not think to carry out a comparison with the right dorsolateral prefrontal cortex.

It is impossible that such a mulitfaceted phenomenon should be associated with one area of the brain, or even with one hemisphere exclusively. However, when one turns to studies which focus on the appreciation, rather than explicit critical judgment, of beauty, the picture leans more clearly (though far from exclusively) towards the right hemisphere. In a particularly revealing study the experience of aesthetic *feeling* was associated with activations that were all right-sided (anterior cingulate, precuneus and inferior frontal gyrus), while aesthetic *judgment* was bilateral.[52] Another study found that as far as aesthetic ratings are concerned, these tend to be more left-sided, while aesthetic responses tend to be more right-sided, which the experimenters describe as 'a confirmation of previous findings', in which the right hemisphere has 'continuously been associated' with beauty appreciation.[53] A study of EEG responses – event-related potentials – found the 'contemplative' rather than judgmental sense of beauty to be right-sided. Another such study found that responses were larger over the right hemisphere than the left during aesthetic 'appreciation' in both experts and non-experts.[54] The right inferior parietal lobule and premotor frontal cortex are activated when watching dance that is found more beautiful.[55] When subjects were scanned while reading a number of proverbs without explicitly evaluating them, and were subsequently asked to rate them for beauty, beauty was found to have activated the right caudate, anterior cingulate, and cerebellum.[56] However, at least one study of beauty appreciation (that did not involve making judgments) found no difference between hemispheres.[57]

One of the best ways to narrow down what it is one is assessing

48 Zeman, Milton, Smith et al 2013.

49 Bohrn, Altmann, Lubrich et al 2013.

50 Cupchik, Vartanian, Crawley et al 2009.

51 Cattaneo, Lega, Flexas et al 2014.

52 Yeh, Lin, Hsu et al 2015.

53 Silveira, Fehse, Vedder et al 2015. See, eg, Lutz, Nassehi, Bao et al 2013.

54 Pang, Nadal, Müller et al 2013.

55 Cross, Kirsch, Ticini et al 2011.

56 Bohrn, Altmann, Lubrich et al 2013.

57 Kawabata & Zeki 2004.

in visual aesthetics is to get subjects to compare an image with the same image after it has been subtly altered in ways designed to reduce its aesthetic appeal, and see which areas of the brain respond differently. This can be done by interfering with the composition of a painting, or adding noise to it, or by changing its proportions. In each case the original is, as expected, preferred to its altered version. In viewing the degraded paintings or images, both right and left-sided areas are activated, but the right and left lingual, and right fusiform, gyri are the most significant cortical areas.[58] In the case of the golden ratio the key area activated is, once again, the right insula.[59] A more recent study by the same investigators has suggested that the aesthetic response of the right insula is specific to art,[60] though other studies, as we have seen, suggest that its role must be more widespread, since it is the most consistent area involved in visual beauty appreciation of every kind.

58 Di Dio, Macaluso & Rizzolatti 2007; Vartanian & Goel 2004.

59 Di Dio *et al, op cit.*

60 Di Dio, Canessa, Cappa *et al* 2011.

⌇ *Appendix 7* · Hemisphere differences and spirituality: review of neuropsychological evidence

This evidence should not be read in isolation from its context in the overall argument of Chapter 28.

In mindfulness meditators, there is 'a pronounced shift away from midline cortices towards a right lateralised network', including the right dorsal and lateral prefrontal cortex, right insula, right secondary somatosensory cortex and right inferior parietal lobule.[1] A meta-analysis of 78 functional neuroimaging studies revealed that 'significant activation clusters for loving-kindness meditation were found in the right somatosensory cortices, the inferior parietal lobule and the right anterior insula.'[2] In satyananda yoga, all the distinctive activity, both alpha and gamma, was found to be in broad cortical networks in the right hemisphere, providing 'evidence for a core right-sided (rather than midline) network'.[3] There is a clear shift toward right hemisphere dominance in altered states of consciousness, such as the shamanic state, induced by meditation;[4] and in the shamanic state there is increased activity across the whole range of electrophysiological frequencies in the right (but not left) frontal region, with, in general,

> a shift from the normally dominant left analytical to the right experiential mode of self-experience, and from the normally dominant anterior prefrontal to the posterior somatosensory mode ... Similar neurophysiological markers have been documented in diverse trance states including yoga and shamanic practices; Zen Buddhist and transcendental meditation ... This may represent a core feature of altered states of consciousness resulting from right-hemispheric activation and homologous trans-callosal inhibition of contralateral left hemispheric functions responsible for ordinary, ego-bound states of consciousness.[5]

The type of meditation may make a difference, in reassuringly predictable ways: visual meditations result in increases in gamma activity mainly in the right posterior region; in meditative self-dissolution the increases are mainly in the right anterior region; but in verbal mantra meditations, they are mainly left central.[6]

Significant both state and trait changes – ie, changes that generalise and persist outside of specific practices – were found in mindfulness

1 Farb, Segal, Mayberg et al 2007.
2 Fox, Dixon, Nijeboer et al 2016.
3 Thomas, Jamieson & Cohen 2014.
4 Krippner & Combs 2002.
5 Flor-Henry, Shapiro & Sombrun 2017.
6 Lehmann, Faber, Achermann et al 2001.

meditation practitioners, marked by increased gamma power in the right posterior cortex when compared with controls.[7] Measures of connectivity within, rather than between, hemispheres are asymmetric in mindfulness meditators, being higher in the right hemisphere and lower in the left hemisphere.[8] And anatomical, as well as neurophysiological, changes have been observed. In meditators, cortical thickening has been observed in the right insula, and the somatosensory cortex of both hemispheres and inferior parietal lobule of the right hemisphere;[9] increased grey matter volumes have been found in the right angular and posterior parahippocampal gyri of loving-kindness meditators;[10] mindfulness practitioners show an increased grey matter volume in the right insula and hippocampus (regions related to intuitive bodily awareness and the regulation of emotion, respectively);[11] and meditators more generally have been shown to have larger right hippocampal volumes than controls.[12] (However, for completeness' sake, and to remind us how complex the subjects of measurement must be, in another study the left hippocampus was found to be larger than controls;[13] and yet another study linked types of meditation in which self-other boundaries are blurred with *decreased* activity in the right inferior parietal lobule, an area known to be critical to the sense of the self.[14])

An exception is the achievement of blissful states, which has repeatedly been associated with activation of the left frontal pole. (In Chapter 6, I discussed the association of the left frontal pole with a positive emotional timbre.) Research by Newberg and colleagues underlines that this excitation of the left frontal pole is strongly correlated with suppression of the superior parietal region in the same (left) hemisphere.[15] In relation to this, it is a possibility that, since meditation of all kinds depends on stilling the normal verbalising activity of the left posterior cortex, the blissful state achieved by some meditators is an (initially incidental) consequence – a 'side-effect' – of learning how maximally to recruit the left frontal pole in its capacity to suppress the posterior cortex of the same hemisphere.

A consensus is emerging from the literature that the phenomena associated with religious experience generally tend to be sustained more by the right hemisphere than by the left. This conclusion is supported by two book-length studies of spirituality and the brain, *The Soul in the Brain* by Michael Trimble,[16] formerly Professor of Behavioural Neurology at the Institute of Neurology, Queen Square, in London, and Patrick McNamara's scholarly *The Neuroscience of Religious Experience*,[17] as well as by an extensive review by Orrin Devinsky and George Lai.[18] Having said all this, it should be emphasised that there is no reductive intent in any of these authors: for instance, McNamara writes that he is 'not interested in debunk-

7 Berkovich-Ohana, Glicksohn & Goldstein 2012.

8 Berkovich-Ohana, Glicksohn & Goldstein 2014.

9 Lazar, Kerr, Wasserman *et al* 2005 ; Lutz, Greischar, Perlman *et al* 2009.

10 Leung, Chan, Yin *et al* 2013.

11 Hölzel, Carmody, Vangel *et al* 2011; Hölzel, Lazar, Gard *et al* 2011; Hölzel, Ott, Hempel *et al* 2007; Luders, Kurth, Mayer *et al* 2012.

12 Luders, Toga, Lepore *et al* 2009; Luders, Clark, Narr *et al* 2011.

13 Hölzel, Carmody, Vangel *et al*, *op cit*.

14 Johnstone & Glass 2008.

15 Newberg, Alavi, Baime *et al* 2001; Newberg, Pourdehnad, Alavi *et al* 2003.

16 Trimble 2007.

17 McNamara 2009.

18 Devinsky & Lai 2008.

ing religion's supposed pretensions or calling it "nothing but ..."[19] McNamara largely implicates right frontotemporal networks, as do others:[20]

> although the range of variance in religious experiences across cultures and time epochs is unknown, I find that changes in religious experiences in the samples of subjects that have been studied with cognitive and neuroscientific techniques are, in fact, reliably associated with a complex circuit of neural structures ... [structures which] nearly always include the key nodes of the amygdala, the right anterior temporal cortex, and the right prefrontal cortex.[21]

Devinsky and Lai, following William James, distinguish the 'religion of the everyday man', with its characteristic ongoing belief pattern and set of convictions, predominantly localised to the frontal region, from ecstatic religious experience, more localised to the temporal region, both in the right hemisphere.[22]

Pathology of a religious nature, by contrast, seems to be associated with the left hemisphere. Religious delusions, like other delusions,[23] are associated with increased left frontal and temporal activity.[24] Hyperreligiosity in subjects with epilepsy has been associated with a decreased volume in the right hippocampus.[25]

As is clear, much depends on what is meant by religion. In some studies what is measured is essentially a pathological hyperreligiosity; in others an openness to ritual; in others the holding of certain beliefs or practices; in others experiences of a 'paranormal' kind. As if to demonstrate difficulties with generalisation, in one study the right hemisphere appeared foundational for aesthetic religious experience, while the left hemisphere was associated with ritual religious experience,[26] a distinction that, however it may work in the laboratory, falls short of the complexity of real-life experience.

19 McNamara *op cit* (xiii).

20 Trimble & Freeman 2006; Devinsky *et al*, *op cit* 2008.

21 McNamara 2009 (xi).

22 It is of interest that Swedenborg's capacity for clairvoyance, hearing angelic voices, and seeing visions all came to a sudden end three months before he died, immediately after he suffered what would appear to have been a non-dominant hemisphere stroke. Though he was still quite able to speak, 'his spiritual sight of nearly 30 years was entirely gone': Wulff 1996 (108). Thanks to David Lorimer for bringing this to my notice.

23 See Chapter 4.

24 Puri, Lekh, Nijran *et al* 2001.

25 Wuerfel, Krishnamoorthy, Brown *et al* 2004.

26 Butler, McNamara & Durso 2011.

~ Appendix 8 · The 'incompatibility' of science and religion, and the harmful nature of religion: some evidence to the contrary

This appendix contains the following sections:

(1) Scientists and religious belief;
(2) Religious institutions and science;
(3) Religion as brainwashing;
(4) Religious belief and the health of individuals and societies.

This evidence should not be read in isolation from its context in the overall argument of Chapter 28.

1 · Scientists and religious belief

Jerry Coyne seems to think it self-evident that science and religion are incompatible. What does he make, then, of the existence of so many great scientists that thought and think that such a view is just as clearly mistaken, not to say naïve? My point here is not that they are right or wrong to accept a religious view: just that they do accept one, in plain contradiction to Coyne's statement that science and religion are incompatible.

When confronted with the overconfident, even contemptuous, pronouncements of some scientists to the effect that God does not exist, I think of Bohr's openness to religion, quoted in Chapter 28. And Heisenberg's. He too saw further – and saw the rather elementary mistake Coyne and his kind are making, to do with different kinds of knowledge expressed in different kinds of language:

> Above all else one must be clear that in religion language is used in an entirely different way from that in which it is used in science. The language of religion is closer to that of poetry than that of science ... one of its most important tasks is, with its language of images and parables, to keep us *mindful of the wider context* ... We know that in religion we are dealing with a language of images and parables, that can never precisely convey what is meant ... At the end of the day, the central Order, or the 'One' as we used to say, with which we commune in the language of religion, must prevail.[1]

The point Heisenberg makes about context seems to me enormously important. As always the tendency of the left hemisphere is to narrow focus so drastically that it fails to see the broad picture at

[1] Heisenberg 2001 (107, 111 & 25; emphasis added): » Aber man muß sich doch vor allem darüber klar sein, daß in der Religion die Sprache in einer ganz anderen Weise gebraucht wird als in der Wissenschaft. Die Sprache der Religion ist mit der Sprache der Dichtung näher verwandt als mit der Sprache der Wissenschaft ... es gehört zu ihren wichtigsten Aufgaben, in ihrer Sprache der Bilder und Gleichniße an den großen Zusammenhang zu erinnern ... Wir wissen, daß es sich bei der Religion um eine Sprache der Bilder und Gleichniße handeln muß, die nie genau das darstellen können, was gemeint ist ... Aber letzten Endes setzt sich doch wohl immer die zentrale Ordnung durch, das ‚Eine', um in der antiken Terminologie zu reden, zu dem wir in der Sprache der Religion in Beziehung treten. «

all, and therefore cannot detect that anything is missing when its impoverished picture becomes a supposed representation of the whole.

It should also be remembered that much in contemporary science itself can be expressed only in poetic language – a point Bohr makes. Moreover, some of its posited entities, such as 'dark matter' (which is thought to constitute about 85% of all matter) and 'dark energy', are neither visible, nor tangible, nor even directly detectable. That's without considering 'multiverses' and 'many worlds'. These are non-empirically verified and in some cases non-empirically verifiable, theoretical constructs, proposed in order to explain what *can* be detected.

I quoted Max Planck on the idea that spirit, not matter, alone exists. 'Visible, perishable matter – that is not what constitutes the real, true and actual', he continued: 'it is the invisible, immortal spirit that is the truth.' And he went on to say that we are compelled to see behind this spirit a creative force:

> I am not afraid to call this mysterious creator, as have all the civilized nations of the earth for thousands of years: God. So you see, my dear friends, how in our days, in which people no longer believe in spirit as the foundation of all creation, and therefore find themselves in bitter estrangement from God, it is precisely the minute and the invisible that leads truth back from the grave of materialist delusion, and opens doors into the lost and forgotten world of the spirit.[2]

Einstein's views on organised religion were largely, though not wholly, dismissive; but he did not deny a religious feeling to the scientist, which 'takes the form of a rapturous amazement at the harmony of natural law, which reveals an intelligence of such superiority that, compared with it, all the systematic thinking and acting of human beings is an utterly insignificant reflection.'[3] Call that what you like, it is not far at all from a sense of a transcendent intelligence revealed in the order of the cosmos that inspires awe and rapture – in fact that is exactly what it is. Someone who had no belief in any kind of God could not say that. Einstein time and again reflected, as incidentally have all the great spiritual masters, that such questions are beyond the capacity of human intelligence to fathom. He further commented: 'in view of such harmony in the cosmos which I, with my limited human mind, am able to recognise, there are yet people who say there is no God. But what makes me really angry is that they quote me for support of such views.'[4]

Einstein's colleague Kurt Gödel, probably the 20th century's most important mathematician, went much further. He was a Lutheran, who read the Bible regularly, believed in an afterlife, and thought that it was impossible to give a credible account of reality without

---

2 Planck 1944: » *Nicht die sichtbare, aber vergängliche Materie ist das Reale, Wahre, Wirkliche, sondern der unsichtbare, unsterbliche Geist ist das Wahre! ... so scheue ich mich nicht, diesen geheimnisvollen Schöpfer so zu benennen, wie ihn alle Kulturvölker der Erde früherer Jahrtausende genannt haben: Gott. So sehen Sie, meine verehrten Freunde, wie in unseren Tagen, in denen man nicht mehr an den Geist als den Urgrund aller Schöpfung glaubt und darum in bitterer Gottesferne steht, gerade das Winzigste und Unsichtbare es ist, das die Wahrheit wieder aus dem Grabe materialistischen Stoffwahnes herausführt und die Türe öffnet in die verlorene und vergessene Welt des Geistes.* «

3 Einstein 1999 (29).

4 Statement to the German anti-Nazi diplomat Hubertus zu Löwenstein around 1941, as quoted in Löwenstein 1968.

5 McElroy 2004 (118).

God.⁵ Of an afterlife, he wrote that it is 'possible today to perceive, by pure reasoning' that it 'is entirely consistent with known facts ... If the world is rationally constructed and has meaning, then there must be such a thing'.⁶ He also thought that rational religion requires 'a direct connection of each individual to God or the whole or what is ultimate';⁷ that 'mechanism in biology' was a fallacy; and that the doctrine that there can be 'no mind separate from matter' would one day be scientifically disproved.⁸

Equally there have been few mathematical minds greater than that of the physicist John von Neumann, who quite apart from his contributions to pure maths, made foundational contributions to quantum mechanics, hydrodynamics, game theory, computing and statistics. 'A mind of von Neumann's inexorable logic had to understand and accept much that most of us do not want to accept and do not even wish to understand', wrote Eugene Wigner.⁹ Neumann's verdict on God? 'There probably is a God. Many things are easier to explain if there is than if there isn't.'¹⁰ Whatever else one might say about that, it would seem that, at least, he saw no incompatibility.

In the words of the Nobel-winning neurophysiologist and philosopher Sir John Eccles: 'Science and religion are very much alike. Both are imaginative and creative aspects of the human mind. The appearance of a conflict is a result of ignorance.'¹¹

Here I have referred to a minute fraction of scientists, all from within the last 100 years, many of them amongst the greatest that ever lived, who demonstrate that the assertion that 'science and religion are incompatible' is simply wrong. I could, of course, have referred to many more: but even one robust case is enough to refute Coyne's argument, let alone more than 95% of Nobel laureate physicists.¹²

6 Wang 1996 (104–5).
7 Wang 1987 (150).
8 *ibid* (2).
9 Wigner 1967 (261).
10 Macrae 1992 (379).
11 Eccles 1984 (50).
12 See p 1278.

2 · *Religious institutions and science*

There has been a very one-sided appraisal of the part played by religious institutions in the history of science. Much – probably a majority – of all important scientific research over human history in the West was actually carried out by clergy. In a world where a large number of highly educated people took holy orders that is perhaps not surprising; but the evidence that these men experienced conflict between pursuing science and sustaining their religion is scant, and they received munificent support from the Church. Until recently scientists saw, not only no conflict, but an obvious synthesis of knowledge between religion and science: in discovering the awe-inspiring coherence and beauty of the world, they felt they were uncovering the work of a great and divine intelligence, and thereby coming into the presence of God. (Newton, a religious man, saw his real contribution to knowledge as his theological writings.)

Natural philosophy (which is what science used to be called), natural history and astronomy, were almost entirely sponsored by the Church; and these disciplines were originally based on the writings of Aristotle and Ptolemy – the best scientific tracts Europe had – as transcribed by Syrian Christians and preserved by Islam. At the fall of Muslim Toledo to the Christians in 1085, the libraries were not ransacked, but rather the texts zealously safeguarded and translated into Latin by European monks.

With the Middle Ages came the era of innovation. Mediaeval science was sophisticated, and laid the foundations for the Newtonian revolution. Robert Grosseteste, who was Bishop of Lincoln in the first half of the 13th century, is credited with being the first person to outline what is now known as the scientific method, and made important advances in astronomy and optics. His pupil Roger Bacon, a Franciscan monk, emphasised empirical methods and made experimental discoveries in physics and chemistry. The polymathic 15th century theologian and cardinal, Nicholas Cusanus, 'anticipated many later ideas in mathematics, cosmology, astronomy and experimental science.'[13] He had insights that took up to four centuries to be proven, but which eventually became established scientific or mathematical truths.[14]

The myth that people believed the earth was flat until Columbus proved them wrong is an example of major, motivated, misdirection. As early as the 6th century BC, Pythagoras – and later Aristotle and Euclid – wrote about the earth as a sphere. Eratosthenes at the beginning of the 2nd century BC devised a method for calculating, fairly accurately, the circumference of the spherical earth.[15] When Ptolemy wrote his *Geography* in the 2nd century AD he considered the idea of a round planet as taken for granted. An important tract of the 13th century, Sacrobosco's *Sphæra Mundi*, widely read across Europe, contains a clear description of the Earth (not just the wider cosmos) as a sphere, in accord with established opinion in Europe at the time. According to Jeffrey Burton Russell, Emeritus Professor of History at the University of California,

> with extraordinarily few exceptions *no* educated person in the history of Western Civilisation from the 3rd century BC onward believed that the earth was flat ... the sphericity of the earth was accepted by all educated Greeks and Romans. Nor did this situation change with the advent of Christianity. A few – at least two and at most five – early Christian fathers denied the sphericity of the earth by mistakenly taking passages such as Psalm 104, verses 2–3, as geographical rather than metaphorical statements. On the other side tens of thousands of Christian theologians, poets, artists, and scientists took the spherical view throughout the early, mediaeval, and modern church. The point is that no educated person believed otherwise.[16]

13 Miller 2017.

14 He demonstrated the impossibility of squaring the circle, finally proven by Ferdinand von Lindemann in 1882 (*De circuli quadratura*, 1450); saw that the earth is not the centre of the universe, is not at rest, but moves round the sun, and that its poles are not fixed, as well as realising that the celestial bodies are not strictly spherical, nor their orbits circular (see Bond 1997); believed the stars were other suns that had other worlds orbiting them, and that space was infinite (*ibid*); published improvements to the Alfonsine Tables which gave a practical method to find the position of the sun, moon and planets; contributed to the development of the concepts of infinitesimal and relative motion – his writings were instrumental for Leibniz's development of calculus (Priday 2019 (88–9)); is credited with having conducted the first modern formal biology experiment, concluding that plants absorb nourishment from the air and water – this some 150 years before van Helmont (Lieth & Whittaker 1975, 8; Krikorian & Steward 1968); and was the first to see the value of counting, rather than merely characterising, the pulse, which he proposed by weighing the quantity of water run out of a water clock while the pulse beat 100 times (Hoff 1964). Although he is sometimes credited with being the first to use concave lenses to correct myopia, this would appear to be an overstatement (*De Beryllo*, 1458).

15 Osserman 1995.

16 Russell 1997 (emphasis in original).

Of the myth of Columbus he writes, 'The courage of the rationalist confronted by the crushing weight of tradition and its cruel institutions of repression is appealing, exciting – and baseless.'[17] The idea that in the Middle Ages people believed the earth was flat was in fact a deliberately engineered 19th century invention designed to bolster (quite unnecessarily, since science needed no bolstering) an attack on religion, seen as an impediment to science:

> the falsehood about the spherical earth became a colourful and unforgettable part of a larger falsehood: the falsehood of the eternal war between science (good) and religion (bad) throughout Western history. This vast web of falsehood was invented and propagated by the influential historian John Draper (1811–1882) and many prestigious followers, such as Andrew Dickson White (1832–1918), the president of Cornell University, who made sure that the false account was perpetrated in texts, encyclopaedias, and even allegedly serious scholarship, down to the present day.[18]

No historian nowadays takes Draper or White seriously, though Coyne, who bases part of his argument on their testimony, would seem not to be aware of this.[19]

Another *locus classicus* for the 'war' between science and religion, according to its proponents, is that of the earth not being at the centre of the universe. First of all, it is a popular misconception that being at the centre of the universe was considered a position of prestige, and that, hence, the disproof of geocentrism by Copernicus was a blow to human dignity. Quite the opposite is the case. The centre was where the lowest parts of creation remained, the more exalted occupying positions in ever higher circles removed from the centre:

> pre-Copernican cosmology pointed not to the metaphysical or axiological 'centrality' but rather to the sheer grossness of humankind and its abode. In this view, the earth appears as a universal pit, figuratively as well as literally the world's low point.[20]

This was a view shared by Christian, Jewish and Islamic writers, with particularly clear expositions by Maimonides in the 12th century ('in the universe, the nearer the parts are to the centre, the greater is their turbidness, their solidity, their inertness, their dimness and darkness, because they are further away from the loftiest element, from the source of light and brightness'),[21] and Al-Biruni in the 11th ('in the centre of the sphere of the moon is the earth, and this centre is in reality the lowest part').[22] 'This negative view encompasses … not only ancient and mediaeval Arabic, Jewish, and Christian writers', writes Dennis Danielson, 'but also many prominent voices that we usually associate with Renaissance humanism, both before and after the time of Copernicus.'[23] A quarter of a century after the

17 Russell 1991 (6).
18 Russell 1997.
19 Numbers 2009.
20 Danielson 2001 (1031).
21 Maimonides 1919 (118–9).
22 Al-Biruni 1934 (45).
23 Danielson *op cit* (1031).

publication of Copernicus' *De Revolutionibus*, Montaigne took up the same theme, declaring that we are 'lodged down here, among the mire and shit of the world, bound and nailed to the deadest, most stagnant part of the universe, in the lowest storey of the building, the farthest from the vault of heaven.'[24]

The Church was committed to astronomical observation: as well as funding virtually all hospitals, schools and universities for over 1,000 years, it funded observatories and meteorological stations since the late Middle Ages. Copernicus was a canon of Frauenburg Cathedral. Nor was his theory rejected, but on the contrary enthusiastically welcomed by a number of cardinals,[25] as well as by Pope Clement VII himself, who invited him to lecture to an assembly of bishops and cardinals in the Vatican, and extravagantly rewarded the scholar Johann Widmanstetter, the Papal secretary who brought Copernicus to his notice.[26] 'As a matter of fact', writes Claudio Ronchi, 'heliocentric doctrines ... already circulated in classical antiquity as well as in Muslim and Christian Middle Ages.'[27] It is often said that Giordano Bruno was somehow a martyr to science, but when he was burned at the stake as a heretic, 'it had nothing to do with his writings in support of Copernican cosmology'.[28] His mistake was his refusal to renounce heterodox religious beliefs, and had nothing to do with science. He was, then, a martyr, not to science, but to his religious beliefs (he did not understand Copernicus's science). A century later, however, interestingly under pressure from the literalistic, left hemisphere-dominated mentality of Reformed theologians, the Roman Catholic Church decided it had better ban Copernicus's teaching, a ban that remained in place for 200 years – indisputably an egregious mistake by a hugely powerful institution. But it does not in itself make science and religion enemies.

And, then, a very large part of scientific progress, up to about 100 years ago, is owed to the work of clergy, often in rural livings – gifted outsiders with intelligent, educated minds, and leisured enough to undertake patient observation and repeat experiments, untrammelled by the conformist pressures of a modern university career. With few exceptions, far from resisting science, as is often implied, they were themselves the makers of science. At the same time scientists have themselves, at times, resisted scientific discovery.[29]

Gregor Mendel, the father of modern genetics, was an Augustinian friar whose experiments took place in a monastery garden. And in the 20th century it was a Catholic priest, the cosmologist Georges Lemaître, who proposed the Big Bang theory. There is no conflict, implicit or explicit, here between religion and science.

Until the 19th century (the high point of belief in mechanistic science), it would never have occurred to anyone that there need be a war between religion and science. It was only then that the term

---

24 Montaigne 1991, 'An apology for Raymond Sebond', Bk II, §12 (505).

25 Eg, letter of Nicholas Schönberg, Cardinal of Capua, to Copernicus, 1 November 1536: 'Some years ago word reached me concerning your proficiency, of which everybody constantly spoke. At that time I began to have a very high regard for you, and also to congratulate our contemporaries among whom you enjoyed such great prestige. For I had learned that you had not merely mastered the discoveries of the ancient astronomers uncommonly well but had also formulated a new cosmology. In it you maintain that the earth moves; that the sun occupies the lowest, and thus the central, place in the universe; that the eighth heaven remains perpetually motionless and fixed; and that, together with the elements included in its sphere, the moon, situated between the heavens of Mars and Venus, revolves around the sun in the period of a year. I have also learned that you have written an exposition of this whole system of astronomy, and have computed the planetary motions and set them down in tables, to the greatest admiration of all. Therefore with the utmost earnestness I entreat you, most learned sir, unless I inconvenience you, to communicate this discovery of yours to scholars, and at the earliest possible moment to send me your writings on the sphere of the universe together with the tables and whatever else you have that is relevant to this subject': Copernicus 1992.

26 Freely 2014 (105–6).

27 Ronchi 2014 (47).

'natural philosophy' was abandoned, and the word 'science', which simply means knowledge, appropriated to that which was practised by a certain caste, with their own rituals, myths – including, prominently, those that denigrated religion, but also those that oversold what science can achieve – and an unproven but historically very unusual materialist creed. The idea of this war between science and religion was bolstered by the generation of a powerful 19th century mythology of science's own fabrication, some of which we have seen, and featuring two cover stories: the supposed 'martyrdom' of Galileo, and in a lesser role, the reported triumph of T. H. Huxley over Bishop Samuel Wilberforce at a debate in Oxford. The science behind Galileo was not universally accepted; there were scientists who opposed him and theologians who were on Galileo's side. His misfortunes had much to do with his personality, and his martyrdom amounted to being allowed to live out his life at his comfortable rural estate on the outskirts of Florence. While the events behind these stories did take place, the realities are complex; the stories often told are tendentious half-truths, and suggestive of the grounding of a cult. Cults motivate themselves by having adversaries.

It is absolutely true that there arose in the middle of the 19th century, at the same time as the attack on religion by science, an authoritarian attitude in the Roman Catholic Church, especially under the papacies of Pius IX and X, to free thinking of any kind, including science. This was a heinous mistake. It is not unreasonable to see the Church, especially at this point, as having constituted an oppressive cultural institution. But its recalcitrance might also be seen as a reaction to the aggressive attitude of those already fanning a 'war' between science and religion at that time, and suggesting that science was incompatible with religion. When such conflicts are invented, both parties inevitably become defensive, polarise their positions, and consequently fulfil their opponents' skewed beliefs. Extreme views are often a reaction to the perception of extreme contrary views. In any case, this Papal posturing had very little influence, if any, on the course of science, and represents only an episode in a long history.

Darwin gets dragged into the unseemly squabble between misunderstood science and misunderstood religion in a way he, in his modest agnosticism, would not have appreciated. Darwin's importance is not in fact rightly understood by many of his disciples. Addressing the topic of what he contributed to our thinking, and as a preamble to celebrating his true greatness, biologist and philosopher of science John Wilkins disposes of some myths:

> Darwin's theories (plural) are not controversial because they imply that species are mutable. This was a widely held view by preachers, moralists, Aristotelians, naturalists, breeders, formalists, folk biology, and even biblical translators ... Darwin was not controversial because

28 This is clearly shown in Finocchiaro's reconstruction of the accusations against Bruno (2002); see also Blumenberg 1987.

29 Barber 1961 (596): 'In the study of the history and sociology of science there has been a relative lack of attention to one of the interesting aspects of the social process of discovery – the resistance on the part of scientists themselves to scientific discovery ... There has been a great deal of attention paid to resistance on the part of economic, technological, religious, and ideological elements and groups outside science itself. Indeed, the tendency of such elements to resist seems sometimes to be emphasized disproportionately as against the support which they also give to science. In the matter of religion, for example, are we not all a little too much aware that religion has resisted scientific discovery, not enough aware of the large support it has given to Western science?'

he thought the age of the earth was large. This preceded him also, and was settled in the late eighteenth century, although the present value wasn't finalised until the 1960s. Darwin was not controversial because his account of humans being animals contradicted the Bible. Linnaeus [a deeply religious man] knew humans were animals a century earlier [as, by the way, did Goethe], and indeed the only issue was whether humans were animals with souls ... Moreover, it was Christians who rejected the literal interpretation of the Bible, long before Darwin (beginning with the Alexandrian school in the second century), and those who realised that the global Flood was a myth (or an allegory) were Christian geologists a half century at least in advance of Darwin.[30]

What can of course be said is that people who believe that the world was literally created in six days are wrong, and that they are wrong to oppose the idea of evolution in its general sense (beliefs about evolution were not carved in tablets of stone, left hemisphere fashion, but evolve, right hemisphere fashion, as all living traditions do: we are learning inevitably that evolution is far more complex than the narrow neo-Darwinian view would hold). But most religious people do not espouse creationism.[31]

As for Darwin's views on the existence of a God, he was far more modest than many of the neo-Darwinians: 'I feel most deeply that the whole subject is too profound for the human intellect. A dog might as well speculate on the mind of Newton.'[32] Elsewhere he wrote: 'In my most extreme fluctuations I have never been an atheist in the sense of denying the existence of a God.'[33]

3 · *Religion as brainwashing*

Religious experience exists across the life-span, from childhood onwards.[34] The literature attests to the existence of profound religious experiences in children.[35] Children's 'intuitive theism' appears to be independent of culture and environment, including parents' beliefs (whether atheistic or theistic), the storybooks that have been read to them or the content of family conversations.[36] Young children whose parents are atheists may have religious experience,[37] and their religiosity may persist without any specific cultural reinforcement.[38] Young children, it turns out, have frequent experience of a naturally 'relational consciousness', as do many non-Western adult cultures.[39] According to David Hay, one of the researchers, this involves two elements:

> an altered state of awareness as compared with other kinds of consciousness, more intense, more serious and more valued; and the experience of being in relationship – with other people, with the environment and with God, and in an important sense, in touch with oneself.[40]

---

30 Wilkins 2012.

31 In a YouGov Poll in 2017, 84% of those who identified as religious/spiritual in the UK rejected creationism (rejected by 92% of the population overall). Slightly confusingly, 3% of the non-religious and non-spiritual nonetheless believed in creationism, and a further 7% of that same group believed in evolution 'guided by God'. The report's authors conclude that 'individuals who find it difficult to accept aspects of evolutionary science overwhelmingly see other sciences as reliable, showing similar trends in attitudes as other groups. Rejection of, or uncertainty about, aspects of human evolution is not necessarily an issue of "religion versus evolutionary science". Universal questions around what it is to be human and about the human experience are implicated in this rejection or uncertainty and affect people of all faiths and none'. See https://sciencereligionspectrum.org/in-the-news/press-release-results-of-major-new-survey-on-evolution/ (accessed 15 March 2021).

32 Letter of Darwin to Asa Gray, dated 22 May 1860; in F Darwin 1887, vol 2 (312).

33 Letter of Darwin to John Fordyce, dated 7 May 1879; in F Darwin 1887, vol 1 (304).

34 Fowler 1981; Oser 1991; Tamminen 1994.

Children are eager to discuss events, experiences and relationships in terms that we would call unambiguously spiritual, and do not shy from talking of what we wrongly call 'the supernatural' (wrongly, since, it seems, there is nothing 'added on', or 'outside' nature, about it). What the research of Nye and her colleague Hay revealed was more like the opposite of the thesis that children would never think in such terms unless they were 'forced' to do so: rather, they naturally thought in such terms until they learnt that it was not considered smart to do so, and an alienated, atomistic, inanimate vision of the world was imposed on them at school.[41] It is at least as arguable that Western children are 'forced' to accept what could be seen as a damaging and unnatural cosmology by teachers cowed by the high-decibel, upper case, pronouncements of people like Richard Dawkins.

A meta-analysis of religious experience tried to disentangle the effects of culture on awareness of God. It used a strict criterion for 'religious experience', specifically excluding experience of supernatural beings or phenomena such as the spirits of dead relatives, other humans, animals or plants, as well as so-called 'extrovertive' experiences[42] involving feelings of unity, knowledge, reality, love, luminosity and so forth in relation to the natural world; and including only experience of something such as a transcendent God or divine realm'.[43]

The study's authors reflect that though culture obviously shapes such experience, it is not obvious that it generates it. They point out that, within the same culture, and given the same religious or non-religious background, only some experiences outwith the 'normal' range are interpreted in a religious way: for example, bizarre experiences in dreams or psychotic phenomena are usually not.[44] Religious experience may arise spontaneously in the most prosaic of circumstances, outside any context suggested by religious doctrine or practice. Equally, within the same culture, not all religious persons claim to have had religious experiences;[45] non-religious persons may report having religious experiences;[46] and, despite the fact that the vast majority of people during the more than 70 years of the Soviet Union were raised and educated in a rigorously anti-religious culture, religious experience was still common.[47]

Many people experience moments of transcendence when the world seems to come more alive and to be transformed: when there is a strong subjective sense of being in the closest touch with something much bigger than oneself from which one had been hitherto somewhat removed. Experience of the divine is not necessarily something at all rarefied or 'super'-natural, but rather a normal and natural phenomenon. 'My chief desire', writes Hart,

> is to show that what is most mysterious and most exalted is also that which, strangely enough, turns out to be most ordinary and nearest

35 See Robinson 1983; Hart & Ailoae 2007; Coles 1990; Stower & Ryan 1998.

36 Evans 2000 & 2001; Barrett & Richert 2003; Bering 2004; Kelemen 2004; Kelemen, Callanan, Casler *et al* 2005; Richert & Barrett 2006; Bloom 2007.

37 Evans 2000 & 2001.

38 Kelemen 2004; Kelemen *et al, op cit*.

39 Nye 1998.

40 Hay 2007 (14).

———

41 Hay & Nye 2006.

42 Marshall 2005 (324).

43 Fingelkurts & Fingelkurts 2009.

44 In this connexion, see Moreira-Almeida & Cardeña 2011. They find that abnormal religious experiences are distinguished from psychotic illness by being 'related to higher levels of spirituality and better mental health, social adjustment, and well-being, rather than to pathology'. Such experiences do not lead to disturbance or suffering, are not associated with social or occupational impediments, involve insight, are compatible with religious traditions, lack psychiatric co-morbidity, and are experienced as leading to personal growth, self-integration and helping others.

45 Saver & Rabin 1997.

46 Hood, Hill & Spilka 2009.

47 Katz 1971.

to hand, and that what is most glorious in its transcendence is also that which is humblest in its wonderful immediacy, and that we know far more than we are usually aware of knowing, in large part because we labour to forget what is laid out before us in every moment, and because we spend so much of our lives wandering in dreams, in a deep but fitful sleep.[48]

This insight into the familiar was given its most famous expression by Blake: 'To see a world in a grain of sand, and a heaven in a wild flower; hold infinity in the palm of your hand, and eternity in an hour.' He it was who also saw in the sunrise not 'a round disc of fire, somewhat like a guinea', but 'an innumerable company of the heavenly host, crying, "Holy, holy, holy is the Lord God Almighty!"'[49]

What have been called somewhat infelicitously 'God encounter' experiences, either with or without the use of drugs, have been studied in a survey of 4,285 subjects:

> About 75 percent of respondents in both the non-drug and psychedelics groups rated their 'God encounter' experience as among the most meaningful and spiritually significant in their lifetime, and both groups attributed to it positive changes in life satisfaction, purpose and meaning.

More than two-thirds of those who said they were atheists before the experience no longer identified as such afterwards. Most participants, in both the non-drug and psychedelics groups, reported vivid memories of the encounter experience, which frequently involved communication with some entity having the attributes of consciousness (approximately 70%), benevolence (approximately 75%), intelligence (approximately 80%), sacredness (approximately 75%) and eternal existence (approximately 70%). Both groups reported a decreased fear of death (70% in the psychedelics group, 57% among non-drug respondents). In the non-drug group, participants were most likely to describe their encounter as with 'God' or 'an emissary of God' (59%), while the psychedelics group were most likely (55%) to describe an encounter with 'ultimate reality'.[50]

'Normal' people report their religious experience as *more* real than 'baseline' reality, even when the experience is later recollected from within that baseline reality.[51] At the same time, 'normal' individuals usually refer to dreams as *less* real than 'baseline' reality when they are recalled. The same is true for the hallucinations accompanying a psychotic illness – they are appraised as less real, once the episode of illness has resolved.[52]

The Upanishads refer to two kinds of knowledge, *apara vidya* (knowledge via the intellect and the senses, limited to a finite world) and *para vidya* (knowledge via the soul, which is not just subjective

---

48 Hart 2013 (84).

49 Blake, 'Auguries of Innocence'; and 'A Vision of the Last Judgment'.

50 Griffiths, Hurwitz, Davis et al 2019.

51 Newberg & d'Aquili 1994; Newberg, d'Aquili & Rause 2001; Newberg, Alavi, Baime et al 2001; and Newberg & Lee 2005.

52 Newberg & Lee *op cit*.

experience of concepts and emotions, but oneness with the infinite). *Apara vidya* presupposes a knower, a thing that is known, and the act of knowing. *Para vidya*, by contrast, is the attainment of knowledge through oneness with Brahman, the creative principle revealed in the whole cosmos.[53] It is thought to be the purpose of life to attain *para vidya*. It is relevant that in the Kabbalah, similarly, 'God, the cosmos, the human soul, and the act of knowledge' are all a single, unified essence or substance.[54]

So what about atheism: is that perhaps – as scientists, remember, we must consider everything – a kind of brainwashing inflicted on us by those in society who want to take control? It's hard to say, but I would think not – at least not in those terms. In a paper entitled 'The origins of religious disbelief', psychologists Ara Norenzayan and Will Gervais suggest four possible paths to atheism.[55] Because there seems to be a (presumably selected for) religious sense in most people, the existence of atheists demands to be explained. They write that 'one widely discussed view holds that disbelief, when it arises, results from significant cognitive effort against these powerful biases [towards religion]... atheism is possible, but requires some hard cognitive work to reject or override the intuitions that nourish religious beliefs'. They reject this view. 'We argue', they say,

> that atheism is more prevalent and enduring than would be expected if it was solely driven by effortful rejection of intuitive theism, that disbelief does not always require hard or explicit cognitive effort, and that rational deliberation is only one of several routes to disbelief.

Of the four paths they consider, two concern aspects of the society, and two concern the cognitive style of the individual. The societal aspects are lack of threat ('where life is safe and predictable, people are less motivated to turn to gods for succour ... even within the same society, religiosity declines over time as conditions become more secure');[56] and a tendency to conform to a culture, often scientific or 'Western liberal', in which belief is stigmatised as unintelligent. On the other hand, the personal cognitive aspects are a tendency to overvalue analytic thinking (more typically atheistic) and undervalue intuitive thinking (more typically theistic); and 'weak mentalizing abilities' (inability to see what the world looks like from another point of view). They conclude:

> So is atheism a 'hard sell', as many evolutionary and cognitive theorists of religion have argued? The answer, as is often the case when asking a complex question about a complex phenomenon, is that it depends ...

> ... on the individual's mental style and on the nature of the society to which they belong.

53 Uebersax 2013.

54 Drob 1997.

55 Norenzayan & Gervais 2013.

56 See eg, Norris & Inglehart 2004.

Since religion is not a matter of ratiocination, primarily – though not of unreason, as I suggested in Chapter 28 – those who find intuitive, emotional and social understanding more difficult are less likely to be religious.[57] The cognitive style with which we approach the world affects what we find there, and this is just as true of the religious sense as of any other. People with certain cognitive abnormalities, such as autistic spectrum conditions, will find religion to be 'largely inscrutable' and 'cognitively challenging'.[58] It turns out that autism-spectrum disorders make belief in God less likely (in a series of four studies, neurotypical ('normal') subjects were 10 times as likely as those on the spectrum strongly to endorse God).[59] According to another study, subjects on the autistic spectrum are more likely than neurotypical subjects to be atheists, and atheists are more likely to be on the autistic spectrum.[60] The elements of autistic cognitive style that most closely correlate with atheism are a preference for systematically logical beliefs, and a distrust of metaphor and figures of speech, along with weaker mentalising tendencies. (Theists with high-functioning autism tend to conceive of God as purely transcendent: Temple Grandin, for example, describes God, in an analogy of the complexity-consciousness theory, as the entanglement of millions of interacting particles.)[61] However I emphasise that this is a tendency only: none of this supports some such simple claim as that atheists 'are' autistic. Atheists are not a uniform group any more than theists, and both groups would, and should, resent being dismissed by association with the softest targets.

Research on atheists and personality characteristics found over 85% of the non-believers sampled to be broadly 'psychologically well-adjusted', and that personality measures have little correlation with varieties of non-belief. However there was one exception. The researchers distinguished six types of atheists, one of which they described as 'anti-theists' (about 14% of atheists). Such persons

> view religion as ignorance and see any individual or institution associated with it as backward and socially detrimental ... view the logical fallacies of religion as an outdated worldview that is not only detrimental to social cohesion and peace, but also to technological advancement and civilised evolution as a whole ... [and believe that] the obvious fallacies in religion and belief should be aggressively addressed in some form or another.[62]

This group of 'anti-theists' were atypical of atheists as a whole, scoring highest of the six groups on four of the personality scales – autonomy, anger, dogmatism and narcissism – and lowest on agreeableness.[63]

Stronger belief in God is linked to more intuitive thinking. This effect of intuitive thinking is not related to intelligence (IQ), educa-

---

[57] But this very incapacity can drive such people into the open arms of demonstratively welcoming religious communities.

[58] McCauley 2011 (253–4).

[59] Norenzayan, Gervais & Trzesniewski 2012.

[60] Caldwell-Harris, Murphy, Velazquez et al 2011.

[61] Bering 2002.

[62] Silver, Coleman, Hood et al 2014.

[63] ibid.

tional level, income, political orientation, or personality.[64] Intuitive thinking style is not correlated with a religious upbringing, but with a growing belief in God since childhood irrespective of upbringing, 'suggesting a causal relationship between cognitive style and change in belief over time.'[65] It may be relevant here that a disposition towards religious thinking is partly innate:[66] genetic variation contributes up to 50% of individual variation in religiosity.[67] Importantly, though, genes cannot determine the situation anywhere near completely.

There is evidence that on average the religious are less intelligent than non-believers.[68] But as we saw in Chapter 28, almost all scientific Nobel Prize winners seem to espouse not just the idea of God, in general terms, but a religion: how is that? There are two different questions here that tend to be conflated. One is horizontal: are people in the top half of the intelligence range more likely to be atheists than believers? The answer is 'no'. The other is vertical: are proportionately more atheists to be found in the top half of the intelligence range than the lower? The answer is 'yes'. These are, of course, entirely compatible findings.

4 · Religious belief and the health of individuals and societies

The evidence is that religion is a human universal – more so than science – and far from being a product of brainwashing. (Whether any given set of religious beliefs is true or not is of course untouched by this observation on its own.) As C. S. Peirce reflected, 'The human mind and the human heart have a filiation to God.'[69] In modern terms, we are often said to be 'hardwired' for God. Again this neither means that our brains make it all up *nor* proves that God exists, though usually our brains have developed faculties of perception in response to elements in the world that are there to be encountered: eg, we are 'hardwired' for sight. Whether or not that be the case, 'scientists who try to dismiss religion as "nothing but" an organic disposition of the evolved human brain', writes Rée, 'forget that the same could be said of the natural sciences.'[70]

That the religious, both communally and individually, are happier, and dramatically healthier, both mentally *and* physically, as well as better adjusted, more resilient and more prosocial in their habits, also does not prove that religion is true. But it suggests that we and our societies function poorly when we neglect it, and that human thriving and fulfilment depend on it to a considerable extent. The evidence on this deserves to be better known.

From a purely evolutionary point of view, we ought to find that religious experience contributes to the survival of those capable of experiencing it.[71] A study of 19th century utopian communities, both religious and secular, demonstrates that religious communities are far more likely to outlast their non-religious counterparts: 'religious

64 See Gervais & Norenzayan 2012; Pennycook, Cheyne, Seli *et al* 2012.

65 Shenhav, Rand & Greene 2012.

66 Waller, Kojetin, Bouchard *et al* 1990; D'Onofrio, Eaves, Murrelle *et al* 1999; Eaves 2004; Eaves, Martin & Heath 1990.

67 Bouchard, Lykken, McGue *et al* 1990.

68 Zuckerman, Silberman & Hall 2013.

69 Peirce 1931–60, vol 8 (262).

70 Rée 2019.

71 Joseph 2001.

communes are more likely than secular communes to survive at every stage of their life course' – four times as likely in any given year.[72]

Religion is a noted source of a sense of purpose. A study relating sense of purpose in life to longevity found that those with the lowest 'life purpose' had a two and a half times greater chance of death in the five-year period of the study than those with the highest. This finding survived adjustments for body mass index, level of physical activity, alcohol consumption, chronic illness, and even smoking status. The finding is in line with those from a number of previous studies in both East and West.[73]

Oxford's *Handbook for Religion and Health* is a thorough undertaking, which examined 3,300 quantitative studies published between 1872 and 2010, two-thirds of them since the year 2000, that address the relationship between health and well-being on the one hand and a religious or spiritual disposition on the other.[74] The religious or spiritual are markedly better at dealing with adversity, whether that be war, natural disaster, disease, bereavement or approaching death. They also enjoy much greater well-being: out of 326 quantitative, peer-reviewed studies, 256 (79%) found significant positive associations only; and only three studies (less than 1%) reported any significant inverse relationship between being religious or spiritual and well-being. These levels are extremely high by comparison with the norms in social psychology. Similar positive correlations with being religious or spiritual were found for a sense of meaning in life, levels of hope, levels of self-esteem, an internal locus of control (roughly a measure of taking responsibility), altruism, and feelings of forgiveness and gratitude. The meta-analysis found that being religious or spiritual has a protective effect against depression, anxiety, substance abuse and suicide.[75] It is correlated strongly with personality traits of conscientiousness, agreeableness, extraversion and openness to experience, and is inversely correlated with psychoticism and neuroticism.

This is in keeping with findings of improved health and happiness among the spiritually or religiously minded young. In their examination of the role of religion and spirituality in the lives of American teenagers, Smith and Denton write: 'the differences between more religious and less religious teenagers in the United States are actually significant and consistent across every outcome measure examined: risk behaviours, quality of family and adult relationships, moral reasoning and behaviour, community participation, media consumption, sexual activity, and emotional well-being'.[76]

In terms of social behaviour, perhaps most predictably, being religious or spiritual is strongly negatively associated with delinquency or crime,[77] alcohol and drug abuse[78] and marital instability; and positively associated with better social supports, and measures of

---

72 Sosis 2000; Sosis & Bressler 2003.

73 Alimujiang, Wiensch, Boss *et al* 2019. See also the meta-analysis of Cohen, Bavishi & Rozanski 2016; as well as Hill & Turiano 2014; Krause 2009; Boyle, Barnes, Buchman *et al* 2009; Gruenewald, Karlamangla, Greendale *et al* 2007; Sone, Nakaya, Ohmori *et al* 2008; Tanno, Sakata, Ohsawa *et al* 2009; Okamoto & Tanaka 2004; Nakanishi, Fukuda & Tatara 2003.

74 Koenig, King & Carson 2012.

75 See also van Praag 2009.

76 Smith & Denton 2005 (218–19).

77 Johnson, Li, Larson *et al* 2000.

78 See also Cook, Goddard & Westall 1997.

'social capital'. It is associated with lower levels of smoking, healthier levels of exercise, and better diet – though not lower weight, the only health measure to be negatively related. In terms of physical health it is associated with lower cholesterol levels; lower incidence of heart disease, hypertension and cerebrovascular disease; improved immune function and endocrine function; and better cognitive function in age.[79]

There are marked effects on longevity, confirmed by four meta-analyses, equivalent in size to the effects of taking cholesterol-lowering drugs, such as statins, or enrolling in exercise-based cardiac rehabilitation regimes.[80] The authors of one meta-analysis found that, 'notably, the protective effect [on mortality] of religiosity/spirituality in the initially healthy population studies was independent of behavioral factors' such as smoking, drinking, or exercising, and of socioeconomic status.[81]

It seems that many of these effects are mediated to a large extent by attendance at a place of worship, and thus attributable to belonging in a group of fellow-worshippers.[82] We know that social connectedness mediates extraordinary benefits in terms both of the social cohesion and well-being of a society, and in terms of individual mental and physical health, the subject of Robert Putnam's classic book, *Bowling Alone: The Collapse and Revival of American Community*.[83] However, regardless of the social aspects of religion, religious experience *per se*, including 'simply' the experience of awe,[84] has positive effects on happiness and prosocial emotions – thus bringing together, the right hemisphere's role in the experience of awe, in prosocial emotions (each discussed in Chapter 26), and in the experience of the divine (see Chapter 28).

While most studies are carried out in North America or Western Europe, similar results have been found for Islam;[85] Judaism;[86] Hinduism;[87] and Buddhism, Shintoism and Confucianism.[88] Finally the literature recognises a category of pathological religiosity associated with certain neuropsychiatric disorders.[89]

79 For this and the previous two paragraphs, see Koenig, King & Carson 2012.

80 Powell, Shahabi & Thoresen 2003; McCullough, Hoyt, Larson et al 2000; Chida, Steptoe & Powell 2009; Koenig 2012; McCullough 2001.

81 Chida et al, op cit.

82 In this connexion, see, eg, Orr, Tobin, Carey et al 2019.

83 Putnam 2000.

84 Sturm, Datta, Roy et al 2020.

85 Abu-Rayya, Abu-Rayya & Khalil 2009.

86 Rosmarin, Pirutinsky, Pargament et al 2009.

87 Tarakeshwar, Pargament & Mahoney 2003.

88 Graham & Crown 2014.

89 See eg Okruszek, Kalinowski & Talarowska 2013.

Bibliography

Aaronson BS, 'Mystic and schizophreniform states and the experience of depth', *Journal for the Scientific Study of Religion*, 1967, 6(2), 246–52

Aarts AA, Anderson JE, Anderson CJ et al, 'Estimating the reproducibility of psychological science', *Science*, 2015, 349(6251), aac4716

Abbie AA, 'The origin of the corpus callosum and the fate of the structures related to it', *Journal of Comparative Neurology*, 1939, 70(1), 9–44

Abbott CC, Jones T, Lemke NT et al, 'Hippocampal structural and functional changes associated with electroconvulsive therapy response', *Translational Psychiatry*, 2014, 4(11), e483

Abdullaev YG & Posner MI, 'Time course of activating brain areas in generating verbal associations', *Psychological Science*, 1997, 8(1), 56–9

Abe K, Oda N, Araki R et al, 'Macropsia, micropsia, and episodic illusions in Japanese adolescents', *Journal of the American Academy of Child & Adolescent Psychiatry*, 1989, 28(4), 493–6

Abe N, Ishii H, Fujii T et al, 'Selective impairment in the retrieval of family relationships in person identification: a case study of delusional misidentification', *Neuropsychologia*, 2007, 45(13), 2902–9

Abecasis D, Brochard R, Del Río D et al, 'Brain lateralization of metrical accenting in musicians', *Annals of the New York Academy of Sciences*, 2009, 1169(1), 74–8

Abele-Brehm A, 'Positive and negative mood influences on creativity: evidence for asymmetrical effects', *Polish Psychological Bulletin*, 1992, 23(3), 203–21

Aberg KC, Doell KC & Schwartz S, 'The "creative right brain" revisited: individual creativity and associative priming in the right hemisphere relate to hemispheric asymmetries in reward brain function', *Cerebral Cortex*, 2017, 27(10), 4946–59

Abert B & Ilsen PF, 'Palinopsia', *Optometry*, 2010, 81(8), 394–404

Aboitiz F, 'Brain connections: interhemispheric fiber systems and anatomical brain asymmetries in humans', *Biological Research*, 1992, 25(2), 51–61

———, 'Evolution of isocortical organization. A tentative scenario including roles of reelin, p35/cdk5 and the subplate zone', *Cerebral Cortex*, 1999, 9(7), 655–61

Aboitiz F & Ide A, 'Anatomical asymmetries in language-related cortex and their relation to callosal function', in E Stemmer & H Whitaker (eds), *Handbook of Neurolinguistics*, Academic Press, 1998, 393–404

Aboitiz F, López J & Montiel J, 'Long distance communication in the human brain: timing constraints for inter-hemispheric synchrony and the origin of brain lateralization', *Biological Research*, 2003, 36(1), 89–99

Aboitiz F & Montiel J, 'One hundred million years of interhemispheric communication: the history of the corpus callosum', *Brazilian Journal of Medical and Biological Research*, 2003, 36(4), 409–20

Aboitiz F, Morales D & Montiel J, 'The inverted neurogenetic gradient of the mammalian isocortex: development and evolution', *Brain Research Reviews*, 2001, 38(1–2), 129–39

Aboitiz F, Scheibel AB, Fisher RS et al, 'Fiber composition of the human corpus callosum', *Brain Research*, 1992, 598(1–2), 143–53

Abrahams S, Pickering A, Polkey CE et al, 'Spatial memory deficits in patients with unilateral damage to the right hippocampal formation', *Neuropsychologia*, 1997, 35(1), 11–24

Abram D, *Becoming Animal: An Earthly Cosmology*, Penguin Random House, 2011

Abramson CI & Chicas-Mosier AM, 'Learning in plants: lessons from *Mimosa pudica*', *Frontiers in Psychology*, 2016, 7, 417

Abu-Rayya HM, Abu-Rayya MH & Khalil M, 'The Multi-Religion Identity Measure: a new scale for use with diverse religions', *Journal of Muslim Mental Health*, 2009, 4(2), 124–38

Abu-Rustum RS, Ziade MF & Abu-Rustum SE, 'Reference values for the right and left fetal choroid plexus at 11 to 13 weeks: an early sign of "developmental" laterality?', *Journal of Ultrasound Medicine*, 2013, 32(9), 1623–39

Acuna BD, Eliassen JC, Donoghue JP et al, 'Frontal and parietal lobe activation during transitive inference in humans', *Cerebral Cortex*, 2002, 12(12), 1312–21

Adachi N, Nagayama M, Anami K et al, 'Asymmetrical blood flow in the temporal lobe in the Charles Bonnet syndrome: serial neuroimaging study', *Behavioural Neurology*, 1994, 7(2), 97–9

Adair JC, Na DL, Schwartz RL et al, 'Caloric stimulation in neglect: evaluation of response as a function of neglect type', *Journal of the International Neuropsychological Society*, 2003, 9(7), 983–8

Ades C & Ramires EN, 'Asymmetry of leg use during prey handling in the spider *Scytodes globula* (Scytodidæ)', *Journal of Insect Behavior*, 2002, 15(4), 563–70

Adler G & Jaffé A (eds), *C. G. Jung Letters*, trans RFC Hull, Princeton University Press, 1973

Adler J & Tso W-W, '"Decision-making" in bacteria: chemotactic response of *Escherichia coli* to conflicting stimuli', *Science*, 1974, 184(4143), 1292–4

Adolphs R, 'The neurobiology of social cognition', *Current Opinion in Neurobiology*, 2001, 11(2), 231–9

Adolphs R, Baron-Cohen S & Tranel D, 'Impaired recognition of social emotions following amygdala damage', *Journal of Cognitive Neuroscience*, 2002, 14(8), 1264–74

Adolphs R, Damasio H, Tranel D et al, 'Cortical systems for the recognition of emotion in facial expressions', *Journal of Neuroscience*, 1996, 16(23), 7678–87

Adolphs R, Tranel D, Bechara A et al, 'Neuropsychological approaches to reasoning and decision-making', in AR Damasio, H Damasio & Y Christen (eds), *Neurobiology of Decision-Making*, Springer-Verlag, 1996, 157–79

Adorno T, *Minima Moralia*, Suhrkamp Verlag, 1944

Adunsky A, 'Early post-stroke parasitic delusions', *Age & Ageing*, 1997, 26(3), 238–9

Aerts D, Gabora L & Sozzo S, 'Concepts and their dynamics: a quantum-theoretic modeling of human thought', *Topics in Cognitive Sciences*, 2013, 5(4), 1–36

Aeschliman MD, *The Restitution of Man: C. S. Lewis and the Case Against Scientism*, Wm B Eerdmans, 1983

Aglioti S, Smania N, Manfredi M et al, 'Disownership of left hand and objects related to it in a patient with right brain damage', *NeuroReport*, 1996, 8(1), 293–6

Aguilar C, Vlamakis H, Losick R et al, 'Thinking about *Bacillus subtilis* as a multicellular organism', *Current Opinion in Microbiology*, 2007, 10(6), 638–43

Agus JB, *Modern Philosophies of Judaism*, Behrman, 1941, 184–94

Ahern GL, Herring AM, Tackenberg JN et al, 'Affective self-report during the intracarotid sodium amobarbital test', *Journal of Clinical and Experimental Neuropsychology*, 1994, 16(3), 372–6

Ahern GL & Schwartz GE, 'Differential lateralization for positive and negative emotion in the human brain: EEG spectral analysis', *Neuropsychologia*, 1985, 23(6), 745–55

Ahn DH, Lee YJ, Jeong JH et al, 'The effect of post-stroke depression on rehabilitation outcome and the impact of caregiver type as a factor of post-stroke depression', *Annals of Rehabilitation Medicine*, 2015, 39(1), 74–80

Ahrens K, Liu HL, Lee CY et al, 'Functional MRI of conventional and anomalous metaphors in Mandarin Chinese', *Brain and Language*, 2007, 100(2), 163–71

Aitchison IJR, 'Nothing's plenty: the vacuum in modern quantum field theory', *Contemporary Physics*, 1985, 26(4), 333–91

Akıncı E, Öncü F & Topçular B, 'Tactile hallucination and delusion following acute stroke: a case report', *Düşünen Adam: The Journal of Psychiatry and Neurological Sciences*, 2016, 29(1), 79–84

Akinola M & Mendes WB, 'The dark side of creativity: biological vulnerability and negative emotions lead to greater artistic creativity', *Personality and Social Psychology Bulletin*, 2008, 34(12), 1677–86

Al-Biruni, *The Book of Instruction in the Elements of Astrology*, trans RR Wright, Luzac, 1934

Alajouanine T, 'Aphasia and artistic realization', *Brain*, 1948, 71(3), 229–41

Alberti LB, *On Painting*, trans R Sinisgalli, Cambridge University Press, 2013

Alberts B, 'Impact factor distortions', *Science*, 2013, 340(6134), 787

Alberts B, Johnson A, Lewis J et al, *Molecular Biology of the Cell*, Garland Science, 5th edn, 2008

Albrecht A & Phillips D, 'Origin of probabilities and their application to the multiverse', *Physical Review D*, 2014, 90(12–15), 123514

Albright L, Kenny DA & Malloy TE, 'Consensus in personality judgments at zero acquaintance', *Journal of Personality and Social Psychology*, 1988, 55(3), 387–95

Alcantara F & Monk M, 'Signal propagation during aggregation in the slime mould *Dictyostelium discoideum*', *Journal of General Microbiology*, 1974, 85(2), 321–34

Alcock KJ, Wade D, Anslow P et al, 'Pitch and timing abilities in adult left-hemisphere-dysphasic and right-hemisphere damaged subjects', *Brain and Language*, 2000, 75(1), 47–65

Alderman MH, 'Reducing dietary sodium: the case for caution', *Journal of the American Medical Association*, 2010, 303(5), 448–9

Alderman MH & Cohen HW, 'Dietary sodium intake and cardiovascular mortality: controversy resolved?', *American Journal of Hypertension*, 2012, 25(7), 727–34

Aldous CR, 'Creativity, problem-solving and innovative science: insights from history, cognitive psychology and neuroscience',

International Education Journal, 2007, 8(2), 176–86

Aldous DE, 'Social, environmental, economic, and health benefits of green spaces', *Acta Horticulturae*, 2007, 762, 171–85

Aleman A, Sommer IE & Kahn RS, 'Efficacy of slow repetitive transcranial magnetic stimulation in the treatment of resistant auditory hallucinations in schizophrenia: a meta-analysis', *Journal of Clinical Psychiatry*, 2007, 68(3), 416–21

Alexander AL, Lee JE, Lazar M et al, 'Diffusion tensor imaging of the corpus callosum in autism', *NeuroImage*, 2007, 34(1), 61–73

Alexander C & Huggins AWF, 'On changing the way people see', *Perceptual & Motor Skills*, 1964, 19(1), 235–53

Alexander MP, Stuss DT & Benson DF, 'Capgras syndrome: a reduplicative phenomenon', *Neurology*, 1979, 29(3), 334–9

Alexander TM, 'The voices of nature: toward a polyphonic conception of philosophy', *Telos*, 2016, 177, 145–67

Alimujiang A, Wiensch A, Boss J et al, 'Association between life purpose and mortality among US adults older than 50 years', *JAMA Network Open*, 2019 May 3, 2(5), e194270

Allé MC, Gandolphe MC, Doba K et al, 'Grasping the mechanisms of narratives' incoherence in schizophrenia: an analysis of the temporal structure of patients' life story', *Comprehensive Psychiatry*, 2016, 69(1), 20–29

Allen FD, 'Etymological and grammatical notes', *American Journal of Philology*, 1880, 1(2), 127–45

Allen JJ, Coan JA & Nazarian M, 'Issues and assumptions on the road from raw signals to metrics of frontal EEG asymmetry in emotion', *Biological Psychology*, 2004, 67(1–2), 183–218

Allison JD, Meador KJ, Loring DW et al, 'Functional MRI cerebral activation and deactivation during finger movement', *Neurology*, 2000, 54(1), 135–42

Allison T, Puce A & McCarthy G, 'Social perception from visual cues: role of the STS region', *Trends in Cognitive Sciences*, 2000, 4(7), 267–78

Allman JM, Tetreault NA, Hakeem AY et al, 'The von Economo neurons in frontoinsular and anterior cingulate cortex in great apes and humans', *Brain Structure & Function*, 2010, 214(5–6), 495–517

———, 'The von Economo neurons in fronto-insular and anterior cingulate cortex', *Annals of the New York Academy of Sciences*, 2011, 1225, 59–71

Allman MJ, Teki S, Griffiths TD et al, 'Properties of the internal clock: first- and second-order principles of subjective time', *Annual Review of Psychology*, 2014, 65, 743–71

Alloy LB & Abramson LY, 'Judgement of contingency in depressed and nondepressed students: sadder but wiser?', *Journal of Experimental Psychology: General*, 1979, 108(4), 441–85

———, 'Depressive realism: four theoretical perspectives', in LB Alloy (ed), *Cognitive Processes in Depression*, Guilford Press, 1988, 223–65

Ally BA, Hussey EP & Donahue MJ, 'A case of hyperthymesia: rethinking the role of the amygdala in autobiographical memory', *Neurocase*, 2013, 19(2), 166–81

Almeida LG, Ricardo-Garcell J, Prado H et al, 'Reduced right frontal cortical thickness in children, adolescents and adults with ADHD and its correlation to clinical variables: a cross-sectional study', *Journal of Psychiatric Research*, 2010, 44(16), 1214–23

Alonso-Alonso M & Pascual-Leone A, 'The right brain hypothesis for obesity', *Journal of the American Medical Association*, 2007, 297(16), 1819–22

Alpherts WCJ, Vermeulen J, Franken MLO et al, 'Lateralization of auditory rhythm length in temporal lobe lesions', *Brain and Cognition*, 2002, 49(1), 114–22

Alvarez P, Puente VM, Blasco MJ et al, 'Concurrent Koro and Cotard syndromes in a Spanish male patient with a psychotic depression and cerebrovascular disease', *Psychopathology*, 2012, 45(2), 126–9

Ambady N & Rosenthal R, 'Thin slices of expressive behavior as predictors of interpersonal consequences: a meta-analysis', *Psychological Bulletin*, 1992, 111(2), 256–74

Ambard L & Beaujard E, 'Causes de l'hypertension arterielle', *Archives générales de médecine*, 1904, série 10, vol 3, 520–33

Ambrosetto G, 'Post-ictal gustatory hallucination, sleep related microspikes and glioma of the sylvian region: report of a case', *Clinical Electroencephalography*, 1986, 17(2), 89–91

Amminger GP, Schlögelhofer M, Lehner T et al, 'Premorbid performance IQ deficit in schizophrenia', *Acta Psychiatrica Scandinavica*, 2000, 102(6), 414–22

Amodio DM, 'The neuroscience of prejudice and stereotyping', *Nature Reviews Neuroscience*, 2014, 15(10), 670–82

Amodio DM, Devine PG & Harmon-Jones E, 'A dynamic model of guilt: implications for motivation and self-regulation in the context of prejudice', *Psychological Science*, 2007, 18(6), 524–30

Amunts K, 'Structural indices of asymmetry', in K Hugdahl & R Westerhausen R (eds), *The*

Two Halves of the Brain, Massachusetts Institute of Technology Press, 2010, 145–76

Amunts K, Jäncke L, Mohlberg H et al, 'Interhemispheric asymmetry of the human motor cortex related to handedness and gender', *Neuropsychologia*, 2000, 38(3), 304–12

Amunts K, Schleicher A, Bürgel U et al, 'Broca's region revisited: cytoarchitecture and inter-subject variability', *Journal of Comparative Neurology*, 1999, 412(2), 319–41

Anagnostopoulos A, Spiegel R, Palmer J et al, 'A left-hand superiority for the implicit detection of a rule', *Cortex*, 2013, 49(2), 582–90

Anaki D, Faust M & Kravetz S, 'Cerebral hemisphere asymmetries in processing lexical metaphors' (1), *Neuropsychologia*, 1998a, 36(4), 353–62

——, 'Cerebral hemisphere asymmetries in processing lexical metaphors' (2), *Neuropsychologia*, 1998b, 36(7), 691–700

Anand KJ & Hickey PR, 'Pain and its effects in the human neonate and fetus', *New England Journal of Medicine*, 1987, 317(21), 1321–9

Ananthaswamy A, 'New quantum paradox clarifies where our views of reality go wrong', 2018: www.quantamagazine.org/frauchiger-renner-paradox-clarifies-where-our-views-of-reality-go-wrong-20181203/ (accessed 23 February 2021)

Anastasopoulos G, 'Zur Frage des pathologischen Erlebens veränderter Körperzustände', *Nervenarzt*, 1954, 25(12), 492–500

Andersch N, 'Gestalt und Gestaltverlust in der Schizophrenie: zur Bedeutung stabilisierender Symbolbildung für unser Bewusstsein und seine Störungen', *Gestalt Theory*, 2016, 38(2–3), 279–96

Andersen BB, Korbo L & Pakkenberg B, 'A quantitative study of the human cerebellum with unbiased stereological techniques', *Journal of Comparative Neurology*, 1992, 326(4), 549–60

Andersen SB, Gerritsma S, Yusah KM et al, 'The life of a dead ant: the expression of an adaptive extended phenotype', *The American Naturalist*, 2009, 174(3), 424–33

Anderson AK, Christoff K, Stappen I et al, 'Dissociated neural representations of intensity and valence in human olfaction', *Nature Neuroscience*, 2003, 6(2), 196–202

Anderson CA, Camp J & Filley CM, 'Erotomania after aneurysmal subarachnoid hemorrhage: case report and literature review', *Journal of Neuropsychiatry and Clinical Neurosciences*, 1998, 10(3), 330–7

Anderson CJ, Bahník Š, Barnett-Cowan M et al, 'Response to Comment on "Estimating the reproducibility of psychological science"', *Science*, 2016, 351(6277), 1037

Anderson PW, 'More is different', *Science*, 1972, 177(4047), 393–96

Anderson SW, Barrash J & Bechara A, 'Impairments of emotion and real-world complex behavior following childhood- or adult-onset damage to ventromedial prefrontal cortex', *Journal of the International Neuropsychological Society*, 2006, 12(2), 224–35

Anderson SW & Rizzo M, 'Hallucinations following occipital lobe damage: the pathological activation of visual representations', *Journal of Clinical and Experimental Neuropsychology*, 1994, 16(5), 651–63

Andino SL, Menendez RG, Khateb A et al, 'Electrophysiological correlates of affective blindsight', *NeuroImage*, 2009, 44(2), 581–9

Ando V, Claridge G & Clark K, 'Psychotic traits in comedians', *British Journal of Psychiatry*, 2014, 204(5), 341–5

Andreasen NC, 'Creativity and mental illness: prevalence rates in writers and their first-degree relatives', *American Journal of Psychiatry*, 1987, 144(10), 1288–92

——, *The Creating Brain: The Neuroscience of Genius*, Dana Press, 2005

——, 'The relationship between creativity and mood disorders', *Dialogues in Clinical Neuroscience*, 2008, 10(2), 251–5

Andreou C, Faber PL, Leicht G et al, 'Resting-state connectivity in the prodromal phase of schizophrenia: insights from EEG microstates', *Schizophrenia Research*, 2014, 152(2–3), 513–20

Andreou C, Nolte G, Leicht G et al, 'Increased resting-state gamma-band connectivity in first-episode schizophrenia', *Schizophrenia Bulletin*, 2014, 41(4), 930–9

Angyal A, 'The experience of the body-self in schizophrenia', *Archives of Neurology and Psychiatry*, 1936, 35(5), 1029–59

Anjum RL & Mumford S, 'Dispositionalism: a dynamic theory of causation', in DJ Nicholson & J Dupré, *Everything Flows: Towards a Processual Philosophy of Biology*, Oxford University Press, 2018, 61–75

Annesley T, Scott M, Bastian H et al, 'Biomedical journals and preprint services: friends or foes?', *Clinical Chemistry*, 2017, 63(2), 453–8

Annett M, *Handedness and Brain Asymmetry: The Right Shift Theory*, Psychology Press, 2002

Annoni JM, Devuyst G, Carota A et al, 'Changes in artistic style after minor posterior stroke', *Journal of Neurology, Neurosurgery, & Psychiatry*, 2005, 76(6), 797–803

Antonova E, Amaratunga K, Wright B et al, 'Schizotypy and mindfulness: magical thinking without suspiciousness characterizes mindfulness meditators', *Schizophrenia Research: Cognition*, 2016, 5, 1–6

Anzellotti F, Onofrj V, Maruotti V et al, 'Autoscopic phenomena: case report and review of literature', *Behavioral and Brain Functions*, 2011, 7(1), 2

Aoki K, Hiroki M, Bando M et al, 'A case of agnosia for streets without visual memory disturbance', *Rinsho Shinkeigaku*, 2003, 43(6), 335–40

Appel K, Haken W & Koch J, 'Every planar map is four colorable. Part II: reducibility', *Illinois Journal of Mathematics*, 1977, 21(3), 491–567

Arai T, Hasegawa Y, Tanaka T et al, 'Transient Charles Bonnet syndrome after excision of a right occipital meningioma: a case report', *No Shinkei Geka*, 2014, 42(5), 445–51

Arai T, Irie K, Akiyama M et al, 'A case of falcotentorial meningioma with visual allesthesia', *Brain and Nerve*, 2002, 54(3), 255–9

Arden R, Chavez RS, Grazioplene R et al, 'Neuroimaging creativity: a psychometric review', *Behavioral Brain Research*, 2010, 214(2), 143–56

Ardila A & Rosseli M, 'Temporal lobe involvement in Capgras syndrome', *International Journal of Neuroscience*, 1988, 43, 219–22

Arendt H, *The Life of the Mind*: vol I: *Thinking*, Harcourt Brace Jovanovich, 1971

Arguin M, Lassonde M, Quattrini A et al, 'Divided visuo-spatial attention systems with total and anterior callosotomy', *Neuropsychologia*, 2000, 38(3), 283–91

Argyriou P, Byfield S & Kita S, 'Semantics is crucial for the right-hemisphere involvement in metaphor processing: evidence from mouth asymmetry during speaking', *Laterality*, 2015, 20(2), 191–210

Ariely D, Gneezy U & Loewenstein G, 'Large stakes and big mistakes', *Review of Economic Studies*, 2009, 76(2), 451–69

Arieti S, 'The microgeny of thought and perception', *Archives of General Psychiatry*, 1962, 6(6), 454–68

Ariew A & Lewontin RC, 'The confusions of fitness', *British Journal for the Philosophy of Science*, 2004, 55(2), 347–63

Armstrong K, *Buddha*, Penguin, 2004

——, *The Case for God: What Religion Really Means*, Bodley Head, 2009

Arnellos A, 'From organizations of processes to organisms and other biological individuals', in DJ Nicholson & J Dupré (eds), *Everything Flows: Towards a Processual Philosophy of Biology*, Oxford University Press, 2018, 199–223

Arnone D, McIntosh AM, Tan GMY et al, 'Meta-analysis of magnetic resonance imaging studies of the corpus callosum in schizophrenia', *Schizophrenia Research*, 2008, 101(1–3), 124–32

Aron AR, 'The neural basis of inhibition in cognitive control', *Neuroscientist*, 2007, 13(3), 214–8

Aron AR, Robbins TW & Poldrack RA, 'Inhibition and the right inferior frontal cortex', *Trends in Cognitive Sciences*, 2004, 8(4), 170–7

——, 'Inhibition and the right inferior frontal cortex: one decade on', *Trends in Cognitive Sciences*, 2014, 18(4), 177–85

Arrington CM, Carr TH, Mayer AR et al, 'Neural mechanisms of visual attention: object-based selection of a region in space', *Journal of Cognitive Neuroscience*, 2000, 12(suppl 2), 106–17

Arseni C & Dănăilă L, 'Logorrhoea syndrome with hyperkinesia', *European Neurology*, 1977, 15(4), 183–7

Arthur WB, 'Economics in nouns and verbs', preprint at: arXiv arxiv.org/abs/2104.01868, 2021

Arzy S, Mohr C, Michel CM et al, 'Duration and not strength of activation in temporoparietal cortex positively correlates with schizotypy', *NeuroImage*, 2007, 35(1), 326–33

Arzy S, Overney LS, Landis T et al, 'Neural mechanisms of embodiment: asomatognosia due to premotor cortex damage', *Archives of Neurology*, 2006, 63(7), 1022–5

Asari T, Konishi S, Jimura K et al, 'Right temporopolar activation associated with unique perception: an event-related analysis', *NeuroImage*, 2008, 41(1), 145–52

Asarnow RF & MacCrimmon DJ, 'Span of apprehension deficits during the postpsychotic stages of schizophrenia: a replication and extension', *Archives of General Psychiatry*, 1981, 38(9), 1006–11

Asch SE, 'Effects of group pressure upon the modification and distortion of judgments', in H Guetzkow (ed), *Groups, Leadership and Men: Research in Human Relations*, Carnegie Press, 1951, 177–90

——, 'Studies of independence and conformity: I. A minority of one against a unanimous majority', *Psychological Monographs: General and Applied*, 1956, 70(9), 1–70

Ashton R & McFarland K, 'A simple dual-task study of laterality, sex differences and handedness', *Cortex*, 1991, 27(1), 105–9

Asimov I, 'How do people get new ideas?', *Massachusetts Institute of Technology Review*, 20 October 2014 [1959]

Asperger H, 'Die "autistischen Psychopathen" im Kindesalter', *Archiv für Psychiatrie und Nervenkrankheiten*, 1944, 117, 76–136

Asri F, Tazi I, Maaroufi K et al, 'Cerebral hydatic cyst and psychiatric disorders: two cases', *L'Encéphale*, 2007, 33(2), 216–9

Assadi M, Baseman S & Hyman D, 'Tc SPECT scan in a patient with occipital lobe infarction and complex visual hallucinations', *Journal of Neuroscience Nursing*, 2003, 35(3), 175–7

Asthana HS & Mandal MK, 'Visual-field bias in the judgment of facial expression of emotion', *Journal of Genetic Psychology*, 2001, 128(1), 21–9

At A, Spierer L & Clarke S, 'The role of the right parietal cortex in sound localization: a chronometric single pulse transcranial magnetic stimulation study', *Neuropsychologia*, 2011, 49(9), 2794–7

Atchley RA, Keeney M & Burgess C, 'Cerebral hemispheric mechanisms linking ambiguous word meaning retrieval and creativity', *Brain and Cognition*, 1999, 40(3), 479–99

Athyros VG, Liberopoulos EN, Mikhailidis DP et al, 'Association of drinking pattern and alcohol beverage type with the prevalence of metabolic syndrome, diabetes, coronary heart disease, stroke, and peripheral arterial disease in a Mediterranean cohort', *Angiology*, 2007, 58(6), 689–97

Atkinson J & Egeth H, 'Right hemisphere superiority in visual orientation matching', *Canadian Journal of Psychology*, 1973, 27, 152–8

Atmanspacher H & Primas H, 'Pauli's ideas on mind and matter in the context of contemporary science', *Journal of Consciousness Studies*, 2006, 13(3), 5–50

Attwood T, *The Complete Guide to Asperger's Syndrome*, Jessica Kingsley, 2006

Augustin J, Guegan-Massardier E, Levillain D et al, 'Musical hallucinosis following infarction of the right middle cerebral artery', *Revue Neurologique (Paris)*, 2001, 157(3), 289–92

Auspitz JL, 'The greatest living American philosopher', *Commentary*, 1983, 76(6), 51–64

Austin G, Hayward W & Rouhe S, 'A note on the problem of conscious man and cerebral disconnection by hemispherectomy', in M Kinsbourne & A Smith (eds), *Hemisphere Disconnection and Cerebral Function*, Charles C Thomas, 1974

Aviv R, 'Which way madness lies', *Harper's Magazine*, December 2010, 35–46

Azevedo FA, Carvalho LR, Grinberg LT et al, 'Equal numbers of neuronal and nonneuronal cells make the human brain an isometrically scaled-up primate brain', *Journal of Comparative Neurology*, 2009, 513(5), 532–41

Aziz-Zadeh L & Damasio A, 'Embodied semantics for actions: findings from functional brain imaging', *Journal of Physiology (Paris)*, 2008, 102(1), 35–9

Aziz-Zadeh L, Kaplan JT & Iacoboni M, '"Aha!": the neural correlates of verbal insight solutions', *Human Brain Mapping*, 2009, 30(3), 908–16

Aziz-Zadeh L, Koski L, Zaidel E et al, 'Lateralization of the human mirror neuron system', *Journal of Neuroscience*, 2006, 26(11), 2964–70

Azoulay P, Fons-Rosen C & Graff Zivin JS, 'Does science advance one funeral at a time?', *National Bureau of Economic Research Working Paper no. 21788*, December 2015

Azriel, 'The explanation of the ten Sefirot', in J Dan (trans RC Kieber), *The Early Kabbalah*, Paulist Press, 1966

Baars BJ, Ramsøy TZ & Laureys S, 'Brain, conscious experience and the observing self', *Trends in Neurosciences*, 2003, 26(12), 671–5

Babcock L & Vallesi A, 'The interaction of process and domain in prefrontal cortex during inductive reasoning', *Neuropsychologia*, 2015, 67, 91–9

Babcock LE & Robison RA, 'Preferences of palaeozoic predators', *Nature*, 1989, 337(6209), 695–6

Babinski J, 'Contribution à l'étude des troubles mentaux dans l'hémiplégie organique cérébrale (anosognosie)', *Revue Neurologique (Paris)*, 1914, 27, 845–7

Baciu M, Koenig O, Vernier MP et al, 'Categorical and coordinate spatial relations: fMRI evidence for hemispheric specialization', *NeuroReport*, 1999, 10(6), 1373–8

Bacon, Sir Francis, *The Advancement of Learning*, Bk II [1605], in *The Works*, trans and ed J Spedding, RL Ellis & DD Heath, Longman, vol 3, 1857, 321–491

——, *Novum Organon* [1620], in *The Works*, trans & ed J Spedding, RL Ellis & DD Heath, Longman, vol 4, 1858a, 39–248

——, *Parasceve ad Historiam Naturalem et Experimentalem* [1620], in *The Works*, trans & ed J Spedding, RL Ellis, DD Heath, Longman, vol 4, 1858b, 249–62

——, *Essays*, Macmillan & Co, 1902 [1597]

Badalian LO, Temin PA, Mukhin KI et al, 'Temporal-lobe epilepsy with psychosensory and gustatory attacks', *Zhurnal Nevrologii i Psikhiatrii Imeni S S Korsakova*, 1993, 93(1), 17–9

Badaway A, *Ancient Egyptian Architectural Design: A Study of the Harmonic System*, University of California Press, 1965

Badzakova-Trajkov G, Häberling IS & Corballis MC, 'Magical ideation, creativity, handedness, and cerebral asymmetries: a combined behavioural and fMRI study', *Neuropsychologia*, 2011, 49(10), 2896–903

Bagesteiro LB & Sainburg RL, 'Nondominant arm advantages in load compensation during rapid elbow joint movements', *Journal of Neurophysiology*, 2003, 90(3), 1503–13

Baier B & Karnath H-O, 'Tight link between our sense of limb ownership and self-awareness of actions', *Stroke*, 2008, 39(2), 486–8

Baillet A, *La Vie de Monsieur Descartes*, Part 1, Daniel Horthemels, 1691

Baird B, Smallwood J, Mrazek MD et al, 'Inspired by distraction: mind wandering facilitates creative incubation', *Psychological Science*, 2012, 23(10), 1117–22

Bak TH, 'Movement disorders: why movement and cognition belong together', *Nature Reviews Neurology*, 2010, 7(1), 10–12

——— TH, 'The neuroscience of action semantics in neurodegenerative brain diseases', *Current Opinion in Neurology*, 2013, 26(6), 671–7

Bak TH & Chandran S, 'What wires together dies together: verbs, actions and neurodegeneration in motor neuron disease', *Cortex*, 2012, 48(7), 936–44

Bak TH, O'Donovan DG, Xuereb JH et al, 'Selective impairment of verb processing associated with pathological changes in the Brodmann areas 44 and 45 in the motor neurone disease/dementia/aphasia syndrome', *Brain*, 2001, 124(1), 103–20

Bak TH, Yancopoulou D, Nestor PJ et al, 'Clinical, imaging and pathological correlates of a hereditary deficit in verb and action processing', *Brain*, 2006, 129(2), 321–32

Baker GP & Hacker PMS, *Wittgenstein: Understanding & Meaning*, vol 1 of *An Analytical Commentary on the Philosophical Investigations, Part II – Exegesis §§1–184*, Blackwell, 2nd edn, 2005

Baker M, 'Is there a reproducibility crisis?', *Nature*, 2016, 533(7604), 452–4

Baker SG, 'A cancer theory kerfuffle can lead to new lines of research,' *Journal of the National Cancer Institute*, 2015, 107(2), dju405

Baldo JV, Kacinik NA, Moncrief A et al, 'You may now kiss the bride: interpretation of social situations by individuals with right or left hemisphere injury', *Neuropsychologia*, 2016, 80, 133–41

Baliunas DO, Taylor BJ, Irving H et al, 'Alcohol as a risk factor for type 2 diabetes: a systematic review and meta-analysis', *Diabetes Care*, 2009, 32(11), 2123–32

Balken ER, 'A delineation of schizophrenic language and thought in a test of imagination', *Journal of Psychology*, 1943, 16(2), 239–71

Ball P, H_2O: *A Biography of Water*, Phoenix, 2000

———, 'A metaphor too far', *Nature*, 23 February 2011: www.nature.com/news/2011/110223/full/news.2011.115.html (accessed 23 February 2021)

———, 'Why the many-worlds interpretation has many problems', *Quanta*, 18 October 2018: www.quantamagazine.org/why-the-many-worlds-interpretation-of-quantum-mechanics-has-many-problems-20181018/ (accessed 23 February 2021)

Balslev AN, 'On abuse of time-metaphors', in S Albeverio & P Blanchard (eds), *Direction of Time*, Springer, 2014, 75–84

Baluška F, Mancuso S, Volkmann D et al, 'The "root-brain" hypothesis of Charles and Francis Darwin: revival after more than 125 years', *Plant Signaling and Behavior*, 2009, 4(12), 1121–7

Bambini V, Gentili C, Ricciardi E et al, 'Decomposing metaphor processing at the cognitive and neural level through functional magnetic resonance imaging', *Brain Research Bulletin*, 2011, 86(3–4), 203–16

Banich MT & Belger A, 'Interhemispheric interaction: how do the hemispheres divide and conquer a task?', *Cortex*, 1990, 26(1), 77–94

Banich MT & Brown WS, 'A life-span perspective on interaction between the cerebral hemispheres', *Developmental Neuropsychology*, 2000, 18(1), 1–10

Banich MT & Depue BE, 'Recent advances in understanding neural systems that support inhibitory control', *Current Opinion in Behavioral Sciences*, 2015, 1(1), 17–22

Banich MT & Federmeier KD, 'Categorical and metric spatial processes distinguished by task demands and practice', *Journal of Cognitive Neuroscience*, 1999, 11(2), 153–66

Banney RM, Harper-Hill K & Arnott WL, 'The Autism Diagnostic Observation Schedule and narrative assessment: evidence for specific narrative impairments in autism spectrum disorders', *International Journal of Speech-Language Pathology*, 2015, 17(2), 159–71

Barbaric I, Miller G & Dear TN, 'Appearances can be deceiving: phenotypes of knockout mice', *Briefings in Functional Genomics and Proteomics*, 2007, 6(2), 91–103

Barber B, 'Resistance by scientists to scientific discovery', *Science*, 1961, 134(3479), 596–602

Barbey AK, Colom R, Paul EJ et al, 'Architecture of fluid intelligence and working memory revealed by lesion mapping', *Brain Structure and Function*, 2014, 219(2), 485–94

Barbey AK & Patterson R, 'Architecture of explanatory inference in the human prefrontal cortex', *Frontiers in Psychology*, 2011, 2, article 162

Barboza RB, De Freitas GR, Tovar-Moll F et al, 'Delayed-onset post-stroke delusional disorder: a case report', *Behavioural Neurology*, 2013, 27(3), 287–91

Bareham CA, Bekinschtein TA, Scott SK et al, 'Does left-handedness confer resistance to spatial bias?', *Scientific Reports*, 2015, 5, article 9162

Bargh JA, 'Reply to the commentaries', in RS Wyer (ed), *The Automaticity of Everyday Life*, Erlbaum, 1997, 231–46

Barkley CL & Gabriel KI, 'Sex differences in cue perception in a visual scene: investigation of cue type', *Behavioral Neuroscience*, 2007, 121(2), 291–300

Barnea-Goraly N, Kwon H, Menon V et al, 'White matter structure in autism: preliminary evidence from diffusion tensor imaging', *Biological Psychiatry*, 2004, 55(3), 323–6

Barnes J, *Early Greek Philosophy*, Penguin, 1987

Barnes LA, 'The fine-tuning of the universe for intelligent life', *Publications of the Astronomical Society of Australia*, 2011, 29(4), 529–64

Barnes LL, De Leon CFM, Bienias JL et al, 'Hostility and change in cognitive function over time in older blacks and whites', *Psychosomatic Medicine*, 2009, 71(6), 652–8

Barnett KJ, 'Colour knowledge: the role of the right hemisphere in colour processing and object colour knowledge', *Laterality*, 2008, 13(5), 456–67

Barnett KJ & Corballis MC, 'Speeded right-to-left information transfer: the result of speeded transmission in right-hemisphere axons?', *Neuroscience Letters*, 2005, 380(1–2), 88–92

Barnett KJ, Corballis MC & Kirk IJ, 'Symmetry of callosal information transfer in schizophrenia: a preliminary study', *Schizophrenia Research*, 2005, 74(2–3), 171–8

Baron-Cohen S, Cox A, Baird G et al, 'Psychological markers in the detection of autism in infancy in a large population', *British Journal of Psychiatry*, 1996, 168(2), 158–63

Baron-Cohen S, Jaffa T, Davies S et al, 'Do girls with anorexia nervosa have elevated autistic traits?', *Molecular Autism*, 2013, 4(1), article 24

Baron-Cohen S, Leslie AM & Frith U, 'Does the autistic child have a "theory of mind"?', *Cognition*, 1985, 21(1), 37–46

Baron-Cohen S, Ring H, Moriarty J et al, 'Recognition of mental state terms: clinical findings in children with autism and a functional neuroimaging study of normal adults', *British Journal of Psychiatry*, 1994, 165(5), 640–9

Baron-Cohen S, Ring HA, Wheelwright S et al, 'Social intelligence in the normal and autistic brain: an fMRI study', *European Journal of Neuroscience*, 1999, 11(6), 1891–8

Barr M, Hamm J, Kirk I et al, 'Early visual evoked potentials in callosal agenesis', *Neuropsychology*, 2005, 19(6), 707–27

Barré M, Hamelin S, Minotti L et al, 'Aura visuelle migraineuse et crise épileptique : la migralepsie revisitée', *Revue Neurologique (Paris)*, 2008, 164(3), 246–52

Barrett D, 'Answers in your dreams', *Scientific American Mind*, 2011, 22(5), 33–9

Barrett JL & Richert RA, 'Anthropomorphism or preparedness? Exploring children's God concepts', *Review of Religious Research*, 2003, 44(3), 300–12

Barron F & Harrington DM, 'Creativity, intelligence, and personality', *Annual Review of Psychology*, 1981, 32, 439–76

Barrós-Loscertales A, González J, Pulvermüller F et al, 'Reading salt activates gustatory brain regions: fMRI evidence for semantic grounding in a novel sensory modality', *Cerebral Cortex*, 2012, 22(11), 2554–63

Barsalou LW, 'Grounded cognition', *Annual Review of Psychology*, 2008, 59(1), 617–45

Barta PE, Pearlson GD, Brill LB et al, 'Planum temporale asymmetry reversal in schizophrenia: replication and relationship to gray matter abnormalities', *American Journal of Psychiatry*, 1997, 154(5), 661–7

Barta PE, Petty RG, McGilchrist I et al, 'Asymmetry of the planum temporale: methodological considerations and clinical associations', *Psychiatry Research*, 1995, 61(3), 137–50

Bartels DM & Pizarro DA, 'The mismeasure of morals: antisocial personality traits predict utilitarian responses to moral dilemmas', *Cognition*, 2011, 121(1), 154–61

Barth KA, Miklosi A, Watkins J et al, '*fsi* zebrafish show concordant reversal of laterality of viscera, neuroanatomy, and a subset of behavioral responses', *Current Biology*, 2005, 15(9), 844–50

Barthelemy S & Boulinguez P, 'Manual asymmetries in the directional coding of reaching: further evidence for hemispatial effects and right hemisphere dominance for movement planning', *Experimental Brain Research*, 2002, 147(3), 305–12

Barthold LS, 'Hans-Georg Gadamer' [undated]: www.iep.utm.edu/gadamer/ (accessed 23 February 2021)

Bartholomeus de Capua, in D Prümmer OP (ed), *Fontes vitæ S Thomæ Aquinatis*, Toulouse [undated: foreword, however, dated 1911]

Bartlett JC, Shastri KK, Abdi H et al, 'Component structure of individual differences in true and false recognition of faces', *Journal of Experimental Psychology: Learning, Memory, and Cognition*, 2009, 35(5), 1207–30

Bartolic EI, Basso MR, Schefft BK et al, 'Effects of experimentally induced emotional states on frontal lobe cognitive task performance', *Neuropsychologia*, 1999, 37(6), 677–83

Bartolomeo P, 'The delusion of the Master: the last days of Henry James', *Neurological Sciences*, 2013, 34(11), 2031–4

Bartolomeo P, De Vito S & Seidel Malkinson T, 'Space-related confabulations after right hemisphere damage', *Cortex*, 2017, 87, 166–73

Barton JJ & Cherkasova M, 'Face imagery and its relation to perception and covert recognition in prosopagnosia', *Neurology*, 2003, 61(2), 220–5

Bashō, trans R Hass, in *The Essential Haiku*, Ecco Press, 1995

Basso A, Capitani E & Laiacona M, 'Progressive language impairment without dementia: a case with isolated category specific semantic defect', *Journal of Neurology, Neurosurgery, and Psychiatry*, 1988, 51(9), 1201–7

Batchelor S, *After Buddhism*, HarperCollins, 2016

Bateson G, *Steps to an Ecology of Mind: Collected Essays in Anthropology, Psychiatry, Evolution and Epistemology*, University of Chicago Press, 2000

Battelli L, Cavanagh P, Intriligator J et al, 'Unilateral right parietal damage leads to bilateral deficit for high-level motion', *Neuron*, 2001, 32(6), 985–95

Battelli L, Pascual-Leone A & Cavanagh P, 'The "when" pathway of the right parietal lobe', *Trends in Cognitive Sciences*, 2007, 11(5), 204–10

Batterman R, 'Autonomy and scales', in B Falkenburg & M Morrison (eds), *Why More is Different: Philosophical Issues in Condensed Matter Physics and Complex Systems*, Springer, 2015, 115–35

Bauer C, Yazzolino L, Hirsch G et al, 'Neural correlates associated with superior tactile symmetry perception in the early blind', *Cortex*, 2015, 63, 104–17

Bauer RM, 'Visual hypoemotionality as a symptom of visual-limbic disconnection in man', *Archives of Neurology*, 1982, 39(11), 702–8

Bauman M & Kemper TL, 'Histoanatomic observations of the brain in early infantile autism', *Neurology*, 1985, 35(6), 866–74

Baumann O, Chan E & Mattingley JB, 'Distinct neural networks underlie encoding of categorical versus coordinate spatial relations during active navigation', *NeuroImage*, 2012, 60(3), 1630–7

Baumeister R, 'The low-down on high self-esteem: you're hot stuff isn't the promised cure-all', *Los Angeles Times*, 25 January 2005

Baumeister RF & Leary MR, 'The need to belong: desire for interpersonal attachments as a fundamental human motivation', *Psychological Bulletin*, 1995, 117(3), 497–529

Baumeister RF, Masicampo EJ & DeWall C, 'Prosocial benefits of feeling free: disbelief in free will increases aggression and reduces helpfulness', *Personality and Social Psychology Bulletin*, 2009, 35(2), 260–8

Baumeister RF, Masicampo EJ & Vohs KD, 'Do conscious thoughts cause behavior?', *Annual Review of Psychology*, 2011, 62, 331–61

Baumeister RF, Vohs KD, Aaker JL et al, 'Some key differences between a happy life and a meaningful life', *The Journal of Positive Psychology*, 2013, 8(6), 505–16

Baumgartner T, Knoch D, Hotz P et al, 'Dorsolateral and ventromedial prefrontal cortex orchestrate normative choice', *Nature Neuroscience*, 2011, 14(11), 1468–74

Bazanova OM & Aftanas LI, 'Individual measures of electro-encephalogram alpha activity and non-verbal creativity', *Neuroscience and Behavioral Physiology*, 2008, 38(3), 227–35

Bäzner H & Hennerici MG, 'Painting after right-hemisphere stroke-case studies of professional artists', *Frontiers of Neurology and Neuroscience*, 2007, 22, 1–13

Beal DN, Hover FS, Triantafyllou MS et al, 'Passive propulsion in vortex wakes', *Journal of Fluid Mechanics*, 2006, 549(1), 385–402

Beall J, 'What I learned from predatory publishers', *Biochemia Medica*, 2017, 27(2), 273–9

Bear DM, 'Hemispheric specialization and the neurology of emotion', *Archives of Neurology*, 1983, 40(4), 195–202

Bear DM & Fedio P, 'Quantitative analysis of interictal behaviour in temporal lobe epilepsy', *Archives of Neurology*, 1977, 34(8), 454–67

Bearden CE, Hoffman KM & Cannon TD, 'The neuropsychology and neuroanatomy of bipolar affective disorder: a critical review', *Bipolar Disorders*, 2001, 3(3), 106–50

Beaty RE & Silvia PJ, 'Metaphorically speaking: cognitive abilities and the production of figurative speech', *Memory & Cognition*, 2013, 41(2), 255–67

Beauchamp MS, Lee KE, Haxby JV et al, 'fMRI responses to video and point-light displays of moving humans and manipulable objects', *Journal of Cognitive Neuroscience*, 2003, 15(7), 991–1001

Beauchamp T & Childress J, *Principles of Biomedical Ethics*, Oxford University Press, 4th edn, 1979

Beaumont JG, 'Lateral organization and aesthetic preference: the importance of peripheral visual asymmetries', *Neuropsychologia*, 1985, 23(1), 103–13

Beauregard M, Trent NL & Schwartz GE, 'Toward a postmaterialist psychology: theory, research, and applications', *New Ideas in Psychology*, 2018, 50, 21–33

Becchetti L, Bachelet M & Pisani F, 'Poor eudaimonic subjective wellbeing as a mortality risk factor', *Economia Politica*, 2019, 36(1), 245–72

Becchio C & Bertone C, 'The ontology of neglect', *Consciousness and Cognition*, 2005, 14(3), 483–94

Bechara A, Damasio H, Tranel D et al, 'Deciding advantageously before knowing the advantageous strategy', *Science*, 1997, 275(5304), 1293–5

Beck HP, *Parametric Studies of the Response of a Protozoan, Spirostomum Ambiguum, to Repetitive Stimulation*, East Carolina University Press, 1975

Beck LW, *Early German Philosophy: Kant and his Predecessors*, Harvard University Press, 1969

Beck U, Aschayeri H & Keller H, 'Prosopagnosie und Farberkennungsstörung bei Rückbildung von Rindenblindheit', *Archiv für Psychiatrie und Nervenkrankheiten*, 1978, 225(1), 55–66

Becker E & Karnath HO, 'Incidence of visual extinction after left versus right hemisphere stroke', *Stroke*, 2007, 38(12), 3172–4

Beekman M & Latty T, 'Brainless but multi-headed: decision making by the acellular slime mould *Physarum polycephalum*', *Journal of Molecular Biology*, 2015, 427(23), 3734–43

Beeman M, 'Semantic processing in the right hemisphere may contribute to drawing inferences from discourse', *Brain and Language*, 1993, 44(1), 80–120

——, 'Coarse semantic coding and discourse comprehension', in M Beeman & C Chiarello (eds), *Right Hemisphere Language Comprehension: Perspectives from Cognitive Neuroscience*, Erlbaum, 1998, 255–84

Beeman MJ & Bowden EM, 'The right hemisphere maintains solution-related activation for yet-to-be-solved problems', *Memory & Cognition*, 2000, 28(7), 1231–41

Beeman M & Chiarello C, *Right Hemisphere Language Comprehension: Perspectives from Cognitive Neuroscience*, Erlbaum, 1998

Beeman M, Friedman RB, Grafman J et al, 'Summation priming and coarse semantic coding in the right hemisphere', *Journal of Cognitive Neuroscience*, 1994, 6(1), 26–45

Been G, Ngo TT, Miller SM et al, 'The use of tDCS and CVS as methods of non-invasive brain stimulation', *Brain Research Reviews*, 2007, 56(2), 346–61

Begbie J, *Resounding Truth*, SPCK, 2008

Begley CG & Ellis LM, 'Drug development: raise standards for preclinical cancer research', *Nature*, 2012, 483(7391), 531–3

Behnke EA, 'Body', in *Encyclopedia of Phenomenology*, Kluver, 1997, 66–71

Behrmann M, Avidan G, Leonard GL et al, 'Configural processing in autism and its relationship to face processing, *Neuropsychologia*, 2006, 44(1), 110–29

Beilock SL, Bertenthal B, McCoy AM et al, 'Haste does not always make waste: expertise, direction of attention, and speed versus accuracy in performing sensorimotor skills', *Psychonomic Bulletin & Review*, 2004, 11(2), 373–9

Beilock SL, Carr TH, MacMahon C et al, 'When paying attention becomes counterproductive: impact on divided versus skill-focused attention on novice and experienced performance of sensorimotor skills', *Journal of Experimental Psychology: Applied*, 2002, 8(1), 6–16

Beisson J & Sonneborn TM, 'Cytoplasmic inheritance of the organization of the cell cortex in Paramecium aurelia', *Proceedings of the National Academy of Sciences of the United States of America*, 1965, 53(2), 275–82

Beißner F, 'Zu den Gedichten der letzten Lebenszeit', *Hölderlin-Jahrbuch*, 1947, 2, 6–10

Bekhtereva NP, Starchenko MG, Klyucharev VA et al, 'Study of the brain organization of creativity: II. Positron-emission tomography data', *Human Physiology*, 2000, 26(5), 516–22

Beking T, Geuze RH, van Faassen M et al, 'Prenatal and pubertal testosterone affect brain lateralization', *Psychoneuroendocrinology*, 2018, 88, 78–91

Belfi AM, Bruss J, Karlan B et al, 'Neural correlates of recognition and naming of musical instruments', *Neuropsychology*, 2016, 30(7), 860–8

Belin P, Zatorre RJ, Lafaille P et al, 'Voice-selective areas in human auditory cortex', *Nature*, 2000, 403(6767), 309–12

Bell EC, Willson MC, Wilman AH et al, 'Males and females differ in brain activation during cognitive tasks', *NeuroImage*, 2006, 30(2), 529–38

Bell JS, 'Against measurement', in *Speakable and Unspeakable in Quantum Mechanics*, Cambridge University Press, 2004, 213–31

Bell V, Reddy V, Halligan P et al, 'Relative suppression of magical thinking: a transcranial magnetic stimulation study', *Cortex*, 2007, 43(4), 551–7

Bellamy KJ & Shillcock R, 'A right hemisphere bias towards false memory', *Laterality*, 2007, 12(2), 154–66

Bellas DN, Novelly RA, Eskenazi B et al, 'The nature of unilateral neglect in the olfactory

sensory system', *Neuropsychologia*, 1988a, 26(1), 45–52
——, 'Unilateral displacement in the olfactory sense: a manifestation of the unilateral neglect syndrome', *Cortex*, 1988b, 24(2), 267–75
Bellgrove MA, Vance A & Bradshaw JL, 'Local-global processing in early-onset schizophrenia: evidence for an impairment in shifting the spatial scale of attention', *Brain and Cognition*, 2003, 51(1), 48–65
Bellugi U, Poizner H & Klima ES, 'Brain organization for language: clues from sign aphasia', *Human Neurobiology*, 1983, 2(3), 155–70
——, 'Language, modality and the brain', *Trends in Neurosciences*, 1989, 12(10), 380–8
Belmonti V, Berthoz A, Cioni G et al, 'Navigation strategies as revealed by error patterns on the Magic Carpet test in children with cerebral palsy', *Frontiers in Psychology*, 2015, 6, article 880
Belyï BI, 'The relationship between the side of the lesion and the psychopathology of focal lesions of the frontal lobes', *Zhurnal Nevrologii i Psikhiatrii Imeni S S Korsakova*, 1975, 75(12), 1793–8
——, 'Mental disorders in patients with unilateral frontal tumors', *Zhurnal Nevrologii i Psikhiatrii Imeni S S Korsakova*, 1985, 85(2), 224–32
Ben-Dov G & Carmon A, 'Rhythm length and hemispheric asymmetry', *Brain and Cognition*, 1984, 3(1), 35–41
Benbow CP, 'Physiological correlates of extreme intellectual precocity', *Neuropsychologia*, 1986, 24(5), 719–25
Benbow CP & Lubinski D, 'Psychological profiles of the mathematically talented: some sex differences and evidence supporting their biological basis', *Ciba Foundation Symposium*, 1993, 178, 44–59
Bender MB & Teuber HL, 'Spatial organization of visual perception following injury to the brain', *Archives of Neurology and Psychiatry*, 1947, 58(6), 721–39
——, 'Spatial organization of visual perception following injury to the brain', *Archives of Neurology and Psychiatry*, 1948, 59(1), 39–62
Bender MB, 'Polyopia and monocular diplopia of cerebral origin', *Archives of Neurology and Psychiatry*, 1945, 54(5), 323–38
Bender MB, Feldman M & Sobin AJ, 'Palinopsia', *Brain*, 1968, 91(2), 321–8
Benedek M, Beaty R, Jauk E et al, 'Creating metaphors: the neural basis of figurative language production', *NeuroImage*, 2014, 90(100), 99–106
Benedek M, Bergner S, Könen T et al, 'EEG alpha synchronization is related to top-down processing in convergent and divergent thinking', *Neuropsychologia*, 2011, 49(12), 3505–11
Benedek M, Schickel RJ, Jauk E et al, 'Alpha power increases in right parietal cortex reflects focused internal attention', *Neuropsychologia*, 2014, 56(100), 393–400
Bengtsson S, Csíkszentmihályi M & Ullén F, 'Cortical regions involved in the generation of musical structures during improvisation in pianists', *Journal of Cognitive Neuroscience*, 2007, 19(5), 830–42
Beniczky S, Kéri S, Vörös E et al, 'Complex hallucinations following occipital lobe damage', *European Journal of Neurology*, 2002, 9(2), 175–6
Benjamin W, 'The Storyteller', *Illuminations*, Schocken, 1969 [*Der Erzähler* 1936]
Benjamin W, 'One-Way Street', in *Reflections, Aphorisms, Autobiographical Writings*, Schocken, 1995 [*Einbahnstrasse* 1955]
Benner P, 'From novice to expert', [undated]: www.currentnursing.com/nursing_theory/Patricia_Benner_From_Novice_to_Expert.html (accessed 6 Mar 2021)
Benner P & Tanner C, 'Clinical judgment: how expert nurses use intuition', *American Journal of Nursing*, 1987, 87(1), 23–31
Benowitz LI, Finkelstein S, Levine DN et al, 'The role of the right cerebral hemisphere in evaluating configurations', in C Trevarthen (ed), *Brain Circuits & Functions of the Mind: Essays in Honor of Roger W Sperry*, Cambridge University Press, 1990, 320–33
Benowitz LI, Moya KL & Levine DN, 'Impaired verbal reasoning and constructional apraxia in subjects with right hemisphere damage', *Neuropsychologia*, 1990, 28(3), 231–41
Benson DF & Barton MI, 'Disturbances in constructional ability', *Cortex*, 1970, 6(1), 19–46
Benson DF, Djenderedjian A, Miller BL et al, 'Neural basis of confabulation', *Neurology*, 1996, 46(5), 1239–43
Benson DF, Gardner H & Meadows JC, 'Reduplicative paramnesia', *Neurology*, 1976, 26(2), 147–51
Benton AL, 'The fiction of the Gerstmann syndrome', *Journal of Neurology, Neurosurgery, and Psychiatry*, 1961, 24(2), 176–81
——, 'Differential behavioral effects in frontal lobe disease', *Neuropsychologia*, 1968, 6(1), 53–60
Benton AL & Hécaen H, 'Stereoscopic vision in patients with unilateral cerebral disease', *Neurology*, 1970, 20(11), 1084–8
Benton AL, Varney NR & Hamsher K, 'Visuospatial judgment', *Archives of Neurology*, 1978, 35(6), 364–7
Berg A, Palomäki H, Lehtihalmes M et al, 'Poststroke depression in acute phase after

stroke', *Cerebrovascular Diseases*, 2001, 12(1), 14–20

Berg JM, Tymoczko JL & Stryer L, *Biochemistry*, WH Freeman, 2002

Berg M & Seeber B, *The Slow Professor: Challenging the Cult of Speed in the Academy*, University of Toronto Press, 2016

Berger PL, *The Desecularization of the World: The Resurgence of Religion in World Politics*, William B Eerdmans, 1999

Bergert S, 'How do our brain hemispheres cooperate to avoid false memories?' *Cortex*, 2013, 49(2), 572–81

Bergson H, *L'évolution créatrice*, Félix Alcan, 1908

——, *Time and Free Will: An Essay on the Immediate Data of Consciousness*, trans FL Pogson, George Allen and Unwin, 1910 [*Essai sur les données immédiates de la conscience* 1889]

——, *Creative Evolution*, trans A Mitchell, Henry Holt, 1911a

——, *La perception du changement, conférences faites à l'Université d'Oxford les 26 et 27 mai 1911*, Clarendon Press, 1911b

——, *Matter and Memory*, trans NM Paul & W Scott Palmer, George Allen & Unwin, 1911c

——, *An Introduction to Metaphysics*, trans TE Hulme, Putnam, 1912 [1903]

——, *Mind-Energy: Lectures and Essays*, trans H Wildon Carr, Henry Holt & Co, 1920

——, *Matière et mémoire : essai sur la relation du corps à l'esprit*, Presses Universitaires de France, 72nd edn, 1965

——, 'Introduction (première partie). Croissance de la vérité. Mouvement rétrograde du vrai', in *La pensée et le mouvant*, Presses Universitaires de France, 1969 [1934]

——, *Essai sur les données immédiates de la conscience*, in A Robinet (ed), *Œuvres*, Presses Universitaires de France, 1970 [1889]

——, 'Introduction (Part I). Growth of truth. Retrograde movement of the true' (1934), in *The Creative Mind: An Introduction to Metaphysics*, trans ML Andison, Dover, 2007a [1946], 1–17

——, 'Introduction (Part II). Stating of problems' (1934), in *The Creative Mind: An Introduction to Metaphysics*, trans ML Andison, Dover, 2007b [1946], 18–72

——, 'Introduction to metaphysics' (1903), in *The Creative Mind: An Introduction to Metaphysics*, trans ML Andison, Dover, 2007c [1946], 133–69

——, 'Philosophical intuition' (1911), in *The Creative Mind: An Introduction to Metaphysics*, trans ML Andison, Dover, 2007d [1946], 87–106

——, 'The perception of change' (1911), in *The Creative Mind: An Introduction to Metaphysics*, trans ML Andison, Dover, 2007e [1946], 107–32

——, 'The possible and the real' (1920), in *The Creative Mind: An Introduction to Metaphysics*, trans ML Andison, Dover, 2007f [1946], 73–86

——, 'Message au congrès Descartes' (1937), in F Worms (ed), *Écrits philosophiques*, Presses Universitaires de France, Paris, 2011

Bering JM, 'The existential theory of mind', *Review of General Psychology*, 2002, 6(1), 3–24

——, 'The evolutionary history of an illusion: religious causal beliefs in children and adults', in BJ Ellis & DF Bjorklund (eds), *Origins of the Social Mind: Evolutionary Psychology and Child Development*, Guilford Press, 2004, 411–37

Berkeley G, *A Treatise Concerning the Principles of Human Knowledge*, Oxford University Press, 2009 [1710]

Berkovich-Ohana A, Glicksohn J & Goldstein A, 'Mindfulness-induced changes in gamma band activity – implications for the default mode network, self-reference and attention', *Clinical Neurophysiology*, 2012, 123(4), 700–10

——, 'Studying the default mode and its mindfulness-induced changes using EEG functional connectivity', *Social Cognitive and Affective Neuroscience*, 2014, 9(10), 1616–24

Berlin B, *The Principles of Ethnobiological Classification*, Princeton University Press, 1992

——, '"Just another fish story?" Size-symbolic properties of fish names', in A Minelli, G Ortalli & G Singa (eds), *Animal Names*, Istituto Veneto di Scienze, Lettere ed Arti, 2005, 9–20

Berlin I, *The Roots of Romanticism*, Princeton University Press, 1999

Berlucchi G & Aglioti S, 'The body in the brain: neural bases of corporeal awareness', *Trends in Neurosciences*, 1997, 20(12), 560–4

Berlyne DE, 'Novelty, complexity, and hedonic value', *Perception & Psychophysics*, 1970, 8(5), 279–86

Berman DD & Anton MT, 'A wilderness therapy program and an alternative to adolescent psychiatric hospitalization', *Residential Treatment For Children & Youth*, 1988, 5(3), 41–53

Berman MG, Jonides J & Kaplan S, 'The cognitive benefits of interacting with nature', *Psychological Science*, 2008, 19(12), 1207–12

Bernard C, *Leçons sur les phénomènes de la vie communs aux animaux et aux végétaux*, Baillière, Paris, 1885

Berntsen S, Kragstrup J, Siersma V et al, 'Alcohol consumption and mortality in patients with mild Alzheimer's disease: a prospective cohort study', *BMJ Open*, 2015, 5, e007851

Berrios GE, 'Musical hallucinations: a statistical analysis of 46 cases', *Psychopathology*, 1991, 24(6), 356–60

Berry W, *Life is a Miracle*, Counterpoint, 2000

Bersanelli M, 'Infinity and the nostalgia of the stars', in M Heller & WH Woodin (eds), *Infinity: New Research Frontiers*, Cambridge University Press, 2011, 193–217

Bersani G, Quartini A, Iannitelli A et al, 'Corpus callosum abnormalities and potential age effect in men with schizophrenia: an MRI comparative study', *Journal of Experimental Child Psychology: Neuroimaging*, 2010, 183, 119–25

Berson RJ, 'Capgras syndrome', *American Journal of Psychiatry*, 1983, 140(8), 969–78

Bertamini M & Makin ADJ, 'Brain activity in response to visual symmetry', *Symmetry*, 2014, 6(4), 975–96

Bertamini M, Silvanto J, Norcia AM et al, 'The neural basis of visual symmetry and its role in mid- and high-level visual processing', *Annals of the New York Academy of Sciences*, 2018, 1426(1), 111–26

Berthier M & Starkstein S, 'Acute atypical psychosis following a right hemisphere stroke', *Acta Neurologica Belgica*, 1987, 87(3), 125–31

Berti A, Bottini G, Gandola M et al, 'Shared cortical anatomy for motor awareness and motor control', *Science*, 2005, 309(5733), 488–91

Bertini M, Violani C, Zoccolotti P et al, 'Right cerebral hemisphere activation in dreaming sleep: evidence from a unilateral tactile recognition test', *Psychophysiology*, 1984, 21(4), 418–23

Berto R, 'Exposure to restorative environments helps restore attentional capacity', *Journal of Environmental Psychology*, 2005, 25(3), 249–59

———, 'The role of nature in coping with psycho-physiological stress: a literature review on restorativeness', *Behavioral Sciences*, 2014, 4(4), 394–409

Bertolaso M & Dupré J, 'A processual perspective on cancer', in DJ Nicholson & J Dupré (eds), *Everything Flows: Towards a Processual Philosophy of Biology*, Oxford University Press, 2018, 321–36

Betsch T, Plessner H, Schwieren C et al, 'I like it but I don't know why: a value-account approach to implicit attitude formation', *Personality and Social Psychology Bulletin*, 2001, 27(2), 242–53

Beulens JWJ, van der Schouw YT, Bergmann MM et al, 'Alcohol consumption and risk of type 2 diabetes in European men and women: influence of beverage type and body size The EPIC-InterAct study', *Journal of Internal Medicine*, 2012, 272(4), 358–70

Beume LA, Klingler A, Reinhard M et al, 'Olfactory hallucinations as primary symptom for ischemia in the right posterior insula', *Journal of the Neurological Sciences*, 2015, 354(1–2), 138–39

Bevan G & Hood C, 'What's measured is what matters: targets and gaming in the English public health care system', *Public Administration*, 2006, 84(3), 517–38

Bever TG & Chiarello RJ, 'Cerebral dominance in musicians and nonmusicians', *Science*, 1974, 185(150), 537–39

Bhagavad Gita, trans J Mascaró, Penguin, 1962

Bhalla M & Proffitt DR, 'Visual-motor recalibration in geographical slant perception', *Journal of Experimental Psychology: Human Perception & Performance*, 1999, 25(4), 1076–96

Bhatia MS, 'Cotard's syndrome in parietal lobe tumor', *Indian Pediatrics*, 1993, 30(8), 1019–21

Bhatia MS, Saha R & Doval N, 'Delusional disorder in a patient with corpus callosum agenesis', *Journal of Clinical and Diagnostic Research*, 2016, 10(12), VD01–2

Bhattacharya J & Petsche H, 'Shadows of artistry: cortical synchrony during perception and imagery of visual art', *Brain Research: Cognitive Brain Research*, 2002, 13(2), 179–86

———, 'Drawing on mind's canvas: differences in cortical integration patterns between artists and non-artists', *Human Brain Mapping*, 2005, 26(1), 1–14

Bhavanani AB, Ramanathan M, Balaji R et al, 'Differential effects of uninostril and alternate nostril pranayamas on cardiovascular parameters and reaction time', *International Journal of Yoga*, 2014, 7(1), 60–5

Bhogal SK, Teasell R, Foley N et al, 'Lesion location and poststroke depression: systematic review of the methodological limitations in the literature', *Stroke*, 2004, 35(3), 794–802

Bianconi E, Piovesan A, Facchin F et al, 'An estimation of the number of cells in the human body', *Annals of Human Biology*, 2013, 40(6), 463–71

Bianki VL, *The Mechanisms of Cerebral Lateralisation*, Gordon & Breach, 1993

Bichat X, *Recherches Physiologiques sur la Vie et la Mort*, 4ᵉ édition, augmentée de notes par F Magendie, Béchet Jeune et Gabon, 1822

Bickhard MH, 'Variations in variation and selection: the ubiquity of the variation-and-selective retention ratchet in emergent organizational complexity, part II: quantum field theory', *Foundations of Science*, 2003, 8(3), 283–93

———, 'The interactivist model', *Synthese*, 2009, 166(3), 547–91

Biermann-Ruben K, Kessler K, Jonas M et al, 'Right hemisphere contributions to imitation tasks', *European Journal of Neuroscience*, 2008, 27(7), 1843–55

Bigio EH, 'Motor neuron disease: the C9orf72 hexanucleotide repeat expansion in FTD and ALS', *Nature Reviews Neurology*, 2012, 8(5), 249–50

Bihrle AM, Brownell HH, Powelson JA et al, 'Comprehension of humorous and nonhumorous materials by left and right brain-damaged patients', *Brain and Cognition*, 1986, 5(4), 399–411

Bilalić M, Kiesel A, Pohl C et al, 'It takes two — skilled recognition of objects engages lateral areas in both hemispheres', *PLoS One*, 2011, 6(1), e16202

Bilder RM & Knudsen KS, 'Creative cognition and systems biology on the edge of chaos', *Frontiers in Psychology*, 2014, 5, article 1104

Bilger RC, Matthies ML, Hammel DR et al, 'Genetic implications of gender differences in the prevalence of spontaneous otoacoustic emissions', *Journal of Speech Language and Hearing Research*, 1990, 33(3), 418–32

Binder J, Marshall R, Lazar R et al, 'Distinct syndromes of hemineglect', *Archives of Neurology*, 1992, 49(11), 1187–94

Binet A, *Psychologie des grands calculateurs et joueurs d'échecs*, Hachette, Paris, 1894

Binford T, 'The machine sees', in M Minsky (ed), *Robotics*, Doubleday, 1985

Binkofski F & Block RA, 'Accelerated time experience after left frontal cortex lesion', *Neurocase*, 1996, 2(6), 485–93

Binswanger L, 'Extravagance, perverseness, manneristic behaviour and schizophrenia', in J Cutting & M Shepherd (eds), *The Clinical Roots of the Schizophrenia Concept: Translations of Seminal European Contributions on Schizophrenia*, Cambridge University Press, 1987, 83–8 [*Drei Formen Missglückten Daseins: Verstiegenheit, Verschrobenheit, Manniertheit*, 1956]

———, *Drei Formen missglückten Daseins: Verstiegenheit, Verschrobenheit, Manieriertheit*, de Gruyter, 2010 [1956]

Birch C, 'The postmodern challenge to biology', in DR Griffin (ed), *The Reenchantment of Science*, State University of New York Press, 1988, 57–68

Birchwood M, Iqbal Z, Chadwick P et al, 'Cognitive approach to depression and suicidal thinking in psychosis. 2. Testing the validity of a social ranking model', *British Journal of Psychiatry*, 2000, 177(6), 522–8

Birnbaum KD & Sánchez Alvarado A, 'Slicing across kingdoms: regeneration in plants and animals', *Cell*, 2008, 132(4), 697–710

Bisazza A, Facchin L, Pignatti R et al, 'Lateralization of detour behaviour in poeciliid fish: the effect of species, gender and sexual motivation', *Behavioural Brain Research*, 1998, 91(1–2), 157–64

Bisazza A, Pignatti R & Vallortigara G, 'Detour tests reveal task- and stimulus-specific behavioural lateralization in mosquitofish (*Gambusia holbrooki*)', *Behavioural Brain Research*, 1997a, 89(1–2), 237–42

———, 'Laterality in detour behaviour: interspecific variation in poeciliid fish', *Animal Behaviour*, 1997b, 54(5), 1273–81

Bishop P, 'Disenchantment in education, or: 'Whither art thou gone, fair world?'— Has the magic gone from the ivory tower?', *International Journal of Jungian Studies*, 2012, 4(1), 55–69

Bisiacchi P, Marzi CA, Nicoletti R et al, 'Left-right asymmetry of callosal transfer in normal human subjects', *Behavioural Brain Research*, 1994, 64(1–2), 173–8

Bisiach E, 'Mental representation in unilateral neglect and related disorders: the twentieth Bartlett Memorial Lecture', *Quarterly Journal of Experimental Psychology*, 1993, 46(3), 435–61

———, 'Understanding consciousness: clues from unilateral neglect and related disorders', in N Block, O Flanagan & G Güzeldere (eds), *The Nature of Consciousness: Philosophical Debates*, Massachusetts Institute of Technology Press, 1997, 237–53

Bisiach E & Luzzatti C, 'Unilateral neglect of representational space', *Cortex*, 1978, 14, 129–33

Bisiach E, Rusconi ML & Vallar G, 'Remission of somatoparaphrenic delusion through vestibular stimulation', *Neuropsychologia*, 1991, 29(10), 1029–31

Bissell MM, Dall'Armellina E & Choudhury RP, 'Flow vortices in the aortic root: in vivo 4D-MRI confirms predictions of Leonardo da Vinci', *European Heart Journal*, 2014, 35(20), 1344

Bitan T, Lifshitz A, Breznitz Z et al, 'Bidirectional connectivity between hemispheres occurs at multiple levels in language processing but depends on sex', *Journal of Neuroscience*, 2010, 30(35), 11576–85

Björkenstam E, Helgesson M, Amin R et al, 'Mental disorders, suicide attempt and suicide: differences in the association in refugees compared with Swedish-born individuals', *British Journal of Psychiatry*, 2020, 217(6), 679–85

Black M, *Models and Metaphors: Studies in Language and Philosophy*, Cornell University Press, 1962

———, *The Prevalence of Humbug and Other Essays*, Cornell University Press, 1983

Blackburn K & Schirillo J, 'Emotive hemispheric differences measured in real-life portraits using pupil diameter and subjective aesthetic preferences', *Experimental Brain Research*, 2012, 219(4), 447–55

Blackiston DJ, Silva Casey E & Weiss MR , 'Retention of memory through metamorphosis: can a moth remember what it learned as a caterpillar?', *PLoS One*, 2008, 3(3), e1736

Blackman L, *Lucy's Story: Autism and Other Adventures*, Book in Hand, 1999

Blackwood NJ, Howard RJ, ffytche DH et al, 'Imaging attentional and attributional bias: an fMRI approach to the paranoid delusion', *Psychological Medicine*, 2000, 30(4), 873–88

Blair C, Gamson D, Thorne S et al, 'Rising mean IQ: cognitive demand of mathematics education for young children, population exposure to formal schooling, and the neurobiology of the prefrontal cortex', *Intelligence*, 2005, 33(1), 93–106

Blair RJR & Coles M, 'Expression recognition and behavioural problems in early adolescence', *Cognitive Development*, 2000, 15(4), 421–34

Blair RJR, Colledge E, Murray L et al, 'A selective impairment in the processing of sad and fearful expressions in children with psychopathic tendencies', *Journal of Abnormal Child Psychology*, 2001, 29(6), 491–8

Blakeslee TR, *The Right Brain*, Macmillan, 1980

Blank H, Wieland N & von Kriegstein K, 'Person recognition and the brain: merging evidence from patients and healthy individuals', *Neuroscience and Biobehavioral Reviews*, 2014, 47, 717–34

Blanke O, 'I and me: self-portraiture in brain damage', *Frontiers of Neurology and Neuroscience*, 2007, 22(1), 14–29

———, 'Brain correlates of the embodied self: neurology and cognitive neuroscience', *Annals of General Psychiatry*, 2008, 7(suppl 1), article S92

Blanke O & Arzy S, 'The out-of-body experience: disturbed self-processing at the temporo-parietal junction', *Neuroscientist*, 2005, 11(1), 16–24

Blanke O & Mohr C, 'Out-of-body experience, heautoscopy, and autoscopic hallucination of neurological origin: implications for neurocognitive mechanisms of corporeal awareness and self-consciousness', *Brain Research Reviews*, 2005, 50(1), 184–99

Blanke O, Morgenthaler FD, Brugger P et al, 'Preliminary evidence for a fronto-parietal dysfunction in able-bodied participants with a desire for limb amputation', *Journal of Neuropsychology*, 2009, 3(2), 181–200

Blanke O, Ortigue S, Landis T et al, 'Stimulating illusory own-body perceptions', *Nature*, 2002, 419(6904), 269–70

Blanke O & Pasqualini I, 'The riddle of style changes in the visual arts after interference with the right brain', *Frontiers in Human Neuroscience*, 2012, 5, article 154

Blankenburg W, 'Ansätze zu einer Psychopathologie des "common sense"', *Confinia Psychiatrica*, 1969, 12, 144–63

———, *Der Verlust der natürlichen Selbstverständlichkeit: ein Beitrag zur Psychopathologie symptomarmer Schizophrenien*, Ferdinand Enke, 1971

———, 'First steps toward a psychopathology of "common sense"', trans AL Mishara, *Philosophy, Psychiatry, & Psychology*, 2001, 8(4), 303–15

Blasco-Fontecilla H, Bragado Jimenez MD, Garcia Santos LM et al, 'Delusional disorder with delusions of parasitosis and jealousy after stroke: treatment with quetiapine and sertraline', *Journal of Clinical Psychopharmacology*, 2005, 25(6), 615–7

Blaxton TA, Bookheimer SY, Zeffiro TA et al, 'Functional mapping of human memory using PET: comparisons of conceptual and perceptual tasks', *Canadian Journal of Experimental Psychology*, 1996, 50(1), 42–56

Bleeker JAC & Sno H, 'Het syndroom van Cotard: een herwaardering aan de hand van twee patiënten met een onsterfelijkheidswaan', *Tijdschrift voor Psychiatrie*, 1983, 25(10), 665–75

Bleuler E, *Dementia Præcox or the Group of Schizophrenias*, trans J Zinkin, International Universities Press, 1950 [*Dementia præcox oder Gruppe der Schizophrenien* 1911]

Bleuler M, *The Schizophrenic Disorders*, Yale University Press, 1978

Block RA, Arnott DP, Quigley B et al, 'Unilateral nostril breathing influences lateralized cognitive performance', *Brain and Cognition*, 1989, 9(2), 181–90

Blom JD, 'Alice in Wonderland syndrome: a systematic review', *Neurology: Clinical Practice*, 2016, 6(3), 259–70

Blom O, Cervenka S, Karabanov A et al, 'Regional dopamine D_2 receptor density and individual differences in psychometric creativity', poster session presented at the 14th Annual Meeting for the Organization for Human Brain Mapping, Melbourne, Australia, June 2008

Blonder LX, Bowers D & Heilman KM, 'The role of the right hemisphere in emotional communication', *Brain*, 1991, 114(3), 1115–27

Blonder LX, Burns AF, Bowers D et al, 'Right hemisphere facial expressivity during natural conversation', *Brain and Cognition*, 1993, 21(1), 44–56

Bloom H, *Genius: A Mosaic of One Hundred Exemplary Creative Minds*, Grand Central Publishing, 2003

———, interviewed by Laura Quinney, 27 November 2005: www.rc.umd.edu/praxis/ bloom_hartman/bloom/bloom.html (accessed 23 February 2021)

Bloom JS & Hynd GW, 'The role of the corpus callosum interhemispheric transfer of information: excitation or inhibition?', *Neuropsychology Review*, 2005, 15(2), 59–71

Bloom P, 'Religion is natural', *Developmental Science*, 2007, 10(1), 147–51

Blount G, 'Dangerousness of patients with Capgras syndrome', *Nebraska Medical Journal*, 1986, 71(6), 207

Blumberg HP, Stern E, Martinez D et al, 'Increased anterior cingulate and caudate activity in bipolar mania', *Biological Psychiatry*, 2000, 48(11), 1045–52

Blumenberg H, 'Not a martyr for Copernicanism: Giordano Bruno', in *The Genesis of the Copernican World*, trans RM Wallace, Massachusetts Institute of Technology Press, 1987, pt 3, ch 5: source: Rabin S, 'Nicolaus Copernicus', *The Stanford Encyclopedia of Philosophy* (fall 2019 edn), EN Zalta (ed), plato.stanford.edu/archives/fall2019/ entries/copernicus/ (accessed 23 February 2021)

Boccia M, Barbetti S, Piccardi L et al, 'Neuropsychology of aesthetic judgment of ambiguous and non-ambiguous artworks', *Behavioral Sciences (Basel)*, 2017, 7(1), pii: E13

Bochkarev VK, Kirenskaya AV, Tkachenko AA et al, 'EEG frequency and regional properties in patients with paranoid schizophrenia: effects of positive and negative symptomatology prevalence', *Zhurnal Nevrologii i Psikhiatrii Imeni S S Korsakova*, 2015, 115(1), 66–74

Boehme J, *Mysterium Pansophicum*, tr JR Earle, Knopf, 1920

Boes AD, Tranel D, Anderson SW et al, 'Right anterior cingulate: a neuroanatomical correlate of aggression and defiance in boys', *Behavioral Neuroscience*, 2008, 122(3), 677–84

Boeschoten MA, Kemner C, Kenemans JL et al, 'The relationship between local and global processing and the processing of high and low spatial frequencies studied by event-related potentials and source modeling', *Brain Research: Cognitive Brain Research*, 2005, 24(2), 228–36

Bogen JE, 'The other side of the brain 1: dysgraphia and dyscopia following cerebral commissurotomy', *Bulletin of the Los Angeles Neurological Society*, 1969, 34(2), 73–105

———, 'The dual brain: some historical and methodological aspects', in DF Benson & E Zaidel (eds), *The Dual Brain: Hemispheric Specialization in Humans*, Guilford Press, 1985, 27–43

Bogen JE & Bogen GM, 'Creativity and the corpus callosum', *Psychiatric Clinics of North America*, 1988, 11(3), 293–301

Boger-Megiddo I, Shaw DW, Friedman SD et al, 'Corpus callosum morphometrics in young children with autism spectrum disorder', *Journal of Autism and Developmental Disorders*, 2006, 36(6), 733–9

Boghossian P ('Peter Boyle') & Lindsay J ('Jamie Lindsay'), 'The conceptual penis as a social construct: a Sokal-style hoax on gender studies', www.skeptic.com/reading_ room/conceptual-penis-social-contruct-sokal-style-hoax-on-gender-studies/ (accessed 23 February 2021); original paper at Lindsay J & Boyle P, 'The conceptual penis as a social construct', *Cogent Social Sciences*, 2017, 3, article 1330439, doi.org/10.1 080/23311886.2017.1330439

Bogousslavsky J, Kumral E, Regli F et al, 'Acute hemiconcern: a right anterior parietotemporal syndrome', *Journal of Neurology, Neurosurgery, and Psychiatry*, 1995, 58(4), 428–32

Bohannon J, 'Who's afraid of peer review?', *Science*, 2013, 342(6154), 60–65

Bohbot VD, Jech R, Růžička E et al, 'Rat spatial memory tasks adapted for humans: characterization in subjects with intact brain and subjects with selective medial temporal lobe thermal lesions', *Physiological Research*, 2002, (51, suppl 1), s 49–65

Bohm D, *Quantum Theory*, Prentice-Hall, 1951

———, 'A suggested interpretation of the quantum theory in terms of "hidden" variables: II', *Physical Review*, 1952, 85(2), 180–92

———, *Causality and Chance in Modern Physics*, University of Pennsylvania Press, 1957

———, 'Some remarks on the notion of order', in CH Waddington (ed), *Towards a Theoretical Biology*, vol 2, Edinburgh Press, 1969, 18–40

———, *Wholeness and the Implicate Order*, Routledge (Ark), 1980

———, in dialogue with philosopher Renée Weber: 'Nature as creativity', *ReVision*, 1982, 5(2), 35–40

———, in dialogue with philosopher Renée Weber: in Weber R, *Dialogues with Scientists*

and Sages: The Search for Unity, Routledge & Kegan Paul, 1986
———, in an interview at the Niels Bohr institute, Copenhagen, 1989, transcribed at www.mindstructures.com/an-interview-with-david-bohm-english-part-3/ (accessed 23 February 2021)
———, *Thought as a System*, Routledge, 1994
———, *The Undivided Universe*, Routledge, 2002
Bohr HH, 'My father', in S Rozental (ed), *Niels Bohr: His Life and Work as Seen by His Friends and Colleagues*, Elsevier, 1967
Bohr N, 'Light and life', *Nature*, 1933, *131*(3308), 457–9
———, *Atomic Theory and the Description of Nature: Four Essays with an Introductory Survey*, Cambridge University Press, 1961 [1934]
———, *Atomic Physics and Human Knowledge*, Dover, 2011 [1932]
Bohrn IC, Altmann U & Jacobs AM, 'Looking at the brains behind figurative language – a quantitative meta-analysis of neuroimaging studies on metaphor, idiom, and irony processing', *Neuropsychologia*, 2012, *50*(11), 2669–83
Bohrn IC, Altmann U, Lubrich O et al, 'When we like what we know – a parametric fMRI analysis of beauty and familiarity', *Brain and Language*, 2013, *124*(1), 1–8
Boldyreva GN & Zhavoronkova LA, 'Interhemispheric asymmetry of EEG coherence as a reflection of different functional states of the human brain', *Biomedical Science*, 1991, *2*(3), 266–70
Boller F, Howes D & Patten DH, 'A behavioral evaluation of brain-scan estimates of lesion size', *Journal of the American Medical Association*, 1970, *20*(9), 852–9
Boller F, Sinforiani E & Mazzucchi A, 'Preserved painting abilities after a stroke: the case of Paul-Elie Gernez', *Functional Neurology*, 2005, *20*(4), 151–5
Bolognini N, Rossetti A, Convento S et al, 'Understanding others' feelings: the role of the right primary somatosensory cortex in encoding the affective valence of others' touch', *Journal of Neuroscience*, 2013, *33*(9), 4201–5
Bolte A & Goschke T, 'On the speed of intuition: intuitive judgments of semantic coherence under different response deadlines', *Memory & Cognition*, 2005, *33*(7), 1248–55
Bölte S, Holtmann M, Poustka F et al, 'Gestalt perception and local-global processing in high-functioning autism', *Journal of Autism and Developmental Disorders*, 2007, *37*(8), 1493–1504
Bona S, Cattaneo Z & Silvanto J, 'The causal role of the occipital face area (OFA) and lateral occipital (LO) cortex in symmetry perception', *Journal of Neuroscience*, 2015, *35*(2), 731–8
Bona S, Herbert A, Toneatto C et al, 'The causal role of the lateral occipital complex in visual mirror symmetry detection and grouping: an fMRI-guided TMS study', *Cortex*, 2014, *51*, 46–55
Bonati B & Csermely D, 'Complementary lateralisation in the exploratory and predatory behaviour of the common wall lizard (*Podarcis muralis*)', *Laterality*, 2011, *16*(4), 462–70
Bonati B, Csermely D & Sovrano VA, 'Advantages in exploring a new environment with the left eye in lizards', *Behavioural Processes*, 2013a, *97*, 80–83
———, 'Looking at a predator with the left or right eye: asymmetry of response in lizards', *Laterality*, 2013b, *18*(3), 329–39
Bonato M, Priftis K, Umiltà C et al, 'Computer-based attention-demanding testing unveils severe neglect in apparently intact patients', *Behavioural Neurology*, 2013, *26*(3), 179–81
Bond HL, introduction to *Nicholas of Cusa: Selected Spiritual Writings*, trans HL Bond, Paulist Press, 1997
Bonda E, Petrides M, Ostry D et al, 'Specific involvement of human parietal systems and the amygdala in the perception of biological motion', *Journal of Neuroscience*, 1996, *16*(11), 3737–44
Bonfils KA, Haas GL & Salyers MP, 'Emotion-specific performance across empathy tasks in schizophrenia: influence of metacognitive capacity', *Schizophrenia Research: Cognition*, 2019, *19*, 100139
Bonfils KA, Lysaker PH, Minor KS et al, 'Affective empathy in schizophrenia: a meta-analysis', *Schizophrenia Research*, 2016, *175*(1–3), 109–17
Bonfils KA, Ventura J, Subotnik KL et al, 'Affective prosody and facial emotion recognition in first-episode schizophrenia: associations with functioning & symptoms', *Schizophrenia Research: Cognition*, 2019, *18*, 100153
Bonin G von, 'Anatomical asymmetries of the cerebral hemispheres', in VB Mountcastle (ed), *Interhemispheric Relations and Cerebral Dominance*, Johns Hopkins Press, 1962, 1–6
Bonnefon J-F, Hopfensitz A & De Neys W, 'The modular nature of trustworthiness detection', *Journal of Experimental Psychology: General*, 2013, *142*(1), 143–50
———, 'Face-ism and kernels of truth in facial inferences', *Trends in Cognitive Sciences*, 2015, *19*(8), 421–2
Bonneh YS, Pavlovskaya M, Ring H et al, 'Abnormal binocular rivalry in unilateral

neglect: evidence for a non-spatial mechanism of extinction', *NeuroReport*, 2004, 15, 473–7

Bonner JT, 'Brainless behavior: a myxomycete chooses a balanced diet', *Proceedings of the National Academy of Sciences of the United States of America*, 2010, 107(12), 5267–8

Bookheimer S, 'Functional MRI of language: new approaches to understanding the cortical organization of semantic processing', *Annual Review of Neuroscience*, 2002, 25, 151–88

Booth JR, Burman DD, Meyer JR et al, 'Functional anatomy of intra- and cross-modal lexical tasks', *NeuroImage*, 2002, 16(1), 7–22

Booth R & Happé F, '"Hunting with a knife and … fork": examining central coherence in autism, attention deficit/hyperactivity disorder, and typical development with a linguistic task', *Journal of Experimental Child Psychology*, 2010, 107(4–5), 377–93

Bora E, Gökçen S & Veznedaroglu B, 'Empathic abilities in people with schizophrenia', *Journal of Experimental Child Psychology*, 2008, 160(1), 23–9

Borah S, McConnell B, Hughes R et al, 'Potential relationship of self-injurious behavior to right temporo-parietal lesions', *Neurocase*, 2016, 22(3), 269–72

Borda JP & Sass LA, 'Phenomenology and neurobiology of self disorder in schizophrenia: primary factors', *Schizophrenia Research*, 2015, 169(1–3), 464–73

Borges JL, *Labyrinths*, trans JE Irby, Penguin, 1964a

———, *Other Inquisitions, 1937–1952*, trans RLC Simms, University of Texas Press, 1964b [*Otras Inquisiciones* 1952]

———, *Selected Poems*, trans S Kessler, Penguin, 1999

———, *Brodie's Report: including the Prose Fictions from In Praise of Darkness*, trans A Hurley, Penguin, 2000

Borgo F & Shallice T, 'When living things and other "sensory quality" categories behave in the same fashion: a novel category specificity effect', *Neurocase*, 2001, 7(3), 201–20

Borgomaneri S, Gazzola V & Avenanti A, 'Transcranial magnetic stimulation reveals two functionally distinct stages of motor cortex involvement during perception of emotional body language', *Brain Structure & Function*, 2015, 220(5), 2765–81

Bornstein B, 'Prosopagnosia', in *Problems in Dynamic Neurology*, ed L Halpern, Grune & Stratton, 1963, 283–318

Bornstein B, Sroka H & Munitz H, 'Prosopagnosia with animal face agnosia', *Cortex*, 1969, 5(2), 164–9

Bornstein RA, King G & Carroll A, 'Neuropsychological abnormalities in Gilles de la Tourette's syndrome', *Journal of Nervous and Mental Disease*, 1983, 171(8), 497–502

Borod JC, 'Cerebral mechanisms underlying facial, prosodic, and lexical emotional expression: a review of neuropsychological studies and methodological issues', *Neuropsychology*, 1993, 7(4), 445–63

Borod JC, Haywood CS & Koff E, 'Neuropsychological aspects of facial asymmetry during emotional expression: a review of the normal adult literature', *Neuropsychology Review*, 1997, 7(1), 41–60

Borod JC & Koff E, 'Hemiface mobility and facial expression asymmetry', *Cortex*, 1983, 19(3), 327–32

Borod JC, Martin CC, Alpert M et al, 'Perception of facial emotion in schizophrenic and right brain-damaged patients', *Journal of Nervous and Mental Disease*, 1993, 181(8), 494–502

Borod JC, Welkowitz J, Alpert M et al, 'Parameters of emotional processing in neuropsychiatric disorders: conceptual issues and a battery of tests', *Journal of Communication Disorders*, 1990, 23(4–5), 247–71

Boroditsky L & Gaby A, 'Remembrances of Times East: absolute spatial representations of time in an Australian Aboriginal community', *Psychological Science*, 2010, 12(11), 1635–9

Borovik AV, *Shadows of the Truth: Meta-mathematics of Elementary Mathematics*, American Mathematical Society, 2014: www.borovik.net/ST_Front_250414.pdf (accessed 12 June 2018)

Borowiecki KJ, 'How are you, my dearest Mozart? Well-being and creativity of three famous composers based on their letters', *The Review of Economics and Statistics*, posted online 8 June 2016

Bosco FM, Berardinelli L & Parola A, 'The ability of patients with schizophrenia to comprehend and produce sincere, deceitful, and ironic communicative intentions: the role of theory of mind and executive functions', *Frontiers in Psychology*, 2019, 10, article 827

Bosley TM, Rosenquist AC, Kushner M et al, 'Ischemic lesions of the occipital cortex and optic radiations: positron emission tomography', *Neurology*, 1985, 35(4), 470–84

Boster JS & Maltseva K, 'A crystal seen from each of its vertices: European views of European national characters', *Cross-Cultural Research*, 2006, 40(1), 47–64

Boswell J, *Life of Johnson*, Oxford University Press, 1953 [1791]

Botez-Marquard T & Botez MI, 'Cognitive behavior in heredodegenerative ataxias', *European Neurology*, 1993, 33(5), 351–7

Bottema-Beutel K & White R, 'By the book: an analysis of adolescents with autism spectrum condition co-constructing fictional narratives with peers', *Journal of Autism and Developmental Disorders*, 2016, 46(2), 361–77

Bottini G, Corcoran R, Sterzi R et al, 'The role of the right hemisphere in the interpretation of figurative aspects of language: a positron emission tomography activation study', *Brain*, 1994, 117(6), 1241–53

Bouchard F & Rosenberg A, 'Fitness, probability and the principles of natural selection', *British Journal for the Philosophy of Science*, 2004, 55(4), 693–712

Bouchard TJ, Lykken DT, McGue M et al, 'Sources of human psychological differences: the Minnesota Study of Twins Reared Apart', *Science*, 1990, 250(4978), 223–8

Bouckoms A, Martuza R & Henderson M, 'Capgras syndrome with subarachnoid hemorrhage', *Journal of Nervous and Mental Disease*, 1986, 174(8), 484–8

Bouillaud M-J, *Traité clinique et physiologique de l'encéphalite, ou inflammation du cerveau, et de ses suites*, J-B Baillière, 1825, 63–6

Boulenger V, Mechtouff L, Thobois S et al, 'Word processing in Parkinson's disease is impaired for action verbs but not for concrete nouns', *Neuropsychologia*, 2008, 46(2), 743–56

Boulenger V, Shtyrov Y & Pulvermüller F, 'When do you grasp the idea? MEG evidence for instantaneous idiom understanding', *NeuroImage*, 2012, 59(4), 3502–13

Bourgeault C, 'From the egoic mind to the mind of the heart: the teaching and lived experience of the Christian contemplative path', *Journal of Consciousness Studies*, 2016, 23(1–2), 45–57

Bourgeois A, Chica AB, Migliaccio R et al, 'Inappropriate rightward saccades after right hemisphere damage: oculomotor analysis and anatomical correlates', *Neuropsychologia*, 2015, 73(1), 1–11

Bourget D & Whitehurst L, 'Capgras syndrome: a review of the neurophysiological correlates and presenting clinical features in cases involving physical violence', *Canadian Journal of Psychiatry*, 2004, 49(11), 719–25

Bourguignon M, De Tiège X, de Beeck MO et al, 'The pace of prosodic phrasing couples the listener's cortex to the reader's voice', *Human Brain Mapping*, 2013, 34(2), 314–26

Bourne VJ, 'Lateralised processing of positive facial emotion: sex differences in strength of hemispheric dominance', *Neuropsychologia*, 2005, 43(6), 953–6

——, 'How are emotions lateralised in the brain? Contrasting existing hypotheses using the Chimeric Faces Test', *Cognition and Emotion*, 2010, 24(5), 903–11

Boutros NN, Lajiness-O'Neill R, Zillgitt A et al, 'EEG changes associated with autism spectrum disorders', *Neuropsychiatric Electrophysiology*, 2015, 1, article 3

Bowden EM & Beeman M, 'Getting the right idea: semantic activation in the right hemisphere may help solve insight problems', *Psychological Science*, 1998, 9(6), 435–40

Bowden EM & Jung-Beeman M, 'Aha! Insight experience correlates with solution activation in the right hemisphere', *Psychonomic Bulletin & Review*, 2003, 10(3), 730–7

Bowden EM, Jung-Beeman M, Fleck J et al, 'New approaches to demystifying insight', *Trends in Cognitive Sciences*, 2005, 9(7), 322–8

Bowen A, McKenna K & Tallis RC, 'Reasons for variability in the reported rate of occurrence of unilateral spatial neglect after stroke', *Stroke*, 1999, 30(6), 1196–1202

Bowers KS, Regehr G, Balthazard C et al, 'Intuition in the context of discovery', *Cognitive Psychology*, 1990, 22(1), 72–110

Bowie A, 'Friedrich Wilhelm Joseph von Schelling', in EN Zalta (ed), *The Stanford Encyclopedia of Philosophy* (Fall 2016 edn), plato.stanford.edu/archives/fall2016/entries/schelling/ (accessed 23 February 2021)

Bowler DE, Buyung-Ali LM, Knight TM et al, 'A systematic review of evidence for the added benefits to health of exposure to natural environments', *BMC Public Health*, 2010, 10, 456

Boyd R, 'On the current status of scientific realism', in R Boyd, P Gaspar & JD Trout (eds), *The Philosophy of Science*, Massachusetts Institute of Technology Press, 1991, 195–222

Boyer P, *The Naturalness of Religious Ideas: A Cognitive Theory of Religion*, University of California Press, 1994

Boyer T, 'The development of risk-taking: a multi-perspective review', *Developmental Review*, 2006, 26(3), 291–345

Boyle D, *The Tyranny of Numbers: Why Counting Can't Make Us Happy*, Flamingo, 2010

Boyle PA, Barnes LL, Buchman AS et al, 'Purpose in life is associated with mortality among community-dwelling older persons', *Psychosomatic Medicine*, 2009, 71(5), 574–9

Braam B, Huang X, Cupples WA et al, 'Understanding the two faces of low-salt intake', *Current Hypertension Reports*, 2017, 19(6), article 49

Bradley FH, *Appearance and Reality*, Oxford University Press, 2nd edn, 1897

Bradley JV, 'Overconfidence in ignorant experts', *Bulletin of the Psychonomic Society*, 1981, 17(2), 82–4

Bradshaw JL & Nettleton NC, 'The nature of hemispheric specialization in man', *Behavioral and Brain Sciences*, 1981, 4(1), 51–91

———, *Human Cerebral Asymmetry*, Prentice-Hall, 1983

Brancucci A, D'Anselmo A, Martello F *et al*, 'Left hemisphere specialization for duration discrimination of musical and speech sounds', *Neuropsychologia*, 2008, 46(7), 2013–9

Brancucci A, Lucci G, Mazzatenta A *et al*, '[Review] Asymmetries of the human social brain in the visual, auditory and chemical modalities', *Philosophical Transactions of the Royal Society of London: Series B, Biological Sciences*, 2009, 364(1519), 895–914

Brand A, Kopmann S, Marbach M *et al*, 'Gestalt problems in schizophrenia include intact and deficient feature fusion in schizophrenia', *European Archives of Psychiatry and Clinical Neuroscience*, 2005, 255(6), 413–8

Brand G & Brisson R, 'Lateralisation in wine olfactory threshold detection: comparison between experts and novices', *Laterality*, 2012, 17(5), 583–96

Brandimonte MA, Hitch GJ & Bishop DVM, 'Influence of short-term memory codes on visual image processing: evidence from image transformation tasks', *Journal of Experimental Psychology: Learning, Memory & Cognition*, 1992a, 18(1), 157–65

———, 'Verbal recoding of visual stimuli impairs mental image transformations', *Memory & Cognition*, 1992b, 20(4), 449–55

Brandt KR, Conway MA, James A *et al*, 'Déjà vu and the entorhinal cortex: dissociating recollective from familiarity disruptions in a single case patient', *Memory*, 2018, 7, 1–10

Brang D, McGeoch PD & Ramachandran VS, 'Apotemnophilia: a neurological disorder', *NeuroReport*, 2008, 19(13), 1305–6

Braque G, 'Le jour et la nuit', *Cahiers:1917–1952*, Gallimard, Paris, 1952

Brass M, Bekkering H & Prinz W, 'Movement observation affects movement execution in a simple response task', *Acta Psychologica (Amsterdam)*, 2001, 106(1–2), 3–22

Brass M & Haggard P, 'To do or not to do: the neural signature of self-control', *Journal of Neuroscience*, 2007, 27(34), 9141–5

Brassaï (Gyula Halász), *Conversations with Picasso*, University of Chicago Press, 2nd edn, 1999 [1964]

Bratsberg B & Rogeberg O, 'Flynn effect and its reversal are both environmentally caused', *Proceedings of the National Academy of Sciences of the United States of America*, 2018, 115(26), 6674–8

Brattico E, Jacobsen T, De Baene W *et al*, 'Electrical brain responses to descriptive *versus* evaluative judgments of music', *Annals of the New York Academy of Sciences*, 2003, 999(1), 155–7

Braun CMJ, 'Estimation of interhemispheric dynamics from simple unimanual reaction time to extrafoveal stimuli', *Neuropsychology Review*, 1992, 3(4), 321–65

———, 'Evolution of hemispheric specialisation of antagonistic systems of management of the body's energy resources', *Laterality*, 2007, 12(5), 397–427

Braun CMJ, Archambault MA, Daigneault S *et al*, 'Right body side performance decrement in congenitally dyslexic children and left body side performance decrement in congenitally hyperactive children', *Neuropsychiatry, Neuropsychology and Behavioural Neurology*, 2000, 13(2), 89–100

Braun CMJ & Chouinard MJ, 'Is anorexia nervosa a neuropsychological disease?', *Neuropsychology Review*, 1992, 3, 171–212

Braun CMJ, Daigneault R, Gaudelet S *et al*, 'Diagnostic and Statistical Manual of Mental Disorders, Fourth Edition symptoms of mania: which one(s) result(s) more often from right than left hemisphere lesions?', *Comprehensive Psychiatry*, 2008, 49(5), 441–59

Braun CMJ, Delisle J, Guimond A *et al*, 'Post unilateral lesion response biases modulate memory: crossed double dissociation of hemispheric specialisations', *Laterality*, 2009, 14(2), 122–64

Braun CMJ, Delisle J, Suffren S *et al*, 'Atypical left-right balance of visuomotor awareness in adult ADHD (combined type) on a test of executive function', *Laterality*, 2013, 18(4), 385–406

Braun CMJ, Desjardins S, Gaudelet S *et al*, 'Psychic tonus, body schema and the parietal lobes: a multiple lesion case analysis', *Behavioural Neurology*, 2007, 18(2), 65–80

Braun CMJ, Dumont M, Duval J *et al*, 'Opposed left and right brain hemisphere contributions to sexual drive: a multiple lesion case analysis', *Behavioural Neurology*, 2003, 14(1–2), 55–61

———, 'Speech rate as a sticky switch: a multiple lesion case analysis of mutism and hyperlalia', *Brain and Language*, 2004, 89(1), 243–52

Braun CMJ, Duval J & Guimond A, 'Auditory hypergnosia as an example of psychic tonus in the temporal lobes: multiple case analyses', *Critical Reviews in Neurobiology*, 2005, 17(3–4), 145–60

Braun CMJ, Larocque C, Daigneault S et al, 'Mania, pseudomania, depression, and pseudodepression resulting from focal unilateral critical lesions', *Neuropsychiatry, Neuropsychology and Behavioral Neurology*, 1999, 12(1), 35–51

Braun CMJ & Suffren S, 'A general neuropsychological model of delusion', *Cognitive Neuropsychiatry*, 2011, 16(1), 1–39

Braver TS, Barch DM & Cohen JD, 'Cognition and control in schizophrenia: a computational model of dopamine and prefrontal function', *Biological Psychiatry*, 1999, 46(3), 312–28

Bray D, 'Genomics: molecular prodigality', *Science*, 2003, 299(5610), 1189–90

Brayne S, Lovelace H & Fenwick P, 'End-of-life experiences and the dying process in a Gloucestershire nursing home as reported by nurses and care assistants', *American Journal of Hospice and Palliative Medicine*, 2008, 25(3), 195–206

Breen N, Caine D & Coltheart M, 'Mirrored-self misidentification: two cases of focal onset dementia', *Neurocase*, 2001, 7(3), 239–54

Breen N, Caine D, Coltheart M et al, 'Towards an understanding of delusions of misidentification: four case studies', *Mind & Language*, 2000, 15(1), 74–110

Brégeat MP, Klein M, Thiébaut F et al, 'Hémi-macropsie homonyme droite et tumeur occipitale gauche', *Revue d'oto-neuro-ophtalmologie*, 1947, 19, 239–40

Breitenfeld T, Solter VV, Breitenfeld D et al, 'Johann Sebastian Bach's strokes', *Acta Clinica Croatica*, 2006, 45, 41–4

Breitenstein C, Daum I & Ackermann H, 'Emotional processing following cortical and subcortical brain damage: contribution of the fronto-striatal circuitry', *Behavioural Neurology*, 1998, 11(1), 29–42

Bremer J, 'New onset auditory hallucinations after right temporal lobectomy', *American Journal of Psychiatry*, 1996, 153(3), 442–3

Brenner S, 'In theory', *Current Biology*, 1997, 7(3), R202

Brent J, *Charles Sanders Peirce: A Life*, Indiana University Press, 2nd edn, 1998

Brewer AA, Liu J, Wade AR et al, 'Visual field maps and stimulus selectivity in human ventral occipital cortex', *Nature Neuroscience*, 2005, 8(8), 1102–9

Brewin CR, Dalgleish T & Joseph S, 'A dual representation theory of posttraumatic stress disorder', *Psychological Review*, 1996, 103(4), 670–86

Bridge H, 'Effects of cortical damage on binocular depth perception', *Philosophical Transactions of the Royal Society of London: Series B, Biological Sciences*, 2016, 371(1697), 20150254

Brink PA, 'Article visibility: journal impact factor and availability of full text in PubMed Central and open access', *Cardiovascular Journal of Africa*, 2013, 24(8), 295–6

Brink TL, 'Idiot savant with unusual mechanical ability: an organic explanation', *American Journal of Psychiatry*, 1980, 137(2), 250–1

Britz J, Landis T & Michel CM, 'Right parietal brain activity precedes perceptual alternation of bistable stimuli', *Cerebral Cortex*, 2009, 19(1), 55–65

Broca P, 'Remarques sur le siège de la faculté du langage articulé, suivies d'une observation d'aphémie', *Bulletin de la Société Anatomique*, 1861, 36, 330–57

Bröder A, 'Decision making with the "adaptive toolbox": influence of environmental structure, intelligence, and working memory load', *Journal of Experimental Psychology: Learning, Memory, and Cognition*, 2003, 29(4), 611–25

Brogaard B, Vanni S & Silvanto J, 'Seeing mathematics: perceptual experience and brain activity in acquired synesthesia', *Neurocase*, 2013, 19(6), 566–75

Broman DA, Olsson MJ & Nordin S, 'Lateralization of olfactory cognitive functions: effects of rhinal side of stimulation', *Chemical Senses*, 2001, 26(9), 1187–92

Bromberger B, Sternschein R, Widick P et al, 'The right hemisphere in esthetic perception', *Frontiers in Human Neuroscience*, 2011, 5, article 109

Bronowski J, 'The creative process' *Scientific American*, 1958, 199(3), 58–65

Brooks AC, 'How loneliness is tearing America apart', *The New York Times*, 23 November 2018, A25

Brooks RA, *Fields of Color: The Theory that Escaped Einstein*, Rodney A Brooks Publishing, 3rd edn, 2016

Brown HD & Kosslyn SM, 'Cerebral lateralization', *Current Opinion in Neurobiology*, 1993, 3(2), 183–6

Brown JW, 'Imagery and the microstructure of perception', *Journal of Neurolinguistics*, 1985, 1(1), 89–141

Brown JW (ed), *Agnosia and Apraxia: Selected Papers of Liepmann, Lange and Pötzl*, Lawrence Erlbaum, 1988

Brown PR, 'Independent auditor judgment in the evaluation of internal audit functions', *Journal of Accounting Research*, 1983, 21(2), 444–55

Brown RG, Lacomblez L, Landwehrmeyer BG et al, 'Cognitive impairment in patients with multiple system atrophy and progressive supranuclear palsy', *Brain*, 2010, 133(8), 2382–93

Brown S, Gao X, Tisdelle L et al, 'Naturalizing aesthetics: brain areas for aesthetic appraisal across sensory modalities,' *NeuroImage*, 2011, 58(1), 250–58

Brown S, Martinez M & Parsons L, 'Music and language side by side: a PET study of the generation of sentences', *European Journal of Neuroscience*, 2006, 23(10), 2791–2803

Brown S & Nicholls ME, 'Hemispheric asymmetries for the temporal resolution of brief auditory stimuli', *Perception & Psychophysics*, 1997, 59(3), 442–7

Brown WA, *The Placebo Effect in Clinical Practice*, Oxford University Press, 2013

Brown WS & Paul LK, 'Cognitive and psychosocial deficits in agenesis of the corpus callosum with normal intelligence', *Cognitive Neuropsychiatry*, 2000, 5(2), 135–57

Brown WS, Symington ['Symigton'] M, VanLancker-Sidtis D et al, 'Paralinguistic processing in children with callosal agenesis: emergence of neurolinguistic deficits', *Brain and Language*, 2005, 93(2), 135–9

Brownell HH, Michel D, Powelson J et al, 'Surprise but not coherence: sensitivity to verbal humor in right-hemisphere patients', *Brain and Language*, 1983, 18(1), 20–7

Brownell HH, Potter HH, Bihrle AM et al, 'Inference deficits in right brain-damaged patients', *Brain and Language*, 1986, 27(2), 310–21

Brownell HH, Potter HH, Michelow D et al, 'Sensitivity to lexical denotation and connotation in brain-damaged patients: a double dissociation?', *Brain and Language*, 1984, 22(2), 253–65

Brownell HH, Simpson TL, Bihrle AM et al, 'Appreciation of metaphoric alternative word meanings by left and right brain-damaged patients', *Neuropsychologia*, 1990, 28(4), 375–83

Brownlee J, 'The golden ratio: design's biggest myth', 13 April 2015: www.fastcompany.com/3044877/the-golden-ratio-designs-biggest-myth (accessed 3 July 2021)

Bruce JP, *Chu Hsi and His Masters*, Probsthain & Co, 1973 [1923]

Bruell JH & Albee GW, 'Higher intellectual functions in a patient with hemispherectomy for tumors', *Journal of Consulting Psychology*, 1962, 26(1), 90–8

Bruen PD, McGeown WJ, Shanks MF et al, 'Neuroanatomical correlates of neuropsychiatric symptoms in Alzheimer's disease', *Brain*, 2008, 131(9), 2455–63

Brugger P, Blanke O, Regard M et al, 'Polyopic heautoscopy: case report and review of the literature', *Cortex*, 2006, 42(5), 666–74

Brugger P, Kollias SS, Müri RM et al, 'Beyond re-membering: phantom sensations of congenitally absent limbs', *Proceedings of the National Academy of Sciences of the United States of America*, 2000, 97(11), 6167–72

Brugger P, Monsch AU & Johnson SA, 'Repetitive behavior and repetition avoidance: the role of the right hemisphere', *Journal of Psychiatry & Neuroscience*, 1996, 21(1), 53–6

Brugger P, Regard M & Landis T, 'Unilaterally felt "presences": the neuropsychiatry of one's invisible *Doppelgänger*', *Neuropsychiatry, Neuropsychology & Behavioral Neurology*, 1996, 9(2), 114–22

——, 'Illusory reduplication of one's own body: phenomenology and classification of autoscopic phenomena', *Cognitive Neuropsychiatry*, 1997, 2(1), 19–38

Brumm K, Walenski M, Haist F et al, 'Functional magnetic resonance imaging of a child with Alice in Wonderland syndrome during an episode of micropsia', *Journal of the American Association for Pediatric Ophthalmology and Strabismus*, 2010, 14(4), 317–22

Brüne M, Schöbel A, Karau R et al, 'Von Economo neuron density in anterior cingulate cortex is reduced in early onset schizophrenia', *Acta Neuropathologica*, 2010, 119(6), 771–8

Brüne M & Schröder SG, 'Erotomania variants in dementia', *Journal of Geriatric Psychiatry and Neurology*, 2003, 16(4), 232–4

Brunia CH & Damen EJ, 'Distribution of slow brain potentials related to motor preparation and stimulus anticipation in a time estimation task', *Electroencephalography and Clinical Neurophysiology*, 1988, 69(3), 234–43

Brunyé TT, Holmes A, Cantelon J et al, 'Direct current brain stimulation enhances navigation efficiency in individuals with low spatial sense of direction', *NeuroReport*, 2014, 25(15), 1175–9

Brust JCM & Behrens MM, '"Release hallucinations" as the major symptom of posterior cerebral artery occlusion: a report of 2 cases', *Annals of Neurology*, 1977, 2(5), 432–6

Bryan KL, 'Assessment of language disorders after right hemisphere damage', *British Journal of Disorders of Communication*, 1988, 23(2), 111–25

Bryden MP, 'Response bias and hemispheric difference in dot localisation', *Perception & Psychophysics*, 1976, 19(1), 23–8

——, 'Evidence for sex differences in cerebral organization', in MA Wittig M & AC Peterson (eds), *Sex-Related Differences in Cognitive Functioning: Developmental Issues*, Academic Press, 1979, 121–43

Brysbaert M, 'Lateral preferences and visual-field asymmetries – appearances may have been overstated', *Cortex*, 1994, *30*(3), 413–29

Bubl E, Kern E, Ebert D et al, 'Seeing gray when feeling blue? Depression can be measured in the eye of the diseased', *Biological Psychiatry*, 2010, *68*(2), 205–8

Bubl E, Tebartz Van Elst L, Gondan M et al, 'Vision in depressive disorder', *World Journal of Biological Psychiatry*, 2009, *10*(4, Pt 2), 377–84

Buchanan DC, Waterhouse GJ & West SC Jr, 'A proposed neurophysiological basis of alexithymia', *Psychotherapy and Psychosomatics*, 1980, *34*(4), 248–55

Buchanan M, 'Another kind of evolution', *New Scientist*, 23 January 2010, 34–7

Buchanan TW, Tranel D & Adolphs R, 'Memories for emotional autobiographical events following unilateral damage to medial temporal lobe', *Brain*, 2006, *129*(1), 115–27

Budson AE, Roth HL, Rentz DM et al, 'Disruption of the ventral visual stream in a case of reduplicative paramnesia', *Annals of the New York Academy of Sciences*, 2000, *911*(1), 447–52

Buhner SH, *Plant Intelligence and the Imaginal Realm*, Bear, 2014

Bullmore E & Sporns O, 'Complex brain networks: graph theoretical analysis of structural and functional systems', *Nature Reviews Neuroscience*, 2009, *10*(3), 186–98

Bullock TH & Horridge GA, *Structure and Function in the Nervous Systems of Invertebrates*, WH Freeman & Co, 1965

Buratto LG, Zimmermann N, Ferré P et al, 'False memories to emotional stimuli are not equally affected in right- and left-brain-damaged stroke patients', *Brain and Cognition*, 2014, *90*, 181–94

Burbach JPH & van der Zwaag B, 'Contact in the genetics of autism and schizophrenia', *Trends in Neurosciences*, 2009, *32*(2), 69–72

Burchard JM, 'Zur Frage nach der Natur von Phantomerlebnissen bei angeborener Gliedmaßenverstümmelung', *Archiv für Psychiatrie und Nervenkrankheiten*, 1965, *207*(4), 360–77

Burchett JN, Elek O, Tejos N et al, 'Revealing the dark threads of the cosmic web', *Astrophysical Journal Letters*, 2020, *891*(2), L35

Burgess N, Maguire EA & O'Keefe J, 'The human hippocampus and spatial and episodic memory', *Neuron*, 2002, *35*(4), 625–41

Burgess PW, 'Strategy application disorder: the role of the frontal lobes in human multitasking', *Psychological Research*, 2000, *63*(3–4), 279–88

Burgess PW, Baxter D, Rose M et al, 'Delusional paramnesic misidentification', in PW Halligan & JC Marshall (eds), *Method in Madness: Case Studies in Cognitive Neuropsychiatry*, Psychology Press, 1996, 51–78

Burgess PW, Veitch E, de Lacy Costello A et al, 'The cognitive and neuroanatomical correlates of multitasking', *Neuropsychologia*, 2000, *38*(6), 848–63

Bürgy M, 'Zur Phaenomenologie der Verzweiflung bei der Schizophrenie', *Zeitschrift für klinische Psychologie, Psychiatrie und Psychotherapie*, 2003, *51*(1), 1–16

Burian RM, 'Reconceiving animals and their evolution: on some consequences of new research on the modularity of development and evolution', in *The Epistemology of Development, Evolution, and Genetics* (Cambridge Studies in Philosophy and Biology), Cambridge University Press, 2005, 234–61

Burke E, *The Works of the Right Hon Edmund Burke*, John C Nimmo (London), 12 vols, 1887

Burr HS & Northrop FSC, 'The electrodynamic theory of life', *Quarterly Review of Biology*, 1935, *10*(3), 322–33

Burrows D, *Handel*, Oxford University Press, 1994

Burt DM & Perrett DI, 'Perceptual asymmetries in judgements of facial attractiveness, age, gender, speech and expression', *Neuropsychologia*, 1997, *35*(5), 685–93

Bury G, García-Huéscar M, Bhattacharya J et al, 'Cardiac afferent activity modulates early neural signature of error detection during skilled performance', *NeuroImage*, 2019, *199*, 704–17

Bushdid C, Magnasco MO, Vosshall LB et al, 'Humans can discriminate more than 1 trillion olfactory stimuli', *Science*, 2014, *343*(6177), 1370–2

Busigny T, Joubert S, Felician O et al, 'Holistic perception of the individual face is specific and necessary: evidence from an extensive case study of acquired prosopagnosia', *Neuropsychologia*, 2010, *48*(14), 4057–92

Busigny T, Van Belle G, Jemel B et al, 'Face-specific impairment in holistic perception following focal lesion of the right anterior temporal lobe', *Neuropsychologia*, 2014, *56*, 312–33

Butler AC, Chapman JE, Forman EM et al, 'The empirical status of cognitive-behavioral therapy: a review of meta-analyses', *Clinical Psychology Review*, 2006, *26*(1), 17–31

Butler PM, McNamara P & Durso R, 'Side of onset in Parkinson's disease and alterations in religiosity: novel behavioral phenotypes', *Behavioural Neurology*, 2011, *24*(2), 133–41

Butler PV, 'Diurnal variation in Cotard's syndrome (copresent with Capgras delusion) following traumatic brain injury', *Australian*

& New Zealand Journal of Psychiatry, 2000, 34(4), 684–7

Butler S, Gilchrist ID, Burt DM et al, 'Are the perceptual biases found in chimeric face processing reflected in eye-movement patterns?', Neuropsychologia, 2005, 43(1), 2–9

Butler SH & Harvey M, 'Does inversion abolish the left chimeric face processing advantage?', NeuroReport, 2005, 16(18), 1991–3

Buttarelli FR, Pellicano C & Pontieri FE, 'Neuropharmacology and behavior in planarians: translations to mammals', Comparative Biochemistry and Physiology Part C: Toxicology & Pharmacology, 2008, 147(4), 399–408

Butters N, Barton M & Brody BA, 'Role of the right parietal lobe in the mediation of cross-modal associations and reversible operations in space', Cortex, 1970, 6, 174–90

Butti C, Sherwood CC, Hakeem AY et al, 'Total number and volume of von Economo neurons in the cerebral cortex of cetaceans', Journal of Comparative Neurology, 2009, 515(2), 243–59

Button KS, Ioannidis JPA, Mokrysz C et al, 'Power failure: why small sample size undermines the reliability of neuroscience', Nature Reviews Neuroscience, 2013, 14(5), 365–76

Buxbaum LJ, Ferraro MK, Veramonti T et al, 'Hemispatial neglect: subtypes, neuroanatomy, and disability', Neurology, 2004, 62(5), 749–56

Buxbaum LJ, Kyle KM & Menon R, 'On beyond mirror neurons: internal representations subserving imitation and recognition of skilled object-related actions in humans', Brain Research: Cognitive Brain Research, 2005, 25(1), 226–39

Buzsáki G, Rhythms of the Brain, Oxford University Press, 2006

Byers W, How Mathematicians Think: Using Ambiguity, Contradiction, and Paradox to Create Mathematics, Princeton University Press, 2007

Bylund E & Athanasopoulos P, 'The Whorfian time warp: representing duration through the language hourglass', Journal of Experimental Psychology: General, 2017, 146(7), 911–6

Byrd W, from the 'Preface to Gradualia I' (1605), in P Brett (ed), The Byrd Edition, vol 5, Stainer & Bell, 1989

Byrne RA, Kuba M & Griebel U, 'Lateral asymmetry of eye use in Octopus vulgaris', Animal Behaviour, 2002, 64(3), 461–8

Cabeza R, 'Hemispheric asymmetry reduction in older adults: the HAROLD model', Psychology and Aging, 2002, 17(1), 85–100

Cacioppo JT, Berntson GG, Lorig TS et al, 'Just because you're imaging the brain doesn't mean you can stop using your head: a primer and set of first principles', Journal of Personality and Social Psychology, 2003, 85(4), 650–61

Caeiro L, Ferro JM & Costa J, 'Apathy secondary to stroke: a systematic review and meta-analysis', Cerebrovascular Diseases, 2013, 35(1), 23–39

Cahill L, 'Why sex matters for neuroscience', Nature Reviews Neuroscience, 2006, 7(6), 477–84

Cahill L, Gorski L, Belcher A et al, 'The influence of sex versus sex-related traits on long-term memory for gist and detail from an emotional story', Consciousness and Cognition, 2004, 13(2), 391–400

Cahill L, Uncapher M, Kilpatrick L et al, 'Sex-related hemispheric lateralization of amygdala function in emotionally influenced memory: an fMRI investigation', Learning & Memory, 2004, 11(3), 261–6

Cahill L & van Stegeren A, 'Sex-related impairment of memory for emotional events with beta-adrenergic blockade', Neurobiology of Learning and Memory, 2003, 79(1), 81–8

Cairns J, Overbaugh J & Miller S, 'The origin of mutants', Nature, 1988, 335(6186), 142–5

Calabrese EJ, Baldwin LA & Holland CD, 'Hormesis: a highly generalizable and reproducible phenomenon with important implications for risk assessment', Risk Analysis, 1999, 19(2), 261–81

Calabrese EJ & Mattson MP, 'How does hormesis impact biology, toxicology, and medicine?', NPJ Aging and Mechanisms of Disease, 2017, 3, article 13

Calabrò RS, Baglieri A, Ferlazzo E et al, 'Neurofunctional assessment in a stroke patient with musical hallucinations', Neurocase, 2012, 18(6), 514–20

Calder AJ, Keane J, Manes F et al, 'Impaired recognition and experience of disgust following brain injury', Nature Neuroscience, 2000, 3(11), 1077–8

Calder T, 'The concept of evil', The Stanford Encyclopedia of Philosophy (Fall 2018 edition), ed EN Zalta: plato.stanford.edu/archives/fall2018/entries/concept-evil/ (accessed 23 February 2021)

Caldwell-Harris C, Murphy CF, Velazquez T et al, 'Religious belief systems of persons with high functioning autism', Proceedings of the Annual Meeting of the Cognitive Science Society, 2011, 3362–6

Caligiuri MP, Brown GG, Meloy MJ et al, 'A functional magnetic resonance imaging study of cortical asymmetry in bipolar disorder', Bipolar Disorders, 2004, 6(3), 183–96

Caliyurt O, Vardar E & Tuglu C, 'Cotard's syndrome with schizophreniform disorder can be successfully treated with electroconvulsive therapy: case report', *Journal of Psychiatry & Neuroscience*, 2004, 29(2), 138–41

Callaway E, 'Faked peer reviews prompt 64 retractions', *Nature*, 2015: doi:10.1038/nature.2015.18202

Calvo-Merino B, Jola C, Glaser DE et al, 'Towards a sensorimotor aesthetics of performing art', *Consciousness and Cognition*, 2008, 17(3), 911–22

Calvo-Merino B, Urgesi C, Orgs G et al, 'Extrastriate body area underlies aesthetic evaluation of body stimuli', *Experimental Brain Research*, 2010, 204(3), 447–56

Camazine S, Deneubourg J-L, Franks NR et al, *Self-Organization in Biological Systems* (Princeton Studies in Complexity), Princeton University Press, 2001, 7–8

Campbell J, in R Walter (ed), *A Joseph Campbell Companion: Reflections on the Art of Living*, Joseph Campbell Foundation, 1991: billmoyers.com/content/ep-2-joseph-campbell-and-the-power-of-myth-the-message-of-the-myth/ (accessed 13 March 2021)

Campbell R, Landis T & Regard M, 'Facial recognition and lip-reading, a neurological dissociation', *Brain*, 1986, 109(3), 509–21

Campos GB & Welker WI, 'Comparisons between brains of a large and a small hystricomorph rodent: capybara, *Hydrochoerus* and guinea pig, *Cavia*; neocortical projection regions and measurements of brain subdivisions', *Brain, Behavior and Evolution*, 1976, 13(4), 243–66

Canales J, 'Einstein's Bergson problem: communication, consensus and good science', in Y Dolev & M Roubach, *Cosmological and Psychological Time*, Springer, 2016a, 53–72

———, *The Physicist and the Philosopher: Einstein, Bergson, and the Debate That Changed Our Understanding of Time*, Princeton University Press, 2016b

Canavero S, Bonicalzi V, Castellano G et al, 'Painful supernumerary phantom arm following motor cortex stimulation for central poststroke pain: case report', *Journal of Neurosurgery*, 1999, 91(1), 121–3

Canfield AR, Eigsti IM, de Marchena A et al, 'Story goodness in adolescents with autism spectrum disorder (ASD) and in optimal outcomes from ASD', *Journal of Speech, Language, and Hearing Research*, 2016, 59(3), 533–45

Canli T, Desmond JE, Zhao Z et al, 'Sex differences in the neural basis of emotional memories', *Proceedings of the National Academy of Sciences of the United States of America*, 2002, 99(16), 10789–94

Canning D, 'Ma' [undated], new.uniquejapan.com/ikebana/ma/ (accessed 23 February 2021)

Cannon WB, *The Way of an Investigator*, Norton, 1945

Cao TY, *Conceptual Foundations of Quantum Field Theory*, Cambridge University Press, 1999

Cao Y, Willett WC, Rimm EB et al, 'Light to moderate intake of alcohol, drinking patterns, and risk of cancer: results from two prospective US cohort studies', *BMJ*, 2015, 351, h4238

Cao Z, Zhao Y, Tan T et al, 'Distinct brain activity in processing negative pictures of animals and objects — the role of human contexts', *NeuroImage*, 2013, 84, 901–10

Capa RL, Duval CZ, Blaison D et al, 'Patients with schizophrenia selectively impaired in temporal order judgments', *Schizophrenia Research*, 2014, 156(1), 51–5

Capaldi CA, Dopko RL & Zelenski JM, 'The relationship between nature connectedness and happiness: a meta-analysis', *Frontiers in Psychology*, 2014, 5, article 976

Capitani E, Laiacona M, Mahon B et al, 'What are the facts of semantic category-specific deficits? A critical review of the clinical evidence', *Cognitive Neuropsychology*, 2003, 20(3), 213–61

Caplan R & Dapretto M, 'Making sense during conversation: an fMRI study', *NeuroReport*, 2001, 12, 3625–32

Caporale LH (ed), *The Implicit Genome*, Oxford University Press, 2006

Capozzoli NJ, 'Why are vertebrate nervous systems crossed?', *Medical Hypotheses*, 1995, 45(5), 471–5

Cappa SF, Perani D, Schnur T et al, 'The effects of semantic category and knowledge type on lexical-semantic access: a PET study', *NeuroImage*, 1998, 8(4), 350–9

Cappa SF & Pulvermüller F, 'Language and the motor system', *Cortex*, 2012, 48(7), 785–7

Cappa S, Sterzi R, Vallar G et al, 'Remission of hemineglect and anosognosia during vestibular stimulation', *Neuropsychologia*, 1987, 25(5), 775–82

Caramazza A, Gordon J, Zurif EB et al, 'Right hemispheric damage and verbal problem solving behavior', *Brain and Language*, 1976, 3(1), 41–6

Caramazza A & Shelton JR, 'Domain-specific knowledge systems in the brain: the animate-inanimate distinction', *Journal of Cognitive Neuroscience*, 1998, 10(1), 1–34

Carassiti D, Altmann DR, Petrova N et al, 'Neuronal loss, demyelination and volume change in the multiple sclerosis neocortex',

Neuropathology and Applied Neurobiology, 2018, 44(4), 377–90

Cardeña E, 'A call for an open, informed study of all aspects of consciousness', Frontiers in Human Neuroscience, 2014, 8, article 17

Cardeña E, Lynn SJ & Krippner S, 'The psychology of anomalous experiences: a rediscovery', Psychology of Consciousness: Theory, Research, and Practice, 2017, 4(1), 4–22

Cardillo ER, Watson CE, Schmidt GL et al, 'From novel to familiar: tuning the brain for metaphors', NeuroImage, 2012, 59(4), 3212–21

Carel HH, 'Illness, phenomenology, and philosophical method', Theoretical Medicine and Bioethics, 2013, 34(4), 345–57

Carl N, 'Does intelligence explain the association between generalized trust and economic development?', Intelligence, 2014, 47, 83–92

Carl N & Billari FC, 'Generalized trust and intelligence in the United States', PLoS One, 2014, 9(3), e91786

Carlsson I, 'Anxiety and flexibility of defense related to high or low creativity', Creativity Research Journal, 2002, 14(3–4), 341–9

Carlsson I, Wendt P & Risberg J, 'On the neurobiology of creativity: differences in frontal lobe activity between high and low creative subjects', Neuropsychologia, 2000, 38(6), 873–85

Carly PG, Golding SJJ & Hall BJD, 'Interrelationships among auditory and visual cognitive tasks: an event-related potential (ERP) study', Intelligence, 1995, 2(3), 297–327

Carmel D, Walsh V, Lavie N et al, 'Right parietal TMS shortens dominance durations in binocular rivalry', Current Biology, 2010, 20(18), R799–800

Carmon A & Bechtoldt HP, 'Dominance of the right cerebral hemisphere for stereopsis', Neuropsychologia, 1969, 7(1), 29–39

Carmon A & Nachshon I, 'Effects of unilateral brain damage on the perception of temporal order', Cortex, 1971, 7(4), 411–8

———, 'Ear asymmetry in perception of emotional non-verbal stimuli', Acta Psychologica, 1973, 37(6), 351–7

Carnap R, 'Überwindung der Metaphysik durch logische Analyse der Sprache', Erkenntnis, 1931, 2(1), 219–41

Carne RP, Vogrin S, Litewka L et al, 'Cerebral cortex: an MRI-based study of volume and variance with age and sex', Journal of Clinical Neuroscience, 2006, 13(1), 60–72

Carper BA, 'Fundamental patterns of knowing in nursing', Advances in Nursing Science, 1978, 1(1), 13–23

Carper RA, Treiber JM, DeJesus SY et al, 'Reduced hemispheric asymmetry of white matter microstructure in autism spectrum disorder', Journal of the American Academy of Child and Adolescent Psychiatry, 2016, 55(12), 1073–80

Carran MA, Kohler CG, O'Connor MJ et al, 'Mania following temporal lobectomy', Neurology, 2003, 61(6), 770–4

Carrera E & Tononi G, 'Diaschisis: past, present, future', Brain, 2014, 137(9), 2408–22

Carrington SJ & Bailey AJ, 'Are there theory of mind regions in the brain? A review of the neuroimaging literature', Human Brain Mapping, 2009, 30(8), 2313–35

Carroll CA, Boggs J, O'Donnell BF et al, 'Temporal processing dysfunction in schizophrenia', Brain and Cognition, 2008, 67(2), 150–61

Carroll JB, 'A survey and analysis of correlational and factor-analytic research on cognitive abilities: overview of outcomes', in Carroll JB, Human Cognitive Abilities: A Survey of Factor-Analytic Studies, Cambridge University Press, 1993a, 115–42

———, Human Cognitive Abilities: A Survey of Factor-Analytic Studies, Cambridge University Press, 1993b

Carroll L, 'What the tortoise said to Achilles', Mind, 1895, 4(14), 278–80

Carroll LS & Owen MJ, 'Genetic overlap between autism, schizophrenia and bipolar disorder', Genome Medicine, 2009, 1(10), article 102

Carse JP, Finite and Infinite Games, Free Press, 2012

Carson AJ, MacHale S, Allen K et al, 'Depression after stroke and lesion location: a systematic review', Lancet, 2000, 356(9224), 122–6

Carson RG, Chua R, Goodman D et al, 'The preparation of aiming movements', Brain and Cognition, 1995, 28(2), 133–54

Carter CS, Robertson LC, Nordahl TE et al, 'Perceptual and attentional asymmetries in schizophrenia: further evidence for a left hemispheric deficit', Journal of Experimental Child Psychology, 1996, 62(2), 111–9

Carter G, Milner A, McGill K et al, 'Predicting suicidal behaviours using clinical instruments: systematic review and meta-analysis of positive predictive values for risk scales', British Journal of Psychiatry, 2017, 210(6), 387–95

Cartwright N, How the Laws of Physics Lie, Oxford University Press, 1983

Casagrande M & Bertini M, 'Night-time right hemisphere superiority and daytime left hemisphere superiority: a repatterning of laterality across wake-sleep-wake states', Biological Psychiatry, 2008, 77(3), 337–42

Casagrande M, Violani C, De Gennaro L et al, 'Which hemisphere falls asleep first?', *Neuropsychologia*, 1995, 33(7), 815–22

Casal J, Struhl G & Lawrence PA, 'Developmental compartments and planar polarity in Drosophila', *Current Biology*, 2002, 12(14), 1189–98

Casasanto D & Boroditsky L, 'Time in the mind: using space to think about time', *Cognition*, 2008, 106(2), 579–93

Casasanto D, Fotakopoulou O & Boroditsky L, 'Space and time in the child's mind: evidence for a cross-dimensional asymmetry', *Cognitive Science*, 2010, 34(3), 387–405

Caspar M, *Kepler*, Dover, 1993

Cassirer E, *An Essay on Man: An Introduction to a Philosophy of Human Culture*, Yale University Press, 1977 [1944]

Cassvan A, Ross PL, Dyer PR et al, 'Lateralization in stroke syndromes as a factor in ambulation', *Archives of Physical Medicine and Rehabilitation*, 1976, 57(12), 583–7

Castelli F, Happé F, Frith U et al, 'Movement and mind: a functional imaging study of perception and interpretation of complex intentional movement patterns', *NeuroImage*, 2000, 12(3), 314–25

Castillo H, Schoderbek D, Dulal S et al, 'Stress induction in the bacteria Shewanella oneidensis and Deinococcus radiodurans in response to below-background ionizing radiation', *International Journal of Radiation Biology*, 2015, 91(9), 749–56

Cattaneo Z, Lega C, Ferrari C et al, 'The role of the lateral occipital cortex in aesthetic appreciation of representational and abstract paintings: a TMS study', *Brain and Cognition*, 2015, 95, 44–53

Cattaneo Z, Lega C, Flexas A et al, 'The world can look better: enhancing beauty experience with brain stimulation', *Social Cognitive and Affective Neuroscience*, 2014, 9(11), 1713–21

Cattell RB, *Abilities: Their Structure, Growth, and Action*, Houghton Mifflin, 1971

Caudill W & Weinstein H, 'Maternal care and infant behavior in Japan and America', *Psychiatry*, 1969, 32(1), 12–43

Cavell S, *Must We Mean What We Say?*, Cambridge University Press, 1976

Cavill P, *Maxims in Old English Poetry*, Boydell & Brewer, 1999

Ceci S & Peters DP, 'Peer review: a study of reliability,' *Change*, 1982, 14(6), 44–8

Cela-Conde CJ, Ayala FJ, Munar E et al, 'Sex-related similarities and differences in the neural correlates of beauty', *Proceedings of the National Academy of Sciences of the United States of America*, 2009, 106(10), 3847–52

Celec P, Ostatníková D & Hodosy J, 'On the effects of testosterone on brain behavioral functions', *Frontiers in Neuroscience*, 2015, 9, article 12

Cereda C, Ghika J, Maeder P et al, 'Strokes restricted to the insular cortex', *Neurology*, 2002, 59(12), 1950–5

Cerf-Ducastel B & Murphy C, 'fMRI activation in response to odorants orally delivered in aqueous solutions', *Chemical Senses*, 2001, 26(6), 625–37

Ceriani F, Gentileschi V, Muggia S et al, 'Seeing objects smaller than they are: micropsia following right temporo-parietal infarction', *Cortex*, 1998, 34(1), 131–8

Černigovskaja [usu Chernigovskaya: vid inf] TV, 'Die Heterogenität des verbalen Denkens als cerebrale Asymmetrie', in P Grzybek (ed), *Psychosemiotik-Neurosemiotik*, Universitätsverlag Dr Norbert Brockmeyer, Bochum, 1993, 15–35

Cerrato P, Imperiale D, Giraudo M et al, 'Complex musical hallucinosis in a professional musician with a left subcortical haemorrhage', *Journal of Neurology, Neurosurgery, and Psychiatry*, 2001, 71(2), 280–1

Cerruti C & Schlaug G, 'Anodal transcranial direct current stimulation of the prefrontal cortex enhances complex verbal associative thought', *Journal of Cognitive Neuroscience*, 2009, 21(10), 1980–7

Chaitin G, 'The limits of reason', *Scientific American*, 2006, 294(3), 74–81

Chalmers D, 'Facing up to the problem of consciousness', *Journal of Consciousness Studies*, 1995, 2(3), 200–19

Chamberlain D, 'Babies don't feel pain: a century of denial in medicine': presentation by David Chamberlain, President of the Association of Pre- and Perinatal Psychology and Health, to the Second International Symposium on Circumcision, San Francisco, California, 2 May 1991

Chambers R, *Rural Development: Putting the Last First*, Longman Scientific and Technical, 1983

Chaminade T & Decety J, 'Leader or follower? Involvement of the inferior parietal lobule in agency', *NeuroReport*, 2002, 13(15), 1975–8

Chamovitz D, interviewed by Gareth Cook, in 'Do plants think?', *Scientific American*, 5 June 2012a

———, *What a Plant Knows: A Field Guide to the Senses*, Farrar, Straus & Giroux, 2012b

Champagne FA, Weaver IC, Diorio J et al, 'Maternal care associated with methylation of the estrogen receptor-alpha1b promoter and estrogen receptor-alpha expression in the medial preoptic area of female offspring', *Endocrinology*, 2006, 147(6), 2909–15

Champagne-Lavau M, Cordonier N, Bellmann A et al, 'Context processing during irony comprehension in right-frontal brain-damaged individuals', Clinical Linguistics & Phonetics, 2018, 32(8), 721–38

Chan D, Anderson V, Pijnenburg Y et al, 'The clinical profile of right temporal lobe atrophy', Brain, 2009, 132(5), 1287–98

Chan W, McCrae RR, De Fruyt F et al, 'Stereotypes of age differences in personality traits: universal and accurate?', Journal of Personality and Social Psychology, 2012, 103(6), 1050–66

Chandler MJ, Lalonde CE, Sokol BW et al, 'Personal persistence, identity development, and suicide: a study of Native and non-Native North American adolescents', Monographs of the Society for Research in Child Development, 2003, 68(2), 1–130

Chandos M, Kosmautikon, Xlibris, 2015

Chanen AM, Velakoulis D, Carison K et al, 'Orbitofrontal, amygdala and hippocampal volumes in teenagers with first-presentation borderline personality disorder', Psychiatry Research, 2008, 163(2), 116–25

Chang CY, Hammitt WE, Chen PK et al, 'Psychophysiological responses and restorative value of natural environments in Taiwan', Landscape and Urban Planning, 2008, 85(2), 79–84

Chao LL, Weisberg J & Martin A, 'Experience-dependent modulation of category-related cortical activity', Cerebral Cortex, 2002, 12(5), 545–51

Chapanis L, 'Language deficits and cross-modal sensory perception', in SJ Segalowitz & FA Gruber (eds), Language Development and Neurological Theory, Academic Press, 1977, 107–20

Chapman J, 'The early symptoms of schizophrenia', British Journal of Psychiatry, 1966, 112(484), 225–51

Chargaff E, 'On the dangers of genetic meddling', Science, 1976, 192(4243), 938 & 940

———, Heraclitean Fire: Sketches from a Life Before Nature, Rockefeller University Press, 1978

Charland-Verville V, Bruno M-A, Bahri MA et al, 'Brain dead yet mind alive: a positron emission tomography case study of brain metabolism in Cotard's syndrome', Cortex, 2013, 49(7), 1997–9

Charles J, Sahraie A & McGeorge P, 'Hemispatial asymmetries in judgment of stimulus size', Perception & Psychophysics, 2007, 69(5), 687–98

Charlier P, Saudamini D, Lippi D et al, 'The cerebrovascular health of Thomas Aquinas', Lancet Neurology, 2017, 16(7), 502

Charlton BG, 'Senescence, cancer and "endogenous parasites": a salutogenic hypothesis', Journal of the Royal College of Physicians of London, 1996, 30(1), 10–12

———, 'Reconceptualizing the metaphysical basis of biology: a new definition based on deistic teleology and an hierarchy of organizing entities', 2016: www.thewinnower.com/papers/3497-reconceptualizing-the-metaphysical-basis-of-biology-a-new-definition-based-on-deistic-teleology-and-an-hierarchy-of-organizing-entities (accessed 23 February 2021)

Chase WG & Simon HA, 'Perception in chess', Cognitive Psychology, 1973, 4(1), 55–81

Chatham CH, Claus ED, Kim A et al, 'Cognitive control reflects context monitoring, not motoric stopping, in response inhibition', PLoS One, 2012, 7(2), e31546

Chávez RA, Graff-Guerrero A, García-Reyna JC et al, 'Neurobiología de la creatividad: resultados preliminares de un estudio de activación cerebral (The neurobiology of creativity: preliminary results of a brain activation study)', Salud Mental, 2004, 27(3), 38–46

Chávez-Eakle R, Graff-Guerrero A, García-Reyna J et al, 'Cerebral blood flow associated with creative performance: a comparative study', NeuroImage, 2007, 38(3), 519–28

Chechlacz M, Mantini D, Gillebert CR et al, 'Asymmetrical white matter networks for attending to global versus local features', Cortex, 2015, 72, 54–64

Chen CC & Liu HC, 'Low-dose aripiprazole resolved complex hallucinations in the left visual field after right occipital infarction (Charles Bonnet syndrome)', Psychogeriatrics, 2011, 11(2), 116–18

Chen IH, Novak V & Manor B, 'Infarct hemisphere and noninfarcted brain volumes affect locomotor performance following stroke', Neurology, 2014, 82(10), 828–34

Chen SX, Lam BCP, Wu WCH et al, 'Do people's world views matter? The why and how', Journal of Personality and Social Psychology, 2016, 110(5), 743–65

Chen X, Bracht JR, Goldman AD et al, 'The architecture of a scrambled genome reveals massive levels of genomic rearrangement during development', Cell, 2014, 158(5), 1187–98

Chen Y, Nakayama K, Levy D et al, 'Processing of global, but not local, motion direction is deficient in schizophrenia', Schizophrenia Research, 2003, 61(2–3), 215–27

Chen Z-X, Huang Y-K & Sun Y, 'The golden ratio and Loshu-Fibonacci Diagram: novel research view on relationship of Chinese medicine and modern biology', Chinese

Journal of Integrative Medicine, 2014, 20(2), 148–54

Chernigovskaya T, 'Neurosemiotic approach to cognitive functions', *Journal of the International Association for Semiotic Studies – Semiotica*, 1999, 127(1), 227–37

Chernigovskaya TV & Deglin VL, 'Brain functional asymmetry and neural organization of linguistic competence', *Brain and Language*, 1986, 29(1), 141–53

Chesterton GK, *Orthodoxy*, John Lane, 1909

Chettih S, Durgin FH & Grodner DJ, 'Mixing metaphors in the cerebral hemispheres: what happens when careers collide?', *Journal of Experimental Psychology: Learning, Memory and Cognition*, 2012, 38(2), 295–311

Cheung V, Chen EYH, Chen RYL et al, 'A comparison between schizophrenia patients and healthy controls on the expression of attentional blink in a rapid serial visual presentation (RSVP) paradigm', *Schizophrenia Bulletin*, 2002, 28(3), 443–58

Chi RP & Snyder AW, 'Facilitate insight by non-invasive brain stimulation', *PLoS One*, 2011, 6(2), e16655

———, 'Brain stimulation enables the solution of an inherently difficult problem', *Neuroscience Letters*, 2012, 515(2), 121–4

Chi RP, Fregni F & Snyder AW, 'Visual memory improved by non-invasive brain stimulation', *Brain Research*, 2010, 1353, 168–75

Chiarello C, 'Lateralization of lexical processes in the normal brain: a review of visual half-field research', in HA Whitaker (ed), *Contemporary Reviews in Neuropsychology*, Springer-Verlag, 1988, 36–76

Chiarello C, Burgess C, Richards L et al, 'Semantic and associative priming in the cerebral hemispheres: some words do, some words don't ... sometimes, some places', *Brain and Language*, 1990, 38(1), 75–104

Chiarello C & Maxfield L, 'Varieties of interhemispheric inhibition, or how to keep a good hemisphere down', *Brain and Cognition*, 1996, 30(1), 81–108

Chiarello C, McMahon MA & Schaefer K, 'Visual cerebral lateralization over phases of the menstrual cycle: a preliminary investigation', *Brain and Cognition*, 1989, 11(1), 18–36

Chiarello C, Vazquez D, Felton A et al, 'Structural asymmetry of the human cerebral cortex: regional and between-subject variability of surface area, cortical thickness, and local gyrification', *Neuropsychologia*, 2016, 93(B), 365–79

Chida Y, Steptoe A & Powell LH, 'Religiosity/spirituality and mortality: a systematic quantitative review', *Psychotherapy and Psychosomatics*, 2009, 78(2), 81–90

Chien H-Y, Lin H-Y, Lai M-C et al, 'Hyperconnectivity of the right posterior temporoparietal junction predicts social difficulties in boys with autism spectrum disorder', *Autism Research*, 2015, 8(4), 427–41

Chien YL, Gau SS, Shang CY et al, 'Visual memory and sustained attention impairment in youths with autism spectrum disorders', *Psychological Medicine*, 2015, 45(11), 2263–73

Chiou WB, Wu WH & Chang MH, 'Think abstractly, smoke less: a brief construal-level intervention can promote self-control, leading to reduced cigarette consumption among current smokers', *Addiction*, 2013, 108(5), 985–92

Chiron C, Jambaqué I, Nabbout R et al, 'The right brain hemisphere is dominant in human infants', *Brain*, 1997, 120(6), 1057–65

Cho H, Jönsson H, Campbell K et al, 'Self-organization in high-density bacterial colonies: efficient crowd control', *PLoS Biology*, 2007, 5(11), e302

Cho J-Y, Moon SY, Hong K-S et al, 'Unilateral prosopometamorphopsia as a dominant hemisphere-specific disconnection sign', *Neurology*, 2011, 76(22), e110

Choe SY & Min KH, 'Who makes utilitarian judgments? The influences of emotions on utilitarian judgments', *Judgment and Decision Making*, 2011, 6(7), 580–92

Choi EJ, Lee JK, Kang JK et al, 'Complex visual hallucinations after occipital cortical resection in a patient with epilepsy due to cortical dysplasia', *Archives of Neurology*, 2005, 62(3), 481–4

Chow HM, Kaup B, Raabe M et al, 'Evidence of fronto-temporal interactions for strategic inference processes during language comprehension', *NeuroImage*, 2008, 40(2), 940–54

Chrispal A, Prabhakar AT, Boorugu H et al, 'Peduncular hallucinosis in "top of the basilar syndrome": an unusual complication following coronary angiography', *National Medical Journal of India*, 2009, 22(5), 240–1

Christ K, Decker B, Koch J et al (eds), *Meister Eckhart: Die lateinischen Werke*, vol 3, Kohlhammer, 1994

Christman S & Pinger K, 'Lateral biases in aesthetic preferences: pictorial dimensions and neural mechanisms', *Laterality*, 1997, 2(2), 155–75

Christman SD & Hackworth MD, 'Equivalent perceptual asymmetries for free viewing of positive and negative emotional expressions in chimeric faces', *Neuropsychologia*, 1993, 31(6), 621–4

Christman SD, Propper RE & Dion A, 'Increased interhemispheric interaction is associated with decreased false memories in a verbal converging associates paradigm', *Brain and Cognition*, 2004, 56(3), 313–9

Christoff K, Gordon AM, Smallwood J et al, 'Experience sampling during fMRI reveals default network and executive system contributions to mind wandering', *Proceedings of the National Academy of Sciences of the United States of America*, 2009, 106(21), 8719–24

Chrysikou EG, Hamilton RH, Coslett HB et al, 'Noninvasive transcranial direct current stimulation over the left prefrontal cortex facilitates cognitive flexibility in tool use', *Cognitive Neuroscience*, 2013, 4(2), 81–9

Chrysikou EG, Weber MJ & Thompson-Schill SL, 'A matched filter hypothesis for cognitive control', *Neuropsychologia*, 2014, 62, 341–55

Chung MK, Dalton KM, Alexander AL et al, 'Less white matter concentration in autism: 2D voxel-based morphometry', *NeuroImage*, 2004, 23(1), 242–51

Chura LR, Lombardo MV, Ashwin E et al, 'Organizational effects of fetal testosterone on human corpus callosum size and asymmetry', *Psychoneuroendocrinology*, 2010, 35(1), 122–32

Ciaramidaro A, Bölte S, Schlitt S et al, 'Transdiagnostic deviant facial recognition for implicit negative emotion in autism and schizophrenia', *European Neuropsychopharmacology*, 2018, 28(2), 264–75

Çiçek M, Gitelman D, Hurley RS et al, 'Anatomical physiology of spatial extinction', *Cerebral Cortex*, 2007, 17(12), 2892–8

Cilliers P, *Complexity and Postmodernism: Understanding Complex Systems*, Routledge, 1998

———, 'Complexity theory as a general framework for sustainability science', in M Burns & A Weaver (eds), *Exploring Sustainability Science: A South African Perspective*, African Sun Media, 2008, 39–57

Cimino CR, Verfaellie M, Bowers D et al, 'Autobiographical memory: influence of right hemisphere damage on emotionality and specificity', *Brain and Cognition*, 1991, 15(1), 106–18

Cioran E, *Précis de décomposition*, Gallimard, 1949

———, *Newsweek*, 1989, 114(23), 52

Cipriani G, Vedovello M, Nuti A et al, 'Othello syndrome and dementia', *Psychiatry and Clinical Neurosciences*, 2012, 66(6), 467–73

Citron FM & Goldberg AE, 'Metaphorical sentences are more emotionally engaging than their literal counterparts', *Journal of Cognitive Neuroscience*, 2014, 26(11), 2585–95

Clapp W, Kirk IJ & Hausmann M, 'Effects of memory load on hemispheric asymmetries of colour memory', *Laterality*, 2007, 12(2), 139–53

Clark HH, 'Space, time, semantics and the child', in TE Moore (ed), *Cognitive Development and the Acquisition of Language*, Academic Press, 1973, 27–63

Clark RW, *Einstein: The Life and Times*, Avon, 1984

Clarke S & Zaidel E, 'Anatomical-behavioral relationships: corpus callosum morphometry and hemispheric specialisation', *Behavioral Brain Research*, 1994, 64(1–2), 185–202

Claxton G, 'Investigating human intuition: knowing without knowing why', *The Psychologist*, 1998, 11(5), 217–20

———, *Intelligence in the Flesh*, Yale University Press, 2016

Clerc O, Nanchen D, Cornuz J et al, 'Alcohol drinking, the metabolic syndrome and diabetes in a population with high mean alcohol consumption', *Diabetic Medicine*, 2010, 27(11), 1241–9

Close F, *Lucifer's Legacy*, Oxford University Press, 2000

Cloud of Unknowing, The, trans C Wolters, Penguin, 1961

Coate M, *Beyond All Reason*, Lippincott, 1965

Coats C, *Living Energies: Viktor Schauberger's Brilliant Work with Natural Energy Explained*, Gill Books, 2001

Cobb JB, 'Ecology, science, and religion: toward a postmodern worldview', in DR Griffin (ed), *The Reenchantment of Science*, State University of New York Press, 1988, 99–114

Cobbett W, 'To the Editor of the Agricultural Magazine: On the Subject of Potatoes', *Cobbett's Weekly Political Register*, vol XXIX, no. 7, Saturday 18 November 1815, 193–4

Coben R, Clarke AR, Hudspeth W et al, 'EEG power and coherence in autistic spectrum disorder', *Clinical Neurophysiology*, 2008, 119(5), 1002–9

Coger RW & Serafetinides EA, 'Schizophrenia, corpus callosum, and interhemispheric communication: a review', *Psychiatry Research*, 1990, 34(2), 163–84

Cohen AS, Iglesias B & Minor KS, 'The neurocognitive underpinnings of diminished expressivity in schizotypy: what the voice reveals', *Schizophrenia Research*, 2009, 109(1–3), 38–45

Cohen BD, 'Referent communication disturbances in schizophrenia', in S Schwartz (ed), *Language and Cognition in Schizophrenia*, Erlbaum, 1978, 1–34

Cohen H & Levy JJ, 'Hemispheric specialization for tactile perception opposed by contralateral noise', *Cortex*, 1988, 24(3), 425–31

Cohen JD & Servan-Schreiber D, 'Context, cortex, and dopamine: a connectionist approach to behavior and biology in schizophrenia', *Psychological Review*, 1992, 99(1), 45–77

Cohen Kadosh R, Bien N & Sack AT, 'Automatic and intentional number processing both rely on intact right parietal cortex: a combined fMRI and neuronavigated TMS study', *Frontiers in Human Neuroscience*, 2012, 6, article 2

Cohen L, Geny C, Hermine O et al, 'Crossed aphasia with visceral situs inversus', *Annals of Neurology*, 1993, 33(2), 215–8

Cohen L, Gray F, Meyrignac C et al, 'Selective deficit of visual size perception: two cases of hemimicropsia', *Journal of Neurology, Neurosurgery, and Psychiatry*, 1994, 57(1), 73–8

Cohen R, Bavishi C & Rozanski A, 'Purpose in life and its relationship to all-cause mortality and cardiovascular events: a meta-analysis', *Psychosomatic Medicine*, 2016, 78(2), 122–33

Cohn F, 'Untersuchungen über Bacterien. IV. Beiträge zur Biologie der Bacillen', *Beiträge zur Biologie der Pflanzen*, 1877, 2, 249–76

Colbert SM & Peters ER, 'Need for closure and jumping-to-conclusions in delusion-prone individuals', *Journal of Nervous and Mental Disease*, 2002, 190(1), 27–31

Cole M, 'When the left brain is not right the right brain may be left: report of personal experience of occipital hemianopia', *Journal of Neurology, Neurosurgery, and Psychiatry*, 1999, 67(2), 169–73

Cole PM, Barrett KC & Zahn-Waxler C, 'Emotion displays in two-year-olds during mishaps', *Child Development*, 1992, 63(2), 314–24

Cole SW & Fredrickson BL, 'Reply to Coyne: genomic analyses are unthwarted', *Proceedings of the National Academy of Sciences of the United States of America*, 2013, 110(45), E4184

Coleman MJ, Cestnick L & Krastoshevsky O, 'Schizophrenia patients show deficits in shifts of attention to different levels of global-local stimuli: evidence for magnocellular dysfunction', *Schizophrenia Bulletin*, 2009, 35(6), 1108–16

Coleman SM, 'Misidentification and non-recognition', *Journal of Mental Science*, 1933, 79, 42–51

Coleridge ST, *Biographia Literaria; or Biographical Sketches of My Literary Life and Opinions*, 2 vols, Rest Fenner, 1817

———, *Seven Lectures on Shakespeare and Milton*, ed JP Collier, Chapman & Hall, 1856

———, in EH Coleridge (ed), *Letters of Samuel Taylor Coleridge*, 2 vols, Heinemann, 1895

———, in K Coburn (ed), *The Notebooks of Samuel Taylor Coleridge*: vol 1, Routledge & Kegan Paul, 1957

———, in K Coburn (ed), *The Notebooks of Samuel Taylor Coleridge*: vol 2, Routledge & Kegan Paul, 1962

———, *Biographia Literaria*, ed J Engell and W Jackson Bate, in *The Collected Works of Samuel Taylor Coleridge*, Princeton University Press, vol 7, 1985

———, *Lectures 1808–1819: On Literature*, ed RA Foakes, 2 vols, 1987

Coles R, *The Spiritual Life of Children*, Houghton Mifflin, 1990

Collerton D, 'Psychotherapy and brain plasticity', *Frontiers in Psychology*, 2013, 4, article 548

Collingwood RG, *An Essay on Metaphysics*, Clarendon Press, 1940

———, *An Autobiography*, Clarendon Press, 1983

Collins F, interviewed by Peter Reuell, *The Harvard Gazette*, 22 November 2017

Colom R & Garcia-Lopez O, 'Secular gains in fluid intelligence: evidence from the Culture-Fair Intelligence Test', *Journal of Biosocial Science*, 2003, 35(1), 33–9

Colom R, Lluis-Font JM & Andrés-Pueyo A, 'The generational intelligence gains are caused by decreasing variance in the lower half of the distribution: supporting evidence for the nutrition hypothesis', *Intelligence*, 2005, 33(1), 83–91

Colombo F & Assal G, 'Persisting aphasia, cerebral dominance, and painting in the famous artist Carl Fredrik Reuterswärd', *Frontiers of Neurology and Neuroscience*, 2007, 22, 169–83

Coltheart M, 'Cognitive neuropsychiatry and delusional belief', *Quarterly Journal of Experimental Psychology*, 2007, 60(8), 1041–62

Coltheart M, 'The neuropsychology of delusions', *Annals of the New York Academy of Sciences*, 2010, 1191(1), 16–26

Coltheart M, Langdon R & McKay R, 'Schizophrenia and monothematic delusions', *Schizophrenia Bulletin*, 2007, 33(3), 642–7

Combettes [no first name or initial given], 'Absence complète du cervelet, des pédoncules postérieurs et de la protubérance cérébrale chez une jeune fille morte dans sa onzième année', *Bulletins de la Société anatomique de Paris*, 1831 (janvier), 5ᵉ année, bulletin 6, 148–57

Compston A, 'Aphasia and artistic realization (being the Harveian lecture of the Harveian Society, delivered March 17 1948) by Th. Alajouanine', *Brain*, 2008, 131(1), 3–5

Coney J & Bruce C, 'Hemispheric processes in the perception of art', *Empirical Studies of the Arts*, 2004, 22(2), 181–200

Connolly AC, Sha L, Guntupalli JS et al, 'How the human brain represents perceived dangerousness or "predacity" of animals', *Journal of Neuroscience*, 2016, 36(19), 5373–84

Conti F & Manzoni T, 'The neurotransmitters and postsynaptic actions of callosally projecting neurons', *Behavioural Brain Research*, 1994, 64(1–2), 37–53

Contreras CM, Dorantes ME, Mexicano G et al, 'Lateralization of spike and wave complexes produced by hallucinogenic compounds in the cat', *Experimental Neurology*, 1986, 92(3), 467–78

Conway BR & Rehding A, 'Neuroaesthetics and the trouble with beauty', *PLoS Biology*, 2013, 11(3), e1001504

Conway CR, Bollini AM, Graham BG et al, 'Sensory acuity and reasoning in delusional disorder', *Comprehensive Psychiatry*, 2002, 43(3), 175–8

Conway MA & Fthenaki A, 'Disruption of inhibitory control of memory following lesions to the frontal and temporal lobes', *Cortex*, 2003, 39(4–5), 667–86

Conway MA & Pleydell-Pearce CW, 'The construction of autobiographical memories in the self-memory system', *Psychological Review*, 2000, 107(2), 261–88

Conway Morris S, 'Aliens like us?', *Astronomy & Geophysics*, 2005, 46(4), 24–6

Cook CCH, Goddard D & Westall R, 'Knowledge and experience of drug use amongst church affiliated young people', *Drug and Alcohol Dependence*, 1997, 46(1–2), 9–17

Cook J, 'From movement kinematics to social cognition: the case of autism', *Philosophical Transactions of the Royal Society of London: Series B, Biological Sciences*, 2016, 371(1693), 20150372

Cook JL, Blakemore S-J & Press C, 'Atypical basic movement kinematics in autism spectrum conditions' *Brain*, 2013, 136(9), 2816–24

Cook ND, 'Callosal inhibition: the key to the brain', *Behavioral Science*, 1984a, 29(2), 98–110

———, 'Homotopic callosal inhibition', *Brain and Language*, 1984b, 23(1), 116–25

Cook TA, *Spirals in Nature and Art*, John Murray, 1902

Cookson C, 'Misconduct pervades UK research', *Financial Times*, 12 January 2012

Coolican J, Eskes GA, McMullen PA et al, 'Perceptual biases in processing facial identity and emotion', *Brain and Cognition*, 2008, 66(2), 176–87

Cooper D, *God is a Verb: Kabbalah and the Practice of Mystical Judaism*, Penguin Putnam, 1997

Cooper NR, Croft RJ, Dominey SJ et al, 'Paradox lost? Exploring the role of alpha oscillations during externally vs. internally directed attention and the implications for idling and inhibition hypotheses', *International Journal of Psychophysiology*, 2003, 47(1), 65–74

Cooper SA, Joshi AC, Seenan PJ et al, 'Akinetopsia: acute presentation and evidence for persisting defects in motion vision', *Journal of Neurology, Neurosurgery, and Psychiatry*, 2012, 83(2), 229–30

Cope LM, Schaich Borg J, Harenski CL et al, 'Hemispheric asymmetries during processing of immoral stimuli', *Frontiers in Evolutionary Neuroscience*, 2010, 2, article 110

Cope MB & Allison DB, 'White hat bias: examples of its presence in obesity research and a call for renewed commitment to faithfulness in research reporting, *International Journal of Obesity*, 2009, 34(1), 84–8

Copernicus, *On the Revolutions: Nicholas Copernicus Complete Works*, trans and ed E Rosen, Johns Hopkins University Press, 1992

Copland A, *Music and Imagination: Charles Eliot Norton Lectures*, Oxford University Press, 1952

Coppo F & Brignolio P, 'Descrizione di un insolito quadro clinico di allucinosi visiva (poliopsia, eritropsia) quale espressione di ischemia temporo-occipitale destra (corteccia dell'associazione visiva, area 37 di Brodman)', *Minerva Medica*, 1990, 81(1–2), 111–4

Corballis MC & Roldan CE, 'On the perception of symmetrical and repeated patterns', *Perception & Psychophysics*, 1974, 16(1), 136–42

Corballis MC & Sergent J, 'Imagery in a commissurotomized patient', *Neuropsychologia*, 1988, 26(1), 13–26

Corballis MC & Sidey S, 'Effects of concurrent memory load on visual-field differences in mental rotation', *Neuropsychologia*, 1993, 31(2), 183–97

Corballis PM, 'Visuospatial processing and the right-hemisphere interpreter', *Brain and Cognition*, 2003, 53(2), 171–6

Corballis PM, Fendrich R, Shapley R et al, 'Illusory contour perception and amodal boundary completion: evidence of a dissociation following callosotomy', *Journal of Cognitive Neuroscience*, 1999, 11(4), 459–66

Corballis PM, Funnell MG & Gazzaniga MS, 'Hemispheric asymmetries for simple visual judgments in the split brain', *Neuropsychologia*, 2002, 40(4), 401–10

Corbetta M, Kincade JM, Ollinger JM et al, 'Voluntary orienting is dissociated from target detection in human posterior parietal cortex', *Nature Neuroscience*, 2000, 3(3), 292–7

Corbetta M & Shulman GL, 'Spatial neglect and attention networks', *Annual Review of Neuroscience*, 2011, 34, 569–99

Corcoran R, Cahill C & Frith CD, 'The appreciation of visual jokes in people with schizophrenia: a study of "mentalizing" ability', *Schizophrenia Research*, 1997, 24(3), 319–27

Corlett PR, Aitken MR, Dickinson A et al, 'Prediction error during retrospective evaluation of causal associations in humans: fMRI evidence in favor of an associative model of learning', *Neuron*, 2004, 44(5), 877–88

Corlett PR, Murray GK, Honey GD et al, 'Disrupted prediction-error signal in psychosis: evidence for an associative account of delusions', *Brain*, 2007, 130(9), 2387–400

Cornford FM, *From Religion to Philosophy*, Arnold, 1912

Cory-Slechta DA, Weston D, Liu S et al, 'Brain hemispheric differences in the neurochemical effects of lead, prenatal stress, and the combination and their amelioration by behavioral experience', *Toxicological Sciences*, 2013, 132(2), 419–30

Coslett HB, Bowers D & Heilman KM, 'Reduction in cerebral activation after right hemisphere stroke', *Neurology*, 1987, 37(6), 957–62

Coslett HB & Heilman KM, 'Hemihypokinesia after right hemisphere stroke', *Brain and Cognition*, 1989, 9(2), 267–78

Cosmides L & Tooby J, 'Cognitive adaptations for social exchange', in J Barkow, L Cosmides & J Tooby (eds), *The Adapted Mind: Evolutionary Psychology and the Generation Culture*, Oxford University Press, 1992, 163–228

Coss RG, 'Evolutionary persistence of memory-like processes', *Concepts in Neuroscience*, 1991, 2(2), 129–68

Costa PT, Terracciano A & McCrae RR, 'Gender differences in personality traits across cultures: robust and surprising findings', *Journal of Personality and Social Psychology*, 2001, 81, 322–31

Costa T, Suardi AC, Diano M et al, 'The neural correlates of hedonic and eudaimonic happiness: an fMRI study', *Neuroscience Letters*, 2019, 712, 13449

Costello JE, Erkanli A & Angold A, 'Is there an epidemic of child or adolescent depression?', *Journal of Child Psychology and Psychiatry*, 2006, 47(12), 1263–71

Cotard J, 'Du délire hypochondriaque dans une forme grave de la mélancolie anxieuse', *Annales médico-psychologiques*, 1880, 4, 168–74

———, 'Du délire de négation', *Archives de Neurologie, revue des maladies nerveuses et mentales*, 1882, 4, 152–70 & 282–6

Cottingham J, '"A brute to the brutes?" Descartes' treatment of animals', *Philosophy*, 1978, 53(206), 551–9

———, *Philosophy and the Good Life*, Cambridge University Press, 1998

———, 'What is humane philosophy and why is it at risk?', in A O'Hear (ed), *Conceptions of Philosophy* (Royal Institute of Philosophy Supplements 65), Cambridge University Press, 2009, 233–56

———, *Philosophy of Religion*, Cambridge University Press, 2014

———, *In Search of the Soul*, Princeton University Press, 2020

Cotton SM, Kiely PM, Crewther DP et al, 'A normative and reliability study for the Raven's Coloured Progressive Matrices for primary school aged children from Victoria, Australia', *Personality and Individual Differences*, 2005, 39(3), 647–59

Coull JT, Vidal F, Nazarian B et al, 'Functional anatomy of the attentional modulation of time estimation', *Science*, 2004, 303(5663), 1506–8

Coulson S & Williams RF, 'Hemispheric asymmetries and joke comprehension', *Neuropsychologia*, 2005, 43, 128–41

Coulson S & Wu YC, 'Right hemisphere activation of joke-related information: an event-related brain potential study', *Journal of Cognitive Neuroscience*, 2005, 17(3), 494–506

Couper J, 'Unilateral musical hallucinations and all that jazz', *Australian & New Zealand Journal of Psychiatry*, 1994, 28(3), 516–9

Courchesne E, Yeung-Courchesne R, Press GA et al, 'Hypoplasia of cerebellar vermal lobules VI and VII in autism', *The New England Journal of Medicine*, 1988, 318(21), 1349–54

Courty A, Maria AS, Lalanne C et al, 'Levels of autistic traits in anorexia nervosa: a comparative psychometric study', *BMC Psychiatry*, 2013, 13(1), article 222

Cousin E, Peyrin C & Baciu M, 'Hemispheric predominance assessment of phonology and semantics: a divided visual field experiment', *Brain and Cognition*, 2006, 61(3), 298–304

Cousins EH, *Bonaventure and the Coincidence of Opposites*, Franciscan Press, 1978

Cowley J, 'Bryan Magee: the restless philosopher', *New Statesman*, 5 April 2018

Coyne JA, *Faith vs. Fact: Why Science and Religion Are Incompatible*, Penguin, 2016

Coyne JC, 'Highly correlated hedonic and eudaimonic well-being thwart genomic analysis', *Proceedings of the National Academy of Sciences of the United States of America*, 2013, 110(45), E4183, and a reply to Coyne: Cole SW & Fredrickson BL, 'Reply to Coyne: genomic analyses are unthwarted', *Proceedings of the National Academy of Sciences of the United States of America*, 2013, 110(45), E4184

Crabb Robinson H, *Diary, Reminiscences, and Correspondence*, Macmillan, 1869

Craig AD, 'How do you feel? Interoception: the sense of the physiological condition of the body', *Nature Reviews Neuroscience*, 2002, 3(8), 655–66

———, 'Forebrain emotional asymmetry: a neuroanatomical basis?', *Trends in Cognitive Sciences*, 2005, 9(12), 566–71

———, 'The sentient self', *Brain Structure and Function*, 2010, 214(5–6), 563–77

Craig AD, Chen K, Bandy D et al, 'Thermosensory activation of insular cortex', *Nature Neuroscience*, 2000, 3(2), 184–90

Craik FIM, Moroz TM, Moscovitch M et al, 'In search of the self: a positron emission tomography study', *Psychological Science*, 1999, 10(1), 26–34

Cramer JG, 'The retarding of science', 1984: www.npl.washington.edu/av/altvw04.html (accessed 23 February 2021)

Crawford MB, *The World Beyond Your Head*, Penguin Random House, 2015

Crespi BJ, 'The evolution of social behaviour in microorganisms', *Trends in Ecology and Evolution*, 2001, 16(4), 178–83

Crichton P & Lewis S, 'Delusional misidentification, AIDS and the right hemisphere', *British Journal of Psychiatry*, 1990, 157(4), 608–10

Crichton-Browne J, 'On the weight of the brain and its component parts in the insane', *Brain*, 1880, 2, 42–67

Critchley HD, Wiens S, Rotshtein P et al, 'Neural systems supporting interoceptive awareness', *Nature Neuroscience*, 2004, 7(2), 189–95

Critchley M, 'Types of visual perseveration: "paliopsia" and "illusory visual spread"', *Brain*, 1951, 74(3), 267–99

———, *The Parietal Lobes*, Hafner, 1953

———, 'Personification of paralysed limbs in hemiplegics', *British Medical Journal*, 1955, 2(4934), 284–6

———, 'The enigma of Gerstmann's syndrome', *Brain*, 1966, 89(2), 183–98

———, 'Misoplegia, or hatred of hemiplegia', *Mount Sinai Journal of Medicine*, 1974, 41(1), 82–7

Cronin-Golomb A, 'Figure-background perception in right and left hemispheres of human commissurotomy subjects', *Perception*, 1986, 15(2), 95–109

Cropley A, 'In praise of convergent thinking', *Creativity Research Journal*, 2006, 18(3), 391–404

Cross ES, Kirsch L, Ticini LF et al, 'The impact of aesthetic evaluation and physical ability on dance perception', *Frontiers in Human Neuroscience*, 2011, 5, article 102

Csaba G, 'Hormesis and immunity: a review', *Acta Microbiologica et Immunologica Hungarica*, 2019, 66(2), 155–68

Csikszentmihályi M, *Flow and the Foundations of Positive Psychology: The Collected Works of Mihaly Csikszentmihalyi*, Springer, 2014

Csiszar A, 'Peer review: troubled from the start', *Nature*, 2016, 532(7599), 306–8

Culumber ZW, Bautista-Hernández CE, Monks S et al, 'Variation in melanism and female preference in proximate but ecologically distinct environments', *Ethology*, 2014, 120(11), 1090–1100

Cummings JL, 'Neuropsychiatric manifestations of right hemisphere lesions', *Brain and Language*, 1997, 57(1), 22–37

Cummings JL & Mendez MF, 'Secondary mania with focal cerebrovascular lesions', *American Journal of Psychiatry*, 1984, 141(9), 1084–7

Cummings JL, Syndulko K, Goldberg Z et al, 'Palinopsia reconsidered', *Neurology*, 1982, 32(4), 444–7

Cunningham MR, Roberts AR, Barbee AP et al, '"Their ideas of beauty are, on the whole, the same as ours": consistency and variability in the cross-cultural perception of female attractiveness', *Journal of Personality and Social Psychology*, 1995, 68(2), 261–79

Cuonzo M, *Paradox*, Massachusetts Institute of Technology Press, Cambridge MA, 2014

Cupchik GC, Vartanian O, Crawley A et al, 'Viewing artworks: contributions of cognitive control and perceptual facilitation to aesthetic experience', *Brain and Cognition*, 2009, 70(1), 84–91

Curie P, 'Sur la symétrie dans les phénomènes physiques, symétrie d'un champ électrique et d'un champ magnétique', *Journal de Physique Théorique et Appliquée*, 1894, 3(1), 393–415

Curran T, Schacter DL, Norman KA et al, 'False recognition after a right frontal lobe infarction: memory for general and specific information', *Neuropsychologia*, 1997, 35(7), 1035–49

Currivan J, *The Cosmic Hologram: In-formation at the Center of Creation*, Inner Traditions, 2017

Curry P, 'The enchantment of learning and "the fate of our times"', in A Voss and S

Wilson (eds), *Re-Enchanting the Academy*, Rubedo Press, 2017, 33–51
——, 'The work of wonder', *Western Humanities Review*, 2018, 72(2), 28–41
——, *Enchantment*, Floris Books, 2019
Cusanus (Nicolas of Cusa), *The Vision of God*, trans EG Salter, Ungar Publishing, 1960 [1453]
——, *De venatione sapientiæ*, trans J Hopkins, Banning Press, 1978a [1462]
——, *Trialogus de possest*, trans J Hopkins, Banning Press, 1978b [1460]
——, *De docta ignorantia*, trans J Hopkins, Banning Press, 1985 [1440]
——, *De visione Dei*, trans J Hopkins, Banning Press, 1988 [1453]
Cushing H, 'Distortions of the visual fields in cases of brain tumour: the field defects produced by temporal lobe lesions', *Brain*, 1922, 44(4), 341–96
Cushman F, Gray K, Gaffey A et al, 'Simulating murder: the aversion to harmful action', *Emotion*, 2012, 12(1), 2–7
Cusinato G, *Person und Selbsttranszendenz: Ekstase und Epoché des Ego als Individuationsprozesse bei Schelling und Scheler*, Königshausen & Neumann, 2012
Custers EJFM, Boshuizen HPA & Schmidt HG, 'The influence of medical expertise, case typicality, and illness script component on case processing and disease probability estimates', *Memory & Cognition*, 1996, 24(3), 384–99
——, 'The role of illness scripts in the development of medical diagnostic expertise: results from an interview study', *Cognition and Instruction*, 1998, 16(4), 367–98
Custers R & Aarts H, 'Learning of predictive relations between events depends on attention, not on awareness', *Consciousness and Cognition*, 2011, 20(2), 368–78
Cutting J, 'Study of anosognosia', *Journal of Neurology, Neurosurgery, and Psychiatry*, 1978, 41(6), 548–55
——, *The Right Cerebral Hemisphere and Psychiatric Disorders*, Oxford University Press, 1990
——, 'Delusional misidentification and the role of the right hemisphere in the appreciation of identity', *British Journal of Psychiatry*, 1991, 159(14, suppl), 70–5
——, 'Eugène Minkowski: La Schizophrénie', in S Crown & H Freeman, *The Book of Psychiatric Books*, Jason Aronson Inc, 1994a, 75–81
——, 'Evidence for right hemisphere dysfunction in schizophrenia', in AS David & JC Cutting (eds), *The Neuropsychology of Schizophrenia*, Erlbaum, 1994b, 231–42
——, *Principles of Psychopathology: Two Worlds – Two Minds – Two Hemispheres*, Oxford University Press, 1997
——, 'Morbid objectivization in psychopathology', *Acta Psychiatrica Scandinavica*, 1999, 99(suppl 395), 30–3
——, *The Living, The Dead, and the Never-Alive*, Forest Publishing Company, 2002
——, *A Critique of Psychopathology*, Forest Publishing Company, 2011
——, 'First rank symptoms of schizophrenia: their nature and origin', *History of Psychiatry*, 2015, 26(2), 131–46
——, 'Max Scheler's phenomenological reduction as the schizophrenic's *modus vivendi*', *Thaumàzein: Rivista di Filosofia*, 2018, 6, 42–65
Cutting J & Dunne F, 'Subjective experience of schizophrenia', *Schizophrenia Bulletin*, 1989, 15(2), 217–31
Cutting J, Moreira V, Naudin J et al, 'The psychopathology of thingness: homage to Arthur Tatossian', *Journal für Philosophie & Psychiatrie*, August 2017: jfpp.org/118.98.html (accessed 23 February 2021)
Cutting J & Murphy D, 'Preference for denotative as opposed to connotative meanings in schizophrenics', *Brain and Language*, 1990, 39(3), 459–68
Cutting J & Musalek M, 'The nature of delusion: psychologically explicable? psychologically inexplicable? philosophically explicable? Part 2', *History of Psychiatry*, 2016, 27(1), 21–37
Cutting J & Shepherd M (eds), *The Clinical Roots of the Schizophrenia Concept: Translations of seminal European contributions on schizophrenia*, Cambridge University Press, 1987
Cutting J & Silzer H, 'Psychopathology of time in brain disease and schizophrenia', *Behavioural Neurology*, 1990, 3(4), 197–215
Cybulska EM, 'The madness of Nietzsche: a misdiagnosis of the millennium?' *Hospital Medicine*, 2000, 61(8), 571–5
Cytowic R, 'Aphasia in Maurice Ravel', *Bulletin of the Los Angeles Neurological Societies*, 1976, 41(3), 109–14
——, *Synesthesia: A Union of the Senses*, Springer, 1989
D'Onofrio BM, Eaves LJ, Murrelle L et al, 'Understanding biological and social influences on religious affiliation, attitudes, and behaviors: a behavior genetic perspective', *Journal of Personality*, 1999, 67(6), 953–84
Dabbs JM, Chang E-L, Strong RA et al, 'Spatial ability, navigation strategy, and geographic knowledge among men and women', *Evolution and Human Behavior*, 1998, 19, 89–98

Dabiri JO, 'Renewable fluid dynamic energy derived from aquatic animal locomotion', *Bioinspiration & Biomimetics*, 2007, 2(3), L1–3

Dagenbach D, Harris LJ & Fitzgerald HE, 'A longitudinal study of lateral biases in parents' cradling and holding of infants', *Infant Mental Health*, 1988, 9(3), 218–34

Dagge M & Hartje W, 'Influence of contextual complexity on the processing of cartoons by patients with unilateral lesions', *Cortex*, 1985, 21(4), 607–16

Dahlgren SO & Gillberg C, 'Symptoms in the first two years of life: a preliminary population study of infantile autism', *European Archives of Psychiatry and Neurological Sciences*, 1989, 283(3), 169–74

Daini R, Facchin A, Bignotti M et al, 'Neuropsychological evidence of high-level processing in binocular rivalry', *Behavioural Neurology*, 2010, 23(4), 233–5

Dainton B, *Self: Philosophy in Transit*, Penguin, 2014

Dakin S, Carlin P & Hemsley D, 'Weak suppression of visual context in chronic schizophrenia', *Current Biology*, 2005, 15(20), R822–4

Dalby JT, Arboleda-Florez J & Seland TP, 'Somatic delusions following left parietal lobe injury', *Neuropsychiatry, Neuropsychology, & Behavioral Neurology*, 1989, 2(4), 306–11

Dalrymple KA, Davies-Thompson J, Oruc I et al, 'Spontaneous perceptual facial distortions correlate with ventral occipitotemporal activity', *Neuropsychologia*, 2014, 59, 179–91

Damasio A, *Descartes' Error: Emotion, Reason and the Human Brain*, Putnam, 1994

Damasio AR, Grabowski TJ, Bechara A et al, 'Subcortical and cortical brain activity during the feeling of self-generated emotions', *Nature Neuroscience*, 2000, 3(10), 1049–56

Damasio H, *Human Brain Anatomy in Computerized Images*, Oxford University Press, 2nd edn, 2005

Damasio H, Grabowski TJ, Tranel D et al, 'A neural basis for lexical retrieval', *Nature*, 1996, 380(6574), 499–505 (erratum: *Nature*, 381(6595), 810)

Damasio H, Tranel D, Grabowski T et al, 'Neural systems behind word and concept retrieval', *Cognition*, 2004, 92, 179–229

Damrad-Frye R & Laird JD, 'The experience of boredom: the role of the self-perception of attention', *Journal of Personality and Social Psychology*, 1989, 57(2), 315–20

Dan J, *The Early Kabbalah*, Paulist Press, 1986

Dan'ko SG, Shemyakina NV, Nagornova ZhV et al, 'Comparison of the effects of the subjective complexity and verbal creativity on the EEG spectral power parameters', *Human Physiology*, 2009, 35(3), 381–3

Dan'ko SG, Starchenko MG & Bechtereva NP, 'EEG local and spatial synchronization during a test on the insight strategy of solving creative verbal tasks', *Human Physiology*, 2003, 29(4), 502–4

Danchin A, 'Bacteria as computers making computers', *FEMS Microbiological Reviews*, 2009, 33(1), 3–26

Danckert J, Ferber S, Pun C et al, 'Neglected time: impaired temporal perception of multisecond intervals in unilateral neglect', *Journal of Cognitive Neuroscience*, 2007, 19(10), 1706–20

Danckert J, Stöttinger E, Quehl N et al, 'Right hemisphere brain damage impairs strategy updating', *Cerebral Cortex*, 2012, 22(12), 2745–60

Danckert JA & Allman A-A, 'Time flies when you're having fun: temporal estimation and the experience of boredom', *Brain and Cognition*, 2005, 59(3), 236–45

Dane E, Rockmann KW & Pratt MG, 'When should I trust my gut? Linking domain expertise to intuitive decision-making effectiveness', *Organizational Behavior and Human Decision Processes*, 2012, 119(2), 187–94

Danek AH, Fraps T, von Müller A et al, 'Aha! experiences leave a mark: facilitated recall of insight solutions', *Psychological Research*, 2013, 77(5), 659–69

Danguecan AN & Smith ML, 'Re-examining the crowding hypothesis in pediatric epilepsy', *Epilepsy & Behavior*, 2019, 94, 281–7

Daniele A, Giustolisi L, Silveri MC et al, 'Evidence for a possible neuroanatomical basis for lexical processing of nouns and verbs', *Neuropsychologia*, 1994, 32(11), 1325–41

Danielmeier C, Eichele T, Forstmann BU et al, 'Posterior medial frontal cortex activity predicts post-error adaptations in task-related visual and motor areas', *Journal of Neuroscience*, 2011, 31(5), 1780–9

Daniels JK, Gaebler M, Lamke JP et al, 'Grey matter alterations in patients with depersonalization disorder: a voxel-based morphometry study', *Journal of Psychiatry & Neuroscience*, 2015, 40(1), 19–27

Danielson DR, 'The great Copernican cliché', *American Journal of Physics*, 2001, 69(10), 1029–35

Danilova MV & Mollon JD, 'The symmetry of visual fields in chromatic discrimination', *Brain and Cognition*, 2009, 69(1), 39–46

Dansereau DF, Knight DK & Flynn PM, 'Improving adolescent judgment and decision making', *Professional Psychology Research and Practice*, 2013, 44(4), 274–82

Danta G, Hilton RC & O'Boyle DJ, 'Hemisphere function and binocular depth perception', *Brain*, 1978, 101(4), 569–89

Darby R & Prasad S, 'Lesion-related delusional misidentification syndromes: a comprehensive review of reported cases', *Journal of Neuropsychiatry and Clinical Neurosciences*, 2016, 28(3), 217–22

Darwin CR, *On the Origin of Species*, John Murray, 1st edn, 1859

———, *The Descent of Man, and Selection in Relation to Sex*, 2 vols, John Murray, 1871

———, *On the Origin of Species*, John Murray, 6th edn, 1872

———, *The Power of Movement in Plants*, John Murray, 1880

———, *The Formation of Vegetable Mould through the Action of Worms with Observations on Their Habits*, John Murray, 1881

———, in F Darwin (ed), *The Life and Letters of Charles Darwin, Including an Autobiographical Chapter*, John Murray, 3 vols, 1887

———, in F Darwin & AC Seward (eds), *More Letters of Charles Darwin: A Record of his Work in a Series of Hitherto Unpublished Letters*, 2 vols, John Murray, 1903

———, in F Burkhardt, J Browne & DM Porter (eds), *The Correspondence of Charles Darwin*, Cambridge University Press, vol 8, 1993

———, *Autobiographies*, Penguin Classics, London, 2002 [1903, 1887]

Darwin E, *Zoonomia, or The Laws of Organic Life*, Thomas & Andrews, 1809

David AS, 'Callosal transfer in schizophrenia: too much or too little?', *Journal of Abnormal Psychology*, 1993, 102(4), 573–9

———, 'Schizophrenia and the corpus callosum: developmental, structural and functional relationships', *Behavioural Brain Research*, 1994, 64(1–2), 203–11

———, 'The clinical importance of insight: an overview', in X Amador & A David (eds), *Insight and Psychosis*, Oxford University Press, 2nd edn, 2004, 359–92

Davidoff JB, 'Hemispheric differences in the perception of lightness', *Neuropsychologia*, 1975, 13(1), 121–4

———, 'Hemispheric sensitivity differences in the perception of colour', *Quarterly Journal of Experimental Psychology*, 1976, 28(3), 387–94

———, 'Hemispheric differences in dot detection', *Cortex*, 1977, 13(4), 434–44

Davidson PA, *Turbulence: An Introduction for Scientists and Engineers*, Oxford University Press, 2004

Davidson RJ, Ekman P, Saron CD et al, 'Approach/withdrawal and cerebral asymmetry: emotional expression and brain physiology', *Journal of Personality and Social Psychology*, 1990, 58(2), 330–41

Davidson RJ, Schwartz GE, Saron C et al, 'Frontal versus parietal EEG asymmetry during positive and negative affect' [SPR Abstracts 1978, Session III], *Psychophysiology*, 1979, 16(2), 202–3

Davidson RJ, Taylor N & Saron C, 'Hemisphericity and styles of information processing: individual differences in EEG asymmetry and their relationship to cognitive performance' [SPR Abstracts 1978, Science Fair II], *Psychophysiology*, 1979, 16(2), 197

Davies P, *God and the New Physics*, Simon & Schuster, 1983

———, 'Particles do not exist', in SM Christensen (ed), *Quantum Theory of Gravity*, Adam Hilger, 1984, 66–77

———, *The Cosmic Blueprint: New Discoveries in Nature's Creative Ability to Order the Universe*, Touchstone, 1988

———, 'The anthropic universe', 18 February 2006, ABC: www.abc.net.au/radionational/programs/scienceshow/the-anthropic-universe/3302686# (accessed 23 February 2021)

Davis KD, Pope GE, Crawley AP et al, 'Perceptual illusion of "paradoxical heat" engages the insular cortex', *Journal of Neurophysiology*, 2004, 92, 1248–51

Davis MT, DellaGioia N, Matuskey D et al, 'Preliminary evidence concerning the pattern and magnitude of cognitive dysfunction in major depressive disorder using cogstate measures', *Journal of Affective Disorders*, 2017, 218, 82–5

Davis PJ & Hersh R, *Descartes' Dream: The World According to Mathematics*, Penguin Press, 1990

Davis TA & Altevogt R, 'Golden mean of the human body', *Fibonacci Quarterly*, 1979, 17(4), 340–4

Dawkins R, *The Selfish Gene*, Oxford University Press, 1976

———, *The Blind Watchmaker*, Norton, 1986

———, 'Universal Darwinism', in DL Hull & M Ruse (eds), *The Philosophy of Biology*, Oxford University Press, 1998, 15–37

———, *A Devil's Chaplain*, Mariner, 2003

Dawson G & Adams A, 'Imitation and social responsiveness in autistic children', *Journal of Abnormal Child Psychology*, 1984, 12(2), 209–25

de Bonald, Louis, vicomte, *Législation primitive*, Adrien le Clere et Cie, 5th edn, 1857 [1802]

de Broglie L, *Matter and Light: The New Physics*, trans WH Johnston, Norton, 1939

———, 'The concepts of contemporary physics and Bergson's ideas on time and motion', in PAY Gunter (ed), *Bergson and the*

Evolution of Physics, University of Tennessee Press, 1969 [1947]

de Clérambault GG, 'Les psychoses passionelles', *Œuvre Psychiatrique*, Presses Universitaires de France, 1921–4, vol 1, 1921, 323–7

de Gelder B, Tamietto M, van Boxtel G et al, 'Intact navigation skills after bilateral loss of striate cortex', *Current Biology*, 2008, 18(24), R1128–9

de Grazia S, *Of Time, Work and Leisure*, Twentieth Century Fund, 1962

De Greck M, Wang G, Yang X et al, 'Neural substrates underlying intentional empathy', *Social Cognitive and Affective Neuroscience*, 2012, 7(2), 135–44

De Groot AD & Gobet F, *Perception and Memory in Chess*, Van Gorcum, 1996

de Haan S & Fuchs T, 'The ghost in the machine: disembodiment in schizophrenia — two case studies', *Psychopathology*, 2010, 43(5), 327–33

de la Fuente J-M, Goldman S, Stanus E et al, 'Brain glucose metabolism in borderline personality disorder', *Journal of Psychiatric Research*, 1997, 31(5), 531–41

De La Garza CO & Worchel P, 'Time and space orientation in schizophrenia', *Journal of Abnormal and Social Psychology*, 1956, 52(2), 191–4

de la Mettrie JO, *Machine Man and Other Writings*, Cambridge University Press, 1996 [1748]

de Leon J, Antelo RE & Simpson G, 'Delusion of parasitosis or chronic tactile hallucinosis: hypothesis about their brain physiopathology', *Comprehensive Psychiatry*, 1992, 33(1), 25–33

De Maeseneire C, Duray MC, Tyberghien A et al, 'Musical hallucinations as a presenting manifestation of a left temporo-insular glioma', *Revue Neurologique (Paris)*, 2014, 170(4), 302–4

De Neys W, 'Dual processing in reasoning: two systems but one reasoner', *Psychological Science*, 2006, 17(5), 428–33

De Neys W, Hopfensitz A & Bonnefon J-F, 'Adolescents gradually improve at detecting trustworthiness from the facial features of unknown adults', *Journal of Economic Psychology*, 2015, 47, 17–22

———, 'Split-second trustworthiness detection from faces in an economic game', *Experimental Psychology*, 2017, 64(4), 231–9

De Neys W, Moyens E & Vansteenwegen D, 'Feeling we're biased: autonomic arousal and reasoning conflict', *Cognitive, Affective & Behavioral Neuroscience*, 2010, 10(2), 208–16

de Pauw KW, Szulecka TK & Poltock TL, 'Fregoli syndrome after cerebral infarction', *Journal of Nervous and Mental Disease*, 1987, 175(7), 433–8

De Quincey C, *Radical Nature: Rediscovering the Soul of Matter*, Park Street Press, 2010

De Renzi E, *Disorders of Space Exploration and Cognition*, Wiley, 1982

———, 'Oculomotor disturbances in hemispheric disease', in CW Johnston & FJ Pirozzolo (eds), *Neuropsychology of Eye Movements*, Erlbaum, 1988, 177–99

De Renzi E & Faglioni P, 'The comparative efficiency of intelligence and vigilance tests in detecting hemispheric cerebral damage', *Cortex*, 1965, 1(4), 410–33

De Renzi E, Faglioni P, Scotti G et al, 'Impairment of color sorting behavior after hemispheric damage: an experimental study with the Holmgren skein test', *Cortex*, 1972, 8(2), 147–63

De Renzi E, Faglioni P & Villa P, 'Sequential memory for figures in brain-damaged patients', *Neuropsychologia*, 1977, 15(1), 43–9

De Renzi E & Lucchelli F, 'Ideational apraxia', *Brain*, 1988, 111, 1173–85

De Renzi E, Perani D, Carlesimo GA et al, 'Prosopagnosia can be associated with damage confined to the right hemisphere: an MRI and PET study and a review of the literature', *Neuropsychologia*, 1994, 32(8), 893–902

De Renzi E & Scotti G, 'Autotopagnosia: fiction or reality? Report of a case', *Archives of Neurology*, 1970, 23(3), 221–7

De Renzi E & Spinnler H, 'Visual recognition in patients with unilateral cerebral disease', *Journal of Nervous and Mental Disease*, 1966, 142(6), 515–25

———, 'Impaired performance on colour tasks in patients with hemispheric damage', *Cortex*, 1967, 3(2), 194–206 & 207–17

de Sélincourt B, 'Music and duration', *Music and Letters*, 1920, 1(4), 286–93

De Simone R, Ranieri A, Marano E et al, 'Migraine and epilepsy: clinical and pathophysiological relations', *Neurological Sciences*, 2007, 28(suppl 2), S150–5

de Tommaso M, Sardaro M & Livrea P, 'Aesthetic value of paintings affects pain thresholds', *Consciousness and Cognition*, 2008, 17(4), 1152–62

de Vito S, Lunven M, Bourlon C et al, 'When brain damage "improves" perception: neglect patients can localize motion-shifted probes better than controls', *Journal of Neurophysiology*, 2015, 114(6), 3351–8

De Vries GJ, 'Sex differences in adult and developing brains: compensation, compensation,

compensation', *Endocrinology*, 2004, 145(3), 1063–8
De Vries M, *The Whole Elephant Revealed: Insights into the Existence and Operation of Universal Laws and the Golden Ratio*, Axis Mundi, 2012
De Zulueta FIS, 'Bilingualism and family therapy', *Journal of Family Therapy*, 1990, 12, 255–65
Deary IJ, *Looking Down on Human Intelligence: From Psychometrics to the Brain*, Oxford University Press, 2000
Deary IJ, Bell JP, Bell AJ et al, 'Sensory discrimination and intelligence: testing Spearman's other hypothesis', *American Journal of Psychology*, 2004, 117(1), 1–18
Deary IJ, Penke L & Johnson W, 'The neuroscience of human intelligence differences', *Nature Reviews Neuroscience*, 2010, 11(3), 201–11
Debruyne H, Portzky M, Peremans K et al, 'Cotard's syndrome: a review', *Mind & Brain*, 2011, 2(1), 67–72
Debruyne H, Portzky M, Van den Eynde F et al, 'Cotard's syndrome: a review', *Current Psychiatry Reports*, 2009, 11(3), 197–202
Decety J & Chaminade T, 'When the self represents the other: a new cognitive neuroscience view on psychological identification', *Consciousness and Cognition*, 2003, 12(4), 577–96
Decety J, Chaminade T, Grèzes J et al, 'A PET exploration of the neural mechanisms involved in reciprocal imitation', *NeuroImage*, 2002, 15(1), 265–72
Decety J & Lamm C, 'The role of the right temporoparietal junction in social interaction: how low-level computational processes contribute to meta-cognition', *The Neuroscientist*, 2007, 13(6), 580–93
Decety J & Sommerville JA, 'Shared representations between self and other: a social cognitive neuroscience view', *Trends in Cognitive Sciences*, 2003, 7(12), 527–33
Deci E, 'The effects of externally mediated rewards on intrinsic motivation', *Journal of Personality and Social Psychology*, 1971, 18(1), 105–15
Deegan G, 'Discovering recovery', *Psychiatric Rehabilitation Journal*, 2003, 26(4), 368–76
DeFelipe J, 'Cortical interneurons: from Cajal to 2001', *Progress in Brain Research*, 2002, 136, 215–38
Deglin VL, 'Our split brain', *The Unesco Courier*, 1976, 29(1), 4–31
——, 'Die paradoxale Mentalität oder warum Fiktionen die Realität ersetzen', in P Grzybek (ed), *Psychosemiotik-Neurosemiotik*, Universitätsverlag Dr N Brockmeyer, 1993, 55–96

Deglin VL & Kinsbourne M, 'Divergent thinking styles of the hemispheres: how syllogisms are solved during transitory hemisphere suppression', *Brain and Cognition*, 1996, 31(3), 285–307
Dehaene S, *The Number Sense: How The Mind Creates Mathematics*, Penguin, 1997
Dehaene S, Spelke E, Pinel P et al, 'Sources of mathematical thinking: behavioral and brain-imaging evidence', *Science*, 1999, 284(5416), 970–4
Dehaene S, Tzourio N, Frak V et al, 'Cerebral activations during number multiplication and comparison: a PET study', *Neuropsychologia*, 1996, 34(11), 1097–1106
Dehon H, Larøi F & Van der Linden M, 'Affective valence influences participant's susceptibility to false memories and illusory recollection', *Emotion*, 2010, 10(5), 627–39
Deicken RF, Calabrese G, Merrin EL et al, '31Phosphorus magnetic resonance spectroscopy of the frontal and parietal lobes in chronic schizophrenia', *Biological Psychiatry*, 1994, 36(8), 503–10
Deisboeck TS, Berens ME, Kansal AR et al, 'Pattern of self-organization in tumour systems: complex growth dynamics in a novel brain tumour spheroid model', *Cell Proliferation*, 2001, 34(2), 115–34
Dekker JM, Crow RS, Folsom AR et al, 'Low heart rate variability in a 2-minute rhythm strip predicts risk of coronary heart disease and mortality from several causes: the ARIC Study', *Circulation*, 2000, 102(11), 1239–44
Delbecq-Derouesne J, Beauvois MF & Shallice T, 'Preserved recall versus impaired recognition — a case-study', *Brain*, 1990, 113(4), 1045–74
Delbrück M, *Mind from Matter?: An Essay on Evolutionary Epistemology*, Blackwell Scientific Publications, 1986
Delgado MG & Bogousslavsky J, 'Misoplegia', *Frontiers of Neurology and Neuroscience*, 2018, 41(1), 23–7
Delis DC, Kiefner MG & Fridlund AJ, 'Visuospatial dysfunction following unilateral brain damage: dissociations in hierarchical and hemispatial analysis', *Journal of Clinical and Experimental Neuropsychology*, 1988, 10(4), 421–31
Delis DC, Robertson LC & Efron R, 'Hemispheric specialization of memory for visual hierarchical stimuli', *Neuropsychologia*, 1986, 24(2), 205–14
della Mirandola P, *Oration on the Dignity of Man*, trans AR Caponigri, Henry Regnery, 1956
DeLong GR & Aldershof AL, 'An association of special abilities with juvenile manic-depressive illness', in Obler LK & Fein D

(eds), *The Exceptional Brain: Neuropsychology of Talent and Special Abilities*, Guilford Press, 1983, 387–95

Demaree HA, Everhart DW, Youngstorm EA et al, 'Brain lateralization of emotional processing: historical roots and a future incorporating dominance', *Behavioral and Cognitive Neuroscience Reviews*, 2005, 4(1), 3–20

Demay G & Renaux J-P, 'Refus d'obéissance d'origine délirante. Conviction de l'irréalité de la guerre chez un combattant', *Annales médico-psychologiques*, 1919, 11, 396–400

DeMiguel V, Garlappi L & Uppal R, 'Optimal versus naive diversification: how inefficient is the 1/N portfolio strategy?', *The Review of Financial Studies*, 2009, 22(5), 2009, 1915–53

Denenberg VH, 'Micro and macro theories of the brain', *Behavioral and Brain Science*, 1983, 6(1), 174–8

Denes G & Dalla Barba G, 'G. B. Vico, precursor of cognitive neuropsychology? The first reported case of noun-verb dissociation following brain damage', *Brain and Language*, 1998, 62(1), 29–33

Denes G, Cappelletti JY, Zilli T et al, 'A category-specific deficit of spatial representation: the case of autotopagnosia', *Neuropsychologia*, 2000, 38(4), 345–50

Denes G & Pizzamiglio L, *Handbook of Clinical and Experimental Neuropsychology*, Psychology Press, 1999

Dennett DC, *Darwin's Dangerous Idea*, Simon & Schuster, 1995

Denny-Brown D, Meyer ST & Horenstein S, 'The significance of perceptual rivalry resulting from parietal lesion', *Brain*, 1952, 75(4), 433–71

Deouell LY, Ivry RB & Knight RT, 'Electrophysiologic methods and transcranial magnetic stimulation in behavioral neurology and neuropsychology', in Feinberg TE & Farah MJ (eds), *Behavioral Neurology and Neuropsychology*, McGraw-Hill, 2nd edn, 2003, 105–34

Der G & Deary IJ, 'The relationship between intelligence and reaction time varies with age: results from three representative narrow-age age cohorts at 30, 50 and 69 years', *Intelligence*, 2017, 64, 89–97

Descartes R, *Discourse on the Method of Rightly Conducting the Reason*, in ES Haldane & GRT Ross (trans), *The Philosophical Works of Descartes*, 2 vols [1911], Cambridge University Press, 1973

——, *The Philosophical Writings of Descartes*, trans J Cottingham, R Stoothoff, D Murdoch et al, Cambridge University Press, 3 vols, 1984–91

——, *Discourse on the Method*, trans G Heffernan, University of Notre Dame Press, 1994

——, *Meditations on First Philosophy: with selections from the Objections and Replies*, ed & trans J Cottingham, Cambridge University Press, 2nd edn, 2017

Desco M, Navas-Sanchez FJ, Sanchez-González J et al, 'Mathematically gifted adolescents use more extensive and more bilateral areas of the fronto-parietal network than controls during executive functioning and fluid reasoning tasks', *NeuroImage*, 2011, 57(1), 281–92

Desmedt JE, 'Active touch exploration of extrapersonal space elicits specific electrogenesis in the right cerebral hemisphere of intact right-handed man', *Proceedings of the National Academy of Sciences of the United States of America*, 1977, 74(9), 4037–40

Desmond A, entry for TH Huxley in the *Oxford Dictionary of National Biography*, Oxford University Press, 2004

Desmurget M, Reilly KT, Richard N et al, 'Movement intention after parietal cortex stimulation in humans', *Science*, 2009, 324(5928), 811–3

Deutsch D, 'The anthropic universe', ABC, 18 February 2006: www.abc.net.au/radionational/programs/scienceshow/the-anthropic-universe/3302686# (accessed 23 February 2021)

Devine MJ, Bentley P, Jones B et al, 'The role of the right inferior frontal gyrus in the pathogenesis of post-stroke psychosis', *Journal of Neurology*, 2014, 261(3), 600–3

Devinsky O, 'Right cerebral hemisphere dominance for a sense of corporeal and emotional self', *Epilepsy & Behavior*, 2000, 1(1), 60–73

——, 'Delusional misidentifications and duplications: right brain lesions, left brain delusions', *Neurology*, 2009, 72(1), 80–7

Devinsky O & D'Esposito M, 'Right hemisphere dominance of the corporeal self', in *Neurology of Cognitive and Behavioral Disorders*, Oxford University Press, 2004, 71–7

Devinsky O, Feldmann E, Burrowes K et al, 'Autoscopic phenomena with seizures', *Archives of Neurology*, 1989, 46(10), 1080–8

Devinsky O & Lai G, 'Spirituality and religion in epilepsy', *Epilepsy and Behavior*, 2008, 12, 636–43

Devinsky O, Putnam F, Grafman J et al, 'Dissociative states and epilepsy', *Neurology*, 1989, 39(6), 835–40

Dewey J, 'The reflex arc concept in psychology', *Psychological Review*, 1896, 3(4), 357–70

——, *Context and Thought* (University of California Publications in Philosophy, vol 12(3)), University of California Press, 1931

———, *The Quest for Certainty*, in JA Boydston (ed), *John Dewey: The Later Works: 1925–1953*, Southern Illinois University Press, vol 4, 1984

———, 'The philosophy of Whitehead', in JA Boydston (ed), *John Dewey: The Later Works, 1925–1953*, Southern Illinois University Press, 1988, vol 14, 123–40

Dharmaretnam M & Rogers LJ, 'Hemispheric specialization and dual processing in strongly versus weakly lateralized chicks', *Behavioural Brain Research*, 2005, 162(1), 62–70

Di Dio C, Canessa N, Cappa SF et al, 'Specificity of esthetic experience for artworks: an fMRI study', *Frontiers in Human Neuroscience*, 2011, 5, article 139

Di Dio C, Macaluso E & Rizzolatti G, 'The golden beauty: brain response to Classical and Renaissance sculptures', *PLoS One*, 2007, 2(11), e1201

di Ludovico A, 'Exploiting perceiving frames in late third millennium Mesopotamia', in P Matthiae, F Pinnock, L Nigro et al (eds), ICAANE 6: *Proceedings of the 6th International Congress of the Archaeology of the Ancient Near East, 5 May–10 May 2009, 'Sapienza', Università di Roma*, Harrassowitz Verlag, 2010

Di Martino A, Ross K, Uddin LQ et al, 'Functional brain correlates of social and nonsocial processes in autism spectrum disorders: an activation likelihood estimation meta-analysis', *Biological Psychiatry*, 2009, 65, 63–74

Di Nicola V, 'Slow thought: a manifesto', *Aeon*, 2018: aeon.co/essays/take-your-time-the-seven-pillars-of-a-slow-thought-manifesto (accessed 23 February 2021)

Di Thiene D, Alexanderson K, Tinghög P et al, 'Suicide among first-generation and second-generation immigrants in Sweden: association with labour market marginalisation and morbidity', *Journal of Epidemiology and Community Health*, 2015, 69(5), 467–73

Diamond MC, Johnson RE & Ingham CA, 'Morphological changes in the young, adult, and aging rate cerebral cortex, hippocampus, and diencephalon', *Behavioral Biology*, 1975, 14(2), 163–74

Dias AM, 'The integration of the glutamatergic and the white matter hypotheses of schizophrenia's etiology', *Current Neuropharmacology*, 2012, 10(1), 2–11

Dias BG & Ressler KJ, 'Parental olfactory experience influences behavior and neural structure in subsequent generations', *Nature Neuroscience*, 2014, 17(1), 89–96

Diaz MT, Barrett KT & Hogstrom LJ, 'The influence of sentence novelty and figurativeness on brain activity', *Neuropsychologia*, 2011, 49(3), 320–30

Dickens C, in J Hartley (ed), *The Selected Letters of Charles Dickens*, Oxford University Press, 2012

Dide M & Guiraud P, *Psychiatrie du médecin praticien*, Masson et Cie, 2nd edn, 1929 [1922]

Diderot D, *Le Rêve de D'Alembert*, in *Œuvres complètes*, ed J Assézat, Garnier, vol 11, 1875, 122–81 [written 1769, published 1830]

Diduca D & Joseph S, 'Schizotypal traits and dimensions of religiosity', *British Journal of Clinical Psychology*, 1997, 36(4), 635–8

Dieguez S, Staub F & Bogousslavsky J, 'Asomatognosia', in O Godefroy & J Bogousslavsky (eds), *The Behavioral and Cognitive Neurology of Stroke*, Cambridge University Press, 2005, 215–53

Diels H, *Die Fragmente der Vorsokratiker*, ed W Kranz, Weidmann Verlag, 6th edn, 1951

Dien J, 'Looking both ways through time: the Janus model of lateralized cognition', *Brain and Cognition*, 2008, 67(3), 292–323

Diesfeldt HF & Troost D, 'Delusional misidentification and subsequent dementia: a clinical and neuropathological study', *Dementia*, 1995, 6(2), 94–8

Dieterich M, Bartenstein P, Spiegel S et al, 'Thalamic infarctions cause side-specific suppression of vestibular cortex activations', *Brain*, 2005, 128(9), 2052–67

Dieterich M & Brandt T, 'Global orientation in space and the lateralization of brain functions', *Current Opinion in Neurology*, 2018, 31(1), 96–104

Dietrich A, 'The wavicle of creativity', *Methods*, 2007, 42(1), 1–2

———, 'You're gonna need a bigger boat', in McLoughlin N & Brien DL, *Text*, 2012 (special issue on Creativity: Cognitive, Social and Cultural Perspectives), 13 [no page numbers provided]

———, *How Creativity Happens in the Brain*, Palgrave Macmillan, 2015

Dietrich A & Kanso R, 'A review of EEG, ERP, and neuroimaging studies of creativity and insight', *Psychological Bulletin*, 2010, 136(5), 822–48

Diette GB, Lechtzin N, Haponik E et al, 'Distraction therapy with nature sights and sounds reduces pain during flexible bronchoscopy: a complementary approach to routine analgesia', *Chest*, 2003, 123(3), 941–8

Dietz MJ, Friston KJ, Mattingley JB et al, 'Effective connectivity reveals right-hemisphere dominance in audiospatial perception: implications for models of spatial neglect', *Journal of Neuroscience*, 2014, 34(14), 5003–11

DiFabio R, Casali C, Giugni E et al, 'Olfactory hallucinations as a manifestation of hidden rhinosinusitis', *Journal of Clinical Neuroscience*, 2009, 16(10), 1353–5

Dijksterhuis A, Bos MW, Nordgren LF et al, 'On making the right choice: the deliberation-without-attention effect', *Science*, 2006, 311(5763), 1005–7

Dijksterhuis A & Meurs T, 'Where creativity resides: the generative power of unconscious thought', *Consciousness and Cognition*, 2006, 15(1), 135–46

Dijksterhuis A & Nordgren LF, 'A theory of unconscious thought', *Perspectives on Psychological Science*, 2006, 1(2), 95–109

Dijksterhuis GB, Møller P, Bredie WL et al, 'Gender and handedness effects on hedonicity of laterally presented odours', *Brain and Cognition*, 2002, 50(2), 272–81

Dimond SJ, 'Depletion of attentional capacity after total commissurotomy in man', *Brain*, 1976, 99(2), 347–56

——, 'Disconnection and psychopathology', in JH Gruzelier & P Flor-Henry (eds), *Hemisphere Asymmetries of Function in Psychopathology*, Elsevier, 1979a, 35–47

——, 'Performance by split-brain humans on lateralized vigilance tasks', *Cortex*, 1979b, 15(1), 43–50

——, 'Symmetry and asymmetry in the vertebrate brain', in DA Oakley & HC Plotkin (eds), *Brain, Behaviour, and Evolution*, Methuen, 1979c, 189–218

——, 'Tactual and auditory vigilance in split-brain man', *Journal of Neurology, Neurosurgery, and Psychiatry*, 1979d, 42(1), 70–4

Dimond S & Beaumont G, 'Use of two cerebral hemispheres to increase brain capacity', *Nature*, 1971, 232(5308), 270–1

DiNicolantonio J, *The Salt Fix: Why the Experts Got it All Wrong and How Eating More Might Save Your Life*, Piatkus, 2017

DiNicolantonio JJ & Lucan SC, 'The wrong white crystals: not salt but sugar as aetiological in hypertension and cardiometabolic disease', *Open Heart*, 2014, 1, e000167

Dionysius the Areopagite, *The Divine Names*, trans CE Rolt, SPCK, 1920

——, *The Mystical Theology of Dionysius the Areopagite*, trans anon, Shrine of Wisdom, 1923

Dirac P, 'The evolution of the physicist's picture of Nature', *Scientific American*, 1963, 208(5), 45–53

Dissanayake E, 'Ancestral minds and the spectrum of symbol', in E Malotki & E Dissanayake, *Early Rock Art of the American West: The Geometric Enigma*, University of Washington Press, 2018, 91–129

Dixon C, 'How Aristotle created the computer', *The Atlantic*, 20 March 2017: www.theatlantic.com/technology/archive/2017/03/aristotle-computer/518697/ (accessed 23 February 2021)

Dixon MJ, Piskopos M & Schweizer TA, 'Musical instrument naming impairments: the crucial exception to the living/nonliving dichotomy in category-specific agnosia', *Brain and Cognition*, 2000, 43(1–3), 158–64

Dobzhansky T, 'Review: Darwinian or "oriented" evolution?', *Evolution*, 1975, 29(2), 376–8

Dodson CS, Johnson MK & Schooler JW, 'The verbal overshadowing effect: why descriptions impair face recognition', *Memory & Cognition*, 1997, 25(2), 129–39

Dodson MJ, 'Vestibular stimulation in mania: a case report', *Journal of Neurology, Neurosurgery, and Psychiatry*, 2004, 75(1), 168–9

Dōgen Zenji, 'Uji' from the *Shōbōgenzō*; translated as 'On "Just for the time being, just for a while, for the whole of time is the whole of existence"', in *The Treasure House of the Eye of the True Teaching*, trans H Nearman, Shasta Abbey Press, 2007

Doherty MJ, Campbell NM, Tsuji H et al, 'The Ebbinghaus illusion deceives adults but not children', *Developmental Science*, 2010, 13(5), 714–21

Doi T, *The Anatomy of Dependence*, trans J Bester, Kodansha International, 1973

Dolev Y, 'Relativity, global tense and phenomenology', in Y Dolev & M Roubach (eds), *Cosmological and Psychological Time*, Springer, 2016, 21–40

Dolina IA, Efimova OI, Kildyushov EM et al, 'Exploring *terra incognita* of cognitive science: lateralisation of gene expression at the frontal pole of the human brain', *Psychology in Russia: State of the Art*, 2017, 10(3), 231–47

Dollfus S, Razafimandimby A, Delamillieure P et al, 'Atypical hemispheric specialisation for language in right-handed schizophrenia patients', *Biological Psychiatry*, 2005, 57(9), 1020–8

Donald M, 'How culture and brain mechanisms interact in decision-making', in C Engel & W Singer (eds), *Better Than Conscious? Decision Making, the Human Mind, and Implications for Institutions* (Strüngmann Forum Reports), Massachusetts Institute of Technology Press, 2008, 191–205

Doniger GM, Foxe JJ, Murray MM et al, 'Impaired visual object recognition and dorsal/ventral stream interaction in schizophrenia', *Archives of General Psychiatry*, 2002, 59(11), 1011–20

Doniger GM, Silipo G, Rabinowicz EF et al, 'Impaired sensory processing as a basis for object-recognition deficits in schizophrenia', *American Journal of Psychiatry*, 2001, 158(11), 1818–26

Donnellan AM, Hill DA & Leary MR, 'Rethinking autism: implications of sensory and movement differences for understanding and support', *Frontiers in Integrative Neuroscience*, 2013, 6, article 124

Dorion A, Chantôme M, Hasboun D et al, 'Hemispheric asymmetry and corpus callosum morphometry: a magnetic resonance imaging study', *Neuroscience Research*, 2000, 36(1), 9–13

Douthat AS, Nagib HM & Fejer AA, 'On the artistic use of fluid flow patterns made visible', *Leonardo*, 1975, 8(1), 7–11

Dowd EC & Barch DM, 'Anhedonia and emotional experience in schizophrenia: neural and behavioral indicators', *Biological Psychiatry*, 2010, 67(10), 902–11

Dowden B, 'Zeno's paradoxes' [undated], www.iep.utm.edu/zeno-par/ (accessed 6 March 2021)

Doyle J, Csete M & Caporale L, 'An engineering perspective: the implicit protocols', in LH Caporale (ed), *The Implicit Genome*, Oxford University Press, 2006, 294–8

Doyon J & Milner B, 'Right temporal-lobe contribution to global visual processing', *Neuropsychologia*, 1991, 29(5), 343–60

Draaisma D, *Why Life Speeds Up As You Get Older*, Cambridge University Press, 2004

Drago V, Foster PS, Okun MS et al, 'Artistic creativity and DBS: a case report', *Journal of Neurological Sciences*, 2009a, 276(1–2), 138–42

——, 'Turning off artistic ability: the influence of left DBS in art production', *Journal of Neurological Sciences*, 2009b, 281(1–2), 116–21

Drago V, Foster PS, Trifiletti D et al, 'What's inside the art? The influence of frontotemporal dementia in art production', *Neurology*, 2006, 67(7), 1285–7

Drake ME, 'Cotard's syndrome and temporal lobe epilepsy', *Psychiatric Journal of the University of Ottawa*, 1988, 13(1), 36–9

Drake RA & Bingham BR, 'Induced lateral orientation and persuasibility', *Brain and Cognition*, 1985, 4(2), 156–64

Drake RA & Seligman ME, 'Self-serving biases in causal attributions as a function of altered activation asymmetry', *International Journal of Neuroscience*, 1989, 45(3–4), 199–204

Drake RA & Ulrich G, 'Line bisecting as a predictor of personal optimism and desirability of risky behaviors', *Acta Psychologica (Amsterdam)*, 1992, 79(3), 219–26

Drane DL, Lee GP, Loring DW et al, 'Time perception following unilateral amobarbital injection in patients with temporal lobe epilepsy', *Journal of Clinical and Experimental Neuropsychology*, 1999, 21(3), 385–96

Dreyfus H & Dreyfus S, *Mind Over Machine: The Power of Human Intuition and Expertise in the Era of the Computer*, Free Press, 1986

Driesch H, 'The potency of the first two cleavage cells in the development of echinoderms: experimental production of partial and double formations' [1892], in BH Willier & JM Oppenheimer (eds), *Foundations of Experimental Embryology*, Hafner, 1964, 38–50

Drieschner M, 'Is time directed?', in S Albeverio & P Blanchard (eds), *Direction of Time*, Springer, 2014, 117–35

Drob SL, 'The Sefirot: kabbalistic archetypes of mind and creation', *Crosscurrents: The Journal of the Association for Religion and Intellectual Life*, 1997, 47(1), 5–29

——, 'The doctrine of *coincidentia oppositorum* in Jewish mysticism', 2000: www.newKabbalah.com (accessed 29 September 2018)

——, *Kabbalah and Postmodernism: A Dialogue*, Peter Lang Publishing, 2009

Druckman DE & Swets JA, *Enhancing Human Performance: Issues, Theories, and Techniques*, National Academies Press, 1988

Dubischar-Krivec AM, Bölte S, Braun C et al, 'Neural mechanisms of savant calendar calculating in autism: an MEG–study of few single cases', *Brain and Cognition*, 2014, 90, 157–64

Dubois J & VanRullen R, 'Visual trails: do the doors of perception open periodically?', *PLoS Biology*, 2011, 9(5), e1001056

Ducasse CJ, *A Critical Examination of the Belief in a Life after Death*, Kessinger Publishing, 2006 [1961]

Ducatez S, Sol D, Sayol F et al, 'Behavioural plasticity is associated with reduced extinction risk in birds', *Nature Ecology & Evolution*, 2020, 4(6), 788–93

Duchêne A, Graves RE & Brugger P, 'Schizotypal thinking and associative processing: a response commonality analysis of verbal fluency', *Journal of Psychiatry and Neuroscience*, 1998, 23(1), 56–60

Dudley R, John C, Young A et al, 'Normal and abnormal reasoning in people with delusions', *British Journal of Clinical Psychology*, 1997, 36(2), 243–58

Dukas H & Hoffmann B (eds), *Albert Einstein, The Human Side: Glimpses from his Archives*, Princeton University Press, 1979

Dumas G, Nadel J, Soussignan R et al, 'Interbrain synchronization during social interaction', *PLoS One*, 2010, 5(8), e12166

Dumont JE, Pécasse F & Maenhaut C, 'Crosstalk and specificity in signalling: are we crosstalking ourselves into general confusion?', *Cellular Signalling*, 2001, 13, 457–63

Dumont S & Prakash M, 'Emergent mechanics of biological structures', *Molecular Biology of the Cell*, 2014, 25(22), 3461–5

Dunayevich E & Keck PE, 'Prevalence and description of psychotic features in bipolar mania', *Current Psychiatry Reports*, 2000, 2, 286–90

Duncan L, Yilmaz Z, Gaspar H et al, 'Significant locus and metabolic genetic correlations revealed in genome-wide association study of anorexia nervosa', *American Journal of Psychiatry*, 2017, 174(9), 850–8

Dunker AK, Lawson JD, Brown CJ et al, 'Intrinsically disordered protein', *Journal of Molecular Graphics and Modelling*, 2001, 19(1), 26–59

Dunn BD, Galton HC, Morgan R et al, 'Listening to your heart. How interoception shapes emotion experience and intuitive decision making', *Psychological Science*, 2010, 21(12), 1835–44

Dupont J (ed), *The Clinical Diary of Sándor Ferenczi*, Harvard University Press, 1995

Duppa R, 'Maxim LIII', *Maxims, Reflections, &c*, Longman & Co, 1830

Dupré J, 'Are there genes?', in A O'Hear (ed), *Philosophy, Biology and Life* (Royal Institute of Philosophy Supplements), Cambridge University Press, 2005, 193–210

———, 'The polygenomic organism', in S Parry & J Dupré (eds), *Nature After the Genome*, Blackwell, 2010, 19–31

———, 'Animalism and the persistence of human organisms', *The Southern Journal of Philosophy*, 2014, 52 (suppl), 6–23

———, 'The metaphysics of evolution', *Interface Focus*, 2017a, 7(5), 20160148

———, 'The metaphysics of metamorphosis': aeon.co/essays/science-and-metaphysics-must-work-together-to-answer-lifes-deepest-questions, 2017b (accessed 23 February 2021)

Dupré J & Nicholson DJ, 'A manifesto for a processual philosophy of biology', in DJ Nicholson & J Dupré (eds), *Everything Flows: Towards a Processual Philosophy of Biology*, Oxford University Press, 2018, 3–47

Dupré J & O'Malley MA, 'Varieties of living things: life at the intersection of lineage and metabolism', *Philosophy, Theory, and Practice in Biology*, 2009, 1(3), 1–25

Durkheim E, *Suicide: A Study in Sociology*, Free Press, 1951 [1897]

Durnford M & Kimura D, 'Right hemisphere specialisation for depth perception reflected in visual field differences', *Nature*, 1971, 231(5302), 394–5

Dusoir H, Owens C, Forbes RB et al, 'Anorexia nervosa remission following left thalamic stroke', *Journal of Neurology, Neurosurgery, & Psychiatry*, 2005, 76(1), 144–5

Dutton E, 'The Savanna-IQ interaction hypothesis: a critical examination of the comprehensive case presented in Kanazawa's *The Intelligence Paradox*', *Intelligence*, 2013, 41(5), 607–14

Dutton E & Lynn R, 'A negative Flynn effect in Finland, 1997–2009', *Intelligence*, 2013, 41(6), 817–20

———, 'A negative Flynn Effect in France, 1999–2008/9', *Intelligence*, 2015, 51, 67–70

Dutton E, van der Linden D & Lynn R, 'The negative Flynn Effect: a systematic literature review', *Intelligence*, 2016, 59, 163–9

Dvořák R, 'The role of vortices in animal locomotion in fluids', *Applied and Computational Mechanics*, 2014, 8, 147–56

Dyson F, obituary of Hermann Weyl, *Nature*, 10 March 1956, 177(4506), 457–8

———, 'Innovation in physics', *Scientific American*, 1958, 199(3), 74–82

———, *Disturbing the Universe*, Basic Books, 1979

———, *Infinite in All Directions*, Harper Perennial, 2004

———, *Dreams of Earth and Sky*, New York Review of Books Publishing, 2015

Dyson J, 'Failure can be an option', *The Guardian*, 22 July 2012

Earle JBB, 'Cerebral laterality and meditation: a review of the literature', *The Journal of Transpersonal Psychology*, 1981, 13(2), 155–73

Earle TC & Cvetkovich GT, *Social Trust: Toward a Cosmopolitan Society*, Praeger, 1995

Earle W, 'Notes on the death of culture', in MR Stein, AJ Vidich & DM White (eds), *Identity and Anxiety*, Free Press of Glencoe, 1960, 367–83

Eaves L, 'Genetic and social influences on religion and values', in M Jeeves (ed), *From Cells to Souls – and Beyond: Changing Portraits of Human Nature*, Eerdmans, 2004, 102–22

Eaves LJ, Martin NG & Heath AC, 'Religious affiliation in twins and their parents: testing a model of cultural inheritance', *Behavioral Genetics*, 1990, 20(1), 1–22

Ebata S, Ogawa M, Tanaka Y et al, 'Apparent reduction in the size of one side of the face associated with a small retrosplenial haemorrhage', *Journal of Neurology, Neurosurgery, and Psychiatry*, 1991, 54(1), 68–70

Ebel H, Gross G, Klosterkotter J et al, 'Basic symptoms in schizophrenic and affective psychoses', *Psychopathology*, 1989, 22(4), 224–32

Eccles J, 'Modern biology and the turn to belief in God', in RA Varghese (ed), *The*

Intellectuals Speak Out About God: A Handbook for the Christian Student in a Secular Society, Regnery, 1984
Eckblad M & Chapman LJ, 'Magical ideation as an indicator of schizotypy', *Journal of Consulting and Clinical Psychology*, 1983, 51(2), 215–25
Eckermann JP, *Gespräche mit Goethe in den letzten Jahren seines Lebens*, III; in GW von Biedermann (ed), *Goethes Gespräche* (vols 1–10), Leipzig, 1889–1896
———, *Conversations with Goethe*, ed JK Moorhead, trans J Oxenford, Dent, 1970
Eckhart, Meister, in F Pfeiffer (ed), 2 vols, *Meister Eckhart*, Göschen, 1857
———, in F Pfeiffer (ed), 2 vols, *Meister Eckhart*, trans C de B Evans, John Watkins, 2 vols, 1931–1947 [1857]
———, in *Meister Eckhart: Selected Treatises and Sermons*, trans JM Clark & JV Skinner, Faber, 1958
———, in M O'C Walshe (trans & ed), *The Complete Mystical Works of Meister Eckhart*, Herder & Herder, 2009
Eddington A, *Space, Time and Gravitation: An Outline of the General Relativity Theory*, Cambridge University Press, 1920
———, *The Nature of the Physical World*, Macmillan, 1928
———, *The Philosophy of Physical Sciences*, Cambridge University Press, 1939
Eddy J, interview conducted for the American Institute of Physics by Spencer Weart (21 April 1999): history.aip.org/climate/eddy_int.htm (accessed 23 February 2021)
Edel L, 'The deathbed notes of Henry James', *The Atlantic Monthly*, 1968, 221(6), 103–5
Edelstyn NM & Oyebode F, 'A review of the phenomenology and cognitive neuropsychological origins of the Capgras syndrome', *International Journal of Geriatric Psychiatry*, 1999, 14(1), 48–59
Edelstyn NM, Oyebode F & Barrett K, 'The delusions of Capgras and intermetamorphosis in a patient with right-hemisphere white-matter pathology', *Psychopathology*, 2001, 34(6), 299–304
Edmonds CJ, Isaacs EB, Visscher PM et al, 'Inspection time and cognitive abilities in twins aged 7 to 17 years: age-related changes, heritability, and genetic covariance', *Intelligence*, 2008, 36(3), 210–25
Edmonds D, 'Reason and romance: the world's most cerebral marriage', *Prospect Magazine*, August 2014
Edwards L, *The Vortex of Life: Nature's Patterns in Space and Time*, Floris Books, 1993
Edwards-Lee T & Cummings JL, 'Focal lesions and psychosis', in J Bogousslavsky & JL Cummings (eds), *Behavior and Mood Disorders in Focal Brain Lesions*, Cambridge University Press, 2000, 419–37
Egaas B, Courchesne E & Saitoh O, 'Reduced size of corpus callosum in autism', *Archives of Neurology*, 1995, 52(8), 794–801
Eggert P, 'CS Peirce, DH Lawrence, and representation: artistic form and polarities', *DH Lawrence Review*, 1999, 28(1–2), 97–113
Egorov AY & Nikolaenko NN, 'Functional brain asymmetry and visuo-spatial perception in mania, depression and psychotropic medication', *Biological Psychiatry*, 1992, 32(5), 399–410
Ehrenwald H, 'Verändertes Erleben des Körperbildes mit konsekutiver Wahnbildung bei linksseitiger Hemiplegie', *Monatsschrift für Psychiatrie und Neurologie*, 1930, 75, 89–97
———, 'Anosognosie und Depersonalisation: ein Beitrag zur Psychologie der linksseitig Hemiplegischen', *Nervenarzt*, 1931, 4(12), 681–8
Ehrlinger J, Johnson K, Banner M et al, 'Why the unskilled are unaware: further explorations of (absent) self-insight among the incompetent', *Organizational Behavior and Human Decision Processes*, 2008, 105(1), 98–121
Ehrsson HH, Holmes NP & Passingham RE, 'Touching a rubber hand: feeling of body ownership is associated with activity in multisensory brain areas', *Journal of Neuroscience*, 2005, 25(45), 10564–73
Einstein A, 'Geometry and experience', a speech delivered to the Prussian Academy of Sciences, Berlin, 27 January 1921
———, interview with Viereck, in 'What life means to Einstein: an interview by George Sylvester Viereck', *The Saturday Evening Post*, 26 October 1929
———, *Cosmic Religion: With Other Opinions and Aphorisms*, 1931
———, obituary for Emmy Noether, *The New York Times*, 5 May 1935
———, 'Physics and reality', *Journal of the Franklin Institute*, 1936a, 221(3), 349–82
———, 'Physik und Realität', *Journal of the Franklin Institute*, 1936b, 221(3), 313–47
———, *Relativity: The Special and General Theory*, Three Rivers Press, 1961
———, *The Evolution of Physics*, Cambridge University Press, 1971
———, *Out of My Later Years*, Kensington Publishing, 1976
———, 'Principles of Research', address on 26 April 1918 to the German Physical Society, Berlin, on the occasion of Max Planck's 60th birthday: collected in HC Meiser (ed), *Ausgewählte Texte*, Goldmann Verlag, 1986
———, 'Religion and science', *The World as I See It*, Citadel Press, 1999, 24–9

Eklund A, Nichols TE & Knutsson H, 'Cluster failure: why fMRI inferences for spatial extent have inflated false-positive rates', *Proceedings of the National Academy of Sciences of the United States of America*, 2016, 113(28), 7900–5

el Gaddal YY, 'De Clérambault's syndrome (erotomania) in organic delusional syndrome', *British Journal of Psychiatry*, 1989, 154(5), 714–6

Eliade M, *A History of Religious Ideas, Vol 1: From the Stone Age to the Eleusinian Mysteries*, University of Chicago Press, 1978

Elias N, *The Society of Individuals*, Blackwell, 1991

Elio R & Scharf PB, 'Modeling novice-to-expert shifts in problem-solving strategy and knowledge organization', *Cognitive Science*, 1990, 14(4), 579–639

Eliot TS, 'Tradition and the individual talent', *Points of View*, Faber, 1941 [1919]

———, 'The three voices of poetry', *On Poetry and Poets*, Faber, 1957

Elkins J & Valiavicharska Z (eds), *Art and Globalization (The Stone Art Theory Institutes Book 1)*, Penn State Press, 2010

Ellena J-C, *The Diary of a Nose: A Year in the Life of a Parfumeur*, Penguin, 2012

Ellenberg L & Sperry RW, 'Capacity for holding sustained attention following commissurotomy', *Cortex*, 1979, 15(3), 421–38

Elliot AJ & Maier MA, 'Color psychology: effects of perceiving color on psychological functioning in humans', *Annual Review of Psychology*, 2014, 65, 95–120

Elliott R & Dolan RJ, 'Activation of different anterior cingulate foci in association with hypothesis testing and response selection', *NeuroImage*, 1998, 8(1), 17–29

Ellis AW, Jordan JL & Sullivan CA, 'Unilateral neglect is not unilateral: evidence for additional neglect of extreme right space', *Cortex*, 2006, 42(6), 861–8

Ellis GFR, 'Does the multiverse really exist?', *Scientific American*, 2011, 305(2), 38–43

———, 'The evolving block universe and the meshing together of times', *Annals of the New York Academy of Sciences*, 2014, 1326, 26–41

Ellis GFR, Kirchner U & Stoeger WR, 'Multiverses and physical cosmology', *Monthly Notices of the Royal Astronomical Society*, 2004, 347(3), 921–36

Ellis HA, *A Study of British Genius*, Hurst & Blackett, 1904

Ellis HD, 'The role of the right hemisphere in the Capgras delusion: review', *Psychopathology*, 1994, 27(3–5), 177–85

Ellis HD & Szulecka TK, 'The disguised lover: a case of Fregoli delusion', in PW Halligan & JC Marshall (eds), *Method in Madness: Case Studies in Cognitive Neuropsychiatry*, Psychology Press, 1996, 39–50

Ellis RD & Newton N, 'Could moving ourselves be the link between emotion and consciousness?' in A Foolen, UM Lüdtke, TP Racine et al (eds), *Moving Ourselves, Moving Others: Motion and Emotion in Intersubjectivity, Consciousness and Language*, John Benjamins, 2012, 57–80

Ellis RM, *The Middle Way Philosophy 1: The Path of Objectivity*, lulu.com, 2015, 316–7

Ellison-Wright I & Bullmore E, 'Meta-analysis of diffusion tensor imaging studies in schizophrenia', *Schizophrenia Research*, 2009, 108(1–3), 3–10

Ellul J, *The Technological Society*, Vintage, 1964

Elmhirst S, 'Is Richard Dawkins destroying his reputation?', *The Guardian*, 9 Jun 2015

Elster J, *Sour Grapes: Studies in the Subversion of Rationality*, Cambridge University Press, 1983

Elvevåg B, McCormack T, Gilbert A et al, 'Duration judgements in patients with schizophrenia', *Psychological Medicine*, 2003, 33(7), 1249–61

Emerson RW, *Nature*, James Munroe & Co, 1836

———, 'Self-reliance', in *Essays: First Series*, Phillips, Sampson & Co, 1854, 37–79

———, in MM Sealts (ed), *Journals and Miscellaneous Notebooks of Ralph Waldo Emerson*, vol 5 (1835–1838), Harvard University Press, 1965

———, in WH Gilman, AR Ferguson, GP Clark et al (eds), *Journals and Miscellaneous Notebooks of Ralph Waldo Emerson*, vol 1 (1819–1822), Harvard University Press, 1960

———, 'Address delivered before the Senior Class in Divinity College, Cambridge 1838', in L Ziff (ed), *Nature and Other Essays*, Penguin, 2003

———, *The Conduct of Life*, in *Collected Works of Ralph Waldo Emerson*, vol 6, Harvard University Press, 2004

Empson W, *Seven Types of Ambiguity*, Chatto & Windus, 1949

———, from *Empson in Granta: The Book, Film & Theatre Reviews of William Empson Originally Printed in the Cambridge Magazine Granta 1927–1929*, The Foundling Press, 1993

Endrass T, Mohr B & Rockstroh B, 'Reduced interhemispheric transmission time in schizophrenia patients: evidence from event-related potentials', *Neuroscience Letters*, 2002, 320, 57–60

Engel C & Singer W (eds), *Better Than Conscious? Decision Making, the Human Mind, and Implications for Institutions* (Strüngmann

Forum Reports), Massachusetts Institute of Technology Press, 2008
Engerth G, 'Zeichenstörungen bei Patienten mit Autotopagnosie', *Zeitschrift für die gesamte Neurologie und Psychiatrie*, 1933, 143(1), 381–402
Engerth G & Hoff H, 'Ein Fall von Halluzinationen im hemianoptischen Gesichtsfeld: Beitrag zur Genese der optischen Halluzinationen', *Monatsschrift für Psychiatrie und Neurologie*, 1929, 74, 246–56
English I, 'Intuition as a function of the expert nurse: a critique of Benner's novice to expert model', *Journal of Advanced Nursing*, 1993, 18(3), 387–93
Enoch MD & Trethowan WH, *Uncommon Psychiatric Syndromes*, Butterfield-Heinemann, 2nd edn, 1979
Eren I, Çivi I & Yıldız M, 'Frontoparietal hypoperfusion in Capgras syndrome: a case report and review', *Türk Psikiyatri Dergisi*, 2005, 16(4), 284–90
Ernout A & Meillet A, *Dictionnaire étymologique de la langue latine: histoire des mots*, Klincksieck, 2001
Ernst B, *The Magic Mirror of M. C. Escher*, Taschen, 2007
Ersche KD, Fletcher PC, Lewis SJG et al, 'Abnormal frontal activations related to decision-making in current and former amphetamine and opiate dependent individuals', *Psychopharmacology*, 2005, 180(4), 612–23
Eslinger P, 'Neurological and neuropsychological bases of empathy', *European Neurology*, 1998, 39(4), 193–9
Etcoff NL, 'Perceptual and conceptual organization of facial emotions: hemispheric differences', *Brain and Cognition*, 1984, 3(4), 385–412
Eurelings-Bontekoe E, Zwinkels K, Schaap-Jonker H et al, 'Formal characteristics of Thematic Apperception Test narratives of adult patients with an autism spectrum disorder: a preliminary study', *Psychology*, 2011, 2(7), 687–93
Evans DW, Orr PT, Lazar SM et al, 'Human preferences for symmetry: subjective experience, cognitive conflict and cortical brain activity', *PLoS One*, 2012, 7(6), e38966
Evans EM, 'The emergence of beliefs about the origin of species in school-age children', *Merrill-Palmer Quarterly*, 2000, 46(2), 221–54
———, 'Cognitive and contextual factors in the emergence of diverse belief systems: creation versus evolution', *Cognitive Psychology*, 2001, 42(3), 217–66
Evans JS, Barston JL & Pollard P, 'On the conflict between logic and belief in syllogistic reasoning', *Memory & Cognition*, 1983, 11(3), 295–306
Evans JSBT & Wason PC, 'Rationalization in a reasoning task', *British Journal of Psychology*, 1976, 67(4), 479–86
Evans KM & Federmeier KD, 'The memory that's right and the memory that's left: event-related potentials reveal hemispheric asymmetries in the encoding and retention of verbal information', *Neuropsychologia*, 2007, 45(8), 1777–90
Evans P, *The Music of Benjamin Britten*, Clarendon Press, 1996
Everhart DE, Demaree HA & Harrison DW, 'The influence of hostility on electroencephalographic activity and memory functioning during an affective memory task', *Clinical Neurophysiology*, 2008, 119(1), 134–43
Evers S, Dannert J, Rodding D et al, 'The cerebral haemodynamics of music perception: a transcranial Doppler sonography study', *Brain*, 1999, 122(1), 75–85
Evers S, Ellger T, Ringelstein EB et al, 'Is hemispheric language dominance relevant in musical hallucinations? Two case reports', *European Archives of Psychiatry and Clinical Neuroscience*, 2002, 252(6), 299–302
Everson SA, Kauhanen J, Kaplan GA et al, 'Hostility and increased risk of mortality and acute myocardial infarction: the mediating role of behavioral risk factors', *American Journal of Epidemiology*, 1997, 146(2), 142–52
Eviatar Z & Just MA, 'Brain correlates of discourse processing: an fMRI investigation of irony and conventional metaphor comprehension', *Neuropsychologia*, 2006, 44(12), 2348–59
Ey H, *Schizophrénie, études cliniques et psychopathologiques*, Synthelabo, 1996
Eysenck MW, *Psychology: A Student's Handbook*, Psychology Press, 2000
Ezura M, Kakisaka Y, Jin K et al, 'A case of focal epilepsy manifesting multiple psychiatric auras', *Brain and Nerve*, 2015, 67(1), 105–9
Fabiani M, Stadler MA & Wessels PM, 'True but not false memories produce a sensory signature in human lateralized brain potentials', *Journal of Cognitive Neuroscience*, 2000, 12(6), 941–9
Fabris F, 'Waddington's processual epigenetics and the debate over cryptic variability', in DJ Nicholson & J Dupré (eds), *Everything Flows: Towards a Processual Philosophy of Biology*, Oxford University Press, 2018, 246–63
Fadiga L & Craighero L, 'Hand actions and speech representation in Broca's area', *Cortex*, 2006, 42(4), 486–90

Falik O, Mordoch Y, Quansah L et al, 'Rumor has it … : relay communication of stress cues in plants', *PLoS One*, 2011, 6, e23625

Falletta N, *The Paradoxicon*, Turnstone Press, 1985

Fallgatter AJ & Strik WK, 'Reduced frontal functional asymmetry in schizophrenia during a cued continuous performance test assessed with near-infrared spectroscopy', *Schizophrenia Bulletin*, 2000, 26(4), 913–9

Farb NAS, Segal ZV, Mayberg H et al, 'Attending to the present: mindfulness meditation reveals distinct neural modes of self-reference', *Social Cognitive and Affective Neuroscience*, 2007, 2(4), 313–22

Farde L, Wiesel F-A, Stone-Elander S et al, 'D$_2$ dopamine receptors in neuroleptic-naive schizophrenic patients: a positron emission tomography study with [^{11}C]raclopride', *Archives of General Psychiatry*, 1990, 47(3), 213–9

Farne A, Buxbaum L, Ferraro M et al, 'Patterns of spontaneous recovery of neglect and associated disorders in acute right brain-damaged patients', *Journal of Neurology, Neurosurgery, and Psychiatry*, 2004, 75(10), 1401–10

Farrer C, Franck N, Georgieff N et al, 'Modulating the experience of agency: a positron emission tomography study', *NeuroImage*, 2003, 18(2), 324–33

Farrera A, Villanueva M, Quinto-Sánchez M et al, 'The relationship between facial shape asymmetry and attractiveness in Mexican students', *American Journal of Human Biology*, 2015, 27(3), 387–96

Farris EA, Ring J, Black J et al, 'Predicting growth in word level reading skills in children with developmental dyslexia using an object rhyming functional neuroimaging task', *Developmental Neuropsychology*, 2016, 41(3), 145–61

Fatemi Y, Boeve BF, Duffy J et al, 'Neuropsychiatric aspects of primary progressive aphasia', *Journal of Neuropsychiatry and Clinical Neurosciences*, 2011, 23(2), 168–72

Faust M & Mashal N, 'The role of the right cerebral hemisphere in processing novel metaphoric expressions taken from poetry: a divided visual field study', *Neuropsychologia*, 2007, 45(4), 860–70

Faust M & Weisper S, 'Understanding metaphoric sentences in the two cerebral hemispheres', *Brain and Cognition*, 2000, 43(1–3), 186–91

Feinberg TE, 'Neuropathologies of the self: clinical and anatomical features', *Consciousness and Cognition*, 2011, 20(1), 75–81

———, 'Neuropathologies of the self and the right hemisphere: a window into productive personal pathologies', *Frontiers in Human Neuroscience*, 2013, 7, article 472

Feinberg TE, Deluca J, Giacino JT et al, 'Right-hemisphere pathology and the self: delusional misidentification and reduplication', in TE Feinberg & JP Keenan (eds), *The Lost Self: Pathologies of the Brain and Identity*, Oxford University Press, 2005, 100–30

Feinberg TE & Keenan JP, 'Where in the brain is the self?', *Consciousness and Cognition*, 2005, 14(4), 661–78

Feinberg TE & Roane DM, 'Anosognosia', in TE Feinberg & MJ Farah (eds), *Behavioral Neurology and Neuropsychology*, McGraw-Hill Professional, 2003

———, 'Delusional misidentification', *Psychiatric Clinics of North America*, 2005, 28(3), 665–83

Feinberg TE, Roane DM & Ali J, 'Illusory limb movements in anosognosia for hemiplegia', *Journal of Neurology, Neurosurgery & Psychiatry*, 2000, 68(4), 511–3

Feinberg TE, Roane DM, Kwan PC et al, 'Anosognosia and visuoverbal confabulation', *Archives of Neurology*, 1994, 51(5), 468–73

Feinberg TE & Shapiro RM, 'Misidentification-reduplication and the right hemisphere', *Neuropsychiatry, Neuropsychology and Behavioral Neurology*, 1989, 2(1), 39–48

Feinberg TE, Venneri A, Simone AM et al, 'The neuroanatomy of asomatognosia and somatoparaphrenia', *Journal of Neurology, Neurosurgery, and Psychiatry*, 2010, 81(3), 276–81

Fellows LK, 'Deciding how to decide: ventromedial frontal lobe damage affects information acquisition in multi-attribute decision making', *Brain*, 2006, 129(4), 944–52

Felsten G, 'Where to take a study break on the college campus: an attention restoration theory perspective', *Journal of Environmental Psychology*, 2009, 29(1), 160–7

Feltovich PJ & Barrows HS, 'Issues of generality in medical problem-solving', in HG Schmidt & ML de Volder (eds), *Tutorials in Problem-Based Learning*, Van Gorcum, 1984, 128–42

Fenwick P, Lovelace H & Brayne S, 'End of life experiences and their implications for palliative care', *International Journal of Environmental Studies*, 2007, 64(3), 315–23

———, 'Comfort for the dying: five year retrospective and one year prospective studies of end of life experiences', *Archives of Gerontology and Geriatrics*, 2010, 51(2), 173–9

Ferenczi S & Rank O, *The Development of Psychoanalysis*, Nervous and Mental Diseases Publishing Co, 1925 [*Entwicklungsziele*

der Psychoanalyse: zur Wechselbeziehung von Theorie und Praxis 1924]
Ferguson C, Marcus A & Oransky I, 'Publishing: the peer-review scam', Nature, 2014, 515(7528), 480–2
Ferman TJ, Primeau M, Delis D et al, 'Global-local processing in schizophrenia: hemispheric asymmetry and symptom-specific interference', Journal of the International Neuropsychological Society, 1999, 5(5), 442–51
Fernald A, 'Intonation and communicative intent in mothers' speech to infants: is the melody the message?', Child Development, 1989, 60(6), 1497–510
Ferrandez AM, Hugueville L, Lehericy S et al, 'Basal ganglia and supplementary motor area subtend duration perception: an fMRI study', NeuroImage, 2003, 19(4), 1532–44
Ferrari C, Lega C, Tamietto M et al, 'I find you more attractive ... after (prefrontal cortex) stimulation', Neuropsychologia, 2015, 72, 87–93
Ferrè ER, Arthur K & Haggard P, 'Galvanic vestibular stimulation increases novelty in free selection of manual actions', Frontiers in Integrative Neuroscience, 2013, 7, article 74
Ferro J, Kertesz A & Black S, 'Subcortical neglect: quantitation, anatomy, and recovery', Neurology, 1987, 37(9), 1487–92
Ferstl EC, 'The functional neuroanatomy of text comprehension: what's the story so far?', in F Schmalhofer & C Perfetti (eds), Higher Level Language Processes in The Brain: Inference and Comprehension Processes, Erlbaum, 2007
Ferstl EC & von Cramon DY, 'The role of coherence and cohesion in text comprehension: an event-related fMRI study', Brain Research: Cognitive Brain Research, 2001, 11(3), 325–40
Fetchenhauer D & Dunning D, 'Why so cynical? Asymmetric feedback underlies misguided skepticism regarding the trustworthiness of others', Psychological Science, 2010, 21(2), 189–93
Feyerabend P, Against Method, New Left Books, 1975
Feynman RP, 'The Value of Science', public address at the meeting of the National Academy of Sciences, Caltech, 2–4 November 1955: https://calteches.library.caltech.edu/40/2/Science.pdf (accessed 3 July 2021)
———, The Character of Physical Law, Penguin, 1992
Feynman RP & Hibbs AR, Quantum Mechanics and Path Integrals, MacGraw-Hill, 1965
Feynman RP, Leighton RB & Sands M, The Feynman Lectures on Physics, Addison-Wesley, 1964

ffytche DH, Lappin JM & Philpot M, 'Visual command hallucinations in a patient with pure alexia', Journal of Neurology, Neurosurgery, and Psychiatry, 2004, 75(1), 80–6
Fichte JG, The Science of Knowledge, trans Heath & Lachs, Appleton-Century-Crofts, 1970 [1794, 1797]
———, Preface to The Foundations of The Entire Science of Knowledge, 1, §87–88, in The Science of Knowledge, ed & trans P Heath & J Lachs, Cambridge University Press, 1991 [1794]
Fiebelkorn IC, Foxe JJ, McCourt ME et al, 'Atypical category processing and hemispheric asymmetries in high-functioning children with autism: revealed through high-density EEG mapping', Cortex, 2013, 49(5), 1259–67
Fiez JA & Raichle ME, 'Linguistic processing', International Review of Neurobiology, 1997, 41, 233–54
Fijalkowska IJ, Schaaper RM & Jonczyk P, 'DNA replication fidelity in Escherichia coli: a multi-DNA polymerase affair', FEMS Microbiology Reviews, 2012, 36(6), 1105–21
Filipowicz A, Anderson B & Danckert J, 'Adapting to change: the role of the right hemisphere in mental model building and updating', Canadian Journal of Experimental Psychology, 2016, 70(3), 201–18
Filley CM & Jarvis PE, 'Delayed reduplicative paramnesia', Neurology, 1987, 37(4), 701–3
Fingelkurts AA & Fingelkurts AA, 'Is our brain hardwired to produce God, or is our brain hardwired to perceive God? A systematic review on the role of the brain in mediating religious experience', Cognitive Processing, 2009, 10(4), 293–326
Fink A, Benedek M, Grabner RH et al, 'Creativity meets neuroscience: experimental tasks for the neuroscientific study of creative thinking', Methods, 2007, 42(1), 68–76
Fink A, Grabner RH, Benedek M et al, 'Short communication: divergent thinking training is related to frontal electroencephalogram alpha synchronization', European Journal of Neuroscience, 2006, 23(8), 2241–6
———, 'The creative brain: investigation of brain activity during creative problem solving by means of EEG and fMRI', Human Brain Mapping, 2009, 30(3), 734–48
Fink A, Graif B & Neubauer AC, 'Brain correlates underlying creative thinking: EEG alpha activity in professional vs novice dancers', NeuroImage, 2009, 46(3), 854–62
Fink A & Neubauer AC, 'EEG alpha oscillations during the performance of verbal creativity tasks: differential effects of sex and verbal intelligence', International Journal of Psychophysiology, 2006, 62(1), 46–53

———, 'Eysenck meets Martindale: the relationship between extraversion and originality from the neuroscientific perspective', *Personality and Individual Differences*, 2008, 44(1), 299–310

Fink GR, Halligan PW, Marshall JC et al, 'Where in the brain does visual attention select the forest and the trees?', *Nature*, 1996, 382(6592), 626–8

Fink GR, Markowitsch HJ, Reinkemeier M et al, 'Cerebral representation of one's own past: neural networks involved in autobiographical memory', *Journal of Neuroscience*, 1996, 16(13), 4275–82

Fink GR, Marshall JC, Halligan PW et al, 'The neural consequences of conflict between intention and the senses', *Brain*, 1999, 122(3), 497–512

Finkelstein Y, Vardi J & Hod I, 'Impulsive artistic creativity as a presentation of transient cognitive alterations', *Behavioral Medicine*, 1991, 17(2), 91–4

Finlay DC & French J, 'Visual field differences in a facial recognition task using signal detection theory', *Neuropsychologia*, 1978, 16(1), 103–7

Finlay J, Franke T, McKay H et al, 'Therapeutic landscapes and wellbeing in later life: impacts of blue and green spaces for older adults', *Health & Place*, 2015, 34, 97–106

Finocchiaro MA, 'Philosophy versus religion and science versus religion: the trials of Bruno and Galileo', in H Gatti (ed), *Giordano Bruno: Philosopher of the Renaissance*, Ashgate, 2002, 51–96

Fiore SM & Schooler JW, 'Right hemisphere contributions to creative problem solving: converging evidence for divergent thinking', in M Beeman & C Chiarello (eds), *Right Hemisphere Language Comprehension: Perspectives from Cognitive Neuroscience*, Erlbaum, 1998, 349–71

Fischer DB, Perez DL, Prasad S et al, 'Right inferior longitudinal fasciculus lesions disrupt visual-emotional integration', *Social Cognitive and Affective Neuroscience*, 2016, 11(6), 945–51

Fischer F, 'Zeitstruktur und Schizophrenie', *Zeitschrift für die gesamte Neurologie und Psychiatrie*, 1929, 121(1), 544–74

———, 'Raum-Zeit-Struktur und Denkstörung in der Schizophrenie: 11. Mitteilung', *Zeitschrift für die gesamte Neurologie und Psychiatrie*, 1930a, 124(1), 241–56

———, 'Weitere Mitteilung über das schizophrene Zeiterleben (zugleich ein Beitrag zum Verlaufsproblem)', *Archiv für Psychiatrie und Nervenkrankheiten*, 1930b, 92(1), 469–71 (part of a much longer report of the 'Wanderversammlung der südwestdeutschen Neurologen und Psychiater am 24. und 25. Mai 1930 in Baden-Baden')

Fischer M & Zwaan R, 'Embodied language: a review of the role of the motor system in language comprehension', *Quarterly Journal of Experimental Psychology*, 2008, 61(6), 825–50

Fish S, 'Does reason know what it is missing?', *New York Times*, 12 April 2010

Fisher CM, 'Disorientation for place', *Archives of Neurology*, 1982, 39(1), 33–36

Fisher JE, Mohanty A, Herrington JD et al, 'Neuropsychological evidence for dimensional schizotypy: implications for creativity and psychopathology', *Journal of Research in Personality*, 2004, 38(1), 24–31

Fisher MC, 'Hunger and the temporal lobe', *Neurology*, 1994, 44(9), 1577–9

Fisher MPA, 'Quantum cognition: the possibility of processing with nuclear spins in the brain', *Annals of Physics*, 2015, 362, 593–602

Fishman MC & Michael P, 'Integration of auditory information in the cat's visual cortex', *Vision Research*, 1973, 13(8), 1415–9

Fitzgerald M, 'Did Ludwig Wittgenstein have Asperger's syndrome?', *European Child & Adolescent Psychiatry*, 2000, 9(1), 61–5

———, 'Was Spinoza autistic?', *The Philosophers' Magazine*, 2001, 14, 15–16

———, *Autism and Creativity: Is there a Link Between Autism in Men and Exceptional Ability?*, Brunner-Routledge, 2004

———, *The Genesis of Artistic Creativity: Asperger's Syndrome and the Arts*, Jessica Kingsley, 2005

Flashman LA & Roth RM, 'Neural correlates of unawareness of illness in psychosis', in X Amador & A David (eds), *Insight and Psychosis*, Oxford University Press, 2nd edn, 2004, 157–76

Fleminger S & Burns A, 'The delusional misidentification syndromes in patients with and without evidence of organic cerebral disorder: a structured review of case reports', *Biological Psychiatry*, 1993, 33(1), 22–32

Fletcher A, *The Art of Looking Sideways*, Phaidon, 2001

Fletcher PC, Anderson JM, Shanks DR et al, 'Responses of human frontal cortex to surprising events are predicted by formal associative learning theory', *Nature Neuroscience*, 2001, 4(10), 1043–8

Fletcher PC, Happé F, Frith U et al, 'Other minds in the brain: a functional imaging study of "theory of mind" in story comprehension', *Cognition*, 1995, 57(2), 109–28

Fletcher PC, Zafiris O, Frith CD et al, 'On the benefits of not trying: brain activity and connectivity reflecting the interactions of explicit and implicit sequence learning', *Cerebral Cortex*, 2005, 15(7), 1002–15

Flöel A, Buyx A, Breitenstein C et al, 'Hemispheric lateralization of spatial attention in right- and left-hemispheric language dominance', *Behavioural Brain Research*, 2005, 158(2), 269–75

Flor-Henry P, 'Psychosis and temporal lobe epilepsy: a controlled investigation', *Epilepsia*, 1969a, 10(3), 363–95

———, 'Schizophrenic-like reactions and affective psychoses associated with temporal lobe epilepsy: etiological factors', *American Journal of Psychiatry*, 1969b, 126(3), 400–4

Flor-Henry P, Lind JC & Koles ZJ, 'A source-imaging (low-resolution electromagnetic tomography) study of the EEGs from unmedicated males with depression', *Psychiatry Research*, 2004, 130(2), 171–90

Flor-Henry P, Shapiro Y & Sombrun C, 'Brain changes during a shamanic trance: altered modes of consciousness, hemispheric laterality, and systemic psychobiology', *Cogent Psychology*, 2017, 4(1), article 1313522

Flor-Henry P, Tomer R, Kumpula I et al, 'Neurophysiological and neuropsychological study of two cases of multiple personality syndrome and comparison with chronic hysteria', *International Journal of Psychophysiology*, 1990, 10(2), 151–61

Florensky P, *The Pillar and Ground of the Truth: An Essay in Orthodox Theodicy in Twelve Letters*, trans B Jakim, Princeton University Press, 2004

Flourens JP, 'Experimental researches on the properties and functions of the nervous system in the vertebrate animal' [1824], in W Dennis (ed & trans), *Readings in the History of Psychology*, Appleton-Century-Crofts, 1948

Flowers JH & Garbin CP, 'Creativity and perception', in JA Glover, RR Ronning & CR Reynolds (eds), *Handbook of Creativity*, Plenum Press, 1989, 147–62

Flynn FG, Cummings JL, Scheibel J et al, 'Monosymptomatic delusions of parasitosis associated with ischemic cerebrovascular disease', *Journal of Geriatric Psychiatry and Neurology*, 1989, 2(3), 134–9

Flynn FG, Cummings JL & Tomiyasu U, 'Altered behavior associated with damage to the ventromedial hypothalamus: a distinctive syndrome', *Behavioural Neurology*, 1988, 1(1), 49–58

Flynn JR, 'The mean IQ of Americans: massive gains 1932 to 1978', *Psychological Bulletin*, 1984, 95(1), 29–51

———, 'Massive IQ gains in 14 nations: what IQ tests really measure', *Psychological Bulletin*, 1987, 101(2), 171–91

———, 'Beyond the Flynn effect: solution to all outstanding problems — except enhancing wisdom', lecture delivered at The Psychometrics Centre, Judge Business School, University of Cambridge, 15 December 2006: www.psychometrics.cam.ac.uk/about-us/directory/beyond-the-flynn-effect (accessed 23 February 2021)

———, 'Requiem for nutrition as the cause of IQ gains: Raven's gains in Britain 1938–2008', *Economics & Human Biology*, 2009, 7, 18–27

Flynn JR & Shayer M, 'IQ decline and Piaget: does the rot start at the top?', *Intelligence*, 2018, 66, 112–21

Fol H, 'Die erste Entwickelung des Geryonideneies', *Jenaische Zeitschrift für Medizin und Naturwissenschaft*, 1873, 7, 471–92

Foldi NS, 'Appreciation of pragmatic interpretations of indirect commands: comparison of right and left hemisphere brain-damaged patients', *Brain and Language*, 1987, 31(1), 88–108

Foldi NS, Cicone M & Gardner H, 'Pragmatic aspects of communication in brain-damaged patients', in SJ Segalowitz (ed), *Language Functions and Brain Organisation*, Academic Press, 1983, 51–86

Folley BS & Park S, 'Verbal creativity and schizotypal personality in relation to prefrontal hemispheric laterality: a behavioral and near-infrared optical imaging study', *Schizophrenia Research*, 2005, 80(2–3), 271–82

Fong CT, 'The effects of emotional ambivalence on creativity', *Academy of Management Journal*, 2006, 49(5), 1016–30

Fontana AF, Rosenberg RL, Marcus JL et al, 'Type A behavior pattern, inhibited power motivation, and activity inhibition', *Journal of Personality and Social Psychology*, 1987, 52(1), 177–83

Fontenot DJ, 'Visual field differences in the recognition of verbal and non-verbal stimuli in man', *Journal of Comparative and Physiological Psychology*, 1973, 85(3), 564–9

Forbes RJ, *Studies in Ancient Technology*, EJ Brill Publications, vol 8, 1971

Ford BJ, 'Cellular intelligence: microphenomenology and the realities of being', *Progress in Biophysics and Molecular Biology*, 2017, 131, 273–87

Forgács B, Lukács A & Pléh C, 'Lateralized processing of novel metaphors: disentangling figurativeness and novelty', *Neuropsychologia*, 2014, 56, 101–19

Forrester GS, Leavens DA, Quaresmini C et al, 'Target animacy influences gorilla handedness', *Animal Cognition*, 2011, 14(6), 903–7

Forrester GS, Quaresmini C, Leavens DA et al, 'Target animacy influences chimpanzee handedness', *Animal Cognition*, 2012, 15(6), 1121–7

Forster B, Corballis PM & Corballis MC, 'Effect of luminance on successiveness discrimination in the absence of the corpus callosum', *Neuropsychologia*, 2000, 38(4), 441–50

Förster J, Friedman RS & Liberman N, 'Temporal construal effects on abstract and concrete thinking: consequences for insight and creative cognition', *Journal of Personality and Social Psychology*, 2004, 87(2), 177–89

Förstl H, 'Capgras' delusion: an example of coalescent psychodynamic and organic factors', *Comprehensive Psychiatry*, 1990, 31(5), 447–9

Förstl H, Almeida OP & Iacoponi E, 'Capgras delusion in the elderly: the evidence for a possible organic origin', *International Journal of Geriatric Psychiatry*, 1991, 6(12), 845–52

Förstl H, Almeida OP, Owen AM et al, 'Psychiatric, neurological and medical aspects of misidentification syndromes: a review of 260 cases', *Psychological Medicine*, 1991, 21(4), 905–10

Förstl H & Beats B, 'Charles Bonnet's description of Cotard's delusion and reduplicative paramnesia in an elderly patient (1788)', *British Journal of Psychiatry*, 1992, 160(3), 416–8

Förstl H, Besthorn C, Burns A et al, 'Delusional misidentification in Alzheimer's disease: a summary of clinical and biological aspects', *Psychopathology*, 1994, 27(3–5), 194–9

Förstl H, Burns A, Jacoby R et al, 'Neuroanatomical correlates of clinical misidentification and misperception in senile dementia of the Alzheimer type', *Journal of Clinical Psychiatry*, 1991, 52(6), 268–71

Fortin A, Ptito A, Faubert J et al, 'Cortical areas mediating stereopsis in the human brain: a PET study', *NeuroReport*, 2002, 13(6), 895–8

Fossati P, Hevenor SJ, Graham SJ et al, 'In search of the emotional self: an fMRI study using positive and negative emotional words', *American Journal of Psychiatry*, 2003, 160(11), 1938–45

Foster C, *Choosing Life, Choosing Death: The Tyranny of Autonomy in Medical Ethics and Law*, Bloomsbury, 2009

———, *In the Hot Unconscious*, Westland Books, 2012

Foster PL, 'Adaptive mutation: has the unicorn landed?', *Genetics*, 1998, 148(4), 1453–9

———, 'Adaptive mutation: implications for evolution', *Bioessays*, 2000, 22(12), 1067–74

Fouad YA & Aanei C, 'Revisiting the hallmarks of cancer', *American Journal of Cancer Research*, 2017, 7(5), 1016–36

Foucher JR, Lacambre M, Pham B-T et al, 'Low time resolution in schizophrenia: lengthened windows of simultaneity for visual, auditory and bimodal stimuli', *Schizophrenia Research*, 2007, 97(1–3), 118–27

Foussias G & Remington G, 'Negative symptoms in schizophrenia: avolition and Occam's razor', *Schizophrenia Bulletin*, 2010, 36(2), 359–69

Fowler JW, *Stages of Faith*, HarperCollins, 1981

Fowler RL, 'Mythos and logos', *Journal of Hellenic Studies*, 2011, 131, 45–66

Fox CJ, Iaria G & Barton JJS, 'Disconnection in prosopagnosia and face processing', *Cortex*, 2008, 44(8), 996–1009

Fox CJ, Moon SY, Iaria G et al, 'The correlates of subjective perception of identity and expression in the face network: an fMRI adaptation study', *NeuroImage*, 2009, 44(2), 569–80

Fox KCR, Dixon ML, Nijeboer S et al, 'Functional neuroanatomy of meditation: a review and meta-analysis of 78 functional neuroimaging investigations', *Neuroscience & Biobehavioral Reviews*, 2016, 65, 208–28

Fox KCR, Yih J, Raccah O et al, 'Changes in subjective experience elicited by direct stimulation of the human orbitofrontal cortex', *Neurology*, 2018, 91(16), e1519–27

Fox M, *Christian Mystics*, New World Library, 2011

Fraguas R, 'Joyas de tinta china: El Conde Duque expone las ilustraciones realizadas por Daniel Vierge para "El Ingenioso Hidalgo Don Quijote"', *El Pais*, 16 October 2005

Frank A, 'Are clocks enough? Science, philosophy, and time', in E Arias, L Combrinck, P Gabor et al (eds), *The Science of Time* (Astrophysics and Space Science Proceedings, vol 50), Springer, 2016, 391–2

———, 'Minding matter', *Aeon*, 13 March 2017

Frank P, 'Contemporary science and the contemporary world view', *Daedalus*, 1958, 87(1), 57–66

Frank W, 'The central problem of modern Man', *Commentary*, 1 June 1946

Frankl VE, *Yes to Life in Spite of Everything*, Penguin Random House, 2019

Franklin A, Catherwood D, Alvarez J et al, 'Hemispheric asymmetries in categorical perception of orientation in infants and adults', *Neuropsychologia*, 2010, 48(9), 2648–57

Franklin A, Clifford A, Williamson E et al, 'Color term knowledge does not affect categorical perception of color in toddlers', *Journal of Experimental Child Psychology*, 2005, 90(2), 114–41

Franklin A, Drivonikou GV, Bevis L et al, 'Categorical perception of color is lateralized to the right hemisphere in infants, but to the left hemisphere in adults', *Proceedings of the*

National Academy of Sciences of the United States of America, 2008, 105(9), 3221–5

Franklin A, Drivonikou GV, Clifford A et al, 'Lateralisation of categorical perception of colour changes with colour term acquisition', Proceedings of the National Academy of Sciences of the United States of America, 2008, 105(47), 18221–5

Franklin B, Autobiography, Dover, 1996 [1791]

———, 'Advice to a Young Tradesman, Written by an Old One', in Franklin: The Autobiography and Other Writings on Politics, Economics and Virtue, Cambridge University Press, 2004 [1748]

Franks F, The Physics and Physical Chemistry of Water, Springer, 1972

Franzon M & Hugdahl K, 'Visual half-field presentations of incongruent color words: effects of gender and handedness', Cortex, 1986, 22(3), 433–5

Frasnelli E, Anfora G, Trona F et al, 'Morpho-functional asymmetry of the olfactory receptors of the honeybee (Apis mellifera)', Behavioural Brain Research, 2010, 209(2), 221–5

Frasnelli E, Iakovlev I & Reznikova Z, 'Asymmetry in antennal contacts during trophallaxis in ants', Behavioural Brain Research, 2012, 232(1), 7–12

Frasnelli E, Vallortigara G & Rogers LJ, 'Left-right asymmetries of behaviour and nervous system in invertebrates', Neuroscience & Biobehavioral Reviews, 2012, 36(4), 1273–91

Frassinetti F, Maini M, Romualdi S et al, 'Is it mine? Hemispheric asymmetries in corporeal self-recognition', Journal of Cognitive Neuroscience, 2008, 20(8), 1507–16

Frassinetti F, Nichelli P & di Pellegrino G, 'Selective horizontal dysmetropsia following prestriate lesion', Brain, 1999, 122(2), 339–50

Frassinetti F, Pavani F, Zamagni E et al, 'Visual processing of moving and static self body-parts', Neuropsychologia, 2009, 47(8–9), 1988–93

Frauchiger D & Renner R, 'Quantum theory cannot consistently describe the use of itself', Nature Communications, 2018, 9(1), 3711

Fredrickson BL, Grewen KM, Coffey KA et al, 'A functional genomic perspective on human well-being', Proceedings of the National Academy of Sciences of the United States of America, 2013, 110(33), 13684–89

Freedman DA & Petitti DB, 'Salt and blood pressure: conventional wisdom reconsidered', Evaluation Review, 2001, 25(3), 267–87

Freedman M, Binns M, Gao F et al, 'Mind-matter interactions and the frontal lobes of the brain: a novel neurobiological model of psi inhibition', Explore, 2018, 14(1), 76–85

Freely J, Celestial Revolutionary: Copernicus, the Man and His Universe, IB Tauris, 2014

Freitag CM, Konrad C, Häberlen M et al, 'Perception of biological motion in autism spectrum disorders', Neuropsychologia, 2008, 46(5), 1480–94

Freud S, 'The Unconscious', in J Strachey (trans & ed), Standard Edition of the Complete Psychological Works, vol 14, Hogarth Press, 1915, 159–215

———, Totem and Taboo, trans AA Brill, Routledge, 1919

———, Lecture XXXI, 'The anatomy of the mental personality' [1932], in New Introductory Lectures on Psycho-analysis, Hogarth Press, 1933

———, in EL Freud (ed), Letters of Sigmund Freud 1873–1939, trans T & J Stern, Hogarth Press, 1961

———, 'Das Unbewußte' (1915), Studienausgabe, vol 3, Fischer Verlag, 1975

———, Three Essays on the Theory of Sexuality, trans J Strachey, Martino Fine Books, 2011 [1905; in this form, in English, 1953]

Freudenreich O, Psychotic Disorders, Lippincott Williams & Wilkins, 2007

Frey SH, Funnell MG, Gerry VE et al, 'A dissociation between the representation of tool-use skills and hand dominance: insights from left- and right-handed callosotomy patients', Journal of Cognitive Neuroscience, 2005, 17(2), 262–72

Frey TS & Hambert G, 'Neuropsychiatric aspects of logorrhoea', Nordsk Psykiatrisk Tidsskrift, 1972, 26(3), 158–73

Fride E & Weinstock M, 'Prenatal stress increases anxiety related behaviour and alters cerebral lateralization of dopamine activity', Life Sciences, 1988, 42(10), 1059–65

Fried C, 'Is liberty possible?', The Tanner Lectures on Human Values, delivered at Stanford University, 14 & 18 May 1981: tannerlectures.utah.edu/_documents/a-to-z/f/fried82.pdf

Friedlander KJ & Fine PA, 'The grounded expertise components approach in the novel area of cryptic crossword solving', Frontiers in Psychology, 2016, 7, article 567

Friedman A & Polson MC, 'Hemispheres as independent resource systems: limited-capacity processing and cerebral specialization', Journal of Experimental Psychology: Human Perception and Performance, 1981, 7(5), 1031–58

Friedman RS & Förster J, 'The effects of approach and avoidance motor actions on the elements of creative insight', Journal of Personality and Social Psychology, 2000, 79(4), 477–92

——, 'The effects of promotion and prevention cues on creativity', *Journal of Personality and Social Psychology*, 2001, 81(6), 1001–13
——, 'The influence of approach and avoidance motor actions on creative cognition', *Journal of Experimental Social Psychology*, 2002, 38(1), 41–55
——, 'Effects of motivational cues on perceptual asymmetry: implications for creativity and analytical problem solving', *Journal of Personality and Social Psychology*, 2005, 88(2), 263–75
Friedrich TE, Harms VL & Elias LJ, 'Dynamic stimuli: accentuating aesthetic preference biases', *Laterality*, 2014, 19(5), 549–59
Frings L, Wagner K, Unterrainer J et al, 'Gender-related differences in lateralization of hippocampal activation and cognitive strategy', *NeuroReport*, 2006, 17(4), 417–21
Frith CD & Corcoran R, 'Exploring "theory of mind" in people with schizophrenia', *Psychological Medicine*, 1996, 26(3), 521–30
Frith CD & Metzinger TK, 'What's the use of consciousness? How the stab of conscience made us really conscious', in AK Engel, KJ Friston & D Kragic (eds), *The Pragmatic Turn: Toward Action-Oriented Views in Cognitive Science*, Massachusetts Institute of Technology Press, 2016, 197–224
Frith CD, Stevens M, Johnstone EC et al, 'Integration of schematic faces and other complex objects in schizophrenia', *Journal of Nervous and Mental Disease*, 1983, 171(1), 34–9
Friz CT, 'The biochemical composition of the free-living amoebae *Chaos chaos*, *Amoeba dubia* and *Amoeba proteus*', *Comparative Biochemistry and Physiology*, 1968, 26(1), 81–90
Fromm E, *Zen Buddhism and Psychoanalysis*, George Allen & Unwin, 1960
——, *The Art of Listening*, Constable, 1994
Fuchs T, 'The tacit dimension', *Philosophy, Psychiatry, & Psychology*, 2001, 8(4), 323–6
——, 'Corporealized and disembodied minds: a phenomenological view of the body in melancholia and schizophrenia', *Philosophy, Psychiatry & Psychology*, 2005, 12(2), 95–107
——, 'Temporality and psychopathology', *Phenomenology and the Cognitive Sciences*, 2013, 12(1), 75–104
Fudali-Czyż A, Francuz P & Augustynowicz P, 'The effect of art expertise on eye fixation-related potentials during aesthetic judgment task in focal and ambient modes', *Frontiers in Psychology*, 2018, 9, article 1972
Fujita K & Carnevale JJ, 'Transcending temptation through abstraction: the role of construal level in self-control', *Current Directions in Psychological Science*, 2012, 21(4), 248–52
Fujito R, Minese M, Hatada S et al, 'Musical deficits and cortical thickness in people with schizophrenia', *Schizophrenia Research*, 2018, 197, 233–9
Fukuhara K, Ogawa Y, Tanaka H et al, 'Impaired interpretation of others' behavior is associated with difficulties in recognizing pragmatic language in patients with schizophrenia', *Journal of Psycholinguistic Research*, 2017, 46(5), 1309–18
Fuller Torrey E & Yolken RH, '*Toxoplasma gondii* and schizophrenia', *Emerging Infectious Diseases*, 2003, 9(11), 1375–80
Funder DC & Colvin CR, 'Friends and strangers: acquaintanceship, agreement, and the accuracy of personality judgment', *Journal of Personality and Social Psychology*, 1988, 55(1), 149–58
Funk AP & Pettigrew JD, 'Does interhemispheric competition mediate motion-induced blindness? A transcranial magnetic stimulation study', *Perception*, 2003, 32(11), 1328–38
Funnell MG, Colvin MK & Gazzaniga MS, 'The calculating hemispheres: studies of a split-brain patient', *Neuropsychologia*, 2007, 45(10), 2378–86
Funnell MG, Corballis PM & Gazzaniga MS, 'Temporal discrimination in the split brain', *Brain and Cognition*, 2003, 53(2), 218–22
Fuqua WC, Winans SC & Greenberg EP, 'Quorum sensing in bacteria: the LuxR-LuxI family of cell density-responsive transcriptional regulators', *Journal of Bacteriology*, 1994, 176(2), 269–75
Furness JB, Callaghan BP, Rivera LR et al, 'The enteric nervous system and gastrointestinal innervation: integrated local and central control', *Advances in Experimental Medicine and Biology*, 2014, 817, 39–71
Furst CJ, 'EEG asymmetry and visuospatial performance', *Nature*, 1976, 260(5548), 254–5
Fusco FB, Gomes DJ, Bispo KCS et al, 'Low-sodium diet induces atherogenesis regardless of lowering blood pressure in hypertensive hyperlipidemic mice', *PLoS One*, 2017, 12(5), e0177086
Futamura A, Katoh H & Kawamura M, 'Successful treatment with anti-epileptic-drug of an 83-year-old man with musical hallucinosis', *Brain and Nerve*, 2014, 66(5), 599–603
Gable PA, Poole BD & Cook MS, 'Asymmetrical hemisphere activation enhances global-local processing', *Brain and Cognition*, 2013, 83(3), 337–41
Gadamer H-G, *Truth and Method*, trans J Weinsheimer & DG Marshall, Crossroad, 2nd edn, 1992

Gagliano M, Grimonprez M, Depczynski M et al, 'Tuned in: plant roots use sound to locate water', *Oecologia*, 2017, 184(1), 151–60

Gagliano M, Mancuso S & Robert D, 'Towards understanding plant bioacoustics', *Trends in Plant Science*, 2012, 17(6), 323–5

Gagliano M, Renton M, Depczynski M et al, 'Experience teaches plants to learn faster and forget slower in environments where it matters', *Oecologia*, 2014, 175(1), 63–72

Gagliano M, Vyazovskiy VV, Borbély AA et al, 'Learning by association in plants', *Scientific Reports*, 2016, 6, 38427

Gainotti G, 'Cognitive and anatomical locus of lesion in a patient with a category-specific semantic impairment for living beings', *Cognitive Neuropsychology*, 1996, 13(3), 357–90

———, 'Components and levels of emotion disrupted in patients with unilateral brain damage', in G Gainotti (ed), *Emotional Behavior and Its Disorders: Handbook of Neuropsychology*, vol 5, Elsevier, 2nd edn, 2001, 161–80

———, 'Emotional disorders in relation to unilateral brain damage', in TE Feinberg & M Farah (eds), *Behavioral Neurology and Neuropsychology*, McGraw-Hill, 2003, 725–35

———, 'Unconscious emotional memories and the right hemisphere', in M Mancia (ed), *Psychoanalysis and Neuroscience*, Springer, Milan, 2006, 151–73

———, 'Face familiarity feelings, the right temporal lobe and the possible underlying neural mechanisms', *Brain Research Reviews*, 2007, 56(1), 214–35

———, 'Unconscious processing of emotions and the right hemisphere', *Neuropsychologia*, 2012, 50(2), 205–18

Gainotti G, Barbier A & Marra C, 'Slowly progressive defect in recognition of familiar people in a patient with right anterior temporal atrophy', *Brain*, 2003, 126(4), 792–803

Gainotti G, D'Erme P & Bartolomeo P, 'Early orientation of attention toward the half space ipsilateral to the lesion in patients with unilateral brain damage', *Journal of Neurology, Neurosurgery, and Psychiatry*, 1991, 54(12), 1082–9

Gainotti G & Marra C, 'Differential contribution of right and left temporo-occipital and anterior temporal lesions to face recognition disorders', *Frontiers in Human Neuroscience*, 2011, 5, article 55

Gaisman J, 'The devout sceptic: a creed for those of little faith', *Standpoint*, September 2018

Galaburda AM, 'Anatomic basis of cerebral dominance', in RJ Davidson & K Hugdahl (eds), *Brain Asymmetry*, Massachusetts Institute of Technology Press, 1995, 51–73

Galaburda AM, Aboitiz F, Rosen GD et al, 'Histological asymmetry in the primary visual cortex of the rat: implications for mechanisms of cerebral asymmetry', *Cortex*, 1986, 22(1), 151–60

Galaburda AM, LeMay M, Kemper TL et al, 'Right-left asymmetrics in the brain', *Science*, 1978, 199(4331), 852–6

Gale CR, Batty GD, McIntosh A et al, 'Is bipolar disorder more common in highly intelligent people? A cohort study of a million men', *Molecular Psychiatry*, 2013, 18(2), 190–4

Gale TM, Done DJ & Frank RJ, 'Visual crowding and category specific deficits for pictorial stimuli: a neural network model', *Cognitive Neuropsychology*, 2001, 18(6), 509–50

Galea LAM & Kimura D, 'Sex differences in route-learning', *Personality and Individual Differences*, 1993, 14(1), 53–65

Galilei G, *Il Saggiatore, nel quale con bilancia esquisita e giusta si ponderano le cose contenute nella Libra Astronomica e Filosofica di Lotario Sarsi Sigensano, scritto in forma di lettera all'Illustrissimo et Reverendissimo Monsignore D. Virginio Cesarini*, G Mascardi, 1623

———, *Le Opere* (Edizione nazionale), ed A Favaro, Barbèra, 1965 [1638], vol 8

Gall FJ, *On the Origin of the Moral Qualities and Intellectual Faculties of Man, and the Conditions of their Manifestations*, 6 vols, trans W Lewis, Marsh, Capen & Lyon, 1835

Gallace A & Spence C, 'The cognitive and neural correlates of "tactile consciousness": a multisensory perspective', *Consciousness and Cognition*, 2008, 17, 370–407

———, 'Touch and the body: the role of the somatosensory cortex in tactile awareness', *Psyche*, 2010, 16(1), 30–67

Gallace A, Tan HZ & Spence C, 'The body surface as a communication system: the state of art after 50 years of research', *Presence: Teleoperators and Virtual Environments*, 2007, 16, 655–76

Gallagher S, 'Self-narrative in schizophrenia', in T Kircher & A David (eds), *The Self in Neuroscience and Psychiatry*, Cambridge University Press, 2003, 336–57

Gallagher S, Butterworth GE, Lew A et al, 'Hand-mouth coordination, congenital absence of limb, and evidence for innate body schemas', *Brain and Cognition*, 1998, 38(1), 53–65

Gallese V, 'Mirror neurons and the social nature of language: the neural exploitation hypothesis', *Social Neuroscience*, 2008, 3(3–4), 317–33

Gallhofer B, Trimble MR, Frackowiak R et al, 'A study of cerebral blood flow and metabolism in epileptic psychosis using positron emission tomography and oxygen', *Journal of*

Neurology, Neurosurgery, and Psychiatry, 1985, 48(3), 201–6

Gallinat J, Winterer G, Herrmann CS et al, 'Reduced oscillatory gamma-band responses in unmedicated schizophrenic patients indicate impaired frontal network processing', *Clinical Neurophysiology*, 2004, 115(8), 1863–74

Galton F, *Hereditary Genius: An Enquiry into its Laws and Consequences*, Macmillan and Company, 1892

Gambetta D & Hertog S, 'Why are so many engineers among Islamic radicals?', *European Journal of Sociology*, 2009, 51(2), 201–30

Gansler DA, McLaughlin NC, Iguchi L et al, 'A multivariate approach to aggression and the orbital frontal cortex in psychiatric patients', *Psychiatry Research*, 2009, 171(3), 145–54

Gao C & Guo C, 'The experience of beauty of Chinese poetry and its neural substrates', *Frontiers in Psychology*, 2018, 9, article 1540

Gao X & Lan C, 'Buddhist metaphors in the *Diamond Sutra* and the *Heart Sutra*: a cognitive perspective', in P Chilton and M Kopytowska (eds), *Religion, Language and the Human Mind*, Oxford University Press, 2018, 229–62

Garcia JM & Stick SL, 'Perceptual feature sorting of brain-injured patients: left versus right hemisphere preferences', *Journal of Communication Disorders*, 1986, 19(5), 395–404

Gardner H, 'The music of the hemispheres', *Psychology Today*, June 1982, 91–2

———, *Frames of Mind: The Theory of Multiple Intelligence*, Basic Books, 1983

Gardner H, Brownell HH, Wapner W et al, 'Missing the point: the role of the right hemisphere in the processing of complex linguistic materials', in E Perecman (ed), *Cognitive Processing in the Right Hemisphere*, Academic Press, 1983, 169–91

Gardner H & Denes G, 'Connotative judgements by aphasic patients on a pictorial adaptation of the semantic differential', *Cortex*, 1973, 9(2), 183–96

Gardner H, Ling PK, Flamm L et al, 'Comprehension and appreciation of humorous material following brain damage', *Brain*, 1975, 98(3), 399–412

Gardner M, *Are Universes Thicker than Blackberries?*, Norton, 2003

Gardner WJ, Karnosh LJ, McClure CC et al, 'Residual function following hemispherectomy for tumour and for infantile hemiplegia', *Brain*, 1955, 78(4), 487–502

Gardner-Thorpe C & Pearn J, 'The Cotard syndrome. Report of two patients: with a review of the extended spectrum of "délire des négations"', *European Journal of Neurology*, 2004, 11, 563–6

Gare A, 'From Kant to Schelling to process metaphysics: on the way to ecological civilization', *Cosmos and History: The Journal of Natural and Social Philosophy*, 2011, 7(2), 26–69

———, 'Chreods, homeorhesis and biofields: finding the right path for science through Daoism', *Progress in Biophysics and Molecular Biology*, 2017, 131, 61–91

———, 'Natural philosophy and the sciences: challenging science's tunnel vision', *Philosophies*, 2018, 3(4), 33

Garety PA & Freeman D, 'Cognitive approaches to delusions: a critical review of theories and evidence', *British Journal of Clinical Psychology*, 1999, 38, 113–54

Garety PA, Freeman D, Jolley S et al, 'Reasoning, emotions and delusional conviction in psychosis', *Journal of Abnormal Psychology*, 2005, 114(3), 373–84

Garety PA, Hemsley DR & Wessely S, 'Reasoning in deluded schizophrenic and paranoid patients: biases in performance on a probabilistic inference task', *Journal of Nervous and Mental Disease*, 1991, 179(4), 194–201

Garety P, Joyce E, Jolley S et al, 'Neuropsychological functioning and jumping to conclusions in delusions', *Schizophrenia Research*, 2013, 150(2–3), 570–4

Garoff RJ, Slotnick SD & Schacter DL, 'The neural origins of specific and general memory: the role of the fusiform cortex', *Neuropsychologia*, 2005, 43(6), 847–59

Garretson HB, Fein D & Waterhouse L, 'Sustained attention in children with autism', *Journal of Autism and Developmental Disorders*, 1990, 20(1), 101–14

Gaukroger S, *The Emergence of a Scientific Culture: Science and the Shaping of Modernity, 1210–1685*, Oxford University Press, 2006

Gaut B, 'The philosophy of creativity', *Philosophy Compass*, 2010, 5(12), 1034–46

Gauthier I, Behrmann M & Tarr MJ, 'Can face recognition really be dissociated from object recognition?', *Journal of Cognitive Neuroscience*, 1999, 11(4), 349–70

Gauthier I, Tarr MJ, Anderson AW et al, 'Activation of the middle fusiform "face area" increases with expertise in recognizing novel objects', *Nature Neuroscience*, 1999, 2(6), 568–73

Gauthier I, Tarr MJ, Moylan J et al, 'Does visual subordinate-level categorisation engage the functionally defined fusiform face area?', *Cognitive Neuropsychology*, 2000, 17(1), 143–64

Gavaler JS & Arria AM, 'Increased susceptibility of women to alcoholic liver disease:

artifactual or real?' in P Hall (ed), *Alcoholic Liver Disease: Pathology and Pathogenesis*, Edward Arnold, 2nd edn, 1995, 123–33

Gazzaniga MS, 'Some effects of cerebral commissurotomy on monkey and man', unpublished PhD thesis, California Institute of Technology, 1965

———, 'The split brain revisited', *Scientific American*, 1998, 279(1), 50–5

———, 'Cerebral specialization and interhemispheric communication: does the corpus callosum enable the human condition?', *Brain*, 2000, 123(7), 1293–326

———, *Who's in Charge?: Free Will and the Science of the Brain*, Ecco, 2011

———, *Tales from Both Sides of the Brain: A Life in Neuroscience*, Ecco, 2015

Gazzaniga MS, Bogen JE & Sperry RW, 'Dyspraxia following division of the cerebral commissures', *Journal of Nervous and Mental Disease*, 1967, 16(6), 606–12

Gazzaniga M, Ivry R & Mangun G, *Cognitive Neuroscience: The Biology of the Mind*, Norton, 2008

Gazzaniga MS & LeDoux JE, *The Integrated Mind*, Plenum Press, 1978

Gazzaniga MS & Miller MB, 'The left hemisphere does not miss the right hemisphere', in S Laureys & G Tononi (eds), *The Neurology of Consciousness*, Academic Press, 2009, 261–70

Gazzaniga MS & Smylie CS, 'Dissociation of language and cognition: a psychological profile of two disconnected right hemispheres', *Brain*, 1984, 107(1), 145–53

GBD 2016 Alcohol Collaborators, 'Alcohol use and burden for 195 countries and territories, 1990–2016: a systematic analysis for the Global Burden of Disease Study 2016', *The Lancet*, 2018, 392(10152), 1015–35

Geddes L, 'Food for thought? French bean plants show signs of intent, say scientists', *The Guardian*, 8 January 2021

Gee DG, Biswal BB, Kelly C et al, 'Low frequency fluctuations reveal integrated and segregated processing among the cerebral hemispheres', *NeuroImage*, 2011, 54(1), 517–27

Geeraerts S, Lafosse C, Vaes N et al, 'Dysfunction of right hemisphere attentional networks in attention deficit hyperactivity disorder', *Journal of Clinical and Experimental Neuropsychology*, 2008, 30(1), 42–52

Geffen G, Bradshaw JL & Wallace G, 'Interhemispheric effects on reaction times to verbal and nonverbal stimuli', *Journal of Experimental Psychology*, 1971, 87(3), 415–22

Geiser E, Zaehle T, Jäncke L et al, 'The neural correlate of speech rhythm as evidenced by metrical speech processing', *Journal of Cognitive Neuroscience*, 2008, 20(3), 541–52

Gelal FM, Kalaycı TÖ, Çelebisoy M et al, 'Clinical and MRI findings of cerebellar agenesis in two living adult patients', *Annals of Indian Academy of Neurology*, 2016, 19(2), 255–7

Gell PGH, 'Destiny and the genes: genetic pathology and the individual', in R Duncan & M Weston-Smith (eds), *The Encyclopædia of Medical Ignorance*, Pergamon Press, 1984, 179–87

Gellner E, *The Legitimation of Belief*, Cambridge University Press, 1975

Gemmell BJ, Costello JH, Colin SP et al, 'Passive energy recapture in jellyfish contributes to propulsive advantage over other metazoans', *Proceedings of the National Academy of Sciences of the United States of America*, 2013, 110(44), 17904–9

Gencoglu EA, Alehan F, Erol I et al, 'Brain SPECT findings in a patient with Alice in Wonderland syndrome', *Clinical Nuclear Medicine*, 2005, 30(11), 758–9

Genthon N, Rougier P, Gissot AS et al, 'Contribution of each lower limb to upright standing in stroke patients', *Stroke*, 2008, 39(6), 1793–9

Gentilucci M, Bernardis P, Crisi G et al, 'Repetitive transcranial magnetic stimulation of Broca's area affects verbal responses to gesture observation', *Journal of Cognitive Neuroscience*, 2006, 18(7), 1059–74

George I, Hara E & Hessler NA, 'Behavioral and neural lateralization of vision in courtship singing of the zebra finch', *Journal of Neurobiology*, 2006, 66(10), 1164–73

George I, Vernier B, Richard JP et al, 'Hemispheric specialization in the primary auditory area of awake and anesthetized starlings (Sturnus vulgaris)', *Behavioral Neuroscience*, 2004, 118(3), 597–610

George MS, Parekh PI, Rosinsky N et al, 'Understanding emotional prosody activates right hemisphere region', *Archives of Neurology*, 1996, 53(7), 665–70

Geroldi C, Akkawi NM, Galluzzi S et al, 'Temporal lobe asymmetry in patients with Alzheimer's disease with delusions', *Journal of Neurology, Neurosurgery, and Psychiatry*, 2000, 69(2), 187–91

Gersztenkorn D & Lee AG, 'Palinopsia revamped: a systematic review of the literature', *Survey of Ophthalmology*, 2015, 60(1), 1–35

Gervais WM & Norenzayan A, 'Analytic thinking promotes religious disbelief', *Science*, 2012, 336(6080), 493–6

Geschwind N, 'The neglect of advances in the neurology of behavior', in R Duncan & M Weston-Smith (eds), *The Encyclopædia of*

Medical Ignorance, Pergamon Press, 1984, 9–15

Geschwind N & Galaburda A, 'Cerebral lateralization: biological mechanisms, associations, and pathology: I. A hypothesis and a program for research', Archives of Neurology, 1985, 42(5), 428–59

Geuter S, Koban L & Wager TD, 'The cognitive neuroscience of placebo effects: concepts, predictions, and physiology', Annual Review of Neuroscience, 2017, 40, 167–88

Ghaemi SN & Rosenquist KJ, 'Insight in mood disorders: an empirical and conceptual review', in X Amador & A David (eds), Insight and Psychosis, Oxford University Press, 2nd edn, 2004, 101–15

Ghaziuddin M & Butler E, 'Clumsiness in autism and Asperger syndrome: a further report', Journal of Intellectual Disability Research, 1998, 42(1), 43–8

Ghiselin B (ed), The Creative Process: Reflections on the Invention in the Arts and Sciences, University of California Press, 1985

Gholipour B, 'A famous argument against free will has been debunked', 10 September 2019 www.theatlantic.com/health/archive/2019/09/free-will-bereitschaftspotential/597736/ (accessed 23 February 2021)

Ghosh GN, Wycoco V & Ghosh S, 'Transient visual hallucinations due to posterior callosal stroke', Journal of Stroke and Cerebrovascular Diseases, 2015, 24(6), e147–8

Ghosh P, Motamedi G, Osborne B et al, 'Reversible blindness: simple partial seizures presenting as ictal and postictal hemianopsia', Journal of Neuro-Ophthalmology, 2010, 30(3), 272–5

Giammattei J & Arndt J, 'Hemispheric asymmetries in the activation and monitoring of memory errors', Brain and Cognition, 2012, 80(1), 7–14

Giannini AJ, Daood J, Giannini MC et al, 'Intellect versus intuition – a dichotomy in the reception of nonverbal communication', Journal of General Psychology, 1978, 99(1), 19–24

Gibson C, Folley BS & Park S, 'Enhanced divergent thinking and creativity in musicians: a behavioral and near-infrared spectroscopy study', Brain and Cognition, 2009, 69(1), 162–9

Gibson JJ, The Ecological Approach to Visual Perception, Houghton Mifflin Harcourt, 1979a

———, The Senses Considered as Perceptual Systems, Greenwood Press, 1979b [1966]

Gidron Y, Gaygısız E, Lajunen T et al, 'Hostility, driving anger, and dangerous driving: the emerging role of hemispheric preference', Accident Analysis and Prevention, 2014, 73, 236–41

Giedd JN, Snell JW, Lange N et al, 'Quantitative magnetic resonance imaging of human brain development: ages 4–18', Cerebral Cortex, 1996, 6(4), 551–60

Giersch A, Lalanne L, Corves C et al, 'Extended visual simultaneity thresholds in patients with schizophrenia', Schizophrenia Bulletin, 2009, 35(4), 816–25

Giersch A, Poncelet PE, Capa RL et al, 'Disruption of information processing in schizophrenia: the time perspective', Schizophrenia Research: Cognition, 2015, 2(2), 78–83

Gigerenzer G, 'From tools to theories: a heuristic of discovery in cognitive psychology', Psychological Review, 1991, 98(2), 254–67

———, 'Why heuristics work', Perspectives on Psychological Science, 2008, 3(1), 20–9

Gigerenzer G & Brighton H, 'Homo heuristicus: why biased minds make better inferences', Topics in Cognitive Science, 2009, 1, 107–43

Gilaie-Dotan S, Saygin AP, Lorenzi LJ et al, 'The role of human ventral visual cortex in motion perception', Brain, 2013, 136(9), 2784–98

Gilbert AL, Regier T, Kay P et al, 'Whorf hypothesis is supported in the right visual field but not the left', Proceedings of the National Academy of Sciences of the United States of America, 2006, 103(2), 489–94

Gilbert AR, Akkal D, Almeida JR et al, 'Neural correlates of symptom dimensions in pediatric obsessive-compulsive disorder: a functional magnetic resonance imaging study', Journal of the American Academy of Child & Adolescent Psychiatry, 2009, 48(9), 936–44

Gilbert DT, King G, Pettigrew S et al, 'Comment on "Estimating the reproducibility of psychological science"', Science, 2016, 351(6277), 1037

Gilhooly KJ & Fioratou E, 'Executive functions in insight versus non-insight problem solving: an individual differences approach', Thinking & Reasoning, 2009, 15(4), 355–76

Gilhooly KJ, McGeorge P, Hunter J et al, 'Biomedical knowledge in diagnostic thinking: the case of electrocardiogram (ECG) interpretation', European Journal of Cognitive Psychology, 1997, 9(2), 199–223

Giliarovsky VA, 1957, quoted in P Ostwald and V Zavarin, 'Studies of language and schizophrenia in the USSR', in RW Rieber (ed), Applied Psycholinguistics and Mental Health, Plenum, 1980, 69–92

Giljov A, Karenina K & Malashichev Y, 'Facing each other: mammal mothers and infants prefer the position favoring right hemisphere processing', *Biology Letters*, 2018, 14(1), 20170707

Gill GJ, O. K., *The Data's Lousy, But It's All We've Got (Being a Critique of Conventional Methods)*, International Institute for Environment and Development (Gatekeeper Series no. 38), 1993

Gillberg C, 'Are autism and anorexia nervosa related?', *British Journal of Psychiatry*, 1983, 142(4), 428

Gillberg IC, Billstedt E, Wentz E et al, 'Attention, executive functions, and mentalizing in anorexia nervosa eighteen years after onset of eating disorder', *Journal of Clinical and Experimental Neuropsychology*, 2010, 32(4), 358–65

Gilmore RL, Heilman KM, Schmidt RP et al, 'Anosognosia during Wada testing', *Neurology*, 1992, 42(4), 925–7

Gilson É, trans J Lyon, *From Aristotle to Darwin and Back Again: A Journey in Final Causality, Species and Evolution*, Ignatius Press, 2009 [*D'Aristote à Darwin … et Retour. Essai sur quelques constantes de la bio-philosophie* 1971]

Giora R, Fein O, Laadan D et al, 'Expecting irony: context versus salience-based effects', *Metaphor and Symbol*, 2007, 22(2), 119–46

Giora R, Zaidel E, Soroker N et al, 'Differential effects of right- and left-hemisphere damage on understanding sarcasm and metaphor', *Metaphor and Symbol*, 2000, 15(1&2), 63–83

Giovanello KS, Alexander M & Verfaellie M, 'Differential impairment of person-specific knowledge in a patient with semantic dementia', *Neurocase*, 2003, 9(1), 15–26

Girkin CA, Perry JD & Miller NR, 'Visual environmental rotation: a novel disorder of visiospatial integration', *Journal of Neuro-Ophthalmology*, 1999, 19, 13–16

Gitelman DR, Alpert NM, Kosslyn S et al, 'Functional imaging of human right hemispheric activation for exploratory movements', *Annals of Neurology*, 1996, 39(2), 174–9

Gittings R (ed), *Letters of John Keats*, Oxford University Press, 1970

Gladwell M, *Blink: The Power of Thinking Without Thinking*, Penguin, 2005

Gläscher J, Rudrauf D, Colom R et al, 'Distributed neural system for general intelligence revealed by lesion mapping', *Proceedings of the National Academy of Sciences of the United States of America*, 2010, 107(10), 4705–9

Glass A & Butler SR, 'Alpha EEG asymmetry and speed of left hemisphere thinking', *Neuroscience Letters*, 1977, 4(3–4), 231–5

Glass H, Shaw G, Ma C et al, 'Agenesis of the corpus callosum in California 1983–2003: a population-based study', *American Journal of Medical Genetics: Part A*, 2008, 146A(19), 2495–500

Gleick J, *Chaos*, Cardinal, 1987

Gleiser M, *The Island of Knowledge: The Limits of Science and the Search for Meaning*, Basic Books, 2014

Glick SD, Ross DA & Hough LB, 'Lateral asymmetry of neurotransmitters in human brain', *Brain Research*, 1982, 234(1), 53–63

Glickstein M, 'Paradoxical inter-hemispheric transfer after section of the cerebral commissures', *Experimental Brain Research*, 2009, 192(3), 425–9

Gigerenzer G & Brighton H, 'Homo heuristicus: why biased minds make better inferences', *Topics in Cognitive Science*, 2009, 1, 107–43

Glöckner A & Betsch T, 'Modeling option and strategy choices with connectionist networks: towards an integrative model of automatic and deliberate decision making', *Judgment & Decision Making*, 2008a, 3(3), 215–28

——, 'Multiple-reason decision making based on automatic processing', *Journal of Experimental Psychology: Learning, Memory, and Cognition*, 2008b, 34(5), 1055–75

Gloning I, Gloning K & Hoff H, *Neuropsychological Symptoms and Syndromes in Lesions of the Occipital Lobe and the Adjacent Areas*, Gauthier-Villars, 1968

Gobet F & Chassy P, 'Towards an alternative to Benner's theory of expert intuition in nursing: a discussion paper', *International Journal of Nursing Studies*, 2008, 45(1), 129–39

Gobet F & Simon HA, 'Recall of rapidly presented random chess positions is a function of skill', *Psychonomic Bulletin & Review*, 1996, 3(2), 159–63

Goble DJ & Brown SH, 'Upper limb asymmetries in the matching of proprioceptive versus visual targets', *Journal of Neurophysiology*, 2008, 99(6), 3063–74

Goble DJ, Noble BC & Brown SH, 'Proprioceptive target matching asymmetries in left-handed individuals', *Experimental Brain Research*, 2009, 197(4), 403–8

Godfrey-Smith P, *Other Minds: The Octopus and the Evolution of Intelligent Life*, William Collins, 2018

Godlee F, Gale CR & Martyn CN, 'Effect on the quality of peer review of blinding reviewers and asking them to sign their reports: a randomized controlled trial', *Journal of the American Medical Association*, 1998, 280(3), 237–40

Goebel C, Linden DEJ, Sireteanu R et al, 'Different attention processes in visual search tasks investigated with functional magnetic resonance imaging', *NeuroImage*, 1997, 5(4, pt 2), s81

Goel V, *Sketches of Thought*, Massachusetts Institute of Technology Press, 1995

———, 'Anatomy of deductive reasoning', *Trends in Cognitive Sciences*, 2007, 11(10), 435–41

———, 'Anatomy of deductive reasoning', *Trends in Cognitive Sciences*, 2007, 11(10), 435–41

Goel V, Buchel C, Frith C et al, 'Dissociation of mechanisms underlying syllogistic reasoning', *NeuroImage*, 2000, 12(5), 504–14

Goel V & Dolan RJ, 'Anatomical segregation of component processes in an inductive inference task', *Journal of Cognitive Neuroscience*, 2000, 12(1), 1–10

———, 'The functional anatomy of humor: segregating cognitive and affective components', *Nature Neuroscience*, 2001, 4(3), 237–8

———, 'Explaining modulation of reasoning by belief', *Cognition*, 2003, 87(1), B11–22

———, 'Differential involvement of left prefrontal cortex in inductive and deductive reasoning', *Cognition*, 2004, 93(3), B109–21

Goel V & Grafman J, 'Role of the right prefrontal cortex in ill-structured planning', *Cognitive Neuropsychology*, 2000, 17(5), 415–36

Goel V, Shuren J, Sheesley L et al, 'Asymmetrical involvement of frontal lobes in social reasoning', *Brain*, 2004, 127(4), 783–90

Goel V, Stollstorff M, Nakic M et al, 'A role for right ventrolateral prefrontal cortex in reasoning about indeterminate relations', *Neuropsychologia*, 2009, 47(13), 2790–7

Goel V, Tierney M, Sheesley L et al, 'Hemispheric specialization in human prefrontal cortex for resolving certain and uncertain inferences', *Cerebral Cortex*, 2007, 17(10), 2245–50

Goel V & Vartanian O, 'Dissociating the roles of right ventral lateral and dorsal lateral prefrontal cortex in generation and maintenance of hypotheses in set-shift problems', *Cerebral Cortex*, 2005, 15(8), 1170–7

Goel V, Vartanian O, Bartolo A et al, 'Lesions to right prefrontal cortex impair real-world planning through premature commitments', *Neuropsychologia*, 2013, 51(4), 713–24

Goethe JW von, *Farbenlehre*, in *Gedenkausgabe der Werke, Briefe und Gespräche*, Artemis Verlag, 1948–

———, *Maximen und Reflektionen* [1833]; in *Sämtliche Werke*, Aufbau-Verlag: *Kunsttheoretische Schriften und Übersetzungen* (vols 17–22), vol 18, 1972

———, 'Significant help given by an ingenious turn of phrase' [*Bedeutende Förderniss durch ein einziges geistreiches Wort*], in *Goethe: Scientific Studies*, trans & ed D Miller, Suhrkamp, 1988

———, *Theory of Colours*, trans CL Eastlake, Dover, 2006

Gogtay N, Lu A, Leow AD et al, 'Three-dimensional brain growth abnormalities in childhood-onset schizophrenia visualized by using tensor-based morphometry', *Proceedings of the National Academy of Sciences of the United States of America*, 2008, 105(41), 15979–84

Goksan S, Hartley C, Emery F et al, 'fMRI reveals neural activity overlap between adult and infant pain', *Elife*, 2015, 4, e06356

Gold JM, Berman KF, Randolph C et al, 'PET validation of a novel prefrontal task: delayed response alternation (DRA)', *Neuropsychology*, 1996, 10(1), 3–10

Goldberg E, *The Executive Brain: Frontal Lobes and the Civilized Mind*, Oxford University Press, 2001

Goldberg E & Costa LD, 'Hemispheric differences in the acquisition and use of descriptive systems', *Brain and Language*, 1981, 14(1), 144–73

Goldberg E, Podell K & Lovell M, 'Lateralization of frontal lobe functions and cognitive novelty. Review', *Journal of Neuropsychiatry and Clinical Neurosciences*, 1994, 6(4), 371–8

Goldberg G, Mayer NH & Toglia JU, 'Medial frontal cortex infarction and the alien hand sign', *Archives of Neurology*, 1981, 38(11), 683–6

Goldberg TE, Maltz A, Bow JN et al, 'Blink rate abnormalities in autistic and mentally retarded children: relationship to dopaminergic activity', *Journal of the American Academy of Child & Adolescent Psychiatry*, 1987, 26(3), 336–8

Golden EC & Josephs KA, 'Minds on replay: musical hallucinations and their relationship to neurological disease', *Brain*, 2015, 138(12), 3793–802

Goldenberg G & Hagmann S, 'Tool use and mechanical problem solving in apraxia', *Neuropsychologia*, 1998, 36(7), 581–9

Goldman-Rakic PS & Selemon LD, 'Functional and anatomical aspects of prefrontal pathology in schizophrenia', *Schizophrenia Bulletin*, 1997, 23(3), 437–58

Goldstein DG & Gigerenzer G, 'Models of ecological rationality: the recognition heuristic', *Psychological Review*, 2002, 109(1), 75–90

Goldstein K, 'Zur Lehre von der motorischen Apraxie', *Journal für Psychologie und Neurologie*, 1908, 11(4–5), 169–87 & 270–83

Goldstein P & Leshem M, 'Dietary sodium, added salt, and serum sodium associations with growth and depression in the US general population', *Appetite*, 2014, 79, 83–90

Goldstein S, Taylor J, Tumulka R et al, 'Are all particles identical?', *Journal of Physics A*, 2005, 38(7), 1567–76

Golgi C, 'Sulla fina struttura dei bulbi olfattorii', *Rivista Sperimentale di Freniatria e Medicina Legale*, 1875, 1, 405–25

Gombrich EH, *Art and Illusion*, Phaidon, 1960

Gómez JC, 'Species comparative studies and cognitive development', *Trends in Cognitive Sciences*, 2005, 9(3), 118–25

Gonçalves LM & Tosoni A, 'Sudden onset of Cotard's syndrome as a clinical sign of brain tumor', *Archives of Clinical Psychiatry*, 2016, 43(2), 35–6

Gonzalez CL, Ganel T & Goodale MA, 'Hemispheric specialization for the visual control of action is independent of handedness', *Journal of Neurophysiology*, 2006, 95(6), 3496–501

González J, Barrós-Loscertales A, Pulvermüller F et al, 'Reading cinnamon activates olfactory brain regions', *NeuroImage*, 2006, 32(2), 906–12

González-Vallejo C, Lassiter GD, Bellezza FS et al, '"Save angels perhaps": a critical examination of unconscious thought theory and the deliberation-without-attention effect', *Review of General Psychology*, 2008, 12(3), 282–96

Gonzalo-Fonrodona I, 'Inverted or tilted perception disorder', *Revue Neurologique (Paris)*, 2007, 44(3), 157–65

Gooding DC, Matts CW & Rollmann EA, 'Sustained attention deficits in relation to psychometrically identified schizotypy: evaluating a potential endophenotypic marker', *Schizophrenia Research*, 2006, 82(1), 27–37

Goodman CP, 'The politics of imperfection', *Polanyiana*, 2008, 17(1–2), 73–8

Goodman RB, 'What Wittgenstein learned from William James', *History of Philosophy Quarterly*, 1994, 11(3), 339–54

Goodrich BG, 'We do, therefore we think: time, motility, and consciousness', *Reviews in the Neurosciences*, 2010, 21(5), 331–61

Goodwin B, 'Towards a science of qualities', in W Harman (ed), *New Metaphysical Foundations of Modern Science*, Institute of Noetic Sciences, 1994, 215–50

Gopnik A & Meltzoff AN, 'Minds, bodies, and persons: young children's understanding of the self and others as reflected in imitation and theory of mind research', in ST Parker, RW Mitchell & ML Boccia (eds), *Self-Awareness in Animals and Humans: Developmental Perspectives*, Cambridge University Press, 2006, 166–86

Gordon EE, Drenth V, Jarvis L et al, 'Neurophysiologic syndromes in stroke as predictors of outcome', *Archives of Physical Medicine and Rehabilitation*, 1978, 59(5), 399–403

Gordon HW, Frooman B & Lavie P, 'Shift in cognitive asymmetry between waking from REM and NREM sleep', *Neuropsychologia*, 1982, 20(1), 99–103

Gordon HW & Sperry RW, 'Lateralization of olfactory perception in the surgically separated hemispheres of man', *Neuropsychologia*, 1969, 7(2), 111–20

Görgülü Y, Alparslan SC & Uygur N, 'Corpus callosum atrophy and psychosis: a case report', *Journal of Psychiatry & Neurological Sciences*, 2010, 23(2), 137–41

Gorynia I & Müller J, 'Hand skill and hand-eye preference in relation to verbal ability in healthy adult male and female right-handers', *Laterality*, 2006, 11(5), 415–35

Goss RJ, *Deer Antlers: Regeneration, Function, and Evolution*, Academic Press, 1983

Gottfredson LS, 'Mainstream science on intelligence: an editorial with 52 signatories, history, and bibliography', *Intelligence*, 1997, 24(1), 13–23

Gottfried JA, O'Doherty J & Dolan RJ, 'Appetitive and aversive olfactory learning in humans studied using event-related functional magnetic resonance imaging', *Journal of Neuroscience*, 2002, 22(24), 10829–37

Gottfries CG, Perris C & Roos BE, 'Visual averaged evoked responses (AER) and monoamine metabolites in cerebrospinal fluid (CSF)', *Acta Psychiatrica Scandinavica*, 1974, 50(255), 135–42

Gotts SJ, Jo HJ, Wallace GL et al, 'Two distinct forms of functional lateralization in the human brain', *Proceedings of the National Academy of Sciences of the United States of America*, 2013, 110(36), E3435–44

Gould SJ, 'Impeaching a self-appointed judge', *Scientific American*, 1992, 267(1), 118–21

Gould SJ, 'Nonoverlapping magisteria', *Natural History*, 1997, 106(2), 16–22 & 60–2

Grabner R, Fink A & Neubauer A, 'Brain correlates of self-rated originality of ideas: evidence from event-related power and phase locking changes in the EEG', *Behavioral Neuroscience*, 2007, 121(1), 224–30

Grabowska A, 'Lateral differences in the detection of stereoscopic depth', *Neuropsychologia*, 1983, 21(3), 249–57

———, 'Visual field differences in the magnitude of the tilt after-effect', *Neuropsychologia*, 1987, 25(6), 957–63

Grabowska A & Nowicka A, 'Visual-spatial-frequency model of cerebral asymmetry: a critical survey of behavioral and electrophysiological studies', *Psychological Bulletin*, 1996, 120(3), 434–49

Graff-Radford J, Whitwell JL, Geda YE et al, 'Clinical and imaging features of Othello's syndrome', *European Journal of Neurology*, 2012, 19(1), 38–46

Graham C & Crown S, 'Religion and wellbeing around the world: social purpose, social time, or social insurance?', *International Journal of Wellbeing*, 2014, 4(1), 1–27

Graham P, *Hackers and Painters: Big Ideas from the Computer Age*, O'Reilly, 2010

Grandin T, *Thinking in Pictures: My Life with Autism*, Vintage, 2006

———, 'How does visual thinking work in the mind of a person with autism? A personal account', *Philosophical Transactions of the Royal Society of London: Series B, Biological Sciences*, 2009, 364(1522), 1437–42

Granholm E, Cadenhead K, Shafer K et al, 'Lateralized perceptual organization deficits on the global-local task in schizotypal personality disorder', *Journal of Abnormal Psychology*, 2002, 111, 42–52

Granholm E, Perry W, Filoteo JV et al, 'Hemispheric and attentional contributions to perceptual organization deficits on the global-local task in schizophrenia', *Neuropsychology*, 1999, 13(2), 271–81

Grann D, 'The mark of a masterpiece', *The New Yorker*, 12 July 2010, 50–61

Grattan LM, Bloomer RH, Archambault FX et al, 'Cognitive flexibility and empathy after frontal lobe lesions', *Neuropsychiatry, Neuropsychology, & Behavioral Neurology*, 1994, 7(4), 251–9

Graudal N, 'Commentary: possible role of salt intake in the development of essential hypertension', *International Journal of Epidemiology*, 2005, 34(5), 972–4

Grawitch MJ, Munz DC & Kramer TJ, 'Effects of member mood states on creative performance in temporary workgroups', *Group Dynamics: Theory, Research, and Practice*, 2003, 7(1), 41–54

Gray A, 'Scientific worthies III: Charles Robert Darwin', *Nature*, 1874, 10(240), 79–81

Gray J, *Seven Types of Atheism*, Penguin, 2018

Gray JA, 'The contents of consciousness: a neuropsychological conjecture', *Behavioral and Brain Sciences*, 1995, 18(4), 659–76

Gray JA, Feldon J, Rawlins NJP et al, 'The neuropsychology of schizophrenia', *Behavioral and Brain Sciences*, 1991, 14(1), 1–20

Grayling AC, 'Probably a ridiculous caveat', *The Guardian*, 7 January 2009

———, 'In two minds', *Literary Review*, December 2009/January 2010, 372, 56–7

Green CD, 'All that glitters: a review of psychological research on the aesthetics of the Golden Section', *Perception*, 1995, 24(9), 937–68

Green M, 'Strings that surprise', 2014: www.cam.ac.uk/research/features/strings-that-surprise-how-a-theory-scaled-up (accessed 23 February 2021)

Greenberg C, 'Modernist Painting', *Forum Lectures*, Voice of America, 1960

Greenberg J, *I Never Promised You a Rose Garden*, St Martin's Press, 2009 (originally published under the pseudonym 'Hannah Green', by The New American Library, 1964)

Greenberg MR, Carey GW & Popper FJ, 'External causes of death among young White Americans', *New England Journal of Medicine*, 1985, 313(23), 1482–3

Greene B, *The Fabric of the Cosmos: Space, Time and the Texture of Reality*, Alfred A Knopf, 2005

Greene J & Haidt J, 'How (and where) does moral judgement work?', *Trends in Cognitive Sciences*, 2002, 6(12), 517–23

Greene JD & Paxton JM, 'Patterns of neural activity associated with honest and dishonest moral decisions', *Proceedings of the National Academy of Sciences of the United States of America*, 2009, 106(30), 12506–11

Greene JD, Nystrom LE, Engell AD et al, 'The neural bases of cognitive conflict and control in moral judgment', *Neuron*, 2004, 44(2), 389–400

Greenman C, Stephens P, Smith R et al, 'Patterns of somatic mutation in human cancer genomes', *Nature*, 2007, 446 (7132), 153–8

Greer G, *The Beautiful Boy*, Rizzoli International, 2003

Greisman HC & Ritzer G, 'Max Weber, critical theory, and the administered world', *Qualitative Sociology*, 1981, 4(1), 34–55

Greyson B, 'Is consciousness produced by the brain?', 2019: medium.com/@richmartini/interview-with-dr-bruce-greyson-is-consciousness-produced-by-the-brain-51894883edca (accessed 23 February 2021)

Griffith J & Hochberg F, 'Anorexia and weight loss in glioma patients', *Psychosomatics*, 1988, 29(3), 335–7

Griffiths P & Stotz K, 'Developmental systems theory as a process theory', in DJ Nicholson & J Dupré (eds), *Everything Flows: Towards a Processual Philosophy of Biology*, Oxford University Press, 2018, 225–45

Griffiths RR, Hurwitz ES, Davis AK et al, 'Survey of subjective "God encounter experiences": comparisons among naturally occurring experiences and those occasioned by the classic psychedelics psilocybin, LSD, ayahuasca, or DMT', *PLoS One*, 2019, 14(4), e0214377

Griffiths TD, Rees A, Witton C et al, 'Spatial and temporal auditory processing deficits following right hemisphere infarction: a psychophysical study', *Brain*, 1997, 120(5), 785–94

Griffiths TD, Warren JD, Dean JL et al, '"When the feeling's gone": a selective loss of musical emotion', *Journal of Neurology, Neurosurgery, and Psychiatry*, 2004, 75(2), 344–5

Grimshaw GM, Bryden MA & Finegan J-AK, 'Relations between prenatal testosterone and cerebral lateralization in children', *Neuropsychology*, 1995, 9(1), 68–79

Griswold A, *Autistic Symphony*, iUniverse, 2007

Gröblacher S, Paterek T, Kaltenbaek R et al, 'An experimental test of non-local realism', *Nature*, 2007, 446(7138), 871–5

Groopman J, *How doctors think*, Houghton Mifflin, Boston, 2007

Gross G & Huber G, 'Sensorische Störungen bei Schizophrenien', *Archiv für Psychiatrie und Nervenkrankheiten*, 1972, 216(2), 119–30

Gross TF, 'The perception of four basic emotions in human and nonhuman faces by children with autism and other developmental disabilities', *Journal of Abnormal Child Psychology*, 2004, 32(5), 469–80

Grossman ED, Battelli L & Pascual-Leone A, 'Repetitive TMS over posterior STS disrupts perception of biological motion', *Vision Research*, 2005, 45(22), 2847–53

Grossman E, Donnelly M, Price R et al, 'Brain areas involved in perception of biological motion', *Journal of Cognitive Neuroscience*, 2000, 12(5), 711–20

Grossman M, 'A bird is a bird is a bird: making reference within and without superordinate categories', *Brain and Language*, 1981, 12(2), 313–31

——, 'Reversal operations after brain damage', *Brain and Cognition*, 1982, 1, 331–59

——, 'Drawing deficits in brain-damaged patients' freehand pictures', *Brain and Cognition*, 1988, 8(2), 189–205

Grossman M & Haberman S, 'The detection of errors in sentences after right hemisphere brain damage', *Neuropsychologia*, 1987, 25(1, pt 2), 163–72

Grossmann I, Brienza JP & Bobocel DR, 'Wise deliberation sustains cooperation', *Nature Human Behaviour*, 2017, 1, article 0061

Grouios G, 'Phantom limb perceptuomotor "memories" in a congenital limb child', *Medical Science Research*, 1996, 24(7), 503–4

Grover S, Kattharaghatta Girigowda V & Kumar V, 'Lilliputian hallucinations in schizophrenia: a case report', *African Journal of Psychiatry*, 2012, 15(5), 311–3

Grover S, Sahoo S & Surendran I, 'Obsessive-compulsive symptoms in schizophrenia: a review', *Acta Neuropsychiatrica*, 2018, 31(2), 63–73

Grow G, *Dancing with a Camera in the Presence of Light*, Longleaf, 2019

Gruber LN, Mangat BS & Abou-Taleb H, 'Laterality of auditory hallucinations in psychiatric patients', *American Journal of Psychiatry*, 1984, 141(4), 586–7

Gruber R, Schiestl M, Boeckle M et al, 'New Caledonian crows use mental representations to solve metatool problems', *Current Biology*, 2019, 29(4), 686–92

Gruenewald TL, Karlamangla AS, Greendale GA et al, 'Feelings of usefulness to others, disability, and mortality in older adults: the MacArthur Study of Successful Aging', *The Journals of Gerontology: Series B, Psychological Sciences and Social Sciences*, 2007, 62(1), 28–37

Grunwald M, Weiss T, Assmann B et al, 'Stable asymmetric interhemispheric theta power in patients with anorexia nervosa during haptic perception even after weight gain: a longitudinal study', *Journal of Clinical and Experimental Neuropsychology*, 2004, 26(5), 608–20

Grüsser O-J, 'Mother-child holding patterns in Western art: a developmental study', *Ethology and Sociobiology*, 1983, 4(2), 89–94

Grüsser O-J & Landis T, *Visual Agnosias and Other Disturbances of Visual Perception and Cognition*, Macmillan, 1991

Grüsser O-J, Selke T & Zynda B, 'Cerebral lateralisation and some implications for art, aesthetic perception and artistic creativity', in I Rentschler, B Herzberger & D Epstein (eds), *Beauty and the Brain: Biological Aspects of Aesthetics*, Birkhauser Verlag, 1988, 257–93

Gruzelier JH, 'Bilateral asymmetry of skin conductance orienting activity and levels in schizophrenics', *Biological Psychology*, 1973, 1(1), 21–41

Gruzelier J & Venables P, 'Bimodality and lateral asymmetry of skin conductance orienting activity in schizophrenics: replication and evidence of lateral asymmetry in patients with depression and disorder of personality', *Biological Psychiatry*, 1974, 8(1), 55–73

Gsponer J & Babu MM, 'The rules of disorder or why disorder rules', *Progress in Biophysics and Molecular Biology*, 2009, 99(2–3), 94–103

Guard O, Delpy C, Richard D et al, 'Une cause mal connue de confusion mentale: le ramollissement temporal droit', *Revue de Médecine*, 1979, 40, 2115–21

Guardini R, *The End of the Modern World*, ISI Books, 1998 [1956]

Guariglia C, Piccardi L, Puglisi Allegra MC et al, 'Is autotopoagnosia real? EC says yes. A case study', *Neuropsychologia*, 2002, 40(10), 1744–9

Guilford JP, *The Nature of Human Intelligence*, 1967

Guillaume S, Jollant F, Jaussent I et al, 'Somatic markers and explicit knowledge are both involved in decision-making', *Neuropsychologia*, 2009, 47(10), 2120–4

Gunaratana H, *Mindfulness in Plain English*, 20th anniversary edn, Wisdom Publications, 2011, 132–46

Gündel H, López-Sala A, Ceballos-Baumann AO et al, 'Alexithymia correlates with the size of right anterior cingulate', *Psychosomatic Medicine*, 2004, 66(1), 132–40

Gunter HL, Ghaziuddin M & Ellis HD, 'Asperger syndrome: tests of right hemisphere functioning and interhemispheric communication', *Journal of Autism and Developmental Disorders*, 2002, 32(4), 263–81

Gunter PAY, *Bergson and the Evolution of Physics*, University of Tennessee Press, 1969

Günther W, Petsch R, Steinberg R et al, 'Brain dysfunction during motor activation and corpus callosum alterations in schizophrenia measured by cerebral blood flow and magnetic resonance imaging', *Biological Psychiatry*, 1991, 29(6), 535–55

Güntürkün O & Bugnyar T, 'Cognition without cortex', *Trends in Cognitive Sciences*, 2016, 20(4), 291–303

Güntürkün O & Ocklenburg S, 'Ontogenesis of lateralization', *Neuron*, 2017, 94(2), 249–63

Güntürkün O, Ströckens F, Scarf D et al, 'Apes, feathered apes, and pigeons: differences and similarities', *Current Opinion in Behavioral Sciences*, 2017, 16, 35–40

Guo K, Meints K, Hall C et al, 'Left gaze bias in humans, rhesus monkeys and domestic dogs', *Animal Cognition*, 2009, 12(3), 409–18

Gupta S & Katsarska M, 'The official record and the receptive field: Zlatyu Boyadzhiev in communist times', *Konsthistorisk Tidskrift/Journal of Art History*, 2010, 79(1), 1–17

Gur RC & Gur RE, 'Hemispheric specialization and regional cerebral blood flow', in A Glass (ed), *Individual Differences in Hemispheric Specialization* (NATO ASI Series A: Life Sciences, vol 130), Plenum Press, 1987, 93–101

Gur RC, Packer IK, Hungerbuhler JP et al, 'Differences in distribution of gray and white matter in human cerebral hemispheres', *Science*, 1980, 207(4436), 1226–8

Gur RE, Resnick SM, Alavi A et al, 'Regional brain function in schizophrenia: I. A positron emission tomography study', *Archives of General Psychiatry*, 1987, 44(2), 119–25

Gurin L & Blum S, 'Delusions and the right hemisphere: a review of the case for the right hemisphere as a mediator of reality-based belief', *Journal of Neuropsychiatry and Clinical Neurosciences*, 2017, 29(3), 225–35

Gurzadyan VG & Penrose R, 'CCC and the Fermi paradox', *European Physical Journal Plus*, 2016, 131, 11

Gustafson K, review of *A World Without Time: The Forgotten Legacy of Gödel and Einstein* by Palle Yourgrau, *Journal of Scientific Exploration*, 2005, 19(2), 274–78

Guttinger S, 'A process ontology for macromolecular biology', in DJ Nicholson & J Dupré, *Everything Flows: Towards a Processual Philosophy of Biology*, Oxford University Press, 2018, 303–20

Guttmann E & Maclay WS, 'Clinical observations on schizophrenic drawings', *British Journal of Medical Psychology*, 1937, 16, 184–205

Guye M, Bettus G, Bartolomei F et al, 'Graph theoretical analysis of structural and functional connectivity MRI in normal and pathological brain networks', *Magnetic Resonance Materials in Physics, Biology and Medicine*, 2010, 23(5), 409–21

Güzelcan Y, Kleinpenning AS & Vuister FM, 'Peduncular hallucinosis caused by a tumour in the right thalamus: a case study', *Tijdschrift voor Psychiatrie*, 2008, 50(1), 65–8

Haack S, 'The Pragmatist theory of truth', *British Journal for the Philosophy of Science*, 1976, 27(3), 231–49

Haag R, 'Quantum events and irreversibility', in S Albeverio & P Blanchard, *Direction of Time*, Springer, 2014, 31–4

Haaga DA & Beck AT, 'Perspectives on depressive realism: implications for cognitive theory of depression', *Behaviour Research and Therapy*, 1995, 33(1), 41–8

Habekost T & Bundesen C, 'Patient assessment based on a theory of visual attention (TVA): subtle deficits after a right frontal-subcortical lesion', *Neuropsychologia*, 2003, 41(9), 1171–88

Habermas J, 'A conversation about God and the world', *Time of Transitions*, Polity Press, 2006

Habib M, 'Visual hypoemotionality and prosopagnosia associated with right temporal lobe isolation', *Neuropsychologia*, 1986, 24(4), 577–82

Hadamard JS, *An Essay on The Psychology of Invention in the Mathematics Field*, Princeton University Press, 1945

Hadhazy A, 'Think twice: how the gut's "second brain" influences mood and well-being', *Scientific American*, 12 February 2010

Hadot P, *The Veil of Isis: An Essay on the History of the Idea of Nature*, trans M Chase, Harvard University Press, 2006

Hadravová T, 'Aesthetic experts', *Espes*, 2019, 8(2), 27–36

Hadžiselimović H & Čuš M, 'The appearance of internal structures of the brain in relation to configuration of the human skull', *Acta Anatomica*, 1966, 63(3), 289–99

Hafiz, in D Ladinsky (trans & ed), *Love Poems from God*, Penguin, 2002

Haga Y & Haidt J, 'The emotional dog and its rational tail: a social intuitionist approach to moral judgment', *Psychological Review*, 1979, 108(4), 814–34

Hagemann D, Naumann E, Lürken A et al, 'EEG asymmetry, dispositional mood and personality', *Personality and Individual Differences*, 1999, 27(3), 541–68

Hagemann D, Naumann E & Thayer JF, 'The quest for the EEG reference revisited: a glance from brain asymmetry research', *Psychophysiology*, 2001, 38(5), 847–57

Haggard MP & Parkinson AM, 'Stimulus and task factors as determinants of ear advantages', *Quarterly Journal of Experimental Psychology*, 1971, 23(2), 168–77

Haidt J, *The Righteous Mind*, Penguin, 2013

Haier RJ, Chueh D, Touchette P et al, 'Brain size and cerebral glucose metabolic rate in nonspecific mental retardation and Down syndrome', *Intelligence*, 1995, 20(2), 191–210

Haier RJ, Siegel BV, MacLachlan A et al, 'Regional glucose metabolism changes after learning a complex visuospatial/motor task: a positron emission tomographic study', *Brain Research*, 1992, 570(1–2), 134–43

Haier RJ, Siegel BV, Nuechterlein KH et al, 'Cortical glucose metabolic rate correlates of abstract reasoning and attention studied with positron emission tomography', *Intelligence*, 1988, 12(2), 199–217

Haisch B, *The God Theory*, Weiser, 2006

Hakim H, Verma NP & Greiffenstein MF, 'Pathogenesis of reduplicative paramnesia', *Journal of Neurology, Neurosurgery, and Psychiatry*, 1988, 51(6), 839–41

Haldane JBS, *The Inequality of Man*, Chatto & Windus, 1932

——, *Keeping Cool and Other Essays*, Chatto & Windus, 1940

——, 'The origin of life', in *Science and Life*, Pemberton, 1968, 1–11 (2–3) [*The Rationalist Annual* 1929]

Haldane JS, *Organism and Environment, as Illustrated by the Physiology of Breathing*, Yale University Press, 1917

——, *The New Physiology and Other Addresses*, Charles Griffin, 1919

Hale MJG & Hale CM, '*I Had No Means to Shout!*', 1st Books, 1999

Hall ET, *The Silent Language*, Anchor, 1973

Hall PM, 'Factors influencing individual susceptibility to alcoholic liver disease', in PM Hall (ed), *Alcoholic Liver Disease: Pathology and Pathogenesis*, Edward Arnold, 2nd edn, 1995, 299–316

Hallak JE, Crippa JA, Pinto JP et al, 'Total agenesis of the corpus callosum in a patient with childhood-onset schizophrenia', *Arquivos de Neuro-Psiquiatria*, 2007, 65(4b), 1216–9

Hallett D, Chandler MJ & Lalonde CE, 'Aboriginal language knowledge and youth suicide', *Cognitive Development*, 2007, 22(3), 392–9

Halligan PW & Marshall JC, 'Toward a principled explanation of unilateral neglect', *Cognitive Neuropsychology*, 1994, 11(2), 167–206

——, 'Supernumerary phantom limb after right hemispheric stroke', *Journal of Neurology, Neurosurgery, and Psychiatry*, 1995, 59(3), 341–2

——, 'Neglect of awareness', *Consciousness and Cognition*, 1998, 7(3), 356–80

Halligan PW, Marshall JC & Ramachandran VS, 'Ghosts in the machine: a case description of visual and haptic hallucinations after right hemisphere stroke', *Cognitive Neuropsychology*, 1994, 11(4), 459–77

Halligan PW, Marshall JC & Wade DT, 'Three arms: a case study of supernumerary phantom limb after right hemisphere stroke', *Journal of Neurology, Neurosurgery, and Psychiatry*, 1993, 56(2), 159–66

——, 'Unilateral somatoparaphrenia after right hemisphere stroke: a case description', *Cortex*, 1995, 31(1), 173–82

Hallowell AI, 'Ojibwa ontology, behavior, and world view', in G Harvey (ed), *Readings in Indigenous Religions*, Continuum, 2002, 17–49

Hallpike CR, *The Foundations of Primitive Thought*, Clarendon Press, 1979

——, *How We Got Here: From Bows and Arrows to the Space Age*, AuthorHouse, 2008

——, *On Primitive Society: And Other Forbidden Topics*, AuthorHouse, 2011

——, 'Constructivism and selection: two opposed theories of social evolution', paper presented to the International Symposium, 'Die Geistes- und Sozialwissenschaften vor der Geschichte: Geschichte in universalhistorischer Perspektivierung', at Freiburg University, 23–25 September 2013

Halmos P, *I Want to be a Mathematician: An Automathography in Three Parts*, Wiley & Sons, 1989

Halpern DF, Benbow CP, Geary DC et al, 'The science of sex differences in science and mathematics', *Psychological Science in the Public Interest*, 2007, 8(1), 1–51

Halpern DF, Straight C & Stephenson CL, 'Beliefs about cognitive gender differences: accurate for direction, underestimated for size', *Sex Roles*, 2011, 64(5–6), 336–47

Halpern ME, Güntürkün O, Hopkins WD et al, 'Lateralization of the vertebrate brain: taking the side of model systems', *Journal of Neuroscience*, 2005, 25(45), 10351–7

Hamilton A, Wolpert D & Frith U, 'Your own action influences how you perceive another person's action', *Current Biology*, 2004, 14(6), 493–8

Hamilton AF, Brindley RM & Frith U, 'Imitation and action understanding in autistic spectrum disorders: how valid is the hypothesis of a deficit in the mirror neuron system?', *Neuropsychologia*, 2007, 45(8), 1859–68

Hamilton AF & Grafton ST, 'Action outcomes are represented in human inferior frontoparietal cortex', *Cerebral Cortex*, 2008, 18(5), 1160–8

Hamilton CR & Vermeire BA, 'Complementary hemispheric specialization in monkeys', *Science*, 1988, 242(4886), 1691–4

Hamilton J, 'Study narrows gap between mind and brain', 29 March 2010: www.npr.org/templates/story/story.php?storyId =125304448 (accessed 23 February 2021)

Hamilton JA, 'Attention, personality and the self-regulation of mood: absorbing interest and boredom', *Progress in Experimental Personality Research*, 1981, 10, 281–315

Hamilton JA, Haier RJ & Buchsbaum MS, 'Intrinsic enjoyment and boredom coping scales: validation with personality, evoked potential and attention measures', *Personality and Individual Differences*, 1984, 5(2), 183–93

Hammond GR & Fox AM, 'Electrophysiological evidence for lateralization of preparatory motor processes', *NeuroReport*, 2005, 16(6), 559–62

Hammond KR, Hamm RM, Grassia J et al, 'Direct comparison of the efficacy of intuitive and analytical cognition in expert judgment', *IEEE Transactions on Systems, Man, and Cybernetics: Systems*, 1987, 17(5), 753–70

Hamzei F, Dettmers C, Rzanny R et al, 'Reduction of excitability ("inhibition") in the ipsilateral primary motor cortex is mirrored by fMRI signal decreases', *NeuroImage*, 2002, 17(1), 490–6

Han Z, Bi Y, Chen J et al, 'Distinct regions of right temporal cortex are associated with biological and human-agent motion: functional magnetic resonance imaging and neuropsychological evidence', *Journal of Neuroscience*, 2013, 33(39), 15442–53

Handberg-Thorsager M, Fernandez E & Salo E, 'Stem cells and regeneration in planarians', *Frontiers in Bioscience – Landmark*, 2008, 13(16), 6374–6394

Handley SJ, Newstead SE & Trippas D, 'Logic, beliefs, and instruction: a test of the default interventionist account of belief bias', *Journal of Experimental Psychology: Learning, Memory, and Cognition*, 2011, 37(1), 28–43

Handy TC, *Event-Related Potentials: A Methods Handbook*, Massachusetts Institute of Technology Press, 2004

Handy TC, Gazzaniga MS & Ivry RB, 'Cortical and subcortical contributions to the representation of temporal information', *Neuropsychologia*, 2003, 41(11), 1461–73

Hankey A, 'A complexity basis for phenomenology: how information states at criticality offer a new approach to understanding experience of self, being and time', *Progress in Biophysics & Molecular Biology*, 2015, 119(3), 288–302

Hankir A, 'Review: bipolar disorder and poetic genius', *Psychiatria Danubina*, 2011, 23(suppl 1), s62–8

Hannay HJ, 'Asymmetry in reception and retention of colours', *Brain and Language*, 1979, 8(2), 191–201

Hannay HJ, Ciaccia PJ, Kerr JW et al, 'Self-report of right-left confusion in college men and women', *Perceptual & Motor Skills*, 1990, 70(2), 451–7

Hansen MM, Jones R & Tocchini K, 'Shinrin-yoku (forest bathing) and nature therapy: a state-of-the-art review', *International Journal of Environmental Research and Public Health*, 2017, 14(8), p ii: E851

Hansen P, Azzopardi P, Matthews P et al, 'Neural correlates of "creative intelligence": an fMRI study of fluid analogies', poster session presented at the annual conference of the Society for Neuroscience, New Orleans, November 2008

Hanson NR, *Patterns of Discovery: An Inquiry into the Conceptual Foundations of Science*, Cambridge University Press, 1958

Happé FGE, 'Studying weak central coherence at low levels: children with autism do not succumb to visual illusions: a research note', *Journal of Child Psychology and Psychiatry, and Allied Disciplines*, 1996, 37(7), 873–7

Happé F & Frith U, 'The weak coherence account: detail-focused cognitive style in autism spectrum disorders', *Journal of Autism and Developmental Disorders*, 2006, 36(1), 5–25

Haramati S, Soroker N, Dudai Y et al, 'The posterior parietal cortex in recognition memory: a neuropsychological study', *Neuropsychologia*, 2008, 46(7), 1756–66

Harasawa M & Shioiri S, 'Asymmetrical brain activity induced by voluntary spatial attention depends on the visual hemifield: A functional near-infrared spectroscopy study', *Brain and Cognition*, 2011, 75(3), 292–8

Harasty J, Double KL, Halliday GM et al, 'Language-associated cortical regions are proportionally larger in the female brain', *Archives of Neurology*, 1997, 54(2), 171–6

Harcombe Z, Baker JS, Cooper SM et al, 'Evidence from randomised controlled trials did not support the introduction of dietary fat guidelines in 1977 and 1983: a systematic review and meta-analysis', *BMJ Open Heart*, 2015, 2, e000196

Hård A & Sivik L, 'A theory of colors in combination – a descriptive model related to the NCS color-order system', *Color Research and Application*, 2001, 26(1), 4–28

Hari R, Hänninen R, Mäkinen T et al, 'Three hands: fragmentation of human bodily awareness', *Neuroscience Letters*, 1998, 240(3), 131–4

Harmon-Jones E, 'Early Career Award: clarifying the emotive functions of asymmetrical frontal cortical activity', *Psychophysiology*, 2003, 40(6), 838–48

———, 'Contributions from research on anger and cognitive dissonance to understanding the motivational functions of asymmetrical frontal brain activity', *Biological Psychology*, 2004, 67(1–2), 51–76

———, 'Trait anger predicts relative left frontal cortical activation to anger-inducing stimuli', *International Journal of Psychophysiology*, 2007, 66(2), 154–60

Harmon-Jones E & Allen JJ, 'Anger and frontal brain activity: EEG asymmetry consistent with approach motivation despite negative affective valence', *Journal of Personality and Social Psychology*, 1998, 74(5), 1310–6

Harmon-Jones E, Gable PA & Price TF, 'Does negative affect always narrow and positive affect always broaden the mind? Considering the influence of motivational intensity on cognitive scope', *Current Directions in Psychological Science*, 2013, 22(4), 301–7

Harmon-Jones E, Peterson CK & Harris CR, 'Jealousy: novel methods and neural correlates', *Emotion*, 2009, 9(1), 113–7

Harmon-Jones E & Sigelman J, 'State anger and prefrontal brain activity: evidence that insult-related relative left-prefrontal activation is associated with experienced anger and aggression', *Journal of Personality and Social Psychology*, 2001, 80(5), 797–803

Harms V, Reese M & Elias LJ, 'Lateral bias in theatre-seat choice', *Laterality*, 2014, 19(1), 1–11

Harmsen IE, 'Empathy in autism spectrum disorder', *Journal of Autism and Developmental Disorders*, 2019, 49(10), 3939–55

Harrington DL, Haaland KY & Knight RT, 'Cortical networks underlying mechanisms of time perception', *Journal of Neuroscience*, 1998, 18(3), 1085–95

Harris A, *Conscious: A Brief Guide to the Fundamental Mystery of the Mind*, Harper, 2019

Harris CP, Townsend JJ, Brockmeyer DL et al, 'Cerebral granular cell tumor occurring with glioblastoma multiforme: case report', *Surgical Neurology*, 1991, 36(3), 202–6

Harris G, 'Doctor admits pain studies were frauds, hospital says', *The New York Times*, 10 March 2009

Harris LJ, 'Early theory and research on hemispheric specialisation', *Schizophrenia Bulletin*, 1999, 25(1), 11–39

Harris LJ, Cárdenas RA, Spradlin MP et al, 'Adults' preferences for side-of-hold as portrayed in paintings of the Madonna and Child', *Laterality*, 2009, 14(6), 590–617

Harris R & de Jong BM, 'Differential parietal and temporal contributions to music perception in improvising and score-dependent musicians, an fMRI study', *Brain Research*, 2015, 1624, 253–64

Harris S, *An Atheist Manifesto*, 2005: samharris.org/an-atheist-manifesto/ (accessed 23 February 2021)

Harrison B & Harrison A, 'Through the looking glass: a literature review of a rare pediatric neuropsychiatric condition: Alice in Wonderland (Todd's) syndrome', *University of Ottawa Journal of Medicine*, 2015, 5(2), 46–9

Hart DB, *The Experience of God: Being, Consciousness, Bliss*, Yale University Press, 2013

———, reviewing Daniel Dennett, 2017: www.thenewatlantis.com/publications/the-illusionist (accessed 23 February 2021)

Hart T & Ailoae C, 'Spiritual touchstones: childhood spiritual experience in the development of influential historic and contemporary figures', *Imagination, Cognition and Personality*, 2007, 26(4), 345–59

Hart Y, report dated August 2008, appended to Woolf J, *The Mystery of Lewis Carroll*, St Martin's Press, 2010, 298–9

Hartig T, Evans GW, Jamner LD et al, 'Tracking restoration in natural and urban field settings', *Journal of Environmental Psychology*, 2003, 23(2), 109–23

Hartmann K, Goldenberg G, Daumüller M et al, 'It takes the whole brain to make a cup of coffee: the neuropsychology of naturalistic actions involving technical devices', *Neuropsychologia*, 2005, 43(4), 625–37

Hartwell LH, Hood L, Goldberg LM et al, *Genetics: From Genes to Genomes*, McGraw-Hill, 2011

Harvey M, Milner AD & Roberts RC, 'An investigation of hemispatial neglect using the Landmark Task', *Brain and Cognition*, 1995, 27(1), 59–78

Harwood DG, Sultzer DL, Feil D et al, 'Frontal lobe hypo-metabolism and impaired insight in Alzheimer disease', *American Journal of Geriatric Psychiatry*, 2005, 13(11), 934–41

Hasegawa Y, *Routledge Course in Japanese Translation*, Routledge, 2012

Haslam J, *Illustrations of Madness: Exhibiting a Singular Case of Insanity, And a No Less Remarkable Difference in Medical Opinions: Developing the Nature of An Assailment, And the Manner of Working Events; with a Description of Tortures Experienced by Bomb-Bursting, Lobster-Cracking and Lengthening the Brain. Embellished with a Curious Plate*, G Hayden, 1810

Hassler M & Nieschlag E, 'Masculinity, femininity, and musical composition: psychological and psychoendocrinological aspects of musical and spatial faculties', *Archives of Psychology*, 1989, 141(11), 71–84

Hassler M, Nieschlag E & de la Motte D, 'Creative musical talent, cognitive functioning, and gender: psychobiological aspects', *Music Perception*, 1990, 8(1), 35–48

Hatazawa J, Brooks RA, di Chiro G et al, 'Glucose utilization rate versus brain size in humans', *Neurology*, 1987, 37(4), 583–8

Hatta T, Ohnishi H & Ogura H, 'Hemispheric asymmetry and sex differences in a comparative judgement task', *International Journal of Neuroscience*, 1982, 16(2), 83–6

Hauk O, Johnsrude I & Pulvermüller F, 'Somatotopic representation of action words in human motor and premotor cortex', *Neuron*, 2004, 41(2), 301–7

Haukkala A, 'Socio-economic differences in hostility measures – a population based study', *Psychology & Health*, 2002, 17(2), 191–202

Haukkala A & Uutela A, 'Cynical hostility, depression, and obesity: the moderating role of education and gender', *International Journal of Eating Disorders*, 2000, 27(1), 106–9

Hauser MD & Akre K, 'Asymmetries in the timing of facial and vocal expressions by rhesus monkeys: implications for hemispheric specialization', *Animal Behaviour*, 2001, 61(2), 391–400

Hausmann M, 'Why sex hormones matter for neuroscience: a very short review on sex, sex hormones, and functional brain asymmetries', *Journal of Neuroscience Research*, 2017, 95(1–2), 40–9

Hausmann M, Ergun G, Yazgan Y et al, 'Sex differences in line bisection as a function of hand', *Neuropsychologia*, 2002, 40(3), 235–40

Hausmann M & Güntürkün O, 'Sex differences in functional cerebral asymmetries in a repeated measures design', *Brain and Cognition*, 1999, 41(3), 263–75

Hausmann M, Waldie KE & Corballis MC, 'Developmental changes in line bisection: a result of callosal maturation?', *Neuropsychology*, 2003, 17(1), 155–60

Hawkes T & Savage M (eds), *Measuring the Mathematics Problem*, Engineering Council, London, 2000: www.engc.org.uk/engcdocuments/internet/Website/Measuring%20the%20Mathematic%20Problems.pdf (accessed 23 February 2021)

Hawking S, *A Brief History of Time*, Bantam, 1989

Haxby JV, Horwitz B, Ungerleider LG et al, 'The functional organization of human extrastriate cortex: a PET-rCBF study of selective attention to faces and locations', *Journal of Neuroscience*, 1994, 14(11 Pt 1), 6336–53

Hay D, *Why Spirituality is Difficult for Westerners*, Societas, 2007

Hay D & Nye R, *The Spirit of the Child*, Jessica Kingsley, revd edn, 2006

Hayashi MJ & Ivry RB, 'Duration selectivity in right parietal cortex reflects the subjective experience of time', *Journal of Neuroscience*, 2020, 40(40), 7749–58

Hayashi R, 'Olfactory illusions and hallucinations after right temporal hemorrhage', *European Neurology*, 2004, 51(4), 240–1

Hayek FA, '"Conscious" direction and the growth of reason', in B Caldwell (ed), *Studies on the Abuse and Decline of Reason*, Routledge, 2010, 149–55

Hayes DP, 'The growing inaccessibility of science', *Nature*, 1992, 356(6372), 739–40

———, 'Nutritional hormesis', *European Journal of Clinical Nutrition*, 2007, 61(2), 147–59

Haznedar MM, Buchsbaum MS, Wei T-C et al, 'Limbic circuitry in patients with autism spectrum disorders studied with positron emission tomography and magnetic resonance imaging', *American Journal of Psychiatry*, 2000, 157(12), 1994–2001

He Q, Duan Y, Karsch K et al, 'Detecting corpus callosum abnormalities in autism based on anatomical landmarks', *Journal of Experimental Child Psychology: Neuroimaging*, 2010, 183(2), 126–32

Heath RL & Blonder LX, 'Spontaneous humor among right hemisphere stroke survivors', *Brain and Language*, 2005, 93(3), 267–76

Heavens A, interview for *Science* magazine, 2016: www.sciencemag.org/news/2016/02/

conditions-life-may-hinge-how-fast-universe-expanding (accessed 23 February 2021)
Hebb DO, *The Organization of Behavior*, Wiley & Sons, 1949
Hécaen H, *Neuropsychologie de la Perception Visuelle*, Masson et Cie, 1972
Hécaen H & Angelergues R, 'Agnosia for faces (prosopagnosia)', *Archives of Neurology*, 1962, 7(2), 92–100
——, *La Cécité Psychique*, Masson et Cie, 1963
Hécaen H & de Ajuriaguerra J, *Méconnaissances et Hallucinations Corporelles: Intégration et Désintégration de la Somatognosie*, Masson et Cie, 1952
Hécaen H, de Ajuriaguerra J & Massonet J, 'Les troubles visuo-constructifs par lésion parieto-occipitale droite: rôle des perturbations vestibulaires', *L'Encéphale*, 1951, 40, 122–79
Hécaen H & Garcia Badaracco J, 'Les hallucinations visuelles au cours des ophthalmopathies et des lésions des nerfs et du chiasma optiques', *L'Évolution Psychiatrique*, 1956a, 21(1), 157–9
——, 'Séméiologie des hallucinations visuelles en clinique neurologique', *Acta Neurológica Latinoamericana*, 1956b, 2(1), 23–57
Hecht D, 'An inter-hemispheric imbalance in the psychopath's brain', *Personality and Individual Differences*, 2011, 51(1), 3–10
——, 'The neural basis of optimism and pessimism', *Experimental Neurobiology*, 2013, 22(3), 173–99
——, 'Cerebral lateralization of pro- and anti-social tendencies', *Experimental Neurobiology*, 2014, 23(1), 1–27
Hegel GWF, *Wissenschaft der Logik* [1812], trans AV Miller, as *The Science of Logic*, George Allen & Unwin, 1969
——, *Encyclopädie der philosophischen Wissenschaften* [1830], Part I, trans W Wallace, as *Part One of the Encyclopædia of Philosophical Sciences*, Clarendon Press, 1975
——, *Phänomenologie des Geistes* [1807], trans AV Miller, as *The Phenomenology of Spirit*, Clarendon Press, 1977
Heidegger M, *Nietzsche I*, Neske, 1961
——, *Being and Time*, trans J Macquarrie & E Robinson, Blackwell, 1962
——, *What is Called Thinking?*, trans FD Wick & JG Gray, Harper & Row, 1968
——, 'The thinker as poet', in *Poetry, Language, Thought*, trans A Hofstadter, Harper-Perennial, 1975a, 1–14
——, 'The origin of the work of art', in *Poetry, Language, Thought*, trans A Hofstadter, HarperPerennial, 1975b, 15–87
——, 'The thing', in *Poetry, Language, Thought*, trans A Hofstadter, Harper-Perennial, 1975c, 163–86
——, '"Only a God can save us now": an interview with Martin Heidegger', trans D Schendler, *Graduate Faculty Philosophy Journal*, 1977, 6(1), 5–27
——, *Sein und Zeit*, Max Niemeyer Verlag, 1986 [1927]
——, in *Martin Heidegger: Gesamtausgabe*, III: *Unveröffentlichte Abhandlungen Vorträge – Gedachtes*, vol 79, *Einblick in das was ist: Bremer Vorträge* [1949]), ed P Jaeger, Klostermann, 1994
——, 'Das Ge-stell', as trans DM Levin in *The Philosopher's Gaze*, University of California Press, 1999
——, in *Martin Heidegger: Gesamtausgabe*, I: *Veröffentlichte Schriften 1910–1976*, vol 16, *Reden und Andere Zeugnisse eines Lebensweges*, ed H Heidegger, Klostermann, 2000
Heider ER, 'Universals in color naming and memory', *Journal of Aesthetic Education*, 1972, 93(1), 10–20
Heilman KM & Acosta LM, 'Visual artistic creativity and the brain', *Progress in Brain Research*, 2013, 204, 19–43
Heilman KM, Bowers D, Speedie L et al, 'Comprehension of affective and nonaffective prosody', *Neurology*, 1984, 34(7), 917–21
Heilman KM, Scholes R & Watson RT, 'Auditory affective agnosia: disturbed comprehension of affective speech', *Journal of Neurology, Neurosurgery, & Psychiatry*, 1975, 38(1), 69–72
Heilman KM, Schwartz HD & Watson RT, 'Hypoarousal in patients with the neglect syndrome and emotional indifference', *Neurology*, 1978, 28(3), 229–32
Heilman KM & van den Abell T, 'Right hemispheric dominance for mediating cerebral activation', *Neuropsychologia*, 1979, 17(3–4), 315–21
——, 'Right hemisphere dominance for attention: the mechanism underlying hemispheric asymmetries of inattention (neglect)', *Neurology*, 1980, 30(3), 327–3
Heilman RM & Miclea M, 'The contributions of declarative knowledge and emotion regulation in the IOWA gambling task', *Cognition, Brain, & Behavior*, 2015, 19(1), 35–53
Hein G & Knight RT, 'Superior temporal sulcus – it's my area: or is it?', *Journal of Cognitive Neuroscience*, 2008, 20(12), 2125–36
Hein G, Lamm C, Brodbeck C et al, 'Skin conductance response to the pain of others predicts later costly helping', *PLoS One*, 2011, 6(8), e22759

Heinik J, Aharon-Peretz J & Hes JP, 'De Clérambault's syndrome in multi-infarct dementia', *Psychiatria Fennica*, 1991, 22, 23–6

Heinlein RA, *Time Enough for Love*, Putnam, 1973

Heinrichs RW & Zakzanis KK, 'Neurocognitive deficit in schizophrenia: a quantitative review of the evidence', *Neuropsychology*, 1998, 12(3), 426–45

Heinze HJ, Hinrichs H, Scholz M et al, 'Neural mechanisms of global and local processing: a combined PET and ERP study', *Journal of Cognitive Neuroscience*, 1998, 10(4), 485–98

Heinze HJ & Münte TF, 'Electrophysiological correlates of hierarchical stimulus processing: dissociation between onset and later stages of global and local target processing', *Neuropsychologia*, 1993, 31(8), 841–52

Heisenberg W, *Physics and Philosophy: The Revolution in Modern Science*, George Allen & Unwin, 1958

——, *Physics and Beyond: Encounters and Conversations*, trans AJ Pomerans, Harper & Row, 1971

——, *Across the Frontiers*, Harper & Row, 1974

——, *Der Teil und das Ganze: Gespräche im Umkreis der Atomphysik*, Piper Taschenbuch, 2001

Heit E, 'Brain imaging, forward inference, and theories of reasoning', *Frontiers in Human Neuroscience*, 2015, 8, article 1056

Heiti W, 'Reading and character: Weil and McDowell on naïve realism and second nature', *Philosophical Investigations*, 2018, 41(3), 267–90

Heldmann B, Kerkhoff G, Struppler A et al, 'Repetitive peripheral magnetic stimulation alleviates tactile extinction', *NeuroReport*, 2000, 11(14), 3193–8

Heller M, 'Infinities in cosmology', in M Heller & WH Woodin (eds), *Infinity: New Research Frontiers*, Cambridge University Press, 2011, 218–29

Hellige JB, Bloch MI, Cowin EL et al, 'Individual variation in hemispheric asymmetry: multitask study of effects related to handedness and sex', *Journal of Experimental Psychology: General*, 1994, 123(3), 235–56

Hellige JB, Cox PJ & Litvac L, 'Information processing in the cerebral hemispheres: selective hemispheric activation and capacity limitations', *Journal of Experimental Psychology: General*, 1979, 108(2), 251–79

Hellige JB & Michimata C, 'Categorization versus distance: hemispheric differences for processing spatial information', *Memory & Cognition*, 1989, 17(6), 770–6

Hemphill RE & Klein R, 'Contribution to dressing disability as focal sign and to the imperception phenomena', *Journal of Mental Science*, 1948, 94(396), 611–22

Hemsley DR, 'Perceptual and cognitive abnormalities as the bases for schizophrenic symptoms', in AS David & JC Cutting (eds), *The Neuropsychology of Schizophrenia*, Erlbaum, 1994, 97–116

Hendry SH, Schwark HD, Jones EG et al, 'Numbers and proportions of GABA-immunoreactive neurons in different areas of monkey cerebral cortex', *Journal of Neuroscience*, 1987, 7(5), 1503–19

Henery CC & Mayhew TM, 'The cerebrum and cerebellum of the fixed human brain: efficient and unbiased estimates of volumes and cortical surface areas', *Journal of Anatomy*, 1989, 167, 167–80

Henkel LA, 'Point-and-shoot memories: the influence of taking photos on memory for a museum tour', *Psychological Science*, 2013, 25(2), 396–402

Hennin B, 'What went wrong on Apollo 11's moon landing', *Peak of Flight*, 2010, 276, 2

Henning BG, 'Of termites and men', in BG Henning & AC Scarfe (eds), *Beyond Mechanism: Putting Life Back into Biology*, Lexington Books, 2013, 233–48

Henrich J, Heine SJ & Norenzayan A, 'The weirdest people in the world?', *Behavioral and Brain Sciences*, 2010, 33(2–3), 61–83

Henriksen MG & Nordgaard J, 'Self-disorders in schizophrenia', in G Stanghellini & M Aragon (eds), *An Experiential Approach to Psychopathology*, Springer, 2016

Henrion M & Fischhoff B, 'Assessing uncertainty in physical constants', *American Journal of Physics*, 1986, 54(9), 791–8

Henriques JB & Davidson RJ, 'Left frontal hypoactivation in depression', *Journal of Abnormal Psychology*, 1991, 100(4), 535–45

Henry PJ & Napier JL, 'Education is related to greater ideological prejudice', *Public Opinion Quarterly*, 2017, 81(4), 930–42

Henry RC, 'The mental universe', *Nature*, 2005, 436(7), 29

——, 'Review of *The God Theory*, by Bernard Haisch', *Journal of Scientific Exploration*, 2008, 22, 266

Henry RC & Palmquist SR, 2007: henry.pha.jhu.edu/aspect.html (accessed 9 March 2021)

Henson R, 'What can functional neuroimaging tell the experimental psychologist?', *Quarterly Journal of Experimental Psychology*, 2005, 58A(2), 193–233

Henson R, Shallice T & Dolan R, 'Neuroimaging evidence for dissociable forms of repetition priming', *Science*, 2000, 287(5456), 1269–72

Heo K, Cho YJ, Lee SK et al, 'Single-photon emission computed tomography in a patient with ictal metamorphopsia', *Seizure*, 2004, 13(4), 250–3

Heraclitus, in Kahn CH, *The Art and Thought of Heraclitus*, Cambridge University Press, 1979

Herbet G, Lafargue G, Moritz-Gasser S et al, 'A disconnection account of subjective empathy impairments in diffuse low-grade glioma patients', *Neuropsychologia*, 2015, 70, 165–76

Herbranson WT, 'Pigeons, humans, and the Monty Hall Dilemma', *Current Directions in Psychological Science*, 2012, 21(5), 297–301

Herbranson WT & Schroeder J, 'Are birds smarter than mathematicians? Pigeons (*Columba livia*) perform optimally on a version of the Monty Hall Dilemma', *Journal of Comparative Psychology*, 2010, 124(1), 1–13

Herculano-Houzel S, 'The human brain in numbers: a linearly scaled-up primate brain', *Frontiers in Human Neuroscience*, 2009, 3, 31

Herder JG, 'Philosophie und Schwärmerei', in J von Müller (ed), *Postscenien zur Geschichte der Menschheit*, Cotta'schen Buchhandlung, 1828, 43–53

———, *Sculpture: Some Observations on Shape and Form from Pygmalion's Creative Dream*, trans & ed J Gaiger, University of Chicago Press, 2002 [*Plastik: Einige Wahrnehmungen über Form und Gestalt aus Pygmalions bildendem Traume* 1778]

Hermle L, Gouzoulis-Mayfrank E & Spitzer M, 'Blood flow and cerebral laterality in the mescaline model of psychosis', *Pharmacopsychiatry*, 1998, (31, suppl 2), 85–91

Herndon RM, 'The fine structure of the Purkinje cell', *Journal of Cell Biology*, 1963, 18(1), 167–80

Herrmann A, Zidansek M, Sprott DE et al, 'The power of simplicity: processing fluency and the effects of olfactory cues on retail sales', *Journal of Retailing*, 2013, 89(1), 30–43

Herrmann CS, Pauen M, Min BK et al, 'Analysis of a choice-reaction task yields a new interpretation of Libet's experiments', *International Journal of Psychophysiology*, 2008, 67(2), 151–7

Hertenstein E, Waibel E, Frase L et al, 'Modulation of creativity by transcranial direct current stimulation', *Brain Stimulation*, 2019, 12(5), 1213–21

Hertwig R & Todd PM, 'More is not always better: the benefits of cognitive limits', in DJ Hardman & L Macchi (eds), *Thinking: Psychological Perspectives on Reasoning, Judgment and Decision Making*, Wiley, 2003, 213–31

Herz RS, McCall C & Cahill L, 'Hemispheric lateralization in the processing of odor pleasantness versus odor names', *Chemical Senses*, 1999, 24(6), 691–5

Herzog G, *Anton Räderscheidt*, DuMont Buchverlag, 1991

Heschel AJ, *God in Search of Man*, Farrar, Straus & Giroux, 1955

———, 'Halakhah and aggadah', in *Between God and Man*, Free Press, 1997

———, *The Sabbath: Its Meaning for Modern Man*, Farrar, Straus & Giroux, 2005 [1951]

Hesling I, Clément S, Bordessoules M et al, 'Cerebral mechanisms of prosodic integration: evidence from connected speech', *NeuroImage*, 2005, 24(4), 937–47

Hess RH, Baker CL & Zihl J, 'The "motion-blind" patient: low-level spatial and temporal filters', *The Journal of Neuroscience*, 1989, 9(5), 1628–40

Hesse H, *If the War Goes On: Reflections on War and Politics*, Canongate, 2018

Hesselmann G, Naccache L, Cohen L et al, 'Splitting of the P3 component during dual-task processing in a patient with posterior callosal section', *Cortex*, 2013, 49(3), 730–47

Hétu S, Taschereau-Dumouchel V & Jackson PL, 'Stimulating the brain to study social interactions and empathy', *Brain Stimulation: Basic, Translational, and Clinical Research in Neuromodulation*, 2012, 5(2), 95–102

Heutink J, Brouwer WH, Kums E et al, 'When family looks strange and strangers look normal: a case of impaired face perception and recognition after stroke', *Neurocase*, 2012, 18(1), 39–49

Heuts BA & Lambrechts DYM, 'Positional biases in leg loss of spiders and harvestmen (Arachnida)', *Entomologische Berichten*, 1999, 59(2), 13–20

Hewitt JNB, 'Iroquoian Cosmology, First Part', *Annual Report of the Bureau of American Ethnology*, 1899–1900, 21, 133–339

———, 'Iroquoian Cosmology, Second Part', *Annual Report of the Bureau of American Ethnology*, 1928, 43, 449–819

Heydrich L & Blanke O, 'Distinct illusory own-body perceptions caused by damage to posterior insula and extrastriate cortex', *Brain*, 2013, 136(3), 790–803

Heydrich L, Dieguez S, Grunwald T et al, 'Illusory own body perceptions: case reports and relevance for bodily self-consciousness', *Consciousness and Cognition*, 2010, 19(3), 702–10

Heydrich L, Lopez C, Seeck M et al, 'Partial and full own-body illusions of epileptic origin in a child with right temporoparietal epilepsy', *Epilepsy & Behavior*, 2011, 20(3), 583–6

Hier D & Kaplan J, 'Verbal comprehension deficits after right hemispheric damage', *Applied Psycholinguistics*, 1980, 1, 279–94

Hilbert D, 'On the infinite', in P Benacerraf & H Putnam (eds), *Philosophy of Mathematics: Selected Readings*, Cambridge University Press, 2nd edn, 1984, 182–201 ['Über das Unendliche', *Mathematische Annalen*, 1926, 95, 161–90]

Hilgard J & Jamieson KH, 'Science as "broken" versus science as "self-correcting": how retractions and peer-review problems are exploited to attack science', in KH Jamieson, D Kahan & DA Scheufele (eds), *The Oxford Handbook of the Science of Science Communication*, Oxford University Press, 2017, 85–92

Hilgetag CC, Theoret H & Pascual-Leone A, 'Enhanced visual spatial attention ipsilateral to rTMS-induced "virtual lesions" of human parietal cortex', *Nature Neuroscience*, 2001, 4(9), 953–7

Hill HM, Guarino S, Calvillo A et al, 'Lateralized swim positions are conserved across environments for beluga (*Delphinapterus leucas*) mother-calf pairs', *Behavioural Processes*, 2017, 138(1), 22–8

Hill PL & Turiano NA, 'Purpose in life as a predictor of mortality across adulthood', *Psychological Science*, 2014, 25(7), 1482–6

Hilliard RD, 'Hemispheric laterality effects on a facial recognition task in normal subjects', *Cortex*, 1973, 9(3), 246–58

Hillis AE & Tippett DC, 'Stroke recovery: surprising influences and residual consequences', *Advances in Medicine*, 2014, 2014, 378263

Hilti LM & Brugger P, 'Incarnation and animation: physical versus representational deficits of body integrity', *Experimental Brain Research*, 2010, 204(3), 315–26

Hilti LM, Hänggi J, Vitacco DA et al, 'The desire for healthy limb amputation: structural brain correlates and clinical features of xenomelia', *Brain*, 2013, 136(1), 318–29

Hiraishi H, Haida M, Matsumoto M et al, 'Differences of prefrontal cortex activity between picture-based personality tests: a near-infrared spectroscopy study', *Journal of Personality Assessment*, 2012, 94(4), 366–71

Hirel C, Lévêque Y, Deiana G et al, 'Amusie acquise et anhédonie musicale', *Revue Neurologique (Paris)*, 2014, 170, 536–40

Hirotani M, Makuuchi M, Rüschemeyer SA et al, 'Who was the agent? The neural correlates of reanalysis processes during sentence comprehension', *Human Brain Mapping*, 2011, 32(11), 1775–87

Hirshfield J, *Nine Gates: Entering the Mind of Poetry*, Harper Perennial, 1998

Hirt ER, Levine GM, McDonald HE et al, 'The role of mood in quantitative and qualitative aspects of performance: single or multiple mechanisms?', *Journal of Experimental Social Psychology*, 1997, 33(6), 602–29

Hirt ER, Melton RJ, McDonald HE et al, 'Processing goals, task interest, and the mood-performance relationship: a mediational analysis', *Journal of Personality and Social Psychology*, 1996, 71(2), 245–61

Hiscock M, Perachio N & Inch R, 'Is there a sex difference in human laterality? IV. An exhaustive survey of dual-task interference studies from six neuropsychology journals', *Journal of Clinical and Experimental Neuropsychology*, 2001, 23(2), 137–48

Hishizawa M, Tachibana N & Hamano T, 'A case of left hemifacial metamorphopsia by a right retrosplenial infarction', *Rinsho Shinkeigaku*, 2015, 55(2), 87–90

Ho M-W, 'Toward an indigenous Western science: causality in the universe of coherent space-time structures', in W Harman (ed), *New Metaphysical Foundations of Modern Science*, Institute of Noetic Sciences, 1994, 179–213

Ho M-W & Popp F-A, 'Bioelectrodynamics and biocommunication: an epilogue', in M-W Ho, F-A Popp & U Warnke (eds), *Bioelectrodynamics and Biocommunication*, World Scientific Publishing, 1994

Hobbes T, *The Questions concerning Liberty and Necessity and Chance*, [1654], in *English Works of Thomas Hobbes*, ed W Molesworth, Bohn, 1841

Hobert O, Johnston RJ & Chang S, 'Left-right asymmetry in the nervous system: the *Cænorhabditis elegans* model', *Nature Reviews Neuroscience*, 2002, 3(8), 629–40

Hobson A, *Physics: Concepts and Connections*, Pearson Prentice Hall, 2007

Hobson RP, *Autism and the Development of Mind*, Psychology Press, 1995

Hobus PP, Schmidt HG, Boshuizen HP et al, 'Contextual factors in the activation of first diagnostic hypotheses: expert-novice differences', *Medical Education*, 1987, 21(6), 471–6

Hochschild J & Einstein KL, *Do Facts Matter? Information and Misinformation in American Politics*, University of Oakland Press, 2016

Hodgson T, Chamberlain M, Parris B et al, 'The role of the ventrolateral frontal cortex in inhibitory oculomotor control', *Brain*, 2007, 130(6), 1525–37

Hoeller K, 'Is Heidegger really a poet?', *Philosophical Topics*, 1981, 12(3), 121–38

Hoenig K, Müller C, Herrnberger B et al, 'Neuroplasticity of semantic representations

for musical instruments in professional musicians', *NeuroImage*, 2011, 56, 1714–25
Hofer S, Merboldt K-D, Tammer R et al, 'Rhesus monkey and human share a similar topography of the corpus callosum as revealed by diffusion tensor MRI in vivo', *Cerebral Cortex*, 2008, 18, 1079–84
Hoff H & Pötzl O, 'Über eine Zeitrafferwirkung bei homonymer linksseitiger Hemianopsie', *Zeitschrift für die gesamte Neurologie und Psychiatrie*, 1934, 151, 599–641
——, 'Über ein neues parieto-occipitales Syndrom: Seelenlähmung des Schauens, Störung des Körperschemas, Wegfall des zentralen Sehens', *Jahrbuch für Psychiatrie und Neurologie*, 1935a, 52, 173–218
——, 'Über Störungen des Tiefensehens bei zerebraler Metamorphopsie', *Monatsschrift für Psychiatrie und Neurologie*, 1935b, 90, 305–26
——, 'Anatomischer Befund eines Falles mit Zeitrafferphänomen', *Deutsche Zeitschrift für Nervenheilkunde*, 1937a, 145 (1–4), 150–78
——, 'Anisotropie des Sehraums bei occipitaler Herderkrankung', *Deutsche Zeitschrift für Nervenheilkunde*, 1937b, 145(1–4), 179–217
——, 'Anatomical findings in a case of time acceleration', in trans G Dean, E Perecman, E Franzen & J Luwisch, *Agnosia and Apraxia: Selected Papers of Liepmann, Lange and Pötzl* (ed JW Brown), Erlbaum, 1988, 231–50
Hoff HE, 'Nicolaus of Cusa, van Helmont, and Boyle: the first experiment of the Renaissance in quantitative biology and medicine', *Journal of the History of Medicine and Allied Sciences*, 1964, 19(2), 99–117
Hoffman D, 'The case against reality', 2016: www.theatlantic.com/science/archive/2016/04/the-illusion-of-reality/479559/ (accessed 23 February 2021)
Hoffman PJ, Slovic P & Rorer LG, 'An analysis-of-variance model for the assessment of configural cue utilization in clinical judgment', *Psychological Bulletin*, 1968, 69(5), 338–49
Hoffmann M, 'Isolated right temporal lobe stroke patients present with Geschwind Gastaut syndrome, frontal network syndrome and delusional misidentification syndromes', *Behavioural Neurology*, 2008, 20(3), 83–9
Hoffmeyer J, *Biosemiotics: An Examination into the Signs of Life and the Life of Signs*, University of Chicago Press, 2008
Hofstadter DR & Dennett DC, *The Mind's I: Fantasies and Reflections on Self and Soul*, Bantam Books, 1982
Hofstede G, 'Dimensionalizing cultures: the Hofstede model in context', *Online Readings in Psychology and Culture*, 2011, 2(1), article 8

Hogan PC, 'Literary aesthetics: beauty, the brain, and Mrs. Dalloway', *Progress in Brain Research*, 2013, 205, 319–37
——, *Beauty and Sublimity: A Cognitive Aesthetics of Literature and the Arts*, Cambridge University Press, 2017
Hogarth RM & Karelaia N, 'Ignoring information in binary choice with continuous variables: when is less "more"?', *Journal of Mathematical Psychology*, 2005, 49(2), 115–24
Hohwy J, *The Predictive Mind*, Oxford University Press, 2014
Hoksbergen I, Pickut BA, Mariën P et al, 'SPECT findings in an unusual case of visual hallucinosis', *Journal of Neurology*, 1996, 243(8), 594–8
Holbrook D, *Education and Philosophical Anthropology*, Fairleigh Dickinson University Press, 1987
Holden JM, Greyson B & James D (eds), *The Handbook of Near-Death Experiences: Thirty Years of Investigation*, Praeger Publishers, 2009
Hölderlin H, in N von Hellingrath, F Seebass & L von Pigenot (eds), *Sämtliche Werke: Historisch-Kritische Ausgabe*, Propyläen-Verlag, 1923
Holdrege C (ed), *The Dynamic Heart and Circulation*, trans K Creeger, AWSNA, 2002
Holland AC & Kensinger EA, 'Emotion and autobiographical memory', *Physics of Life Reviews*, 2010, 7(1), 88–131
Holland J, 'In the spirit of "clever inventions and constellations": the mechanics of Romantic systems', 2016: www.rc.umd.edu/praxis/systems/praxis.systems.2016.holland.html (accessed 23 February 2021)
Hollander AC, Pitman A, Sjöqvist H et al, 'Suicide risk among refugees compared with non-refugee migrants and the Swedish-born majority population', *British Journal of Psychiatry*, 2020, 217(6), 686–92
Holm-Hadulla RM, Roussel M & Hofmann FH, 'Depression and creativity – the case of the German poet, scientist and statesman J. W. v. Goethe', *Journal of Affective Disorders*, 2010, 127(1–3), 43–9
Holmes E & Gruenberg G, 'Learning in plants', *Worm Runner's Digest*, 1965, 7(1), 9–12
Holmes G & Horrax G, 'Disturbances of spatial orientation and visual attention, with loss of stereoscopic vision', *Archives of Neurology and Psychiatry*, 1919, 1(4), 385–407
Holst C, Becker U, Jørgensen ME et al, 'Alcohol drinking patterns and risk of diabetes: a cohort study of 70,551 men and women from the general Danish population', *Diabetologia*, 2017, 60(10), 1941–50
Holster A, 'Time symmetry in physics', *Principles of Physical Time Directionality*

Hof–Hol

Hol–Hou

and *Fallacies of the Conventional View*, ch 2, 2014: philarchive.org/archive/HOLTTF-6 (accessed 23 February 2021)

Holt J, 'Numbers Guy: are our brains wired for math?', *The New Yorker*, 3 March 2008, 42–7

Holtzman JD & Gazzaniga MS, 'Enhanced dual task performance following corpus commissurotomy in humans', *Neuropsychologia*, 1985, 23(3), 315–21

Hölzel BK, Carmody J, Vangel M et al, 'Mindfulness practice leads to increases in regional brain gray matter density', *Journal of Experimental Child Psychology: Neuroimaging*, 2011, 191(1), 36–43

Hölzel BK, Lazar SW, Gard T et al, 'How does Mindfulness meditation work? Proposing mechanisms of action from a conceptual and neural perspective', *Perspectives on Psychological Science*, 2011, 6(6), 537–59

Hölzel BK, Ott U, Hempel H et al, 'Differential engagement of anterior cingulate and adjacent medial frontal cortex in adept meditators and non-meditators', *Neuroscience Letters*, 2007, 421(1), 16–21

Hood RW, Hill PC & Spilka B, *The Psychology of Religion: An Empirical Approach*, Guilford, 4th edn, 2009

Hooghe M, Marien S & de Vroome T, 'The cognitive basis of trust: the relation between education, cognitive ability, and generalized and political trust', *Intelligence*, 2012, 40(6), 604–13

Hook S, Gordon E, Lazzaro I et al, 'Regional differentiation of cortical activity in schizophrenia: a complementary approach to conventional analysis of regional cerebral blood flow', *Journal of Experimental Child Psychology*, 1995, 61(2), 85–93

Hooper R, 'The man who records his entire life', BBC World Service, 3 November 2016: www.bbc.co.uk/news/magazine-37631646 (accessed 27 March 2021)

Hopfinger JB, Woldorff MG, Fletcher EM et al, 'Dissociating top-down attentional control from selective perception and action', *Neuropsychologia*, 2001, 39(12), 1277–91

Hopkins GM, 'On the origin of beauty: a Platonic dialogue' [1865]; in *Poems and Prose*, ed WH Gardner, Penguin, 1963

Hopkins J, '*Coincidentia Oppositorum* in Nicholas of Cusa's Sermons' [undated]: www.jasper-hopkins.info/CusaOnCoincidencePlusNotes.pdf (accessed 23 February 2021)

Hoppe KD, 'Split brain and psychoanalysis', *Psychoanalytic Quarterly*, 1977, 46(2), 220–44

Hoppe KD & Bogen JE, 'Alexithymia in 12 commissurotomized patients', *Psychotherapy and Psychosomatics*, 1977, 28(1–4), 148–55

Hoppe KD & Kyle NL, 'Dual brain, creativity, and health', *Creativity Research Journal*, 1990, 3(2), 150–7

Höppner J, Kunesch E, Buchmann J et al, 'Demyelination and axonal degeneration in corpus callosum assessed by analysis of transcallosally mediated inhibition in multiple sclerosis', *Clinical Neurophysiology*, 1999, 110(4), 748–56

Hoptman MJ & Davidson RJ, 'How and why do the two cerebral hemispheres interact?' *Psychological Bulletin*, 1994, 116(2), 195–219

Hori H, Nagamine M, Soshi T et al, 'Schizotypal traits in healthy women predict prefrontal activation patterns during a verbal fluency task: a near-infrared spectroscopy study', *Neuropsychobiology*, 2008, 57(1–2), 61–9

Hori H, Ozeki Y, Terada S et al, 'Functional near-infrared spectroscopy reveals altered hemispheric laterality in relation to schizotypy during verbal fluency task', *Progress in Neuro-Psychopharmacology & Biological Psychiatry*, 2008, 32(8), 1944–51

Horikawa H, Monji A, Sasaki M et al, 'Different SPECT findings before and after Capgras' syndrome in interictal psychosis', *Epilepsy & Behavior*, 2006, 9(1), 189–92

Hornak J, 'Ocular exploration in the dark by patients with visual neglect', *Neuropsychologia*, 1992, 30, 547–52

Horrax G, 'Visual hallucinations as a cerebral localizing phenomenon with especial reference to their occurrence in tumors of the temporal lobes', *Archives of Neurology & Psychiatry*, 1923, 10(5), 532–47

Horrobin FD, 'The philosophical basis of peer review and the suppression of innovation', *Journal of the American Medical Association*, 1990, 263(10), 1438–41

Hortensius R, Schutter DJ & Harmon-Jones E, 'When anger leads to aggression: induction of relative left frontal cortical activity with transcranial direct current stimulation increases the anger-aggression relationship', *Social Cognitive and Affective Neuroscience*, 2012, 7(3), 342–7

Horton R, 'African traditional thought and western science', *Africa: Journal of the International African Institute*, 1967, 37(2), 155–87

Horváth RA, Schwarcz A, Aradi M et al, 'Lateralisation of non-metric rhythm', *Laterality*, 2011, 16(5), 620–35

Hou C, Miller BL, Cummings JL et al, 'Autistic savants' [corrected by the authors from published title, 'Artistic savants'], *Neuropsychiatry, Neuropsychology & Behavioral Neurology*, 2000, 13(1), 29–38

Houdé O & Tzourio-Mazoyer N, 'Neural foundations of logical and mathematical cognition', *Nature Reviews Neuroscience*, 2003, 4(6), 507–14

Houdé O, Zago L, Crivello F et al, 'Access to deductive logic depends on a right ventromedial prefrontal area devoted to emotion and feeling: evidence from a training paradigm', *NeuroImage*, 2001, 14(6), 1486–92

Houdé O, Zago L, Mellet E et al, 'Shifting from the perceptual brain to the logical brain: the neural impact of cognitive inhibition training', *Journal of Cognitive Neuroscience*, 2000, 12(5), 721–8

Hough MS, 'Narrative comprehension in adults with right and left hemisphere brain-damage: theme organization', *Brain and Language*, 1990, 38(2), 253–77

House A & Hodges JR, 'Persistent denial of handicap after infarction of the right basal ganglia: a case study', *Journal of Neurology, Neurosurgery, and Psychiatry*, 1988, 51(1), 112–5

Houzel JC, Carvalho ML & Lent R, 'Interhemispheric connections between primary visual areas: beyond the midline rule', *Brazilian Journal of Medical and Biological Research*, 2002, 35(12), 1441–53

Howard RJ, ffytche DH, Barnes J et al, 'The functional anatomy of imagining and perceiving colour', *NeuroReport*, 1998, 9(6), 1019–23

Howard SR, Avarguès-Weber A, Garcia JE et al, 'Numerical cognition in honeybees enables addition and subtraction', *Science Advances*, 2019, 5(2), eaav0961

Howard-Jones PA, Blakemore S-J, Samuel EA et al, 'Semantic divergence and creative story generation: an fMRI investigation', *Brain Research: Cognitive Brain Research*, 2005, 25(1), 240–50

Howes D & Boller F, 'Simple reaction time: evidence for focal impairment from lesions of the right hemisphere', *Brain*, 1975, 98(2), 317–32

Hoy JA, Hatton C & Hare D, 'Weak central coherence: a cross-domain phenomenon specific to autism?', *Autism*, 2004, 8(3), 267–81

Hoyle F, 'The universe: past and present reflections', *Annual Review of Astronomy and Astrophysics*, 1982, 20, 1–35

Hsu YT, Duann JR, Chen CM et al, 'Anatomical and electrophysiological manifestations in a patient with congenital corpus callosum agenesis', *Brain Topography*, 2013, 26(1), 171–6

Huang P, Huang H, Luo Q et al, 'The difference between aesthetic appreciation of artistic and popular music: evidence from an fMRI study', *PLoS One*, 2016, 11(11), e0165377

Huang P, Qiu L, Shen L et al, 'Evidence for a left-over-right inhibitory mechanism during figural creative thinking in healthy non-artists', *Human Brain Mapping*, 2013, 34(10), 2724–32

Huang S, Ernberg I & Kauffman S, 'Cancer attractors: a systems view of tumors from a gene network dynamics and developmental perspective', *Seminars in Cell and Developmental Biology*, 2009, 20(7), 869–76

Huber G & Gross G, 'The concept of basic symptoms in schizophrenic and schizoaffective psychoses', *Recenti Progressi in Medicina*, 1989, 80(12), 646–52

Huber M, Karner M, Kirchler E et al, 'Striatal lesions in delusional parasitosis revealed by magnetic resonance imaging', *Progress in Neuro-Psychopharmacology & Biological Psychiatry*, 2008, 32(8), 1967–71

Huber-Okrainec J, Blaser SE & Dennis M, 'Idiom comprehension deficits in relation to corpus callosum agenesis and hypoplasia in children with spina bifida meningomyelocele', *Brain and Language*, 2005, 93(3), 349–68

Hübner R, 'Hemispheric differences in global/local processing revealed by same-different judgements', *Visual Cognition*, 1998, 5(4), 457–78

Hucker S, Langevin R, Dickey R et al, 'Cerebral damage and dysfunction in sexually aggressive men', *Annals of Sex Research*, 1988, 1(1), 33–47

Hudry J, Ryvlin P, Saive AL et al, 'Lateralization of olfactory processing: differential impact of right and left temporal lobe epilepsies', *Epilepsy & Behavior*, 2014, 37, 184–90

Hudson D, *Ludwig Wittgenstein*, John Knox Press, 1968

Hugdahl K, Iversen PM & Johnsen BH, 'Laterality for facial expressions: does the sex of the subject interact with the sex of the stimulus face?', *Cortex*, 1993, 29(2), 325–31

Hugdahl K, Løberg EM, Jørgensen HA et al, 'Left hemisphere lateralisation of auditory hallucinations in schizophrenia: a dichotic listening study', *Cognitive Neuropsychiatry*, 2008, 13(2), 166–79

Huggett N, 'Philosophical foundations of quantum field theory', *British Journal for the Philosophy of Science*, 2000, 51(suppl), 617–37

Hughes JR, 'The savant syndrome and its possible relationship to epilepsy', *Advances in Experimental Medicine and Biology*, 2012, 724, 332–43

Hughes R, *The Shock of the New*, Knopf, 1991

Hughlings Jackson J, 'Clinical remarks on cases of defects of expression (by words, writing, signs, etc.) in diseases of the nervous system.

(Under the care of Dr. Hughlings Jackson.)', *The Lancet*, 1864, 2, 604–5 (604)

——, 'On the nature of the duality of the brain', *Medical Press and Circular*, 1874, 1, 19–25, 41–49 & 63–70; reprinted in *Brain*, 1915, 38, 80–103

Hugo V, Preface to *Cromwell*, Ambroise Dupont et Cie, 1827

Huke V, Turk J, Saeidi S *et al*, 'Autism spectrum disorders in eating disorder populations: a systematic review', *European Eating Disorders Review*, 2013, 21(5), 345–51

Hulbert AJ, 'Metabolism and the development of endothermy', in CH Tyndale-Briscoe & PA Janssens (eds), *The Developing Marsupial: Models for Biomedical Research*, Springer-Verlag, 1988, 148–61

Hull DL, *Science as a Process: An Evolutionary Account of the Social and Conceptual Development of Science*, University of Chicago Press, 1990

Hume D, *A Treatise of Human Nature*, ed LA Selby-Bigge, Clarendon Press, 1896 [1739]

Humphrey NK, 'The illusion of beauty', *Perception*, 1973, 2(4), 429–39

——, *Seeing Red: A Study in Consciousness*, Harvard University Press, 2006

Hundertwasser F, 'Über das durch die gerade Linie zerstörte Paradies', 1985: translation at hundertwasser.com/en/texts/ueber_das_durch_die_gerade_linie_zerstoerte_paradies (accessed 23 February 2021)

Hunt GR & Gray RD, 'Direct observations of pandanus-tool manufacture and use by a New Caledonian crow (*Corvus moneduloides*)', *Animal Cognition*, 2004, 7(2), 114–20

Hunt GR, 'Human-like, population-level specialisation in the manufacture of pandanus tools by New Caledonian crows *Corvus moneduloides*', *Proceedings of the Royal Society of London, Series B: Biological Sciences*, 2000, 267(1441), 403–13

Hunt GR, Corballis MC & Gray RD, 'Laterality in tool manufacture by crows – neural processing and not ecological factors may influence "handedness" in these birds', *Nature*, 2001, 414(6865), 707

Hunt T, 'The hippies were right: it's all about vibrations, man! A new theory of consciousness', *Scientific American*, 5 December 2018

Huq S, Garety P & Hemsley D, 'Probabilistic judgements in deluded and non-deluded subjects', *Quarterly Journal of Experimental Psychology A*, 1988, 40(4), 801–12

Hurschler MA, Liem F, Jäncke L *et al*, 'Right and left perisylvian cortex and left inferior frontal cortex mediate sentence-level rhyme detection in spoken language as revealed by sparse fMRI', *Human Brain Mapping*, 2013, 34(12), 3182–92

Hurschler MA, Liem F, Oechslin M *et al*, 'fMRI reveals lateralized pattern of brain activity modulated by the metrics of stimuli during auditory rhyme processing', *Brain & Language*, 2015, 147, 41–50

Husain M & Rorden C, 'Non-spatially lateralized mechanisms in hemispatial neglect', *Nature Reviews Neuroscience*, 2003, 4(1), 26–36

Husain M, Shapiro K, Martin J *et al*, 'Abnormal temporal dynamics of visual attention in spatial neglect patients', *Nature*, 1997, 385(6612), 154–6

Hut P, 'There are no things' [undated]: www.edge.org/response-detail/11514 (accessed 23 February 2021)

Hutcheon L & Hutcheon M, 'Creativity, productivity, aging: the case of Benjamin Britten', *Age Culture Humanities*, 2014, 1: ageculturehumanities.org/WP/creativity-productivity-aging-the-case-of-benjamin-britten/ (accessed 23 February 2021)

Hutchison J, Thomas NA & Elias L, 'Leftward lighting in advertisements increases advertisement ratings and purchase intention', *Laterality*, 2011, 16(4), 423–32

Hutner N & Liederman J, 'Right hemisphere participation in reading', *Brain and Language*, 1991, 41(4), 475–95

Hutsler J & Galuske RA, 'Hemispheric asymmetries in cerebral cortical networks', *Trends in Neurosciences*, 2003, 26(8), 429–35

Hutsler JJ & Gazzaniga MS, 'Acetylcholinesterase staining in human auditory and language cortices: regional variation of structural features', *Cerebral Cortex*, 1996, 6(2), 260–70

Hutson M, 'Magical thinking', *Psychology Today*, March/April 2008, 89–95

Huxley A, 'Pascal', in *Do What You Will*, Watts & Co, 1936a, 181–246

Huxley A, 'Wordsworth in the Tropics', in *Do What You Will*, Watts & Co, 1936b, 90–103

Huxley A, *The Divine Within: Selected Writings on Enlightenment*, Harper Collins, 1992

Huxley JS, 'The biologist looks at man', *Fortune*, 1942, 26(6), 139–52

Huxley JS, preface to Teilhard de Chardin P, *The Phenomenon of Man*, trans B Wall, Harper & Row, 1959a

Huxley JS, *Religion without Revelation*, Max Parrish, 1959b

Huxley TH, 'On the hypothesis that animals are automata, and its history [1874]', in *Method and Results: Essays*, Macmillan, 1893, 199–250

Huxley TH, *Autobiography and Selected Essays*, CreateSpace, 2015 [1909]

Hyde KL, Lerch JP, Zatorre RJ *et al*, 'Cortical thickness in congenital amusia: when less is better than more', *Journal of Neuroscience*, 2007, 27(47), 13028–32

Hyde KL, Peretz I & Zatorre RJ, 'Evidence for the role of the right auditory cortex in fine pitch resolution', *Neuropsychologia*, 2008, 46(2), 632–9

Hyde KL, Zatorre RJ, Griffiths TD et al, 'Morphometry of the amusic brain: a two-site study', *Brain*, 2006, 129, 2562–70

Hyde KL, Zatorre RJ & Peretz I, 'Functional MRI evidence of an abnormal neural network for pitch processing in congenital amusia', *Cerebral Cortex*, 2010, 21(2), 292–9

I Ching, or Book of Changes, trans R Wilhelm & CF Baynes, Routledge & Kegan Paul, 3rd edn, 1968

Iacoboni M, Rayman J & Zaidel E, 'Left brain says yes, right brain says no: normative duality in the split brain', in SR Hameroff, AW Kasniak & AC Scott (eds), *Toward a Scientific Basis of Consciousness*, Massachusetts Institute of Technology Press, 1996, 197–202

Iacoboni M, Woods RP, Brass M et al, 'Cortical mechanisms of human imitation', *Science*, 1999, 286(5449), 2526–8

Ichheiser G, *Appearances and Realities*, Josey-Bass, 1970

Ideno Y, Hayashi K, Abe Y et al, 'Blood pressure-lowering effect of Shinrin-yoku (forest bathing): a systematic review and meta-analysis', *BMC Complementary and Alternative Medicine*, 2017, 17(1), 409

Ido F, Badran R, Dmytruk B et al, 'Auditory hallucinations as a rare presentation of occipital infarcts', *Case Reports in Neurological Medicine*, 2018, article 1243605

Iglói K, Doeller CF, Berthoz A et al, 'Lateralized human hippocampal activity predicts navigation based on sequence or place memory', *Proceedings of the National Academy of Sciences of the United States of America*, 2010, 107(32), 14466–71

Igou ER & Bless H, 'On undesirable consequences of thinking: framing effects as a function of substantive processing', *Journal of Behavioral Decision Making*, 2007, 20(2), 125–42

Ihara A, Hirata M, Fujimaki N et al, 'Neuroimaging study on brain asymmetries in situs inversus totalis', *Journal of the Neurological Sciences*, 2010, 288(1–2), 72–8

Ikejima K, Enomoto N, Iimuro Y et al, 'Estrogen increases sensitivity of hepatic Kupffer cells to endotoxin', *American Journal of Physiology*, 1998, 274(4), G669–76

Ikemi Y & Nakagawa S, 'A psychosomatic study of contagious dermatitis', *Kyushu Journal of Medical Science*, 1962, 13, 335–50

Ilomaki J, Jokanovic N, Tan EC et al, 'Alcohol consumption, dementia and cognitive decline: an overview of systematic reviews', *Current Clinical Pharmacology*, 2015, 10(3), 204–12

Imai N, Nohira O, Miyata K et al, 'A case of metamorphopsia caused by a very localized spotty infarct', *Rinsho Shinkeigaku*, 1995, 35(3), 302–5

Inafuku T, Sakai F, Sakamoto T et al, 'Visual hallucination in the hemianopic field caused by dural arteriovenous malformation', *Rinsho Shinkeigaku*, 1994, 34(5), 484–8

Indersmitten T & Gur RC, 'Emotion processing in chimeric faces: hemispheric asymmetries in expression and recognition of emotions', *Journal of Neuroscience*, 2003, 23(9), 3820–5

Ingalhalikar M, Smith A, Parker D et al, 'Sex differences in the structural connectome of the human brain', *Proceedings of the National Academy of Sciences of the United States of America*, 2014, 111(2), 823–8

Inglis J & Lawson JS, 'Sex differences in the effects of unilateral brain damage on intelligence', *Science*, 1981, 212(4495), 693–5

Ingold T, 'Rethinking the animate, re-animating thought', *Ethnos*, 2006, 71(1), 9–20

Iniesta I, 'Tomas Tranströmer's stroke of genius: language but no words', *Progress in Brain Research*, 2013, 206, 157–67

Innes BR, Burt DM, Birch YK et al, 'A leftward bias however you look at it: revisiting the emotional chimeric face task as a tool for measuring emotion lateralization', *Laterality*, 2016, 21(4–6), 643–61

Innocenti GM, 'General organization of callosal connections in the cerebral cortex', in EG Jones & AA Peters (eds), *Cerebral Cortex*, vol 5, Plenum Press, 1986, 291–353

——, 'The primary visual pathway through the corpus callosum: morphological and functional aspects in the cat', *Archives Italiennes de Biologie*, 1980, 118(2), 124–88

Intersalt Cooperative Research Group, 'Intersalt: an international study of electrolyte excretion and blood pressure: results for 24 hour urinary sodium and potassium excretion', *BMJ*, 1988, 297(6644), 319–28

Inutsuka M, Ogino T, Yoshinaga H et al, 'A child with ictal fear as the primary epileptic manifestation', *No To Hattatsu*, 2003, 35(4), 336–41

Ioannidis JP, 'Why most published research findings are false', *PLoS Medicine*, 2005, 2(8), e124

Ioannadis JP & Trikalinos TA, 'Early extreme contradictory estimates may appear in published research: the Proteus phenomenon in molecular genetics research and randomised trials', *Journal of Clinical Epidemiology*, 2005, 58(6), 543–9

Ionta S, Heydrich L, Lenggenhager B et al, 'Multisensory mechanisms in temporo-parietal cortex support self-location and first-person perspective', *Neuron*, 2011, 70(2), 363–74

Iosa M, Morone G, Bini F et al, 'The connection between anthropometry and gait harmony unveiled through the lens of the golden ratio', *Neurosci Lett*, 2016, 612, 138–44

Iosa M, Morone G & Paolucci S, 'Review article: Phi in physiology, psychology and biomechanics: the Golden Ratio between myth and science', *BioSystems*, 2018, 165(1), 31–9

Iparraguirre J, 'Socioeconomic determinants of risk of harmful alcohol drinking among people aged 50 or over in England', *BMJ Open*, 2015, 5, e007684

Irani F, Platek SM, Panyavin IS et al, 'Self-face recognition and theory of mind in patients with schizophrenia and first-degree relatives', *Schizophrenia Research*, 2006, 88(1–3), 151–60

Irle E, Lange C & Sachsse U, 'Reduced size and abnormal asymmetry of parietal cortex in women with borderline personality disorder', *Biological Psychiatry*, 2005, 57(2), 173–82

Irle E, Lange C, Weniger G et al, 'Size abnormalities of the superior parietal cortices are related to dissociation in borderline personality disorder', *Journal of Experimental Child Psychology*, 2007, 156(2), 139–49

Isen AM, Daubman KA & Nowicki GP, 'Positive affect facilitates creative problem solving', *Journal of Personality and Social Psychology*, 1987, 52(6), 1122–3

Isenman L, 'Understanding unconscious intelligence and intuition: "Blink" and beyond', *Perspectives in Biology and Medicine*, 2013, 56(1), 148–66

Isern RD, 'Family violence and the Klüver-Bucy syndrome', *Southern Medical Journal*, 1987, 80(3), 373–7

Ishizu T & Zeki S, 'A neurobiological enquiry into the origins of our experience of the sublime and beautiful', *Frontiers in Human Neuroscience*, 2014, 8, 891

Isolan GR, Bianchin MM, Bragatti JA et al, 'Musical hallucinations following insular glioma resection', *Neurosurgical Focus*, 2010, 28(2), E9

Ito J, Yamane Y, Suzuki M et al, 'Switch from ambient to focal processing mode explains the dynamics of free viewing eye movements', *Scientific Reports*, 2017, 7, 1082

Ito Y, 'Hemispheric asymmetry in the induction of false memories', *Laterality*, 2001, 6(4), 337–46

Iturria-Medina Y, Fernández AP, Morris DM et al, 'Brain hemispheric structural efficiency and interconnectivity rightward asymmetry in human and nonhuman primates', *Cerebral Cortex*, 2011, 21(1), 56–67

Ives CD, Abson DJ, von Wehrden H et al, 'Reconnecting with nature for sustainability', *Sustainability Science*, 2018, 13(5), 1389–97

Ivry RB & Keele SW, 'Timing functions of the cerebellum', *Journal of Cognitive Neuroscience*, 1989, 1(2), 136–52

Ivry RB & Robertson LC, *The Two Sides of Perception*, Massachusetts Institute of Technology Press, 1998

Iwamura Y, 'Bilateral receptive field neurons and callosal connections in the somatosensory cortex', *Philosophical Transactions of the Royal Society of London, Series B: Biological Sciences*, 2000, 355, 267–73

Iwasaki M, Noguchi Y & Kakigi R, 'Two-stage processing of aesthetic information in the human brain revealed by neural adaptation paradigm', *Brain Topography*, 2018, 31(6), 1001–13

Iwata K, *Kami to Kami: Animizumu Uchu no Tabi (Spirits and Gods: Travel in the Universe of Animism)*, Kodansha, 1989

Iyer J, Singh MD, Jensen M et al, 'Pervasive genetic interactions modulate neurodevelopmental defects of the autism-associated 16p11.2 deletion in Drosophila melanogaster', *Nature Communications*, 2018, 9(1), 2548

Jabbi M, Kippenhan J S, Kohn P et al, 'The Williams syndrome chromosome 7q11.23 hemideletion confers hypersocial, anxious personality coupled with altered insula structure and function', *Proceedings of the National Academy of Sciences of the United States of America*, 2012, 109(14), E860–6

Jablonka E & Lamb MJ, *Evolution in Four Dimensions*, Massachusetts Institute of Technology Press, 2005

Jablonka E & Lamb MJ, 'The inheritance of acquired epigenetic variations', *International Journal of Epidemiology*, 2015, 44(4), 1094–103

Jack CV, Cruz C, Hull RM et al, 'Regulation of ribosomal DNA amplification by the TOR pathway', *Proceedings of the National Academy of Sciences of the United States of America*, 2015, 112(3), 9674–9

Jackson CJ, Hobman EV, Jimmieson NL et al, 'Do left and right asymmetries of hemispheric preference interact with attention to predict local and global performance in applied tasks?', *Laterality*, 2012, 17(6), 647–72

Jackson PL, Brunet E, Meltzoff AN et al, 'Empathy examined through the neural mechanisms involved in imagining how I feel versus how you feel pain', *Neuropsychologia*, 2006, 44(5), 752–61

Jackson PM, 'Another case of lycanthropy', *Am J Psychiatry*, 1978, 135(1), 134–5

Jackson WS, *Melancholia and Depression: From Hippocratic Times to Modern Times*, Yale University Press, 1986

Jacob F, 'Evolution and tinkering', *Science*, 1977, 196(4295), 1161–6

Jacob R, Schall M & Scheibel AB, 'A quantitative dendritic analysis of Wernicke's area in humans. II. Gender, hemispheric, and environmental factors', *Journal of Comparative Neurology*, 1993, 327(1), 97–111

Jacobs A & Shiffrar M, 'Walking perception by walking observers', *Journal of Experimental Psychology: Human Perception and Performance*, 2005, 31(1), 157–69

Jacobs J, Korolev IO, Caplan JB et al, 'Right-lateralized brain oscillations in human spatial navigation', *Journal of Cognitive Neuroscience*, 2010, 22(5), 824–36

Jacobs L, Feldman M & Bender MB, 'The persistence of visual or auditory percepts as symptoms of irritative lesions of the cerebrum of man', *Zeitschrift für Neurologie*, 1972, 203(3), 211–8

Jacobs N, van Os J, Derom C et al, 'Heritability of intelligence', *Twin Research and Human Genetics*, 2007, 10(1), 11–14

Jacobsen T & Höfel L, 'Aesthetics electrified: an analysis of descriptive symmetry and evaluative aesthetic judgment processes using event-related brain potentials', *Empirical Studies of the Arts*, 2001, 19(2), 177–90

———, 'Descriptive and evaluative judgment processes, behavioral and electrophysiological indices of processing symmetry and aesthetics', *Cognitive, Affective and Behavioural Neuroscience*, 2003, 3(4), 289–99

Jacobsen T, Schubotz RI, Höfel L et al, 'Brain correlates of aesthetic judgment of beauty', *NeuroImage*, 2006, 29(1), 276–85

Jacome DE, 'Aphasia with elation, hypermusia, musicophilia and compulsive whistling', *Journal of Neurology, Neurosurgery, and Psychiatry*, 1984, 47(3), 308–10

Jafaria Z, Esmaili M, Delbari A et al, 'Post-stroke acquired amusia: a comparison between right- and left-brain hemispheric damages', *NeuroRehabilitation*, 2017, 40(2), 233–41

Jahn O, *W. A. Mozart*, Breitkopf & Härtel, 4 vols, 1856–59

Jaillais Y & Chory J, 'Unraveling the paradoxes of plant hormone signaling integration', *Nature Structural & Molecular Biology*, 2010, 17(6), 642–5

Jakobson R & Lübbe-Grothues G, 'The language of schizophrenia: Hölderlin's speech and poetry', in Jakobson R (ed), *Verbal Art, Verbal Sign, Verbal Time*, University of Minnesota Press, 1985, 133–40

Jalili M, 'Hemispheric asymmetry of electroencephalography-based functional brain networks', *NeuroReport*, 2014, 25(16), 1266–71

Jalili M, Meuli R, Do KQ et al, 'Attenuated asymmetry of functional connectivity in schizophrenia: a high-resolution EEG study', *Psychophysiology*, 2010, 47(4), 706–16

James H (ed), *The Letters of William James*, Atlantic Monthly Press, Boston, 2 vols, 1920

James I, *Asperger's Syndrome and High Achievement: Some Very Remarkable People*, Jessica Kingsley, 2005

James W, *The Principles of Psychology*, Henry Holt & Co, 2 vols, 1890

———, *The Will to Believe and Other Essays in Popular Philosophy*, Longmans, Green & Co, 1897

———, *The Varieties of Religious Experience*, Longmans, Green & Co, 1902

———, *A Pluralistic Universe*, Longmans, Green & Co, 1909a

———, 'Confidences of a "psychical researcher"', *The American Magazine*, 1909b, 68, 580–9

———, *Some Problems of Philosophy*, Longmans, Green & Co, 1911a

———, *The Meaning of Truth*, Longmans, Green & Co, 1911b

———, 'The Function of Cognition', Address to the Aristotelian Society, 1 December 1884; in *Pragmatism: And Four Essays from The Meaning of Truth*, Meridian Books, 1943

———, *Pragmatism*, Meridian Books, 1960 [1907]

———, 'On human immortality' [1898], in EM Gerald (ed), *William James: Writings 1878–1899*, The Library of America, 1992, 1100–27

James W, 'The will', *Talks to Teachers on Psychology and to Students on Some of Life's Ideals*, Cosimo, 2008 [1899]

Jamison KR, 'Mood disorders and patterns of creativity in British writers and artists', *Psychiatry*, 1989, 52(2), 125–34

———, *Touched with Fire: Manic-Depressive Illness and the Artistic Temperament*, Simon & Schuster, 1993

Jammer M, *The Philosophy of Quantum Mechanics: The Interpretations of Quantum Mechanics in Historical Perspective*, Wiley, 1974

Jang JW, Youn YC, Seok JW et al, 'Hypermetabolism in the left thalamus and right inferior temporal area on positron emission tomography-statistical parametric mapping

(PET-SPM) in a patient with Charles Bonnet syndrome resolving after treatment with valproic acid', *Journal of Clinical Neuroscience*, 2011, 18(8), 1130–2

Janka Z, 'Artistic creativity and bipolar mood disorder', *Orvosi Hetilap*, 2004, 145(33), 1709–18

Jansiewicz EM, Goldberg MC, Newschaffer CJ et al, 'Motor signs distinguish children with high functioning autism and Asperger's syndrome from controls', *Journal of Autism and Developmental Disorders*, 2006, 36(5), 613–21

Jarosz AF, Colflesh GJ & Wiley J, 'Uncorking the muse: alcohol intoxication facilitates creative problem-solving', *Consciousness and Cognition*, 2012, 21(1), 487–93

Jarvis ED, Güntürkün O, Bruce L et al, 'Avian brains and a new understanding of vertebrate brain evolution', *Nature Reviews Neuroscience*, 2005, 6(2), 151–9

Jasmin KM, McGettigan C, Agnew ZK et al, 'Cohesion and joint speech: right hemisphere contributions to synchronized vocal production', *Journal of Neuroscience*, 2016, 36(17), 4669–80

Jasper JD, Kunzler JS, Prichard EC et al, 'Individual differences in information order effects: the importance of right-hemisphere access in belief updating', *Acta Psychologica (Amsterdam)*, 2014, 148, 115–22

Jaspers K, *General Psychopathology*, trans J Hoenig & MW Hamilton, Manchester University Press, 1963 [1946, *Allgemeine Psychopathologie* 1913]

Jauk E, Benedek M, Dunst B et al, 'The relationship between intelligence and creativity: new support for the threshold hypothesis by means of empirical breakpoint detection', *Intelligence*, 2013, 41(4), 212–21

Jaušovec N, 'Affect in analogical transfer', *Creativity Research Journal*, 1989, 2(4), 255–66

——, 'Differences in cognitive processes between gifted, intelligent, creative, and average individuals while solving complex problems: an EEG study', *Intelligence*, 2000, 28(3), 213–37

Jaušovec N & Jaušovec K, 'Differences in resting EEG related to ability', *Brain Topography*, 2000a, 12(3), 229–40

——, 'EEG activity during the performance of complex mental problems', *International Journal of Psychophysiology*, 2000b, 36(1), 73–88

Javitt DC, 'Glutamate and schizophrenia: phencyclidine, N-methyl-D-aspartate receptors, and dopamine-glutamate interactions', *International Review of Neurobiology*, 2007, 78, 69–108

Jaynes J, *The Origins of Consciousness in the Breakdown of the Bicameral Mind*, Penguin, 1993 [1976]

Jeannerod M, *Motor Cognition: What Actions Tell the Self*, Oxford University Press, 2006

Jeans J, 'The new world-picture of modern physics', Presidential Address delivered at Aberdeen, 5 September 1934: in *Nature*, 1934, 134, 355–65

——, *Physics and Philosophy*, Cambridge University Press, 1942

——, *The Mysterious Universe*, Cambridge University Press, 1930

Jefferson T, Alderson P, Wager E et al, 'Effects of editorial peer review: a systematic review', *Journal of the American Medical Association*, 2002, 287(21), 2784–6

Jelinek E, 'Ich möchte seicht sein' (trans J Bramann as 'I want to be shallow'), in P von Becker, M Merschmeier & H Rischbieter (eds), *Theater Heute Jahrbuch 1983*, Orell Füssli & Friedrich Verlag, 1983

Jensen AR, 'How much can we boost IQ and scholastic achievement?', *Harvard Educational Review*, 1969, 39(1), 1–123

——, 'Individual differences in the Hick paradigm', in PA Vernon (ed), *Speed of Information Processing and Intelligence*, Ablex, 1987, 101–75

——, *The g Factor: The Science of Mental Ability*, Praeger, 1998

——, 'The theory of intelligence and its measurement', *Intelligence*, 2011, 39(4), 171–7

Jenner AR, Rosen GD & Galaburda AM, 'Neuronal asymmetries in primary visual cortex of dyslexic and nondyslexic brains', *Annals of Neurology*, 1999, 46(2), 189–96

Jensen I & Larsen JK, 'Mental aspects of temporal lobe epilepsy: follow-up of 74 patients after resection of a temporal lobe', *Journal of Neurology, Neurosurgery, and Psychiatry*, 1979, 42(3), 256–65

Jensen O, Gelfand J, Kounios J et al, 'Oscillations in the alpha band (9–12 Hz) increase with memory load during retention in a short-term memory task', *Cerebral Cortex*, 2002, 12(8), 877–82

Jerison HJ, 'Vigilance: biology, psychology, theory and practice', in Mackie RR (ed), *Vigilance* (NATO conference series), Plenum Press, 1977, 27–40

Jiang Y & Han S, 'Neural mechanisms of global/local processing of bilateral visual inputs: an ERP study', *Clinical Neurophysiology*, 2005, 116(6), 1444–54

Jibiki I, Matsuda H, Yamaguchi N et al, 'Acutely administered haloperidol-induced pattern changes of regional cerebral blood flow in schizophrenics', *Neuropsychobiology*, 1992, 25(4), 182–7

Jin S, Kwon Y, Jeong J et al, 'Differences in brain information transmission between gifted and normal children during scientific hypothesis generation', *Brain and Cognition*, 2006, 62(3), 191–7

Jirak D, Menz MM, Buccino G et al, 'Grasping language – a short story on embodiment', *Consciousness and Cognition*, 2010, 19(3), 711–20

Joanette Y, Goulet P & Hannequin D, *Right Hemisphere and Verbal Communication*, Springer-Verlag, 1990

John LK, Loewenstein G & Prelec D, 'Measuring the prevalence of questionable research practices with incentives for truth telling', *Psychological Science*, 2012, 23(5), 524–32

John S & Ovsiew F, 'Erotomania in a brain-damaged male', *Journal of Intellectual Disability Research*, 1996, 40(3), 279–83

Johnson BR, Li S de, Larson DB et al, 'A systematic review of the religiosity and delinquency literature: a research note', *Journal of Contemporary Criminal Justice*, 2000, 16(1), 32–52

Johnson BW, McKenzie KJ & Hamm JP, 'Cerebral asymmetry for mental rotation: effects of response hand, handedness and gender', *NeuroReport*, 2002, 13(15), 1929–32

Johnson DL, 'The moral judgment of schizophrenics', *The Journal of Nervous and Mental Disease*, 1960, 130, 278–85

Johnson D & Myklebust H, *Learning Disabilities: Educational Principles and Practices*, Grune and Stratton, 1967

Johnson J & Raab M, 'Take the first: option generation and resulting choices', *Organizational Behavior and Human Decision Processes*, 2003, 91(2), 215–29

Johnson RC, 'Are human brains quantum computers?', 2018: cacm.acm.org/news/227590-are-human-brains-quantum-computers/fulltext (accessed 23 February 2021)

Johnson SS, Hebsgaard MB, Christensen TR et al, 'Ancient bacteria show evidence of DNA repair', *Proceedings of the National Academy of Sciences of the United States of America*, 2007, 104(36), 14401–5

Johnson W, Bouchard TJ, McGue M et al, 'Genetic and environmental influences on the Verbal-Perceptual-Image Rotation (VPR) model of the structure of mental abilities in the Minnesota Study of Twins Reared Apart', *Intelligence*, 2007, 35, 542–62

Johnson-Laird PN, 'Mental models, deductive reasoning, and the brain', in MS Gazzaniga (ed), *The Cognitive Neurosciences*, Massachusetts Institute of Technology Press, 1995

———, *Mental Models: Towards a Cognitive Science of Language, Inference and Consciousness*, Harvard University Press, 1983

Johnsrude IS, Zatorre RJ, Milner BA et al, 'Left-hemisphere specialization for the processing of acoustic transients', *NeuroReport*, 1997, 8(7), 1761–5

Johnstone B, Cohen D, Bryant KR et al, 'Functional and structural indices of empathy: evidence for self-orientation as a neuropsychological foundation of empathy', *Neuropsychology*, 2015, 29(3), 463–72

Johnstone B & Glass BA, 'Support for a neuropsychological model of spirituality in persons with traumatic brain injury', *Zygon*, 2008, 43(4), 861–74

Jokic T, Zakay D & Wittmann M, 'Individual differences in self-rated impulsivity modulate the estimation of time in a real waiting situation', *Timing & Time Perception*, 2018, 6(1), 71–89

Jonas H, *The Phenomenon of Life: Toward a Philosophical Biology*, Northwestern University Press, 2001 [1966]

Jonas J, Descoins M, Koessler L et al, 'Focal electrical intracerebral stimulation of a face-sensitive area causes transient prosopagnosia', *Neuroscience*, 2012, 222, 281–8

Jonas J, Frismand S, Vignal JP et al, 'Right hemispheric dominance of visual phenomena evoked by intracerebral stimulation of the human visual cortex', *Human Brain Mapping*, 2014, 35(7), 3360–71

Jonas J, Rossion B, Brissart H et al, 'Beyond the core face-processing network: intracerebral stimulation of a face-selective area in the right anterior fusiform gyrus elicits transient prosopagnosia', *Cortex*, 2015, 72, 140–55

Jones E, *Free Associations: Memoirs of a Psychoanalyst*, Basic Books, 1959

Jones EG, 'The origins of cortical interneurons: mouse versus monkey and human', *Cerebral Cortex*, 2009, 19(9), 1953–6

Joo SW, Yoon W, Shon S-H et al, 'Altered white matter connectivity in patients with schizophrenia: an investigation using public neuroimaging data from SchizConnect', *PLoS One*, 2018, 13(10), e0205369

Jordan JT, 'The rodent hippocampus as a bilateral structure: a review of hemispheric lateralization', *Hippocampus*, 2020, 30(3), 278–92

Jordania J [Zhordania IM], *Who Asked the First Question? The Origins of Human Choral Singing, Intelligence, Language and Speech*, Logos, 2006: citeseerx.ist.psu.edu/viewdoc/download?doi=10.1.1.470.3973&rep=rep1&type=pdf (accessed 23 February 2021)

Joseph AB, 'Koro: computed tomography and brain electrical activity mapping in two patients', *Journal of Clinical Psychiatry*, 1986a, 47(8), 430–2

———, 'Cotard's syndrome with coexistent Capgras' syndrome, syndrome of subjective doubles, and palinopsia', *Journal of Clinical Psychiatry*, 1986b, 47(12), 605–6

Joseph AB & O'Leary DH, 'Brain atrophy and interhemispheric fissure enlargement in Cotard's syndrome', *Journal of Clinical Psychiatry*, 1986, 47(10), 518–20

Joseph KA, 'Capgras syndrome and its relationship to neurodegenerative disease', *Archives of Neurology*, 2007, 64, 1762–6

Joseph R, 'The neuropsychology of development: hemispheric laterality, limbic language, and the origin of thought', *Journal of Clinical Psychology*, 1982, 38(1), 4–33

———, 'Confabulation and delusional denial: frontal lobe and lateralized influences', *Journal of Clinical Psychology*, 1986, 42(3), 507–20

———, 'The right cerebral hemisphere: emotion, music, visual-spatial skills, body-image, dreams, and awareness', *Journal of Clinical Psychology*, 1988, 44(5), 630–73

———, *The Right Brain and the Unconscious*, Plenum Press, 1992

———, 'The limbic system and the soul: evolution and the neuroanatomy of religious experience', *Zygon*, 2001, 36(1), 105–36

Jostmann NB, Lakens D & Schubert TW, 'Weight as an embodiment of importance', *Psychological Science*, 2009, 20(9), 1169–74

Joyce J, *A Portrait of the Artist as a Young Man*, Penguin, 2000

Jozet-Alves C, Viblanc VA, Romagny S et al, 'Visual lateralization is task and age dependent in cuttlefish, *Sepia officinalis*', *Animal Behaviour*, 2012, 83, 1313–8

Juárez J & Corsi-Cabrera M, 'Sex differences in interhemispheric correlation and spectral power of EEG activity', *Brain Research Bulletin*, 1995, 38(2), 149–51

Juda A, 'The relationship between highest mental capacity and psychic abnormalities', *American Journal of Psychiatry*, 1949, 106(4), 296–307

Judd DB, *Color in Business, Science and Industry*, John Wiley & Sons, 1952

Judd T, 'A neuropsychologist looks at music behavior', in FL Hohmann & FR Wilson (eds), *The Biology of Music Making* (Proceedings of the 1984 Denver Conference), MMB Music, 1988a, 57–76

———, 'The varieties of musical talent', in LK Obler & D Fein (eds), *The Exceptional Brain: Neuropsychology of Talent and Special Abilities*, Guilford Press, 1988b, 127–55

Judd TL, Arslanian A, Davidson L et al, 'A right hemisphere stroke in a composer', paper presented to the International Neuropsychological Society Seventh Annual Conference, New York, 2 February 1979

Jung CG, *The Collected Works of CG Jung* (Bollingen Series XX), 20 vols, trans RFC Hull, ed H Read, M Fordham & G Adler, Princeton University Press, 1953–79

———, *Memories, Dreams, Reflections*, Random House, 1961

———, *Modern Man in Search of a Soul*, trans WS Dell & CF Baynes, Routledge & Kegan Paul, 1962

———, *Letters*, ed G Adler & A Jaffé (eds), Princeton University Press, 2 vols, 1973–76

Jung R, 'Neuropsychologie und Neurophysiologie des Kontur- und Formsehens in Zeichnung und Malerei', in HH Wieck (ed), *Psychopathologie musischer Gestaltungen*, FK Schattauer, 1974, 27–88

Jung RE, Gasparovic C, Chavez RS et al, 'Biochemical support for the "threshold" theory of creativity: a magnetic resonance spectroscopy study', *Journal of Neuroscience*, 2009, 29(16), 5319–25

Jung RE, Segall JM, Bockholt HJ et al, 'Neuroanatomy of creativity', *Human Brain Mapping*, 2010, 31(3), 398–408

Jung-Beeman M, 'Bilateral brain processes for comprehending natural language', *Trends in Cognitive Sciences*, 2005, 9(11), 512–8

Jung-Beeman M, Bowden EM, Haberman J et al, 'Neural activity when people solve verbal problems with insight', *PLoS Biology*, 2004, 2(4), e97

Juniper A, *Wabi Sabi: The Japanese Art of Impermanence*, Tuttle Publishing, 2003

Juolasmaa A, Outakoski J, Hirvenoja R et al, 'Effect of open heart surgery on intellectual performance', *Journal of Clinical Neuropsychology*, 1981, 3(3), 181–97

Jussim L, *Social Perception and Social Reality: Why Accuracy Dominates Bias and Self-Fulfilling Prophecy*, Oxford University Press, 2012a

———, 'Stereotype inaccuracy: extraordinary scientific delusions and the blindness of psychologists', *Psychology Today*, 25 October 2012b: www.psychologytoday.com/blog/rabble-rouser/201210/stereotype-inaccuracy (accessed 23 February 2021)

———, interviewed by Kate Hardiman in *The College Fix*, 31 May 2016: www.thecollegefix.com/post/27657/ (accessed 23 February 2021)

Jussim L, Cain TR, Crawford JT et al, 'The unbearable accuracy of stereotypes', in TD Nelson (ed), *Handbook of Prejudice, Stereotyping, and Discrimination*, Psychology Press, 2010, 199–225

Jussim L, Crawford JT, Anglin SM et al, 'Stereotype accuracy: one of the largest and most replicable effects in all of social psychology', in TD Nelson (ed), *Handbook of Prejudice,*

Stereotyping, and Discrimination, Psychology Press, new edn, 2016, 31–59

Just MA, Carpenter PA, Keller TA et al, 'Brain activation modulated by sentence comprehension', *Science*, 1996, 274(5284), 114–6

Just MA, Cherkassky VL, Keller TA et al, 'Functional and anatomical cortical underconnectivity in autism: evidence from an fMRI study of an executive function task and corpus callosum morphometry', *Cerebral Cortex*, 2007, 17(4), 951–61

Kaas JH, 'The organization of neocortex in mammals: implications for theories of brain function', *Annual Review of Psychology*, 1987, 38, 129–51

Kabadayi C & Osvath M, 'Ravens parallel great apes in flexible planning for tool-use and bartering', *Science*, 2017, 357(6347), 202–4

Kaessmann H, 'Origins, evolution, and phenotypic impact of new genes', *Genome Research*, 2010, 20(10), 1313–26

Kagan J, *The Three Cultures: Natural Sciences, Social Sciences and the Humanities in the Twenty-First Century*, Cambridge University Press, 2009

Kagerer FA, Wittmann M, Szelag E et al, 'Cortical involvement in temporal reproduction: evidence for differential roles of the hemispheres', *Neuropsychologia*, 2002, 40(3), 357–66

Kahan D, Braman D & Jenkins-Smith H, 'Cultural cognition of scientific consensus', *Journal of Risk Research*, 2011, 14(2), 147–74

Kahan DM, Peters E, Wittlin M et al, 'The polarizing impact of science literacy and numeracy on perceived climate change risks', *Nature Climate Change*, 2012, 2, 732–5

Kahane G, Everett JAC, Earp BD et al, '"Utilitarian" judgments in sacrificial moral dilemmas do not reflect impartial concern for the greater good', *Cognition*, 2015, 134, 193–209

Kahn CH, *The Art and Thought of Heraclitus*, Cambridge University Press, 1979

Kahneman D, *Thinking, Fast and Slow*, Penguin, 2012

Kaiser D, 'Building a multicellular organism', *Annual Review of Genetics*, 2001, 35(1), 103–23

Kakolewski KE, Crowson JJ, Sewell KW et al, 'Laterality, word valence, and visual attention: a comparison of depressed and non-depressed individuals', *International Journal of Psychophysiology*, 1999, 34(3), 283–92

Kamikubo T, Abo M & Yatsuzuka H, 'Case of long-term metamorphopsia caused by multiple cerebral infarction', *Brain and Nerve*, 2008, 60(6), 671–5

Kaminski J, Tempelmann S, Call J et al, 'Domestic dogs comprehend human communication with iconic signs', *Developmental Science*, 2009, 12(6), 831–7

Kaminsky I & Towler K (eds), *A God in the House: Poets Talk about Faith*, Tupelo Press, 2011

Kamoto-Barth S, Call J & Tomasello M, 'Great apes' understanding of other individuals' line of sight', *Psychological Science*, 2007, 18(5), 462–8

Kan R, Mori Y, Suzuki S et al, 'A case of temporal lobe astrocytoma associated with epileptic seizures and schizophrenia-like psychosis', *Psychiatry and Clinical Neurosciences*, 1989, 43(1), 97–103

Kanazawa S & Perina K, 'Why more intelligent individuals like classical music', *Journal of Behavioral Decision Making*, 2012, 25(3), 264–75

Kane J, 'Poetry as right-hemispheric language', *Journal of Consciousness Studies*, 2004, 11(5–6), 21–59

Kanemoto K, 'Peri-ictal Capgras syndrome after clustered ictal fear: depth-electroencephalogram study', *Epilepsia*, 1997, 38, 847–50

Kanner L, 'Autistic disturbances of affective contact', *Nervous Child*, 1943, 2, 217–50

——, 'Autistic disturbances of affective contact', *Acta Paedopsychiatrica*, 1968, 35(4), 100–36

Kant I, 'Vom Ideale der Schönheit', in *Kritik der Urteilskraft* [1790], 1: 'Kritik der ästhetischen Urteilskraft', sect 1, bk 1, i, §5, [*Gesammelte Schriften* (Akademie-Ausgabe), De Gruyter, 1962, vol 5]; trans WS Pluhar, as 'Critique of aesthetic judgement', in *Critique of Judgment*, Hackett, 1987

——, *Critique of Judgment*, trans JC Meredith, Oxford University Press, 2007 [*Kritik der Urteilskraft* 1790]

Kantrowitz JT, Butler PD, Schechter I et al, 'Seeing the world dimly: the impact of early visual deficits on visual experience in schizophrenia', *Schizophrenia Bulletin*, 2009, 35(6), 1085–94

Kantrowitz JT, Scaramello N, Jakubovitz A et al, 'Amusia and protolanguage impairments in schizophrenia', *Psychological Medicine*, 2014, 44(13), 2739–48

Kanwisher N, 'Domain specificity in face perception', *Nature Neuroscience*, 2000, 3(8), 759–63

Kanwisher N, Chunn MM, McDermott J et al, 'Functional imaging of human visual recognition', *Cognitive Brain Research*, 1996, 5(1–2), 55–67

Kanwisher N, McDermott J & Chun MM, 'The fusiform face area: a module in human extrastriate cortex specialized for face perception', *Journal of Neuroscience*, 1997, 17(11), 4302–11

Kanwisher N, Tong F & Nakayama K, 'The effect of face inversion on the human

fusiform face area', *Cognition*, 1998, 68(1), B1–11
Kaplan JA, Brownell HH, Jacobs JR et al, 'The effects of right hemisphere damage on the pragmatic interpretation of conversational remarks', *Brain and Language*, 1990, 38(2), 315–33
Kaplan JA & Gardner H, 'Artistry after unilateral brain disease', in F Boller & J Grafman (eds), *Handbook of Neuropsychology*, vol 2, Elsevier, 1989, 141–55
Kaplan JT, Aziz-Zadeh L, Uddin LQ et al, 'The self across the senses: an fMRI study of self-face and self-voice recognition', *Social Cognitive and Affective Neuroscience*, 2008, 3(3), 218–23
Kaplan R & Kaplan S, *The Experience of Nature: A Psychological Perspective*, Cambridge University Press, 1989
Kaplan S, 'The restorative benefits of nature: toward an integrative framework', *Journal of Environmental Psychology*, 1995, 15(3), 169–82
Kaplan-Solms K & Solms M (eds), *Clinical Studies in Neuro-Psychoanalysis: Introduction to a Depth Neuropsychology*, Karnac Books, 2000
Kapogiannis D, Sutin A, Davatzikos C et al, 'The five factors of personality and regional cortical variability in the Baltimore longitudinal study of aging', *Human Brain Mapping*, 2013, 34(11), 2829–40
Kapp RO, 'Living and lifeless machines', *British Journal for the Philosophy of Science*, 1954, 5(18), 91–103
Kapur N & Coughlan AK, 'Confabulation and frontal lobe dysfunction', *Journal of Neurology, Neurosurgery, and Psychiatry*, 1980, 43(5), 461–3
Kapur N, Turner A & King C, 'Reduplicative paramnesia: possible anatomical and neuropsychological mechanisms', *Journal of Neurology, Neurosurgery, and Psychiatry*, 1988, 51(4), 579–81
Karbe H, Herholz K, Halber M et al, 'Collateral inhibition of transcallosal activity facilitates functional brain asymmetry', *Journal of Cerebral Blood Flow and Metabolism*, 1998, 18(10), 1157–61
Karim AA, Schneider M, Lotze M et al, 'The truth about lying: inhibition of the anterior prefrontal cortex improves deceptive behavior', *Cerebral Cortex*, 2010, 20(1), 205–13
Karlsson JL, 'Genetic association of giftedness and creativity with schizophrenia', *Hereditas*, 1970, 66(2), 177–82
Karmanova IG, *Evolution of Sleep: Stages of the Formation of the 'Wakefulness-Sleep' Cycle in Vertebrates*, trans AI Koryushkin & OP Uchastkin, Karger, 1982
Karnath H-O & Baier B, 'Right insula for our sense of limb ownership and self-awareness of actions', *Brain Structure & Function*, 2010, 214(5–6), 411–7
Karnath H-O, Baier B & Nägele T, 'Awareness of the functioning of one's own limbs mediated by the insular cortex?', *Journal of Neuroscience*, 2005, 25(31), 7134–8
Karnath H-O & Fetter M, 'Ocular space exploration in the dark and its relation to subjective and objective body orientation in neglect patients with parietal lesions', *Neuropsychologia*, 1995, 33, 371–7
Karnath H-O & Rorden C, 'The anatomy of spatial neglect', *Neuropsychologia*, 2012, 50(6), 1010–7
Karpiński S & Szechyńska-Hebda M, 'Secret life of plants: from memory to intelligence', *Plant Signaling and Behavior*, 2010, 5(11), 1391–4
Karson CN, 'Spontaneous eye-blink rates and dopaminergic systems', *Brain*, 1983, 106(3), 643–53
Karson CN, Dykman RA & Paige SR, 'Blink rates in schizophrenia', *Schizophrenia Bulletin*, 1990, 16(2), 345–54
Kartsounis LD & Warrington EK, 'Failure of object recognition due to a breakdown of figure-ground discrimination in a patient with normal acuity', *Neuropsychologia*, 1991, 29(10), 969–80
Kas A, Lavault S, Habert MO et al, 'Feeling unreal: a functional imaging study in patients with Kleine-Levin syndrome', *Brain*, 2014, 137(7), 2077–87
Kasai K, Asada T, Yumoto M et al, 'Evidence for functional abnormality in the right auditory cortex during musical hallucinations', *Lancet*, 1999, 354(9191), 1703–4
Kasof J, 'Creativity and breadth of attention', *Creativity Research Journal*, 1997, 10, 303–15
Kass LR, *Toward a More Natural Science*, Free Press, 1988
Kassubek J, Otte M, Wolter T et al, 'Hemimicropsia in a patient with a cavernoma of the visual association cortex: imaging analysis utilizing a new 3D brain atlas system and fMRI', *NeuroImage*, 1998, 7(4, pt 2), S358
———, 'Brain imaging in a patient with hemimicropsia', *Neuropsychologia*, 1999, 37(12), 1327–34
Kastner RE, 'Taking Heisenberg's "potentia" seriously', 2017: transactionalinterpretation.org/2017/10/02/taking-heisenbergs-potentia-seriously-featured-on-science-news-blog/ (accessed 23 February 2021)
Kastrup B, *Why Materialism is Baloney*, Iff Books, 2014
Kattah JC, Luessenhop AJ, Kolsky M et al, 'Removal of occipital arteriovenous

malformations with sparing of visual fields', *Archives of Neurology*, 1981, 38(5), 307–9

Katz AN, 'Creativity and individual differences in asymmetric cerebral hemispheric functioning', *Empirical Studies of the Arts*, 1983, 1(1), 3–16

———, 'Setting the record right: comments on creativity and hemispheric functioning', *Empirical Studies of the Arts*, 1985, 3(1), 109–13

Katz Z, 'Sociology of religion in the USSR: a beginning?', *Slavic Review*, 1971, 30(4), 870–5

Kauffman S, *At Home in the Universe: The Search for the Laws of Self-Organization and Complexity*, Oxford University Press, 1995

Kaufman JC, 'The door that leads into madness: Eastern European poets and mental illness', *Creativity Research Journal*, 2005, 17(1), 99–103

———, 'The Sylvia Plath effect: mental illness in eminent creative writers', *Journal of Creative Behavior*, 2001, 35(1), 37–50

Kaufman JC & Baer J, 'I bask in the dreams of suicide: mental illness, poetry, and women', *Review of General Psychology*, 2002, 6(3), 271–86

Kaufmann G & Vosburg SK, '"Paradoxical" mood effects on creative problem-solving', *Cognition and Emotion*, 1997, 11(2), 151–70

Kaushall P, 'Functional asymmetries of the human visual system as revealed by binocular rivalry and binocular brightness matching', *American Journal of Optometry and Physiological Optics*, 1975, 52(8), 509–20

Kavcic V, Fei R, Hu S et al, 'Hemispheric interaction, metacontrol, and mnemonic processing in split-brain macaques', *Behavioural Brain Research*, 2000, 111(1–2), 71–82

Kawabata H & Zeki S, 'Neural correlates of beauty', *Journal of Neurophysiology*, 2004, 91(4), 1699–705

Kawaguchi Y, 'Receptor subtypes involved in callosally-induced postsynaptic potentials in rat frontal agranular cortex in vitro', *Experimental Brain Research*, 1992, 88(1), 33–40

Kawasaki K, 'Kansatsu no kenkyu' [An epistemological study on 'kansatsu' believed to be a precise equivalent for 'observation'], *Nihon Rika Kyoiku Gakkai Kenkyu Kiyo* [Bulletin of the Society of Japanese Teaching], 1992, 33, 71–80

———, 'A cross-cultural comparison of Japanese and English linguistic assumptions influencing pupil learning of science', *Canadian and International Education*, 2002, 31(1), 19–51

Kay J, *Obliquity*, Profile Books, 2011a

———, 'Economics: rituals of rigour', *The Financial Times*, 25 August 2011b

———, 'Beware of Franklin's Gambit in making decisions', *Financial Times*, 17 April 2012

Kay P & Kempton W, 'What is the Sapir-Whorf hypothesis?', *American Anthropologist*, 1984, 86, 65–79

Kazin A, *Journals*, ed RM Cook, Yale University Press, 2011

Ke M, Zou R, Shen H et al, 'Bilateral functional asymmetry disparity in positive and negative schizophrenia revealed by resting-state fMRI', *Journal of Experimental Child Psychology: Neuroimaging*, 2010, 182(1), 30–9

Kean C, 'Silencing the self: schizophrenia as a self-disturbance', *Schizophrenia Bulletin*, 2009, 35(6), 1034–6

Kearl MC & Harris R, 'Individualism and the emerging "modern" ideology of death', *Omega: Journal of Dying and Death*, 1981, 12(3), 269–80

Keats J, in R Gittings (ed), *Letters of John Keats*, Oxford University Press, 1970

Keefover RT, Ringel R & Roy EP, 'Negative hallucinations: an ictal phenomenon of partial complex seizures', *Journal of Neurology, Neurosurgery, and Psychiatry*, 1988, 51(3), 454–5

Keenan JP, Gallup GG Jr & Falk D, *The Face In The Mirror: The Search For The Origins Of Consciousness*, Harper Collins, 2003

Keenan JP, McCutcheon B, Freund S et al, 'Left hand advantage in a self-face recognition task', *Neuropsychologia*, 1999, 37(12), 1421–5

Keenan JP, Nelson A, O'Connor M et al, 'Self recognition and the right hemisphere', *Nature*, 2001, 409(6818), 305

Keenan JP, Rubio J, Racioppi C et al, 'The right hemisphere and the dark side of consciousness', *Cortex*, 2005, 41(5), 695–704

Keenan JP, Wheeler MA, Gallup GG Jr et al, 'Self-recognition and the right prefrontal cortex', *Trends in Cognitive Sciences*, 2000, 4(9), 338–44

Kelemen D, 'Are children "intuitive theists"? Reasoning about purpose and design in nature', *Psychological Science*, 2004, 15(5), 295–301

Kelemen D, Callanan MA, Casler K et al, 'Why things happen: teleological explanation in parent-child conversations', *Developmental Psychology*, 2005, 41(1), 251–64

Kelemen D, Rottman J & Seston R, 'Professional physical scientists display tenacious teleological tendencies: purpose-based reasoning as a cognitive default', *Journal of Experimental Psychology: General*, 2013, 142(4), 1074–83

Keller EF, 'Genes as difference makers', in S Krimsky & J Gruber (eds), *Genetic Explanations: Sense and Nonsense*, Harvard University Press, 2014, 34–42

Kelley CM & Jacoby LL, 'The construction of subjective experience: memory attributions', *Mind & Language*, 1990, 5(1), 49–68

Kelly T, 'Following the argument where it leads', *Philosophical Studies*, 2011, 154(1), 105–24

Kempler D, Van Lancker D, Marchman V et al, 'Idiom comprehension in children and adults with unilateral brain damage', *Developmental Neuropsychology*, 1999, 15(3), 327–49

Kenemans JL, Baas JMP, Mangun GR et al, 'On the processing of spatial frequencies as revealed by evoked potential source modeling', *Clinical Neurophysiology*, 2000, 111, 1113–23

Kenkō, *Essays in Idleness: the Tsurezuregusa of Kenkō*, trans D Keene, Columbia University Press, 1967

Kennedy DN, O'Craven KM, Ticho BS et al, 'Structural and functional brain asymmetries in human situs inversus totalis', *Neurology*, 1999, 53(6), 1260–5

Kennedy M, *Britten*, Dent, 1993

Kensinger EA, Garoff-Eaton RJ & Schacter DL, 'How negative emotion enhances the visual specificity of a memory', *Journal of Cognitive Neuroscience*, 2007, 19(11), 1872–87

Kenyon EE, *The American Weekly*, 30 September 1956

Kepler J, *Mysterium Cosmographicum: The Secret of the Universe*, trans AM Duncan, Abaris Books, 1981

Kerestes R, Chase HW, Phillips ML et al, 'Multimodal evaluation of the amygdala's functional connectivity', *NeuroImage*, 2017, 148, 219–29

Kerkhoff G, in R Götze & B Höfer (eds), *AOT – Alltagsorientierte Therapie bei Patienten mit erworbener Hirnschädigung*, Thieme, 1999, 102–16

Kerkhoff G, Schindler I, Keller I et al, 'Visual background motion reduces size distortion in spatial neglect', *NeuroReport*, 1999, 10, 319–23

Kertesz A, Polk M, Black SE et al, 'Anatomical asymmetries and functional laterality', *Brain*, 1992, 115(2), 589–605

Keshavan MS, Kahn EM & Brar JS, 'Musical hallucinations following removal of a right frontal meningioma', *Journal of Neurology, Neurosurgery, and Psychiatry*, 1988, 51(9), 1235–6

Keune PM, van der Heiden L, Várkuti B et al, 'Prefrontal brain asymmetry and aggression in imprisoned violent offenders', *Neuroscience Letters*, 2012, 515(2), 191–5

Keysar B, Hayakawa SL & An SG, 'The foreign-language effect: thinking in a foreign tongue reduces decision biases', *Psychological Science*, 2012, 23(6), 661–8

Khateb A, Pegna AJ, Landis T et al, 'Rhyme processing in the brain: an ERP mapping study', *International Journal of Psychophysiology*, 2007, 63(3), 240–50

Kiefer M & Dehaene S, 'The time course of parietal activation in single-digit multiplication: evidence from event-related potentials', *Mathematical Cognition*, 1997, 3(1), 1–30

Kiefer M, Sim EJ, Herrnberger B et al, 'The sound of concepts: four markers for a link between auditory and conceptual brain systems', *Journal of Neuroscience*, 2008, 28(47), 12224–30

Kiefer M, Weisbrod M, Kern I et al, 'Right hemisphere activation during indirect semantic priming: evidence from event-related potentials', *Brain and Language*, 1998, 64(3), 377–408

Kierkegaard S, in A Dru (trans & ed), *The Journals of Kierkegaard*, Fontana, 1958

——, *Johannes Climacus or De omnibus dubitandum est*, trans TH Croxall, Stanford University Press, 1967 [1842]

——, 'Diapsalmata', *Either/Or*, Princeton University Press, 1971

——, 'The absolute paradox: a metaphysical caprice', *Philosophical Fragments: Johannes Climacus*, trans HV Hong & EH Hong, Princeton University Press, 1985

——, 'Some reflections on marriage' [1845], in HV Hong & EH Hong (eds), *Collected Writings*, Princeton University Press, vol 11, 1988

——, *The Concept of Anxiety* [1844], in HV Hong & EH Hong (eds), *The Essential Kierkegaard*, Princeton University Press, 2000

Killgore WD & Yurgelun-Todd DA, 'Sex differences in amygdala activation during the perception of facial affect', *NeuroReport*, 2001, 12, 2543–7

Kim E, 'A post-ictal variant of Capgras syndrome in a patient with a frontal meningioma: case report', *Psychosomatics*, 1991, 32(4), 448–51

Kim HS, 'We talk, therefore we think? A cultural analysis of the effect of talking on thinking', *Journal of Personality and Social Psychology*, 2002, 83(4), 828–42

Kim J, Park S & Blake R, 'Perception of biological motion in schizophrenia and healthy individuals: a behavioral and fMRI study', *PLoS One*, 2011, 6(5), e19971

Kim SM, Park CH, Intenzo CM et al, 'Brain SPECT in a patient with post-stroke hallucination', *Clinical Nuclear Medicine*, 1993, 18(5), 413–6

Kim W, Lim SK, Chung EJ et al, 'The effect of cognitive behavior therapy-based psychotherapy applied in a forest environment on physiological changes and remission of

major depressive disorder', *Psychiatry Investigation*, 2009, 6(4), 245-54
Kim Y, Takemoto K, Mayahara K et al, 'An analysis of the subjective experience of schizophrenia', *Comprehensive Psychiatry*, 1994, 35(6), 430-6
Kim YM & Garrett K, 'On-line and memory-based: revisiting the relationship between candidate evaluation processing models', *Political Behavior*, 2012, 34(2), 345-68
Kimura B, 'Psychopathologie der Zufälligkeit oder Verlust des Aufenthaltsortes beim Schizophrenen', *Daseinsanalyse*, 1994, 11(3), 192-204
———, 'Cogito and I: a bio-logical approach', *Philosophy, Psychiatry, & Psychology*, 2001, 8(4), 331-6
Kimura D, 'Right temporal-lobe damage: perception of unfamiliar stimuli after damage', *Archives of Neurology*, 1963a, 8(3), 264-71
———, 'Speech lateralisation in young children as determined by an auditory test', *Journal of Comparative and Physiological Psychology*, 1963b, 56(5), 899-902
———, 'Spatial localization in left and right visual fields', *Canadian Journal of Psychology*, 1969, 23(6), 445-58
———, 'Sex differences in the brain', *Scientific American*, 1992, 267(3), 118-25
King D, Dockrell J & Stuart M, 'Constructing fictional stories: a study of story narratives by children with autistic spectrum disorder', *Research in Developmental Disabilities*, 2014, 35(10), 2438-49
King DG, 'Genetic variation among developing brain cells', *Science*, e-letters, 16 May 2011: science.sciencemag.org/content/332/6027/300/tab-e-letters (accessed 23 February 2021)
King DG & Kashi Y, 'Mutability and evolvability: indirect selection for mutability', *Heredity*, 2007, 99, 123-4
King FL & Kimura D, 'Left-ear superiority in dichotic perception of vocal nonverbal sounds', *Canadian Journal of Psychology*, 1972, 26(2), 111-6
King L & Appleton JV, 'Intuition: a critical review of the research and rhetoric', *Journal of Advanced Nursing*, 1997, 26(1), 194-202
Kingsland W, *An Anthology of Mysticism and Mystical Philosophy*, Methuen, 1927
Kingstone A, 'Covert orienting in the split brain: right hemisphere specialization for object-based attention', *Laterality*, 2016, 21(4-6), 732-44
Kinney DK, Richards R, Lowing PA et al, 'Creativity in offspring of schizophrenic and control patients: an adoption study', *Creativity Research Journal*, 2001, 13(1), 17-25

Kinnier Wilson SA & Walshe FMR, 'The phenomenon of "tonic innervation" and its relation to motor apraxia', *Brain*, 1914, 37, 199-246
Kinno R, Kawamura M, Shioda S et al, 'Neural correlates of noncanonical syntactic processing revealed by a picture-sentence matching task', *Human Brain Mapping*, 2008, 29(9), 1015-27
Kinomura S, Kawashima R, Yamada K et al, 'Functional anatomy of taste perception in the human brain studied with positron emission tomography', *Brain Research*, 1994, 659(1-2), 263-6
Kinsbourne M, 'Mechanisms of hemisphere interaction in man', in M Kinsbourne & WL Smith (eds), *Hemispheric Disconnection and Cerebral Function*, Thomas, 1974, 260-85; & 'Cerebral control and mental evolution', 286-289
Kinsbourne M, 'The mechanisms of hemispheric control of the lateral gradient of attention', in PMA Rabbitt & J Dornic (eds), *Attention and Performance v*, Academic Press, 1975
Kinsbourne M, 'Evolution of language in relation to lateral action', in M Kinsbourne (ed), *Asymmetrical Function of the Brain*, Cambridge University Press, 1978, 553-65
Kinsbourne M, 'Hemisphere interactions in depression', in M Kinsbourne (ed), *Cerebral Hemisphere Function in Depression*, American Psychiatric Press, 1988, 133-62
Kinsbourne M, 'Orientational bias model of unilateral neglect: evidence from attentional gradients within hemispace', in IH Robertson & JC Marshall (eds), *Unilateral Neglect: Clinical and Experimental Studies*, Lawrence Erlbaum, 1993
Kinsbourne M, 'Somatic twist: a model for the evolution of decussation', *Neuropsychology*, 2013, 27(5), 511-5
Kinsbourne M & Bemporad B, 'Lateralization of emotion: a model and the evidence', in NA Fox & RJ Davidson (eds), *The Psychobiology of Affective Development*, Erlbaum, 1984, 259-91
Kircher TT, Brammer M, Tous Andreu N et al, 'Engagement of right temporal cortex during processing of linguistic context', *Neuropsychologia*, 2001, 39(8), 798-809
Kircher TT, Seiferth NY, Plewnia C et al, 'Self-face recognition in schizophrenia', *Schizophrenia Research*, 2007, 94(1-3), 264-72
Kircher TT, Senior C, Phillips ML et al, 'Recognising one's own face', *Cognition*, 2001, 78(1), B1-15
Kirk U, 'The neural basis of object-context relationships on aesthetic judgment', *PLoS One*, 2008, 3(11), e3754

Kirk U, Skov M, Christensen MS et al, 'Brain correlates of aesthetic expertise: a parametric fMRI study', *Brain and Cognition*, 2009, 69(2), 306–15

Kirk U, Skov M, Hulme O et al, 'Modulation of aesthetic value by semantic context: an fMRI study', *NeuroImage*, 2009, 44(3), 1125–32

Kirsner K, 'Hemisphere-specific processes in letter matching', *Journal of Experimental Psychology: Human Perception and Performance*, 1980, 6(1), 167–79

Kitamura T & Kumar R, 'Time passes slowly for patients with depressed state', *Acta Psychiatrica Scandinavica*, 1982, 65(6), 415–20

Kitterle FL, Christman S & Hellige JB, 'Hemispheric differences are found in the identification, but not the detection, of low versus high spatial frequencies', *Perception & Psychophysics*, 1990, 48(4), 297–306

Kitterle FL & Selig LM, 'Visual field effects in the discrimination of sine-wave gratings', *Perception & Psychophysics*, 1991, 50(1), 15–18

Klaczynski PA & Lavallee KL, 'Domain-specific identity, epistemic regulation, and intellectual ability as predictors of belief-based reasoning: a dual-process perspective', *Journal of Experimental Child Psychology*, 2005, 92(1), 1–24

Klaczynski PA & Robinson B, 'Personal theories, intellectual ability, and epistemological beliefs: adult age differences in everyday reasoning tasks', *Psychology and Aging*, 2000, 15(3), 400–16

Klar AJS, 'Fibonacci's flowers', *Nature*, 2002, 417(6889), 595

Klein C, Keller B, Silver C et al, 'Positive adult development and "spirituality": psychological well-being, generativity, and emotional stability', in H Streib & RW Hood (eds), *Semantics and Psychology of 'Spirituality': A Cross-cultural Analysis*, Springer, 2016, 401–36

Klein G, 'Critical thoughts about critical thinking', *Theoretical Issues in Ergonomics Science*, 2011, 12(3), 201–24

Klein G & Jarosz A, 'A naturalistic study of insight', *Journal of Cognitive Engineering and Decision Making*, 2011, 5(4), 335–51

Klein TA, Endrass T, Kathmann N et al, 'Neural correlates of error awareness', *NeuroImage*, 2007, 34(4), 1774–81

Kleinman JT, Sepkuty JP, Hillis AE et al, 'Spatial neglect during electrocortical stimulation mapping in the right hemisphere', *Epilepsia*, 2007, 48(12), 2365–8

Kleinschmidt A, Buchel C, Hutton C et al, 'The neural structures expressing perceptual hysteresis in visual letter recognition', *Neuron*, 2002, 34(4), 659–66

Kleiter I, Luerding R, Diendorfer G et al, 'A lightning strike to the head causing a visual cortex defect with simple and complex visual hallucinations', *Journal of Neurology, Neurosurgery, and Psychiatry*, 2007, 78(4), 423–6

Kleven MS & Koek W, 'Differential effects of direct and indirect dopamine agonists on eye blink rate in cynomolgus monkeys', *Journal of Pharmacology and Experimental Therapeutics*, 1996, 279(3), 1211–19

Kliewer EV & Ward RH, 'Convergence of immigrant suicide rates to those in the destination country', *American Journal of Epidemiology*, 1988, 127(3), 640–53

Klimesch W, Doppelmayr M, Röhm D et al, 'Simultaneous desynchronization and synchronization of different alpha responses in the human electroencephalograph: a neglected paradox?', *Neuroscience Letters*, 2000, 284(1–2), 97–100

Klimesch W, Doppelmayr M, Schwaiger J et al, '"Paradoxical" alpha synchronization in a memory task', *Brain Research: Cognitive Brain Research*, 1999, 7(4), 493–501

Klimesch W, Sauseng P & Hanslmayr S, 'EEG alpha oscillations: the inhibition-timing hypothesis', *Brain Research Reviews*, 2007, 53(1), 63–88

Klin A, Volkmar FR, Sparrow SS et al, 'Validity and neuropsychological characterization of Asperger syndrome: convergence with nonverbal learning disabilities syndrome', *Journal of Child Psychology and Psychiatry*, 1995, 36(7), 1127–40

Kline M, *Mathematical Thought from Ancient to Modern Times*, Oxford University Press, 3 vols, 1972–90

Kloos G, 'Störungen des Zeiterlebens in der endogenen Depression', *Nervenarzt*, 1938, 11, 225–44

Knauff M, Fangmeier T, Ruff CC et al, 'Reasoning, models, and images: behavioral measures and cortical activity', *Journal of Cognitive Neuroscience*, 2003, 15(4), 559–73

Knight RA, Manoach DS, Elliott DS et al, 'Perceptual organization in schizophrenia: the processing of symmetrical configurations', *Journal of Abnormal Psychology*, 2000, 109(4), 575–87

Knoblich G, Ohlsson S, Haider H et al, 'Constraint relaxation and chunk decomposition in insight problem solving', *Journal of Experimental Psychology: Learning, Memory, and Cognition*, 1999, 25(6), 1534–55

Knobloch E, 'Galileo and Leibniz: different approaches to infinity', *Archive for History of Exact Sciences*, 1999, 54, 87–99

Knoch D & Fehr E, 'Resisting the power of temptations: the right prefrontal cortex and self-control', *Annals of the New York Academy of Sciences*, 2007, 1104(1), 123–34

Knoch D, Gianotti LR, Baumgartner T et al, 'A neural marker of costly punishment behavior', *Psychological Science*, 2010, 21(3), 337–42

Knoch D, Nitsche MA, Fischbacher U et al, 'Studying the neurobiology of social interaction with transcranial direct current stimulation – the example of punishing unfairness', *Cerebral Cortex*, 2008, 18(9), 1987–90

Knoch D, Pascual-Leone A, Meyer K et al, 'Diminishing reciprocal fairness by disrupting the right prefrontal cortex', *Science*, 2006, 314(5800), 829–32

Knorr KD, 'The scientist as an analogical reasoner: a critique of the metaphor theory of innovation', in KD Knorr, R Krohn & RP Whitley (eds), *The Social Process of Scientific Investigation*, Sociology of the Sciences Yearbook, vol IV, D Reidel Publishing, 1980, 25–52

Knott C, Bell S & Britton A, 'Alcohol consumption and the risk of type 2 diabetes: a systematic review and dose-response meta-analysis of more than 1.9 million individuals from 38 observational studies', *Diabetes Care*, 2015, 38(9), 1804–12

Knyazeva MG, 'Splenium of corpus callosum: patterns of interhemispheric interaction in children and adults', *Neural Plasticity*, 2013, 2013, article 639430

Ko Y-G & Kim J-Y, 'Scientific geniuses' psychopathology as a moderator in the relation between creative contribution types and eminence', *Creativity Research Journal*, 2008, 20(3), 251–61

Kobayashi M, 'Functional organization of the human gustatory cortex', *Journal of Oral Biosciences*, 2006, 48, 244–60

Kobayashi Y, 'A case of traumatic brain injury presenting with musical hallucinations', *Case Reports in Neurology*, 2018, 10(1), 7–11

Koch C, *The Quest for Consciousness*, Roberts & Co, 2004

Koch G, Oliveri M, Carlesimo GA et al, 'Selective deficit of time perception in a patient with right prefrontal cortex lesion', *Neurology*, 2002, 59(10), 1658–9

Koch K, *Locus Solus II*, Kraus, 1971

Koch SV, Larsen JT, Mouridsen SE et al, 'Autism spectrum disorder in individuals with anorexia nervosa and in their first- and second-degree relatives: Danish nationwide register-based cohort-study', *British Journal of Psychiatry*, 2015, 206(5), 401–7

Kociba RJ, Keyes DG, Beyer JE et al, 'Results of a two-year chronic toxicity and oncogenicity study of 2,3,7,8-tetrachlorodibenzo-p-dioxin in rats', *Toxicology and Applied Pharmacology*, 1978, 46(2), 279–303

Koechlin E, Ody C & Kouneiher F, 'The architecture of cognitive control in the human prefrontal cortex', *Science*, 2003, 302(5648), 1181–5

Koeda M, Takahashi H, Yahata N et al, 'A functional MRI study: cerebral laterality for lexical-semantic processing and human voice perception', *American Journal of Neuroradiology*, 2006, 27(7), 1472–9

———, 'Neural responses to human voice and hemisphere dominance for lexical-semantic processing: an fMRI study', *Methods of Information in Medicine*, 2007, 46(2), 247–50

Koedinger KR & Anderson JR, 'Abstract planning and perceptual chunks: elements of expertise in geometry', *Cognitive Science*, 1990, 14(4), 511–50

Koelkebeck K, Miyata J, Kubota M et al, 'The contribution of cortical thickness and surface area to gray matter asymmetries in the healthy human brain', *Human Brain Mapping*, 2014, 35(12), 6011–22

Koelsch S, Kasper E, Sammler D et al, 'Music, language, and meaning: brain signatures of semantic processing', *Nature Neuroscience*, 2004, 7(3), 302–7

Koenig HG, 'Religion, spirituality, and health: the research and clinical implications', *International Scholarly Research Network Psychiatry*, 2012, 2012, article 278730

Koenig HG, King DE & Carson VB, *Handbook of Religion and Health*, Oxford University Press, 2012

Koenigs M, Kruepke M, Zeier J et al, 'Utilitarian moral judgment in psychopathy', *Social Cognitive and Affective Neuroscience*, 2012, 7(6), 708–14

Koenigs M, Young L, Adolphs R et al, 'Damage to the prefrontal cortex increases utilitarian moral judgements', *Nature*, 2007, 446(7138), 908–11

Kõgesaar M, 'Flynni efekti esinemine Eesti abiturientide seas Raveni testi põhjal' [The Flynn effect in Estonian adolescents based on Raven's testing], Seminar series, Tartu University, 2013: hdl.handle.net/10062/30644 (accessed 23 February 2021)

Kohler CG, Walker JB, Martin EA et al, 'Facial emotion perception in schizophrenia: a meta-analytic review', *Schizophrenia Bulletin*, 2010, 36(5), 1009–19

Kohno T, Shiga T, Kusumi I et al, 'Left temporal perfusion associated with suspiciousness score on the Brief Psychiatric Rating Scale in schizophrenia', *Journal of Experimental Child Psychology*, 2006, 147(2–3), 163–71

Kok P, Jehee JF & de Lange FP, 'Less is more: expectation sharpens representations in the primary visual cortex', *Neuron*, 2012, 75(2), 265–70

Kokoschka O, *Mein Leben*, GeraNova Bruckmann, 1971

———, *My Life*, trans D Britt, Macmillan, 1974 [*Mein Leben* 1971]

Kolakowski L, *Bergson*, St Augustine's Press, 1985

Kolenbrander PE, 'Oral microbial communities: biofilms, interactions, and genetic systems', *Annual Review of Microbiology*, 2000, 54(1), 413–37

Koles ZJ, Lind JC & Flor-Henry P, 'A source-imaging (low-resolution electromagnetic tomography) study of the EEGs from unmedicated men with schizophrenia', *Journal of Experimental Child Psychology*, 2004, 130(2), 171–90

———, 'Gender differences in brain functional organization during verbal and spatial cognitive challenges', *Brain Topography*, 2010, 23(2), 199–204

Köllner MG & Schultheiss OC, 'Meta-analytic evidence of low convergence between implicit and explicit measures of the needs for achievement, affiliation, and power', *Frontiers in Psychology*, 2014, 5, 826

Kölmel HW, 'Complex visual hallucinations in the hemianopic field', *Journal of Neurology, Neurosurgery, and Psychiatry*, 1985, 48(1), 29–38

Kömpf D, Piper HF, Neundörfer B et al, 'Palinopsie (visuelle Perseveration) und zerebrale Polyopie – klinische Analyse und computertomographische Befunde', *Fortschritte der Neurologie-Psychiatrie*, 1983, 51(8), 270–81

Kong XZ, Mathias SR, Guadalupe T et al, 'Mapping cortical brain asymmetry in 17,141 healthy individuals worldwide via the ENIGMA Consortium', *Proceedings of the National Academy of Sciences of the United States of America*, 2018, 115(22), E5154–63

Konishi S, Hayashi T, Uchida I et al, 'Hemispheric asymmetry in human lateral prefrontal cortex during cognitive set shifting', *Proceedings of the National Academy of Sciences of the United States of America*, 2002, 99(11), 7803–8

Konrath SH, O'Brien EH & Hsing C, 'Changes in dispositional empathy in American college students over time: a meta-analysis', *Personality and Social Psychology Review*, 2011, 15(2), 180–98

Koonin EV, 'The cosmological model of eternal inflation and the transition from chance to biological evolution in the history of life', *Biology Direct*, 2007, 2, article 15

———, *The Logic of Chance: The Nature and Origin of Biological Evolution*, FT Press, 2011

———, 'The meaning of biological information', *Philosophical Transactions Series A: Mathematical, Physical, and Engineering Sciences*, 2016, 374(2063), 20150065

Kopala LC, Good KP & Honer WG, 'Olfactory hallucinations and olfactory identification ability in patients with schizophrenia and other psychiatric disorders', *Schizophrenia Research*, 1994, 12(3), 205–11

Koren L, *Wabi-Sabi: for Artists, Designers, Poets & Philosophers*, Imperfect Publishing, 2008

Korkman M, Granström M-L, Kantola-Sorsa E et al, 'Two-year follow-up of intelligence after pediatric epilepsy surgery', *Pediatric Neurology*, 2005, 33(3), 173–8

Korn H & Faure P, 'Is there chaos in the brain? II. Experimental evidence and related models', *Comptes Rendus Biologies*, 2003, 326(9), 787–840

Kornbrot DE, Msetfi RM & Grimwood MJ, 'Time perception and depressive realism: judgment type, psychophysical functions and bias', *PLoS One*, 2013, 8(8), e71585

Kornhuber HH & Deecke L, 'Hirnpotentialänderungen bei Willkürbewegungen und passiven Bewegungen des Menschen: Bereitschaftspotential und reafferente Potentiale', *Pflügers Archiv – European Journal of Physiology*, 1965, 284(1), 1–17

Kosillo P & Smith AT, 'The role of the human anterior insular cortex in time processing', *Brain Structure and Function*, 2010, 214(5–6), 623–8

Koski L & Petrides M, 'Time-related changes in task performance after lesions restricted to the frontal cortex', *Neuropsychologia*, 2001, 39(3), 268–81

Kosslyn SM, 'Seeing and imagining in the cerebral hemispheres: a computational approach', *Psychological Review*, 1987, 94(2), 148–75

Kosslyn SM, Chabris CF, Marsolek CJ et al, 'Categorical versus coordinate spatial relations: computational analyses and computer simulations', *Journal of Experimental Psychology: Human Perception and Performance*, 1992, 18(2), 562–77

Kosslyn SM, Koenig O, Barrett A et al, 'Evidence for two types of spatial representations: hemispheric specialization for categorical and coordinate relations', *Journal of Experimental Psychology: Human Perception and Performance*, 1989, 15(4), 723–35

Kosslyn SM, Thompson WL, Costantini-Ferrando MF et al, 'Hypnotic visual illusion alters color processing in the brain', *American Journal of Psychiatry*, 2000, 157(8), 1279–84

Kounios J & Beeman M, 'The cognitive neuroscience of insight', *Annual Review of Psychology*, 2014, 65, 71–93

Kounios J, Fleck JI, Green DL et al, 'The origins of insight in resting-state brain activity', *Neuropsychologia*, 2008, 46(1), 281–91

Kounios J, Frymiare JL, Bowden EM et al, 'The prepared mind: neural activity prior to problem presentation predicts subsequent solution by sudden insight', *Psychological Science*, 2006, 17(10), 882–91

Koutstaal W, Wagner AD, Rotte M et al, 'Perceptual specificity in visual object priming: functional magnetic resonance imaging evidence for a laterality difference in fusiform cortex', *Neuropsychologia*, 2001, 39(2), 184–99

Kovács I, 'Human development of perceptual organization', *Vision Research*, 2000, 40(10–12), 1301–10

Kovács-Bálint Z, Bereczkei T & Hernádi I, 'The telltale face: possible mechanisms behind defector and cooperator recognition revealed by emotional facial expression metrics', *British Journal of Psychology*, 2013, 104(4), 563–76

Kowatari Y, Lee SH, Yamamura H et al, 'Neural networks involved in artistic creativity', *Human Brain Mapping*, 2009, 30(5), 1678–90

Kramer JH, Ellenberg L, Leonard J et al, 'Developmental sex differences in global-local perceptual bias', *Neuropsychology*, 1996, 10(3), 402–7

Krause N, 'Meaning in life and mortality', *The Journals of Gerontology: Series B, Psychological Sciences and Social Sciences*, 2009, 64B(4), 517–27

Kravariti E, Toulopoulou T, Mapua-Filbey F et al, 'Intellectual asymmetry and genetic liability in first-degree relatives of probands with schizophrenia', *British Journal of Psychiatry*, 2006, 188(2), 186–7

Kretschmer E, *Physique and Character: An Investigation of the Nature of Constitution and of the Theory of Temperament*, trans WJH Sprott, Kegan Paul, Trench, Trubner & Co, 2nd edn, 1925 [*Körperbau und Charakter* 1921]

Krikorian AD & Steward FC, 'Water and solutes in plant nutrition: with special reference to van Helmont and Nicholas of Cusa', *BioScience*, 1968, 18(4), 286–92

Krippner SC & Combs A, 'The neurophenomenology of shamanism: an essay review', *Journal of Consciousness Studies*, 2002, 9(3), 77–82

Kroger JK, Nystrom LE, Cohen JD et al, 'Distinct neural substrates for deductive and mathematical processing', *Brain Research*, 2008, 1243, 86–103

Krug R, Mölle M, Dodt C et al, 'Acute influences of estrogen and testosterone on divergent and convergent thinking in postmenopausal women', *Neuropsychopharmacology*, 2003, 28(8), 1538–45

Kruger J & Dunning D, 'Unskilled and unaware of it: how difficulties in recognizing one's own incompetence lead to inflated self-assessments', *Journal of Personality and Social Psychology*, 1999, 77(6), 1121–34

Kruglanski AW & Gigerenzer G, 'Intuitive and deliberate judgments are based on common principles', *Psychological Review*, 2011, 118(1), 97–109

Krumbholz K, Eickhoff SB & Fink GR, 'Feature- and object-based attentional modulation in the human auditory "where" pathway', *Journal of Cognitive Neuroscience*, 2007, 19(10), 1721–33

Krumbholz K, Schonwiesner M, Rubsamen R et al, 'Hierarchical processing of sound location and motion in the human brainstem and planum temporale', *European Journal of Neuroscience*, 2005, 21(1), 230–38

Krutetskii VA, *The Psychology of Mathematical Abilities in Schoolchildren*, University of Chicago Press, 1976

Krystal JH, Karper LP, Seibyl JP et al, 'Subanesthetic effects of the noncompetitive NMDA antagonist, ketamine, in humans: psychotomimetic, perceptual, cognitive, and neuroendocrine responses', *Archives of General Psychiatry*, 1994, 51(3), 199–214

Kubicki M, McCarley RW & Shenton ME, 'Evidence for white matter abnormalities in schizophrenia', *Current Opinion in Psychiatry*, 2005, 18(2), 121–34

Kubicki M, Westin C-F, Maier SE et al, 'Uncinate fasciculus findings in schizophrenia: a magnetic resonance diffusion tensor imaging study', *American Journal of Psychiatry*, 2002, 159(5), 813–20

Kubo K, Okanoya K & Kawai N, 'Apology isn't good enough: an apology suppresses an approach motivation but not the physiological and psychological anger', *PLoS One*, 2012, 7(3), e33006

Kuck H, Grossbach M, Bangert M et al, 'Brain processing of meter and rhythm in music: electrophysiological evidence of a common network', *Annals of the New York Academy of Sciences*, 2003, 999(1), 244–53

Kudlur SNC, George S & Jaimon M, 'An overview of the neurological correlates of Cotard syndrome', *European Journal of Psychiatry*, 2007, 21(2), 99–116

Kuehni RG, 'How many object colors can we distinguish?', *Color Research & Application*, 2016, 41(5), 439–44

Kuhl J & Kazén M, 'Motivation, affect, and hemispheric asymmetry: power versus affiliation', *Journal of Personality and Social Psychology*, 2008, 95(2), 456–69

Kuhn A, *Lovis Corinth*, Propyläen Verlag, 1925

Kuhn R, 'Daseinsanalytische Studie über die Bedeutung von Grenzen im Wahn', *Monatsschrift für Psychiatrie und Neurologie*, 1952, 124(4–6), 354–83

Kuhn T, *The Structure of Scientific Revolutions*, University of Chicago Press, 1962

Kulisevsky J, Berthier ML & Pujol J, 'Hemiballismus and secondary mania following a right thalamic infarction', *Neurology*, 1993, 43(7), 1422–4

Kumar S, Sedley W, Barnes GR et al, 'A brain basis for musical hallucinations', *Cortex*, 2014, 52, 86–97

Kumar V & Takahashi JS, 'PARP around the clock', *Cell*, 2010, 142(6), 841–3

Kumral E & Öztürk Ö, 'Delusional state following acute stroke', *Neurology*, 2004, 62(1), 110–3

Kumral E, Uluakay A & Dönmez İ, 'Charles Bonnet syndrome in a patient with right medial occipital lobe infarction: epileptic or deafferentation phenomenon?', *The Neurologist*, 2015, 20(1), 13–15

Kunda Z & Thagard P, 'Forming impressions from stereotypes, traits, and behaviors: a parallel-constraint-satisfaction theory', *Psychological Review*, 1996, 103, 284–308

Kundel HL & Nodine CF, 'A visual concept shapes image perception', *Radiology*, 1983, 146(2), 363–8

Kundel HL, Nodine CF, Krupinski EA et al, 'Using gaze-tracking data and mixture distribution analysis to support a holistic model for the detection of cancers on mammograms', *Academic Radiology*, 2008, 15(7), 881–6

Kunkel TA, 'DNA replication fidelity', *Journal of Biological Chemistry*, 2004, 279(17), 16895–8

Kuo FE & Sullivan WC, 'Aggression and violence in the inner city: impacts of environment via mental fatigue', *Environment and Behavior*, 2001, 33(4), 543–71

Kuo YT, Chiu NC, Shen EY et al, 'Cerebral perfusion in children with Alice in Wonderland syndrome', *Pediatric Neurology*, 1998, 19(2), 105–8

Kuperberg GR, Lakshmanan BM, Caplan DN et al, 'Making sense of discourse: an fMRI study of causal inferencing across sentences', *NeuroImage*, 2006, 33(1), 343–61

Kuperberg GR, McGuire PK & David AS, 'Reduced sensitivity to context in schizophrenic thought disorder: evidence from online monitoring of words in linguistically anomalous sentences', *Journal of Abnormal Psychology*, 1998, 107(3), 423–34

Kuppens T & Spears R, 'You don't have to be well-educated to be an aversive racist, but it helps', *Social Science Research*, 2014, 45, 211–23

Kurakin A, 'The self-organizing fractal theory as a universal discovery method: the phenomenon of life', *Theoretical Biology and Medical Modelling*, 2011, 8, article 4

Kurata A, Miyasaka Y, Yoshida T et al, 'Venous ischemia caused by dural arteriovenous malformation: case report', *Journal of Neurosurgery*, 1994, 80(3), 552–5

Kurlansky M, *Salt: A World History*, Penguin, 2003

Kürnberger F, *Amerika-Müde, amerikanisches Kulturbild*, von Meidinger, 1855

Kurth F, Zilles K, Fox PT et al, 'A link between the systems: functional differentiation and integration within the human insula revealed by meta-analysis', *Brain Structure and Function*, 2010, 214(5–6), 519–34

Kyaga S, Landén M, Boman M et al, 'Mental illness, suicide and creativity: 40-year prospective total population study', *Journal of Psychiatric Research*, 2013, 47(1), 83–90

Kyaga S, Lichtenstein P, Boman M et al, 'Creativity and mental disorder: family study of 300,000 people with severe mental disorder', *British Journal of Psychiatry*, 2011, 199(5), 373–9

Labate A, Cerasa A, Mumoli L et al, 'Neuroanatomical differences among epileptic and non-epileptic déjà-vu', *Cortex*, 2015, 64, 1–7

Labounsky A, *Jean Langlais: The Man and His Music*, Amadeus Press, 2000

Lacey S, Hagtvedt H, Patrick VM et al, 'Art for reward's sake: visual art recruits the ventral striatum', *NeuroImage*, 2011, 55(1), 420–33

Lachower F & Tishby I (eds), *Wisdom of the Zohar: an Anthology of Texts*, trans D Goldstein, Liverpool University Press, 3 vols, 1991

Lacroix D, Chaput Y, Rodriguez JP et al, 'Quantified EEG changes associated with a positive clinical response to clozapine in schizophrenia', *Progress in Neuro-Psychopharmacology & Biological Psychiatry*, 1995, 19(5), 861–76

Laeng B, 'Lateralization of categorical and coordinate spatial functions: a study of unilateral stroke patients', *Journal of Cognitive Neuroscience*, 1994, 6(3), 189–203

Laeng B, Shah J & Kosslyn S, 'Identifying objects in conventional and contorted poses: contributions of hemisphere-specific mechanisms', *Cognition*, 1999, 70(1), 53–85

Laeng B, Zarrinpar A & Kosslyn SM, 'Do separate processes identify objects as exemplars versus members of basic-level categories?

Evidence from hemispheric specialization', *Brain and Cognition*, 2003, 53(1), 15–27

Laforgue R & Allendy R, *La Psychanalyse et les Névroses*, Payot, 1924

Lag T, Hveem K, Ruud KPE et al, 'The visual basis of category effects in object identification: evidence from the visual hemifield paradigm', *Brain and Cognition*, 2006, 60(1), 1–10

Lahat E, Berkovitch M, Barr J et al, 'Abnormal visual evoked potentials in children with "Alice in Wonderland" syndrome due to infectious mononucleosis', *Journal of Child Neurology*, 1999, 14(11), 732–5

Lai VT, van Dam W, Conant LL et al, 'Familiarity differentially affects right hemisphere contributions to processing metaphors and literals', *Frontiers in Human Neuroscience*, 2015, 9, 44

Laing RD, 'Minkowski and schizophrenia', *Review of Existential Psychology and Psychiatry*, 1963, 3(3), 195–207

——, *The Divided Self*, Penguin, 1990 [1959]

Lakatos I, *Proofs and Refutations*, Cambridge University Press, 1976

Lakoff G & Johnson M, *Metaphors We Live By*, University of Chicago Press, 1980

Lakoff G & Johnson M, *Philosophy in the Flesh: The Embodied Mind and Its Challenge to Western Thought*, Basic Books, 1999

Lalanne L, van Assche M & Giersch A, 'When predictive mechanisms go wrong: disordered visual synchrony thresholds in schizophrenia', *Schizophrenia Bulletin*, 2012, 38(3), 506–13

Lalanne L, van Assche M, Wang W et al, 'Looking forward: an impaired ability in patients with schizophrenia?', *Neuropsychologia*, 2012, 50(12), 2736–44

Lamantia AS & Rakic P, 'Cytological and quantitative characteristics of four cerebral commissures in the rhesus monkey', *Journal of Comparative Neurology*, 1990, 291(4), 520–37

Lamarche B & Couture P, 'It is time to revisit current dietary recommendations for saturated fat', *Applied Physiology, Nutrition, and Metabolism*, 2014, 39(12), 1409–11

Lamb MR, Robertson LC & Knight RT, 'Component mechanisms underlying the processing of hierarchically organized pattern: inferences from patients with unilateral cortical lesions', *Journal of Experimental Psychology: Learning, Memory, and Cognition*, 1990, 16(3), 471–83

Lambert N, Chen Y-N, Cheng Y-C et al, 'Quantum biology', *Nature Physics*, 2013, 9, 10–18

Lamendella J, 'General principles of neurofunctional organization and their manifestation in primary and secondary language acquisition', *Language Learning*, 1977, 27(1), 155–96

Lampl Y, Lorberboym M, Gilad R et al, 'Auditory hallucinations in acute stroke', *Behavioural Neurology*, 2005, 16(4), 211–6

Lance JW, 'Simple formed hallucinations confined to the area of a specific visual field defect', *Brain*, 1976, 99(4), 719–34

Lance JW, Cooper B & Misbach J, 'Visual hallucinations as a symptom of right parieto-occipital lesions', *Proceedings of the Australian Association of Neurologists*, 1974, 11, 209–17

Landau B, Gleitman H & Spelke E, 'Spatial knowledge and geometric representation in a child blind from birth', *Science*, 1981, 213(4513), 1275–8

Landis T, Assal G & Perret E, 'Opposite cerebral hemispheric superiorities for visual associative processing of emotional facial expressions and objects', *Nature*, 1979, 278(5706), 739–40

Landis T, Cummings JL, Benson DF et al, 'Loss of topographic familiarity: an environmental agnosia', *Archives of Neurology*, 1986, 43(2), 132–6

Landis T, Cummings JL, Christen L et al, 'Are unilateral right posterior cerebral lesions sufficient to cause prosopagnosia? clinical and radiological findings in six additional patients', *Cortex*, 1986, 22(2), 243–52

Landis T & Regard M, 'The right hemisphere's access to lexical meaning: a function of its release from left-hemisphere control?', in C Chiarello (ed), *Right Hemisphere Contributions to Lexical Semantics*, Springer-Verlag, 1988, 33–46

Landtblom AM, Dige N, Schwerdt K et al, 'A case of Kleine-Levin syndrome examined with SPECT and neuropsychological testing', *Acta Neurologica Scandinavica*, 2002, 105(4), 318–21

Lane RD, Novelly R, Cornell C et al, 'Asymmetric hemispheric control of heart rate', *Psychophysiology*, 1988, 25(4), 464

Lane SM & Schooler JW, 'Skimming the surface: verbal overshadowing of analogical retrieval', *Psychological Science*, 2004, 15(11), 715–9

Lang CJ, Kneidl O, Hielscher-Fastabend M et al, 'Voice recognition in aphasic and non-aphasic stroke patients', *Journal of Neurology*, 2009, 256(8), 1303–6

Lang S, Kanngieser N, Jaśkowski P et al, 'Precursors of insight in event-related brain potentials', *Journal of Cognitive Neuroscience*, 2006, 18(12), 2152–66

Langan J, Peltier SJ, Bo J et al, 'Functional implications of age differences in motor system connectivity', *Frontiers in Systems Neuroscience*, 2010, 4, 1–11

Langdon D & Warrington EK, 'The role of the left hemisphere in verbal and spatial reasoning tasks', *Cortex*, 2000, 36(5), 691–702

Langdon R, Coltheart M, Ward PB et al, 'Disturbed communication in schizophrenia: the role of poor pragmatics and poor mind-reading', *Psychological Medicine*, 2002, 32(7), 1273–84

Langdon R, McKay R & Coltheart M, 'The cognitive neuropsychological understanding of persecutory delusions', in D Freeman, P Garety & R Bentall (eds), *Persecutory Delusions: Assessment, Theory and Treatment*, Oxford University Press, 2008, 221–36

Lange J, 'Agnosien und Apraxien', in O Bumke & O Förster (eds), *Handbuch der Neurologie*, vol 6, Springer, 1936, 807–960

———, 'Agnosia and apraxia', in JW Brown (ed), and G Dean et al (trans), *Agnosia and Apraxia: Selected Papers of Liepmann, Lange, and Pötzl*, Lawrence Erlbaum, 1988, 43–226

Lange-Eichbaum W & Paul ME, *The Problem of Genius*, trans E & C Paul, Kegan Paul & Co, 1931 [*Das Genie-Problem: Eine Einführung* 1931]

Langer S, *Philosophy in a New Key: A Study in the Symbolism of Reason, Rite, and Art*, Harvard University Press, 1942

———, *Feeling and Form: A Theory of Art*, Scribner, 1953

Langevin R, Ben-Aron MH, Coulthard R et al, 'Sexual aggression: constructing a predictive equation – a controlled pilot study', in R Langevin (ed), *Erotic Preference, Gender Identity, and Aggression in Men: New Research Studies*, Lawrence Erlbaum, 1985

Langford CH, 'The notion of analysis in Moore's philosophy', in PA Schilpp (ed), *The Philosophy of GE Moore*, Northwestern University Press, 1942

Langlois JH, Kalakanis L, Rubenstein AJ et al, 'Maxims or myths of beauty? A meta-analytic and theoretical review', *Psychological Bulletin*, 2000, 126(3), 390–423

Langlois JH, Ritter JM, Roggman LA et al, 'Facial diversity and infant preference for attractive faces', *Developmental Psychology*, 1991, 27(1), 79–84

Langlois JH, Roggman LA, Casey RJ et al, 'Infant preferences for attractive faces: rudiments of a stereotype', *Developmental Psychology*, 1987, 23(3), 363–9

Lansdell H, 'The effects of neurosurgery on a test of proverbs', *The American Psychologist*, 1961, 16(7), 448

———, 'Verbal and nonverbal factors in right-hemisphere speech: relation to early neurological history', *Journal of Comparative and Physiological Psychology*, 1969, 69(4, pt 1), 734–8

———, 'Relation of extent of temporal removals to closure and visuomotor factors', *Perceptual & Motor Skills*, 1970, 31(2), 491–8

LaPlante E, *Seized*, HarperCollins, 1993

Larson EB & Brown WS, 'Bilateral field interactions, hemispheric specialisation and evoked potential interhemispheric transmission time', *Neuropsychologia*, 1997, 35(5), 573–81

Lassonde M, Sauerwein HC & Lepore F, 'Extent and limits of callosal plasticity: presence of disconnection symptoms in callosal agenesis', *Neuropsychologia*, 1995, 33(8), 989–1007

Lattner S, Meyer ME & Friederici AD, 'Voice perception: sex, pitch, and the right hemisphere', *Human Brain Mapping*, 2005, 24(1), 11–20

Latty T & Beekman M, 'Irrational decision-making in an amoeboid organism: transitivity and context-dependent preferences', *Proceedings of the Royal Society B: Biological Sciences*, 2011a, 278(1703), 307–12

———, 'Speed-accuracy trade-offs during foraging decisions in the acellular slime mould *Physarum polycephalum*', *Proceedings of the Royal Society B: Biological Sciences*, 2011b, 278(1705), 539–45

Laughlin PR & Ellis AL, 'Demonstrability and social combination processes on mathematical intellective tasks', *Journal of Experimental Social Psychology*, 1986, 22(3), 177–89

Laughlin PR, Hatch EC, Silver JS et al, 'Groups perform better than the best individuals on letters-to-numbers problems: effects of group size', *Journal of Personality and Social Psychology*, 2006, 90(4), 644–51

Laurent JP, Denhières G, Passerieux C et al, 'On understanding idiomatic language: the salience hypothesis assessed by ERPs', *Brain Research*, 2006, 1068(1), 151–60

Laures-Gore JS & Defife LC, 'Perceived stress and depression in left and right hemisphere post-stroke patients', *Neuropsychological Rehabilitation*, 2013, 23(6), 783–97

Laurian S, Gaillard J-M, Le PK et al, 'Topographic aspects of EEG profile of some psychotropic drugs', in C Perris, D Kemali & M Koukkou-Lehmann (eds), *Neurophysiological Correlates of Normal Cognition and Psychopathology: Advances in Biological Psychiatry* [series], vol 13, Karger, Basel, 1983, 165–71

Lavell JC, '"The power of naming": co-option in fine art practice', 2011, unpublished PhD thesis, University of Northumbria

Lavelle L, *L'Erreur de Narcisse*, Floch, 1939

———, *The Dilemma of Narcisse*, tr WT Gairdner, George Allen & Unwin, 1973

Lavoie S, Bartholomeuz CF, Nelson B et al, 'Sulcogyral pattern and sulcal count of the orbitofrontal cortex in individuals at ultra high risk for psychosis', *Schizophrenia Research*, 2014, 154(1–3), 93–9

Lavric A, Forstmeier S & Rippon G, 'Differences in working memory involvement in analytical and creative tasks: an ERP study', *NeuroReport*, 2000, 11(8), 1613–8

Lawlor L & Moulard-Leonard V, 'Henri Bergson', *Stanford Encyclopedia of Philosophy*, 2004

Lawrence DH, 'Sex versus loveliness', *Selected Essays*, Penguin, 1950, 13–18

———, 'The real thing', in ED McDonald (ed), *Phoenix: The Posthumous Papers of DH Lawrence*, Heinemann, 1961 [1936], 196–203

———, *The Lost Girl*, Penguin, 1980 [1920]

———, 'The painted tombs of Tarquinia: II', in S de Filippis (ed), *Sketches of Etruscan Places and Other Essays*, Cambridge University Press, 1992 [1932]

———, 'Chaos in Poetry' (1928), introduction to Harry Crosby's, *Chariot of the Sun*; in NH Reeve & J Worthen (eds), *The Works of DH Lawrence: Introductions and Reviews*, Cambridge University Press, 2005, 107–16

Lawrence PA, 'The mismeasurement of science', *Current Biology*, 2007, 17(15), 583–85

Lawton CA, Charleston SI & Zieles AS, 'Individual- and gender-related differences in indoor wayfinding', *Environment and Behavior*, 1996, 28(2), 204–19

Lazar SW, Kerr CE, Wasserman RH et al, 'Meditation experience is associated with increased cortical thickness', *NeuroReport*, 2005, 16(17), 1893–7

Lazazzera BA, 'Lessons from DNA microarray analysis: the gene expression profile of biofilms', *Current Opinion in Microbiology*, 2005, 8(2), 222–7

Lazure CL & Persinger MA, 'Right hemisphericity and low self-esteem in high school students: a replication', *Perceptual & Motor Skills*, 1992, 75(3, pt 2), 1058

Lazzarini A, 'The case of Shebalin: comparative analysis of musical scores composed before and after stroke', paper presented at the 17th Annual Meeting of the International Society for the History of the Neurosciences (ISHN), 19 June 2012

Le Bourg E, 'Hormesis, aging and longevity', *Biochimica et Biophysica Acta*, 2009, 1790(10), 1030–9

Le Corbusier [Charles-Édouard Jeanneret], 'Le chemin des ânes, le chemin des hommes' [The path for donkeys, the path for men], *L'Esprit Nouveau*, 1922, 17, 33–6

Le Guen O & Balam LI, 'No metaphorical timeline in gesture and cognition among Yucatec Mayas', *Frontiers in Psychology*, 2012, 3, article 271

Le Guin UK, 'The ones who walk away from Omelas', *New Dimensions*, 3, 1973

Lea RB, 'On-line evidence for elaborative logical inferences in text', *Journal of Experimental Psychology: Learning, Memory, and Cognition*, 1995, 21(6), 1469–82

Lea RB, O'Brien DP, Fisch SM et al, 'Predicting propositional logic inferences in text comprehension', *Journal of Memory and Language*, 1990, 29(3), 361–87

Leach S & Weick M, 'Can people judge the veracity of their intuitions?', *Social Psychological and Personality Science*, 2018, 9(1), 40–9

Leader D, *What is Madness?*, Penguin, 2012

Leavens DA & Hopkins WD, 'The whole-hand point: the structure and function of pointing from a comparative perspective', *Journal of Comparative Psychology*, 1999, 113(4), 417–25

Lebel C, Gee M, Camicioli R et al, 'Diffusion tensor imaging of white matter tract evolution over the lifespan', *NeuroImage*, 2012, 60(1), 340–52

Lechevalier B, 'Perception of musical sounds: contributions of positron emission tomography', *Bulletin of the Academy of National Medicine*, 1997, 181(6), 1191–9

Leclerc I, *The Nature of Physical Existence*, George Allen & Unwin, 1972

Leclercq M, 'Theoretical aspects of the main components and functions of attention', in M Leclercq & P Zimmerman (eds), *Applied Neuropsychology of Attention*, Psychology Press, 2002, 3–55

Lee GP, Strauss E, Loring DW et al, 'Sensitivity of figural fluency on the five-point test to focal neurological dysfunction', *The Clinical Neuropsychologist*, 1997, 11(1), 59–68

Lee J, Chung D, Chang S et al, 'Gender differences revealed in the right posterior temporal areas during Navon letter identification tasks', *Brain Imaging and Behavior*, 2012, 6(3), 387–96

Lee K-H, Bhaker RS, Mysore A et al, 'Time perception and its neuropsychological correlates in patients with schizophrenia and in healthy volunteers', *Journal of Experimental Child Psychology*, 2009, 166(2–3), 174–83

Lee K, Shinbo M, Kanai H et al, 'Reduplicative paramnesia after a right frontal lesion', *Cognitive and Behavioral Neurology*, 2011, 24(1), 35–9

Lee M, Martin GE, Hogan A et al, 'What's the story? A computational analysis of narrative competence in autism', *Autism*, 2018, 22(3), 335–44

Lee S, Kim DY, Kim JS et al, 'Visual hallucinations following a left-sided unilateral tuberothalamic artery infarction', *Innovations in Clinical Neuroscience*, 2011, 8(5), 31–4

Lee SM, Gao T & McCarthy G, 'Attributing intentions to random motion engages the posterior superior temporal sulcus', *Social Cognitive and Affective Neuroscience*, 2014, 9(1), 81–7

Lee SS & Dapretto M, 'Metaphorical vs literal word meanings: fMRI evidence against a selective role of the right hemisphere', *NeuroImage*, 2006, 29(2), 536–44

Lee TM, Liu HL, Hoosain R et al, 'Gender differences in neural correlates of recognition of happy and sad faces in humans assessed by functional magnetic resonance imaging', *Neuroscience Letters*, 2002, 333(1), 13–16

Lehman Blake M, 'Affective language and humor appreciation after right hemisphere brain damage', *Seminars in Speech and Language*, 2003, 24(2), 107–19

Lehmann D, Faber PL, Achermann P et al, 'Brain sources of EEG gamma frequency during volitionally meditation-induced, altered states of consciousness, and experience of the self', *Journal of Experimental Child Psychology: Neuroimaging*, 2001, 108(2), 111–21

Lehmann G, Bremond J, Rabaud C et al, 'Space-occupying lesions of the occipital lobe of the cerebral cortex', *Neurochirurgie*, 1975, 21(1), 55–79

Leibniz GW, *La monadologie*, 1714

———, in *Die philosophischen Schriften von Gottfried Wilhelm Leibniz*, ed CI Gerhardt, vol 3, Weidmannsche Buchhandlung, 1887

———, *De quadratura arithmetica circuli ellipseos et hyperbolæ cujus corollarium est trigonometria sine tabulis*, ed E Knobloch, Vandenhoeck & Ruprecht, 1993

———, *New Essays on Human Understanding*, trans & ed P Remnant & J Bennett, Cambridge University Press, 1996 [publ posth 1764]

———, *De arte characteristica ad perficiendas scientias ratione nitentes*, in *Sämtliche Schriften und Briefe*, series 6, vol 4, part A, Akademie Verlag, 1999, 909–15

Leinonen E, Tuunainen A & Lepola U, 'Postoperative psychoses in epileptic patients after temporal lobectomy', *Acta Neurologica Scandinavica*, 1994, 90(6), 394–9

Leiserson WM, Bonini NM & Benzer S, 'Transvection at the eyes absent gene of Drosophila', *Genetics*, 1994, 138(4), 1171–9

LeMay M, 'Morphological aspects of human brain asymmetry: an evolutionary perspective', *Trends in Neurosciences*, 1982, 5(8), 273–5

Lemos GB, Borish V, Cole GD et al, 'Quantum imaging with undetected photons', *Nature*, 2014, 512(7515), 409–12

Lempert H & Kinsbourne M, 'Effects of laterality of orientation on verbal memory', *Neuropsychologia*, 1982, 20(2), 211–4

Lengger P, Fischmeister FP, Leder H et al, 'Functional neuroanatomy of the perception of modern art: a DC-EEG study on the influence of stylistic information on aesthetic experience', *Brain Research*, 2007, 1158, 93–102

Lennox JC, *God and Stephen Hawking: Whose Design Is It Anyway?*, Lion Hudson, 2011

Leo PD & Greene AJ, 'Is awareness necessary for true inference?', *Memory & Cognition*, 2008, 36(6), 1079–86

Leonhard D & Brugger P, 'Creative, paranormal, and delusional thought: a consequence of right hemisphere semantic activation?', *Neuropsychiatry, Neuropsychology & Behavioral Neurology*, 1998, 11(4), 177–83

Lepper MP, Greene D & Nisbett RE, 'Undermining children's intrinsic interest with extrinsic reward: a test of the "overjustification" hypothesis', *Journal of Personality and Social Psychology*, 1973, 28(1), 129–37

Lerouge D, 'Evaluating the benefits of distraction on product evaluations: the mind-set effect', *Journal of Consumer Research*, 2009, 36(3), 367–79

Lessing GE, *Anti-Goeze: Eine Duplik* [1778], in *Werke*, ed HG Göpfert, Hanser Verlag, 1979, vol 8, 32–3; trans HB Garland, in *Lessing: the Founder of Modern German Literature*, Bowes & Bowes, 1937

Lester D, 'Note on Mohave theory of suicide', *Cross-Cultural Research*, 1997, 31(3), 268–72

Letzner S, Güntürkün O & Beste C, 'How birds outperform humans in multi-component behavior', *Current Biology*, 2017, 27(18), R996-8

Leung M, Chan C, Yin J et al, 'Increased gray matter volume in the right angular and posterior parahippocampal gyri in loving-kindness meditators', *Social Cognitive and Affective Neuroscience*, 2013, 8(1), 34–9

Leutmezer F, Podreka I, Asenbaum S et al, 'Postictal psychosis in temporal lobe epilepsy', *Epilepsia*, 2003, 44(4), 582–90

Lévi-Strauss C, *Tristes Tropiques*, Plon, 1955

———, *Mythologiques, 1: Le cru et le cuit*, Plon, 1964

———, *Myth and Meaning* (The 1977 Massey Lectures), Routledge & Kegan Paul, 2001 [1978]

Levin DM, *The Philosopher's Gaze: Modernity in the Shadows of the Enlightenment*, University of California Press, 1999

Levin M, 'Morphogenetic fields in embryogenesis, regeneration, and cancer: non-local control of complex patterning', *Biosystems*, 2012, 109(3), 243–61

Levine B, Black SE, Cabeza R et al, 'Episodic memory and the self in a case of isolated retrograde amnesia', *Brain*, 1998, 121(10), 1951–73

Levine DN & Finklestein S, 'Delayed psychosis after right temporoparietal stroke or trauma: relation to epilepsy', *Neurology*, 1982, 32(3), 267–73

Levine DN & Grek A, 'The anatomic basis of delusions after right cerebral infarction', *Neurology*, 1984, 34(5), 577–82

Levine GM, Halberstadt JB & Goldstone RL, 'Reasoning and the weighting of attributes in attitude judgments', *Journal of Personality and Social Psychology*, 1996, 70(2), 230–40

Levine J, Toder D, Geller V et al, 'Beneficial effects of caloric vestibular stimulation on denial of illness and manic delusions in schizoaffective disorder: a case report', *Brain Stimulation*, 2012, 5(3), 267–73

Levins R & Lewontin R, *The Dialectical Biologist*, Harvard University Press, 1985

Levitan DR, 'Do sperm really compete and do eggs ever have a choice? Adult distribution and gamete mixing influence sexual selection, sexual conflict, and the evolution of gamete recognition proteins in the sea', *The American Naturalist*, 2017, 191(1), 88–105

Levitt MD, Li R, DeMaster EG et al, 'Use of measurements of ethanol absorption from stomach and intestine to assess human ethanol metabolism', *American Journal of Physiology: Gastrointestinal and Liver Physiology*, 1997, 3(4), G951–7

Lévy ED, Landry CR & Michnick SW, 'Signaling through cooperation', *Science*, 2010, 328(5981), 983–4

Levy J, 'Information processing and higher psychological functions in the disconnected hemispheres of human commissurotomy patients', unpublished thesis, California Institute of Technology, 1969a

——, 'Possible basis for the evolution of lateral specialisation of the human brain', *Nature*, 1969b, 224(5219), 614–5

——, 'Psychobiological implications of bilateral asymmetry', in SJ Dimond & JG Beaumont (eds), *Hemisphere Function in the Human Brain*, Elek Science, 1974, 121–83

——, 'Interhemispheric collaboration: single mindedness in the asymmetrical brain', in CT Best (ed), *Hemispheric Function and Collaboration in the Child*, Academic Press, 1985, 11–32

Levy J, Heller W, Banich MT et al, 'Are variations among right-handed individuals in perceptual asymmetries caused by characteristic arousal differences between hemispheres?', *Journal of Experimental Psychology: Human Perception and Performance*, 1983a, 9(3), 329–59

——, 'Asymmetry of perception in free viewing of chimeric faces', *Brain and Cognition*, 1983b, 2(4), 404–19

Levy J & Trevarthen C, 'Color-matching, color-naming and color-memory in split-brain patients', *Neuropsychologia*, 1981, 19(4), 523–41

Levy J, Trevarthen C & Sperry RW, 'Perception of bilateral chimeric figures following hemisphere deconnexion', *Brain*, 1972, 95(1), 61–78

Lewin JS, Friedman L, Wu D et al, 'Cortical localisation of human sustained attention: detection with functional MR using a visual vigilance paradigm', *Journal of Computer Assisted Tomography*, 1996, 20(5), 695–70

Lewin R, 'Is your brain really necessary?', *Science*, 1980, 210(4475), 1232–4

——, 'Why is development so illogical?', *Science*, 1984, 224(4655), 1327–9

Lewis A, 'The experience of time in mental disorder', *Proceedings of the Royal Society of Medicine*, 1932, 25(5), 611–20

Lewis CS, *The Four Loves*, Harcourt, 1988 [1960]

——, 'Myth became fact', from *God in the Dock*, Wm B Eerdmans, 2014 [1970]

——, *The Problem of Pain*, Harper Collins, 2015 [1940]

——, *Surprised by Joy: The Shape of My Early Life*, HarperCollins, 2017 [1955]

Lewis G & Barnes LA, *A Fortunate Universe: Life in a Finely-Tuned Cosmos*, Cambridge University Press, 2016

Lewis GJ, Kanai R, Rees G et al, 'Neural correlates of the "good life": eudaimonic well-being is associated with insular cortex volume', *Social Cognitive and Affective Neuroscience*, 2014, 9(5), 615–8

Lewis PA & Miall RC, 'A right hemispheric prefrontal system for cognitive time measurement', *Behavioural Processes*, 2006, 71, 226–34

Lewis RD, *When Cultures Collide: Leading Across Cultures*, Nicholas Brealey International, 2nd edn 2006

Lewis SW, 'Brain imaging in a case of Capgras syndrome', *British Journal of Psychiatry*, 1987, 150(1), 117–21

Lewontin RC, 'The corpse in the elevator', *New York Review of Books*, 20 January 1983, 34–7

——, foreword to S Oyama (ed), *The Ontogeny of Information*, Duke University Press, 2nd edn, 2000

Lewontin RC, Rose S & Kamin LJ, *Not in Our Genes: Biology, Ideology and Human Nature*, Pantheon, 1984

Ley RG & Bryden MP, 'Hemispheric differences in processing emotions and faces', *Brain and Language*, 1979, 7(1), 127–38

Leys S, 'Chinese shadows: bureaucracy, happiness, history', 1977: www.chinafile.com/chinese-shadows-bureaucracy-happiness-history (accessed 11 March 2021)

Lhermitte F, Chedru F & Chain F, 'À propos d'un cas d'agnosie visuelle', *Revue Neurologique (Paris)*, 1973, 128(5), 301–22

Lhermitte J, *L'image de notre corps*, L'Harmattan, 1998 [1939]

Li R, 'Why women see differently from the way men see? A review of sex differences in cognition and sports', *Journal of Sport and Health Science*, 2014, 3(3), 155–62

Li R & Bowerman B, 'Symmetry breaking in biology', *Cold Spring Harbor Perspectives in Biology*, 2010, 2, a003475

Li Y & Du S, *Chinese Mathematics: A Concise History*, Clarendon Press, 1987

Liao JC, 'Neuromuscular control of trout swimming in a vortex street: implications for energy economy during the Kármán gait', *Journal of Experimental Biology*, 2004, 207(20), 3495–506

Liaw S-B & Shen E-Y, 'Alice in Wonderland syndrome as a presenting symptom of EBV infection', *Pediatric Neurology*, 1991, 7(6), 464–6

Libet B, 'Unconscious cerebral initiative and the role of conscious will in voluntary action', *Behavioral and Brain Sciences*, 1985, 8(4), 529–39

Lichtenberg GC, *Sudelbuch F*, in *Georg Christoph Lichtenberg: Schriften und Briefe*, vol 1, Hanser, 1967–72a [1776–1779]

———, *Sudelbuch K*, in *Georg Christoph Lichtenberg: Schriften und Briefe*, vol 2, Hanser, 1967–72b [1799]

Liederman J, 'Subtraction in addition to addition: dual task performance improves when tasks are presented to separate hemispheres', *Journal of Clinical and Experimental Neuropsychology*, 1986, 8(5), 486–502

Liederman J & Meehan P, 'When is between-hemisphere division of labor advantageous?', *Neuropsychologia*, 1986, 24(6), 863–74

Liepelt R, Von Cramon DY & Brass M, 'How do we infer others' goals from non-stereotypic actions? The outcome of context-sensitive inferential processing in right inferior parietal and posterior temporal cortex', *NeuroImage*, 2008, 43(4), 784–92

Liepmann H, 'Das Krankheitsbild der Apraxie ("motorische Asymbolie") auf Grund eines Falles von einseitiger Apraxie', *Monatsschrift für Psychiatrie und Neurologie*, 1900, 8(3), 182–97

Liester MB, 'Personality changes following heart transplantation: the role of cellular memory', *Medical Hypotheses*, 2020, 135, 109468

Lieth H & Whittaker RH (eds), *Primary Productivity of the Biosphere*, Springer-Verlag, 1975

Likitcharoen Y & Phanthumchinda K, 'Environmental reduplication in a patient with right middle cerebral artery occlusion', *Journal of the Medical Association of Thailand*, 2004, 87(12), 1526–9

Lim CY, Park JY, Kim DY et al, 'Terminal lucidity in the teaching hospital setting', *Death Studies*, 2020, 44(5), 285–91

Lin Z, Chuah A, Mohan T et al, 'Movement disorder as prodrome of schizophrenia', *Australian & New Zealand Journal of Psychiatry*, 2011, 45(10), 904

Lincoln B, *Theorizing Myth: Narrative, Ideology, and Scholarship*, University of Chicago Press, 1999

Lincoln GA, 'Biology of antlers', *Journal of Zoology*, 1992, 226(3), 517–28

Lindell AK, 'Lateral thinkers are not so laterally minded: hemispheric asymmetry, interaction, and creativity', *Laterality*, 2011, 16(4), 479–98

———, 'Continuities in emotion lateralization in human and non-human primates', *Frontiers in Human Neuroscience*, 2013, 7, article 464

Lindsay J & Boyle P, 'The conceptual penis as a social construct', *Cogent Social Sciences*, 2017, 3, article 1330439

Lindsey DT & Brown AM, 'Color naming and the phototoxic effects of sunlight on the eye', *Psychological Science*, 2002, 13(6), 506–12

Linhares JM, Pinto PD & Nascimento SM, 'The number of discernible colors in natural scenes', *Journal of the Optical Society of America, A: Optics, Image Science and Vision*, 2008, 25(12), 2918–24

Linnæus C, 'Glömska af alla Substantiva och i synnerhet namn', *Kungliga Svenska Vetenskaps-Akademiens Handlingar*, 1745, 6, 116–7 (translated in Viets HR, 'Aphasia as described by Linnæus and as painted by Ribera', *Bulletin for the History of Medicine*, 1943, 13, 328–9)

Linscott RJ, Marie D, Arnott KL et al, 'Over-representation of Maori New Zealanders among adolescents in a schizotypy taxon', *Schizophrenia Research*, 2006, 84(2–3), 289–96

Liotti M & Tucker DM, 'Right hemisphere sensitivity to arousal and depression', *Brain and Cognition*, 1992, 18(2), 138–51

——, 'Emotion in asymmetric corticolimbic networks', in RJ Davidson & K Hugdahl (eds), *Human Brain Laterality*, Oxford University Press, 1994, 389–424

Lippman CW, 'Certain hallucinations peculiar to migraine', *Journal of Nervous and Mental Disease*, 1952, 116(4), 346–51

Lipton B, *The Biology of Belief*, Hay House, 2005

Litman LC, 'A case of erotic violence syndrome', *Canadian Journal of Psychiatry*, 2004, 49(3), 217–8

Little AC, Jones BC & DeBruine LM, 'Facial attractiveness: evolutionary based research', *Philosophical Transactions of the Royal Society of London: Series B, Biological Sciences*, 2011, 366(1571), 1638–59

Little AC, Jones BC, DeBruine L et al, 'Accuracy in discrimination of self-reported cooperators using static facial information', *Personality and Individual Differences*, 2013, 54(4), 507–12

Litwin-Kumar A, Harris KD, Axel R et al, 'Optimal degrees of synaptic connectivity', *Neuron*, 2017, 93(5), 1153–64.e7

Liu C, Brattico E, Abu-Jamous B et al, 'Effect of explicit evaluation on neural connectivity related to listening to unfamiliar music', *Frontiers in Human Neuroscience*, 2017, 11, article 611

Liu H, Radisky DC, Yang D et al, 'MYC suppresses cancer metastasis by direct transcriptional silencing of (alpha)v and (beta)3 integrin subunits', *Nature Cell Biology*, 2012, 14(6), 567–74

Liu H, Stufflebeam SM, Sepulcre J et al, 'Evidence from intrinsic activity that asymmetry of the human brain is controlled by multiple factors', *Proceedings of the National Academy of Sciences of the United States of America*, 2009, 106(48), 20499–503

Liu J, 'What is nature? – ziran in early Daoist thinking', *Asian Philosophy*, 2016, 26(3), 265–79

Liu S, Li A, Liu Y et al, 'Polygenic effects of schizophrenia on hippocampal grey matter volume and hippocampus-medial prefrontal cortex functional connectivity', *British Journal of Psychiatry*, 2020, 216(5), 267–74

Livio M, *The Accelerating Universe: Infinite Expansion, the Cosmological Constant, and the Beauty of the Cosmos*, Wiley, 2000

Livni E, 'A debate over plant consciousness is forcing us to confront the limitations of the human mind', *Quartz*, 3 June 2018: qz.com/1294941/a-debate-over-plant-consciousness-is-forcing-us-to-confront-the-limitations-of-the-human-mind/ (accessed 23 February 2021)

Llinás R, *I of the Vortex: From Neurons to Self*, Massachusetts Institute of Technology Press, 2001

Lobry C, Oh P, Mansour MR et al, 'Notch signaling: switching an oncogene to a tumor suppressor', *Blood*, 2014, 123(16), 2451–9

Lock S, *A Difficult Balance: Editorial Peer Review in Medicine*, Nuffield Provincials Hospital Trust, 1985

Locke J, *An Essay Concerning Human Understanding*, W Tegg & Co, 1849 [1690]

Löckenhoff CE, Chan W, McCrae RR et al, 'Gender stereotypes of personality: universal and accurate?', *Journal of Cross-Cultural Psychology*, 2014, 45(5), 675–94

Lockhart P, *A Mathematician's Lament*, Bellevue, 2009

Loehlin JC, 'Dysgenesis and IQ: what evidence is relevant?', *The American Psychologist*, 1997, 52(11), 1236–9

Loetscher T, Regard M & Brugger P, 'Misoplegia: a review of the literature and a case without hemiplegia', *Journal of Neurology, Neurosurgery, and Psychiatry*, 2006, 77(9), 1099–1100

Lohr VI & Pearson-Mims CH, 'Responses to scenes with spreading, rounded, and conical tree forms', *Environment and Behavior*, 2006, 38(5), 667–88

Lombardo M, Ashwin E, Auyeung B et al, 'Fetal testosterone influences sexually dimorphic gray matter in the human brain', *Journal of Neuroscience*, 2012, 32(2), 674–80

Lombardo MV, Chakrabarti B, Bullmore ET et al, 'Specialization of right temporo-parietal junction for mentalizing and its relation to social impairments in autism', *NeuroImage*, 2011, 56(3), 1832–8

Lombroso C, *The Man of Genius*, Walter Scott, 1891

Long JA, Mark-Kurik E, Johanson Z et al, 'Copulation in antiarch placoderms and the origin of gnathostome internal fertilization', *Nature*, 2015, 517(7533), 196–9

Longden K, Ellis C & Iverson SD, 'Hemispheric differences in the discrimination of curvature', *Neuropsychologia*, 1976, 14(2), 195–202

Longo MR, Trippier S, Vagnoni E et al, 'Right hemisphere control of visuospatial attention in near space', *Neuropsychologia*, 2015, 70, 350–7

Lopera F & Ardila A, 'Prosopamnesia and visuolimbic disconnection syndrome: a case study', *Neuropsychology*, 1992, 6(1), 3–12

Lorberbaum JP, Newman JD, Horwitz AR et al, 'A potential role for thalamocingulate circuitry in human maternal behavior', *Biological Psychiatry*, 2002, 51(6), 431–45

Lord CG, Ross L & Lepper MR, 'Biased assimilation and attitude polarization: the effects of prior theories on subsequently considered evidence', *Journal of Personality and Social Psychology*, 1979, 37(11), 2098–109

Lorenz K, *Studies in Animal and Human Behaviour*, Methuen, 2 vols, 1971

Lorenz-Spreen P, Mønsted BM, Hövel P et al, 'Accelerating dynamics of collective attention', *Nature Communications*, 2019, 10, article 1759

Lotman Yu & Nikolaenko N, 'The "golden section" and problems of intracerebral dialogue', *Decorative Arts USSR*, 1983, 9, 31–4 [Lotman Y & Nikolaenko N, '"Zolotoye secheniye" i problemy vnutrimozgovogo dialoga', *Dekorativnoye Iskusstvo SSSR*, 1983, 9, 31–4]

Lötsch J, Reither N, Bogdanov V et al, 'A brain-lesion pattern based algorithm for the diagnosis of posttraumatic olfactory loss', *Rhinology*, 2015, 53(4), 365–70

Lötsch J, Ultsch A, Eckhardt M et al, 'Brain lesion-pattern analysis in patients with olfactory dysfunctions following head trauma', *NeuroImage: Clinical*, 2016, 11, 99–105

Lou HC, Luber B, Crupain M et al, 'Parietal cortex and representation of the mental Self', *Proceedings of the National Academy of Sciences of the United States of America*, 2004, 101(17), 6827–32

Loughry CW, Sheffer DB, Price TE et al, 'Breast volume measurement of 598 women using biostereometric analysis', *Annals of Plastic Surgery*, 1989, 22(5), 380–5

Löwenstein, Prince Hubertus zu, *Towards the Further Shore: An Autobiography*, Victor Gollancz, 1968

Lu LW & Chiang W, 'Emptiness we live by: metaphors and paradoxes in Buddhism's Heart Sutra', *Metaphor and Symbol*, 2007, 22(4), 331–55

Luauté J & Saladini O, 'Neuroimaging correlates of chronic delusional jealousy after right cerebral infarction', *Journal of Neuropsychiatry and Clinical Neurosciences*, 2008, 20(2), 245–7

Lucas JR, *The Future: An Essay on God, Temporality and Truth*, Blackwell, 1990

Lucas P & Sheeran A, 'Asperger's syndrome and the eccentricity and genius of Jeremy Bentham', *Journal of Bentham Studies*, 8, 2006

Luchins AS, 'Mechanization of problem solving: the effect of Einstellung', *Psychological Monographs*, 1942, 54(6), 1–95

Luchins AS & Luchins EH, 'New experimental attempts at preventing mechanization in problem solving', *Journal of General Psychology*, 1950, 42(2), 279–97

Luck SJ, *An Introduction to the Event-Related Potential Technique*, Massachusetts Institute of Technology Press, 2005

Luck SJ, Hillyard SA, Mangun GR et al, 'Independent hemispheric attentional systems mediate visual search in split-brain patients', *Nature*, 1989, 342(6249), 543–5

———, 'Independent attentional scanning in the separated hemispheres of split-brain patients', *Journal of Cognitive Neuroscience*, 1994, 6(1), 84–91

Luckey TD, *Radiation Hormesis*, CRC Press, 1991

Luders E, Clark K, Narr KL et al, 'Enhanced brain connectivity in long-term meditation practitioners', *NeuroImage*, 2011, 57(4), 1308–16

Luders E, Kurth F, Mayer EA et al, 'The unique brain anatomy of meditation practitioners: alterations in cortical gyrification', *Frontiers in Human Neuroscience*, 2012, 6, article 34

Luders E, Toga AW, Lepore N et al, 'The underlying anatomical correlates of long-term meditation: larger hippocampal and frontal volumes of gray matter', *NeuroImage*, 2009, 45(3), 672–8

Ludwig AM, 'Creative achievement and psychopathology – comparison among professions', *American Journal of Psychotherapy*, 1992, 46(3), 330–56

———, 'Mental illness and creative activity in female writers', *American Journal of Psychiatry*, 1994, 151(11), 1650–6

Ludz U (ed), *Letters 1925–1975: Hannah Arendt and Martin Heidegger*, trans A Shields, Harcourt, 2004

Lueken U, Schwarz M, Hertel F et al, 'Impaired performance on the Wisconsin Card Sorting Test under left- when compared to right-sided deep brain stimulation of the subthalamic nucleus in patients with Parkinson's disease', *Journal of Neurology*, 2008, 255(12), 1940–8

Luft CDB, Zioga I, Banissy MJ et al, 'Relaxing learned constraints through cathodal tDCS on the left dorsolateral prefrontal cortex', *Nature Research: Scientific Reports*, 2017, 7, article 2916

Lugli E, 'Watery manes: reversing the stream of thought about quattrocento Italian heads', *Internet Archaeology*, 2016, 42, dx.doi.org/10.11141/ia.42.6.11

Luh KE, 'Line bisection and perceptual asymmetries in normal individuals: what you see is not what you get', *Neuropsychology*, 1995, 9(4), 435–48

Luh KE, Rueckert LM & Levy J, 'Perceptual asymmetries for free viewing of several types of chimeric stimuli', *Brain and Cognition*, 1991, 16(1), 83–103

Lumer ED, Friston KJ & Rees G, 'Neural correlates of perceptual rivalry in the human brain', *Science*, 1998, 280(5371), 1930–4

Lunardi P, Tacconi L, Missori P et al, 'Palinopsia: unusual presenting symptom of a cerebral abscess in a man with Kartagener's syndrome', *Clinical Neurology and Neurosurgery*, 1991, 93(4), 337–9

Lunn V, 'Om legemsbevidstheden, belyst ved Nogle forstyrrelser af den normale oplevelsesmaade', *Ugeskrift vor laeger*, 1948, 110(7), 178–80

——, 'Autoscopic phenomena', *Acta Psychiatrica Scandinavica*, 1970, 46, 118–25

Luo J & Niki K, 'Function of hippocampus in "insight" of problem solving', *Hippocampus*, 2003, 13(3), 316–23

Luo J, Niki K & Knoblich G, 'Perceptual contributions to problem solving: chunk decomposition of Chinese characters', *Brain Research Bulletin*, 2006, 70(4–6), 430–43

Luo Z-X, Ruf I, Schultz JA et al, 'Fossil evidence on evolution of inner ear cochlea in Jurassic mammals', *Proceedings of the Royal Society B: Biological Sciences*, 2011, 278(1702), 28–34

Luria AR, Tsvetkova LS & Futer DS, 'Aphasia in a composer (V. G. Shebalin)', *Journal of the Neurological Sciences*, 1965, 2(3), 288–92

Lustenberger C, Boyle MR, Foulser AA et al, 'Functional role of frontal alpha oscillations in creativity', *Cortex*, 2015, 67, 74–82

Lutz A, Greischar LL, Perlman D et al, 'BOLD signal in insula is differentially related to cardiac function during compassion meditation in experts vs. novices', *NeuroImage*, 2009, 47(3), 1038–1046

Lutz A, Nassehi A, Bao Y et al, 'Neurocognitive processing of body representations in artistic and photographic images', *NeuroImage*, 2013, 66, 288–92

Lux S, Marshall JC, Ritzl A et al, 'A functional magnetic resonance imaging study of local/global processing with stimulus presentation in the peripheral visual hemifields', *Neuroscience*, 2004, 124(1), 113–20

Luzzati F, 'A hypothesis for the evolution of the upper layers of the neocortex through co-option of the olfactory cortex developmental program', *Frontiers in Neuroscience*, 2015, 9, 162

Lykouras L, Typaldou M, Mourtzouchou P et al, 'Neuropsychological relationships in paranoid schizophrenia with and without delusional misidentification syndromes: a comparative study', *Progress in Neuro-Psychopharmacology & Biological Psychiatry*, 2008, 32(6), 1445–8

Lynn R, *Dysgenics: Genetic Deterioration in Modern Populations*, Ulster Institute for Social Research, revised edn, 2011

Lyons V & Fitzgerald M, *Asperger Syndrome – A Gift or a Curse?*, Nova Biomedical, 2006

——, 'Did Hans Asperger (1906–1980) have Asperger syndrome?', *Journal of Autism and Developmental Disorders*, 2007, 37(10), 2020–1

——, 'Atypical sense of self in autism spectrum disorders: a neuro-cognitive perspective', in M Fitzgerald (ed), *Recent Advances in Autism Spectrum Disorders*, vol 1, InTech, 2013, 753–4

Ma X-S, Herbst T, Scheidl T et al, 'Quantum teleportation using active feed-forward between two Canary Islands', *Nature*, 2012, 489(7415), 269–73

Ma X-S, Kofler J, Qarry A et al, 'Quantum erasure with causally disconnected choice', *Proceedings of the National Academy of Sciences of the United States of America*, 2013, 110(4), 1221–6

Ma X-S, Zotter S, Kofler J et al, 'Experimental delayed-choice entanglement swapping', *Nature Physics*, 2012, 8, 479–84

MacCabe JH, Lambe MP, Cnattingius S et al, 'Excellent school performance at age 16 and risk of adult bipolar disorder: national cohort study', *British Journal of Psychiatry*, 2010, 196(2), 109–15

MacCabe JH, Sariaslan A, Almqvist C, 'Artistic creativity and risk for schizophrenia, bipolar disorder and unipolar depression: a Swedish population-based case-control study and sib-pair analysis', *British Journal of Psychiatry*, 2018, 212(6), 370–6

Mace CJ & Trimble MR, 'Psychosis following temporal lobe surgery: a report of six cases', *Journal of Neurology, Neurosurgery, and Psychiatry*, 1991, 54(7), 639–44

MacGregor N, *Living With the Gods*, Penguin Random House, 2018

Machado C, Estévez M & Leisman G, 'QEEG spectral and coherence assessment of autistic children in three different experimental conditions', *Journal of Autism and Developmental Disorders*, 2015, 45, 406–24

MacIntyre A, *Whose Justice? Which Rationality?*, Duckworth, 1988

Mack JE, Meltzer-Asscher A, Barbieri E et al, 'Neural correlates of processing passive sentences', *Brain Sciences*, 2013, 3(3), 1198–214

MacKay DM & MacKay V, 'Explicit dialogue between left and right half-systems of split brains', *Nature*, 1982, 295(5851), 690–1

Mackenzie C, Begg T, Brady M et al, 'The effects on verbal communication skills of right hemisphere stroke in middle age', *Aphasiology*, 1997, 11(10), 929–45

Mackenzie C, Begg T, Lees KR et al, 'The communication effects of right brain damage on the very old and the not so old', *Journal of Neurolinguistics*, 1999, 12(1), 79–93

Mackert A, Flechtner K-M, Woyth C et al, 'Increased blink rates in schizophrenics: influences of neuroleptics and psychopathology', *Schizophrenia Research*, 1991, 4(1), 41–7

Mackintosh FC & Schmidt CF, 'Active cellular materials', *Current Opinion in Cell Biology*, 2010, 22(1), 29–35

Mackworth-Young CG, 'Sequential musical symptoms in a professional musician with presumed encephalitis', *Cortex*, 1983, 19(3), 413–9

Macleod AD, 'Lightening up before death', *Palliative and Supportive Care*, 2009, 7(4), 513–6

Macrae D & Trolle E, 'The defect of function in visual agnosia', *Brain*, 1956, 79(1), 94–110

MacRae N, *John Von Neumann: The Scientific Genius Who Pioneered the Modern Computer, Game Theory, Nuclear Deterrence and Much More*, American Mathematical Society, 1992

Madjar N & Oldham GR, 'Preliminary tasks and creative performance on a subsequent task: effects of time on preliminary tasks and amount of information about the subsequent task', *Creativity Research Journal*, 2002, 14(2), 239–51

Madoz-Gúrpide A & Hillers-Rodríguez R, 'Capgras delusion: a review of aetiological theories', *Revista de Neurologia*, 2010, 50(7), 420–30

Madzharov AV, Block LG & Morrin M, 'The cool scent of power: effects of ambient scent on consumer preferences and choice behavior', *Journal of Marketing*, 2015, 79(1), 83–96

Maeda K, Yamamoto Y, Yasuda M et al, 'Delusions of oral parasitosis', *Progress in Neuropsychopharmacology and Biological Psychiatry*, 1998, 22(1), 243–8

Magee B, *Confessions of a Philosopher: A Journey through Western Philosophy*, Phoenix, 1998a

——, *The Philosophy of Schopenhauer*, Oxford University Press, 1998b

Magee GA, 'The recovery of myth and the *sensus communis*', in GA Magee (ed), *Philosophy and Culture: Essays in Honor of Donald Phillip Verene*, Philosophy Documentation Center, 2002

Magnan V, 'Des hallucinations bilatérales de caractère différent suivant le côté affecté', *Archives de Neurologie*, 1883, 6, 336–55

Maguire EA, Burgess N, Donnett JG et al, 'Knowing where and getting there: a human navigation network', *Science*, 1998, 280(5365), 921–4

Maguire EA, Burgess N & O'Keefe J, 'Human spatial navigation: cognitive maps, sexual dimorphism, and neural substrates', *Current Opinion in Neurobiology*, 1999, 9(2), 171–7

Maguire EA, Frackowiak RS & Frith CD, 'Recalling routes around London: activation of the right hippocampus in taxi drivers', *Journal of Neuroscience*, 1997, 17(18), 7103–10

Maguire EA, Gadian DG, Johnsrude IS et al, 'Navigation-related structural change in the hippocampi of taxi drivers', *Proceedings of the National Academy of Sciences of the United States of America*, 2000, 97(8), 4398–403

Mahood E, *The Primordial Leap and the Present: The Ever-Present Origin – An Overview of the Work of Jean Gebser*, 1996: www.gaiamind.org/Gebser (accessed 2 March 2021)

Mai X, Luo J, Wu J-H et al, '"Aha!" effects in a guessing riddle task: an event related potential study', *Human Brain Mapping*, 2004, 22(4), 261–70

Mai X, Zhang W, Hu X et al, 'Using tDCS to explore the role of the right temporoparietal junction in theory of mind and cognitive empathy', *Frontiers in Psychology*, 2016, 7, article 380

Maidhof C, 'Error monitoring in musicians', *Frontiers in Human Neuroscience*, 2013, 7, article 401

Maier M, Mellers J, Toone B et al, 'Schizophrenia, temporal lobe epilepsy and psychosis: an in vivo magnetic resonance spectroscopy and imaging study of the hippocampus/amygdala complex', *Psychological Medicine*, 2000, 30(3), 571–81

Mailo J & Tang-Wai R, 'Insight into the precuneus: a novel seizure semiology in a child with epilepsy arising from the right posterior precuneus', *Epileptic Disorders*, 2015, 17(3), 321–7

Maimonides, *A Guide for the Perplexed*, trans M Friedländer, Dutton, 2nd edn, 1919

Majima Y, 'Belief in pseudoscience, cognitive style and science literacy', *Applied Cognitive Psychology*, 2015, 29(4), 552–9

Makarov VA, Schmidt KE, Castellanos NP et al, 'Stimulus-dependent interaction between the visual areas 17 and 18 of the 2 hemispheres of the ferret (*Mustela putorius*)', *Cerebral Cortex*, 2008, 18(8), 1951–60

Maki RH, Grandy CA & Hauge G, 'Why is telling right from left more difficult than telling above from below?', *Journal of Experimental Psychology: Human Perception and Performance*, 1979, 5(1), 52–67

Makin ADJ, Rampone G, Wright A et al, 'Visual symmetry in objects and gaps', *Journal of Vision*, 2014, 14(3), article 12

Makinson DC, 'The paradox of the preface', *Analysis*, 1965, 25(6), 205–7

Makris N, Biederman J, Valera EM et al, 'Cortical thinning of the attention and executive function networks in adults with attention-deficit/hyperactivity disorder', Cerebral Cortex, 2007, 17(6), 1364–75

Malaspina D, Bruder G, Furman V et al, 'Schizophrenia subgroups differing in dichotic listening laterality also differ in neurometabolism and symptomatology', Journal of Neuropsychiatry and Clinical Neurosciences, 2000, 12(4), 485–92

Malaspina D, Simon N, Mujica-Parodi L et al, 'Using figure ground perception to examine the unitary and heterogeneity models for psychopathology in schizophrenia', Schizophrenia Research, 2003, 59(2–3), 297–9

Malhotra P, Coulthard EJ & Husain M, 'Role of right posterior parietal cortex in maintaining attention to spatial locations over time', Brain, 2009, 132(3), 645–60

Malinowski P, Hübner R, Keil A et al, 'The influence of response competition on cerebral asymmetries for processing hierarchical stimuli revealed by ERP recordings', Experimental Brain Research, 2002, 144(1), 136–9

Malkoc SA, Zauberman G & Bettman JR, 'Unstuck from the concrete: carryover effects of abstract mindsets in intertemporal preferences', Organizational Behavior and Human Decision Processes, 2010, 113(2), 112–26

Mallet L, Schüpbach M, N'Diaye K et al, 'Stimulation of subterritories of the subthalamic nucleus reveals its role in the integration of the emotional and motor aspects of behavior', Proceedings of the National Academy of Sciences of the United States of America, 2007, 104(25), 10661–6

Mallon EB & Franks NR, 'Ants estimate area using Buffon's needle', Proceedings of the Royal Society of London: Series B, Biological Sciences, 2000, 267, 765–70

Malloy P, Cimino C & Westlake R, 'Differential diagnosis of primary and secondary Capgras delusions', Neuropsychiatry, Neuropsychology & Behavioral Neurology, 1992, 5(2), 83–96

Malloy PF & Richardson ED, 'The frontal lobes and content-specific delusions', Journal of Neuropsychiatry and Clinical Neurosciences, 1994, 6(4), 455–66

Manchanda R, Miller H & Mclachlan RS, 'Post-ictal psychosis after right temporal lobectomy', Journal of Neurology, Neurosurgery, and Psychiatry, 1993, 56(3), 277–9

Mancing H, The Cervantes Encyclopedia, Greenwood, 2 vols, 2004

Mancini M, Presenza S, Di Bernardo L et al, 'The life-world of persons with schizophrenia: a panoramic view', Journal of Psychopathology, 2014, 20(4), 423–34

Mandal MK & Ambady N, 'Laterality of facial expressions of emotion: universal and culture-specific influences', Behavioural Neurology, 2004, 15(1–2), 23–34

Mandell J, Schulze K & Schlaug G, 'Congenital amusia: an auditory-motor feedback disorder?', Restorative Neurology and Neuroscience, 2007, 25(3–4), 323–34

Manes F, Piven J, Vrancic D et al, 'An MRI study of the corpus callosum and cerebellum in mentally retarded autistic individuals', Journal of Neuropsychiatry and Clinical Neurosciences, 1999, 1(4)1, 470–4

Manes F, Sahakian B, Clark L et al, 'Decision-making processes following damage to the prefrontal cortex', Brain, 2002, 125(3), 624–39

Manning AG, Khakimov RI, Dall RG et al, 'Wheeler's delayed-choice gedanken experiment with a single atom', Nature Physics, 2015, 11(7), 539–42

Manning JT & Chamberlain A, 'Left-side cradling and brain lateralization', Ethology and Sociobiology, 1991, 12(3), 237–44

Manoach D, Sandson TA & Weintraub S, 'The developmental social-emotional processing disorder is associated with right hemisphere abnormalities', Neuropsychiatry, Neuropsychology and Behavioral Neurology, 1995, 8(2), 99–105

Manzotti R & Parks T, Dialogues on Consciousness, OR Books, 2020

Maoz U, Rutishauser U, Kim S et al, 'Predeliberation activity in prefrontal cortex and striatum and the prediction of subsequent value judgment', Frontiers in Neuroscience, 2013, 7, article 225

Marcel AJ, Tegnér R & Nimmo-Smith I, 'Anosognosia for plegia: specificity, extension, partiality and disunity of bodily unawareness', Cortex, 2004, 40(1), 19–40

Marchewka A, Brechmann A, Nowicka A et al, 'False recognition of emotional stimuli is lateralised in the brain: an fMRI study', Neurobiology of Learning and Memory, 2008, 90(1), 280–4

Marchewka A, Jednorog K, Nowicka A et al, 'Grey-matter differences related to true and false recognition of emotionally charged stimuli – a voxel based morphometry study', Neurobiology of Learning and Memory, 2009, 92(1), 99–105

Marcos LR, 'Bilinguals in psychotherapy: language as an emotional barrier', American Journal of Psychotherapy, 1976, 30(4), 552–60

—, 'Effects of interpreters on the evaluation of psychopathology in non-English-speaking patients', American Journal of Psychiatry, 1979, 136(2), 171–4

Marcus A & Oransky I, 'How the biggest fabricator in science got caught', 21 May 2015: nautil.us/issue/24/Error/how-the-biggest-fabricator-in-science-got-caught

Marder M, 'Plant intentionality and the phenomenological framework of plant intelligence', *Plant Signaling and Behavior*, 2012, 7(11), 1365–72

———, 'Plant intelligence and attention', *Plant Signaling & Behavior*, 2013, 8(5), e23902

Marianetti M, Mina C, Marchione P et al, 'A case of visual hypoemotionality induced by interferon alpha-2b therapy in a patient with chronic myeloid leukemia', *Journal of Neuropsychiatry and Clinical Neurosciences*, 2011, 23(3), E34–5

Marin OSM & Perry DW, 'Neurological aspects of music perception and performance', in D Deutsch (ed), *The Psychology of Music*, Academic Press, 2nd edn, 1999, 653–724

Marinkovic K, Baldwin S, Courtney MG et al, 'Right hemisphere has the last laugh: neural dynamics of joke appreciation', *Cognitive, Affective, & Behavioral Neuroscience*, 2011, 11(1), 113–30

Marino G, *Kierkegaard in the Present Age*, Marquette University Press, 2001

Marinsek N, Turner BO, Gazzaniga M et al, 'Divergent hemispheric reasoning strategies: reducing uncertainty versus resolving inconsistency', *Frontiers in Human Neuroscience*, 2014, 8, article 839

Mariotti P, Iuvone L, Torrioli MG et al, 'Linguistic and non-linguistic abilities in a patient with early left hemispherectomy', *Neuropsychologia*, 1998, 36(12), 1303–12

Markowitsch HJ, 'Which brain regions are critically involved in the retrieval of old episodic memory?', *Brain Research: Brain Research Reviews*, 1995, 21(2), 117–27

Markowitsch HJ, Calabrese P, Fink GR et al, 'Impaired episodic memory retrieval in a case of probable psychogenic amnesia', *Journal of Experimental Child Psychology*, 1997, 74(2), 119–26

Markowitsch HJ, Calabrese P, Haupts M et al, 'Searching for the anatomical basis of retrograde amnesia', *Journal of Clinical and Experimental Neuropsychology*, 1993, 15(6), 947–67

Markowitsch HJ, Calabrese P, Neufeld H et al, 'Retrograde amnesia for world knowledge and preserved memory for autobiographic events', *Cortex*, 1999, 35(2), 243–52

Marryat F, *Mr Midshipman Easy*, Saunders and Otley, 1836

Marsh A, 'Visual hallucinations during hallucinogenic experience and schizophrenia', *Schizophrenia Bulletin*, 1979, 5(4), 627–30

Marshall P, *Mystical Encounters with the Natural World: Experiences and Explanations*, Oxford University Press, 2005

Marsolek CJ, 'Dissociable neural subsystems underlie abstract and specific object recognition', *Psychological Science*, 1999, 10, 111–8

Marsolek CJ, Kosslyn SM & Squire LR, 'Form-specific visual priming in the right cerebral hemisphere', *Journal of Experimental Psychology: Learning, Memory, and Cognition*, 1992, 18, 492–508

Martin A & Weisberg J, 'Neural foundations for understanding social and mechanical concepts', *Cognitive Neuropsychology*, 2003, 20(3–6), 575–87

Martin A, Wiggs CL, Ungerleider LG et al, 'Neural correlates of category-specific knowledge', *Nature*, 1996, 379(6566), 649–52

Martin A, Wiggs CL & Weisberg J, 'Modulation of human medial temporal lobe activity by form, meaning and experience', *Hippocampus*, 1997, 7(6), 587–93

Martin J, *The Education of John Dewey: A Biography*, Columbia University Press, 2003

Martin JR & Pacherie E, 'Out of nowhere: thought insertion, ownership and context-integration', *Consciousness and Cognition*, 2013, 22(1), 111–22

Martin M, 'Hemispheric specialization for local and global processing', *Neuropsychologia*, 1979, 17(1), 33–40

Martín-María N, Miret M, Caballero FF et al, 'The impact of subjective well-being on mortality: a meta-analysis of longitudinal studies in the general population', *Psychosomatic Medicine*, 2017, 79(5), 565–75

Martinaud O, Pouliquen D, Gérardin E et al, 'Visual agnosia and posterior cerebral artery infarcts: an anatomical-clinical study', *PLoS One*, 2012, 7(1), e30433

Martindale C, 'Biological bases of creativity', in RJ Sternberg (ed), *Handbook of Creativity*, Cambridge University Press, 1999, 137–51

Martindale C & Hasenfus N, 'EEG differences as a function of creativity, stage of the creative process, and effort to be original', *Biological Psychology*, 1978, 6(3), 157–67

Martindale C & Hines D, 'Creativity and cortical activation during creative, intellectual and EEG feedback tasks', *Biological Psychology*, 1975, 3(2), 91–100

Martindale C, Hines D, Mitchell L et al, 'EEG alpha asymmetry and creativity', *Personality and Individual Differences*, 1984, 5(1), 77–86

Martindale C & Moore K, 'Priming, prototypicality, and preference', *Journal of Experimental Psychology: Human Perception and Performance*, 1988, 14(4), 661–70

Martinez A, Moses P, Frank L et al, 'Hemispheric asymmetries in global and local processing: evidence from fMRI', *NeuroReport*, 1997, 8(7), 1685–9

Martins AT, Faísca LM, Esteves F et al, 'Atypical moral judgment following traumatic brain injury', *Judgment and Decision Making*, 2012, 7(4), 478–87

Martinson BC, Anderson MS & de Vries R, 'Scientists behaving badly', *Nature*, 2005, 435(7043), 737–8

Marzi CA, 'Asymmetry of interhemispheric communication', *Wiley Interdisciplinary Reviews: Cognitive Science*, 2010, 1(3), 433–8

Marzi CA & Berlucchi G, 'Right visual field superiority for accuracy of recognition of famous faces in normals', *Neuropsychologia* 1977, 15(6), 751–6

Marzi CA, Bisiacchi P & Nicoletti R, 'Is interhemispheric transfer of visuomotor information asymmetric? Evidence from a meta-analysis', *Neuropsychologia*, 1991, 29(12), 1163–77

Marzluff JM, Miyaoka R, Minoshima S et al, 'Brain imaging reveals neuronal circuitry underlying the crow's perception of human faces', *Proceedings of the National Academy of Sciences of the United States of America*, 2012, 109(39), 15912–7

Mash EJ & Barkley RA (eds), *Child Psychopathology*, Guilford Press, 1996

Mashal N & Faust M, 'Right hemisphere sensitivity to novel metaphoric relations: application of the signal detection theory', *Brain and Language*, 2008, 104(2), 103–12

Mashal N, Faust M & Hendler T, 'The role of the right hemisphere in processing nonsalient metaphorical meanings: application of principal components analysis to fMRI data', *Neuropsychologia*, 2005, 43(14), 2084–100

Mashal N, Faust M, Hendler T et al, 'An fMRI investigation of the neural correlates underlying the processing of novel metaphoric expressions', *Brain and Language*, 2007, 100(2), 115–26

——, 'Hemispheric differences in processing the literal interpretation of idioms: converging evidence from behavioral and fMRI studies', *Cortex*, 2008, 44(7), 848–60

——, 'An fMRI study of processing novel metaphoric sentences', *Laterality*, 2009, 14(1), 30–54

Mason A & Blankenship V, 'Power and affiliation motivation, stress, and abuse in intimate relationships', *Journal of Personality and Social Psychology*, 1987, 5(1)2, 203–10

Masserman JH, 'Ethology, comparative biodynamics and psychoanalytic research', in JH Masserman (ed), *Science and Psychoanalysis*, Grune & Stratton, 1960

Masserman JH, Wechkin S & Terris W, '"Altruistic" behavior in rhesus monkeys', *American Journal of Psychiatry*, 1964, 121(6), 584–5

Massman PJ, Delis DC, Filoteo JV et al, 'Mechanisms of spatial impairment in Alzheimer's disease subgroups: differential breakdown of directed attention to global-local stimuli', *Neuropsychology*, 1993, 7(2), 172–81

Mastria G, Mancini V, Viganò A et al, 'Alice in Wonderland syndrome: a clinical and pathophysiological review', *Biomed Research International*, 2016, 2016, 8243145

Masuda T & Nisbett RE, 'Attending holistically versus analytically: comparing the context sensitivity of Japanese and Americans', *Journal of Personality and Social Psychology*, 2001, 81(5), 922–34

Masuda T, Wang H, Ito K et al, 'Culture and the mind: implications for art, design and advertisement', in S Okazaki (ed), *Handbook of Research on International Advertising*, Edward Elgar Publishing, 2012, 109–33

Masullo C, Piccininni C, Quaranta D et al, 'Selective impairment of living things and musical instruments on a verbal "Semantic Knowledge Questionnaire" in a case of apperceptive visual agnosia', *Brain and Cognition*, 2012, 80(1), 155–9

Mataix-Cols D, Nakatani E, Micali N et al, 'Structure of obsessive-compulsive symptoms in pediatric OCD', *Journal of the American Academy of Child & Adolescent Psychiatry*, 2008, 47(7), 773–8

Mathis KI, Wynn JK, Breitmeyer B et al, 'The attentional blink in schizophrenia: isolating the perception/attention interface', *Journal of Psychiatric Research*, 2011, 45(10), 1346–51

Mathis KI, Wynn JK, Jahshan C et al, 'An electrophysiological investigation of attentional blink in schizophrenia: separating perceptual and attentional processes', *International Journal of Psychophysiology*, 2012, 86(1), 108–13

Matlack S, 'Quantum poetics', *The New Atlantis*, 2017, 53, 46–67: www.thenewatlantis.com/authors/samuel-matlack (accessed 23 February 2021)

Matsuda I, Nittono H & Allen JJ, 'Detection of concealed information by P3 and frontal EEG asymmetry', *Neuroscience Letters*, 2013, 537, 55–9

Matsue Y & Okuma T, 'Relative advance of eye movement to the target in the rightward tracking in schizophrenics', *Tohoku Journal of Experimental Medicine*, 1984, 143(3), 345–59

Matsunaga H, Kiriike N, Iwasaki Y et al, 'Clinical characteristics in patients with anorexia nervosa and obsessive-compulsive disorder', *Psychological Medicine*, 1999, 29(2), 407–14

Matsunaga H, Miyata A, Iwasaki Y et al, 'A comparison of clinical features among

Japanese eating-disordered women with obsessive-compulsive disorder', *Comprehensive Psychiatry*, 1999, 40(5), 337–42

Matsuo R, Kawaguchi E, Yamagishi M et al, 'Unilateral memory storage in the procerebrum of the terrestrial slug *Limax*', *Neurobiology of Learning and Memory*, 2010, 93(3), 337–42

Matsuura M, 'Psychosis of epilepsy, with special reference to anterior temporal lobectomy', *Epilepsia*, 1997, 38(suppl 6), 32–4

Matt DC, *The Essential Kabbalah*, HarperOne, 2009

Matteis M, Silvestrini M, Troisi E et al, 'Transcranial Doppler assessment of cerebral flow velocity during perception and recognition of melodies', *Journal of the Neurological Sciences*, 1997, 149(1), 57–61

Matthews D, 'Britten's Third Quartet', *Tempo*, 1978, 25, 21–24

——, *Britten*, Haus, 2013

Matthysse S, 'A theory of the relation between dopamine and attention', *Journal of Psychiatric Research*, 1978, 14(1–4), 241–8

Mattingley JB, Berberovic N, Corben L et al, 'The greyscales task: a perceptual measure of attentional bias following unilateral hemispheric damage', *Neuropsychologia*, 2004, 42(3), 387–94

Mattingley JB, Bradshaw JL, Nettleton NC et al, 'Can task specific perceptual bias be distinguished from unilateral neglect?', *Neuropsychologia*, 1994, 32(7), 805–17

Matussek P, 'Studies in delusional perception', in J Cutting & M Shepherd (eds), *The Clinical Roots of the Schizophrenia Concept: Translations of Seminal European Contributions on Schizophrenia*, Cambridge University Press, 1987 [1952], 89–103

Maudlin T, 'Part and whole in quantum mechanics', in E Castellani (ed), *Interpreting Bodies: Classical and Quantum Objects in Modern Physics*, Princeton University Press, 1998, 46–60

Maunsell JHR & Newsome WT, 'Visual processing in monkey extrastriate cortex', *Annual Review of Neuroscience*, 1987, 10, 363–401

Maximus the Confessor, in GEH Palmer, P Sherrard & Fr Kallistos Ware (eds), *The Philokalia: The Complete Text*, Faber, 1990–2020

Maxwell CR, Villalobos ME, Schultz RT et al, 'Atypical laterality of resting gamma oscillations in autism spectrum disorders', *Journal of Autism and Developmental Disorders*, 2015, 45(2), 292–7

Maxwell JK, Tucker DM & Townes BD, 'Asymmetric cognitive function in anorexia nervosa', *International Journal of Neuroscience*, 1984, 24(1), 37–44

Maxwell N, *Is Science Neurotic?*, Imperial College Press, 2004

Mayer BJ, Blinov ML & Loew LM, 'Molecular machines or pleiomorphic ensembles: signaling complexes revisited', *Journal of Biology*, 2009, 8(9), 81

Mayr E, 'Teleological and teleonomic: a new analysis', in *Evolution and the Diversity of Life: Selected Essays*, Harvard University Press, 1976, 383–404

——, *Toward a New Philosophy of Biology: Observations of an Evolutionist*, Harvard University Press, 1989

Mayseless N & Shamay-Tsoory SG, 'Enhancing verbal creativity: modulating creativity by altering the balance between right and left inferior frontal gyrus with tDCS', *Neuroscience*, 2015, 291, 167–76

Mazières D & Kohler E, 'Get me off your fucking mailing list', *The Guardian*, 25 November 2014: www.theguardian.com/australia-news/2014/nov/25/journal-accepts-paper-requesting-removal-from-mailing-list (accessed 6 March 2021)

Mazzoni M, Moretti P, Pardossi L et al, 'A case of music imperception', *Journal of Neurology, Neurosurgery, and Psychiatry*, 1993, 56(3), 322–4

Mazzucchi A, Marchini C, Budai R et al, 'A case of receptive amusia with prominent timbre perception defect', *Journal of Neurology, Neurosurgery, and Psychiatry*, 1982, 45(7), 644–7

Mazzucchi A, Sinforiani E & Boller F, 'Focal cerebral lesions and painting abilities', *Progress in Brain Research*, 2013, 204, 71–98

McBride AJ, 'Comedians: fun and dysfunctionality', *British Journal of Psychiatry*, 2004, 185(2), 177

McCabe H, *God and Evil in the Philosophy of Thomas Aquinas*, Continuum, 2010 [1957]

McCammon JM, Blaker-Lee A, Chen X et al, 'The 16p11.2 homologs fam57ba and doc2a generate certain brain and body phenotypes', *Human Molecular Genetics*, 2017, 26(19), 3699–712

McCarthy RA & Warrington EK, *Cognitive Neuropsychology: A Clinical Introduction*, Academic Press, 1990

McCarty CW, Gordon GM, Walker A et al, 'Prosopometamorphopsia and alexia following left splenial corpus callosum infarction: case report and literature review', *eNeurologicalSci*, 2017, 6, 1–3

McCauley C & Stitt CL, 'An individual and quantitative measure of stereotypes', *Journal of Personality and Social Psychology*, 1978, 36(9), 929–40

McCauley RN, *Why Religion Is Natural and Science Is Not*, Oxford University Press, 2011

McClelland DC, 'Some reflections on the two psychologies of love', *Journal of Personality*, 1986, 54(2), 334–53

McClelland DC & Burnham DH, 'Power is the great motivator', *Harvard Business Review*, 2003, 81(1), 117–26 &142

McClelland DC, Koestner R & Weinberger J, 'How do self-attributed and implicit motives differ?', *Psychological Review*, 1989, 96(4), 690–702

McClintock B, 'The significance of responses of the genome to challenge', *Science*, 1984, 226(4676), 792–801

McCormack B, 'Intuition: concept analysis and application to curriculum development. II. Application to curriculum development', *Journal of Clinical Nursing*, 1993, 2(1), 11–17

McCormack P, *Vortex, Molecular Spin and Nanovorticity: An Introduction*, Springer, 2012

McCosker P, 'Parsing paradox, analysing "and": Christological configurations of theological paradox in some mystical theologies', unpublished PhD thesis, submitted to the University of Cambridge, October 2008

McCourt ME, Shpaner M, Javitt DC et al, 'Hemispheric asymmetry and callosal integration of visuospatial attention in schizophrenia: a tachistoscopic line bisection study', *Schizophrenia Research*, 2008, 102(1–3), 189–96

McCrae RR, Chan W, Jussim L et al, 'The inaccuracy of national character stereotypes', *Journal of Research in Personality*, 2013, 47(6), 831–42

McCullough M, 'Religious involvement and mortality: answers and more questions', in TG Plante & AC Sherman (eds), *Faith and Health: Psychological Perspectives*, Guilford Press, 2001, 53–74

McCullough ME, Hoyt WT, Larson DB et al, 'Religious involvement and mortality: a meta-analytic review', *Health Psychology*, 2000, 19(3), 211–22

McCutcheon HHI & Pincombe J, 'Intuition: an important tool in the practice of nursing', *Journal of Advanced Nursing*, 2001, 35(3), 342–8

McDaniel MA, 'Big-brained people are smarter: a meta-analysis of the relationship between in vivo brain volume and intelligence', *Intelligence*, 2005, 33(4), 337–46

McDermott KB, Szpunar KK & Christ SE, 'Laboratory-based and autobiographical retrieval tasks differ substantially in their neural substrates', *Neuropsychologia*, 2009, 47(11), 2290–8

McDine DA, Livingston IJ, Thomas NA et al, 'Lateral biases in lighting of abstract artwork', *Laterality*, 2011, 16(3), 268–79

McElroy T, *A to Z of Mathematicians*, Facts on File, 2004

McElroy T & Stroh N, 'Making estimates and sensitivity to anchors: exploring the role of hemispheric processing', *Laterality*, 2013, 18(3), 294–302

McEvoy JP, Hartman M, Gottlieb D et al, 'Common sense, insight, and neuropsychological test performance in schizophrenia patients', *Schizophrenia Bulletin*, 1996, 22(4), 635–41

McFie J, Piercy MF & Zangwill OL, 'Visual-spatial agnosia associated with lesions of the right cerebral hemisphere', *Brain*, 1950, 73(2), 167–90

McGee A & Foster C, *Intuitively Rational*, forthcoming

McGeoch PD, Brang D & Song T, 'Xenomelia: a new right parietal lobe syndrome', *Journal of Neurology, Neurosurgery, & Psychiatry*, 2011, 82(12), 1314–9

McGeorge P, Beschin N, Colnaghi A et al, 'A lateralized bias in mental imagery: evidence for representational pseudoneglect', *Neuroscience Letters*, 2007, 421(3), 259–63

McGhie A & Chapman J, 'Disorders of attention and perception in early schizophrenia', *British Journal of Medical Psychology*, 1961, 34(2), 103–16

McGilchrist I, 'A problem of symmetries', *Philosophy, Psychiatry, & Psychology*, 2009a, 16(2), 161–9

———, *The Master and his Emissary: The Divided Brain and the Making of the Western World*, Yale University Press, 2009b

———, *LA Review of Books*, 25 September 2013: lareviewofbooks.org/article/can-this-couple-work-it-out/

———, '"Selving" and union', *Journal of Consciousness Studies*, 2016, 23(1–2), 196–213

———, 'Depression is not like anything on earth', in *Depression: Law and Ethics*, ed C Foster and J Herring, Oxford University Press, 2017, 2–5

———, 'God, metaphor and the language of the hemispheres', in P Chilton & M Kopytowska (eds), *Religion, Language, and the Human Mind*, Oxford University Press, 2018, 135–68

———, 'A response to commentators', *Religion, Brain and Behavior*, 2019, 9(4), 399–422

McGilchrist I & Cutting J, 'Somatic delusions in schizophrenia and the affective psychoses', *British Journal of Psychiatry*, 1995, 167(3), 350–61

McGilchrist IK, Jadresic D, Goldstein LH et al, 'Thalamo-frontal psychosis', *British Journal of Psychiatry*, 1993, 163(1), 113–5

McGilchrist I, Wolkind SN & Lishman WA, '"Dyschronia" in a patient with Tourette's syndrome presenting as maternal neglect', *British Journal of Psychiatry*, 1994, 164(2), 261–3

McGinn B, *The Mystical Thought of Meister Eckhart*, Herder & Herder, 2001

McGinn C, 'Consciousness and cosmology: hyperdualism ventilated', in M Davies & GW Humphreys (eds), *Consciousness*, Blackwell, 1993, 55–77

——, 'Consciousness and space', *Journal of Consciousness Studies*, 1995, 2(3), 220–30

McGlone J & Davidson W, 'The relation between cerebral speech laterality and spatial ability with special reference to sex and hand preference', *Neuropsychologia*, 1973, 11(1), 105–13

McGlone J & Kertesz A, 'Sex differences in cerebral processing of visuospatial tasks', *Cortex*, 1973, 9(3), 313–20

McGlone J, 'Sex differences in the cerebral organization of verbal functions in patients with unilateral brain lesions', *Brain*, 1977, 100(4), 775–93

——, 'Sex differences in functional brain asymmetry', *Cortex*, 1978, 14(1), 122–8

——, 'Sex differences in functional brain asymmetry: a critical survey', *Behavioral and Brain Sciences*, 1980, 3(2), 215–27

McGonigle DJ, Hänninen R, Salenius S et al, 'Whose arm is it anyway? An fMRI case study of supernumerary phantom limb', *Brain*, 2002, 125(6), 1265–74

McGrath J, Johnson K, O'Hanlon E et al, 'White matter and visuospatial processing in autism: a constrained spherical deconvolution tractography study', *Autism Research*, 2013, 6(5), 307–19

McGrath JA, Avramopoulos D, Lasseter VK et al, 'Familiality of novel factorial dimensions of schizophrenia', *Archives of General Psychiatry*, 2009, 66(6), 591–600

McGraw M, 'Neural maturation as exemplified in the changing reactions of the infant to pin prick', *Child Development*, 1941, 12(1), 31–42

McGuire J, Langdon R & Brüne M, 'Moral cognition in schizophrenia', *Cognitive Neuropsychiatry*, 2014, 19(6), 495–508

McIntosh AR & Gonzalez-Lima F, 'Structural equation modeling and its application to network analysis in functional brain imaging', *Human Brain Mapping*, 1994, 2(1–2), 2–22

McKay R, Arciuli J, Atkinson A et al, 'Lateralisation of self-esteem: an investigation using a dichotically presented auditory adaptation of the Implicit Association Test', *Cortex*, 2010, 46(3), 367–73

McKay R, Tamagni C, Palla A et al, 'Vestibular stimulation attenuates unrealistic optimism', *Cortex*, 2013, 49(8), 2272–5

McKenna F, Babb J, Miles L et al, 'Reduced microstructural lateralization in males with chronic schizophrenia: a diffusional kurtosis imaging study', *Cerebral Cortex*, 2020, 30(4), 2281–94

McKinley J, Dempster M & Gormley GJ, '"Sorry I meant the patient's left side": impact of distraction on right/left discrimination', *Medical Education*, 2015, 49(4), 427–35

McKinnell S, *Choices: One Man's Spiritual Journey*, Amazon Media, 2017

McKinnon MC & Schellenberg EG, 'A left-ear advantage for forced-choice judgments of melodic contour', *Canadian Journal of Experimental Psychology*, 1997, 51(2), 171–5

McNabb AW, Carroll WM & Mastaglia FL, '"Alien hand" and loss of bimanual coordination after dominant anterior cerebral artery territory infarction', *Journal of Neurology, Neurosurgery, and Psychiatry*, 1988, 51(2), 218–22

McNamara P, *The Neuroscience of Religious Experience*, Cambridge University Press, 2009

McNeill D, *Hand and Mind: What Gestures Reveal about Thought*, University of Chicago Press, 1992, 345–52

Mead AM & McLaughlin JP, 'The roles of handedness and stimulus asymmetry in aesthetic preference', *Brain and Cognition*, 1992, 20(2), 300–7

Meador KJ, Loring DW, Feinberg TE et al, 'Anosognosia and asomatognosia during intracarotid amobarbital inactivation', *Neurology*, 2000, 55(6), 816–20

Meador KJ, Ray PG, Day L et al, 'Physiology of somatosensory perception: cerebral lateralization and extinction', *Neurology*, 1998, 51(3), 721–7

Meadows JC, 'The anatomical basis of prosopagnosia', *Journal of Neurology, Neurosurgery, and Psychiatry*, 1974, 37(5), 489–501

Meadows JC & Munro SSF, 'Palinopsia', *Journal of Neurology, Neurosurgery, and Psychiatry*, 1977, 40(1), 5–8

Meares R, Schore A & Melkonian D, 'Is borderline personality a particularly right hemispheric disorder? A study of P3a using single trial analysis', *Australian & New Zealand Journal of Psychiatry*, 2011, 45(2), 131–9

Mechthilde of Magdeburg: *The Flowing Light of The Godhead*, Paulist Press, 2012

Meck WH, 'Neuropsychology of timing and time perception', *Brain and Cognition*, 2005, 58(1), 1–8

Medawar PB, 'Is the scientific paper fraudulent?', *The Saturday Review*, 1 August 1964, 43

———, *Induction and Intuition in Scientific Thought*, Methuen, 1969

———, *Advice to a Young Scientist*, Harper & Row, 1979

———, *The Limits of Science*, Oxford University Press, 1988

Medvedev AV, 'Does the resting state connectivity have hemispheric asymmetry? A near-infrared spectroscopy study', *NeuroImage*, 2014, 85(1), 400–7

Mega LF, Gigerenzer G & Volz KG, 'Do intuitive and deliberate judgments rely on two distinct neural systems? A case study in face processing', *Frontiers in Human Neuroscience*, 2015, 9, 456

Mega M & Cummings JL, 'Frontal subcortical circuits: anatomy and function', in SP Salloway, PF Malloy & JD Duffy (eds), *The Frontal Lobes and Neuropsychiatric Illness*, American Psychiatric Publishing, 2001, 15–32

Megidish E, Halevy A, Shacham T et al, 'Entanglement swapping between photons that have never coexisted', *Physical Review Letters*, 2013, 110(21), 210403

Megumi F, Bahrami B, Kanai R et al, 'Brain activity dynamics in human parietal regions during spontaneous switches in bistable perception', *NeuroImage*, 2015, 107, 190–7

Mehta N & Myrskylä M, 'The population health benefits of a healthy lifestyle: life expectancy increased and onset of disability delayed', *Health Affairs*, 2017, 36(8), doi: 10.1377/hlthaff.2016.1569

Meier CA (ed), *Atom and Archetype: The Pauli/Jung Letters 1932–1958*, Princeton University Press, 2001

Meijer DKF & Geesink HJH, 'Favourable and unfavourable EMF frequency patterns in cancer: perspectives for improved therapy and prevention', *Journal of Cancer Therapy*, 2018, 9(3), 188–230

Meinschaefer J, Hausmann M & Güntürkün O, 'Laterality effects in the processing of syllable structure', *Brain and Language*, 1999, 70(2), 287–93

Meisner G, 'The golden ratio, beauty and design: it's time to "face" the facts', 9 November 2015: www.goldennumber.net/golden-ratio-design-beauty-face-evidence-facts/ (accessed 12 March 2021)

Meissner CA & Brigham JC, 'A meta-analysis of the verbal overshadowing effect in face identification', *Applied Cognitive Psychology*, 2001, 15(6), 603–16

Melcher JM & Schooler JW, 'Perceptual and conceptual training mediate the verbal overshadowing effect in an unfamiliar domain', *Memory & Cognition*, 2004, 32(4), 618–31

Mele AR, *Effective Intentions: The Power of Conscious Will*, Oxford University Press, 2009

Melhuish G, *The Paradoxical Nature of Reality*, St Vincent's Press, 1973

Mell J, Howard S & Miller B, 'Art and the brain: the influence of frontotemporal dementia on an accomplished artist', *Neurology*, 2003, 60(10), 1707–10

Meltzoff AN & Brooks R, '"Like me" as a building block for understanding other minds: bodily acts, attention and intention', in DA Baldwin, LJ Moses & BF Malle (eds), *Intentions and Intentionality: Foundations of Social Cognition*, Massachusetts Institute of Technology Press, 2001, 171–92

Meltzoff AN & Moore MK, 'Newborn infants imitate adult facial gestures', *Child Development*, 1983, 54(3), 702–9

Melzack R, 'Phantom limbs', *Scientific American*, 1992, 266(4), 120–6

Melzack R, Israel R, Lacroix R et al, 'Phantom limbs in people with congenital limb deficiency or amputation in early childhood', *Brain*, 1997, 120(9), 1603–20

Men A, *A History of Religion: In Search of the Way, the Truth and the Life*, Slovo, 2 vols, 1997

Mendez MF, 'Dementia as a window to the neurology of art', *Medical Hypotheses*, 2004, 63(1), 1–7

———, 'What frontotemporal dementia reveals about the neurobiological basis of morality', *Medical Hypotheses*, 2006, 67(2), 411–8

———, 'The neurobiology of moral behavior: review and neuropsychiatric implications', *CNS Spectrums*, 2009, 14(11), 608–20

Mendez MF, Chen AK, Shapira JS et al, 'Acquired sociopathy and frontotemporal dementia', *Dementia and Geriatric Cognitive Disorders*, 2005, 20(2–3), 99–104

Mendez MF & Fras IA, 'The false memory syndrome: experimental studies and comparison to confabulations', *Medical Hypotheses*, 2011, 76(4), 492–6

Mendez MF & Lim GT, 'Alterations of the sense of "humanness" in right hemisphere predominant frontotemporal dementia patients', *Cognitive and Behavioral Neurology*, 2004, 17(3), 133–8

Mendez MF & Perryman KM, 'Neuropsychiatric features of frontotemporal dementia: evaluation of consensus criteria and review', *Journal of Neuropsychiatry and Clinical Neurosciences*, 2002, 14(4), 424–9

Mendez MF & Shapira JS, 'Altered emotional morality in frontotemporal dementia', *Cognitive Neuropsychiatry*, 2009, 14(3), 165–79

Mendola JD, Rizzo JF, Cosgrove GR et al, 'Visual discrimination after anterior temporal lobectomy in humans', *Neurology*, 1999, 52(5), 1028–37

Menenti L, Petersson KM, Scheeringa R et al, 'When elephants fly: differential sensitivity of right and left inferior frontal gyri to discourse and world knowledge', *Journal of Cognitive Neuroscience*, 2009, 21(12), 2358–68

Menon M, Mizrahi R & Kapur S, '"Jumping to conclusions" and delusions in psychosis: relationship and response to treatment', *Schizophrenia Research*, 2008, 98(1–3), 225–31

Mensh IN, Schwartz HG, Matarazzo RG et al, 'Psychological functioning following cerebral hemispherectomy in man', *Archives of Neurology and Psychiatry*, 1952, 67(6), 787–96

Menshutkin VV & Nikolaenko NN, 'The role of the right hemisphere in size constancy', *Fiziologiia Cheloveka (Human Physiology)*, 1987, 13(2), 324–6

Mente A, O'Donnell MJ & Yusuf S, 'The population risks of dietary salt excess are exaggerated', *Canadian Journal of Cardiology*, 2014, 30(5), 507–12

——, 'How robust is the evidence for recommending very low salt intake in entire populations?', *Journal of the American College of Cardiology*, 2016, 68(15), 1618–21

Mercier H & Sperber D, 'Why do humans reason? Arguments for an argumentative theory', *Behavioral and Brain Sciences*, 2011, 34(2), 57–74; discussion 74–111

Merker B, 'Consciousness without a cerebral cortex: a challenge for neuroscience and medicine', *Behavioral and Brain Sciences*, 2007, 30(1), 63–81; discussion 81–134

Merleau-Ponty M, *Phénoménologie de la perception*, Gallimard, 1945

——, *Phenomenology of Perception*, trans C Smith, Routledge, 1962 [*Phénoménologie de la perception* 1945]

——, *The Primacy of Perception*, Northwestern University Press, 1964

——, *La prose du monde*, Gallimard, 1969

Mermin ND, *Boojums All the Way Through: Communicating Science in a Prosaic Age*, Cambridge University Press, 1990

——, 'What is quantum mechanics trying to tell us?', *American Journal of Physics*, 1998, 66, 753–67

Merola JL & Liederman J, 'Developmental changes in hemispheric independence', *Child Development*, 1985, 56(5), 1184–94

Merrifield C, Hurwitz M & Danckert J, 'Multimodal temporal perception deficits in a patient with left spatial neglect', *Cognitive Neuroscience*, 2010, 1(4), 244–53

Mesulam J, 'Possible basis for the evolution of lateral specialisation of the human brain', *Nature*, 1969, 224(5219), 614–5

Mesulam M-M, 'Behavioral neuroanatomy: large-scale net-works, association cortex, frontal syndromes, the limbic system and hemispheric specialization', in M-M Mesulam (ed), *Principles of Behavioral and Cognitive Neurology*, Oxford University Press, 2nd ed, 2000a, 1–120

—— (ed), *Principles of Behavioral and Cognitive Neurology*, Oxford University Press, 2nd ed, 2000b

Mesulam M-M, Waxman SG, Geschwind N et al, 'Acute confusional states with right middle cerebral artery infarction', *Journal of Neurology, Neurosurgery, and Psychiatry*, 1976, 39(1), 84–9

Metcalfe J, 'Premonitions of insight predict impending error', *Journal of Experimental Psychology: Learning, Memory, and Cognition*, 1986, 12(4), 623–34

Metcalfe J, Funnell M & Gazzaniga MS, 'Right-hemisphere memory superiority: studies of a split-brain patient', *Psychological Science*, 1995, 6(3), 157–64

Metter EJ, Riege WH, Hanson WR et al, 'Comparison of metabolic rates, language, and memory in subcortical aphasias', *Brain and Language*, 1983, 19(1), 33–47

——, 'Correlations of glucose metabolism and structural damage to language function in aphasia', *Brain and Language*, 1984, 21(2), 187–207

Metuki N, Sela T & Lavidor M, 'Enhancing cognitive control components of insight problems solving by anodal tDCS of the left dorsolateral prefrontal cortex', *Brain Stimulation*, 2012, 5(2), 110–5

Meudell PR & Greenhalgh M, 'Age related differences in left and right hand skill and in visuo-spatial performance: their possible relationships to the hypothesis that the right hemisphere ages more rapidly than the left', *Cortex*, 1987, 23(3), 431–45

Mevorach C, Humphreys GW & Shalev L, 'Attending to local form while ignoring global aspects depends on handedness: evidence from TMS', *Nature Neuroscience*, 2005, 8(3), 276–7

Meyer B-U, Röricht S, Gräfin von Einsiedel H et al, 'Inhibitory and excitatory interhemispheric transfers between motor cortical areas in normal subjects and patients with abnormalities of the corpus callosum', *Brain*, 1995, 118(2), 429–40

Meyer CS, Hagmann-von Arx P, Lemola S et al, 'Correspondence between the general ability to discriminate sensory stimuli and general intelligence', *Journal of Individual Differences*, 2010, 31(1), 46–56

Meyer M, Alter K, Friederici AD et al, 'fMRI reveals brain regions mediating slow prosodic modulations in spoken sentences', *Human Brain Mapping*, 2002, 17(2), 73–88

Meyer M, Baumann S, Wildgruber D et al, 'How the brain laughs: comparative evidence from behavioral, electrophysiological and neuroimaging studies in human and monkey', *Behavioural Brain Research*, 2007, 182(2), 245–60

Meyer M, Steinhauer K, Alter K et al, 'Brain activity varies with modulation of dynamic pitch variance in sentence melody', *Brain and Language*, 2004, 89(2), 277–89

Meyer M, Zysset S, von Cramon DY et al, 'Distinct fMRI responses to laughter, speech, and sounds along the human perisylvian cortex', *Brain Research: Cognitive Brain Research*, 2005, 24(2), 291–306

Meyers MA, *Happy Accidents: Serendipity in Modern Medical Breakthroughs*, Arcade, 2007

Michel EM & Troost BT, 'Palinopsia: cerebral localization with CT', *Neurology*, 1980, 30(8), 887–9

Michelson M, 'Convergent evolution in the genes', California Academy of Sciences Scientific News, 12 September 2013: www.calacademy.org/explore-science/convergent-evolution-in-the-genes (accessed 23 February 2021)

Midgley M, *Wisdom, Information, and Wonder: What is Knowledge For?*, Routledge, 1989

———, *Science and Poetry*, Routledge, 2006

Midorikawa A & Kawamura M, 'The emergence of artistic ability following traumatic brain injury', *Neurocase*, 2015, 21(1), 90–4

Midrash: from *Forms of Prayer for Jewish Worship*, The Reform Synagogues of Great Britain, 7th edn, 1977

Mihov KM, Denzler M & Förster J, 'Hemispheric specialization and creative thinking: a meta-analytic review of lateralization of creativity', *Brain and Cognition*, 2010, 72(3), 442–8

Mihrshahi R, 'The corpus callosum as an evolutionary innovation', *Journal of Experimental Zoology, Part B: Molecular & Developmental Evolution*, 2006, 306(1), 8–17

Miklósi Á & Soproni K, 'Review: a comparative analysis of animals' understanding of the human pointing gesture', *Animal Cognition*, 2006, 9(2), 81–93

Mikulecky DC, 'Robert Rosen: the well posed question and its answer – why are organisms different from machines?', *Systems Research & Behavioral Science*, 2000, 17(5), 419–32

Milano N, Goldman A, Woods A et al, 'The influence of right and left frontotemporal stimulation on visuospatial creativity', *Neurology*, 2016, 86(16, suppl), P4.051

Mileaf MI & Byne W, 'Neuronal deficit in medial pulvinar from right but not left hemisphere in schizophrenia', *Schizophrenia Research*, 2012, 134(2–3), 291–2

Miles LK, Tan L, Noble GD et al, 'Can a mind have two time lines? Exploring space-time mapping in Mandarin and English speakers', *Psychonomic Bulletin & Review*, 2011, 18(3), 598–604

Mill JS, 'Coleridge', *Mill on Bentham and Coleridge*, Chatto & Windus, 1950

———, 'Essay on Bentham', in *Utilitarianism and On Liberty, including Mill's 'Essay on Bentham' and selections from the writings of Jeremy Bentham and John Austin*, ed M Warnock, Blackwell, Oxford, 2003, 52–87

Miller AC, Hickman LC & Lemasters GK, 'A distraction technique for control of burn pain', *Journal of Burn Care & Rehabilitation*, 1992, 13(5), 576–80

Miller BL, Boone K, Cummings J et al, 'Functional correlates of musical and visual ability in frontotemporal dementia', *British Journal of Psychiatry*, 2000, 176(5), 458–63

Miller BL, Chang L, Mena I et al, 'Progressive right frontotemporal degeneration: clinical, neuropsychological and SPECT characteristics', *Dementia*, 1993, 4(3–4), 204–13

Miller BL, Cummings J, Mishkin F et al, 'Emergence of artistic talent in frontotemporal dementia', *Neurology*, 1998, 51(4), 978–82

Miller CL, entry for Cusanus in the *Stanford Encyclopedia of Philosophy*, 2017: plato.stanford.edu/entries/cusanus/ (accessed 23 July 2019)

Miller G, 'The magical number seven, plus or minus two: some limits on our capacity for processing information', *Psychological Review*, 1956, 63(2), 81–97

Miller JW, Jayadev S, Dodrill CB et al, 'Gender differences in handedness and speech lateralisation related to early neurologic insults', *Neurology*, 2005, 65(12), 1974–5

Miller LA & Tippett LJ, 'Effects of focal brain lesions on visual problem-solving', *Neuropsychologia*, 1996, 34, 387–98

Miller MB & Gazzaniga MS, 'Creating false memories for visual scenes', *Neuropsychologia*, 1998, 36(6), 513–20

Miller MB, Sinnott-Armstrong W, Young L et al, 'Abnormal moral reasoning in complete and partial callosotomy patients', *Neuropsychologia*, 2010, 48(7), 2215–20

Miller RE, Giannini AJ & Levine JM, 'Nonverbal communication in man with a cooperative conditioning task', *Journal of Social Psychology*, 1977, 103(1), 101–13

Miller W, 'Death of a genius', *Time*, 2 May 1955, 64

Milleret C & Houzel JC, 'Visual interhemispheric transfer to areas 17 and 18 in cats with convergent strabismus', *European Journal of Neuroscience*, 2001, 13, 137–52

Mills L & Rollman GB, 'Hemispheric asymmetry for auditory perception of temporal order', *Neuropsychologia*, 1980, 18(1), 41–8

Mills M, Alwatban M, Hage B et al, 'Cerebral hemodynamics during scene viewing: hemispheric lateralization predicts temporal gaze behavior associated with distinct modes of visual processing', *Journal of Experimental Psychology: Human Perception and Performance*, 2017, 43(7), 1291–302

Millstein RL & Skipper RA, 'Population genetics', in DL Hull & M Ruse (eds), *The Cambridge Companion to the Philosophy of Biology*, Cambridge University Press, 2007

Milner AD & Harvey M, 'Distortion of size perception in visuospatial neglect', *Current Biology*, 1995, 5(1), 85–9

Milner B, 'Interhemispheric differences in the localisation of physiological processes in man', *British Medical Bulletin*, 1971, 27(3), 272–7

———, 'Functional recovery after lesions of the nervous system. 3. Developmental processes in neural plasticity. Sparing of language functions after early unilateral brain damage', *Neurosciences Research Program Bulletin*, 1974, 12(2), 213–7

Milner B & Taylor L, 'Right hemisphere superiority in tactile pattern-recognition after cerebral commissurotomy: evidence for non-verbal memory', *Neuropsychologia*, 1971, 10(1), 1–15

Miłosz C, *The Captive Mind*, trans J Zielonko, Vintage Books, 1955

Mimura M, Nakagome K, Hirashima N et al, 'Left frontotemporal hyperperfusion in a patient with post-stroke mania', *Journal of Experimental Child Psychology*, 2005, 139(3), 263–7

Minkowska F, 'Troubles essentiels de la schizophrénie dans leurs rapports avec les données de la psychologie et de la biologie modernes', *Évolution Psychiatrique*, 1925, 1, 127–41

Minkowski E, 'Étude psychologique et analyse phénoménologique d'un cas de mélancolie schizophrénique', *Journal de Psychologie normale et pathologique*, 1923, 20(10), 543–58, trans B Bliss, as 'Findings in a case of schizophrenic depression', in R May, E Angel & HF Ellenberger (eds), *Existence*, Rowman & Littlefield, 2004 [1958], 127–38

———, *La notion de perte de contact vital avec la réalité et ses applications en psychopathologie*, Jouve et Cie, 1926

———, *Le temps vécu*, Quadrige, 1968

———, *Lived Time*, trans N Metzel, Northwestern University Press, 1970 [1933]

———, 'The essential disorder of schizophrenia' [1927], in J Cutting & M Shepherd (eds), *The Clinical Roots of the Schizophrenia Concept*, Cambridge University Press, 1987, 188–212 (translation of *La schizophrénie* ch 2)

———, *Au-delà du rationalisme morbide*, L'Harmattan, 1997

———, *Traité de psychopathologie*, Institut Sythelabo, 1999 [1968]

———, *La schizophrénie: psychopathologie des schizoïdes et des schizophrènes*, Payot, 2002 [1927]

Mintz M, Tomer R & Myslobodsky MS, 'Neuroleptic-induced lateral asymmetry of visual evoked potentials in schizophrenia', *Biological Psychiatry*, 1982, 17(7), 815–28

Misselbrook D, 'Virtue ethics – an old answer to a new dilemma? Part 2. The case for inclusive virtue ethics', *Journal of the Royal Society of Medicine*, 2015, 108(3), 89–92

Mitchell JP, Ames DL, Jenkins AC et al, 'Neural correlates of stereotype application', *Journal of Cognitive Neuroscience*, 2009, 21(3), 594–604

Mitchell RL & Crow TJ, 'Right hemisphere language functions and schizophrenia: the forgotten hemisphere?', *Brain*, 2005, 128(5), 963–78

Mitchell RL, Elliott R, Barry M et al, 'The neural response to emotional prosody, as revealed by functional magnetic resonance imaging', *Neuropsychologia*, 2003, 41(10), 1410–21

———, 'Neural response to emotional prosody in schizophrenia and in bipolar affective disorder', *British Journal of Psychiatry*, 2004, 184(3), 223–30

Mitchley NJ, Barber J, Gray JM et al, 'Comprehension of irony in schizophrenia', *Cognitive Neuropsychiatry*, 1998, 3(2), 127–38

Mithen SJ, *Singing Neanderthals: the Origin of Music, Language, Mind and Body*, Phoenix, 2005

Mittelstrass J, 'Nature and science in the Renaissance', in RS Woolhouse (ed), *Metaphysics and Philosophy of Science in the Seventeenth and Eighteenth Centuries*, Kluwer, 1988

Mitterschiffthaler MT, Fu CHY, Dalton JA et al, 'A functional MRI study of happy and sad affective states induced by classical music', *Human Brain Mapping*, 2007, 28(11), 1150–62

Mittleman MA, Maclure M, Sherwood JB et al, 'Triggering of acute myocardial infarction onset by episodes of anger: Determinants of

Myocardial Infarction Onset Study Investigators', *Circulation*, 1995, 92(7), 1720–5

Miwa H & Kondo T, 'Metamorphopsia restricted to the right side of the face associated with a right temporal lobe lesion', *Journal of Neurology*, 2007, 254(12), 1765–7

Miyamoto Y, Nisbett RE & Masuda T, 'Culture and the physical environment: holistic versus analytic perceptual affordances', *Psychological Science*, 2006, 17(2), 113–9

Mize KD & Jones NA, 'Infant physiological and behavioral responses to loss of maternal attention to a social-rival', *International Journal of Psychophysiology*, 2012, 83(1), 16–23

Mizobuchi M, Ito N, Tanaka C et al, 'Unidirectional olfactory hallucination associated with ipsilateral unruptured intracranial aneurysm', *Epilepsia*, 1999, 40(4), 516–9

Mizokami Y, Terao T, Hatano K et al, 'Difference in brain activations during appreciating paintings and photographic analogs', *Frontiers in Human Neuroscience*, 2014, 8, 478

Mizukami K, Yamakawa Y, Yokoyama H et al, 'A case of psychotic disorder associated with a right temporal lesion: a special reference to magnetic resonance imaging and single photon emission computed tomography findings', *Psychiatry and Clinical Neurosciences*, 1999, 53(5), 603–6

Mizuno A, Liu Y, Williams DL et al, 'The neural basis of deictic shifting in linguistic perspective-taking in high-functioning autism', *Brain*, 2011, 134(8), 2422–35

Mizuno M, 'Neuropsychological characteristics of right hemisphere damage: investigation by attention tests, concept formation and change test, and self-evaluation task', *Keio Journal of Medicine*, 1991, 40(4), 221–34

Mobbs D, Hassabis D, Seymour B et al, 'Choking on the money: reward-based performance decrements are associated with midbrain activity', *Psychological Science*, 2009, 20(8), 955–62

Moes PE, Brown WS & Minnema MT, 'Individual differences in interhemispheric transfer time (IHTT) as measured by event related potentials', *Neuropsychologia*, 2007, 45(11), 2626–30

Molfese DL, 'Left hemisphere sensitivity to consonant sounds not displayed by the right hemisphere: electrophysiological correlates', *Brain and Language*, 1984, 22(1), 109–27

Moll J, de Oliveira-Souza R, Eslinger PJ et al, 'The neural correlates of moral sensitivity: a functional magnetic resonance imaging investigation of basic and moral emotions', *Journal of Neuroscience*, 2002, 22(7), 2730–36

Mölle M, Marshall L, Wolf B et al, 'EEG complexity and performance measures of creative thinking', *Psychophysiology*, 1999, 36(1), 95–104

Møller P & Husby R, 'The initial prodrome in schizophrenia: searching for naturalistic core dimensions of experience and behavior', *Schizophrenia Bulletin*, 2000, 26(1), 217–32

Moltmann J, *God in Creation*, Fortress Press, 1993

Monbiot G, 'Hopeless realism', *The Guardian*, 14 November 2018

Mondloch CJ, Geldart S, Maurer D et al, 'Developmental changes in the processing of hierarchical shapes continue into adolescence', *Journal of Experimental Child Psychology*, 2003, 84(1), 20–40

Monk CS, Weng SJ, Wiggins JL et al, 'Neural circuitry of emotional face processing in autism spectrum disorders', *Journal of Psychiatry & Neuroscience*, 2010, 35(2), 105–14

Monk R, in 'How the untimely death of RG Collingwood changed the course of philosophy forever', *Prospect Magazine*, 5 September 2019: www.prospectmagazine.co.uk/magazine/how-the-untimely-death-of-rg-collingwood-changed-the-course-of-philosophy-forever-gilbert-ryle-ray-monk-analytic-continental (accessed 23 February 2021)

Monod J, *Chance and Necessity*, trans A Wainhouse, Collins/Fount, 1977 [*Le hasard et la nécessité* 1970]

Montague PR & Chiu PH, 'For goodness' sake', *Nature Neuroscience*, 2007, 10(2), 137–8

Montaigne M de, *The Complete Essays*, trans M Screech, Penguin, 1991 [*Essais* 1580–95]

Montalvo MJ & Khan MA, 'Clinicoradiological correlation of macropsia due to acute stroke: a case report and review of the literature', *Case Reports in Neurological Medicine*, 2014, 2014, article 272084

Monti MM, Parsons LM & Osherson DN, 'The boundaries of language and thought in deductive inference', *Proceedings of the National Academy of Sciences of the United States of America*, 2009, 106(30), 12554–9

———, 'Thought beyond language: neural dissociation of algebra and natural language', *Psychological Science*, 2012, 23(8), 914–22

Moore DW, Bhadelia RA, Billings RL et al, 'Hemispheric connectivity and the visual-spatial divergent-thinking component of creativity', *Brain and Cognition*, 2009, 70(3), 267–72

Moore EO, 'A prison environment's effect on health care service demands', *Journal of Environmental Systems*, 1981, 11(1), 17–34

Moore LL, Singer MR, Bradlee ML et al, 'Low sodium intakes are not associated with lower blood pressure levels among Framingham Offspring Study adults', *Federation of*

American Societies for Experimental Biology Journal, 2017, 31(1 Suppl), 446
Moore R, *Niels Bohr: The Man, His Science, and the World They Changed*, Knopf, 1966
Moore W, *Schrödinger: Life and Thought*, Cambridge University Press, 1992
Moran M, Seidenberg M, Sabsevitz D et al, 'The acquisition of face and person identity information following anterior temporal lobectomy', *Journal of the International Neuropsychological Society*, 2005, 11(3), 237–48
Moreira-Almeida A & Cardeña E, 'Differential diagnosis between non-pathological psychotic and spiritual experiences and mental disorders: a contribution from Latin American studies to the ICD-11', *Revista Brasileira de Psiquiatria*, 2011, 33(suppl 1), S21–36
Moreno CR, Borod JC, Welkowitz J et al, 'Lateralization for the expression and perception of facial emotion as a function of age', *Neuropsychologia*, 1990, 28(2), 199–209
Morgan CJ & Curran HV, 'Acute and chronic effects of ketamine upon human memory: a review', *Psychopharmacology*, 2006, 188(4), 408–24
Morgan KD & David AS, 'Neuropsychological studies of insight in patients with psychotic disorders', in X Amador & A David (eds), *Insight and Psychosis*, Oxford University Press, 2nd edn, 2004, 177–93
Morgan KD, Dazzan P, Morgan C et al, 'Insight, grey matter and cognitive function in first-onset psychosis', *British Journal of Psychiatry*, 2010, 197(2), 141–8
Mori M, 'The uncanny valley', *Energy*, 1970, 7(4), 33–5
Morishima Y, Schunk D, Bruhin A et al, 'Linking brain structure and activation in temporoparietal junction to explain the neurobiology of human altruism', *Neuron*, 2012, 75(1), 73–9
Morita T, Itakura S, Saito DN et al, 'The role of the right prefrontal cortex in self-evaluation of the face: a functional magnetic resonance imaging study', *Journal of Cognitive Neuroscience*, 2008, 20(2), 342–55
Moritz S & Woodward TS, 'Jumping to conclusions in delusional and non-delusional schizophrenic patients', *British Journal of Clinical Psychology*, 2005, 44(2), 193–207
Morlaas J, *Contribution à l'étude de l'apraxie*, Legrand, 1928: in JW Brown (ed), *Agnosia and Apraxia: Selected Papers of Liepmann, Lange and Pötzl*, Erlbaum, 1988, 182–3
Morland D, Wol V, Dietemann J-L et al, 'Robin Hood caught in Wonderland: brain SPECT findings', *Clinical Nuclear Medicine*, 2013, 38(12), 979–81
Mormann F, Dubois J, Kornblith S et al, 'A category-specific response to animals in the right human amygdala', *Nature Neuroscience*, 2011, 14(10), 1247–9
Moro A, Tettamanti M, Perani D et al, 'Syntax and the brain: disentangling grammar by selective anomalies', *NeuroImage*, 2001, 13(1), 110–8
Morrow L & Ratcliff G, 'The disengagement of covert attention and the neglect syndrome', *Psychobiology*, 1988, 16(3), 261–9
Morrow L, Vrtunski PB, Kim Y et al, 'Arousal responses to emotional stimuli and laterality of lesion', *Neuropsychologia*, 1981, 19(1), 65–71
Morsanyi K & Handley SJ, 'How smart do you need to be to get it wrong? The role of cognitive capacity in the development of heuristic-based judgment', *Journal of Experimental Child Psychology*, 2008, 99(1), 18–36
——, 'Logic feels so good – I like it! Evidence for intuitive detection of logicality in syllogistic reasoning', *Journal of Experimental Psychology: Learning, Memory, and Cognition*, 2012, 38(3), 596–616
Mortensen EL, Barefoot JC & Avlund K, 'Do depressive traits and hostility predict age-related decline in general intelligence?', *Journal of Aging Research*, 2012, 2012, article 973121
Morylowska-Topolska J, Zieminski R, Molas A et al, 'Schizophrenia and anorexia nervosa – reciprocal relationships: a literature review', *Psychiatria Polska*, 2017, 51(2), 261–70
Moscovitch M, 'Information processing and the cerebral hemispheres', in MS Gazzaniga (ed), *Handbook of Behavioral Neurobiology, vol 2: Neuropsychology*, Plenum, 1979, 379–446
Moscovitch M & Winocur G, 'The frontal cortex and working with memory', in DT Stuss & RT Knight (eds), *Principles of Frontal Lobe Function*, Oxford University Press, 2002, 188–209
Moseley RL & Pulvermüller F, 'Nouns, verbs, objects, actions, and abstractions: local fMRI activity indexes semantics, not lexical categories', *Brain & Language*, 2014, 132(1), 28–42
Moser DJ, Cohen RA, Malloy PF et al, 'Reduplicative paramnesia: longitudinal neurobehavioral and neuroimaging analysis', *Journal of Geriatric Psychiatry and Neurology*, 1998, 11(4), 174–80
Moshman D & Geil M, 'Collaborative reasoning: evidence for collective rationality', *Thinking & Reasoning*, 1998, 4(3), 231–48
Mosidze VM, Mkheidze RA & Makashvili MA, 'Disorders of visuo-spatial attention in patients with unilateral brain damage', *Behavioural Brain Research*, 1994, 65(1), 121–2

Moss AD & Turnbull OH, 'Hatred of the hemiparetic limbs (misoplegia) in a 10 year old child', *Journal of Neurology, Neurosurgery, and Psychiatry*, 1996, 61(2), 210–1

Mossbridge J, Tressoldi PE & Utts J, 'Predictive physiological anticipation preceding seemingly unpredictable stimuli: a meta-analysis', *Frontiers in Psychology*, 2012, 3, article 390

Mostofsky E, Maclure M, Sherwood JB et al, 'Risk of acute myocardial infarction after death of a significant person in one's life: The Determinants of Myocardial Infarction Onset Study', *Circulation*, 2012, 125(3), 491–6

Motomura N, Satani S & Inaba M, 'Monozygotic twin cases of the agenesis of the corpus callosum with schizophrenic disorder', *Psychiatry and Clinical Neurosciences*, 2002, 56(2), 199–202

Möttönen R & Watkins KE, 'Motor representations of articulators contribute to categorical perception of speech sounds', *Journal of Neuroscience*, 2009, 29(31), 9819–25

Moulin CJA, Conway MA, Thompson RG et al, 'Disorders of memory awareness: recollective confabulation in two cases of persistent déjà vécu', *Neuropsychologia*, 2005, 43(9), 1362–78

Mounk Y, 'What an audacious hoax reveals about academia: three scholars wrote 20 fake papers using fashionable jargon to argue for ridiculous conclusions', *The Atlantic*, 5 October 2018: www.theatlantic.com/ideas/archive/2018/10/new-sokal-hoax/572212/ (accessed 23 February 2021)

Mouren P & Tatossian A, 'Les illusions visuo-spatiales: étude clinique', *L'Encéphale*, 1963, 52, 439–80

Moya KL, Benowitz LI, Levine DN et al, 'Covariant defects in visuospatial abilities and recall of verbal narrative after right hemisphere stroke', *Cortex*, 1986, 22(3), 381–97

Mu B, *Chinese Philosophy A–Z*, Edinburgh University Press, 2009

Mucci A, Galderisi S, Bucci P et al, 'Hemispheric lateralization patterns and psychotic experiences in healthy subjects', *Journal of Experimental Child Psychology*, 2005, 139(2), 141–54

Mueser KT, Sayers SL, Schooler NR et al, 'A multisite investigation of the reliability of the Scale for the Assessment of Negative Symptoms', *American Journal of Psychiatry*, 1994, 151(10), 1453–62

Mugford ST, Mallon EB & Franks NR, 'The accuracy of Buffon's needle: a rule of thumb used by ants to estimate area', *Behavioral Ecology*, 2001, 12(6), 655–8

Muir J, 'Wild Wool', *Overland Monthly*, April 1875, 14, 361–6

———, *Travels in Alaska*, Houghton Mifflin, 1915

Mullan S & Penfield W, 'Illusions of comparative interpretation and emotion; production by epileptic discharge and by electrical stimulation in the temporal cortex', *AMA Archives of Neurology & Psychiatry*, 1959, 81(3), 269–84

Müller F, Lenz C, Dolder P et al, 'Increased thalamic resting-state connectivity as a core driver of LSD-induced hallucinations', *Acta Psychiatrica Scandinavica*, 2017, 136(6), 648–57

Müller FM, 'Lectures on Mr. Darwin's philosophy of language' [Part 3], *Fraser's Magazine*, 1873, 8(43), 1–24

———, *Lectures on the Science of Language*, Longman, Green & Co, 6th edn, 1885 [1861]

Müller JL, Gänssbauer S, Sommer M et al, 'Gray matter changes in right superior temporal gyrus in criminal psychopaths: evidence from voxel-based morphometry', *Journal of Experimental Child Psychology*, 2008, 163(3), 213–22

Müller JL, Sommer M, Döhnel K et al, 'Disturbed prefrontal and temporal brain function during emotion and cognition interaction in criminal psychopathy', *Behavioral Sciences & the Law*, 2008, 26(1), 131–50

Muller JZ, *The Tyranny of Metrics*, Princeton University Press, 2018

Muller RA, *Now: The Physics of Time*, Norton, 2016

Müller T, Büttner T, Kuhn W et al, 'Palinopsia as sensory epileptic phenomenon', *Acta Neurologica Scandinavica*, 1995, 91(6), 433–6

Müller WA, *Developmental Biology*, Springer, 1997

Müller-Oehring EM, Schulte T, Raassi C et al, 'Local-global interference is modulated by age, sex and anterior corpus callosum size', *Brain Research*, 2007, 1142(1), 189–205

Mullin T, 'Turbulent times for fluids', *New Scientist*, 11 November 1989, 52

Mullins RT, *The End of the Timeless God*, Oxford University Press, 2016

Mummery CJ, Patterson K, Hodges JR et al, 'Generating "tiger" as an animal name or a word beginning with T: differences in brain activation', *Proceedings of the Royal Society of London, Series B: Biological Sciences*, 1996, 263(1373), 989–95

———, 'Functional neuroanatomy of the semantic system: divisible by what?', *Journal of Cognitive Neuroscience*, 1998, 10(6), 766–77

Munafò MR & Flint J, 'How reliable are scientific studies?', *British Journal of Psychiatry*, 2010, 197(4), 257–58

Munafò MR, Stothart G & Flint J, 'Bias in genetic association studies and impact factor', *Molecular Psychiatry*, 2009, 14(2), 119–20

Mundt C, Richter P, van Hees H et al, 'Zeiterleben und Zeitschätzung depressiver Patienten', *Nervenarzt*, 1998, 69(1), 38–45

Mundy P & Crowson M, 'Joint attention and early social communication: implications for research on intervention with autism', *Journal of Autism and Developmental Disorders*, 1997, 27(6), 653–76

Murai T & Fukao K, 'Paramnesic multiplication of autobiographical memory as a manifestation of interictal psychosis', *Psychopathology*, 2003, 36(1), 49–51

Murai T, Toichi M, Sengoku A et al, 'Reduplicative paramnesia in patients with focal brain damage', *Neuropsychiatry, Neuropsychology & Behavioral Neurology*, 1997, 10(3), 190–6

Murata J, 'Colors in the life-world', *Continental Philosophy Review*, 1998, 31(3), 293–305

Muratori F, Cesari A & Casella C, 'Autism and cerebellum: an unusual finding with MRI', *Panminerva Medica*, 2001, 43(4), 311–5

Murdoch I, 'On "God" and "good"', in *Existentialists & Mystics*, Penguin, 1999

Murphy AP, Leopold DA, Humphreys GW et al, 'Lesions to right posterior parietal cortex impair visual depth perception from disparity but not motion cues', *Philosophical Transactions of the Royal Society of London: Series B, Biological Sciences*, 2016, 371(1697), 20150263

Muscatello CF & Giovanardi Rossi P, 'Perdita della visione mentale e patologia dell'esperienza temporale', *Giornale di Psichiatria e di Neuropatologia*, 1967, 95(4), 765–90

Musser G, *Spooky Action at a Distance*, Farrar, Straus and Giroux, 2015

——, 'What is spacetime?', *Nature*, 2018, 557(7704), s3–6

Must O & Must A, 'Changes in test-taking patterns over time', *Intelligence*, 2013, 41(6), 780–90

Must O, Must A & Raudik V, 'The secular rise in IQs: in Estonia, the Flynn effect is not a Jensen effect', *Intelligence*, 2003, 31(5), 461–71

Muter V, Taylor S & Vargha-Khadem F, 'A longitudinal study of early intellectual development in hemiplegic children', *Neuropsychologia*, 1997, 35(3), 289–98

Mychack P, Kramer JH, Boone KB et al, 'The influence of right frontotemporal dysfunction on social behavior in frontotemporal dementia', *Neurology*, 2001, 56(11 suppl 4), s11–5

Myers PS, 'Profiles of communication deficits in patients with right cerebral hemisphere damage: implications for diagnosis and treatment', *Aphasiology*, 2005, 19(12), 1147–60 (reprinted from RH Brookshire (ed), *Clinical Aphasiology: Proceedings of the Conference*, vol 9, BRK Publishers, 1979, 27–36)

Myers PS & Linebaugh CW, 'Comprehension of idiomatic expressions by right-hemisphere-damaged adults', in RH Brookshire (ed), *Clinical Aphasiology: Proceedings of the Conference*, vol 11, BRK Publishers, 1981, 254–61

Nachev P, Roberts R, Husain M et al, 'The neural basis of meta-volition', *Communications Biology*, 2019, 2, article 101

Nadal M, Munar E, Capó MA et al, 'Towards a framework for the study of the neural correlates of aesthetic preference', *Spatial Vision*, 2008, 21(3–5), 379–96

Nadal M, Schiavi S & Cattaneo Z, 'Hemispheric asymmetry of liking for representational and abstract paintings', *Psychonomic Bulletin & Review*, 2018, 25(5), 1934–42

Nadig AS, Ozonoff S, Young GS et al, 'A prospective study of response to name in infants at risk for autism', *Archives of Pediatrics and Adolescent Medicine*, 2007, 161(4), 378–83

Naeser MA, Martin PI, Ho M et al, 'Transcranial magnetic stimulation and aphasia rehabilitation', *Archives of Physical Medicine and Rehabilitation*, 2012, 93(1 Suppl), s26–34

Nagaishi A, Narita T, Gondo Y et al, 'Left-sided metamorphopsia of the face and simple objects caused by an infarction at the right side of the splenium of the corpus callosum', *Rinsho Shinkeigaku*, 2015, 55(7), 465–71

Nagaoka K, Ookawa S & Maeda K, 'A case of corticobasal degeneration presenting with visual hallucination', *Rinsho Shinkeigaku*, 2004, 44(3), 193–7

Nagara T, Ohara H & Yano K, 'Disappearance of hallucinations and delusions following left putaminal hemorrhage in a case of schizophrenia', *Seishin Shinkeigaku Zasshi*, 1996, 98, 498

Nagaratnam N & O'Neile L, 'Delusional parasitosis following occipito-temporal cerebral infarction', *General Hospital Psychiatry*, 2000, 22(2), 129–32

Nagaratnam N, Virk S & Brdarevic O, 'Musical hallucinations associated with recurrence of a right occipital meningioma', *British Journal of Clinical Practice*, 1996, 50(1), 56–7

Nagel BJ, Herting MM, Maxwell EC et al, 'Hemispheric lateralization of verbal and spatial working memory during adolescence', *Brain and Cognition*, 2013, 82(1), 58–68

Nagel T, *The View from Nowhere*, Oxford University Press, 1986

——, *Mind and Cosmos: Why the Materialist Neo-Darwinian Conception of Nature is*

Almost Certainly False, Oxford University Press, 2012

Nagy M, Ákos Z, Biro D et al, 'Hierarchical group dynamics in pigeon flocks', *Nature*, 2010, 464(7290), 890–3

Nahm M & Greyson B, 'Terminal lucidity in patients with chronic schizophrenia and dementia: a survey of the literature', *Journal of Nervous and Mental Disease*, 2009, 197(12), 942–4

Nahm M, Greyson B, Kelly EW et al, 'Terminal lucidity: a review and a case collection', *Archives of Gerontology and Geriatrics*, 2012, 55(1), 138–42

Naito E, Roland PE, Grefkes C et al, 'Dominance of the right hemisphere and role of area 2 in human kinesthesia', *Journal of Neurophysiology*, 2005, 93(2), 1020–34

Nakajima K, 'Visual hallucination associated with anterior cerebral artery occlusion', *Brain and Nerve*, 1991, 43(1), 71–6

Nakamura M, Nestor PG, McCarley RW et al, 'Altered orbitofrontal sulcogyral pattern in schizophrenia', *Brain*, 2007, 130(3), 693–707

Nakamura RK & Mishkin M, 'Blindness in monkeys following non-visual cortical lesions', *Brain Research*, 1980, 188(2), 572–7

———, 'Chronic "blindness" following lesions of nonvisual cortex in the monkey', *Experimental Brain Research*, 1986, 63(1), 173–84

Nakanishi N, Fukuda H & Tatara K, 'Changes in psychosocial conditions and eventual mortality in community-residing elderly people', *Journal of Epidemiology*, 2003, 13(2), 72–9

Nakano S, Yamashita F, Matsuda H et al, 'Relationship between delusions and regional cerebral blood flow in Alzheimer's disease', *Dementia and Geriatric Cognitive Disorders*, 2006, 21(1), 16–21

Narayan VM, Narr KL, Kumari V et al, 'Regional cortical thinning in subjects with violent antisocial personality disorder or schizophrenia', *American Journal of Psychiatry*, 2007, 164(9), 1418–27

Narr KL, Bilder RM, Kim S et al, 'Abnormal gyral complexity in first-episode schizophrenia', *Biological Psychiatry*, 2004, 55(8), 859–67

Narumoto J, Nakamura K, Kitabayashi Y et al, 'Othello syndrome secondary to right orbitofrontal lobe excision', *Journal of Neuropsychiatry and Clinical Neurosciences*, 2006, 18(4), 560–1

Narumoto J, Okada T, Sadato N et al, 'Attention to emotion modulates fMRI activity in human right superior temporal sulcus', *Cognitive Brain Research*, 2001, 12(2), 225–31

Narumoto J, Ueda H, Tsuchida H et al, 'Regional cerebral blood flow changes in a patient with delusional parasitosis before and after successful treatment with risperidone: a case report', *Progress in Neuropsychopharmacology and Biological Psychiatry*, 2006, 30(4), 737–40

Nasar S, *A Beautiful Mind*, Simon & Schuster, 2011

Nasrallah HA, 'The unintegrated right cerebral hemispheric consciousness as alien intruder: a possible mechanism for Schneiderian delusions in schizophrenia', *Comprehensive Psychiatry*, 1985, 26(3), 273–82

Nass R, Sinha S & Solomon G, 'Epileptic facial metamorphopsia', *Brain and Development*, 1985, 7(1), 50–52

Nasseri K & Moulton LH, 'Patterns of death in the first and second generation immigrants from selected Middle Eastern countries in California', *Journal of Immigrant and Minority Health*, 2011, 13(2), 361–70

Nathaniel-James DA & Frith CD, 'Confabulation in schizophrenia: evidence of a new form?', *Psychological Medicine*, 1996, 26(2), 391–9

Nathanson M, Bergman PS & Gordon GG, 'Denial of illness; its occurrence in one hundred consecutive cases of hemiplegia', *AMA Archives of Neurology and Psychiatry*, 1952, 68(3), 380–7

National Institute on Alcohol Abuse and Alcoholism, *Alcohol Alert*, no. 46, December 1999

Navon D, 'Forest before trees: the precedence of global features in visual perception', *Cognitive Psychology*, 1977, 9(3), 353–83

Nebes RD, 'Investigation on lateralization of function in the disconnected hemispheres of man', unpublished thesis, California Institute of Technology, 1971

Nebes RD & Sperry RW, 'Hemispheric deconnection with cerebral birth injury in the dominant arm area', *Neuropsychologia*, 1971, 9, 247–59

Needham J, *Science and Civilisation in China*, Cambridge University Press, 1954–98

———, *The Grand Titration: Science and Society in East and West*, George Allen & Unwin, 1969

Needham T, *Visual Complex Analysis*, Oxford University Press, 1999

Needleman J, *The American Soul*, Tarcher/Putnam, 2003

———, *I Am Not I*, North Atlantic Books, 2016

Neetesh B & Yogaratnam J, 'Organic Othello syndrome following a stroke a rare complication', *German Journal of Psychiatry*, 2012, 15(1), 41–3

Negoianu D & Goldfarb S, 'Just add water', *Journal of the American Society of Nephrology*, 2008, 19(6), 1041–3

Negri A, *Subversive Spinoza*, Manchester University Press, 2004

Negri GA, Rumiati RI, Zadini A et al, 'What is the role of motor simulation in action and object recognition? Evidence from apraxia', *Cognitive Neuropsychology*, 2007, 24(8), 795–816

Neihart M, 'The impact of giftedness on psychological well-being: what does the empirical literature say?', *Roeper Review*, 1999, 22(1), 10–17

Neisser U, 'Rising scores on intelligence tests', *American Scientist*, 1997, 85(5), 440–7

Nejad AG & Toofani K, 'Co-existence of lycanthropy and Cotard's syndrome in a single case', *Acta Psychiatrica Scandinavica*, 2005, 111(3), 250–2

Nelsen EM, Frankel J & Jenkins LM, 'Nongenetic inheritance of cellular handedness', *Development*, 1989, 105(3), 447–56

Nelson E, 'Warning signs of a possible collapse of contemporary mathematics', in Heller M, *Infinity: New Research Frontiers*, Cambridge University Press, 2014, 76–85

Nelson S, Dosenbach N, Cohen A et al, 'Role of the anterior insula in task-level control and focal attention', *Brain Structure & Function*, 2010, 214(5–6), 669–80

Nelson XJ & Jackson RR, 'The role of numerical competence in a specialized predatory strategy of an araneophagic spider', *Animal Cognition*, 2012, 15(4), 699–710

Nemeroff C & Rozin P, 'The contagion concept in adult thinking in the United States: transmission of germs and interpersonal influence', *Ethos*, 1994, 22(2), 158–86

———, 'The makings of the magical mind: the nature and function of sympathetic magical thinking', in KS Rosengren, CN Johnson & PL Harris (eds), *Imagining the Impossible: Magical, Scientific and Religious Thinking in Children*, Cambridge University Press, 2000, 1–34

Nestle M, 'Food industry funding of nutrition research: the relevance of history for current debates', *JAMA Internal Medicine*, 2016, 176(11), 1685–6

Nestor PG, Nakamura M, Niznikiewicz M et al, 'In search of the functional neuroanatomy of sociality: MRI subdivisions of orbital frontal cortex and social cognition', *Social Cognitive and Affective Neuroscience*, 2013, 8(4), 460–7

Nettl B, *The Study of Ethnomusicology: Twenty-Nine Issues and Concepts*, University of Illinois Press, 1983

Nettle D, *Strong Imagination: Madness, Creativity and Human Nature*, Oxford University Press, 2002

Newberg AB, Alavi A, Baime M et al, 'The measurement of regional cerebral blood flow during the complex cognitive task of meditation: a preliminary SPECT study', *Journal of Experimental Child Psychology: Neuroimaging*, 2001, 106(2), 113–22

Newberg AB & d'Aquili EG, 'The near death experience as archetype: a model for "prepared" neurocognitive processes', *Anthropology of Consciousness*, 1994, 5(4), 1–15

Newberg AB, d'Aquili EG & Rause VP, *Why God Won't Go Away: Brain Science and the Biology of Belief*, Ballantine, 2001

Newberg AB & Lee BY, 'The neuroscientific study of religious and spiritual phenomena: or why God doesn't use biostatistics', *Zygon*, 2005, 40(2), 469–90

Newberg AB, Pourdehnad M, Alavi A et al, 'Cerebral blood flow during meditative prayer: preliminary findings and methodological issues', *Perceptual & Motor Skills*, 2003, 97(2), 625–30

Newell A & Simon HA, 'Computer science as empirical inquiry: symbols and search', 1975 ACM Turing Award Lecture: citeseerx.ist.psu.edu/viewdoc/download?doi=10.1.1.334.2089&rep=rep1&type=pdf (accessed 23 February 2021)

Newman JH, Cardinal, 'Sermon 13: Implicit and Explicit Reason', *Fifteen Sermons Preached Before the University of Oxford between A.D. 1826–1843*, University of Notre Dame Press, 1998 [1909], 261–2

Newman SA, *Notre Dame Philosophical Reviews*, 6 May 2014: ndpr.nd.edu/news/in-search-of-mechanisms-discoveries-across-the-life-sciences-2/ (accessed 23 February 2021)

Newton I, *Philosophiæ Naturalis Principia Mathematica*, 3 vols, 1687

Newton N, 'Emergence and the uniqueness of consciousness', *Journal of Consciousness Studies*, 2001, 8(9–10), 47–59

Nguyen T-V, McCracken J, Ducharme S et al, 'Testosterone-related cortical maturation across childhood and adolescence', *Cerebral Cortex*, 2013, 23(6), 1424–32

Nichelli P, Rinaldi M & Cubelli R, 'Selective spatial attention and length representation in normal subjects and in patients with unilateral spatial neglect', *Brain and Cognition*, 1989, 9(1), 57–70

Nichol L (ed), *The Essential David Bohm*, Routledge, 2003

Nicholls ME, 'Hemispheric asymmetries for temporal resolution: a signal detection analysis of threshold and bias', *Quarterly Journal of Experimental Psychology A: Human Experimental Psychology*, 1994, 47(2), 291–310

Nicholls ME, Bradshaw JL & Mattingley JB, 'Free-viewing perceptual asymmetries for the judgement of brightness, numerosity and size', *Neuropsychologia*, 1999, 37(3), 307–14

Nicholls ME, Ellis BE, Clement JG et al, 'Detecting hemifacial asymmetries in emotional expression with three-dimensional computerized image analysis', *Proceedings of the Royal Society B: Biological Sciences*, 2004, 271(1540), 663–8

Nicholls ME, Orr CA & Lindell AK, 'Magical ideation and its relation to lateral preference', *Laterality*, 2005, 10(6), 503–15

Nicholls MER, Wolfgang BJ, Clode D et al, 'The effect of left and right poses on the expression of facial emotion', *Neuropsychologia*, 2002, 40, 1662–5

Nicholson DJ, 'Organisms ≠ machines', *Studies in History and Philosophy of Biological and Biomedical Sciences*, 2013, 44(4), 669–78

——, 'The machine conception of the organism in development and evolution: a critical analysis', *Studies in History and Philosophy of Biological and Biomedical Sciences*, 2014, 48(B), 162–74

——, 'Reconceptualizing the organism: from complex machine to flowing stream', in DJ Nicholson & J Dupré (eds), *Everything Flows: Towards a Processual Philosophy of Biology*, Oxford University Press, 2018, 139–66

Nicholson DJ & Dupré J (eds), *Everything Flows: Towards a Processual Philosophy of Biology*, Oxford University Press, 2018

Nickerson D & Newhall SM, 'A psychological color solid', *Journal of the Optical Society of America*, 1943, 33(7), 419–22

Nicolis SC, Zabzina N, Latty T et al, 'Collective irrationality and positive feedback', *PLoS One*, 2011, 6(4), e18901

Niebauer CL, 'A possible connection between categorical and coordinate spatial relation representations', *Brain and Cognition*, 2001, 47(3), 434–45

Nielsen M & Collison P, 'Science is getting less bang for its buck', 2018: www.theatlantic.com/science/archive/2018/11/diminishing-returns-science/575665/ (accessed 23 February 2021)

Nietzsche F, *Götzen-Dämmerung, oder, Wie man mit dem Hammer philosophiert*, Naumann Verlag, 1889

——, *The Wanderer and his Shadow*, trans PV Cohn, MacMillan, 1913 [*Der Wanderer und Sein Schatten* 1880]

——, *The Birth of Tragedy* and *The Genealogy of Morals*, trans F Golffing, Doubleday, 1956 [*Zur Genealogie der Moral: Eine Streitschrift* 1887]

——, *Philosophy in the Tragic Age of the Greeks*, trans M Cowan, Regnery, 1962 [*Philosophie im tragischen Zeitalter der Griechen* unpublished, written c 1873]

——, *The Will to Power*, trans W Kaufmann & RJ Hollingdale, Vintage Books, 1967 [*Der Wille zur Macht* 1901]

——, 'Der Don Juan der Erkenntniß', *Morgenröthe: Gedanken über die moralischen Vorurtheile*, Bk IV, §327; in G Colli & M Montinari (eds), *Werke: Kritische Gesamtausgabe*, Walter de Gruyter & Co, Section 5, vol 1, 1970 [1881]

——, *Beyond Good and Evil*, trans RJ Hollingdale, Penguin, 1974a [*Jenseits von Gut und Böse* 1886]

——, *The Gay Science*, trans W Kaufmann, Vintage, 1974b [*Die fröhliche Wissenschaft* 1882]

——, *Twilight of the Idols*, trans RJ Hollingdale, Penguin, 1990 [*Götzen-Dämmerung* 1889]

——, *Human, All Too Human*, trans M Faber & S Lehmann, Penguin, 1994 [*Menschliches, Allzumenschliches* 1878]

——, *The Birth of Tragedy and other writings*, ed R Geuss & R Spiers, trans R Spiers, Cambridge University Press, 1999 [*Die Geburt der Tragödie aus dem Geiste der Musik* 1872]

——, in B Williams (ed), *The Gay Science*, trans J Nauckhoff, Cambridge University Press, 2001 [*Die fröhliche Wissenschaft* 1882]

——, *Daybreak*, ed M Clark & B Leiter, trans RJ Hollingdale, Cambridge University Press, 2003a [*Morgenröthe* 1881]

——, in R Bittner (ed), *Writings from the Late Notebooks*, trans K Sturge, Cambridge University Press, 2003b

——, *Also Sprach Zarathustra*, Nikol Verlagsgesellschaft, 2011 [1883–5]

——, *On Truth and Lies in a Nonmoral Sense*, Theophania, 2012 [*Über Wahrheit und Lüge im aussermoralischen Sinne* 1896]

Nightingale S, 'Somatoparaphrenia: a case report', *Cortex*, 1982, 18(3), 463–7

Nijboer TC & Jellema T, 'Unequal impairment in the recognition of positive and negative emotions after right hemisphere lesions: a left hemisphere bias for happy faces', *Journal of Neuropsychology*, 2012, 6(1), 79–93

Nijboer TC, Nys GM, van der Smagt MJ et al, 'A selective deficit in the appreciation and recognition of brightness: brightness agnosia?', *Cortex*, 2009, 45(7), 816–24

Nijboer TC, Ruis C, van der Worp HB et al, 'The role of *Funktionswandel* in metamorphopsia', *Journal of Neuropsychology*, 2008, 2(1), 287–300

Nijhout HF, 'Metaphors and the role of genes in development', *Bioessays*, 1990, 12(9), 441–6

Niklas KJ, 'The role of phyllotactic pattern as a "developmental constraint" on the interception of light by leaf surfaces', *Evolution*, 1988, 42(1), 1–16

Nikolaenko NN, 'Representation activity of the right and left hemispheres of the brain', *Acta Neuropsychologica*, 2003, 1(1), 34–47

——, 'Sex differences and activity of the left and right brain hemispheres', *Journal of Evolutionary Biochemistry and Physiology*, 2005, 41(6), 689–99

Nikolaenko NN & Egorov AY, 'The role of the right and left cerebral hemispheres in depth perception', *Fiziologiia Cheloveka (Human Physiology)*, 1998, 24(6), 21–31

Nikolaenko NN, Egorov AY & Freiman EA, 'Representation activity of the right and left hemispheres of the brain', *Behavioural Neurology*, 1997, 10(1), 49–59

Nili U, Goldberg H, Weizman A et al, 'Fear thou not: activity of frontal and temporal circuits in moments of real-life courage', *Neuron*, 2010, 66(6), 949–62

Nisbet H, *Herder and the Philosophy and History of Science*, Modern Humanities Research Association, 1970

Nisbett RE & Wilson TD, 'Telling more than we can know: verbal reports on mental processes', *Psychological Review*, 1977, 84(3), 231–59

Nishio Y & Mori E, 'Delusions of death in a patient with right hemisphere infarction', *Cognitive and Behavioral Neurology*, 2012, 25(4), 216–23

Njemanze PC, Gomez CR & Horenstein S, 'Cerebral lateralization and color perception: a transcranial Doppler study', *Cortex*, 1992, 28(1), 69–75

Nobile M, Perego P, Piccinini L et al, 'Further evidence of complex motor dysfunction in drug naive children with autism using automatic motion analysis of gait', *Autism*, 2011, 15(3), 263–83

Noble D, 'Evolution viewed from physics, physiology and medicine', *Interface Focus*, 2017, 7(5), 20160159

——, [undated]: www.musicoflife.website/pdfs/The%20Dance%20Sourcebook.pdf (accessed 23 February 2021)

Noë A, 'Art and the Limits of Neuroscience', *The New York Times*, 4 December 2011

Noguchi Y & Murota M, 'Temporal dynamics of neural activity in an integration of visual and contextual information in an esthetic preference task', *Neuropsychologia*, 2013, 51(6), 1077–84

Nomura K, Kazui H, Wada T et al, 'Classification of delusions in Alzheimer's disease and their neural correlates', *Psychogeriatrics*, 2012, 12(3), 200–10

Norenzayan A & Gervais WM, 'The origins of religious disbelief', *Trends in Cognitive Sciences*, 2013, 17(1), 20–5

Norenzayan A, Gervais WM & Trzesniewski KH, 'Mentalizing deficits constrain belief in a personal God', *PLoS One*, 2012, 7(5), e36880

Norman DA, Coblentz CL, Brooks LR et al, 'Expertise in visual diagnosis: a review of the literature', *Academic Medicine*, 1992, 67(10 suppl), s78–83

Norredam M, Olsbjerg M, Petersen JH et al, 'Are there differences in injury mortality among refugees and immigrants compared with native-born?', *Injury Prevention*, 2013, 19(2), 100–5

Norris P & Inglehart R, *Sacred and Secular: Religion and Politics Worldwide*, Cambridge University Press, 2004

Noterdaeme M, Mildenberger K, Minow F et al, 'Evaluation of neuromotor deficits in children with autism and children with a specific speech and language disorder', *European Child & Adolescent Psychiatry*, 2002, 11(5), 219–25

Nottale L, 'Scale relativity and fractal space-time: applications to quantum physics, cosmology and chaotic systems', *Chaos, Solitons and Fractals*, 1996, 7(6), 877–938

Nottebohm F, 'The ontogeny of bird song', *Science*, 1970, 167(3920), 950–6

Novalis (Friedrich Leopold, Freiherr von Hardenberg), 'Aus den Fragmentensammlungen', *Gesammelte Werke*, Sigbert Mohn Verlag, 1967

Nowell-Smith G, *Luchino Visconti*, British Film Institute, London, 2003

Nowicka A & Fersten E, 'Sex-related differences in interhemispheric transmission time in the human brain', *NeuroReport*, 2001, 12(18), 4171–5

Nowicka A & Tacikowski P, 'Transcallosal transfer of information and functional asymmetry of the human brain', *Laterality*, 2011, 16(1), 35–74

Nozaki S, Kato M, Takano H et al, 'Regional dopamine synthesis in patients with schizophrenia using L-[beta-^{11}C]DOPA PET', *Schizophrenia Research*, 2009, 108(1–3), 78–84

Nuechterlein KH & Dawson ME, 'Information processing and attentional functioning in the developmental course of schizophrenic disorders', *Schizophrenia Bulletin*, 1984, 10(2), 160–203

Numbers RL (ed), *Galileo Goes to Jail and Other Myths about Science and Religion*, Harvard University Press, 2009

Nuñez PL, *Neocortical Dynamics and Human EEG Rhythms*, Oxford University Press, 1995

———, *Brain, Mind, and the Structure of Reality*, Oxford University Press, 2010

Núñez R, Cooperrider K, Doan D et al, 'Contours of time: topographic construals of past, present, and future in the Yupno valley of Papua New Guinea', *Cognition*, 2012, 124(1), 25–35

Núñez RE & Sweetser E, 'With the future behind them: convergent evidence from Aymara language and gesture in the crosslinguistic comparison of spatial construals of time', *Cognitive Science*, 2006, 30(3), 401–50

Nuske HJ & Bavin EL, 'Narrative comprehension in 4–7-year-old children with autism: testing the weak central coherence account', *International Journal of Language & Communication Disorders*, 2011, 46(1), 108–19

Nutt AE, *Shadows Bright as Glass*, Free Press, 2011

Nuttall AD, *New Mimesis: Shakespeare and the Representation of Reality*, Yale University Press, 2007

Nye E & Arendts G, 'Intracerebral haemorrhage presenting as olfactory hallucinations', *Emergency Medicine Australasia*, 2002, 14(4), 447–9

Nye R, 'Psychological perspectives on children's spirituality', PhD thesis, University of Nottingham, 1998

O'Boyle EH, Forsyth D, Banks GC et al, 'A meta-analytic review of the Dark Triad-intelligence connection', *Journal of Research in Personality*, 2013, 47(6), 789–94

O'Boyle MW, Alexander JE & Benbow CP, 'Enhanced right hemisphere activation in the mathematically precocious: a preliminary EEG investigation', *Brain and Cognition*, 1991, 17(2), 138–53

O'Boyle MW & Benbow CP, 'Enhanced right hemisphere involvement during cognitive processing may relate to intellectual precocity', *Neuropsychologia*, 1990, 28(2), 211–6

O'Boyle MW, Benbow CP & Alexander JE, 'Sex differences, hemispheric laterality, and associated brain activity in the intellectually gifted', *Developmental Neuropsychology*, 1995, 11(4), 415–43

O'Boyle MW, Cunnington R, Silk TJ et al, 'Mathematically gifted male adolescents activate a unique brain network during mental rotation', *Brain Research: Cognitive Brain Research*, 2005, 25(2), 583–7

O'Boyle MW & Sanford M, 'Hemispheric asymmetry in the matching of melodies to rhythm sequences tapped in the right and left palms', *Cortex*, 1988, 24(2), 211–21

O'Brien G, 'The behavioral and developmental consequences of callosal agenesis', in M Lassonde & MA Jeeves (eds), *Callosal agenesis: a natural split brain?*, Plenum Press, 1994

O'Connor LE, Berry JW, Lewis T et al, 'Empathy and depression: the moral system on overdrive', in TFD Farrow & PWR Woodruff (eds), *Empathy in Mental Illness*, Cambridge University Press, 2007, 49–75

O'Donnell MJ, Yusuf S, Mente A et al, 'Urinary sodium and potassium excretion and risk of cardiovascular events', *Journal of the American Medical Association*, 2011, 306(20), 2229–38

O'Mahony P, *Irish Times*, 12 August 2014

O'Neill D, Macsweeney CA, Cornell IA et al, 'Stravinsky syndrome: giving a voice to chronic stroke disease', *Quarterly Journal of Medicine*, 2014, 107(6), 489–93

O'Neill O, *A Question of Trust*, Cambridge University Press, 2002

O'Regan JK, 'Solving the "real" mysteries of visual perception: the world as an outside memory', *Canadian Journal of Psychology*, 1992, 46, 461–88

O'Regan JK & Noë A, 'A sensorimotor account of vision and visual consciousness', *Behavioral and Brain Sciences*, 2001, 24(5), 939–1031

Obama B, speech to the Obama Foundation summit in Chicago, 29 October 2019

Oberhummer H, Csótó A & Schlattl H, 'Fine-tuning carbon-based life in the universe by the triple-alpha process in Red Giant stars', *Science*, 2000, 289(5476), 88–90

Oelman H & Loeng B, 'A validation of the emotional meaning of single intervals according to classical and Indian music theory', *Proceedings of the 5th Triennial ESCOM Conference, Hanover University of Music & Drama*, 2003, 393–6

Oepen G, Harrington A, Spitzer M et al, '"Feelings" of conviction: on the relation of affect and thought disorder', in M Spitzer, FA Uehlein & G Oepen (eds), *Psychopathology and Philosophy*, Springer, 1989, 43–55

Oertel V, Knöchel C, Rotarska-Jagiela A et al, 'Reduced laterality as a trait marker of schizophrenia – evidence from structural and functional neuroimaging', *Journal of Neuroscience*, 2010, 30(6), 2289–99

Oertel-Knöchel V, Knöchel C, Matura S et al, 'Reduced functional connectivity and asymmetry of the planum temporale in patients with schizophrenia and first-degree relatives', *Schizophrenia Research*, 2013, 147(2–3), 331–8

Oerter R, *The Theory of Almost Everything: The Standard Model, the Unsung Triumph of Modern Physics*, Plume, 2006

Ogawa M, 'A cultural history of science education in Japan: an epic description', in WW Cobern (ed), *Socio-Cultural Perspectives on Science Education: An International Dialogue*, Kluwer, 1998, 139–61

Ohara K, Nishii R, Nakajima T et al, 'Alterations of symptoms with borderline personality disorder after fronto-temporal traumatic brain injury: a case study', *Seishin Shinkeigaku Zasshi*, 2004, 106(4), 458–66

Ohloff G, *Scent and Fragrances*, Springer-Verlag, 1994

Okada T, Sato W, Kubota Y et al, 'Right hemispheric dominance and interhemispheric cooperation in gaze-triggered reflexive shift of attention', *Psychiatry and Clinical Neurosciences*, 2012, 66(2), 97–104

Okada T, Sato W & Toichi M, 'Right hemispheric dominance in gaze-triggered reflexive shift of attention in humans', *Brain and Cognition*, 2006, 62(2), 128–33

Okamoto K & Tanaka Y, 'Subjective usefulness and 6-year mortality risks among elderly persons in Japan', *The Journals of Gerontology: Series B, Psychological Sciences and Social Sciences*, 2004, 59(5), 246–9

Okamoto-Barth S, Call J & Tomasello M, 'Great apes' understanding of other individuals' line of sight', *Psychological Science*, 2007, 18(5), 462–8

Oke A, Keller R, Mefford I et al, 'Lateralization of norepinephrine in human thalamus', *Science*, 1978, 200(4348), 1411–3

Okike K, Hug KT, Kocher MS et al, 'Single-blind vs double-blind peer review in the setting of author prestige', *Journal of the American Medical Association*, 2016, 316(12), 1315–6

Okri B, 'Plato's dream', *A Time for New Dreams*, Rider, 2009

Okruszek Ł, Kalinowski K & Talarowska M, 'Religiosity as a symptom of selected neuropsychiatric disorders', *Medical Science and Technology*, 2013, 54, 136–40

Olausson H, Charron J, Marchand S et al, 'Feelings of warmth correlate with neural activity in right anterior insular cortex', *Neuroscience Letters*, 2005, 389, 1–5

Olby R, 'Quiet debut for the double helix', *Nature*, 2003, 421(6921), 402–5

Oleksiak A, Postma A, van der Ham IJM et al, 'A review of lateralization of spatial functioning in nonhuman primates', *Brain Research Reviews*, 2011, 67(1–2), 56–72

Olivares R, Michalland S & Aboitiz F, 'Cross-species and intraspecies analysis of the corpus callosum', *Brain, Behavior and Evolution*, 2000, 55(1), 37–43

Olivares R, Montiel J & Aboitiz F, 'Species differences and similarities in the fine structure of the mammalian corpus callosum', *Brain, Behavior and Evolution*, 2001, 57(2), 98–105

Oliveira-Pinto AV, Andrade-Moraes CH, Oliveira LM et al, 'Do age and sex impact on the absolute cell numbers of human brain regions?', *Brain Structure & Function*, 2016, 221(7), 3547–59

Oliver DA, 'Autonomous quanta: on the dynamism of nature and the nature of dynamism', 2019: www.davidanthonyoliver.net/wp-content/uploads/2019/10/AutonomousQuanta_10-09-2019-1.pdf (accessed 23 February 2021)

Oliveri M, Rossini PM, Filippi MM et al, 'Time-dependent activation of parieto-frontal networks for directing attention to tactile space: a study with paired transcranial magnetic stimulation pulses in right-brain-damaged patients with extinction', *Brain*, 2000, 123, 1939–47

Olivola CY, Funk F & Todorov A, 'Social attributions from faces bias human choices', *Trends in Cognitive Sciences*, 2014, 18(11), 566–70

Olkowicz S, Kocourek M, Lučan RK et al, 'Birds have primate-like numbers of neurons in the forebrain', *Proceedings of the National Academy of Sciences of the United States of America*, 2016, 113(26), 7255–60

Olsen L, 'The ironic dialectic in Yeats', *Colby Quarterly*, 1983, 19(4), 215–20

Olson IR, von der Heide RJ, Alm KH et al, 'Development of the uncinate fasciculus: Implications for theory and developmental disorders', *Developmental Cognitive Neuroscience*, 2015, 14, 50–61

Omnès R, 'The direction of time in quantum mechanics', in S Albeverio & P Blanchard, *Direction of Time*, Springer, 2014, 49–56

Ōno S, *Nihongo no bunpō o kangaeru*, Iwanami, 1978

Oomen PMF, 'God's power and almightiness in Whitehead's thought', *Open Theology*, 2015, 1, 277–92

Oparin AI, *Life: Its Nature, Origin and Development*, Oliver & Boyd, 1961

Open Science Collaboration, 'Estimating the reproducibility of psychological science', *Science*, 2015, 349(6251), 943

Oppenheimer JR, *Science and the Common Understanding*, Simon & Schuster, 1954

Oppenheimer SM, Gelb A, Girvin JP et al, 'Cardiovascular effects of human insular cortex stimulation', *Neurology*, 1992, 42(9), 1727–32

Oreshkov O, Costa F & Brukner C, 'Quantum correlations with no causal order', *Nature Communications*, 2012, 3, article 1092

Oriel E, 'Whom would animals designate as "persons"? On avoiding anthropocentrism and including others', *Journal of Evolution and Technology*, 2014, 24(3), 44–59

Orjuela-Rojas JM, Sosa-Ortiz AL, Díaz-Victoria AR et al, 'The painter from Sinaloa: artistic analysis of a case of spatial agnosia and neglect of visual shapes', *Neurocase*, 2017, 23(5–6), 304–13

Orme L & Maggs C, 'Decision-making in clinical practice: how do expert nurses, midwives and health visitors make decisions?', *Nurse Education Today*, 1993, 13(4), 270–6

Ornstein R, Herron J, Johnstone J et al, 'Differential right hemisphere involvement in two reading tasks', *Psychophysiology*, 1979, 16(4), 398–401

Orr HA, 'Awaiting a new Darwin', *New York Review of Books*, 7 February 2013

Orr J, Tobin K, Carey D et al, 'Religious attendance, religious importance, and the pathways to depressive symptoms in men and women aged 50 and over living in Ireland', *Research on Aging*, 2019, 41(9), 891–911

Ortega y Gasset J, 'The barbarism of "specialisation"', in *The Revolt of the Masses*, Norton, 1994 [*La Rebelión de las Masas* 1930], 107–14

Ortigue S & Bianchi-Demicheli F, 'Intention, false beliefs, and delusional jealousy: insights into the right hemisphere from neurological patients and neuroimaging studies', *Medical Science Monitor*, 2011, 17(1), RA1–11

Ortigue S, King D, Gazzaniga M et al, 'Right hemisphere dominance for understanding the intentions of others: evidence from a split-brain patient', *BMJ Case Reports*, 2009, 2009, bcr07.2008.0593

Ortigue S, Sinigaglia C, Rizzolatti G et al, 'Understanding actions of others: the electrodynamics of the left and right hemispheres. A high-density EEG neuroimaging study', *PLoS One*, 2010, 5(8), e12160

Ortigue S, Thompson JC, Parasuraman R et al, 'Spatio-temporal dynamics of human intention understanding in temporo-parietal cortex: a combined EEG/fMRI repetition suppression paradigm', *PLoS One*, 2009, 4(9), e6962

Ortuño F, Guillén-Grima F, López-García P et al, 'Functional neural networks of time perception: challenge and opportunity for schizophrenia research', *Schizophrenia Research*, 2011, 125(2–3), 129–35

Oser FK, 'The development of religious judgement', *New Directions for Child and Adolescent Development*, 1991, 52, 5–25

Osherson D, Perani D, Cappa S et al, 'Distinct brain loci in deductive versus probabilistic reasoning', *Neuropsychologia*, 1998, 36(4), 369–76

Oskamp S, 'Overconfidence in case-study judgments', *Journal of Consulting Psychology*, 1965, 29(3), 261–5

Oskarsson S, Dawes C, Johannesson M et al, 'The genetic origins of the relationship between psychological traits and social trust', *Twin Research and Human Genetics*, 2012, 15(1), 21–33

Osserman R, *Poetry of the Universe: A Mathematical Exploration of the Cosmos*, Anchor, 1995

Ossola M, Romani A, Tavazzi E et al, 'Epileptic mechanisms in Charles Bonnet syndrome', *Epilepsy & Behavior*, 2010, 18(1–2), 119–22

Östberg P, '18th century cases of noun-verb dissociation: the contribution of Carl Linnæus', *Brain and Language*, 2003, 84(3), 448–50

Osterling J & Dawson G, 'Early recognition of children with autism: a study of first birthday home videotapes', *Journal of Autism and Developmental Disorders*, 1994, 24(3), 247–57

Ostrom TM, Carpenter SL, Sedikides C et al, 'Differential processing of in-group and out-group information', *Journal of Personality and Social Psychology*, 1993, 64(1), 21–34

Otsuka-Hirota N, Yamamoto H, Miyashita K et al, 'Invisibility of moving objects: a core symptom of motion blindness', *BMJ Case Reports*, 2014, 2014, pii: bcr2013201233

Ott BR, Lafleche G, Whelihan WM et al, 'Impaired awareness of deficits in Alzheimer disease', *Alzheimer Disease and Associated Disorders*, 1996, 10(2), 68–76

Otto R, *The Idea of the Holy*, trans JW Harvey, Oxford University Press, 1917

Ovsiew F, 'The *Zeitraffer* phenomenon, akinetopsia, and the visual perception of speed of motion: a case report', *Neurocase*, 2014, 20(3), 269–72

Owen GS, Cutting J & David AS, 'Are people with schizophrenia more logical than healthy volunteers?', *British Journal of Psychiatry*, 2007, 191(5), 453–4

Owens D & Kelley R, 'Predictive properties of risk assessment instruments following self-harm', *British Journal of Psychiatry*, 2017, 210(6), 384–5

Oyama S, 'The nurturing of natures', in A Grunwald, M Gutmann & EM Neumann-Held (eds), *On Human Nature: Anthropological, Biological and Philosophical Foundations* (Studienreihe der Europäischen Akademie), Springer, 2002, 163–70

Oyebode F, 'The neurology of psychosis', *Medical Principles and Practice*, 2008, 17(4), 263–9

Ozsarac M, Aksay E, Kiyan S et al, 'De novo cerebral arteriovenous malformation: Pink Floyd's song "Brick in the Wall" as a warning sign', *Journal of Emergency Medicine*, 2012, 43(1), e17–20

Pace-Schott EF, Nave G, Morgan A et al, 'Sleep-dependent modulation of affectively guided decision-making', *Journal of Sleep Research*, 2012, 21(1), 30–9

Pachalska M, Grochmal-Bach B, Wilk M et al, 'Rehabilitation of an artist after right-hemisphere stroke', *Medical Science Monitor*, 2008, 14(10), 110–24

Pack AA & Herman LM, 'The dolphin's (*Tursiops truncatus*) understanding of human gazing and pointing: knowing what and where', *Journal of Comparative Psychology*, 2007, 121(1), 34–45

Padgett J, in interview reported by Susannah Cahalan in the *New York Post*, 20 April 2014a: | nypost.com/2014/04/20/how-a-brain-injury-turned-a-college-dropout-into-a-genius/ (accessed 23 February 2021)

Padgett J, in interview reported by Amanda Cochran for *CBS News*, 24 April 2014b: www.cbsnews.com/news/jason-padgett-changes-into-mathematical-prodigy-after-violent-assault/ (accessed 23 February 2021)

Padgett J, in interview reported by Tanya Lewis, 5 May 2014c: www.livescience.com/45349-brain-injury-turns-man-into-math-genius.html (accessed 23 February 2021)

Padgett J, in interview reported by Tanya Lewis in *The Washington Post*, 12 May 2014d: www.washingtonpost.com/national/health-science/a-man-became-a-math-wiz-after-suffering-brain-injuries-researchers-think-they-know-why/2014/05/12/88c4738e-d613-11e3-95d3-3bcd77cd4e11_story.html (accessed 23 February 2021)

Padgett J & Seaberg M, *Struck by Genius: How a Brain Injury Made Me a Mathematical Marvel*, Headline, 2014

Pai MC, 'Topographic disorientation: two cases', *Journal of the Formosan Medical Association*, 1997, 96(8), 660–3

Paine RT, 'Food web complexity and species diversity', *The American Naturalist*, 1966, 100(910), 65–75

Pais A, *Subtle is the Lord: The Science and the Life of Albert Einstein*, Clarendon Press, 1982

Pakkenberg B & Gundersen HJ, 'Neocortical neuron number in humans: effect of sex and age', *Journal of Comparative Neurology*, 1997, 384(2), 312–20

Pal PK, Hanajima R, Gunraj CA et al, 'Effect of low-frequency repetitive transcranial magnetic stimulation on interhemispheric inhibition', *Journal of Neurophysiology*, 2005, 94(3), 1668–75

Palaniyappan L & Liddle PF, 'Aberrant cortical gyrification in schizophrenia: a surface-based morphometry study', *Journal of Psychiatry & Neuroscience*, 2012, 37(6), 399–406

Palermo L, Ranieri G, Boccia M et al, 'Map-following skills in left and right brain-damaged patients with and without hemineglect', *Journal of Clinical and Experimental Neuropsychology*, 2012, 34(10), 1065–79

Pallasmaa J, *The Eyes of the Skin*, Wiley, 2005

Pallis CA, 'Impaired identification of faces and places with agnosia for colours; report of a case due to cerebral embolism', *Journal of Neurology, Neurosurgery, and Psychiatry*, 1955, 18(3), 218–24

Palmer AH, *The Life and Letters of Samuel Palmer, Painter & Etcher*, Seeley & Co, 1892

Palmerini F & Bogousslavsky J, 'Right hemisphere syndromes', in M Paciaroni, G Agnelli, V Caso et al (eds), *Manifestations of Stroke* (*Frontiers of Neurology & Neuroscience, vol 30*), Karger Medical and Scientific Publishers, 2012, 61–4

Palminteri S, Serra G, Buot A et al, 'Hemispheric dissociation of reward processing in humans: insights from deep brain stimulation', *Cortex*, 2013, 49(10), 2834–44

Pang C-Y, Nadal M, Müller J et al, 'Electrophysiological correlates of looking at paintings and its association with art expertise', *Biological Psychology*, 2013, 93(1), 246–54

Panikkar R, *The Vedic Experience: Mantramañjari*, Motilal Banarsidass, 2001

———, *Christophany: The Fullness of Man*, Orbis, 2004

Panksepp J, *Affective Neuroscience: the Foundations of Human and Animal Emotions*, Oxford University Press, 1998

———, 'At the interface of the affective, behavioral, and cognitive neurosciences: decoding the emotional feelings of the brain', *Brain and Cognition*, 2003, 52(1), 4–14

———, 'Emotional feelings originate below the neocortex: toward a neurobiology of the soul', *Behavioral and Brain Sciences*, 2007, 30(1), 101–3

Panksepp J, Burgdorf J, Turner C et al, 'Modeling ADHD-type arousal with unilateral frontal cortex damage in rats and beneficial effects of play therapy', *Brain and Cognition*, 2003, 52(1), 97–105

Pantelis PC, Byrge L, Tyszka JM et al, 'A specific hypoactivation of right temporo-parietal junction/posterior superior temporal sulcus in response to socially awkward situations in autism', *Social Cognitive and Affective Neuroscience*, 2015, 10(10), 1348–56

Papageorgiou C, Ventouras E, Lykouras L et al, 'Psychophysiological evidence for altered information processing in delusional misidentification syndromes', *Progress in Neuro-Psychopharmacology & Biological Psychiatry*, 2003, 27(3), 365–72

Papagno C, Fogliata A, Catricalà E et al, 'The lexical processing of abstract and concrete nouns', *Brain Research*, 2009, 1263, 78–86

Papagno C, Martello G & Mattavelli G, 'The neural correlates of abstract and concrete words: evidence from brain-damaged patients', *Brain Sciences*, 2013, 3(3), 1229–43

Paquier P, van Vugt P, Bal P et al, 'Transient musical hallucinosis of central origin: a review and clinical study', *Journal of Neurology, Neurosurgery, and Psychiatry*, 1992, 55(11), 1069–73

Pardo JV, Fox PT & Raichle ME, 'Localization of a human system for sustained attention by positron emission tomography', *Nature*, 1991, 349(6304), 61–4

Paré PW & Tumlinson JH, 'Plant volatiles as a defense against insect herbivores', *Plant Physiology*, 1999, 121(2), 325–32

Parente R & Tommasi L, 'A bias for the female face in the right hemisphere', *Laterality*, 2008, 13(4), 374–86

Parfit D, *Reasons and Persons*, Clarendon Press, 1984

Pariyadath V & Eagleman D, 'The effect of predictability on subjective duration', *PLoS One*, 2007, 2(11), e1264

Park G, Lubinski D & Benbow CP, 'Contrasting intellectual patterns predict creativity in the arts and sciences: tracking intellectually precocious youth over 25 years', *Psychological Science*, 2007, 18(11), 948–52

———, 'Ability differences among people who have commensurate degrees matter for scientific creativity', *Psychological Science*, 2008, 19(10), 957–61

Park H-J, Westin C-F, Kubicki M et al, 'White matter hemisphere asymmetries in healthy subjects and in schizophrenia: a diffusion tensor MRI study', *NeuroImage*, 2004, 23(1), 213–23

Park M-G, Choi K-D, Kim J-S et al, 'Hemimacropsia after medial temporo-occipital infarction', *Journal of Neurology, Neurosurgery, and Psychiatry*, 2007, 78(5), 546–8

Park RL, *Superstition: Belief in the Age of Science*, Princeton University Press, 2009

Parker J, Tsagkogeorga G, Cotton JA et al, 'Genome-wide signatures of convergent evolution in echolocating mammals', *Nature*, 2013, 502(7470), 228–31

Parkin AJ, Bindschaedler C, Harsent L et al, 'Pathological false alarm rates following damage to the left frontal cortex', *Brain and Cognition*, 1996, 32(1), 14–27

Parks NE, Rigby HB, Gubitz GJ et al, 'Dysmetropsia and Cotard's syndrome due to migrainous infarction – or not?', *Cephalalgia*, 2014, 34(9), 717–20

Parks RW, Loewenstein DA, Dodrill KL et al, 'Cerebral metabolic effects of a verbal fluency test – a PET scan study', *Journal of Clinical and Experimental Neuropsychology*, 1988, 10(5), 565–75

Parks T, *Out of My Head: On the Trail of Consciousness*, Harvill Secker, 2018

Parnas J, 'The self and intentionality in the pre-psychotic stages of schizophrenia', in D Zahavi (ed), *Exploring the Self*, John Benjamins, 2000, 115–74

Parnas J & Bovet P, 'Autism in schizophrenia revisited', *Comprehensive Psychiatry*, 1991, 32(1), 1–15

Parnas J & Handest P, 'Phenomenology of anomalous self-experience in early schizophrenia', *Comprehensive Psychiatry*, 2003, 44(2), 121–34

Parnas J & Korsgaard S, 'Epilepsy and psychosis', *Acta Psychiatrica Scandinavica*, 1982, 66(2), 89–99

Parnas J, Møller P, Kircher T et al, 'EASE: examination of anomalous self-experience', *Psychopathology*, 2005, 38(5), 236–58

Parnas J, Vianin P, Saebye D et al, 'Visual binding abilities in the initial and advanced stages of schizophrenia', *Acta Psychiatrica Scandinavica*, 2001, 103(3), 171–80

Parry GA, *Original Thinking: A Radical Revisioning of Time, Humanity, and Nature*, North Atlantic Books, 2015

Parsons BD, Gandhi S, Aurbach EL et al, 'Lengthened temporal integration in schizophrenia', *Neuropsychologia*, 2013, 51(2), 372–6

Parsons LM & Osherson D, 'New evidence for distinct right and left brain systems for deductive versus probabilistic reasoning', *Cerebral Cortex*, 2001, 11(10), 954–65

Pascal B, 'Discourse on the passion of love', in CW Eliot (ed), WF Trotter, ML Booth & OW Wight (trans), *Blaise Pascal: Thoughts, Letters & Minor Works*, PF Collier & Son, 1910

———, *De l'Esprit géométrique*, 2 ('De l'art de persuader'), § 3–5, in J Mesnard (ed), *Œuvres Complètes*, Desclée De Brouwer, 1964, III, 413–4

———, *Pensées*, édition Brunschvicg, Garnier Flammarion, 1976 [1670]

Passeri A, Capotosto P & Di Matteo R, 'The right hemisphere contribution to semantic categorization: a TMS study', *Cortex*, 2015, 64, 318–26

Passmore J, *A Hundred Years of Philosophy*, Duckworth, 1968

Passynkova N, Neubauer H & Scheich H, 'Spatial organization of EEG coherence during listening to consonant and dissonant chords', *Neuroscience Letters*, 2007, 412(1), 6–11

Pasteur L, Inaugural lecture as professor and dean of the faculty of science, University of Lille, Douai, 7 December 1854

———, 'Observations sur les forces dissymétriques', *Comptes rendus de l'Académie des* sciences, 1874, 78, 1515–8 (in P Vallery-Radot (ed), Œuvres de Pasteur, vol 1: Dissymétrie Moléculaire, Masson et Cie, 1922)

Patel VL, Groen GJ & Arocha JF, 'Medical expertise as a function of task difficulty', Memory & Cognition, 1990, 18(4), 394–406

Paterna S, Gaspare P, Fasullo S et al, 'Normal-sodium diet compared with low-sodium diet in compensated congestive heart failure: is sodium an old enemy or a new friend?', Clinical Science, 2008, 114(3), 221–30

Paterson A & Zangwill OL, 'A case of topographical disorientation associated with a unilateral cerebral lesion', Brain, 1945, 68(3), 188–211

Patterson MB & Mack JL, 'Neuropsychological analysis of a case of reduplicative paramnesia', Journal of Clinical and Experimental Neuropsychology, 1985, 7(1), 111–21

Paul GL, 'The production of blisters by hypnotic suggestion: another look', Psychosomatic Medicine, 1963, 25(3), 233–44

Paul LK, Corsello C, Kennedy DP et al, 'Agenesis of the corpus callosum and autism: a comprehensive comparison', Brain, 2014, 137(6), 1813–29

Paul LK, Schieffer B & Brown WS, 'Social processing deficits in agenesis of the corpus callosum: narratives from the Thematic Apperception Test', Archives of Clinical Neuropsychology, 2004, 19(2), 215–25

Paul LK, VanLancker-Sidtis D, Schieffer B et al, 'Communicative deficits in agenesis of the corpus callosum: nonliteral language and affective prosody', Brain and Language, 2003, 85(2), 313–24

Pauli W, 'Die philosophische Bedeutung der Idee der Komplementarität', Experientia, 1950, 6(2), 72–81

———, 'Der Einfluss archetypischer Vorstellungen auf die Bildung naturwissenschaftlicher Theorien bei Kepler', in CG Jung & W Pauli, Naturerklärung und Psyche, Rascher Verlag, 1952, 109–94

———, 'Matter', in H Muschel (ed), Man's Right to Knowledge, Columbia University Press, 1954, 10–18

———, 'The influence of archetypal ideas on the scientific theories of Kepler', in CG Jung & W Pauli, Interpretation of Nature and the Psyche, trans P Silz, Pantheon Books, 1955

———, 'Phänomen und physikalische Realität', Dialectica, 1957, 11(1–2), 36–48

———, Aufsätze und Vorträge über Physik und Erkenntnistheorie, Springer Verlag, 1961

———, 'Probability and physics', in CP Enz & K von Meyenn (eds), Wolfgang Pauli: Writings on Physics and Philosophy, Springer Verlag, 1994, 43–8

Paumgarten N, 'We are a camera', The New Yorker, 22 September 2014

Paxton JM, Ungar L & Greene JD, 'Reflection and reasoning in moral judgment', Cognitive Science, 2012, 36(1), 163–77

Pazart L, Comte A, Magnin E et al, 'An fMRI study on the influence of sommeliers' expertise on the integration of flavor', Frontiers in Behavioral Neuroscience, 2014, 8, article 358

Pazzaglia M, Smania N, Corato E et al, 'Neural underpinnings of gesture discrimination in patients with limb apraxia', Journal of Neuroscience, 2008, 28(12), 3030–41

Peach A, 'Richard Duppa', Oxford Dictionary of National Biography, 1885–1900

Pearce JM, 'Misoplegia: review', European Neurology, 2007, 57(1), 62–4

Pearsall P, Schwartz GER & Russek LGS, 'Changes in heart transplant recipients that parallel the personalities of their donors', Journal of Near-Death Studies, 2002, 20(3), 191–206

Pearson H, 'What is a gene?', Nature, 2006, 441(7092), 398–401

Peirce CS, 'The fixation of belief', Popular Science Monthly, November 1877, 12, 1–15

———, in C Hartshorne, P Weiss & AW Burks (eds), The Collected Papers of Charles Sanders Peirce, Harvard University Press, 8 vols, 1931–1960

———, 'The law of mind', in J Buchler (ed), Philosophical Writings of Peirce, Dover, 1955

———, 'Some consequences of four incapacities', in N Houser & C Kloesel (eds), The Essential Peirce, Volume 1: Selected Philosophical Writings (1867–1893), Indiana University Press, 1992, 28–55

Peirce JW, Leigh AE & Kendrick KM, 'Configurational coding, familiarity and the right hemisphere advantage for face recognition in sheep', Neuropsychologia, 2000, 38(4), 475–83

Pelak VS & Hoyt WF, 'Symptoms of akinetopsia associated with traumatic brain injury and Alzheimer's disease', Neuro-Ophthalmology, 2005, 29(4), 137–42

Pellicano E, Gibson L, Maybery M et al, 'Abnormal global processing along the dorsal visual pathway in autism: a possible mechanism for weak visuospatial coherence?', Neuropsychologia, 2005, 43(7), 1044–53

Pelly DF, Ukkusiksalik: The People's Story, Dundurn, 2016

Pelphrey K, Morris JP & McCarthy G, 'Grasping the intentions of others: the perceived intentionality of an action influences activity in the superior temporal sulcus during social perception', Journal of Cognitive Neuroscience, 2004, 16(10), 1706–16

Pelphrey K, Singerman JD, Allison T et al, 'Brain activation evoked by perception of gaze shifts: the influence of context', *Neuropsychologia*, 2003, 41(2), 156–70

Pelvig DP, Pakkenberg H, Stark AK et al, 'Neocortical glial cell numbers in human brains', *Neurobiology of Aging*, 2008, 29(11), 1754–62

Penfield W, 'Functional localisation in temporal and deep sylvian areas', *Research Publications – Association for Research in Nervous and Mental Disease*, 1958, 36, 210–26

———, 'The anatomy of temporal lobe seizures', in L van Bogaert & J Radermecker (eds), *First International Congress of Neurological Science*, vol 3, Pergamon Press, 1959, 513–27

———, in the Foreword to the Tercentenary edition of Thomas Willis's *The Anatomy of the Brain and Nerves*, ed W Feindel, McGill University Press, 1965–6

———, *Mystery of the Mind: A Critical Study of Consciousness and the Human Brain*, Princeton University Press, 1975

Penfield W & Perot P, 'The brain's record of auditory and visual experience: a final summary and discussion', *Brain*, 1963, 86(4), 595–696

Pennal BE, 'Human cerebral asymmetry in colour discrimination', *Neuropsychologia*, 1977, 15(4–5), 563–68

Pennell J, *Pen Drawing and Pen Draughtsmen*, Macmillan, 1889

Pennisi E, 'Microbial genomes: sequences reveal borrowed genes', *Science*, 2001, 294(5547), 1634–35

———, 'Flea boasts whopper gene count', *Science*, 2009, 324(5932), 1252

Pennycook G, Cheyne JA, Seli P et al, 'Analytic cognitive style predicts religious and paranormal belief', *Cognition*, 2012, 123(3), 335–46

Penrose R, *Shadows of the Mind: A Search for the Missing Science of Consciousness*, Oxford University Press, 1994

Penrose R & Clark J, 'Discussion of "Shadows of the Mind"', *Journal of Consciousness Studies*, 1994, 1(1), 17–24

Penttonen M & Buzsáki G, 'Natural logarithmic relationship between brain oscillators', *Thalamus & Related Systems*, 2003, 2(2), 145–52

Péran P, Rascol O, Démonet JF et al, 'Deficit of verb generation in nondemented patients with Parkinson's disease', *Movement Disorders*, 2003, 18(2), 150–6

Percy W, *Lost in the Cosmos*, Farrar Straus & Giroux, 1983

———, 'The fateful rift: the San Andreas fault in the modern mind', *Design For Arts in Education*, 1990, 91(3), 2–7 & 51–3

Pérennou D, 'Postural disorders and spatial neglect in stroke patients: a strong association', *Restorative Neurology and Neuroscience*, 2006, 24(4–6), 319–34

Pérennou D, Bénaïm C, Rouget E et al, 'Postural balance following stroke: towards a disadvantage of the right brain-damaged hemisphere', *Revue Neurologique (Paris)*, 1999, 155(4), 281–90

Pérennou DA, Mazibrada G, Chauvineau V et al, 'Lateropulsion, pushing and verticality perception in hemisphere stroke: a causal relationship?', *Brain*, 2008, 131(9), 2401–13

Peres A, 'Einstein, Podolsky, Rosen, and Shannon', *Foundations of Physics*, 2005, 35(3), 511–4

Peres JF, Moreira-Almeida A, Caixeta L et al, 'Neuroimaging during trance state: a contribution to the study of dissociation', *PLoS One*, 2012, 7(11), e49360

Peretz I, 'Processing of local and global musical information in unilateral brain damaged patients', *Brain*, 1990, 113(4), 1185–205

Peretz I, Cagnon L & Bouchard B, 'Music and emotion: perceptual determinants, immediacy, and isolation after brain damage', *Cognition*, 1998, 68(2), 111–41

Peretz I, Kolinsky R, Tramo M et al, 'Functional dissociations following bilateral lesions of auditory cortex', *Brain*, 1994, 117(6), 1283–1301

Perez DL, Fuchs BH & Epstein J, 'A case of Cotard syndrome in a woman with a right subdural hemorrhage', *Journal of Neuropsychiatry & Clinical Neurosciences*, 2014, 26(1), E29–30

Perez MM, Trimble MR, Murray NM et al, 'Epileptic psychosis: an evaluation of PSE profiles', *British Journal of Psychiatry*, 1985, 146(2), 155–63

Peroutka SJ, Sohmer BH, Kumar AJ et al, 'Hallucinations and delusions following a right temporoparieto-occipital infarction', *Johns Hopkins Medical Journal*, 1982, 151(4), 181–85

Perreault A, Gurnsey R, Dawson M et al, 'Increased sensitivity to mirror symmetry in autism', *PLoS One*, 2011, 6(4), e19519

Perry DW, Zatorre RJ & Evans AC, 'Co-variation of CBF during singing with vocal fundamental frequency', *NeuroImage*, 1996, 3(3), S315

Perry DW, Zatorre RJ, Petrides M et al, 'Localization of cerebral activity during simple singing (1)', *NeuroReport*, 1999a, 10(16), 3453–8

———, 'Localization of cerebral activity during simple singing (2)', *NeuroReport*, 1999b, 10(18), 3979–84

Perry RB, *The Thought and Character of William James*, 2 vols, Little Brown, 1935

Perry RJ, Rosen HR, Kramer JH et al, 'Hemispheric dominance for emotions, empathy and social behaviour: evidence from right and left handers with frontotemporal dementia', *Neurocase*, 2001, 7(2), 145–60

Perry S, *Coleridge and the Uses of Division*, Oxford University Press, 1999

Perry S, 7 August 2017: www.minnpost.com/second-opinion/2017/08/think-your-intuition-helps-you-make-wiser-decisions-probably-not-say-research (accessed 23 February 2021)

Persinger MA & Makarec K, 'Greater right hemisphericity is associated with lower self-esteem in adults', *Perceptual & Motor Skills*, 1991, 73(3, pt 2), 1244–6

Peru A, Leder M & Aglioti S, 'Right-sided neglect following a left subcortical lesion', *Revue Neurologique (Paris)*, 2000, 156(5), 475–80

Pesenti M, Zago L, Crivello F et al, 'Mental calculation in a prodigy is sustained by right prefrontal and medial temporal areas', *Nature Neuroscience*, 2001, 4(1), 103–7

Pessoa L, 'To what extent are emotional visual stimuli processed without attention and awareness?', *Current Opinion in Neurobiology*, 2005, 15(2), 188–96

Pessoa L, Japee S & Ungerleider LG, 'Visual awareness and the detection of fearful faces', *Emotion*, 2005, 5(2), 243–7

Pessoa L, Thompson E & Noë A, 'Finding out about filling-in: a guide to perceptual completion for visual science and the philosophy of perception', *Behavioral and Brain Sciences*, 1998, 21(6), 723–48 (peer commentary 748–802)

Peters D & Ceci S, 'Peer-review practices of psychological journals: the fate of published articles, submitted again', *Behavioral and Brain Sciences*, 1982, 5(2), 187–255

Petersen A, 'The philosophy of Niels Bohr', *Bulletin of the Atomic Scientists*, 1963, 19(7), 8–14

Peterson CK, Gravens LC & Harmon-Jones E, 'Asymmetric frontal cortical activity and negative affective responses to ostracism', *Social Cognitive and Affective Neuroscience*, 2011, 6(3), 277–85

Peterson CK, Shackman AJ & Harmon-Jones E, 'The role of asymmetrical frontal cortical activity in aggression', *Psychophysiology*, 2008, 45(1), 86–92

Peterson JB, *12 Rules for Life: An Antidote to Chaos*, Random House, 2018

Petit L, Simon G, Joliot M et al, 'Right hemisphere dominance for auditory attention and its modulation by eye position: an event related fMRI study', *Restorative Neurology and Neuroscience*, 2007, 25(3–4), 211–25

Petracca G, Migliorelli R, Vázquez S et al, 'SPECT findings before and after ECT in a patient with major depression and Cotard's syndrome', *Journal of Neuropsychiatry and Clinical Neurosciences*, 1995, 7(4), 505–7

Pettigrew JD, 'Searching for the switch: neural bases for perceptual rivalry alternations', *Brain and Mind*, 2001, 2(1), 85–118

Petty RE & Cacioppo JT, 'Issue involvement can increase or decrease persuasion by enhancing message-relevant cognitive responses', *Journal of Personality and Social Psychology*, 1979, 37(10), 1915–26

Petty RG, 'Structural asymmetries of the human brain and their disturbance in schizophrenia', *Schizophrenia Bulletin*, 1999, 25(1), 121–40

Petty RG, Barta PE, Pearlson GD et al, 'Reversal of asymmetry of the planum temporale in schizophrenia', *American Journal of Psychiatry*, 1995, 152(5), 715–21

Peuskens H, Vanrie J, Verfaillie K et al, 'Specificity of regions processing biological motion', *European Journal of Neuroscience*, 2005, 21(10), 2864–75

Peyrin C, Baciu M, Segebarth C et al, 'Cerebral regions and hemispheric specialization for processing spatial frequencies during natural scene recognition: an event-related fMRI study', *NeuroImage*, 2004, 23(2), 698–707

Peyrin C, Mermillod M, Chokron S et al, 'Effect of temporal constraints on hemispheric asymmetries during spatial frequency processing', *Brain and Cognition*, 2006, 62(3), 214–20

Peyrin C, Schwartz S, Seghier M et al, 'Hemispheric specialization of human inferior temporal cortex during coarse-to-fine and fine-to-coarse analysis of natural visual scenes', *NeuroImage*, 2005, 28(2), 464–73

Pfeffer W, *Physiologische Untersuchungen*, Engelmann, 1873

———, *The Physiology of Plants*, in 3 vols, trans AJ Ewart, Clarendon Press, 1900

Pfennig DW, Wund MA, Snell-Rood EC et al, 'Innovation and robustness in complex regulatory gene networks', *Proceedings of the National Academy of Sciences of the United States of America*, 2007, 104(34), 13591–6

Phelps EA & Gazzaniga MS, 'Hemispheric differences in mnemonic processing: the effects of left hemisphere interpretation', *Neuropsychologia*, 1992, 30(3), 293–7

Phelps EA, Hyder F, Blamire AM et al, 'fMRI of the prefrontal cortex during overt verbal fluency', *NeuroReport*, 1997, 8(2), 561–5

Philip M, Kornitzer J, Marks D et al, 'Alice in Wonderland Syndrome associated with a temporo-parietal cavernoma', *Brain Imaging and Behavior*, 2015, 9(4), 910–2

Phillips WA & Silverstein SM, 'Convergence of biological and psychological perspectives on cognitive coordination in schizophrenia: a physiological, computational, and psychological perspective', *Behavioural and Brain Sciences*, 2003, 26(1), 65–138

Phillipson TO & Harris JP, 'Perceptual changes in schizophrenia: a questionnaire survey', *Psychological Medicine*, 1985, 15(4), 859–66

Phua C, Bhaskar S & Calic Z, 'Hemiprosopometamorphopsia: a case of impaired facial perception restricted to the eye', *SN Comprehensive Clinical Medicine*, 2019, 1, 931–3

Pia L, Neppi-Modona M, Ricci R et al, 'The anatomy of anosognosia for hemiplegia: a meta-analysis', *Cortex*, 2004, 40(2), 367–77

Piazza L, Lummen TT, Quiñonez E et al, 'Simultaneous observation of the quantization and the interference pattern of a plasmonic near-field', *Nature Communications*, 2015, 6, 6407

Pichler E, 'Über Störungen des Raum- und Zeiterlebens bei Verletzungen des Hinterhauptlappens', *Zeitschrift für die gesamte Neurologie und Psychiatrie*, 1943, 176(1), 434–64

Pickup GJ & Frith CD, 'Theory of mind impairments in schizophrenia: symptomatology, severity and specificity', *Psychological Medicine*, 2001, 31(2), 207–20

Pieper J, 'Learning how to see again', in *Only the Lover Sings: Art and Contemplation*, Ignatius Press, 1990 [1952], 29–36

———, *Leisure: The Basis of Culture*, trans G Malsbary, St Augustine's Press, 1998

Pierre D & Hübler A, 'A theory for adaptation and competition applied to logistic map dynamics', *Physica D: Nonlinear Phenomena*, 1994, 75(1–3), 343–60

Pietschnig J & Gittler G, 'A reversal of the Flynn effect for spatial perception in German-speaking countries: evidence from a cross-temporal IRT-based meta-analysis', *PLoS One*, 2015, 5, e14406

Pietschnig J & Voracek M, 'One century of global IQ gains: a formal meta-analysis of the Flynn Effect (1909–2013)', *Perspectives on Psychological Science*, 2015, 10(3), 282–306

Pietschnig J, Voracek M & Formann AK, 'Pervasiveness of the IQ rise: a cross-temporal meta-analysis', *PLoS One*, 2010, 5(12), e14406

Pile J, *A History of Interior Design*, Laurence King Publishing, 2005

Pinkard T, *Hegel: A Biography*, Cambridge University Press, 2001

Pinsent A, *The Second-Person Perspective in Aquinas's Ethics: Virtues and Gifts*, Routledge, 2012

Pinsk MA, DeSimone K, Moore T et al, 'Representations of faces and body parts in macaque temporal cortex: a functional MRI study', *Proceedings of the National Academy of Sciences of the United States of America*, 2005, 102(19), 6996–7001

Piran T, Jimenez R, Cuesta AJ et al, 'Cosmic explosions, life in the universe, and the cosmological constant', *Physical Review Letters*, 2016, 116(8), 081301

Pirot M, Pulton TW & Sutker LW, 'Hemispheric asymmetry in reaction time to color stimuli', *Perceptual & Motor Skills*, 1977, 45(3, pt 2), 1151–5

Pitcher D, Charles L, Devlin JT et al, 'Triple dissociation of faces, bodies, and objects in extrastriate cortex', *Current Biology*, 2009, 19(4), 319–24

Piven J, Bailey J, Ranson BJ et al, 'An MRI study of the corpus callosum in autism', *American Journal of Psychiatry*, 1997, 154(8), 1051–6

Pizzagalli D, Lehmann D & Brugger P, 'Lateralized direct and indirect semantic priming effects in subjects with paranormal experiences and beliefs', *Psychopathology*, 2001, 34(2), 75–80

Place EJ & Gilmore GC, 'Perceptual organization in schizophrenia', *Journal of Abnormal Psychology*, 1980, 89(3), 409–18

Plailly J, Bensafi M, Pachot-Clouard M et al, 'Involvement of right piriform cortex in olfactory familiarity judgments', *NeuroImage*, 2005, 24(4), 1032–41

Plaisted K, O'Riordan M & Baron-Cohen S, 'Enhanced visual search for a conjunctive target in autism: a research note', *Journal of Child Psychology and Psychiatry*, 1998, 39(5), 777–83

Plaisted KC, Swettenham J & Rees L, 'Children with autism show local precedence in a divided attention task and global precedence in a selective attention task', *Journal of Child Psychology and Psychiatry*, 1999, 40(5), 733–42

Planck M, interview in *The Observer*, 25 January 1931a, 17 (column 3)

———, *The Universe in the Light of Modern Physics*, George Allen & Unwin, 1931b

———, *Where is Science Going?*, WW Norton, 1932

———, *The Philosophy Of Physics*, trans WH Johnston, George Allen & Unwin, 1936

———, *Das Wesen der Materie* [The Nature of Matter], speech delivered in Florence in 1944, Archiv zur Geschichte der Max-Planck-Gesellschaft, Abt Va, Rep. 11 Planck, Nr. 1797

———, *Wissenschaftliche Selbstbiographie: mit einem Bildnis und der von Max von Laue gehaltenen Traueransprache*, Johann Ambrosius Barth Verlag, 1948

———, *Scientific Autobiography and Other Papers*, trans F Gaynor, Philosophical Library, 1949

———, *A Survey of Physical Theory*, tr R Jones & DH Williams, Dover, 1960

Plant GT, Laxer KD, Barbaro NM et al, 'Impaired visual motion perception in the contralateral hemifield following unilateral posterior cerebral lesions in humans', *Brain*, 1993, 116(6), 1303–35

Platek SM, Thomson JW & Gallup GG Jr, 'Cross-modal self-recognition: the role of visual, auditory, and olfactory primes', *Consciousness and Cognition*, 2004, 13(1), 197–210

Platel H, Price C, Baron J-C et al, 'The structural components of music perception: a functional anatomical study', *Brain*, 1997, 120(2), 229–43

Pletzer B, 'Sex-specific strategy use and global-local processing: a perspective toward integrating sex differences in cognition', *Frontiers in Neuroscience*, 2014, 8, article 425

Pletzer B, Kronbichler M, Nuerk H-C et al, 'Sex differences in the processing of global vs. local stimulus aspects in a two-digit number comparison task – an fMRI study', *PLoS One*, 2013, 8(1), e53824

Pletzer B, Petasis O & Cahill L, 'Switching between forest and trees: opposite relationship of progesterone and testosterone to global-local processing', *Hormones and Behavior*, 2014, 66(2), 257–66

Plomin R, DeFries JC, Knopik VS et al (eds), *Behavioral Genetics*, Worth Publishers, 6th edn, 2013

Ploran E, Nelson SM, Velanova K et al, 'Evidence accumulation and moment of recognition: dissociating perceptual recognition processes using fMRI', *Journal of Neuroscience*, 2007, 27(44), 11912–24

Plotinus, *Enneads*, trans S MacKenna & BS Page, The Medici Society, 1921

———, *Enneads*, trans & ed AH Armstrong, Loeb Classical Library, 1995

Pluchon C, Salmon F, Houeto JL et al, 'Paramnésie de réduplication d'évènement après hémorragie du noyau caudé droit', *Canadian Journal of Neurological Sciences*, 2010, 37(4), 468–72

Pobric G & Hamilton AF de C, 'Action understanding requires the left inferior frontal cortex', *Current Biology*, 2006, 16(5), 524–9

Pobric G, Mashal N, Faust M et al, 'The role of the right cerebral hemisphere in processing novel metaphoric expressions: a transcranial magnetic stimulation study', *Journal of Cognitive Neuroscience*, 2008, 20(1), 170–81

Podell K, Lovell M, Zimmerman M et al, 'The Cognitive Bias Task and lateralised frontal lobe functions in males', *Journal of Neuropsychiatry and Clinical Neurosciences*, 1995, 7(4), 491–501

Podoll K & Robinson D, 'Lewis Carroll's migraine experiences', *Lancet*, 1999, 353(9161), 1366

Poeck K, 'Phantoms following amputation in early childhood and in congenital absence of limbs', *Cortex*, 1964, 1(3), 269–75

Poeppel D, 'The analysis of speech in different temporal integration windows: cerebral lateralization as "asymmetric sampling in time"', *Speech Communication*, 2003, 41(1), 245–55

Poincaré H, *Science and Hypothesis*, in *The Foundations of Science*, trans GB Halsted, Cambridge University Press, 1913a [*La Science et l'hypothèse* 1902]

———, *The Value of Science*, in *The Foundations of Science*, trans GB Halsted, Cambridge University Press, 1913b [*La Valeur de la science* 1905]

———, *Science and Method*, in *The Foundations of Science*, trans GB Halsted, Cambridge University Press, 1913c [*Science et Méthode* 1908]

———, *Science and Method*, trans F Maitland, Nelson & Sons, 1914

Pol H, Cohen-Kettenis P, van Haren N et al, 'Changing your sex changes your brain: influences of testosterone and estrogen on adult human brain structure', *European Journal of Endocrinology*, 2006, 155(suppl 1), S107–14

Polányi M, *Personal Knowledge: Towards a Post-Critical Philosophy*, Routledge & Kegan Paul, 1958

———, *The Tacit Dimension*, University of Chicago Press, 1966

Polderman TJ, Benyamin B, de Leeuw CA et al, 'Meta-analysis of the heritability of human traits based on fifty years of twin studies', *Nature Genetics*, 2015, 47(7), 702–9

Polenz D & Rubitz F, 'Staff perceptions of the effect of therapeutic camping upon psychiatric patients' affect', *Therapeutic Recreation Journal*, 1977, 11(2), 70–3

Polge J, 'Critical thinking: the use of intuition in making clinical nursing judgments', *Journal of the New York State Nurses Association*, 1995, 26(1), 4–9

Politis M & Loa C, 'Reduplicative paramnesia: a review', *Psychopathology*, 2012, 45(6), 337–43

Polkinghorne J, 'Science and theology: parallelisms', in G Tanzella-Nitti, I Colagé & A Strumia (eds), *Interdisciplinary Encyclopedia of Religion and Science*, 2008, inters.org/science-theology

Pollatos O & Schandry R, 'Accuracy of heartbeat perception is reflected in the amplitude of the heartbeat-evoked brain potential', *Psychophysiology*, 2004, 41(3), 476–82

Pope SM, Fagot J, Meguerditchian A et al, 'Enhanced cognitive flexibility in the semi-nomadic Himba', *Journal of Cross-Cultural Psychology*, 2018, 50(1), 47–62

Popescu M, Otsuka A & Ioannides AA, 'Dynamics of brain activity in motor and frontal cortical areas during music listening: a magnetoencephalographic study', *NeuroImage*, 2004, 21(4), 1622–38

Pophristic V & Goodman L, 'Hyperconjugation not steric repulsion leads to the staggered structure of ethane', *Nature*, 2001, 411(6837), 565–8

Poppelreuter W, *Die Psychische Schädigungen durch Kopfschuß im Kriege 1914–16*, Leopold Voss, 2 vols, 1917

Popper KR, 'Natural selection and the emergence of mind', *Dialectica*, 1978, 32(3–4), 339–55

———, *The Open Society and its Enemies*, University of Klagenfurt Press, 1994 [1945]

Poreh AM, Whitman D & Ross TP, 'Creative thinking abilities and hemispheric asymmetry in schizotypal college students', *Current Psychology*, 1993, 12(4), 344–52

Porete M, *The Mirror of Simple Souls*, trans 'MN' & ed C Kirchberger, Burns Oates & Washbourne, 1927

Porter S, Spencer L & Birt AR, 'Blinded by emotion? Effect of the emotionality of a scene on susceptibility to false memories', *Canadian Journal of Behavioural Science*, 2003, 35(3), 165–75

Porto P, Oliveira L, Mari J et al, 'Does cognitive behavioral therapy change the brain? A systematic review of neuroimaging in anxiety disorders', *The Journal of Neuropsychiatry and Clinical Neurosciences*, 2009, 21(2), 114–25

Posner GP, 'Nation's mathematicians guilty of "innumeracy"', *Skeptical Inquirer*, 1991, 15, 342–9

Posner MI, 'Attention in cognitive neuroscience: an overview', in MS Gazzaniga (ed), *The Cognitive Neurosciences*, Massachusetts Institute of Technology Press, Cambridge MA, 1995, 615–24

Posner MI & Petersen SE, 'The attention system of the human brain', *Annual Review of Neuroscience*, 1990, 13(1), 25–42

Posner MI, Walker JA, Friedrich FJ et al, 'Effects of parietal injury on covert orienting of attention', *Journal of Neuroscience*, 1984, 4(7), 1863–74

Post F, 'Creativity and psychopathology: a study of 291 world-famous men', *British Journal of Psychiatry*, 1994, 165(1), 22–34

Postal KS, 'The mirror sign delusional misidentification symptom', in TE Feinberg & JP Keenan (eds), *The Lost Self: Pathologies of the Brain and Identity*, Oxford University Press, 2005, 131–46

Postema MC, van Rooij D, Anagnostou E et al, 'Altered structural brain asymmetry in autism spectrum disorder in a study of 54 datasets', *Nature Communications*, 2019, 10(1), 4958

Poston C (ed), *The Making of a Mystic: New and Selected Letters of Evelyn Underhill*, University of Illinois Press, 2010

Pötzl O, 'Über Störungen der Selbstwahrnehmung bei linksseitiger Hemiplegie', *Zeitschrift für die gesamte Neurologie und Psychiatrie*, 1924, 93, 117–68

Pötzl O & Redlich E, 'Bilaterale Affektion beider Okzipitallappen' [the case report has no title as such in the original], *Wiener Klinische Wochenschrift*, 1911, 24, 517–8

Poundstone W, *Prisoner's Dilemma*, Doubleday, 1992

Powell LH, Shahabi L & Thoresen CE, 'Religion and spirituality: linkages to physical health', *The American Psychologist*, 2003, 58(1), 36–52

Powell WR & Schirillo JA, 'Asymmetrical facial expressions in portraits and hemispheric laterality: a literature review', *Laterality*, 2009, 14(6), 545–72

Power M, *The Audit Society: Rituals of Verification*, Oxford University Press, 1997

Power RA, Steinberg S, Bjornsdottir G et al, 'Polygenic risk scores for schizophrenia and bipolar disorder predict creativity', *Nature Neuroscience*, 2015, 18(7), 953–5

Powers C, Bencic R, Horton WS et al, 'Hemispheric inference priming during comprehension of conversations and narratives', *Neuropsychologia*, 2012, 50(11), 2577–83

Poynter W & Roberts C, 'Hemispheric asymmetries in visual search', *Laterality*, 2012, 17(6), 711–26

Požgain I, Požgain Z & Degmečić D, 'Placebo and nocebo effect: a mini-review', *Psychiatria Danubina*, 2014, 26(2), 100–7

Prado J, Chadha A & Booth JR, 'The brain network for deductive reasoning: a quantitative meta-analysis of 28 neuroimaging studies', *Journal of Cognitive Neuroscience*, 2011, 23(11), 3483–97

Prado J & Noveck IA, 'Overcoming perceptual features in logical reasoning: a parametric functional magnetic resonance imaging study', *Journal of Cognitive Neuroscience*, 2007, 19(4), 642–57

Prat CS, Mason RA & Just MA, 'An fMRI investigation of analogical mapping in metaphor comprehension: the influence of context and individual cognitive capacities on processing demands', *Journal of Experimental Psychology: Learning, Memory, and Cognition*, 2012, 38(2), 282–94

Preisler A, Gallasch E & Schulter G, 'Hemispheric asymmetry and the processing of harmonies in music', *International Journal of Neuroscience*, 1989, 47(1–2), 131–40

Prescott JW, 'Early somatosensory deprivation as an ontogenetic process in the abnormal development of the brain and behavior', in EI Goldsmith & J Moor-Jankowski (eds), *Medical Primatology*, Karger, 1970, 356–75

Prescott J, Gavrilescu M, Cunnington R et al, 'Enhanced brain connectivity in math-gifted adolescents: an fMRI study using mental rotation', *Cognitive Neuroscience*, 2010, 1(4), 277–88

Prete G, Fabri M, Foschi N et al, 'Asymmetry for symmetry: right-hemispheric superiority in bi-dimensional symmetry perception', *Symmetry*, 2017, 9(5), 76

Pretty J, 'How nature contributes to mental and physical health', *Spirituality & Health International*, 2004, 5(2), 68–78

Pribram KH, 'Emotions', in SB Filskov & TJ Boll (eds), *Handbook of Clinical Neuropsychology*, Wiley, 1981, 102–34

Price BH & Mesulam M, 'Psychiatric manifestations of right hemisphere infarctions', *Journal of Nervous and Mental Disease*, 1985, 173(10), 610–4

Price CJ & Friston KJ, 'Degeneracy and cognitive anatomy', *Trends in Cognitive Sciences*, 2002a, 6(10), 416–21

———, 'Functional imaging studies of category specificity', in EME Forde & GW Humphreys (eds), *Category Specificity in Brain and Mind*, Psychology Press, 2002b, 427–47

Priday H, *Seizing the Essence: A Value Cosmology for the Modernist*, Global Summit House, 2019

Priftis K, Rusconi E, Umiltà C et al, 'Pure agnosia for mirror stimuli after right inferior parietal lesion', *Brain*, 2003, 126(4), 908–19

Prigogine I, *The End of Certainty: Time, Chaos and the New Laws of Nature*, The Free Press, 1997

Pronin E, Wegner DM, McCarthy K et al, 'Everyday magical powers: the role of apparent mental causation in the overestimation of personal influence', *Journal of Personality and Social Psychology*, 2006, 91(2), 218–31

Pross A, *What is Life? How Chemistry Becomes Biology*, Oxford University Press, 2012

———, on his thesis, in the *Huffington Post*, 2014: www.huffingtonpost.com/addy-pross/what-is-life_b_4992980.html (accessed 23 February 2021)

Proust M, trans CK Scott Moncrieff, revised DJ Enright, *Remembrance of Things Past*, vol 1, *Swann's Way*, Chatto & Windus, 1992

Proverbio AM, Brignone V, Matarazzo S et al, 'Gender and parental status affect the visual cortical response to infant facial expression', *Neuropsychologia*, 2006, 44(14), 2987–99

Proverbio AM, Crotti N, Zani A et al, 'The role of left and right hemispheres in the comprehension of idiomatic language: an electrical neuroimaging study', *BMC Neuroscience*, 2009, 10(1), 116

Proverbio AM, Minniti A & Zani A, 'Electrophysiological evidence of a perceptual precedence of global vs local visual information', *Cognitive Brain Research*, 1998, 6(4), 321–34

Proverbio AM, Riva F, Martin E et al, 'Face coding is bilateral in the female brain', *PLoS One*, 2010, 5(6), e11242

Pucarin-Cvetković J, Zuskin E, Mustajbegović J et al, 'Known symptoms and diseases of a number of classical European composers during 17th and 20th century in relation with their artistic musical expressions', *Collegium Antropologicum*, 2011, 35(4), 1327–31

Puccetti R, 'Two brains, two minds? Wigan's theory of mental duality', *The British Journal for the Philosophy of Science*, 1989, 40(2), 137–44

Puce A, Allison T, Gore JC et al, 'Face-sensitive regions in human extrastriate cortex studied by functional MRI', *Journal of Neurophysiology*, 1995, 74(3), 1192–9

Pugh GE, *The Biological Origin of Human Values*, Routledge & Kegan Paul, 1978

Pugliese L, Catani M, Ameis S et al, 'The anatomy of extended limbic pathways in Asperger syndrome: a preliminary diffusion tensor imaging tractography study', *NeuroImage*, 2009, 47(2), 427–34

Pujol J, Deus J, Losilla JM et al, 'Cerebral lateralization of language in normal left-handed people studied by functional MRI', *Neurology*, 1999, 52(5), 1038–43

Pullman P, *Daemon Voices*, David Fickling Books, 2017

Pullum G, 'The great Eskimo vocabulary hoax', *Natural Language and Linguistic Theory*, 1989, 7, 275–81

Pulsifer MB, Brandt J, Salorio CF et al, 'The cognitive outcome of hemispherectomy in 71 children', *Epilepsia*, 2004, 45(3), 243–54

Pulvermüller F, 'Brain mechanisms linking language and action', *Nature Reviews Neuroscience*, 2005, 6(7), 576–82

Pun C, Adamo M, Weger UW et al, 'The right time and the left time: spatial associations of temporal cues affect target detection in right brain-damaged patients', *Cognitive Neuroscience*, 2010, 1(4), 289–95

Puri BK, Lekh SK, Nijran KS et al, 'SPECT neuroimaging in schizophrenia with religious delusions', *International Journal of Psychophysiology*, 2001, 40(2), 143–8

Putnam R, *Bowling Alone*, Simon & Schuster, 2000

Pyles SH & Stern PN, 'Discovery of nursing Gestalt in critical care nursing: the importance of the Gray Gorilla Syndrome', *Image – The Journal of Nursing Scholarship*, 1983, 15(2), 51–7

Pynte J, Besson M, Robichon FH et al, 'The time-course of metaphor comprehension: an event-related potential study', *Brain and Language*, 1996, 55(3), 293–316

Qiu J, Li H, Jou J et al, 'Spatiotemporal cortical activation underlies mental preparation for successful riddle solving: an event-related potential study', *Experimental Brain Research*, 2008, 186(4), 629–34

Qiu J, Li H, Yang D et al, 'The neural basis of insight problem solving: an event-related potential study', *Brain and Cognition*, 2008, 68(1), 100–6

Quadflieg S & Macrae CN, 'Stereotypes and stereotyping: what's the brain got to do with it?', *European Review of Social Psychology*, 2011, 22(1), 215–73

Querné L, Eustache F & Faure S, 'Interhemispheric inhibition, intrahemispheric activation, and lexical capacities of the right hemisphere: a tachistoscopic, divided visual-field study in normal subjects', *Brain and Language*, 2000, 74(2), 171–90

Querné L & Faure S, 'Activating the right hemisphere by a prior spatial task: equal lexical decision accuracy in left and right visual fields in normal subjects', *Brain and Cognition*, 1996, 32(2), 142–6

Quine W van O, 'Two dogmas of empiricism', *Philosophical Review*, 1951, 60(1), 20–43

———, *Philosophy of Logic*, Prentice Hall, 1970

Quinlivan L, Cooper J, Meehan D et al, 'Predictive accuracy of risk scales following self-harm: multicentre, prospective cohort study', *British Journal of Psychiatry*, 2017, 210(6), 429–36

Quinn D, *Iris Exiled: A Synoptic History of Wonder*, University Press of America, 2002

Quinn PC, Kelly DJ, Lee K et al, 'Preference for attractive faces in human infants extends beyond conspecifics', *Developmental Science*, 2008, 11(1), 76–83

Quint J, *Meister Eckehart ['Eckehart']: Deutsche Predigten und Traktate*, Carl Hanser, 1955

Quinton AM, 'The foundations of knowledge', in B Williams & A Montefiore (eds), *British Analytical Philosophy*, Routledge & Kegan Paul, 1966, 55–86

Quirin M, Gruber T, Kuhl J et al, 'Is love right? Prefrontal resting brain asymmetry is related to the affiliation motive', *Frontiers in Human Neuroscience*, 2013, 7, article 902

Quirin M, Meyer F, Heise N et al, 'Neural correlates of social motivation: an fMRI study on power versus affiliation', *International Journal of Psychophysiology*, 2013, 88(3), 289–95

Qureshy A, Kawashima R, Imran MB et al, 'Functional mapping of human brain in olfactory processing: a PET study', *Journal of Neurophysiology*, 2000, 84(3), 1656–66

Rabinowicz EF, Opler LA, Owen DR et al, 'Dot Enumeration Perceptual Organization Task (DEPOT): evidence for a short-term visual memory deficit in schizophrenia', *Journal of Abnormal Psychology*, 1996, 105(3), 336–48

Rabinowicz T, de Courten-Myers GM, Petetot JM et al, 'Human cortex development: estimates of neuronal numbers indicate major loss late during gestation', *Journal of Neuropathology & Experimental Neurology*, 1996, 55(3), 320–8

Rabinowicz T, Dean DE, Petetot JM et al, 'Gender differences in the human cerebral cortex: more neurons in males; more processes in females', *Journal of Child Neurology*, 1999, 14(2), 98–107

Rabinowicz T, Petetot JM, Gartside PS et al, 'Structure of the cerebral cortex in men and women', *Journal of Neuropathology & Experimental Neurology*, 2002, 61(1), 46–57

Radak Z, Chung HY, Koltai E et al, 'Exercise, oxidative stress and hormesis', *Ageing Research Reviews*, 2008, 7(1), 34–42

Radden J (ed), *The Nature of Melancholy: from Aristotle to Kristeva*, Oxford University Press, 2000

Rader AW & Sloutsky VM, 'Processing of logically valid and logically invalid conditional inferences in discourse comprehension', *Journal of Experimental Psychology: Learning, Memory, and Cognition*, 2002, 28(1), 59–68

Radin D, *Real Magic*, Harmony Books, 2018

Radin D, Michel L & Delorme A, 'Psychophysical modulation of fringe visibility in a distant double-slit optical system', *Physics Essays*, 2016, 29(1), 14–22

Radler C, '"In love I am more God": the centrality of love in Meister Eckhart's mysticism', *The Journal of Religion*, 2010, 90(2), 171–98

Rafique SA, Richards JR & Steeves JKE, 'Altered white matter connectivity associated with visual hallucinations following occipital stroke', *Brain and Behavior*, 2018, 8(6), e01010

Rahnev D, Lau H & de Lange FP, 'Prior expectation modulates the interaction between sensory and prefrontal regions in the human brain', *Journal of Neuroscience*, 2011, 31(29), 10741–8

Raine A, Andrews H, Sheard C et al, 'Interhemispheric transfer in schizophrenics, depressives, and normals with schizoid tendencies', *Journal of Abnormal Psychology*, 1989, 98(1), 35–41

Raja V, Silva PL, Holghoomi R et al, 'The dynamics of plant nutation', *Nature: Science Reports*, 2020, 10(1), 19465

Rajagopal S, 'The impact factor and psychiatry journals: an international perspective', *British Journal of Psychiatry International*, 2017, 14(1), 15–18

Rakic P, 'Evolution of the neocortex: perspective from developmental biology', *Nature Reviews Neuroscience*, 2009, 10(10), 724–35

Rakison DH & Poulin-Dubois D, 'Developmental origin of the animate-inanimate distinction', *Psychological Bulletin*, 2001, 127(2), 209–28

Ramachandran VS, 'Behavioral and magnetoencephalographic correlates of neural plasticity in the adult human brain', *Proceedings of the National Academy of Sciences of the United States of America*, 1993, 90(22), 10413–20

——, 'Phantom limbs, neglect syndromes, repressed memories, and Freudian psychology', *International Review of Neurobiology*, 1994, 37, 291–333

——, 'Anosognosia in parietal lobe syndrome', *Consciousness and Cognition*, 1995, 4(1), 22–51

——, 'The evolutionary biology of self-deception, laughter, dreaming and depression: some clues from anosognosia', *Medical Hypotheses*, 1996, 47(5), 347–62

——, *Phantoms in the Brain: Human Nature and the Architecture of the Mind*, HarperCollins, 2005

Ramachandran VS, Altschuler EL & Hillyer S, 'Mirror agnosia', *Proceedings of the Royal Society B: Biological Sciences*, 1997, 264, 645–7

Ramachandran VS & Blakemore C, 'Consciousness', in C Blakemore & S Jennett, *The Oxford Companion to the Body*, Oxford University Press, 2001

Ramachandran VS & Hirstein W, 'The science of art: a neurological theory of aesthetic experience', *Journal of Consciousness Studies*, 1999, 6(6–7), 15–31

Ramachandran VS & Hubbard EM, 'Synaesthesia: a window into perception, thought and language', *Journal of Consciousness Studies*, 2001, 8(12), 3–34

Ramani V, 'Cortical blindness following ictal nystagmus', *Archives of Neurology*, 1985, 42(2), 191–2

Ramón y Cajal S, 'Origen y terminación de las fibras nerviosas olfatorias' (1), *Gaceta Sanitaria de Barcelona*, 11 octubre 1890, 133–9

——, *Die Struktur des Chiasma opticum nebst einer allgemeinen Theorie der Kreuzung der Nervenbahnen*, trans J Bresler, JA Barth Verlag, 1899

——, *Texture of the Nervous System of Man and the Vertebrates*, ed P Pasik & T Pasik, Springer 2002 [1899–1903]

Ramos-Loyo J & Sanchez-Loyo LM, 'Gender differences in EEG coherent activity before and after training navigation skills in virtual environments', *Fiziologiia Cheloveka*, 2011, 37(6), 68–75

Rampone G, O'Sullivan N & Bertamini M, 'The role of visual eccentricity on preference for abstract symmetry', *PLoS One*, 2016, 11(4), e0154428

Rand DG, Greene JD & Nowak MA, 'Spontaneous giving and calculated greed', *Nature*, 2012, 489(7416), 427–30

Ranganath C & Rainer G, 'Neural mechanisms for detecting and remembering novel events', *Nature Reviews Neuroscience*, 2003, 4(3), 193–202

Rangaswami K, 'Indian system of psychotherapy', *Indian Journal of Clinical Psychotherapy*, 1996, 23(1), 62–75

Rankin KP, Gorno-Tempini ML, Allison SC et al, 'Structural anatomy of empathy in neurodegenerative disease', *Brain*, 2006, 129(11), 2945–56

Rankin KP, Rosen HJ, Kramer JH et al, 'Right and left medial orbitofrontal volumes show an opposite relationship to agreeableness in FTD', *Dementia and Geriatric Cognitive Disorders*, 2004, 17(4), 328–32

Rao SM, Mayer AR & Harrington DL, 'The evolution of brain activation during temporal processing', *Nature Neuroscience*, 2001, 4, 317–23

Rao TSS & Andrade C, '*Indian Journal of Psychiatry*: changes in instructions to contributors', *Indian Journal of Psychiatry*, 2014, 56(4), 319–20

Rapcsak SZ, Nielsen L, Littrell LD et al, 'Face memory impairments in patients with frontal lobe damage', *Neurology*, 2001, 57, 1168–75

Rapcsak SZ, Polster MR, Comer JF et al, 'False recognition and misidentification of faces following right hemisphere damage, *Cortex*, 1994, 30(4), 565–83

Rapcsak SZ, Polster MR, Glisky ML et al, 'False recognition of unfamiliar faces following right hemisphere damage:

neuropsychological and anatomical observations', *Cortex*, 1996, 32(4), 593–611

Rapcsak SZ, Reminger SL, Glisky EL et al, 'Neuropsychological mechanisms of false facial recognition following frontal lobe damage', *Cognitive Neuropsychology*, 1999, 16(3–5), 267–92

Rapp AM, Leube DT, Erb M et al, 'Neural correlates of metaphor processing', *Brain Research: Cognitive Brain Research*, 2004, 20(3), 395–402

———, 'Laterality in metaphor processing: lack of evidence from functional magnetic resonance imaging for the right hemisphere theory', *Brain and Language*, 2007, 100(2), 142–9

Rapp AM, Mutschler DE & Erb M, 'Where in the brain is nonliteral language? A coordinate-based meta-analysis of functional magnetic resonance imaging studies', *NeuroImage*, 2012, 63(1), 600–10

Rasmjou, Hausmann & Güntürkün, 'Hemispheric dominance and gender in the perception of an illusion', *Neuropsychologia*, 1999, 37(9), 1041–7

Rasmussen T & Milner B, 'The role of early left-brain injury in determining lateralization of cerebral speech functions', *Annals of the New York Academy of Sciences*, 1977, 299(1), 355–69

Rassoulzadegan M, Grandjean V, Gounon P et al, 'RNA-mediated non-mendelian inheritance of an epigenetic change in the mouse', *Nature*, 2006, 441(7092), 469–74

Rastelli F, Funes MJ, Lupiáñez J et al, 'Left visual neglect: is the disengage deficit space- or object-based?', *Experimental Brain Research*, 2008, 187(3), 439–46

Ratcliffe M, '"Folk psychology" is not folk psychology', *Phenomenology and the Cognitive Sciences*, 2006, 5(1), 31–52

———, *Experiences of Depression: A Study in Phenomenology*, Oxford University Press, 2015

Ratinckx E, Brysbaert M & Vermeulen E, 'CRT screens may give rise to biased estimates of interhemispheric transmission time in the Poffenberger paradigm', *Experimental Brain Research*, 2001, 136(3), 413–6

Raud R, 'The existential moment: rereading Dōgen's theory of time', *Philosophy East and West*, 2012, 62(2), 153–73

Rausch R, 'Cognitive strategies in patients with unilateral temporal lobe excisions', *Neuropsychologia*, 1977, 15(3), 385–95

———, 'Differences in cognitive function with left and right temporal lobe dysfunction', in DF Benson & E Zaidel (eds), *The Dual Brain*, Guilford Press, 1985, 247–61

Rauscher FH, Krauss RM & Chen Y, 'Gesture, speech and lexical access: the role of lexical movements in speech production', *Psychological Science*, 1996, 7(4), 226–31

Rawls J, *A Theory of Justice*, Harvard University Press, 1971

Rawolle M, Schultheiss M & Schultheiss OC, 'Relationships between implicit motives, self-attributed motives, and personal goal commitments', *Frontiers in Psychology*, 2013, 4, article 923

Ray WJ & Cole HW, 'EEG alpha activity reflects attentional demands, and beta activity reflects emotional and cognitive processes', *Science*, 1985, 228(4700), 750–2

Rayner A, *Origin of Life Patterns: In the Natural Inclusion of Space in Flux* (Springer Briefs in Psychology), Springer, 2017

Raysor TM (ed), *Coleridge's Miscellaneous Criticism*, Harvard University Press, 1936

Razumnikova OM, 'Functional organization of different brain areas during convergent and divergent thinking: an EEG investigation', *Cognitive Brain Research*, 2000, 10(1–2), 11–18

———, 'Gender differences in hemispheric organization during divergent thinking: an EEG investigation in human subjects', *Neuroscience Letters*, 2004a, 362(3), 193–5

———, *Myshlenie i funktsional'naya asimmetriya mozga* (*Thinking and Functional Asymmetry of the Brain*), Siberian Division of the Russian Academy of Medical Sciences, Novosibirsk, 2004b

———, 'Hemispheric activity during creative thinking: role of gender factor', in *KORUS 2005: Proceedings of the 9th Russian-Korean International Symposium on Science and Technology*, 2005, 1027–31

———, 'Creativity related cortex activity in the remote associates task', *Brain Research Bulletin*, 2007, 73(1–3), 96–102

Razumnikova OM & Vol'f NV, 'Selection of visual hierarchical stimuli between global and local aspects in men and women', *Fiziologiia Cheloveka*, 2011, 37, 14–19

Razumnikova OM, Vol'f N & Tarasova IV, 'Strategy and results: sex differences in electrographic correlates of verbal and figural creativity', *Human Physiology*, 2009, 35(3), 285–94

Read DB, 'Solving deductive-reasoning problems after unilateral temporal lobectomy', *Brain and Language*, 1981, 12(1), 116–27

Read R, 'Religion as sedition: on liberalism's intolerance of real religion', *Ars Disputandi*, 2011, 11, 83–100

———, *Wittgenstein's Liberatory Philosophy*, Routledge, 2021

Reber AS, 'Implicit learning of artificial grammars', *Journal of Verbal Learning & Verbal Behavior*, 1967, 6(6), 855–63

———, *Implicit Learning and Tacit Knowledge: an Essay on the Cognitive Unconscious*, Oxford University Press, 1996

Reber J & Tranel D, 'Sex differences in the functional lateralization of emotion and decision making in the human brain', *Journal of Neuroscience Research*, 2017, 95(1–2), 270–8

Rechavi O, Minevish G & Hobert O, 'Transgenerational inheritance of an acquired small RNA-based antiviral response in C. elegans', *Cell*, 2011, 147(6), 1248–56

Redcay E, 'The superior temporal sulcus performs a common function for social and speech perception: implications for the emergence of autism', *Neuroscience & Biobehavioral Reviews*, 2008, 32(1), 123–42

Reddien PW & Sánchez Alvarado A, 'Fundamentals of planarian regeneration', *Annual Review of Cell and Developmental Biology*, 2004, 20, 725–57

Reder M & Schmidt J, 'Habermas and religion', in J Habermas et al, *An Awareness of What is Missing: Faith and Reason in a Post-Secular Age*, Polity Press, 2010, 15–23

Redmond T & Taniguchi T, 'The Buddhist priest who became a billionaire snubbing investors', *Bloomberg*, 4 November 2015

Rée J, on *The Varieties of Religious Experience*, speaking on BBC 4's programme *In Our Time*, 13 May 2010

———, 'Certainty is uncertain', *The Guardian*, 7 March 2013

———, '1901: Intelligent love', in *Witcraft: The Invention of Philosophy in English*, Yale University Press, 2019, 369–468

Rees M, *Just Six Numbers*, Weidenfeld & Nicolson, 2015

Regard M, Cook ND, Wieser HG et al, 'The dynamics of cerebral dominance during unilateral limbic seizures', *Brain*, 1994, 117(1), 91–104

Regard M & Landis T, 'The "smiley": a graphical expression of mood in right anterior cerebral lesions', *Neuropsychiatry, Neuropsychology & Behavioral Neurology*, 1994, 7(4), 303–7

———, '"Gourmand syndrome": eating passion associated with right anterior lesions', *Neurology*, 1997, 48(5), 1185–90

Regolin L, Marconato F, Tommasi L et al, 'Do chicks complete partly occluded objects only with their right hemisphere?', *Behavioural Pharmacology*, 2001, 12(suppl 1), S82

Regolin L, Marconato F & Vallortigara G, 'Hemispheric differences in the recognition of partly occluded objects by newly-hatched domestic chicks (*Gallus gallus*)', *Animal Cognition*, 2004, 7(3), 162–70

Regolin L & Vallortigara G, 'Perception of partly occluded objects by young chicks', *Perception & Psychophysics*, 1995, 57(7), 971–6

Reiber C, Shattuck EC, Fiore S et al, 'Change in human social behavior in response to a common vaccine', *Annals of Epidemiology*, 2010, 20(10), 729–33

Reich SS & Cutting J, 'Picture perception and abstract thought in schizophrenia', *Psychological Medicine*, 1982, 12(1), 91–6

Reid CR, Latty T, Dussutour A et al, 'Slime mold uses an externalized spatial "memory"', *Proceedings of the National Academy of Sciences of the United States of America*, 2012, 109(43), 17490–4

Reid I, Young AW & Hellawell DJ, 'Voice recognition impairment in a blind Capgras patient', *Behavioural Neurology*, 1993, 6(4), 225–8

Reinvang I, 'Crosssed aphasia and apraxia in an artist', *Aphasiology*, 1987, 1(5), 423–4

Reis J, Swayne OB, Vandermeeren Y et al, 'Contribution of transcranial magnetic stimulation to the understanding of cortical mechanisms involved in motor control', *Journal of Physiology*, 2008, 586(2), 325–51

Remedios JD, Chasteen AL, Rule NO et al, 'Impressions at the intersection of ambiguous and obvious social categories: does gay + Black = likable?', *Journal of Experimental Social Psychology*, 2011, 47(6), 1312–5

Reniers RL, Völlm BA, Elliott R et al, 'Empathy, ToM, and self-other differentiation: an fMRI study of internal states', *Social Neuroscience*, 2014, 9(1), 50–62

Rennie D, 'Misconduct and journal peer review', in F Godlee & T Jefferson (eds), *Peer Review in Health Sciences*, BMJ Books, 2nd edn, 2003, 118–29

Rentería ME, 'Cerebral asymmetry: a quantitative, multifactorial, and plastic brain phenotype', *Twin Research and Human Genetics*, 2012, 15(3), 401–13

Rescher N, *Paradoxes: Their Roots, Range, and Resolution*, OpenCourt, 2001

Restak RM, *Brain: The Last Frontier*, Warner Books, 1979

Rettew DC, Billman D & Davis RA, 'Inaccurate perceptions of the amount others stereotype: estimates about stereotypes of one's own group and other groups', *Basic and Applied Social Psychology*, 1993, 14(2), 121–42

Reusser K, 'Problem solving beyond the logic of things: textual and contextual effects on understanding and solving word problems', paper presented at the Annual Meeting of

the American Educational Research Association, San Francisco, 1986
Reverberi C, Toraldo A, D'Agostini S et al, 'Better without (lateral) frontal cortex? Insight problems solved by frontal patients', Brain, 2005, 128(12), 2882–90
Rew L, 'Intuition: concept analysis of a group phenomenon', Advances in Nursing Science, 1986, 8(2), 21–8
Reyna VF, 'A theory of medical decision making and health: fuzzy trace theory', Medical Decision Making, 2008, 28(6), 850–65
Reyna VF & Farley F, 'Risk and rationality in adolescent decision making: implications for theory, practice, and public policy', Psychological Science in the Public Interest, 2006, 7(1), 1–44
Reynolds GP, 'Increased concentrations and lateral asymmetry of amygdala dopamine in schizophrenia', Nature, 1983, 305(5934), 527–9
Rezlescu C, Pitcher D & Duchaine B, 'Acquired prosopagnosia with spared within-class object recognition but impaired recognition of degraded basic-level objects', Cognitive Neuropsychology, 2012, 29(4), 325–47
Rhees R, Recollections of Wittgenstein, Oxford University Press, 1984
Ribeiro AF, Mansur LL & Radanovic M, 'Impairment of inferential abilities based on pictorial stimuli in patients with right-hemisphere damage', Applied Neuropsychology: Adult, 2015, 22(3), 161–9
Richard FD, Bond CF, Stokes-Zoota JJ, 'One hundred years of social psychology quantitatively described', Review of General Psychology, 2003, 7(4), 331–63
Richards IA, The Philosophy of Rhetoric, Oxford University Press, 1965
Richards RL, Kinney DK, Lunde I et al, 'Creativity in manic-depressives, cyclothymes, their normal relatives and control subjects', Journal of Abnormal Psychology, 1988, 97(3), 281–8
Richardson ED, Malloy PF & Grace J, 'Othello syndrome secondary to right cerebrovascular infarction', Journal of Geriatric Psychiatry and Neurology, 1991, 4(3), 160–5
Richardson JK, 'Psychotic behavior after right hemispheric cerebrovascular accident: a case report', Archives of Physical Medicine and Rehabilitation, 1992, 73(4), 381–4
Richardson M, Cormack A, McRobert L et al, '30 days wild: development and evaluation of a large-scale nature engagement campaign to improve well-being', PLoS One, 2016, 11(2), e0149777
Richert R & Barrett JL, 'The child's God', in EM Dowling & WJ Scarlett (eds), The Encyclopædia of Spiritual Development, Sage Publishing, 2006, 70–2
Ricoeur P, 'The model of the text: meaningful action considered as a text', New Literary History, 1973, 5(1), 91–117
———, La métaphore vive, Éditions du Seuil, 1975
Ridley M, Origins of Virtue: Human Instincts and the Evolution of Cooperation, Viking, 1996
Riege WH, Klane LT, Metter EJ et al, 'Decision speed and bias after unilateral stroke', Cortex, 1982, 18(3), 345–55
Riege WH, Metter EJ & Hanson WR, 'Verbal & nonverbal recognition memory in aphasic and non-aphasic stroke patients', Brain and Language, 1980, 10(1), 60–70
Riehemann S, Volz HP, Stützer P et al, 'Hypofrontality in neuroleptic-naive schizophrenic patients during the Wisconsin Card Sorting Test – an fMRI study', European Archives of Psychiatry and Clinical Neuroscience, 2001, 251(2), 66–71
Rigosi E, Haase A, Rath L et al, 'Asymmetric neural coding revealed by in vivo calcium imaging in the honey bee brain', Proceedings of the Royal Society B: Biological Sciences, 2015, 282(1803), 20142571
Rikers RMJP, Schmidt HG, Boshuizen HPA et al, 'The robustness of medical expertise: clinical case processing by medical experts and subexperts', American Journal of Psychology, 2002, 115(4), 609–29
Rilea SL, Roskos-Ewoldsen B & Boles D, 'Sex differences in spatial ability: a lateralization of function approach', Brain and Cognition, 2004, 56(3), 332–43
Riley EN & Sackeim HA, 'Ear asymmetry in recognition of unfamiliar voices', Brain and Cognition, 1982, 1(3), 245–58
Rilke RM, Letters to a Young Poet, trans C Louth, Penguin, 2011
Rilling JK & Insel TR, 'Differential expansion of neural projection systems in primate brain evolution', NeuroReport, 1999, 10(7), 1453–9
Rimé B, Schiaratura L, Hupet M et al, 'Effects of relative immobilisation on the speaker's nonverbal behaviour and on the dialogue imagery level', Motivation and Emotion, 1984, 8(4), 311–25
Rinaldi MC, Marangolo P & Baldassarri F, 'Metaphor comprehension in right brain-damaged patients with visuo-verbal and verbal material: a dissociation (re)considered', Cortex, 2004, 40(3), 479–90
Rinehart NJ, Bradshaw JL, Moss SA et al, 'Atypical interference of local detail on global processing in high-functioning autism and Asperger's disorder', Journal of Child

Psychology and Psychiatry, 2000, 41(6), 769–78

Rinehart NJ, Tonge BJ, Iansek R et al, 'Gait function in newly diagnosed children with autism: cerebellar and basal ganglia related motor disorder', *Developmental Medicine and Child Neurology*, 2006, 48(10), 819–24

Ring K & Cooper S, 'Near-death and out-of-body experiences in the blind: a study of apparent eyeless vision', *Journal of Near-death Studies*, 1997, 16(2), 101–47

Ringbauer M, Duffus B, Branciard C et al, 'Measurements on the reality of the wavefunction', *Nature Physics*, 2015, 11, 249–54

Ringo JL, 'Neuronal interconnection as a function of brain size', *Brain, Behavior and Evolution*, 1991, 38(1), 1–6

Ringo JL, Doty RW, Demeter S et al, 'Time is of the essence: a conjecture that hemispheric specialization arises from interhemispheric conduction delay', *Cerebral Cortex*, 1994, 4(4), 331–43

Rissman J, Eliassen JC & Blumstein SE, 'An event-related fMRI investigation of implicit semantic priming', *Journal of Cognitive Neuroscience*, 2003, 15(8), 1160–75

Ritblatt SN, 'Children's level of participation in a false-belief task, age, and theory of mind', *Journal of Genetic Psychology*, 2000, 161(1), 53–64

Ritchie S, *Intelligence: All That Matters*, John Murray, 2015

Rivera LO, Arms-Chavez CJ & Zárate MA, 'Resource-dependent effects during sex categorisation', *Journal of Experimental Social Psychology*, 2009, 45(4), 908–12

Rizzo M, Akutsu H & Dawson J, 'Increased attentional blink after focal cerebral lesions', *Neurology*, 2001, 57(5), 795–800

Rizzolatti G & Buchtel HA, 'Hemispheric superiority in reaction time to faces: a sex difference', *Cortex*, 1977, 13(3), 300–5

Rizzolatti G & Sinigaglia C, *Mirrors in the Brain: How Our Minds Share Actions and Emotions*, Oxford University Press, 2008

Roalf D, Lowery N & Turetsky BI, 'Behavioral and physiological findings of gender differences in global-local visual processing', *Brain and Cognition*, 2006, 60(1), 32–42

Robbins M, *The Primordial Mind in Health and Illness: A Cross-Cultural Perspective*, Routledge, 2011

Robbins SE, 'Bergson, perception and Gibson', *Journal of Consciousness Studies*, 2000, 7(5), 23–45

—— , 'On time, memory and dynamic form', *Consciousness and Cognition*, 2004, 13(4), 762–88

Roberson D & Davidoff J, 'The categorical perception of colors and facial expressions: the effect of verbal interference', *Memory & Cognition*, 2000, 28(6), 977–86

Roberts NA, Beer JS, Werner KH et al, 'The impact of orbital prefrontal cortex damage on emotional activation to unanticipated and anticipated acoustic startle stimuli', *Cognitive, Affective and Behavioral Neuroscience*, 2004, 4(3), 307–16

Robertson CE, Thomas C, Kravitz DJ et al, 'Global motion perception deficits in autism are reflected as early as primary visual cortex', *Brain*, 2014, 137(9), 2588–99

Robertson DA, Gernsbacher MA, Guidotti SJ et al, 'Functional neuroanatomy of the cognitive process of mapping during discourse comprehension', *Psychological Science*, 2000, 11(3), 255–60

Robertson LC & Lamb MR, 'Neuropsychological contributions to theories of part/whole organization', *Cognitive Psychology*, 1991, 23(2), 299–330

Robertson LC, Lamb MR & Knight RT, 'Effects of lesions of temporal-parietal junction on perceptual and attentional processing in humans', *Journal of Neuroscience*, 1988, 8(10), 3757–69

Robinson E, *The Original Vision: A Study of the Religious Experience of Childhood*, Seabury Press, 1983

Robinson I, 'Can an atheist have a religious experience?', *The Australian Rationalist*, 2000, 54, 5–16

Robinson R, 'Different paths, same structure: "developmental systems drift" at work', *PLoS Biology*, 2011, 9(7), e1001113

Robinson RG, Boston JD, Starkstein SE et al, 'Comparison of mania and depression after brain injury: causal factors', *American Journal of Psychiatry*, 1988, 145(2), 172–8

Robinson RG & Price TR, 'Post-stroke depressive disorders: a follow-up study of 103 patients', *Stroke*, 1982, 13(5), 635–41

Rochester S & Martin JR, *Crazy Talk: A Study of the Discourse of Schizophrenic Speakers*, Springer, 1979

Rockstroh B & Lutzenberger W, 'Differences between anhedonic and control subjects in brain hemispheric specialization as revealed by brain potentials', in A Glass (ed), *Individual Differences in Hemispheric Specialization* (NATO ASI Series A: Life Sciences, vol 130), Plenum Press, 1987, 183–94

Rode G, Charles N, Perenin MT et al, 'Partial remission of hemiplegia and somatoparaphrenia through vestibular stimulation in a case of unilateral neglect', *Cortex*, 1992, 28, 203–8

Rode G, Vallar G, Chabanat E et al, 'What do spatial distortions in patients' drawing after right brain damage teach us about space representation in art?', *Frontiers in Psychology*, 2018, 9, article 1058

Rodríguez RL, Briceño RD, Briceño-Aguilar E et al, '*Nephila clavipes* spiders (Araneæ: Nephilidæ) keep track of captured prey counts: testing for a sense of numerosity in an orb-weaver', *Animal Cognition*, 2015, 18, 307–14

Rodríguez-Ferreiro J, Menéndez M, Ribacoba R et al, 'Action naming is impaired in Parkinson disease patients', *Neuropsychologia*, 2009, 47(14), 3271–4

Rodway P, Schepman A, Crossley B et al, 'A leftward perceptual asymmetry when judging the attractiveness of visual patterns', *Laterality*, 2019, 24(1), 1–25

Roediger HL & McDermott KB, 'Two types of event memory', *Proceedings of the National Academy of Sciences of the United States of America*, 2013, 110(52), 20856–7

Rogers B, *A. J. Ayer: A Life*, Chatto & Windus, 1999

Rogers LJ & Andrew RJ, *Comparative Vertebrate Lateralization*, Cambridge University Press, 2002

Rogers LJ, Zucca P & Vallortigara G, 'Advantages of having a lateralized brain', *Proceedings of the Royal Society of London, Series B: Biological Sciences*, 2004, 271(suppl 6), S420–2

Rogers RD, Everitt BJ, Baldacchino A et al, 'Dissociable deficits in the decision-making cognition of chronic amphetamine abusers, opiate abusers, patients with focal damage to prefrontal cortex, and tryptophan-depleted normal volunteers: evidence for monoaminergic mechanisms', *Neuropsychopharmacology*, 1999, 20(4), 322–39

Rogers RD, Owen AM, Middleton HC et al, 'Choosing between small, likely rewards and large, unlikely rewards activates inferior and orbital prefrontal cortex', *Journal of Neuroscience*, 1999, 19(20), 9029–38

Rojo V, Caballero L, Iruela LM et al, 'Capgras' syndrome in a blind patient', *American Journal of Psychiatry*, 1991, 148(9), 1271–2

Roland PE, Larsen B, Lassen NA et al, 'Supplementary motor area and other cortical areas in organization of voluntary movements in man', *Journal of Neurophysiology*, 1980, 43(1), 118–36

Roland PE, Skinhøj E & Lassen NA, 'Focal activation of human cerebral cortex during auditory discrimination', *Journal of Neurophysiology*, 1981, 45(6), 1139–51

Ronchi C, *The Tree of Knowledge: The Bright and the Dark Sides of Science*, Springer, 2014

Rorden C & Karnath H-O, 'Using human brain lesions to infer function: a relic from a past era in the fMRI age?', *Nature Reviews Neuroscience*, 2004, 5(10), 813–9

Röricht S, Irlbacher K, Petrow E et al, 'Normwerte transkallosal und kortikospinal vermittelter Effekte einer hemisphärenselektiven elektromyographischer magnetischen Kortexreizung beim Menschen', *Zeitschrift für Elektroenzephalographie, Elektromyographie und Verwandte Gebiete*, 1997, 28(1), 34–8

Rosa C, Lassonde M, Pinard C et al, 'Investigations of hemispheric specialization of self-voice recognition', *Brain and Cognition*, 2008, 68(2), 204–14

Rosen HJ, Allison SC, Schauer GF et al, 'Neuroanatomical correlates of behavioural disorders in dementia', *Brain*, 2005, 128(11), 2612–25

Rosen R, *Essays on Life Itself*, Columbia University Press, 2000

Rosenberg A, 'Defending information-free genocentrism', *History and Philosophy of the Life Sciences*, 2005, 27(3–4), 345–59

Rosenblueth A & Wiener N, 'The role of models in science', *Philosophy of Science*, 1945, 2(4), 316–21

Rosenfeld L, 'Niels Bohr's contribution to epistemology', *Physics Today*, 1963, 16(10), 47–54

Rosenthal R & Bigelow LB, 'Quantitative brain measurements in chronic schizophrenia', *British Journal of Psychiatry*, 1972, 121(562), 259–64

Rosenthal R, Hall JA, Di Matteo MR et al, *Sensitivity to Nonverbal Communication: The PONS Test*, Johns Hopkins University Press, 1979

Rosenthal SB, 'Continuity, contingency, and time: the divergent intuitions of Whitehead and Pragmatism', *Transactions of the Charles S Peirce Society*, 1996, 32(4), 542–67

Roser ME, Fugelsang JA, Dunbar KN et al, 'Dissociating processes supporting causal perception and causal inference in the brain', *Neuropsychology*, 2005, 19(5), 591–602

Rosmarin DH, Pirutinsky S, Pargament KI et al, 'Are religious beliefs relevant to mental health among Jews?', *Psychology of Religion and Spirituality*, 2009, 1(3), 180–90

Ross ED, 'The aprosodias: functional-anatomic organization of the affective components of language in the right hemisphere', *Archives of Neurology*, 1981, 38(9), 561–9

Ross ED, Homan RW & Buck R, 'Differential hemispheric lateralization of primary and social emotions: implications for developing a comprehensive neurology for emotions, repression, and the subconscious', *Neuropsychiatry, Neuropsychology, and Behavioral Neurology*, 1994, 7(1), 1–19

Ross ED, Orbelo DM, Cartwright J et al, 'Affective-prosodic deficits in schizophrenia: profiles of patients with brain damage and comparison with relation to schizophrenic symptoms', *Journal of Neurology, Neurosurgery, and Psychiatry*, 2001, 70(5), 597–604

Ross ED, Thompson RD & Yenkosky J, 'Lateralisation of affective prosody in the brain and callosal integration of hemispheric language functions', *Brain and Language*, 1977, 56(1), 27–54

Rossell SL & Van Rheenen TE, 'Theory of mind performance using a story comprehension task in bipolar mania compared to schizophrenia and healthy controls', *Cognitive Neuropsychiatry*, 2013, 18(5), 409–21

Rosser R, 'The psychopathology of feeling and thinking in a schizophrenic', *International Journal of Psychoanalysis*, 1979, 60(2), 177–88

Rossion B, 'Understanding face perception by means of prosopagnosia and neuroimaging', *Frontiers in Bioscience (Elite Edition)*, 2014, 6, 258–307

Rossion B, Caldara R, Seghier M et al, 'A network of occipito-temporal face-sensitive areas besides the right middle fusiform gyrus is necessary for normal face processing', *Brain*, 2003, 126(11), 2381–95

Rotarska-Jagiela A, van de Ven V, Oertel-Knöchel V et al, 'Resting-state functional network correlates of psychotic symptoms in schizophrenia', *Schizophrenia Research*, 2010, 117(1), 21–30

Roth G & Dicke U, 'Evolution of the brain and intelligence', *Trends in Cognitive Sciences*, 2005, 9(5), 250–7

Rothbart MK, Ahadi SA & Hershey KL, 'Temperament and social behavior in childhood', *Merrill-Palmer Quarterly*, 1994, 40(1), 21–39

Rothman S, *Lessons from the Living Cell: The Limits of Reductionism*, McGraw Hill, 2002

Rothmayr C, Baumann O, Endestad T et al, 'Dissociation of neural correlates of verbal and non-verbal visual working memory with different delays', *Behavioral and Brain Functions*, 2007, 3, article 56

Rothschild FS, 'Über Links und Rechts: eine erscheinungswissenschaftliche Untersuchung', *Zeitschrift für die gesamte Neurologie und Psychiatrie*, 1930, 124(3–4), 451–511

Rourke BP, *Nonverbal Learning Disabilities: The Syndrome and the Model*, Guilford Press, 1989

Rousseaux M, Debrock D, Cabaret M et al, 'Visual hallucinations with written words in a case of left parietotemporal lesion', *Journal of Neurology, Neurosurgery, and Psychiatry*, 1994, 57(10), 1268–71

Rousseaux M, Fimm B & Cantagallo A, 'Attention disorders in cerebrovascular diseases', in M Leclercq & P Zimmerman (eds), *Applied Neuropsychology of Attention*, Psychology Press, 2002, 280–304

Rowe TB, Macrini TE & Luo Z-X, 'Fossil evidence on origin of the mammalian brain', *Science*, 2011, 332(6032), 955–7

Rowland LM, Spieker EA, Francis A et al, 'White matter alterations in deficit schizophrenia', *Neuropsychopharmacology*, 2009, 34(6), 1514–22

Rowson J, *Chess for Zebras*, Gambit, 2005

———, *Spiritualise: revitalising spirituality to address 21st century challenges*, RSA publications, December 2014: www.thersa.org/globalassets/pdfs/reports/spiritualise-report.pdf (accessed 15 March 2021)

Roy O, *The Failure of Political Islam*, Harvard University Press, 1994

Royet J-P & Plailly J, 'Lateralization of olfactory processes', *Chemical Senses*, 2004, 29(8), 731–45

Rozenblit L & Keil F, 'The misunderstood limits of folk science: an illusion of explanatory depth', *Cognitive Science*, 2002, 26(5), 521–62

Rozensky RH & Gomez MY, 'Language switching in psychotherapy with bilinguals: two problems and case examples', *Psychotherapy: Theory, Research and Practice*, 1983, 20(2), 152–60

Rozin P & Nemeroff C, 'Sympathetic magical thinking: the contagion and similarity "heuristics"', in T Gilovich, D Griffin & D Kahneman (eds), *Heuristics and Biases: The Psychology of Intuitive Judgment*, Cambridge University Press, 2002, 201–16

Rozin P, Markwith M & Ross B, 'The sympathetic magical law of similarity, nominal realism and neglect of negatives in response to negative labels', *Psychological Science*, 1990, 1(6), 383–4

Rubens AB, 'Caloric stimulation and unilateral visual neglect', *Neurology*, 1985, 35(7), 1019–24

Rubenzer R, 'The role of the right hemisphere in learning and creativity implications for enhancing problem solving ability', *Gifted Child Quarterly*, 1979, 23(1), 78–100

Rubin AN, Espiridion ED & Lofgren DH, 'A sub-acute cerebral contusion presenting with medication-resistant psychosis', *Cureus*, 2018, 10(7), e2938

Rubinsztein JS, Fletcher PC, Rogers RD et al, 'Decision-making in mania: a PET study', *Brain*, 2001, 124(12), 2550–63

Rueckert L & Grafman J, 'Sustained attention deficits in patients with right frontal lesions', *Neuropsychologia*, 1996, 34(10), 953–63

Ruff RM, Allen CC, Farrow CE et al, 'Figural fluency: differential impairment in patients with left versus right frontal lobe lesions',

Archives of Clinical Neuropsychology, 1994, 9(1), 41–55
Rugani R, Vallortigara G, Priftis K et al, 'Number-space mapping in the newborn chick resembles humans' mental number line', *Science*, 2015, 347(6221), 534–6
Ruiz MH, Strübing F, Jabusch H-C et al, 'EEG oscillatory patterns are associated with error prediction during music performance and are altered in musician's dystonia', *NeuroImage*, 2011, 55(4), 1791–1803
Rumbelow H, 'How healthy is thin? The big BMI debate', *The Times*, 12 May 2016
Rumi, *Mathnawi*, from trans and ed C & K Helminski, *Daylight: A Daybook of Spiritual Guidance*, Shambhala, 1999
Rumiati RI, 'Right, left or both? Brain hemispheres and apraxia of naturalistic actions', *Trends in Cognitive Sciences*, 2005, 9(4), 167–9
Rumiati RI, Zanini S, Vorano L et al, 'A form of ideational apraxia as a selective deficit of contention scheduling', *Cognitive Neuropsychology*, 2001, 18(7), 617–42
Rushton JP, 'Creativity, intelligence, and psychoticism', *Personality and Individual Differences*, 1990, 11(12), 1291–8
——, 'Secular gains in IQ not related to the g factor and inbreeding depression – unlike Black-White differences: a reply to Flynn', *Personality and Individual Differences*, 1999, 26(2), 381–9
Rushton JP & Jensen AR, 'The rise and fall of the Flynn effect as a reason to expect a narrowing of the Black-White IQ gap', *Intelligence*, 2010, 38(2), 213–9
Ruskin J, *Modern Painters*, John Wiley, 6 vols, 1863 [1843–60].
Russell B, *Mysticism and Logic: And Other Essays*, Longman, 1919
——, *An Outline of Philosophy*, George Allen & Unwin, 1927
——, *The Scientific Outlook*, WW Norton, 1931
——, 'Is there a God?' (1952), in JG Slater (ed), *The Collected Papers of Bertrand Russell*, Routledge, 1943–68, vol 11, 542–8
——, *History of Western Philosophy*, George Allen & Unwin, 1961
Russell JB, *Inventing the Flat Earth: Columbus and Modern Historians*, Praeger, 1991
——, 'The myth of the flat earth', 1997: www.veritas-ucsb.org/library/russell/FlatEarth.html (accessed 22 February 2019)
Russo P, Persegani C, Papeschi LL et al, 'Sex differences in hemisphere preference as assessed by a paper-and-pencil test', *International Journal of Neuroscience*, 2000, 100(1–4), 29–37
Rutledge R & Hunt GR, 'Lateralised tool use in New Caledonian crows', *Animal Behaviour*, 2004, 67(2), 327–32
Ruysbroeck J, 'De calculo', *The Sparkling Stone*, c1340
Ryan CS, 'Stereotype accuracy', *European Review of Social Psychology*, 2002, 13(1), 75–109
Rybash JM & Hoyer WJ, 'Hemispheric specialization for categorical and coordinate spatial representations: a reappraisal', *Memory & Cognition*, 1992, 20(3), 271–6
Sá W, Kelley C, Ho C et al, 'Thinking about personal theories: individual differences in the coordination of theory and evidence', *Personality and Individual Differences*, 2005, 38(5), 1149–61
Saadah ESM & Melzack R, 'Phantom limb experiences in congenital limb-deficient adults', *Cortex*, 1994, 30(3), 479–85
Saban-Bezalel R & Mashal N, 'Comprehension and hemispheric processing of irony in schizophrenia', *Frontiers in Psychology*, 2017, 8, article 943
Sack AT, Camprodon JA, Pascual-Leone A et al, 'The dynamics of interhemispheric compensatory processes in mental imagery', *Science*, 2005, 308(5722), 702–4
Sack AT & Schuhmann T, 'Hemispheric differences within the fronto-parietal network dynamics underlying spatial imagery', *Frontiers in Psychology*, 2012, 3, 214
Sackeim HA, Gur RC & Saucy MC, 'Emotions are expressed more intensely on the left side of the face', *Science*, 1978, 202(4366), 434–6
Sacks O, *The Man Who Mistook His Wife for a Hat*, Picador, 1986
——, 'The landscape of his dreams', *The New Yorker*, 27 July 1992
——, *Musicophilia*, Knopf, 2007
Safer DL, Wenegrat B & Roth WT, 'Risperidone in the treatment of delusional parasitosis: a case report', *Journal of Clinical Psychopharmacology*, 1997, 17(2), 131–2
Sagan C, *The Varieties of Scientific Experience*, Penguin, 2006
Saggar M, Quintin EM, Kienitz E et al, 'Pictionary-based fMRI paradigm to study the neural correlates of spontaneous improvisation and figural creativity', *Scientific Reports*, 2015, 5, 10894
Sahoo A & Josephs KA, 'A neuropsychiatric analysis of the Cotard delusion', *Journal of Neuropsychiatry & Clinical Neuroscience*, 2018, 30(1), 58–65
Saigusa T, Tero A, Nakagaki T et al, 'Amoebae anticipate periodic events', *Physical Review Letters*, 2008, 100(1), 018101
Sainburg RL, 'Evidence for a dynamic-dominance hypothesis of handedness',

Experimental Brain Research, 2002, 142(2), 241–58

Saito H, Kanayama S & Takahashi T, 'Right angular lesion and selective impairment of motion vision in left visual field', *Tohoku Journal of Experimental Medicine*, 1992, 166(2), 229–38

Saito Y, 'The Japanese aesthetics of imperfection and insufficiency', *Journal of Aesthetics and Art Criticism*, 1997, 55(4), 377–85

Saito Y, Matsunaga A, Yamamura O et al, 'A case of left hemi-facial metamorphopsia induced by infarction of the right side of the splenium of the corpus callosum', *Rinsho Shinkeigaku*, 2014, 54(8), 637–42

Saj A, Fuhrman O, Vuilleumier P et al, 'Patients with left spatial neglect also neglect the "left side" of time', *Psychological Science*, 2013, 25(1), 207–14

Sakai K, Hikosaka O, Miyauchi, S et al, 'Neural representation of a rhythm depends on its interval ratio', *Journal of Neuroscience*, 1999, 19(22), 10074–81

Sakai M, Hishii T, Takeda S et al, 'Laterality of flipper rubbing behaviour in wild bottlenose dolphins (*Tursiops aduncus*): caused by asymmetry of eye use?', *Behavioural Brain Research*, 2006, 170(2), 204–10

Sakai T, Kondo M & Tomimoto H, 'Complex partial status epilepticus with recurrent episodes of complex visual hallucinations: study by using 123I-IMP-SPECT, brain MRI and EEG', *Rinsho Shinkeigaku*, 2015, 55(8), 580–4

Sakhno S & Tersis N, 'Is a "friend" an "enemy"? Between "proximity" and "opposition"', in M Vanhove (ed), *From Polysemy to Semantic Change*, John Benjamins, 2008, 317–40

Saks ER, *The Center Cannot Hold: My Journey Through Madness*, Hyperion, 2007

Sakurai K, Kurita T, Takeda Y et al, 'Akinetopsia as epileptic seizure', *Epilepsy & Behavior Case Reports*, 2013, 1, 74–6

Salam A, *Unification of Fundamental Forces*, Cambridge University Press, 1990

Salas CE, Radovic D, Yuen KSL et al, '"Opening an emotional dimension in me": changes in emotional reactivity and emotion regulation in a case of executive impairment after left fronto-parietal damage', *Bulletin of the Menninger Clinic*, 2014, 78(4), 301–34

Salazar-Ciudad I, Solé RV & Newman SA, 'Phenotypic and dynamical transitions in model genetic networks. II. Application to the evolution of segmentation mechanisms', *Evolution & Development*, 2001, 3(2), 95–103; erratum at *Evolution & Development*, 2001, 3(5), 371

Salk J, interviewed for the Academy of Achievement in May 1991, after winning the Congressional Gold Medal: achievement.org/achiever/jonas-salk-m-d/#interview (accessed 23 February 2021)

Salk L, 'The role of the heartbeat in the relations between mother and infant', *Scientific American*, 1973, 228(5), 24–9

Salmon W, *Causality and Explanation*, Oxford University Press, 1998

Salva OR, Regolin L, Mascalzoni E et al, 'Cerebral and behavioural asymmetries in animal social recognition', *Comparative Cognition & Behavior Reviews*, 2012, 7, 110–38

Salvi C, Bricolo E, Bowden E et al, 'Insight solutions are correct more often than analytic solutions', *Thinking and Reasoning*, 2016, 22(4), 443–60

Salvi C, Bricolo E, Franconeri S et al, 'Sudden insight is associated with shutting out visual inputs', *Psychonomic Bulletin & Review*, 2015, 22(6), 1814–9

Sample I, 'Group of biologists tries to bury the idea that plants are conscious', *The Guardian*, 3 July 2019

Samson D, Apperly IA, Braithwaite JJ et al, 'Seeing it their way: evidence for rapid and involuntary computation of what other people see', *Journal of Experimental Psychology: Human Perception and Performance*, 2010, 36(5), 1255–66

Samson S & Zatorre R, 'Neuropsychologie de la musique: approche anatomo-fonctionnelle', in A Zenatti (ed), *Psychologie de la musique*, Presses Universitaires de France, 1994, 291–316

Samuels CA, Butterworth G, Roberts T et al, 'Facial aesthetics: babies prefer attractiveness to symmetry', *Perception*, 1994, 23(7), 823–31

Samuels CA & Ewy R, 'Aesthetic perception of faces during infancy', *British Journal of Developmental Psychology*, 1985, 3(3), 221–8

Sanches RF, de Lima Osório F, Dos Santos RG et al, 'Antidepressant effects of a single dose of ayahuasca in patients with recurrent depression: a SPECT study', *Journal of Clinical Psychopharmacology*, 2016, 36(1), 77–81

Sanchez F, *The Master Illusionist: Principles of Neuropsychology*, Xlibris, 1990

Sander K & Scheich H, 'Auditory perception of laughing and crying activates human amygdala regardless of attentional state', *Brain Research: Cognitive Brain Research*, 2001, 12(2), 181–98

Sanders LD & Poeppel D, 'Local and global auditory processing: behavioral and ERP evidence', *Neuropsychologia*, 2006, 45(6), 1172–86

Sanders RJ, 'Sentence comprehension following agenesis of the corpus callosum', *Brain and Language*, 1989, 37(1), 59–72

Sandkühler S & Bhattacharya J, 'Deconstructing insight: EEG correlates of insightful problem solving', *PLoS One*, 2008, 3(1), e1459

Sandson J & Albert ML, 'Perseveration in behavioral neurology', *Neurology*, 1987, 37(11), 1736–41

Sandson TA, Manoach DS, Price BH et al, 'Right hemisphere learning disability associated with left hemisphere dysfunction: anomalous dominance and development', *Journal of Neurology, Neurosurgery, and Psychiatry*, 1994, 57(9), 1129–32

Sandstrom NJ, Kaufman J & Huettel SA, 'Males and females use different distal cues in a virtual environment navigation task', *Brain Research: Cognitive Brain Research*, 1998, 6(4), 351–60

Sandyk R, 'Improvement of right hemispheric functions in a child with Gilles de la Tourette's syndrome by weak electromagnetic fields', *International Journal of Neuroscience*, 1995, 81(3–4), 199–213

Saniga M, 'Geometry of psychological time', in S Albeverio & P Blanchard (eds), *Direction of Time*, Springer, 2014, 171–86

Sar V, Unal SN & Ozturk E, 'Frontal and occipital perfusion changes in dissociative identity disorder', *Journal of Experimental Child Psychology*, 2007, 156(3), 217–23

Sarfati Y, Hardy-Baylé MC, Besche C et al, 'Attribution of intentions to others in people with schizophrenia: a non-verbal exploration with comic strip', *Schizophrenia Research*, 1997, 25(3), 199–209

Särkämö T, Tervaniemi M, Soinila S et al, 'Auditory and cognitive deficits associated with acquired amusia after stroke: a magnetoencephalography and neuropsychological follow-up study', *PLoS One*, 2010, 5(12), e15157

Sarnat HB & Netsky MG, *Evolution of the Nervous System*, Oxford University Press, 1981

———, 'The brain of the planarian as the ancestor of the human brain', *Canadian Journal of Neurological Sciences*, 1985, 12(4), 296–302

Saron CD & Davidson RJ, 'Visual evoked potential measures of interhemispheric transfer time in humans', *Behavioral Neuroscience*, 1989, 103(5), 1115–38

Saron CD, Foxe JJ, Schroeder CE et al, 'Complexities of interhemispheric communication in sensorimotor tasks revealed by high-density event-related potential mapping', in K Hugdahl & RJ Davidson (eds), *The Asymmetrical Brain*, Massachusetts Institute of Technology Press, 2003, 341–408

Saron CD, Foxe JJ, Simpson GV et al, 'Interhemispheric visuomotor activation: spatiotemporal electrophysiology related to reaction time', in E Zaidel & M Iacoboni (eds), *The Parallel Brain: The Cognitive Neuroscience of the Corpus Callosum*, Massachusetts Institute of Technology Press, 2002, 171–219

Sartre J-P, *Genet: Actor and Martyr*, trans B Frechtman, University of Minnesota Press, 1963

Sasaki H, Morimoto A, Nishio A et al, 'Right hemisphere specialization for color detection', *Brain and Cognition*, 2007, 64(3), 282–9

Sasaki M, Kamei A & Chida S, 'Abnormal magnetic resonance imaging in a child with Alice in Wonderland syndrome following Epstein-Barr virus infection', *Brain and Development*, 2002, 34(4), 348–52

Sass LA, *The Paradoxes of Delusion: Wittgenstein, Schreber, and the Schizophrenic Mind*, Cornell University Press, 1994

———, 'Surface and depth: Wittgenstein's reflections on psychoanalysis', *Partisan Review*, 1998, 65(4), 590–614

———, 'Self and world in schizophrenia: three classic approaches', in *Philosophy, Psychiatry, & Psychology*, 2001, 8(4), 251–70

———, 'Self-disturbance in schizophrenia: hyperreflexivity and diminished self-affection', in T Kircher & A David (eds), *The Self in Neuroscience and Psychiatry*, Cambridge University Press, 2003, 242–71

———, 'Schizophrenia: a disturbance of the thematic field', in L Embree (ed), *Gurwitsch's Relevancy for Cognitive Science*, Springer, 2004, 59–78

———, *Madness and Modernism: Insanity in the Light of Modern Art, Literature and Thought*, revised edn, Oxford University Press, 2017 [1992]

Sass LA & Borda JP, 'Phenomenology and neurobiology of self disorder in schizophrenia: secondary factors', *Schizophrenia Research*, 2015, 169(1–3), 474–82

Sass LA & Byrom G, 'Phenomenological and neurocognitive perspectives on delusions: a critical overview', *World Psychiatry*, 2015, 14(2), 164–73

Sass LA & Parnas J, 'Phenomenology of self-disturbances in schizophrenia: some research findings and directions', *Philosophy, Psychiatry, & Psychology*, 2001, 8(4), 347–56

———, 'Schizophrenia, consciousness, and the self', *Schizophrenia Bulletin*, 2003, 29(3), 427–44

Sass LA & Pienkos E, 'Delusion: the phenomenological approach', in KWM Fulford & M Davies (eds), *The Oxford Handbook of Philosophy & Psychiatry*, Oxford University Press, 2013a, 632–57

———, 'Varieties of self-experience: a comparative phenomenology of melancholia,

mania, and schizophrenia (Part I)', *Journal of Consciousness Studies*, 2013b, 20(7–8), 103–30
——, 'Space, time, and atmosphere: a comparative phenomenology of melancholia, mania, and schizophrenia (Part II)', *Journal of Consciousness Studies*, 2013c, 20(7–8), 131–52
——, 'Beyond words: linguistic experience in melancholia, mania, and schizophrenia', *Phenomenology and the Cognitive Sciences*, 2015a, 14(3), 475–95
——, 'Faces of intersubjectivity: a phenomenological study of interpersonal experience in melancholia, mania, and schizophrenia', *Journal of Phenomenological Psychology*, 2015b, 46, 1–32
Satoh M, Kato N, Tabei K et al, 'A case of musical anhedonia due to right putaminal hemorrhage: a disconnection syndrome between the auditory cortex and insula', *Neurocase*, 2016, 22(6), 518–25
Satoh M, Nakase T, Nagata K et al, 'Musical anhedonia: selective loss of emotional experience in listening to music', *Neurocase*, 2011, 17(5), 410–7
Satoh M, Suzuki K, Miyamura M et al, 'Metamorphopsia and transient increase in the cerebral blood flow of the left occipital pole on 123I-IMP SPECT: a case report', *Rinsho Shinkeigaku*, 1997, 37(7), 631–5
Sauerwein H & Lassonde MC, 'Intra- and interhemispheric processing of visual information in callosal agenesis', *Neuropsychologia*, 1983, 21(2), 167–71
Sauseng P, Klimesch W, Doppelmayr M et al, 'EEG alpha synchronization and functional coupling during top-down processing in a working memory task', *Human Brain Mapping*, 2005, 26(2), 148–55
Saver JL & Rabin J, 'The neural substrates of religious experience', *Journal of Neuropsychiatry and Clinical Neurosciences*, 1997, 9(3), 498–510
Savic I & Lindström P, 'PET and MRI show differences in cerebral asymmetry and functional connectivity between homo- and heterosexual subjects', *Proceedings of the National Academy of Sciences of the United States of America*, 2008, 105(27), 9403–8
Saxe R, Carey S & Kanwisher N, 'Understanding other minds: linking developmental psychology and functional neuroimaging', *Annual Review of Psychology*, 2004, 55, 87–124
Saxe R & Wexler A, 'Making sense of another mind: the role of the right temporo-parietal junction', *Neuropsychologia*, 2005, 43(10), 1391–9
Saxe R, Xiao D-K, Kovacs G et al, 'A region of right posterior superior temporal sulcus responds to observed intentional actions', *Neuropsychologia*, 2004, 42(11), 1435–46
Sayen J, *Einstein in America*, Crown, 1985
Sayers SL, Curran PJ & Mueser KT, 'Factor structure and construct validity of the Scale for the Assessment of Negative Symptoms', *Psychological Assessment*, 1996, 8(3), 269–80
Schaadt AK, Brandt SA, Kraft A et al, 'Holmes and Horrax (1919) revisited: impaired binocular fusion as a cause of "flat vision" after right parietal brain damage – a case study', *Neuropsychologia*, 2015, 69, 31–8
Schaaper RM, 'Base selection, proofreading, and mismatch repair during DNA replication in Escherichia coli', *Journal of Biological Chemistry*, 1993, 268(32), 23762–5
Schachter M, 'Erotomania or the delusional conviction of being loved: contribution to the psychopathology of the love life', *Annales Médico-psychologiques (Paris)*, 1977, 1, 729–47
Schacter DL, Curran T, Galluccio L et al, 'False recognition and the right frontal lobe: a case study', *Neuropsychologia*, 1996, 34(8), 793–808
Schacter DL, Glisky EL & McGlynn SM, 'Impact of memory disorder on everyday life: awareness of deficits and return to work', in D Tapper & K Cicerone (eds), *The Neuropsychology of Everyday Life*, Kluwer, 1990, 231–57
Schaffer CE, Davidson RJ & Saron C, 'Frontal and parietal electroencephalogram asymmetry in depressed and nondepressed subjects', *Biological Psychiatry*, 1983, 18(7), 753–62
Scharf C, 'Cosmic (in)significance', *Scientific American*, 2014, 311(2), 74–7
Schechter B, *My Brain is Open: The Mathematical Journeys of Paul Erdös*, Simon & Schuster, 2000
Schechter I, Butler PD, Jalbrzikowski M et al, 'A new dimension of sensory dysfunction: stereopsis deficits in schizophrenia', *Biological Psychiatry*, 2006, 60(11), 1282–4
Scheibel AB, Fried I, Paul L et al, 'Differentiating characteristics of the human speech cortex: a quantitative Golgi study', in DF Benson & E Zaidel (eds), *The Dual Brain: Hemispheric Specialization in Humans*, Guilford, 1985, 65–74
Scheid W, 'Der Zeiger der Schuld in seiner Bedeutung für die Prognose involutiver Psychosen', *Zeitschrift für die gesamte Neurologie und Psychiatrie*, 1934, 150(1), 528–55
Scheler M, *Der Formalismus in der Ethik und die materiale Wertethik*, Niemeyer Verlag, 1916
——, *Vom Ewigen im Menschen*, Francke Verlag, 1954
——, in Maria Scheler (ed), *Schriften aus dem Nachlass*, vol 1: *Zur Ethik und Erkenntnislehre*, Francke Verlag, 2nd edn, 1957

———, *On the Eternal in Man*, trans B Noble, SCM Press, 1960
———, *Selected Philosophical Essays*, trans & ed DR Lachterman, Northwestern University Press, 1973 [1927]
———, *On Feeling, Knowing and Valuing*, ed HJ Bershady, University of Chicago Press, 1992
———, *The Constitution of the Human Being*, from *Schriften aus dem Nachlass*, vols 11 & 12, trans J Cutting, Marquette University Press, 2008
———, *The Human Place in the Cosmos*, tr MS Frings, Northwestern University Press, 2009 [1929]
Schelling FWJ von, *Von der Weltseele: eine Hypothese der höhern Physik zur Erklärung des allgemeinen Organismus*, Friedrich Perthes, 1798
———, *Einleitung zu seinem Entwurf eines Systems der Naturphilosophie*, Christian Ernst Gabler Verlag, 1799a
———, *Erster Entwurf eines Systems der Naturphilosophie*, Christian Ernst Gabler Verlag, 1799b
———, *Darlegung des wahren Verhältnisses der Naturphilosophie, zu der verbesserten Fichte'schen Lehre*, Cotta'sche Buchhandlung, 1806
———, *Vorlesungen über die Methode des akademischen Studium*, Cotta'sche Buchhandlung, 1830
———, *Die Weltalter* [1815]; in *Sämmtliche Werke: 1811–1815*, vol 1, Cotta'scher Verlag, 1856–61a
———, *Philosophie der Mythologie* [1842]; in *Sämmtliche Werke: 1811–1815*, vol 2, Cotta'scher Verlag, 1856–61b
———, *Zur Geschichte der neueren Philosophie*, Philipp Reclam jun Verlag, 1966 [1836–37]
———, *Aphorismen zur Einleitung in die Naturphilosophie*, in *Ausgewählte Schriften*, Suhrkamp Verlag, 6 vols, 1985– [1805]
———, *Ideas for a Philosophy of Nature*, trans EE Harris & P Heath, Cambridge University Press, 1988
———, *On the History of Modern Philosophy*, trans A Bowie, Cambridge University Press, 1994 [1833–34]
———, *The Ages of the World*, trans JM Wirth, State University of New York Press, 2000 [1815]
———, *First Outline of a System of the Philosophy of Nature*, trans KR Peterson, State University of New York Press, 2004 [1799]
Scherf KS, Behrmann M, Kimchi R et al, 'Emergence of global shape processing continues through adolescence', *Child Development*, 2009, 80(1), 162–77
Schettino A, Lauro LR, Crippa F et al, 'The comprehension of idiomatic expressions in schizophrenic patients', *Neuropsychologia*, 2010, 48(4), 1032–40
Scheuringer A & Pletzer B, 'Sex differences in the Kimchi-Palmer task revisited: global reaction times, but not number of global choices differ between adult men and women', *Physiology & Behavior*, 2016, 165, 159–65
Schiffer F, Zaidel E, Bogen J et al, 'Different psychological status in the two hemispheres of two split-brain patients', *Neuropsychiatry, Neuropsychology and Behavioral Neurology*, 1998, 11(3), 151–6
Schilder P, *The Image and Appearance of the Human Body*, Kegan Paul, Trench, Trubner, 1935
Schiller FCS ['A Troglodyte'], *Riddles of the Sphinx: a Study in the Philosophy of Evolution*, Swan Sonnenschein, 1891
———, *Logic For Use: An Introduction to the Voluntarist Theory of Knowledge*, G Bell & Sons, 1929
———, *Must philosophers disagree?*, Macmillan, 1933
Schilpp PA (ed), *The Philosophy of Rudolf Carnap: Intellectual Autobiography*, Open Court, 1963
Schlegel KWF, 'Athenæum Fragments', in *Philosophical Fragments*, trans P Firchow, University of Minnesota Press, 1991 [*Athenäums-Fragmente* 1797–8]
Schleiermacher F, *On Religion: Speeches to its Cultured Despisers*, trans J Oman, Westminster/John Knox Press, 1994 [*Über die Religion* 1799, 3rd edn 1831]
Schlosser R, Hutchinson M, Joseffer S et al, 'Functional magnetic resonance imaging of human brain activity in a verbal fluency task', *Journal of Neurology, Neurosurgery, and Psychiatry*, 1998, 64(4), 492–8
Schlosshauer M, Kofler J & Zeilinger A, 'A snapshot of foundational attitudes toward quantum mechanics', *Studies in History and Philosophy of Modern Physics*, 2013, 44(3), 222–30
Schlottmann A & Tring J, 'How children reason about gains and losses: framing effects in judgment and choice', *Swiss Journal of Psychology*, 2005, 64(3), 153
Schlumpf YR, Reinders AATS, Nijenhuis ERS et al, 'Dissociative part-dependent resting-state activity in dissociative identity disorder: a controlled fMRI perfusion study', *PLoS One*, 2014, 9(6), e98795
Schmahmann J, 'An emerging concept. The cerebellar contribution to higher function', *Archives of Neurology*, 1991, 48(11), 1178–87
———, 'Review: the role of the cerebellum in cognition and emotion: personal reflections since 1982 on the dysmetria of thought

hypothesis, and its historical evolution from theory to therapy', *Neuropsychology Review*, 2010, 20(3), 236–60

Schmahmann JD & Pandya DN, 'Disconnection syndromes of basal ganglia, thalamus, and cerebrocerebellar systems', *Cortex*, 2008, 44(8), 1037–66

Schmemann A, in *The Journals of Father Alexander Schmemann 1973–1983*, trans & ed J Schmemann, St Vladimir's Seminary Press, 2000

Schmidt A, Müller F, Lenz C et al, 'Acute LSD effects on response inhibition neural networks', *Psychological Medicine*, 2018, 48(9), 1464–73

Schmidt B & Hanslmayr S, 'Resting frontal EEG alpha-asymmetry predicts the evaluation of affective musical stimuli', *Neuroscience Letters*, 2009, 460(3), 237–40

Schmidt G, 'A review of the German literature on delusion between 1914 and 1939', in J Cutting & M Shepherd (eds), *The Clinical Roots of the Schizophrenia Concept*, Cambridge University Press, 1987, 104–33

Schmidt GL, DeBuse CJ & Seger CA, 'Right hemisphere metaphor processing? Characterizing the lateralization of semantic processes', *Brain and Language*, 2007, 100(2), 127–41

Schmidt GL & Seger CA, 'Neural correlates of metaphor processing: the roles of figurativeness, familiarity and difficulty', *Brain and Cognition*, 2009, 71(3), 375–86

Schmidt H, McFarland J, Ahmed M et al, 'Low-level temporal coding impairments in psychosis: preliminary findings and recommendations for further studies', *Journal of Abnormal Psychology*, 2011, 120(2), 476–82

Schmidt HG & Boshuizen HP, 'On the origin of intermediate effects in clinical case recall', *Memory & Cognition*, 1993, 21(3), 338–51

Schmucker D, Clemens JC, Shu H et al, 'Drosophila Dscam is an axon guidance receptor exhibiting extraordinary molecular diversity', *Cell*, 2000, 101(6), 671–84

Schneider F, Gur RE, Alavi A et al, 'Cerebral blood flow changes in limbic regions induced by unsolvable anagram tasks', *American Journal of Psychiatry*, 1996, 153(2), 206–12

Schneider MS, *A Beginner's Guide to Constructing the Cosmos: the Mathematical Archetypes of Nature, Art, and Science*, Harper, 1995

Schneider S, Peters J, Bromberg U et al, 'Boys do it the right way: sex-dependent amygdala lateralization during face processing in adolescents', *NeuroImage*, 2011, 56(3), 1847–53

Schneiderman EI, Murasugi KG & Saddy JD, 'Story arrangement ability in right-brain damaged patients', *Brain and Language*, 1992, 43(1), 107–20

Schnell AK, Hanlon RT, Benkada A et al, 'Lateralization of eye use in cuttlefish: opposite direction for anti-predatory and predatory behaviors', *Frontiers in Physiology*, 2016, 7, article 620

Schnider A, Regard M, Benson DF et al, 'Effects of a right-hemisphere stroke on an artist's performance', *Neuropsychiatry, Neuropsychology, & Behavioural Neurology*, 1993, 6(4), 249–55

Schofield K & Mohr C, 'Schizotypy and hemispheric asymmetry: results from two Chapman scales, the O-LIFE questionnaire, and two laterality measures', *Laterality*, 2014, 19(2), 178–200

Scholem G, *Origins of the Kabbalah*, trans A Arkush, ed RJ Zwi Werblowski, Princeton University Press, 1987 [*Ursprung und Anfänge der Kabbala* 1962]

Scholz J, Triantafyllou C, Whitfield-Gabrieli S et al, 'Distinct regions of right temporoparietal junction are selective for theory of mind and exogenous attention', *PLoS One*, 2009, 4(3), e4869

Schooler JW & Engstler-Schooler TY, 'Verbal overshadowing of visual memories: some things are better left unsaid', *Cognitive Psychology*, 1990, 22(1), 36–71

Schooler JW, Fiore SM & Brandimonte MA, 'At a loss for words: verbal overshadowing of non-verbal memories', in DL Medin (ed), *The Psychology of Learning and Motivation*, vol 37, Academic Press, 1997, 291–340

Schooler JW, Ohlsson S & Brooks K, 'Thoughts beyond words: when language overshadows insight', *Journal of Experimental Psychology: General*, 1993, 122(2), 166–83

Schopenhauer A, *Parerga and Paralipomena*, trans EFJ Payne, Clarendon Press, 1974

———, *The World as Will and Representation*, trans EFJ Payne, 2 vols, Dover, 2000 [*Die Welt als Wille und Vorstellung* 1819]

———, *Essays and Aphorisms*, trans RJ Hollingdale, Penguin, 2004

Schor JB, *The Overworked American: The Unexpected Decline in Leisure*, Basic Books, 1992

Schore AN, *Affect Regulation and the Origin of the Self: The Neurobiology of Emotional Development*, Erlbaum, 1994

———, 'The experience-dependent maturation of a regulatory system in the orbital prefrontal cortex and the origin of developmental psychopathology', *Development and Psychopathology*, 1996, 8(1), 59–87

———, *Affect Dysregulation and Disorders of the Self*, WW Norton, 2003

———, *The Science of the Art of Psychotherapy*, WW Norton, 2012

—, 'Early interpersonal neurobiological assessment of attachment and autistic spectrum disorders', *Frontiers in Psychology*, 2014a, 5, article 1049

—, 'The right brain is dominant in psychotherapy', *Psychotherapy*, 2014b, 51(3), 388–97

—, *The Development of the Unconscious Mind*, Norton, 2019

—, 'Forging connections in group psychotherapy through right brain-to-right brain emotional communications. Part 1: theoretical models of right brain therapeutic action. Part 2: clinical case analyses of group right brain regressive enactments', *International Journal of Group Psychotherapy*, 2020, 70(1), 29–88

Schore AN & Marks-Tarlow T, 'How love opens creativity, play, and the arts through early right brain development', in AN Schore (ed), *Right Brain Psychotherapy*, Norton, 2019, 155–79

Schott BH, Voss M, Wagner B et al, 'Frontolimbic novelty processing in acute psychosis: disrupted relationship with memory performance and potential implications for delusions', *Frontiers in Behavioral Neuroscience*, 2015, 9, article 144

Schott GD, 'Pictures as a neurological tool: lessons from enhanced and emergent artistry in brain disease', *Brain*, 2012, 135(6), 1947–63

Schou M, 'Artistic productivity and lithium prophylaxis in manic-depressive illness', *British Journal of Psychiatry*, 1979, 135(2), 56–65

Schraeder BD & Fischer DK, 'Using intuitive knowledge to make clinical decisions', *Journal of Maternal-Child Nursing*, 1986, 11, 161–2

—, 'Using intuitive knowledge in the neonatal intensive care nursery', *Holistic Nursing Practice*, 1987, 1(3), 45–51

Schreber D, *Denkwürdigkeiten eines Nervenkranken*, Oswald Mutze, 1903

—, *Memoirs of My Nervous Illness*, ed RA Hunter and trans I Macalpine, Harvard University Press, 1955

Schröder H, 'Ueber eine optische Inversion bei Betrachtung verkehrter, durch optische Vorrichtung entworfener, physischer Bilder', *Annalen der Physik*, 1858, 181, 298–311

Schröder KA, 'Nähe und Ferne: Faktur und Ausdruck im Schaffen Lovis Corinths', in KA Schröder (ed), *Lovis Corinth*, Prestel, 1992, 8–35

Schrödinger E, *Science and Humanism*, Cambridge University Press, 1951

—, *My View of the World*, Cambridge University Press, 1964

—, *Mind and Matter* [1958], in *What is Life? and Mind and Matter*, Cambridge University Press, 1967a

—, *What is Life?* [1944], in *What is Life? and Mind and Matter*, Cambridge University Press, 1967b

—, 'General scientific and popular papers', in *Collected Papers*, vol 4, Austrian Academy of Sciences and Vieweg & Sohn, 1984

—, *Nature and the Greeks*, in *Nature and the Greeks and Science and Humanism*, Cambridge University Press, 1996 [1954]

Schroeder C, Schneider-Gold C, Behrendt V et al, 'Unilateral right prosopometamorphopsia with positive "half-face-covering-test" after small occipitotemporal stroke', *Journal of the Neurological Sciences*, 2017, 379, 247–8

Schroter S, Black N, Evans S et al, 'Effects of training on quality of peer review: randomised controlled trial', *BMJ*, 2004, 328(7441), 673

Schuepbach D, Skotchko T, Duschek S et al, 'Gender and rapid alterations of hemispheric dominance during planning', *Neuropsychobiology*, 2012, 66(3), 149–57

Schuler A-L, Kasprian G, Schwartz E et al, 'Mens inversus [sic] in corpore inverso? Language lateralization in a boy with situs inversus totalis', *Brain and Language*, 2017, 174, 9–15

Schultheiss OC, 'Implicit motives', in OP John, RW Robins & LA Pervin (eds), *Handbook of Personality: Theory and Research*, Guilford Press, 3rd edn, 2008, 603–33

—, 'Implicit motives and hemispheric processing differences are critical for understanding personality disorders: a commentary on Hopwood', *European Journal of Personality*, 2018, 32(5), 580–2

Schultheiss OC & Brunstein JC, 'Inhibited power motivation and persuasive communication: a lens model analysis', *Journal of Personality*, 2002, 70(4), 553–82

Schultheiss OC, Riebel K & Jones NM, 'Activity inhibition: a predictor of lateralized brain function during stress?', *Neuropsychology*, 2009, 23(3), 392–404

Schultheiss OC, Yankova D, Dirlikov B et al, 'Are implicit and explicit motive measures statistically independent? A fair and balanced test using the Picture Story Exercise and a cue- and response-matched questionnaire measure', *Journal of Personality Assessment*, 2009, 91(1), 72–81

Schultz G, 'Bericht über die Feier der Deutschen Chemischen Gesellschaft zu Ehren August Kekulé's', *European Journal of Inorganic Chemistry*, 1890, 23(1), 1265–312

Schultz RT, 'Developmental deficits in social perception in autism: the role of the amygdala and fusiform face area', *International Journal of Developmental Neuroscience*, 2005, 23(2–3), 125–41

Schultze-Lutter F, 'Subjective symptoms of schizophrenia in research and the clinic: the basic symptom concept', *Schizophrenia Bulletin*, 2009, 35(1), 5–8

Schulz SM, 'Neural correlates of heart-focused interoception: a functional magnetic resonance imaging meta-analysis', *Philosophical Transactions of the Royal Society of London: Series B, Biological Sciences*, 2016, 371(1708), 20160018

Schumacher J, 'Ich wurde Christ, weil ich Wissenschaftler bin', *Christliches Medienmagazin pro*, 2017, 6, 16ff

Schutz LE, 'Broad-perspective perceptual disorder of the right hemisphere', *Neuropsychology Review*, 2005, 15(1), 11–27

Schüz A & Preißl H, 'Basic connectivity of the cerebral cortex and some considerations on the corpus callosum', *Neuroscience & Biobehavioral Reviews*, 1996, 20(4), 567–70

Schwabl H & Klima H, 'Spontaneous ultraweak photon emissions from biological systems and the endogenous light field', *Forschende Komplementärmedizin und Klassische Naturheilkunde*, 2005, 12(2), 84–9

Schwanberg JS, 'Does language of retrieval affect the remembering of trauma?', *Journal of Trauma and Dissociation*, 2010, 11, 44–56

Schwartz AS, Marchok PL & Flynn RE, 'A sensitive test for tactile extinction: results in patients with parietal and frontal lobe disease', *Journal of Neurology, Neurosurgery, and Psychiatry*, 1977, 40, 228–33

Schwartz AS, Marchok PL, Kreinick CJ et al, 'The asymmetric lateralization of tactile extinction in patients with unilateral cerebral dysfunction', *Brain*, 1979, 102(4), 669–84

Schwartz B, *The Paradox of Choice: Why More is Less*, HarperCollins, 2005

Schwartz M, Creasey H, Grady CL et al, 'Computed tomographic analysis of brain morphometrics in 30 healthy men, aged 21 to 81 years', *Annals of Neurology*, 1985, 17(2), 146–57

Schwartz M & Smith ML, 'Visual asymmetries with chimeric faces', *Neuropsychologia*, 1980, 18(1), 103–6

Schwartz MA, Wiggins OP, Naudin J et al, 'Rebuilding reality: a phenomenology of aspects of chronic schizophrenia', *Phenomenology and the Cognitive Sciences*, 2005, 4(1), 91–115

Schwartz MF, Buxbaum LJ, Montgomery MW et al, 'Naturalistic action production following right hemisphere stroke', *Neuropsychologia*, 1999, 37(1), 51–66

Schwartz Place & Gilmore 1980: *see* Place & Gilmore 1980

Schwartz R, Shipkin D & Cermak LS, 'Verbal and non-verbal memory abilities of adult brain-damaged patients', *American Journal of Occupational Therapy*, 1979, 33(2), 79–83

Schwarzschild L, *The Red Prussian: the Life and Legend of Marx*, trans M Wing, Hamish Hamilton, 1948

Schweitzer L, 'Evidence of right cerebral hemisphere dysfunction in schizophrenic patients with left hemisphere overactivation', *Biological Psychiatry*, 1982, 17(6), 655–73

Schwenk T, *Sensitive Chaos: The Creation of Flowing Forms in Water and Air*, Schocken, 1976

Sciama DW, 'The physical significance of the vacuum state of a quantum field', in S Saunders & HR Brown (eds), *The Philosophy of Vacuum*, Clarendon Press, 1991, 137–58

Scopelliti I, Morewedge CK, McCormick E et al, 'Bias blind spot: structure, measurement, and consequences', *Management Science*, 2015, 61(10), 2468–86

Scotti G & Spinnler H, 'Colour imperception in unilateral hemisphere-damaged patients', *Journal of Neurology, Neurosurgery, and Psychiatry*, 1970, 33(1), 22–8

Scruton R, *Sexual Desire: A Philosophical Investigation*, Weidenfeld & Nicolson, 1986

———, *An Intelligent Person's Guide to Philosophy*, Duckworth, 1996

———, *Modern Philosophy*, Arrow, 1997

———, *Beauty*, Oxford University Press, 2009

———, 'Scientism in the arts and humanities', *The New Atlantis*, 2013, 40, 33–46

———, 'Parfit the perfectionist', *Philosophy*, 2014, 89(4), 621–34

Seashore CE & Metfessel M, 'Deviation from the regular as an art principle', *Proceedings of the National Academy of Sciences of the United States of America*, 1925, 11(9), 538–42

Sechehaye M, *Autobiography of a Schizophrenic Girl: The True Story of 'Renee'*, trans G Rubin-Rabson, Meridian, 1994 [1951]

Secret of the Golden Flower, The, trans T Cleary, Harper Collins, 1991

Sedivy J, 'The strange persistence of first languages', *Nautilus*, 5 November 2015

Seeley WW, Matthews BR, Crawford RK et al, 'Unravelling Boléro: progressive aphasia, transmodal creativity and the right posterior neocortex', *Brain*, 2008, 131(1), 39–49

Segal E, 'Incubation in insight problem solving', *Creativity Research Journal*, 2004, 16(1), 141–8

Segall M, 'Panpsychist Physicalism', 3 May 2018: footnotes2plato.com/2018/05/03/panpsychist-physicalism/ (accessed 23 February 2021)

———, 'Why German idealism matters', 4 October 2018: thesideview.co/tag/philosophy-of-nature/ (accessed 23 February 2021)

Seger CA, Desmond JE, Glover GH et al, 'Functional magnetic resonance imaging evidence for right-hemisphere involvement in processing unusual semantic relationships', *Neuropsychology*, 2000, 14(3), 361–9

Seger CA, Poldrack RA, Prabhakaran V et al, 'Hemispheric asymmetries and individual differences in visual concept learning as measured by functional MRI', *Neuropsychologia*, 2000, 38(9), 1316–24

Séglas J, 'Diagnostic des délires de persécution systématisés', *La Semaine médicale*, 1890, 10, 419

Segré D & Lancet D, 'Composing life', *EMBO Reports*, 2000, 1(3), 217–22

Seidenwurm DJ & Devinsky O, 'Neuroradiology in the humanities and social sciences', *Radiology*, 2006, 239(1), 13–17

Seife C, 'For sale: "Your name here" in a prestigious science journal: an investigation into some scientific papers finds worrying irregularities', *Nature News*, 19 December 2014

Seldon HL, 'Structure of human auditory cortex. I. Cytoarchitectonics and dendritic distributions', *Brain Research*, 1981, 229(2), 277–94

Sellal F, Renaseau-Leclerc C & Labrecque R, 'The man with 6 arms: an analysis of supernumerary phantom limbs after right hemisphere stroke', *Revue Neurologique (Paris)*, 1996, 152(3), 190–5

Selz J, *Munch*, Bonfini Press, 1991

Semendeferi K & Horton CF, 'Evolution of the cerebral cortex', in AW Toga (ed), *Brain Mapping: An Encyclopedic Reference*, vol 2, Academic Press (Elsevier), 2015, 1–10

Semendeferi K, Lu A, Schenker et al, 'Humans and great apes share a large frontal cortex', *Nature Neuroscience*, 2002, 5(3), 272–6

Semenza C, 'Impairment in localization of body parts following brain damage', *Cortex*, 1988, 24(3), 443–9

Semmes J, 'Hemispheric specialisation: a possible clue to mechanism', *Neuropsychologia*, 1968, 6(1), 11–26

Senter HJ, Lieberman AN & Pinto R, 'Cerebral manifestations of ergotism: report of a case and review of the literature', *Stroke*, 1976, 7(1), 88–92

Serafetinides EA, 'The significance of the temporal lobes and of hemispheric dominance in the production of the LSD-25 symptomatology in man: a study of epileptic patients before and after temporal lobectomy', *Neuropsychologia*, 1965, 3(1), 69–79

Sergent J, 'About face: left-hemisphere involvement in processing physiognomies', *Journal of Experimental Psychology: Human Perception and Performance*, 1982a, 8(1), 1–14

———, 'The cerebral balance of power: confrontation or cooperation?', *Journal of Experimental Psychology: Human Perception and Performance*, 1982b, 8(2), 253–72

———, 'Cognitive and neural structures in face processing', in A Kertesz (ed), *Localization and Neuro-imaging in Neuropsychology*, Academic Press, 1994, 473–94

Sergent J, Ohta S & MacDonald B, 'Functional neuroanatomy of face and object processing: a positron emission tomography study', *Brain*, 1992, 115(1), 15–36

Sergent J & Signoret JL, 'Varieties of functional deficits in prosopagnosia', *Cerebral Cortex*, 1992, 2(5), 375–88

Seron X, Mataigne F, Coyette F et al, 'Étude d'un cas de métamorphopsie limitée aux visages et à certains objets familiers', *Revue Neurologique (Paris)*, 1995, 151(12), 691–8

Serra Catafau J, Rubio F & Peres Serra J, 'Peduncular hallucinosis associated with posterior thalamic infarction', *Journal of Neurology*, 1992, 239(2), 89–90

Sévigné, Marie de Rabutin-Chantal, marquise de, *Lettres de Mme de Sévigné*, ed M Suard, Firmin Didot frères, 1846

Sfard A, 'Reification as the birth of metaphor', *For the Learning of Mathematics*, 1994, 14(1), 44–55

Shadmehr R & Holcomb HH, 'Neural correlates of motor memory consolidation', *Science*, 1997, 277(5327), 821–5

Shah A & Frith U, 'An islet of ability in autistic children: a research note', *Journal of Child Psychology and Psychiatry*, 1983, 24(4), 613–20

Shah C, Erhard K, Ortheil HJ et al, 'Neural correlates of creative writing: an fMRI study', *Human Brain Mapping*, 2013, 34(5), 1088–101

Shah I, 'Isa and the doubters', *Tales of the Dervishes*, ISF Publishing, 2016

Shahani B, Burrows PT & Whitty CWM, 'The grasp reflex and perseveration', *Brain*, 1970, 93(1), 181–92

Shaku S, *Sermons of a Buddhist Abbot*, trans DT Suzuki, Pinnacle Press, 2017 [1906]

Shakya DR, Shyangwa PM, Pandey AK et al, 'Self injurious behavior in temporal lobe epilepsy', *Journal of the Nepal Medical Association*, 2010, 49(179), 239–42

Shalev BA, *100 Years of Nobel Prizes*, Americas Group Publications, 2002

Shallis T, *On Time*, Pelican, 1982

Shamay-Tsoory SG, Adler N, Aharon-Peretz J et al, 'The origins of originality: the neural bases of creative thinking and originality', *Neuropsychologia*, 2011, 49(2), 178–85

Shamay-Tsoory SG, Tomer R & Aharon-Peretz J, 'The neuroanatomical basis of understanding sarcasm and its relationship to social cognition', *Neuropsychology*, 2005, 19(3), 288–300

Shamay-Tsoory SG, Tomer R, Berger BD et al, 'Characterization of empathy deficits following prefrontal brain damage: the role of the right ventromedial prefrontal cortex', *Journal of Cognitive Neuroscience*, 2003, 15(3), 324–37

Shammi P & Stuss DT, 'Humour appreciation: a role of the right frontal lobe', *Brain*, 1999, 122(4), 657–66

Shannahoff-Khalsa D & Golshan S, 'Nasal cycle dominance and hallucinations in an adult schizophrenic female', *Journal of Experimental Child Psychology*, 2015, 226(1), 289–94

Shanteau J, 'Psychological characteristics of expert decision makers', *Acta Psychologica*, 1988, 68(1–3), 203–15

Shapiro JA, 'Thinking about bacterial populations as multicellular organisms', *Annual Review of Microbiology*, 1998, 52(1), 81–104

———, 'Revisiting the central dogma in the 21st century', *Annals of the New York Academy of Sciences*, 2009, 1178(1), 6–28

———, *Evolution: A View from the 21st Century*, Financial Times/ Prentice Hall, 2011

———, 'Review: how life changes itself: the Read-Write (RW) genome', *Physics of Life Reviews*, 2013, 10(3), 287–323

Shapleske J, Rossell SL, Chitnis XA et al, 'A computational morphometric MRI study of schizophrenia: effects of hallucinations', *Cerebral Cortex*, 2002, 12(12), 1331–41

Sharma K, *Interdependence*, Fordham University Press, 2015

Sharot T, 'The optimism bias', *Current Biology*, 2011, 21(23), R941–5

Sharot T, Kanai R, Marston D et al, 'Selectively altering belief formation in the human brain', *Proceedings of the National Academy of Sciences of the United States of America*, 2012, 109(42), 17058–62

Sharp DJ, Scott SK & Wise RJ, 'Monitoring and the controlled processing of meaning: distinct prefrontal systems', *Cerebral Cortex*, 2004, 14(1), 1–10

Shattuck R, *The Banquet Years: The Origins of the Avant Garde in France, 1885 to World War 1*, Random House, 1988

Shaw P, Lalonde F, Lepage C et al, 'Development of cortical asymmetry in typically developing children and its disruption in attention-deficit/hyperactivity disorder', *Archives of General Psychiatry*, 2009, 66(8), 888–96

Shaw P, Mellers J, Henderson M et al, 'Schizophrenia-like psychosis arising de novo following a temporal lobectomy:

timing and risk factors', *Journal of Neurology, Neurosurgery, and Psychiatry*, 2004, 75(7), 1003–8

Shayer M, 'Cognitive acceleration through science education II: its effects and scope', *International Journal of Science Education*, 1999, 21(8), 883–902

Shayer M & Ginsburg D, 'Thirty years on — a large anti-Flynn effect? The Piagetian test *Volume & Heaviness* norms 1975–2003', *British Journal of Educational Psychology*, 2007, 77, 25–41

———, 'Thirty years on – a large anti-Flynn effect? (II) 13- and 14-year-olds. Piagetian tests of formal operations norms 1976–2006/7', *British Journal of Educational Psychology*, 2009, 79(3), 409–18

Shayer M, Ginsburg D & Coe R, 'Thirty years on – a large anti-Flynn effect? The Piagetian test *Volume & Heaviness* norms 1975–2003', *British Journal of Educational Psychology*, 2007, 77(1), 25–41

Shaywitz BA, Shaywitz SE, Pugh KR et al, 'Sex differences in the functional organization of the brain for language', *Nature*, 1995, 373(6515), 607–9

Shebani Z & Pulvermüller F, 'Moving the hands and feet specifically impairs working memory for arm and leg-related action words', *Cortex*, 2013, 49(1), 222–31

Sheehan W, 'WW Morgan and the discovery of the spiral arm structure of our galaxy', *Journal of Astronomical History & Heritage*, 2008, 11(1), 3–21

Sheldrake R, *A New Science of Life: The Hypothesis of Morphic Resonance*, Park Street Press, 1995 [1981]

Shelley PB, *A Defence of Poetry* [1821, publ 1840], in HFB Brett-Smith (ed), *Peacock's Four Ages of Poetry, Shelley's Defence of Poetry, Browning's Essay on Shelley*, Blackwell, Oxford, 1921

Shemyakina NV & Dan'ko SG, 'Changes in the power and coherence of the β_2 EEG band in subjects performing creative tasks using emotionally significant and emotionally neutral words', *Human Physiology*, 2007, 33(1), 20–6

Shenefelt PD, 'Use of hypnosis, meditation, and biofeedback in dermatology', *Clinics in Dermatology*, 2017, 35(3), 285–91

Shenhav A, Rand DG & Greene JD, 'Divine intuition: cognitive style influences belief in God', *Journal of Experimental Psychology: General*, 2012, 141(3), 423–8

Shepard RN, 'Recognition memory for words, sentences and pictures', *Journal of Verbal Learning and Verbal Behavior*, 1967, 6(1), 156–63

Shepherd M, 'Morbid jealousy: some clinical and social aspects of a psychiatric symptom', *Journal of Mental Science*, 1961, 107(449), 687–753

Sheppard DM, Bradshaw JL & Mattingley JB, 'Abnormal line bisection judgments in children with Tourette's syndrome', *Neuropsychologia*, 2002, 40(3), 253–9

Sheppard DM, Bradshaw JL, Mattingley JB et al, 'Effects of stimulant medication on the lateralisation of line bisection judgements of children with attention deficit hyperactivity disorder', *Journal of Neurology, Neurosurgery, and Psychiatry*, 1999, 66(1), 57–63

Shermer M, 'The brain is not modular: what fMRI really tells us', *Scientific American Mind*, 2008, 19(5), 66–71

Sherover CM, '*Res cogitans*: the time of mind', in JT Fraser (ed), *Time and Mind: Interdisciplinary Issues*, vol 6 ('The Study of Time'), International Universities Press, 1989

Sherrington C, *Man on His Nature*, Cambridge University Press, 2009 [1940]

Sherwin I, 'Psychosis associated with epilepsy: significance of the laterality of the epileptogenic lesion', *Journal of Neurology, Neurosurgery, and Psychiatry*, 1981, 44(1), 83–5

Sherwin I, Peron-Magnan P, Bancaud J et al, 'Prevalence of psychosis in epilepsy as a function of the laterality of the epileptogenic lesion', *Archives of Neurology*, 1982, 39(10), 621–5

Sherwood K, 'How a cerebral hemorrhage altered my art', *Frontiers in Human Neuroscience*, 2012, 6, article 55

Sheth BR, Sandkühler S & Bhattacharya J, 'Posterior beta and anterior gamma oscillations predict cognitive insight', *Journal of Cognitive Neuroscience*, 2009, 21(7), 1269–79

Shewmon DA, Holmes GL & Byrne PA, 'Consciousness in congenitally decorticate children: developmental vegetative state as self-fulfilling prophecy', *Developmental Medicine & Child Neurology*, 1999, 41(6), 364–74

Shi S, Kumar P & Lee KF, 'Generation of photonic entanglement in green fluorescent proteins', *Nature Communications*, 2017, 8(1), article 1934

Shibahara N & Lucero-Wagoner B, 'Hemispheric asymmetry in accessing word meanings: concrete and abstract nouns', *Perceptual & Motor Skills*, 2002, 94(3, pt 2), 1292–300

Shiffrar M, 'People watching: visual, motor, and social processes in the perception of human movement', *Wiley Interdisciplinary Reviews: Cognitive Science*, 2011, 2(1), 68–78

Shiga K, Makino M, Ueda Y et al, 'Metamorphopsia and visual hallucinations restricted to the right visual hemifield after a left putaminal haemorrhage', *Journal of Neurology, Neurosurgery, and Psychiatry*, 1996, 61(4), 420–1

Shiloah A, 'Music and religion in Islam', *Acta Musicologica*, 1997, 69(2), 143–55

Shimamura AP, 'The role of the prefrontal cortex in dynamic filtering', *Psychobiology*, 2000, 28(2), 207–18

Shimkunas A, 'Conceptual deficit in schizophrenia: a reappraisal', *British Journal of Medical Psychology*, 1972, 45(2), 149–57

——, 'Hemispheric asymmetry and schizophrenic thought disorder', in S Schwartz (ed), *Language and Cognition in Schizophrenia*, Psychology Press, 1978, 193–235

Shinbrot T & Young W, 'Why decussate? Topological constraints on 3D wiring', *The Anatomical Record*, 2008, 291(10), 1278–92

Shiraishi H, Ito M, Hayashi H et al, 'Sulpiride treatment of Cotard's syndrome in schizophrenia', *Progress in Neuro-Psychopharmacology & Biological Psychiatry*, 2004, 28(3), 607–9

Shlain L, *Leonardo's Brain: Understanding da Vinci's Creative Genius*, Lyons Press, 2014

Shomrat T & Levin M, 'An automated training paradigm reveals long-term memory in planaria and its persistence through head regeneration', *Journal of Experimental Biology*, 2013, 216(20), 3799–810

Shortt R, *Does Religion Do More Harm Than Good?*, SPCK, 2019a

——, *Outgrowing Dawkins: God for Grown-ups*, SPCK, 2019b

Shrestha R, 'Clinical lycanthropy: delusional misidentification of the "self"', *Journal of Neuropsychiatry and Clinical Neurosciences*, 2014, 26(1), E53–4

Shulman GL, Pope DLW, Astafiev SV et al, 'Right hemisphere dominance during spatial selective attention and target detection occurs outside the dorsal fronto-parietal network', *Journal of Neuroscience*, 2010, 30(10), 3640–51

Shultz S, Lee SM, Pelphrey K et al, 'The posterior superior temporal sulcus is sensitive to the outcome of human and non-human goal-directed actions', *Social Cognitive and Affective Neuroscience*, 2011, 6(5), 602–11

Shultz S & McCarthy G, 'Goal-directed actions activate the face-sensitive posterior superior temporal sulcus and fusiform gyrus in the absence of human-like perceptual cues', *Cerebral Cortex*, 2012, 22(5), 1098–106

——, 'Perceived animacy influences the processing of human-like surface features in the fusiform gyrus', *Neuropsychologia*, 2014, 60, 115–20

Sichart U & Fuchs T, 'Visuelle Halluzinationen bei älteren Menschen mit reduziertem

Visus: das Charles Bonnet-Syndrom', *Klinische Monatsblätter für Augenheilkunde*, 1992, 200(3), 224-7

Sieböger FT, Ferstl EC & von Cramon DY, 'Making sense of nonsense: an fMRI study of task induced inference processes during discourse comprehension', *Brain Research*, 2007, 1166, 77-91

Siegal M, Carrington J & Radel M, 'Theory of mind and pragmatic understanding following right hemisphere damage', *Brain and Language*, 1996, 53(1), 40-50

Siegrist M, Cvetkovich G & Roth C, 'Salient value similarity, social trust, and risk/benefit perception', *Risk Analysis*, 2000, 20(3), 353-62

Siéroff E, 'Focussing on/in visual-verbal stimuli in patients with parietal lesions', *Cognitive Neuropsychology*, 1990, 7(5-6), 519-54

———, 'Les mécanismes attentionnels', in X Seron & M Jeannerod (eds), *Neuropsychologie humaine*, Mardaga, 1994, 127-51

Siéroff E, Decaix C, Chokron S et al, 'Impaired orienting of attention in left unilateral neglect: a componential analysis', *Neuropsychology*, 2007, 21(1), 94-113

Sierra M, Lopera F, Lambert MV et al, 'Separating depersonalisation and derealisation: the relevance of the "lesion method"', *Journal of Neurology, Neurosurgery, and Psychiatry*, 2002, 72(4), 530-2

Sierra M, Nestler S, Jay EL et al, 'A structural MRI study of cortical thickness in depersonalisation disorder', *Journal of Experimental Child Psychology*, 2014, 224(1), 1-7

Signer SS & Cummings JL, 'Erotomania and cerebral dysfunction', *British Journal of Psychiatry*, 1987, 151(2), 275

Sikka S, 'Heidegger's appropriation of Schelling', *The Southern Journal of Philosophy*, 1994, 32(4), 421-48

Silva JA & Leong GB, 'A case of organic Othello syndrome', *Journal of Clinical Psychiatry*, 1993, 54(7), 277

Silva JA, Leong GB & Wine DB, 'Misidentification delusions, facial misrecognition, and right brain injury', *Canadian Journal of Psychiatry*, 1993, 38(4), 239-41

Silva JA, Tekell JL, Leong GB et al, 'Delusional misidentification of the self associated with nondominant cerebral pathology', *Journal of Clinical Psychiatry*, 1995, 56(4), 171

Silva JR, Sakamoto AC, Thomé Ú et al, 'Left hemispherectomy in older children and adolescents: outcome of cognitive abilities', *Child's Nervous System*, 2020, 36(6), 1275-82

Silveira S, Fehse K, Vedder A et al, 'Is it the picture or is it the frame? An fMRI study on the neurobiology of framing effects', *Frontiers in Human Neuroscience*, 2015, 9, article 528

Silver CF, Coleman TJ, Hood RW et al, 'The six types of nonbelief: a qualitative and quantitative study of type and narrative', *Mental Health, Religion & Culture*, 2014, 17(10), 990-1001

Silveri MC & Ciccarelli N, 'The deficit for the word-class "verb" in corticobasal degeneration: linguistic expression of the movement disorder?', *Neuropsychologia*, 2007, 45(11), 2570-9

Silverman IW, 'Simple reaction time: it is not what it used to be', *American Journal of Psychology*, 2010, 123(1), 39-50

Silverstein SM & Keane BP, 'Perceptual organization impairment in schizophrenia and associated brain mechanisms: review of research from 2005 to 2010', *Schizophrenia Bulletin*, 2011, 37(4), 690-9

Silverstein SM & Palumbo DR, 'Nonverbal perceptual organization output disability and schizophrenia spectrum symptomatology', *Psychiatry*, 1995, 58(1), 66-81

Silverstein SM, Bakshi S, Chapman RM et al, 'Perceptual organization of configural and visual patterns in schizophrenia: effects of repeated exposure', *Cognitive Neuropsychiatry*, 1998, 3(3), 209-23

Silverstein SM, Keane BP, Wang Y et al, 'Effects of short-term inpatient treatment on sensitivity to a size contrast illusion in first-episode psychosis and multiple-episode schizophrenia', *Frontiers in Psychology*, 2013, 4, article 466

Silverstein SM, Knight RA, Schwarzkopf SB et al, 'Stimulus configuration and context effects in perceptual organization in schizophrenia', *Journal of Abnormal Psychology*, 1996, 105(3), 410-20

Silverstein SM, Matteson S & Knight R, 'Reduced top-down influences in auditory perceptual organization in schizophrenia', *Journal of Abnormal Psychology*, 1996, 105(4), 663-7

Silverstein SM, Schenkel LS, Valone C et al, 'Cognitive deficits and psychiatric rehabilitation outcomes in schizophrenia', *Psychiatric Quarterly*, 1998, 69(3), 169-91

Simard S, 'How trees talk to each other', 2016: www.ted.com/talks/suzanne_simard_how_trees_talk_to_each_other (accessed 3 July 2021)

Simeon D, Guralnik O, Hazlett EA et al, 'Feeling unreal: a PET study of depersonalization disorder', *American Journal of Psychiatry*, 2000, 157(11), 1782-8

Simeonova DI, Chang KD, Strong C et al, 'Creativity in familial bipolar disorder', *Journal of Psychiatric Research*, 2005, 39(6), 623-31

Simkin MV & Roychowdhury VP, 'Read before you cite!', *Complex Systems*, 2003, 14, 269-74

Simon HA, 'Discovery, invention and development: human creative thinking', *Proceedings of the National Academy of Sciences of the United States of America*, 1983, 80(14), 4569–71

Simonis M, Klous P, Splinter E et al, 'Nuclear organization of active and inactive chromatin domains uncovered by chromosome conformation capture-on-chip (4C)', *Nature Genetics*, 2006, 38(11), 1348–54

Simons DJ & Chabris CF, 'Gorillas in our midst: sustained inattentional blindness for dynamic events', *Perception*, 1999, 28(9), 1059–74

Simons DJ & Levin DT, 'Failure to detect changes to people during a real-world interaction', *Psychonomic Bulletin & Review*, 1998, 5(4), 644–9

Simons JS, Koutstaal W, Prince S et al, 'Neural mechanisms of visual object priming: evidence for perceptual and semantic distinctions in fusiform cortex', *NeuroImage*, 2003, 19(3), 613–26

Simons P, 'Processes and precipitates', in DJ Nicholson & J Dupré (eds), *Everything Flows: Towards a Processual Philosophy of Biology*, Oxford University Press, 2018, 49–60

Simonton DK, 'The mad (creative genius): what do we know after a century of historiometric research?', in JC Kaufman (ed), *Creativity and Mental Illness*, Cambridge University Press, 2014a, 25–41

——, 'The mad-genius paradox: can creative people be more mentally healthy but highly creative people more mentally ill?', *Perspectives on Psychological Science*, 2014b, 9(5), 470–80

Simpson EM, *A Study of the Prose Works of John Donne*, Clarendon Press, 1948

Simpson GG, 'Biology and the nature of science', *Science*, 1963, 139(3550), 81–8

Simpson J & Done DJ, 'Elasticity and confabulation in schizophrenic delusions', *Psychological Medicine*, 2002, 32(3), 451–8

Sinclair, Marie, Countess of Caithness, *Old Truths in a New Light: An Earnest Endeavour to Reconcile Material Science with Spiritual Science and with Scripture*, Chapman & Hall, 1876

Singal J, 'The case of the amazing gay-marriage data: how a graduate student reluctantly uncovered a huge scientific fraud', *The Cut*, 29 May 2015: nymag.com/scienceofus/2015/05/how-a-grad-student-uncovered-a-huge-fraud.html (accessed 6 March 2021)

Singer T, Seymour B, O'Doherty JP et al, 'Empathic neural responses are modulated by the perceived fairness of others', *Nature*, 2006, 439(7075), 466–9

Singh B, *Indian Metaphysics*, Humanities Press, 1987

Singh P, 'The so-called Fibonacci numbers in ancient and medieval India', *Historia Mathematica*, 1985, 12(3), 229–44

Singh PB, Iannilli E & Hummel T, 'Segregation of gustatory cortex in response to salt and umami taste studied through event-related potentials', *NeuroReport*, 2011, 22(6), 299–303

Sinha JB, 'Collectivism, social energy, and mental health', *Dynamische Psychiatrie*, 1988, 21(1–2), 110–8

Siniscalchi M, Pergola G & Quaranta A, 'Detour behaviour in attack-trained dogs: left-turners perform better than right-turners', *Laterality*, 2013, 18(3), 282–93

Siniscalchi M, Sasso R, Pepe AM et al, 'Catecholamine plasma levels following immune stimulation with rabies vaccine in dogs selected for their paw preferences', *Neuroscience Letters*, 2010, 476(3), 142–5

—— 'Sniffing with right nostril: lateralisation of response to odour stimuli by dogs', *Animal Behaviour*, 2011, 82(2), 399–404

Sitting Bull, Chief, addressing the Dawes Commission, 1883: quoted at *Native American Rights Fund Legal Review*, 1973, 2(1), 2

Skaf CR, Yamada A, Garrido GE et al, 'Psychotic symptoms in major depressive disorder are associated with reduced regional cerebral blood flow in the subgenual anterior cingulate cortex: a voxel-based single photon emission computed tomography (SPECT) study', *Journal of Affective Disorders*, 2002, 68(2–3), 295–305

Skinner M, 'Unified theory of evolution: Darwin's theory that natural selection drives evolution is incomplete without input from evolution's anti-hero: Lamarck', *Aeon*, 2016: aeon.co/essays/on-epigenetics-we-need-both-darwin-s-and-lamarck-s-theories (accessed 23 February 2021)

Sklar S, 'How beauty will save the world: William Blake's prophetic vision', *Spiritus*, 2007, 7(1), 30–9

Skrbina D, *Panpsychism in the West*, Massachusetts Institute of Technology Press, 2017

Skyrms B, 'EPR: lessons for metaphysics', in PA French, TE Uehling & HK Wettstein (eds), *Causation and Causal Theories*, University of Minnesota Press, 1984

Slagter HA, Davidson RJ & Tomer R, 'Eye-blink rate predicts individual differences in pseudoneglect', *Neuropsychologia*, 2010, 48(5), 1265–8

Slater A, Bremner G, Johnson SP et al, 'Newborn infants' preference for attractive faces: the role of internal and external facial features', *Infancy*, 2000, 1(2), 265–74

Slater A, Quinn PC, Hayes R et al, 'The role of orientation in newborn infants' preference for attractive faces', *Developmental Science*, 2000, 3(2), 181–5

Slepian ML & Ambady N, 'Fluid movement and creativity', *Journal of Experimental Psychology: General*, 2012, 141(4), 625–9

Slepian ML, Weisbuch M, Pauker K et al, 'Fluid movement and fluid social cognition: bodily movement influences essentialist thought', *Personality and Social Psychology Bulletin*, 2014, 40(1), 111–20

Slotnick SD, 'Resting-state fMRI data reflects default mode network activity rather than null data: a defense of commonly employed methods to correct for multiple comparisons', *Cognitive Neuroscience*, 2017, 8(3), 141–3

Slusher MP & Anderson CA, 'Using causal persuasive arguments to change beliefs and teach new information: the mediating role of explanation availability and evaluation bias in the acceptance of knowledge', *Journal of Educational Psychology*, 1996, 88(1), 110–22

Small DM, Jones-Gotman M, Zatorre RJ et al, 'A role for the right anterior temporal lobe in taste quality recognition', *Journal of Neuroscience*, 1997, 17(13), 5136–42

Small DM & Prescott J, 'Odor/taste integration and the perception of flavor', *Experimental Brain Research*, 2005, 166(3–4), 345–57

Smith A, 'Differing effects of hemispherectomy in children and adults', paper presented at APA 84th Annual Convention, Washington DC, 1976

Smith A & Sugar O, 'Development of above normal language and intelligence 21 years after left hemispherectomy', *Neurology*, 1975, 25(9), 813–8

Smith C & Denton ML, *Soul Searching: The Religious and Spiritual Lives of American Teenagers*, Oxford University Press, 2005

Smith DJ, Anderson J, Zammit S et al, 'Childhood IQ and risk of bipolar disorder in adulthood: prospective birth cohort study', *British Journal of Psychiatry Open*, 2015, 1(1), 74–80

Smith R, 'Peer review: a flawed process at the heart of science and journals', *Journal of the Royal Society of Medicine*, 2006, 99(4), 178–82

Smith SD, Dixon MJ, Tays WJ et al, 'Anomaly detection in the right hemisphere: the influence of visuospatial factors', *Brain and Cognition*, 2004, 55(3), 458–62

Smith SD, Tays WJ, Dixon MJ et al, 'The right hemisphere as an anomaly detector: evidence from visual perception', *Brain and Cognition*, 2002, 48(2–3), 574–9

Smith TW, 'Hostility and health: current status of a psychosomatic hypothesis', *Health Psychology*, 1992, 11(3), 139–50

Smith WC, Crombie IK, Tavendale RT et al, 'Urinary electrolyte excretion, alcohol consumption, and blood pressure in the Scottish heart health study', *BMJ*, 1988, 297(6644), 329–30

Smolin L, *The Life of the Cosmos*, Oxford University Press, 1999

———, *The Trouble With Physics: The Rise of String Theory, the Fall of Science and What Comes Next*, Penguin, 2008 [2006]

———, *Time Reborn*, Knopf, 2013

Sneddon L, 'Pain perception in fish: evidence and implications for the use of fish', *Journal of Consciousness Studies*, 2011, 18(9–10), 209–29

Snell B, *The Discovery of the Mind*, Harper & Row, 1960 [*Die Entdeckung des Geistes: Studien zur Entstehung des europäischen Denkens bei den Griechen* 1946]

Snider MJ & Wolfenden R, 'The rate of spontaneous decarboxylation of amino acids', *Journal of the American Chemical Society*, 2000, 122(46), 11507–8

Sniekers S, Stringer S, Watanabe K et al, 'Genome-wide association meta-analysis of 78,308 individuals identifies new loci and genes influencing human intelligence', *Nature Genetics*, 2017, 49(7), 1107–12

Snyder AW, Mulcahy E, Taylor JL et al, 'Savant-like skills exposed in normal people by suppressing the left fronto-temporal lobe', *Journal of Integrative Neuroscience*, 2003, 2(2), 149–58

Snyder JJ & Chatterjee A, 'Spatial-temporal anisometries following right parietal damage', *Neuropsychologia*, 2004, 42(12), 1703–8

Soares JS & Storm BC, 'Forget in a flash: a further investigation of the photo-taking-impairment effect', *Journal of Applied Research in Memory & Cognition*, 2018, 7(1), 154–60

Sokal AD, 'A physicist experiments with cultural studies', *Lingua Franca*, 2016. The original article is published in *Social Text*, 1996, 46/47, 217–52: www.physics.nyu.edu/sokal/transgress_v2/transgress_v2_singlefile.html (accessed 23 February 2021)

Sol D, 'Behavioural innovation: a neglected issue in the ecological and evolutionary literature?', in SM Reader & KN Laland (eds), *Animal Innovation*, Oxford University Press, 2003, 63–82

Sollberger M, Stanley CM, Wilson SM et al, 'Neural basis of interpersonal traits in neurodegenerative diseases', *Neuropsychologia*, 2009, 47(13), 2812–27

Solon I, 'How intelligence mediates liberalism and prosociality', *Intelligence*, 2014, 47, 44–53

Solso R, 'Brain activities in a skilled versus a novice artist: an fMRI study', *Leonardo*, 2001, 34(1), 31–4

Solzhenitsyn A, *One Word of Truth* (Nobel Prize Lecture), trans N Bethell, Stenvalley Press, 1973

———, 'Godlessness: the first step to the Gulag', Templeton Prize Lecture, delivered London, 10 May 1983

Somers D & Sheremata SL, 'Attention maps in the brain', *Wiley Interdisciplinary Reviews: Cognitive Science*, 2013, 4(4), 327–40

Sommer IE, Aleman A, Bouma A et al, 'Do women really have more bilateral language representation than men? A meta-analysis of functional imaging studies', *Brain*, 2004, 127(8), 1845–52

Sommer IEC, Ramsey NF & Kahn RS, 'Language lateralisation in schizophrenia: an fMRI study', *Schizophrenia Research*, 2001, 52(1–2), 57–67

Sommer IEC, Ramsey NF, Mandl RCW et al, 'Language lateralisation in female patients with schizophrenia: an fMRI study', *Schizophrenia Research*, 2003, 60(2–3), 183–90

Son TG, Camandola S & Mattson MP, 'Hormetic dietary phytochemicals', *Neuromolecular Medicine*, 2008, 10(4), 236–46

Sone T, Nakaya N, Ohmori K et al, 'Sense of life worth living (*ikigai*) and mortality in Japan: Ohsaki Study', *Psychosomatic Medicine*, 2008, 70(6), 709–15

Sontag S (ed), *Antonin Artaud: Selected Writings*, University of California Press, 1976

Soon CS, Brass M, Heinze H-J et al, 'Unconscious determinants of free decisions in the human brain', *Nature Neuroscience*, 2008, 11(5), 543–9

Sosis R, 'Religion and intragroup cooperation: preliminary results of a comparative analysis of utopian communities', *Cross-Cultural Research*, 2000, 34(1), 70–87

Sosis R & Bressler ER, 'Co-operation and commune longevity: a test of the costly signaling theory of religion', *Cross-Cultural Research*, 2003, 37(2), 211–39

Sotillo M, Carretié L, Hinojosa JA et al, 'Neural activity associated with metaphor comprehension: spatial analysis', *Neuroscience Letters*, 2005, 373(1), 5–9

Southgate V & Hamilton AF, 'Unbroken mirrors: challenging a theory of autism', *Trends in Cognitive Sciences*, 2008, 12(6), 225–9

Soyka M, 'Delusional jealousy and localised cerebral pathology', *Journal of Neuropsychiatry and Clinical Neuroscience*, 1998, 10(4), 472

Spangenberg KB, Wagner MT & Bachman DL, 'Neuropsychological analysis of a case of abrupt onset mirror sign following a hypotensive crisis in a patient with vascular dementia', *Neurocase*, 2008, 4(2), 149–54

Speck O, Ernst T, Braun J et al, 'Gender differences in the functional organization of the brain for working memory', *NeuroReport*, 2000, 11(11), 2581–5

Spelke E, 'Infants' intermodal perception of events', *Cognitive Psychology*, 1976, 8(4), 553–60

Spencer-Brown G, *The Laws of Form*, Dutton, 1969

Sperry RW, 'Neurology and the mind-brain problem', *American Scientist*, 1952, 40(2), 291–312

———, 'Hemisphere deconnection and unity in conscious awareness', *The American Psychologist*, 1968, 23(10), 723–33

———, 'Perception in the absence of the neocortical commissures', *Research Publications – Association for Research in Nervous and Mental Disease*, 1970, 48, 123–38

———, 'Lateral specialization of cerebral function in the surgically separated hemispheres', in FJ McGuigan and RA Schoonover (eds), *The Psychophysiology of Thinking: Studies of Covert Processes*, Academic Press, 1973, 209–29

———, 'Lateral specialization in the surgically separated hemispheres', in FO Schmitt & FG Worden (eds), *The Neurosciences Third Study Program*, Massachusetts Institute of Technology Press, 1974a, 5–19

———, 'Science and the problem of values', *Zygon*, 1974b, 9(1), 7–21

———, 'Consciousness, personal identity and the divided brain', in DF Benson & E Zaidel (eds), *The Dual Brain: Hemispheric Specialization in Humans*, Guilford Press, 1985, 11–26

Sperry RW, Gazzaniga MS & Bogen JE, 'Interhemispheric relationships: the neocortical commissures; syndromes of hemisphere disconnection', in PJ Vinken & GW Bruyn (eds), *Handbook of Clinical Neurology*, vol 4, North Holland, 1969, 273–90

Spierer L, De Lucia M, Bernasconi F et al, 'Learning-induced plasticity in human audition: objects, time, and space', *Hear Res*, 2010, 271(1–2), 88–102

Spiers HJ, Burgess N, Maguire EA et al, 'Unilateral temporal lobectomy patients show lateralized topographical and episodic memory deficits in a virtual town', *Brain*, 2001, 124(12), 2476–89

Spilianakis CG, Lalioti MD, Town T et al, 'Interchromosomal associations between

alternatively expressed loci', *Nature*, 2005, 435(7042), 637-45
Spinella M, 'A relationship between smell identification and empathy', *International Journal of Neuroscience*, 2002, 112(6), 605-12
Spinelli DN, Starr A & Barrett TW, 'Auditory specificity in unit recordings from cat's visual cortex', *Experimental Neurology*, 1968, 22(1), 75-84
Spinoza B, *Ethics*, Dutton, trans A Boyle, 1910
Spira R, interviewed by Daphne Astor, 'Consciousness and the role of the artist', 2002: non-duality.rupertspira.com/read/interview_with_daphne_astor_consciousness_and_the_role_of_the_artist_2002 (accessed 23 February 2021)
Spironelli C, Angrilli A & Stegagno L, 'Failure of language lateralisation in schizophrenia patients: an ERP study on early linguistic components', *Journal of Psychiatry & Neuroscience*, 2008, 33(3), 235-43
Spitzer C, Willert C, Grabe HJ et al, 'Dissociation, hemispheric asymmetry, and dysfunction of hemispheric interaction: a transcranial magnetic stimulation approach', *Journal of Neuropsychiatry and Clinical Neurosciences*, 2004, 16(2), 163-9
Spitzer M, 'Ich-Störungen: in search of a theory', in M Spitzer, FA Uehlein & G Oepen (eds), *Psychopathology and Philosophy*, Springer, 1988, 167-83
———, 'On defining delusions', *Comprehensive Psychiatry*, 1990, 31(5), 377-97
———, 'The phenomenology of delusions', *Psychiatric Annals*, 1992, 22(5), 252-9
Spitzer M, Breuckers J, Beyer S et al, 'Contextual insensitivity in thought-disordered schizophrenic patients: evidence from pauses in spontaneous speech', *Language and Speech*, 1994, 37(2), 171-85
Spitzer NC, 'Neurotransmitter switching in the developing and adult brain', *Annual Review of Neuroscience*, 2017, 40(1), 1-19
Sporns O, Tononi G & Edelman GM, 'Theoretical neuroanatomy: relating anatomical and functional connectivity in graphs and cortical connection matrices', *Cerebral Cortex*, 2000, 10(2), 127-41
Spyrou L & Sanei S, 'Source localization of event-related potentials incorporating spatial notch filters', *IEEE Transactions on Biomedical Engineering*, 2008, 55(9), 2232-9
Squeri V, Sciutti A, Gori M et al, 'Two hands, one perception: how bimanual haptic information is combined by the brain', *Journal of Neurophysiology*, 2012, 107(2), 544-50
Squire LR, Ojemann JG, Miezin FM et al, 'Activation of the hippocampus in normal humans: a functional anatomical study of memory', *Proceedings of the National Academy of Sciences of the United States of America*, 1992, 89(5), 1837-41
Sri Aurobindo, *The Human Cycle*: in *The Complete Works of Sri Aurobindo*, vol 25, Sri Aurobindo Ashram Press, 1997
Sridharan D, Levitin DJ & Menon V, 'A critical role for the right fronto-insular cortex in switching between central-executive and default-mode networks', *Proceedings of the National Academy of Sciences of the United States of America*, 2008, 105(34), 12569-74
Sritharan A, Line P, Sergejew A et al, 'EEG coherence measures during auditory hallucinations in schizophrenia', *Journal of Experimental Child Psychology*, 2005, 136(2-3), 189-200
Srivastava A, Taly AB, Gupta A et al, 'Stroke with supernumerary phantom limb: case study, review of literature and pathogenesis', *Vision Research*, 2008, 20(5), 256-64
Srygley RB & Thomas ALR, 'Unconventional lift-generating mechanisms in free-flying butterflies', *Nature*, 2002, 420(6916), 660-4
St George M, Kutas M, Martinez A et al, 'Semantic integration in reading: engagement of the right hemisphere during discourse processing', *Brain*, 1999, 122(7), 1317-25
Stack S, 'The effect of modernization on suicide in Finland: 1750-1986', *Sociological Perspectives*, 1993, 36(2), 137-48
Stadler MA, 'On learning complex procedural knowledge', *Journal of Experimental Psychology: Learning, Memory, and Cognition*, 1989, 15(6), 1061-9
Staff RT, Shanks MF, Macintosh L et al, 'Delusions in Alzheimer's disease: SPET evidence of right hemispheric dysfunction', *Cortex*, 1999, 35(4), 549-60
Staff RT, Venneri A, Gemmell HG et al, 'HMPAO SPECT imaging of Alzheimer's disease patients with similar content-specific autobiographic delusion: comparison using statistical parametric mapping', *Journal of Nuclear Medicine*, 2000, 41(9), 1451-5
Stafford T, 'What's the evidence on using rational argument to change people's minds?', *Contributoria*, May 2014: www.contributoria.com/issue/2014-05/5319c4add63a707e780000cd.html (accessed 23 February 2021)
Stagno SJ & Gates TJ, 'Palinopsia: a review of the literature', *Behavioural Neurology*, 1991, 4(2), 67-74
Standing L, 'Learning 10,000 pictures', *Quarterly Journal of Experimental Psychology*, 1973, 25(2), 207-22
Standing L, Conezio J & Haber RN, 'Perception and memory for pictures: single-trial

learning of 2500 visual stimuli', *Psychonomic Science*, 1970, 19(2), 73–4

Stanescu-Cosson R, Pinel P, van de Moortele P-F et al, 'Understanding dissociations in dyscalculia: a brain imaging study of the impact of number size on the cerebral networks for exact and approximate calculation', *Brain*, 2000, 123(11), 2240–55

Stangeland H, Orgeta V & Bell V, 'Poststroke psychosis: a systematic review', *Journal of Neurology, Neurosurgery, and Psychiatry*, 2018, 89(8), 879–85

Stanghellini G, *Disembodied Spirits and Deanimated Bodies: The Psychopathology of Common Sense*, Oxford University Press, 2004

Stanghellini G & Ballerini M, 'Autism: disembodied existence', *Philosophy, Psychiatry & Psychology*, 2004, 11(3), 259–68

—, 'Values in persons with schizophrenia', *Schizophrenia Bulletin*, 2007, 33(1), 131–41

Stanghellini G, Ballerini M, Blasi S et al, 'The bodily self: a qualitative study of abnormal bodily phenomena in persons with schizophrenia', *Comprehensive Psychiatry*, 2014, 55(7), 1703–11

Stanghellini G, Ballerini M, Presenza S et al, 'Psychopathology of lived time: abnormal time experience in persons with schizophrenia', *Schizophrenia Bulletin*, 2016, 42(1), 45–55

Stanovich KE & West RF, 'Natural myside bias is independent of cognitive ability', *Thinking & Reasoning*, 2007, 13(3), 225–47

—, 'On the failure of intelligence to predict myside bias and one-sided bias', *Thinking & Reasoning*, 2008, 14(2), 129–67

Stanovich KE, West RF & Toplak ME, 'Myside bias, rational thinking, and intelligence', *Current Directions in Psychological Science*, 2013, 22(4), 259–64

Stapp HP, *Mindful Universe: Quantum Mechanics and the Participating Observer*, Springer, 2007

—, *Mind, Matter, and Quantum Mechanics*, Springer, 3rd edn, 2009

—, *Quantum Theory and Free Will: How Mental Intentions Translate into Bodily Actions*, Springer, 2017

Stark DE, Margulies DS, Shehzad ZE et al, 'Regional variation in interhemispheric coordination of intrinsic hemodynamic fluctuations', *Journal of Neuroscience*, 2008, 28(51), 13754–64

Stark F, *Perseus in the Wind*, IB Tauris, 2013 [1948]

Stark-Adamec C & Adamec R, 'Breaking into the grant proposal market', *International Journal of Women's Studies*, 1981, 4(2), 105–17

Starkstein SE, Berthier ML, Fedoroff P et al, 'Anosognosia and major depression in 2 patients with cerebrovascular lesions', *Neurology*, 1990, 40(9), 1380–2

Starkstein SE, Boston JD & Robinson RG, 'Mechanisms of mania after brain injury: 12 case reports and a review of the literature', *Journal of Nervous and Mental Disease*, 1988, 176(2), 87–100

Starkstein SE, Fedoroff P, Berthier ML et al, 'Manic-depressive and pure manic states after brain lesions', *Biological Psychiatry*, 1991, 29(2), 149–58

Starkstein SE & Robinson RG, 'Mechanism of disinhibition after brain lesions: a review', *Journal of Nervous and Mental Disease*, 1997, 185(2), 108–14

Starkstein SE, Robinson RG, Honig MA et al, 'Mood changes after right-hemisphere lesions', *British Journal of Psychiatry*, 1989, 155(1), 79–85

Staton RD, Brumback RA & Wilson H, 'Reduplicative paramnesia: a disconnection syndrome of memory', *Cortex*, 1982, 18(1), 23–35

Stavrova O & Ehlebracht D, 'Cynical beliefs about human nature and income: longitudinal and cross-cultural analyses', *Journal of Personality and Social Psychology*, 2016, 110(1), 116–32

—, 'Education as an antidote to cynicism', *Social Psychological and Personality Science*, 2018, 9(1), 59–69

—, 'The cynical genius illusion: exploring and debunking lay beliefs about cynicism and competence', *Personality and Social Psychology Bulletin*, 2019, 45(2), 254–69

Stavy R, Goel V, Critchley H et al, 'Intuitive interference in quantitative reasoning', *Brain Research*, 2006, 1073–1074, 383–88

Steane A, *Science and Humanity: A Humane Philosophy of Science and Religion*, Oxford University Press, 2018

Steckler CM, Hamlin JK, Miller MB et al, 'Moral judgement by the disconnected left and right cerebral hemispheres: a split-brain investigation', *Royal Society Open Science*, 2017, 4(7), 170172

Steer G, Klimanek W & Löser F (eds), *Meister Eckhart: Die deutschen und lateinischen Werke* (*Die deutschen Werke: Meister Eckharts Predigten*), vol 4(1), pts 5–8, Kohlhammer, 2002

Steffen Moritz S & Woodward TS, 'Jumping to conclusions in delusional and non-delusional schizophrenic patients', *British Journal of Clinical Psychology*, 2005, 44(2), 193–207

Stein RL, 'Towards a process philosophy of chemistry', *HYLE – International Journal for Philosophy of Chemistry*, 2004, 10(1), 5–22

Steiner G, *Has Truth A Future?*, BBC Publications, 1978a

———, *Martin Heidegger*, Viking, 1978b

———, *Real Presences*, University of Chicago Press, 1991

———, *Nostalgia for the Absolute*, Anansi, 2004

Steinhardt PJ, 'The inflation debate: is the theory at the heart of modern cosmology deeply flawed?', *Scientific American*, 2011, 304(4), 36

Steinhauser G, Adlassnig W, Risch JA et al, 'Peer review versus editorial review and their role in innovative science', *Theoretical Medicine and Bioethics*, 2012, 33(5), 359–76

Stenger VJ, 'The universe shows no evidence for design', in JP Moreland, C Meister & KA Sweis (eds), *Debating Christian Theism*, Oxford University Press, 2013, 47–58

Stent GS, 'That was the molecular biology that was', *Science*, 1968, 160(3826), 390–5

Stephan KE, Fink GR & Marshall JC, 'Mechanisms of hemispheric specialization: insights from analyses of connectivity', *Neuropsychologia*, 2007, 45(2–4), 209–28

Stephan KE, Penny WD, Marshall JC et al, 'Investigating the functional role of callosal connections with dynamic causal models', *Annals of the New York Academy of Sciences*, 2005, 1064, 16–36

Stephens T, 'Astronomers use slime mold model to reveal dark threads of the cosmic web', 2020: news.ucsc.edu/2020/03/cosmic-web.html (accessed 23 February 2021)

Sternberg RJ & Davidson JE (eds), *The Nature of Insight*, Massachusetts Institute of Technology Press, 1996

Sterzer P & Kleinschmidt A, 'A neural basis for inference in perceptual ambiguity', *Proceedings of the National Academy of Sciences of the United States of America*, 2007, 104(1), 323–8

Stevens D, Charman T & Blair RJ, 'Recognition of emotion in facial expressions and vocal tones in children with psychopathic tendencies', *Journal of Genetic Psychology*, 2001, 162(2), 201–11

Stevens SS & Davis H, *Hearing: Its Psychology and Physiology*, John Wiley, 1938

Stevenson B & Wolfers J, 'The paradox of declining female happiness', Working Paper 14969, National Bureau of Economic Research, May 2009

Stewart I, 'Thinking about patterns', interview with Richard Bright, *Interalia*, March 2018: www.interaliamag.org/interviews/ian-stewart/ (accessed 23 February 2021)

Stewart I & Golubitsky M, *Fearful Symmetry: Is God a Geometer?*, Blackwell, 1992

Stewart J & Kolb BE, 'The effects of neonatal gonadectomy and prenatal stress on cortical thickness and asymmetry in rats', *Behavioral and Neural Biology*, 1988, 49(3), 344–60

Stewart JL, Bismark AW, Towers DN et al, 'Resting frontal EEG asymmetry as an endophenotype for depression risk: sex-specific patterns of frontal brain asymmetry', *Journal of Abnormal Psychology*, 2010, 119(3), 502–12

Stewart JL, Levin-Silton R, Sass SM et al, 'Anger style, psychopathology, and regional brain activity', *Emotion*, 2008, 8(5), 701–13

Stewart L, von Kriegstein K, Warren JD et al, 'Music and the brain: disorders of musical listening', *Brain*, 2006, 129(10), 2533–53

Stigler SM, 'A historical view of statistical concepts in psychology and educational research', *American Journal of Education*, 1992, 101(1), 60–70

Stillman TF, Baumeister RF, Vohs KD et al, 'Personal philosophy and personnel achievement: belief in free will predicts better job performance', *Social Psychological and Personality Science*, 2010, 1(1), 43–50

Stoeckel MC, Weder B, Binkofski F et al, 'Left and right superior parietal lobule in tactile object discrimination', *European Journal of Neuroscience*, 2004, 19(4), 1067–72

Stoeger WR, Ellis GFR & Kirchner U, 'Multiverses and cosmology: philosophical issues', 2004: arXiv:astro-ph/0407329

Stolarz-Skrzypek K, Kuznetsova T, Thijs L et al, 'Fatal and nonfatal outcomes, incidence of hypertension, and blood pressure changes in relation to urinary sodium excretion', *Journal of the American Medical Association*, 2011, 305(17), 1777–85

Stollstorff M, Vartanian O & Goel V, 'Levels of conflict in reasoning modulate right lateral prefrontal cortex', *Brain Research*, 2012, 1428, 24–32

Stone JM, Morrison PD & Pilowsky LS, 'Glutamate and dopamine dysregulation in schizophrenia – a synthesis and selective review', *Journal of Psychopharmacology*, 2007, 21(4), 440–52

Stone SP, Halligan PW, Greenwood RJ et al, 'The incidence of neglect phenomena and related disorders in patients with an acute right or left hemisphere stroke', *Age and Aging*, 1993, 22(1), 46–52

Stotz K & Griffiths P, 'Genes: philosophical analyses put to the test', *History & Philosophy of the Life Sciences*, 2004, 26(1), 5–28

Stout JC, Busemeyer JR, Lin A et al, 'Cognitive modeling analysis of the decision-making processes used by cocaine abusers', *Psychonomic Bulletin & Review*, 2004, 11(4), 742–7

Stove D, *Darwinian Fairytales: Selfish Genes, Errors of Heredity and Other Fables of Evolution*, Encounter, 2006

Stower L & Ryan MA, 'A vision of the whole child: the significance of religious experiences in early childhood', *Australian Journal of Early Childhood*, 1998, 23(1), 1–4

Strasburger H & Waldvogel B, 'Sight and blindness in the same person', *PsyCh Journal*, 2015, 4(4), 178–85

Strathern P, *Mendeleyev's Dream: The Quest for the Elements*, St Martin's Press, 2000

Strauss E & Moscovitch M, 'Perception of facial expressions', *Brain and Language*, 1981, 13(2), 308–32

Strawson G, 'Realistic monism: why physicalism entails panpsychism', in G Strawson (ed), *Consciousness and Its Place in Nature: Does Physicalism Entail Panpsychism?*, Academic, 2006, 3–31

———, *Real Materialism and Other Essays*, Clarendon Press, Oxford, 2008

———, 'Real naturalism', *London Review of Books*, 2013, 35(18), 28–30

Strick M, Dijksterhuis A & van Baaren RB, 'Unconscious-thought effects take place offline, not on-line', *Psychological Science*, 2010, 21(4), 484–8

Stringaris AK, Medford NC, Giampietro V et al, 'Deriving meaning: distinct neural mechanisms for metaphoric, literal, and non-meaningful sentences', *Brain and Language*, 2007, 100(2), 150–62

Stringaris AK, Medford N, Giora R et al, 'How metaphors influence semantic relatedness judgments: the role of the right frontal cortex', *NeuroImage*, 2006, 33(2), 784–93

Stroganova TA, Nygren G, Tsetlin MM et al, 'Abnormal EEG lateralization in boys with autism', *Clinical Neurophysiology*, 2007, 118(8), 1842–54

Strough J, Karns TE & Schlosnagle L, 'Decision-making heuristics and biases across the life span', *Annals of the New York Academy of Sciences*, 2011, 1235(1), 57–74

Struck PT, *Divination and Human Nature: A Cognitive History of Intuition in Classical Antiquity*, Princeton University Press, 2016

Stump E, *Wandering in Darkness: Narrative and the Problem of Suffering*, Oxford University Press, 2010

Sturgis P, Read S & Allum N, 'Does intelligence foster generalized trust? An empirical test using the UK birth cohort studies', *Intelligence*, 2010, 38(1), 45–54

Sturm VE, Datta S, Roy ARK et al, 'Big smile, small self: awe walks promote prosocial positive emotions in older adults', *Emotion*, 21 September 2020 [advance online publication: doi.org/10.1037/emo0000876] (accessed 23 February 2021)

Sturm W & Willmes K, 'On the functional neuroanatomy of intrinsic and phasic alertness', *NeuroImage*, 2001, 14(1 Pt 2), S76–84

Sturm W, de Simone A, Krause BJ et al, 'Functional neuroanatomy of intrinsic alertness: evidence for a fronto-parietal-thalamic-brainstem network in the right hemisphere', *Neuropsychologia*, 1999, 37(7), 797–805

Sturm W, Fimm B, Cantagallo A et al, 'Computerized training of specific attention deficits in stroke and traumatic brain-injured patients: a multicentric efficacy study', in M Leclercq & P Zimmerman (eds), *Applied Neuropsychology of Attention*, Psychology Press, 2002, 365–80

Stuss DT, 'Disturbance of self-awareness after frontal system damage', in GP Prigatano & DL Schacter (eds), *Awareness of Deficit after Brain Injury: Clinical and Theoretical Issues*, Oxford University Press, 1991a, 66–83

———, 'Self, awareness, and the frontal lobes: a neuropsychological perspective', in J Strauss & GR Goethals (eds), *The Self: Interdisciplinary Approaches*, Springer-Verlag, 1991b, 255–78

Stuss DT & Alexander MP, 'Executive functions and the frontal lobes: a conceptual view', *Psychological Research*, 2000, 63(3–4), 289–98

Stuss DT, Gallup GG, Jnr & Alexander MP, 'The frontal lobes are necessary for "theory of mind"', *Brain*, 2001, 124(2), 279–86

Stuss DT, Picton TW & Alexander MP, 'Consciousness, self-awareness and the frontal lobes', in S Salloway, P Malloy & J Duffy (eds), *The Frontal Lobes and Neuropsychiatric Illness*, American Psychiatric Press, 2001, 101–9

Subbotsky EV, 'Causal explanations of events by children and adults: can alternative causal modes coexist in one mind?', *British Journal of Developmental Psychology*, 2001, 19(1), 23–46

———, 'The permanence of mental objects: testing magical thinking on perceived and imaginary realities', *Developmental Psychology*, 2005, 41(2), 301–18

———, 'Children's and adults' reactions to magical and ordinary suggestion: are suggestibility and magical thinking psychologically close relatives?', *British Journal of Psychology*, 2007, 98(4), 547–74

———, 'Curiosity and exploratory behaviour towards possible and impossible events in children and adults', *British Journal of Psychology*, 2010, 101(3), 481–501

———, 'The belief in magic in the age of science', *SAGE Open*, 2014, 1–17

Subbotsky E & Quinteros G, 'Do cultural factors affect causal beliefs? Rational and magical thinking in Britain and Mexico', *British Journal of Psychology*, 2002, 93(4), 519–43

Suberi M & McKeever WF, 'Differential right hemispheric memory storage of emotional and non-emotional faces', *Neuropsychologia*, 1977, 15(6), 757–68

Subramaniam K, Kounios J, Parrish TB et al, 'A brain mechanism for facilitation of insight by positive affect', *Journal of Cognitive Neuroscience*, 2009, 21(3), 415–32

Suchan J & Karnath H-O, 'Spatial orienting by left hemisphere language areas: a relict from the past?', *Brain*, 2011, 134(10), 3059–70

Suda M, Brooks SJ, Giampietro V et al, 'Provocation of symmetry/ordering symptoms in anorexia nervosa: a functional neuroimaging study', *PLoS One*, 2014, 9(5), e97998

Sugiura M, Sassa Y, Jeong H et al, 'Face specific and domain-general characteristics of cortical responses during self-recognition', *NeuroImage*, 2008, 42(1), 414–22

Sullivan PF, Magnusson C, Reichenberg A et al, 'Family history of schizophrenia and bipolar disorder as risk factors for autism', *Archives of General Psychiatry*, 2012, 69(11), 1099–103

Sullivan RM, 'Hemispheric asymmetry in stress processing in rat prefrontal cortex and the role of mesocortical dopamine', *Stress*, 2004, 7(2), 131–43

Sullivan RM & Dufresne MM, 'Mesocortical dopamine and HPA axis regulation: role of laterality and early environment', *Brain Research*, 2006, 1076(1), 49–59

Sullivan RM & Gratton A, 'Behavioral effects of excitotoxic lesions of ventral medial prefrontal cortex in the rat are hemisphere-dependent', *Brain Research*, 2002, 927(1), 69–79

Sultzer DL, Brown CV, Mandelkern MA et al, 'Delusional thoughts and regional frontal/temporal cortex metabolism in Alzheimer's disease', *American Journal of Psychiatry*, 2003, 160(2), 341–9

Summerfield C & de Lange FP, 'Expectation in perceptual decision making: neural and computational mechanisms', *Nature Reviews Neuroscience*, 2014, 15(11), 745–56

Summers RL, Dave A, Dolstra TJ et al, 'Diverse mutational pathways converge on saturable chloroquine transport via the malaria parasite's chloroquine resistance transporter', *Proceedings of the National Academy of Sciences of the United States of America*, 2014, 111(17), E1759–67

Sun Y, Chen Y, Collinson SL et al, 'Reduced hemispheric asymmetry of brain anatomical networks is linked to schizophrenia: a connectome study', *Cerebral Cortex*, 2017, 27(1), 602–15

Sun YT & Lin CC, 'Sequential appearance and disappearance of hemianopia, palinopsia and metamorphopsia: a case report and literature review', *Acta Neurologica Taiwanica*, 2004, 13(2), 77–83

Sundararajan L & Raina MK, 'Revolutionary creativity, East and West: a critique from indigenous psychology', *Journal of Theoretical and Philosophical Psychology*, 2015, 35(1), 3–19

Sundet JM, Barlaug DG & Torjussen TM, 'The end of the Flynn effect? A study of secular trends in mean intelligence test scores of Norwegian conscripts during half a century', *Intelligence*, 2004, 32(4), 349–62

Surawicz FG & Banta R, 'Lycanthropy revisited', *Canadian Psychiatric Association Journal*, 1975, 20(7), 537–42

Sutterer MJ, Koscik TR & Tranel D, 'Sex-related functional asymmetry of the ventromedial prefrontal cortex in regard to decision-making under risk and ambiguity', *Neuropsychologia*, 2015, 75, 265–73

Suzuki M, Okamura N, Kawachi Y et al, 'Discrete cortical regions associated with the musical beauty of major and minor chords', *Cognitive, Affective and Behavioral Neuroscience*, 2008, 8(2), 126–31

Suzuki S, *Nurtured by Love: A New Approach to Education*, trans W Suzuki, Exposition Press, 1969

———, *Branching Streams: Zen Talks on the Sandokai*, University of California Press, 1999

———, *Not Always So: Practicing the True Spirit of Zen*, HarperCollins, 2002

———, 'Transiency', *Zen Mind, Beginner's Mind*, Shambhala Publications, 2011 [1973]

Swaddle JP & Cuthill IC, 'Asymmetry and human facial attractiveness: symmetry may not always be beautiful', *Proceedings: Biological Sciences*, 1995, 261(1360), 111–16

Swanson LW, Sporns O & Hahn JD, 'Network architecture of the cerebral nuclei (basal ganglia) association and commissural connectome', *Proceedings of the National Academy of Sciences of the United States of America*, 2016, 113(40), E5972–81

Swash M, 'Visual perseveration in temporal lobe epilepsy', *Journal of Neurology, Neurosurgery, and Psychiatry*, 1979, 42(6), 569–71

Swayze VW, Andreasen NC, Erhardt JC et al, 'Developmental abnormalities in the corpus callosum in schizophrenia', *Archives of Neurology*, 1990, 47(7), 805–8

Swift MW, van de Walle CG & Fisher MPA, 'Posner molecules: from atomic structure to nuclear spins', *Physical Chemistry Chemical Physics*, 2018, 20(18), 12373–80

Swim JK, 'Perceived versus meta-analytic effect sizes: an assessment of the accuracy of gender stereotypes', *Journal of Personality and Social Psychology*, 1994, 66(1), 21–36

Swisher L & Hirsh IJ, 'Brain damage and the ordering of two temporally successive stimuli', *Neuropsychologia*, 1972, 10(2), 137–52

Sylva D, Rieger G, Linsenmeier JA et al, 'Concealment of sexual orientation', *Archives of Sexual Behavior*, 2010, 39(1), 141–52

Symington SH, Paul LK, Symington MF et al, 'Social cognition in individuals with agenesis of the corpus callosum', *Social Neuroscience*, 2010, 5(3), 296–308

Symons A, *William Blake*, Constable & Co, 1907

Szatkowska I, Grabowska A & Nowicka A, 'Hemispheric asymmetry in stimulus size evaluation', *Acta Neurobiologiae Experimentalis*, 1993, 53(1), 257–62

Szechyńska-Hebda M, Kruk J, Górecka M et al, 'Evidence for light wavelength-specific photoelectrophysiological signaling and memory of excess light episodes in Arabidopsis', *Plant Cell*, 2010, 22(7), 2201–18

Szent-Györgyi A, *Introduction to a Submolecular Biology*, Academic Press, 1960

Szilard L, *The Voice of the Dolphins*, Gollancz, 1961

———, *The Collected Works of Leo Szilard: Scientific Papers*, ed BT Feld & GW Szilard, Massachusetts Institute of Technology Press, 1972

Tabibnia G, Monterosso JR, Baicy K et al, 'Different forms of self-control share a neurocognitive substrate', *Journal of Neuroscience*, 2011, 31(13), 4805–10

Taiz L, Alkon D, Draguhn A et al, 'Plants neither possess nor require consciousness', *Trends in Plant Science*, 2019, 24(8), 677–87

Takahashi T, Ozawa H, Inuzuka S et al, 'Sulpiride for treatment of delusion of parasitosis', *Psychiatry and Clinical Neuroscience*, 2003, 57(5), 552–3

Takahashi T, Wood SJ, Yung AR et al, 'Insular cortex gray matter changes in individuals at ultra-high-risk of developing psychosis', *Schizophrenia Research*, 2009, 111(1–3), 94–102

Takahata K, Saito F, Muramatsu T et al, 'Emergence of realism: enhanced visual artistry and high accuracy of visual numerosity representation after left prefrontal damage', *Neuropsychologia*, 2014, 57(1), 38–49

Takemoto T & Brinthaupt TM, 'We imagine therefore we think: the modality of self and thought in Japan and America', *Yamaguchi Journal of Economics, Business Administrations & Laws*, 2017, 65(7–8), 1–29

Talamini LM, de Haan L, Nieman DH et al, 'Reduced context effects on retrieval in first-episode schizophrenia', *PLoS One*, 2010, 5(4), e10356

Talamini LM & Meeter M, 'Dominance of objects over context in a mediotemporal lobe model of schizophrenia', *PLoS One*, 2009, 4(8), e6505

Talbott SL, 'Getting over the code delusion', *The New Atlantis*, Summer 2010a, 3–27

———, 'The unbearable wholeness of beings', *The New Atlantis*, Fall 2010b, 27–51

———, 'A modest champion of the whole organism: Paul Weiss, scientist of distinction', *In Context*, 2011a, 25, 4–8

———, 'Evolution and the illusion of randomness', *The New Atlantis*, Fall 2011b, 37–64

———, 'What do organisms mean?', *The New Atlantis*, Winter 2011c, 24–49

Taleb NN, *The Black Swan*, Random House, 2008

———, *Antifragile: Things that Gain from Disorder*, Penguin, 2013

Tamminen K, 'Religious experiences in childhood and adolescence: a viewpoint of religious development between the ages of 7 and 20', *International Journal for the Psychology of Religion*, 1994, 4(2), 61–85

Tanabe H, Nakagawa Y, Ikeda M et al, 'Selective loss of semantic memory for words', in K Ishikawa, J McGauch & H Sakata (eds), *Brain Processes and Memory*, Elsevier, 1996, 141–52

Tanaka H, Hachisuka K & Ogata H, 'Sound lateralisation in patients with left or right cerebral hemispheric lesions: relation with unilateral visuospatial neglect', *Journal of Neurology, Neurosurgery, and Psychiatry*, 1999, 67(4), 481–6

Tanizaki J, *In Praise of Shadows*, trans TJ Harper & EG Seidensticker, Leete's Island Books, 1977

Tankersley D, Stowe CJ & Huettel SA, 'Altruism is associated with an increased neural response to agency', *Nature Neuroscience*, 2007, 10(2), 150–1

Tanno K, Sakata K, Ohsawa M et al, 'Associations of *ikigai* as a positive psychological factor with all-cause mortality and cause-specific mortality among middle-aged and elderly Japanese people: findings from the Japan Collaborative Cohort Study', *Journal of Psychosomatic Research*, 2009, 67(1), 67–75

Tanriover N, Kacira T, Ulu MO et al, 'Epidermoid tumour within the collateral sulcus: a rare location and atypical presentation', *Journal of Clinical Neuroscience*, 2008, 15(8), 950–4

Tao Te Ching, trans JCH Wu, *Tao Teh Ching*, St John's University Press, 1961

———, trans G-F Feng & J English, Wildwood House, 1973

———, trans Y Wu, *The Book of Lao Tzu (The Tao Te Ching)*, Great Learning Publishing, 1989

———, trans S Senudd, *Tao Te Ching: The Taoism of Lao Tzu Explained*, Arriba, 2011

Tapp PD, Siwak CT, Gao FQ et al, 'Frontal lobe volume, function, and beta-amyloid pathology in a canine model of aging', *Journal of Neuroscience*, 2004, 24(38), 8205–13

Tarakeshwar N, Pargament KI & Mahoney A, 'Measures of Hindu pathways: development and preliminary evidence of reliability and validity', *Cultural Diversity & Ethnic Minority Psychology*, 2003, 9, 316–32

Tarkovsky A, speaking in the film *A Poet in the Cinema: Andrey Tarkovsky*, 1983

———, *Sculpting in Time*, 1986

Tarnas R, *Cosmos and Psyche: Intimations of a New World View*, Viking, 2006

Tassy S, Oullier O, Duclos Y et al, 'Disrupting the right prefrontal cortex alters moral judgement', *Social Cognitive and Affective Neuroscience*, 2012, 7(3), 282–8

Tatossian A, 'La phénoménologie des psychoses', *L'Art du comprendre*, juillet 1997 [1979], double hors série, 5–151

———, 'Étude phénoménologique d'un cas de schizophrénie paranoïde' [1957], in *Psychiatrie phénoménologique*, MJW Fédition, 2014a, 25–128

———, 'Approche phénoménologique du fait hallucinatoire' [1958], in *Psychiatrie phénoménologique*, MJW Fédition, 2014b, 137–42

Tautz D & Domazet-Lošo T, 'The evolutionary origin of orphan genes', *Nature Reviews Genetics*, 2011, 12(10), 692–702

Tavano A, Grasso R, Gagliardi C et al, 'Disorders of cognitive and affective development in cerebellar malformations', *Brain*, 2007, 130(10), 2646–60

Tavano A, Sponda S, Fabbro F et al, 'Specific linguistic and pragmatic deficits in Italian patients with schizophrenia', *Schizophrenia Research*, 2008, 102(1–3), 53–62

Tavare A, 'Scientific misconduct is worryingly prevalent across the UK, shows BMJ survey', *BMJ*, 2012, 344(7840), e377

Taylor AF, Kuo FE & Sullivan WC, 'Views of nature and self-discipline: evidence from inner city children', *Journal of Environmental Psychology*, 2002, 22(1–2), 49–63

Taylor AH, Elliffe D, Hunt GR et al, 'Complex cognition and behavioural innovation in New Caledonian crows', *Proceedings of the Royal Society B: Biological Sciences*, 2010, 277(1694), 2637–43

Taylor JB, *My Stroke of Insight: A Brain Scientist's Personal Journey*, Hodder & Stoughton, 2009

Taylor LJ, Brown RG, Tsermentseli S et al, 'Is language impairment more common than executive dysfunction in amyotrophic lateral sclerosis?', *Journal of Neurology, Neurosurgery, and Psychiatry*, 2013, 84(5), 494–8

Taylor RW, Hsieh YW, Gamse JT et al, 'Making a difference together: reciprocal interactions in C. elegans and zebrafish asymmetric neural development', *Development*, 2010, 137(5), 681–91

Taylor S, 'How a flawed experiment "proved" that free will doesn't exist', *Scientific American*, 6 December 2019: blogs.scientificamerican.com/observations/how-a-flawed-experiment-proved-that-free-will-doesnt-exist/

Taylor TB, Mulley G, Dills AH et al, 'Evolution. Evolutionary resurrection of flagellar motility via rewiring of the nitrogen regulation system', *Science*, 2015, 347(6225), 1014–7

Tchanturia K, Smith E, Weineck F et al, 'Exploring autistic traits in anorexia: a clinical study', *Molecular Autism*, 2013, 4(1), article 44

te Nijenhuis J, De Jong MJ, Evers A et al, 'Are cognitive differences between immigrant and majority groups diminishing?', *European Journal of Personality*, 2004, 18(5), 405–34

te Nijenhuis J & van der Flier H, 'The secular rise in IQs in the Netherlands: is the Flynn effect on g?', *Personality and Individual Differences*, 2007, 43(5), 1259–65

———, 'Is the Flynn Effect on g? a meta-analysis', *Intelligence*, 2013, 41(6), 802–7

te Nijenhuis J, van Vianen AEM & van der Flier H, 'Score gains on g-loaded tests: no g', *Intelligence*, 2007, 35(3), 283–300

Teasdale TW & Owen DR, 'Secular declines in cognitive test scores: a reversal of the Flynn effect', *Intelligence*, 2008, 36(2), 121–6

Tedlock D (ed & trans), *Popul Vuh: The Mayan Book of the Dawn of Life*, Simon & Schuster, 1996

Tegmark M, 'Parallel universes', *Scientific American*, 2003, 288(5), 40–51

Teicholz N, 'The scientific report guiding the US dietary guidelines: is it scientific?', *BMJ*, 2015, 351, h4962

Teilhard de Chardin P, *The Phenomenon of Man*, trans B Wall, Harper & Row, 1959 [*Le phénomène humain* 1955]

———, *Le Milieu Divin: An Essay on the Interior Life*, trans B Wall, Harper & Row, 1960 [1957]

Tellenbach H, 'Die Räumlichkeit der Melancholischen. I. Mitteilung', *Nervenarzt*, 1956a, 27, 12–18

Teller C & Dennis C, 'The effect of ambient scent on consumers' perception, emotions and behaviour: a critical review', *Journal of Marketing Management*, 2012, 28(1–2), 14–36

Temple CM & Ilsley J, 'Phonemic discrimination in callosal agenesis', *Cortex*, 1993, 29(2), 341–8

Temple CM, Jeeves MA & Vilarroya O, 'Ten pen men: rhyming skills in two children with callosal agenesis', *Brain and Language*, 1989, 37(4), 548–64

——, 'Reading in callosal agenesis', *Brain and Language*, 1990, 39(2), 235–53

Templeton JJ, McCracken BG, Sher M et al, 'An eye for beauty: lateralized visual stimulation of courtship behavior and mate preferences in male zebra finches, *Taeniopygia guttata*', *Behavioural Processes*, 2014, 102, 33–9

Templeton JJ, Mountjoy DJ, Pryke SR et al, 'In the eye of the beholder: visual mate choice lateralization in a polymorphic songbird', *Biology Letters*, 2012, 8(6), 924–7

Ten Donkelaar HJ, 'Reptiles', in R Nieuwenhuys, HJ Ten Donkelaar & C Nicholson (eds), *The Central Nervous System of Vertebrates*, Springer-Verlag, 1998, 1315–524

TenHouten WD, Hoppe KD, Bogen JE et al, 'Alexithymia and the split brain III. Global-level content analysis of fantasy and symbolization', *Psychotherapy and Psychosomatics*, 1985, 44(2), 89–94

Tennekes H & Lumley JL, *A First Course in Turbulence*, Massachusetts Institute of Technology Press, 1999

Tennessen CM & Cimprich B, 'Views to nature: effects on attention', *Journal of Environmental Psychology*, 1995, 15(1), 77–85

Terao T & Tani Y, 'Carbamazepine treatment in a case of musical hallucinations with temporal lobe abnormalities', *Australian & New Zealand Journal of Psychiatry*, 1998, 32(3), 454–6

Terracciano A & McCrae RR, 'Perceptions of Americans and the Iraq invasion: implications for understanding national character stereotypes', *Journal of Cross-Cultural Psychology*, 2007, 38(6), 695–710

Tervaniemi M & Hugdahl K, 'Lateralization of auditory-cortex functions', *Brain Research: Brain Research Reviews*, 2003, 43(3), 231–46

Tettamanti M, Rotondi I, Perani D et al, 'Syntax without language: neurobiological evidence for cross-domain syntactic computations', *Cortex*, 2009, 45(7), 825–38

Teuber H-L, 'Why two brains?', in FO Schmitt & FG Worden (eds), *The Neurosciences: Third Study Program*, Massachusetts Institute of Technology Press, 1974, 71–4

Teuber H-L, Battersby WS & Bender MB, *Visual Field Defects After Penetrating Missile Wounds To The Brain*, Harvard University Press, 1960

Thackeray WM, *Roundabout Papers*, Smith, Elder & Company, 1879

Thatcher RW, Biver CJ & North D, 'Spatial-temporal current source correlations and cortical connectivity', *Clinical EEG and Neuroscience*, 2007, 38(1), 35–48

Thatcher RW, North D & Biver C, 'EEG and intelligence: relations between EEG coherence, EEG phase delay and power', *Clinical Neurophysiology*, 2005, 116(9), 2129–41

Theise ND, 'Now you see it, now you don't', *Nature*, 2005, 435(7046), 1165

——, 'Implications of "post-modern biology" for pathology: the cell doctrine', *Laboratory Investigation*, 2006, 86(4), 335–44

Theise ND & Kafatos MC, 'Fundamental awareness: a framework for integrating science, philosophy and metaphysics', *Communicative & Integrative Biology*, 2016, 9(3), e1155010

Thellier M & Lüttge U, 'Plant memory: a tentative model', *Plant Biology (Stuttgart)*, 2013, 15(1), 1–12

Thibodeau R, Jorgensen RS & Kim S, 'Depression, anxiety, and resting frontal EEG asymmetry: a meta-analytic review', *Journal of Abnormal Psychology*, 2006, 115(4), 715–29

Thode KI, Faber RA & Chaudhuri TK, 'Delusional misidentification syndrome: right-hemisphere findings on SPECT', *Journal of Neuropsychiatry and Clinical Neurosciences*, 2012, 24(1), E22–3

Thomas C, *Diversity and Fairness in the Jury System*, Ministry of Justice Research Series 2/07 June 2007, Crown Copyright

——, *Are Juries Fair?*, 2010: www.justice.gov.uk/downloads/publications/research-and-analysis/moj-research/are-juries-fair-research.pdf (accessed 8 March 2021)

Thomas C, Humphreys K, Jung K-J et al, 'The anatomy of the callosal and visual-association pathways in high-functioning autism: a DTI tractography study', *Cortex*, 2010, 47(4), 863–73

Thomas F, Ulitsky P, Augier R et al, 'Biochemical and histological changes in the brain of the cricket *Nemobius sylvestris* infected by the manipulative parasite *Paragordius tricuspidatus* (Nematomorpha)', *International Journal for Parasitology*, 2003, 33(4), 435–43

Thomas J, Jamieson G & Cohen M, 'Low and then high frequency oscillations of distinct right cortical networks are progressively enhanced by medium and long term Satyananda Yoga meditation practice', *Frontiers in Human Neuroscience*, 2014, 8, article 197

Thomas NA, Wignall SJ, Loetscher T et al, 'Searching the expressive face: evidence for both the right hemisphere and valence-specific hypotheses', *Emotion*, 2014, 14(5), 962–77

Thomas P, Barrès P & Chatel M, 'Complex partial status epilepticus of extratemporal origin: report of a case', *Neurology*, 1991, 41(7), 1147–9

Thomas R & Forde E, 'The role of local and global processing in the recognition of living and nonliving things', *Neuropsychologia*, 2006, 44(6), 982–6

Thompson D'AW, *On Growth and Form: A New Edition*, Dover, 1992 [1942, original edn 1917]

Thompson MI, Silk KR & Hover GL, 'Misidentification of a city: delimiting criteria for Capgras syndrome', *American Journal of Psychiatry*, 1980, 137(10), 1270–2

Thompson S, *The Folktale*, Holt, Rinehart & Winston, 1946

Thomson JF, 'Tasks and super-tasks', *Analysis*, 1954, 15(1), 1–13

Thoreau HD, in B Torrey (ed), *The Writings of Henry David Thoreau*, 14 vols, Houghton Mifflin, 1906

Thorup A, Petersen L, Jeppesen P et al, 'Frequency and predictive values of first rank symptoms at baseline among 362 young adult patients with first-episode schizophrenia Results from the Danish OPUS study', *Schizophrenia Research*, 2007, 97(1–3), 60–7

Thurman J, 'On the weight of the brain and the circumstances affecting it', *Journal of Mental Science*, 1866, 12, 1–43

Thybony S, *Burntwater*, University of Arizona Press, 1997

Tian L, Wang J, Yan C et al, 'Hemisphere- and gender-related differences in small-world brain networks: a resting-state functional MRI study', *NeuroImage*, 2011, 54(1), 191–202

Tibber MS, Anderson EJ, Bobin T et al, 'Visual surround suppression in schizophrenia', *Frontiers in Psychology*, 2013, 4, article 88

Tiedt HO, Weber JE, Pauls A et al, 'Sex-differences of face coding: evidence from larger right hemispheric M170 in men and dipole source modelling', *PLoS One*, 2013, 8(7), e69107

Tiihonen J, Katila H, Pekkonen E et al, 'Reversal of cerebral asymmetry in schizophrenia measured with magnetoencephalography', *Schizophrenia Research*, 1998, 30(3), 209–19

Tillich P, 'The lost dimension in religion', *Saturday Evening Post*, 14 June 1958, 230(50), 28–9 & 76–9

Ting WK, Fischer CE, Millikin CP et al, 'Grey matter atrophy in mild cognitive impairment/early Alzheimer disease associated with delusions: a voxel-based morphometry study', *Current Alzheimer Research*, 2015, 12(2), 165–72

Titianova EB & Tarkka IM, 'Asymmetry in walking performance and postural sway in patients with chronic unilateral cerebral infarction', *Journal of Rehabilitation Research and Development*, 1995, 32(3), 236–44

Todd J, 'The syndrome of Alice in Wonderland', *Canadian Medical Association Journal*, 1955, 73(9), 701–4

Todorov A, Olivola CY & Funk F, 'Reply to Bonnefon et al: Limited "kernels of truth" in facial inferences', *Trends in Cognitive Sciences*, 2015, 19(8), 422–3

Todorova GK, Hatton REM & Pollick FE, 'Biological motion perception in autism spectrum disorder: a meta-analysis', *Molecular Autism*, 2019, 10, article 49

Toga AW & Thompson PM, 'Mapping brain asymmetry', *Nature Reviews: Neuroscience*, 2003, 4(1), 37–48

Tognetti A, Berticat C, Raymond M et al, 'Is cooperativeness readable in static facial features? An intercultural approach', *Evolution and Human Behavior*, 2013, 34(6), 427–32

Tokida H, Takeshima S, Takeshita J et al, 'A case of various illusion, and hallucination caused by occipital lobe infarction', *Rinsho Shinkeigaku*, 2018, 58(9), 556–9

Toller G, Adhimoolam B, Rankin KP et al, 'Right fronto-limbic atrophy is associated with reduced empathy in refractory unilateral mesial temporal lobe epilepsy', *Neuropsychologia*, 2015, 78, 80–7

Tolstoy, Count Leo, *Resurrection*, trans L Maude, Dover, 2004 [1899]

——, 'The Kreutzer Sonata'; in *The Kreutzer Sonata and Other Stories*, Oxford University Press, 2009

Tomarken AJ, Davidson RJ, Wheeler RE et al, 'Individual differences in anterior brain asymmetry and fundamental dimensions of emotion', *Journal of Personality and Social Psychology*, 1992, 62(4), 676–87

Tomasello M, *Why We Cooperate*, Massachusetts Institute of Technology Press, 2009

Tomer R & Flor-Henry P, 'Neuroleptics reverse attention asymmetries in schizophrenic patients', *Biological Psychiatry*, 1989, 25(7), 852–60

Tomer R, Mintz M, Levy A et al, 'Smooth pursuit pattern in schizophrenic patients during cognitive task', *Biological Psychiatry*, 1981, 16(2), 131–44

Tomiyama AJ, Hunger JM, Nguyen-Cuu J et al, 'Misclassification of cardiometabolic health when using body mass index categories in NHANES 2005–2012', *International Journal of Obesity*, 2016, 40(5), 883–6

Tomlinson SP, Davis NJ, Morgan HM et al, 'Hemispheric specialisation in haptic processing', *Neuropsychologia*, 2011, 49(9), 2703–10

Tommasi L & Vallortigara G, 'Hemispheric processing of landmark and geometric information in male and female domestic chicks (*Gallus gallus*)', *Behavioural Brain Research*, 2004, 155(1), 85–96

Tong D, 'The Unquantum Quantum', *Scientific American*, 2012, 307(6), 46–9

———, 'The real building blocks of the universe', lecture delivered at the Royal Institution, London, 25 November 2016: www.artandeducation.net/classroom/video/197535/david-tong-quantum-fields-the-real-building-blocks-of-the-universe (accessed 23 February 2021)

Tonna M, Ottoni R, Paglia F et al, 'Obsessive-compulsive symptoms in schizophrenia and in obsessive-compulsive disorder: differences and similarities', *Journal of Psychiatric Practice*, 2016, 22(2), 111–6

Tononi G & Koch C, 'Consciousness: here, there and everywhere?', *Philosophical Transactions of the Royal Society B: Biological Sciences*, 2015, 370(1668), 20140167

Tootell RB, Mendola JD, Hadjikhani NK et al, 'The representation of the ipsilateral visual field in human cerebral cortex', *Proceedings of the National Academy of Sciences of the United States of America*, 1998, 95(3), 818–24

Topiwala A, Allan CL, Valkanova V et al, 'Moderate alcohol consumption as risk factor for adverse brain outcomes and cognitive decline: longitudinal cohort study', *BMJ*, 2017, 357, j2353

Tordjman S, 'Représentations et perceptions du temps', *L'Encéphale*, 2015, 41(4, suppl 1), S1–14

Torrey EF, 'Schizophrenia and the inferior parietal lobule', *Schizophrenia Research*, 2007, 97(1–3), 215–25

Townes C, interviewed by Bonnie Azab Powell of *UC Berkeley News* on 17 June 2005: www.berkeley.edu/news/media/releases/2005/06/17_townes.shtml (accessed 23 February 2021)

Townsend H, *An Account of the Visit of Handel to Dublin*, James McGlashan, 1852

Toynbee A, *A Study of History*, 12 vols, Oxford University Press, 1934–1961: vol 12, *Reconsiderations*, 1961

Tracy JI, Monaco C, McMichael H et al, 'Information-processing characteristics of explicit time estimation by patients with schizophrenia and normal controls', *Perceptual & Motor Skills*, 1998, 86(2), 515–26

Trahan LH, Stuebing KK, Fletcher JM et al, 'The Flynn effect: a meta-analysis', *Psychological Bulletin*, 2014, 140(5), 1332–60

Tramo MJ & Bharucha JJ, 'Musical priming by the right hemisphere post-callosotomy', *Neuropsychologia*, 1991, 29(4), 313–25

Tranel D, Bechara A & Denburg NL, 'Asymmetric functional roles of right and left ventromedial prefrontal cortices in social conduct, decision-making, and emotional processing', *Cortex*, 2002, 38(4), 589–612

Tranel D, Damasio H, Denburg NL et al, 'Does gender play a role in functional asymmetry of ventromedial prefrontal cortex?', *Brain*, 2005, 128(12), 2872–81

Treasure J, 'Coherence and other autistic spectrum traits and eating disorders: building from mechanism to treatment', *Nordic Journal of Psychiatry*, 2013, 6(1)7, 38–42

Treffert DA, 'The savant syndrome: an extraordinary condition. A synopsis: past, present, future', *Philosophical Transactions of the Royal Society of London B: Biological Sciences*, 2009, 364(1522), 1351–7

———, 'Accidental genius', *Scientific American*, 2014, 311(2), 52–7

Trelawny EJ, *Recollections of the Last Days of Shelley and Byron*, Ticknor & Fields, 1858

Tremblay C, Monetta L, Langlois M et al, 'Intermittent theta-burst stimulation of the right dorsolateral prefrontal cortex to promote metaphor comprehension in Parkinson disease: a case study', *Archives of Physical Medicine and Rehabilitation*, 2016, 97(1), 74–83

Trevena J & Miller J, 'Brain preparation before a voluntary action: evidence against unconscious movement initiation', *Consciousness and Cognition*, 2010, 19(1), 447–56

Trewavas A, 'Aspects of plant intelligence', *Annals of Botany*, 2003, 92(1), 1–20

Trimble MR, 'Post-operative psychosis', in MR Trimble (ed), *The Psychoses of Epilepsy*, Raven Press, 1991, 91–108

———, *The Soul in the Brain: The Cerebral Basis of Language, Art, and Belief*, Johns Hopkins University Press, 2007

Trimble M & Freeman A, 'An investigation of religiosity and the Gastaut-Geschwind syndrome in patients with temporal lobe epilepsy', *Epilepsy and Behavior*, 2006, 9(3), 407–14

Troiani I & Campbell H, 'Orchestrating spatial continuity in the urban realm', *Architecture and Culture*, 2015, 3(1), 7–16

Trojano L, Conson M, Salzano S et al, 'Unilateral left prosopometamorphopsia: a neuropsychological case study', *Neuropsychologia*, 2009, 47(3), 942–8

Troup GA, Bradshaw JL & Nettleton NC, 'The lateralization of arithmetic and number processing: a review' *International Journal of Neuroscience*, 1983, 19(1–4), 231–42

True JR & Haag ES, 'Developmental system drift and flexibility in evolutionary trajectories', *Evolution & Development*, 2001, 3(2), 109–19

Tsai MH, Hsu SP, Huang CR et al, 'Transient attenuation of visual evoked potentials during focal status epilepticus in a patient with occipital lobe epilepsy', *Acta Neurologica Taiwanica*, 2010, 19(2), 131–6

Tsakiris M, Costantini M & Haggard P, 'The role of the right temporo-parietal junction in maintaining a coherent sense of one's body', *Neuropsychologia*, 2008, 46(12), 3014–8

Tsakiris M, Hesse MD, Boy C et al, 'Neural signatures of body ownership: a sensory network for bodily self-consciousness', *Cerebral Cortex*, 2007, 17(10), 2235–44

Tse PU, Intriligator J, Rivest J et al, 'Attention and the subjective expansion of time', *Perception & Psychophysics*, 2004, 66(7), 1171–89

Tseng A & Levin M, 'Cracking the bioelectric code: probing endogenous ionic controls of pattern formation', *Communicative & Integrative Biology*, 2013, 6(1), e22595

Tskhay KO & Rule NO, 'Accuracy in categorizing perceptually ambiguous groups: a review and meta-analysis', *Personality and Social Psychology Review*, 2013, 17(1), 72–86

Tsujii T, Masuda S, Akiyama T et al, 'The role of inferior frontal cortex in belief-bias reasoning: an rTMS study', *Neuropsychologia*, 2010, 48(7), 2005–8

Tsujii T, Okada M & Watanabe S, 'Effects of aging on hemispheric asymmetry in inferior frontal cortex activity during belief-bias syllogistic reasoning: a near-infrared spectroscopy study', *Behavioural Brain Research*, 2010, 210(2), 178–83

Tsujii T & Watanabe S, 'Neural correlates of dual-task effect on belief-bias syllogistic reasoning: a near-infrared spectroscopy study', *Brain Research*, 2009, 1287, 118–25

———, 'Neural correlates of belief-bias reasoning under time pressure: a near-infrared spectroscopy study', *NeuroImage*, 2010, 50(3), 1320–6

Tsukiura T & Cabeza R, 'Remembering beauty: roles of orbitofrontal and hippocampal regions in successful memory encoding of attractive faces', *NeuroImage*, 2011, 54(1), 653–60

Tsunetsugu Y, Park BJ, Ishii H et al, 'Physiological effects of Shinrin-yoku (taking in the atmosphere of the forest) in an old-growth broadleaf forest in Yamagata Prefecture, Japan', *Journal of Physiological Anthropology*, 2007, 26(2), 135–42

Tucker DM, 'Lateral brain function, emotion, and conceptualization', *Psychological Bulletin*, 1981, 89(1), 19–46

———, 'Developing emotions and cortical networks', in MR Gunnar & CA Nelson (eds), *Minnesota Symposium on Child Psychology*, vol 24, Developmental Behavioral Neuroscience, Lawrence Erlbaum, 1992, 75–128

Tucker DM, Roth DL & Bair TB, 'Functional connections among cortical regions: topography of EEG coherence', *Electroencephalography and Clinical Neurophysiology*, 1986, 63(3), 242–50

Tucker DM, Stenslie CE, Roth RS et al, 'Right frontal lobe activation and right hemisphere performance: decrement during a depressed mood', *Archives of General Psychiatry*, 1981, 38(2), 169–74

Tucker DM, Watson RT & Heilman KM, 'Discrimination and evocation of affectively intoned speech in patients with right parietal disease', *Neurology*, 1977, 27(10), 947–50

Tucker DM & Williamson PA, 'Asymmetric neural control systems in human self-regulation [review]', *Psychological Review*, 1984, 91(2), 185–215

Tudge C, *The Secret Life of Trees: How They Live and Why They Matter*, Penguin, 2006

———, *Why Genes are Not Selfish and People are Nice*, Floris, 2013

Tullett AM, Harmon-Jones E & Inzlicht M, 'Right frontal cortical asymmetry predicts empathic reactions: support for a link between withdrawal motivation and empathy', *Psychophysiology*, 2012, 49(8), 1145–53

Tulving E, Kapur S, Craik FI et al, 'Hemispheric encoding/retrieval asymmetry in episodic memory: positron emission tomography findings [review]', *Proceedings of the National Academy of Sciences of the United States of America*, 1994, 91(6), 2016–20

Tulving E, Markowitsch HJ, Craik FE et al, 'Novelty and familiarity activations in PET studies of memory encoding and retrieval', *Cerebral Cortex*, 1996, 6(1), 71–9

Tunç S & Başbuğ HS, 'Alice in Wonderland syndrome: a strange visual perceptual disturbance', *Psychiatry and Clinical Psychopharmacology*, 2017, 27(4), 412–5

Tuomisto J, Pekkanen J, Kiviranta H et al, 'Dioxin cancer risk – example of hormesis?', *Dose-Response*, 2006, 3(3), 332–41

Turk A, Brown WS, Symington M et al, 'Social narratives in agenesis of the corpus callosum: linguistic analysis of the Thematic Apperception Test', *Neuropsychologia*, 2010, 48(1), 43–50

Turnbull OH, Worsey RB & Bowman CH, 'Emotion and intuition: does Schadenfreude make interns poor learners?', *Philoctetes*, 2007, 1, 5–43

Turner DC, Aitken MR, Shanks DR et al, 'The role of the lateral frontal cortex in causal associative learning: exploring preventative and super-learning', *Cerebral Cortex*, 2004, 14(8), 872–80

Turner JS, *The Extended Organism: The Physiology of Animal-Built Structures*, Harvard University Press, 2000

——, *The Tinkerer's Accomplice: How Design Emerges From Life Itself*, Harvard University Press, 2007

Tuyns AJ & Pequignot G, 'Greater risk of ascitic cirrhosis in females in relation to alcohol consumption', *International Journal of Epidemiology*, 1984, 13(1), 53–7

Tversky A & Kahneman D, 'The framing of decisions and the psychology of choice', *Science*, 1981, 211(4481), 453–8

Twenge JM, Gentile B, DeWall CN et al, 'Birth cohort increases in psychopathology among young Americans, 1938–2007: a cross-temporal meta-analysis of the MMPI', *Clinical Psychology Review*, 2010, 30(2), 145–54

Tymms P & FitzGibbon CT, 'Standards, achievement and educational performance: a cause for celebration?', in R Philips & J Furlong (eds), *Education, Reform and The State*, Routledge, 2001

Tymoczko T, 'The four-color problem and its philosophical significance', *The Journal of Philosophy*, 1979, 76(2), 57–83

Tysk L, 'A longitudinal study of time estimation in psychotic disorders', *Perceptual & Motor Skills*, 1984, 59(3), 779–89

Tzavaras A, Merienne L & Masure MC, 'Prosopagnosie, amnésie et troubles de langage par lésion temporale gauche chez un sujet gaucher', *L'Encéphale*, 1973, 62(4), 382–94

Uca AU & Kozak HH, 'The Alice in Wonderland syndrome: a case of aura accompanying cluster headache', *Balkan Medical Journal*, 2015, 32(3), 320–2

Udalova GP, 'Hemispheric specialization of recognition of textured images in the rat', *Zhurnal Vyssheĭ Nervnoĭ Deiatelnosti Imeni I P Pavlova*, 1984, 34(1), 53–61

Uddin LQ, Davies MS, Scott AA et al, 'Neural basis of self and other representation in autism: an fMRI study of self-face recognition', *PLoS One*, 2008, 3(10), e3526

Uddin LQ, Kaplan JT, Molnar-Szakacs I et al, 'Self-face recognition activates a frontoparietal "mirror" network in the right hemisphere: an event-related fMRI study', *NeuroImage*, 2005, 25(3), 926–35

Uddin LQ, Molnar-Szakacs I, Zaidel E et al, 'rTMS to the right inferior parietal lobule disrupts self-other discrimination', *Social Cognitive and Affective Neuroscience*, 2006, 1(1), 65–71

Udell MAR & Wynne CDL, 'A review of domestic dogs' (*Canis familiaris*) human-like behaviors: or why behavior analysts should stop worrying and love their dogs', *Journal of the Experimental Analysis of Behavior*, 2008, 89(2), 247–61

Uebersax JS, 'Higher reason and lower reason', 2013: philpapers.org/rec/UEBHRA (accessed 23 February 2021)

Uher R, Murphy T, Friederich H-C et al, 'Functional neuroanatomy of body shape perception in healthy and eating-disordered women', *Biological Psychiatry*, 2005, 58(12), 990–7

Uher R & Treasure J, 'Brain lesions and eating disorders', *Journal of Neurology, Neurosurgery, & Psychiatry*, 2005, 76(6), 852–7

Uhlhaas PJ, Phillips WA, Mitchell G et al, 'Perceptual grouping in disorganized schizophrenia', *Journal of Experimental Child Psychology*, 2006, 145(2–3), 105–17

Uhlhaas PJ, Phillips WA, Schenkel LS et al, 'Theory of mind and perceptual context-processing in schizophrenia', *Cognitive Neuropsychiatry*, 2006, 11(4), 416–36

Uhlhaas PJ & Silverstein SM, 'Perceptual organization in schizophrenia spectrum disorders: empirical research and theoretical implications', *Psychological Bulletin*, 2005, 131(4), 618–32

Uhls YT, Michikyan M, Morris J et al, 'Five days at outdoor education camp without screens improves preteen skills with nonverbal emotion cues', *Computers in Human Behavior*, 2014, 39, 387–92

Ulinski PS, 'The cerebral cortex of reptiles', in EG Jones & A Peters (eds), *Comparative Structure and Evolution of Cerebral Cortex, Pt 1*, Plenum, 1990, 139–215

Ulla M, Thobois S, Lemaire JJ et al, 'Manic behaviour induced by deep-brain stimulation in Parkinson's disease: evidence of substantia nigra implication?', *Journal of Neurology, Neurosurgery, & Psychiatry*, 2006, 77(12), 1363–6

Ulrich R, Nitschke J & Rammsayer T, 'Perceived duration of expected and unexpected stimuli', *Psychological Research*, 2006, 70(2), 77–87

Ulrich RS, 'Visual landscape and psychological well-being', *Landscape Research*, 1979, 4(1), 17–23

——, 'View through a window may influence recovery from surgery', *Science*, 1984, 224(4647), 420–1

———, 'Biophilia, biophobia, and natural landscapes', in SA Kellert & EO Wilson, (eds), *The Biophilia Hypothesis*, Island Press, 1993, 73–137

Ulrich RS, Simons RF, Losito BD et al, 'Stress recovery during exposure to natural and urban environments', *Journal of Environmental Psychology*, 1991, 11(3), 201–30

Umiltà C, Bagnara S & Simion F, 'Laterality effects for simple and complex geometrical figures and nonsense patterns', *Neuropsychologia*, 1978, 16(1), 43–9

Unamuno M de, *The Tragic Sense of Life*, trans JEC Flitch, Dover, 1954 [1921]

Underhill E, *Mysticism: A Study in the Nature and Development of Spiritual Consciousness*, Dutton, 1911

———, *Practical Mysticism: A Little Book for Normal People*, JM Dent & Sons, 1914

Unema P, Pannasch S, Joos M et al, 'Timecourse of information processing during scene perception: the relationship between saccade amplitude and fixation duration', *Visual Cognition*, 2005, 12(3), 473–94

Unkelbach C, 'Reversing the truth effect: learning the interpretation of processing fluency in judgments of truth', *Journal of Experimental Psychology: Learning, Memory, and Cognition*, 2007, 33(1), 219–30

Unnithan SB, David AS & Cutting JC, 'Magnetic attraction of gaze: further evidence of hemispheric imbalance in schizophrenia?', *Behavioural Neurology*, 1991, 4(2), 63–6

Upanishads, trans S Radhakrishnan, Allen & Unwin, 1953

Urfer A, 'Phenomenology and psychopathology of schizophrenia: the views of Eugène Minkowski', in *Philosophy, Psychiatry, & Psychology*, 2001, 8(4), 279–89

Urgesi C, Aglioti SM, Skrap M et al, 'The spiritual brain: selective cortical lesions modulate human self-transcendence', *Neuron*, 2010, 65(3), 309–19

Usher M, Russo Z, Weyers M et al, 'The impact of the mode of thought in complex decisions: intuitive decisions are better', *Frontiers in Psychology*, 2011, 2, article 37

Uversky VN, 'The mysterious unfoldome: structureless, underappreciated, yet vital part of any given proteome', *Journal of Biomedicine and Biotechnology*, 2010, article 568068

Uylings HB & van Eden CG, 'Qualitative and quantitative comparison of the prefrontal cortex in rat and in primates, including humans', *Progress in Brain Research*, 1990, 85, 31–62

Vaihinger H, *The Philosophy of 'As If'*, trans CK Ogden, Routledge & Kegan Paul, 2nd edn, 1935 [1911]

Vaina LM, 'Selective impairment of visual motion interpretation following lesions of the right occipito-parietal area in humans', *Biological Cybernetics*, 1989, 61(5), 347–59

Vaiserman AM, 'Radiation hormesis: historical perspective and implications for low-dose cancer risk assessment', *Dose-Response*, 2010, 8(2), 172–91

Valentin G, 'Ueber die subjectiven Gefühle von Personen, welche mit mangelhaften Extremitäten geboren sind', *Repertorium für Anatomie und Physiologie*, 1836, 1, 328–37

Valentine L & Gabbard GO, 'Can the use of humor in psychotherapy be taught?', *Academic Psychiatry*, 2014, 38(1), 75–81

Valentine M & Robin AA, 'Aspects of thematic apperception testing: paranoid schizophrenia', *Journal of Mental Science*, 1950, 96(405), 869–88

Valins S, 'Emotionality and information concerning internal reactions', *Journal of Personality & Social Psychology*, 1967, 6(4), 458–63

Valladares F & Brites D, 'Leaf phyllotaxis: does it really affect light capture?', *Plant Ecology*, 2004, 174(1), 11–17

Vallar G, 'A hemispheric asymmetry in somatosensory processing', *Behavioral and Brain Sciences*, 2007, 30(2), 223–4

Vallar G & Ronchi R, 'Somatoparaphrenia: a body delusion. A review of the neuropsychological literature', *Experimental Brain Research*, 2009, 192(3), 533–51

Vallortigara G & Andrew RJ, 'Lateralization of response to change in a model partner by chicks', *Animal Behaviour*, 1991, 41(2), 187–94

Valsangkar-Smyth MA, Donovan CL, Sinnett S et al, 'Hemispheric performance in object-based attention', *Psychonomic Bulletin & Review*, 2004, 11(1), 84–91

Valtin H, '"Drink at least eight glasses of water a day." Really? Is there scientific evidence for "8 x 8"?', *American Journal of Physiology – Regulatory, Integrative and Comparative Physiology*, 2002, 283(5), R993–1004

van Dam WO, Decker SL, Durbin JS et al, 'Resting state signatures of domain and demand-specific working memory performance', *NeuroImage*, 2015, 118, 174–82

van den Bos R, Homberg J & de Visser L, 'A critical review of sex differences in decision-making tasks: focus on the Iowa Gambling Task', *Behavioural Brain Research*, 2013, 238(1), 95–108

van der Ham IJM & Borst G, 'The nature of categorical and coordinate spatial relation processing: an interference study', *Journal of Cognitive Psychology*, 2011, 23(8), 922–30 van der Ham IJM, van Strien JW, Oleksiak A et al, 'Temporal characteristics of working memory for spatial relations: an ERP study', *International Journal of Psychophysiology*, 2010, 77(2), 83–94

van der Knaap LJ & van der Ham IJM, 'How does the corpus callosum mediate interhemispheric transfer? A review', *Behavioural Brain Research*, 2011, 223(1), 211–21

van der Merwe WL & Voestermans PP, 'Wittgenstein's legacy and the challenge to psychology', *Theory & Psychology*, 1995, 5(1), 27–48

van der Wall EE, 'Journal impact factor: holy grail?', *Netherlands Heart Journal*, 2012, 20(10), 385–6

Van Dongen S, 'Associations among facial masculinity, physical strength, fluctuating asymmetry and attractiveness in young men and women', *Annals of Human Biology*, 2014, 41(3), 205–13

Van Duuren M, Kendell-Scott L & Stark N, 'Early aesthetic choices: infant preference for attractive premature infant faces', *International Journal of Behavioral Development*, 2003, 27(3), 212–19

van Eden CG, Uylings HB & van Pelt J, 'Sex-difference and left-right asymmetries in the prefrontal cortex during postnatal development in the rat', *Brain Research*, 1984, 314(1), 146–53

Van Essen DC & Maunsell JHR, 'Hierarchical organization and functional streams in the visual cortex', *Trends in Neurosciences*, 1983, 6(9), 370–75

van Kemenade BM, Muggleton N, Walsh V et al, 'Effects of TMS over premotor and superior temporal cortices on biological motion perception', *Journal of Cognitive Neuroscience*, 2012, 24(4), 896–904

van Kleeck MH, 'Hemispheric differences in global versus local processing of hierarchical visual stimuli by normal subjects: new data and a meta-analysis of previous studies', *Neuropsychologia*, 1989, 27(9), 1165–78

Van Lancker DR & Canter GJ, 'Impairment of voice and face recognition in patients with hemispheric damage', *Brain and Cognition*, 1982, 1(2), 185–95

Van Lancker D & Kreiman J, 'Voice discrimination and recognition are separate abilities', *Neuropsychologia*, 1987, 25(5), 829–34

Van Lancker DR, Kreiman J & Cummings J, 'Voice perception deficits: neuroanatomical correlates of phonagnosia', *Journal of Clinical and Experimental Neuropsychology*, 1989, 11(5), 665–74

Van Noorden R, 'Publishers withdraw more than 120 gibberish papers', *Nature News*, 24 February 2014: www.nature.com/news/publishers-withdraw-more-than-120-gibberish-papers-1.14763 (accessed 6 March 2021)

van Praag HM, 'The role of religion in suicide prevention', in D Wasserman & C Wasserman (eds), *Oxford Textbook of Suicidology and Suicide Prevention*, Oxford University Press, 2009

van Rooyen S, Godlee F, Evans S et al, 'Effect of open peer review on quality of reviews and on reviewers' recommendations: a randomised trial', *BMJ*, 1999, 318(7175), 23–7

van Stralen HE, van Zandvoort MJ & Dijkerman HC, 'The role of self-touch in somatosensory and body representation disorders after stroke', *Philosophical Transactions of the Royal Society of London: Series B, Biological Sciences*, 2011, 366(1581), 3142–52

van Veelen NM, Vink M, Ramsey NF et al, 'Left dorsolateral prefrontal cortex dysfunction in medication-naive schizophrenia', *Schizophrenia Research*, 2010, 123, 22–9

van't Wout M, Kahn RS, Sanfey AG et al, 'Repetitive transcranial magnetic stimulation over the right dorsolateral prefrontal cortex affects strategic decision-making', *NeuroReport*, 2005, 16(16), 1849–52

Vancleef K, Wagemans J & Humphreys GW, 'Impaired texture segregation but spared contour integration following damage to right posterior parietal cortex', *Experimental Brain Research*, 2013, 230(1), 41–57

Vandenberg LN, Adams DS & Levin M, 'Normalized shape and location of perturbed craniofacial structures in the *Xenopus* tadpole reveal an innate ability to achieve correct morphology', *Developmental Dynamics*, 2012, 241(5), 863–78

Varela FJ & Shear J, 'First-person methodologies: what, why, how?', *Journal of Consciousness Studies*, 1999, 6(2–3), 1–14

Varney NR & Benton AL, 'Tactile perception of direction in relation to handedness and familial handedness', *Neuropsychologia*, 1975, 13(4), 449–54

Vartanian O & Goel V, 'Neuroanatomical correlates of aesthetic preference for paintings', *NeuroReport*, 2004, 15(5), 893–7

——, 'Task constraints modulate activation in right ventral lateral prefrontal cortex', *NeuroImage*, 2005, 27(4), 927–33

Vaskinn A, Sundet K, Østefjells T et al, 'Reading emotions from body movement: a generalized impairment in schizophrenia', *Frontiers in Psychology*, 2016, 6, article 2058

Vauvenargues, Luc de Clapiers, marquis de, *Réflexions et Maximes* [1746], in *Œuvres de Vauvenargues*, ed D-L Gilbert, Furne, 1859

Vazsonyi A, 'Which door has the Cadillac?', *Decision Line*, Dec/Jan 1999, 17–19

Vecera SP & Behrmann M, 'Spatial attention does not require preattentive grouping', *Neuropsychology*, 1997, 11(1), 30–43

Vega WA, Kolody B, Aguilar-Gaxiola S et al, 'Lifetime prevalence of DSM-III-R psychiatric disorders among urban and rural Mexican Americans in California', *Archives of General Psychiatry*, 1998, 55(9), 771–8

Velay JL, Daffaure V, Raphael N et al, 'Hemispheric asymmetry and interhemispheric transfer in pointing depend on the spatial components of the movement', *Cortex*, 2001, 37(1), 75–90

Veldhuizen MG, Albrecht J, Zelano C et al, 'Identification of human gustatory cortex by activation likelihood estimation', *Human Brain Mapping*, 2011, 32(12), 2256–66

Velichkovsky BM, Korosteleva AN, Pannasch S et al, 'Two visual systems and their eye movements: a fixation-based event-related experiment with ultrafast fMRI reconciles competing views', *Sovremennye Tehnologii v Medicine*, 2019, 11(4), 7

Velichkovsky BM, Krotkova OA, Kotov AA et al, 'Consciousness in a multilevel architecture: evidence from the right side of the brain', *Consciousness and Cognition*, 2018, 64, 227–39

Velichkovsky BM, Krotkova OA, Sharaev MG et al, 'In search of the "I": neuropsychology of lateralized thinking meets Dynamic Causal Modeling', *Psychology in Russia: State of the Art*, 2017, 10(3), 7–27

Velichkovsky BM, Rothert A, Kopf M et al, 'Towards an express diagnostics for level of processing and hazard perception', *Transportation Research, Part F: Traffic Psychology and Behaviour*, 2002, 5(2), 145–56

Velioğlu SK, Kuzeyli K & Özmenoğlu M, 'Cerebellar agenesis: a case report with clinical and MR imaging findings and a review of the literature', *European Journal of Neurology*, 1998, 5(5), 503–6

Venneri A & Shanks MF, 'Belief and awareness: reflections on a case of persistent anosognosia', *Neuropsychologia*, 2004, 42(2), 230–8

Ventura P, Delgado J, Ferreira M et al, 'Hemispheric asymmetry in holistic processing of words', *Laterality*, 2019, 24(1), 98–112

Vergely B, *Retour à l'émerveillement*, Éditions Albin Michel, 2010

Verhaeghen P, Joorman J & Khan R, 'Why we sing the blues: the relation between self-reflective rumination, mood, and creativity', *Emotion*, 2005, 5(2), 226–32

Verleger R, Sprenger A, Gebauer S et al, 'On why left events are the right ones: neural mechanisms underlying the left-hemifield advantage in rapid serial visual presentation', *Journal of Cognitive Neuroscience*, 2009, 21(3), 474–88

Verma A & Brysbaert M, 'A right visual field advantage for tool-recognition in the visual half-field paradigm', *Neuropsychologia*, 2011, 49(9), 2342–8

Verma A, van der Haegen L & Brysbaert M, 'Symmetry detection in typically and atypically speech lateralized individuals: a visual half-field study', *Neuropsychologia*, 2013, 51(13), 2611–9

Vermeire BA & Hamilton CR, 'Effects of facial identity, facial expression, and subject's sex on laterality in monkeys', *Laterality*, 1998, 3(1), 1–19

Vermeire BA, Hamilton CR & Erdmann AL, 'Right-hemispheric superiority in split-brain monkeys for learning and remembering facial discrimination', *Behavioral Neuroscience*, 1998, 112(5), 1048–61

Verona E, Sadeh N & Curtin JJ, 'Stress-induced asymmetric frontal brain activity and aggression risk', *Journal of Abnormal Psychology*, 2009, 118(1), 131–45

Verosky SC & Turk-Browne NB, 'Representations of facial identity in the left hemisphere require right hemisphere processing', *Journal of Cognitive Neuroscience*, 2012, 24(4), 1006–17

Verster AJ, Ramani AK, McKay SJ et al, 'Comparative RNAi screens in C. elegans and C. briggsæ reveal the impact of developmental system drift on gene function', *PLoS Genetics*, 2014, 10(2), e1004077

Vessel EA, Starr GG & Rubin N, 'The brain on art: intense aesthetic experience activates the default mode network', *Frontiers in Human Neuroscience*, 2012, 6, article 66

Vickers A, Goyal N, Harland R et al, 'Do certain countries produce only positive results? A systematic review of controlled trials', *Controlled Clinical Trials*, 1998, 19(2), 159–66

Vico GB, *The New Science of Giambattista Vico*, trans TG Bergin & MH Fisch, Cornell University Press, 1948

——, *Principi di Scienza Nuova*, in P Cristofolini (ed), *Vico: Opere Filosofiche*, Sansoni, Firenze, 1971 [1744]

Vidal CN, Nicolson R, DeVito TJ et al, 'Mapping corpus callosum deficits in autism: an index of aberrant cortical connectivity', *Biological Psychiatry*, 2006, 60(3), 218–25

Videler JJ, Stamhuis EJ & Povel GDE, 'Leading-edge vortex lifts swifts', *Science*, 2004, 306(5703), 1960–2

Viets HR, 'Aphasia as described by Linnæus and as painted by Ribera', *Bulletin for the History of Medicine*, 1943, 13, 328–9

Vigliocco G, Vinson DP, Druks J et al, 'Nouns and verbs in the brain: a review of behavioural, electrophysiological, neuropsychological and imaging studies', *Neuroscience & Biobehavioral Reviews*, 2011, 35(3), 407–26

Vignal J-P, Maillard L, McGonigal A et al, 'The dreamy state: hallucinations of autobiographical memory evoked by temporal lobe stimulations and seizures', *Brain*, 2007, 130(1), 88–99

Vigouroux RA, Bonnefoi B & Khalil R, 'Réalisations picturales chez un artiste peintre présentant une héminégligence gauche', *Revue Neurologique (Paris)*, 1990, 146(11), 665–70

Vikingstad EM, George KP, Johnson AF et al, 'Cortical language lateralization in right handed normal subjects using functional magnetic resonance imaging', *Journal of the Neurological Sciences*, 2000, 175(1), 17–27

Villarejo A, Martin VP, Moreno-Ramos T et al, 'Mirrored-self misidentification in a patient without dementia: evidence for right hemispheric and bifrontal damage', *Neurocase*, 2011, 17(3), 276–84

Villarreal MF, Cerquetti D, Caruso S et al, 'Neural correlates of musical creativity: differences between high and low creative subjects', *PLoS One*, 2013, 8(9), e75427

Vincent M, 'Cancer: a de-repression of a default survival program common to all cells?: a life-history perspective on the nature of cancer', *Bioessays*, 2012, 34(1), 72–82

Virtue S, Haberman J, Clancy Z et al, 'Neural activity of inferences during story comprehension', *Brain Research*, 2006, 1084, 104–14

Virtue S, Parrish T & Beeman M, 'Inferences during story comprehension: cortical recruitment affected by predictability of events and working memory capacity', *Journal of Cognitive Neuroscience*, 2008, 20(12), 2274–84

Vitruvius, *The Ten Books on Architecture*, trans MH Morgan, Harvard University Press, 1914

Vitz PC, 'The use of stories in moral development', *The American Psychologist*, 1990, 45(6), 709–20

Vivanti G, Fanning PAJ, Hocking DR et al, 'Social attention, joint attention and sustained attention in autism spectrum disorder and Williams syndrome: convergences and divergences', *Journal of Autism and Developmental Disorders*, 2017, 47(6), 1866–77

Vladusich T, Olu-Lafe O, Kim DS et al, 'Prototypical category learning in high-functioning autism', *Autism Research*, 2010, 3(5), 226–36

Vlasov PN, Chervyakov AV & Gnezditskii VV, 'Déjà vu phenomenon-related EEG pattern: case report', *Epilepsy & Behavior Case Reports*, 2013, 1, 136–41

———, 'Electroencephalographic characteristics of the déjà vu phenomenon', *Neuroscience and Behavioral Physiology*, 2014, 44(7), 754–60

Vogel D & Dussutour A, 'Direct transfer of learned behaviour via cell fusion in non-neural organisms', *Philosophical Transactions of the Royal Society of London: Series B, Biological Sciences*, 2016, 283(1845), pii: 20162382

Vogeley K & Fink GR, 'Neural correlates of the first-person-perspective', *Trends in Cognitive Sciences*, 2003, 7(1), 38–42

Vogeley K, Tepest R, Pfeiffer U et al, 'Right frontal hypergyria differentiation in affected and unaffected siblings from families multiply affected with schizophrenia: a morphometric MRI study', *American Journal of Psychiatry*, 2001, 158(3), 494–6

Vohs KD & Baumeister RF, 'Addiction and free will', *Addiction Research & Theory*, 2009, 17(3), 231–5

Vohs KD & Schooler JW, 'The value of believing in free will: encouraging a belief in determinism increases cheating, *Psychological Science*, 2008, 19(1), 49–54

Voineskos AN, Farzan F, Barr MS et al, 'The role of the corpus callosum in transcranial magnetic stimulation induced interhemispheric signal propagation', *Biological Psychiatry*, 2010, 68(9), 825–31

Volberg G, 'Right-hemisphere specialization for contour grouping', *Experimental Psychology*, 2014, 61(5), 331–9

Volberg G & Hübner R, 'On the role of response conflicts and stimulus position for hemispheric differences in global/local processing: an ERP study', *Neuropsychologia*, 2004, 42(13), 1805–13

Volz HP, Nenadic I, Gaser C et al, 'Time estimation in schizophrenia: an fMRI study at adjusted levels of difficulty', *NeuroReport*, 2001, 12(2), 313–6

von Balthasar HU, *The Word Made Flesh*, Ignatius Press, 1989

von Bertalanffy L, *Modern Theories of Development: An Introduction to Theoretical Biology*, Oxford University Press, 1933

———, *Problems of Life: An Evaluation of Modern Biological and Scientific Thought*, Harper & Brothers, 1952

———, *General System Theory*, George Braziller Inc, 2003 [1969]

von Helmholtz H, 'The modern development of Faraday's conception of electricity', Faraday Lecture delivered before the Fellows of the Chemical Society in London on 5 April 1881: www.chemteam.info/Chem-History/Helmholtz-1881.html (accessed 15 July 2017)

von Kriegstein K, Eger E, Kleinschmidt A et al, 'Modulation of neural responses to speech by directing attention to voices or verbal content', Brain Research: Cognitive Brain Research, 2003, 17(1), 48–55

von Meyenn K (ed), *Wolfgang Pauli: Wissenschaftlicher Briefwechsel*, vol IV, part I (1950–1952), trans H Atmanspacher & H Primas, Springer, 1996

——, *Wolfgang Pauli: Wissenschaftlicher Briefwechsel*, vol IV, part III (1955–1956), trans H Atmanspacher & H Primas, Springer, 2001

von Monakow C, *Die Lokalisation im Grosshirn und Abbau der Funktion durch kortikale Herde*, Bergmann, 1914

von Neumann J, 'Method in the physical sciences', in LG Leary (ed), *The Unity of Knowledge*, Doubleday, 1955a

——, *Mathematical Foundations of Quantum Mechanics*, trans RT Beyer, Princeton University Press, 1955b

von Weizsäcker CF, *Die Geschichte der Natur: Zwölf Vorlesungen*, Vandenhoeck & Ruprecht, 1948

——, introduction to Krishna G, *The Biological Basis of Religion and Genius*, Harper & Row, 1972

——, *The Unity of Nature*, Farrar, Straus, Giroux, 1980

Vörös V, Tényi T, Simon M et al, 'Clonal pluralization of the self': a new form of delusional misidentification syndrome', Psychopathology, 2003, 36(1), 46–48

Vos L & Whitman D, 'Maintaining perceptual constancy while remaining vigilant: left hemisphere change blindness and right hemisphere vigilance', Laterality, 2014, 19(2), 129–45

Voskoboynik A, Simon-Blecher N, Soen Y et al, 'Striving for normality: whole body regeneration through a series of abnormal generations', The Federation of American Societies for Experimental Biology Journal, 2007, 21(7), 1335–44

Voss JL & Paller KA, 'An electrophysiological signature of unconscious recognition memory', Nature Neuroscience, 2009, 12(3), 349–55

Vowles D, 'Neuroethology, evolution and grammar', in R Aronson, E Tobach, D Lehrman et al (eds), *Development and Evolution of Behavior: Essays in Memory of TC Schneirla*, WH Freeman & Co, 1970, 194–215

Voyer D, 'On the magnitude of laterality effects and sex differences in functional lateralities', Laterality, 1996, 1(1), 51–83

——, 'Sex differences in dichotic listening', Brain and Cognition, 2011, 76(2), 245–55

Voyer D, Bowes A & Techentin C, 'On the perception of sarcasm in dichotic listening', Neuropsychology, 2008, 22(3), 390–9

Voyer D & Bryden MP, 'Gender, level of spatial ability, and lateralization of mental rotation', Brain and Cognition, 1990, 13(1), 18–29

Vuilleumier P, Ghika-Schmid F, Bogousslavsky J et al, 'Persistent recurrence of hypomania and prosopoaffective agnosia in a patient with right thalamic infarct', Neuropsychiatry, Neuropsychology and Behavioral Neurology, 1998, 11(1), 40–4

Vuilleumier P, Henson RN, Driver J et al, 'Multiple levels of visual object constancy revealed by event-related fMRI of repetition priming', Nature Neuroscience, 2002, 5(5), 491–9

Vyas A, Kim S-K, Giacomini N et al, 'Behavioral changes induced by *Toxoplasma* infection of rodents are highly specific to aversion of cat odors', Proceedings of the National Academy of Sciences of the United States of America, 2007, 104(15), 6442–7

Vygotsky LS, *Thought and Language*, trans A Kozulin, Massachusetts Institute of Technology Press, 1986

Waber DP, 'Cognitive abilities and sex-related variations in the maturation of cerebral cortical functions', in MA Wittig & AC Petersen (eds), *Sex-related Differences in Cognitive Functioning: Developmental Issues*, Academic Press, 1979, 161–86

Wacker J, Heldmann M & Stemmler G, 'Separating emotion and motivational direction in fear and anger: effects on frontal asymmetry', Emotion, 2003, 3(2), 167–93

Wada T, Kawakatsu S, Komatani A et al, 'Possible association between delusional disorder, somatic type and reduced regional cerebral blood flow', Progress in Neuropsychopharmacology and Biological Psychiatry, 1999, 23(2), 353–7

Waddington CH, 'Genetic assimilation of an acquired character', Evolution, 1953, 7(2), 118–26

——, *The Strategy of the Genes: A Discussion of Some Aspects of Theoretical Biology*, George Allen & Unwin, 1957

——, *The Nature of Life*, George Allen & Unwin, 1961

Wade D, *Li: Dynamic Form in Nature*, Wooden Books, 2007

Wager TD, Phan KL, Liberzon I et al, 'Valence, gender, and lateralization of functional brain anatomy in emotion: a meta-analysis of findings from neuroimaging', NeuroImage, 2003, 19(3), 513–31

Wager TD & Smith EE, 'Neuroimaging studies of working memory: a meta-analysis', Cognitive, Affective & Behavioral Neuroscience, 2003, 3(4), 255–74

Wagner HN, Burns HD, Dannals RF et al, 'Imaging dopamine receptors in the human brain by positron emission tomography', *Science*, 1983, 221(4617), 1264–6

Wagner PS & Spiro CS, *Divided Minds: Twin Sisters and their Journey through Schizophrenia*, St Martin's Press, 2008

Wagner U, Gais S, Haider H et al, 'Sleep inspires insight', *Nature*, 2004, 427(6972), 352–5

Wagner U, N'Diaye K, Ethofer T et al, 'Guilt-specific processing in the prefrontal cortex', *Cerebral Cortex*, 2011, 21(11), 2461–70

Wagoner B, 'There is more to memory than inaccuracy and distortion', *Behavioral and Brain Sciences*, 2017, 40, E16

Wahl OF & Sieg D, 'Time estimation among schizophrenics', *Perceptual & Motor Skills*, 1980, 50(2), 535–41

Wai J, Lubinski D & Benbow CP, 'Creativity and occupational accomplishments among intellectually precocious youth: an age 13 to age 33 longitudinal study', *Journal of Educational Psychology*, 2005, 97(3), 484–92

Waismann F, *How I See Philosophy*, Macmillan, 1968

———, *The Principles of Linguistic Philosophy*, Palgrave Macmillan, 2nd edn, 1997 [1965]

Wald C, 'Why red means red in almost every language: the confounding consistency of color categories', *Nautilus*, 26, 30 July 2015

Wald G, 'Life and mind in the universe', *International Journal of Quantum Chemistry: Quantum Biology Symposium*, 1984, 11(1), 1–15

———, 'The cosmology of life and mind,' *Noetic Sciences Review*, 1989, 10, 10

Waldman P, 'Tragedy turns a right-handed artist into a lefty – and a star in art world', *Wall Street Journal*, 12 May 2000

Waley A, *The Analects of Confucius*, George Allen & Unwin, 1938

Wallace BA, *Hidden Dimensions: The Unification of Physics and Consciousness*, Columbia University Press, 2007

Wallace C, *Portrait of a Schizophrenic Nurse*, Hammond & Hammond, 1965

Wallace JC, Vodanovich SJ & Restino R, 'Predicting cognitive failures from boredom proneness and daytime sleepiness scores: an investigation within military and undergraduate samples', *Personality and Individual Differences*, 2003, 34(4), 635–44

Wallas G, *The Art of Thought*, Harcourt Brace, 1926

Waller NG, Kojetin BA, Bouchard TJ et al, 'Genetic and environmental influences on religious interests, attitudes, and values: a study of twins reared apart and together', *Psychological Science*, 1990, 1(2), 138–42

Walls RM, Hockberger RS & Gausche-Hill M, 'How this medical textbook should be viewed by the practising clinician and judicial system', *Rosen's Emergency Medicine: Concepts and Clinical Practice*, Elsevier, 9th edn, 2018

Walser R, *Looking at Pictures*, trans S Bernofsky, L Davis & C Middleton, New Directions, 2015

Walsh DM, 'Objectcy and agency: towards a methodological vitalism', in DJ Nicholson & J Dupré (eds), *Everything Flows: Towards a Processual Philosophy of Biology*, Oxford University Press, 2018, 167–85

Walsh S, *Stravinsky: The Second Exile: France and America, 1934–1971*, Jonathan Cape, 2006

Walterfang M, Wood AG, Reutens DC et al, 'Morphology of the corpus callosum at different stages of schizophrenia: cross-sectional study in first-episode and chronic illness', *British Journal of Psychiatry*, 2008, 192(6), 429–34

———, 'Corpus callosum size and shape in first-episode affective and schizophrenia-spectrum psychosis', *Psychiatry Research*, 2009, 173(1), 77–82

Walters RP, Harrison DW, Williamson J et al, 'Lateralized visual hallucinations: an analysis of affective valence', *Applied Neuropsychology*, 2006, 13(3), 160–5

Wang F, Peng K, Chechlacz M et al, 'The neural basis of independence versus interdependence orientations: a voxel-based morphometric analysis of brain volume', *Psychological Science*, 2017, 28(4), 519–29

Wang H, *Beyond Analytic Philosophy*, Massachusetts Institute of Technology Press, 1986

———, *Reflections on Kurt Gödel*, Massachusetts Institute of Technology Press, 1987

———, *A Logical Journey: From Gödel to Philosophy*, 1996

———, 'On ge wu: recovering the way of the "Great Learning"', *Philosophy East and West*, 2007, 57(2), 204–26

Wang J, Barstein J, Ethridge LE et al, 'Resting state EEG abnormalities in autism spectrum disorders', *Journal of Neurodevelopmental Disorders*, 2013, 5(1), 24

Wang X & Sommer RJ, 'Antagonism of LIN-17/Frizzled and LIN-18/Ryk in nematode vulva induction reveals evolutionary alterations in core developmental pathways', *PLoS Biology*, 2011, 9(7), e1001110

Wang YX, He GX, Tong GH et al, 'Cerebral asymmetry in a selected Chinese population', *Australasian Radiology*, 1999, 43(3), 321–4

Wangh M, 'Boredom in psychoanalytic perspective', *Social Research*, 1975, 42(3), 538–50

Wapner W, Hamby S & Gardner H, 'The role of the right hemisphere in the apprehension of complex linguistic materials', *Brain and Language*, 1981, 14(1), 15–33

Waragai M, Takaya Y & Hayashi M, 'Complex visual hallucinations in the hemianopic field following an ischemic lesion of the occipito-temporal base – confirmation of the lesion by MRI and speculations on the pathophysiology', *Brain and Nerve*, 1996, 48(4), 371–6

Warman DM, Lysaker PH, Martin JM et al, 'Jumping to conclusions and the continuum of delusional beliefs', *Behaviour Research and Therapy*, 2007, 45(6), 1255–69

Waroquier L, Marchiori D, Klein O et al, 'Is it better to think unconsciously or to trust your first impression? A reassessment of unconscious thought theory', *Social Psychological and Personality Science*, 2010, 1(2), 111–8

Warrington EK & Rudge P, 'A comment on apperceptive agnosia', *Brain and Cognition*, 1995, 28(2), 173–7

Warrington EK & Shallice T, 'Category specific semantic impairments', *Brain*, 1984, 107(3), 829–54

Warrington EK & Taylor AM, 'The contribution of the right parietal lobe to object recognition', *Cortex*, 1973, 9(2), 152–64

———, 'Two categorical stages of object recognition', *Perception*, 1978, 7(6), 695–705

Wason PC, 'Reasoning about a rule', *Quarterly Journal of Experimental Psychology*, 1968, 20(3), 273–81

Watanabe H, Kuhn A, Fushiki M et al, 'Sequential actions of β-catenin and Bmp pattern the oral nerve net in *Nematostella vectensis*', *Nature Communications*, 2014, 5, 5536

Watanabe S, 'The concept of time in modern physics and Bergson's pure duration', in PAY Gunter (ed), *Bergson and the Evolution of Physics*, University of Tennessee Press, 1969, 62–76

Waters F, Collerton D, ffytche DH et al, 'Visual hallucinations in the psychosis spectrum and comparative information from neurodegenerative disorders and eye disease', *Schizophrenia Bulletin*, 2014, 40(suppl 4), S233–45

Waters F & Jablensky A, 'Time discrimination deficits in schizophrenia patients with first-rank (passivity) symptoms', *Psychiatry Research*, 2009, 167(1–2), 12–20

Watkins KE, Paus T, Lerch JP et al, 'Structural asymmetries in human brain: a voxel-based statistical analysis of 142 brains', *Cerebral Cortex*, 2001, 11(9), 868–77

Watling D & Bourne VJ, 'Sex differences in the relationship between children's emotional expression discrimination and their developing hemispheric lateralization', *Developmental Neuropsychology*, 2013, 38(7), 496–506

Watson D, 'Strangers' ratings of the five robust personality factors: evidence of a surprising convergence with self-report', *Journal of Personality and Social Psychology*, 1989, 57(1), 120–8

Watson G, *A Philosophy of Emptiness*, Reaktion Press, 2014

Watt RJ & Phillips WA, 'The function of dynamic grouping in vision', *Trends in Cognitive Sciences*, 2000, 4(12), 447–54

Watts A, *The Two Hands of God*, Rider & Co, 1963

———, *Tao: The Watercourse Way*, Pantheon, 1975

———, *The Book: On the Taboo Against Knowing Who You Are*, Random House, 1989 [1966]

———, 'Philosophy of nature', in *Eastern Wisdom, Modern Life: Collected Talks: 1960–1969*, New World Library, 2006, 123–38

———, 'Taoism' [undated]: terebess.hu/english/watts4.html#tao (accessed 23 February 2021)

Way JC, Collins JJ, Keasling JD et al, 'Integrating biological redesign: where synthetic biology came from and where it needs to go', *Cell*, 2014, 157(1), 151–61

Weber EU & Johnson EJ, 'Mindful judgment and decision making', *Annual Review of Psychology*, 2009, 60, 53–85

Webster G & Goodwin B, *Form and Transformation: Generative and Relational Principles in Biology*, Cambridge University Press, 2011

Webster Rudmin F, Ferrada-Noli M & Skolbekken J-A, 'Questions of culture, age and gender in the epidemiology of suicide', *Scandinavian Journal of Psychology*, 2003, 44(4), 373–81

Wechsler AF, 'The effect of organic brain disease on recall of emotionally charged versus neutral narrative texts', *Neurology*, 1973, 23(2), 130–5

Weckowicz TE, 'Size constancy in schizophrenic patients', *Journal of Mental Science*, 1957, 103(432), 475–86

Wei N, Yong W, Li X et al, 'Post-stroke depression and lesion location: a systematic review', *Journal of Neurology*, 2015, 262(1), 81–90

Weick M, speaking about his research on 'All in the Mind', BBC Radio 4, 12 December 2017 (at 04:30 min)

Weijers NR, Rietveld A, Meijer FJA et al, 'Macrosomatognosia in frontal lobe infarct: a case report', *Journal of Neurology*, 2013, 260(3), 925–6

Weilnhammer V, Stuke H, Hesselmann G et al, 'A predictive coding account of bistable perception – a model-based fMRI', *PLoS Computational Biology*, 2017, 13(5), e1005536

Weinand ME, Hermann B, Wyler AR et al, 'Long-term subdural strip

electrocorticographic monitoring of ictal déjà vu', *Epilepsia*, 1994, 35(5), 1054–9
Weinstein EA, Kahn RL, Malitz S et al, 'Delusional reduplication of parts of the body', *Brain*, 1954, 77(1), 45–60
Weinstein S & Graves RE, 'Are creativity and schizotypy products of a right hemisphere bias?', *Brain and Cognition*, 2002, 49(1), 138–51
Weinstein S & Sersen EA, 'Phantoms in cases of congenital absence of limbs', *Neurology*, 1961, 11, 905–11
Weintraub S & Mesulam M-M, 'Developmental learning disabilities of the right hemisphere: emotional, interpersonal, and cognitive components', *Archives of Neurology*, 1983, 40(8), 463–8
Weis S, Haug H, Holoubek B et al, 'The cerebral dominances: quantitative morphology of the human cerebral cortex', *International Journal of Neuroscience*, 1989, 47(1–2), 165–8
Weismann-Arcache C, 'L'adolescent savant: penser la mort pour rêver d'amour', *Adolescence*, 2010, 28(2), 347–60
Weismann-Arcache C & Tordjman S, 'Relationships between depression and high intellectual potential', *Depression Research and Treatment*, 2012, 2012, article 567376
Weiss EM, Hofer A, Golaszewski S et al, 'Language lateralisation in unmedicated patients during an acute episode of schizophrenia: a functional MRI study', *Psychiatry Research*, 2006, 146(2), 185–90
Weiss M, 'Is there a link between music and language? How loss of language affected the compositions of Vissarion Shebalin', 2013: www.sylff.org/news_voices/12911/ (accessed 23 February 2021)
Weiss P, 'The prediction paradox', *Mind*, 1952, 61(242), 265–9
Weiss PA, 'From cell to molecule' [1960], in JM Allen (ed), *The Molecular Control of Cellular Activity*, McGraw Hill, 1962, 1–72
———, 'The living system: determinism stratified', in A Koestler & JR Smythies (eds), *Beyond Reductionism: New Perspectives in the Life Sciences*, Macmillan, 1970, 361–400
———, 'The basic concept of hierarchic systems', in PA Weiss (ed), *Hierarchically Organized Systems in Theory and Practice*, Hafner, 1971, 1–44
Weissman MM, Bland RC, Canino GJ et al, 'Cross-national epidemiology of major depression and bipolar disorder', *Journal of the American Medical Association*, 1996, 276(4), 293–9
Weller JA, Levin IP, Shiv B et al, 'Neural correlates of adaptive decision making for risky gains and losses', *Psychological Science*, 2007, 18(11), 958–64

Wells CE, 'Transient ictal psychosis', *Archives of General Psychiatry*, 1975, 32(9), 1201–3
Wells DS & Leventhal D, 'Perceptual grouping in schizophrenia: a replication of Place and Gilmore', *Journal of Abnormal Psychology*, 1984, 93(2), 231–4
Welsch W, 'Animal aesthetics', *Contemporary Aesthetics*, 2004, 2(1) [no page numbers]
Wende KC, Nagels A, Blos J et al, 'Differences and commonalities in the judgment of causality in physical and social contexts: an fMRI study', *Neuropsychologia*, 2013, 51(13), 2572–80
Wendell Holmes O, Jnr: in MD Howe (ed), *Holmes-Pollock Letters: The Correspondence of Mr. Justice Holmes and Sir Frederick Pollock, 1874–1932*, Harvard University Press, 2nd edn, 1961
Werner H, *Comparative Psychology of Mental Development*, International Universities Press, 1957
Wertheimer M, 'Experimentelle Studien über das Sehen von Bewegung', *Zeitschrift für Psychologie*, 1912, 61, 161–265
———, *Productive Thinking*, Harper & Row, 2nd edn, 1959 [1945]
Wessels T, *The Myth of Progress*, University Press of New England, revised edn, 2013 [2006]
West RF, Meserve RJ & Stanovich KE, 'Cognitive sophistication does not attenuate the bias blind spot', *Journal of Personality & Social Psychology*, 2012, 103(3), 506–19
Westerberg CE & Marsolek CJ, 'Hemispheric asymmetries in memory processes as measured in a false recognition paradigm', *Cortex*, 2003, 39(4–5), 627–42
Westlake RJ & Weeks SM, 'Pathological jealousy appearing after cerebrovascular infarction in a 25-year-old woman', *Australian & New Zealand Journal of Psychiatry*, 1999, 33(1), 105–7
Westphal [ACO], 'Über einen Fall von motorischer Apraxie', *Allgemeine Zeitschrift für Psychiatrie und psychisch-gerichtliche Medizin*, 1907, 64(2–3), 452–9
Wever EG & Lawrence M, *Physiological Acoustics*, Princeton University Press, 1954
Wexler BE, 'Alterations in cerebral laterality during acute psychotic illness', *British Journal of Psychiatry*, 1986, 149(2), 202–9
———, 'Regional brain dysfunction in depression', in M Kinsbourne (ed), *Cerebral Hemisphere Function in Depression*, American Psychiatric Press, 1988, 65–78
Weyl H, *Philosophy of Mathematics and Natural Science*, Princeton University Press, 2nd edn, 1950 [1949]
———, *Unterrichtsblätter für Mathematik und Naturwissenschaften*, 1932, 38, 177–88; trans

A Shenitzer, 'Part I. Topology and abstract algebra as two roads of mathematical comprehension', *The American Mathematical Monthly*, 1995, 102(5), 453–60; and 'Part II. Topology and abstract algebra as two roads of mathematical comprehension', *The American Mathematical Monthly*, 1995, 102(5), 646–51

———, 'The open world: three lectures on the metaphysical implications of science', in P Pesic (ed), *Mind and Nature: Selected Writings on Philosophy, Mathematics, and Physics*, Princeton University Press, 2010 [1932]

Wheeler JA, '"No fugitive and cloistered virtue" – a tribute to Niels Bohr', *Physics Today*, 1963, 16(1), 30–2

———, *The Intellectual Digest*, June 1973

———, interviewed by MR Gearhart, in 'From the Big Bang to the Big Crunch', *Cosmic Search*, 1979, 1(4), 2–8

———, 'Information, physics, quantum: the search for links', in WH Zurek (ed), *Complexity, Entropy and the Physics of Information*, Addison-Wesley, 1990, 3–28

———, 'The anthropic universe', 18 February 2006: www.abc.net.au/radionational/programs/scienceshow/the-anthropic-universe/3302686#transcript (accessed 23 February 2021)

Wheeler MA, Stuss DT & Tulving E, 'Toward a theory of episodic memory: the frontal lobes and autonoetic consciousness [review]', *Psychological Bulletin*, 1997, 121(3), 331–54

Wheeler RE, Davidson RJ & Tomarken AJ, 'Frontal brain asymmetry and emotional reactivity: a biological substrate of affective style', *Psychophysiology*, 1993, 30(1), 82–9

Whitehead AN, *An Introduction to Mathematics*, Williams & Norgate, 1911

——— *An Enquiry Concerning The Principles of Natural Knowledge*, Cambridge University Press, 1919

———, 'Theories of the bifurcation of Nature', in *The Concept of Nature*, Cambridge University Press, 1920, 26–48

———, 'The philosophical aspects of the principle of relativity', *Proceedings of the Aristotelian Society, New Series*, vol 22, Williams & Norgate, 1922, 215–23

———, 'The education of an Englishman', *The Atlantic Monthly*, 1926a, 138, 192–8

———, *Science and the Modern World*, Cambridge University Press, 1926b

———, *Process and Reality: An Essay in Cosmology*, ed DR Griffin & DW Sherburne, Macmillan, 1929a

———, *The Function of Reason*, Princeton University Press, 1929b

———, *Adventures of Ideas*, Cambridge University Press, 1933

———, *Nature and Life*, Cambridge University Press, 1934

———, 'Remarks: analysis of meaning', *The Philosophical Review*, 1937, 46(2), 178–86

———, *Modes of Thought*, Macmillan, 1938

———, *Dialogues of Alfred North Whitehead*, (ed L Price), Little, Brown & Co, 1954

———, *The Principle of Relativity*, Cosimo, 2007 [1922]

Whitehead R, 'Right hemisphere processing superiority during sustained visual attention', *Journal of Cognitive Neuroscience*, 1991, 3(4), 329–34

Whiteley AM & Warrington EK, 'Prosopagnosia: a clinical, psychological, and anatomical study of three patients', *Journal of Neurology, Neurosurgery, and Psychiatry*, 1977, 40(4), 395–403

Whitfield J, 'A brain in doubt leaves it out: sometimes the brain ignores what the eyes tell it', *Nature News*, 14 June 2001, doi:10.1038/news010614-9

Whitfield-Gabrieli S, Thermenos HW, Milanovic S et al, 'Hyperactivity and hyperconnectivity of the default network in schizophrenia and in first-degree relatives of persons with schizophrenia', *Proceedings of the National Academy of Sciences of the United States of America*, 2009, 106(4), 1279–84

Whitford F, *Oskar Kokoschka: A Life*, Atheneum, 1984

Whitford TJ, Kubicki M, Ghorashi S et al, 'Predicting inter-hemispheric transfer time from the diffusion properties of the corpus callosum in healthy individuals and schizophrenia patients: a combined ERP and DTI study', *NeuroImage*, 2011, 54(3), 2318–29

Whitney C, Huber W, Klann J et al, 'Neural correlates of narrative shifts during auditory story comprehension', *NeuroImage*, 2009, 47(1), 360–6

Whittle S, Chanen AM, Fornito A et al, 'Anterior cingulate volume in adolescents with first-presentation borderline personality disorder', *Psychiatry Research*, 2009, 172(2), 155–60

Whittle S, Yücel M, Fornito A et al, 'Neuroanatomical correlates of temperament in early adolescents', *Journal of the American Academy of Child & Adolescent Psychiatry*, 2008, 47(6), 682–93

Wicherts JM, Dolan CV, Hessen DJ et al, 'Are intelligence tests measurement invariant over time? Investigating the nature of the Flynn effect', *Intelligence*, 2004, 32(5), 509–37

Wicks R, 'The idealization of contingency in traditional Japanese aesthetics', *Journal of Aesthetic Education*, 2005, 39(3), 88–101

Wiech K, Kahane G, Shackel N et al, 'Cold or calculating? Reduced activity in the subgenual cingulate cortex reflects decreased emotional aversion to harming in counterintuitive utilitarian judgment', *Cognition*, 2013, 126(3), 364–72

Wieser HG, 'Music and the brain. Lessons from brain diseases and some reflections on the "emotional" brain', *Annals of the New York Academy of Sciences*, 2003, 999, 76–94

Wiggins OP & Schwartz MA, 'Schizophrenia: a phenomenological-anthropological approach', in MD Chung, KWM Fulford & G Graham (eds), *Reconceiving Schizophrenia*, Oxford University Press, 2007, 113–27

Wigner E, *Symmetries and Reflections: Scientific Essays of Eugene P Wigner*, Indiana University Press, 1967

———, 'Remarks on the mind-body question' (1961), reprinted in JA Wheeler & WH Zurek (eds), *Quantum Theory and Measurement*, Princeton University Press, 1983, 168–81

Wilber K, *No Boundary: Eastern and Western Approaches to Personal Growth*, Shambhala, 2001

Wilczek F, 'On Absolute Units, III: Absolutely Not?', *Physics Today*, 2006, 59(5), 10–11

———, *The Lightness of Being: Mass, Ether, and the Unification of Forces*, Basic Books, 2008

———, *A Beautiful Question: Finding Nature's Deep Design*, Allen Lane, 2015

Wildgruber D, Pihan H, Ackermann H et al, 'Dynamic brain activation during processing of emotional intonation: influence of acoustic parameters, emotional valence, and sex', *NeuroImage*, 2002, 15(4), 856–69

Wilhelm R & Wilhelm H, *Understanding the I Ching: The Wilhelm Lectures on The Book of Changes*, trans Baynes CF, Princeton University Press, 1995

Wilkins AJ, Shallice T & McCarthy R, 'Frontal lesions and sustained attention', *Neuropsychologia*, 1987, 25(2), 359–65

Wilkins JS, 'Why is Darwin's theory so controversial?', 24 November 2012: evolvingthoughts.net/2012/11/24/why-is-darwins-theory-so-controversial (accessed 23 February 2021)

Wilkins MC, 'The effect of changed material on the ability to do formal syllogistic reasoning', *Archives of Psychology*, 1928, 102(16), 5–83

Wilkinson DT & Halligan PW, 'The effects of stimulus symmetry on landmark judgments in left and right visual fields', *Neuropsychologia*, 2002, 40(7), 1045–58

Willems RM, Labruna L, D'Esposito M et al, 'A functional role for the motor system in language understanding: evidence from theta-burst transcranial magnetic stimulation', *Psychological Science*, 2011, 22(7), 849–54

Williams D, 'The candle eater', *Breaking the Ice: Essays About Understanding*, Gronow Press, 1991, 97–113

Williams D, *Autism: An Inside-out Approach*, Jessica Kingsley Publishing, 1996

———, *Like Colour to the Blind: Soul Searching and Soul Finding*, Jessica Kingsley Publishing, 1999

———, *Exposure Anxiety – The Invisible Cage: An Exploration of Self-Protection Responses in the Autism Spectrum and Beyond*, Jessica Kingsley Publishing, 2003

Williams EF, Dunning D & Kruger J, 'The hobgoblin of consistency: algorithmic judgment strategies underlie inflated self-assessments of performance', *Journal of Personality and Social Psychology*, 2013, 104(6), 976–94

Williams VG, Tremont G & Blum AS, 'Musical hallucinations after left temporal lobectomy', *Cognitive and Behavioral Neurology*, 2008, 21(1), 38–40

Williamson T, *Vagueness*, Routledge, 1994

Willis J & Todorov A, 'First impressions: making up your mind after a 100-ms exposure to a face', *Psychological Science*, 2006, 17(7), 592–8

Wilson DS, Near D & Miller RR, 'Machiavellianism: a synthesis of the evolutionary and psychological literatures', *Psychological Bulletin*, 1996, 119, 285–99

Wilson TD, *Strangers to Ourselves: Discovering the Adaptive Unconscious*, Cambridge University Press, 2002

Wilson TD, Hodge SD & LaFleur SJ, 'Effects of introspecting about reasons: inferring attitude from accessible thoughts', *Journal of Personality and Social Psychology*, 1995, 69(1), 16–28

Wilson TD, Lisle DJ, Schooler JW et al, 'Introspecting about reasons can reduce postchoice satisfaction', *Personality and Social Psychology Bulletin*, 1993, 19(3), 331–9

Wilson TD & Schooler JW, 'Thinking too much: introspection can reduce the quality of preferences and decisions', *Journal of Personality and Social Psychology*, 1991, 60(2), 181–92

Winkielman P, Schwarz N & Belli RF, 'The role of ease of retrieval and attribution in memory judgments: judging your memory as worse despite recalling more events', *Psychological Science*, 1998, 9(2), 124–6

Winner E, 'The origins and ends of giftedness', *The American Psychologist*, 2000, 55(1), 159–69

Winner E, Brownell H, Happé F et al, 'Distinguishing lies from jokes: theory of mind deficits and discourse interpretation in right hemisphere brain-damaged patients', *Brain and Language*, 1998, 62(1), 89–106

Winner E & Gardner H, 'The comprehension of metaphor in brain-damaged patients', *Brain*, 1977, 100(4), 717–29

Wintersgill P, 'Music and melancholia', *Journal of the Royal Society of Medicine*, 1994, 87(12), 764–6

Wisdom JO, *The Metamorphosis of Philosophy*, Blackwell, 1947

Wise J, 'Boldt the great pretender', *BMJ*, 2013, 346:f1738

Wisniewski AB, 'Sexually-dimorphic patterns of cortical asymmetry, and the role for sex steroid hormones in determining cortical patterns of lateralization', *Psychoneuroendocrinology*, 1998, 23(5), 519–47

Witelson SF, 'Sex and the single hemisphere: specialization of the right hemisphere for spatial processing', *Science*, 1976, 193(4251), 425–7

Witelson SF, Beresh H & Kigar DL, 'Intelligence and brain size in 100 postmortem brains: sex, lateralization and age factors', *Brain*, 2006, 129(2), 386–98

Wittgenstein L, *Tractatus Logico-Philosophicus*, trans DF Pears & BF McGuinness, Routledge & Kegan Paul, 1961

———, *On Certainty*, ed GEM Anscombe & GH von Wright, trans D Paul & GEM Anscombe, Blackwell, 1975

———, *Philosophical Investigations*, Blackwell, 1976a

———, *Wittgenstein's Lectures on Foundations of Mathematics, Cambridge, 1939*, Cornell University Press, 1976b

———, *Notebooks 1914–1916*, Wiley, 1981

———, *Culture and Value*, ed GH von Wright & H Nyman, trans P Winch, University of Chicago Press, 1984 [*Vermischte Bemerkungen* 1977]

———, *Culture and Value*, ed GH von Wright & H Nyman, trans P Winch, revised edn, Wiley, 1998

Wittling W & Pflüger M, 'Neuroendocrine hemispheric asymmetries: salivary cortisol secretion during lateralized viewing of emotion-related and neutral films', *Brain and Cognition*, 1990, 14(2), 243–65

Wittling W, Block A, Genzel S et al, 'Hemisphere asymmetry in parasympathetic control of the heart', *Neuropsychologia*, 1998, 36(5), 461–8

Woese CR, 'A new biology for a new century', *Microbiology and Molecular Biology Reviews*, 2004, 68(2), 173–86

Wohlleben P, *The Hidden Life of Trees: What They Feel, How They Communicate*, William Collins, 2017

Wohlschläger A, 'Visual motion priming by invisible actions', *Vision Research*, 2000, 40(8), 925–30

Wolańczyk T, Komender J & Brzozowska A, 'Catatonic syndrome preceded by symptoms of anorexia nervosa in a 14-year-old boy with arachnoid cyst', *European Child and Adolescent Psychiatry*, 1997, 6(3), 166–9

Wolf YI & Koonin EV, 'On the origin of the translation system and the genetic code in the RNA world by means of natural selection, exaptation, and subfunctionalization', *Biology Direct*, 2007, 2, article 14

Wolff C, *Bach: Essays on His Life and Music*, Harvard University Press, 1994

Wolff G & McKenzie K, 'Capgras, Fregoli and Cotard's syndromes and Koro in folie à deux', *British Journal of Psychiatry*, 1994, 165(6), 842

Wolff S, '"Schizoid" personality in childhood and adult life, 1: The vagaries of diagnostic labelling', *British Journal of Psychiatry*, 1991, 159(5), 615–20

———, *Loners: The Life Path of Unusual Children*, Routledge, 1995

Wolford G, Miller MB & Gazzaniga MS, 'The left hemisphere's role in hypothesis formation', *Journal of Neuroscience*, 2000, 20: RC64, 1–4

Wong CW, Ko SF & Wai YY, 'Arachnoid cyst of the lateral ventricle manifesting positional psychosis', *Neurosurgery*, 1993, 32(5), 841–3

Woo PY, Leung LN, Cheng ST et al, 'Monoaural musical hallucinations caused by a thalamocortical auditory radiation infarct: a case report', *Journal of Medical Case Reports*, 2014, 8, 400

Wood AM & Joseph S, 'The absence of positive psychological (eudemonic) well-being as a risk factor for depression: a ten year cohort study', *Journal of Affective Disorders*, 2010, 122(3), 213–7

Wood AM, Kaptoge S, Butterworth AS et al, 'Risk thresholds for alcohol consumption: combined analysis of individual-participant data for 599 912 current drinkers in 83 prospective studies', *The Lancet*, 2018, 391(10129), 1513–23

Woodger JH, *Biological Principles*, Routledge, 1929

Woodley of Menie MA & Fernandes HBF, 'Showing their true colours: possible secular declines and a Jensen effect on colour acuity — more evidence for the weaker variant of Spearman's Other Hypothesis', *Personality and Individual Differences*, 2016, 88, 280–4

Woodley of Menie MA, Fernandes HBF, José Figueredo A et al, 'By their words ye shall know them: evidence of genetic selection against general intelligence and concurrent environmental enrichment in vocabulary usage since the mid 19th century', *Frontiers in Psychology*, 2015, 6, article 361

Woodley of Menie MA & Figueredo AJ, *Historical Variability in Heritable General Intelligence: Its Evolutionary Origins and Socio-Cultural Consequences*, The University of Buckingham Press, 2013

Woodley of Menie MA & Meisenberg G, 'In the Netherlands the anti-Flynn effect is a Jensen effect', *Personality and Individual Differences*, 2013, 54(8), 871–6

Woodley of Menie MA, te Nijenhuis J & Murphy R, 'Were the Victorians cleverer than us? The decline in general intelligence estimated from a meta-analysis of the slowing of simple reaction time', *Intelligence*, 2013, 41(6), 843–50

———, 'Is there a dysgenic secular trend towards slowing simple reaction time? Responding to a quartet of critical commentaries', *Intelligence*, 2014, 46(1), 131–47

Woodman GF, 'A brief introduction to the use of event-related potentials (ERPs) in studies of perception and attention', *Attention, Perception & Psychophysics*, 2010, 72(8), 2031–46

Woodruff PW, McManus IC & David AS, 'Meta-analysis of corpus callosum size in schizophrenia', *Journal of Neurology, Neurosurgery, and Psychiatry*, 1995, 58(4), 457–61

Woodruff PW, Pearlson GD, Geer MJ et al, 'A computerized magnetic resonance imaging study of corpus callosum morphology in schizophrenia', *Psychological Medicine*, 1993, 23(1), 45–56

Woods AJ, Hamilton RH, Kranjec A et al, 'Space, time, and causality in the human brain', *NeuroImage*, 2014, 92, 285–97

Wordsworth W, *The Prose Works of William Wordsworth*, Edward Moxon, Son & Co, 3 vols, 1876

Workman L, Chilvers L, Yeomans H et al, 'Development of cerebral lateralisation for recognition of emotions in chimeric faces in children aged 5 to 11', *Laterality*, 2006, 11(6), 493–507

Workman L, Peters S & Taylor S, 'Lateralisation of perceptual processing of pro- and anti-social emotions displayed in chimeric faces', *Laterality*, 2000, 5(3), 237–49

Worley MM & Boles DB, 'The face is the thing: faces, not emotions, are responsible for chimeric perceptual asymmetry', *Laterality*, 2016, 21(4–6), 672–88

Wotherspoon T & Hübler A, 'Adaptation to the edge of chaos with random-wavelet feedback', *Journal of Physical Chemistry A*, 2009, 113(1), 19–22

Wraw C, Deary IJ, Der G et al, 'Intelligence in youth and mental health at age 50', *Intelligence*, 2016, 58, 69–79

Wright AA, Magnotti JF, Katz JS et al, 'Corvids outperform pigeons and primates in learning a basic concept', *Psychological Science*, 2017, 28(4), 437–44

Wright C, 'Narrative of the image: a correspondence with Charles Simic', *Quarter Notes: Improvisations and Interviews*, University of Michigan Press, 1995

Wright D, Makin ADJ & Bertamini M, 'Right-lateralised alpha desynchronization during regularity discrimination: hemispheric specialization or directed attention?', *Psychophysiology*, 2015, 52(5), 638–47

———, 'Electrophysiological responses to symmetry presented in the left or in the right visual hemifield', *Cortex*, 2017, 86, 93–108

Wright NT, 'Can a scientist trust the New Testament?', The James Gregory Lecture, 17 February 2014: www.jamesgregory.org.uk/wp-content/uploads/2013/11/James-Gregory-Feb-17.pdf

Wright S, Young AW & Hellawell DJ, 'Sequential Cotard and Capgras delusions', *British Journal of Clinical Psychology*, 1993, 32(3), 345–9

Wu L, Wang D & Evans JA, 'Large teams develop and small teams disrupt science and technology', *Nature*, 2019, 566(7744), 378–82

Wu TQ, Miller ZA, Adhimoolam B et al, 'Verbal creativity in semantic variant primary progressive aphasia', *Neurocase*, 2015, 21(1), 73–8

Wuerfel J, Krishnamoorthy ES, Brown RJ et al, 'Religiosity is associated with hippocampal but not amygdala volumes in patients with refractory epilepsy', *Journal of Neurology, Neurosurgery, and Psychiatry*, 2004, 75(4), 640–2

Wulff DH, *Psychology of Religion: Classical and Contemporary*, John Wiley and Sons, 1996

Wyman MA, 'Whitehead's philosophy of science in the light of Wordsworth's poetry', *Philosophy of Science*, 1956, 23(4), 283–96

Wyndham Lewis P, *Tarr*, Egoist Press, 1918

Wyrick RA & Wyrick LC, 'Time experience during depression', *Archives of General Psychiatry*, 1977, 34(12), 1441–3

Wyrsch J, 'Klinik der Schizophrenie', in E Gruhle, R Jung, R Mayer-Gross et al (eds), *Psychiatrie der Gegenwart*, Springer, 1949

Xu J, Kemeny S, Park G et al, 'Language in context: emergent features of word, sentence, and narrative comprehension', *NeuroImage*, 2005, 25(3), 1002–15

Yalin Ş, Taş FV & Güvenir T, 'The coexistence of Capgras, Fregoli and Cotard's syndromes in an adolescent case', *Archives of Neuropsychiatry*, 2008, 45, 149–51

Yamadori A, Mori E, Tabuchi M et al, 'Hypergraphia: a right hemisphere syndrome', *Journal of Neurology, Neurosurgery, & Psychiatry*, 1986, 49(10), 1160–4

Yamaguchi S, Yamagata S & Kobayashi S, 'Cerebral asymmetry of the "top-down" allocation of attention to global and local features', *The Journal of Neuroscience*, 2000, 20(9), RC72

Yamakawa H, Takenaka K, Sumi Y et al, 'Intracranial bullet retained since the Sino-Japanese war manifesting as hallucination – case report', *Neurologia Medico-Chirurgica (Tokyo)*, 1994, 34(7), 451–4

Yamamoto T, Koashi M, Ozdemir SK et al, 'Experimental extraction of an entangled photon pair from two identically decohered pairs', *Nature*, 2003, 421(6921), 343–46

Yamazaki Y, Aust U, Huber L et al, 'Lateralized cognition: asymmetrical and complementary strategies of pigeons during discrimination of the "human concept"', *Cognition*, 2007, 104(2), 315–44

Yanagisawa K, Masui K, Furutani K et al, 'Does higher general trust serve as a psychosocial buffer against social pain? An NIRS study of social exclusion', *Social Neuroscience*, 2011, 6(2), 190–7

Yang J, 'The role of the right hemisphere in metaphor comprehension: a meta-analysis of functional magnetic resonance imaging studies', *Human Brain Mapping*, 2014, 35(1), 107–22

Yang J, Bellgowan PS & Martin A, 'Threat, domain-specificity and the human amygdala', *Neuropsychologia*, 2012, 50(11), 2566–72

Yang X, 'A wheel of time: the circadian clock, nuclear receptors, and physiology', *Genes and Development*, 2010, 24(8), 741–7

Yang Y, Raine A, Colletti P et al, 'Abnormal temporal and prefrontal cortical gray matter thinning in psychopaths', *Molecular Psychiatry*, 2009, 14(6), 561–2

Yang Z, Oathes DJ, Linn KA et al, 'Cognitive behavioral therapy is associated with enhanced cognitive control network activity in major depression and posttraumatic stress disorder', *Biological Psychiatry: Cognitive Neuroscience and Neuroimaging*, 2017, 3(4), 311–9

Ye Z & Zhou X, 'Conflict control during sentence comprehension: fMRI evidence', *NeuroImage*, 2009, 48(1), 280–90

Yeh Y, Lin C-W, Hsu W-C et al, 'Associated and dissociated neural substrates of aesthetic judgment and aesthetic emotion during the appreciation of everyday designed products', *Neuropsychologia*, 2015, 73, 151–60

Yeh ZT & Tsai CF, 'Impairment on theory of mind and empathy in patients with stroke', *Psychiatry and Clinical Neurosciences*, 2014, 68(8), 612–20

Yehuda R, Daskalakis NP, Bierer LM et al, 'Holocaust exposure induced intergenerational effects on FKBP5 methylation', *Biological Psychiatry*, 2015, 80(5), 372–80

Yellott, JI, 'Probability learning with noncontingent success', *Journal of Mathematical Psychology*, 1969, 6(3), 541–75

Yin J, Cao Y, Yong H-L et al, 'Lower bound on the speed of nonlocal correlations without locality and measurement choice loopholes', *Physical Review Letters*, 2013, 110(26), 260407

Yirmiya N & Charman T, 'The prodrome of autism: early behavioral and biological signs, regression, peri- and post-natal development and genetics', *Journal of Child Psychology & Psychiatry*, 2010, 51(4), 432–58

Yokoyama S, Miyamoto T, Riera J et al, 'Cortical mechanisms involved in the processing of verbs: an fMRI study', *Journal of Cognitive Neuroscience*, 2006, 18, 1304–13

Yokoyama S, Okamoto H, Miyamoto T et al, 'Cortical activation in the processing of passive sentences in L1 and L2: an fMRI study', *NeuroImage*, 2006, 30, 570–9

Yokoyama S, Watanabe J, Iwata K et al, 'Is Broca's area involved in the processing of passive sentences? An event-related fMRI study', *Neuropsychologia*, 2007, 45(5), 989–96

Yolken RH, Dickerson FB & Fuller Torrey E, 'Toxoplasma and schizophrenia', *Parasite Immunology*, 2009, 31(11), 706–15

Yomogida Y, Sugiura M, Watanabe J et al, 'Mental visual synthesis is originated in the fronto-temporal network of the left hemisphere', *Cerebral Cortex*, 2004, 14(12), 1376–83

Yong E, 'Let slime moulds do the thinking!' *The Guardian*, 8 September 2010: www.theguardian.com/science/blog/2010/sep/08/slime-mould-physarum (accessed 23 February 2021)

———, 'A brainless slime that shares memories by fusing', *The Atlantic*, 21 December 2016: www.theatlantic.com/science/archive/2016/12/the-brainless-slime-that-can-learn-by-fusing/511295/?utm_source=eb (accessed 23 February 2021)

Yoo SD, Kim DH, Jeong YS et al, 'Atypical supernumerary phantom limb and pain in two patients with pontine hemorrhage', *Journal of Korean Medical Science*, 2011, 26(6), 844–7

Yoruk S & Runco MA, 'The neuroscience of divergent thinking', *Activitas Nervosa Superior*, 2014, 56(1/2), 1–16

Yoshii F, Barker WW, Chang JY et al, 'Sensitivity of cerebral glucose metabolism to age, gender, brain volume, brain atrophy, and cerebrovascular risk factors', *Journal of Cerebral Blood Flow and Metabolism*, 1988, 8(5), 654–61

Yoshimura N & Otsuki M, 'A case of topographic disorientation without right occipital lesion', *Brain and Nerve*, 2002, 54(7), 601–4

Yoshimura M, Uchiyama Y, Kaneko A et al, 'Formed visual hallucination after excision of the right temporo parietal cystic meningioma – a case report', *Brain and Nerve*, 2010, 62(8), 893–7

Young AW, 'Wondrous strange: the neuropsychology of abnormal beliefs', *Mind & Language*, 2000, 15(1), 47–73

Young AW, Bion PJ & McWeeny KH, 'Age and sex differences in lateral asymmetries to visual and tactile stimuli', in A Glass (ed), *Individual Differences in Hemispheric Specialisation*, Plenum Press, 1987, 215–31

Young AW & Ellis AW, 'Perception of numerical stimuli felt by fingers of the left and right hands', *Quarterly Journal of Experimental Psychology*, 1979, 31(2), 263–72

Young AW, Flude BM & Ellis AW, 'Delusional misidentification incident in a right hemisphere stroke patient', *Behavioural Neurology*, 1991, 4, article 316241

Young AW & Leafhead KM, 'Betwixt life and death: case studies of the Cotard delusion', in PW Halligan & JC Marshall (eds), *Method in Madness: Case Studies in Cognitive Neuropsychiatry*, Psychology Press, 1996, 147–71

Young AW, Robertson IH, Hellawell DJ et al, 'Cotard delusion after brain injury', *Psychological Medicine*, 1992, 22(3), 799–804

Young CE, 'Intuition and nursing process', *Holistic Nursing Practice*, 1987, 1(3), 52–62

Young L, Camprodon JA, Hauser M et al, 'Disruption of the right temporoparietal junction with transcranial magnetic stimulation reduces the role of beliefs in moral judgments', *Proceedings of the National Academy of Sciences of the United States of America*, 2010, 107(15), 6753–8

Young L, Cushman F, Adolphs R et al, 'Does emotion mediate the effect of an action's moral status on its intentional status? Neuropsychological evidence', *Journal of Cognition and Culture*, 2006, 6(1–2), 291–304

Young L, Cushman F, Hauser M et al, 'The neural basis of the interaction between theory of mind and moral judgment', *Proceedings of the National Academy of Sciences of the United States of America*, 2007, 104(20), 8235–40

Younger SD & Zongo J-B, 'West Africa: the Onchocerciasis Control Program', in *Successful Development in Africa* (EDI Development Policy Case Series, Analytical Case Studies, 1), World Bank, 1989

Yourgrau P, *A World Without Time: The Forgotten Legacy of Gödel and Einstein*, Basic Books, 2005

Yu F, Jiang Q-J, Sun X-Y et al, 'A new case of complete primary cerebellar agenesis: clinical and imaging findings in a living patient', *Brain*, 2015, 138(6), e353

Yuasa S, Kurachi M, Suzuki M et al, 'Clinical symptoms and regional cerebral blood flow in schizophrenia', *European Archives of Psychiatry and Clinical Neuroscience*, 1995, 246(1), 7–12

Yücel M, Fornito A, Youssef G et al, 'Inhibitory control in young adolescents: the role of sex, intelligence, and temperament', *Neuropsychology*, 2012, 26(3), 347–56

Yuvaraj R, Yuvaraj R, Murugappan M et al, 'Review of emotion recognition in stroke patients', *Dementia and Geriatric Cognitive Disorders*, 2013, 36(3–4), 179–96

Zagvazdin Y, 'Stroke, music, and creative output: Alfred Schnittke and other composers', in *Music, Neurology, and Neuroscience: Historical Connections and Perspectives (Progress in Brain Research)*, Elsevier, 2015, 149–65

Zahn R, Garrido G, Moll J et al, 'Individual differences in posterior cortical volume correlate with proneness to pride and gratitude', *Social Cognitive and Affective Neuroscience*, 2014, 9(11), 1676–83

Zahn-Waxler C & Robinson J, 'Empathy and guilt: early origins of feelings of responsibility', in JP Tangney & KW Fischer (eds), *Self-conscious Emotions*, Guilford Press, 1995, 143–73

Zaidel DW, 'View of the world from a split-brain perspective', in E Critchley (ed), *Neurological Boundaries of Reality*, Farrand Press, 1994, 161–74

———, 'Split-brain, the right hemisphere, and art: fact and fiction', *Progress in Brain Research*, 2013, 204, 3–17

———, *The Neuropsychology of Art: Neurological, Cognitive, and Evolutionary Perspectives*, Psychology Press, 2nd edn, 2015

———, 'Braque and Kokoschka: brain tissue injury and preservation of artistic skill', *Behavioral Sciences (Basel)*, 2017, 7(3), 56

Zaidel DW, Aarde SM & Baig K, 'Appearance of symmetry, beauty, and health in human faces', *Brain and Cognition*, 2005, 57(3), 261–3

Zaidel DW & Cohen JA, 'The face, beauty, and symmetry: perceiving asymmetry in beautiful faces', *International Journal of Neuroscience*, 2005, 115(8), 1165–73

Zaidel DW & Deblieck C, 'Attractiveness of natural faces compared to computer constructed perfectly symmetrical faces',

International Journal of Neuroscience, 2007, 117(4), 423–31

Zaidel DW & Hessamian M, 'Asymmetry and symmetry in the beauty of human faces', *Symmetry*, 2010, 2(1), 136–49

Zaidel DW & Kasher A, 'Hemispheric memory for surrealistic versus realistic paintings', *Cortex*, 1989, 25(4), 617–41

Zaidel E, Clarke J & Suyenbu B, 'Hemispheric independence: a paradigm case for cognitive neuroscience', in AB Scheibel & AF Wechsler (eds), *Neurobiology of Higher Cognitive Function*, Guilford Press, 1990, 297–355

Zaidel E, Iacoboni M, Zaidel DW et al, 'The callosal syndromes', in KM Heilman & E Valenstein (eds), *Clinical Neuropsychology*, Oxford University Press, 4th edn, 2003, 347–403

Zaimov K, Kitov D & Kolev N, 'Aphasie chez un peintre', *L'Encéphale* 1969, 58(5), 377–417

Zajkowski WK, Kossut M & Wilson RC, 'A causal role for right frontopolar cortex in directed, but not random, exploration', *eLife*, 2017, 6, e27430

Zajonc RB, 'On the primacy of affect', *The American Psychologist*, 1984, 39(2), 117–23

Zajonc RB, Pietromonaco P & Bargh J, 'Independence and interaction of affect and cognition', in MS Clark and S Fiske (eds), *Affect and Cognition: the 17th Annual Carnegie Symposium on Cognition*, Lawrence Erlbaum, 1982

Zaki J & Mitchell JP, 'Intuitive prosociality', *Current Directions in Psychological Science*, 2013, 22(6), 466–70

Zald DH & Pardo JV, 'Emotion, olfaction, and the human amygdala: amygdala activation during aversive olfactory stimulation', *Proceedings of the National Academy of Sciences of the United States of America*, 1997, 94(8), 4119–24

——, 'Functional neuroimaging of the olfactory system in humans', *International Journal of Psychophysiology*, 2000, 36(2), 165–81

Zammit S, Allebeck P, David AS et al, 'A longitudinal study of premorbid IQ score and risk of developing schizophrenia, bipolar disorder, severe depression, and other nonaffective psychoses', *Archives of General Psychiatry*, 2004, 61(4), 354–60

Zander T, Öllinger M & Volz KG, 'Intuition and insight: two processes that build on each other or fundamentally differ?', *Frontiers in Psychology*, 2016, 7, article 1395

Zárate MA, Stoever CJ, MacLin MK et al, 'Neurocognitive underpinnings of face perception: further evidence of distinct person and group perception processes', *Journal of Personality and Social Psychology*, 2008, 94(1), 108–15

Zaretskaya N, Thielscher A, Logothetis NK et al, 'Disrupting parietal function prolongs dominance durations in binocular rivalry', *Current Biology*, 2010, 20(23), 2106–11

Zatorre RJ, Belin P & Penhune VB, 'Structure and function of auditory cortex: music and speech', *Trends in Cognitive Sciences*, 2002, 6(1), 37–46

Zatorre RJ & Jones-Gotman M, 'Right-nostril advantage for discrimination of odors', *Perception & Psychophysics*, 1990, 47(6), 526–31

Zatorre RJ, Jones-Gotman M, Evans AC et al, 'Functional localization and lateralization of human olfactory cortex', *Nature*, 1992, 360(6402), 339–40

Zatorre RJ, Jones-Gotman M & Rouby C, 'Neural mechanisms involved in odor pleasantness and intensity judgments', *NeuroReport*, 2000, 11(12), 2711–6

Zedelius CM & Schooler JW, 'The richness of inner experience: relating styles of daydreaming to creative processes', *Frontiers in Psychology*, 2016, 6, article 2063

Zeh HD, 'Remarks on the compatibility of opposite arrows of time' and 'Remarks on the compatibility of opposite arrows of time II', in S Albeverio & P Blanchard, *Direction of Time*, Springer, 2014, 265–73 & 287–92

Zeki S, 'Artistic creativity and the brain', *Science*, 2001, 293(5527), 51–2

Zeki S & Shipp S, 'The functional logic of cortical connections', *Nature*, 1988, 335(6188), 311–7

Zeman A, *Consciousness: A User's Guide*, Yale University Press, 2004

Zeman A, Milton F, Smith A et al, 'By heart: an fMRI study of brain activation by poetry and prose', *Journal of Consciousness Studies*, 2013, 20(9–10), 132–58

Zervos C, 'Conversation avec Picasso', *Cahiers d'Art*, 1935, trans and ed B Ghiselin, *The Creative Process*, 48–53

Zetzsche T, Preuss UW, Frodl T et al, 'White matter alterations in schizophrenic patients with pronounced negative symptomatology and with positive family history for schizophrenia', *European Archives of Psychiatry and Clinical Neuroscience*, 2008, 258(5), 278–84

Zhang L, Shu H, Zhou F et al, 'Common and distinct neural substrates for the perception of speech rhythm and intonation', *Human Brain Mapping*, 2010, 31(7), 1106–16

Zhao D, Wang Y, Han K et al, 'Does target animacy influence manual laterality of monkeys? First answer from northern pig-tailed macaques (*Macaca leonina*)', *Animal Cognition*, 2015, 18(4), 931–6

Zhao Q, Zhou Z, Xu H et al, 'Neural pathway in the right hemisphere underlies verbal insight problem solving', *Neuroscience*, 2014, 256, 334–41

Zhaoping L & Guyader N, 'Interference with bottom-up feature detection by higher-level object recognition', *Current Biology*, 2007, 17(1), 26–31

Zhavoronkova LA & Trofimova EV, 'Coherence dynamics of EEG and motor reactions while falling asleep in dextrals and sinistrals: communication. I. Analysis of interhemispheric correlations', *Human Physiology*, 1997, 23(6), 18–26

———, 'Dynamics of EEG coherence in right-handers and left-handers when falling asleep. II. An analysis of interhemispheric relations', *Human Physiology*, 1998, 24(1), 32–9

Zhu X, Luo Y, Huang Y et al, 'Unlocking the underlying information of golden ratio in the Hogarth curves', *International Journal of Clothing Science and Technology*, 2018, 30(6), 738–46

Zihl J, von Cramon D & Mai N, 'Selective disturbance of movement vision after bilateral brain damage', *Brain*, 1983, 106(2), 313–40

Zilles K, Dabringhaus A, Geyer S et al, 'Structural asymmetries in the human forebrain and the forebrain of non-human primates and rats', *Neuroscience and Biobehavioral Reviews*, 1996, 20(4), 593–605

Zingerle H, 'Ueber Störungen der Wahrnehmung des eigenen Körpers bei organischen Gehirnerkrankungen', *Monatsschrift für Psychiatrie und Neurologie*, 1913a, 34(1), Pt 1, 13–24

———, 'Ueber Störungen der Wahrnehmung des eigenen Körpers bei organischen Gehirnerkrankungen', *Monatsschrift für Psychiatrie und Neurologie*, 1913b, 34(1), Pt 2, 25–36

Zmigrod L, Rentfrow PJ & Robbins TW, 'The partisan mind: is extreme political partisanship related to cognitive inflexibility?', *Journal of Experimental Psychology: General*, 2020, 149(3), 407–18

Zmigrod S, Colzato LS & Hommel B, 'Stimulating creativity: modulation of convergent and divergent thinking by transcranial direct current stimulation (tDCS)', *Creativity Research Journal*, 2015, 27(4), 353–60

Zohar: see Lachower & Tishby 1991

Zorzi M, Priftis K & Umiltà C, 'Brain damage: neglect disrupts the mental number line', *Nature*, 2002, 417(6885), 138–9

Zubiaurre-Elorza L, Junque C, Gómez-Gil E et al, 'Effects of cross-sex hormone treatment on cortical thickness in transsexual individuals', *Journal of Sexual Medicine*, 2014, 11(5), 1248–61

Zubler F, Seeck M, Landis T et al, 'Contralateral medial temporal lobe damage in right but not left temporal lobe epilepsy: a ^{1}H magnetic resonance spectroscopy study', *Journal of Neurology, Neurosurgery & Psychiatry*, 2003, 74(9), 1240–4

Zucco GM & Tressoldi PE, 'Hemispheric differences in odour recognition', *Cortex*, 1989, 25(4), 607–15

Zucker NL, Losh M, Bulik CM et al, 'Anorexia nervosa and autism spectrum disorders: guided investigation of social cognitive endophenotypes', *Psychological Bulletin*, 2007, 133(6), 976–1006

Zuckerman M, Silberman J & Hall JA, 'The relation between intelligence and religiosity: a meta-analysis and some proposed explanations', *Personality and Social Psychology Review*, 2013, 17(4), 325–54

Zukav G, *The Dancing Wu Li Masters*, Rider & Co, 1979

Zurif E, Caramazza A, Myerson R et al, 'Semantic feature representations for normal and aphasic language', *Brain and Language*, 1974, 1(2), 167–87

Zwaan R, 'The immersed experiencer: toward an embodied theory of language comprehension', in BH Ross (ed), *Psychology of Learning and Motivation*, vol 44, Academic Press, 2004, 35–62

Zwicky J, *Lyric Philosophy*, Brush Education, 2014a

———, *Wisdom & Metaphor*, Brush Education, 2014b

———, *The Experience of Meaning*, McGill Queen's University Press, 2019

Zwijnenburg PJG, Wennink JMB, Laman DM et al, 'Alice in Wonderland syndrome: a clinical presentation of frontal lobe epilepsy', *Neuropediatrics*, 2002, 33(1), 53–5

Index of Topics

abstraction, from context 116, 339, 371–3, 571–9, 643, 749, 851, 869, 903; *and see* embodiment
affordances 452–3
aggadah and *halakhah* 1229–30, 1266
akinetopsia, *see* palinopsia
akrasia 837
alexia 190
Alice in Wonderland syndrome 118, 125–128, 129 n 169
Alzheimer's type dementia 116, 137, 141, 802, 1087
anamnesis 607, 683, 1227, 1269
animacy/inanimacy; *see* devitalisation
anomaly detector, RH as 28, 255, 260
anorexia nervosa, RH deficits 145; and autism 321, 1011
anosodiaphoria 89, 138
anosognosia 89, 93, 138, 149–51
antifragility 828, 1034
Apollo 11 mission and creativity 282–3, 288–9
apotemnophilia, *see* xenomelia
appetite, disorders of 145; gourmand syndrome 145–6
apraxia, lateralisation in ideomotor 182–3; ideational 183; dressing 183; constructional 183, 212, 269
archetypes 596, 599, 626, 683–4, 713; *and see* uniqueness and generality
asomatognosia 92–3, 138
Asperger's syndrome 616–7, 988; RH deficit syndromes and 322; *see also* autism
asymmetry, as foundational 840, 1028–35 and *passim*; in living world 837–40; cerebral 25ff and *passim*: pathological absence of 26, 317; *coincidentia oppositorum* asymmetry of 832–7, 841; asymmetry of symmetry and asymmetry 1034
attention 17–27; lateralisation in animals 23–6; RH dominance for attention in general 68; hierarchy

of 98; nature of RH 68–9, nature of LH 69–73, and *passim*; local processing, also carried on in RH 258; detail-focussed attention, distinguished from awareness of anomalous detail 259; attentional switching 114; meta-control centre in midbrain 32 n 76; contrast between attentional and perceptual deficits 73ff; attention deficit and hyperactivity disorder 155, 319 n 82
autism
—not a unitary condition 307 n 7
—in relation to RH deficits: phenomenology 199–203, 321–2, 328–330, 350–1, 354, 760; brain correlates 101, 222, 307–8, 323–9, 983; corpus callosum abnormalities, including agenesis in 323–4; RH release syndromes in 274–5; *and see* savants
—lateralising characteristics: literal-mindedness in 350–1, 354; devitalisation in 101, 321, 351, 354; narrative difficulties in 365; philosophical style and 616–8, 662; lack of emotional depth 616; quantification, excessive in 364–5; systemisation, excessive in 350–1; utility, focus on 615; loss of fluidity in time, cognition and motion 588, 986–90, 1010; atheism and 1279, 1372; third-person perspective 322
—features in common with schizophrenia 305–370 *passim*, esp 305–7; relation to anorexia nervosa 321
autoscopy 127, 146
autotopagnosia 98, 188–90

Bǎojìng Sānmèi (*Hōkyō Zanmai*) 1219
beauty 596, 1146–64, lateralisation and 1349–56; guide to truth in maths and science 609, 754, 757–60, 845–7, 977; and *phi* (ϕ), *see*

golden ratio; 1121–9 *passim*, 1146–65; 'useless' 1168–72; Darwin and 1184; and the divine 1259–60, 1289; and the *coincidentia* 1299–1300, 1323; lateralisation in 1349–56
'beginner's mind' 562
Begriffsschrift 571
belief bias, laterality in 171–2
Bereitschaftspotential 1082–3
'betweenness' 333–6, 367, 402, 459, 550–1, 635, 847, 994, 999, 1004, 1009, 1027, 1039, 1063, 1112, 1119 n 379, 1211, 1219, 1237
Bhagavad Gita 878, 1231
bi-stable percepts 114
binocular rivalry 114–5
blindsight 697–8
Buddha 1000, 1208, 1283; Buddhism: 604, 875, 996, 999, 1146, 1201, 1209, 1213, 1234, 1237, 1246, 1252, 1255, 1278 n 297, 1283, 1297, 1302, 1333, 1375; *and see* Zen

calculation *vs* judgment, *see under* judgment
callosal agenesis 110, 281, 319 n 77; and autism 323–4
Capgras' syndrome 97, 140, 143, 144 nn62, 65, 193–4, 309, 366, 655, 657, 844, 904
causation in scientific thought 252; distinct meanings 420; in organisms 447–9, 471, 475–83 *passim*; LH and attribution of 152; linear 601–2; top-down 103, 440, 455, 465, 1175; significance of scale in 916–8; and time 913–8
cells, cognition in 435–6, 468, 1068–74; systemic behaviour in 482–3
cerebellum 278, 983–6, 1067; abnormalities in autism 983
cerebral cortex, mammalian, evolution of 54–66
chengyu riddles 257
chess, unconscious processing in 252, 695, 700–703, 728
Christianity 1193–1304 *passim*; 1201, 1268–9, 1290–8, 1301;

1561

[Christianity, *cont'd*] Orthodox 1220; mysticism in 1234ff
cognitive biases 729–32
commissures, anterior and posterior 56–8, 324
common sense, LH and failure of 313, 351–2, 619, 746–8, 754; implying a sense of proportion, and capacity to accept uncertainty 746–7; *see also* humour
complexity, as prior to simplicity 8; intuition better than rationality in 674, 692–3, 750; RH better at handling 216
confabulation 91–3, 154–8, 698
conformism, LH and 172, 540
Confucianism 860, 973, 1375; *and see* Confucius *in the Index of Names*
consciousness 1037–1120 *passim*; ontologically primary 1039–47, 1059–62; relationship to matter 1047–59, 1098ff; and life 1062; and neurones 1066–85; free will 1082–5; 'permission' and 1085–9; open and closed systems and 1094–8; individual and universal 1102–4; hemisphere hypothesis and 1104–19
consistency *vs* experience 400–1; *and see* perfection
context *passim* throughout; lateralisation in appreciation of 178, 216, 625; science and neglect of 418, 510; in organism 438–66, 500; reason and 548, 572, 577, 586, 624; intuition and 698; 1272, 1307, 1312
corpus callosum, origins of in mammals 57–63
Cotard's syndrome 122, 135 n 2, 140, 142–4, 149, 309
cradling bias 209
creativity 158, 239–304; problems in measuring 240; ways of analysing 240; as a process 815–6; temporal phases 240–2; generative factors 242–5; permissive factors 246–8; translational factors 248–50; lateralisation in 245–9, 781–91; part played by neuronal architecture in 255; relationship between hemispheres in 280; associations, impoverishment of in RH lesions 213; divergent thinking in 242ff; Gestalt perception in 250; intelligence and 245; imagination and 250–4, 764ff; insight and 254–7, 753–7; role of analysis in 250; the unconscious in 252–8; posteriority *vs* anteriority in 254; in healthy artists and composers 277–8; effect of lateralised suppression or stimulation 274–6; after lateralised brain insult, in artists 261–71, composers 271–3, and poets 273–4; mental illness and 292–303; lesion studies in 259–276; EEG and imaging studies in 279–80, 781–791; cerebellum in 278; Goethe on 251
crying, laterality in sensitivity to 204
cryptic crosswords and insight 256

~

de Clérambault's syndrome 142
Deborah number 490 n 285
decussation 21 n 20
déjà vu, déjà vécu, lateralisation in 196
delusions, lateralisation of 117–8, 135–49; somatic delusions in schizophrenia and depression, laterality in 314–5; *and see individual syndromes*
denial 88–9, 172, 474, 530–1, 540, 619, 1315; *see also* anosognosia *and* somatoparaphrenia
depression, lateralisation in 90, 135 n 2, 151–2; time perception in 152, 1337–41
devitalisation, in RH dysfunction 99–103, 209; in schizophrenia and autism 101, 336–345, 350, 360; *see also* model, machine
dhat 149
diaschisis 325
dipoles 64 and *passim*
DNA, a resource not a deterministic agent 437–457
dyschronia 79
dysmegalopsia, *see* metamorphopsia
dysmelia, *see* xenomelia
dysmetropsia, *see* metamorphopsia

~

Einstellung effect 169, 611
élan vital 344, 350 n 306; misunderstandings of 892, 923, 974
embodiedness: embodiment *vs* abstraction 571–82, 769–70; lateralisation of: in general 28–9, awareness of body 205, facial expressivity and receptivity 205–9, bodily gestures 207, body schema 145
'emergence' 440–1, 1044–5 and n 30, 1067, 1075, 1198
emotion: and the RH 87–91; approach/withdrawal model, alternatives to 204–7; emotional depth 87–92, 206, loss of 213; emotional intelligence, lateralisation in 197ff, incl animals 199; emotional receptivity and expressivity, lateralisation in 203–15; disgust, lateralisation in 147, 204, 206, 215, 301; anger, lateralisation in 197–9, 204, 206, 223, 475, 531; empathy, lateralisation in 68, 201ff; empathy *vs* rivalry 206; sadness and relation to empathy 197; sadness, loss of capacity for 215, in modernity 1322–3
entanglement 913–16, 1026, 1077; over time 928
Epicurus, on will 1060; Epicureanism 1140
epigenetics 161, 438–9, 451–61, 630, 681–2, 1176, 1182
epileptic abnormalities, interpretation of 194
erotomania, *see* de Clérambault's syndrome
expertise and *Gestalt* perception 698–706; adverse effects of explicit procedures on 702–5; misrepresentation of 724–9, 742–3
exploration, *vs* grasping, lateralisation evidenced in release phenomena 183–6

~

fact-value distinction 416–9

Fibonacci series 1030–1, 1159, pl 22[d–e]
flexibility *vs* rigidity, lateralisation in 44, 72, 173, 178, 212–4, 260–1, 274, 311, 645, 789; in reason 548, 698; in creativity 242–60, 277
flow and motion 945–96 *passim*; foundational nature 8, 978–80, 993–6; dependency for appreciation on RH 83–7, 945–52, 989–90; correlate of consciousness 953–5; conceptual thought and resistance to 954; effect of language on 955ff; stream of life model, 431, 487–99; flow as metaphor of life, music as 968–71, water as 967–75, *li* as 973–5; cognition and 980–5, disturbance in schizophrenia and autism 986–9; flow *vs* linearity 976–8 and *passim*; effects of scale and perspective on reception of 490–3; necessary resistance and, see separate entry
form, nature of 8–9, 998–1006; the *Gestalt* 13, 29, 45, 203, 243, 250 and *passim*; role of the *Gestalt* in creativity 250–4; effects of linearity 601–12, 824–8, 887, 976–8; as 'the downfall of mankind' 824
Fregoli's syndrome 97, 140, 844
frontal lobes, evolution of 63–6

≈

GABA neurones 60–1
gaze, leftward-bias 207; mutual 207; eye contact 213
Gerstmann's syndrome 188
glutamatergic transmission 60, 318
Golden Mean, nature of 821
golden ratio 596 & n 91, 977, 999 n 17, 1030–1, 1159–61, 1356
'grand illusion' 18
groupthink 714–5

≈

halakhah, see *aggadah*
hallucinations, lateralisation of, in general 117–34; visual 118–29; auditory 129–31 (musical 131); tactile/haptic 131; olfactory and gustatory 131; Lilliputian 123, 128
harmony 596–7, 706, 846–7; *harmoniē* in Heraclitus 817–21; lateralisation in appreciation of 110–1, 970; harmony of difference and sameness 838; Einstein on 1362
heautoscopy 146
hemimacropsia, see metamorphopsia
hemimicropsia, see metamorphopsia
hemineglect, lateralisation in 70–8, 81–3, 88–9, 138–9, 261–9; in schizophrenia 320
hemiprosopometamorphopsia, see prosopometamorphopsia
hemisphere differences: introductory outline 28–31, 44; hemisphere hypothesis, core ideas 19–28; hemisphere hypothesis, common misunderstandings 32–43; in intra- and inter-hemispheric communication 62–3; hemisphere specialisation, evolution of 59; hemispheres and personality 39, 68, 194, 212; neuropsychological analysis at hemisphere level, advantages of 325–6
heuristics 729–40
Hinduism 1200–1, 1220, 1232, 1283, 1375; Tantric 835, Vedic 1200, 1244
Homeric Hymns 626
hormesis 827–9
humour, sense of, lateralisation in 29, 210–12, 217, 222, 298, 309, 627, 659, 1257, 1267, 1313, 1325–7; fate of in modernity 746–7; relationship to depression 298

≈

I Ching 420 n 35, 1303 n 373, 1309, 1315–6
Ich-Spaltung 366
illness, as a way of exploring human experience 305ff
imagination, as foundational 239, 768–76; operating unconsciously 753–67; distinction from fantasy 768–773; role in science and maths 421ff and *passim*; co-creation of the world 767–78; transcending the subject-object divide 763ff; impairment of, consequences 212–4; lateralisation in 772 and *passim*; ability to synthesise opposites 773; philosophy and 773ff; effects of specialisation and technicalisation on 559–62; analysis *vs* synthesis in 563–4
implicit, capacity to comprehend, and lateralization 28, 309; importance of 8, 167, 178–9, 210, 214, 340, 354–5, 365, 409, 564–6, 586–9, 603, 632–6, 700, 751
individuality, different conceptions of 393–4: see also uniqueness
Indra's net 1224, 1231, 1244, 1297, 1307
inference-making, LH and excessive 136, 177–9; RH and logical 175–7, 210–11
infinity 665–8, 878; as qualitative, not quantitative 929; a process 1241; potential and actual 1247, 1259; limitation conveying infinity 830; infinity within the finite 878, 1251
innovation, in animals 239
insight *passim*; insight (creative), lateralisation of 254–7, 667; role of intelligence in 256–7; *Gestalt* in 757–9; role of beauty in 758–60; unconscious in 760ff; insight (into own motivations), RH and 157; insight (into illness) 150–5, 212; see also anosognosia and denial
instrumentality, lateralisation in 182, 202–3, 870
intelligence, lateralisation in 225–8; Piagetian stage assessments 230–2; Raven's matrices and 234; simple reaction times and 227–8, 234; colour discrimination and 228, 234; role in creativity 245, 256, 784; Flynn effect 229–31; reverse Flynn effect 231–7
intelligent design 538, 1192
interdependence of genome and environment 441–3, 451–4

interregional connectivity in brain 106, 200–1
intuition 554, 673–776 *passim*; superior to explicit reasoning in complex cases 674–5; capacity to make complex and subtle judgements 692; accepting of apparent opposites 693–4; as guiding reason 723; part played in scientific and mathematical discoveries 754–60; relationship to skill 694–706, 724–9; interpreting of somatic indicators 687–94; embodied cognition and 694–706; unconscious memory and pattern recognition in 695–7; instinctual understanding and 679–84, incl acquired instincts 681; prejudices, distinct from bias 710–20; precluded by excess consciousness 673–94, 760–2; sleep and 760–3; an affront to the desire for control 723–4; perils of mistrusting 740–5, failure of common sense 746–51; Bergson's distinction between intuition and intellect 894–6, and *passim*
ipseity, loss of in schizophrenia 337, 352, 749
irony, lateralisation in 29, 210, 222, 627
Iroquois legend of two brothers, see Onondaga
Islam 1200–1, 1268–9, 1283, 1364, 1269, 1375; *and see* Sufism

~

jamais vu 196
jealousy, lateralisation in 141
Judaism 1200–1, 1220, 1234, 1269, 1283, 1375; *and see* Kaballah
judgment *vs* calculation 593–601, 744–5
judiciousness, RH and 154, 156, 170–80

~

Kabbalah 825–6, 835, 1242, 1302–3, 1371; in Renaissance thought 825; *chochmah* and *binah* 826; *da'at* 826; *Keter* 1246, 1260; *Ein-Sof* and creation 1248–9, 1255–7 1302; *sefirot* 1260
kenosis 1251

keystone species 832 n 69
kinaesthesia, lateralisation in 202
kintsugi (*kintsukuroi*) 1161, 1303 n 371, *pl* 22[c]
Kleine-Levin syndrome 194
knowledge, Aristotle's kinds 698
koro 149
kouros, the Getty 692–3

~

language: gesture and 106, 181; manipulation and 184ff; representation and 185–8; aspects of lateralisation in 186–7; second language, effects of speaking 715–6; dominance by 1318
Lao Tzu 181, 974, 1197, 1200–1, 1209, 1289
lateralisation, see *under individual phenomena*; differences between sexes 162–7
laughter, lateralisation in 204
lǐ 860, 968, 973–5, 999, 1154, 1200, 1206, 1223
limit case 7ff
logorrhoea, LH and 214, 339
logos vs mythos, see *under mythos*
love and strife, see *under* form
lycanthropy 140, 143

~

ma 1004–5
Machiavellianism 1346
macropsia, see metamorphopsia
magical thinking 158–62
manipulation, see utilisation
Marx, Karl 1127–8; Marxism-Leninism 1292
matter, RH and LH in comprehension of 1016–8, 1021–8; fields *vs* particles 1018–21; itself an abstraction 1047; consciousness and 1037–1120 *passim*
Maxims (Anglo-Saxon) 860–2
meaning, denotative and connotative, lateralisation in 217; gist, understanding of 210–2
meditation and mindfulness 117, 159, 195, 922, 1213–4, 1357–9
memories, false 155–7
metamorphopsia 120–9
metaphor *vs* literalism 45, 186–7, 210, 216–22; importance of metaphor 8, 291,

409–11, 494, 542, 578, 603, 621, 630–5, 757–8, 936, 1125; lateralisation in 214–22; importance of distinction from cliché 218–9
micropsia, see metamorphopsia
Midrash 829, 1229
mirror agnosia 142
mirror neurones 201–2
misidentification, delusional 97, 140; *and see* Capgras' *and* Fregoli's syndromes
misoplegia 147–9
model
—affecting what is found 409–12
—the machine model: biology contrasted with physics 431–3, 494; language in 433–6; attraction of 472–5; LH and 475–7; problems in: organisms not machine-like 431ff, esp 443ff; genes not robots 436–40, but a resource used by the organism in context 441–2, 454–8; 'black boxes' 440–1; open *vs* closed systems 443–7; non-linearity 447–451; mutual constitution of cell, genome and environment 451–4; illusion of 'parts' 454–70; imprecise boundaries 470–2; boot-strapping 471–3; attempts to save model 484–7; effect on humanity 495–500
— stream of life model: see *under* flow
Monty Hall dilemma 708–9
morality, lateralisation and 1343–7; goodness 1132–46
mother-infant bonding 201
motion, see flow
motion-induced blindness 172
music, experience as having the nature of 11–16, 1032
mythos vs logos 621–39

~

narrative, lateralisation in generating and comprehending 29, 210–11, 309; *and see under* schizophrenia *and* autism
neglect, see hemineglect

~

objectivity 393–4; nature of 413ff; subject-object divide

303, 413–7, 550–1; impersonal *vs* personal 612–621
Onondaga legend of two brothers 813–6, 824–5, 833, 841, 877, 1303–4, 1308, 1319
opponent processors 54–61
opposites: coincidence of 706, 813–841: as foundational 825; generative power of 816–21, 825ff; not irrational 821–3; asymmetry of 832–7, 841; enantiodromia 832; 'dark side' 837
order, as prior to randomness 8; and randomness together 455–7, 839–40; *rubato*, and irregularity of the heart 838–840
Othello syndrome 141
ouroboros 1298

~

palinopsia 79–80, 124, pl 3[b]
panentheism 1231–9, 1248–50, 1267
panpsychism/panexperientialism 1044, 1059–61
paradoxes: Achilles and the Tortoise 647–8; Amphibius 651; Antinomy of Change 648; Arrow 649; Baldness 651; 'Blub' 1211; Bradleys' Paradox of Relations 654; Carroll's Paradox of Entailment 652–3; Dichotomy 645–7; Dollar Cost Auction 651; Einstein-Podolsky-Rosen 668; Euathlus 660; Flint's 664; Grandfather 664; Grelling's 660; Growing 662–4; Kripke's Pierre 655; Liar 658–9; McTaggart's 650; Moore's 656, 665; Morning Star 655; Negation 656; Newcomb's 664–5; Place 654; Prediction 664; Preface 657; Prisoner's Dilemma 652; Quine-Duhem 642–3; Russell's Barber 661; Sceptical 643–4; Self-deception 657; Ship of Theseus 661–3; Sorites 650–1; Third Man 653; Thomson's Lamp 665–6; Unexpected Examination 644–5
paranoia 90, 96, 140–1, 154–5, 745
Parkinson's disease 221, 276, 982

passivity phenomena 90
passivity, and LH 309, 342
pelopsia, *see* metamorphopsia
perception, a reciprocal process 106; lateralisation of, in general 106ff; perception, local *vs* global 112ff,116; visual 107–10; auditory 110–11 tactile 112; olfactory and gustatory 112; of the body and embodied self 116ff; changes with age 115
perceptual rivalry 114–5
perfection, imperfection of 840, 926, 1159, 1161–2, 1247; role of inconsistency in triggering insights 255; in creative evolution 840, 1175
perseverance 213–4, 240, 248
perseveration 213, 984, 988
personal *vs* impersonal, *see under* objectivity
personality disorder, borderline 334, 336 n 203; lateralisation in 194
persons, not brains 27–8
phantom limb phenomenon 145–7, 181–2, 264
Pirkei Avot 46 n 94
plainsong 838, 847
Potemkin villages 357
potential 8, 834; types of 910–3; *potentia vs potestas* 856
pragmatics 29, 187, 215, 594
Pragmatism 390–1, 402, 420, 427, 495, 577
precision *vs* accuracy 174, 392, 582–93, 642–5, 734
prejudice *vs* bias 710–22
prepotency of LH 214, 324
presence *vs* re-presentation, hemisphere difference in 7, 11, 105, 336–7, 366, 373, 380–3, 387, 393, 632, 1112
probability matching 709–10
prosody 29, 187; loss of in RH damaged patients and schizophrenia 204, 207, 211, 213, 309
prosopagnosia 97, 120, 139, 206–7, 868
prosopometamorphopsia 121–2, 125, pl 5[a]
Psalms 1258
psychopathy 197, 200–1, 1126, 1133–4, 1142; lateralisation in 200–1, 1346
psychosis, *de novo* lateralisation of 118; *and see* schiz-

ophrenia *and individual syndromes*
public health policy 530–1, 797–807
Puritanism 1268, 1327
purpose 1100–2, 1167–92 *passim*; purposiveness 421, in organisms 477; non-instrumental 1168–70; open nature 1170–1; scale and 1171–2; Darwin and 1172–4; compatibility with randomness 1174–7; unadulterated chance unlikely 1177–80; misunderstandings of 1180–4; evolution towards complexity 1184–6; thisness and 1186–8; in mind and time 1188–92

~

realism, lateralisation in 152–4, 172–3; undue optimism 152–3; 177–80
reality, sense of 118, 124, 186, 193–6; essential hiddenness of 562, 1245–6; Goethe's 'holy open secret' 1245
reason: different meanings 397–8, 547–9, 552–4; ratiocentrism 555–67; role in forming judgements 167–80; induction and deduction, lateralisation in 168–70; necessary assumptions for 552; dependence on intuition 554–5, 563; incompleteness of 567–9; awakening, not compelling, an understanding 552–3; reason beyond reasoning 552–5; linearity *vs* the *Gestalt* 554, 601–12; explicit *vs* implicit 564–9; abstraction *vs* embodiment 571–582; precision *vs* accuracy 582–93; calculation *vs* judgment 593–601, 740; impersonal *vs* personal 612–21; *logos vs mythos* 621–40; not necessarily in the service of truth 721–2
recognition, lateralisation in 106–7
reductionism 5–7, 45, 448; fearful of intuition 724, of the body 1054–5; Simpson on 477; Salk on 508; schizophrenia simulating 344; self-defeating 1054;

[reductionism, cont'd] consciousness and 1080, 1094; beauty and 1148, 1158; open systems and 1095–6; cynicism and 1143; engineering and 1269 n 258; modern culture 851–2, 1306, 1323, 1331
reduplicative paramnesia 96–7, 140, 1118
relationship, as prior to *relata* 6–7, 11, 459, 1004–6; in physics 1006, 1027–8, 1119 n 379; in theology 1264, 1307; *and see* Indra's net
resistance: as necessary to creation 60, 63–5, 766–7, 775–6, 818–21, 832, 957–67, 1260–6 and *passim*; cortical inhibition: increasing in mammals, primates and humans 61–2; mutual inhibition of hemispheres 214; RH inhibitory control over emotional arousal 209, 223, 721, 740; effects of deficits in 223; impulsivity and LH 154, 170, 176; Scheler and 1091–2; *see also tzimtzum, śūnyatā,* kenosis
response-to-next-patient syndrome 143–5
responsibility, sense of, lateralisation 89–90, 150–2, 197, 336, 1332, 1340; *Zeiger der Schuld* 90, 1346
Rey-Osterrieth figure 152–3, 164–5
rhyme and half-rhyme 110, 838, 847
rhythm, lateralisation in 29, 110, 113 n 59, 298, 596, 628, 975; living rhythms flexible 838 & n 84; foundational 926; as image of life in Whitehead 975–6; *rubato* in 976
'rules of thumb', *see* heuristics
Ryōan-ji 550, 1058, 1104

~

sacred and divine, the: the ground of being 1193–1200; recalcitrance to language 1200–1207, 1223–8; dispositions *vs* propositions 1207–23; active receptivity 1213 and *passim*; need for transcending LH constraints 1207ff; later reintegration of LH 1228ff, *aufgehoben* by RH 1248ff; processual in nature 1234–41; as *coincidentia oppositorum* 1246–8; negation of negation 1255–6; how knowable 1256–60; omnipotence, omniscience, and hemispheric differences in understanding 1251ff; LH usurpation 1266ff; dogmatism 1266–70; fundamentalism in theism and atheism 1270ff; religion and spirituality 1290–8; evil 1298–1304; religion and science, compatibility of 1361–8; religion as brainwashing 1368–73; religious belief and health 1373–5; spirituality, lateralisation and 1357–9
sarcasm 210, 222, 309
savants 274–5, 364, 1088–9 n 273, 1255
scale, changing quality 490–1, 499, 589, 835–6, 916–8, 1058, 1168, 1171–2, 1175, 1243–4
schizophrenia 140, 167, 223
—in relation to RH deficits: phenomenology 306–15; brain correlates 315–20; cerebral lateralisation in 90, 98, 101
—lateralising characteristics: abstraction in 336–40, 353, 357; contextual understanding, failure of 313, 328; aversion from the natural in 349–50; 'betweenness', loss of 333ff; body, alienation from 332–3, 337–8, 341, 345–354; common sense and sense of humour, loss of 747–9; depth, loss of: in emotion 309–11, 337; in space 358–9; in time 333–5, 351, 356, 359ff; devitalisation in 336–345, 350–1, 354, 360; focus on utility in 349–51; fragmentation in 329–332; hallucinations in: visual 122–5; auditory 129; tactile/haptic 131; olfactory & gustatory 131; hyper-consciousness in 355ff; hyper-rationalism in 333, 350–3; hyper-reflexivity in 367; immobility and 336–9; 'ipseity', loss of 352, 749; literal-mindedness in 354; mechanisation in 345ff; 'morbid geometrism' in 352–3; narrative difficulties in 365; re-presentation and 193, 356–66; rigidity in 352–3; schematisation in 349–53; self, loss of 333–6; sense of motion, loss of 336–9; sense of reality, loss of 193, 366–8; third-person perspective in 337, 352
—relation to other conditions: anorexia nervosa 145; autism 305–370 *passim*, esp 305–7
schizotypy 158–9, 300, 307 n 7, 312 n 41
science: necessary assumptions in 419ff; scientific method, part played by 421–3; role of imagination in 423; important role of chance in 425; divorce from philosophy 507–8; effects of overspecialisation 502–8; technicalisation and 507–8; neuroimaging and 509–513; science, institution of: effects on originality 502–8, 531–542; reliability 509; replication 513–6; pressures that distort 516ff; peer review 523–30
Secret of the Golden Flower, The 1310–11, 1318
Sefer yetzirah 1298
self *vs* ego, lateralisation and 874–8; sense of self 116–7; *and see* ipseity
self-esteem, lateralisation in 152; cult of 1140–1
self-recognition, lateralisation in 142
set-shifting, lateralisation in 72 n 28, 173, 193, 197, 260, 274, 789
shevirat ha-kelim 1249–50, 1303
sign language, lateralisation in 187
skill 703; unconscious processing in 604–5, 674–8, 694–5, 700, 728, 743, 746, 1041; *and see under* intuition
somatoparaphrenia 90–6, 138–9
space 997–1035 *passim*; lateralisation and 81–3, 1006–8; depth in 1000–4; loss of depth, effects of 1011–2; effects of schizophrenia on 1008–11; modernity and

1012–15; primacy of relation over *relata* 1004–6, 1027; *and see* 'betweenness'
stereotype accuracy 717–9
stereotypy 213
Sufism 835, 1091, 1202, 1221, 1268
śūnyatā 1209–10
syllapsies 817–8
syllogisms, hemisphere difference in evaluating 399–401
symbols, in evolution of hemisphere difference 105, 185–7
symmetry, morbid 352–3, 1010–11, 1162; an abstraction 1031

≈

tacit knowledge, importance of 337, 416, 584 n 52, 639, 698, 748–50; in organisms 461
taijitu 36
Talmud 734 n 99, 833 n 76, 1221, 1267
Taoism 745, 820, 971, 1234, 1302, 1311–3; *and see* Lao Tzu *and* Chuang Tzu *in the Index of Names*
teleology *vs* genetic programming 440, 477–83; rapid adaptations of organisms 451–2, 466–8
teleopsia, *see* metamorphopsia
theory of mind, lateralisation 29, 199–200; in schizophrenia and autism 311, 322
theory over experience, LH adherence to 172–3, 317, 399–402, 573 and *passim*
things, origins of concept 884–5; secondary to processes 886, 1005–6
tikkun 1243, 1249–50, 1284, 1303
time 881–944 *passim*; foundational nature 888ff, 925–8, 932–6, 943–4; not a thing 883–4, but a flow 896–901; lateralisation in appreciation of 78–81, 901–9; spatialisation, effects of 886–9; representation, effects of 889–96; shaped by value 920–2; flow, and clock *vs* lived time 922–5; determinism and the misconception of time 909–18; the present having extension 918–20; the 'growing block' 930–2; eternity 928–30; time in depression 1337–41; in schizophrenia 333ff; in modernity 936–43
tools, lateralisation in appreciation of 29, 101 *&* n 161, 102, 110, 183, 870
'top-down' constraints 103, 112, 440, 453, 1081
Torah 1253
Truman Show, The 357
truth: perils of postmodern approaches 10, 390, 4011, 565; hemisphere differences in 379–403, 1122–32; presence *vs* re-presentation 380–1, 393–4; imagination a requirement for 768; science and 407–429; thing *vs* process 382–7, 402; potential and actual 1241–3; holarchies and 1243–5; hidden *vs* manifest 1245–6
TT races 684–6
tzimtzum 1249–1260 *passim*, 1302–3

≈

uncertainty, intrinsic, lateralisation in appreciation of 108, 154, 169, 179, 384, 1211, 1236; in creativity 909, 1078; in physics 417, 839, 881–2, 894, 1017, 1078–9, 1110, 1186; in time and motion 909–10, 914, 1027; in relation to human meaning 174, 389, 541, 734, 910, 939, 1004, 1135, 1145, 1271; in relation to the divine 1269, 1285; antipathy to in autism and schizophrenia 352, 616–7; 'the salt of life' 775
understanding, *vs* explanation 409–13, 499–500; *vs* knowledge 407–8
union and division: union of 837ff and *passim* throughout; Goethe and 837; nervous system as an embodiment of 839, 858–60; architective *vs* connective interactions 834–6; competition and co-operation, each needed 497ff; whole, as subsuming tension between parts and whole 825
uniqueness and generality: One and Many, the 817, 825, 843–880 *passim*; generalities and particulars, 572–6, 844–6; *hæcceitas* 855, 1186, 1236; process of individuation 848–77; part played by time in 925–8; uniqueness and levels of categorisation, lateralisation in 864–6; parts and wholes, lateralisation in 848–9, 1243–5; equality, reducibility, utility 849–56; integration of unique and general 856–858; beauty and the coincidence of One and Many 846–8, archetypes as instances of, *see separate entry*
Upanishads 820, 1370
'uselessness', supreme value of 745, 1168, 1171
utilisation, lateralisation in 181–4, 209, 215; language and 181, 185ff; schizophrenia 313, 349; *see also* tools
utilitarianism 1132–9, 1144, 1170; brain correlates of 1133–4, 1146, 1343–7; and modernity 1314, 1324–5

value 1121–65 *passim*
vestibular stimulation 71, 93, 276, 325
voice perception, lateralisation in 204

wabi-sabi 1161, pl 22[c]
Wason tests 176, 731, pl 16[c–d]
Williams syndrome 222
working memory, hemispheres difference in 80, 153, 178, 221–3

≈

xenomelia 149

≈

Zeitraffer phenomenon 79–80
Zen 16, 46, 562, 606, 823, 835, 839, 875, 929, 933, 955, 1149, 1161, 1197, 1200, 1208, 1219, 1221, 1234, 1283, 1310–11, 1316, 1357
zero, invention of 656
Zohar 1235, 1298, 1302 n 369
zombies 1113–4
Zwischensein 334

Index of Names

Abram, David 1103
Abramowitz, Mike 833–5, 972
Adams, Anne 268, 273
Adorno, Theodor 1286–7
Aeschylus 1002
Agathon 935
Agus, Rabbi Jacob 379
Aha, Rabbi Isaac bar 734
Aherne, Caroline 298
Ajuriaguerra, Julian de 93, 147, 904–5
Al-Biruni 1365
Al-Khalili, Jim 1015
Alajouanine, Théophile 264–5
Alberti, Leon Battista 977
Albrecht, Andreas 1078
Aldous, Carol 788 n 33
Alexander, Christopher 611
Allen, Woody 298, 843
Alzheimer's type dementia 116, 137, 141, 802, 1087
Amodio, David 721
Anaxagoras 978
Anaximander 935, 977–8, 1059, 1299
Andersen, Hans Christian 630
Anderson, Philip 1035, 1306
Andreasen, Nancy 296–9
Angelergues, René 139
Angelico, Fra 1259–60
Anjum, Rani Lill 448, 915
'Anne' [pt of Blankenburg] 333, 345, 348, 353
Annoni, Jean-Marie 267
Anselm, Saint 1129
Aquinas, Saint Thomas 559, 1226, 1242
Arai, Takao 80
Archilochus 859
Arendt, Hannah 581 n 42, 615, 1104, 1301–2
Aristotle 65, 109, 291–2, 400, 488, 544, 552, 577, 582–3, 648, 653, 698, 758, 817, 826, 837, 885, 917, 930, 935, 979, 987, 992, 995, 998, 1051, 1184, 1186, 1203–4, 1226, 1364
Armstrong, Karen 625–6
Arnellos, Argyris 470
Artaud, Antonin 341, 344
Asimov, Isaac 243–4, 247–8
Assal, Gil 266
Attwood, Tony 616
Augustine of Hippo, Saint 881, 933, 943, 1129, 1201, 1210

Aurobindo, Sri 749
Auspitz, JL 526
Austen, Jane 793
Averroes (Ibn Rushd) 1102
Avicenna (Ibn Sina) 197
Ayer, AJ 590 n 76, 617
Azriel of Gerona 1253–4, 1256

Bach, JS 272, 113 n 59, 282–5, 493, 838, 847, 968, 1193, 1289
Bacon, Roger 544, 1364
Bacon, Sir Francis, Lord Verulam 48, 425, 562, 811, 1191, 1271, 1275
Bak, Thomas 982–3
Balanchine, George 271
Ball, Philip 410–1, 971–3, 1118
Balthasar, Hans Urs von 1163
Baluška, František 1071–3
Banich, Marie 1344
Barber, Bernard 537
Barbour, Ginny 516
Barnes, Julian 939
Barnes, Luke 1116
Barrett, Deirdre 762
Barthold, Lauren 712
Bartneck, Christoph 521
Basaldella, Afro 266
Bashō 363, 393, 764
Bate, Walter Jackson 772
Bateson, Gregory 635–6
Baumeister, Roy 1081–2, 1084, 1140–1
Bayes, Thomas 643
Bäzner, Hansjörg 264, 269
Beall, Jeffrey 520–1
Beauchamp, T 1145
Beaumont, Francis 1238
Beauregard, Mario 1081
Becchio, Cristina 78
Bechly, Günther 538
Beck, LW 739
Beeman (also Jung-Beeman), Mark 257, 277
Beethoven, Ludwig van 271, 299, 1215–6
Beißner, Friedrich 339
Belfi, Amy 102
Bell, John Stewart 1026, 1047
Bellini, Vincenzo 1152
Bene, Gyula 1006
Benedek, Matthias 787
Benjamin, Walter 360, 1309
Benner, Patricia 704–5
Bentham, Jeremy 617

Benton, AL 188
Berenson, Bernard 693
Berg, Maggie 937
Berger, Peter 1281–2
Bergson, Henri 31, 78, 305, 328, 343–9, 381, 441, 573, 577, 607, 647–50, 653, 663–9, 723, 753, 766, 774, 822, 843, 877, 891–930 passim, 943–58, 967–9, 977–9, 994–6, 1002, 1018, 1021, 1028, 1061, 1086, 1088–90, 1110, 1131, 1172–4, 1191, 1212, 1234–5, 1247; distinction between 'intuition' and 'intellect' 343–5
Berkeley, George 74, 77, 575, 611
Berlin, Sir Isaiah 859, 1000–1
Bernard, Claude 1185
Berrios, German 131
Berry, Wendell 430, 496, 1210
Bersanelli, Marco 1179
Bertone, Cesare 78
Bethe, Hans 551
Bethlem Royal Hospital 125, 360, 618
Bevan, Gwyn 745
Bickhard, Mark 489, 494
Binet, Alfred 980
Binswanger, Ludwig 351–9, 905
Birch, Charles 447
Birnbaum, Johann 285
Bishop, Paul 1286
Bisiach, Edoardo 76–7, 88, 93
Black, Max 632
Blake, William 67, 363, 541, 571, 607, 753, 764, 821, 830, 852–3, 855, 857–61, 888, 923, 929, 1000, 1085, 1193, 1224, 1284, 1293, 1299, 1308, 1370, pl 14
Blakemore, Colin 1046
Blanke, Olaf 262
Blankenburg, Wolfgang 333, 348, 351–3, 618, 747–8
Bleuler, Eugen 321, 332–7, 343, 1340
Bloom, Harold 16 n 15
Boehme, Jakob 1198, 1237, 1255
Bogen, Joseph 40
Boghossian, Peter 795
Bogousslavsky, Julien 143
Bohm, David 403, 432, 450, 474, 494, 562, 682, 957,

1022–7, 1062, 1093–1105, 1241, 1244, 1315–9
Bohr, Hans Henrik 641
Bohr, Niels 7, 42, 419, 424, 473, 490, 535, 612, 620, 641, 657, 727, 811, 813, 816, 894, 1006, 1016–20, 1026, 1046–51, 1057, 1064, 1105–7, 1115, 1119, 1177, 1225–31, 1274–5, 1307, 1361–2
Bohrn, Isabel 1355
Boiyadjiev, Zlatyu 265
Boldt, Joachim 522
Bonald, Louis de 591
Bond, HL 1242, 1246
Boon, Wouter 284
Borges, Jorge Luis 912, 917, 923, 927, 929
Borovik, Alexandre 965
Borowiecki, Karol Jan 299
Bosch, Hieronymus 314, pl 4
Boswell, James 298
Bourgeault, Cynthia 876
Bowerman, Bruce 1029
Bowie, Andrew 1092
Boyer, Pascal 420
Boyle, David 425
Bradley, FH 612, 653–4
Brahe, Tycho 415, 537
Brahmagupta 656, 819
Braque, Georges 270, 589
'Brassaï' [Gyula Halász] 288
'Brassau, Pierre' 793–4
Braun, Claude 40, 117, 136–7, 156, 178
Brenner, Sydney 448, 1077
'Brights', the 1213, 1272
Brinthaupt, Thomas 557
Britten, Benjamin 271
Broca, Paul 106
Brogaard, Berit 946
Bronowski, Jacob 283, 499
Brooks, Arthur 1321
Brooks, Rodney 1015, 1019–25, 1109
Brouwer, LEJ 742
Brown, Paul R 726
Bruce, Christine 1354
Bruce, Lenny 298
Brugger, Peter 162
Brüne, Martin 142
Brunelleschi, Filippo 1012
Bruno, Giordano 1366
Brunswik, Egon 734
Bulgakov, Sergei 822
Burgess, Paul 140
Burian, Richard 441
Burke, Edmund 384, 557, 583, 1151

Buxton, Jedediah 364, 597, 599
Buzsáki, György 839–40, 992
Byers, William 1255
Byrom, Greg 372
Byron, George Gordon, Lord 763

≈

Calderón de la Barca, Pedro 643
Callas, Maria 1152
Calvo, Paco 1071
Campbell, Joseph 831
Canales, Jimena 892–3
Cantor, Georg 667
Čapek, Milič 893
Carel, Havi 305
Carnap, Rudolf 558, 818 n 17, 931 n 181
Carr, Nicholas 1003
Carroll, Lewis 126–7, 603, 652–3, 659
Carse, James 396, 505, 554, 606, 1168, 1280
Cassirer, Ernst 562
Cavell, Stanley 569
Cavill, Paul 862
Ceci, Stephen 526–7, 1152
Cézanne, Paul 343
Chaitin, Greg 568
Chambers, Robert 425, 594–5
Chaminade, Thierry 201
Chapman, Graham 298
Chapman, James 122 n 130, 335, 1008
Chargaff, Erwin 412–5, 419, 422, 505–6, 775–6, 1213, 1319–20
Charlton, Bruce 1188–90
Chartier, Tim 946
Chernigovskaya, Tatyana 869–70
Chesterton, GK 167
Childress, J 1145
Chomsky, Noam 441
Chrysikou, Evangelia 278
Chuang Tzu 606, 934
Cicero 829
Cilliers, Paul 1299
Cimabue 1013, 1149
Cioran, Emil 968, 1289
Clare, John 1216
Claxton, Guy 697, 981
Cleary, Thomas 1310–1
Cleese, John 298
Clement VII, Pope 1366
Close, Frank 1032
Cloud of Unknowing, The 1193
Cobbett, William 622
Coleman, Stanley 127–8

Coleridge, Samuel Taylor 488, 588, 635, 768–76, 828, 846, 854, 912, 1234, 1238–40
Collingwood, RG 407, 615, 856, 1203
Collini, Stefan 937
Collins, Francis 433
Colombo, Françoise 266
Coltheart, Max 137
Columbus 1278, 1364–5
Combettes [first name unknown] 984
Coney, Jeffrey 1354
Confucius 385, 1197, 1200, 1208, 1220
Conway Morris, Simon 478, 1190
Cook, Peter 298, 1182
Cooper, Tommy 298
Cope, David 285
Copernicus, Nicolaus 395, 537, 1278, 1365–6
Copland, Aaron 16 n 15
Corballis, Michael 168
Corinth, Lovis 262–3, pll 8–9
Cornford, Francis 612
Corot, Jean-Baptiste-Camille 869
Coslett, HB 990
Cottingham, John 556, 559, 562–3, 581, 1304
Coyne, Jerry 1273–82, 1361–5
Craig, Bud 1164
Cratylus 953
Crawford, Matt 981
Crick, Francis 412, 505, 531
Critchley, Macdonald 79–80, 188, 195
Cropley, Arthur 243
Csikszentmihályi, Mihály 921
Cuonzo, Margaret 642, 666
Curie, Pierre 833, 1033
Curry, Patrick 707, 1288
Cusanus (Nicholas of Cusa) 667, 821, 948–9, 1201, 1241–59 passim, 1265, 1364
Cushing, Harvey 195
Cusinato, Guido 1131
Cutting, John 79–80, 97, 117–8, 122, 195–6, 311, 314, 327–8, 334–44 passim, 845, 870, 990, 1340–1
Cytovic, Richard 273

≈

Dalí, Salvador 122, pl 6[a]
Damasio, Antonio 224, 549, 737
Dandy, Walter 536
Danielson, Dennis 1365

Darwin, Charles 19, 42, 243–4, 299, 433–4, 453, 467–8, 506, 538–40, 619–20, 636, 734, 1071–4, 1127–8, 1142–50 *passim*, 1170–4, 1184, 1187, 1277, 1367–8; Darwin-Eigen spiral 1178
Darwin, Erasmus 462
Darwin, Francis 1073, 1173
Davidson, Richard 204
Davies, Dame Sally 802
Davies, Paul 1048, 1115–6, 1179
Dawkins, Richard 434, 436, 453, 456, 484–5, 498, 538, 1093, 1170, 1174, 1180, 1277–9, 1282, 1369
de Broglie, Louis 535, 669, 759, 892–4, 914, 1018, 1109
de Chirico, Giorgio 340
de Haan, Sanneke 357
de la Mettrie, Julien Offray 347, 484, 487
de Quincey, Christian 1044
de Volder, Burcher 1185
Decety, Jean 201
Dee, Jack 298
Deecke, Lüder 1083
Deglin, Vadim 170–3, 399, 474
Demay, G 357
Demetrius of Phalerum 661
Democritus 945, 978
Denes, Gianfranco 217
Dennett, Daniel 1043, 1093
Denton, Melinda 1292, 1374
Descartes, René 11, 248, 251, 332, 348–66 *passim*, 413, 421, 432, 494, 556–63 *passim*, 617, 619, 662, 712–3, 749, 769, 897, 903–4, 954, 994–5, 1040, 1049, 1052, 1076, 1107, 1113, 1192
Deutsch, David 1093, 1164, 1193
Devinsky, Orrin 116, 136, 1358–9
Devlin, Keith 1160
Dewey, John 390, 418, 504, 551, 577, 582, 613, 617, 892, 970, 991, 1061, 1167, 1285
Dickens, Charles 765
Dickinson, Emily 1121, 1125–6, 1151, 1156
Dide, Maurice 350
Diderot, Denis 1097–8
Dietrich, Arne 282–92, 302, 781–91
Dimond, Stuart 69, 215, 359, 1318
DiNicolantonio, James 798–9

Dionysius the Areopagite 1202, 1246
Dirac, Paul 535, 759, 1153, 1172
Dissanayake, Ellen 760
Divided Brain, The (film) 197, 507
Dix, Otto 147, 263–4, *pl 7*; obtrusive, polydactylous hand in 264
Dixon, Chris 571
Dobzhansky, Theodosius 480 *& n* 234
'Dr P' [pt of Sacks] 207, 871–4, 1351
Dodge, Wag 753–4
Dōgen 930, 933, 1208
Dolev, Yuval 933
Donald, Merlin 684
Donne, John 778, 818, 1000, 1216, 1251
Draper, John 1365
Drell, Daniel 471
Dreyfus, Hubert *and* Dreyfus, Stuart 605, 703–4, 742–50 *passim*
Driesch, Hans 468–9
Drieschner, Michael 934
Drob, Sanford 1235, 1243, 1249, 1254
Dryden, John 292
Ducasse, CJ 1081
Duhem, Pierre 540, 642–3
Dumont, Sophie 460
Dunne, Francis 311, 327
Dunning-Kruger effect 1315
Duppa, Richard 588
Dupré, John 431, 440–57 *passim*, 470, 490–8 *passim*, 1190
Durkheim, Émile 1322
Dyson, Freeman 531, 558, 608, 755, 1024, 1058–9, 1227
Dyson, James 283

Eagleman, David 1112
Earle, William 560–1
Eccles, Sir John 1363
Eckhart, Meister 943, 1138, 1211–64 *passim*, 1283–4
Eddington, Sir Arthur 563, 997, 1054–6, 1061, 1100, 1110
Eddy, Jack 503–4
Edison, Thomas 282–4
Edwards, Lawrence 466
Edwards, Oliver 298
Ehrenwald, Hans 100–1, 146, 343
Einstein, Albert 254, 288, 415, 420, 424, 488, 506, 525, 535, 540, 562, 612, 624, 650, 666, 746, 754–5, 758, 836, 883,

892–3, 922, 931, 997, 1019, 1027, 1053, 1094–5, 1102, 1115, 1193–4, 1227, 1273, 1275, 1333, 1362
Eliade, Mircea 1297
Elias, Norbert 1316–7
Eliot, TS 158, 348, 607, 912, 1239
'Elliot' [pt of Damasio] 224, 549, 737
Ellis, George 931, 1116
Ellis, Henry Havelock 295
Ellis, Robert 740
Ellul, Jacques 923, 942
Elster, Jon 605
Emerson, Ralph Waldo 617, 767, 877, 1127, 1156, 1238
Empedocles 641, 824, 826, 841, 978, 1060
Empson, William 588, 818, 831
Engel, Christoph 674, 692
Engell, James 772
Engels, Friedrich 1128
Engerth, Gottfried 189
Epimenides the Cretan 658–9
Eratosthenes 1364
Erdös, Paul 709
Escher, MC 1139, 1302, *pl 24*
Euathlus 660
Eucherius 1149
Euclid 1020, 1364
Euler, Johann 423, 1149

Faraday, Michael 424, 1019
Farrer, TH 1174
Fasce, Gianfranco 266
Feinberg, Todd 92–6
Felig, Philip 528
Fellini, Federico 264
Fellows, Graham 298
Ferenczi, Sándor 616, 1128
Feyerabend, Paul 333, 618
Feynman, Richard 535, 540, 753, 776, 914, 965, 1024–5, 1172
Fichte, Johann Gottlieb 16, 613, 768, 993–4
Fierz, Lukas 1327
Fink, Andreas 785 n 18
Fiore, Stephen 225
Fischer, David 1349
Fischer, Franz 345, 358, 905
Fischhoff, Baruch 514
Fish, Stanley 549, 567, 1291
Fisher, Matthew 1077
Fitzgerald, Michael 322
Flaubert, Gustave 992
Fleming, Alexander 425
Fletcher, Alan 1005

Fletcher, John 1238
Flint, Jonathan 514
Fløistad, Guttorm 937
Flourens, Jean Pierre 858
Flynn, James 229–37 *passim*;
  Flynn effect 229–31; reverse
  Flynn effect 231–7
Forbes, Robert James 1150
Ford, Brian 228–9, 451, 461,
  464, 481–2, 486, 1066
Ford, Henry 386
Förster, Jens 257
Förstl, Hans 140
Foster, Charles 707, 1137, 1141,
  1269
Fox, Robbie 526
Fox, William 1128
Fraguas, Rafael 265
Frank, Adam 1046, 1048–9
Frank, Joseph 942
Frank, Philipp 1017, 1020
Frank, Waldo 1317
Frankl, Viktor 832, 1297, 1317
Franklin, Benjamin 65, 613,
  936
Franks, Felix 971
Frassinetti, Francesca 120
Frauchiger-Renner
  experiment 1052
Freedman, David 799–800
Frege, Gottlob 571–2
Freud, Sigmund 42, 88, 249,
  328, 334, 339–40, 506, 540,
  616, 626, 636, 697, 1127–8,
  1287
Fried, Charles 573
Friedman, Ronald 257
Fromm, Erich 572, 1130
Fry, Stephen 298
Fuchs, Thomas 354, 357, 907,
  921, 923, 969
Fujii Yoshitaka 522
Funk, Agnes 172

Gabbard, Glen 702
Gadamer, Hans-Georg 712
Gagliano, Monica 1069,
  1072–3
Gainotti, Guido 204
Gaisman, Jonathan 1211, 1280,
  1292–3
Galilei, Galileo 42, 667, 1000,
  1367
Gall, Franz 328
Galton, Sir Francis 295, 980,
  992
Gamble, Clive 1220
Gardner, Howard 217, 272, 372
Gardner, Martin 1117

Gare, Arran 488, 554, 774,
  1099–1100
Garrick, David 364
Garrow, JS 806
Gauss, Carl Friedrich 423
Gazzaniga, Michael 40, 91–2,
  107, 136, 155, 157, 179, 203, 365,
  373–4, 507–8
Gebser, Jean 1201, 1234
Gell-Mann, Murray 1038
Gell, Philip 448
Gellner, Ernest 549, 577–8
Gendlin, Eugene 1248
Gernez, Paul-Élie 264
Gersztenkorn, David 80
Gervais, William 1371
Geschwind, Norman 535–6,
  539
Ghiselin, Brewster 248, 288
Gibson, James 452, 1130
Gigerenzer, Gerd 696, 706,
  732–3
Gikatilla, Joseph 1257
Gill, Gerald 594
Gilson, Étienne 1184, 1192
Giotto 1012–3
Gladwell, Malcolm 692
Gleiser, Marcelo 48, 417, 448,
  755, 928, 1015, 1026–8, 1120,
  1281
Gödel, Kurt 568, 756, 931,
  1362–3
Godfrey-Smith, Peter 1074–5
Goel, Vinod 167, 171, 174, 260
Goethe, Johann Wolfgang von
  3, 31, 105, 251, 301–2, 363, 631,
  747, 774, 795, 818, 837–8, 841,
  902, 1058, 1060, 1064, 1126–7,
  1204, 1234, 1244–5, 1254,
  1293, 1297, 1300, 1339, 1368
Goldberg, Elkhonon 510, 866
Goldstein, Dan 696
Goldstein, Kurt 536
Golgi, Camillo 858, pl 17[a]
Goodman, CP 416
Goodrich, Barbara 839–40
Goodwin, Brian 469
Gottfredson, Linda 225
Gould, Stephen Jay 593, 1277,
  1281
Grafman, Jordan 260
Graham, Paul 1211
Graham, William 1174
Grandin, Temple 329, 760,
  1372
Graves, Robert 969
Gray, Asa 1173
Gray, John 1202, 1290
Grayling, AC 1270–1
Green, Michael 1015, 1020

Greenberg, Clement 1013
Greene, Brian 427, 1015
Greene, Joshua 1142–3
Greenfield, Susan 1003
Grelling, Kurt 660
Greyson, Bruce 1088
Griffiths, Paul 439, 471
Groopman, Jeremy 701–2
Gross, Gisela 125
Grosseteste, Robert 1364
Grow, Gerald 414 n 13
Gruenberg, G 1069
Guigó, Roderic 439
Guilford, JP 242–3, 282–3, 285
Guiraud, Paul 350
Gunaratana, Bhante Henepola
  1213–4
Gunter, Peter 893, 1028
Güntürkün, Onur 24, 40
Gurzadyan, Vahe 931
Gustafson, Karl 742
Guttinger, Stephan 470

Haack, Susan 390–1, 402
Habermas, Jürgen 637,
  1220 n 89, 1290–1
Habura, Krystyna 261
Hadamard, Jacques 288, 610,
  758
Hafiz 1251
Haisch, Bernard 1056
Haldane, JBS 428, 432, 443,
  477, 488, 499–500, 1045,
  1062
Haldane, John Scott [father of
  JBS] 432, 444, 461, 489
Halligan, Peter 74
Hallowell, Peter 102
Hallpike, Christopher 235,
  471, 473, 1150
Halvorson, Hans 1080
Hancock, Tony 298
Handel, George Frideric 272,
  288
Handley, Simon 738
Hankey, Alex 451
Hanson, Norwood 415
Hardy, GH 759
Harris, Annaka 1067
Harris, Sam 1271, 1280
Hart, David Bentley 1158,
  1168–9, 1199, 1276, 1282,
  1369–70
Hartshorne, Charles 892, 1061
Haslam, John 360, 362
Hawking, Stephen 1114 n 360,
  1115
Hay, David 1368–9
Hayek, Friedrich 567
Healy, David 531

Hebb, Donald 465, 983
Hécaen, Henri 93, 138–9, 147, 904–5
Hecht, David 1138, 1343, 1346
Hegel, Georg Wilhelm Friedrich 64, 584, 617, 768, 773, 819–20, 829, 840, 847, 878, 900, 935, 957–8, 994, 1227, 1234, 1250, 1256, 1260
Heidegger, Martin 247, 340, 381, 386, 390, 419, 535, 547, 562, 628–9, 641, 713, 773, 818–9, 848, 878, 892, 902, 905, 911, 926–7, 974, 992, 1012, 1091, 1104, 1204, 1206, 1227, 1260, 1264, 1311
Heilman, Kenneth 268, 990
Heinlein, Robert 501
Heisenberg, Werner 424, 535, 575, 839, 894, 914, 995, 1016, 1027, 1035, 1046, 1051–2, 1056–7, 1064, 1105, 1107, 1186, 1361
Heiti, Warren 1130
Helmholtz, Hermann von 424, 539
Hennerici, Michael 264, 269
Hennin, Bart 289
Henning, Brian 469
Henrion, Max 514
Henry, Richard Conn 6, 1055–6
Heraclitus 244, 431, 562, 565, 611, 627, 641, 773, 816–24 *passim*, 843, 857, 892, 900, 903, 935, 953, 973, 977, 996, 1059–60, 1105, 1200, 1234, 1260, 1299
Herbert, George 1216, 1298
Herder, Johann Gottfried 184, 193, 768, 977, 1060
Herschel, John 971
Herzog, Gunter 263
Heschel, Abraham 14 n 12, 884–5, 1207–8, 1229–30, 1266, 1333
Hesiod 626
Hibbs, Albert 914
Hilbert, David 609, 667, 1116
Hilgard, Joseph 529
Hirshfield, Jane 875
Ho, Mae-Wan 1038, 1054, 1090
Hobbes, Thomas 557
Hobson, Art 1021
Höfel, L 1157
Hoff, Hans 82, 85
Hoffman, Donald 431, 1090
Hoffmeyer, Jesper 461
Hofstadter, Douglas 624, 711

Hofstede, Geert 1320–1
Hogan, Patrick Colm 1164
Hogarth, William 977
Hokusai 965, *pl 20*
Hölderlin, Friedrich 339, 368, 813
Holdrege, Craig 445
Holmes, E 1069
Holmes, Jane 164
Holmes, Oliver Wendell 831
Holstein, Thomas 55
Holster, Andrew 931
Homer 626, 698, 939
Hood, Christopher 745
Hopkins, Gerard Manley 844–7, 854–5, 874, 1040, 1186, 1216, 1259
Hoppe, Klaus 214
Horace 435
Horkheimer, Max 1286
Horton, Robin 159
Hoyle, Sir Fred 1115
Huang, Sui 455
Huber, Gerd 125
Hughlings Jackson, John 51, 106, 512
Hugo, Victor 818
Hull, David 533
Hume, David 394–5, 449 n 85, 555, 559, 575, 740
Humperdinck, Engelbert 272
Hundertwasser, Friedensreich 824, 949, 977
Husserl, Edmund 340, 562, 750, 892
Hut, Piet 1006
Huxley, Aldous 588, 875, 1011
Huxley, Sir Julian 1061, 1092, 1281
Huxley, Thomas 244, 620, 1139, 1173, 1273, 1367

≈

Ichheiser, Gustav 717
Inamori, Kazuo 604
Ingold, Tim 1232–3
Ioannidis, John 515, 528

≈

Jablonka, Eva 467
Jacob, François 480
Jacob, Hélène [pt of Tatossian] 343–9 *passim*, 358, 362, 366
Jacobsen, T 1157
Jahn, Otto 287 n 234
James, Henry 96 n 143
James, William 48, 247, 305, 381, 390, 414, 501, 539–40, 574, 576–7, 580, 613–4, 617, 643, 653–4, 673, 680, 769–70, 829, 831–2, 836, 848, 857, 865, 877, 879, 892, 903, 908–9, 933, 935, 949; 953–8, 970, 978, 993–4, 996, 999, 1002–3, 1018, 1021, 1037, 1040–1, 1045, 1061, 1085–6, 1089, 1103, 1110, 1120, 1144, 1170, 1190, 1200, 1212, 1219–21, 1234, 1247, 1251, 1272, 1293, 1359
Jamieson, Kathleen Hall 529
Jamison, Kay Redfield 296–7
Jarry, Alfred 936
Jaspers, Karl 343, 360, 986, 1009
Jaynes, Julian 886–7, 891
'Jean Paul' (Richter) 768
Jeans, Sir James 1052, 1055, 1107–8
Jefferson, Thomas *pl 23*[*b*]
Jelinek, Elfriede 1013
Jensen, Arthur 233
Jobs, Steve 244
Johnson-Laird, Philip 175
Johnson, Mark 542, 578, 632, 635, 936, 981, 1090
Johnson, Samuel 298, 552
Jonas, Hans 5–6, 476
Jones, Ernest 616
Jordan, Jake 81
Jordan, Pascual 1016
Jordania, Joseph 628 n 193
Joyce, James 1231
Judd, Tedd 272
Jung, Carl Gustav 595, 614, 626, 630, 680, 683–4, 713, 821, 831–2, 837, 876, 1065, 1102, 1107, 1140, 1242, 1287, 1300, 1302, 1327
Jung, Richard 263
Jussim, Lee 716–20

≈

Kaessmann, Henrik 480
Kafatos, Menas 1058, 1121, 1243
Kafka, Franz 1005
Kahn, Charles 817, 1059
Kahneman, Daniel 698, 722–7, 729, 731, 737; 'system 1 & 2 thinking' 722, 738–40; supposed critique of expertise 724ff
Kamo no Chōmei 996
Kanner, Leo 321–2, 988
Kanso, Riam 282–3, 292, 302, 781–91
Kant, Immanuel 31, 459 n 141, 549, 559, 568, 581, 616–7, 666, 739, 822, 1013, 1124, 1151, 1238, 1240; antinomies 568, 666, 822

Kapp, Reginald 487
Kárpinski, Stanisław 1069
Kass, Leon 1187–8
Kastner, Ruth 953
Kastrup, Bernardo 1087
'Katharina' [pt] 346
Kaufman, James 297
Kaufmann, Stuart 458
Kay, John 579, 588, 598, 603–5, 613, 694, 736
Kazén, Miguel 1345
Kazin, Alfred 823, 831
Keasling, Jay 499
Keats, John 589, 1154, 1261, 1293
Keenan, Julian 203
Keil, Frank 422
Kekulé, August 424
Keller, Evelyn Fox 452
Kelly, Thomas 553
Kenkō [Yoshida Kenkō] 939
Kepler, Johannes 415, 596, 756, 949
Kerkhoff, Georg 73, 88
Kierkegaard, Søren 574, 617, 642, 647, 930, 955, 1270
Kimura Bin 334, 366, 1118
Kimura, Doreen 163
Kinnier Wilson, Samuel 536
Kinsbourne, Marcel 170–3, 181, 325, 399, 474
Klein, Yves 292
Kline, Morris 608
Knorr, Karin 424
Koch, Christof 1101, 1110, 1112
Koch, Robert 537
Koestler, Arthur 1243
Kogon, Eugen 1286
Kokoschka, Oskar 270–1
Kolakowski, Leszek 843, 895, 1172, 1176, 1187–8, 1191, 1201, 1228, 1261
Kömpf, Detlef 80
Koonin, Eugene 1177–8
Kornhuber, Hans Helmut 1083
Kosslyn, Stephen 110, 865
Kounios, John 257, 277
Kraus, Karl 614
Kretschmer, Ernst 350, 367–8, 988, 1145
Kripke, Saul 655
Krutetsky, Vadim 302, 759
Kuhl, Julius 1345
Kuhn, Alfred 262
Kuhn, Roland 329, 341, 351, 995
Kuhn, Thomas 536
Kunda, Ziva 719

Kurelek, William 349, *pl 11*
Kürnberger, Ferdinand 360, 1286, 1319

~

Labbé, Cyril 522
Lacan, Jacques 354
Lai, George 1358–9
Laing, RD 328, 1113
Lakatos, Imre 424
Lakoff, George 542, 578, 632, 635, 936, 981, 1090
Lamarck, Jean-Baptiste 681
Lamb, Marion 467
Lamb, Sir Horace 965
Landis, Theodor 80, 139, 868
Lange-Eichbaum, Wilhelm 295
Lange, Joseph 190
Langer, Suzanne 251, 968–9
Langlais, Jean 271
Langland, William 941
Lao Tzu 181, 974, 1197, 1200–1, 1209, 1289
Laplace, Pierre-Simon, marquis de 909, 1078, 1170
Lassman, David 793
Laurent, Eric 348
Lavelle, Louis 17, 928
Lawrence, DH 684, 768, 958, 1037, 1149, 1319
Lawrence, Peter 517–8, 528, 534
Le Guin, Ursula 1299–1300
Leader, Darian 354
LeDoux, Joseph 92
Lee, Andrew 80
Lee, Tsung-Dao 1033
Lefschetz, Solomon 424
Leibniz, Gottfried Wilhelm 418 n 21, 431, 564, 599, 667, 843 n 3, 915, 978, 994, 1060, 1107, 1151, 1185, 1245
Leifer, Matthew 1052
Lemaître, Georges 1366
Lenin, Vladimir 1290, 1292
Leonardo da Vinci 281, 431, 964, 1159–60, 1196, *pl 18–19*
Leopardi, Giacomo 830
Lessing, Gotthold Ephraim 1126
Lévi-Strauss, Claude 584, 630
Levin, DM 561, 995
Levin, Michael 462
Levins, Richard 441
Levy, Jerre 430, 761–2
Lewin, Roger 1067
Lewis, CS 636, 638–9, 1284, 1289

Lewontin, Richard 437, 441, 451, 466, 500
Lhermitte, François 96
Li, Jet 889
Li, Rong 1029
Libet, Benjamin 696–7, 1082–4
Lichtenberg, Georg Christoph 239, 505, 630, 1040
Liebig, Justus von 537
Liepmann, Hugo Karl 186, 536, 982 n 153
Life of Brian, The 1267
Lincoln, Bruce 626
Lindell, Annukka 270, 279–80
Linnæus, Carl 983, 1368
Lippman, Caro 127–8
Lipton, Bruce 468, 1065–6, 1073, 1085
Liszt, Franz 299
Livio, Mario 1153
Llinás, Rodolfo 992
Locke, John 611, 623, 632
Löckenhoff, Corinna 718
Lockhart, Paul 283, 1153
Lombroso, Cesare 295
Lorber, John 1067
Lorentz, Hendrik 892
Lorenz, Konrad 253
Lucas, JR 931, 1236
Luchins, Abraham 169
Lucretius 1037
Ludwig, Arnold 297
Luria, Aleksandr 234–5, 872
Luria, Isaac 1248–50, 1254, 1303
Luzzatti, Claudio 77, 88
Lyons, Viktoria 322

~

MacGregor, Neil 1150
MacIntyre, Alasdair 623
Macrae, D 873
Maddox, Sir John 795
Magee, Bryan 250, 419, 556, 585, 590, 633, 1090 n 276, 1119, 1196
Magendie, François 539
Magritte, René 660–1, *pl 15*
Maimonides 1201, 1365
Mallarmé, Stéphane 48, 1005
'Malley, Ern' 793–4
Malthus, Thomas 42, 244
Manzotti, Riccardo 1112
Mao Tse-Tung 1290
Marcus Aurelius 1151
Marder, Michael 1070–2
Marianetti, Massimo 1350
Marinsek, Nikki 136, 153–4, 512, 602

Markosian, Ned 892
Markowitz, Harry 734
Marshall, John 74
Marshall, Julian 1162
Martindale, Colin 240
Masuda, Takahiko 1265
Matisse, Henri 288 n 239
Matlack, Sam 1119
Matthews, David 271
Matthews, James Tilly 360–2
Maudlin, Tim 6
Maximus the Confessor 431 n 4
Maxwell, James Clark 1019
Maxwell, Nicholas 419
Mayr, Ernst 477, 1192
Mazzucchi, Anna 262, 268
McAuley, James 793–4
McCabe, Herbert 1203, 1206
McClelland, David 158
McClintock, Barbara 239, 436, 460, 463–4, 479, 486, 774
McCosker, Philip 1258
McCrae, Robert 718
McGee, Andrew 707, 1137
McGinn, Colin 1008, 1045, 1053
McGraw, Myrtle 1076
McKinnell, Stephen 196, 205, 1213
McNamara, Patrick 1358–9
McTaggart, JME 650, 652, 654, 901–2
Mechthild of Magdeburg 1237
Medawar, Sir Peter 225, 253, 423, 1277
Medvedev, Andrei 62
Meisner, Gary 1160
Melhuish, George 648
Mendel, Gregor 539, 1366
Mendeleev, Dmitri 424
Mendez, Mario 259
Mercier, Hugo 171, 557, 626, 721–2
Merleau-Ponty, Maurice 106, 135, 333, 338, 377, 381, 390–1, 571, 617, 713, 749, 892, 901–2, 919–20, 923, 927–8, 934, 993, 1008, 1189, 1339
Mermin, David 496, 1006, 1024, 1027, 1186, 1243
Metcalfe, Janet 756
Michelangelo 776, 1085, 1188
Michelson, Albert 892
Midgley, Mary 1193 n 5, 1287–8
Mihov, Konstantin 258, 270
Mikulecky, Don 1006
Mill, John Stuart 417–8
Miller, CL 1244, 1265
Milligan, Spike 298

Miłosz, Czesław 1269
Milton, John: *Paradise Lost* 622, 1294, 1302
Minkowska (*née* Brokman), Françoise 331, 337–8, 348, 355, 357–8, 365
Minkowski, Eugène 167, 328–30, 336–8, 342–3, 345, 347, 352, 355–6, 359, 618, 652, 749, 904–5, 923, 986–8, 1010, 1351
Minsky, Marvin 624
Misselbrook, David 1144–5
Mitchell, Jason 1142
Molière 441
Moltmann, Jürgen 1266–7
Monet, Claude 869, 1349
Monod, Jacques 477
Montaigne, Michel de 657, 829, 837, 1208, 1325, 1366
Moore, Dudley 298
Moore, GE 656, 665
Morgan, William Wilson 756
Morlaas, Joseph 183
Morland, George 869
Mouren, Pierre 118, 121
Mourier, Franck (& Capucine) 674–8, 698, 728, 740, 753, 762
Mozart, Wolfgang Amadeus 251, 284–7, 299, 622, 1084, 1216; *Don Giovanni* 849
Muir, John 843, 1121, 1233
Mullan, Sean 86, 194–5
Müller, Friedrich Max 105, 575, 980, 1284
Muller, Jerry 744–5
Muller, Richard 909, 916, 1280
Müller, Werner 437
Mumford, Stephen 448, 915
Munafò, Marcus 514
Munch, Edvard 292, 349, pl 10[b]
Murdoch, Iris 614, 620, 1258
Musser, George 1020, 1026
Myers, Penelope 212

Nadal, Marcos 1157, 1354
Nagel, Thomas 612, 1037, 1044 n 30, 1093–4, 1099–1100, 1121, 1191
Nägeli, Karl von 539
Nasar, Sylvia 314
Nash, John 314
Needham, Joseph 973
Needham, Tristan 633
Needleman, Jacob 823
Negri, Antonio 856
Nelson, Edward 1247

Neumann, John von 421, 543, 1046, 1078, 1363
Newberg, Andrew 1358
Newcomb, William 664–5
Newman, John Henry, Cardinal 250–1, 566–7
Newton, Nakita 1045
Newton, Sir Isaac 7, 42, 391–2, 420, 420 n 36, 423, 431, 450, 535, 667, 755, 857, 881, 913, 916, 922, 932, 945, 997, 1010, 1017, 1079, 1110, 1185–6, 1199, 1266, 1363–4
Nicholas of Cusa, *see* Cusanus
Nicholson, Daniel 436–7, 440, 443–4, 460, 470, 484–7, 490
Nietzsche, Friedrich 31, 350, 508, 556, 558 n 44, 572, 576–7, 614, 616, 629, 706, 817, 827, 850–2, 867, 885, 968, 992, 1013, 1058, 1129, 1167, 1197, 1264, 1299
Nijboer, Tanja 120
Nijhout, Fred 453
Nikolaenko, Nikolai 82
Nikolayeva, Tatiana 285 n 233
Nisbett, Richard 1265
Noble, Denis 440, 463–4, 468, 839, 1078, 1175–6
Noë, Alva 27–8, 638, 992
Norenzayan, Ara 1371
Novalis (Georg Philipp Friedrich Freiherr von Hardenberg) 444, 662, 954, 1289
Nowell-Smith, Geoffrey 263
Nowicka, Anna 109
Nuñez, Paul 1081, 1085
Nuttall, AD 1239
Nye, Rebecca 1369

O'Malley, Maureen 497–8
O'Regan, Kevin 992
Occam (Ockham), William of 1077, 1099, 1188, 1236, 1239
Oerter, Robert 1025
Oliver, David A 473, 839, 909, 1095
Onasander 678
Ōno Susumu 886
Oppenheimer, Robert 929
Orr, Allen 1180
Orr, Gregory 1259
Ortega y Gasset, José 430 n 72, 583, 633
Ostwald, Wilhelm 1019
Ovid 935
Ovsiew, Fred 79
Owen, Wilfred 838
Oyama, Susan 471

INDEX OF NAMES

Padgett, Jason 945–52
Paine, Robert 832 n 69
Pais, Abraham 1023, 1050
Paley, William 347, 485, 1170
Palladio, Andrea pl 23[a]
Pallasmaa, Juhani 888
Pallis, CA 870
Panikkar, Raimon 1206
Panksepp, Jaak 153, 193, 203
Paracelsus (Theophrastus von Hohenheim) 806, 827
Parfit, Derek 580, 617, 663, 954, 1259
Park, Robert 1181
Parker, Joe 478
Parks, Tim 922, 1112
Parmenides 611, 648, 656, 892, 903
Parnas, Josef 353
Parry, Glenn 921
Parry, Sir Hubert 285 n 233
Pascal, Blaise 31, 51, 368, 547, 565, 568–9, 583, 617–8, 822, 1129–30, 1208, 1294; wager 1262–3
Pasqualini, Isabella 262
Pasteur, Louis 425, 499, 537, 539, 1028, 1032
Patmore, Coventry 1125–6
Patočka, Jan 1284
Patsev, Atanas 265
Paul, Saint 1208
Pauli, Wolfgang 423, 535, 1023, 1040, 1050–2, 1056, 1064–5, 1105–7, 1115, 1177, 1241
Peirce, CS 281, 381, 390, 601, 617, 630, 680, 740, 813, 1022, 1061, 1096, 1154, 1172, 1373
Penfield, Wilder v, 86, 118, 194–5, 508, 1089
Pennell, Joseph 265
Penrose, Sir Roger 611, 931, 1055
Penttonen, Markku 839
Percy, Walker 1309, 1324
Pessoa, Luiz 769
Peters, Douglas 526
Petersen, Aage 473
Peterson, Jordan 1253
Petitti, Diana 800
Petrarch 1216
Pettigrew, Jack 172
Pfeffer, Wilhelm 1069
Pfister, Oskar 1127–8
Phillips, Daniel 1078
Piaget, Jean 230–2, 1101
Picasso, Pablo 122, 288, pl 5[b]
Pichler, Ernst 86
Pick, Arnold 182

Pico della Mirandola, Giovanni 825, 1091
Pienkos, Elizabeth 167, 987
Pieper, Josef 940, 968, 1204, 1209
Piperno, Reginald of 1226
Pitt, Brad 889–90
Pius IX & X, Popes 1367
Planck, Max 414, 420, 426, 535, 667, 1019, 1052, 1054–6, 1064, 1106–7, 1126, 1362
Plato 12, 549, 551–3, 559, 574, 577, 607, 625, 627, 648, 653, 837, 885–6, 903, 925–7, 930, 936, 953, 973, 996, 998, 1000, 1051, 1204, 1216, 1226, 1227, 1269
Pletzer, Belinda 164
Pliny 799
Plotinus v, 4–5, 430, 431 n 4, 495, 500, 508, 560, 1105–6, 1204, 1206, 1228, 1293
Plutarch 661, 818
Poincaré, Henri 244, 248, 250–2, 283, 288–9, 609, 759, 892, 1005, 1007–8, 1084
Pol Pot 1290
Polányi, Michael 254 n 45, 395, 416
Polkinghorne, Sir John 1034, 1182, 1197
Pólya, George 424
Pope, Alexander 630, 1156
Poppelreuter, Walther 211
Popper, Sir Karl 1170, 1175, 1275–7
Porete, Marguerite 1227
Post, Felix 295, 299
Pötzl, Otto 80 n 61, 82, 85–6
Prakash, Manu 460
Prescott, James 984, 993
Prigogine, Ilya 422, 489, 932–3, 1191
'Professor F' [pt] 205–6
Pross, Addy 416, 420, 496–7
Protagoras 660
Proust, Marcel 888
Pryor, Richard 298
Ptolemy 1364
Pugh, Emerson 49
Pushkin, Alexander 763–4, 1261
Putnam, Robert 1375
Pythagoras 596, 1154, 1364
Pythia of Delphi, the 679, 687, 707

≈

Quijada, John 629
Quine, Willard v O 540, 558, 582, 642–3, 655, 668, 756

Quinn, Dennis 1287
Quinton, Anthony (Baron Quinton) 388

≈

Rachmaninov, Sergei 1352
Räderscheidt, Anton 263
Ramachandran, VS 28, 82–3, 93–4, 172, 181, 255, 260, 1046
Ramdas, Swami 1208
Ramón y Cajal, Santiago 61, 858, pl 17[b]
Rank, Otto 616
Ratcliffe, Matthew 201, 1337
Ravel, Maurice 272–3
Rawls, John 1144, 1290
Rayleigh, Lord (John William Strutt) 532
Read, Rupert 1290–1
Reber, Arthur 689
Redlich, Emil 80 n 61
Rée, Jonathan 1220–2, 1270, 1373
Regard, Marianne 868
Regli, Franco 143
Renaux, J-P 357
'Renée' [pt of Sechehaye] 340, 345, 348, 357, 367
Rennie, Drummond 528
Rensch, Bernhard 1062
Rescher, Nicholas 641
Reuben, Scott 522
Reuterswärd, Carl Fredrik 266–7
Reynolds, Sir Joshua 859
Ribeiro, Ariella 177
Richards, IA 634
Richardson, Ralph 1005
Ricoeur, Paul 384, 1225
Rilke, Rainer Maria 943–4, 1152, 1164, 1197, 1226, 1318–9
Ripa, Cesare 293–4
Ritchie, Stuart 226
'RL' [Polish painter] 266–7
Robbins, Stephen 914–5
Ronchi, Claudio 407, 995, 1366
Rosen, Robert 476, 1094–7
Rosenblueth, Arturo 411
Rosenfeld, Léon 1051
Rosenthal, Sandra 902
Rosenzweig, Franz 379 n 1
Ross, David 265
Rosser, Rachel 344
Rothschild, Friedrich 181
Rovelli, Carlo 1006, 1015
Rozenblit, Leonid 422
Rumi 1202, 1317
Rusconi, Maria 93
Ruskin, John 1058, 1163, 1299

Russell, Bertrand 429, 506, 572–3, 617, 661, 759, 1060–1, 1063, 1272
Russell, Jeffrey Burton 1364
Rutherford, Ernest 531–2
Ruysbroeck, Jan van 1247, 1289
Ryle, Gilbert 615

Sacks, Oliver 197, 272, 374–5, 871–4, 1351
Sacks, Rabbi Jonathan 833
Sacrobosco, Johannes de 1364
Sagan, Carl 1276
Saks, Elyn 330
'Salamoun' [pt of Pick] 182
Salk, Jonas 504, 508, 722–3, 740
Salvi, Carola 756
Sappho 1216
Sarkin, Jon 268
Sartre, Jean-Paul 338, 381, 1145
Sass, Louis 167, 308, 313, 328, 332, 339, 341, 346–7, 352–3, 357, 365, 368–70, 372, 615, 619, 636, 749, 936, 942, 986, 987, 1338
Saussure, Ferdinand de 186
Savant, Marilyn vos 709
Schauberger, Viktor 966 n 77
Scheid, Werner 90, 1346
Scheler, Max 31, 381, 390, 562, 713, 832, 845, 892, 990, 997, 1091, 1103, 1126–7, 1130–1, 1155, 1252–3, 1283–4, 1323–4
Schelling, Friedrich Joseph Wilhelm von 444 n 58, 488, 576, 764, 768, 772–4, 825, 840, 878, 917–8, 948–9, 958–61, 967, 971, 992–4, 1021, 1064, 1091–1100 *passim*, 1121–2, 1205, 1228, 1234, 1238, 1240, 1247, 1256, 1265, 1328
Schilder, Paul 334, 348
Schiller, Ferdinand CS 917, 1038, 1040, 1047–8, 1079, 1086, 1089
Schiller, Friedrich 1293
Schlegel, Friedrich 559–60, 1276, 1293
Schleiermacher, Friedrich 821, 826
Schmahmann, Jeremy 985
Schmemann, Alexander 929
Schneier, Bruce 523
Schnider, Armin 261–2
Schnittke, Alfred 272
Schooler, Jonathan 225
Schopenhauer, Arthur 31, 243, 250, 628, 673, 773–4, 881, 896, 932, 968, 1049–50, 1060–1
Schor, Juliet 941
Schore, Allan 201, 322, 1345
Schott, Geoffrey 268
Schreber, Daniel 328, 337, 357, 360, 362, 615
Schröder, Heinrich 114
Schröder, Stefan 142
Schrödinger, Erwin 4, 5, 393, 428, 430, 451, 474, 494, 500, 502, 560, 609, 759, 919, 928, 952–3, 998, 1043, 1053, 1064, 1078, 1094–5, 1102–4, 1117, 1121, 1316
Schubart, Christian 285
Schubert, Franz 1149, 1215–6
Schultheiss, Oliver 1347
Schwinger, Julian 535, 1024
Schwitzgebel, Eric 1112
Scotus, Duns 855 n 34, 1236
Scotus, Michael 488
Scruton, Sir Roger 584, 599–600, 616, 932, 1155–6, 1247–8, 1277
Seaberg, Maureen 946
Searle, John 590 n 76
Sechehaye, Marguerite 357, 367
Sedivy, Julie 716
Seeber, Narnara 937
Segall, Matt 1093, 1240
Seife, Charles 521–2
Sélincourt, Basil de 968–9
Sellers, Peter 298
Seneca 1151
Serafetinides, EA 133
Sergent, Justine 513
Sévigné, Marie de Rabutin-Chantal, marquise de 606
Sfard, Anna 302, 423, 757–9
Shakespeare, William 16 n 15, 284, 386, 588–9, 596, 765, 854, 1002, 1163, 1239; Coleridge on 854, 1238–9; *Richard III* 364; *Merchant of Venice* 596; *King Lear* 623; *Winter's Tale* 864; *As You Like It* 1298; *Hamlet* 1338
Shallis, Michael 924
Shamay-Tsoory, Simone 259
Shanteau, James 725–7
Shapiro, James 435, 438, 446, 451, 461, 468, 1066, 1188
Sharma, Kriti 453–4, 459
Shattuck, Roger 936
Shayer, Michael 232
Shear, Jonathan 551
Shebalin, Vissarion 271
Sheldrake, Rupert 681–2, 794–5, 1079 n 226, 1189
Shelley, Percy Bysshe 46, 239, 760, 764, 857, 888, 1086, 1254
Shepard, Roger 696
Sherover, Charles 903, 925
Sherrington, Sir Charles 54, 858, 1038–9, 1046
Sherwood, Katherine 265
Shishkin, Ivan 869
Shortt, Rupert 1199
Shostakovich, Dmitri 271–2
Silesius, Angelus 1333
Silverman, Sarah 298
Simard, Suzanne 1072–3
Simons, Peter 489, 956
Simonton, Dean Keith 302
Simpson, George Gaylord 422, 432, 477, 504
Singh, Balbir 993
Sitting Bull, Chief 855
Skell, Philip 539
Skinner, Michael 1182
Skyrms, Brian 420 n 32
Smith, Christian 1292, 1374
Smith, Richard 517, 524–8
Smolin, Lee 360, 426–7, 499, 504, 533–4, 843, 853, 896, 909–10, 913, 916, 922, 932, 934, 1006, 1114, 1184, 1259, 1305
Snell, Bruno 624–5, 648
Socrates 553, 556, 766–7, 1208
Sokal, Alan 794–5
Solzhenitsyn, Aleksandr 852, 1292
Soman, Vijay 528
Soros, George 728
Spelke, Elizabeth 1007, 1101
Spencer-Brown, George 414 n 13
Sperber, Dan 171, 557, 626, 722
Sperry, Roger 39–40, 53, 214, 430, 761, 1332
Spiegelhalter, Sir David 804–5
Spinoza, Baruch 31, 488, 559, 576, 616–7, 856, 1060, 1151, 1238, 1247–8, 1267
Spyra, Robert [pt of Poppelreuter] 211
Ssu-ma Ch'ien 840
St Victor, Richard of 1284
Stalin, Joseph 1290
Standing, Lionel 696
Stanghellini, Giovanni 326–7, 338, 344, 619, 905, 1113
Stapp, Henry Pierce 1015, 1056–7, 1059, 1063–4, 1078–9, 1089, 1092
Stark-Adamec, Cannie 527

INDEX OF NAMES

Stark, Freya 1253
Steane, Andrew 477, 1122, 1171, 1179
Stein, Ross 446, 459, 479, 489
Steiner, George 247, 992, 1168, 1260, 1272–3
Steinhardt, Paul 1117
Stent, Gunther 1045–6
Stephen, Sir Leslie 585
Stern, Isaac 1005
Sterne, Laurence 977
Stewart, Harold 793
Stewart, Ian 1031–2
Stotz, Karola 438, 471
Stove, David 1143
Strauss, Richard 1163
Stravinsky, Igor 271–2
Strawson, Galen 1042–4, 1048–9, 1051, 1059, 1085, 1089, 1120
Struck, Peter 678, 687
Suffren, Sabrina 40, 117, 136–7, 178
Sutherland, Stuart 1120
Suzuki, Shinichi 754
Suzuki, Shunryū 823, 875, 996, 1197, 1201, 1213, 1283
Szechyńska-Hebda, Magdalena 1069–70
Szilard, Leo 525, 551

~

Tacikowski, Pawel 109
Tagore, Rabindranath 589
Takemoto, Timothy 557
Talbott, Stephen 433, 435, 442, 445, 476, 480
Taleb, Nassim Nicholas 169, 418, 457, 828, 1034, 1188
Tallis, Thomas 847, 1215
Tanizaki, Jun'ichiro 1163
Tarkovsky, Andrei 414, 847, 1156–7
Tarnas, Richard 67
Tarski, Alfred 558
Tatossian, Arthur 118, 121, 340–1, 343, 349, 358, 362, 1009
Taverner, John 1259
Taylor, Charles 887
Taylor, Jill Bolte 85–7, 196–7, 268, 924, 1213
Tegmark, Max 1118
Teilhard de Chardin, Pierre 1061, 1092, 1261
Tesla, Nikola 1084
Thackeray, William Makepeace 764
Thagard, Paul 719
Thales 977
Theise, Neil 1058, 1121, 1243

Theocritus 698
Thomas, Cheryl 719
Thompson, D'Arcy 483, 840, 999, 1154
Thomson, JF 665–6
Thoreau, Henry David 67, 1128, 1209, 1331
Thybony, Scott 1155
Tillich, Paul 1003, 1202, 1218, 1286
Todd, John 126–8
Tolkien, JRR 767
Tolstoy, Count Leo 405, 832, 1154
Tomonaga, Shin'ichirō 1024
Tong, David 493, 997, 1015–20, 1023, 1027, 1033, 1199
Tononi, Giulio 1101, 1110
Townes, Charles 1192
Toynbee, Arnold 502–3
Tranströmer, Tomas 273
Treffert, Donald 274–5
Trimble, Michael 1358
Trojano, Luigi 122
Trolle, E 873
Tudge, Colin 1142, 1145, 1288, 1294
Turing, Alan 568
Turner, J Scott 441, 458, 474 n 218, 481, 483, 485–6
Tversky, Amos 722, 729
Twenge, Jean 1321
Tynyanov, Yuri 398
Tzu-kung 385

~

Unamuno, Miguel 563 n 62
Underhill, Evelyn 381, 637, 845, 1194, 1209–10, 1284

~

Vaihinger, Hans 563 n 62
Valentine, Lisa 702
Valins, Stuart 688
Vallar, Giuseppe 93
Vamplew, Peter 519–20
Varela, Francisco 551
Vaughan, Henry 1246
Vauvenargues, Luc de Clapiers, marquis de 741
Velichkovsky, Boris 116, 154
Vergely, Bertrand 1287
Verster, Adrian 456
Vial de Saint-Bel, Charles 445 n 62
Vico, Giambattista 747, 982–3
Viereck, George 754
Vierge, Daniel 265
Visconti, Luchino 262–4
Voltaire 53
von Bertalanffy, Ludwig 432, 474, 490, 1188

von Economo, Constantin 223
von Monakow, Constantin 325
von Neumann, John 421, 543, 1046, 1078, 1363
Vuilleumier, Patrik 109

~

Waber, Deborah 164
Waddington, Conrad Hal 432, 451–2, 463, 488, 1175–6
Wade, David 974–5
Wagner, Moritz 453
Wagoner, Brady 710
Waismann, Friedrich 6, 12, 251, 281, 386–7, 389 n 12, 555–7, 585–6, 592, 600–1, 673, 765–6, 770, 776, 811, 883, 924
Wald, George 1102
Wall, Patrick 1068
Walliams, David 298
Walser, Robert 343
Walsh, Denis 457, 459
Wang, Hao 738, 756
Ward, Keith 1232
Watanabe, Satosi 931, 1185
Watson, James 412, 505, 531
Watson, Mark 298
Watts, Alan 449, 819, 973–4, 1285
Wax, Ruby 298
Webb, James 303
Weber, Franz [pt of Kuhn] 351, 995
Weber, Max 549, 1293
Webster, JD 806
Weil, Simone 1130
Weinstein, Edwin 146
Weisberg, Robert 285
Weismann-Arcache, Catherine 303
Weiss, Paul 457, 479, 644
Weizsäcker, Carl von 288, 1049, 1111, 1193
Wertheimer, Max 80 n 61, 254, 754
West, Richard 716
Wexler, Bruce 510
Weyl, Hermann 488, 608–10, 668
Wheeler, John Archibald 503, 1050, 1052, 1056–7, 1172
White, Andrew Dickson 1365
Whitehead, Alfred North 42, 342, 377, 407, 411–2, 418, 428, 431–4, 470, 499, 506, 540, 547–89 passim, 621–9 passim, 700, 727, 751, 823–52 passim, 881, 892, 975–6, 998, 1000, 1041, 1049, 1061–5, 1093–7,

[Whitehead, AN, *cont'd*] 1167, 1172, 1187–97 *passim*, 1204, 1225, 1234, 1240–4, 1247, 1254, 1264, 1308
Whitfield, John 172
Whitman, Walt 273, 657
Wicks, Robert 1161
Widmanstetter, Johann 1366
Wiener, Norbert 411
Wigner, Eugene 1049, 1054, 1121, 1363
Wilberforce, Samuel 1367
Wilczek, Frank 1021, 1024, 1115
Wilkins, John 1367
Williams, Denis 145
Williams, Kenneth 298
Williams, Robin 298
Williams, William Carlos 273
Wisser, Richard 419 n 27
Wittgenstein, Ludwig 3, 29, 105, 325, 381, 390, 506, 552, 556, 585, 587, 614–7, 628, 632, 636, 658, 990, 1064, 1140, 1194, 1199, 1203, 1222, 1267, 1273, 1286
Woese, Carl 433, 460–1, 479, 494
Wolff, Christoph 285
Wolff, Patrick 695
Wood, Victoria 298
Woodger, Joseph Henry 432, 479
Woodley of Menie, Michael 233
Woodman, Geoffrey 781
Woolf, Virginia 368
Wordsworth, William 241, 288, 381, 397, 550, 765–6, 770, 845–6, 1104, 1216, 1231, 1234, 1238, 1261, 1290
Wotton, Sir Henry 818
Wright, Charles 1208
Wright, NT 624
Wu, Chien-Shiung 1033
Wyman, Jeffries 1173
Wyndham Lewis, Percy 345

≈

Yang Chen-Ning 1033
Yeats, William Butler 13, 897, 935
Yukawa Hideki 1051, 1102

≈

Zagvazdin, Yuri 272
Zaidel, Dahlia 214–5, 269–71, 1349
Zaki, Jamil 1142
'Zasetsky' [pt of Luria] 872
Zeh, Dieter 931
Zeilinger, Anton 428
Zeno of Elea 645–54, 664, 883, 896, 979, 994, 1256
Zihl, Josef 86
Zingerle, Hermann 75–6, 87–8, 90, 197, 1339
Zweig, George 1038
Zwicky, Jan 348, 397, 547, 554, 558, 608, 611, 633–4, 853–4, 997, 1319

CREDITS FOR THE COLOUR PLATES IN VOLUME II

17 (a) Golgi 1875; (b) Ramón y Cajal 1890;

18 Royal Collection Trust © Royal Collection / Royal Collection Trust © Her Majesty Queen Elizabeth II, 2021 / Bridgeman Images;

19 (a) Royal Collection Trust © Royal Collection / Royal Collection Trust © Her Majesty Queen Elizabeth II, 2021 / Bridgeman Images; (b) Royal Collection Trust © Royal Collection / Royal Collection Trust © Her Majesty Queen Elizabeth II, 2021 / Bridgeman Images;

20 Rijksmuseum, Amsterdam, The Netherlands / Bridgeman Images;

21 Shutterstock;

22 (a) public domain; (b) Robert Bringhurst; (c) Shutterstock; (d) photo: © Ron Knott, www.maths.surrey.ac.uk/hosted- sites/R.Knott/Fibonacci/copyright.html; (e) photo: Paul Tomkiss;

23 (a) www.theartpostblog.com: Flavia Sciortino, 'Andrea Palladio e il Palladianesimo', 19 August 2020, courtesy of Caterina Stringhetta; (b) photo: courtesy of Filippo Romano;

24 M. C. Escher's 'Circle Limit IV' © 2021 The M. C. Escher Company – The Netherlands. www.mcescher.com.

¶ *Volumes I & II of Iain McGilchrist's* THE MATTER WITH THINGS *were designed and set in type by Robert Bringhurst Ltd on Quadra Island, British Columbia. The text face (roman, italic, Greek, and Cyrillic) is* ARNO, *designed by Robert Slimbach for Adobe Systems and issued in 2007. Other faces used here and there include* CENTAUR, *Bruce Rogers's early 20th-century interpretation of a roman face cut in Venice in 1469 by the French printer Nicolas Jenson; John Peters's* CASTELLAR, *first cut in metal by the Monotype Corporation, Redhill, Surrey, in 1957; and Hermann Zapf's* PALATINO SANS, *which was first sketched in Darmstadt in 1973 and completed, with the assistance of Akira Kobayashi, in Bad Homberg in 2006.*